METHODS OF SOIL ANALYSIS

Part 1, Second Edition

AGRONOMY

A Series of Monographs

The American Society of Agronomy and Academic Press published the first six books in this series. The General Editor of Monographs 1 to 6 was A. G. Norman. They are available through Academic Press, Inc., 111 Fifth Avenue, New York, NY 10003.

1. C. EDMUND MARSHALL: The Colloid Chemical of the Silicate Minerals, 1949
2. BYRON T. SHAW, *Editor*: Soil Physical Conditions and Plant Growth, 1952
3. K. D. JACOB, *Editor*: Fertilizer Technology and Resources in the United States, 1953
4. W. H. PIERRE and A. G. NORMAN, *Editors*: Soil and Fertilizer Phosphate in Crop Nutrition, 1953
5. GEORGE F. SPRAGUE, *Editor*: Corn and Corn Improvement, 1955
6. J. LEVITT: The Hardiness of Plants, 1956

The Monographs published since 1957 are available from the American Society of Agronomy, 677 S. Segoe Road, Madison, WI 53711.

7. JAMES N. LUTHIN, *Editor*: Drainage of Agricultural Lands, 1957 *General Editor,* D.E. Gregg
8. FRANKLIN A. COFFMAN, *Editor*: Oats and Oat Improvement
   Managing Editor, H. L. Hamilton
9. A. KLUTE, *Editor*: Methods of Soil Analysis, 1986
   Part 1—Physical and Mineralogical Methods, Second Edition *Managing Editor,* R. C. Dinauer
   A. L. PAGE, *Editor*: Methods of Soil Analysis, 1982
   Part 2—Chemical and Microbiological Properties, Second Edition *Managing Editor*, R. C. Dinauer
10. W. V. BARTHOLOMEW and F. E. CLARK, *Editors*: Soil Nitrogen, 1965
    (Out of print; replaced by no. 22) *Managing Editor*, H. L. Hamilton
11. R. M. HAGAN, H. R. HAISE, and T. W. EDMINSTER, *Editors*: Irrigation of Agricultural Lands, 1967 *Managing Editor,* R. C. Dinauer
12. FRED ADAMS, *Editor*: Soil Acidity and Liming, Second Edition, 1984
    Managing Editor, R. C. Dinauer
13. K. S. QUISENBERRY and L. P. REITZ, *Editors*: Wheat and Wheat Improvement, 1967
    Managing Editor, H. L. Hamilton
14. A. A. HANSON and F. V. JUSKA, *Editors*: Turfgrass Science, 1969
    Managing Editor, H. L. Hamilton
15. CLARENCE H. HANSON, *Editor*: Alfalfa Science and Technology, 1972
    Managing Editor, H. L. Hamilton
16. B. E. CALDWELL, *Editor*: Soybeans: Improvement, Production, and Use, 1973
    Managing Editor, H.L. Hamilton
17. JAN VAN SCHILFGAARDE, *Editor*: Drainage for Agriculture, 1974
    Managing Editor, R. C. Dinauer
18. GEORGE F. SPRAGUE, *Editor*: Corn and Corn Improvement, 1977
    Managing Editor, D. A. Fuccillo
19. JACK F. CARTER, *Editor*: Sunflower Science and Technology, 1978
    Managing Editor, D. A. Fuccillo
20. ROBERT C. BUCKNER and L. P. BUSH, *Editors*: Tall Fescue, 1979
    Managing Editor, D. A. Fuccillo
21. M. T. BEATTY, G. W. PETERSEN, and L. D. SWINDALE, *Editors*: Planning the Uses and Management of Land, 1979 *Managing Editor*, R. C. Dinauer
22. F. J. STEVENSON, *Editor*: Nitrogen in Agricultural Soils, 1982
    Managing Editor, R. C. Dinauer
23. H. E. DREGNE and W. O. WILLIS, *Editors*: Dryland Agriculture, 1983
    Managing Editor, D. A. Fuccillo
24. R. J. KOHEL and C. F. LEWIS, *Editors*: Cotton, 1984
    Managing Editor, D. A. Fuccillo
25. N. L. TAYLOR, *Editor*: Clover Science and Technology, 1985
    Managing Editor, D. A. Fuccillo
26. D. C. RASMUSSON, *Editor*: Barley, 1985
    Managing Editor, D. A. Fuccillo
27. M. A. TABATABAI, *Editor*: Sulfur in Agriculture, 1986
    Managing Editor, R. C. Dinauer

METHODS OF SOIL ANALYSIS

Part 1

Physical and Mineralogical Methods

Second Edition

Arnold Klute, ***Editor***

Number 9 (Part 1) in the series

AGRONOMY

American Society of Agronomy, Inc.
Soil Science Society of America, Inc.
Publisher
Madison, Wisconsin USA

1986

American Society of Agronomy, Inc.
Soil Science Society of America, Inc.
677 South Segoe Road, Madison, Wisconsin 53711 USA

Library of Congress Cataloging in Publication Data

(Revised for no. 9, pt. 1)

Methods of soil analysis.

(Agronomy; no. 9)
Includes bibliographies and indexes.
Contents: pt. 1. Physical and mineralogical methods—pt. 2. Chemical and microbiological properties.
1. Soils—Analysis—Collected works. I. Page, A. L. (Albert Lee), 1927- . II. Miller, R. H., 1933- . III. Keeney, Dennis R. IV. Series.
S593.M4453 1982 631.4'1'0287 82-22630
ISBN 0-89118-088-5 (pt. 1)
ISBN 0-89118-072-9 (pt. 2)

Printed in the United States of America

CONTENTS

13 Bulk Density

G. R. BLAKE AND K. H. HARTGE

14 Particle Density

G. R. BLAKE AND K. H. HARTGE

15 Particle-size Analysis

G. W. GEE AND J. W. BAUDER

16 Specific Surface

D. L. CARTER, M. M. MORTLAND, AND W. D. KEMPER

17 Aggregate Stability and Size Distribution

W. D. KEMPER AND R. C. ROSENAU

18 Porosity

R. E. DANIELSON AND P. L. SUTHERLAND

19 Penetrability

J. M. BRADFORD

20 Compressibility

J. M. BRADFORD AND S. C. GUPTA

21 Water Content

WALTER H. GARDNER

22 Water Potential: Piezometry

R. C. REEVE

38 Heat Capacity and Specific Heat

STERLING A. TAYLOR AND RAY D. JACKSON

39 Thermal Conductivity and Diffusivity

RAY D. JACKSON AND STERLING A. TAYLOR

40 Heat Flux

M. FUCHS

41 Heat of Immersion

D. M. ANDERSON

42 Solute Content

J. D. RHOADES AND J. D. OSTER

43 Solute Diffusivity

W. D. KEMPER

DEDICATION

Charles A. Black

It is truly fitting that *Methods of Soil Analysis,* Part 1, Second Edition, be dedicated to Dr. Charles A. Black. Dr. Black was editor-in-chief of the 1965 *Methods of Soil Analysis,* Parts 1 and 2, one of the most successful and widely acclaimed of the Society's monograph series. His dedicated efforts were largely responsible for the overall high quality of the first edition of the monograph. It is also fitting to recognize Dr. Black for his contributions to research and teaching and for his current role as one of the chief spokespersons for agriculture.

Dr. Black was born 22 January 1916 in Lone Tree, Iowa. He received B.S. degrees in chemistry and soil science from Colorado State University in 1937, and the M.S. and Ph.D. degrees in soil fertility from Iowa State University in 1938 and 1942.

He began his professional career as a research fellow in the Department of Agronomy, Iowa State University in 1937, and in 1939 joined that faculty as instructor in soils. Except for service with the U.S. Navy during World War II, a visiting professorship at Cornell University in 1955–56, and a NSF Fellowship at UC-Davis in 1964–65, Dr. Black has remained at Iowa State. He retired as distinguished professor in 1979 to devote full time to his duties with the Council for Agricultural Science and Technology (CAST).

Dr. Black's research and teaching career has had a major influence on the discipline of soil science, particularly soil fertility and soil chemistry. He has contributed much to our knowledge of phosphate reactions in soils, uptake by plants, and interpretation of yield curves. He is author or co-author of approximately 100 research papers, has written two editions of a widely used textbook entitled *Soil-Plant Relationships,* and several editions of a laboratory manual on soil chemistry. He has also served as associate editor for the *SSSA Journal*; as consulting editor for *Soil Science,* and as editor of more than 100 publications issued by CAST. He has served the ASA and SSSA as a member of numerous committees, as SSSA president in 1961, and as ASA president in 1971. He has received numerous awards and honors, including the ASA Soil Science Award (1957), ASA Fellow (1962), Fellow of the American Institute of Chemists (1969), Honorary Member of SSSA (1975) and ASA (1981), AAAS Fellow (1976), the Henry A. Wallace Award from Iowa State University for Distinguished Service to Agriculture (1981), and the Bouyoucos Soil Science Distinguished Career Award, SSSA (1981).

Dr. Black's critical and forthright evaluation of research findings, coupled with a warm personality and a dry sense of humor, have made him a much sought-after counselor by students and colleagues. His graduate level soil-plant relationship courses at Iowa State were especially popular. Those privileged to learn under Dr. Black gained the type of knowledge and philosophy which has served them well in their varied careers.

Dr. Black's career took on a new dimension in 1970 when, largely under his direction, CAST was developed. He was the president of CAST in 1973 and then served as the executive vice-president of this innovative, independent association of agricultural science societies from 1973 through April 1985. Since May 1985 he has served as executive chairman of the board of directors of CAST.

He is providing invaluable service to the community of food and agricultural scientists through his dedicated efforts on the behalf of CAST. Through the Council, the scientific societies and the scientists they represent, can make an input into the development of national policies on food and agriculture by supplying scientific information to decision makers and opinion leaders.

FOREWORD

Characterization and, hence, our understanding of soils requires that they be precisely and reproducively analyzed or described. The parameters generated by such analyses are needed to generalize hypotheses for differences among soils as well as observations on the same soil under different circumstances of time and manipulation.

Mineralogical characteristics of soils simultaneously reflect the parent material as well as the processes which formed the soil from the geologic matrix. Knowing and understanding the mineralogical composition provides an insight into the behavior of soils under different temporal conditions and their usefulness or suitability for various purposes. To a great extent, the chemical properties of soils depend upon soil mineralogy.

The physical properties of soils, perhaps, more than the chemical properties, determine their adaptability to cultivation for food and fiber production—the most important use of soils on a densely populated planet. It is the physical properties which are most prominent in determining the adaptability of soils to other civilized activities including housing, communications, and recreation.

Great strides have been made in the conception of physical and mineralogical characteristics of soils and how they relate to each other and to chemical properties. These have been stimulated by and, in turn, have encouraged better methods of analysis. It is appropriate from time to time to record those analytical procedures that seem best to serve the scientific community in its understanding of soils.

The methods of analyses included in this volume provide a uniform set of procedures which can be used by the majority of scientists and engineers working with soil and, in that way, will improve their ability to communicate about their observations with each other. This volume also provides a launching point from which others might depart in an effort to refine methods and procedures or develop new ones. Given our rapidly changing understanding of soils and our ability to make new, different, and more precise measurements, it is certain that other volumes such as this will follow.

June 1986

JOHN PESEK, *president*
Soil Science Society of America

DALE M. MOSS, *president*
American Society of Agronomy

PREFACE

It has been more than 20 years since the first edition of *Methods of Soil Analysis, Part 1, Physical and Mineralogical Properties* was published under the able editorship of Dr. C. A. Black. The first edition was extremely well received. More than 13 000 copies have been sold, and sales have continued up to the present time at more than 500 copies per year.

Since the publication of the first edition there has been substantial progress in the development of improved physical and mineralogical measurements. The study of transport processes in soil in relation to environmental quality concerns has brought about an increased interest in the application of methods of physical measurement to field situations. More emphasis has been placed on the development of methods to cope with the inherent spatial variability of natural soils. In addition, techniques of measurement of physical and mineralogical properties of soils have generally been improved. Following a recommendation by the ASA Monographs Committee, publication of the second edition of the monograph was approved by the Executive Committee of ASA, and an editorial committee was appointed.

Major changes have been made in the subject coverage of Part 1, as compared to the first edition. The book consists of 50 chapters prepared by 71 authors and coauthors. Four chapters deal with statistical subject matter, including a new chapter on geostatistical methods applied to measurements in soils. Eight chapters focus on various mineralogical methods, and eight chapters describe methods for evaluation of the soil matrix and its structure. Methods for assessing the energy status of soil water, hydraulic conductivity and diffusivity of soils, intake rate, and water retention of soils are described in 16 chapters. There are five chapters on methods for measurement of thermal properties of soil, four chapters on methods for determining the concentration and flux of soil solutes, and five chapters on methods for study of the soil gas phase.

Members of the editorial committee who participated in the planning and development of the book are:

A. Klute, editor, Colorado State University, and USDA-ARS, Ft. Collins, CO

G. S. Campbell, Washington State University, Pullman, WA

R. D. Jackson, U. S. Water Conservation Laboratory, Tempe, AZ

M. M. Mortland, Michigan State University, East Lansing, MI

D. R. Nielsen, University of California, Davis, CA

To the many authors, who drew from their expertise to prepare descriptions of the many methods, I extend my thanks and appreciation. I also wish to express my appreciation to the members of the editorial committee, and to many other anonymous reviewers who provided their time and talents to help produce the monograph. Mr. R. C. Dinauer, and the ASA Headquarters staff are to be given special thanks for their diligence, and competence in handling the many mechanical details of production of the book.

This second edition of Part 1 is dedicated to Dr. C. A. Black in recognition of his efforts as editor of the first edition, and its success as a publication of the American Society of Agronomy. I and the other members of the editorial committee hope that the revised edition will be as successful, and that it will be found useful by those who need information on physical and mineralogical methods.

April 1986

A. KLUTE, *editor*
Agricultural Research Service, USDA, and
Colorado State University, Ft. Collins, CO

CONTRIBUTORS

Lajpat R. Ahuja Soil Physicist, Water Quality and Watershed Research Laboratory, Agricultural Research Service, U.S. Department of Agriculture, Durant, Oklahoma

William R. Allardice Staff Research Associate, Department of Land, Air, and Water Resources, University of California, Davis, California

R. R. Allmaras Soil Scientist, Agricultural Research Service, U.S. Department of Agriculture, St. Paul, Minnesota

A. Amoozegar Assistant Professor, Soil Science Department, North Carolina State University, Raleigh, North Carolina

Duwayne M. Anderson Associate Provost for Research, Texas A&M University, College Station, Texas

Isaac Barshad Soil Chemist, Soil and Plant Biology Department, University of California, Berkeley, California

J. W. Bauder Professor of Soil Science, Cooperative Extension, Montana State University, Bozeman, Montana

George R. Blake Professor Emeritus of Soil Physics, Soil Science Department, University of Minnesota, St. Paul, Minnesota

Charles W. Boast Associate Professor of Soil Physics, Department of Agronomy, University of Illinois, Urbana, Illinois

Herman Bouwer Director, U.S. Water Conservation Laboratory, Agricultural Research Service, U.S. Department of Agriculture, Phoenix, Arizona

J. M. Bradford Soil Scientist, National Soil Erosion Laboratory, Agricultural Research Service, U.S. Department of Agriculture, Purdue University, West Lafayette, Indiana

R. R. Bruce Soil Scientist, Southern Piedmont Conservation Research Center, Agricultural Research Service, U.S. Department of Agriculture, Watkinsville, Georgia

G. D. Bubenzer Professor, Department of Agricultural Engineering, University of Wisconsin, Madison, Wisconsin

J. G. Cady Lecturer in Pedology, Department of Geography and Environmental Engineering, The Johns Hopkins University, Baltimore, Maryland

Lyle D. Calvin Professor of Statistics, Department of Statistics, Oregon State University, Corvallis, Oregon

Gaylon S. Campbell Professor of Soils, Department of Agronomy and Soils, Washington State University, Pullman, Washington

D. L. Carter Supervisory Soil Scientist, Snake River Conservation Research Center, Agricultural Research Service, U.S. Department of Agriculture, Kimberly, Idaho

D. K. Cassel Professor of Soil Science, Department of Soil Science, North Carolina State University, Raleigh, North Carolina

She-Kong Chong Associate Professor of Soil Physics and Hydrology, Plant and Soil Science Department, Southern Illinois University, Carbondale, Illinois. Formerly Associate Professor of Forest Hydrology, Department of Forestry, Southern Illinois University, Carbondale, Illinois

Arthur T. Corey Professor Emeritus, Department of Agricultural and Chemical Engineering, Colorado State University, Fort Collins, Colorado

Robert E. Danielson Professor, Department of Agronomy, Colorado State University, Fort Collins, Colorado

C. Dirksen Senior Scientist, Department of Soil Science and Plant Nutrition, Agricultural University, Wageningen, The Netherlands

Joe B. Dixon Professor of Soil Mineralogy, Department of Soil and Crop Sciences, Texas A&M University, College Station, Texas

W. J. Dixon Professor of Biomathematics, Department of Biomathematics, School of Medicine, University of California, Los Angeles, California

L. R. Drees Research Associate, Department of Soil and Crop Sciences, Texas A&M University, College Station, Texas

H. Flühler Professor of Soil Physics, Swiss Federal Institute of Technology, Zurich, Switzerland

Marcel Fuchs Scientist, Agricultural Research Organization, Institute of Soils and Water, Bet Dagan, Israel

Walter H. Gardner Professor of Soils Emeritus, Department of Agronomy and Soils, Washington State University, Pullman, Washington

G. W. Gee Staff Scientist, Battelle Pacific Northwest Laboratories, Richland, Washington

Richard E. Green Professor of Soil Science, Department of Agronomy and Soil Science, University of Hawaii, Honolulu, Hawaii

S. C. Gupta Associate Professor of Soil Physics, Department of Soil Science, University of Minnesota, St. Paul, Minnesota

B. F. Hajek Professor, Agronomy and Soil Department, Auburn University, Auburn, Alabama

K. H. Hartge Professor of Soil Science, University of Hannover, Federal Republic of Germany

Marion L. Jackson The F. H. King Professor of Soil Science, Department of Soil Science, University of Wisconsin, Madison, Wisconsin

Ray D. Jackson Research Physicist, U.S. Water Conservation Laboratory, Agricultural Research Service, U.S. Department of Agriculture, Phoenix, Arizona

W. Doral Kemper Director, Snake River Conservation Research Center, Agricultural Research Service, U.S. Department of Agriculture, Kimberly, Idaho

Oscar Kempthorne Professor of Statistics, Department of Statistics, Iowa State University, Ames, Iowa

Dennis C. Kincaid Agricultural Engineer, Snake River Conservation Research Center, Agricultural Research Service, U.S. Department of Agriculture, Kimberly, Idaho

A. Klute Soil Scientist and Professor, Agricultural Research Service, U.S. Department of Agriculture, and Agronomy Department, Colorado State University, Fort Collins, Colorado

George W. Kunze Professor Emeritus of Soil Mineralogy, Department of Soil and Crop Sciences, Texas A&M University, College Station, Texas

Chin H. Lim Formerly Post-doctoral Fellow, Department of Soil Science, University of Wisconsin, Madison, Wisconsin. Now at Guthrie Research Chemara, Negri Sembilan, Malaysia

R. J. Luxmoore Soil and Plant Scientist, Oak Ridge National Laboratory, Oak Ridge, Tennessee

Murray B. McBride Associate Professor of Soil Chemistry, Department of Agronomy, Cornell University, Ithaca, New York

Max M. Mortland Professor, Department of Crop and Soil Sciences, Michigan State University, East Lansing, Michigan

Yechezkel Mualem Soil Scientist, Seagram Centre for Soil and Water Sciences, The Hebrew University of Jerusalem, Rehovot, Israel

Donald E. Myers Professor of Mathematics, Department of Mathematics, University of Arizona, Tucson, Arizona

D. R. Nielsen Professor, Department of Land, Air and Water Resources, University of California, Davis, California

J. D. Oster Extension Soil and Water Specialist, Department of Soil and Environmental Sciences, University of California, Riverside, California

A. J. Peck Chief Research Scientist, Division of Groundwater Research, CSIRO, Wembley, Western Australia. Formerly Senior Principal Research Scientist

Roger G. Petersen Professor of Statistics, Department of Statistics, Oregon State University, Corvallis, Oregon

Arthur E. Peterson Professor, Department of Soil Science, University of Wisconsin, Madison, Wisconsin

Claude J. Phene Supervisory Soil Scientist, Water Management Research Laboratory, Agricultural Research Service, U.S. Department of Agriculture, Fresno, California

Stephen L. Rawlins Soil Physicist, National Program Staff, Agricultural Research Service, U.S. Department of Agriculture, Beltsville, Maryland

Ronald C. Reeve Civil Engineer, Irrigation and Drainage, Agricultural Research Service, U.S. Department of Agriculture, Columbus, Ohio

J. D. Rhoades Supervisory Soil Scientist, U.S. Salinity Laboratory, Agricultural Research Service, U.S. Department of Agriculture, Riverside, California

D. E. Rolston Professor of Soil Science, Department of Land, Air and Water Resources, University of California, Davis, California

Russell C. Rosenau Technician, Snake River Conservation Research Center, Agricultural Research Service, U.S. Department of Agriculture, Kimberly, Idaho

Charles B. Roth Professor of Soil Mineralogy/Chemistry, Department of Agronomy, Purdue University, West Lafayette, Indiana

Brij L. Sawhney Soil Chemist, The Connecticut Agricultural Experiment Station, New Haven, Connecticut

Lewis H. Stolzy Professor of Soil Physics, Department of Soil and Environmental Sciences, University of California, Riverside, California

P. Lorenz Sutherland Assistant Professor, Department of Agronomy, Southeast Colorado Research Center, Colorado State University, Lamar, Colorado

Kim H. Tan Professor of Agronomy, Department of Agronomy, University of Georgia, Athens, Georgia

Sterling A. Taylor Professor of Soil Physics, Department of Soils and Biometeorology, Utah State University, Logan, Utah. Deceased 8 June 1967

M. Th. van Genuchten Research Soil Scientist, U.S. Salinity Laboratory, Agricultural Research Service, U.S. Department of Agriculture, Riverside, California

R. J. Wagenet Associate Professor, Department of Agronomy, Cornell University, Ithaca, New York

A. W. Warrick Professor, Department of Soil and Water Science, University of Arizona, Tucson, Arizona

Joe L. White Professor of Soil Mineralogy, Department of Agronomy, Purdue University, West Lafayette, Indiana

Lynn D. Whittig Professor of Soil Science, Department of Land, Air and Water Resources, University of California, Davis, California

P. J. Wierenga Professor of Soil Physics, Department of Crop and Soil Sciences, New Mexico State University, Las Cruces, New Mexico

L. P. Wilding Professor of Pedology, Soil and Crop Sciences Department, Texas A&M University, College Station, Texas

Lucian W. Zelazny Professor of Soil Mineralogy, Agronomy Department, Virginia Polytechnic Institute and State University, Blacksburg, Virginia

Conversion Factors for SI and non-SI Units

To convert Column 1 into Column 2, multiply by	Column 1 SI Unit	Column 2 non-SI Unit	To convert Column 2 into Column 1 multiply by
		Length	
0.621	kilometer, km (10^3 m)	mile, mi	1.609
1.094	meter, m	yard, yd	0.914
3.28	meter, m	foot, ft	0.304
1.0	micrometer, μm (10^{-6} m)	micron, μ	1.0
3.94×10^{-2}	millimeter, mm (10^{-3} m)	inch, in	25.4
10	nanometer, nm (10^{-9} m)	Angstrom, Å	0.1
		Area	
2.47	hectare, ha	acre	0.405
247	square kilometer, km^2 $(10^3\ m)^2$	acre	4.05×10^{-3}
0.386	square kilometer, km^2 $(10^3\ m)^2$	square mile, mi^2	2.590
2.47×10^{-4}	square meter, m^2	acre	4.05×10^3
10.76	square meter, m^2	square foot, ft^2	9.29×10^{-2}
1.55×10^{-3}	square millimeter, mm^2 $(10^{-6}\ m)^2$	square inch, in^2	645
		Volume	
6.10×10^4	cubic meter, m^3	cubic inch, in^3	1.64×10^{-5}
2.84×10^{-2}	liter, L (10^{-3} m^3)	bushel, bu	35.24
1.057	liter, L (10^{-3} m^3)	quart (liquid), qt	0.946
3.53×10^{-2}	liter, L (10^{-3} m^3)	cubic foot, ft^3	28.3
0.265	liter, L (10^{-3} m^3)	gallon	3.78
33.78	liter, L (10^{-3} m^3)	ounce (fluid), oz	2.96×10^{-2}
2.11	liter, L (10^{-3} m^3)	pint (fluid), pt	0.473
9.73×10^{-3}	$meter^3$, m^3	acre-inch	102.8
35.3	$meter^3$, m^3	cubic foot, ft^3	2.83×10^{-2}

continued on next page

Conversion Factors for SI and non-SI Units

To convert Column 1 into Column 2, multiply by	Column 1 SI Unit	Column 2 non-SI Unit	To convert Column 2 into Column 1 multiply by
	Mass		
2.20×10^{-3}	gram, g (10^{-3} kg)	pound, lb	454
3.52×10^{-2}	gram, g	ounce (avdp), oz	28.4
2.205	kilogram, kg	pound, lb	0.454
10^{-2}	kilogram, kg	quintal (metric), q	10^{2}
1.10×10^{-3}	kilogram, kg	ton (2000 lb), ton	907
1.102	megagram, Mg (tonne)	ton (U.S.), ton	0.907
	Yield and Rate		
0.893	kilogram per hectare, kg ha^{-1}	pound per acre, lb $acre^{-1}$	1.12
7.77×10^{-2}	kilogram per cubic meter, kg m^{-3}	pound per bushel, lb bu^{-1}	12.87
1.49×10^{-2}	kilogram per hectare, kg ha^{-1}	bushel per acre, 60 lb	67.19
1.59×10^{-2}	kilogram per hectare, kg ha^{-1}	bushel per acre, 56 lb	62.71
1.86×10^{-2}	kilogram per hectare, kg ha^{-1}	bushel per acre, 48 lb	53.75
0.107	liter per hectare, L ha^{-1}	gallon per acre	9.35
893	megagram per hectare, Mg ha^{-1}	pound per acre, lb $acre^{-1}$	1.12×10^{-3}
0.446	megagram per hectare, Mg ha^{-1}	ton (2000 lb) per acre, ton $acre^{-1}$	2.24
2.24	meter per second, m s^{-1}	mile per hour	0.447
	Specific Surface		
10	square meter per kilogram, m^2 kg^{-1}	square centimeter per gram, cm^2 g^{-1}	0.1
10^{3}	square meter per kilogram, m^2 kg^{-1}	square millimeter per gram, mm^2 g^{-1}	10^{-3}

	Pressure		
9.90	megapascal, MPa (10^6 Pa)	atmosphere	0.101
10	megapascal, MPa (10^6 Pa)	bar	0.1
1.00	megagram per cubic meter, Mg m^{-3}	gram per cubic centimeter, g cm^{-3}	1.00
2.09×10^{-2}	pascal, Pa	pound per square foot, lb ft^{-2}	47.9
1.45×10^{-4}	pascal, Pa	pound per square inch, lb in^{-2}	6.90×10^3
	Temperature		
1.00 (K − 273)	Kelvin, K	Celsius, °C	1.00 (°C + 273)
(9/5 °C) + 32	Celsius, °C	Fahrenheit, °F	5/9 (°F −32)
	Energy, Work, Quantity of Heat		
9.52×10^{-4}	joule, J	British thermal unit, Btu	1.05×10^3
0.239	joule, J	calorie, cal	4.19
10^7	joule, J	erg	10^{-7}
0.735	joule, J	foot-pound	1.36
2.387×10^{-5}	joule per square meter, J m^{-2}	calorie per square centimeter (langley)	4.19×10^4
10^5	newton, N	dyne	10^{-5}
1.43×10^{-3}	watt per square meter, W m^{-2}	calorie per square centimeter minute (irradiance), cal cm^{-2} min^{-1}	698
	Transpiration and Photosynthesis		
3.60×10^{-2}	milligram per square meter second, mg m^{-2} s^{-1}	gram per square decimeter hour, g dm^{-2} h^{-1}	27.8
5.56×10^{-3}	milligram (H_2O) per square meter second, mg m^{-2} s^{-1}	micromole (H_2O) per square centimeter second, μmol cm^{-2} s^{-1}	180
10^{-4}	milligram per square meter second, mg m^{-2} s^{-1}	milligram per square centimeter second, mg cm^{-2} s^{-1}	10^4
35.97	milligram per square meter second, mg m^{-2} s^{-1}	milligram per square decimeter hour, mg dm^{-2} h^{-1}	2.78×10^{-2}
	Angle		
57.3	radian, rad	degrees (angle), °	1.75×10^{-2}

continued on next page

Conversion Factors for SI and non-SI Units

To convert Column 1 into Column 2, multiply by	Column 1 SI Unit	Column 2 non-SI Unit	To convert Column 2 into Column 1 multiply by
	Electrical Conductivity		
10	siemen per meter, S m^{-1}	millimho per centimeter, mmho cm^{-1}	0.1
	Water Measurement		
9.73×10^{-3}	cubic meter, m^3	acre-inches, acre-in	102.8
9.81×10^{-3}	cubic meter per hour, m^3 h^{-1}	cubic feet per second, ft^3 s^{-1}	101.9
4.40	cubic meter per hour, m^3 h^{-1}	U.S. gallons per minute, gal min^{-1}	0.227
8.11	hectare-meters, ha-m	acre-feet, acre-ft	0.123
97.28	hectare-meters, ha-m	acre-inches, acre-in	1.03×10^{-2}
8.1×10^{-2}	hectare-centimeters, ha-cm	acre-feet, acre-ft	12.33
	Concentrations		
1	centimole per kilogram, cmol kg^{-1} (ion exchange capacity)	milliequivalents per 100 grams, meq 100 g^{-1}	1
0.1	gram per kilogram, g kg^{-1}	percent, %	10
1	megagram per cubic meter, Mg m^{-3}	gram per cubic centimeter, g cm^{-3}	1
1	milligram per kilogram, mg kg^{-1}	parts per million, ppm	1
	Plant Nutrient Conversion		
	Elemental	***Oxide***	
2.29	P	P_2O_5	0.437
1.20	K	K_2O	0.830
1.39	Ca	CaO	0.715
1.66	Mg	MgO	0.602

23 April 1986

1 Errors and Variability of Observations[1]

OSCAR KEMPTHORNE

Iowa State University
Ames, Iowa

R. R. ALLMARAS

Agricultural Research Service, USDA
St. Paul, Minnesota

1-1 INTRODUCTION

The experimenter is confronted nearly every day with the examination of results from his or her own experiments as well as those of others. There is a need to know the methods by which the data were obtained and what confidence can be placed in the numerical results. Significant aspects of these matters involve the principles and methods of statistics, and the objective of this chapter is to describe the basic ideas of statistics that are relevant to errors of observations and numbers derived from these observations.

A measurement is a quantification of an attribute of the material under investigation, directed to the answering of a specific question in an experiment. The quantification implies a sequence of operations or steps that yields the resultant measurement. Thus, the concept of measurement may be said to include not only the steps used to obtain the measurement but also use of the measurement by the experimenter to draw conclusions.

The measurement process provides a result which serves as part of the basis upon which an experimenter makes a judgment about the attribute under investigation. Some judgments may require a more reliable basis than others. Desired reliability in the measurements will depend on the purpose for which the measurements are to be used, but the degree of reliability may be limited by resources available to the experimenter. The experimenter may control reliability by choosing a measurement process making use of a number of relevant scientific principles, by con-

[1]Contribution from the Department of Statistics, Iowa State University, Ames, IA, and the Agricultural Research Service, USDA, St. Paul, MN. Journal Paper no. J-11311 of the Iowa Agricultural & Home Economics Experiment Station, Ames, IA. Project no. 890.

trolling attributes of the environment in which measurement is made, and by repeating the measurements. Different combinations of these control alternatives may be suitable for a given measurement process; the suitability of a particular combination of control alternatives varies among measurement processes, and depends on the reliability desired.

The number, which the experimenter uses to judge about the substance under investigation, may not be a single measurement but may be a derived number, that is, some function of several measurements utilizing the same or different measurement processes. The best function will be dictated by the scientific nature of the investigation as well as by the theory of combination of observations and related statistical concepts.

1–2 CLASSIFICATION OF MEASUREMENT ERRORS

From the standpoint of errors of measurement, the simplest situation is that of obtaining an attribute of an object which is not affected by the measuring process, so that the measuring process can be applied again and again to the unchanged object, as, for example, determining the length of a bar of steel. Repeated application of the measuring process to this unchanged object, following the directions laid down in the specifications, will yield a sequence of nonidentical numbers. Variation in the numbers arises because no sequence of operations or "state of nature" is perfectly reproducible. In other words, no human or machine can do exactly the same sequence of operations again and again, and the identical circumstances and object of measurement cannot be achieved perfectly. If the measurement is coarse, the lack of reproducibility of the measuring process may have no effect, as for instance would be the case with most adults measuring the length of a 14.6 cm rod to the nearest centimeter with a ruler graduated in centimeters. In general, however, the lack of reproducibility of operations will produce variability. For instance the instruction to bring the pH of a solution to 6 by adding 1.0 *M* HCl can be performed only to a certain degree of correctness, depending on the indicator and the operator. Supposedly simple operations like weighing a precipitate will not lead to the same answer on repetition with a sufficiently sensitive balance. Of course, much of the training in elementary analytical sciences is directed to the performance of operations in a manner sufficiently exact to achieve negligible variability in results and conclusions, but this ideal can rarely be achieved. The variability among results of a measuring process applied to a constant object may be called measurement error or, for emphasis, pure measurement error.

In contrast to the above situation, we may imagine that the object of measurement, which may be a batch of material, is heterogeneous and can be measured only in parts which are unlike, but that the measurement process is perfectly reproducible for each part. If we could apply the measuring process to all parts of the whole under these ideal circumstances, there would be no error in the final result, but the circumstances under which one can process the whole are very rare. To give an obvious example, suppose we wish to characterize the potassium status of plots of land in connection with a study of potassium-fertilizer uptake. We cannot process all the soil of the plot for obvious reasons, of which perhaps the most compelling reason is that if we did so we would not then have the plot of land to experiment on. We, therefore, must determine the status of the land by drawing samples and applying a measurement process to these samples. The samples will not be equal with regard to the attributes under examination, and the variation among the sample results will lead to uncertainty in the final result. Variation of this type, in which it is supposed that the measuring process itself is exactly reproducible, is called pure sampling error.

We can, of course, take the view that what we called pure measurement error arises from sampling a population of repetitions of the measurement process.

In practice few measurement situations lie at either of the two extremes we have mentioned. In the majority of cases the measurement process destroys the object being measured, so that repetition of the measuring process on the object of measurement cannot be performed. A common way out of this difficulty in soil analysis is to homogenize the sample by passing it through a fine sieve and mixing it thoroughly. The homogenized sample is then subdivided, and one performs repeated measurements on subsamples, which one has good reason to regard as identical. Where analyses are made in this way, the total error of an observation on a subsample will include both measurement error and sampling error, the latter because the total sample from which the subsamples are derived is only a small part of the whole for which information is desired. In soil analytical practice, of course, there is always a lower limit to the subsample size that can be taken "without sampling error" from even a sample that has been finely ground and mixed. This problem is of considerable importance in some kinds of soil analysis.

The measurement process consists of a sequence of operations, and at each step in the sequence there will be a certain lack of perfect repeatability of the operation. The measurement error may, therefore, have a structure, in the sense that part arises from step 1 of the sequence of operations, part from step 2, and so on. Similarly the total operation of sampling may in each particular situation be broken down into distinct steps, perhaps according to the sampling design, and each step will introduce a part of the sampling error.

The total error is that arising from measurement and sampling. In practice we must attempt to control this total error, so that conclusions based on results will not differ in any important respect from conclusions that would be made in the absence of error, or so that the effect of the error on subsequent conclusions can be assessed.

In general, error in the sense discussed above is *not* the result of incorrect procedure; also, it is to be distinguished from mistakes in following the directions, which will lead to large deviations of single measurements from other measurements made with the same measurement process. The detection of gross mistakes is discussed in chapter 4.

1-3 SCIENTIFIC VALIDITY OF MEASUREMENTS

In general, the measurement process is aimed at the characterization of an object of measurement in terms of scientific concepts. We might, for instance, wish to measure the total content of combined nitrogen in a soil that contains appreciable amounts of NO_3^-. We can imagine a highly reproducible Kjeldahl analysis wherein no additives were used to assure conversion of the NO_3^- to NH_4^+ in the digestion. The measurement would not give the desired answer, because the total content of combined nitrogen is the answer sought, while the measurement process does not measure all the nitrogen present as NO_3^-. The occurrence of such a defect in the measurement process can be found only on the basis of scientific principles or by special tests of validity.

Underlying any measurement process applied to an object of measurement is what one may term the *scientific true value* which can be visualized as being approached more and more closely by refinement of measurement operations and by appropriate changes in the process in accordance with scientific principles. For example, in the instance just considered, the scientific true value would be approached more closely if the process were changed to include NO_3^-.

Scientific relevance of the measurement process is not a matter of the theory of "error," which involves the "wandering" of the results. Clearly, however, the theory of "error" enters into what we may call the validation of a measurement process, because such validation will require the comparison of results obtained in different ways, each way having its own peculiar error characteristics. It is because the scientific validation of a measurement process involves such comparisons that the experimenter must have some awareness of the concept of "error" and of procedures for drawing conclusions in the presence of "error."

It is, one supposes, obvious that examination of the results given by a single measurement process cannot per se lead to scientific validation of the measurement process, but it is equally obvious that validation involves consideration of the variability exhibited by the results of applying the measurement process. We, therefore, take up the problem of characterizing the numbers one will obtain by repeating the measurement process.

1-4 CHARACTERIZATION OF VARIABILITY

We suppose that we have at hand a large sample of soil which can be subdivided into a very large number of smaller samples, to each of which we can apply the measurement process. We suppose that a large number of the smaller samples have been analyzed. Under these circumstances the variability among analytical values may arise from pure measurement error, sampling (of the large sample) error, or both. We can then construct a frequency distribution or histogram of the resulting numbers, which could have the appearance of Fig. 1–1, in which the area of the block between a and b is the relative frequency of numbers between a and b, and the units of area are chosen so that the total area of all the blocks is unity. This histogram is an empirical or observed distribution.

We have supposed that we could have obtained an indefinitely large number of results, and we can imagine making successive histograms in which the intervals become progressively narrower until we have essentially a continuous curve rather than a curve which proceeds by steps. The distribution which we would so obtain may be called the true distribution underlying the whole measurement operation, and it may take a particular mathematical form, of which examples will be given later.

The mean of this essentially infinite population, which will be hypothetical in all cases (that is, the arithmetic average of an indefinitely large number of results), may be called the *statistical true value* or *limiting mean* associated with the measurement process and the object of measurement. It may also be called the *operational true value,* in the sense that it is a mean value associated with operations of the measurement process.

This statistical true value may be different from the *scientific true value* which is being sought, for one or more of the following reasons:

1. The specifications of the measurement process may be inadequate. The deficiency may be one of selectivity, in which the process does not permit inclusion of all that is desired; or it may include more than is intended. Alternatively, the specifications may not provide for the problem encountered with some measurements, in that the quantitative expression of the property being measured is not entirely independent of the nature of the material on which the analysis is performed. Inadequate specification of a measurement process may be characterized as a scientific deficiency of the process, and the resulting

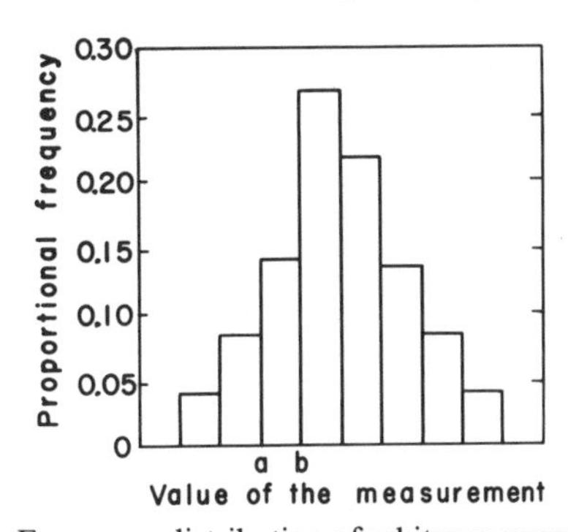

Fig. 1–1. Frequency distribution of arbitrary measurements.

deviation of the statistical true value from the scientific true value may be termed *scientific bias.*

2. The auxiliary apparatus or materials used in making the measurement may be faulty, so that the results tend to be too high or too low. Differences between the scientific and statistical true values arising in this way may be called *measurement bias.*
3. The process for obtaining the samples to be measured may result in selection of samples that are not representative of the whole. A difference between the scientific and statistical true values arising in this way may be called *sampling bias.*

These are very generalized statements of what may go wrong with a measurement process. Another source of discrepancy arises from the human element. We can imagine two different operators analyzing comparable samples of the same soil by the same method but obtaining different distributions because of differences in performance such as filling pipettes to consistently different levels. The purpose of training is, of course, to attempt to eliminate this type of personal effect as much as possible, but one cannot assume that the training has been effective or has not been forgotten.

The spread or dispersion of the distribution of Fig. 1–1 must now be considered. One can imagine two measurement processes with distributions (*1*) and (*2*) in Fig. 1–2, in which the means of the two distributions are the same, but distribution (*2*) is clearly more spread out than distribution (*1*). If the sampling procedure in the two processes is the same, so that for instance the two distributions were estimated by first drawing 2000 samples and then partitioning these into two sets of 1000 samples, one set for each measurement process, then process (*1*) may be said to give more *precise* measurements than process (*2*).

Precision is inversely related to the variability among results obtained by applying a measurement process again and again to an object of measurement or to samples from a population that is the object of measurement. This variability can be represented by the construction of a histogram or frequency distribution of the results, and we can imagine a mathematically defined form for the distribution that would result from

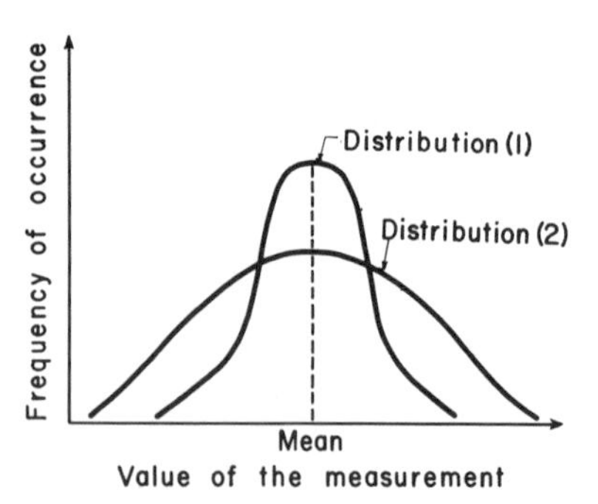

Fig. 1–2. Frequency distributions expected in two different arbitrary situations of (a) reproducibility of the measurement process or (b) sample homogeneity.

an infinite number of repetitions. Usually one is satisfied with a mathematically defined distribution that is developed from a model of the measurement process.

The most common distribution in measurement work is the "normal" distribution, and most of our ideas on precision and modes of handling imprecision are based on the normal distribution. Let us suppose that the total error of an observation, say e, is made up additively of components, say, e_1 arising from taking a prescribed aliquot imperfectly, e_2 arising from not adding the prescribed amount of one reagent, and so on. Then

$$e = e_1 + e_2 + \ldots + e_n .$$

It is a mathematical fact that if these component errors are not associated, e.g., an extreme value of e_1 does not induce an extreme value for e_2, and if the number of constituent errors is large, the total error will follow a distribution close to the normal distribution. If $f(x)dx$ denotes the relative frequency that e has in the interval x to $x + dx$, then

$$f(x) = [1/\sigma(2\pi)^{1/2}] \exp(-x^2/2\sigma^2) . \qquad [1]$$

Classical writers liked to write this in another form, obtained by writing h in place of $1/\sigma$, so that

$$f(x) = [h/(2\pi)^{1/2}] \exp(-h^2x^2/2) . \qquad [2]$$

This gives the familiar bell-shaped curve of error which is characterized by one parameter σ, which is called the standard deviation, or by h. We give in Fig. 1–3 the two curves for:

(a) $\sigma = 1$, or $h = 1$, and

(b) $\sigma = 1/2$, or $h = 2$.

A numerical quantity for which the relative frequency of possible values

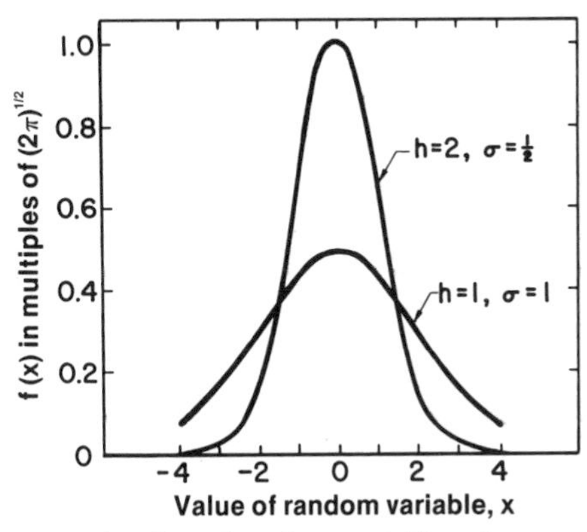

Fig. 1–3. Frequency distribution of random variable x for two values of h and σ.

is specified is called a random variable (actually, a real-valued random variable). It is a purely mathematical exercise to determine the frequency with which errors lie in particular ranges, and the most useful symmetrical ones are given in Table 1–1. To facilitate presentation, the situation is considered where $\sigma = 1$ and Eq. [1] is integrated between specified limits corresponding to particular segments of the random variable. An empirical distribution which is virtually "normal" in appearance is presented in Fig. 1–4. In practice one may not get a normal distribution because sets of observations have common errors, and for other reasons.

The fact that the normal distribution of error is determined by one parameter, σ or h, makes the definition of a measure of precision for such errors easy to specify, and the idea that less variability should be associated with higher precision led classical writers to use the quantity h as a measure of precision. If then we have normally distributed errors, the matter of quantifying precision is simple. It is more usual nowadays to use σ, which is called the standard deviation, so that the lower the value of σ the higher the precision.

The concept of *probable error* was frequently used in the past. Its meaning is given in Table 1–1. Instead of considering one standard deviation, only 0.674 σ is considered. The frequency of occurrence of errors within the range $\pm 0.674\ \sigma$ is equal to the frequency of occurrence of errors outside this range, so that the confined proportional frequency is 0.50.

Table 1–1. Frequencies of errors in particular segments for a normal distribution with zero mean and unit standard deviation.

Segment of the random variable	Confined proportion of the total frequency	Usual name of segment denoted
−0.674 to 0.674	0.50	Probable error
−1.0 to 1.0	0.67	Standard deviation
−1.96 to 1.96	0.95	"95% confidence region"
−2.58 to 2.58	0.99	"99% confidence region"

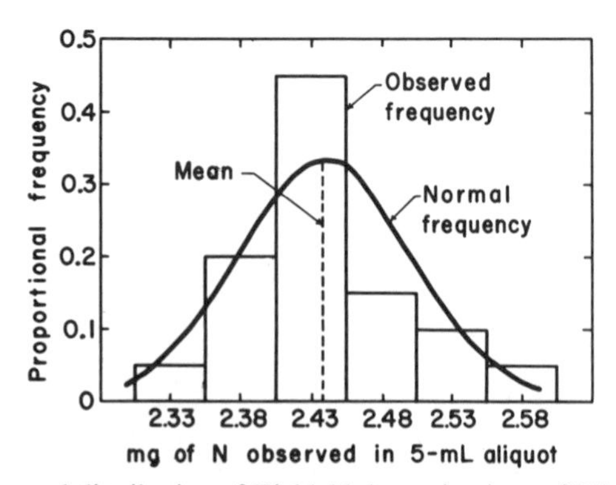

Fig. 1–4. Histogram and distribution of Kjeldahl determinations of *N* in 20 aliquots taken from a homogeneous $(NH_4)_2SO_4$ solution. (Personal communication, D. R. Timmons, Ames, Iowa.)

Frequently the experimenter may observe that a plot of the frequency versus the magnitude of the random variable reveals a positive skew such as in the log-normal distribution of pore-water velocity in Fig. 1–5(A). This type of distribution can be detected by methods such as illustrated in Table 1–2 and Fig. 1–5(B). After the observations are listed in increasing order, their cumulative probability is plotted versus the obser-

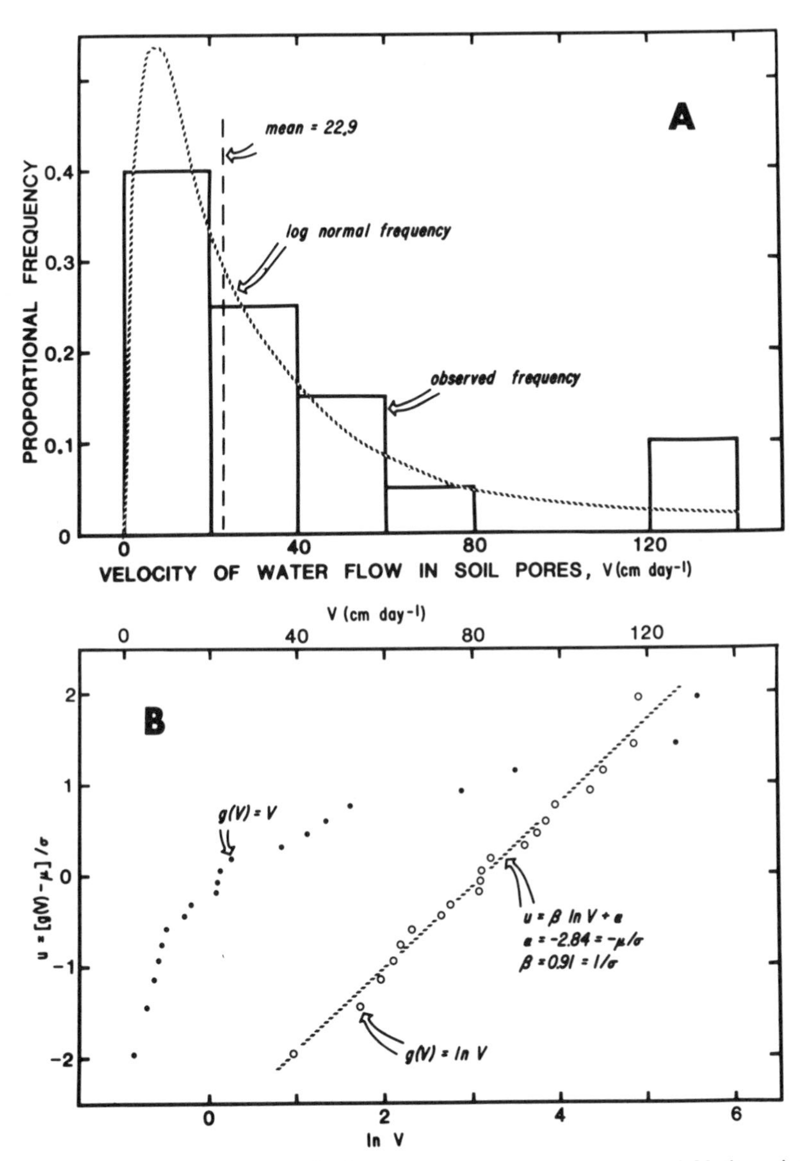

Fig. 1–5. (A) Histogram and distribution of pore-water velocity measured 20 times in a Panoche silt loam, and (B) graphical determination of transformation to achieve normality. (Data from Biggar and Nielsen, 1976.)

Table 1-2. Twenty observations of pore-water velocity, V in cm day^{-1}, and corresponding statistical parameters (data from Biggar and Nielsen, 1976).

i	V_i†	$(\ln V)_i$†	$(i - 0.5)/n$‡	u_i
1	2.6	0.96	0.025	−1.96
2	5.6	1.72	0.075	−1.44
3	7.0	1.95	0.125	−1.15
4	8.0	2.08	0.175	−0.93
5	8.7	2.16	0.225	−0.76
6	9.8	2.28	0.275	−0.60
7	14.0	2.64	0.325	−0.45
8	15.5	2.74	0.375	−0.32
9	21.4	3.06	0.425	−0.19
10	21.6	3.07	0.475	−0.06
11	22.1	3.10	0.525	0.06
12	24.8	3.21	0.575	0.19
13	36.3	3.59	0.625	0.32
14	42.2	3.74	0.675	0.45
15	46.5	3.84	0.725	0.60
16	52.0	3.95	0.775	0.76
17	77.4	4.35	0.825	0.93
18	89.7	4.50	0.875	1.15
19	126.6	4.84	0.925	1.44
20	131.4	4.88	0.975	1.96

† When $X = g(V)$, then a linear fit of u versus $g(V)$ can be used to determine what form of $g(V)$ is normally distributed. These constants may then be used to reconstruct the theoretical relation.

‡ $(i - 0.5)/n$ approximates the area under the cumulative probability function $P\{u\}$, and $u = (x - \mu)/\sigma$ is the corresponding upper limit in:

$$P\{u\} = 1/(2\pi)^{1/2} \int_{-\infty}^{u} \exp(-X^2/2)\, dX$$

vation (V) or a function of the observation [$g(V) = \ln V$], both as in Fig. 1-5(B). A normal distribution of the observation or of a function of the observation produces a linear plot. The parameters computed in Fig. 1-5(B) are the same as those computed in the usual manner in Table 1-2; moreover, the observed goodness of linear fit in Fig. 1-5(B) gives confidence in use of the distribution parameters. An approximation for the density of the log-normal frequency is

$$f(x) = [\Delta x/x]\,[1/\sigma(2\pi)^{1/2}]\exp\{-(\ln x - \mu)^2/2\sigma^2\} \qquad [3]$$

where μ and σ are computed using x as the variate, and Δx is the class interval selected, such as 10 cm day^{-1}, in Fig. 1-5(A).

Many soil properties are log-normally distributed. Nielsen et al. (1973) found that saturated hydraulic conductivity, pore-water velocity, and a scaling coefficient that relates horizontal heterogeneity of soil water flow properties all were log-normally distributed. Meanwhile, soil water content and dry bulk density were normally distributed. The apparent diffusion coefficient for Cl^- is also log-normally distributed (Van de Pol et al., 1977). The log-normal distribution is often observed for size (diameter) distributions of aggregates and primary particles (Gardner, 1956), dispersed clay size (Austin, 1939), exchangeable calcium and strontium

(Menzel and Heald, 1959), and heights of constant length pins impinging on a newly tilled soil surface (Allmaras et al., 1966). All of these distributions, whether normal or log-normal, may be determined by using the methods illustrated in Table 1–2 and Fig. 1–5(B).

Analogous to the situation in the normal case where the magnitude of the random variables is the additive effect of a large number of small independent causes, the arithmetic magnitude of the log-normally distributed random variable is the multiplicative effect of a large number of independent causes. This analogy is discussed by Aitchison and Brown (1957) and Gaddum (1945). As an example, we may consider the diameter of an aggregate of soil particles in relation to two causes that may bring about a change in diameter. The proportionate-effect hypothesis predicts that the diameter change, resulting from the action of a cause, is some proportion of the initial diameter. Let the initial diameter be d_1, the diameter after the first cause acts be d_2, and the final diameter after the second cause acts be d_3. The effect resulting from the first cause is $(d_2 - d_1)/d_1 = e_1$ and that after the second cause acts is $(d_3 - d_2)/d_2 = e_2$. Hence

$$d_3 = d_1(1 + e_1)(1 + e_2)$$

or

$$\ln d_3 = \ln d_1 + \ln (1 + e_1) + \ln (1 + e_2)$$

or

$$\ln d_3 = \ln d_1 + f_1 + f_2$$

where f_1 is $\ln (1 + e_1)$ and f_2 is $\ln (1 + e_2)$. With a large number of independent causes, $\ln d$ will have an error which is the sum of a large number of independent errors and will tend to have a normal distribution.

The Poisson distribution has a position of some importance, especially in count or enumeration data. This distribution arises where discrete events occur at random over a long time or a large area, and where the random variable is the frequency of occurrence of these events in any small time interval or small area chosen at random. The total number of possible events should be large (infinite, theoretically), but the probability of occurrence of any individual event in the time interval or area considered should be small. The underlying random variable can then have the integral values 0, 1, 2, . . . , and the probability that x will be observed is

$$f(x) = e^{-m}(m^x/x!); \quad 0 \leq x < \infty \qquad [4]$$

where m is the limiting mean or statistical true value.

The Poisson distribution arises, for example, in counts of disintegrations of radioactive elements. Experimentally, the Poisson distribution of radioactive counts per unit time may be verified by repeatedly counting for short periods of time a nuclide having a long half-life. An example is the number of alpha particles emitted per unit time from a polonium source, shown in Table 1–3. The observed frequencies and the fitted

Table 1–3. Number of α-particle emissions per unit of time from polonium observed by scintillation counting, and the χ^2 goodness of fit to the Poisson distribution (data from Rutherford and Geiger, 1910).

Number of α-particles observed unit time x	Number of times observed in 2 608 trials f_0	$f_0 x$	Expected number of times based on Poisson distribution† f_ϵ	$\frac{(f_0 - f_\epsilon)^2}{f_\epsilon}$
0	57	0	54	0.17
1	203	203	210	0.23
2	383	766	407	1.42
3	525	1 575	525	0.00
4	532	2 128	508	1.13
5	408	2 040	394	0.50
6	273	1 638	254	1.42
7	139	973	141	0.03
8	45	360	68	7.78
9	27	243	29	0.14
10	10	100	11	0.09
11	4	44	4	0.00
12	0)	0	2)	0.00
13	1)	13	)	
14	1)	14	)	
Sum	2 608	10 097	χ^2 (11 df) = 12.91	

† Estimates from assumed Poisson distribution:

$m = 10\ 097/2608 = 3.87$

$f(0) = 2608\{[(3.87)^0/0!]\exp(-3.87)\} = 54$

$f(1) = 3608\{[(3.87)^1/1!]\exp(-3.87)\} = 210$

⋮

$f(14) = 2608\{[(3.87)^{14}/14!]\exp(-3.87)\} = 0.11.$

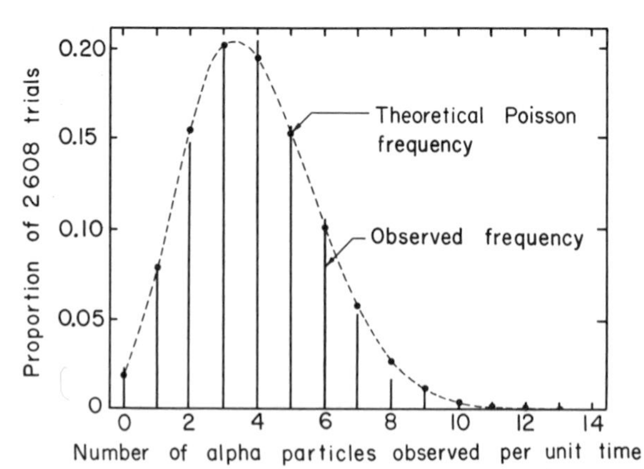

Fig. 1–6. Observed frequency distribution of alpha-particle emission per unit time from polonium, and the theoretical Poisson distribution. (Data from Rutherford and Geiger, 1910.)

Poisson distribution are shown in Fig. 1–6. Under the null hypothesis, which states that there is no difference between paired f_0 and f_ϵ in Table 1–3, the quantity $\Sigma\ [(f_0-f_\epsilon)^2/f_\epsilon]$ belongs to the χ^2 distribution with 11 degrees of freedom. Because the probability $(\chi^2 \geq 12.9) = 0.30$, there is no evidence that the distribution was not properly identified. The distribution could also have been evaluated using the normal approximation that $\mu = (x-m)/m^{1/2}$ and the methods in Table 1–2 and Fig. 1–5(B).

In passing, it should be noted that there are instances where frequencies of radioactive counts do not assume the Poisson form. Counts of nuclides having short half-lives tend to show distribution forms more closely approximating the normal in general shape. A tendency for the distribution to shift toward the normal form is caused also by the internal and external conditions playing upon the measurement process. For example, the scalers commonly used in measuring radioactivity consist of many electronic components, each of which conceivably could add some directional effect to the count. The observed count thus contains not only the factors determined by sampling from the Poisson population, but also a number of other factors. As the latter become determinant relative to the former, the distribution should show increasing normality of form. Other situations to which the Poisson distribution applies include the number of colonies of a fungal pathogen on an agar plate from a dilution of soil (Nash and Snyder, 1962). The Poisson frequency is expected in this instance because a volume of soil contains many organisms, only a small proportion of which are the test species. Statistical treatment of data that follow a Poisson distribution is discussed in most texts on statistical methods. A common procedure based on section 1–6.1, Case 1 is to transform the counts y to $y^{1/2}$, because the variance of a Poisson variable is equal to the mean value.

Of other mathematical distributions which will occur occasionally in measurement work, we mention only the binomial distribution and the multinomial distribution. The binomial distribution relates to outcomes that are dichotomous, such as success or failure, or presence or absence. With n trials the probability of r "successes" is

$$n!/[r!(n - r)!]p^r(1 - p)^{n-r}$$

where p is the probability of "success" on any one trial. This distribution is interesting per se, and also in that the more important distributions in measurement work, the normal and the Poisson, arise as limiting forms of the binomial distribution. If the number of trials n increases with p remaining constant, the quantity $(r - np)/[np(1 - p)]^{1/2}$ tends to have a normal distribution with zero mean and unit standard deviation. If the number of trials n increases and p decreases, so that np remains constant, the number of "successes" tends to have a Poisson distribution with mean np. The multinomial distribution arises when the outcome is multichotomous, such as would arise when judgments are made about the semiquantitative or qualitative occurrence of some attribute. A reaction might be categorized as "absent," "weak," or "strong," to give a multinomial with three classes. The binomial distribution is used in mineralogical analysis; the chance that a given mineral grain in a microscopic field belongs to a given mineral species is dichotomous, and such a dichotomous outcome applies to each of the n different mineral species in the thin section (Brewer, 1964).

In general, measures of precision and procedures for handling imprecision of data are based on an assumed mathematical distribution. There are, however, basic measures and concepts which can be applied to any distribution.

The variance of any distribution that can occur in a measurement situation is defined as the average of squared deviations of results from the mean of an essentially infinite number of measurements by a given process on a given object (known as the statistical true value). The variance is generally denoted by μ_2, or by σ^2. The square root of the variance is the standard deviation. If the distribution is normal it is specified completely by the statistical true value and the standard deviation, as we have seen; but of course, the distribution may not always be normal. The standard deviation σ is useful for any distribution because of a fundamental concept known as Tchebychef's Theorem, which states that the relative frequency with which an error exceeds $t\sigma$ in absolute magnitude is less than or equal to $1/t^2$ for any distribution (t is an arbitrary constant). In symbols, using probability in place of relative frequency, as is usually done.

$$\text{Probability } (|\epsilon| > t\sigma) \leq 1/t^2$$

where $|\epsilon|$ is the absolute magnitude of the error. For example, the probability that an error greater than 2σ will arise is less than 1/4 for any distribution of error. Note that if the distribution of error were normal, this statement could be replaced by the much stronger one

$$\text{Probability } (|\epsilon| > 2\sigma) = 1/20 \text{ (approximately)}.$$

Thus we pay for ignorance of the nature of the distribution by being able only to make a weaker probability statement.

There are several other measures of variability. The interquartile range is the interval such that 25% of the observations are above the interval and 25% are below the interval. Other percentile ranges can be envisaged. In a normal distribution these ranges are simply related to the standard deviation, as shown in Table 1–1. The variance measures the second moment of inertia around the mean. Other moments may also be considered, such as the third moment, which is the average value of the cube of deviations of observations from the mean. Usually the second moment will be adequate to characterize the variability.

It is very useful to transform observations so that the transforms are approximately normally distributed. This general approach has great utility for a number of reasons: (i) the normal distribution of error occurs frequently in an ordinary measurement process, (ii) the normal distribution has a mathematical role for expressing the distribution of a total error made up from a large number of small independent errors, as will often occur in a measurement process, and (iii) the normal distribution of errors allows the use of a wide variety of numerical processes for analyzing variability and specifying limits of error of an observation. The analysis of variance and associated statistical tests are an analysis of variability. Thus the use of transforms often enables the stronger normal law probability to be used, as elaborated upon in section 1–6.1.

1–5 THE ESTIMATION OF PRECISION

In practice we do not have an infinite number of samples, so that we must estimate the unknown true properties of the distributions we meet. We can give only introductory ideas with regard to the general problem, which forms a substantial part of the body of statistical theory.

We shall assume that the distributions encountered are such that they may reasonably be characterized by a mean and a variance, and that they are approximately normal in form, so that mathematically developed procedures for normal distributions can be used.

To estimate the variance, a standard procedure with n results x_1, $x_2, \ldots, x_n$ is to calculate

$$\Sigma(x_i - \bar{x})^2/(n - 1)$$

where $\bar{x}$ is the arithmetic average. The result is taken as an estimate of μ_2 or σ^2, and is commonly denoted by s^2. It is an unbiased estimate, so that if one made a large number of estimates from independent samples, the arithmetic average of the s^2 values obtained would approach the true variance. One can take s^2 to be the estimated variance of a single observation. The positive square root of s^2, which is usually denoted by s, is called the estimated standard deviation of a single observation, or the "standard error" of a single observation. It is a measure of the precision of a single observation. This measure of precision is easily calculated, and gives a measure of precision for the observed arithmetic mean (which is the common "estimate" of the statistical true value) by forming $s/(n)^{1/2}$. This quantity is the estimated standard deviation of the observed mean or the standard error of the observed mean. One can say forthrightly that any proposed method of analysis should be accompanied by an estimated standard deviation or standard error of a single result, with a statement of the number of observations on which this estimate is based. More commonly, rather than giving the sample size, it is appropriate to give the *degrees of freedom* on which the standard error is based, which for a simple random sample of size n is equal to $n - 1$.

Presentation of results from any use of the measurement process must contain the number of observations, the mean, and the standard error of the mean. This guideline will allow the scientist or reader to obtain intervals of uncertainty for the unknown or true value.

A particularly simple estimation of variance arises when only two observations are obtained, because the formula given above reduces to one-half the squared difference of the two observations. This measure of variability has only one degree of freedom and is the absolute minimum which can be regarded as scientific in nature.

Rather than use the formula above for variance, one can estimate the standard deviation directly from the range, i.e., the difference between the smallest and largest observation. To convert the range to an estimate of standard deviation, one must divide the range by a number which depends on the sample size. This number is given in many statistical texts, e.g., Snedecor and Cochran (1980, Table 2.4.1). For sample sizes of 2, 3, 4, 5, and 6, the divisors are 1.13, 1.69, 2.06, 2.33, and 2.53, respectively. This estimate of standard deviation is not as good as the earlier estimate given if the distribution is normal, but is probably good enough for most routine assessments of precision.

A rather commonly used measure of precision is the coefficient of variation (CV). For a population this is defined as σ/μ, where σ is the true standard deviation and μ is the statistical true value. This is estimated with a sample by $s/\bar{x}$, where s is the estimate of standard deviation. It measures, in a rough way, the relative variability of observations. For example, if one adopted as rough limits of uncertainty for a particular observation

$$x - 2s \text{ to } x + 2s$$

then the relative spread is approximately

$$1 - (2s/x) \text{ to } 1 + (2s/x)$$

which is approximated by the interval

$$1 - 2\text{CV} \text{ to } 1 + 2\text{CV} .$$

An approximate interval of uncertainty of a sample mean based on n observations can then be taken as

$$\bar{x} - [2s/(n)^{1/2}] \text{ to } \bar{x} + [2s/(n)^{1/2}]$$

so that the relative spread of uncertainty is

$$1 - 2\text{CV}/(n)^{1/2} \text{ to } 1 + 2\text{CV}/(n)^{1/2} .$$

The coefficient of variation may therefore be used to judge what sample size is desirable to have reasonable relative uncertainty on the sample mean.

We have used the term *approximate interval of uncertainty* without specifying what is meant. This term and the computational formula are based on the idea of *confidence intervals,* as related to normal distributions. If the observations are from a normal distribution with statistical true value μ, then the quantity

$$(\bar{x} - \mu)/[s/(n)^{1/2}]$$

is a random variable with a distribution that depends only on sample size, or rather on the degrees of freedom on which s is based, which in the case of a simple random sample is one less than the sample size. This quantity follows Student's t distribution. From this distributional fact, one can assert that if one determines from a table of the t distribution (which is given in essentially every book on statistical methods) the quantity $t_{\alpha,k}$ such that

$$\text{Probability } (|t| \leq t_{\alpha,k}) = 1 - \alpha$$

where the subscript k indicates that degrees of freedom, then one can assert that the interval

$$\bar{x} - t_{\alpha,n-1}[s/(n)^{1/2}] \text{ to } \bar{x} + t_{\alpha,n-1}[s/(n)^{1/2}]$$

contains the unknown statistical true value with the probability $1 - \alpha$ that the statement is true. Conventionally $1 - \alpha$ is taken to be 0.95 or 0.99, and the resulting interval is called the 95% confidence interval or the 99% confidence interval on the unknown statistical true value.

In writing an approximate interval of uncertainty as

$$\bar{x} - 2s/(k)^{1/2} \text{ to } \bar{x} + 2s/(k)^{1/2}$$

we are using a t value of 2, which corresponds to a probability of about 0.95 if the sample size is 40 or more. If the sample size is as low as 5 the associated probability is about 0.90; and if the sample size is only 2, so that s is based on only one degree of freedom, the associated probability is 0.50.

1-6 PRECISION OF DERIVED OBSERVATIONS

A derived number or observation is considered to be some function of a single attribute of the substance investigated by the measurement process, or a function of single attributes, each observed from a different measurement process. The objective of this section is to consider the precision of these derived numbers.

This topic is treated by Topping (1957) and Beers (1957) by essentially the same method. Another excellent but somewhat different treatment of errors is given by Schenck (1979). The advantage of an approximate method for estimating precision of derived numbers is that it is not necessary to assume that the attribute belongs to any particular parent distribution, i.e., normally distributed or otherwise. Its disadvantage is that, in the absence of an assumed parent distribution and by virtue of the assumptions, the method is an approximation. Exposition of this method of estimating variance of derived numbers will help to extend the experimenter's objective evaluation of data throughout the continuum from measurement to reporting of results. Two cases of the general problem are treated.

1–6.1 Case 1: A Single-Valued Function of an Observation

Suppose we have an observation x, with mean μ and variance σ^2, and we wish to assess the precision of a single-valued function of x, say, $f(x)$. Examples are $f(x) = \ln x, f(x) = x^2, f(x) = \sin x$. The approximate procedure is to note that

$$\begin{aligned} f(x) &= f(\mu + x - \mu) \\ &= f(\mu + \delta), \text{ where } \delta = x - \mu\,, \end{aligned}$$

and to suppose that this function can be expanded by a Taylor series:

$$f(\mu + \delta) = f(\mu) + \delta f'(\mu) + [\delta^2/(2!)][f''(\mu)] + \text{remainder}\,.$$

Now suppose that the remainder can be ignored, and consider the average value in repetitions, or the "expectation," denoted by E, of $f(x)$. We have

$$E[f(x)] = f(\mu) + [f''(\mu)/2]\sigma^2$$

since

$$E(\delta) = 0, \text{ and } E(\delta^2) = \sigma^2\,.$$

Hence we note that, approximately,

$$f(x) - [f''(x)/2]\sigma^2$$

is an unbiased estimate of $f(\mu)$. Here, of course, σ^2 must be replaced by an estimate. A first approximation to $f(\mu)$ would be to take $f(x)$.

Turning now to precision, we note that if $f''(\mu)$ is small,

$$\text{Variance } [f(x)] = \text{Variance } [f(\mu) + \delta f'(\mu)]$$
$$= [f'(\mu)]^2\sigma^2$$

because variance $= E\{f(x) - E[f(x)]\}^2$ and variance $(cx) = c^2\sigma^2$. In practice one would use the approximation

$$\text{Variance } [f(x)]^2 s^2$$

where s^2 is an estimate of σ^2. When $f(\bar{x})$ is desired and $\bar{x}$ is the mean of n observations, the quantities s^2 and σ^2 are replaced by s^2/n and σ^2/n, respectively. Simple examples are as follows:

$f(x)$	Estimated variance of $f(x)$
$\ln x$	s^2/x^2
$1/x$	s^2/x^4
$\sin x$	$(\cos^2 x)s^2$
x^p	$p^2 x^{2(p-1)} s^2$

In the case when a function of $\bar{x}$, the mean of n observations, is desired, the quantity s^2 in the above formulas is replaced by s^2/n.

The importance of adjustment of the variance of the mean of the measurements when a derived number is calculated may be appreciated by considering several examples. The first example concerns the precision of the estimate of the area A of a circle (a derived number) when the diameter d is measured. In this example $f(\bar{d}) = \pi(\bar{d})^2/4$. The area, $\bar{A} = f(\bar{d})$, is approximated as $\pi(\Sigma_{i=1}^n d_i/n)^2/4$, where the d_i are the individual measurements of diameter, and there are n of these measurements. The variance $\sigma_{\bar{A}}^2$ of the estimate of area of the circle is approximated by

$$\sigma_{\bar{A}}^2 = [\pi(\Sigma_1^n d_i)/2n]^2 \, \sigma_{\bar{d}}^2$$

because $f'(\bar{d}) = (\pi\bar{d}/2)$. The disperson of the mean area is, therefore, inflated with respect to the disperson of the diameter. Furthermore, with a lower specified limit of precision on the area, the required lower limit of precision of the mean diameter can be calculated. In practice this application is important because the variance of the mean diameter can be controlled not only by the measurement process itself but also by the number of diameter measurements.

A second example concerns the situation where x is measured, and it is desired to estimate $f(\bar{x}) = \ln(\bar{x})$ and the precision of $f(\bar{x})$. The quantity $f(\bar{x})$ would be estimated approximately by $\ln(\Sigma_{i=1}^n x_i/n)$, and its variance is observed to be $(n/\Sigma_{i=1}^n x_i)^2\sigma_{\bar{x}}^2$. This example is mentioned because of the possibility of a misleading implication about the precision involved in measuring the arithmetic value when the variance of $\ln x$ or $\ln(\bar{x})$ is reported.

A very simple example is the expression of a derived number in different units of measurement than the unit of measurement for the measurements themselves. Thus $f(\bar{x}) = k\bar{x}$, where k is a constant related to the change in units of measurement. The estimate of $f(\bar{x})$ is then $k(\Sigma_1^n x_i/n)$ and the variance of $f(\bar{x})$ is $k^2\sigma_{\bar{x}}^2$ which is an exact result because there is no error in the Taylor expansion. Incidentally, this result agrees with the result obtained by first changing the measurements to the unit of measurement for the derived number and then considering the variance of the derived number. The derived number would be kx_i. Its mean would be $k\ \Sigma_{i=1}^n x_i/n$ and its variance,

$$\text{Variance } (k\ \Sigma_1^n x_i/n) \qquad \text{is } k^2 n\sigma_x^2/n^2 = k^2\sigma_{\bar{x}}^2 .$$

1–6.2 Case 2: A Number Derived from Measurements of More Than One Attribute on the Same Sample

This case is more complex than Case 1, but utilizes some of its properties. We illustrate the situation with the case of a number based on two attributes, x and y. Consider the function $f(x,y)$, and suppose it is a "smooth" function. Then it can be expanded around μ_x,μ_y, the limiting means, in a Taylor series expansion

$$f(x,y) = f(\mu_x,\mu_y) + (x - \mu_x)(\partial f/\partial\mu_x) + (y - \mu_y)(\partial f/\partial\mu_y) + \text{remainder} .$$

If the remainder is small, the expectation of $f(x,y)$ is nearly equal to $f(\mu_x,\mu_y)$, so that $f(x,y)$ is nearly unbiased. We could obtain a more nearly unbiased estimate by taking account of second derivatives as in the case of a single attribute. The variance of $f(x,y)$ is approximately the expectation of

$$[(x - \mu_x)(\partial f/\partial\mu_x) + (y - \mu_y)(\partial f/\partial\mu_y)]^2$$

which is equal to

$$(\partial f/\partial\mu_x)^2\sigma_x^2 + (\partial f/\partial\mu_y)^2\sigma_y^2 + 2(\partial f/\partial\mu_x)(\partial f/\partial\mu_y)\sigma_{xy}$$

where σ_{xy} is the expectation of $(x - \mu_x)(y - \mu_y)$, that is, the covariance of x and y. In using this relationship, the variances, σ_x^2 and σ_y^2, and the covariance σ_{xy} are estimated by

$$s_x^2 = \Sigma(x - \bar{x})^2/(n - 1)$$
$$s_y^2 = \Sigma(y - \bar{y})^2/(n - 1)$$
$$s_{xy} = \Sigma(x - \bar{x})(y - \bar{y})/(n - 1) .$$

Also, to use the formula one must insert $\bar{x}$ and $\bar{y}$ for μ_x and μ_y. The whole procedure is approximate, but if the standard deviations of x and y are small relative to x and y, it will give a reasonable guide to the precision of the estimate $f(x,y)$. To base the results on means instead of individual observations, the estimated variances, s_x^2 and s_y^2, and estimated covariance s_{xy} would be divided by n, the sample size.

We give in Table 1-4 some elementary functions which arise with some frequency. The formulas shown in the table were derived in the manner described above. Rao et al. (1977) applied these concepts to a water flow problem in a Panoche clay loam.

The argument given above is applicable also in the case where x and y are obtained from independent samples, with the modification that s_{xy} would be zero. The type of argument above can be extended easily to the case of a function of several means.

The case of a function of m measurements $x_1, x_2, \ldots, x_m$ and a single function $f(x_1, x_2, \ldots, x_m)$ is treated similarly, in that approximately

$$f(x_1, x_2, \ldots, x_m) = f(\mu_1, \mu_2, \ldots, \mu_m) + \sum_{i=1}^{m} (x_i - \mu_i)(\partial f/\partial \mu_i)$$

where derivatives are evaluated at μ_i. Thus $f(x_1, x_2, \ldots, x_m)$ is approximately a linear function of the errors in $x_1, x_2, \ldots, x_m$. One therefore uses the formulas in Table 1-4, which apply for functions of errors that are linear.

Two particular cases with many attributes have moderate frequency of occurrence. Let $Z_1, Z_2, \ldots, Z_n$ be random variables, and $a_1, a_2, \ldots, a_n$ be constants. Then the variance of a linear function

$$a_1 Z_1 + a_2 Z_2 + \ldots + a_n Z_n$$

is equal to

$$a_1^2 \sigma_{Z1}^2 + a_2^2 \sigma_{Z2}^2 + \ldots + a_n^2 \sigma_{Zn}^2 + 2a_1 a_2 \sigma_{Z1Z2} + 2a_1 a_3 \sigma_{Z1Z3} + \text{etc}\,.$$

In the particular case when the Z's are uncorrelated so that $\sigma_{Z_iZ_j}$ is zero, the variance function reduces to the simple form

Table 1-4. Approximate precision of $f(x,y)$ measured by variance in terms of variability of x and y.

Estimate of $f(x,y)$	Variance of $f(x,y)$
$x \pm y$	$s_x^2 + s_y^2 \pm 2s_{xy}$
xy	$y^2 s_x^2 + x^2 s_y^2 + 2xy\, s_{xy}$
x/y	$\{s_x^2 + (x/y)^2 s_y^2 - 2(x/y)s_{xy}\}/y^2$
$\ln(x/y)$	$(1/x^2)\, s_x^2 + (1/y^2)\, s_y^2 - (2/xy)\, s_{xy}$
$\ln(xy)$	$(1/x^2)\, s_x^2 + (1/y^2)\, s_y^2 + (2/xy)\, s_{xy}$

$$a_1^2\sigma_{Z1}^2 + a_2^2\sigma_{Z2}^2 + \ldots + a_n^2\sigma_{Zn}^2 ,$$

of which simpler forms with equal values for a_i can be recognized. Consider also a product function such as

$$P = Z_1^{a1} Z_2^{a2}, \ldots, Z_n^{an}$$

in which $a_1, a_2, \ldots, a_n$ are constants. Then

$$\log P = a_1 \log Z_1 + a_2 \log Z_2 + \ldots + a_n \log Z_n$$

and

$$(1/P)\Delta P = (a_1/Z_1)\Delta Z_1 + (a_2/Z_2)\Delta Z_2 + \ldots + (a_n/Z_n)\Delta Z_n$$

where $\Delta P, \Delta Z_1, \ldots, \Delta Z_n$ are small errors in the respective variables. If now the errors in $Z_1, Z_2, \ldots, Z_n$ are uncorrelated, we obtain by squaring, replacing $Z_1, Z_2, \ldots, Z_n$ by their expected values, and taking expectations,

$$\mathrm{CV}^2(P) = a_1^2\mathrm{CV}^2(Z_1) + a_2^2\mathrm{CV}^2(Z_2) + \ldots + a_n^2\mathrm{CV}^2(Z_n)$$

where CV (P), CV (Z_1), etc. are coefficients of variation of P, Z_1, etc. In the simple case of a pure product, all a_i being equal to unity, or if the a_i are all plus or minus unity, this function would reduce to

$$\mathrm{CV}^2 \text{ (Product of uncorrelated variables)} = \text{Sum of } \mathrm{CV}^2 \text{ (each variable)} .$$

This expression serves for products the same role as variance does for sums of uncorrelated variables in that

$$\text{Variance (sum of uncorrelated variables)} = \text{Sum of variances of each variable} .$$

1–7 THE ROLES OF BIAS AND PRECISION

We have defined bias as the deviation of the statistical true value (limiting mean of repetitions) from the scientific true value, and precision has been defined as a measure of variability of an observation around the statistical true value. Bias and precision have intrinsically different roles. One can note, for instance, that precision can be increased merely by making more repetitions, since the precision of the mean of n independent measurements (i.e., independent samples and independent application of the unchanged measurement process) is equal to the precision of a single observation multiplied by n, with almost any reasonable definition. Hence with sufficient expenditure of resources in sampling and measuring, the precision can be made arbitrarily high (unless the material to be measured is used up). Thus in a general way a choice between processes of different precisions must be made not only on the basis of the precision of a single observation with each process but also on the relative costs of observations with the different processes. If for example a measurement with process *A* gives an unbiased answer with a standard deviation of 1 unit at a cost of $9, while with process *B* one gets an unbiased number with a standard deviation of 2 units at a cost of $1, process *B* is the better one for the research worker with a shortage of money but plenty of time. Four repetitions with process *B* would lead to a standard deviation of 1 unit at a cost of $4, which is clearly cheaper than what process *A* can achieve.

To examine the bias of a measurement process, we must consider the precision of observations so that we can form judgments as to whether the underlying statistical true value differs from the scientific true value, and by how much. Consequently the logical process of examining experimental results per se must start first with estimation of precision and lead up to the comparison of means and variances.

Repetition achieves nothing with respect to additive bias because every repeated measurement (and hence the average of repeated measurements) contains a constant bias. One's attitude to bias is, however, based on its magnitude relative to the precision as measured by the standard deviation. For example, suppose we have two processes, *A* and *B*, with distributions as in Fig. 1–7. Here it is supposed that *A* gives unbiased results while *B* does not, because each *B* observation contains a constant positive quantity relative to the *A* observation. It is clear, however, that process *B* will give answers which deviate from the scientific true value by less than one standard deviation with much higher frequency than *A*. In the comparison of methods a composite measure of goodness combining effects of both bias and precision is the root-mean-square error, defined as

$$(\text{Bias}^2 + \text{Standard Deviation}^2)^{1/2}.$$

If one used n independent measurements with the same bias, the root-mean-square error of the average would be

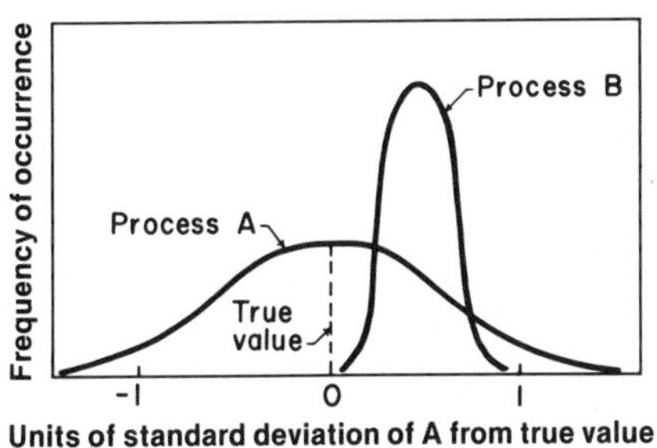

Fig. 1–7. Frequency distributions of two processes differing both in precision and in statistical true value.

$$(\text{Bias}^2 + \text{Standard Deviation}^2/n)^{1/2}\,.$$

This measure shows how the comparison of processes A and B depends on the two aspects, precision and bias of the additive type.

When bias is proportional, i.e.

$$y = cu + \text{error}$$

and c is not unity as with additive bias, processes such as A and B in Fig. 1–7 are much more difficult to compare visually or by using techniques like the root-mean-square error. Our assumption of a simple additive or proportional bias component is an oversimplification of how bias generally occurs and the attendant problems confronted when using statistical techniques to determine (or detect) bias and to make inferences in its presence. The interested reader should consult Cochran (1968) for a comprehensive review and analysis of this problem.

The attitude of the experimental scientist to bias should also be based on the type of investigation pursued. In a general sense of purpose for which measurements are taken, two types of experimentation may be defined. With absolute experimentation, measurements are made to assess the absolute magnitude of the attribute of a particular object; but, in comparative experimentation, comparative magnitudes of the attribute are measured when specific environmental changes have been imposed to bring about a quantitative change in the attribute. The distinction between absolute and comparative experimentation may be illustrated by considering, respectively, the conductivity of the saturation extract of soil for specification of a critical level, and the comparative conductivity of the saturation extract of soil from control and irrigated plots in the same field. In the first case, specification of a critical level implies that other laboratories may use this conductivity, in which case absolute values are important since measurement processes may not be the same among laboratories. Youden (1959, 1961) should be consulted for special techniques to study bias when absolute values are sought. Some of these techniques are elaborate interlaboratory tests, such as conducted by Shaw and Claire (1956) and Attoe (1959) to describe bias in the flame photometric determination of magnesium and calcium. In the second case, absolute values are not so important if the comparative conductivities are determined under comparable conditions by the same analyst, because consistent bias may not seriously interfere with certain types of inferences about the effect of irrigation on the conductivity so long as blocking, randomization, and replication are used to protect against unexpected bias (see Section 1–8). Precision, of course, must be considered in both absolute and comparative experimentation.

An experimenter may wish to reduce or remove bias when precision is sufficiently good or some other cause demands. While statistical approaches are essential as we have explained, reduction or elimination of bias ultimately depends upon the particular measurement process and how specialized experiments may be used to understand the bias component(s). An experimenter may then use an empirical mathematical adjustment, alter the original measurement process, or select a different measurement process. No brief generalization can be given that will include the approaches in specialized experiments to deal with bias—these specialized experiments are the essence of scientific inquiry.

For instance, approaches in soil chemistry may be unsuitable or nongermane in soil physics, soil microbiology, or soil mineralogy because of different natural phenomena in the measurement process. Described methods of analysis should include precautions against bias, based on knowledge of the natural phenomena involved in the measurement process. Likewise, the skilled experimenter will carefully evaluate the natural phenomena involved.

1-8 HOW TO STUDY ERRORS OF OBSERVATION

Clearly any proposed method of observation should be examined per se, to understand and quantify variability arising from error of observation.

The simplest situation is that in which the measurement process does not affect the entity being observed, as for instance repeated planimeter measurements of an area on a soil survey map. Even this should be done with care, as when repeated measurements are made on a single entity there is an unavoidable tendency to "read the same result" on repetitions because the experimenter knows the same entity is being measured. A useful way to overcome such a tendency is to take a number (e.g., 6) of the entities to which the measurement process is to be applied. Suppose that five observations are to be made on each entity. Then one may use a simple experimental design with randomization. The general idea is as follows. We are to make 30 (6×5) observations in all. Then obtain a random permutation of the numbers 1, 2, . . . , 30. Suppose this permutation turns out to be 3, 14, 26, 15, 6, ... Then the first observation is made on entity 3, the second on entity 2 ($14 - 2 \times 6$), the third on entity 2 ($26 - 4 \times 6$), the fourth on entity 3 ($15 - 2 \times 6$), the fifth on entity 6, and so on. Furthermore, the entities should have identities that are known to the scientist who will examine the data but have not been revealed to the observer. This scientist should first obtain the mean square among readings for the same entity (there are six in this case) and check for homogeneity as described in Snedecor and Cochran (1980). Next the scientist should consider the possibility that the variability of measuring an entity depends on the mean for the entity, and if this happens, a transform of the measurements may be necessary to avoid heterogeneity of measurement-error variance.

The question may arise whether there is observer variability; that is, whether some observers obtain results that average higher than other observers, or whether some observers are more variable than others. This problem again involves ideas about experimental design. In the simple case of an entity for which the measurement process has negligible effect, one should construct a factorial design with observers and entities as factors and with repetitions of observations of each entity by each observer. Such a study should also include randomization and concealment of the identity of the entity as it is being observed, so that the observer does not know values he or she has obtained on previous observations of any one of the entities. This procedure will lead to mean values and errors of observation (estimated as variances) for each operator-entity combination.

If the observation process is destructive, one will necessarily start with a batch of material that can be divided into portions. One will then involve a more complex experimental design than when the entity is not destroyed by the measurement process. Such designs and associated statistical analyses are indicated in such texts as Snedecor and Cochran (1980) or Ostle and Mensing (1979).

Studies of methods and the accuracy of observations derived therefrom cannot be ignored, and should be a matter for scientific study per se when a new method of measurement is being considered. We trust that our brief exposition of the value of experimental design will encourage its proficient use in the study of errors of observation.

1–9 ROLE OF ERRORS OF OBSERVATION IN THE STUDY OF RELATIONSHIPS

Suppose two attributes, y and x, are measured on each entity. We shall be interested in the relationships between these two attributes and how errors of observation may affect our evaluation of these relationships. To illustrate, we must describe the situation formally. Suppose we have n plots of land; for each plot we measure y and x, giving observations (y_i, x_i), $i = 1, 2, \ldots, n$. Then one can suppose that

$$y_i = \mu_i + e_i \text{, and } x_i = \gamma_i + f_i$$

where μ_i, γ_i are the true values, and e_i, f_i are independent errors of observation. Obviously, we prefer to form ideas about the relation of μ_i to γ_i, but in practice we must examine the relations of y_i to x_i. One might think that if the relation of μ_i to γ_i is

$$\mu_i = \alpha + \beta \gamma_i,$$

then the relation of y_i to x_i will be

$$y_i = \alpha + \beta x_i + h_i,$$

where h_i is an error of the relationship. Then one would estimate β by the usual linear regression formula:

$$\hat{\beta} = \Sigma(y_i - \bar{y})(x_i - \bar{x})/\Sigma(x_i - \bar{x})^2 .$$

The situation is very subtle however, and the suggested estimate will be biased. To understand this subtlety, we use reasoning that is highly approximative. Consider the numerator of $\hat{\beta}$. It is equal to

$$\Sigma\,(\mu_i - \bar{\mu} + e_i - \bar{e})(\gamma_i - \bar{\gamma} + f_i - \bar{f}) .$$

If the errors e_i and f_i have an average value of zero so that the measurements of μ_i and γ_i are unbiased, then this quantity will be approximately

$$\Sigma(\mu_i - \bar{\mu})(\gamma_i - \bar{\gamma}) .$$

Now the denominator is estimated as

$$\Sigma\,(\gamma_i - \bar{\gamma} + f_i - \bar{f})^2$$

which will be approximately

$$\Sigma\,(\gamma_i - \bar{\gamma})^2 + (n - 1)\sigma_f^2 ,$$

where σ_f^2 is the variance of the error of measurement in the x variable. We see that $\hat{\beta}$ is approximately

$$\Sigma\,(\mu_i - \bar{\mu})(\gamma_i - \bar{\gamma})/[\Sigma(\gamma_i - \bar{\gamma})^2 + (n - 1)\sigma_f^2] .$$

but the value we seek for the regression of μ_i on γ_i is

$$\Sigma\,(\mu_i - \bar{\mu})(\gamma_i - \bar{\gamma})/\Sigma(\gamma_i - \bar{\gamma})^2 .$$

So we see that if there is a linear relation of μ_i to γ_i with slope β, the coefficient of regression of y_i on x_i will be $c\beta$, with c less than 1. If the numbers γ_i have variance σ_γ^2 and the measurement errors of γ_i have variance σ_f^2, then the slope from regressing y_i on x_i will be less than the true slope in the ratio $\sigma_\gamma^2/(\sigma_\gamma^2 + \sigma_f^2)$. All of the formulas given here are approximate, but it may be concluded generally that errors in the observation of the independent, or explanatory, variable will lead to a regression or slope that is smaller than would be obtained if this independent variable were measured without error. This phenomenon is called *attenuation* and is of critical importance in all branches of science.

A simple example is the interpretation of nitrogen uptake by paired measurements of nitrogen assimilated and available soil nitrogen on n plots of land. If the available soil nitrogen is measured with error, we must be aware that the linear regression of assimilated nitrogen on available soil nitrogen will have a lower slope than if the available soil nitrogen in each plot were measured without error. A consequence of this problem is that when one is studying dependence of one variable y on another variable x, one must make a determination of the variance describing the error of measuring x (which is σ_f^2) and of the variance (σ_γ^2) of the true value, γ in our example. One can then correct for the attenuation using an obvious algorithm.

1–10 A NOTE ON TERMINOLOGY

It appears that everyone who writes on the topic "errors of observation" uses slightly different terminology, and this will be a source of confusion to the reader. The following terms frequently arise in the various descriptions: random error, systematic error, precision, bias, accuracy. We shall give our understanding of these terms.

The taking of a measurement is the application of a process, with more or less human intervention, which leads to a number. One can imagine applying this process to a constant object of measurement or to independent samples of an object of measurement. In certain cases this is easily done, as for instance when we make up, say, 10 L of solution and draw out successively 1-mL samples to which we apply the measurement process. We can do this again and again, being careful to assure no carry-over of information or of variations in process conditions from one sample to the next. We can then consider the sequence of results and can examine this sequence for structure or system; in other words we can check for randomness. If we can find no system in the sequence of results, we say that the fluctuations of the results are random, and that each result contains a *random error.* We could estimate the random error of any particular result exactly if we knew the statistical true value, which would be the mean of infinitely many results—the limiting mean.

Insofar as a sequence of results obtained in the above manner does exhibit some system, we should properly describe that system of deviation of the results from the limiting mean as systematic error of the process. More commonly a *systematic error* is thought of as the difference between the limiting mean obtained by the process in question and the limiting mean obtained by some other process which is deemed to give the correct result, perhaps by edict, or perhaps by calculation when samples are made up chemically to check the measurement process. We have mentioned that one individual may consistently underevaluate a titration or a reading, this presumably being relative to some highly trained individual whose results can be accepted as being the correct ones, subject only to lack of precision because of finiteness of number of results. Such a difference would be classified as a systematic error. If, however, we had in mind a population of possible individuals of which we were going to use one at random, this systematic error would then become a random error, in relation to the results of other individuals.

We have already discussed in nontechnical terms what we mean by *precision,* which we can summarize as the inverse of the degree of variability among results which arise in a random sequence.

We have also discussed the matter of *bias,* a concept which can be analyzed extensively. Initially it seemed appropriate to us to mention scientific bias of a method, which relates to deviation of a limiting mean (the statistical true value) from a scientifically defined true value, which may not be easily obtained. This scientific bias may be referred to as a systematic error of the measurement method. Also we have mentioned *measurement bias,* an example of which would be a tendency to obtain high values as a result of use of an impure substance as a primary standard. We have also mentioned what may be termed *sampling bias,* which is attributable to nonrandom sampling of material to be measured.

The term *accuracy* has been used very loosely. It appears that a method is described as "accurate" if it has no biases of any sort and has high precision. It appears, therefore, to be a summary type of categorization, including many identifiable or partially identifiable facets of the method. Under some circumstances a reasonable measure of accuracy would be the root-mean-square error, defined as average (observation-scientific true value)^2.

1-11 STATISTICAL PROBLEMS AND TECHNIQUES IN GENERAL

The presentation of detailed statistical techniques for the vast variety of situations that can occur is neither appropriate nor possible in this book. We can refer the worker to several standard texts on statistical methods, such as Fisher (1958), Snedecor and Cochran (1980), Steel and Torrie (1980), Ostle and Mensing (1979), Brownlee (1965), Finney (1972), and Schenck (1979).

It does, however, seem appropriate to discuss briefly the simpler problems the experimenter will meet. We have discussed briefly the precision of a single mean. Frequently the experimenter will have to compare two or more means, for instance, to check one method of analysis against another. Statistical texts contain procedures for comparing means. The comparison of methods of analysis may also involve comparisons of precision. This would ordinarily be done by comparing estimated variances. In some cases the experimenter will wish to check whether data are satisfactorily represented by a particular mathematical distribution, and must use some goodness-of-fit test. Problems of relationships between methods of analysis will involve the fitting and testing of mathematical models, using what is commonly called regression analysis (see, e.g., Kempthorne, 1952; Kempthorne and Folks, 1971; Draper and Smith,

1981). An excellent treatment of problems of straight line data is given by Acton (1959).

It will be more common than not that observations have a structure, such as methods, samples for each method, and aliquots for each sample. The treatment of such data is usually best handled by the analysis of variance, which serves as a procedure for estimating variance arising from different origins. Extensive treatment of analysis-of-variance procedures is given in chapter 2 on sampling.

Also, mention should be made of the existence of what are called nonparametric tests of significance, as for instance the comparison of two means. The word nonparametric means that the test procedure does not depend on an assumed mathematical form for the underlying distribution and requires only that the observations be a random sample from some arbitrary distribution.

1–12 REFERENCES

Acton, F. S. 1959. Analysis of straight-line data. John Wiley and Sons, Inc., New York. Reprinted 1966, Dover Publications Inc., New York.

Aitchison, J., and J. A. C. Brown. 1957. The lognormal distribution. Cambridge University Press, New York.

Allmaras, R. R., R. E. Burwell, W. E. Larson, R. F. Holt, and W. W. Nelson. 1966. Total porosity and random roughness of the interrow zones as influenced by tillage. Conservation Res. Rep. 7. USDA, Washington, DC.

Attoe, O. J. 1959. Report on flame photometric determination of exchangeable magnesium in soils. Soil Sci. Soc. Am. Proc. 23:460–462.

Austin, J. B. 1939. Methods of representing distribution of particle size. Ind. Eng. Chem., Anal. Ed. 11:334–339.

Beers, Y. 1957. Introduction to the theory of error. Rev. Ed. Addison-Wesley Publishing Co., Inc., Reading, MA.

Biggar, J. W., and D. R. Nielsen. 1976. Spatial variability of the leaching characteristics of a field soil. Water Resources Res. 12:78–84.

Brewer, R. 1964. Fabric and mineral analysis of soils. John Wiley and Sons, Inc., New York.

Brownlee, K. A. 1965. Statistical theory and methodology in science and engineering. 2nd ed. John Wiley and Sons, Inc., New York.

Cochran, W. G. 1968. Errors of measurement in statistics. Technometrics 10:637–666.

Draper, N., and H. Smith. 1981. Applied regression analysis. 2nd ed. John Wiley and Sons, Inc., New York.

Finney, D. J. 1972. An introduction to statistical science in agriculture. 4th ed. Halsted Press, New York.

Fisher, R. A. 1958. Statistical methods for research workers. 13th ed. Oliver and Boyd, Ltd., London.

Gaddum, J. G. 1945. Lognormal distributions. Nature (London) 156:463–466.

Gardner, W. R. 1956. Representation of soil aggregate-size distribution by logarithmic-normal distribution. Soil Sci. Soc. Am. Proc. 20:151–153.

Kempthorne, O. 1952. The design and analysis of experiments. John Wiley & Sons, Inc., New York. Reprinted 1973, Krieger, Huntington, NY.

Kempthorne, O., and L. Folks. 1971. Probability, statistics and data analysis. Iowa State Univ. Press, Ames, IA.

Menzel, R. G., and W. R. Heald. 1959. Strontium and calcium contents of crop plants in relation to exchangeable strontium and calcium of the soil. Soil Sci. Soc. Am. Proc. 23:110–112.

Nash, S. M., and W. C. Snyder. 1962. Quantitative estimations by plate counts of propagules of the bean root rot *Fusarium* in field soils. Phytopathology. 52:567–572.

Nielsen, D. R., J. W. Biggar, and K. T. Erh. 1973. Spatial variability of field measured soil-water properties. Hilgardia 42:215–260.

Ostle, B., and R. W. Mensing. 1979. Statistics in research. 3rd ed. Iowa State Univ. Press, Ames, IA.

Rao, P. S. C., P. V. Rao, and J. M. Davidson. 1977. Estimation of the spatial variability of the soil-water flux. Soil Sci. Soc. Am. J. 41:1208–1209.

Rutherford, E., and H. Geiger. 1910. The probability variations in the distribution of α particles. Phil. Mag. J. Sci. Ser. 6, 20:698–704.

Schenck, H. 1979. Theories of engineering experimentation. McGraw Hill, New York.

Shaw, W. M., and V. N. Claire. 1956. Flame photometric determination of exchangeable calcium and magnesium in soils. Soil Sci. Soc. Am. Proc. 20:328–333.

Snedecor, G. W., and W. G. Cochran. 1980. Statistical methods. 7th ed. Iowa State Univ. Press, Ames, IA.

Steel, R. G. D., and J. H. Torrie. 1980. Principles and procedures of statistics: a biometrical approach. 2nd ed. McGraw Hill, New York.

Topping, J. 1957. Errors of observation and their treatment. Rev. ed. The Institute of Physics, 47 Belgrave Square, London.

Van de Pol, R. M., P. J. Wierenga, and D. R. Nielsen. 1977. Solute movement in a field soil. Soil Sci. Soc. Am. J. 41:10–13.

Youden, W. J. 1959. Accuracy and precision: evaluation and interpretation of analytical data. *In* I. M. Kolthoff and P. J. Elving, eds. Treatise on analytical chemistry. Vol. 1, pp. 47–66. The Interscience Encyclopedia Inc., New York.

Youden, W. J. 1961. Systematic errors in physical constants. Physics Today 14 No (9):32–34, 36, 38, 40–43.

2 Sampling

R. G. PETERSEN AND L. D. CALVIN

Oregon State University
Corvallis, Oregon

2-1 INTRODUCTION

The purpose of any soil sample is to obtain information about a particular soil. The sample itself is seldom, if ever, the entire soil mass in which we are interested. This larger aggregate of material, in which we are ultimately interested, is called the "population" or "universe." Information from the sample is of interest only insofar as it yields information about the population, and the information may or may not be representative, depending on how the sample is selected.

The population itself may be large or small, or even a part of what would ordinarily be considered a larger population. For example, a population might be all the soil in a field to a depth of 1 m, the clay portion in the top 15 cm of a field plot, or the organic matter in the B horizon over several fields of a common soil type. It is that portion of the soil for which we want additional information.

For any population, there are certain characteristics which describe it. For soils, these may include the thickness of each horizon in a soil, the percent organic matter, the amount of soluble salts, or the pH. The true value of each such characteristic in the population is called a parameter. The purpose of sampling is to estimate these parameters with an accuracy that will meet our needs at the lowest possible cost. All practical studies are limited in funds, and sampling therefore becomes a necessity in nearly all scientific investigations. If the population is relatively homogeneous, a very small sample may tell us all we desire to know about the population. With soils, however, variation and heterogeneity seem the rule rather than the exception.

2-2 VARIATION OF SOILS

Soils are characterized by several types of variation, and if we are to do an adequate job of representing a particular soil population by means of a sample, it is necessary that we consider the nature of these types of variation. The soil is not a homogeneous mass but a rather heterogeneous body of material. And because of this heterogeneity, systems have been set up that attempt to delineate soil classification units which approach homogeneity within themselves, but which, at the same time, are distinctly different from all other units. Thus, one type of soil variation is the variation among the several units which have been classified as homogeneous. Differences among these units may be large or small depending, among other things, on the differential effect of the factors which formed the soils. Poorly drained soils formed from recent alluvium, for example, are usually different in most of their properties from well-drained soils formed from residual parent material. The variation in properties among soils formed from the same parent material under similar conditions, on the other hand, may be rather small even though the soils be classified as different soils.

Because of the nature of soil-forming processes, distinct boundaries between soil classification units are rare. Although the modal profiles of two adjacent soils series may be distinctly different, there is usually a gradual transition, in the field, between one series and another. Superimposed on this pattern of slowly changing characteristics, however, we may find rather marked local variations.

These local variations may result from natural causes, such as sharp vegetative or topographic variations, or from man-made variation. Available phosphorus, for example, may vary widely within a series because of differential fertilization or liming practices in the past. A similar pattern of variation is found in the subsoil.

Soil properties vary not only from one location to another but also among the horizons of a given profile. The horizon boundaries may be more distinct than are the surface boundaries of a soil classification unit. Here, also, however, zones of transition are found between adjacent horizons. Furthermore, considerable local variation may occur within a particular horizon.

These characteristics should be kept in mind when sampling soils. The soil population to be sampled should be subdivided, both horizontally and vertically, into sampling strata which are as homogeneous as possible, and the several sources of variation within the population should be sampled if valid inferences are to be made about the population from the sample.

The intensity with which a soil must be sampled to estimate with given accuracy some characteristic will depend on the magnitude of the variation within the soil population under consideration. The more heterogeneous the population, the more intense must be the sampling rate to attain a given precision. Few data are available, however, from which the magnitudes of the several sources of variation in soils may be estimated. In general, although differences have been found to exist among soil classification units, considerable variation may be expected within the units for such characteristics as pH, available P, exchangeable K, exchangeable Na, conductivity, volume weight, permeability, and porosity (Allmaras and Gardner, 1956; Mason et al., 1957; Olson et al., 1958; Sayegh et al., 1958). In some instances, the variation within contiguous classification units is so great that it is not feasible to estimate differences among the units with any satisfactory degree of precision. For most characteristics, the variation, both within and among units, decreases, with increasing depth in the profile. Hence, fortunately, it is usually necessary to sample the subsoil at a lower rate than the topsoil to attain comparable accuracy.

2–3 SAMPLING PLANS

When a sample is actually drawn from the population, it is necessary to think of the population as composed of a number of separate units. These may be the number of cores, 7 cm square and 15 cm long, that can be taken from a field, or the number of spade loads which would comprise the area under study. They may be natural units of the population, such as an individual plant; or they may be artificial ones, such as fields within a soil series. The essential feature is that they be separate and distinct units and that their total number comprise the entire population.

The sampling plan designates which units of the population are to be included in the sample. There are always a number of different plans or designs which can be used. Some are more precise than others (provide a smaller error), and some may be carried out at a much lower cost. In general, the best design is one that provides the maximum precision at a given cost or that provides a specified precision (error) at the lowest cost.

The principles of soil sampling were outlined by Cline (1944) and have not changed materially. There has been a greater appreciation of the systematic sample since then, and new methods have been suggested for estimating sampling error; however, very few studies have been made upon which to base sound sampling procedures. Perhaps the cost of making intensive studies has discouraged such work, but if progress is to be forthcoming, further investigation of the reliability and efficiency of different sampling plans is needed.

2–3.1 Judgment Sample

The research worker ordinarily knows something about his population and would like to use this information in obtaining a representative sample. Effort has sometimes been directed toward the use of his judgment in selecting the most "typical" sites from which to draw the sample. These "typical" sites are obtained by selecting the sites that, in the sampler's judgment, present a representative picture of the population. Because the sampling units are selected with different but unknown probabilities, samples selected in this manner are biased. As an example, some workers are very careful to include the extremes of the population (and may therefore oversample the extremes), while others attempt to exclude all extremes as being unrepresentative of the bulk of the population.

Unfortunately, unless additional data are available or the true characteristics are known for the population, there is no way of assessing the accuracy of the results from such a sample. The sample may represent the population very well or it may not; any confidence in the results must rest entirely on faith in the sampler's judgment.

If a small sample is to be taken, a judgment sample may have a lower error than a random sample. As sample size increases, however, the error with a random sample becomes progressively smaller, whereas with a judgment sample, the error drops much more slowly and quickly becomes larger than with a random sample. Unless the research worker is particularly skillful in selecting "typical" sites, the error with a random sample becomes smaller than that with a judgment sample at a relatively small sample size. As sample size increases, the selection of "typical" sites also becomes more difficult and time-consuming. However, if low accuracy is satisfactory and no estimate of precision needed, a judgment sample may be satisfactory for the purpose.

2–3.2 Simple Random Sample

If n units are to be selected from the population, a simple random sample is defined as a sample obtained in such a manner that each possible combination of n units has an equal chance of being selected. In practice, it is usually drawn by selecting each unit separately, randomly, and independently of any units previously drawn. This, of course, requires that each unit can be listed or potentially listed. With a field, for example, it is not necessary to list all possible core samples which can be taken, although it could be done if necessary.

In soil investigations, the unit to be included in the sample is usually a volume of soil, although it may be an area of ground, a plant, or a volume of water. If the units are listed, a random sample can easily be taken by the use of a table of random numbers (Fisher and Yates, 1967). Often, however, it is more convenient to spot the field location by selecting random distances on a coordinate system and using the intersection of the two random distances as the point at which the unit is to be taken. This system works well for fields of both regular and irregular shape, since the points outside the area of interest are merely discarded and only the points inside the area are used in the sample.

The location of the point at which each sampling unit is to be taken is often reached in two steps. The two steps provide a fairly simple method to use and also ensure objectivity, to prevent a selection bias which can occur when only a single step is used. The first step is to reach an approximate point in the field. This can be accomplished by (i) mapping the area and establishing two base lines at right angles to each other, which intersect at an arbitrarily selected origin, say the southwest corner;

(ii) establishing a scale interval to be used in locating the selected point, e.g., a pace, yard, chain, or 0.1 mile; (iii) drawing two random numbers representing the number of basic scaling units to be measured along the two directions; and (iv) locating the intersection of two distances in the field as an approximate point for sampling. Using this point as the origin of a new coordinate system, a second random coordinate is measured off to provide an exact point for the sampling unit. The sampling unit to be taken is then defined as the soil unit for which the selected point is some prescribed point, say the center of the southwest corner of the sampling tool. The scale interval for the second random coordinate should be of the same dimensions as the sampling tool, with the maximum value of the second coordinate equal to the scale interval of the first coordinate system. For example (see Fig. 2–1), if the first scale interval is a pace and a 10- by 10-cm sampling unit is to be used, then the second set of random distances should be selected from numbers 0 to 9 (9 10-cm units per pace). This procedure provides an unbiased method of selecting the sample, since every unit has an equal chance of being included in the sample. With large fields or areas, it is sometimes advisable to use a three-step procedure as an easier method of locating the sampling unit. In this case the first scale interval may be in kilometers. For very small areas, a one-step procedure may be used. When a circular sampling tool is used, this procedure could conceivably introduce some bias into the estimates; however, this is not thought to be important in practice.

Sometimes a sample is taken from a one-dimensional population, such as a list of cards, a row of pots, or distances along a line. A simple random sample of this type can be taken in the same manner as the sample of distances in one direction only in the coordinate or two-dimensional system. Ordinarily only a single-step procedure is used, drawing one random number for each unit to be sampled. Sampling is usually without replacement, i.e., no unit is used more than once in the sample.

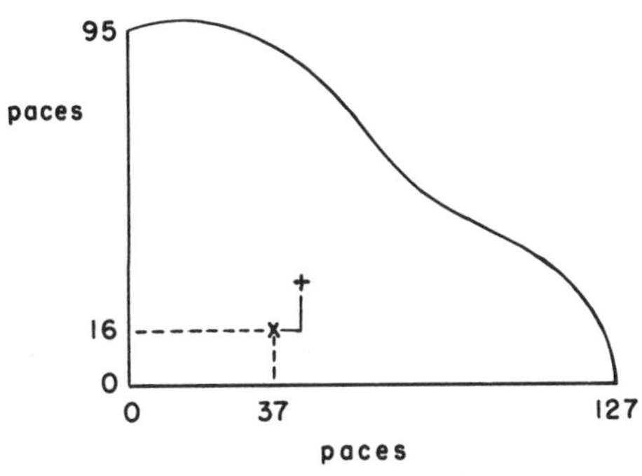

Fig. 2–1. Selection of randomly selected points in a field by a two-step procedure. Random lengths were 37 and 16 paces along the east and north axes, respectively, to establish the approximate point for sampling (*x*). Random 10-cm units drawn were 2 and 8 units to the east and north, respectively, to establish an exact point (+) for locating the southwest corner of the 10- by 10-cm sampling tool. Subsequent sampling units are selected in a like manner. *Note:* The 10-cm units are not drawn to scale, but are exaggerated to illustrate the procedure.

If more than one sampling unit is included in the sample, the random sample provides an estimate of the sampling error. Estimates of the mean $\bar{y}$ and variance $V(\bar{y})$ of the mean are given by

$$\bar{y} = (\Sigma_{i=1}^{n} y_i)/n \qquad [1]$$

$$V(\bar{y}) = \Sigma_{i=1}^{n} (y_i - \bar{y})^2/n(n - 1) = s^2/n \qquad [2]$$

where y_i is the value observed for the ith sampling unit and n is the number of sampling units in the sample. If more than 10% of the population is included in the sample, an adjustment should be made in the estimate of the variance of the mean. Discussions of this correction for sampling from finite populations and other details of analysis are ably presented by several authors (Cochran, 1977; Deming, 1950; Hansen et al., 1953; Hendricks, 1956).

Once the variance of the mean has been estimated, the usual confidence limits may be placed around the mean by the relationship

$$L = \bar{y} \pm t_{\alpha}(s^2/n)^{1/2} \qquad [3]$$

in which L is the confidence limit, t_{α} is the Student's t with $(n - 1)$ degrees of freedom (which is tabulated in most textbooks on statistics) at the α probability level, and s^2/n is as previously defined.

If an estimate of the variance is available from previous samples of the population or can be arrived at from knowledge of the population, then an estimate of the number of samples necessary in future sampling to obtain a given precision with a specified probability may be obtained in the following manner from Eq. [2] and [3]:

$$n = t_{\alpha}^2 s^2/D^2 \qquad [4]$$

where D is the specified limit.

As an example of the application of these equations, suppose we have taken 10 cores at random from the 0- to 15-cm layer of soil in a particular field and have obtained the following estimates of exchangeable potassium (K) in ppm:

59, 47, 58, 80, 57, 58, 62, 52, 50, 47 .

The mean exchangeable K level, estimated using Eq. [1], would be

$$\bar{y} = (59 + 47 + \ldots + 47)/10 = 570/10 = 57.0 \text{ ppm} .$$

The variance of the mean would be, from Eq. [2],

$$V(\bar{y}) = (834.00)/(10)(9) = 92.67/10 = 9.267 .$$

The 95% confidence limits are then obtained using Eq. [3]:

$$L = 57.0 \pm 2.26(9.267)^{1/2} = 57.0 \pm (2.26)(3.04)$$
$$L = 57.0 \pm 6.87 .$$

Thus, we can say that unless a 1 in 20 chance has occurred in sampling, the true population mean lies within the range 50.13 to 63.87 ppm K; that is, the probability is 0.95 that the true mean lies within the range 50.13 to 63.87 ppm K.

Suppose we are interested in determining the number of observations required in future sampling of the population to estimate the mean within ±5.00 ppm, at the 95% probability level. The necessary number of samples would be estimated using Eq. [4]:

$$n = (2.26)^2\,(92.67)/(5.00)^2 = (5.11)(92.67)/25.00$$
$$n = 18.94 .$$

Hence, to obtain the desired precision it is estimated that the sampling rate should be increased from 10 to 19 in future sampling.

2–3.3 Stratified Random Sample

With stratified random sampling the population is broken into a number of subpopulations, and a simple random sample is taken from each subpopulation; e.g., a field may be mapped by series (subpopulations) and a random sample taken from each series. The reasons for sampling soils in this manner include the desire (i) to make statements about each of the subpopulations separately and (ii) to increase the precision of estimates over the entire population. For the first reason, the sampling can be considered a series of separate sampling studies on a number of different populations. The second reason, however, raises some special considerations, both from the standpoint of design and analysis.

If the stratified random sample is to have greater precision than the simple random sample, stratification must eliminate some of the variation from the sampling error. Since every subpopulation, or stratum, is sampled, the differences among the stratum means are eliminated from the error, and only the within-stratum variation contributes to the sampling error. If strata are constructed in such a manner as to make the units within each stratum more homogeneous than the variation among the stratum means, the stratified random sample will have greater precision than the simple random sample. The basis for making effective stratification may be another variable related to or correlated with the characteristic of interest, previous knowledge about the distribution of the characteristic, or merely geographic proximity. Any prior information which will aid in making homogeneous groups of units of the population can be used.

In general, the more stratification the greater the increase in precision. There are several precautions that should be considered, though, to prevent excessive stratification. Precision increases at a decreasing rate as strata are divided more and more, until a point is reached where no further gain in precision is obtained. The additional strata also complicate the analysis so that the gain in precision must be considered in relation to the effort required to obtain it. Since at least two units must be sampled in each stratum to have an estimate of the sampling error from that stratum, it is usually advisable to keep the number of strata small enough to allow satisfactory estimates of error.

The total number of sampling units is often allocated to the strata proportionately; e.g., if a stratum contains 20% of the population then 20% of the sampling units will be taken from that stratum. This allocation is not optimum in the sense that the variance of the mean will be a minimum; however, unless the variation within the strata differs markedly from stratum to stratum it will be nearly as good as optimum allocation. Optimum allocation, both with equal and unequal costs per unit in different strata, is discussed at length by Cochran (1977) and Hansen et al. (1953).

The estimate of the mean over all strata, $\overline{\overline{y}}$, and the variance of this mean, $V(\overline{\overline{y}})$, are given by

$$\overline{\overline{y}} = \left(\sum_{h=1}^{L} N_h \overline{y}_h\right) \Big/ N \qquad [5]$$

$$V(\overline{\overline{y}}) = (1/N^2) \sum_{h=1}^{L} N_h^2 (S_h^2/n_h) \qquad [6]$$

where N_h is the total number of units in the hth stratum, L is the total number of strata, N is the total number of units in all strata, and

$$s_h^2/n_h = V(\overline{y}_h)$$

as given in section 2–3.2. It may be noted that N_h/N is the proportion of all sampling units in the hth stratum; hence the proportion of the population in the hth stratum may be used in the equation rather than the total number.

When the allocation is proportional and all strata have a common variance, which is a common occurrence in soil sampling, then the estimation of the mean and variance is simplified:

$$\overline{\overline{y}} = \left(\sum_{i=1}^{n} y_i\right) \Big/ n \qquad [7]$$

$$V(\overline{\overline{y}}) = s_p^2/n \qquad [8]$$

where n is the total number of units in the sample and s_p^2 is the pooled variance within strata (s_p^2 measures the variation among samples within strata).

If more than 10% of the units in any stratum are included in the sample, use should be made of the finite population correction referred to in section 2–3.2.

To illustrate the computations involved in stratified random sampling, suppose we are interested in estimating the cation-exchange capacity of the surface soil in a field which contains three soil types: *A*, *B* and *C*. Suppose further that type *A* represents 1/6 of the total area of the field, type *B* 1/3, and *C* 1/2 of the total area. We might stratify the field with respect to soil type and draw a simple random sample of cores from each stratum, the number of cores from each stratum being proportional to the fraction of the total area occupied by each soil type. That is, we might take 2 samples from type *A*, 4 from type *B*, and 6 from type *C*. The results of such a sampling plan are shown in Table 2–1.

The mean exchange capacity would be estimated using Eq. [7] as

$$\bar{\bar{y}} = (11.6 + 13.4 + \cdots + 17.1 + 19.5)/12 = 16.62 \text{ meq per 100 g.}$$

Assuming a common within-stratum variance, the variance of the mean could be estimated using an analysis of variance of the sample data, shown in Table 2–2.

The variance of the mean $V(\bar{\bar{y}})$, would then be estimated using the "within-strata" mean square and Eq. [8]:

$$V(\bar{\bar{y}}) = 2.94/12 = 0.24.$$

And, by analogy to simple random sampling, we might place the 95% confidence limits about the mean using Eq. [3]:

Table 2–1. Cation-exchange capacity of soil samples from a field stratified according to soil type.

	Cation-exchange capacity (meq/100 g) of samples from indicated type		
	A	B	C
	11.6	19.0	14.2
	13.4	21.4	16.5
		18.1	15.7
		17.8	15.2
			17.1
			19.5
Total	25.0	76.3	98.2
Mean	12.50	19.08	16.37

Table 2–2. Analysis of variance of cation-exchange capacity (meq/100 g) of soil samples in Table 2–1.

Source of variation	Degrees of freedom	Mean square
Among strata	2	29.22
Among samples within strata	9	2.94

$$L = 16.62 \pm 2.26(0.24)^{1/2} = 16.62 \pm (2.26)(.49) = 16.62 \pm 1.11\,.$$

Thus we can say that the true population mean lies within the range of 15.51 to 17.73 meq per 100 g with probability 0.95.

2–3.4 Systematic Sample

As the stratified random sample is an attempt to ensure that better coverage of the population is obtained than with a simple random sample, so also is the use of a systematic sample a further step in this same direction. It is a natural extension and one that is receiving increasing attention. The fact that it is easy to use in practice has undoubtedly been a factor in its wide usage.

The definition of a systematic sample that is used here covers samples in which the selected units are at regular distances from each other, either in one or two dimensions. If the population is of one dimension (e.g., units along a line, in a list, or in any linear order), the first unit is assumed to be selected at random from the first k units and subsequent units at each kth interval (Cochran, 1977). If the population is of two dimensions, e.g., plots in a field or sections of an area, the surface can be considered to be composed of a number of strata of common size and shape. In one stratum a unit is selected at random, and all units in comparable positions in all strata are included in the sample. Alternatively, the stratum pattern can be imposed at random on the surface and the center unit from each stratum included in the sample (see Fig. 2–2). In practice, it is not necessary that the stratum outlines actually be designated, although the procedure must permit them to be. This definition includes the type of systematic sample most common in practice, although many others could be designed. This type of systematic sample can also be considered as a single complex sampling unit selected at random from such k units in the population.

For the one-dimensional case, systematic samples have been compared with random and stratified random samples, both theoretically

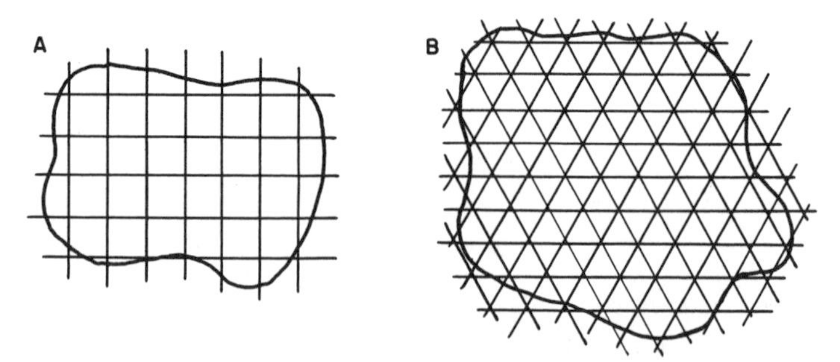

Fig. 2–2. Systematic samples in two dimensions. Design A is composed of all units at the intersections of equidistant parallel lines at right angles to each other. Design B is composed of all units at the intersections of equidistant parallel lines, each set at 60° from vertical, Triangles are formed by drawing horizontal lines through the intersections.

(Cochran, 1946; Madow and Madow, 1944; Madow 1949, 1953; Yates, 1948) and empirically (Finney, 1948; Hasel, 1938; Osborne, 1942; Williams, 1956; Yates, 1948). The results have favored systematic samples in nearly all cases. An exception pointed out by several workers (Cochran, 1953; Finney, 1948; Madow and Madow, 1944; Yates, 1948) may occur when the population has a periodic trend, such as might be experienced by a line across hills and valleys in a forested area, or across a field with row effects caused from previous fertilizer treatment. When this occurs, the systematic sample is much less precise than random sampling if the interval between the sampling units in the systematic sample is equal to an integral multiple of the period, e.g., sampling units taken on each row and none between rows. If the interval is equal to an odd multiple of the half period, systematic sampling will be more precise. Milne (1959) states that such periodicities seldom occur in nature; however, Matérn (1960) gives several examples but indicates that a considerable loss of precision occurs only when the interval coincides closely with the periodicity.

The two-dimensional case, applicable for most field sampling, has been studied theoretically (Das, 1950; Matérn, 1960; Quenouille, 1949; Zubrzycki, 1958) for several types of populations. Empirical studies have been very limited and confined to crops and forested areas. The theoretical studies have all assumed that a correlation exists between units close to each other and that this correlation decreases exponentially with the distance between the units. This seems to be a realistic assumption, and it has been supported by empirical studies (Finney, 1948; Milne, 1959; Rigney and Reed, 1946). The results are similar to the one-dimensional case in that the systematic sample has greater precision than random or stratified random samples, if the correlation between plots decreases exponentially with distance between the plots in both directions. T. Dalenius et al. (unpublished) have also shown that the optimum pattern for several classes of populations is the triangular grid shown in Fig. 2–2B, although the square grid is almost as precise.

The effect of additional trends in the population has not yet been fully studied, although such trends may be a more important factor than the correlation in the precision of systematic sampling. It has been pointed out (Madow and Madow, 1944; Yates, 1948) that the systematic grid pattern is inefficient when there is a fertility gradient along rows or columns of a field. Since this may be the case in fields which have a slope or drainage problem (or other not such obvious conditions imposing a gradient effect), it is apparent that systematic designs cannot be used without consideration of the form of the population distribution.

The empirical comparisons have shown somewhat conflicting results. In sampling for crop yield on 14 fields, Haynes (1948) found that the systematic square grid pattern (see Fig. 2–2A) had about the same precision as the simple random sample and less precision than the stratified random sample. Matérn (1960) found that square and triangular grids gave greater precision than the stratified random and considerably greater precision than the random sample when sampling for percent forested area and percent lake area in a region. Haynes hypothesized that a linear trend might be the cause of his results and showed on the basis of theory that the systematic sample is indeed less precise in the presence of such a trend. It can be shown that the precision also depends on relative gradients in the two directions. In general, when linear trends are present, more sampling units should be selected in the direction of the strongest gradient.

One of the main problems with the systematic sample has been in the estimate of sampling error from the sample itself. A number of methods have been suggested (Hansen et al., 1953; Madow, 1953; Milne, 1959; Osborne, 1942; Smith, 1938; Williams, 1956), but those most commonly used in practice are the following:

1. Assume that the population was in random order before the systematic sample was drawn and estimate the sampling error as with a simple random sample;
2. Block or stratify the sample, assuming that the variation among units within a block or stratum is only sampling variation, and estimate the sampling error as with a stratified random sample; and
3. Take a number of separate systematic samples, each drawn at random from all possible systematic samples of the same type, i.e., each systematic sample is drawn with a separate randomly selected starting point.

The sample mean is calculated from each sample, and these means are treated as the data from a simple random sample of size equal to the number of separate systematic samples. Method 3 is the only unbiased method of assessing the sampling error without making some assumptions about the form of the population; however, if the assumptions made under methods 1 and 2 are reasonably valid, these methods give greater precision.

2–4 SOURCES OF ERRORS

Errors arising from soil sampling (Das, 1950) fall into three general categories, viz., sampling error, selection error, and measurement error. Each contributes to the total error, and consideration of each of them is necessary to ensure a satisfactory sampling procedure.

Sampling error is the error arising from the fact that the sample includes only the selected sampling units rather than the entire population. It is caused by the inherent variation among the units of the population and can be avoided completely only by including all of the population in the sample.

Selection error arises from any tendency to select some units of the population with a greater or lesser probability than was intended, e.g., a tendency to avoid rocky sites in a field or to oversample the borders of a field. There is some tendency for selection errors to cancel each other;

and yet, as sample size increases, bias resulting from the selection procedure may become an increasing portion of the total error. The two-step procedure outlined in section 2–3.2 is a method designed to minimize selection error, and analogous procedures can usually be devised.

Measurement error is that error caused by the failure of the observed measurement to be the true value for the unit. It includes both the random errors of measurement, which tend to cancel out with increased sample size, and biases, which are usually independent of sample size. Examples of random errors are those caused by assuming constant weights for cores of soil which have variable weights and those caused by chance variations of technique in the analytical procedure. Biases, independent of sample size, would result if tare weights were ignored or if a calibration curve were offset to one side of the appropriate curve.

If a biased technique is used, such as subsampling from the top of a container of improperly mixed soil (selection error) or using an analytical test which gives a reading too high (measurement error), then this error, or bias, would not be included in the computed sampling error. Only constant attention to technique can hold these biases to a minimum, and even then no estimate of their magnitude is ordinarily available.

Although considerable study is needed on the sources of error in practical surveys, some evidence would point to sampling error generally being larger than random measurement error (Cline, 1944; Hammond et al., 1958; Rigney and Reed, 1946).

2–5 SUBSAMPLING

In many types of soil investigation, the use of subsampling, or multistage sampling, is advantageous. With this technique, the sampling unit, selected by one of the previously described methods, is divided into a number of smaller elements. The characteristic under consideration is then measured on a sample of these elements drawn at random from the unit. For example, a sample of cores may be taken from a field plot, and a number of small samples taken from each core for chemical analysis.

The primary advantage of subsampling is that it permits the estimation of some characteristic of the larger sampling unit without the necessity of measuring the entire unit. Hence, by using subsampling, the cost of the investigation might be considerably reduced. At the same time, however, subsampling will usually decrease the precision with which the characteristic is estimated. At each stage of sampling, an additional component of variation, the variation among smaller elements within the larger units, is added to the sampling error. Thus, the efficient use of subsampling depends on striking a balance between cost and precision.

To illustrate the statistical considerations involved in subsampling, suppose we draw n sampling units at random from a population, and from each unit we select, at random, m elements. The mean, on a per element basis, may be estimated from

$$\overline{y} = \left(\sum_{i=1}^{n} \sum_{j=1}^{m} y_{ij} \right) \Big/ nm \quad [9]$$

in which y is the sample mean, y_{ij} is the observation on the jth element in the ith unit, n is the number of units in the sample, and m is the number of elements sampled per unit.

The variance of the mean, $V(y)$, on a per element basis, is given by

$$V(\overline{y}) = \sigma_b^2/n + \sigma_w^2/mn \quad [10]$$

where σ_b^2 is the variation among units, σ_w^2 is the variation among elements within units, and n and m are as previously defined.

The components of variance, σ_b^2 and σ_w^2, may be estimated from an analysis of variance of the sample data, shown in Table 2–3.

The variation among elements, σ_w^2, is estimated by the mean square among elements within units, s_w^2. The variation among units, σ_b^2, is estimated from

$$\sigma_b^2 = (s_b^2 - s_w^2)/m\,. \quad [11]$$

If the sample includes > 10% of the units in the population, an adjustment should be made in the estimate of the variance of the mean. The proper adjustment is adequately discussed by several authors (Cochran, 1977; Deming, 1950; Hansen et al., 1953).

If estimates of σ_w^2 and σ_b^2 are available from previous samples or can be arrived at from a knowledge of the population, we can use equation [10] to predict the variance of the mean in future sampling from the same type of population. The proposed sampling and subsampling rates, n and m, and the estimates of the variance components, σ_w^2 and σ_b^2, are substituted into the appropriate equation.

It is apparent from Eq. [10] that the variance of the mean may always be reduced at a more rapid rate by increasing the number of units sampled than by increasing the sampling rate within units. This may not be the most efficient procedure, however, if the cost of taking the sample is considered. The optimum sampling and subsampling rate will depend on the relative cost of sampling the unit and that of sampling the element

Table 2–3. Analysis of variance of the sample (on a per element basis).

Source of variation	Degrees of freedom	Mean square†	Mean square is an estimate of
Among units	$(n - 1)$	s_b^2	$\sigma_w^2 + m\sigma_b^2$
Among elements within units	$n(m - 1)$	s_w^2	σ_w^2

† $s_b^2 = m\Sigma_i(\bar{y}_i - \bar{\bar{y}})^2/(n - 1)$
$s_w^2 = \Sigma_i\Sigma_j(y_{ij} - \bar{y}_i)^2/n(m - 1)$
where $\bar{y}_i$ = the mean of the ith unit $= (\Sigma_{j=1}^{m} y_{ij})/m$.

within the unit. A cost relationship which has been useful in many types of soil investigation is given by

$$C = nc_b + nmc_w \qquad [12]$$

in which C is the total cost of obtaining an estimate of the mean, c_b is the cost per unit, directly assignable to the unit and independent of the number of elements per units, c_w is the cost per element, and n and m are the number of units and the number of elements per unit, respectively, in the sample.

When the variance of the mean is of the form given by Eq. [10] and the cost relationship is that given by Eq. [12], it can be shown (Cochran, 1977) that the optimum subsampling rate, m', may be obtained from

$$m' = [(c_b\sigma_w^2)/(c_w\sigma_b^2)]^{1/2} . \qquad [13]$$

That is, m' can be shown to give the smallest variance for a given cost, or alternatively, the least cost for a given variance. The sample rate, n, is found by solving either Eq. [10] or Eq. [12], depending on whether the variance or the cost has been specified.

In practice, m' will usually not be an integer, and the nearest integer should be selected. In addition, the variance usually changes rather slowly for values of m in the region of the optimum. Thus some latitude in the selection of m' can be tolerated.

As an example of the computations involved in subsampling, suppose we were interested in estimating the effect of continuous grazing on the bulk density of a pasture soil. And suppose that we drew a random sample of five 3- by 3-m sampling units, and within each unit we selected, at random, three subsamples in the form of 7.5-cm cores upon which the bulk density measurements were made. The resultant observations might appear as in Table 2–4.

The mean bulk density may be estimated from Eq. [9] as

Table 2–4. Bulk density of three subsamples of soil in each of five sampling units.

	Bulk density of soil in indicated sampling unit				
Subsample	1	2	3	4	5
	g/cm³				
1	1.53	1.63	1.47	1.59	1.64
2	1.48	1.58	1.41	1.36	1.58
3	1.64	1.67	1.50	1.41	1.50

Table 2–5. Analysis of variance of bulk density data in Table 2–4.

Source of variation	Degrees of freedom	Mean square
Among units	4	0.0168
Among cores within units	10	0.0061

$$\overline{y} = \frac{(1.53 + 1.48 + \cdots + 1.58 + 1.50)}{5 \times 3} = \frac{22.99}{15} = 1.53.$$

The analysis of variance of the data is shown in Table 2–5.

The variance components may then be estimated as

$$\sigma_w^2 = 0.0061$$

$$\sigma_b^2 = (0.0168 - 0.0061)/3 = 0.0036\,.$$

Thus from Eq. [10] the variance of the mean, on a core basis, is found to be

$$V(\overline{y}) = 0.0036/5 + 0.0061/5 \times 3 = 0.0007 + 0.0004 = 0.0011.$$

Now suppose we are interested in determining the effect on the variance of reducing the number of sampling units to 3 and increasing the number of cores per unit to 5. Again we would use eq. [10] to obtain

$$V(\overline{y}) = 0.0036/3 + 0.0061/3 \times 5 = 0.0012 + 0.0004 = 0.0016.$$

Now suppose we wish to estimate what the optimum sampling rate would be in future sampling from the same type of population, and suppose that it costs five times as much to sample an additional unit as it does to take an additional core in each unit. From Eq. [13] we would find the optimum sampling rate to be

$$m' = [(5 \times 0.0061)/(1 \times 0.0036)]^{1/2} = (0.0305/0.0036)^{1/2} = (8.47)^{1/2}$$
$$m' = 2.91.$$

Thus, in sampling from a similar population, we would expect the maximum precision for a given cost by taking 3 cores per unit.

It should be pointed out that subsampling need not be restricted to a two-stage procedure. In some instances, the sample may be taken in three or more stages. A core of soil, for example, might be subsampled to provide small quantities of soil for chemical analysis which, in turn, might be subsampled to provide aliquots for the final measurement. The principles involved in multistage sampling are essentially extensions of the principles discussed here. The statistical considerations are adequately presented by several authors (Cochran, 1977; Deming, 1950; Hansen et al., 1953; Yates, 1953).

2–6 COMPOSITE SAMPLES

In many soil investigations, a substantial saving in total cost can result if laboratory analyses are performed on a composite of the field samples rather than on the individual samples. The procedure consists of taking a number of field samples adequate to represent the population in question, thoroughly mixing these samples to form one composite or bulk sample, and performing the laboratory analyses on the composite sample or on a subsample of the composite. The assumption, of course, is that a valid estimate of the mean of some characteristic of the population may be obtained from this single analysis of the composite sample. This assumption is true only under certain conditions. All samples which form the composite must be drawn from the population under consideration, and each sample must contribute the same amount to the composite.

For example, suppose a study is initiated to determine the changes in available soil phosphorus which have taken place in a field experiment which involved the annual application of varying increments of phosphorus to certain of the plots in the experiment. A composite sample for each plot could be obtained by pooling a number of borings taken at random from the plot. Similarly, if no estimate of precision is needed, the mean available phosphorus for a given treatment would be obtained from a composite of equal numbers of samples from all plots with a common treatment. Forming a composite by pooling samples taken from several treatments, however, is seldom if ever justified.

It should be pointed out that the composite samples provide only an estimate of the mean of the population from which the samples forming the composite are drawn. No estimate of the variance of the mean and hence the precision with which the mean is estimated can be obtained from a single composite sample. It is not sufficient to analyze two or more subsamples from the same composite to obtain an estimate of the variation within the population. Such a procedure would permit the estimation of variation among subsamples within the composite, but not the variation among samples in the field. Similarly, if composites are formed from samples within different parts of a population, the variability among the parts, but not the variability within the parts, can be estimated. If an estimate of the variability among sampling units within the population is required, two or more samples taken at random within the population may be analyzed separately.

The accuracy with which a population mean is estimated from a composite is dependent on the variability among sampling units within the population and the number of such units included in the composite. Because of the difference in variability among the several characteristics of a soil and among the several types of sampling units, no generalizations can be made regarding the number of sampling units required for a composite sample. If an estimate of the variability among units is available, it can be used to determine the number of units to include in the composite to attain a given precision. The procedure is the same as that given in section 2–3.2 for estimating the size of a simple random sample. If no estimate of variability is available, compositing should be avoided if possible.

Subject to the foregoing restrictions, compositing may prove valuable in soil investigation. It permits the precision with which the mean is estimated to be increased by increasing the number of units included in the sample without, at the same time, increasing the cost of analysis.

2-7 REFERENCES

Allmaras, R. R., and C. O. Gardner. 1956. Soil sampling for moisture determination in irrigation experiments. Agron. J. 48:15–17.

Cline, Marlin D. 1944. Principles of soil sampling. Soil Sci. 58:275–288.

Cochran, W. G. 1946. Relative accuracy of systematic and stratified samples for a certain class of populations. Ann. Math. Statist. 17:164–177.

Cochran, W. G. 1977. Sampling techniques. 3rd ed. John Wiley and Sons, Inc., New York.

Das, A. C. 1950. Two-dimensional systematic sampling and associated stratified and random sampling. Sankhya 10:95–108.

Deming, W. E. 1950. Some theory of sampling. John Wiley and Sons, Inc., New York.

Finney, D. J. 1948. Random and systematic sampling in timber surveys. J. Forestry 22:64–99.

Fisher, R. A., and F. Yates. 1967. Statistical tables for biological, agricultural, and medical research. 6th ed. Oliver and Boyd, Edinburgh.

Hammond, L. C., W. L. Pritchett, and V. Chew. 1958. Soil sampling in relation to soil heterogeneity. Soil Sci. Soc. Am. Proc. 22:548–552.

Hansen, M. H., W. N. Hurwitz, and W. G. Madow. 1953. Sample survey methods and theory, Vols. I and II. John Wiley and Sons, New York.

Hasel, A. A. 1938. Sampling errors in timber surveys. J. Agric. Res. 57:713–736.

Haynes, J. D. 1948. An empirical investigation of sampling methods for an area. M.S. thesis, North Carolina State College, Raleigh, NC.

Hendricks, W. A. 1956. The mathematical theory of sampling. Scarecrow Press, New Brunswick, NJ.

Madow, W. G. 1949. On the theory of systematic sampling: II. Ann. Math. Statist. 20:333–354.

Madow, W. G. 1953. On the theory of systematic sampling: III. Ann. Math. Statist. 24:101–106.

Madow, W. G., and L. H. Madow. 1944. On the theory of systematic sampling: I. Ann. Math. Statist. 15:1–24.

Mason, D. D., J. F. Lutz, and R. G. Petersen. 1957. Hydraulic conductivity as related to certain soil properties in a number of great soil groups—sampling errors involved. Soil Sci. Soc. Am. Proc. 21:554–560.

Matérn, Bertel. 1960. Spatial variation. Stochastic models and their application to some problems in forest surveys and other sampling investigations. Meddelanden Fran Statens Skogsforskningsinstitut. Band. 49, NR. 5.

Milne, A. 1959. The centric system area-sample treated as a random sample. Biometrics 15:270–297.

Olson, R. A., A. F. Drier, and R. C. Sorensen. 1958. The significance of subsoil and soil series in Nebraska soil testing. Agron. J. 50:185–188.

Osborne, J. G. 1942. Sampling errors of systematic and random surveys of cover-type areas. J. Am. Statist. Assoc. 37:256–264.

Quenouille, M. H. 1949. Problems in plane sampling. Ann. Math. Statist. 20:355–375.

Rigney, J. A., and J. F. Reed. 1946. Some factors affecting the accuracy of soil sampling. Soil Sci. Soc. Am. Proc. 10:257–259.

Sayegh, A. H., L. A. Alban, and R. G. Petersen. 1958. A sampling study in a saline and alkali area. Soil Sci. Soc. Am. Proc. 22:252–254.

Smith, H. F. 1938. An empirical law governing soil heterogeneity. J. Agric. Sci. 28:1–23.

Williams, R. M. 1956. The variance of the mean of systematic samples. Biometrika 43:137–148.

Yates, F. 1948. Systematic sampling. Phil. Trans. Roy. Soc. London, Ser. A. 241:345–377.

Yates, F. 1953. Sampling methods for censuses and surveys, 2nd ed. Charles Griffen and Co. Ltd., London.

Zubrzycki, S. 1958. Remarks on random, stratified and systematic sampling in a plane. Colloquium Mathematicum 6:251–264.

3 Geostatistical Methods Applied to Soil Science

A. W. WARRICK AND D. E. MYERS

University of Arizona
Tucson, Arizona

D. R. NIELSEN

University of California
Davis, California

3-1 INTRODUCTION

Variations in soil properties tend to be correlated over space—both vertically and horizontally. That is, two values taken close together tend to be more alike than two samples far apart. Most often, the classical approach in the field is to group soils together in like units or lay out small plots and assume variability within the plots or units is purely random. Conceptually, geostatistics offers an alternative approach in that spatial correlations are quantified. Estimates for a property at an unsampled location will be principally determined by measurements made close by, rather than by assuming a class (or plot) average.

Historically, the methodology for geostatistics began in mining engineering for assessment of ore bodies by D. G. Krige, for whom "kriging" is named. Matheron (1973) provided a sound theoretical basis by the formation of random functions. The approach is obviously not a panacea; nevertheless, some very difficult concepts are addressed. For example, spatial and inter-variable correlation is quantified, optimum interpolation schemes are designed, the scale of the sample is considered, and samples for different support volumes can be included. Additionally, new sampling locations can be defined in order to best improve estimates for a total population or location. When sampling locations are far apart, the approach reduces to classical random fields.

Dimensionally, applications of geostatistics could be for distances of a few molecules or kilometers. Methods can be used to analyze any number of soil properties (physical, chemical, biological) and can be extended to include plant response and crop yields. The development of the techniques was for application to very practical problems—e.g., optimizing the selection of blocks of ore to be processed on a sliding economic scale

 Methods of Soil Analysis, Part 1. Physical and Mineralogical Methods–Agronomy Monograph no. 9 (2nd Edition)

according to the market price of the end product. So far, applications to soil problems are somewhat embryonic, with versatility not yet fully exploited. Obvious choices include interpolation for preparing maps, for either transient or invariant properties. Also, sampling to attain a given precision or locating new sampling points is a logical choice. The theory for conditional simulation has not been applied to any extent and neither have the interrelationships of correlated properties.

A comprehensive treatment of the subject is by Journel and Huijbregts (1978). Other readily available texts are by Clark (1979), David (1977) and Rendu (1978). In addition, there has been an explosion of articles in soil science and hydrology journals within the past few years.

3–2 QUANTIFICATION OF SPATIAL INTERDEPENDENCE

Spatial variations with interdependence are commonly described with a correlogram or a variogram. In either case, we consider a set of values $Z(x_1)$, $Z(x_2)$, . . ., $Z(x_n)$ at x_1, x_2. . . ., x_n where each location defines a point in 1-, 2-, or 3-dimensional space. It is not a requirement that the value be for an exact point, but rather that each value is for a defined *support volume* which is centered at x. For the correlogram $\rho(h)$ (which we will define shortly), strong stationarity is required, that is

Strong stationarity (stationarity of order 2)

1. $E[Z(x)]$ exists and is equal to the same constant value for all x.
2. The covariance exists and is a unique function of separation distance h.

A weaker assumption is sufficient for the variogram function $\gamma(h)$ to be defined, namely

Weak stationarity (the intrinsic hypothesis)

1. $E[Z(x)]$ exists as above
2. For all vectors h the variance of $Z(x + h) - Z(x)$ is defined and is a unique function of h.

A system which satisfies the strong stationarity requirements also satisfies the intrinsic hypothesis, but the converse is not true.

3–2.1 Correlograms

The correlogram $\rho(h)$ of the regionalized variable Z is defined by

$$\rho(h) = \text{Cov}[Z(x),Z(x + h)]/\sigma^2 \quad [1]$$

The covariance "Cov" is for any two values of Z at a distance h apart and σ^2 is the variance of Z. Thus, the correlogram is a series of correlations for a common variable where each couple is separated by distance h. In general, x and h are vector quantities and ρ will depend on the direction as well as the magnitude of h. The correlogram can have possible values from -1 to 1 just as can an ordinary correlation coefficient.

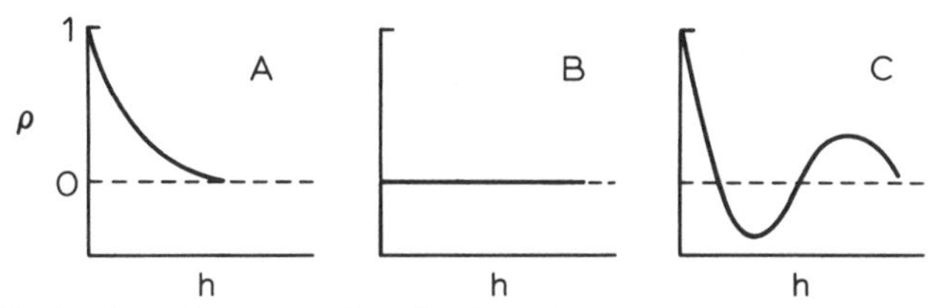

Fig. 3–1. Idealized correlograms: A is well-behaved and decreases monotonically; B is for independent or random values: and C is for a cyclical system.

Some "typical" correlograms are shown in Fig. 3–1. In Fig. 3–1A the maximum value of $\rho(h)$ is at $h = 0$ for which Eq. [1] trivially is 1. As h increases, the covariance decreases gradually until eventually, for large distances, ρ is zero and no correlation or spatial dependence exists. Thus, samples close together are alike; samples somewhat separated are less alike; and samples remote from each other are not correlated at all.

Figure 3–1B is for a variable which is not correlated over space. The value $\rho(0)$ is 1 as before, but for all other points $\rho(h)$ is zero, indicating the results are independent of separation distance. This is a purely random system. Fig. 3–1C shows results for a system showing a cyclical effect. As the distance h increases, the correlogram becomes alternatively positive and negative, and at large distances eventually approaches zero. Physical systems which may exhibit such a correlogram include sediment deposits for periodic flooding or compacted-noncompacted patterns due to wheel traffic in a row crop.

To estimate the correlogram function, let us assume $n(h)$ is the number of pairs of sample points a distance h apart. When dealing with a one-dimensional transect, perhaps all pairs of points will be in discrete classes; however, in general, a class size would be defined and $n(h)$ would refer to all pairs of points within that class. The sample covariance function is

$$C(h) = \left[\frac{1}{n(h) - 1} \right] \sum_{i=1}^{n(h)} \left[Z(x_i) - \overline{Z} \right] \left[Z(x_i + h) - \overline{Z} \right] \qquad [2]$$

and the sample correlogram is

$$r(h) = C(h)/s^2 \qquad [3]$$

with $\overline{Z}$ and s^2 as estimates of the mean and variance. Ideally, the number of data pairs used would be very large for each value of h, but in practice this is normally not the case, especially for extremely small and extremely large h. Following is an example illustrating a sample correlogram.

3–2.1.1 EXAMPLE 1. A SAMPLE CORRELOGRAM FOR A 100-POINT TRANSECT OF 1.5 MPa (15-BAR) WATER VALUES.

Gajem (1980) and Gajem et al. (1981) reported values for the water retained in soil samples collected at the 50-cm depth of a Pima clay loam

Table 3–1. Sequence of 100 values for percent water (g/g) retained at 1.5 MPa (15 bars)† (after Gajem, 1980).

16.1	18.0	16.7	19.7	15.3	15.4	15.2	14.7
17.0	16.9	16.8	18.4	17.7	18.5	18.3	18.4
17.5	20.9	15.9	19.7	16.4	17.7	17.6	17.1
18.9	18.3	17.9	19.9	18.7	17.8	20.4	18.3
17.5	17.7	19.6	16.9	17.0	17.3	17.2	18.1
15.7	15.8	17.8	17.3	17.2	16.7	16.9	16.0
17.3	18.5	17.7	17.5	18.0	15.9	13.9	19.5
17.1	20.0	19.9	18.8	19.4	16.9	21.2	19.5
16.5	16.8	19.3	16.5	16.1	16.0	16.4	16.6
14.0	16.9	16.3	16.9	17.6	17.3	17.7	17.9
18.7	19.9	23.1	20.9	20.4	24.8	20.3	21.4
21.4	21.2	21.6	21.7	22.3	20.9	19.8	20.8
20.8	21.4	21.2	22.6				

† Samples were at 50-cm depth, 20 cm apart, on Pima clay loam. The sequence goes from left to right.

(fine silty, mixed thermic family of Typic Torrifluvents). In Table 3–1 are 100 values of water content at 15-bar suctions. The 100 soil samples were collected on a 20-cm spacing along a 2000-cm transect with a 7.5-cm diameter bucket auger. The depth increment was 40 to 60 cm; the experimental mean and standard deviations were 18.3 and 2.1.

For n points equally spaced at Δh, Eq. [2] may equivalently be written as

$$C\,(k\Delta h) = \frac{(n-k)\,\Sigma Z_i Z_{i+k} - \Sigma Z_i \Sigma Z_{i+k}}{(n-k)(n-k-1)} \qquad [4]$$

where $Z_i = Z(i\Delta h)$ and the summations are taken from $i = 1$ to $n - k$.

If we choose for example Row 5 of Table 3–1 and take $n = 8$ and $k = 1$, the corresponding sums in Eq. [4] are

$$\sum_{i=1}^{7} Z_i = 123.2 \qquad \sum_{i=1}^{7} Z_{i+1} = 123.8$$

$$\sum_{i=1}^{7} Z_i Z_{i+1} = 2178.19$$

Therefore the sample covariance for 20 cm ($k = 1$ and $\Delta h = 20$) is

$$\mathrm{C}(20) = -4.83/42 = -0.115\,.$$

The corresponding estimate of the sample autocorrelation 1 for this row is

$$r(20) = \mathrm{C}(20)/(s^*)^2 = -0.151$$

with $s^* = 0.873$ based on the eight values. Similarly for 40 cm ($k = 2$) and the same row:

$$\sum_{i=1}^{6} Z_i = 106.0 \qquad \sum_{i=1}^{6} Z_{i+2} = 106.1$$

$$\sum_{i=1}^{6} Z_i Z_{i+2} = 1873.23$$

resulting in

$$C(40) = -7.22/30 = -0.241$$

$$r(40) = -0.241/(0.873)^2 = -0.316$$

As demonstrated, the sample values are erratic for small series. There is no exact minimum number of points necessary for estimating $r(h)$, but generally 100 points or so will suffice.

Table 3–2 shows values of $r(h)$ for h = 20 to 500 cm. The value 1.0 at h = 0 is included for completeness. Values are also plotted as Fig. 3–2A. The $r(h)$ values begin at about 0.6 and gradually decrease towards zero at about 250 cm. The estimated values are small negative values for h = 300 to 480 cm, but can be interpreted as insignificant correlations. Fig. 3–2 compares well with the idealized Type A of Fig. 3–1, with the

Table 3–2. Autocorrelation values for moisture content values of Table 3–1.

Lag	h	$r(h)$	Lag	h	$r(h)$
0	cm			cm	
0	0	1.00	13	260	0.01
1	20	0.58	14	280	0.06
2	40	0.61	15	300	−0.07
3	60	0.58	16	320	0.00
4	80	0.49	17	340	−0.11
5	100	0.45	18	360	−0.07
6	120	0.38	19	380	−0.08
7	140	0.37	20	400	−0.12
8	160	0.28	21	420	−0.10
9	180	0.17	22	440	−0.07
10	200	0.11	23	460	−0.04
11	220	0.12	24	480	−0.04
12	240	0.01	25	500	0.06

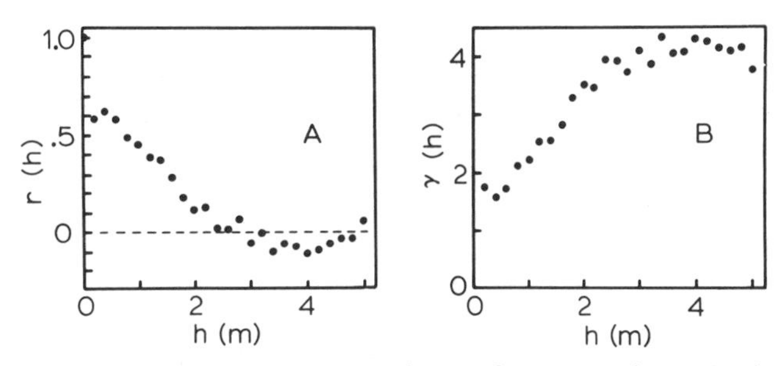

Fig. 3–2. (A) Sample correlogram and (B) variogram for the experimental values of Table 3–1.

exception that the values for small distance do not approach 1 but rather 0.6. This is not surprising, in view of the measurement error involved. On split samples from the same general area, the estimated standard deviation was determined to be approximately 0.009 (g/g) water, which would be about 45% of the estimate of the standard deviation. This will be addressed in more detail later with Example 2.

3–2.2 Variograms

The variogram $\gamma(h)$ is defined as

$$\gamma(h) = (1/2)\,\mathrm{Var}[Z(x) - Z(x + h)] \qquad [5]$$

with "Var" the variance of the argument. As for the correlogram, x and h are, in general, vectors. For 2 or 3 dimensions, both $\gamma(h)$ and $r(h)$ can be directionally dependent. Under the zero drift assumption $E[Z(x + h)] = E[Z(x)]$, and Eq. [5] is equivalent to

$$\gamma(h) = E[Z(x + h) - Z(x)]^2 \,. \qquad [6]$$

An estimate of γ is γ^* given by

$$\gamma^*(h) = \left[\frac{1}{2n(h)}\right] \sum_{i=1}^{n(h)} [Z(x_i + h) - Z(x_i)]^2 \qquad [7]$$

with $n(h)$ the number of pairs separated by a distance h.

If the strong stationarity conditions are met, then both $\rho(h)$ and $\gamma(h)$ exist and Eq. [1] and [6] may be used to show that

$$\gamma(h) = \sigma^2\,[1 - \rho(h)], \text{ (strong stationarity)} \qquad [8]$$

Examples of $\gamma(h)$ are given in Fig. 3–3. Fig. 3–3A shows a typical linear variogram starting at $\gamma(0) = 0$ and reaching a maximum or "sill" value $\gamma(h) = C$ for $h \geq a$. The value $h = a$ is called the *range* and is the maximum separation distance for which sample pairs remain correlated. In some cases, $\gamma(h)$ will remain nonzero as h approaches zero, as shown in Fig. 3–3B. This limiting value of $\gamma = C_0$ is called the *nugget*, so named because of the analogy in mining where a pure metal nugget exists and at any finite distance away a much lower concentration is found. For linear models having a sill, $\gamma(h)$ is

$$\begin{aligned} \gamma(h) &= C_0 + C(h/a), & h &< a \\ &= C_0 + C, & h &\geq a \,. \end{aligned} \qquad [9]$$

Another useful model is the spherical model illustrated by Fig. 3–3C and given by

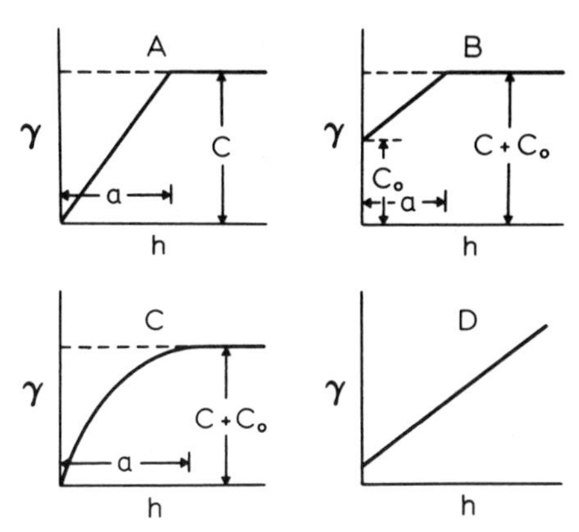

Fig. 3–3. Idealized variograms; A is a linear model with range a and sill C; B is a linear model with range a, sill C and nugget C_0; C is a spherical model with range a; and D is a linear model without a sill.

$$\gamma(h) = C_0 + C\,[(3/2)(h/a) - (1/2)(h/a)^3], \qquad 0 < h < a$$
$$= C_0 + C, \qquad h \geq a \qquad [10]$$

where, as before, C_0, $C_0 + C$, and a are the nugget, sill, and range, respectively.

The preceding examples of variograms assume that $\gamma(h)$ reaches a constant maximum value for h large. Such is not always the case. Figure 3–3D is for such a system and shows $\gamma(h)$ continuing to increase with h, at least on the scale over which the figure is drawn. For such a system, the intrinsic hypothesis is satisfied, but the strong stationarity conditions are not met. Consequently there can be no correlogram, as the variance is unbounded. However, the variogram function remains defined and is of value.

The choice of valid variogram models is restricted such that the negative of $\gamma(h)$ is a *positive-definite* function (Journel and Huijbregts, 1978; Armstrong and Jabin, 1981). It is best to use models known to behave properly. When a sill exists, these include a spherical model, an exponential model $\gamma(h) = 1 - \exp(-h/a)$, and a Guassian model $\gamma(h) = 1 - \exp(-h^2/a^2)$. When a sill does not exist, included are a power model $\gamma(h) = h^\alpha$ $(0 < \alpha < 2)$ and a logarithmic model $\gamma(h) = \log h$. Fortunately, linear combinations of the above models are also acceptable. The linear model with a finite sill of Eq. [9] is not a valid variogram in the true sense, and when applied can only be used with confidence in relating points within the range.

Spatial dependence may depend on separation distance only or on both distance and direction. If the variogram is a function of distance only, it is called isotropic; otherwise, anisotropic. All the models described above are isotropic.

The choice of the model used to approximate the sample variogram is often somewhat subjective. No fool-proof procedure exists for best-fitting all situations. Some guidelines however are to (i) use a valid functional form (at least within the range of application), (ii) lend more credibility to points with large numbers of pairs $n(h)$ in the determination, and (iii) pay more attention to short distances. As the behavior at short distances is critical, automatic curve fitting based on minimization of sums of squares of errors in general is not appropriate. The adequacy of a given variogram model may, to some extent, be judged by *jack-knifing*, to be discussed later.

3–2.2.1 EXAMPLE 2. A SAMPLE VARIOGRAM

Returning to Row 5 of Table 3–1, we now calculate a sample variogram value. For equally spaced observations, Eq. [7] reduces to

$$\gamma^*(k\Delta h) = (1/2n)\sum_{i=1}^{n}(Z_{i+k} - Z_i)^2.$$

With $k = 1$ and $n = 7$, the result for Row 5 is

$$\gamma^*(20) = (1/14)[(17.7 - 17.5)^2 + (19.6 - 17.7)^2 + \ldots + (18.1 - 17.2)^2] = 0.85\,.$$

Similarly with $k = 2$ and $n = 6$, the result for $\gamma(40)$ is

$$\gamma^*(40) = (1/2)[(19.6 - 17.5)^2 + (16.9 - 17.7)^2 + \ldots + (18.1 - 17.3)^2] = 1.05\,.$$

Of course, for such a short series, the values are not reliable.

For the total 100-point transect of Table 3–1, sample variogram values were calculated for each lag and given as Fig. 3–2B. The value for small distances is about 1.7 and increases to a value of about 4 at h = 250 cm. The values are shown up to about 500 cm or 1/4 of the total transect, roughly the limit of reliability. The results can be reasonably modeled by a linear or a spherical model (e.g. Fig. 3–3B, 3–3C) provided a nugget is included. Whether a sill exists is not clear; i.e. whether the variance would remain finite if the transect were much longer is not clear, but if it were, then Eq. [8] would relate $\rho(h)$ to $\gamma(h)$ and vice versa. We can compare for small h the approximate nugget value 1.7 to the sample $(2.1)^2[1 - (0.6)^2] = 2.8$. Similarly, the largest value shown on Fig. 3–2B is about 4 and can be compared to the sample variance $(2.1)^2$. If $\gamma(h)$ is extended beyond 500 cm with this data set, it decreases and then increases to about 6 at 1000 cm, but the results are not reliable for the longer distances.

3–2.3 Range of Influence and Integral Scale

A natural question to ask is "What is the range of influence beyond which values are independent of each other?" Such a limit exists provided that strong stationarity conditions are satisfied, or equivalently that the covariance function is defined for all h. Otherwise a degree of interdependence exists for all samples, although uncertainty regarding large distances can be circumvented somewhat by considering *moving neighborhoods* only. The distance for which $\rho(h)$ approaches zero in Fig. 3–1A is the range of influence for that example; for Fig. 3–3A, 3–3B or 3–3C, it will be the range of the variogram.

The *integral scale* (cf. Lumley and Panofsky, 1964; Bakr et al., 1978; Russo and Bresler, 1981) is defined by

$$\lambda = \int_0^\infty \rho(h)dh \qquad \text{(1-dimensional)} \qquad [11]$$

or

$$\lambda = [2\int_0^\infty h\rho(h)dh]^{0.5} \qquad \text{(2-dimensional)}\,. \qquad [12]$$

The estimated value depends to some extent on the choice of sampling points. There is a tendency for samples over larger regions to result in larger integral scales. Dependence of integral scale to measuring area is a complex problem involving sample variation, scale of measurement and stationarity.

3–2.3.1 EXAMPLE 3. REPORTED VALUES FOR RANGE OF INFLUENCE AND INTEGRAL SCALE

Some reported values of range of influence and integral scales are reported in Table 3–3 for a number of different parameters and locations. The definition and method of determination of range or scale differ somewhat. Overall generalizations are difficult, but, for the most part, the larger the area sampled, the larger is the range. For example, on the same site Gajem et al. (1981) found ranges of 1.5, 21, and 260 m for pH values of 100-member transects spaced at 0.2, 2, and 20 m. Campbell (1978) found random values on 8 × 20 grids at 10-m spacings, and Yost et al. (1982) obtained values of 14.3 km on long transects in Hawaii. Interpretations and comparisons are extremely difficult, but the overall scale of the experiment is a determining factor on the ranges. Simple explanations such as row directions, changing soil types, and topography offer a deterministic answer for some of the results but will not explain many others. The question of whether homogeneity and proper stationarity exist is always valid. Fortunately for many applications, the exact range is a moot consideration, as observations at closer distances dominate the end results—e.g. kriging estimates.

3–3 PUNCTUAL KRIGING

A primary application of geostatistics is for estimating values at locations where measurements have not been made. The most common

Table 3–3. Reported values of integral scale or range.

Source	Parameter	Range or scale	Site
		m	
Al-Sanabani (1982)	Log of saturated EC	>90	Tucson fine loam, Typic Haplargids (Arizona) 10 ha, 101 random samples, 0–30 cm in depth
Burgess and Webster (1980a)	Sodium	61	Approx. 50 ha, Plas Gogerddan (Gr. Britain), 440 samples, 0–15 cm depth
	Depth cover loam	100	Approx. 18 ha, Hole Farm (Gr. Britain), 450 observations
Campbell (1978)	Sand content	30	Ladysmith series, mesic Pachic arguistolls (Kansas), 8 × 20 grid at 10-m spacing in B2 horizons
	Sand content	40	Pawnee series, mesic Aquic Argiudoll (Kansas) (as above)
	Soil pH	Random	Pawnee and Ladysmith
Clifton and Neuman (1982)	Log of transmissivity	9 600	Avra Valley (Arizona), about 15 × 50 km, 148 wells
Folorunso and Rolston (1984)	Flux of N_2 and N_2O at surface	<1	Yolo loam, Typic Xerorthents (California) 100- by 100-m area
Gajem et al. (1981)	Sand content	>5	Pima clay loam, Typic torrifluvents (Arizona), 20-m transect, 20-cm spaces, 50-cm depth
	Soil pH	1.5	Pima, as above, 4 transects
		21	Pima, as above but 4 transects, 2-m spacing
		260	Pima, as above, 1 transect, 20-m spacing, 100 points
	1.5 MPa	0.6	Pima, 20-cm spacing, 4 transects, each 100 points
		>32	As above, 2-m spacing
		150	As above, 20-m spacing
Hajrasuliha et al. (1980)	Saturated EC	<80	Clay loam to loam, Haft Tappeh Plantation (Iran). 0–1m depth, 150 ha, 232 points, (Site 1)
	Saturated EC	>1 200	As above, 455 ha, 710 points (Site 3)
Kachanoski et al. (1985)	Depth of A-horizon and mass of A-horizon	<2	Mix of Typic Haploborolls and Typic Argiborolls (Saskatchewan)
Liss (1983)	Water-soluble organic carbon	<8	Yolo loam, Typic Xerorthents (California) 100- by 100-m area
	Soil water content of 0–10 cm soil depth	<16	As above

(continued on next page)

scenario would assume that $Z(x_1)$, $Z(x_2)$, . . ., $Z(x_n)$ are known at locations x_1, x_2, . . ., x_n and that the variogram (or correlogram) is determined as in the previous section. The question that remains is to estimate the value Z^* at position x_0. The procedure leads to not only an optimal

Table 3-3. Continued.

Source	Parameter	Range or scale	Site
		m	
Russo and Bresler (1981a, 1981b)	Saturated conductivity	34	Surface, Harma Red Mediterranean, Rhodoxeralf (Israel). 30 random sites in 0.8 ha.
		14	90-cm depth, as above.
	Saturated water content	76	Surface, as above.
		28	90 cm, as above.
	Sorptivity	37	Surface
		39	90 cm, as above.
	Wetting front	16–30	Simulated for above site, 1 to 12.5 h
Sisson and Wierenga (1981)	Steady-state infiltration	0.13	Sandy clay loam, Typic torrifluvent (New Mexico). 6.4- by 6.4-m plot, transect of 125 contiguous 5-cm rings
van Kuilenberg et al. (1982)	Moisture supply capacity	600	Cover sand, 30 mapping units, 9 soil types including Haplaquods, Humaquepts, and Psammaquents (Netherlands). 2 by 2 km, 1191 borings
Vauclin et al. (1983)	Sand content	35	Sandy clay loam (Tunisia). 7 × 4 grid at 10-m spacing, 20–40 cm depth
	pF 2.5	25	Same
Vauclin et al. (1982)	Surface soil temperature	8–21	Yolo loam clay, Typic Xerorthents (California). 60 and 100 m transects, 1-m spacing
Vieira et al. (1981)	Steady-state infiltration	50	Yolo loam, Typic Xerorthents (California). 55- × 160-m area.
Wollum and Cassel (1984)	Log of most probable number of *Rhizobium japonicum*		Pocalla loamy sand, thermic Arenis Plinthic (N. Carolina),
		1	0°, 3-m spacing
		>12	0°, 20-cm spacing
		Random	90°, 3-m spacing
		>12	90°, 20-cm spacing
Yost et al. (1982)	Soil pH	14 000–32 000	Various transects on Island of Hawaii at 1- to 2-km intervals, 0–15 cm depth.
	Phosphorus sorbed at 0.02 mg P/L	32 000	As above
	Phosphorus sorbed at 0.2 mg P/L	58 000	As above

solution for Z^*, but also an estimate of $\mathrm{Var}(Z - Z^*)$ which indicates the reliability of the result.

An estimate of Z^* is assumed to be a linear function (nonlinear estimates are rarely used) of known values:

$$Z^*(x_0) = \sum_{i=1}^{n} \lambda_i Z(x_i). \qquad [13]$$

The best linear estimate is found by choosing the weight factors λ_i such that the expected value and variance of $Z^*(x_0) - Z(x_0)$ are 0 and a minimum, respectively, that is,

$$E[Z^*(x_0) - Z(x_0)] = 0 \quad [14]$$

$$\text{Var}[Z^*(x_0) - Z(x_0)] = \text{a minimum}. \quad [15]$$

Of course, the "true value" $Z(x_0)$ is not known. If we assume the expected value of Z is not known, then the first condition (Eq. [14]) which guarantees $Z^*(x_0)$ to be an unbiased estimate results in

$$\sum_{i=1}^{n} \lambda_i = 1. \quad [16]$$

This leaves $\text{Var}[Z^*(x_0) - Z(x_0)]$ to be minimized subject to the constraint that the λ_i's sum to 1. This is done by introducing a Lagrangian multiplier -2μ and minimizing

$$\text{Var}\left[Z^*(x_0) - Z(x_0)\right] - 2\mu\left(\sum_{i=1}^{n} \lambda_i - 1\right). \quad [17]$$

By definition and by Eq. [13], it follows that

$$\text{Var}\left[Z^*(x_0) - Z(x_0)\right] = -\sum_{i=1}^{n}\sum_{j=1}^{n} \lambda_i\lambda_j\gamma_{ij} + 2\sum_{j=1}^{n} \lambda_j\gamma_{0j} \quad [18]$$

where γ_{ij} is defined by

$$\gamma_{ij} = \gamma(x_i - x_j). \quad [19]$$

Substitution of the right side of Eq. [18] for the "Var" term of Eq. [17] and taking the partial derivatives of the result with respect to each λ_i gives the set of linear equations (cf. Burgess and Webster, 1980a, esp. p. 310–321):

$$\sum_{j=1}^{n} \lambda_j\gamma_{ij} + \mu = \gamma_{i0}, \qquad i = 1, 2, \ldots, n. \quad [20]$$

In matrix notation, the equivalence is

$$A\begin{bmatrix} \lambda \\ \mu \end{bmatrix} = b \quad [21]$$

where

$$
A = \begin{bmatrix} \gamma_{12}\gamma_{21} & \cdots & \gamma_{n1} & 1 \\ \gamma_{21}\gamma_{22} & \cdots & \gamma_{n2} & 1 \\ \vdots & & \vdots & \vdots \\ \gamma_{n1}\gamma_{n2} & \cdots & \gamma_{nn} & 1 \\ 1 & \cdots & 1 & 0 \end{bmatrix} \quad [22]
$$

and the column matrices given by (the transposes of)

$$
\begin{bmatrix} \lambda \\ \mu \end{bmatrix}^{T} = [\lambda_1\lambda_2 \ldots \lambda_n\mu] \quad [23]
$$

$$
b^{T} = [\gamma_{10}\ \gamma_{20} \cdots \gamma_{n0}\ 1] \quad [24]
$$

Thus, the solutions for the λ's and μ are

$$
\begin{bmatrix} \lambda \\ \mu \end{bmatrix} = A^{-1}b \quad [25]
$$

Furthermore, by Eq. [18], the minimum estimation error is

$$
\sigma_E^2 = b^{T}\begin{bmatrix} \lambda \\ \mu \end{bmatrix} \quad [26]
$$

When the nugget is nonzero, Eq. [21] may be evaluated with γ_{ii} (for $i = 1$ to n) equal to the nugget or taken as 0. In either case, the weights λ_i are algebraically the same. The Lagrange multiplier when γ_{ii} is taken as 0 is $\mu + C_o$ where C_o is the nugget and μ the value when $\gamma_{ii} = C_o$.

An alternative to Eq. [21] is the "covariance" form where the γ_{ij} entries are replaced by the corresponding covariance (for i and j from 1 to n). The elements $a_{i,n+1}$ are changed from 1 to -1 for $i = 1$ to n. In place of Eq. [26], the kriging variance is $\sigma^2 - \mu - \Sigma\,\lambda_i C_{oj}$, where C_{oj} is the covariance corresponding to the distance between points "o" and "j".

The most tedious calculation in the procedure is the inversion of the matrix A. If the number of points used for the estimator is large, A and the necessary machine operations become unwieldy. Fortunately, the number of points necessary for the estimate may be relatively small (10 or less), as the inclusion of other weighting factors affects the results only negligibly. Another labor-saving feature, especially for data on regular grids, is that the A matrix is dependent upon the sampled locations only and not on the values of the Z_i's themselves. Thus, a separate inversion is not required for each $Z(x_0)$ of interest.

The adequacy of the estimates $Z^*(x_0)$ are described to some extent by the values of $\sigma_E{}^2$ (note that $\sigma_E{}^2$ varies throughout the domain of estimation). "Error maps" can be prepared to reveal areas where the variance is the highest and presumably any additional sampling should be made. A nicety is that the error maps depend on the variograms and locations only; thus alternative sampling grids or locations can be evaluated a priori (McBratney and Webster, 1981, 1983b).

An additional technique to test the adequacy of the estimates is by jack-knifing. This is accomplished by systematically evaluating $Z^*(x_i)$ for each data point as if it were unknown and then comparing the estimate to the known values. We will illustrate in the following example. This technique can be used to evaluate alternative variogram models, since generally the more acceptable model would give a lower variance between the *kriged* and actual value.

3–3.1 Example 4. Punctual Kriging–One-dimensional

We illustrate the calculations necessary for kriging by a simple one-dimensional case. Suppose the estimate is to be made for a missing point from Row 4 of Table 3–1. Specifically, assume that the fifth value (18.7) was missing and the estimate is on the basis of the nearest neighbors (19.9 to the left and 17.8 to the right). In order to use Eq. [21], $\gamma(0)$, $\gamma(20)$, and $\gamma(40)$ are required, which we estimate from Fig. 3–2B as 1.5, 1.6, and 1.7 for illustration. Thus, the system to be solved is

$$\begin{bmatrix} 1.5 & 1.7 & 1 \\ 1.7 & 1.5 & 1 \\ 1 & 1 & 0 \end{bmatrix} \begin{bmatrix} \lambda_1 \\ \lambda_2 \\ \mu \end{bmatrix} = \begin{bmatrix} 1.6 \\ 1.6 \\ 1 \end{bmatrix}$$

The solution is $\lambda_1 = 0.5$, $\lambda_2 = 0.5$ and $\mu = 0$. Thus, the estimated value is the simple arithmetic average

$$Z_0^* = \lambda_1 Z_1 + \lambda_2 Z_2 = 18.8$$

with a "kriging variance"

$$\sigma_E^2 = \mu + \sum_{i=1}^{2} \lambda_i \gamma_{i0} = 1.6.$$

That the λ's are 0.5 is not surprising, in that the neighbors are equidistant from the point to be estimated and the system is isotropic.

3–3.2 Example 5. Punctual Kriging–Two-dimensional

Suppose Z_0 is to be estimated from its three nearest points, using the hypothetical values below, and with $\gamma(h) = 4h$. (The grid spacing is 1 unit.)

·	·	·	·	·
	36			
	Z_3			
·	·	·	·	·
	33			
		Z_0?	Z_1	
·	·	·	·	·
35			42	33
			Z_2	
·	·	·	·	·
			39	

The closest three points are Z_1, Z_2, and Z_3. If relevant values for the variogram are

$$\gamma_{12} = \gamma_{10} = 4;\ \gamma_{13} = 4\sqrt{5};\ \gamma_{23} = 4\sqrt{8} = 8\sqrt{2}$$

$$\gamma_{20} = \gamma_{30} = 4\sqrt{2}$$

the system of Eq. [21] becomes

$$\begin{bmatrix} 0 & 4 & 8.9 & 1 \\ 4 & 0 & 11.3 & 1 \\ 8.9 & 11.3 & 0 & 1 \\ 1 & 1 & 1 & 0 \end{bmatrix} \begin{bmatrix} \lambda_1 \\ \lambda_2 \\ \lambda_3 \\ \mu \end{bmatrix} = \begin{bmatrix} 4 \\ 5.66 \\ 5.66 \\ 1 \end{bmatrix}$$

which has the solution

$$\lambda_1 = 0.398 \qquad \lambda_2 = 0.215$$
$$\lambda_3 = 0.387 \qquad \mu = -0.308.$$

Thus, the estimated value is

$$Z_0^* = \sum_{i=1}^{3} \lambda_i Z_i = 37.9$$

with a kriging variance of

$$\sigma_E^2 = \mu + \sum_{i=1}^{3} \lambda_i \gamma_{i0} = 4.69.$$

Note that the diagonal entries of A are all zero for this example, as the nugget is zero.

3–3.3 Example 6. Kriging Map for Salinity

Al-Sanabani (1982) sampled 101 random sites in a 10-ha field of a typic haplargid soil in southern Arizona. The soil samples were from the 0-to 30-cm depth and were analyzed for the electrical conductivity (EC) of the saturated extract. Values of EC ranged from 0.6 to 32 dS/m and were found to follow approximately a log-normal frequency distribution with a mean of 1.4 and variance of 0.70 for ln EC.

Figure 3–4 shows the estimated variogram for the 0 to 125 m. Values were calculated for a 10-m lag, with values also shown for 5-m lags out to h = 20 m. Also shown on the figure was $\gamma(h)$, given by the spherical model

$$\begin{aligned} \gamma(h) &= 0.3 + 0.6\,[1.5(r/160) - 0.5(r/160)^3], \qquad && r \leq 160 \\ &= 0.9 && r > 160\,. \end{aligned} \qquad [27]$$

(This model was verified to give low error between measured and kriged estimates, as discussed later.) Directional variograms were also calculated, but little if any difference by direction was in evidence. The values are not shown beyond 125 m, as the field was only 300 by 350 m. The relationship shows considerable dependence out to at least 100 m.

A map was prepared using the variogram model and the 10 closest points on a 15- by 15-m grid, using the computer algorithm of Carr et al. (1983). Contours were drawn with the results shown in Fig. 3–5A. Generally, a low salt region exists through the center of the field from southwest to northeast with ln EC < 1. A high salt area is to the east with a sizeable area in excess of ln EC = 2 to the east. The kriging variance σ_E^2 is mapped as Fig. 3–5B. The kriging variance indicates which regions

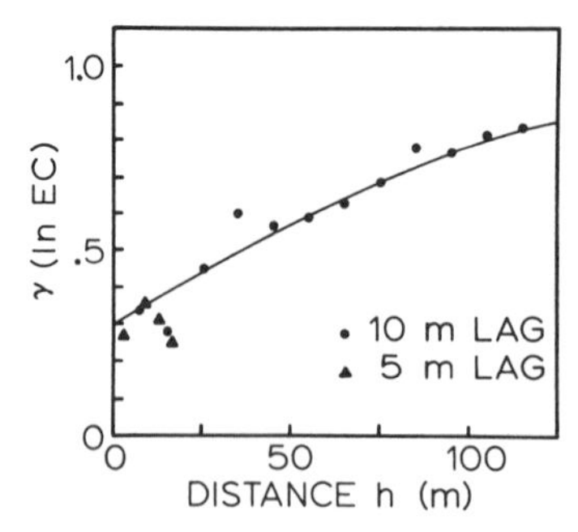

Fig. 3–4. Variogram for EC measurements. Solid line is spherical model with nugget, range, and sill as 0.3, 160, and 0.9, respectively. (Data from Al-Sanabi, 1982.)

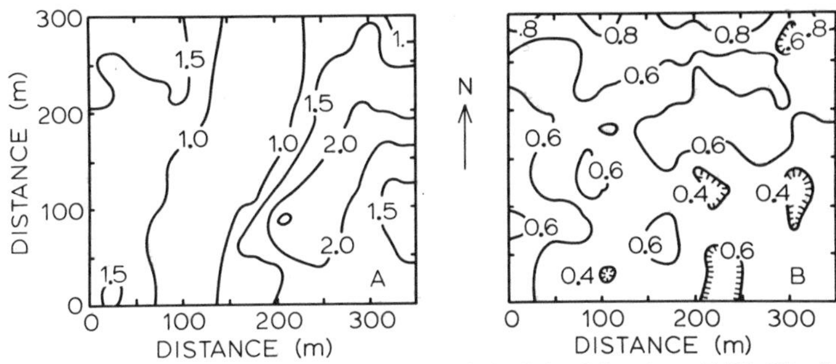

Fig. 3–5. Map of ln EC for a 10-ha field (A) and the kriged "variance" (B). Hatched line indicates interior depression.

tend to be known with the greatest confidence and is a function of the sampling pattern and variogram model.

An indication of the quality of the kriging estimates is by a "cross-validation" or "jack-knifing" technique. In this case one measured point at a time is excluded and a kriged estimate is made and compared with that measurement. Ideally, the estimates should be close to the experimental points and indicate no bias. Doing this for the 101 experimental points of the EC measurements and the variogram of Eq. [27] resulted in an average absolute difference of 0.010 with a variance of 0.57. As the difference is small, the model is judged to be adequate. The variance value of 0.60 is reasonably close to the average σ_E^2 of 0.49.

3–3.4 Example 7. Efficiency of Sampling—Infiltration

Vieira et al. (1981) measured limiting infiltration rates over a 160- by 55-m area of Yolo loam (fine-silty, mixed, nonacid, thermic Typic Xerothents). A total of 1280 measurements were made in eight columns of 160 measurements. Water was ponded in 46-cm diameter single rings for 36 hours, at which time steady rates were measured.

The 1280 values were approximately normally distributed with a mean value of 7.0 mm h^{-1} and variance of 7.8 mm^2h^{-2}. The sample values were highly dependent on position, as evidenced by the sample variogram of Fig. 3–6A. They addressed the question of what minimum number of samples would give results similar to the true 1280 (or more measured values). For a first trial, only 16 measured values are used for the entire field and the remaining 1264 positions was found by kriging. The 1264 kriged values resulted in a correlation coefficient (r) of about 0.16 with the actually measured points. The correlation coefficient is plotted by the first point on Fig. 3–6B. Repetition of the process was done assuming 32, then 64, 128, and 256 points were known resulted in increasing r values as shown in the figure. If an r^2 of 0.8 between measured and estimated values is acceptable, then 128 sampling points give similar results to the 1280 measured values. Figure 3–6C is a scatter diagram showing measured vs. kriged values based on 256 samples.

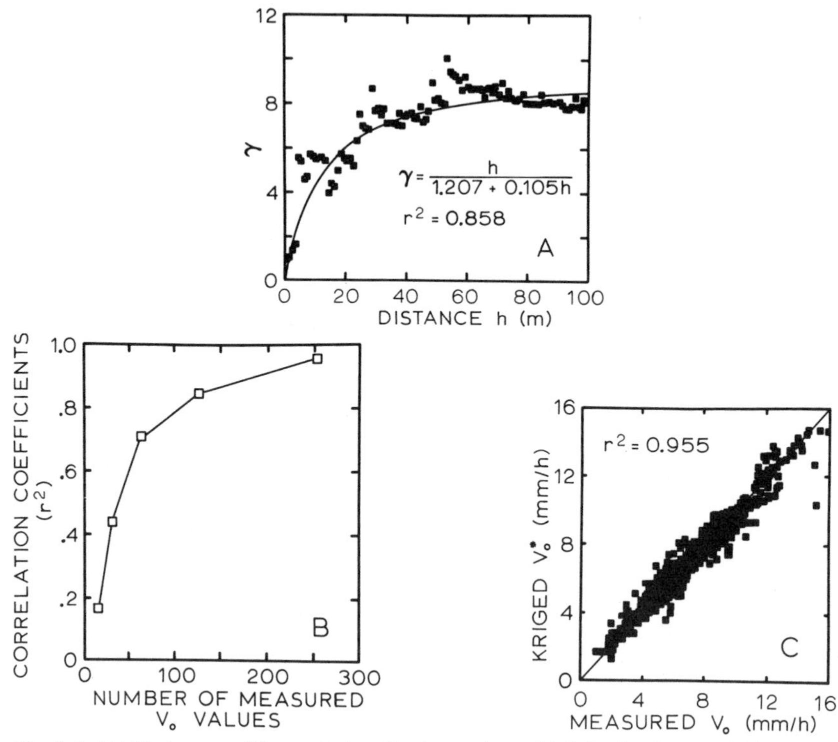

Fig. 3–6. (A) Variogram; (B) correlation for data points with kriged values; and (C) scatter diagram for 1024 data points with kriged values. (After Vieira et al., 1981.)

3–4 BLOCK KRIGING

Not only punctual (point) values are of interest, but also averages over finite regions. The average value $\overline{Z}(x_0)$ over the region V_0 is

$$\overline{Z}(x_0) = (1/V_0)\int_{V_0} Z(x)dx \qquad [28]$$

where V_0 is centered at x_0. The *block* V_0 can be a finite line, an area, or a volume, depending on whether Z is defined in one, two, or three dimensions. Block averages are smoother than point values and regional effects are displayed more clearly.

The kriging estimate $\overline{Z}^*(x_0)$ is

$$\overline{Z}^*(x_0) = \sum_{i=1}^{n} \lambda_i Z(x_i) \qquad [29]$$

(The support of the sample itself could be formally introduced into the discussion, but for simplicity we assume samples are taken at points.) Proceeding as before, Eq. [16] is still valid ($\Sigma\ \lambda_i = 1$) if the estimate is to be unbiased. Minimizing the variance, Var$[\overline{Z}^*(x_0) - \overline{Z}(x_0)]$, leads to (cf. Burgess and Webster, 1980b, esp. p. 335)

$$A\begin{bmatrix}\lambda\\ \mu\end{bmatrix} = s \qquad [30]$$

where the column vector s is the transpose of

$$s^T = [\bar{\gamma}(x_1, V_0), \bar{\gamma}(x_2, V_0), \ldots, \bar{\gamma}(x_n, V_0), 1] \qquad [31]$$

with $\bar{\gamma}$ the average of the variogram over the block, V_0, i.e.

$$\bar{\gamma}(x_i, V_0) = (1/V_0)\int_{V0} \gamma(x_i, x)dx \qquad [32]$$

The solution of the linear system of Eq. [32] is as for the punctual case. The resulting kriging variance, however, is

$$\bar{\sigma}^2 = \sum_{i=1}^{n} \lambda_i \bar{\gamma}(x_i, V_0) + \mu - (1/V_0)\int_{V_0} \bar{\gamma}(x, V_0)\,dx \qquad [33]$$

In general, the variance for block kriging is smaller than for point estimates by an amount corresponding to the last term. This last term of Eq. [33] is a "within block" term analogous to "within" effects of classical statistics. This is a special case of the extension variance of V by a support value v (cf. Journel and Huijbregts, 1978, p. 54):

$$\bar{\sigma}_E^2 = 2\bar{\gamma}(V,v) - \bar{\gamma}(V,V) - \bar{\gamma}(v,v) \qquad [34]$$

where

$$\bar{\gamma}(V,v) = [1/(vV)]\int\int_{v,V} \gamma(x_i, x_j)dvdV \qquad [35]$$

3-4.1. Example 8. Block Kriging for Salinity

The EC values of Example 6 were used with block kriging, resulting in the map of Fig. 3-7A. The kriging estimates are for the center of 50- by 50-m blocks based on the closest 10 points and were from the computer

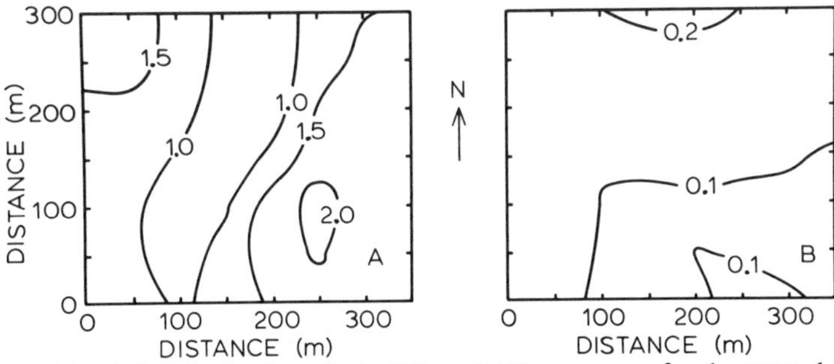

Fig. 3-7. (A) Block kriged values for ln EC, and (B) error map for the same data and variogram as for the punctual results in Fig. 3-5A.

code of Biafi (1982). The result is a smoothing of values compared to punctual kriging. The high value from southwest to northeast through the center is evident, but the boundaries are smoothed. The high value region (ln EC > 2) is reduced to a single area to the lower right. The error map for the block kriging is given as Fig. 3–7B. The values are much smaller than the values for punctual kriging (about 0.1, compared to 0.4 to 0.8). This is evidence of the error reduction analogous to the "within block" component.

3–5 SAMPLING STRATEGIES FOR SPECIFIED ESTIMATION ERROR

Scientists almost without exception are confronted again and again with the question of how many locations to sample and how to best locate them. The best known relationship is that the sampling number n necessary to be within a specified value of the population mean is

$$n = z_{\alpha}{}^2\sigma^2/(x - \mu)^2 \qquad [36]$$

where σ^2 and μ are the population variance and mean values, $|x - \mu|$ the allowable deviation to be attained $(1 - \alpha)$ (100)% of the time, and z_α the two-tailed normalized deviate ($z_{0.05} = 1.96$; $z_{0.1} = 1.645$; $z_{0.5} = 0.842$). The assumptions are (i) independence of samples and (ii) n sufficiently large that the "central limit theorem" applies. For application, the best estimate of σ^2 and μ would be used. (For estimating confidence limits when s^2 and n are already specified, the appropriate Student t value is inserted for z_α, but it should not be used to estimate a sampling number required.) The problem of sampling strategies and estimation variance has recently been addressed by Burgess et al. (1981), McBratney et al. (1981), and McBratney and Webster (1983a).

The classical approach to reducing the sampling size required is to logically break the area (or appropriate population) into classes. The appropriate model is

$$Z_{ij} = \mu_j + \epsilon_{ij} \qquad [37]$$

where μ_j is the class average and ϵ_{ij} is taken to be spatially uncorrelated. The best estimate at an unvisited site is

$$\bar{Z}_{ij} = \bar{Z}_j = [1/n(j)] \sum_{i=1}^{n(j)} Z_{kj} \qquad [38]$$

(assuming we know which class it belongs to). The estimation variance σ_E^2 is

$$\sigma_E^2 = \mathrm{Var}(\overline{Z}_{ij} - Z_{ij}) = \mathrm{Var}(\overline{Z}_j) + \mathrm{Var}(\epsilon_{ij})$$

or

$$\sigma_E^2 = \sigma^2 + \sigma^2/[n(j)] \quad [39]$$

The σ^2 is the within-class variance and $\sigma^2/[n(j)]$ is the variance of the class mean. By increasing $n(j)$, better and better estimates of the class mean are found but the estimation variance at an unvisited site can never be less than the estimation variance σ^2.

The above approach implies that the initial subdivision accounts for all spatial variations and that ϵ_{ij} is purely random. In addition, it assumes that class lines are sharp. This may be the best approach if only very limited data are available, but leads to overly conservative estimates of precision. An alternate is to use Eq. [13] for punctual kriging or Eq. [29] for block kriging. Example 4 illustrates this.

3–5.1 Example 9. Sampling Error

Burgess et al. (1981) and McBratney et al. (1981) considered the design of optimal sampling schemes using regionalized variables. Square and triangular sampling grids are considered. The maximum distance between an interpolated point and the nearest observation will be 0.62 for unit triangles and $1/(2)^{1/2} = 0.71$ for unit squares at x_0. The maximum estimation variance will correspond to these points. The kriged variance at the maximum point can be calculated a priori by Eq. [26] and will depend on $\gamma(h)$ and the spacing. In general, the triangular (hexagonal) geometry is more efficient (although not by much, since x_0 for the square grid has four close neighbors rather than 3). The slight advantage of the triangular pattern is largely mitigated by the simplicity of sampling on a square grid.

Firstly, linear variograms are considered with unit slope and nugget of 0, 1, and 2. The estimation was by the 25 nearest neighbors (a computer code OSSFIM is given by McBratney and Webster, 1981). Resulting estimation variance as a function of sample spacing is in Fig. 3–8A. For zero nugget (C_0 of Eq. [9]), the estimation variance increases from zero for very close spacings to about 1.8 for sampling at 4 units. The dashed lines correspond to the triangular grids and show only negligible improvement over the square grids. When a nugget exists, the estimation variance is correspondingly higher and cannot be reduced below the nugget value.

Also, results are for maximum estimation variance for punctual kriging of sodium at Plas Gogerddon. The field site is described in detail by Burgess and Webster (1980a). The variogram is assumed isotropic and linear up to $h = 15.2$ m, with γ as

$$\gamma(h) = 8.7 + (1.69/15.1)h, \qquad h < 60.1 . \quad [40]$$

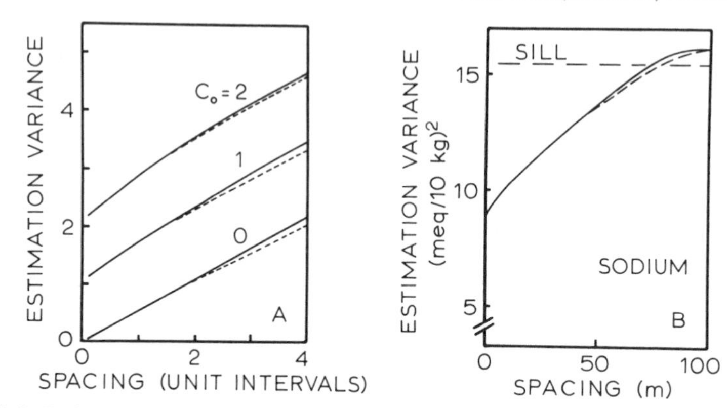

Fig. 3–8. Estimation of variance as a function of spacing. Effects of varying nuggets are shown in A. Sodium results at Plas Gogerddan are in B. Solid lines are for square patterns, dashed for triangular. (After Burgess et al., 1981.)

The resulting maximum estimation variance is given as Fig. 3–8B, with results again calculated by the nearest 25 neighbors. The value starts at 8.7 (the nugget) for very small spacings and increases with distance until it levels off at about 60 m which was the maximum distance over which dependence is assumed. The estimated variance overshoots the sill value for dependent samples. This is an anomaly of the calculations, in that 25 points were chosen, giving a long-range estimation of $[8.7 + (1.69/15.1)(60.1)](1 + 1/25) = 16.0$. Had more points been chosen, this would be reduced to the sill value of about 15.5. This reiterates that for spacings beyond the range of influence, kriging estimates reduce to the same results as for random sampling. Several more examples, including those for block kriging, are given in the above references.

3–6 FURTHER APPLICATIONS

3–6.1 Universal Kriging

Both the strong stationarity and intrinsic hypotheses imply an assumption of zero drift, that is,

$$E[Z(x + h) - Z(x)] = 0 \qquad [41]$$

In practice this assumption is scale-related and may be satisfied by partitioning the region into sub-regions. In other instances there may be a strong or pronounced drift, and Eq. [20] and [21] must be changed. Since this drift is generally unknown it is necessary to model or estimate it. The simplest model is given by

$$E[Z(x)] = \sum_{i=0}^{p} a_i f_i(x) \qquad [42]$$

where $f_0, f_1, \ldots, f_p$ are known, linearly independent functions, but $a_0, \ldots, a_p$ are unknown. While knowledge of these coefficients is necessary to estimate the variogram, it is not necessary for kriging. The estimator

$$Z^*(x_0) = \sum_{i=1}^{n} \lambda_i Z(x_i) \qquad [43]$$

is of the same form as Eq. [13] but Eq. [20] becomes

$$\sum_{i=1}^{n} \lambda_i \gamma(x_i, x_j) + \sum_{k=0}^{p} \mu_k f_k(x_j) = \gamma(x_0, x_j) \qquad [44]$$

and Eq. [42] becomes

$$\sum \lambda_i f_j(x_i) = f_j(x_0) \qquad j = 0, \ldots, p. \qquad [45]$$

The kriging variance is given by

$$\sigma_{UK}^2 = \sum \lambda_i \gamma(x_i, x_0) + \sum_{k=0}^{p} \mu_k f_k(x_0). \qquad [46]$$

The μ_k's are Lagrange multipliers corresponding to the $p + 1$ constraints which are required to insure that the estimator is unbiased. The kriging variance is in general larger due to the uncertainty associated with modelling the drift.

The simplest choices for the drift functions are polynomials, [e.g., in one dimension $f_0(x) = 1, f_1(x) = x, f_2(x) = x^2, \ldots$].

The difficulty in utilizing universal kriging is the circular nature of the problem. Estimation/fitting of $\gamma(h)$ requires the drift function, but the coefficients in Eq. [42] can only be estimated optimally if the variogram is known. Some authors have used least squares to fit the drift. This will result in a bias in estimating the variogram. If a linear variogram is used this bias is known. Burgess and Webster (1980b) have given an example of universal kriging and show the computation of the bias.

Even in mining applications, universal kriging has received limited application because of these difficulties. There are several possible solutions. If the drift is at most second order, the samples are taken on a regular grid or transect, and a linear variogram is postulated, then least squares fitting can be used to model the drift and the bias in the variogram compensated. If the data are in two or three dimensions, the drift may be only in one direction and the variogram can be modeled using only data in other directions. In this case universal kriging may be used without difficulty. More recently a different formulation using generalized covariances has been developed. As yet there are few if any examples of applications to problems in soil science. There are several computer programs commercially available such as BLUEPACK (available through

Geomath, Inc., 4891 Independence Street, Suite 250, Wheatridge, CO 80023). The method is described in greater detail in Delfiner (1976).

3–6.2 Co-Regionalization

Kriging utilizes the spatial dependence of a particular soil characteristic. Some attributes such as clay and wilting percentage are dependent. This dependence can be used in estimation as well as the spatial dependence. When one or more variables are estimated by a linear combination using both the spatial and inter-variable dependence, the technique is known as *co-kriging* or *co-regionalization.*

Co-kriging is utilized for several kinds of problems. For example, Vauclin et al. (1983) reported available water content (AWC), pF 2.5 (pF = log of soil matric potential as cm of water), and sand values sampled on a regular grid at 10-m intervals. The AWC and pF 2.5 values were then co-kriged on a 5-m grid. It is also possible to replace missing or insufficient data for one variable by data on other variables by utilizing the intervariable dependence. This is known as the "undersampled problem." In both forms of co-kriging the objective is to reduce the kriging variance.

The most compact form of the co-kriging equations is given in matrix form. Let $Z_1(x), \ldots, Z_m(x)$ be the random functions representing the soil attributes where x is a location in 1-, 2-, or 3-space. If

$$\overline{Z}(x) = [Z_1(x), \ldots, Z_m(x)] \qquad [47]$$

and $x_1, \ldots, x_n$ are sample locations, the estimator for $\overline{Z}(x)$ is written as

$$Z^*(x) = \sum_{j=1}^{n} Z(x_j)\Gamma_j \qquad [48]$$

where $\Gamma_1, \ldots, \Gamma_n$ are $m \times m$ matrices with entries λ^i_{jk}. The λ^i_{jk} is the weight for location i given to Z_j in estimating Z_k.

The system of equations used to obtain the Γ_i's is given by

$$\begin{aligned} &\sum \overline{\gamma}(x_i - x_j)\Gamma_i + \overline{\mu} = \overline{\gamma}(x_i - x_0) \\ &\sum \Gamma_i = I \end{aligned} \qquad [49]$$

Each $\overline{\gamma}$ is an $m \times m$ matrix whose entries are variograms (on the diagonal) and cross-variograms. For the undersampled problem the $\overline{\gamma}$'s must be modified by inserting zeros in appropriate places.

McBratney and Webster (1983b) and Vauclin et al. (1983) did not use the matrix form, since only two variables were considered. The details of the matrix form are found in Myers (1982, 1983).

To apply co-kriging it is necessary to model variograms for each variable separately as well as cross-variograms for all pairs. Examples are found in McBratney and Burgess (1983a, 1983b) and Vauclin et al. (1983), with theoretical results given in Myers (1982). A computer program utilizing an iterative method for solving the large system of equations as well as the undersampled option is given in Carr et al. (1983).

3–6.3 Conditional Simulation

Kriging is formulated in terms of random functions, although only one realization of the random function is available. Often another realization is sought which exhibits the spatial variability and known values at the sample locations.

By the use of a random number generator one can generate many values of a random variable with a desired distribution. However, for one value of each of many dependent random variables a different technique is necessary. Generation of the set of values, one each for a set of spatially dependent random variables, is called simulation. Conditional simulation produces a simulation such that at the sample locations the estimate coincides with the sample value.

Unlike some laboratory or field experiments, a new value cannot be obtained by repeating the experiment. If the same locations are sampled then the same results (except for instrument or analysis error) should be obtained. Sampling new locations provides additional data on the same realization. For another, realization simulation is necessary. Delhomme (1979) has used conditional simulation to study the Bathonian aquifer in France. In particular, simulated values were obtained for the log of the transmissivity.

The spatial dispersion in soils of nutrients, water, or pollutants could be studied by producing many simulations. Conditioning or matching the simulation to the data at sample locations is accomplished by using kriging. With any location x where a simulated value is to be made, then

$$Z(x) = Z^*(x) + [Z(x) - Z^*(x)]$$

where Z^* is the kriging estimator. Moreover, the $Z(x) - Z^*(x)$ has a mean of zero; in fact, at a sample location $Z(x) - Z^*(x) = 0$. The procedure then is to simulate values of $Z(x) - Z^*(x)$ (except at sample locations) and add to $Z^*(x)$.

To produce simulations in 2- or 3-space, the *turning bands method* was developed by Matheron (1973) and described in more detail in Journel and Huijbrets (1978). The simulated value at a point in 3-space is the sum of 15 simulations on equally spaced lines. These lines are determined by the edges on an iscosahedron, which provides the optimal polygonal approximation to a sphere.

The simulations on lines are produced by a *moving average* as described in Box and Jenkins (1970). It is necessary to relate covariances in 3-space to a covariance in 1-space and then represent this covariance as a convolution. This decomposition is known for the standard covariance models used in geostatistics. Lenton and Rodriquez-Iturbe (1977) and Smith and Freeze (1979b) have given examples of other methods for simulation of rainfall and groundwater flow.

3–6.4 Miscellaneous Comments and Notes

Variograms—As noted above, valid models must satisfy the positive-definite condition. The use of an invalid model can give rise to a negative estimated variance. The collection of valid models can be enlarged by nesting, that is, by using positive linear combinations of valid models. The shape of the basic models can then be modified substantially by nesting and hence match the sample variogram more closely. Nesting is also a method for constructing an anisotropic model from isotropic models. It should also be noted that replacing γ by $a + \gamma$ or $a\gamma$ has no effect on the kriged values, but the kriging variance is changed correspondingly.

Unique vs. Moving Neighborhoods—The kriging estimator utilizes spatial correlation in such a way that sample locations close to the location for which the estimate is desired receive greater weights and those farther away lesser weights. The sum of the weights is 1.0, but the weights are not constrained to be non-negative. Thus, negative weights can lead to negative estimated values. There is no theoretical contradiction, but for physical phenomena negative values are often not realistic. One solution is to use only those locations that are close; that is, for each location to be estimated, a different neighborhood is used. Another advantage of using moving neighborhoods is that the coefficient matrix for the kriging equations is much smaller. One disadvantage when contouring is that the moving neighborhoods lead to discontinuities in the plot.

Computing—One of the advantages of kriging is its simplicity of application. Computing sample variograms requires finding average squared distances only, and the kriged values require solving only a linear system. In addition to the complexities of data handling, inverting large matrices (i.e., solving large systems) can create problems. In general, the coefficient matrix is not positive definite when variograms are used and pivotal methods do not work well. There are several ways to avoid this problem. Frequently the variogram is replaced by the corresponding covariance. This avoids all the zeros on the diagonal. Alternatively the matrix can be partitioned; the upper left block containing the variogram values is positive definite. Finally, iterative methods such as the projection method are useful. A version of this method is embedded in the co-kriging program of Carr et al. (1983). One advantage of the projection method is that the universality conditions are checked at each iteration. The subroutine used to solve linear systems is a major component of any kriging program.

Screening, Declustering and Smoothing—Because spatial correlation is incorporated in kriging, several sample locations in close proximity constitute related information. The weights assigned by kriging compensate for this clustering. With this in mind, it is seen that cluster sampling is not efficient, although there may be other factors such as screening: locations screened by other locations will receive lesser weights. Finally it should be noted that the kriged random function Z^* is smoother than the original function Z; that is, the variance of Z^* is smaller than the variance of Z.

3–7 DISCUSSION

Several points regarding geostatistics should be reiterated or stated more explicitly:

1. The applications are relatively new, especially to soil science.
2. The applications are quite general in terms of processes, scale of measurement, and type of application.
3. The results very often give the same answer as more conventional statistics.
4. There are many "gray" areas, partially because of (1) but also because of general uncertainty in the assumptions.

Most applications to this date emphasize mapping and contouring. These include efforts towards general surveying as well as a number of physical and chemical parameters. Although these are of obvious benefit, they need not be the final product. In fact, the background of geostatistics is heavily rooted in practical economics of mining; thus possibilities for use in operational analysis in soil science seem feasible.

The parameters addressed have been strongly slanted towards chemical and physical properties. There are no inherent reasons to eliminate any spatially variable property—e.g., yields, plant nutrients, or microbes. Likewise applications have been mostly for a few centimeters to kilometers, but any meaningful scale can be used.

In many cases, results are trivial. In Example 2, after a lengthy analysis, the unknown value was estimated by the average of its two neighbors—hardly surprising in retrospect, but pointing out that the results should be reasonable and must be dependent on available information and data. In general, as spacings between samples get larger, the potential advantages of regionalized variables are less; results approach those based on independence of samples and agree fully with other statistical approaches.

There are many "gray" areas which underscore basic uncertainties regarding natural systems. The form of the variogram influences the analysis; yet it cannot be known with certainty, and its best approximation cannot be given with confidence. Whether and to what degree stationarity exists is usually not really known. In the case of universal kriging, the relationship between estimators of the drift and bias in the results becomes clumsily entangled. The complications, however, are mostly a consequence of a more detailed analysis of the system—the absolute status, which cannot be totally established.

The challenge then is to use geostatistics as a tool where favorable and advantageous. Recognition of best-suited situations hopefully will come into focus with time, experience, and further development. How to best complement existing knowledge and techniques is a key ingredient. Examples are how to best interface with existing soil survey and descriptions, or how to best address contemporary problems such as raising or maintaining productivity, protecting the environment, or utilizing less energy and water—all under difficult economic and/or social constraints.

3-8 REFERENCES

Agterberg, F. P. 1974. Geomathematics. Elsevier, New York.

Al-Sanabani, M. 1982. Spatial variability of salinity and sodium adsorption ratio in a typic haplargid soil. M.S. Thesis, The Univ. of Arizona, Tucson.

Armstrong, M., and R. Jabin. 1981. Variogram models must be positive definite. Math. Geol. 13:455–459.

Bakr, A. A., L. W. Gelhar, A. L. Gutjahr, and H. R. MacMillan. 1978. Stochastic analysis of spatial variability of subsurface flow. I. Comparison of one and three dimensional flows. Water Resour. Res. 14:263–272.

Biafi, E. Y. 1982. Program user's manual for basis geostatistics systems. Dept. of Mining and Geol. Engr., Univ. of Arizona, Tucson.

Box, G. P., and G. M. Jenkins. 1970. Time series analysis forecasting and control. Holden-Day, San Francisco.

Bresler, E., and G. Dagan. 1983. Unsaturated flow in spatially variable fields. 2. Application of water flow models to various fields. Water Resour. Res. 19:421–428.

Burgess, T. M., and R. Webster. 1980a. Optimal interpolation and isarithmic mapping of soil properties I. The variogram and punctual kriging. J. Soil Sci. 31:315–331.

Burgess, T. M., and R. Webster. 1980b. Optimal interpolation and isarithmic mapping of soil properties II. Block kriging. J. Soil Sci. 31:333–341.

Burgess, T. M., R. Webster, and A. B. McBratney. 1981. Optimal interpolation and isarithmic mapping of soil properties VI. Sampling strategy. J. Soil Sci. 31:643–659.

Campbell, J. B. 1978. Spatial variability of sand content and pH within continuous delineations of two mapping units. Soil Sci. Soc. Am. J. 42:460–464.

Carr, J., D. Myers, and C. Glass. 1983. Co-kriging: A computer program. Dept. of Mathematics, Univ. of Arizona, Tucson.

Clark, I. 1979. Practical geostatistics. Applied Sci. Pub., Ltd., London.

Clifton, P.M., and S.P. Neuman. 1982. Effects of kriging and inverse modeling on conditional simulation of the Avra Valley in southern Arizona. Water Resour. Res. 18:1215–1234.

Dagan, G., and E. Bresler. 1983. Unsaturated flow in spatially variable fields. 1. Derivation of models of infiltration and redistribution. Water Resour. Res. 19:413–420.

David, M. 1977. Geostatistical ore reserve estimation. Elsevier, New York.

Davis, J. C. 1973. Statistics and data analysis in geology. John Wiley and Sons, Inc., New York.

Delfiner, P. 1976. Linear estimation of non-stationary phenomena, *In* M. Guarascio, M. David, and Ch. Huijbregts (eds.) Advanced geostatistics in the mining industry. Reidel. NATO Symposium. pp. 49–68.

Delhomme, J. P. 1979. Spatial variability and uncertainty in groundwater flow parameters: a geostatistical approach. Water Resour. Res. 15:269–280.

DuBrule, O. 1983. Two methods with different objectives: splines and kriging. Math. Geol. 15:245–258.

Folorunso, O. A., and D. E. Rolston. 1984. Spatial variability of field measured denitrification gas fluxes. Soil Sci. Soc. Am. J. 48:1214–1219.

Gajem, Y. M. 1980. Spatial structure of physical properties of a typic Torrifluvent. M.S. Thesis, Univ. of Arizona, Tucson.

Gajem, Y. M., A. W. Warrick, and D. E. Myers. 1981. Spatial dependence of physical properties of a typic Torrifluvent soil. Soil Sci. Soc. Am. J. 46:709–715.

Hajrasuliha, S., N. Baniabbassi, J. Matthey, and D. R. Nielsen. 1980. Spatial variability of soil sampling for salinity studies in southwest Iran. Irrig. Sci. 1:197–208.

Hassan, H. M., A. W. Warrick, and A. Amoozegar-Fard. 1983. Sampling volume effects on determining salt in a soil profile. Soil Sci. Soc. Am. J. 47: 1265–1267.

Hawley, M. E., H. M. Richard, and J. J. Thomas. 1982. Volume-accuracy relationship in soil moisture sampling. J. Irrig. and Drain. Proc., ASCE 108:1–11.

Journel, A. G., and Ch. Huijbregts. 1978. Mining geostatistics. Academic Press, London.

Kachanoski, R. G., D. E. Rolston, and E. deJong. 1985. Spatial and spectral relationships of soil properties and microtopagraphy: I. Density and thickness of A horizon. Soil Sci. Soc. Am. J. 49:804–812.

Lenton, R. L. and I. Rodriquez-Iturbe. 1977. A multidimensional model for the synthesis of processes of areal rainfall averages. Water Resour. Res. 13:605–612.

Liss, H. F. 1983. Spatial and temporal variability of water soluble organic carbon for a cropped soil. Ph.D. Dissertation. Univ. of California, Davis.

Lumley, J. L., and A. Panofsky. 1964. The structure of atmospheric disturbance. John Wiley and Sons, Inc., New York.

Luxmoore, R. J., B. P. Spalding, and I. M. Munro. 1981. Areal variation and chemical modification of weathered shale infiltration characteristics. Soil Sci. Soc. Am. J. 45:687–691.

Matheron, G. 1973. The Intrinsic random functions and their applications. Adv. Appl. Prob. 5:239–465.

McBratney, A. B., and R. Webster. 1981. The design of optimal sampling schemes for local estimating and mapping of regionalized variables. II. Program and examples. Computers Geosci. 7:335–365.

McBratney, A. B., and R. Webster. 1983a. How many observations are needed for regional estimation of soil properties? Soil Sci. 135:177–183.

McBratney, A. B., and R. Webster. 1983b. Optimal interpolation and isorithmic mapping of soil properties. V. Co-regionalization and multiple sampling strategies. J. Soil Sci. 34:137–162.

McBratney, A. B., R. Webster, and T. M. Burgess. 1981. The design of optimal sampling schemes for local estimating and mapping of regionalized variables. I. Theory and method. Computers Geosci. 7:331–334.

Myers, D. E. 1982. Matrix formulation of co-kriging. Math. Geol. 14:248–257.

Myers, D. E. 1983. Estimation of linear combinations and co-kriging. Math. Geol. 15:633–637.

Myers, D. E. 1984. Co-kriging: new developments. p. 295–305. *In* G. Verly, M. David, A. Journel, and A. Marechal (ed.) Geostatistics for natural resource characterization. D. Reidel Publ. Co., Dordrecker, The Netherlands.

Rendu, J. M. 1978. An introduction to geostatistical methods of mineral evaluation. S. Afr. Inst. Mining and Metallurgy, Johannesburg.

Russo, D., and E. Bresler. 1981a. Effect of field variability in soil hydraulic properties on solutions of unsaturated water and salt flows. Soil Sci. Soc. Am. J. 45:675–681.

Russo, D., and E. Bresler. 1981b. Soil hydraulic properties as stochastic processes: I. An analysis of field spatial variability. Soil Sci. Soc. Am. J. 45:682–687.

Russo, D., and E. Bresler. 1982. Soil hydraulic properties as stochastic processes: II. Errors of estimates in a heterogeneous field. Soil Sci. Soc. Am. J. 46:20–26.

Sharma, M. L., G. A. Grander, and C. G. Hunt. 1980. Spatial variability of infiltration in a watershed. J. Hydrol. 45:101–122.

Sisson, J. B., and P. J. Wierenga. 1981. Spatial variability of steady-state infiltration rates as a stochastic process. Soil Sci. Soc. Am. J. 45:699–704.

Smith, L. 1981. Spatial variability flow parameters in a stratified sand. Math. Geol. 13:1–21.

Smith, L., and R. A. Freeze. 1979a. Stochastic analysis of steady state groundwater flow in a bounded domain, I. One-dimensional simulations. Water Resour. Res. 15:521–528.

Smith, L., and R. A. Freeze. 1979b. Stochastic analysis of steady state groundwater flow in a bounded domain, II. Two-dimensional simulations. Water Resour. Res. 15:1543–1559.

van Kuilenburg, J., J. J. DeGruijter, B. A. Marsma, and J. Bouma. 1982. Accuracy of spatial interpretation between point data on soil moisture supply capacity compared with estimations from mapping units. Geoderma 27:311–325.

Vauclin, M., S. R. Vieira, R. Bernard, and J. L. Hatfield. 1982. Spatial variability of surface temperature along two transects of a bare soil. Water Resour. Res. 18:1677–1686.

Vauclin, M., S. R. Vieira, G. Vachaud, and D. R. Nielsen. 1983. The use of co-kriging with limited field soil observations. Soil Sci. Soc. Am. J. 47:175–184.

Vieira, S. R., D. R. Nielsen, and J. W. Biggar. 1981. Spatial variability of field-measured infiltration rate. Soil Sci. Soc. Am. J. 45:1040–1048.

Webster, R. 1977. Quantitative and numerical methods in soil classification and survey. Oxford Univ. Press.

Webster, R., and T. M. Burgess. 1980. Optimal interpolation and isarithmic mapping of soil properties. III. Changing drift and universal kriging. J. Soil Sci. 31:505–524.

Wilding, L. P., and L. R. Drees. 1978. Spatial variability: a pedologist's viewpoint. pp. 1–12. *In* Diversity of Soils in the Tropics. Special Pub. 34. Am. Soc. of Agronomy, Soil Sci. Soc. Am., 677 S. Segoe Rd., Madison, WI 53711.

Wollum, A. G., II, and D. K. Cassel. 1984. Spatial variability of *Rhizobium japonicum* in two North Carolina soils. Soil Sci. Soc. Am. J. 48:1082–1086.

Yost, R. S., G. Uehara, and R. L. Fox. 1982. Geostatistical analysis of soil chemical properties of large land areas. I. Variograms. Soil Sci. Soc. Am. J. 46:1028–1032.

4 Extraneous Values

W. J. DIXON

University of California
Los Angeles, California

4-1 INTRODUCTION

Every experimenter has at some time or other faced the problem of whether certain of his observations properly belong in his presentation of measurements obtained. He must decide whether these observations are valid. If they are not valid the experimenter will wish to discard them, or at least treat his data in a manner which will minimize their effect on his conclusions. Frequently, interest in this topic arises only in the final stages of data processing. It is the author's view that a consideration of this sort is more properly made at the recording stage or perhaps at the stage of preliminary processing.

This problem will be discussed in terms of the following general models: We assume that observations are independently drawn from a particular distribution; alternatively, we assume that an observation is occasionally obtained from some other population, and that there is nothing in the experimental situation to indicate that this has happened except what may be inferred from the observational reading itself.

We assume that if no extraneous observations occur, the observations (or some transformation of them, such as logs) follow a normal distribution. The rules to be recommended are based on the assumption that the occasional extraneous observations are either from a population with a shifted mean, or from a population with the same mean and a larger variance. These assumptions may not be completely realistic, but procedures developed for these alternatives should be helpful.

If one is taking observations where either of these models applies, there remain two distinct problems.

First, one may attempt to designate the particular observation or observations which are from the different populations. One may be interested in this selection either to decide that something has gone wrong with the experimental procedure resulting in this observation (in which case he will not wish to include the result), or that this observation gives an indication of some unusual occurrence which the investigator may wish to explore further.

The second problem is not concerned with designating the particular observation that is from a different population, but with obtaining a procedure of analysis not appreciably affected by the presence of such observations. This second problem is of importance whenever one wishes to *estimate* the mean or variance of the basic distribution or to test hypotheses on the mean in a situation where unavoidable contamination occasionally occurs.

The first problem—designating the particular observation—is of importance in looking for "gross errors" or outliers, or the best or largest of several different products. Frequently the analysis of variance test for difference in means is used in the latter case. This is not a particularly good procedure, since many types of inequality of means have the same chance of being discovered. It should be noted that the power of the analysis of variance test decreases as more products are considered when testing a situation of one product different from others that are all alike.

The problem of testing particular apparently divergent observations (outliers) was discussed in a paper by Dixon (1953). The power of numerous criteria was investigated, and recommendations were made for various circumstances.

4–2 THE PROBLEM OF ESTIMATION (USE OF THE MEDIAN AND RANGE)

The median M of a small number of observations can usually be determined by inspection. Although it is less efficient than the average $\bar{x}$ if the population is normally distributed, M may be more efficient than the average $\bar{x}$ if gross errors are present. A chemist is frequently faced with the problem of deciding whether or not to reject an observation that deviates greatly from the rest of the data. If the observation is actually a gross error, it will have an undesirable effect on estimates using that value. The median is obviously less influenced by a gross error than is the average. It may be desirable to use the median to avoid deciding whether a gross error is present. It has been shown by Lieblein (1952) that, for three observations from a normal population, the median is better than the "best two out of three," i.e., the average of the two closest observations. It has also been shown (Dixon, 1953) that the median is better than the mean if as many as 3 to 5% of the observations are displaced by as much as 3 or 4 standard deviations from the mean.

No attempt is made to give a complete treatment of the problem of gross errors here, but an approach based on a fairly complete summary of this problem (Dixon, 1950; Dixon, 1953; Dixon & Massey, 1983) is given in section 4–5.

The following series of observations from Dean and Dixon, 1953, represents calculated percentages of sodium oxide in soda ash. The data have been arranged in order of magnitude, and are presented graphically in Fig. 4–1 and 4–2.

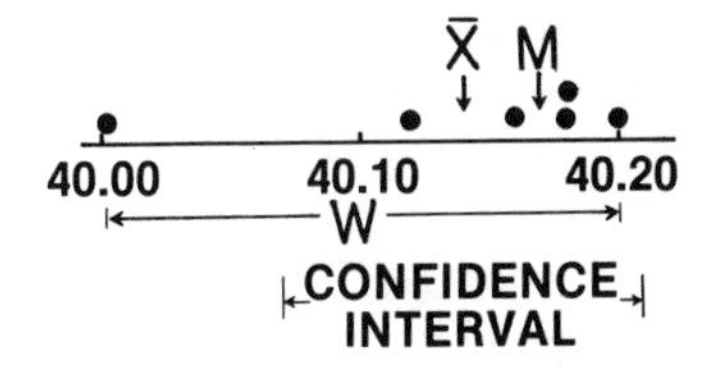

Fig. 4–1. Graphical presentation of data chosen in example.

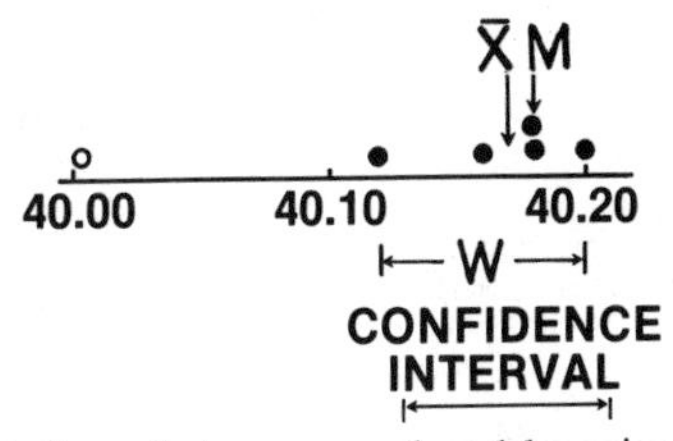

Fig. 4–2. Graphical presentation of changes produced by rejection of a questionable observation (compared with Fig. 1).

40.02	x_1
40.12	x_2
40.16	x_3
40.18	x_4
40.18	x_5
40.20	x_6
$\bar{x} = 40.14$	$M = 40.17$

The first value may be considered doubtful. The average is 40.14, and the median is 40.17. We may wish to place more confidence in the median 40.17 than in the average 40.14. (If the series were obtained in the order given, we might justifiably rule out the first observation as the result of unfamiliarity with the techniques involved.)

The range of the observations w is the difference between the greatest and least value; $w = x_n - x_1$. The range w is a convenient measure of the dispersion. It is highly efficient for ten or fewer observations, as is evident from column 3 of Table 4–1. This high relative efficiency arises in part from the fact that the standard deviation is a poor estimate of the dispersion for a small number of observations, even though it is the best known estimate for a given set of data. The range is also more efficient than the average deviation for fewer than eight observations. To convert the range to a measure of dispersion independent of the number of observations, we must multiply by the factor K_w, which is tabulated in column 2 of Table 3–1. This factor adjusts the range w so that on the average we estimate the standard deviation of the population. The product $wK_w = s_w$ is therefore an estimate of the standard deviation which can be obtained from the range. In the series presented above, the range is 0.18. From the table we find that K_w for six observations is 0.40, and so $s_w = 0.072$. The standard deviation s, calculated according to the

Table 4–1. Numerical values for various statistical parameters for numbers of observations from 2 to 10.

1	2	3	4	5	6	7	8	9	10	11
n	K_w	E_w	$t_{0.95}$	$t_{0.99}$	$t_{w_{0.95}}$	$t_{w_{0.99}}$	$Q_{0.90}$	$Q_{0.80}$	$(w/\sigma)_{0.95}$	$(w/\sigma)_{0.99}$
2	0.886	1.00	12.7	63.7	6.35	31.8	--	--	2.77	3.64
3	0.591	0.99	4.30	9.92	1.30	3.01	0.941	0.886	3.31	4.12
4	0.486	0.98	3.18	5.84	0.717	1.32	0.765	0.679	3.63	4.40
5	0.430	0.96	2.78	4.60	0.507	0.843	0.642	0.557	3.86	4.60
6	0.395	0.93	2.57	4.03	0.399	0.628	0.560	0.482	4.03	4.76
7	0.370	0.91	2.45	3.71	0.333	0.507	0.507	0.434	4.17	4.88
8	0.351	0.89	2.36	3.50	0.288	0.429	0.554*	0.479*	4.29	4.99
9	0.337	0.87	2.31	3.36	0.255	0.374	0.512*	0.441*	4.39	5.08
10	0.325	0.85	2.26	3.25	0.230	0.333	0.477*	0.409*	4.47	5.16

* These critical values are for a modified Q ratio: $Q^1 = (x_2 - x_1)/(x_{n-1} - x_1)$ or $(x_n - x_{n-1})/(x_n - x_2)$.

following equation, equals 0.066.

$$s = [\Sigma(x - \bar{x})^2/(n - 1)]^{1/2}$$

As the number of observations n increases, the efficiency of the range decreases. If the data are randomly presented, such as in order of production rather than in order of size, the average of the ranges of successive subgroups of 6 or 8 is more efficient than a single range. The same table of multipliers is used, the appropriate K_w being determined by the subgroup size.

4–3 CONFIDENCE LIMITS AS ESTIMATES

Although s and s_w are useful measures of the dispersion of the original data, we are usually more interested in the *confidence interval* or *confidence limits.* By the confidence interval, we mean the distance on either side of $\bar{x}$ in which we would expect to find, with a given probability, the "true" central value. For example, we would expect the true average to be covered by the 95% confidence limits 95% of the time. By taking wider confidence limits, say 99% limits, we can increase our chances of covering the "true" average, but the interval will necessarily be longer. The shortest interval for a given probability corresponds to the "t test" of Student (Fisher & Yates, 1957; Merrington, 1942). $\bar{x} \pm ts/(n)^{1/2}$ is the confidence interval, where the quantity t varies with the number of observations and the degree of confidence desired. For convenience, t is tabulated for 95 and 99% confidence values (Table 4–1 columns 4 and 5).

Confidence limits might be calculated in a similar manner, using s_w obtained from the range and a corresponding but different table for t. However, it is more convenient to calculate the limits directly from the range as $\bar{x} \pm wt_w$. The factor for converting w to s_w has been included in the quantity t_w, which is tabulated in column 6 of Table 4–1 for 95%

confidence and in column 7 for 99% confidence (Dixon & Massey, 1983, p. 538; Lord, 1947, 1950).

For a set of six observations, t_w is 0.40, and the range of the set previously listed is 0.18; therefore, wt_w equals 0.072. Hence, we can report as a 95% confidence interval, 40.14 ± 0.072. If we calculate a confidence interval using the tables of t and the calculated value of s, we find t = 2.6 at the 95% level and $ts/(6)^{1/2}$ = 0.070. We would report 40.14 ± 0.070, a result substantially the same as that obtained from the range of this particular sample.

If the standard deviation of a given population is known or assumed from previous data, we can use the normal curve to calculate the confidence limits. This situation may arise when a given analysis has been used for 50 or so sets of analyses of similar samples. The standard deviation of the population may be estimated by averaging the variance s^2 of the sets of observations, or from K_w times the average range of the sets of observations. The interval $\bar{x} \pm 1.96s/(m)^{1/2}$, where x is computed from a new set of m observations, can be expected to include the population average 95% of the time.

4–4 THE PROBLEM OF DESIGNATING EXTRANEOUS VALUES

Simplified statistics have been presented which enable one to obtain estimates of a central value and to set confidence limits on the result. Let us consider now the problem of extraneous values. The use of the median eliminates a large part of the effect of extraneous values on the estimate of the central value. The range, on the other hand, obviously gives unnecessary weight to an extraneous value in an estimate of the dispersion; for this reason, we may wish to eliminate values which fail to pass a screening test.

One very simple test, the Q test, is as follows:

Calculate the distance of a doubtful observation from its nearest neighbor; then divide this distance by the range. The ratio is Q where

$$Q = (x_2 - x_1)/w$$

or

$$Q = (x_n - x_{n-1})/w\,.$$

If Q exceeds the tabulated values (see Table 4–1), the questionable observations may be rejected with 90% confidence (Dixon, 1950; Dixon, 1953; Dixon and Massey, 1983). In the example cited, 40.02 is the questionable value and

$$Q = (40.12 - 40.02)/(40.20 - 40.02) = 0.56$$

This value of Q just equals the tabulated value of 0.56 for 90% confidence.

If we had decided to reject an extreme low value if Q was as large or larger than would occur 90% of the time in sets of observations from a normal population, we would now reject the observation 40.02. In other words, a deviation this great or greater would occur by chance only 10% of the time at one or the other end of a set of observations from a normally distributed population.

By rejecting the first value, we increase the median from 40.17 to 40.18 and the average from 40.14 to 40.17 (see Fig. 4–2). The standard deviation s falls from 0.066 to 0.030 (it might be > 0.030 if we have erred in rejecting the value 40.02), and s_w falls from 0.072 to 0.034. The 95% confidence interval corresponding to the t test is now 40.17 ± 0.038 and, from the median and range, is 40.18 ± 0.040, a reduction of about one-half in the length of the interval. (The 95% is only approximate, as we have performed an intermediate statistical test.)

4–5 RECOMMENDED RULES FOR DESIGNATING EXTRANEOUS VALUES

The problem of a test of significance for designating an extraneous value is straightforward. We choose a level of significance, using the standard considerations, and make a test on the set of observations we are processing. If a significant ratio is obtained, we declare the extreme value to be from a population differing from that of the remaining observations. Depending on the practical situation, we then declare the apparently extraneous value to represent a gross error or an exceptional individual. The best[1] statistic for this test, if the standard deviation σ is known from extensive data, is the range over σ for divergent values in either direction (see Table 4–1 for critical values) or the ratio $(x_n - \bar{x})/\sigma$ for a one-sided test. The largest observation is represented by x_n. For a one-sided test in the other direction, we substitute $\bar{x} - x_1$ for $x_n - \bar{x}$. Here x_1 represents the smallest observation. The power of these tests is discussed by Dixon (1950). Critical values for range over σ are given by Dixon (1950), and for $(x_n - \bar{x})/\sigma$ by Pearson and Hartley (1958).

If an independent estimate of σ is available, the best tests for extraneous values are the same as above, with s replacing σ. Critical values for these tests are given by Pearson and Hartley (1958). If no external estimate of σ is available, the best statistic is the Q-ratio given here. Critical values for these ratios are given in Table 4–1.

[1]Best is used here in the sense of power greater than or equal to all other tests investigated by Dixon (1950).

4-6 RECOMMENDED RULES FOR ESTIMATION IN THE PRESENCE OF EXTRANEOUS VALUES

Now let us suppose that in place of designating an individual value as an extraneous observation from some different distribution, we wish to estimate the parameters of the basic distribution free from this contaminating effect. How might we process the data to come closer to the mean and variance of this basic distribution?

If little is known about the contamination to be expected, about the best one can do is to label particular observations as extraneous values, as described in section 4–5, and remove them from estimates of the mean and standard deviation.

If even a moderate amount of information about the type of contamination to be expected is available, a process can be prescribed which will minimize the efects of contamination on the estimates of mean and dispersion in small samples. Dixon (1953) gives detailed rules. However, if more than 10% of the observations deviate from the mean by 3σ or more, one should test for and remove the extraneous values and then use the median and range. If somewhat fewer observations are expected to deviate to that extent, one should test for and remove the extraneous values and use the mean and range. In either case, it seems advantageous to use a fairly large α; i.e., in general one would use $Q_{0.80}$ in place of $Q_{0.90}$.

4-7 TEST OF HYPOTHESES (TRIMMED AND WINSORIZED *t*-TESTS)

If we wish to test hypotheses on a mean value or to test a hypothesis on the difference of two means in the presence of extraneous values, we can use either a trimmed or Winsorized *t*-test. The trimmed mean is the mean of the observations omitting the largest and smallest, i.e., $(x_2 + x_3 + \ldots x_{n-1})/(\mathrm{n} - 2)$, where the subscript indicates the observations ordered by magnitude. The Winsorized mean is the mean of n observations, except that the smallest is replaced by the next smallest and the largest is replaced by the next largest. The Winsorized standard deviation is the standard deviation of the n items as modified for the Winsorized mean. The ratio of the Winsorized mean to the Winsorized standard deviation if multiplied by $(n - 1)/[(n - 3)(n)^{1/2}]$ can be referred to the standard *t*-table with $n-3$ degrees of freedom.

This procedure may be used for two-sample tests and for confidence interval estimates. The 95% Winsorized confidence limits for the example in section 4–2 is obtained by computing the Winsorized mean $(40.12 + 40.12 + 40.16 + 40.18 + 40.18 + 40.18)/6 = 40.1567$. The Winsorized standard deviation $= 0.0294$ with the confidence interval

$$40.157 \pm 3.182(0.0294)(5/3)/(6)^{1/2}$$

or

$$40.16 \pm 0.064\,.$$

4–8 REFERENCES

Dean, R. B., and W. J. Dixon. 1951. Simplified statistics for small numbers of observations. Anal. Chem. 23:636–638.

Dixon, W. J. 1950. Ratios involving extreme values. Ann. Math. Statist. 21:488–506.

Dixon, W. J. 1953. Processing data for outliers. Biometrics 9:74–89.

Dixon, W. J., and F. J. Massey. 1983. Introduction to statistical analysis. 4th ed. pp. 538, 548. McGraw-Hill Book Co., New York.

Fisher, R. A., and F. Yates. 1963. Statistical tables for biological, agricultural and medical research. 6th ed. Oliver and Boyd, Edinburgh.

Lieblein, Julius. 1952. Properties of certain statistics involving the closest pair in a sample of three observations. Natl. Bur. Std. J. Res. 48:255–268.

Lord, E. 1947. The use of range in place of standard deviation in the t-test. Biometrika 34:41–67.

Lord, E. 1950. Power of the modified t-test (u-test) based on range. Biometrika 37:64–77.

Merrington, M. 1942. Table of percentage points of the t-distribution. Biometrika 32:300.

Pearson, E. S., and H. O. Hartley. 1958. Biometrika tables for statisticians. Vol 1. Cambridge Univ. Press, Cambridge.

5 Pretreatment for Mineralogical Analysis

G. W. KUNZE AND J. B. DIXON

Texas A&M University
College Station, Texas

5–1 GENERAL INTRODUCTION

The use of pretreatments involves the risk of altering or destroying fractions of the soil other than those for which the treatments are intended. In many cases, however, the investigator has very little choice; either pretreatments are utilized or the obtainable data are very much limited. The pretreatments are designed to have a minimum effect on constituents other than those to be eliminated. Even so, the philosophy of using a minimum of pretreatments remains sound.

The removal of free iron oxides was initially employed to clean the coarser fractions of the soil for study with the petrographic microscope and to eliminate aggregates of particles resulting from the cementing action of these materials. Iron-rich soils often require removal of free iron oxides for studies of particle size distribution. Removal of free iron oxides is in many cases desirable and sometimes necessary to obtain definitive x-ray diffraction patterns. Differential thermal, infrared, and other mineralogical analyses are many times simplified and aided by the removal of free iron oxides. The removal of organic matter, soluble salts, and calcium carbonate has also proven advantageous for most mineralogical studies.

Since the 1965 edition of this book, two significant developments have made mineralogical analysis of soil samples possible with less chemical pretreatment. Sonic and ultrasonic methods effect mechanical dispersion of soil particles with little or no prior chemical treatment (Edwards and Bremner, 1967). The time of sample exposure to such vibrations without damage to the particles requires selection dictated by the objectives of the investigations, the sample composition, and the vibrational intensity and frequency. Dispersion of soils with an ultrasonic probe apparently has not been tested as a standard method of preparation for mineralogical analysis, but it has been employed for special studies (Schulze and Dixon, 1979) where 15-min ultrasonic probe treatment was employed to disperse soil clay samples. A 15-min treatment may be too long for routine use, at least where labile samples are to be investigated. Brief (0.5- to 1-min) exposure of clay samples to ultrasonic vibrations is useful for dispersion in many mineralogical preparative procedures. The second recent advance is the crystal monochromator; it is employed to separate Kα x-rays from fluorescent x-rays and other x-ray spectral interferences in diffraction analysis. The graphite monochromator eliminates most of the background from iron minerals where Cu Kα radiation is employed. This innovation permits identification of iron oxides in soil clays where only a few percent free Fe_2O_3 is present (Schulze and Dixon, 1979).

5–2 REMOVAL OF SOLUBLE SALTS AND CARBONATES

5–2.1 Introduction

Soil samples containing soluble salts, including gypsum, may be difficult or impossible to disperse and fractionate as a result of the flocculating action of the salts. If the nature of the salts is such that the soil suspension is alkaline, hydrogen peroxide will decompose readily, and removal of organic matter will be difficult or impossible. The presence of soluble salts also makes it impossible to saturate the exchange complex with a specific cation for the purpose of determining cation-exchange capacity. Removal of these soluble constituents will simplify x-ray diffraction and differential thermal analyses as well as other mineralogical analyses. Since these soluble constituents may be readily identified and measured quantitatively (Rhoades, 1982), removal does not constitute destroying a portion of the sample which cannot be accounted for.

The carbonates commonly encountered in soils in quantities sufficient to cause difficulties are those of calcium and magnesium. As in the case of soluble salts, the carbonates may be accounted for quantitatively (Nelson, 1982); thus their removal does not detract from characterization of the sample. If positive identification of the carbonates is desired through x-ray diffraction or other mineralogical techniques, this may be accomplished with a sample prepared specifically for this purpose.

The more crystalline, concretionary forms of the carbonates are less troublesome than the poorly crystalline, finely divided forms. It is not uncommon to find horizons, particularly the Ck horizon, containing 50% or more of finely divided carbonates. Unless these are removed, it is impossible to achieve any meaningful separation of silt and clay by centrifugation (Chapter 12), due to continual breakdown of the carbonate particles. Furthermore, in many cases the presence of carbonates causes a great deal of scatter and, in general, a poor x-ray diffraction pattern. In the case of x-ray diffraction samples, the degree of orientation of clay particles is reduced in the presence of significant quantities of finely divided carbonates, resulting in a less definitive diffraction pattern. The presence of carbonates also needlessly increases the complexity of differential thermal analysis patterns. Cation-exchange capacity and surface area measurements for genetic studies can be made more meaningful through removal of carbonates and soluble salts. Finally the efficiency of hydrogen peroxide is extremely low in an alkaline medium, which requires that carbonates be removed if organic matter is to be oxidized by this reagent.

Soluble salts are most simply removed by dissolution in water. Carbonates may be removed by treatment with acid sodium acetate, acid, or disodium dihydrogen ethylenediaminetetraacetate. Soluble salts may be removed as part of the same treatment.

5–2.2 Methods

5–2.2.1 TEST FOR GYPSUM AND SOLUBLE SALTS (U.S. Salinity Laboratory Staff, 1954, modified by authors)

5–2.2.1.1 Reagent

1. Acetone, reagent grade.

5–2.2.1.2 Procedure. To test initially for the presence of gypsum, place 10 to 20 g of air-dried soil into an 8-ounce bottle, 250-mL centrifuge tube, or 250-mL Erlenmeyer flask, and then add 100 to 150 mL of distilled water. Stopper the container, and shake it by hand six times at 15-min intervals or agitate it for 15 min in a mechanical shaker. Filter the extract through a paper of medium porosity. Place about 5 mL of the extract in a test tube, add an approximately equal volume of acetone, and mix the solutions. The formation of a white (gel-like) precipitate indicates the presence of gypsum in the soil.

In the absence of gypsum, check another portion of the supernatant liquid for soluble salts with a conductivity bridge (Rhoades, 1982) to determine the need for pretreatment.

5–2.2.1.3 Comments. Because of the variations in the properties of the soluble salts found in soils, it is not feasible to suggest a lower limit below which removal is not necessary. For example, a soil concentration of 500 to 1000 ppm of NaCl would have little if any effect on dispersion or oxidation of organic matter with hydrogen peroxide, but the same would not be true for a similar concentration of $CaCl_2$. To convert conductivity values to quantities of salt, reference should be made to Figures 2 to 4 in *Agricultural Handbook 60* (U. S. Salinity Laboratory Staff, 1954). In the absence of knowledge of the behavior of the samples under investigation, it is good routine practice to treat all samples to remove excess soluble salts as described in the next section.

5–2.2.2 DISSOLUTION OF SOLUBLE SALTS (method after Kunze & Rich, 1959)

5–2.2.2.1 Reagent

1. Magnesium chloride ($MgCl_2 \cdot 6H_2O$), 0.5*M*, 102 g/L.

5–2.2.2.2 Procedure. Place the required amount of sample, ground to pass a 2-mm sieve, into a beaker of sufficient size to allow for the addition of water at the rate of 100 mL of distilled water to 1 g of soil. Cover the beaker with a watchglass and stir the suspension at intervals of 15 to 20 min over a period of 2 hours. Allow the suspended material to settle. If clay remains suspended after 2 or more hours, add a minimum of 0.5*M* $MgCl_2$ to flocculate the clay. Decant or siphon off the supernatant liquid, being careful not to lose any soil material.

Repeat the above procedure as many times as necessary to dissolve the gypsum and/or other soluble salts. Check for dissolution and removal of gypsum by treating an aliquot of the supernatant solution according to the procedure described in section 5–2.2.1.2.

Following dissolution of the gypsum and the subsequent final decantation, transfer the sample with the aid of a powder funnel and wash bottle to a centrifuge tube of appropriate size for washing to remove the soluble salts. Balance pairs of tubes with distilled water, centrifuge the tubes at 1600 to 2200 rpm for 5 to 10 min, and decant the supernatant liquid. Fill the tube half full with distilled water, stopper it tightly, jar the soil loose from the walls by striking the bottom of the tube on a large rubber stopper, and then shake the tube for 5 min in a reciprocating shaker. Wash adhering soil particles from the stopper and walls of the tube with a fine jet of distilled water, balance pairs of tubes, centrifuge the tubes for 10 to 15 min at 1600 to 2200 rpm, and decant the supernatant liquid. Continue to wash the sample in this manner until clay remains suspended after centrifuging for 10 to 15 min at 1600 to 2200 rpm. Mix several drops of 0.5*M* $MgCl_2$ *with the supernatant* liquid to flocculate the suspended clay without disturbing the sedimented material. Repeat the centrifuging, and decant the supernatant liquid. Without drying, transfer the sample with a minimum of distilled water to a beaker of appropriate size for removal of organic matter.

5–2.2.3 DISSOLUTION OF CARBONATES AND SOLUBLE SALTS WITH SODIUM ACETATE BUFFER (method after Grossman & Millet, 1961)

5–2.2.3.1 Reagent.

1. Sodium acetate ($NaC_2H_3O_2 \cdot 3H_2O$), 0.5*M*, 136 g/L, adjusted to pH 5 with acetic acid.

5–2.2.3.2 Procedure. Place a suitable amount of sample, ground to pass a 2-mm sieve, into a dialysis membrane, one end of which is tied with a rubber band. Add several hundred milliliters (depending upon size of sample and size of casing used) of Na-acetate buffer solution. Tie the top of the dialysis membrane around a glass "breather" tube (approximately 10 cm long), and hang the sample in a reservoir of the Na-acetate buffer solution contained in a plastic or glass container. (A 20-gallon plastic garbage can containing 60 L of buffer solution serves well for samples of 1 kg or larger.)

Knead the membrane after several days. If carbonates are still being dissolved, bubbles of CO_2 will be released. (The time required for dissolution of carbonates is dependent upon particle size, percentage and type of carbonate, and sample size.) When bubbles of CO_2 are no longer evident on kneading, open the dialysis membrane and check some of the coarser particles with strong acid for the presence of carbonates. When the sample is free of carbonates, transfer it (still in the dialysis membrane) to another container, and desalt it against tap water flowing continuously through the container. Check the ionic concentration inside the membrane by conductivity measurements on a small volume of the supernatant liquid poured out through the breather tube. Continue dialysis until the salt concentration drops below 10 meq/L. Remove excess water in the membrane with a filter candle.

As an alternative procedure for removing excess salt, and if the sample size permits, transfer the sample to one or more centrifuge tubes and wash the sample as outlined in section 5–2.2.2.2.

Following removal of excess salts, transfer the sample without drying and with a minimum of distilled water to a beaker for removal of organic matter.

5–2.2.3.3 Comments. Grossman and Millet (1961) reported that holding noncalcareous samples in contact with the buffer for 9 weeks did not affect the particle-size distribution, cation-exchange capacity, organic carbon, nitrogen, and free iron values.

The rate of carbonate removal is strongly affected by the concentration of alkaline earth ions in the buffer solution. By changing the buffer in the reservoir well before the buffer capacity has been exhausted, and thereby keeping the alkaline-earth ion concentration low, the rate of carbonate removal can be markedly increased.

Jackson (1969) described basically the same method for removal of carbonates, except that it is carried out in beakers, and the samples are heated in a boiling water bath to expedite dissolution.

5–3 REMOVAL OF ORGANIC MATTER

5–3.1 Introduction

Organic matter has an aggregating effect; hence removal is necessary if the analysis requires dispersion of the sample. Removal of organic matter is required for (or at least expedites) most mineralogical analyses, such as differential thermal, x-ray diffraction, and infrared.

Hydrogen peroxide, first used by Robinson (1922), is now generally used to oxidize organic matter. Efficient use of H_2O_2 requires an acid medium, which means that the soil needs prior treatment for removal of carbonates and soluble salts imparting an alkaline reaction to the soil suspension. To overcome the effects of the manganese oxides, Olmstead et al. (1930) proposed the addition of a small amount of glacial acetic acid, while Jackson (1969) suggested that the reaction be carried out in the presence of a sodium acetate solution adjusted to pH 5.

Atterberg (1912) and Troell (1931) proposed the use of sodium hypobromite to oxidize organic matter. With this reagent, it is not necessary to remove carbonates.

5–3.2 **Method** (modified from Kunze & Rich, 1959)

5–3.2.1 REAGENTS

1. Hydrogen peroxide (H_2O_2), 30%.
2. Magnesium chloride ($MgCl_2 \cdot 6H_2O$), 0.5*M*, 102 g/L.

5–3.2.2 PROCEDURE

If the sample was treated previously to remove soluble salts and/or carbonates, transfer it to a beaker with a minimum of distilled water. The beaker should be of sufficient size to eliminate loss of sample resulting from moderate to strong frothing. Cover the beaker with a ribbed watchglass, place it on a steam bath or hot plate, and allow excess water to evaporate until a soil-to-water ratio of 1:1 to 1:2 is obtained. Remove the beaker, and allow it to cool.

If the sample required no other treatment prior to removal of organic matter, grind it to pass a 2-mm sieve and place the required amount into a beaker. The beaker should be of sufficient size to eliminate loss of sample resulting from moderate to strong frothing. Add distilled water to the sample to give a 1:1 to 1:2 soil-to-water ratio, and cover the beaker with a ribbed watchglass.

If necessary, make the suspension acid to litmus paper with a few drops of 1M HCl. Initially add 30% H_2O_2 in increments of 5 to 10 mL or less, stir the suspension, and allow time for any strong effervescence or frothing to subside. Control reactions that are too vigorous by cooling the beaker in a water bath. Continue adding H_2O_2 in small amounts until the sample ceases to froth; then transfer it to a steam bath or hot plate at low heat (65° to 70°C) and observe it closely for 10 to 20 min, or until danger of any further strong reaction has passed. Add additional H_2O_2 in amounts to give approximately a 10% solution. Evaporate excess liquid between additions of H_2O_2 to maintain a soil-to-water ratio of 1:1 to 1:2. Do not allow the sample to evaporate to dryness.

The reaction of soil with H_2O_2 is essentially complete when the soil sample loses its dark color or when conspicuous effervescence ceases. Some effervescence will always be present due to the decomposition of the H_2O_2. The majority of soils will show some color as a result of highly colored mineral particles and free iron oxides.

Transfer the sample with the aid of a powder funnel and wash bottle to a centrifuge tube of appropriate size, balance pairs of tubes, centrifuge the tubes at 1600 to 2200 rpm for 10 to 15 min, and decant and discard the supernatant liquid. If clay remains suspended, add a few drops of 0.5*M* $MgCl_2$, mix the suspension without disturbing the sedimented material, centrifuge the tubes, and decant the supernatant liquid. Further washing is not necessary. If, however, a clear supernatant liquid is obtained without the addition of any flocculating agent, fill the tube approximately one-third or less with distilled water, stopper it tightly, jar the soil loose from the walls of the tube by striking the bottom of the tube on a large rubber stopper, and then shake the tube for 5 min in a reciprocating shaker. Wash adhering particles from the stopper and walls of the tube with a fine jet of distilled water. Balance pairs of tubes, centrifuge them for 10 to 15 min at 1600 to 2200 rpm and decant the supernatant liquid. Flocculate suspended clay by the addition of a few drops of 0.5*M* $MgCl_2$ as outlined above. The sample is now ready for removal of free iron oxides, dispersion, et cetera. If free iron oxides are not to be removed, flocculation, if required, should be effected by making the supernatant 0.5 *M* NaCl and warming it. Introduction of Mg prior to raising the pH to 9.5 would risk precipitation of a new phase.

5-3.2.3 COMMENTS

Hydrogen peroxide is a strong oxidizing agent, and contact with the skin should be avoided. If contact is made, affected parts should be washed with copious quantities of water.

Elevated temperatures result in decomposition of the H_2O_2, which is the reason for not heating the sample in excess of 70°C.

Normally the deeper samples in the profile contain very little organic matter. However, it is here that the manganese oxides, if present, occur in greatest abundance. Therefore, any vigorous reaction of H_2O_2 with samples from the lower portion of the profile is very likely due to something other than organic matter.

If H_2O_2 is present in the sample at the time of the washing and centrifuging operation, it will be difficult to obtain a clear supernatant liquid to decant because of a slow, continuous release of oxygen resulting from decomposition of the H_2O_2. To dispose of an excess of H_2O_2, heat the sample to boiling for a few minutes.

5-4 REMOVAL OF FREE IRON OXIDES

5-4.1 Introduction

Free iron oxides, occurring as discrete particles or as coatings, are frequently removed from soil or its fractions to expedite mineralogical studies. Observation of the optical properties with a petrographic microscope is greatly facilitated if the particles are clean or free of coatings. Iron-containing samples, such as Oxisols and some Ultisols, are generally very difficult to disperse unless the free iron oxides are removed. Particles coated with iron oxides will not allow for a satisfactory heavy mineral separation. Iron oxides fluoresce when a copper target is used for x-ray diffraction analyses, resulting in an increased background count and a general decrease in the quality or clarity of the diffraction pattern. This, however, does not mean that each sample requires treatment for iron oxide removal to obtain a satisfactory pattern. It has been the writers' experience in working with soil profile samples that $< 10\%$ required removal of free iron oxides for the purpose of obtaining satisfactory x-ray diffraction patterns. Removal of iron oxides would, of course, allow for a greater degree of parallel orientation of the clay particles. Electron microscope observations and differential thermal analysis could possibly be improved. Hence, individual judgment needs to be exercised concerning the necessity or desirability for removing free iron oxides.

The most successful methods for removing free iron oxides, though none remove magnetite and ilmenite (Jackson, 1969), involve the chemical reduction of the iron to the ferrous form (Deb, 1950; Jeffries, 1947). Other methods have utilized biological reduction (Allison and Scarseth, 1942), or solution of the iron with oxalic acid (Robinson and Holmes, 1924; Schofield, 1949) or one of its salts (Drosdoff, 1935; Tamm, 1922). The earlier methods required subjecting the sample to highly acid conditions in the presence of high concentrations of cations such as potassium, ammonium, or aluminum, which are fixed rather strongly by certain of the clay minerals. The sodium dithionite-citrate procedure (Mehra and Jackson, 1960) which follows, overcomes these objectionable features and removes free iron oxides with a minimum of destructive action to the clay minerals. Even so, silica and alumina are common components of the extract and are thought to be contributed primarily by the amorphous and poorly crystalline fractions.

Sodium citrate serves as the chelating agent for ferrous and ferric forms of iron. The sodium bicarbonate buffers the solution, while the sodium dithionite (also known as sodium hydrosulfite and sodium hyposulfite) reduces the iron.

5–4.2 **Method** (after Mehra & Jackson, 1960, with minor modifications by authors—GWK & JBD)

5–4.2.1 REAGENTS

1. Sodium-citrate dihydrate, 0.3*M*, 88 g/L.
2. Sodium bicarbonate ($NaHCO_3$), 0.5*M*, 84 g/L.
3. Sodium dithionite ($Na_2S_2O_4$).
4. Sodium chloride (NaCl) solution, saturated.
5. Acetone, reagent grade.

5–4.2.2 PROCEDURE

Transfer to a 100-mL centrifuge tube a suitable amount of sample (4 g of many soils or 1 g of clay) that has been treated to remove organic matter, soluble salts, and preferably carbonates. The sample should contain no more than 0.5 g of extractable Fe_2O_3. Add 40 mL of 0.3*M* Na-citrate solution and 5 mL of 0.5*M* $NaHCO_3$ solution to the sample. Warm the suspension to 80°C in a water bath, and then add 1 g of solid $Na_2S_2O_4$ (0.5 g suffices for clays low in free iron oxides); stir the suspension constantly for 1 min and occasionally for a total of 15 min. Avoid heating above 80°C because FeS forms. Following the 15-min digestion period, add 10 mL of a saturated NaCl solution. If the suspension fails to flocculate with the NaCl, add 10 mL of acetone. Mix the suspension, warm it in a water bath if needed to expedite flocculation, and centrifuge the tube for 10 to 15 min at 1600 to 2200 rpm. If Si, Fe, and Al determinations are to be made, decant the clear supernatant liquid into a 500- or 1000-mL volumetric flask; otherwise discard it. (Hallmark et al., 1982; Barnhisel & Bertsch, 1982; Olson & Ellis, 1982).

Repeat the above treatment once or twice for samples which originally contained $>$ 5% extractable Fe_2O_3 (the sample may be combined into fewer tubes for the second treatment). Combine extracting and washing solutions with those of the first treatment if Fe, Al, and Si are to be determined. Wash (two or more times for samples of $>$ 1 g of residue) the sample finally with the Na-citrate solution (with NaCl and acetone if necessary for flocculation) and combine the washings with the previous decantates. Avoid HCl and $CaCl_2$ solutions as flocculants. (Caution: exercise care to prevent solutions containing acetone from boiling.) The sample, freed of extractable Fe_2O_3 but not dried at any time during the procedure, is now ready for further processing as prescribed by the procedure for the specific analysis to be made.

5–4.2.3 COMMENTS

Because many minerals are rather strongly colored and because it is impossible to remove the last trace of organic matter with H_2O_2, samples freed of extractable Fe_2O_3 generally do not have a pure white color.

5–5 PARTICLE-SIZE SEPARATIONS

Mineralogical analyses are normally performed on specific size fractions of the soil such as 2 to 0.2μm (coarse clay). For mineralogical work, the particle-size separations are usually made after application of pretreatments such as those described in this chapter. The separation of sands, silt, and clay by a combination of sieving and sedimentation under gravity is discussed in chapter 15. The use of a centrifuge to separate different particle-size fractions of clay is described in chapter 12.

5–6 REFERENCES

Allison, L. E., and G. D. Scarseth. 1942. A biological reduction method for removing free iron oxides from soils and colloidal clay. J. Am. Soc. Agron. 34:616–623.

Atterberg, A. 1912. Die mechanische Bodenanalyse und die Klassifikation der Mineralböden Schwedens. Int. Mitt. Bodenk. 2:312–342.

Barnhisel, R., and P. M. Bertsch. 1982. Aluminum. *In* A. L. Page et al. (ed.) Methods of soil analysis, Part 2. 2nd ed. Agronomy 9:275–300.

Deb, B. C. 1950. The estimation of free iron oxides in soils and clays and their removal. J. Soil Sci. 1:212–220.

Drosdoff, M. 1935. The separation and identification of the clay mineral constituents of colloidal clay. Soil Sci. 39:463–478.

Edwards, A. P., and J. M. Bremner. 1967. Dispersion of soil particles by sonic vibration. J. Soil Sci. 18:47–63.

Grossman, R. B., and J. C. Millet. 1961. Carbonate removal from soils by a modification of the acetate buffer method. Soil Sci. Soc. Am. Proc. 25:325–326.

Hallmark, C. T., L. P. Wilding, and N. E. Smeck. 1982. Silicon. *In* A. L. Page et al. (ed.) Methods of soil analysis, Part 2. 2nd ed. Agronomy 9:263–273.

Jackson, M. L. 1969. Soil chemical analysis—advanced course. 2nd ed. 8th printing, 1973 Publ. by the author, Dept. of Soils, Univ. of Wis., Madison, WI 53706.

Jeffries, C. D. 1947. A rapid method for the removal of free iron oxides in soil prior to petrographic analysis. Soil Sci. Soc. Am. Proc. 11:211–212.

Kunze, G. W., and C. I. Rich. 1959. Mineralogical methods. *In* C. I. Rich, L. F. Seatz, and G. W. Kunze, ed. Certain properties of selected southeastern United States soils and mineralogical procedures for their study. Southern Coop. Series Bul. 61:135–146.

Mehra, O. P., and M. L. Jackson. 1960. Iron oxide removal from soils and clays by a dithionite-citrate system buffered with sodium bicarbonate. *In* Clays and Clay Minerals, Proc. 7th Conf., pp. 317–327. Natl. Acad. Sci.-Natl. Res. Council Publ., Washington, DC.

Nelson, R. E. 1982. Carbonate and gypsum. *In* A. L. Page et al. (ed.) Methods of soil analysis, Part 2. 2nd ed. Agronomy 9:181–197.

Olmstead, L. B., L. T. Alexander, and N. E. Middleton. 1930. A pipette method of mechanical analysis of soils based on an improved dispersion procedure. U. S. Dep. Agric. Tech. Bull. 170.

Olson, R. V., and R. Ellis, Jr. 1982. Iron. *In* A. L. Page et al. (ed.) Methods of soil analysis, Part 2. 2nd ed. Agronomy 9:301–312.

Rhoades, J. D. 1982. Soluble salts. *In* A. L. Page et al. (ed.) Methods of soil analysis, Part 2. 2nd ed. Agronomy 9:167–179.

Robinson, G. W. 1922. Note on the mechanical analysis of humus soils. J. Agric. Sci. 12:287–291.

Robinson, W. O., and R. S. Holmes. 1924. The chemical composition of soil colloids. U. S. Dep. Agric. Bull. 1311.

Schofield, R. K. 1949. Effect of pH on electric charges carried by clay particles. J. Soil Sci. 1:1–8.

Schulze, D. G., and J. B. Dixon. 1979. High gradient magnetic separation of iron oxides and other magnetic minerals from soil clays. Soil Sci. Soc. Am. J. 43:793–799.

Tamm, O. 1922. Eine Methode zur Bestimmung der anorganischen Komponenten des Gelkomplexes in Boden. Meddel. Statens Skogsförsöksanst (Sweden). 19:385–404.

Troell, E. 1931. The use of sodium hypobromite for the oxidation of organic matter in mechanical analyses of soils. J. Agric. Sci. 21:476–484.

U. S. Salinity Laboratory Staff. 1954. Diagnosis and improvement of saline and alkali soils. U. S. Dep. Agric. Handb. 60.

6 Oxides, Hydroxides, and Aluminosilicates[1]

MARION L. JACKSON AND CHIN H. LIM

University of Wisconsin
Madison, Wisconsin

LUCIAN W. ZELAZNY

Virginia Polytechnic Institute and State University
Blacksburg, Virginia

6–1 INTRODUCTION

A series of chemical techniques greatly assist the instrumental methods covered in other chapters (7 through 12) on quantitative soil mineral evaluation. Free oxides and hydroxides of soils include crystalline and noncrystalline (amorphous) compounds of silicon (e.g., quartz), aluminum (e.g., gibbsite), iron (e.g., hematite, goethite), titanium (e.g., anatase) and manganese (e.g., pyrolusite). Small amounts of various other elements are usually substituted in free oxides and hydroxides of soils—for example, Al for Fe. The noncrystalline and finely divided crystalline compounds are hydrous. The term *free* refers to compounds having mainly a single species of coordinating cation such as oxides of silicon or aluminum. Hydrous oxides containing two or more coordinating cation species are spoken of as *combined* oxides, with the most abundant kinds in soils being aluminosilicates or ferrialuminosilicates. Many combined oxides can be differentiated quantitatively by selective dissolution analysis (SDA), cation exchange properties, and thermal techniques.

The term *noncrystalline* (Wada, 1977) is used in preference to the commonly used term *amorphous*. The aluminosilicates of soils include the broad spectrum of soil constituents, ranging from noncrystalline materials (exhibiting local and nonrepetitive short-range order) to paracrystalline materials (intermediate-range order) to crystalline phyllosilicates characterized by three-dimensional periodicity over appreciable distances (long-range order) (Table 6–1). Noncrystalline aluminosilicates

[1]This contribution was supported in part by the Department of Soil Science, College of Agricultural and Life Sciences, University of Wisconsin-Madison; in part by the National Science Foundation EAR-8405422-JACKSON; and in part by the Department of Agronomy, Virginia Polytechnic Institute and State University, Blacksburg, VA.

Table 6-1. Order-disorder in oxides, hydroxides, and aluminosilicates.

Continuum of mineral structures			
Crystalline,		Paracrystalline, (Imogolite†, 1-dim.)	Noncrystalline "amorphous"
Long-range	→	(Smectite†, 2-dim.)	Short-range (Allophane†)
Ordered	→		Disordered
Atomic: positional‡; substitutional or "strange."			
Layer: shifts‡; defective; curvature (halloysite).			
Mixed: regular; irregular§ or "random."			
Sample purity		(Confusion of properties) ←	Sample impurities¶

† Wada (1977).
‡ Brindley (1977).
§ Lim and Jackson (1980).
¶ Lim et al. (1980).

of interest include gel-like compounds containing mainly aluminum, iron, manganese, silicon, oxygen, hydroxyl, and water, but usually also containing more or less magnesium, occasionally phosphate, and minor amounts of other ions. The poorly crystalline as well as noncrystalline hydrous oxides and aluminosilicates of soils reflect soil genesis and mineral weathering. They also take part in many important chemical reactions, such as liming responses, anion retention, and cation retention.

Some of the noncrystalline aluminosilicate compounds, associated mainly with weathered volcanic ash, have been designated *allophane.* A definition of allophane was proposed, in accordance with Ross and Kerr (1934), at an international seminar on amorphous clays in Japan in 1969 (van Olphen, 1971):

> Allophanes are members of a series of naturally occurring minerals which are hydrous aluminum silicates of widely varying chemical composition, characterized by short range order, by the presence of Si–O–Al bonds, and by a differential thermal analysis curve displaying a low temperature endotherm and a high temperature exotherm with no intermediate endotherm.

These criteria limit allophane to a small sector of the total spectrum of noncrystalline and paracrystalline aluminosilicates developed by weathering of volcanic ash and pumice and other materials of soils and deposits. *Imogolite*, a mineral closely associated with allophane, is a hydrated aluminosilicate having a thread-like morphology. It consists of paracrystalline cylindrical assemblies of a one-dimensional structure unit (Cradwick et al., 1972). For the other noncrystalline aluminosilicates, it is recommended strongly that specific names not be given, but they be described so far as possible in terms of their chemical composition (Brindley and Pedro, 1972). Hence, the term *noncrystalline aluminosilicates* will be used herein, in order to address the whole problem of short-range order of weathered parent materials and soils.

Crystalline phyllosilicates, also called *layer silicates*, are defined by the AIPEA Nomenclature Committee (Bailey, 1980) as containing "continuous two-dimensional tetrahedral sheets of composition T_2O_5 (T = Si, Al, Be . . . .) with tetrahedra linked by sharing three corners of each, and with the fourth corner pointing in any direction. The tetrahedral sheets are linked in the unit structure to octahedral sheets, or to groups of coordinated cations, or individual cations." The AIPEA classification scheme for phyllosilicates (Table 1, Bailey, 1980) has been adopted. *Smectite*, in lieu of montmorillonite–saponite, is the group name for clay minerals with layer charge between 0.2 and 0.6 per formula unit. Chlorite is treated as consisting of a 2:1 plus an interlayer hydroxide sheet, rather than as a 2:1:1 or 2:2 layer type. In soils and sediments, the phyllosilicates of interest include the 1:1 layer types (kaolinite, halloysite) and the 2:1 layer types (smectite, vermiculite, mica, and chlorite).

The ubiquity in soils of various noncrystalline ferrialuminosilicates that have properties distinct from those of allophane and phyllosilicates has also been recognized. One class of such compounds is thought to be interlayer and surface coatings of vermiculites, smectites, and other minerals. In addition, highly disordered (three-dimensionally), subcrystalline, clay-like, and clay-relic materials, transitional from "short-range order" to crystalline clay minerals, are being deionized or synthesized by weathering. A highly precise analytical distinction between noncrystalline and crystalline materials of soils is at present impossible because of the transitional nature of the boundary between the two categories (Table 6–1). There is virtually a continuum from perfect crystallinity (long-range order) to disorder constrained only by alternating linkage of cations and anions as radial distribution functions (short-range order). The current approach is mainly an operational one of SDA by a series of reagents and procedures, each appropriate for a given objective or material. The specific sequences of layer-silicate structural disorder (a kind of noncrystallinity) along the *Z*-axis involved in random interstratification or mixed layering (Table 6–1) is excluded from the present consideration and left to the x-ray diffraction analysis approach (chapter 12).

Selective dissolution analysis methods are needed for independent determinations of various inorganic constituents of soils because difficulty is experienced with many physical analytical methods in estimating or even recognizing the presence of noncrystalline and paracrystalline free oxides or aluminosilicates mixed with crystalline soil components. The reaction of noncrystalline compounds of soils with neutral fluoride as KF (Huang & Jackson, 1965) or NaF (Fieldes & Perrott, 1966) complexes Al and Fe as the corresponding fluorates and yields KOH or NaOH giving pH 11.6 (Huang & Jackson, 1965). Specificity of this test is hindered at pH 11.2 to 11.4 with aluminous chlorite and halloysite (Huang & Jackson, 1965) and by $CaCO_3$ (El-Attar et al., 1972).

The crystalline free oxides and phyllosilicates of soils can be identified qualitatively and estimated semiquantitatively by x-ray diffraction analysis (chapter 12). Those containing hydroxyl can sometimes be determined fairly quantitatively by differential thermal analysis (DTA), differential scanning calorimetry (DSC), and thermalgravimetric analysis (chapter 7). The diagnostic properties of the hydrous oxides, as measured by the above techniques, vary a great deal with variation in particle size and degree of crystallinity of the oxides. For example, fine-grained goethite (limonite) has its differential endotherm in the range of 300° to 350°C, overlapping the endotherm of gibbsite (Jackson, 1979, p. 267), instead of 400°C, which is typical of highly crystalline goethite. Thus the selective dissolution of free iron oxides from a ferruginous gibbsite sample provides for more accurate differential thermal analysis for gibbsite. This example illustrates one general objective of SDA methods, namely, to free the remaining mineral materials of certain noncrystalline or free oxide material.

Coarser particles in the sand and silt fractions can be counted by petrographic methods (chapter 8). Quartz and cristobalite can be isolated (section 6-2) in monomineralic form, weighed, and analyzed, by for example, the mass spectrometric method for oxygen isotopic ratio (Clayton et al., 1972). Free MnO_2 (pyrolusite, etc.) is selectively dissolved in buffered (pH 5) H_2O_2 (Jackson, 1979, p. 31) during the usual preparations of soil for mineralogical analysis (chapter 5). Free oxides, for which no specific selective dissolution methods are available, include $ZrSiO_4$ (zircon), $FeO \cdot TiO_2$ (ilmenite), Fe_3O_4 (magnetite), AlOOH (boehmite, diaspore), and Al_2O_3 (corundum). If the presence of these minerals is determined by x-ray diffraction (XRD), petrographic, or other methods, then appropriate percentages can be estimated from total elemental analysis of the residues after removal of substances for which SDA methods are available. Some of the trace element oxides have been fractionated by specific dissolution methods (Viets, 1962; Hodgson, 1963; LeRiche & Weir, 1963; McLaren & Crawford, 1973; Shuman, 1979). Poorer XRD patterns from soil clays may be preferred (Brewster, 1980) to sharper ones that result from SDA.

The methods to be presented include (i) the $Na_2S_2O_7$–H_2SiF_6 method for quartz, cristobalite, and feldspar isolation (section 6–2), (ii) the acid ammonium oxalate extraction in the dark (AOD), for noncrystalline aluminosilicates (section 6–3), (iii) the citrate–bicarbonate–dithionite method for crystalline hematite (Fe_2O_3) and goethite (FeOOH) and associated Al, Si, and Mn (section 6–4), (iv) the selective dissolution (flash heating in 0.5 *M* NaOH) methods for poorly crystalline and/or fine aluminosilicates (section 6–5), (v) the cation exchange methods for smectite and vermiculite and exchange-capacity hysteresis for noncrystalline clays (section 6–6), and (vi) the H_2TiF_6 method for anatase and rutile isolation (section 6–7).

6–2 QUARTZ AND FELDSPARS

6–2.1 Introduction

Crystalline SiO_2 occurs in most soils mainly as the mineral quartz (SiO_2). Opal ($SiO \cdot nH_2O$) is common in many soils. Feldspars ($KAlSi_3O_8$

and $NaAlSi_3O_8$) are common in the sand and silt of many soils. The polymorphs of quartz, cristobalite and tridymite (also crystalline SiO_2), make up appreciable percentages of some soils and, together with quartz, can dominate (95%) even the clay fraction in strongly podzolized soils (Swindale & Jackson, 1956, 1960). Coesite and stishovite are high-density crystalline polymorphs of SiO_2 found around meteor craters (Sclar et al., 1962). The resistance of coarser particles of quartz and opal (Hallmark et al., 1982—section 15–1.2) to chemical weathering results in their common occurrence in soils.

Work on selective dissolution of silica in laboratories concerned with ceramics, geochemistry, silicosis, and soils extends back many decades. In early work, free silica was selectively dissolved away from ferrialuminosilicates by fusion of the sample in equal parts of $KHCO_3$ and KCl, or $NaHCO_3$ and NaCl (Polezhaev, 1958); the Si from free SiO_2 subsequently was determined colorimetrically.

Quartz, cristobalite, and feldspars can be estimated by the internal-standard x-ray diffractometry (chapter 12). A high degree of variability is experienced in the diffraction intensity of standard quartz (Pollack et al., 1954; Nagelschmidt, 1956; Brindley, 1961). The reasons probably arise (i) from the great variations in the degree of crystallinity of quartz specimens, and (ii) from the effect of the noncrystalline surface layer which forms on quartz particles (Henderson et al., 1970; Sayin & Jackson, 1979). Quartz also can be estimated by infrared absorption (Tuddenham & Lyon, 1960) or from its alpha-beta inversion differential endotherm at 573°C (Grim, 1953; Sysoeva, 1958). Quartz and feldspars in sand and silt sizes can be determined by count in a petrographic microscope (chapter 8).

Quartz and feldspars determination as a residue from acid decomposition of more labile aluminosilicates is based on their relatively slow solubility in acid. The metallic cations are liberated from phyllosilicates by digestion in mineral acids. Hot aqueous acids that have been used for decomposition of the labile minerals include a mixture of concentrated hydrochloric, sulfuric, and nitric acids (2:4:1 by volume) (Hardy & Follett-Smith, 1931); phosphoric acid (Talvitie, 1951; Jophcott & Wall, 1955); perchloric acid (Medicus, 1955); 4.5M H_2SO_4 (Shaw, 1934; Nagelschmidt, 1956); and concentrated HCl followed by boiling in a solution of Na_2S and digestion in 6 M HCl–HNO_3 (Shchekaturina & Petrashen, 1958). The tri-acid mixture of HCl–HNO_3–H_2SO_4 (above) has greater effectiveness than $HClO_4$ in decomposing some layer silicates, particularly unheated kaolinite (Corey, 1952); anorthoclase feldspar resists decomposition during digestions in acids (Hashimoto, 1961). The use of a hot $HClO_4$ digestion in later stages of an acid digestion procedure has the advantage that the acid-released silica is quantitatively dehydrated; the metallic cations then can be washed out with HCl, and finally, the dissolved silicon from the decomposed aluminosilicate can be dissolved in dilute NaOH for determination.

Talvitie (1951) found that the minerals albite, pyrophyllite, stillimanite, kyanite, tourmaline, beryl, and topaz resisted decomposition in H_3PO_4. Appreciable dissolution of quartz in H_3PO_4 was reported (Jophcott & Wall, 1955). Phosphoric acid also interferes with dissolved silicate determination with molybdate. Digestion of silicate dusts in pyrophosphoric acid ($H_4P_2O_7$; mp., 61°C; digestion temperature, 250°C) decomposed most aluminosilicates (Dobrovol'skaya, 1958), leaving beryl, topaz, tourmaline, and zircon along with the quartz little attacked. Previous ignition of some dust samples was required to obtain the desired aluminosilicate decomposition by $H_4P_2O_7$ digestion (Bulycheva & Mel'nikova, 1958), and even then subsequent treatment with concentrated HCl was required to decompose some rock powders.

The soil contents of feldspars and quartz are used as indexes of physical weathering, as in glacial "rock flour" (Jackson et al., 1948; Bockheim, 1982), and of quartz as a reference index to chemical weathering of other minerals (Barshad, 1955). Moreover, the presence and amounts of quartz are often considered in ratio to feldspars as a measure of weathering of soils. The persistent interest in and improvements of chemical determination of quartz (and its polymorphs) have resulted in the development of methods with a fairly high degree of accuracy. Chemical removal of aluminosilicates by acids has been used, for example, to identify crystalline SiO_2 polymorphs in the Mt. St. Helens ashfall of May-June 1980, in relation to possible silicosis hazard (Anon., 1980).

6–2.2 Principles

Dehydroxylation of kaolinite and dioctahedral micas by heating has been found to ensure their complete decomposition and dissolution in the several acid-NaOH dissolution routines that preserve the quartz and feldspars. An efficacious way to dehydroxylate layer silicates, providing at the same time a rigorous acid treatment to which quartz and feldspars are resistant, is fusion of the sample in $Na_2S_2O_7$ (Kiely and Jackson, 1964, 1965). Fusion of mineral samples in $Na_2S_2O_7$ or $K_2S_2O_7$ has been employed for the analytical decomposition of certain minerals for over 100 years (Smith, 1865; Hillebrand & Lundell, 1929, p. 705), and $Na_2S_2O_7$ was given preference. The $Na_2S_2O_7$ fusion begins at about 300°C, and the dehydroxylation is promoted by the presence of a hot acid flux consisting of free SO_3 (sulfuric acid anhydride) liberated as the temperature is slowly elevated to full red heat in a covered silica crucible. Gibbsite and various oxides, which tend to be rendered acid- and alkali-insoluble by dry heating (preignition, frequently recommended, is avoided), are dissolved in the fusion. The pyrosulfate fusion method of dehydroxylation is not only more effective but also is several steps simpler than the various aqueous and above-mentioned low-temperature acid procedures, which call for extraction of the sample by aqueous acid, drying the residue, heating the residue for dehydroxylation, and re-extraction with acid.

Fusion in $NaHSO_4$ (which forms $Na_2S_2O_7$) has been employed for quartz and opal determination in rocks (Astaf'ev, 1958); pretreatments consisting of digestion in concentrated HCl, washing, drying, and ignition were given before the fusion. Trostel and Wynne (1940) employed $K_2S_2O_7$ fusion for refractory clays, materials which are high in alumina and kaolinite (and which these authors considered usually to be free of feldspars), to determine the quartz in the insoluble residue. The $K_2S_2O_7$ fusion cake was slaked in hot water, and the resulting suspension was made alkaline by the addition of NaOH pellets. After digestion for 0.5 hr at 90°C in this approximately 2 *M* NaOH, the suspension was filtered through paper. The residue on the paper was washed successively with H_2O, 6 *M* HCl, and H_2O, and then dried and recovered by ignition of the paper. Incomplete removal of iron oxides by this procedure was experienced by Florentin and Heros (1947) when applying the procedure to quartz determination in rocks rich in iron, and these authors consequently employed two successive pyrosulfate fusions and digestions of the residue in concentrated HCl. These difficulties were also experienced for soils in the authors' laboratory; also, the filtration of an alkaline suspension on paper was found to be unsatisfactory for fine-grained rocks and soils.

The procedure subsequent to the $Na_2S_2O_7$ fusion was therefore completely redesigned as follows: (i) the fusion cake is taken up in 3 *M* HCl (instead of NaOH) so as to keep in solution the Fe, Mg, and other metallic cations liberated by the fusion; (ii) the residue is washed with 3 *M* HCl to free the residue of Fe and Mg, which would precipitate in the subsequently used NaOH solutions; (iii) centrifugation is used instead of filtration on paper; and (iv) the residue is digested in 0.5 *M* NaOH for 2.5 min (Hashimoto & Jackson, 1960) ample for dissolution of amorphous silica in the absence of precipitated Fe and Mg (which, if present, inhibit the dissolution of amorphous silica and alumina). This NaOH treatment is less destructive to fine-grained quartz and feldspars than the longer digestion in stronger solutions employed previously.

Potassium and sodium feldspars are relatively resistant to both $Na_2S_2O_7$ and $K_2S_2O_7$ fusion and to 3 *M* HCl and 0.5 *M* NaOH washings. Thus use of the $Na_2S_2O_7$–HCl–NaOH procedure facilitates the differential determination of K in mica or illite and of K feldspars, which is one important objective of the method.

The resistance of quartz sand and silt to oxygen isotopic exchange during weathering makes possible the use of quartz in these soil fractions for determining provenance of soil materials by the oxygen isotopic ratios (Jackson et al., 1971; Clayton et al., 1972; Jackson, 1981). Quantitative isolation of monomineralic quartz from soils and sediments is therefore frequently wanted. The hexafluorosilicic acid (H_2SiF_6) treatment to remove feldspars after $Na_2S_2O_7$ fusion has been developed (Syers et al., 1968; Chapman et al., 1969; Jackson et al., 1976; Jackson, 1979). The H_2SiF_6 is first completely saturated with SiO_2 and then the F complexes the Al as soluble AlF_6^{3-}. Opal and polymorphs of quartz are subsequently removed by heavy liquid separation.

6–2.3 Method (Modified from Kiely & Jackson, 1964, 1965; Jackson, 1964; Jackson et al., 1976; and Jackson, 1979).

6–2.3.1 SPECIAL APPARATUS

1. Heavy walled 70-mL pointed centrifuge tubes and special rubber seats for 100-mL International centrifuge cups.
2. Vitreous silica (fused quartz) crucibles: 50 mL, with closely fitting covers.
3. Nickel or stainless steel beakers, 500 mL.
4. Water bath: 18 ± 2°C (cool tap water often is satisfactory).

6–2.3.2 REAGENTS

1. Sodium pyrosulfate ($Na_2S_2O_7$): Use powdered reagent-grade $Na_2S_2O_7$, or begin with reagent-grade $NaHSO_4$ and let the conversion occur during the fusion.
2. Hydrochloric acid (HCl): Dilute concentrated HCl (approximately 12 *M*) with water to give stocks of approximately 6 *M*, 3 *M*, and 0.05 *M* HCl.
3. Sodium hydroxide (NaOH): Dissolve 2 g of reagent-grade NaOH pellets in 100 mL of water to give approximately 0.5 *M* NaOH.
4. Perchloric acid ($HClO_4$): Use reagent-grade 60% $HClO_4$.
5. Sulfuric acid (H_2SO_4): Cautiously add 50 mL of reagent-grade concentrated H_2SO_4, a few drops at a time with constant stirring, to 50 mL of water to give approximately 9 *M* H_2SO_4.
6. Nitric acid (HNO_3): Dilute 20 mL of reagent-grade HNO_3 to 100 mL to give 3 *M* HNO_3.
7. Hydrofluoric acid (HF): Use reagent-grade 48% HF; dilute 1 mL to 250 mL in H_2O to give 0.1 *M* HF.
8. Hexafluorosilicic acid (H_2SiF_6): Treat commercial 30% H_2SiF_6 with excess commercial finely ground quartz (57 mesh/cm and sized at 1-100 μm) for 3 days at 4°C. Shake intermittently each day. Centrifuge in a plastic tube and decant the supernatant H_2SiF_6 through a Whatman no. 50 filter paper just before each use (section 6–2.3.3.1).
9. Methanol: technical 99% methanol.
10. Boric acid (H_3BO_3): Saturate 100 mL of H_2O with H_3BO_3.
11. Tetrabromoethane and nitrobenzene.

6–2.3.3 PROCEDURE

The sample should be a powder, not coarse aggregates. To prepare a soil sample, boil 2 g in 6 *M* HCl for 10 min, then wash it with 0.5 *M* HCl, and 99% methanol by centrifugation. Dry the soil in a centrifuge tube, then powder the soil with a spatula and a rubber-tipped rod. Silt or fine sand fractions may be used directly for the isolation procedure. Disaggregate strongly aggregated materials by chemical agents and soft (rubber, plastic) instruments, if the size distribution of the quartz and feldspars is to be taken into account. Grind sieve-separated gravel or coarse sand (>1 mm) in an agate mortar so that the powder passes a 25-mesh (per cm) sieve, and keep it separate from the finer sample fractions.

Weigh approximately 0.5 g of dark-colored sedimentary clay, silt, or sand into a 250-mL beaker, add 30 mL of 6 *M* HCl and 10 mL of 3 *M* HNO_3, and digest at 80°C for 30 min to remove carbonates, some hydrous oxides, and other finely divided materials. Transfer to a centrifuge tube, centrifuge, and discard the supernatant liquid. Wash twice with 3 *M* HCl and then with water and 99% methanol. At this point the material is ordinarily almost free of organic matter, except for black shales. The latter would liberate CO_2 during the $Na_2S_2O_7$ fusion and cause excess frothing. To prevent this, return a sample dark with organic matter to the beaker. Add 10 drops of 9 *M* H_2SO_4 and 1 mL of HNO_3, and boil. Then add 1 mL of $HClO_4$. Bring to fumes over a low flame (hood) as the sample turns white and dries. *Caution:* Do not add any organic solvent for washing while $HClO_4$ is present! Cool; wash once with water and centrifuge, and dry at 105°C. Powder the sample with a plastic rod and mix.

Weigh 0.2 g of powdered clay, silt, or sand fraction (dried at 105°C) and transfer the sample into a 50-mL vitreous silica crucible containing 15 g of $NaHSO_4$ (or 12 g of $Na_2S_2O_7$ powder), and mix the sample with the reagent by means of a glass rod. Cover the crucible with a tight-fitting vitreous silica cover. Working with the crucible in a fume hood, fuse the $NaHSO_4$ slowly (converting it to $Na_2S_2O_7$) over a Meker burner, using a low flame at first until vigorous bubbling ceases and then gradually increasing the flame. The reactions are:

$$2\ NaHSO_4 \rightarrow Na_2S_2O_7 + H_2O \qquad [1]$$

$$Na_2S_2O_7 \rightarrow Na_2SO_4 + SO_3\ . \qquad [2]$$

The objective is to confine and conserve the SO_3 within the crucible in a prolonged fusion, and to avoid rapid accumulation of Na_2SO_4. If the mixture begins to lose fluidity, cool and add more $NaHSO_4$. The fusion should be at the full heat of the Meker burner for 2 h. The endpoint is the loss of nearly all the liquid; however, heating without liquid present can etch fused quartz crucibles.

Cool the crucible for 30 min. Transfer the salt with the aid of 3 *M* HCl to a 250-mL beaker, and dissolve it in about 50 mL of 3 *M* HCl on a hot plate at 80°C. (The whole crucible may be placed in the beaker if the cake does not drop out readily.)

Slake the fusion cake in about 50 mL of 3 *M* HCl, and heat the suspension just to boiling. When the fusion cake has disintegrated, transfer the resulting suspension to a 70-mL pointed centrifuge tube. Centrifuge the tube at 1800 rpm for 4 min, or longer if necessary to clear the supernatant liquid. Decant the supernatant solution and discard it. Wash the crucible with 3 *M* HCl to complete the transfer of the residue (consisting of quartz, feldspars, and amorphous silica), and break up the residue in the tube with a glass rod. Centrifuge the residue, and decant the solution as before. Give the residue a third washing with 3 *M* HCl as before. Transfer the residue from the tube into a 500-mL Ni or stainless steel beaker with the aid of a little 0.5 *M* NaOH, and add 0.5 *M* NaOH to a total volume of 100 mL. Bring the suspension rapidly to boiling over a Meker burner, and boil it for exactly 2.5 min to dissolve amorphous silica (and a little alumina). Cool the solution by placing the beaker in a water bath. Transfer the solution to centrifuge tubes, scrub and wash the beaker with 0.5 *M* NaOH to ensure complete transfer of the residue from beaker to tube, and centrifuge. Discard the supernatant solution. Wash the residue (usually mostly quartz and feldspars) and tube thoroughly four times with 3 *M* HCl to remove soluble Na and other soluble components. At this point, the procedure divides, according to objective: (i) to isolate monomineralic quartz, or (ii) to determine individual feldspars. Fuse a separate sample in $Na_2S_2O_7$ and bring to this point for either objective.

6–2.3.3.1 Quartz Isolation.

Add 10 mL of the filtered H_2SiF_6 to the residue in the polyethylene tube and shake. Hold a series of such samples for 3 days in a bath at 18 ± 2°C, with shaking to remix at least twice daily. Centrifuge the tubes containing the suspensions; wash once with 0.1 *M* HF and four times with H_2O. Transfer a portion of the residue to a glass slide, dry, and x-ray (chapter 12) to check the purity of the quartz (0.334, 0.425, and 0.245 nm peaks) and absence of feldspars (0.318 and/or 0.324 nm peaks). If feldspars are still present (seldom the case), return the XRD sample to the tube and repeat the treatment with H_2SiF_6. Wash with H_2O and repeat the XRD test. Repeat as required for quartz purity. A peak at 0.405 to 0.410 nm would indicate cristobalite.

To effect the heavy liquid separation, wash the residue with acetone to remove H_2O. Transfer a 1- to 10-μm diameter quartz fraction with a minimum of acetone to the surface (in a pointed centrifuge tube) of a tetrabromoethane–nitrobenzene mixture of 2.33 g cm^{-3} density (previously tested by weight in a volumetric flask). Rotate the tube gently to mix the acetone suspension slightly into the surface of the heavy liquid and centrifuge at 500 rpm for 15 min. Opal and cristobalite float and quartz and heavy glass sink. Draw off the top fraction onto another tube of the heavy liquid for a second centrifugation. Stir the original bottom fraction lightly with a pointed rod and recentrifuge in the original tube. Combine the two light (floating) fractions. Separately, combine the two heavy fractions using a glass rod with a stopper of suitable size to trap the bottom fraction while the heavy liquid is poured into a bottle for re-use. Wash the fractions with acetone, dry, and check for purity by XRD. Scanning electron microscopy assists in identifying glass and (or) opal.

Remove either with suitable heavy liquid density (e.g., 2.38 g cm^{-3} floats basic volcanic glass). Each quartz size fraction has a different density range (e.g., 2.28 g cm^{-3} for 2- to 5-μm crystals; Henderson et al., 1972). When the sample is free of contaminating substances, treat the quartz with 5 mL of saturated H_3BO_3 overnight to remove any fluorates in interstices. Wash four times with H_2O. Weigh the fractions to calculate the percentage of the original fused fraction. The quartz is ready for the determination of oxygen isotopic ratios.

6–2.3.3.2 Feldspar Determinations.

Under option (ii), weigh the isolate in Section 6–2.3.3, which contains quartz and feldspars, in a platinium crucible and analyze it for K_2O, Na_2O, and CaO by HF dissolution (Lim & Jackson, 1982; Baker & Suhr, 1982; Soltanpour et al., 1982; Helmke, 1982; Jones, 1982; Knudsen et al., 1982; and Lanyon & Heald, 1982—sections 1 through 5, 13–3.3.2, 13–3.3.3, and 14–3.3).

6–2.3.4 CALCULATION OF RESULTS

Under option (i), calculate the quartz content as the percentage of the size fraction from which it was isolated. Also calculate the percentage of the size fraction of the total sample.

Under option (ii), calculate the K, Na, and Ca content found in the residue as oxide percentage of the original fraction (sample) employed. Calculate weight ratios of Na_2O/CaO. Then the percentages of feldspars in the fraction are given by:

$$\text{Microcline–orthoclase content} = \%\ K_2O \times X \qquad [3]$$

$$\text{Albite content} = \%\ Na_2O \times Y - Y' \times \%\ \text{Microcline} \qquad [4]$$

$$\text{Anorthite content} = \%\ CaO \times Z \qquad [5]$$

in which values for X, Y, Y', and Z for different size fractions are given in Table 6–2.

Select the values of Y and Z according to the weight ratio ranges for Na_2O/CaO (Table 6–2). The albite and anorthite contents obtained represent the equivalent end-members of the high Na plagioclase present, since high anorthite feldspars are rare in soils and sediments and also would be largely dissolved by the fusion.

Calculate the residue weight percentage of the original sample from the weighing before the HF treatment. This is "% residue." When only quartz and feldspar are present in the residue, as is substantially the case in soils from acid rocks and most sedimentary rocks, the following relation holds:

Table 6-2. Summary of factors† for conversion of residue K_2O, Na_2O, and CaO to the respective endmember equivalent feldspars.

	Residue K_2O to		Residue Na_2O to albite, Y		Residue CaO to anorthite, Z		
Particle size, μm	Microcline X	Feldspar K_2O, X'	Na_2O/CaO >0.82‡	Na_2O/CaO 0.82–0.37	Na_2O/CaO >0.82	Na_2O/CaO 0.82–0.37	Y'
2000–500	6.05	1.026					
500–50	6.1	1.036	8.9	8.3	5.2	5.9	0.0
50–20	6.5	1.042	9.1	8.3	5.2	6.0	0.02
20–5	7.0	1.10	9.5	8.8	5.5	6.3	0.04
5–2	8.4	1.23	10.3	10.5	6.0	8.5	0.12
2–0.2	13.2	1.73	13.7	19.6	8.1	17.6	0.20

† These factors differ from the theoretical ($X = 5.9$, $Y = 8.5$, $Z = 4.95$) for 16.9% K_2O in microcline-orthoclase, 11.8% Na_2O in albite, and 20.2% CaO in anorthite, respectively, because of dissolution of feldspars, surface loss of K, Na and Ca, and replacement of Na for K and Ca during the $Na_2S_2O_7$ fusion. Y' is used in the estimation of albite to correct and residue Na_2O for the Na uptake by microcline during the fusion. X' is used in the mica determination to correct for microcline dissolution during the fusion.

‡ Ratio of the percentage by weight of the oxides present in the original sample.

Table 6-3. Factors for Eq. [6] to obtain the quartz or cristobalite content.

	Percentage weight recovery						
		Albite, B		Anorthite, C			
Size fraction, μm	Microcline, A	Na_2O/CaO >0.82†	Na_2O/CaO 0.82–0.37	Na_2O/CaO >0.82	Na_2O/CaO 0.82–0.37	Quartz, D	Cristobalite, D'
500–50	96.5	95.2	93.7	95.2	93.7	99.6	98
50–20	96.0	94.5	90.9	94.5	90.9	99.4	97
20–5	93.6	92.7	77.4	92.7	77.4	99.1	93
5–2	84.4	86.2	41.9	86.2	41.9	98.2	77
2–0.2	64.0	68.3	16.4	68.3	16.4	96.5	53

† Ratio of the percentage of weight of the oxides present in the original sample.

$$\%\ \text{quartz} = \frac{100}{D}\left[\%\ \text{residue} - \frac{A \times \%\ \text{K feldspar}}{100} - \frac{B \times \%\ \text{Na feldspar}}{100} - \frac{C \times \%\ \text{Ca feldspar}}{100}\right] \quad [6]$$

in which values for A,B,C, and D for the different particle-size fractions are given a Table 6–3. Substitute D' for D if x-ray diffraction evidence establishes the presence of cristobalite and the absence of quartz.

Calculate the K-mica content of the fraction from the equation:

$$\%\ \text{mica} = 10\ (\text{total initial}\ \%K_2O - \text{feldspar}\ \%K_2O) \times X' \quad [7]$$

in which X' is from Table 6–2. Na-mica can be derived in a similar calculation, using 7% Na_2O content. NH_4^+-mica may be appreciable (Cooper & Evans, 1983).

6–2.3.5 COMMENTS

The chief positive error in the quartz determination arises from the presence of nonquartz silicates other than feldspars (for example, zircon, $ZrSiO_4$) that resist the $Na_2S_2O_7$–HCl–NaOH treatments. The quantities of such minerals present in most soils are inappreciable, but the possibility of their being present must be considered for a given sample. Tremolite (rare) is fairly resistant to the treatment; its Ca content makes it appear mainly as "feldspar" by the Ca analysis after the HF treatment of the residue. Trostel and Wynne (1940) recommended examination of the residue with a microscope prior to the HF treatment. The author has found examination by x-ray diffraction to be highly useful. Quartz and feldspar peaks are the only peaks present in most of the residues from soils (Sayin & Jackson, 1979).

The chief negative errors for feldspars and quartz arise from variations of the determined minerals from the standards used in determining the factors given in Tables 6–2 and 6–3. Excessive losses during fusion may occur with altered feldspars. The losses may represent dissolution of alteration products such as sericite, but may reflect excessive porosity in the macrocrystal. The correction factors for quartz represent chiefly dissolved silica but include slight manipulative losses. Other forms of SiO_2 such as cristobalite (rare in soils except certain ones derived from some rhyolites) are less resistant and more variable from specimen to specimen. Quartz is the form of SiO_2 commonly found in soils and a wide variety of rocks, and the correction factors given for quartz should invariably be employed unless positive evidence of other crystalline forms of SiO_2 is obtained by x-ray diffraction. The overall error in the determination mainly reflects variations from those represented in the correction factors and those of the spectrophotometry (Knudsen et al., 1982; Lanyon & Heald, 1982.) The overall error of the quartz determination is generally of the order of 1 to 5% of the sample (increasing with decreasing particle size) when only quartz and feldspars are present in the residue.

6–3 NONCRYSTALLINE ALUMINOSILICATES AND HYDROUS OXIDES BY ACID AMMONIUM OXALATE IN THE DARK

6–3.1 Introduction

The term *noncrystalline* is employed instead of "amorphous" and "allophane" in accord with Wada (1977) and will be understood generally to include *paracrystalline*, which includes somewhat ordered ("short-range" ordered) materials such as the tubular mineral imogolite (Cradwick et al., 1972). The term allophane is restricted (van Olphen, 1971) to the materials having DTA characteristics of the materials of Ross and Kerr (1934), while noncrystalline and paracrystalline cover a broader range of soil materials (Table 6–1).

Although most soils consist primarily of crystalline minerals, many contain appreciable amounts of noncrystalline inorganic material. Those soils derived from volcanic ash and weathered pumice may primarily consist of "allophane" and imogolite (Thorp & Smith, 1949; Mitchell et al., 1964) or other noncrystalline Al, Fe, or Si materials as shown for various Hydrandepts (Wada & Wada, 1976). The amounts of noncrystalline materials, even when small, can contribute significantly to the physical and chemical properties of soils (Fey & LeRoux, 1977), since they may have (i) high cation-exchange capacity, which may be influenced by pH (Wada & Ataka, 1958; Aomine & Jackson, 1959); (ii) high surface area (Aomine & Otsuka, 1968); and (iii) high reactivity with phosphate (Saunders, 1964) and organics (Inoue & Wada, 1968). Thus it is desirable to characterize and quantify the amount of noncrystalline materials present in soils as well as the amount of the crystalline components.

6–3.2 Principles

Selective dissolution analysis (SDA) has been extensively used in the study of the noncrystalline material content of soils and sediments. There are limitations, however, which must be considered in using SDA. First, a continuum of crystalline order exists, ranging from no long-range order to paracrystalline to poorly crystalline to well crystalline (Follet et al., 1965). It is difficult to assess adequately the portion of this continuum that is extracted by a particular reagent. Acid ammonium oxalate allowed to react in darkness (AOD) for 2 h has been shown to be a selective reagent for dissolution of noncrystalline materials (Schwertmann, 1959, 1973; McKeague & Day, 1966; Higashi & Ikeda, 1974; Fey & LeRoux, 1977; Hodges & Zelazny, 1980).

The AOD procedure has been shown to remove most noncrystalline and paracrystalline materials ("allophane" and imogolite) from volcanic ash soils (Higashi & Ikeda, 1974, Hodges & Zelazny, 1980). This reagent also removes some short-range-ordered oxides and hydroxides of Al, Fe, and Mn (Schwertmann, 1959, 1964; McKeague & Day, 1966; McKeague et al., 1971; Fey & LeRoux, 1977) and Ti (Fitzpatrick et al., 1978). Noncrystalline silica is not dissolved by this method (Wada, 1977).

The AOD treatment has been reported to dissolve very little <200-μm hematite and goethite and only minor amounts (9%) of <200-μm magnetite (Baril & Bitton, 1969; McKeague et al., 1971). The severity of this treatment, however, increases with a decrease in particle size (McKeague et al., 1971).

There are conflicting reports as to the susceptibility of clay minerals to AOD treatment. McKeague and Day (1966) reported that the AOD treatment has little effect on kaolinite, montmorillonite, and illite; and Hodges and Zelazny (1980) reported similar results for gibbsite, kaolinite, and montmorillonite. However, Arshad et al. (1972) found that AOD treatment caused considerable decomposition of finely ground trioctahedral minerals (biotite and chlorite), as measured by solubilization of Fe, Al, and Mg but with lesser amounts of Si. Pawluk (1972) observed that prolonged extraction with AOD treatment resulted in slight dissolution of hydrous mica and trioctahedral chlorite, but Hodges and Zelazny (1980) found that $<0.5\%$ of a trioctahedral vermiculite dissolved during a 2-h AOD treatment. Iyengar et al. (1981), reported that an AOD treatment revealed no x-ray diffraction detectable alteration to mineral phases present in the clay fraction of soils dominated by kaolinite or hydroxy-interlayered vermiculite. McKeague and Day (1966) found that some interlayer Al was removed from artificially prepared Al-chloritized bentonite during a 4-h treatment, but also noted that interlayer materials of natural Al-chloritized clays were only slightly affected by this treatment. Some highly weathered soils of a humid tropical region released a fraction of one percent each of Fe_2O_3, Al_2O_3, and SiO_2 with the AOD method (Fey & LeRoux, 1977; Makumbi & Jackson, 1977).

Another problem in using SDA is the assignment of water content to the oxides of Si, Al, and Fe determined by chemical analysis. Noncrystalline materials are noted for their variable chemical composition (van Olphen, 1971) and thus may be expected to have variable water contents. A constant water content has often been assumed for the noncrystalline components, resulting in assignment of these values without regard for sample composition (Jackson, 1979; Alexiades & Jackson, 1966; de Villiers, 1971). However, SiO_2/Al_2O_3 molar ratios have also been used to assign water contents within specified ranges (Jackson, 1979).

Fey and LeRoux (1976) assigned water content by comparing the SiO_2:Al_2O_3 ratio of the sample with those of noncrystalline, synthetic aluminosilicate gels of known water content. This method assumes that synthetic materials are adequate models for predicting the characteristics of natural materials. A more satisfactory method would be to measure the weight loss of a given sample after SDA. Hodges and Zelazny (1980) made a comparison of measured weight loss to chemical determination with water content assigned on the basis of the SiO_2/Al_2O_3 molar ratio, according to the procedure of Jackson (1979) on AOD-treated soils. As would be expected from heterogeneous composition, the values were not in complete agreement, although they were remarkably similar. It therefore seems possible to quantify the noncrystalline material content of soils without time-consuming chemical determinations or assuming sample homogeneity and water content. However, used in conjunction with chemical analysis, more accurate characterization seems possible in that the water content may be measured, instead of estimated.

6–3.3 Method (Schwertmann, 1964; Fey & LeRoux, 1977; Hodges & Zelazny, 1980)

6–3.3.1 SPECIAL APPARATUS

1. Polypropylene centrifuge tubes, 100-mL.
2. Centrifuge.
3. Aluminum foil.
4. Analytical balance: 0.00001 g readability. A wire hook attached to the frame of the balance pan is used to hold centrifuge tubes in a vertical position, to reduce weighing errors resulting from variable orientation of the tubes on the balance pan.
5. Shaker.
6. Vacuum desiccator containing dry P_2O_5.
7. Oven: set to 110° ± 2°C.

6–3.3.2 REAGENTS

1. Ammonium oxalate $[(NH_4)_2C_2O_4 \cdot H_2O]$, approximately 0.2 *M* at pH 3.0: dissolve 28.4 g of reagent-grade ammonium oxalate monohydrate in 900 mL of distilled water, adjust pH to 3.0 using NH_4OH or HCl, and dilute to 1 L.
2. Ammonium carbonate $[(NH_4)_2CO_3]$, approximately 0.5 *M*: dissolve 47.0 g of reagent-grade ammonium carbonate in 1 L of distilled water.
3. Phosphorus pentoxide (P_2O_5): use reagent-grade powder.

6–3.3.3 PROCEDURE

The sample should be a powder, not coarse aggregates. It can have any cation saturation, although NH_4 saturation would provide the closest weight comparison after SDA. The sample should have prior treatments to remove carbonates, soluble salts, and organic matter (chapter 5). Generally, clay-sized or whole soil samples are treated, although any sized fraction of interest could be examined. The sample should be dried in a vacuum desiccator over P_2O_5, with an aliquot dried at 110°C to determine sample moisture content; or the entire sample may be oven dried at 110°C, depending on the drying characteristics of the sample.

Operationally, a known amount of sample of approximately 250 mg is weighed into a preweighed (to 0.00001 g) and predried 100-mL polypropylene centrifuge tube. The initial sample weight should be based on a 110°C oven-dried weight basis. A set of blank tubes should be carried through each procedure to account for any weight loss by the centrifuge tubes upon drying. Care should be exercised in reducing weighing errors by only handling the tubes with forceps, drying all tubes in a 110°C oven, and cooling in a vacuum desiccator containing P_2O_5, duplicating the time of weighing and exact position of the centrifuge tubes in a vertical position at the center of the balance pan.

To the weighed centrifuge tube containing the sample, add 50 mL of

0.2 *M* ammonium oxalate solution adjusted to pH 3.0, stopper the centrifuge tube, immediately wrap in aluminum foil to eliminate light, and shake for 2 h on a reciprocating shaker. After the designated time, centrifuge the sample and decant the supernatant solution from the sample. Although not essential for quantification of noncrystalline material, more chemical information can be obtained by saving the supernatant solution in a plastic container and analyzing for at least Al, Fe, and Si.

Methods for the analysis of single dilutions of AOD extracts of soils for Al, Fe, and Mn by flame emission spectrometry and Si by atomic absorption spectrometry have been established (Searle & Daly, 1977). Methods have also been developed to reduce the tendency of the nebulizer or burner slot to clog when solutions containing high salt concentrations or salts of low solubility are aspirated (Simmons & Plues-Foster, 1977). Alternatively, the Fe may be determined by orthophenanthroline or KSCN, the Si by the molybdosilicate-blue colorimetric method (Weaver et al., 1968; Hallmark et al., 1982, section 15–3.4.3) after overcoming interference arising from Mo complexation by oxalate (Fey & LeRoux, 1976) and Al by the colorimetric Aluminon method (Jackson, 1979; Hsu, 1963; Barnhisel & Bertsch, 1982, section 16–5.2) after destroying the oxalate. Alternatively, determine Fe and Al by the ferron method after digestion of the oxalate extract with H_2SO_4–H_2O_2 (Tokashiki & Wada, 1972).

The residues in the tubes should be washed three times with 0.5 *M* $(NH_4)_2CO_3$ and once with an equal volume of distilled water to remove remaining dissolution treatment chemicals. The residues in the tube should be dried overnight in an oven at 110°C, to volatize excess $(NH_4)_2CO_3$ as NH_3, CO_2, and H_2O. Inspection of the oven-dried samples should reveal no visible salt residues. The tubes should be allowed to cool in a P_2O_5 vacuum desiccator and weighed for the final time to determine amount of material dissolved by the AOD treatment.

6–3.3.4 CALCULATION OF RESULTS

The noncrystalline content of a sample is calculated as the percent weight loss resulting from the AOD treatment. All calculations are based on weights after or adjusted to oven drying at 110°C:

$$\%\ \text{noncrystalline material} = \frac{(110°\text{C sample weight before AOD} - 110°\text{C sample weight after AOD}) \times 100}{110°\text{C sample weight before AOD}}.$$

From chemical analysis, convert Al, Fe, and Si to the percentage of the respective oxides Al_2O_3, Fe_2O_3, and SiO_2; sum the oxides, and calculate an SiO_2/Al_2O_3 molar ratio. A percent water content of the noncrystalline material may be calculated as a percentage of material unaccounted for as oxides:

% H_2O of noncrystalline material =

$$\frac{(\%\ \text{noncrystalline material} - \%\ \text{oxide sum}) \times 100}{\%\ \text{noncrystalline material}}.$$

Compare results to published values given by Jackson (1979) on the basis of SiO_2/Al_2O_3 molar ratios section 6–5.3.4).

6–3.3.5 COMMENTS

Inherent sources of error in this technique result from weighing errors, presence of salts, and sample loss from decantation. These errors are minimized if procedures are followed carefully. Greater sources of error result from not excluding light, or assuming that the noncrystalline content of a sample is primarily present in the clay-sized fraction.

The addition of light, especially ultraviolet, to the ammonium oxalate treatment removes both crystalline and noncrystalline oxides from soils (DeEndredy, 1963; Schwertmann, 1964). The photolytic ammonium oxalate treatment largely dissolves hematite, goethite, and magnetite and does degrade nontronite and hydroxy-interlayered vermiculite from soils (DeEndredy, 1963; LeRiche & Weir, 1963; Chao & Theobald, 1976; Iyengar et al., 1981).

Many studies dealing with noncrystalline materials have been conducted using only the clay size fractions obtained by dispersion procedures. This results from the general assumption that noncrystalline materials occur only in this fraction and that efficient dispersion of the clay fraction results by the methods employed. Hodges and Zelazny (1980) determined that the noncrystalline material analyses on the clay fractions underestimated the noncrystalline material content for selected whole soils by from 0 to 34%. Much of this difference is undoubtedly due to incomplete dispersion of coatings of noncrystalline materials from sand and silt particles. The dissolution of crystalline particles and aggregates may also have contributed to the observed differences; however, Follett et al. (1965) and Jones and Uehara (1973) showed that most noncrystalline soil materials exist as coatings that bind aggregates of minerals together, rather than as separate particles.

If the percent water content of the noncrystalline material is calculated, it should be compared to published values given by Alexiades and Jackson (1966). Water contents much greater than published values may indicate sources of error, especially the presence of undetermined metal oxides and (or) hydroxides. The AOD solution should then be examined for Mn, Ti, or any other metal suspected of contributing to the loss in sample weight.

6–4 FREE IRON-ALUMINUM OXIDES AND HYDROXIDES

6–4.1 Introduction

Pedogensis results in the accumulation of soil colloids with compositions that are enriched in hydrous oxides of Al, Fe, and Mn relative to Si. In soils and sediments, these pedogenic oxides and hydroxides frequently occur in amorphous surface coatings (Follett et al., 1965; Jones & Uehara, 1973), as hydrous metal oxide coatings on mineral surfaces (Davidtz & Sumner, 1965; Roth et al., 1967, 1968, 1969; Greenland et al., 1968; Jackson & Frolking, 1982), as discrete crystalline minerals such as goethite and hematite which may have varying amount of Al substitution (Correns & von Engelhardt, 1941; Callière et al., 1960; Norrish & Taylor, 1961; Schwertmann et al., 1977; Jackson & Frolking, 1982), and as concretions or nodules (Winters, 1938; Drosdoff & Nikiforoff, 1940).

A dithionite method employing sodium dithionite ($Na_2S_2O_4$, "hydrosulphite") solution at about 40°C in the absence of chelating agent was proposed by Galabutskaya and Govorova (1934) for removal of free iron oxides from kaolins. A similar but more rapid extraction of free iron oxides was obtained by the dithionite solution in acid (pH 3.5 to 6) systems (Mitchell and Mackenzie, 1954). Dithionite was further used with 0.1 *M* tartrate as a chelator and 1 *M* sodium acetate as a buffer at 40°C (Deb, 1950). The amount of free iron oxides removed was independent of pH in the range of 2.9 to 6.0. The latter procedure was found to be slow in removing the free iron oxides from Latosols (Oxisols) and to result in the precipitation of unwanted FeS and S (Jackson, 1979, p. 45). For complete and rapid removal of iron oxides from soils, Aguilera and Jackson (1953) proposed the use of sodium dithionite with sodium citrate (with or without Fe^{3+}-specific Versene), with pH adjusted to 7.3 from moment to moment by dropwise additions of 10% NaOH at 80 to 90°C. As much as 20% of free iron oxides was removed in 15 min without the precipitation of either FeS or elemental S. The removal of free iron oxides and hydroxides is more complete and is free of effects on cation exchange capacity (CEC) in systems using citrate as the chelating agent (Aguilera & Jackson, 1953; Coffin, 1963). The incorporation of sodium bicarbonate as a buffer at pH 7.3 in the citrate–dithionite method resulted in the widely adopted buffered neutral citrate–bicarbonate–dithionite (CBD) system (Jackson, 1979, p. 44–51; Mehra & Jackson, 1960) for the removal of reluctant soluble iron oxides and phosphate (Jackson, 1958). Calcite of soils was selectively removed from dolomite by citrate and dithionite (Petersen et al., 1966) at pH 5.85 (Jackson, 1979, p. 522).

The original objectives of the CBD method are primarily the determination of free iron oxides and the removal of amorphous coatings and crystals of free iron oxide, acting as cementing agents, for subsequent physical and chemical analysis of soils, sediments, and clay minerals. In view of the adsorption of silica by iron and aluminum oxides (Beckwith and Reeve, 1964; Weaver et al., 1968) and the substitution of iron by aluminum in soil goethite and hematite, extractable Al, Si, and Mn are determined by the CBD procedure (Follett et al., 1965; Weaver et al., 1968; Blume & Schwertmann, 1969; Tokashiki & Wada, 1972; Makumbi & Jackson, 1977). Free MnO_2 is made soluble by any H_2O_2 pretreatment (Jackson, 1979).

A knowledge of the nature and distribution of pedogenic oxides and hydroxides in the soil profile aids in documenting the pedological processes of soil formation (Oades, 1963) and in soil classification (McKeague & Day, 1966; Soil Survey Staff, 1975). In *Soil Taxonomy* (Soil Survey Staff, 1975), the CBD-extractable Fe plus Al is used as one of the diagnostic characteristics of the spodic horizon. The ratio of oxalate-extractable (Fe_o) to CBD-extractable (Fe_d) iron oxides has been used as a relative measure of the degree of aging or crystallinity of free iron oxides (Schwertmann, 1964) and as a supplementary criterion for deciding the age sequence of associated but noncontinguous Quaternary deposits (Alexander, 1974). In addition, the removal of amorphous and crystalline pedogenic oxides and hydroxides aids in dispersion and preconcentration of the silicate fractions. The CBD treatment greatly enhances the degree of parallel orientation of the layer silicates, with a consequent increase in the X-ray diffraction intensity (Jackson, 1979, p. 44). The CBD treatment in essence helps to "clean up" samples for mineralogical analysis by both physical and chemical techniques.

Apart from the use of dithionite, various other procedures for removal of free iron oxides from soils and sediments have been employed. These procedures may be classified according to Jackson (1979, p. 52–55) as (i) those without reductants, including acid ammonium oxalate (section 6–3), oxalic acid at elevated temperature, sodium acid oxalate at room temperature for several days, alkaline Tiron solution at pH 10.5, and citric acid, and (ii) those with reductants, such as oxalate with exposure to sunlight, with H_2S reduction, nascent hydrogen reduction, and biological reduction carried out in the presence of sucrose.

Up to 100 times as much SiO_2 is made soluble by CBD as by AOD (section 6–3). The volcanic ashes (Jackson, 1959) as well as opal (Huang & Vogler, 1972; Sayin & Jackson, 1979) and phyllosilicate clays in porous soils are subject to intense depletion of structural cations, including aluminum (Schwertmann & Jackson, 1963), and noncrystalline siliceous residues are accumulated (Dyal, 1953; Jackson, 1959).

6–4.2 Principles

Prerequisite to a good method for removal of free iron hydrous oxides and hydroxides is a reagent with a high oxidation potential one that has a high tendency to become oxidized, and thus is a good reducing agent (Jackson, 1979, p. 651). A chelating agent is also required for sequestering Fe^{2+} and Fe^{3+} ions. The CBD system employs sodium citrate as a chelating agent, sodium bicarbonate ($NaHCO_3$, pH 7.3) as a buffer, and sodium dithionite ($Na_2S_2O_4$) for the reduction (Jackson, 1979, p. 44–51; Mehra & Jackson, 1960).

The citrate anion, a tridentate ligand, forms relatively more stable Fe and Al chelates than either oxalate or tartrate anion. Coordination of metallic cations with the citrate ligand can occur with the formation of two chelate rings (one five-membered ring and one six-membered ring) involving two carboxyl groups and one hydroxyl group (Chaberek &

Martell, 1959, p. 308). Performance of the CBD treatment after carbonate and organic matter removal and the excess of citrate ligand ensure optimum chelation of Fe, Al, and Mn ions. The strong chelating action of citrate prevents the precipitation of FeS at the near-neutral pH of extraction. The $NaHCO_3$ strongly buffers the CBD system at pH 7.3, thereby increasing the oxidation potential of the system because of the supply of four OH^- groups used in the oxidation of each dithionite molecule. The basic system has a theoretical redox potential of -1.12 V, while that of the acid system is -0.08 V (Jackson, 1979, p. 45):

$$\text{Basic system } 4OH^- + S_2O_4^{2-} \rightarrow 2SO_3^{2-} + 2H_2O + 2_e^- \quad [8]$$

$$\text{Acid system } 2H_2O + HS_2O_4^- \rightarrow 2H_2SO_3 + H^+ + 2_e^- \quad [9]$$

A value of the redox potential at pH 7.3 of -0.7 V was measured for the CBD system (Mehra & Jackson, 1960). The earlier dithionite methods (Deb, 1950; MacKenzie, 1954), which used a lower pH (acid system), did not give effective removal of free oxides from soils high in iron oxides because of the low oxidation potential of acid systems. Nonprecipitation of sulfur, rapid reaction rate, and little removal of Fe from silicates or effect on CEC are benefits derived from the extraction with dithionite at a near-neutral pH of 7.3 rather than at an acid pH. Because the solubility of Fe and Al hydroxy oxides decreases as the third power of OH^- concentration, the pH of extraction is maintained at 7.3. Sodium dithionite has sometimes been designated in the literature as sodium "hydrosulphite" or even "dithionate," though these names are incongruous with the accepted nomenclature of the sulfur–oxygen salts.

In addition to free iron oxides, Al-substituted crystalline hematite and goethite are dissolved by the dithionite treatment (Norrish & Taylor, 1961; Fey & LeRoux, 1976, 1977). Magnetite and ilmenite, however, are not extracted by CBD. Reduction of combined Fe^{3+} to Fe^{2+} (after oxidation with H_2O_2) can be effected in micaceous vermiculite (Roth et al., 1968, 1969; Veith & Jackson, 1974). Appreciable dissolution of structural Fe, change in oxidation state of Fe^{3+} to Fe^{2+}, and change in Fe coordination environment in nontronite can be brought about by the CBD treatment (Dudas & Harward, 1971; Stucki et al., 1976; Russell et al., 1979). A considerable amount of Al from hydroxy aluminum interlayers of vermiculitic chlorite is also extracted, but gibbsite is little attacked (Dixon & Jackson, 1962). The infrared spectra of CBD-soluble fractions of volcanic ash soils suggest dissolution of allophane-like materials with a low SiO_2/Al_2O_3 molar ratio ranging from 0.2 to 0.4 and additional alumina-rich components with a ratio less than 0.2 (Wada & Greenland, 1970; Wada & Tokashiki, 1972; Tokashiki & Wada, 1975). High CBD-extracted Al related well to phosphate adsorption in ash soils (Shoji & Ono, 1978).

6–4.3 Method (Aguilera & Jackson, 1953; Mehra & Jackson, 1960; Jackson, 1979).

6–4.3.1 REAGENTS

1. Sodium citrate, approximately 0.3 *M*: Dissolve 88 g of trisodium citrate per liter of solution in water.
2. Sodium bicarbonate ($NaHCO_3$), approximately 1 *M*: Dissolve 8.4 g of reagent-grade $NaHCO_3$ per 100 mL of solution in water.
3. Sodium dithionite ($Na_2S_2O_4$): Use freshly opened, purest grade.
4. Sodium chloride (NaCl), saturated solution: Shake 80 g of reagent-grade NaCl in 200 mL of water.
5. Acetone, reagent-grade.

6–4.3.2 PROCEDURE

Transfer to a 100-mL centrifuge tube a suitable amount of sample (5 g of many soils or 1 g of clay), containing 0.5 g of extractable free Fe_2O_3 or less, that has been treated to remove carbonates, soluble salts, and organic matter (Chapter 5). Add 40 mL of 0.3 *M* sodium citrate and 5 mL of 1 *M* $NaHCO_3$ solution (these solutions can be combined in the same proportion ahead of time; Kittrick & Hope, 1963), shake the tube to mix the contents, and heat the tube in a water bath at 75 to 80°C for several minutes, while stirring the suspension with a glass rod. When the temperature of the soil or clay suspension rises to the range of 75 to 80°C, add about 1 g of $Na_2S_2O_4$ powder with a calibrated spoon; immediately stir thoroughly for 1 min, and then intermittently for 5 min. Add a second 1-g portion of $Na_2S_2O_4$ and continue occasional stirring for another 10 min. After the 15-min digestion period, add 10 mL of saturated NaCl solution to promote flocculation, and centrifuge. If the suspension fails to flocculate with the NaCl, add 10 mL of acetone (particularly needed for volcanic ash soils), mix the contents, warm in a water bath, and centrifuge for 5 min at 1600 to 2200 rpm. Decant the clear supernatant into a 500-mL volumetric flask (or a 1000-mL flask, if the volume of extractant exceeds 500 mL). Repeat the procedure for samples in which a brown or red coloration persists and those containing more than 5% Fe_2O_3. Combine extraction and washing solutions with those of the first treatment. Wash (two or more times for samples with more than 1 g of residue) the sample with sodium citrate solution (with NaCl and acetone if necessary for flocculation), combine the washings with the previous decantates, dilute with water to the marked volume, and mix. Add H_2O_2 dropwise if black FeS appears.

Determine the Fe by atomic absorption spectrometry (AAS) by direct aspiration of a 1:10 dilution of the CBD extract solution (Jenne et al., 1974; Olson & Ellis, 1982). Alternatively, the Fe may be determined by orthophenanthroline or KSCN, the Si by the molybdosilicate blue colorimetric method (Weaver et al., 1968; Hallmark et al., 1982, section 15–

3.4.3), and the Al by the colorimetric Aluminon method (Jackson, 1979; Hsu, 1963; Barnhisel & Bertsch, 1982, section 16–5.2.3) after destroying the citrate. Alternatively, determine Fe and Al by the ferron method after digestion of the CBD extract with H_2SO_4–H_2O_2 (Tokashiki & Wada, 1972).

The sample, freed of extractable oxides and hydroxides but not dried during the procedure, is now ready for aluminosilicate dissolution (section 6–5); smectite, vermiculite, and CEC hysteresis determinations (section 6–6); x-ray diffraction analysis (chapter 12); thermal analysis (Chapter 7); infrared analysis (chapter 11); total elemental analysis (Lim and Jackson, 1982); or petrographic (chapter 8) and electron microscopic examination or other procedures.

6–4.3.3 CALCULATION OF RESULTS

Calculate the percentages of Fe_2O_3, Al_2O_3, and SiO_2 dissolved in the extraction. Alternatively, calculate the total amount of free oxides and hydroxides from the dry weight difference between the sample before and after extraction, as for amorphous materials (section 6–3).

6–4.3.4 COMMENTS

Avoid $CaCl_2$ as a flocculant, because $CaCO_3$ would form. Avoid adding HCl, because this would promote sulfur precipitation.

A maximum of about 0.5 g of Fe_2O_3 can be dissolved in a 40-mL portion of the citrate reagent, which is relatively high compared to the amount of free iron oxides removed by most methods. The sample size of a given soil (per unit volume of reagent) is adjusted in accordance with the expected percentage of extractable Fe_2O_3 present. For most soils of the temperate regions, a 10-g sample of soil is employed. Samples high in concretionary crystalline iron oxides such as hematite, goethite, or limonite are first crushed moderately well in an agate mortar, and a sample of approximately 0.2 g is used. For samples containing a high but unknown quantity of iron oxides, a 0.1-g pilot sample is taken through one treatment of the procedure, and the iron is determined.

Heating above 80°C is avoided because FeS forms at higher temperature. Ordinarily, little sulfur forms unless too high a temperature or too low a pH value is employed. To remove the sulfur if it does form, the sample is washed twice with 80 to 100% acetone and once with carbon tetrachloride (CCl_4) at 30 to 50°C. The CCl_4 is finally removed by washing with methanol or acetone. Alternatively, sulfur can be effectively removed by carbon disulfide (CS_2) washings. The sample is washed twice with 95% ethanol saturated with NaCl, and then three times with CS_2 reagent (consisting of 1 volume of CS_2 to 2 volumes of 95% ethanol saturated with NaCl). Finally the sample is washed four to five times with 95% ethanol saturated with NaCl to remove the CS_2.

A pure white color of the residue after CBD treatment should not be expected, as many soils and colloids contain greenish, bluish, cream-colored clays arising because of the presence of structural iron. Black minerals, such as ilmenite or magnetite, which are not dissolved, may also be present and leave a black color.

6–5 POORLY CRYSTALLINE ALUMINOSILICATES

6–5.1 Introduction

Noncrystalline mineral colloids occur extensively in soils (Aomine & Yoshinaga, 1955; Briner & Jackson, 1969; Kanehiro & Whittig, 1961; Mitchell & Farmer, 1962; Schwertmann, 1964; Hodges & Zelazny, 1980). The commonly used method of characterizing them is to determine their relative resistance to dissolution in various solutions (for example, allophane and imogolite in acid ammonium oxalate in the dark (AOD, section 6–3). An older method was to use the acid salt (pH 3.5) of oxalate in the presence of light for SDA of the "inorganic gel complex" of soils (Tamm, 1922).

Selective dissolution analysis (SDA) with alkaline solutions has been employed with whole soils or clays for a long time as a series of operational procedures, each for specific dissolution objectives (Table 6–4). Percentages dissolved from volcanic ash soils by alternating treatments with 8 *M* HCl for 30 min and warm 0.5 *M* NaOH for 5 min (Segalen, 1968) were correlated well with heating weight losses at 200°C (Kitagawa, 1977). A kinetic dissolution study mainly of silica and iron oxides was carried out on "sensitive" ("quick," subject to landslides) clays of Champlain Sea sediments of Ontario and Quebec (McKyes et al., 1974) by a modified Segalen method. The XRD intensity for the crystalline constituents increased when 11 to 12% of soluble material was extracted by alternate 0.5-h treatments at room temperature. Extensive treatment of clays in 6 *M* NaOH decomposed most of the layer silicate clay minerals except mica, which became enriched in the sample (Reynolds, 1960). Stringent SDA treatments have continued to have a role for soil mineralogical purposes. An example is the Bayer process for dissolution of Al from bauxite in strong NaOH.

Low concentrations of Na_2CO_3 and NaOH solutions have been employed for selective dissolution of various soil constituents, dissolving silica and alumina cements, cleaning coarser mineral grains, and dispersing clays. Repeated alkaline extractions with Na_2CO_3 solutions, for example, 5% Na_2CO_3 solution (Van Bemmelen in 1877, according to Hillebrand and Lundell, 1929), have a moderate solvent effect for finely divided noncrystalline silica and some alumina. A 2% Na_2CO_3 solution has a pH of about 10.7, which is about the same as that of 0.0003 *M* NaOH (Jackson, 1979, p. 72); it is buffered at this pH by hydrolysis. A 2% solution of Na_2CO_3, though dissolving only small amounts of silica, has a remarkably favorable dispersion effect on soils and clays (Jackson et al., 1950) when used after the removal of exchangeable Ca and Mg (Jackson, 1979). It is ineffective in simultaneously dissolving appreciable

Table 6–4. Treatment reagents for dissolution of Al, Fe, and Si in various clay and organic constituents.

Element in specified complex or component	0.5 *N* NaOH†	2% $NaCO_3$‡	$NH_4HC_2O_4$ or $NaHC_2O_4$ (pH 3.0–3.5)§	CBD¶	0.1 *M* $Na_2P_2O_7$ or $K_2P_2O_7$#	8 *N* HCl and 0.5 *N* NaOH††	0.1 *M* alkaline tiron solution‡‡	NaF or KF§§
	(+ + Good, + Poor, 0 No, -- Not tested)							
Al:								
Organic	+ +	+ +	+ +	+ +	+ +	+ +	+ +	--
Hydrous oxides								
Noncrystalline	+ +	+ +	+ +	+ +	+	+ +	+ +	+ +
Crystalline	+ +	+	0	+	0	+	+	+
Fe:								
Organic	0	0	+ +	+ +	+ +	+ +	+ +	--
Hydrous oxides								
Amorphous	0	0	+ +	+ +	+	+ +	+ +	+ +
Crystalline	0	0	0	+ +	0	+ +	+	+
Si:								
Opaline	+ +	+	0	0	0	+ +	--	+ +
Crystalline	+	0	0	0	0	--	--	0
Al and Si:								
Amorphous	+ +	+ +	+ +	+ +	+	+ +	+ +	+ +
Allophane	+ +	+	+ +	+	+	+ +	+	+ +
Imogolite	+ +	+	+	+	+	+ +	--	+ +
Layer silicates	+	0	+	0	0	+	+	+

† Hashimoto and Jackson (1960); Follett et al. (1965); Wada and Greenland (1970).
‡ Jackson et al. (1950); Jackson (1979); Tokashiki and Wada (1975).
§ Schwertmann (1964); McKeague and Day (1966); Higashi and Ikeda (1974); Fey and LeRoux (1976); Wada and Wada (1976).
¶ Mehra and Jackson (1960); Wada and Greenland (1970); Tokashiki and Wada (1975).
# McKeague et al. (1971); Arshad et al. (1972); Wada and Higashi (1976).
†† Segalen (1968); Kitagawa (1977).
‡‡ Biermans and Baert (1977).
§§ Huang and Jackson (1965); Fieldes and Perrott (1966); Perrott et al. (1976).

amounts of both free silica and alumina from soils (Hashimoto and Jackson, 1960) or the noncrystalline aluminosilicates of dehydroxylated kaolinite (Hislop, 1944).

A boiling solution of 0.5 *M* NaOH was used for the dissolution of noncrystalline silica from soils (Hardy & Follett-Smith, 1931). This solution was used in a 4-h digestion at 100°C to dissolve free silica or alumina in montmorillonitic samples (Foster, 1953) and in soils (Dyal, 1953; Whittig et al., 1957). Gibbsite was dissolved from soil samples high in goethite and halloysite by digestion of the samples in 1.25 *M* NaOH on a steam bath for 20 min (Muñoz Taboadela, 1953; Mackenzie & Robertson, 1961). Treatment with 0.5 *M* NaOH dissolved 2 to 15 times more "free Al_2O_3" than "free SiO_2" from a <0.3 μm soil fraction high in both aluminum phosphate and noncrystalline silicate (Dyal, 1953).

The dissolution of free hydrous oxides, such as fine-grained opaline silica (Jones, 1969; Tokashiki & Wada, 1975), cristobalite (Higashi & Ikeda, 1974), and some gibbsite (Wada & Greenland, 1970; Hodges & Zelazny, 1980) is effected by the 0.5 *M* NaOH SDA treatment (Hashimoto & Jackson, 1960) following the CBD extraction (section 6–4). The NaOH treatment also dissolves fine-grained phyllosilicate clays such as halloysite (Langston & Jenne, 1964; Wada & Tokashiki, 1972; Higashi & Ikeda, 1974), some poorly crystalline kaolinite (Hashimoto and Jackson, 1960; Wada & Greenland, 1970; Fey & Le Roux, 1977; Hodges & Zelazny, 1980), and some poorly crystalline 2:1 phyllosilicates (Hashimoto & Jackson, 1960; Follett et al., 1965). Aluminosilicates having SiO_2/Al_2O_3 molar ratio of 1.5 to 2.3 (Fieldes & Furkert, 1966) dissolved from weathered volcanic ash soils, although over 90% of these materials did not give the allophane DTA (Bracewell et al., 1970). Some poorly crystalline aluminosilicates and glass may have been extracted.

Clay fractions of soils derived from volcanic materials may contain from 30% (Tamura et al., 1953) to nearly 100% (Aomine & Yoshinaga, 1955; Aomine & Jackson, 1959) of alkali-extractable clay materials, and therefore should be considered in SDA methods. Extraction of an imogolite or allophane suspension at 1 mg/mL in 5% $NaCO_3$ solution (Mitchell et al., 1968) at 20°C for 16 h (Farmer et al., 1977) removes small percentages of Si and Al (on the order of 10% or less of the sample). The remaining materials retain their original infrared spectra. Hot 2% (Jackson, 1979) or 5% (Follett et al., 1965) Na_2CO_3 solution dissolves an indeterminant quantity of these materials and alters the infrared spectra. Use of the 1 mg/mL concentrations in 0.5 *M* NaOH during a boiling period of 2.5 min (Hashimoto and Jackson, 1960) results in complete dissolution of imogolite and allophane (Farmer et al., 1977), similar to the action of ammonium oxalate (section 6–3). The oxalate procedure provides an option, therefore, to remove these entities prior to extraction of poorly crystalline aluminosilicates by the flash 0.5 *M* NaOH extraction.

Scarcely recognizable 1.8-nm montmorillonite diffraction peaks of clays from Tama soil (Glenn et al., 1960) and Dodge soil (Weaver et al., 1971) developed from loess of southern Wisconsin became clearly resolved after 0.5 *M* NaOH extraction of quickly soluble constituents from the clay. The mole fraction of SiO_2 in extracted ($SiO_2+Al_2O_3$) of the poorly crystalline "reactive" components extracted was related, through the equation of Paces (1973), to the geomorphic-geochemical soil site characteristics (Weaver et al., 1976).

Only fractional percentages of Al_2O_3, Fe_2O_3, and SiO_2 were dissolved by the acid ammonium oxalate procedure (section 6–3) from kaolinitic soil clays of central Zaire (Makumbi & Jackson, 1977). In contrast, 13 to 42% of the unheated clay fractions of 22 kaolinitic soils developed on phyllite, gneiss, or metabasalt of southern India was quickly soluble in the flash 0.5 *M* NaOH procedure (Rao & Krishna Murti, 1982). The poorly crystalline ferrialuminosilicate fractions of soil clays, soluble in 0.5 *M* NaOH, are clearly both abundant and reactive in the soil system.

6–5.2 Principles

Use of a high ratio of 0.5 *M* NaOH volume to sample weight (Hashimoto & Jackson, 1960) (so as to avoid saturation of the solution with respect to silica and alumina) brings about an entirely different SDA result as compared to results with higher sample-to-solution ratios, which have often been employed. Treatment with a small volume of NaOH solution may dissolve all of the free noncrystalline silica or free alumina but not both; noncrystalline aluminosilicates are dissolved only to a limited extent in concentrated suspensions. Boiling a soil or clay for only 2.5 min with a low ratio of sample to 0.5 *M* NaOH (1 mg/mL) causes dissolution of some noncrystalline silica (but not opal), free alumina, and large percentages of noncrystalline combined aluminosilicates including allophane and imogolite (which optionally may be removed by acid oxalate, section 6–3).

The large dissolution capacity factor of this flash SDA treatment with 0.5 *M* NaOH is important. Prior removal of iron oxides by CBD (section 6–4) is also important because such oxides partially block the reaction. Selective dissolution of hydrous silica and alumina, noncrystalline or poorly crystalline ferrialuminosilicates, depends upon the fact that these materials have higher specific surface (giving a higher dissolution rate) than less reactive crystalline clays. Careful limitation of treatment time is essential because the well crystallized clays are appreciably soluble during an hour or more of treatment with dilute suspensions in NaOH (Hashimoto & Jackson, 1960). Crystalline quartz and silt-sized opal are little dissolved in the brief extraction time (2.5 min of boiling) given in the procedure and are determined otherwise (section 6–2). After the treatment, well crystallized kaolinite, smectite, vermiculite, chlorite, mica, quartz, feldspars, and many other minerals remain for other analyses.

6–5.3 Method[2]

6–5.3.1 SPECIAL APPARATUS

1. Nickel or stainless steel beakers, 500 mL.
2. Muffle furnace with temperature regulater.

6–5.3.2 REAGENTS

1. Sodium hydroxide (NaOH), approximately 0.5 *M*: Prepare a fresh solution for each use by dissolving 4 g of reagent-grade NaOH pellets in 100 mL of water in a nickel or stainless steel beaker.
2. Acetone, 99%, reagent grade.
3. 99% methanol, technical grade; 80 methanol, by dilution with H_2O.
4. Sodium chloride (NaCl) saturated solution: Shake 80 g of reagent-grade NaCl in 200 mL of water.
5. Hydrochloric acid (HCl), approximately 0.05 *M*: Dilute 4 mL of concentrated HCl to 1 liter with water.

6–5.3.3 PROCEDURE

Transfer a volume of clay suspension, from a soil from which the free iron oxides have been removed (section 6–4.3), containing 200 mg of Na-saturated sample (base the weight on drying a separate aliquot) into a 500-mL nickel or stainless steel beaker. Alternatively, crush the aggregates of dried NH_4^+-saturated clay, silt, soil, or deposit sample, from which the noncrystalline aluminosilicates (section 6–3.3) have been removed, in a small mortar, or dry the sample from 99% methanol. Then weigh a 0.200-g sample, transfer it to a 500-mL nickel or stainless steel beaker, and rub it with a rubber-tipped rod in 1 to 2 ml of 0.5 *M* NaOH to facilitate the dispersion of the sample. Place the beaker with the sample on an electric hot plate, promptly add 200 mL of boiling 0.5 *M* NaOH (prepared beforehand in another beaker), and continue the boiling for 2.5 min. Rapidly cool the suspension in a cold water bath and remove the supernatant liquid by centrifugation. Immediately determine the dissolved Si and Al by atomic absorption (Hallmark et al., 1982; Barnhisel & Bertsch, 1982, sections 15–3.3 and 16–6.4) or colorimetrically as molybdosilicate (Hallmark et al., 1982, section 15–3.4), and as the salt of the aurin tricarboxylic acid (Aluminon) (Barnhisel & Bertsch, 1982, section 16–5.2).

Extract the iron oxide released (now giving a brown stain to the suspended residue) by the CBD method (section 6–4) and determine Fe by atomic absorption (Olson & Ellis, 1982, section 17–2.3.2) or by orthophenanthroline or KSCN. Examine the undissolved residue by means of XRD (chapter 12), since the removal of noncrystalline and (or) poorly

[2]Hashimoto and Jackson (1960). Use of a Na-saturated sample instead of a dried H-saturated sample is less destructive of fine-clay layer silicates (A. B. Hanna and M. L. Jackson, unpublished).

crystalline materials frequently discloses interesting crystalline components in the residue, such as mica, montmorillonite, or chlorite, which were not clearly revealed in the diffractogram of the material before the above treatments were given.

To determine the remaining kaolinite plus halloysite content, heat a Na-saturated sample, preferably one already given the above NaOH–CBD treatments to remove gibbsite and silica-rich materials, in a muffle furnace at 550°C for 4 h, and allow it to cool in a desiccator. Then lightly crush the sample in an agate mortar to break up the aggregates. Weigh a 0.100-g sample and repeat the above NaOH dissolution procedure, using 100 mL of 0.5 *M* NaOH. Repeat the CBD extraction of Fe (section 6–4). Determine Si, Al, and Fe on the extracts, as given above. Wash the residue three times with 1 *M* $(NH_4)_2CO_3$ and once with 80% methanol (Lim & Jackson, 1982, section 1–5.3) to remove the CBD reagents. Alternatively, dry the residue at 110°C and weigh the residue (section 6–3) to determine by difference the content of kaolinite plus halloysite dissolved.

To determine noncrystalline interlayer material alumina, silica and iron (Dixon and Jackson, 1959), H-saturate a second sample by washing it with 0.05 *M* HCl, acetone, and methanol. Dry the sample, heat it to 400°C for 4 h, and allow it to cool in a desiccator. Then powder it, and weigh a 0.100-g sample for extraction with 100 mL of 0.5 *M* NaOH by the same procedure as given above. Repeat the CBD procedure (section 6–4) for extraction of Fe. Determine Si, Al, and Fe as in the above. Alternatively, dry the residue at 110°C and weigh to determine the quantity of interlayer material dissolved.

6–5.3.4 CALCULATION OF RESULTS

Calculate the percentages of SiO_2, Al_2O_3, and Fe_2O_3 dissolved in the extractions. Alternatively, calculate the total amount of dissolved materials from the dry weight difference between the sample before and after extraction. If other evidence, such as XRD (chapter 12), indicates the presence of gibbsite, then subtract the percentage of gibbsite-derived Al_2O_3, obtained from DTA or DSC (chapter 7) before and after NaOH treatment, from the Al_2O_3 dissolved from the unheated sample. Some free silica (uncombined) and the poorly-crystallized and fine-sized ("reactive") phyllosilicates will be represented in the dissolved constituents, if the option of preceding the NaOH extraction by the AOD extraction (section 6–3) was elected. To the sum of SiO_2, Al_2O_3, and Fe_2O_3 weight dissolved, the approximate water content may be added according to the SiO_2/Al_2O_3 molar ratio (Jackson, 1979): 0.8 to 1.4, 21%; 1.5 to 2.8, 15%; 2.9 to 4.5, 10% H_2O.

For the kaolinite plus halloysite content, convert the percentages of the NaOH–CBD dissolved SiO_2, Al_2O_3, and Fe_2O_3 (obtained after heating to 550°C) to mol/100 g of sample. When the SiO_2/R_2O_3 molar ratio is

2 to 4, correct for the dissolution of some Fe-containing 2:1 phyllosilicates (Briner & Jackson, 1970; El-Attar & Jackson, 1973) by the equation:

$$\% \text{ kaolinite plus halloysite} = \frac{[4(B + C) - A]\ 60}{0.465} \quad [10]$$

in which

A = moles of SiO_2 dissolved in NaOH per 100 g;
B = moles of Al_2O_3 dissolved in NaOH per 100 g; and
C = moles of Fe_2O_3 dissolved by CBD, subsequent to the 550°C heating-NaOH extraction, per 100 g.

When the SiO_2/R_2O_3 molar ratio is <2, correct for the dissolution of some interlayer alumina and/or crystalline alumina (Alexiades and Jackson, 1966) by the equation:

$$\% \text{ kaolinite plus halloysite} = (\% \ SiO_2/46.5) \times 100\,. \quad [11]$$

When the SiO_2/R_2O_3 molar ratio exceeds 4, correct for the dissolution of some silica by the equation:

$$\% \text{ kaolinite plus halloysite} = (\% \ Al_2O_3/39.5) \times 100\,. \quad [12]$$

For the noncrystalline interlayer material, multiply the difference in percent Al_2O_3 extracted before and after heating at 400°C by 4.4 (Dixon & Jackson, 1962) to obtain the theoretical chlorite equivalent of the interlayer alumina extracted.

6–5.3.5 COMMENTS

Interlayers are usually discussed as hydroxy aluminum (Rich, 1968). An increment of alumina, silica, and iron oxide often becomes soluble as a result of heating to 400°C, suggesting that ferrialuminosilicate interlayer and surface coatings may be characteristic of expanded layer silicates of soils. A pretreatment with NaOH–CBD is necessary for the determination of kaolinite plus halloysite in gibbsite-containing samples. The heat treatment at 550°C transforms gibbsite to a corundum-like product (Al_2O_3) which is only slightly soluble in the NaOH (Hashimoto & Jackson, 1960). In gibbsite-free samples, the pretreatment may be omitted; however, its inclusion will provide some evaluation of the size, crystallinity, and other characteristics of the most "reactive" fraction of the soil, related to soil genetic properties (Weaver et al., 1976). After discounting for the gibbsite and free silica in the dissolved constituents, the excess SiO_2, Al_2O_3, and Fe_2O_3 can be used for the calculation of fine-sized or poorly-crystallized kaolinite plus halloysite. This value can then be added to the kaolinite plus halloysite content obtained from the second extraction of the sample dehydroxylated at 550°C. For samples containing intergradient chlorite-expandable 2:1 phyllosilicates, the kaolinite can be calculated from the difference between percent SiO_2 or percent Al_2O_3

dissolved by the 2.5-min boiling in 0.5 *M* NaOH following the 400°C and 550°C heating (Dixon & Jackson, 1962).

The extraction of the additional amounts of Al_2O_3, SiO_2, and Fe_2O_3 after the 400°C heating treatment (in excess of amounts extracted before the heating treatment) causes the expanded layer silicates of many soils to have changed expandable characteristics from those clays not subjected to thermal plus NaOH treatment. The treated clays undergo more complete thermal collapse at 300°C, have an increase in the measurable interlayer specific surface, have an increased CEC capacity, and may expand to 1.8 nm with glycerol (Dixon & Jackson, 1959). Soil genetic and soil acidity properties are related to interlayer aluminum (Jackson, 1960, 1963a, 1963b) as are polymorphic phases of $Al(OH)_3$ (Violante & Jackson, 1981).

An aluminosilicate precipitate may form in the NaOH solution if the solution after the extraction is allowed to stand for an appreciable time prior to the Si and Al determinations. Also, serious contamination with glass may occur before or after the extraction, and therefore plastic ware should be used to contain the extracts.

Highly crystalline gibbsite may require two or more successive NaOH treatments for its complete dissolution (Hashimoto and Jackson, 1960). Magnesium hydroxide is, of course, not dissolved by NaOH. Magnesium-containing silicate minerals such as mafic chlorite are protected by $Mg(OH)_2$ formation at surfaces during treatment with NaOH, and therefore they are relatively stable to the NaOH treatment. This protection is indicated by the remarkable resistance of chlorite to dissolution (Hashimoto & Jackson, 1960). Since kaolinite and halloysite are converted to noncrystalline aluminosilicates by heating to 550°C (Hashimoto & Jackson, 1960), they can be dissolved in NaOH after such heating, and can be sharply differentiated from mafic chlorite minerals. The percentage of NaOH-dissolved noncrystalline product which results from kaolinite plus halloysite decomposition between 550° and 400°C checks well (Andrew et al., 1960) with the structural-expansion method for these minerals. This concordance identifies the NaOH-dissolution SDA method with short-range order and poor crystallinity, small size, and weathered relicts of some layer silicates, in the mid-range of crystallinity (Table 6–1). The selective acid oxalate (AOD) selective dissolution (section 6–3) of noncrystalline aluminosilicates, silica, and alumina, helps to give a fairly good differentiation of them from the poorly crystalline and well developed crystalline states throughout the transitional range (Table 6–1). The foregoing separation of noncrystalline, poorly crystalline, and coarsely crystalline phyllosilicates is probably accurate to within about 5% of the amounts of each present in a given soil.

6–6 SMECTITE, VERMICULITE, AND CEC HYSTERESIS

6–6.1 Introduction

The quantitative analysis of complex soil mineralogy has long been considered an objective that is better pursued by multiple physical and chemical techniques (XRD, chapter 12; thermal, chapter 7; infrared, chapter 11; and chemical analysis, chapter 6) than by any one or two techniques (Kelley & Jenny, 1936). After the early examination of soil clays by XRD (Hendricks and Fry, 1930; Kelley et al., 1931), through 50 years of developments, G. W. Brindley (personal communication, 1980) still stated that "at best" XRD requires the most sophisticated procedures and the most favorable circumstances for reasonable [semiquantitative] results. Chemical analyses of clays, which suggested the possibility of mixtures of crystals (Robinson & Holmes, 1924), have become increasingly useful for assisting the quantification of various mineral types in soils. For example, smectites have a fairly high CEC charge (Marshall, 1935) and are not prone to cation fixation. Vermiculites have still higher CEC charge, related to their biotite origin (Barshad, 1948), and characteristically fix cations such as NH_4^+, K^+, and Cs^+ (Alexiades & Jackson, 1965; Lim et al., 1980). The CEC charges arise from a net equivalent of isomorphous substitution of cations in the structural end member, pyrophyllite–$Si_8Al_4O_{20}(OH)_4$ (MacEwan, 1961). The CEC is of the "permanent" charge.

A fairly large variation in CEC occurs with some soils or clays, depending on the pH at which it is measured. For soils, the "pH variable" charge is large when the noncrystalline silicates or organic matter is high (Clark, 1965) or when content of hydrous metal oxides of iron and aluminum is appreciable (Sawhney et al., 1970; section 6–3; Rhoades, 1982, section 8–4). The CEC at the pH of an acid soil can be measured with 0.1*M* $BaCl_2$ (Rhoades, 1982, section 8–4). It has been measured by washing with the unbuffered salt 2 *M* NaCl and determining replaced basic cations (McKeague, 1978). Determination of exchanged Al^{3+} of acid soils additionally provides the permanent CEC as the cation sum. To measure the CEC to include the pH variable charge to pH 7, Clark et al. (1966) used washing with 0.45 *M* $Ca(OAc)_2$–0.05 *M* $CaCl_2$ of pH 7, water washing to freedom from chloride, and replacement of Ca in 2*M* NaCl. The CEC increment over the cation sum is the variable charge (McKeague, 1978). A measure of the CEC to pH 8.2 is done with NaOAc in 60% ethanol (Rhoades, 1982), although 15 or 20% of the CEC may arise from $CaCO_3$, if present (van Bladel et al., 1975). Historically, the reference pH for soil CEC has varied up to pH 11 or more. The proton donor-acceptor phenomenon of silinol (–SiOH), aquo-alumina (–Al–OH_2^+), and aluminol (–AlOH) groups from mineral matter and carboxyl (RCOOH) and phenolic (ROH) groups of humus can be activated at varying pH values. Measurement of the reversible pH variable charge is largely excluded from this section. The emphasis is on hysteretic CEC of crystalline phyllosilicates (particularly of vermiculite) and of noncrystalline (Wada, 1977) clays (e.g. imogolite and "allophane") involving short-range order. Noncrystalline clays are included to the extent that the short-range order is so interpreted.

6–6.2 Principles

Hysteresis of CEC refers to a change in CEC measured by a buffered solution at pH 7 induced by a specified pretreatment of the sample. Effectively, for hysteresis, the pretreatment must cause changes in the samples which are not reversed in the buffered solution at pH 7 (Tabikh et al., 1960). The noncrystalline or paracrystalline exchangers of volcanic ash soils (mainly "allophane" and imogolite: Ross and Kerr, 1934; Aomine and Yoshinaga, 1955; Wada, 1977) were found to exhibit hysteresis when exposed to heated acid (pH 3.5) or alkaline pH (10.7), which cause infrared changes (section 6–5). These noncrystalline and paracrystalline materials mainly in volcanic soils can be dissolved (section 6–3), but also can be characterized by their hysteretic CEC properties (section 6–6.4). Moreover, interlayer aluminum (section 6–5; Barnhisel & Bertsch, 1982, section 16–4) in expandable layer silicates can be dissolved in acid ammonium oxalate (section 6–3) in ultraviolet light; it shows a small, characteristic CEC hysteresis (de Villiers & Jackson, 1967; Sawhney et al., 1970; Sawhney & Norrish, 1971). Vermiculite fixes K^+ against replacement by NH_4^+ and the amount present in clays has been determined by this hysteretic property (Alexiades & Jackson, 1965). Smectite clays show virtually no CEC hysteresis, but its CEC is found after the vermiculite CEC loss by K fixation. Organic matter, $CaCO_3$, and free iron oxides are removed from the sample to avoid ordinary pH variable CEC. The weighing method is employed to measure the amount of 0.02 *M* KCl, thus avoiding problems with clay dispersion during washing out of salts and hydrolysis (Birrell & Gradwell, 1956) of the exchangeable cations even in aqueous alcohol.

6–6.3 Method for Vermiculite[3]

6–6.3.1 SPECIAL APPARATUS

1. Ultrasonic probe.
2. Vortex mixer for tubes.

6–6.3.2. REAGENTS

1. Acetate buffer, pH 5: dissolve 82 g of NaOAc salt in water, add 27 mL of glacial HOAc, and dilute the solution to a volume of 1 L to give a solution approximately 1 *M* with respect to NaOAc and of pH 5.
2. Hydrogen peroxide (H_2O_2), 27 to 35%, reagent grade.
3. Calcium chloride ($CaCl_2$), approximately 0.25 *M* and 0.005 *M*: dissolve 56 g of reagent grade $CaCl_2$ in H_2O and dilute to 1 L for 0.25 *M*. Dilute 20 mL of this solution to 1 L for 0.005 *M*.

[3]Alexiades and Jackson (1965) and Jackson (1979). Appreciation to S. L. Chapman for testing the weighing method as being more consistent than the washing out of excess salts and to F. H. Abdel-Kader for introducing the use of 0.02 *M* KCl in 80% methanol.

4. Magnesium chloride ($MgCl_2$), approximately 0.25 *M*: dissolve 48 g of reagent grade $MgCl_2$ in H_2O and dilute to 1 L.
5. Potassium chloride (KCl), approximately 0.5 *M* and 0.02 *M*: dissolve 37 g of reagent grade KCl in H_2O and dilute to 1 L. Dilute 40 mL of this solution to 1 L for 0.02 *M*. Dissolve 1.5 g of reagent grade KCl in 1 L of 80% methanol (v/v) for 0.02 *M*.
6. Ammonium chloride (NH_4Cl), approximately 0.5 *M*: dissolve 27 g of reagent grade NH_4Cl in H_2O and dilute to 1 L.
7. Ammonium acetate (NH_4OAc), approximately 0.5 *M*: to 28 mL of glacial HOAc, add 500 mL of H_2O and then NH_4OH to obtain pH 7. Dilute to 1 L.

6–6.3.3 PROCEDURE

Weigh out a 100-mg sample (200 mg for low CEC clays) of air-dried or freeze-dried clay or take an aliquot of clay suspension into a clean, weighed (to the nearest 0.0001 g) 15-mL centrifuge tube. (In all cases, the sample will have been pretreated with NaOAc of pH 5, H_2O_2, CBD, and preferably with AOD, and brought to neutrality by washing with NaOAc of pH 7).

Wash the sample three times with 0.25 *M* $CaCl_2$, with thorough mixing with an ultrasonic probe for a few seconds, and then as required with a vortex mixer. Follow each time by centrifugation; decant the supernatant liquid carefully and discard it. Then wash the sample five times with 0.005 *M* $CaCl_2$. Discard the supernatant liquid each time. Wipe the tube externally. Weigh the tube with sample and entrained liquid to the nearest 0.001 g.

Replace the Ca by five washings with 0.25 *M* $MgCl_2$ and collect the washings in a 100-mL volumetric flask. Fill the flask to 100 mL with 0.25 *M* $MgCl_2$ solution. Determine the Ca (Lanyon & Heald, 1982, section 14–3.1) including the 0.25 *M* $MgCl_2$ in the blank and standards. Save the sample for determination of potassium exchange capacity (KEC) with a drying step.

Wash the same sample three times with 0.5 *M* KCl, once with aqueous 0.02 *M* KCl and four times with 0.02 *M* KCl in 80% methanol. Wipe the tube externally and weigh the tube, sample, and entrained KCl to the nearest 0.001 g. Dry the tube at 110°C overnight and weigh to the nearest 0.0001 g.

Wash the sample five times with 10-mL portions of 0.5 *M* NH_4Cl and collect the supernatant solution in a 100-mL volumetric flask. Use the ultrasonifier as necessary for dispersion in the first washing (usually 1 min) and the vortex mixer for each washing thereafter. Make up the remaining 250 mL with 0.5 *M* NH_4OAc. Determine by flame emission (Knudsen et al., 1982, section 13–3.3.2).

6–6.3.4 CALCULATIONS

Subtract entrained cations (Ca, K):

Entrained cations (meq) = solution $N \times$ g of entrained solution

$$\text{CEC} = \frac{100\ (\text{total cations, meq} - \text{entrained cations, meq})}{\text{sample weight, g}}. \quad [13]$$

This calculation of CEC (meq/100 g) is done for Ca to obtain CaEC and for K to obtain K/EC, in which the slash signifies the 110°C drying step.

$$\%\ \text{vermiculite} = (\text{CaEC} - \text{K/EC})/1.54 \quad [14]$$

in which 1.54 represents 159 ± 20 for the CaEC of vermiculites, from which 5 meq is subtracted for the K/EC of external faces and edges.

$$\%\ \text{smectite} = (\text{K/EC} - 5)/1.05 \quad [15]$$

in which 1.05 represents 110 ± 30 meq/100 g of CEC for various smectites and 5 has been subtracted for the CEC of the external surfaces. This formula assumes that noncrystalline materials have been removed by AOD (section 6–3) and that poorly crystalline and interlayer materials have not been removed (section 6–5 was not used).

6–6.3.5 COMMENTS

The smectite interlayers reopen in NH_4Cl solution and the CEC is therefore measured as K/EC. The vermiculite interlayers are kept closed in NH_4Cl, but would open materially with divalent salts such as Mg and Ca (Alexiades & Jackson, 1965).

Chlorite can be estimated rather well from thermal analysis (chapter 7), assuming weight losses between 300 and 950°C of 5% for smectites and vermiculites, 16.3% for halloysites, 3% for gibbsite (remaining after 300°C heating), and 14% for kaolinite (section 6–5) and chlorite (Alexiades & Jackson, 1966, 1967). Serpentines (rare in most soils) would fall in with chlorites. Noncrystalline materials (section 6–3) and gibbsite (chapter 7) will have been estimated separately. The poorly crystalline fraction has been removed (Rengasamy et al., 1975) by the flash 0.5 *M* NaOH procedure prior to the use of the Alexiades and Jackson (1966) system applied to the residual larger size and more crystalline minerals. Although this seems to be an attractive alternative, it is not without difficulties with mineralogical interpretations of the mica and vermiculite contents (Jackson & Abdel-Kader, 1977).

6–6.4 CEC Hysteresis of Noncrystalline Aluminosilicates (Aomine & Jackson, 1959)

6–6.4.1 SPECIAL APPARATUS

1. Nickel or stainless steel beakers, 500 mL.
2. Weighed 100-mL porcelain dishes.

3. A vortex centrifuge tube agitator.

6–6.4.2 REAGENTS

1. Sodium carbonate (Na_2CO_3), approximately 2% (or 0.2 *M*): dissolve 20 g of reagent-grade Na_2CO_3 in H_2O and dilute to 1 L.
2. Acetate buffer of pH 3.5: to 57 mL of reagent-grade glacial HOAc, add 900 mL of water and then add 60 mL of 1 *M* NaOH. Adjust the pH to 3.5 (glass electrode) by addition of HOAc or NaOH as needed, and dilute the solution to 1 L.
3. Potassium acetate (KOAc), approximately 1*M*: dissolve 98 g of reagent-grade KOAc in 900 mL of H_2O. Adjust the pH to 7 with glacial HOAc, and dilute to 1 L.
4. Ammonium acetate (NH_4OAc), approximately 1 *M*: to 57 mL of glacial HOAc, add 500 mL of water and then NH_4OH to obtain pH 7. Dilute the solution to 1 L.
5. Methanol and acetone: dilute 200 mL of 99% acetone to 50% (v/v) with methanol. Dilute a second 200 mL of acetone to 75% (v/v) with methanol.
6. Sodium chloride (NaCl), saturated solution: shake 80 g of reagent-grade NaCl in 200 mL of water.

6–6.4.3 PROCEDURE

The determination is usually made on whole soil samples, but may be done with separated size fractions. The sample will have been treated with H_2O_2 in NaOAc of pH 5 to remove organic matter (section 6–6.3.3) and CBD to remove free iron oxides (section 6–4). To assist in dispersion, transfer the sample from which the free iron oxides have been removed to a 500-mL nickel or stainless steel beaker with 200 mL of 2% Na_2CO_3 solution, stir the suspension thoroughly, cover the beaker with a watch-glass, and boil the contents for 5 min. Recover the residual sample by centrifuging, and wash it successively with methanol-diluted 50% and 75% acetone and with 99% acetone. Transfer the residue to a tared porcelain dish and dry the dish and contents in an oven at 110°C. Determine the weight of the dried residue.

Mix the dried sample with a spatula and transfer 0.1-g subsamples into each of four 15-mL centrifuge tubes. (Alternatively, use 0.3-g subsamples of soil low in noncrystalline clays.) Add 5 mL of 2% Na_2CO_3 solution to each of two tubes and 5 mL of acetate buffer of pH 3.5 to each of the other two tubes. Place all tubes in a boiling water bath and mix intermittently by a vortex stirrer. After 15 min, remove the tubes containing the pH 3.5 buffer, and after 1 hour, remove the tubes containing the Na_2CO_3 solution. Mix 1 mL of saturated NaCl into the contents of each tube, and centrifuge the tubes to sediment the solids.

Determine the usual (pH 7) CEC by the K method, washing 5 times with neutral 1 *M* KOAc and successively with 5 mL of each of the following solvents, to remove salts: once with 99% methanol, once with 50%

acetone (diluted with 99% methanol), once with 75% acetone (diluted with 99% methanol), and once with 99% acetone. Then replace the exchangeable K by 5 washings, each consisting of 8 mL of neutral 1 *M* NH_4OAc, and collect the decantate. Determine K in the decantate (Knudsen et al., 1982, section 13–3.3.2).

6–6.4.4 CALCULATION OF RESULTS

The difference (in meq/100 g) obtained by subtraction of the CEC of the samples treated at pH 3.5 from that of the samples treated in 2% Na_2CO_3 is the cation exchange capacity hysteresis ("delta value") on the dried sample weight basis. The Ross and Kerr (1934) "allophane" percentage, in the first approximation, equals the CEC delta value. Subtraction of CEC delta values attributable to other minerals improves the result. To do this, allocate the observed CEC delta value to allophane at 100 meq/100 g (it varies from 90 to 110, averaging about 100), to halloysite at 18, to montmorillonite and vermiculite at 10, and to illite and chlorite at 5 meq/100 g. To make this calculation, the contents of these minerals must be known from other types of analysis.

6–6.4.5 COMMENTS

Dispersion during washing is more of a problem with paracrystalline clays of soils developed from volcanic ash than with those containing crystalline clays. For this reason, acetone solutions are employed in the washings. The CEC delta value for separated noncrystalline clay from a given locality can be standardized. The range of values given covers two allophanes of Ross and Kerr (1934) and several ash-derived soils of Kyushu Island, Japan. A few highly leached soils of the USA, including a sandy soil of Wisconsin (Whittig & Jackson, 1955), have been found to give CEC hysteresis indicative of a considerable paracrystalline content in the clay fraction.

It was noticed (Aomine & Jackson, 1959) that the CEC of clay separates, measured by using washing solutions having a pH value of 7, varied a great deal according to the pH of the dispersion reagents employed for separation. The clay fraction originally separated in an alkaline medium (about pH 10.7) showed a high (±150 meq/100 g) exchange capacity, while the clay fraction from the same soil originally separated with an acid dispersion medium (about pH 3.5) had an exchange capacity of only about one-third that of the sample receiving alkaline dispersion. The CEC of the acid-dispersed sample was increased to equal that of the former on treatment with a mildly alkaline buffer. The Kyushu soil clays dispersed in acid were known to have thread-like morphology in electron micrographs (Aomine & Yoshinaga, 1955; Aomine & Jackson, 1959). This morphology was characterized as having an aluminol external sheath on a paracrystalline tube (Cradwick et al., 1972) and is known as imogolite (Wada, 1977).

Use of K as the saturating cation provides a sensitive measure of

CEC, although K fixation may occur in vermiculitic clays. This fixation is not a problem, since it occurs in both the acid- and alkali-treated samples. The use of Ca gives a less sensitive measure and requires washing of the Na_2CO_3-treated samples with neutral 1 *M* NaOAc prior to $Ca(OAc)_2$ washing to prevent the precipitation of $CaCO_3$.

The determination of hysteretic (delta value) CEC gives duplicates falling within about 10%, and the mean values agree well with the percentage of noncrystalline clay derived from thermal dehydration data (Aomine & Jackson, 1959). Although not a highly accurate method, the CEC delta value method contributes a moderately rapid and reliable quantitative estimate of these materials, which may otherwise be missed altogether or merely suspected to be present in a mixture with crystalline minerals and noncrystalline free oxides of aluminum or silicon. Many short-range order materials have been termed noncrystalline (Bunn, 1961; Wada, 1977) herein, although some of them might be termed paracrystalline. Separation of allophane in alkaline suspension (Yoshinaga & Aomine, 1962a) and imogolite in acid suspension (Yoshinaga & Aomine, 1962b) led to their subsequent discovery and characterization in many countries. A hollow spheroidal paracrystalline structure has been proposed for allophane (Wada, 1979).

6–7 RUTILE AND ANATASE

6–7.1 Introduction

The titanium oxide minerals rutile and anatase (TiO_2), often termed "accessory minerals," are usually present in soils and sediments and because of their high resistance to weathering are used as index minerals. Their content in soils may vary widely, from 0.5% in young or little weathered soils to about 4% in many soils of the tropical and temperate regions; unusually high contents of TiO_2 up to 25% have been reported in the highly weathered Humic Ferruginous Latosols of Hawaii (Sherman, 1952). The usual method employed for their identification is XRD; but detection of their presence is often missed in diffraction patterns, as their contents are relatively low and the X-ray peaks (0.351 nm for anatase and 0.325 nm for rutile) fall very close to those of various silicates in soils. Isolation of the free oxide minerals, rutile and anatase, is necessary to verify that the Ti they contain is present in the form of the oxide and not in the octahedral position in layer silicates. In the absence of such information, Mankin and Dodd (1963) allocated the entire Ti content of their proposed reference Blaylock illite to octahedral positions. Analysis by XRD and electron microscope methods has shown, however (Raman & Jackson, 1965), that almost the entire Ti content is present in this sample as rutile needles. Use of HF to concentrate the rutile from the Blaylock and anatase in the Cashel and Moorepark silts and clays (soils from Ireland) was demonstrated by XRD. The ratio of HF and HCl used was evidently sufficiently small to produce H_2SiF_6 as the silicate dissolution reagent, since excess HF-HCl dissolves TiO_2 minerals (Campbell, 1973; Sayin & Jackson, 1975). Differentiation of silicate-combined Ti from free oxide TiO_2 (variously substituted with Fe) was carried out by H_2TiF_6 dissolution of silicate minerals (Dolcater et al., 1970).

The noncrystalline ("amorphous") forms of synthetic Ti/Ti+Fe systems were studied by selective dissolution in acid NH_4-oxalate (pH 3) in the dark (AOD) for possible interpretation of pedogenesis (Fitzpatrick et al., 1978). Crystallization of the oxides occurred on aging at pH 5.5 at 70°C for 70 days. The assessment of noncrystalline Ti (e.g., in leucoxene, $TiO_2 \cdot n\ H_2O$) and Fe by AOD is treated in section 6–3.

6–7.2 Principles

Qualitative isolation of rutile and anatase generally occurring in small amounts in soils is made possible by the selective dissolution of the siliceous materials in the sample by treatment with hexafluorotitanic acid (H_2TiF_6). These crystalline minerals are relatively unattacked. The presence of rutile and anatase is detected by XRD. Exchangeable bases are replaced with hydrogen, and free (Ca, Mg)CO_3 is dissolved prior to the treatment to eliminate the formation of insoluble Ca and Mg fluorides. These fluorides would coat the siliceous minerals and impede dissolution of the latter by H_2TiF_6. Anatase shows a diffraction peak at 0.351 nm and rutile at 0.325 nm. These spacings change slightly with isomorphous substitutions in the structures. The minerals rutile and brookite (TiO_2), sphene ($CaSiTiO_5$), and ilmenite ($FeTiO_3$) are largely detrital, occur in the coarser fractions of soils, and are subject to heavy mineral fraction identification (chapter 8).

6–7.3 Method (Sayin & Jackson, 1975; Dolcater et al., 1970. Tiron method: Jackson, 1958; Yoe & Armstrong, 1947)

6–7.3.1 SPECIAL APPARATUS

1. Ice-water bath: to hold a 1-L polyethylene bottle with room for vigorous swirling movement.
2. Polyethylene centrifuge tubes, 100 mL.
3. Water bath at 45 ± 2°C.
4. Vortex mixer.
5. Scanning electron microscope (chapter 10).

6–7.3.2 REAGENTS

1. Hydrofluoric acid (HF): use reagent grade HF 49% (aq)(1.2 g/mL).
2. Hydrochloric acid (HCl): dilute concentrated HCl (approximately 12 *M*) with water to give a 500-mL stock of approximately 1 *M* and a second stock of approximately 0.1 *M*.

3. Hexafluorotitanic acid (H_2TiF_6): synthesize from 49% (aq) hydrofluoric acid (HF) to 16.5% Ti by weight. To do this, place 400 mL of 49% HF(d = 1.2 g/mL) in a 1-L polyethylene bottle. Place the bottle in an ice bath. With a scoop, add about 10 g from a weighed quantity of 200 g of reagent-grade TiO_2, with care that no TiO_2 sticks to the neck of the bottle. Cap the bottle loosely and swirl it rapidly to keep the mixture at 50 to 70° (below 80°C), to avoid softening the polyethylene bottle. The reaction is highly exothermic. Add a 10-g increment of TiO_2 every 2 h, with swirling every 30 min, until 50 g of TiO_2 has been added. The exothermic reaction will have largely subsided:

$$6\ HF(aq) + TiO_2 \rightleftharpoons H_2TiF_6(aq) + 2\ H_2O\ . \qquad [16]$$

   Tighten the cap and use a slow mechanical shaker, with further 10-g additions of TiO_2 and a vigorous shaking each hour until the 200 g of TiO_2 has been added. Then allow the suspension to settle for 10 min, filter 10 mL through a fine filter paper (Whatman no. 50), and weigh approximately 5 mL for dilution. Dilute (1:160 000) with enough 1 M HCl and distilled water to give 0.4 M HCl, and immediately analyze for Ti with Tiron, keeping 0.4 M HCl in the final dilution as with the standard (below). Depending on the fineness of the reagent-grade TiO_2, more TiO_2 and shaking may be needed to reach 16.5% Ti in the H_2TiF_6 solution. Excess coarse solid TiO_2 settles out. Cool the supernatant solution to room temperature, decant into 100-mL polyethylene centrifuge tubes, centrifuge, and decant through filter paper as before. Store in a tightly capped polyethylene bottle.
4. Tiron reagent: prepare the reagent daily by dissolving 4 g of Tiron in 75 mL of distilled water and diluting the solution to 100 mL.
5. Buffer solution, pH 4.7: mix equal volumes of 1 M HOAc (60 mL of glacial acetic acid per liter) and 1 M NaOAc (82 g of anhydrous NaOAc per liter) and adjust to pH 4.7 using a glass electrode.
6. Standard Ti solution: prepare the solution by fusion of 0.1668 g of standard TiO_2 in $K_2S_2O_7$, taking up the melt with 10 mL of 6 M HCl. Dilute the solution to 1 liter with 50 mL of 6 M HCl and water. The concentration is 100 mg of elemental Ti per liter of 0.4 M HCl. After thorough mixing, dilute 10 mL of this solution to 100 mL with 0.4 M HCl, to give a standard Ti solution of 10 ppm. Aliquots (2,4,6, 8, and 10 mL) of this standard are taken for the standard curve, giving 20 to 100 μg of Ti per 50 mL.

6–7.3.3 PROCEDURE

The sample should be free of $CaCO_3$ (pH 5 NaOAc buffer treatment), organic matter (H_2O_2 treatment), and CBD soluble iron oxides (section 6–4, with a Ti determination to monitor Ti). Weigh out a 1-g sample of clay or soil into a 100-mL polyethylene centrifuge tube. Add 30 mL of H_2TiF_6 solution and cap the tube. Vent the tube slightly from time to time to release SiF_4 formed (hood). Mix the suspension vigorously each few minutes to 30 min with a vortex mixer. After 3 h, place the tube in a water bath at 45 ± 2°C for 2 days, but mix thoroughly seven times a day. Keep stoppered to avoid evaporation of H_2TiF_6 around the mouth of the tube.

Centrifuge and decant the reagent and soluble product. Quickly wash the residue three times with 30 mL of 1 *M* HCl, twice with 0.1 *M* HCl, and five times with H_2O, to avoid drying any residual HCl. Dry at 110°C overnight and weigh the residue. Evaluate the mineralogy by XRD (chapter 12). If fluorates have formed, wash the residue with saturated H_3BO_3 solution (section 6–2.3.3.1, paragraph 2). Analyze the residue for Ti (and other elements if desired).

6–7.3.4 CALCULATION OF RESULTS

Calculate the weight percentage of residue. Approximately 95% of the Ti from rutile and anatase is recovered by this procedure. The total TiO_2 content determined is therefore multiplied by 1.05 to obtain the total that was present. The 1 *M* and 0.1 *M* HCl treatments are given to insure the removal of any possible TiO_2 precipitate from the reagent, but these treatments also may be responsible for some TiO_2 loss from the sample.

The isolates are predominantly $(TiFe)O_2$ because the oxide of Ti is generally partially substituted with Fe, with 0.5 to 3% Fe_2O_3 equivalent (Sayin & Jackson, 1975) or more. Quartz, muscovite, and cristobalite may not be removed entirely from some samples. These are evaluated by elemental and XRD analyses (section 6–6; chapter 12).

6–7.3.5 COMMENTS

The 1:30 solid to solution ratio is maintained to insure an excess of the reagent. Only approximately 5% of the H_2TiF_6 is consumed in forming SiF_4 (gas), H_2SiF_6, H_3AlF_6, H_2FeF_5 and other complexes. The large excess of H_2TiF_6 is required to prevent further oxygenation and consequent precipitation of TiO_2.

The definitive x-ray diffraction peaks for anatase and rutile are prominent in the residue. In addition, scanning electron microscopy (chapter 10) reveals irregular detrital rutile and occasionally euhedral anatase and rutile crystals. Rutile and anatase have been found by this method in the clay fraction of several soils and clay sediments in which they are not detected in diffraction patterns of the bulk sample. Electronoscopic observations indicate the presence of euhedral crystals of different sizes in some samples before the HF treatment, though they are diluted by siliceous components. The morphology of the crystals was not changed by H_2TiF_6 treatment, indicating nonsynthesis by this treatment.

6-8 REFERENCES

Aguilera, N. H., and M. L. Jackson. 1953. Iron oxide removal from soils and clays. Soil Sci. Soc. Am. Proc. 17:359-364 (also 18:223 and 350, 1954).

Alexander, E. B. 1974. Extractable iron in relation to soil age on terraces along the Truckee river, Nevada. Soil Sci. Soc. Am. Proc. 38:121-124.

Alexiades, C. A., and M. L. Jackson. 1965. Quantitative determination of vermiculite in soils. Soil Sci. Soc. Am. Proc. 29:522-527.

Alexiades, C. A., and M. L. Jackson. 1966. Quantitative clay mineralogical analysis of soils and sediments. Clays Clay Miner. 14:35-52.

Alexiades, C. A., and M. L. Jackson. 1967. Chlorite determination in clays of soils and mineral deposits. Am. Miner. 52:1855-1873.

Andrew, R. W., M. L. Jackson, and K. Wada. 1960. Intersalation as a technique for differentiation of kaolinite from chloritic minerals by X-ray diffraction. Soil Sci. Soc. Am. Proc. 24:422-424.

Anonymous. 1980. Controversy erupts over Mt. St. Helens ash. Anal. Chem. 52:1136A-1140A.

Aomine, S., and M. L. Jackson. 1959. Allophane determination in Ando soils by cation-exchange capacity delta value. Soil Sci. Soc. Am. Proc. 23:210-214.

Aomine, S., and H. Ostuka. 1968. Surface of soil allophane clays. Trans. Int. Congr. Soil Sci., 9th 1:731-737.

Aomine, S., and N. Yoshinaga. 1955. Clay minerals of some well-drained volcanic ash soils in Japan. Soil Sci. 79:349-358.

Arshad, M. A., R. J. St. Arnaud, and P. M. Huang. 1972. Dissolution of trioctahedral layer silicates by ammonium oxalate, sodium dithionite-citrate-bicarbonate, and potassium pyrophosphate. Can. J. Soil Sci. 52:19-26.

Astaf'ev, V. P. 1958. Method of determination of quartz and opal in rocks. Opredelenie Svobodnoĭ Dvuokisi Kremniya v Gorn. Prodokh i Rudn. Pyli, Akad. Nauk SSSR, Inst. Gorn. Dela, Sbornik Stateĭ. 1958:51-53.

Bailey, S. W. 1980. Summary of recommendations of AIPEA nomenclature committee on clay minerals. Am. Mineral. 65:1-7.

Baker, D. E., and N. H. Suhr. 1982. Atomic adsorption and flame emission spectrometry. *In* A. L. Page et al. (ed.) Methods of soil analysis, part 2. 2nd ed. Agronomy 9:13-27.

Baril, R., and G. Bitton. 1969. Teneurs élevées de fer libre et identification taxonomique de certain sols due Québec contenant de la magnetite. Can. J. Soil Sci. 49:1-9.

Barnhisel, R., and P. M. Bertsch. 1982. Aluminum. *In* A. L. Page et al. (ed.) Methods of soil analysis, part 2. 2nd ed. Agronomy 9:275-300.

Barshad, I. 1948. Vermiculite and its relation to biotite as revealed by base exchange reactions, X-ray analyses, differential thermal curves, and water content. Am. Mineral. 33:655-678.

Barshad, I. 1955. Soil development. p. 1-52. *In* F. E. Bear, ed. Chemistry of the soil. Reinhold Publishing Corp., New York.

Beckwith, R. S., and R. Reeve. 1964. Studies on soluble silica in soils. II. The release of monosilicic acid from soils. Aust. J. Soil Res. 2:33-45.

Biermans, V., and L. Baert. 1977. Selective extraction of the amorphous Al, Fe and Si oxides using an alkaline Tiron solution. Clay Miner. 12:127-134.

Birrell, K. S., and M. Gradwell. 1956. Ion-exchange phenomena in some soils containing amorphous mineral constituents. J. Soil Sci. 7:130-147.

Blume, H. P., and U. Schwertmann. 1969. Genetic evaluation of profile distribution of aluminum, iron, and manganese oxides. Soil Sci. Soc. Am. Proc. 33:438-444.

Bockheim, J. G. 1982. Properties of a chronosequence of ultraxerous soils in the trans-Antarctic mountains. Geoderma 28:239-255.

Bracewell, J. M., A. C. Campbell, and B. D. Mitchell. 1970. An assessment of some thermal and chemical techniques used in the study of the poorly-ordered aluminosilicates in soil clays. Clay Miner. 8:325-335.

Brewster, G. R. 1980. Effect of chemical pretreatment on X-ray powder diffraction characteristics of clay minerals derived from volcanic ash. Clays Clay Miner. 28:303-310.

Brindley, G. W. 1961. Quantitative analysis of clay mixture. p. 489-516. *In* G. Brown, ed. The X-ray identification and crystal structures of clay minerals. Mineralogical Society, London.

Brindley, G. W. 1977. Aspects of order-disorder in clay minerals—a review. Clay Science 5:103–112.

Brindley, G. W., and G. Pedro. 1972. Report of the AIPEA nomenclature committee. Assoc. Int. Etud. Argiles (AIPEA) Newsletter 7:8–13.

Briner, G. P., and M. L. Jackson. 1969. Allophane material in soils derived from Pleistocene basalt. Aust. J. Soil Res. 7:163–169.

Briner, G. P., and M. L. Jackson. 1970. Mineralogical analysis of clays in soils developed from basalt in Australia. Israel J. Chem. 8:487–500.

Bulycheva, A. I., and P. A. Mel'nikova. 1958. Determination of free silica in the presence of silicates with pyrophosphoric acid. Opredelenie Svobodnoi Dvuokisi Kremniya v Gorn. Porodakh i Rudn. Pylĭ, Akad. Nauk SSSR, Inst. Gorn. Dela, Sbornik Stateĭ. 1958:23–32.

Bunn, C. W. 1961. Chemical crystallography: Clarendon Press, Oxford.

Callière, S., L. Gatineau, and St. Hénin. 1960. Preparation à base témperature d'hématite alumineuse. Compt. Rend. Acad. Sci. Seance, Mai 1960:3677–3679.

Campbell, A. S. 1973. Anatase and rutile determination in soil clays. Clay Miner. 10:57–58.

Chaberek, S., and A. E. Martell. 1959. Organic sequestering agents. John Wiley & Sons, Inc., New York.

Chao, T. T., and P. K. Theobald, Jr. 1976. The significance of secondary iron and manganese oxides in geochemical exploration. Econ. Geol. 71:1560–1569.

Chapman, S. L., J. K. Syers, and M. L. Jackson. 1969. Quantitative determination of quartz in soils, sediments, and rocks by pyrofulfate fusion and hydrofluosilicic acid treatment. Soil Sci. 107:348–355.

Clark, J. S. 1965. The extraction of exchangeable cations from soils. Can. J. Soil Sci. 45:311–322.

Clark, J. S., J. A. McKeague, and W. E. Nichol. 1966. The use of pH-dependent cation-exchange capacity for characterizing the B horizons of Brunisolic and Podsolic soils. Can. J. Soil Sci. 46:161–166.

Clayton, R. N., R. W. Rex, J. K. Syers, and M. L. Jackson. 1972. Oxygen isotope abundance in quartz from Pacific pelagic sediments. J. Geophys. Res. 77:3907–3915.

Coffin, D. E. 1963. A method for the determination of free iron in soils and clays. Can. J. Soil Sci. 43:7–17.

Cooper, J. E., and W. S. Evans. 1983. Ammonium-nitrogen in Green River formation oil shale. Science 219:492–493.

Corey, R. B. 1952. Allocation of elemental constituents to mineral species in polycomponents colloids of soils. Ph.D. thesis, University of Wisconsin Library, Madison (no. 0262).

Correns, C. W., and W. von Engelhardt. 1941. Röntgenographische Untersuchungen über den Mineralbestand sedimentärer Eisenerze. Nachr. Akad. Wiss. Göttingen, Math.-Phys. Klasse 213:131–137.

Cradwick, P. D. G., V. C. Farmer, J. D. Russell, C. R. Masson, K. Wada, and N. Yoshinaga. 1972. Imogolite, a hydrated aluminum silicate of tubular structure. Nature Phys. Sci. 240:187–189.

Davidtz, J. C., and M. E. Sumner. 1965. Blocked charges on clay minerals in subtropical soils. J. Soil Sci. 16:270–274.

Deb, B. C. 1950. The estimations of free iron oxide in soils and clays and their removal. J. Soil Sci. 1:212–220.

DeEndredy, A. S. 1963. Estimation of free iron oxides in soils and clays by a photolytic method. Clay Miner. Bull. 9:209–217.

de Villiers, J. M. 1971. The problem of quantitative determination of allophane in soil. Soil Sci. 112:2–7.

de Villiers, J. M., and M. L. Jackson. 1967. Cation exchange capacity variation with pH in soil clays. Soil Sci. Soc. Am. Proc. 31:473–476.

Dixon, J. B., and M. L. Jackson. 1959. Dissolution of interlayers from intergradient soil clays after preheating at 400°C. Science 129:1616–1617.

Dixon, J. B., and M. L. Jackson. 1962. Properties of intergradient chlorite-expansible layer silicates in soils. Soil Sci. Soc. Am. Proc. 26:358–362.

Dobrovol'skaya, V. V. 1958. Determination of free silica in dust by a method with pyrophosphoric acid. Opredelenie Svobodnoĭ Dvuokisi Kremniya v Gorn. Porodakh i Rudn. Pyli, Akad. Nauk SSSR, Inst. Gorn. Dela Sbornick Stateĭ. 1958:15–22.

Dolcater, D. L., J. K. Syers, and M. L. Jackson. 1970. Titanium as free oxide and substituted forms in kaolinite and other soil minerals. Clays Clay Miner. 18:71–79.

Drosdoff, M., and C. C. Nikiforoff. 1940. Iron-manganese concretions in Dayton soils. Soil Sci. 49:333–345.

Dudas, M. J., and M. E. Harward. 1971. Effect of dissolution treatment on standard and soil clays. Soil Sci. Soc. Am. Proc. 35:134–140.

Dyal, R. S. 1953. Mica leptyls and wavellite content of clay fraction from Gainesville loamy fine sand of Florida. Soil Sci. Soc. Am. Proc. 17:55–58.

El-Attar, H. A., and M. L. Jackson. 1973. Montmorillonitic soils developed in Nile River sediments. Soil Sci. 116:191–201.

El-Attar, H. A., M. L. Jackson, and V. V. Volk. 1972. Fluorine loss from silicates on ignition. Am. Mineral. 57:246–252.

Farmer, V. C., B. F. L. Smith, and J. M. Tait. 1977. Alteration of allophane and imogolite by alkaline digestion. Clay Miner. 12:195–198.

Fey, M. V., and J. LeRoux. 1976. Quantitative determinations of allophane in soil clays. Proc. Int. Clay Conf. 1975 (Mexico City). 5:451–463. Applied Publishing, Ltd., Wilmette IL.

Fey, M. V., and J. LeRoux. 1977. Properties and quantitative estimation of poorly crystalline components in sesquioxidic soil clays. Clays Clay Miner. 25:285–294.

Fieldes, M., and R. J. Furkert. 1966. The nature of allophane in soils Part 2—differences in composition. N. Z. J. Sci. 9:608–622.

Fieldes, M., and K. W. Perrott. 1966. The nature of allophane in soils Part 3—rapid field and laboratory test for allophane. N. Z. J. Sci. 9:623–629.

Fitzpatrick, R. W., J. LeRoux, and U. Schwertmann. 1978. Amorphous and crystalline titanium and iron-titanium oxides in synthetic preparations, at near ambient conditions, and in soil clays. Clays Clay Miner. 26:189–201.

Florentin, M. D., and M. Heros. 1947. Dosage de la silice libre (quartz) dans les silicates. Bull. Soc. Chim. France. 1947M:213–215.

Follett, E. A. C., W. J. McHardy, B. D. Mitchell, and B. F. L. Smith. 1965. Chemical dissolution techniques in the study of soil clays. I. & II. Clay Miner. 6:23–43.

Foster, M. D. 1953. Geochemical studies of clay minerals. III. The determination of free silica and free alumina in montmorillonite. Geochim. Cosmochim. Acta 3:143–154.

Galabutskaya, E., and R. Govorova. 1934. Bleaching of kaolin. Min. Suir'e. 9:27.

Glenn, R. C., M. L. Jackson, F. D. Hole, and G. B. Lee. 1960. Chemical weathering of layer silicate clays in loess-derived Tama silt loam of southwestern Wisconsin. Clays Clay Miner. 8:63–83.

Greenland, D. J., J. M. Oades, and T. W. Sherwin. 1968. Electron microscope observations of iron oxides in some red soils. J. Soil Sci. 19:123–126.

Grim, R. E. 1953. Clay mineralogy. McGraw-Hill Book Co., New York.

Hallmark, C. T., L. P. Wilding, and N. E. Smeck. 1982. Silicon. *In* A. L. Page et al. (ed.) Methods of soil analysis, part 2. 2nd ed. Agronomy 9:263–273.

Hardy, F., and R. R. Follett-Smith. 1931. Studies in tropical soils. II. Some characteristic igneous rock soil profiles in British Guiana, South America. J. Agric. Sci. 21:739–761.

Hashimoto, I. 1961. Differential dissolution analysis of clays and its application to Hawaiian soils. Ph.D. thesis, University of Wisconsin library, Madison. Diss. Abstr. No. 61-2957.

Hashimoto, I., and M. L. Jackson. 1960. Rapid dissolution of allophane and kaolinite-halloysite after dehydration. Clays Clay Miner. 7:102–113.

Helmke, P. A. 1982. Neutron activation analysis. *In* A. L. Page et al. (ed.) Methods of soil analysis, part 2. 2nd ed. Agronomy 9:67–84.

Henderson, J. H., R. N. Clayton, M. L. Jackson, J. K. Syers, R. W. Rex, J. L. Brown, and I. B. Sachs. 1972. Cristobalite and quartz isolation from soils and sediments by hydrofluosilicic acid treatment and heavy liquid separation. Soil Sci. Soc. Am. Proc. 36:830–835.

Henderson, J. H., J. K. Syers, and M. L. Jackson. 1970. Quartz dissolution as influenced by pH and the presence of a disturbed surface layer. Israel J. Chem. 8:357–372.

Hendricks, S. B., and W. H. Fry. 1930. The results of X-ray and microscopical examinations of soil colloids. Soil Sci. 29:457–479.

Higashi, T., and H. Ikeda. 1974. Dissolution of allophane by acid oxalate solution. Clay Science 4:205–211.

Hillebrand, W. F., and G. E. F. Lundell. 1929. Applied inorganic analysis. p. 705 and 715, 10th printing (1948). John Wiley & Sons, New York.

Hislop, J. F. 1944. The decomposition of clay by heat. Trans. Br. Ceram. Soc. 43:49–51.

Hodges, S. C., and L. W. Zelazny. 1980. Determination of noncrystalline soil components by weight difference after selective dissolution. Clays Clay Miner. 28:35–42.

Hodgson, J. R. 1963. Chemistry of the micronutrient elements in soils. Adv. Agron. 15:119–159.

Hsu, P. H. 1963. Effect of initial pH, phosphate, and silicate on the determination of aluminum with Aluminon. Soil Sci. 96:230–238.

Huang, P. M., and M. L. Jackson. 1965. Mechanism of reaction of neutral fluoride solution with layer silicates and oxides of soils. Soil Sci. Soc. Am. Proc. 29:661–665.

Huang, W. H., and D. C. Vogler. 1972. Dissolution of opal in water and its water contents. Nature Phys. Sci. 235:157–158.

Inoue, T., and K. Wada. 1968. Adsorption of humified clover extracts by various clays. Trans Int. Congr. Soil., 9th 3:289–298.

Iyengar, S. S., L. W. Zelazny, and D. C. Martens. 1981. Effect of photolytic oxalate treatment on soil hydroxy-interlayered vermiculites. Clays Clay Miner. 29:429–434.

Jackson, M. L. 1958. Soil chemical analysis. Prentice-Hall, Inc., Englewood Cliffs, New Jersey. Available from the author Madison, WI 53705.

Jackson, M. L. 1959. Frequency distribution of clay minerals in major great soil groups as related to the factors of soil formation. Clays Clay Miner. 6:133–143.

Jackson, M. L. 1960. Structural role of hydronium in layer silicates during soil genesis. Trans. Int. Congr. Soil Sci., 7th 2:445–455.

Jackson, M. L. 1963a. Interlaying of expansible layer silicates in soils by chemical weathering. Clays Clay Miner. 11:29–46.

Jackson, M. L. 1963b. Aluminum bonding in soils: A unifying principle in soil science. Soil Sci. Soc. Am. Proc. 27:1–10.

Jackson, M. L. 1964. Soil clay mineralogical analysis. p. 245–294. *In* C. I. Rich and G. W. Kunze, ed. Soil clay mineralogy, University of North Carolina Press, Chapel Hill, NC.

Jackson, M. L. 1979. Soil chemical analysis-advanced course. 2nd ed., 11th Printing. Published by the author, Madison, WI 53705.

Jackson, M. L. 1981. Oxygen isotopic ratios in quartz as an indicator of provenance of dust. p. 27–36, Ch. 3. *In* T. L. Péwé, (ed.). Desert dust: origin, characteristics, and effect on man. Special Paper 186 of the Geological Society of America, P.O. Box 9140, 3300 Penrose Place, Boulder, CO.

Jackson, M. L., and F. H. Abdel-Kader. 1977. Effect on CEC of soil clays after removal of amorphous materials by KOH or NaOH. Agron. Abstr., Am. Soc. Agron. Madison, WI. p. 188.

Jackson, M. L. and T. A. Frolking. 1982. Mechanism of terra rossa red coloration. Clay Research 1(1):1–5.

Jackson, M. L., T. W. M. Levelt, J. K. Syers, R. W. Rex, R. N. Clayton, G. D. Sherman, and G. Uehara. 1971. Geomorphological relationships of tropospherically derived quartz in the soils of the Hawaiian Islands. Soil Sci. Soc. Am. Proc. 35:515–525.

Jackson, M. L., M. Sayin, and R. N. Clayton. 1976. Hexafluorosilicic acid reagent modification for quartz isolation. Soil Sci. Soc. Am. J. 40:958–960.

Jackson, M. L., S. A. Tyler, A. L. Willis, G. A. Bourbeau, and R. P. Pennington. 1948. Weathering sequence of clay-size minerals in soils and sediments. I. J. Phys. Chem. 52:1237–1260.

Jackson, M. L., L. D. Whittig, and R. P. Pennington. 1950. Segregation procedure for the mineralogical analysis of soils. Soil Sci. Soc. Am. Proc. 14:77–81.

Jenne, E. A., J. W. Ball, and C. Simpson. 1974. Determination of trace metals in sodium dithionite–citrate extracts of soils and sediments by atomic absorption. J. Environ. Qual. 3:281–287.

Jones, A. A. 1982. X-ray fluorescence spectrometry. *In* A. L. Page et al. (ed.) Methods of soil analysis, part 2. 2nd ed. Agronomy 9:85–121.

Jones, R. L. 1969. Determination of opal in soil by alkali dissolution analysis. Soil Sci. Soc. Am. Proc. 33:976–978.

Jones, R. C., and G. Uehara. 1973. Amorphous coating on mineral surfaces. Soil Sci. Soc. Am. Proc. 37:792–798.

Jophcott, C. M., and H. F. V. Wall. 1955. Determination of quartz of various particle sizes in quartz-silicate mixture. Arch. Ind. Health 11:425–430.

Kanehiro, Y., and L. D. Whittig. 1961. Amorphous mineral colloids of soils of the Pacific region and adjacent areas. Pacific Sci. 40:477–482.

Kelley, W. P., W. M. Dore, and S. M. Brown. 1931. The nature of base exchange material of bentonite, soils, and zeolites, as revealed by chemical investigation and X-ray analysis. Soil Sci. 31:25–55.

Kelley, W. P., and H. Jenny. 1936. The relation of crystal structure to base exchange and its bearing in base exchange in soils. Soil Sci. 41:367–382.

Kiely, P. V., and M. L. Jackson. 1964. Selective dissolution of micas from potassium feldspars by sodium pyrosulfate fusion of soils and sediments. Am. Mineral. 49:1648–1659.

Kiely, P. V., and M. L. Jackson. 1965. Quartz, feldspar, and mica determination for soils by sodium pyrosulfate fusion. Soil Sci. Soc. Am. Proc. 29:159–163.

Kitagawa, Y. 1977. Determination of allophane and amorphous inorganic matter in clay fraction of soils. II. Soil clay fractions. Soil Sci. Plant Nutr. 23:21–31.

Kittrick, J. A., and E. W. Hope. 1963. A procedure for the particle-size separation of soils for X-ray diffraction analysis. Soil Sci. 96:319–325.

Knudsen, D., G. A. Peterson, and P. F. Pratt. 1982. Lithium, sodium, and potassium. *In* A. L. Page et al. (ed.) Methods of soil analysis, part 2. 2nd ed. Agronomy 9:225–246.

Langston, R. B., and E. A. Jenne. 1964. NaOH dissolution of some oxide impurities from kaolins. Clays Clay Miner. 12:633–647.

Lanyon, L. E., and W. R. Heald. 1982. Magnesium, calcium, strontium, and barium. *In* A. L. Page et al. (ed.) Methods of soil analysis, part 2. 2nd ed. Agronomy 9:247–262.

LeRiche, H. H., and A. H. Weir. 1963. A method for studying trace elements in soil fractions. J. Soil Sci. 14:225–235.

Lim, C. H., and M. L. Jackson. 1980. Polycomponent interstratified phyllosilicates in dolomite residuum and sandy till of central Wisconsin. Soil Sci. Soc. Am. J. 44:868–872.

Lim, C. H., and M. L. Jackson. 1982. Dissolution for total elemental analysis. *In* A. L. Page et al. (ed.) Methods of soil analysis, part 2. 2nd ed. Agronomy 9:1–12.

Lim, C. H., M. L. Jackson, R. D. Koons, and P. A. Helmke. 1980. Kaolins: sources of differences in cation-exchange capacities and cesium retention. Clays Clay Miner. 28:223–229.

MacEwan, D. M. C. 1961. Montmorillonite minerals. p. 143–207. *In* G. Brown (ed.) The X-ray identification and crystal structures of clay minerals. Mineralogical Society, London.

Mackenzie, R. C. 1954. Free iron-oxide removal from soils. J. Soil Sci. 5:167–172.

Mackenzie, R. C., and R. H. S. Robertson. 1961. The quantitative determination of halloysite, goethite, and gibbsite. Acta Universtatis Carolinae—Geologica Supplementum. 1:139–149.

Makumbi, M. N., and M. L. Jackson. 1977. Weathering of Karroo argillite under equatorial conditions. Geoderma 19:181–197.

Mankin, C. J., and C. G. Dodd. 1963. Proposed reference illite from the Ouachita Mountains. Clays Clay Miner. 10:372–379.

Marshall, C. E. 1935. Layer lattices and the base-exchange of clays. Ztschr. Krist. 91:433–449.

McKeague, J. A. 1978. Manual on soil sampling and methods of analysis. 2nd ed. pp. 72–81. Can. Soc. Soil Sci., Ottawa.

McKeague, J. A., J. E. Brydon, and N. M. Miles. 1971. Differentiation of forms of extractable iron and aluminum in soils. Soil Sci. Soc. Am. Proc. 35:33–38.

McKeague, J. A., and J. H. Day. 1966. Dithionite- and oxalate-extractable Fe and Al as aids in differentiating various classes of soils. Can. J. Soil Sci. 46:13–22.

McKyes, E., A. Sethi, and R. N. Yong. 1974. Amorphous coatings on particles of sensitive clay soils. Clays Clay Miner. 22:427–433.

McLaren, R. G., and D. V. Crawford. 1973. Studies on soil copper: I. Fractionation of copper in soils. J. Soil Sci. 24:172–181.

Medicus, K. 1955. Schnellbestimmung der Kieselsäure im Bauxite nach der Perchlorsäure method. Zeit. Anal. Chemie. 145:337–338.

Mehra, O. P., and M. L. Jackson. 1960. Iron oxide removal from soils and clays by a dithionite-citrate system buffered with sodium bicarbonate. Clays Clay Miner. 7:317–327.

Mitchell, B. D., J. M. Bracewell, A. S. de Endredy, W. J. McHardy and B. F. L. Smith. 1968. Mineralogical and chemical characteristics of a gley soil from north-east Scotland. Trans. Int. Congr. Soil Sci., 9th 3:67–77.

Mitchell, B. D., and V. C. Farmer. 1962. Amorphous clay minerals in some Scottish soil profiles. Clay Miner. Bull. 5:128–144.

Mitchell, B. D., V. C. Farmer, and W. J. McHardy. 1964. Amorphous inorganic materials in soils. Adv. Agron. 16:327–383.

Mitchell, B. D., and R. C. Mackenzie. 1954. Removal of free iron oxide from clays. Soil Sci. 77:173–184.

Muñoz Taboadela, M. 1953. The clay mineralogy of some soils from Spain and from Rio Muni (West Africa). J. Soil Sci. 4:48–55.

Nagelschmidt, G. 1956. Inter-laboratory trials on the determination of quartz in dusts of respirable size. Analyst 81:210–219.

Norrish, K., and R. M. Taylor. 1961. The isomorphous replacement of iron by aluminum in soil goethites. J. Soil Sci. 12:294–306.

Oades, J. M. 1963. The nature and distribution of iron compounds in soils. Soils Fert. 26:69–80.

Olson, R. V., and R. Ellis, Jr. 1982. Iron. *In* A. L. Page et al. (ed.) Methods of soil analysis, part 2. 2nd ed. Agronomy 9:301–312.

Paces, T. 1973. Steady-state kinetics and equilibrium between ground water and granitic rock. Geochim. Cosmochim. Acta 37:2641–2663.

Pawluk, S. 1972. Measurement of crystalline and amorphous iron removal in soils. Can. J. Soil Sci. 52:119–123.

Perrott, K. W., B. F. L. Smith, and B. D. Michell. 1976. Effect of pH on the reaction of sodium fluoride with hydrous oxides of silicon, aluminum, and iron, and poorly ordered aluminosilicates. J. Soil Sci. 27:348–356.

Peterson, G. W., G. Chesters, and G. B. Lee. 1966. Quantitative determination of calcite and dolomite in soils. J. Soil Sci. 17:328–338.

Polezhaev, N. G. 1958. New method of determination of free silica in presence of silicates. Opredelenie Svobodnoĭ Dvuokisi Kremniya v Gorn. Porodakh i Rudn. Pyli, Akad. Nauk SSSR, Inst. Gorn. Dela, Sbornik Stateĭ. 1958:33–43.

Pollack, S. S., E. P. Whiteside, and D. E. Van Varowe. 1954. X-ray diffraction of common silica minerals and possible applications to studies of soil genesis. Soil Sci. Soc. Am. Proc. 18:268–272.

Raman, K. V., and M. L. Jackson. 1965. Rutile and anatase determinations in soils and sediments. Am. Mineral. 50:1086–1092.

Rao, T. V., and G. S. R. Krishna Murti. 1982. Clay mineralogy of some laterite and associated soils of Goa, India. Clay Research 1:6–16.

Rengasamy, P., V. A. K. Sarma, and G. S. R. Krishna Murti. 1975. Quantitative mineralogical analysis of soil clays containing amorphous materials: a modification of the Alexiades and Jackson procedure. Clays Clay Miner. 23:78–79.

Reynolds, R. C., Jr. 1960. Separating illite for geochemical analysis. U. S. Patent 2,946,657.

Rhoades, J. D. 1982. Cation exchange capacity. *In* A. L. Page et al. (ed.) Methods of soil analysis, part 2. 2nd ed. Agronomy 9:149–157.

Rich, C. I. 1968. Hydroxy interlayers in expansible layer silicates. Clays Clay Miner. 16:119–123.

Robinson, W. O., and R. S. Holmes. 1924. The chemical composition of soil colloids. U. S. Dep. Agric. Bull. 1311.

Ross, C. S., and P. F. Kerr. 1934. Halloysite and allophane. U. S. Geol. Survey, Prof. Paper 185 G:135–148.

Roth, C. B., M. L. Jackson, J. M. de Villiers, and V. V. Volk. 1967. Surface colloids on micaceous vermiculite. p. 217–221. *In* G. V. Jacks (ed.) Soil chemistry and soil fertility, Trans. Comm. II and IV, Int. Soc. Soil Sci. 1966. (Aberdeen).

Roth, C. B., M. L. Jackson, E. G. Lotse, and J. K. Syers. 1968. Ferrous-ferric ratio and CEC changes on deferration of weathered micaceous vermiculite. Israel J. Chem. 6:261–273.

Roth, C. B., and M. L. Jackson, and J. K. Syers. 1969. Deferration effect on structural ferrous-ferric iron ratio and CEC of micaceous vermiculites and soils. Clays Clay Miner. 17:253–264.

Russell, J. D., B. A. Goodman, and A. R. Fraser. 1979. Infrared and Mössbauer studies of reduced nontronites. Clays Clay Miner. 27:63–71.

Saunders, W. M. H. 1964. Phosphate retention by New Zealand soils and its relationship to free sesquioxides, organic matter and other soil properties: N.Z.J. Agric. Res. 8:30–57.

Sawhney, B. L., C. R. Frink, and D. E. Hill. 1970. Components of pH dependent cation exchange capacity. Soil Sci. 109:272–278.

Sawhney, B. L., and K. Norrish. 1971. pH dependent cation exchange capacity minerals and soils of tropical regions. Soil Sci. 112:213–215.

Sayin, M., and M. L. Jackson. 1975. Anatase and rutile determination in kaolinite deposits. Clays Clay Miner. 23:437–443.

Sayin, M., and M. L. Jackson. 1979. Size and shape of fine quartz in the clay fraction of soils and geological materials. Z. Pflanzenernähr. Bodenk. 142:865–873.

Schwertmann, U. 1959. Die fraktionierte Extraktion der freien Eisenoxide in Boden, ihre mineralogischen Formen und ihre Entstehungsweisen. Z. Pflanzenernähr. Dueng. Bodenk. 84:194–204.

Schwertmann, U. 1964. Differenzierung der Eisenoxide des Bodens durch photochemische Extraktion mit saurer Ammoniumoxalate-Lösung. Z. Pflanzenernähr. Dueng. Bodenk. 105:194–202.

Schwertmann, U. 1973. Use of oxalate for Fe extraction from soils. Can. J. Soil Sci. 53:244–246.

Schwertmann, U., R. W. Fitzpatrick, and J. LeRoux. 1977. Al substitution and differential disorder in soil hematites. Clays Clay Miner. 25:373–374.

Schwertmann, U., and M. L. Jackson. 1963. A third buffer range in the potentiometric titration of H-Al-clays. Science 139:1052–1054.

Sclar, C. B., L. C. Carrison, and C. M. Schwartz. 1962. Relation of infrared spectra to coordination in quartz and two high-pressure polymorphs of SiO_2. Science 138:525–526.

Searle, P. L., and B. K. Daly. 1977. The determination of aluminum, iron, manganese and silicon in acid oxalate soil extracts by flame emission and atomic absorption spectrometry. Geoderma. 19:1–10.

Segalen, P. 1968. Note sur une méthod de determination des produits minéraux amorphes dans certains sols a hydroxydes tropicaux. Cah. ORSTOM. Sér. Pédol. 6:105–125.

Shaw, A. 1934. The determination of free silica in coal-measure rocks. Analyst 59:446–461.

Shchekaturina, L. G., and V. I. Petrashen. 1958. Determination of free silica in coal dust. Opredelenie Svobodnoĭ Dvuokisi Kremniya v Gorn. Porodakh i Rudn. Pyli, Akad. Nauk SSSR, Inst. Gorn. Dela, Sbornik Stateĭ. 1958:54–57.

Sherman, G. D. 1952. The titanium oxide content of Hawaiian soils and its significance. Soil Sci. Soc. Am. Proc. 16:15–18.

Shoji, S., and T. Ono. 1978. Physical and chemical properties and clay mineralogy of andosols from Kitakami, Japan. Soil Sci. 126:297–312.

Shuman, L. M. 1979. Zinc, manganese, and copper in soil fractions. Soil Sci. 127:10–17.

Simmons, W. L. and L. A. Plues-Foster. 1977. Improved method of analysing difficult soil extracts by flame atomic absorption spectrometry—Application to measurement of copper in ammonium oxalate extracts. Aust. J. Soil Res. 15:171–175.

Smith, J. L. 1865. On the use of the bisulphate of soda as a substitute for the bisulphate potash in the decomposition of minerals, especially the aluminous minerals. Am. J. Sci. Arts 40:248–249.

Soil Survey Staff. 1975. Soil taxonomy. A basic system of soil classification for making and interpreting soil surveys. U. S. Gov. Printing Office, Washington, DC.

Soltanpour, P. N., J. B. Jones, Jr., and S. M. Workman. 1982. Optical emission spectrophotometry. *In* A. L. Page et al. (ed.) Methods of soil analysis, part 2. 2nd ed. Agronomy 9:29–65.

Stucki, J. W., C. B. Roth, and W. E. Baitinger. 1976. Analysis of iron-bearing clay minerals by electron spectroscopy for chemical analysis (ESCA). Clays Clay Miner. 24:289–292.

Swindale, L. D., and M. L. Jackson. 1956. Genetic processes in some residual podzolised soils of New Zealand. Trans. Int. Congr. Soil Sci., 6th 5:233–239.

Swindale, L. D., and M. L. Jackson. 1960. A mineralogical study of soil formation in four rhyolite-derived soils from New Zealand. N.Z.J. Geol. Geophys. 3:141–183.

Syers, J. K., S. L. Chapman, R. W. Rex, M. L. Jackson, and R. N. Clayton. 1968. Quartz isolation from rocks, sediments, and soils for the determination of oxygen isotopic composition. Geochim. Cosmochim. Acta 32:1022–1024.

Sysoeva, R. S. 1958. Test of parallel determination of the free silica in the dust of a crushing mill by chemical, petrographic, X-ray spectral and thermal methods. Opredelenie Svobodnoĭ Dvuokisi Kremniya v Gorn. Porodakh i Rudn. Pyli, Akad. Nauk SSSR, Inst. Gorn. Dela, Sbornik Stateĭ. 1958:103–110.

Tabikh, A. A., I. Barshad, and R. Overstreet. 1960. Cation-exchange hysteresis in clay minerals. Soil Sci. 90:219–226.

Talvitie, N. A. 1951. Determination of quartz in presence of silicate using phosphoric acid. Anal. Chem. 23:623–626.

Tamm, O. 1922. Eine Methode zur Bestimmung der anorganischen Komponenten des Gel-Komplex in Boden. Medd. Statens Skogforsokanst. 19:385–404.

Tamura, T., M. L. Jackson, and G. D. Sherman. 1953. Mineral content of low humic, humic, and hydrol humic latosols of Hawaii. Soil Sci. Soc. Am. Proc. 17:343–346.

Thorp, J., and G. D. Smith. 1949. Higher categories of soil classification: order, suborder, and great soil group. Soil Sci. 67:117–126.

Tokashiki, Y., and K. Wada. 1972. Determination of silicon, aluminum and iron dissolved by successive and selective dissolution treatments of volcanic ash soil clays. Clay Sci. 4:105–114.

Tokashiki, Y., and K. Wada. 1975. Weathering implications of the mineralogy of clay fractions of two ando soils, Kyushu. Geoderma 14:47–62.

Trostel, L. J., and D. J. Wynne. 1940. Determination of quartz (free silica) in refractory clays. J. Am. Ceram. Soc. 23:18–22.

Tuddenham, W. M., and R. J. P. Lyon. 1960. Infrared techniques in the identification and measurement of minerals. Anal. Chem. 32:1630–1634.

van Bladel, R., R. Frankart, and H. R. Gheyi. 1975. A comparison of three methods of determining the cation exchange capacity of calcareous soils. Geoderma 13:289–298.

van Olphen, H. 1971. Amorphous clay materials. Science 171:91–92.

Veith, J. A., and M. L. Jackson. 1974. Iron oxidation and reduction effects on structural hydroxyl and layer charge in aqueous suspensions of micaceous vermiculites. Clays Clay Miner. 22:345–353.

Viets, F. G., Jr. 1962. Chemistry and availability of micronutrients in soils. J. Agric. Food Chem. 10:174–178.

Violante, A., and M. L. Jackson. 1981. Clay influence on the crystallization of aluminum hydroxide polymorphs in the presence of citrate, sulfate or chloride. Geoderma 25:199–214.

Wada, K. 1977. Allophane and imogolite. p. 603–638. *In* J. B. Dixon and S. B. Weed (eds.) Minerals in soil environments. Soil Sci. Soc. Am., Madison, WI 53711.

Wada, K. 1979. Structural formula of allophanes. *In* M. M. Mortland and V. C. Farmer (eds.) Proc. Int. Clay Conf. 1978. 6:537–545. Elsevier Sci. Pub. Co., Amsterdam.

Wada, K., and H. Ataka. 1958. The ion-uptake mechanism of allophane. Soil Plant Food (Tokyo) 4:12–18.

Wada, K., and D. J. Greenland. 1970. Selective dissolution and differential infrared spectroscopy for characterization of 'amorphous' constituents in soil clays. Clay Miner. 8:241–254.

Wada, K., and T. Higashi. 1976. The categories of aluminum- and iron-humus complexes in ando soils determined by selective dissolution. J. Soil Sci. 27:357–368.

Wada, K., and Y. Tokashiki. 1972. Selective dissolution and difference infrared spectroscopy in quantitative mineralogical analysis of volcanic-ash soil clays. Geoderma 7:199–213.

Wada, K., and S. Wada. 1976. Clay mineralogy of the B horizons of two Hydrandepts, a Torrox and a Humitropept in Hawaii. Geoderma 16:139–157.

Weaver, R. M., M. L. Jackson, and J. K. Syers. 1971. Magnesium and silicon activities in matrix solutions of montmorillonite-containing soils in relation to clay mineral stability. Soil Sci. Soc. Am. Proc. 35:823–830.

Weaver, R. M., M. L. Jackson, and J. K. Syers. 1976. Clay mineral stability as related to activities of aluminum, silicon, and magnesium in matrix solution of montmorillonite-containing soils. Clays Clay Miner. 24:246–252.

Weaver, R. M., J. K. Syers, and M. L. Jackson. 1968. Determination of silica in citrate-bicarbonate-dithionite extracts of soils. Soil Sci. Soc. Am. Proc. 32:497–501.

Whittig, L. D., and M. L. Jackson. 1955. Interstratified layer silicates in some soils of northern Wisconsin. Clays Clay Miner. 2:322–336.

Whittig, L. D., V. J. Kilmer, R. C. Roberts, and J. G. Cady. 1957. Characteristics and genesis of Cascade and Powell soils of northwestern Oregon. Soil Sci. Soc. Am. Proc. 21:226–232.

Winters, E. 1938. Ferromanganiferous concretions from some podzolic soils. Soil Sci. 46:33–40.

Yoe, J. H., and A. R. Armstrong. 1947. Colorimetric determination of titanium with disodium-1,2-dihydroxybenzene-3,5-disulfonate. Anal. Chem. 19:100–102.

Yoshinaga, N., and S. Aomine. 1962a. Allophane in some ando soils. Soil Sci. Plant Nutr. (Tokyo) 8:6–13.

Yoshinaga, N., and S. Aomine. 1962b. Imogolite in some ando soils. Soil Sci. Plant Nutr. (Tokyo) 8:22–29.

7 Thermal Analysis Techniques

K. H. TAN

University of Georgia
Athens, Georgia

B. F. HAJEK

Auburn University
Auburn, Alabama

I. BARSHAD

University of California
Berkeley, California

7–1 INTRODUCTION

Thermal analysis is a term covering a group of analyses that determine some physical parameter, such as energy, weight, dimension, and evolved volatiles, as a dynamic function of temperature. The measured parameter changes with temperature. Many soil constituents will undergo several thermal reactions upon heating, which can serve as a diagnostic property for the qualitative and quantitative identification of the substances. Some reactions, such as evaporation of adsorbed water, occur at low temperature, whereas other reactions, such as oxidation of organic compounds and metallic ions in a reduced state, occur at intermediate temperatures. Many of the reactions may occur only at high temperatures, such as loss of crystal-lattice OH as water and CO_3^{2-} as CO_2. The reactions may be exothermic or endothermic, and phase changes or crystal changes may take place in the course of reactions. Whatever the reactions are, when properly measured, they yield useful information for the identification of the substances under investigation.

A large number of techniques, related to each other, are available (Fig. 7–1). For detailed information reference is made to Wendlandt (1974), Smothers and Chiang (1966), Schultze (1969) and Mackenzie (1970). Among the many methods listed in Fig. 7–1, three of the major types that find frequent application in soil research, will be discussed: differential thermal analysis (DTA), differential scanning calorimetry (DSC), and thermogravimetry (TG).

 Methods of Soil Analysis, Part 1. Physical and Mineralogical Methods—Agronomy Monograph no. 9 (2nd Edition)

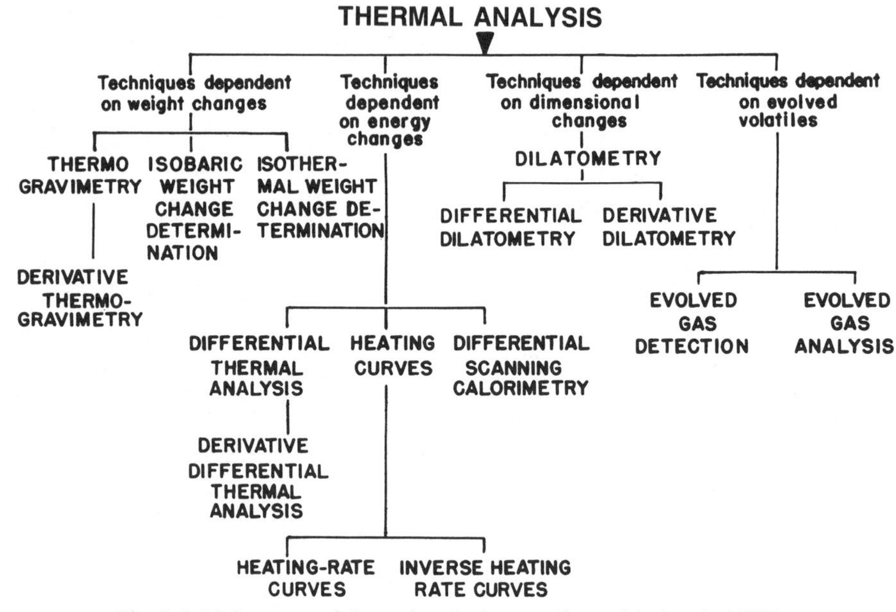

Fig. 7–1. Major types of thermal analysis according to Mackenzie (1970).

7–2 PRINCIPLES OF REACTIONS AND ANALYSIS.

7–2.1 General Principles

Water is driven off by heating a clay. This water is present in the clay in two distinct forms: OH^- ions and H_2O molecules. Customarily, the OH^- ions are referred to as *crystal-lattice water*, and their removal from the mineral is termed *dehydroxylation.* The H_2O molecules are referred to as *water of hydration* and *water of adsorption.,* and their removal is called *desorption* or *dehydration.*

The OH^- ions are present in fixed positions, either as individual ions among the oxygens at the apices of the tetrahedra or octahedra, or as continuous sheets as in the 1:1 and 2:2 mineral species. Their removal from the clay, as a rule, is irreversible and is accompanied by an irreversible change or a complete destruction of the clay mineral structure. The water molecules, on the other hand, are present either on the external surfaces of the clay particles only, as in kaolinite, or on both external and internal surfaces, as in montmorillonite, vermiculite, and hydrated halloysite. Their removal from a clay particle may or may not cause a reversible change in the crystal lattice, depending on the clay mineral. For example, upon removal and readsorption of H_2O molecules, the crystal lattice is not changed in kaolinite but is changed in a reversible manner in montmorillonite and vermiculite. Most of the adsorbed water molecules tend to be organized into sheets of monomolecular layers on the oxygen surfaces and to be grouped around the exchangeable ions. This form is termed here *layer-water*. Some water molecules, however, are present as individual molecules inside the cavities of the oxygen surfaces which form the bases of the tetrahedra. This water is termed here *cavity water*.

The ratio of crystal-lattice water to crystal-lattice oxygen is constant for each clay mineral, but the ratio of adsorbed water to crystal-lattice oxygen varies with the vapor pressure at which the mineral is equilibrated. At a given vapor pressure, the amount of adsorbed water depends on the total surface area, on the ratio of internal to external surface area, on the nature and amount of the exchangeable ions, and on the history of the clay with respect to wetting and drying (Barshad, 1955).

The loss of water, as a rule, is measured by loss in weight of the clay mineral upon heating. With soil clays, gross errors may occur in this measurement unless organic matter and carbonates are eliminated. In minerals which contain reduced ions, such as ferrous iron, some of the loss in weight from loss of water is counterbalanced by a gain of oxygen upon oxidation of the reduced element. In certain temperature ranges, both crystal-lattice water and adsorbed water may be loss simultaneously. The temperature at which the major amount of the crystal-lattice water is lost differs greatly among the different species and is the most singular property for the identification of the species. Because organic matter interferes in the determination of the crystal-lattice water, it is important to ascertain that it is absent from the sample to be analyzed. Absence of organic matter can be verified by DTA—even traces can be detected by this method. Differential thermal analysis is useful also for ascertaining the critical temperature range within which water losses occur. It is recommended, therefore, that DTA precede TG analysis.

The pretreatment of a soil clay required before thermal analysis depends on the objective of the analysis and on the nature of the soil in which the clay occurs. To obtain qualitative information, it may be possible to analyze the whole soil by DTA, even without removal of organic matter, particularly if the soil is high in clay and low in organic matter. However, for quantitative analysis of all constituents present in a clay, as for the purpose of evaluation of soil profile development and for the purpose of soil classification, the sample must be treated in various ways prior to the analysis. These treatments are described elsewhere in the book in detail in the chapters on particle size analysis (chapter 15), pretreatment for mineralogical analysis (chapter 5), and oxides, hydroxides and aluminosilicates (chapter 6).

7–2.2 Differential Thermal Analysis

Differential thermal analysis determines energy changes between a sample and reference material as the two are heated at a controlled rate. When the sample undergoes a transformation, the heat effect causes a difference in temperature between the sample and reference material. This difference in temperature is measured and recorded as a function of temperature.

The method, usually called DTA, is probably the most important and widely used technique (Mackenzie, 1970). It originated with Le Chatelier in 1887, and its usefulness was ignored for some time. However, recently it has been employed in geology, in ceramic, glass, polymer, cement, and plaster industries, and in research in chemistry and catalysis. Currently DTA also finds application in studies of organic matter, explosives, and radioisotopes. In soil research, it was first employed by Matejka (1922) for determination of kaolinite. Russell and Haddock (1940) and Hendricks and Alexander (1940) were perhaps among the first American scientists to recognize it as an important tool for analysis of clays and clay minerals in soils. A review of developments in DTA during 1940 to 1960 was published by Murphy (1962). A large proportion of the information in the literature deals with thermal analysis of clays from clay deposits and not of clays from soils. Indications are currently available that soil clays differ from clay deposits in many respects, as will be discussed in the following sections.

In DTA the sample and reference material are heated side by side at a controlled rate, usually from 0 to 1000°C. The sample temperature is continuously measured and compared with the temperature of the reference material by means of a set of thermocouples. When the sample undergoes a transformation, the heat effect causes a difference in temperature between the sample and reference material. This difference in temperature is recorded and usually plotted against the temperature at which the difference occurs (Fig. 7–2). If the temperature of the sample falls below that of the reference material (ΔT is negative) an endothermic peak develops. When the temperature of the sample rises above that of the reference material (ΔT is positive) an exothermic peak develops. The portion of the curve for which ΔT is approximately zero is considered to be the base line (Mackenzie, 1969; 1972). Endothermic and exothermic reactions are attributed to reactions, such as dehydration, dehydroxyla-

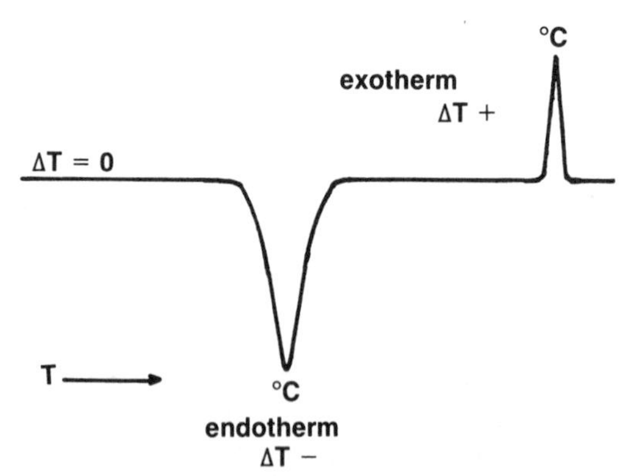

Fig. 7–2. Idealized DTA curve showing endothermic and exothermic peaks.

tion, decomposition, boiling, sublimation, vaporization, oxidation, fusion, destruction of crystal structure, and crystalline structure inversion. Dehydration, dehydroxylation, fusion, evaporation, and sublimation yield endothermic reactions. On the other hand, oxidation, crystalline structure formation, and some decomposition reactions yield exothermic effects.

The points on DTA curves that indicate maximum temperature differences are called *peak temperatures.* They are, of course, situated within the "thermal breaks" (regions in the curves which show the differences in temperature). The shape, size, and temperature of the thermal breaks, or peaks, as they are termed in DTA, are affected not only by the heats of reaction, but also by various instrumental factors, such as rate of heating, nature of thermocouples, size, shape, and nature of sample holder, position of the thermocouple in the sample hole, nature of the recording instrument, and other details of the instrument itself (Arens, 1951; Mackenzie, 1969).

The temperature differences discussed above are records of net heat or enthalpy changes, and they do not exclude the possibility that both exothermic and endothermic reactions are occurring simultaneously. Hence the absence of a peak (break or loop) in a DTA curve does not necessarily indicate the absence of a thermal reaction. It is conceivable that an endothermic and exothermic reaction with equal heat changes might occur simultaneously, canceling each other. Such a cancelation is thought to occur in DTA curves of vermiculite in the temperature region between 700 and 900°C, and in some forms of chlorites in the temperature region between 700 and 800°C (Martin, 1955).

7–2.3 Differential Scanning Calorimetry

This technique, also known as DSC, belongs to a different group of analyses called *calorimetry.* Calorimetry measures specific heat or thermal capacity of a substance, and among the many methods available, DSC exhibits the closest relationship with DTA (Mackenzie, 1970). Differential scanning calorimetry measures the amount of energy required to establish zero temperature difference between sample and reference material as the two are heated side by side at a controlled rate. The measurements are recorded as a function of time or temperature. When a thermal reaction, or transition, occurs in the sample, thermal energy is substracted from or added to the sample or reference containers in order to maintain the sample at the same temperature as the reference material or furnace block. The level of energy input is equivalent in amount to the energy evolved or absorbed as a result of the transition reaction in the sample. A recording of this balancing energy yields calorimetric measurements of the transition energy, but the technique of heating samples side by side is a principle in DTA.

As indicated above, in DSC the sample and reference material are heated side by side at a constant rate, and the differential heat flow (in millicalories per second) between the sample and reference is recorded as a function of a linearly programmed temperature (Brennan, 1974; Flynn, 1974; Skinner, 1969). DSC curves look similar to those of DTA. During the heating process endothermic and exothermic changes may occur in the sample. In order to maintain the sample temperature equal to that of the reference material or the furnace block, thermal energy is added to the sample when an endothermic change occurs, and substracted from the sample when an exothermic transition occurs. As indicated earlier, the energy input is exactly equivalent in magnitude to the energy absorbed or evolved in the particular transition. The amount of thermal energy required to balance the temperature differences is recorded.

7–2.4 Thermogravimetry

This is a technique whereby a sample is continuously weighed as it is being heated or cooled at a controlled rate. The weight changes are recorded as a function of temperature (T) or time (t), i.e.,

$$\text{weight} = f(T \text{ or } t)\,.$$

The recording is called a *thermogravigram* or a *thermogravimetric curve* (Mackenzie et al., 1972). In derivative thermogravimetry, the time derivative of the weight (w) is recorded as a function of temperature (T) or time (t):

$$dw/dt = f(T \text{ or } t)\,.$$

The change in weight is recorded against the temperature and yields information on the thermal stability and composition of the material under investigation. Weight changes are attributed to rupture and formation of various physical and chemical bonds at elevated temperatures, leading to the release of volatile substances or formation of heavier reaction products. For example, loss in weight as a result of heating is due to dehydration, dehydroxylation, or decomposition.

Thermogravimetry apparently started with the work of K. Honda at Tohoku University in 1915, who developed the first thermobalance (Saito, 1969; Wendlant, 1974). Much of the information obtained at that period was, however, empirical in nature. Only since the introduction of the Chevenard thermobalance in 1945 has the method obtained a better scientific foundation (Redfern, 1970; Wendlandt, 1974). Two major techniques are recognized in this category of analyses: (i) *thermogravimetry* (TG), also known as *thermogravimetric analysis* (TGA), *integral thermal analysis* (Jackson, 1956) or *dynamic thermogravimetric analysis*, and (ii) *derivative thermogravimetry* (DTG), which in the past was called *differential thermogravimetry* or *differential thermogravimetric analysis*. The International Committee for Standardization of Thermal Analysis (ICTA) proposed to replace the term thermogravimetric analysis with thermogravimetry (Redfern, 1970, McAdie, 1969; Mackenzie, 1969). As pointed out by Mackenzie et al. (1972), this type of thermal analysis (DTG) should be called *derivative thermogravimetry*, instead of *differential thermogravimetry*. Use of the latter term can result in confusion, since the term "differential" has a different meaning in thermogravimetry and in DTA.

7-3 METHODS

7-3.1 Differential Thermal Analysis

7-3.1.1 APPARATUS

The DTA instrument can be very simple or very complex. For detailed descriptions of commercially available DTA instruments see Mackenzie (1970), Wendlandt (1974) and Smothers and Chiang (1966). All of the instruments have in common the following basic components: (i) a furnace or heating unit, (ii) a specimen holder, and (iii) a temperature regulating and measuring system (Fig. 7-3).

Different kinds of furnaces, operable from 0 to 2800°C, have been used in DTA, with the heating furnished by infrared radiation, by high-frequency rf oscillation, by a coiled tubing through which heated liquid or gas is circulated, or by resistance elements (Wendlandt, 1974). Furnaces heated by resistance elements, such as the Hoskins furnace, are the most common.

The sample holders consist of either a rectangular or circular nickel

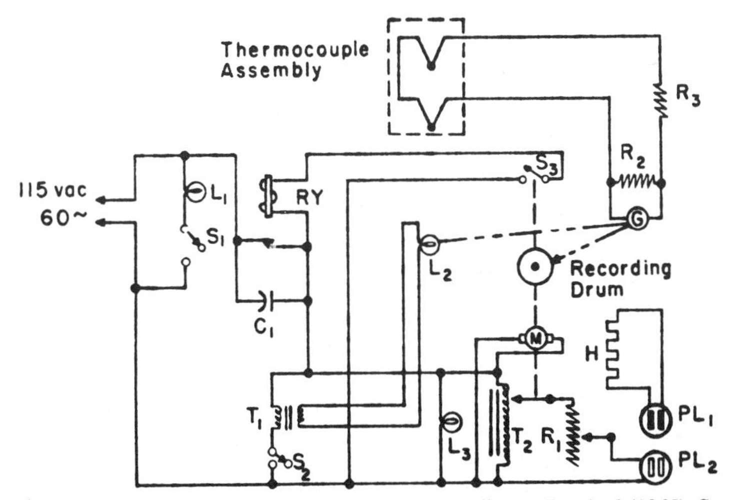

Fig. 7-3. Schematic diagram for a DTA apparatus according to Barshad (1965). S_1, safety lamp switch; S_2, Galvanometer lamp switch; S_3, shut-off switch; L_1, safety lamp; L_2, galvanometer lamp; L_3, pilot lamp; C_1, capacitor; R_1, rheostat; R_2, galvanometer shunt resistance; R_3, galvanometer sensitivity resistance; RY, mercury relay (latching type); T_1, filament transformer; T_2, variac transformer; M, timing motor; G, galvanometer; H, furnace; PL_1, male plug; PL_2, female plug.

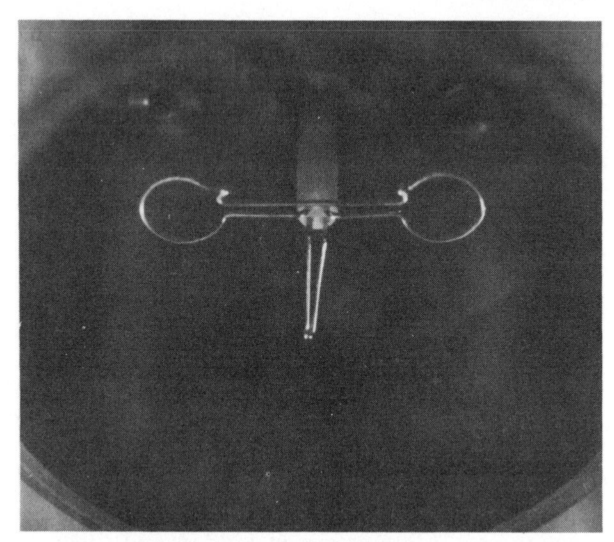

Fig. 7–4. Nickel alloy sample holder of Tracor-Stone DTA instrument with ring type Platinum-Pt/Rhodium thermocouples (5 mm in diameter).

or ceramic block. They are either equipped with cylindrical holes, called well-holders, in which the samples and the thermocouples are placed, or they are designed to hold ring-type thermocouples (Fig. 7–4) upon which a small platinum pan can be seated. In case of well-holders, it is important that the thermocouple be placed exactly in the center of the hole.

Thermocouple elements in most common use are made from platinum and platinum (90%)–rhodium (10%), or from chromel-P and alumel. For use with soil clays, the former elements are preferable, because they are more resistant to corrosion and consequently last longer. The size of the two thermocouples in the DTA circuit must be identical; otherwise the DTA curves tend to drift upward or downward from the baseline. The thermocouple is placed in the hole either from its side or from its bottom. The thermocouple's wires should be cemented in place, so that the junction cannot be displaced easily from a fixed position. It is for this reason that fairly sturdy thermocouple wires of about no. 22 gauge should be used, even though such wire is not as sensitive to temperature differences as finer wires.

The recording of the temperature differences and furnace temperature is accomplished by placing in the thermocouple circuit either a reflecting galvanometer, which records the variation in emf on photographic paper fastened to a rotating drum, or a pen-and-ink recorder. In many instruments, the temperature differences and the furnace temperature are recorded separately (Barshad, 1952).

7–3.1.2 REFERENCE MATERIAL

The reference material, sometimes called standard material, is a known substance that is thermally inert over the temperature range under investigation (Mackenzie et al., 1972). It is a "neutral" body or comparison standard against which the temperature of the sample is measured. For a detailed listing of preferred properties that a reference material should have, see the progress report of the ICTA by McAdie (1969). A number of compounds, both organic and inorganic, such as octanol, benzol, tributyrin, Al_2O_3, clay, quartz, NaCl, KCl, MgO, NH_4NO_3, $NaNO_3$, AgCl, glass beads, and aluminum foil have been used as reference material (Schultze, 1969; Smothers & Chiang, 1966). Among the compounds listed, the most commonly used reference material is α- or γ-alumina, or alumina calcined to 1200°C. For the latter, Al-oxide powder is ignited in a platinum crucible, cooled, and stored in a desiccator. Clay, especially kaolinite, preheated this way is also used as a reference in DTA. However, caution is required in the use of α- or γ-alumina. It has been observed (Arens, 1951) that calcined alumina may become hygroscopic after use in DTA, and may need to be replaced after two or three runs. Calcined clay creates a similar problem and may contain components with reversible thermal reactions. According to Arens (1951), the large differences in thermal conductivity between kaolinite (0.72×10^{-3} cal-cm $°C^{-1}$ s^{-1}) and calcined kaolinite (1.6×10^{-3} cal-cm $°C^{-1}$ s^{-1}) is a decisive factor for rejection of calcined kaolinite as reference material.

7-3.1.3 SAMPLE PREPARATION

7-3.1.3.1 Soil Fractions. Generally, DTA can be performed on any liquid or soil sample. Soil is complex because it is a mixture of mineral and organic matter, varying widely in particle size and in physical, chemical, and biological properties. Soils also vary in mineralogical composition. One question of concern is which fraction of soil should be analyzed. Should the "whole" soil be analyzed, since this is the medium to which plants and management practices react; should the organic matter be destroyed by H_2O_2 prior to DTA; or is it sufficient to analyze only individual mineral and organic fractions? Particle size separates can be obtained by fractionation procedures for other types of mineralogical investigation such as x-ray analysis (Jackson, 1956). However, if size separates are to be analyzed by DTA, one must remember that particle size and crystallinity have a pronounced influence on size, shape, and temperature of endothermic peaks (Smothers & Chiang, 1966).

7-3.1.3.2 Whole Soils. No pertinent data are available in the literature concerning DTA of whole soil samples, perhaps because soil organic matter may obscure characteristic endothermic peaks by strong exothermic reactions. DTA curves of whole soils (untreated and H_2O_2 pretreated), as shown in Fig. 7-5, show that organic matter caused a strong exothermic reaction culminating into a peak at approximately 300°C, which obscured any endothermic peak that may have developed between 250 and 400°C. In soils with low organic matter content, the latter is not a problem (Fig. 7-5, no. 2). Removal of organic matter with H_2O_2 or

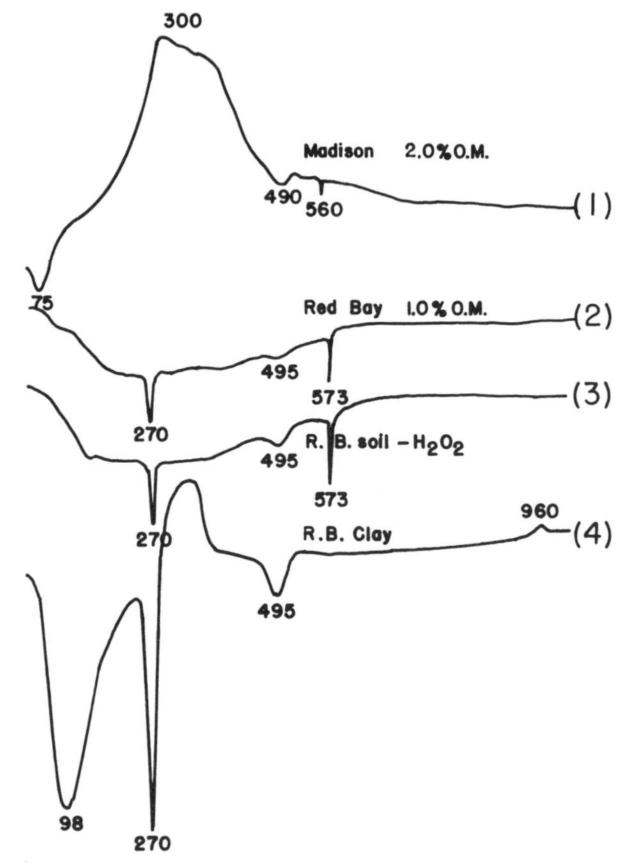

Fig. 7–5. DTA curves of whole soil and soil clay: (1) nontreated Madison surface soil (2.0% organic matter); (2) nontreated Red Bay surface soil (1.0% organic matter); (3) Red Bay surface soil pretreated with 30% H_2O_2; (4) clay (<2μm) of Red Bay surface soil. Tracor-Stone instrument, 3 mg of sample in Pt cups.

analysis in an inert N_2 atmosphere usually increases the intensity of endothermic peaks at 270°C, 490°C, and 573°C for gibbsite, kaolinite, and quartz, respectively (Fig. 7–5, no. 3). In addition to organic matter effects, peaks may also be obscured because of dilution by high quartz content, characteristic of many soils of the USA.

When whole soils are analyzed, the <2-mm fraction should be treated with 30% H_2O_2, washed with distilled water, dried, and ground again to pass a 2 mm or smaller sieve. In general, the analysis of whole soils gives peaks of low intensity. These same peaks are very large and intense if only the clay fractions are analyzed (Fig. 7–5, no. 4); however, the quartz $\alpha - \beta$ inversion peak is often absent.

7–3.1.3.3 Sand Fraction. For use in DTA, the composite sand fraction, 2.0 to 0.05 mm, can be used directly or separated further into very course sand (2.0–1.00 mm), coarse sand (1.00–0.50 mm), medium sand

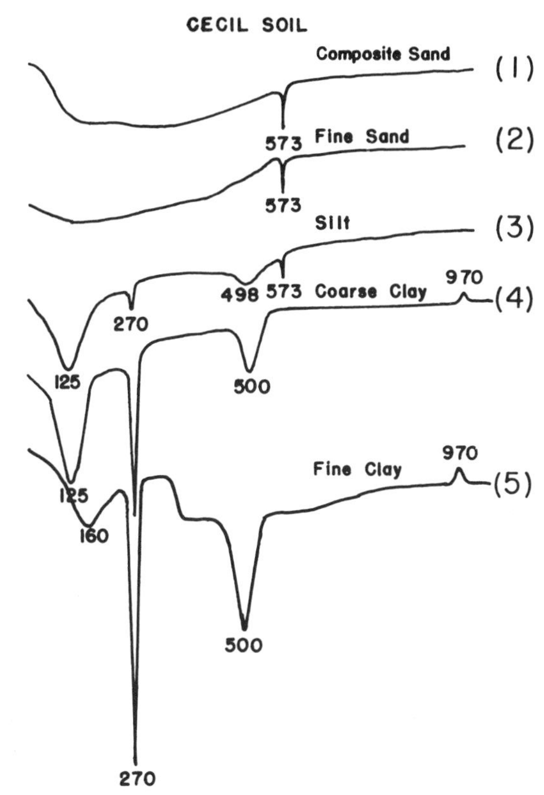

Fig. 7–6. DTA curves of sand fraction separated from Cecil soil AB horizon: (1) composite sand fraction, 2.0–0.05 mm; (2) fine sand, 0.25–0.10 mm; (3) silt fraction, 0.05–0.002 mm; (4) coarse clay, 2.0–0.2 μm; (5) fine clay, <0.2 μm. Tracor-Stone instrument, 3 mg of sample in Pt cups.

(0.50–0.25 mm), fine sand (0.25–0.10 mm), and very fine sand (0.10–0.05 mm). If small sample sizes are required the finer sand fractions are more suitable. When a relatively large amount of sample can be used, as is the case with DTA instruments equipped with well-type sample holders, it is immaterial whether coarse or fine sand is used, since particle size has little influence on the DTA curve of sand. As can be noticed from Fig. 7–6 (no. 1 and 2), the DTA curve of the composite sand fraction (2.0–0.05 mm) was identical to that of the fine sand. In both cases a strong endothermal peak of quartz at 573°C was the only characteristic of the curve. Analysis of the sand fraction by DTA is only of importance in investigation of primary minerals and/or Fe-Mn concretions. In soils of the southeastern USA and in other parts of continental USA, quartz is the dominant component of the sand fraction, as indicated earlier. Therefore, DTA of the sand fraction of these soils results in featureless DTA curves, except for the quartz peak.

7–3.1.3.4 Silt Fraction. The silt fraction of soils (0.05–0.002 mm) can be analyzed directly and usually yields curves showing more complexity than that of sand (Fig. 7–6, no. 3). Often the amount of sesquioxides and kaolinite in the silt size fraction is large enough to yield detectable endothermic peaks at 270 and 498°C, respectively, in addition to the quartz peak at 573°C. Generally the DTA curve of silt resembles that of whole soils. However, the peak intensities are less than in whole soil.

7–3.1.3.5 Clay Fraction. The clay fraction of soils ($< 2\ \mu m$) can be used directly or can be separated into coarse clay (2.0–0.20 μm) and fine clay ($< 0.2\ \mu m$) fractions. The choice of size fraction to be used depends on many factors, e.g., the purpose and objective of the study, the desired precision, and the type and quantity of minerals present. A number of investigators have preferred the use of clay separated into various fractions (Jackson, 1956), while for the analysis of amorphous minerals such as allophane, the use of fine clay has been suggested. However, the present authors found that for general purposes the clay fraction ($< 2\ \mu m$) gives results that are satisfactory for qualitative and quantitative interpretations, although it was noted that the finer clay fraction exhibited DTA curves with more intense peaks (Fig. 7–6, no. 4 and 5).

7–3.1.3.6 Organic Fraction. The application of DTA in the investigation of soil organic matter has received increasing research attention with the increased knowledge on extraction and purification of humic and fulvic acids and related compounds from soils. Isolation of humic fractions contained in soils is accomplished by extraction of soil samples with 0.1 *M* NaOH, using a soil/extractant ratio of 1:5. Humic acid is separated from fulvic acid by acidifying the extract with HCl to pH 1.5. For detailed extraction procedures reference is made to Tan (1975) and Tan and Clark (1969). The humic fractions obtained can be used directly in DTA or TG analysis, or they can be saturated first with different kinds of cations before analysis (Tan, 1978; Schnitzer & Kodama, 1972).

In general it can be observed that DTA is able to distinguish between humic and fulvic acid. Humic acid is characterized by a DTA curve with a strong exothermic reaction at 400°C, whereas fulvic acid exhibits a curve with an exothermic peak at 500°C (Fig. 7–7, no. 1 and 2). Indications have also been obtained by the present authors that different kinds of soils may contain humic acids with different kinds of DTA features. The main decomposition peak of humic acid at 400°C may shift to lower or higher temperatures depending on the different cations used for saturation (Fig. 7–7, no. 3—5). The cations Ca^{2+}, Ba^{2+}, Fe^{3+}, Cu^{2+}, and Mn^{2+} usually decrease the thermal stability of humic acid (Tan, 1978).

7–3.1.3.7 Cation Saturation. Saturation of the sample with a known cation prior to DTA has been proposed for clay or material with cation exchange properties. Clay in soils is expected to be saturated with a variety of cations (Na^+, K^+, Ca^{2+}, Mg^{2+}, Al^{3+}). Hydration properties of these cations, which vary considerably, markedly affect the results of DTA. Mackenzie (1970) recommended that for comparative work each sample should have received identical pretreatment. The authors have found

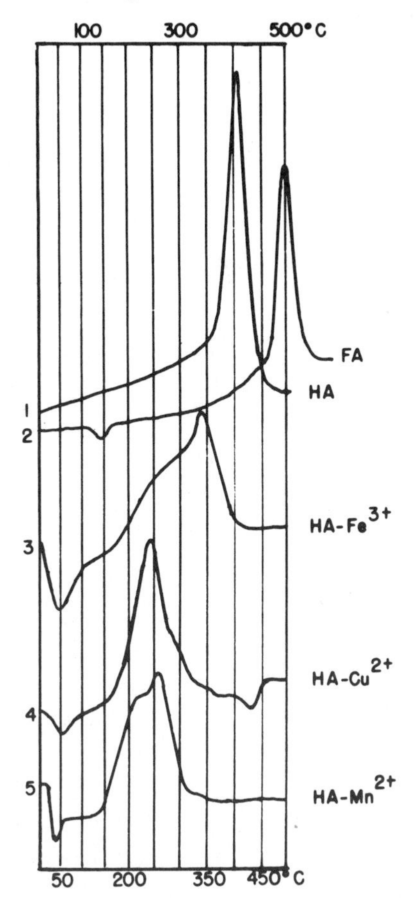

Fig. 7–7. DTA curves of soil humic fractions and metal-humic acid complexes: (1) humic acid (HA); (2) fulvic acid (FA); (3), (4) and (5) are iron-, copper- and manganese-humic acid complexes, respectively. Tracor-Stone instrument, 0.5 mg of sample in Pt cups (Tan, 1978).

that Ca^{2+} saturation of the samples is usually satisfactory. To determine in detail the effect of different cation saturation on DTA curves, kaolinite and bentonite were saturated with Na^{+}, Ca^{2+}, Al^{3+}, and H^{+} by shaking for 1 h with 0.1 *M* NaCl, $CaCl_2$, $AlCl_3$, and HCl solutions and allowing them to stand overnight. After washing with distilled water the samples were dried at 45°C, ground, and stored in a desiccator over $CaCl_2$. DTA curves of these clay samples are shown in Fig. 7–8 and 7–9. The curves in Fig. 7–8 show that different cations affected both the size and shape of endothermic and exothermic peaks for kaolinite. However, no change in peak temperature occurred. Calcium-kaolinite had the lowest peak intensity for both the 530°C and 1000°C peaks. Peak intensity increased from Na-kaolinite to H-kaolinite to Al-kaolinite (Fig. 7–8, no. 2–4). Saturation with Al^{3+} made the main endothermic peak at 530°C very sharp and slender and resulted in an additional strong endothermal peak at

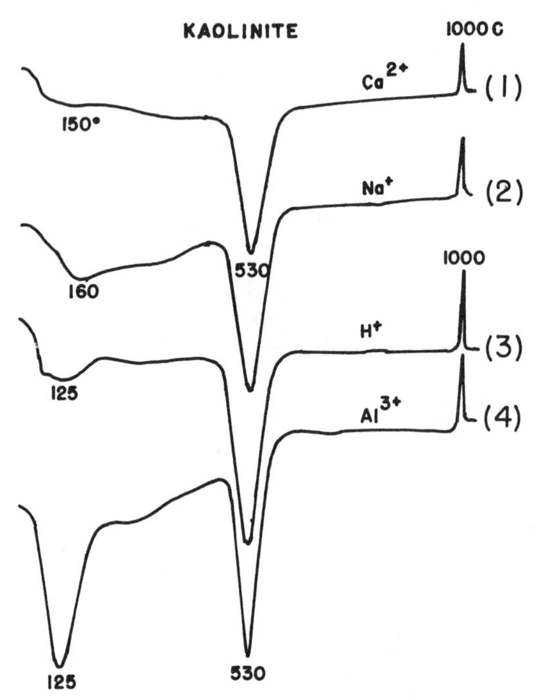

Fig. 7–8. DTA curves of kaolinite (no. 2, Birch Pit, Macon GA) saturated with different cations: (1) Ca-kaolinite; (2) Na-kaolinite; (3) H-kaolinite; (4) Al-kaolinite. Tracor-Stone instrument, 3 mg of sample in Pt cups.

125°C. The latter suggests that Al-kaolinite retains considerable amounts of water even after overnight storage in a desiccator over $CaCl_2$.

Saturation of bentonite with H^+ ions gave curves with a broad peak at 675°C and an S-shape curve at 950°C (Fig. 7–9, no. 1). Calcium-bentonite resulted in a curve with its main endothermic peak shifted to 700°C (Fig. 7–9, no. 2). The presence of two peaks at 75 and 140°C for Ca-bentonite is in agreement with reports in the literature for bentonite treated with Ca^{2+} (Mackenzie, 1970; Barshad, 1965). Curve no. 3 of Fig. 7–9 shows DTA features of a soil smectite for comparison. This sample was obtained from a Houston Black soil (Vertisol). The curve resembles that exhibited by Ca-bentonite, with the difference that the main endothermal peak at 695°C was very pronounced.

7.3.1.3.8 Hydration and Solvation. As discussed in the preceeding section, hydration of samples prior to DTA may result in changes in the low temperature endothermic peaks at 0 to 200°C. Soil colloids are very reactive due to large specific surface and electric charge and may adsorb considerable amounts of water. Moreover, the various types of soil colloids are known to have different capacities for adsorption of moisture; differences in these factors may result in different DTA curve features. Samples should be equilibrated at constant relative humidity to insure that amorphous material and 2:1 lattice-type clays exhibit low temper-

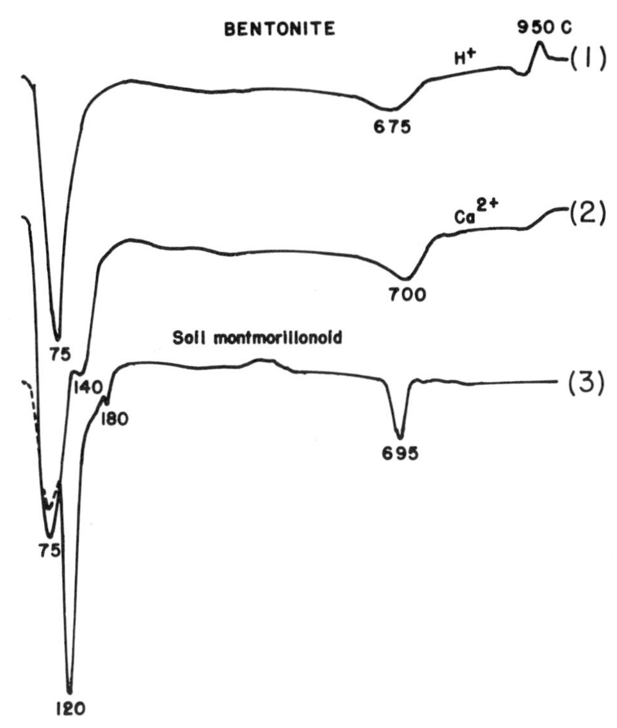

Fig. 7–9. DTA curves of montmorillonite (Oklahoma Geological Survey) saturated with different cations: (1) H-montmorillonite; (2) Ca-montmorillonite; (3) calcium-clay (< 2 μm) from a Houston Black soil. Tracor-Stone instrument, 3 mg of sample in Pt-cups.

ature endothermic curves that can be used for meaningful comparisons (Jackson, 1956). Equilibration is usually carried out in a desiccator over a compound developing a stable relative humidity. Compounds such as $Mg(NO_3)_2 \cdot 6H_2O$, $Mg(NO_3)_2$, and H_2SO_4 have been used (Barshad, 1965; Jackson, 1956). The authors have found that keeping samples overnight over $CaCl_2$ produces satisfactory results and will not lead to disappearance of characteristic low temperature endothermic peaks (Fig. 7–10, no. 3). Keeping samples for prolonged periods over $CaCl_2$ may, however, prove to be a disadvantage. Fig. 7–10 also shows DTA curves of Ca-kaolinite kept overnight over $CaCl_2$ and kept open in contact with air in the laboratory (air-dry). The curve of $CaCl_2$-equilibrated Ca-kaolinite shows practically no low temperature reaction, which is, of course, normal. However, air-dry Ca-kaolinite gave a DTA curve with a strong endothermic peak at 125°C (Fig. 7–10, no. 1 and 2).

7-3.1.4 SAMPLE SIZE

The size of sample to be analyzed is in many cases dictated by the type of DTA instrument available. Most instruments, especially those equipped with well-type sample holders, require sizeable amounts of sam-

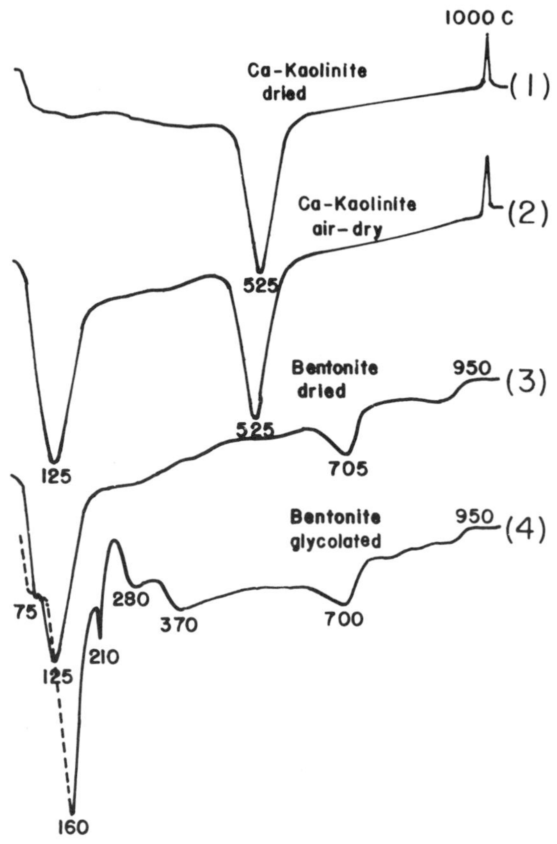

Fig. 7-10. DTA curves of: (1) Ca-kaolinite, dried overnight over $CaCl_2$; (2) air-dry Ca-kaolinite; (3) nontreated montmorillonite, stored overnight over $CaCl_2$; (4) montmorillonite, glycolated with ethylene-glycol. Tracor-Stone instrument, 3 mg of sample in Pt cups.

ple (> 100 mg) to fill the hole. Analysis of a large amount of sample yields extremely large peaks, which are difficult to record on a normal piece of paper. In this case dilution, by mixing the sample with an inactive material, becomes beneficial. Dilution is also suggested to reduce baseline drift and/or increase accuracy. For the merits of and objections to dilution, reference is made to Mackenzie (1970).

Instruments such as the Tracor-Stone apparatus equipped with ring-type thermocouples or the Dupont 990 Thermal Analyzer (plate-type, Al-pans) require very small samples. Although the amount required for an analysis will vary according to type and thermal characteristic of the material, the size is usually in the order of 1 to 10 mg; as little as 0.5 mg of sample can be used, especially in the analysis of soil humic compounds. These small samples approach the ideal postulated by Mackenzie (1970); i.e., the ideal sample should be an infinitely small sphere surrounding the thermocouple junction.

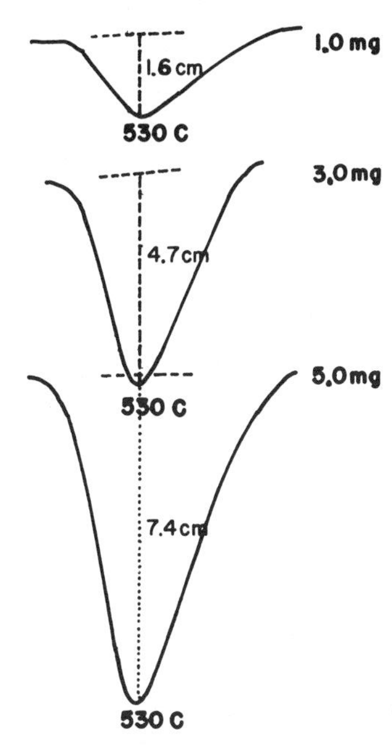

Fig. 7–11. Main endothermic peaks of kaolinite (Mesa Alta, New Mexico) obtained by analysis using different sample sizes: 1 mg, 3 mg, and 5 mg. Tracor-Stone instrument in Pt cups, heating rate of 15°C/min.

In qualitative analysis, it is not necessary to weigh the sample for DTA, although comparison of curves should be made with curves obtained from identical amounts of samples. However, in quantitative DTA analysis, the sample must be weighed accurately. As can be seen in Fig. 7–11, the main endothermic peak height (or area) of kaolinite increases proportionally with sample size.

7–3.1.5 SAMPLE PACKING.

Packing the sample is required when instruments with well-type sample holders are used. A number of methods have been proposed, including hand tapping the sample or the block and layer packing (Arens, 1951; Schultze, 1969; Smothers & Chiang, 1966; Mackenzie, 1970). In the latter, the sample is packed around the thermocouple junction as a sandwich between two layers of reference material. The sandwich method may yield peaks of approximately one-half the intensity of those obtained by filling the whole cavity with the sample.

Reproducibility in packing is of importance, since differences in packing may create differences in sample density, leading in turn to differences in heat conductivity and thermal diffusivity of the sample. The latter is a problem in the low temperature ranges, where heat transfer is mainly controlled by conduction. In the high temperature region, the effects of packing are less apparent, because heat transfer is mostly through radiation. Both reference and sample materials should be of similar aggregate size and should be packed to a similar bulk density in order to avoid base-line drift. In plate or ring-type instruments, where a small amount of sample is placed in an Al or Pt cup, packing effects are negligible. Minimum zero drift is usually attained by carefully balancing the relative amounts of reference and/or test sample in the cups.

7–3.1.6 FURNACE ATMOSPHERE

Some control of furnace atmosphere is desirable and often necessary. The furnace atmosphere affects DTA through one or a combination of the following reactions (Schultze, 1969): (i) a change in furnace atmosphere creates a change in partial pressure of the gas atmosphere inducing a shift of peak temperature, (ii) interactions occurring between gas used and gaseous products of the sample may change furnace atmosphere and obscure the appearance of characteristic DTA peaks, or (iii) peaks are enhanced because oxidation reactions can be eliminated. A number of methods, such as use of vacuum, static air, static gas, dynamic gas, gas-flow over the sample, gas-flow through the sample, and self-generated gas have been tested. For the advantages and disadvantages of one method over the other, reference is made to Mackenzie (1970). When the furnace is evacuated or when an atmosphere of nitrogen is introduced during an analysis, the curve of a sample containing organic matter does not have the exothermic break associated with its oxidation; hence, for qualitative DTA the elimination of organic matter is unnecessary. The use of a furnace similar to that employed for total carbon, but modified to the extent that the gases evolved during heating can be trapped at various temperature ranges, enables the measurement of the actual losses of water and CO_2 during DTA. The modification consists of attaching to the outlet of the furnace pyrex glass tubes arranged in series for trapping quantitatively the water and CO_2 evolved during heating.

For most analyses, it is sufficient to run DTA in the presence of a self-generated gas atmosphere. However, it is necessary to place a close-but loose-fitting lid on top of the sample holder to maintain uniform pressure and thermal conditions around sample and reference material. By using a lid, direct radiation of both the specimens by the furnace will also be avoided.

7–3.1.7 HEATING RATE

The heating rate is defined as the rate of temperature increase, expressed in degrees centigrade per minute. Correspondingly, the cooling rate is the rate of temperature decrease. The heating or cooling rate is constant when the temperature/time curve is linear (Mackenzie et al., 1972). The heating must be controlled at a uniform and steady rate through the analysis. Heating rates used vary from 0.1 to 200°C per minute. For

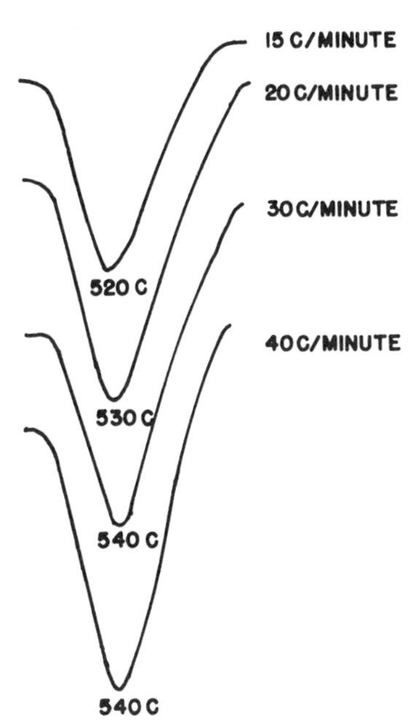

Fig. 7–12. Main endothermic peaks of 3 mg kaolinite (Mesa Alta, New Mexico) obtained by analysis using different heating rates: 15°C/min, 20°C/min, 30°C/min, and 40°C/min. Tracor-Stone instrument in Pt cups.

a review of the effects of slow and fast heating rates on DTA curves see Mackenzie (1970). For most purposes heating rates of 5 to 20°C/min are used.

Although many investigators have stressed the fact that heating rates affect peak height, peak width, peak temperature, and peak area, the authors have found that heating rate is seldom a serious problem. The differences in DTA curves obtained by different heating rates, as shown in Fig. 7–12, indicated that peak height increases gradually and consistently, when heating rates are increased from 15 to 20, 30, and 40°C/min. The most significant difference is obtained between heating at 15° and 40°C/min. Kaolinite heated at a rate of 15°C per minute yielded a broad and relatively shallower endothermic peak at 520°C. When heated at a rate of 40°C/min a higher and slender endothermic peak was obtained, with the peak temperature shifted to 540°C. The peak temperatures, at 520 or 540°C, are well within the range for kaolinite. Since the shape and size of endothermic peaks obtained by heating at 15 or 40°C/min are well within detection limits, the choice of a heating rate between 15 and 40°C/min makes little difference.

7–3.1.8 INTERPRETATION OF RESULTS

7–3.1.8.1 Qualitative Interpretation. Qualitative identification of soil minerals can be achieved by using the DTA curves as fingerprints and comparing or matching them with DTA curves of standard minerals or well-known established minerals. Although within a mineral species the DTA features may vary somewhat according to origin, each mineral exhibits thermal reactions, as reflected by endothermic and exothermic peak temperatures, within specific or well-defined limits (Fig. 7–13).

7–3.1.8.1.1 Kaolinite

DTA curves of kaolinite are generally characterized by a strong endothermic peak at 450 to 600°C and by a strong exothermic peak at 900 to 1000°C. The endothermic peak is caused by dehydroxylation, whereas the exothermic peak is attributed to formation of γ-alumina and/or mullite.

7–3.1.8.1.2 Halloysite

The curve of halloysite is similar to that of kaolinite; however, it has in addition a low temperature (100-200°C) endothermic peak of medium

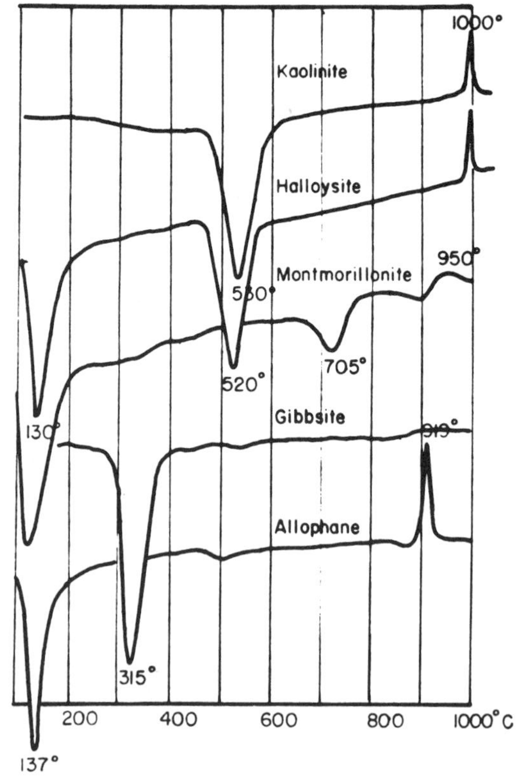

Fig. 7–13. DTA curves of selected major clay minerals: Kaolinite (Macon, GA); Halloysite (Bedford, Indiana); Montmorillonite (Oklahoma Geol. Survey); Gibbsite (Brazil); Allophane (< 2-μm fraction of soil extracted from a tropical Andept). The last was analyzed using 100 mg of sample and a manually operated instrument equipped with a well-holder; the others were analyzed with the Tracor-Stone employing 3 mg of sample in Pt cups.

to strong intensity for loss of adsorbed water. Some investigators are of the opinion that the shape of the main endothermic peak between 450 and 600°C can be used to distinguish between kaolinite and halloysite. This endothermic peak is symmetrical for kaolinite, whereas for halloysite it is asymmetrical. However, the present authors have noted cases in which endothermic peaks of kaolinite were also asymmetrical.

7-3.1.8.1.3 Montmorillonite

This mineral exhibits a DTA curve characterized by a low-temperature (100-200°C) endothermic peak, an endothermic peak between 600 and 700°C, followed by a weak exothermic peak between 900 and 1000°C.

7-3.1.8.1.4 Vermiculite

The DTA curve (not shown) of this mineral resembles closely that of montmorillonite, except for a stronger intensity of the endothermic break at 800 to 900°C and the absence of the main endothermic reaction between 600 and 700°C.

7-3.1.8.1.5 Gibbsite and Goethite

These minerals are usually characterized by DTA curves with only a strong endothermic peak between 290 and 350°C. Often goethite and the other iron oxide minerals have their endothermic peaks at higher temperature ranges than gibbsite. Differential dissolution is frequently applied to distinguish between gibbsite and goethite. The mineral mixture is shaken in 0.5 *M* NaOH for 4 h prior to DTA. This treatment will dissolve the gibbsite component, leaving goethite unaffected for DTA.

7-3.1.8.1.6 Allophane

Allophane exhibits DTA features with a strong low temperature (50–150°C) endothermic peak and a strong exothermic peak between 900 and 1000°C. The low temperature endothermic reaction is attributed to loss of adsorbed water, whereas the main exothermic peak is caused by γ-alumina formation.

A list of characteristic endothermic and exothermic peaks of major clay minerals is provided in Table 7–1.

7-3.1.8.2 Quantitative Interpretation. The use of DTA for quantitative determination of soil clays requires the samples to be weighed accurately, as indicated before. Particle size, packing of samples, and heating rate must be kept similar to maintain uniformity and reproducibility. The availability of automated and programmed control in many of the modern instruments has simplified quantitative analysis with DTA. Since the analysis is made on the basis of height, area, or width of the main endothermic peak, this procedure can be applied only to soil minerals yielding endothermic peaks of sizes sufficiently large to be measured

Table 7-1. DTA endothermic and exothermic peak temperatures of major clay minerals and the reactions causing the peaks (Mackenzie, 1970; Mackenzie and Caillere, 1975; Tan and Hajek, 1977).

Mineral	Endothermic peak temp.	Main reaction	Exothermic peak temp.	Main reaction
	°C		°C	
Kaolinite	500–600	Dehydroxylation	900–1000	γ-alumina formation
Dickite	500–700	Dehydroxylation	900–1000	γ-alumina formation
Nacrite	500–700	Dehydroxylation	900–1000	γ-alumina formation
Montmorillonite	100–250	Loss of adsorbed water	900–1000	recrystallization
	600–750	Dehydroxylation		
Beidellite	100–250	Loss of adsorbed water	900–1000	
	500–600	Dehydroxylation		
Nontronite	100–200	Loss of adsorbed water	900–1000	
	500	Dehydroxylation		
Vermiculite	150	Loss of adsorbed water	800–900	recrystallization
	850	Dehydroxylation		
Illite	100–200	Loss of adsorbed water	920–950	recrystallization
	600	Dehydroxylation		
	900–920	Dehydroxylation		
Chlorite	500–600	Dehydroxylation	800	
Halloysite	100–200	Loss of adsorbed water	900–1000	γ-alumina formation
	500–600	Dehydroxylation		
Gibbsite	250–350	Dehydroxylation		
Boehmite	570	Dehydroxylation		
Diaspore	400–500	Dehydroxylation		
Goethite	300–400	Dehydroxylation		
Quartz	573	α to β inversion		
Allophane	50–150	Loss of adsorbed water	800–900	γ-alumina formation

precisely and accurately. Kaolinite and the other 1:1 type of minerals, gibbsite and goethite, or their mixtures exhibit well-developed endothermic peaks which lend themselves to quantitative analysis. However, montmorillonite and vermiculite have poorly developed endothermic peaks, making a quantitative assessment of these minerals very difficult.

To correct for variations attributed to particle size and crystal imperfections, Jackson (1956) suggested the use of slope ratio and/or ratio of peak area to peak width. The slope ratio was defined as the ratio of slope of the low temperature to high temperature side of the peak, measured by the angles formed by lines running from the steepest tangent to the peak sides. This ratio was reported to reflect the particle size effect and contents of kaolinite and halloysite.

The present authors have successfully used peak heights of main

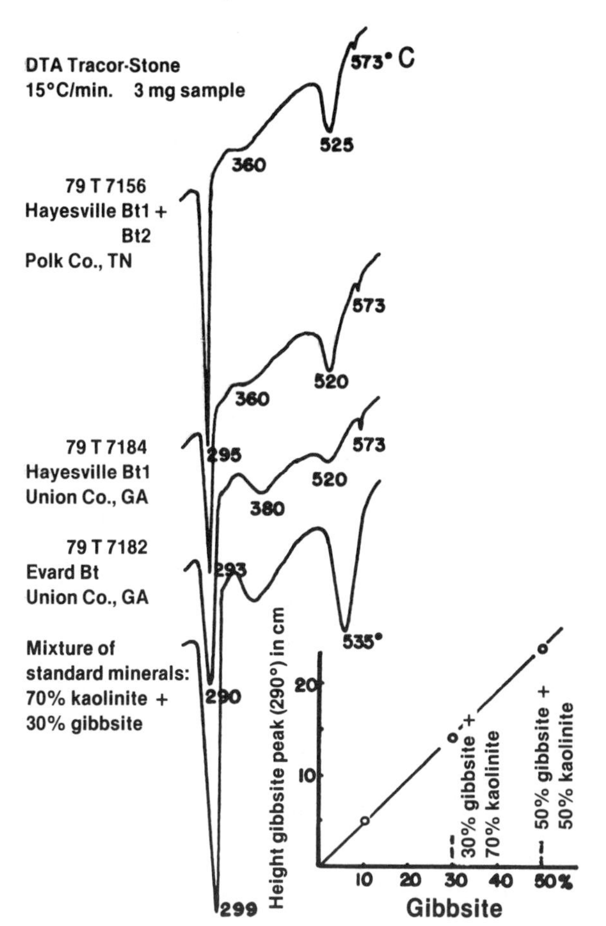

Fig. 7–14. Quantitative interpretation of DTA curves of clay fractions isolated from soil and of selected mixtures of reference (standard) kaolinite and gibbsite.

endothermic reactions for the quantitative determination of gibbsite and kaolinite present as a mixture in the clay fractions of soils. For this purpose a standard curve had to be constructed, from which the unknown content of the respective mineral could be extrapolated. Standard (reference) gibbsite and kaolinite minerals were ground and sieved to pass a 50-μm sieve. Mixtures made of (i) 90% kaolinite + 10% gibbsite, (ii) 70% kaolinite + 30% gibbsite, and (iii) 50% kaolinite + 50% gibbsite were analyzed by DTA. The height of the main endothermic peak of gibbsite at 290 to 299°C plotted against the gibbsite concentration yielded a perfect regression (Fig. 7–14). According to the authors the use of peak height is simpler than that of area. A standard curve for kaolinite can be constructed in a similar way employing the height of endothermic peaks of kaolinite at 520 to 525°C, with which kaolinite concentrations in soil clays can be estimated.

7-3.2 Differential Scanning Calorimetry

7-3.2.1 APPARATUS

There are two types of instruments of completely different design called "differential scanning calorimeter." Wendlandt (1974) describes these as (i) differential scanning calorimeters which are heat-flow-recording instruments, such as the Perkin Elmer instrument, and (ii) differential scanning calorimeters which are actually differential-temperature-recording instruments such as the Dupont and Stone instruments.[1]

Basically, a DSC instrument consists of two matching covered calorimeter cups to hold the sample and the reference material, respectively. Each cup has a heater and a thermocouple. A container of high thermal conductivity, holding the sample, is placed in the "sample" cup in good thermal contact with its surface, whereas an empty container is placed in the "reference" cup. The enthalpic change in the sample is measured with respect to the empty pan, usually serving as the reference material. The temperature sensors in each cup are controlled to read equal temperatures by apportioning the electric power to their heaters. The total electricity to the heater bridge is increased or decreased to keep the temperature of the sample and reference cup in correspondence with the programmed temperature.

7-3.2.2 REFERENCE, SAMPLE SIZE AND HEATING RATE

DSC curves (Fig. 7-15) are very similar to DTA curves. The area under the DSC curve is proportional to the enthalpy, and the ordinate

[1]Trade names are used for the convenience of the readers and do not imply endorsement or preference by the Soil Science Society of America or the American Society of Agronomy to the exclusion of other comparable products that might also be available.

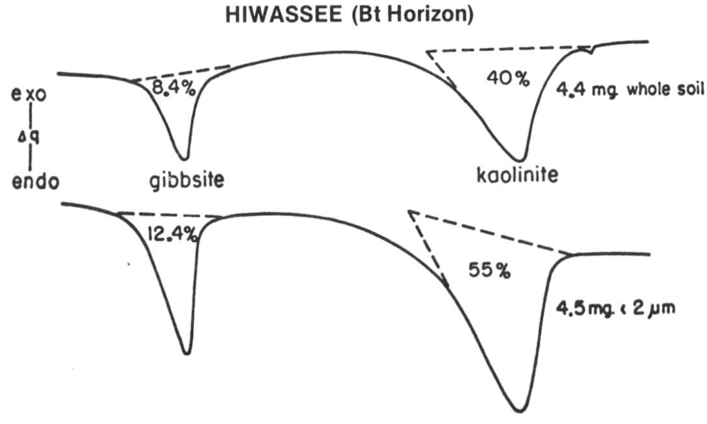

Fig. 7-15. DSC curves for whole soil and the clay fraction of soil from the Bt horizon of a Hiwassee loam. Gibbsite and kaolinite percents on a whole-soil basis from the <2-μm fraction are 8.5 and 38%, respectively. Dupont 990 Thermal Analyzer in Al pans.

value at any given temperature (or time) is directly proportional to the differential heat flow between a sample and reference material. Thus the ordinate is in terms of Δq (cal or mcal/in, equivalent to 0.0017 J/cm) recorded in relation to time since heating started.

Reference material, sample size and heating rate effects are similar to those discussed for DTA of soils and clays. The most commonly used references for DSC analysis are either Al_2O_3 or an empty sample pan. Empty Al-pan references have given satisfactory DSC curves on the DuPont 990 instrument. A baseline adjustment is required due to the mass-heat capacity difference between the sample and reference. For most quantitative determinations, 1- to 5-mg samples give DSC curves with peak areas that can be measured accurately. Larger samples result in some sensitivity loss because of heat conductivity effects of the sample. Heating rates of 10 to 20°C/min give satisfactory curves for soil minerals that can be quantitatively determined by DSC.

7–3.2.3 CALORIMETRIC AND QUANTITATIVE DETERMINATIONS

The determination of ΔH (heat of reaction) and the mass of the reacting part of the sample is a commonly used procedure for DSC. The exact equation used for ΔH depends on the instrument being used. The simple form is

$$\Delta H m = KA$$

in which m is the reactive mass, K is a constant, and A is the peak area (Wendlandt 1974, Mackenzie 1970, Barshad 1965). In the past the equation was further simplified for m by including ΔH with K, yielding,

$$m = KA\,.$$

A standard curve of m vs. A was then developed for comparison with unknowns.

Differential scanning calorimetry can be used in the same manner; however, ΔH or m is usually obtained by using the peak area and substituting into an equation of the form:

$$\Delta H - (A/m)(BE\Delta qs)$$

in which ΔH, A, and m are as defined previously, B is the chart speed in cm/min or mm/s, E is the cell calibration coefficient, and Δqs is the instrument sensitivity setting in cal/s per inch (=1.68 J/s per cm). The cell coefficient E is somewhat dependent on temperature for the DuPont DSC; consequently E must be determined as a function of temperature by analyzing samples of known ΔH and solving for E. Only a single temperature calibration is required for instruments such as the Perkin-Elmer (Wendlandt 1974). To find A (peak area) a line is drawn from the point where the thermogram departs from the baseline to the point where it returns. The area is measured with a planimeter for best results.

As pointed out by Mackenzie (1970) for DTA, quantitative DSC is most useful for soils that have undergone intensive weathering. Usually kaolinite, gibbsite, and hydrated iron minerals dominate the clay fraction, and quartz is predominant in the sand. Fig. 7–15 shows whole soil and clay fraction DSC curves for the Bt horizon of a Hiwassee soil. Soils in this series are clayey and representative of highly weathered soils in the southeastern USA. The points of departure and return to baseline were used for gibbsite determination. However because of the presence of highly chloritized vermiculite, the area for kaolinite was drawn to exclude the endothermic effect of interlayer hydroxyl loss. When the mineral quantities are compared on a whole soil basis the difference between clay and whole soil are well within the limits of error. For routine analysis, if minerals can be measured quantitatively by DSC and the soils contain > 35% clay, fractionation is seldom necessary. Iron removal may sometimes be required (Mackenzie 1970). The quartz α-β inversion peak is easily excluded from the peak area of kaolinite. In many soils montmorillonite will give endothermic peaks in the kaolinite region, thus preventing quantitative determination of kaolinite. In addition, some highly hydroxy interlayered vermiculite clays can interfere.

Fig. 7–16 shows DSC and the derivation curve (ΔT sample/ΔT) of untreated air-dry whole soils. The low-temperature peaks allow positive identification of montmorillonite. Kaolinite can be identified but cannot be estimated quantitatively.

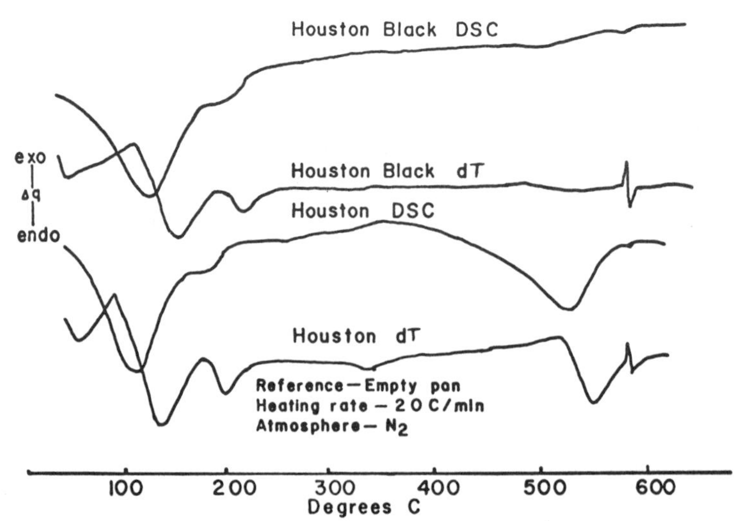

Fig. 7–16. DSC and derivative dry curves for whole soil samples of the AC horizon of a Houston clay (Alabama) and A horizon of a Houston Black clay (Texas). Dupont 990 Thermal Analyzer in Al pans.

7–3.3 Thermogravimetry

7–3.3.1 APPARATUS

The basic components of a thermogravimetric instrument also consists of a furnace, a temperature programmer, and a recording balance called the thermobalance. The thermobalance is an essential part for weighing the sample continuously during the analysis as a function of temperature or time. Manually operated and automatic recording thermobalances are currently available. A comprehensive coverage of theory, instruments, and application of TG is given by Wendlandt (1974). The current availability of TG instruments with sensitive solid state automatic recording capability and low mass cooling furnaces have made TG a rapid, accurate, and relatively simple analytical technique.

Since DTA or DSC alone is often not adequate for analysis of clay minerals, TG and derivative thermogravimetry are valuable complementary techniques (Mackenzie, 1970). In soil studies TG is essential for the determination of adsorbed water and crystal-lattice water. The exact temperature at which soil minerals lose water varies. In addition, sample pretreatment and heating rate will affect the weight-loss region. The latter variables may or may not be serious, depending on the objectives of the study, type of minerals present, and capability of the thermal instruments used. The temperatures of dehydroxylation and desorption and theoretical weight losses are given by Jackson (1956) and Barshad (1965) for most minerals found in soils.

7–3.3.2 SAMPLE PREPARATION AND SAMPLE SIZE

As with DTA and DSC analysis, untreated whole soils can often be used for qualitative and semi-quantitative analysis. The whole-soil curves for the highly weathered Hiwassee soil (Fig. 7–17) and the two Vertisols in Fig. 7–18 (Houston from Alabama and Houston Black from Texas) were obtained by grinding the air-dry whole soil to pass a 140-mesh sieve 106 μm) and using a 10-mg sample for analysis. The clay fraction of Hiwassee was obtained by fractionation methods given by Jackson (1956). All analyses were made at a heating rate of 10°C/min in an N_2 gas atmosphere.

The TG curve for the Hiwassee soil confirms the quantitative DSC analysis for gibbsite and kaolinite for both the whole soil and clay fraction. No quantitative attempt was made for the Vertisols. Montmorillonite, kaolinite, and calcite can be identified in the Houston soil. Montmorillonite and calcite can be identified in the Houston Black; however, additional analysis of the clay fraction would be required to confirm the presence of kaolinite.

7–3.3.3 COMMENTS

To determine the temperature range over which adsorbed water is completely eliminated and the loss of crystal-lattice water begins, it is

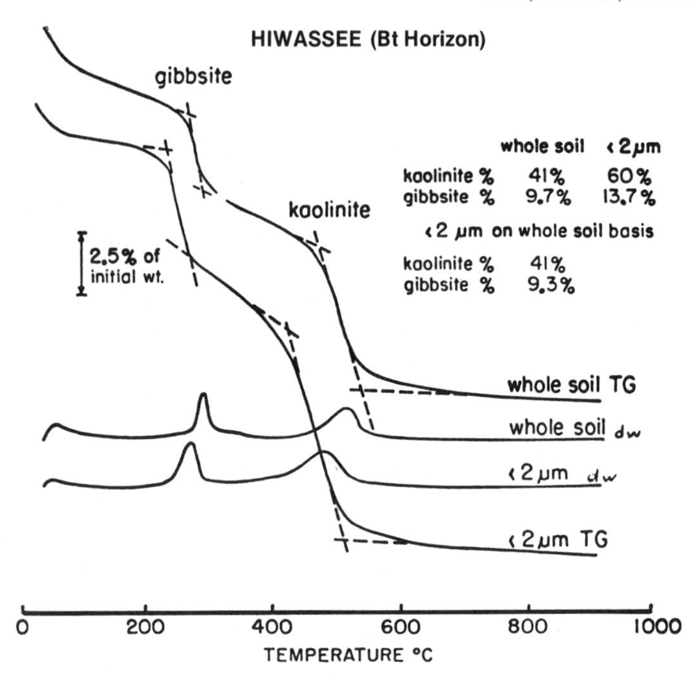

Fig. 7–17. TG and derivative curves for whole soil and the clay fraction of soil from the Bt horizon of a Hiwassee loam. Dupont 990 Thermal Analyzer.

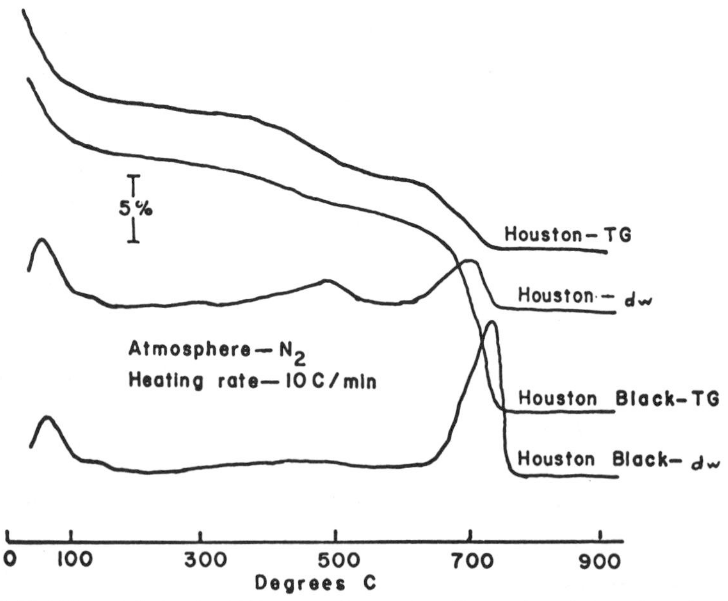

Fig. 7–18. TG and derivative TG curves for whole soil samples of the AC horizon of a Houston clay (Alabama) and A horizon of a Houston Black clay (Texas). Dupont 990 Thermal Analyzer.

necessary to heat individual samples of clay to constant weight at successively increasing temperatures. The temperature intervals should be of the order of 25 to 50°C in the critical range. From such an analysis of pure clay species, it is possible to determine this range by comparing the accumulated water losses with the theoretical crystal-lattice water that should be present.

To calculate the theoretical amount of crystal-lattice water of a given pure species, it is necessary first to determine the chemical composition of the species and then to calculate its formula and molecular weight on a water-free basis. The theoretical number of moles of water is then added to the formula, and the composition is recalculated on a basis of percentage by weight, either on the formula weight with water or without water (on the ignited basis). For determining the temperature range above which all crystal-lattice water is removed, the ignited basis is the more convenient; this is the basis used in Table 7–2.

Two significant facts are brought out by this table: (i) that the crystal-lattice water for all the important silicate clay minerals is lost by heating between about 150 or 350 and 1000°C, and (ii) that the clay minerals can be divided into two distinct groups on the basis of the total amount of crystal-lattice water, namely, the 2:1 minerals and the 1:1 and 2:2 minerals. Among the 2:1 minerals, i.e., montmorillonites, vermiculites, and micas, the water content lies in the range of 4.2 to 5.1%. Among the 1:1 and 2:2 minerals, it lies in the range of 13.4 to 16.2%. It may readily be seen, therefore, that on the basis of only a crystal-lattice water analysis, i.e., total water loss by heating between 150 or 350 and 1000°C, it is possible to determine approximately the relative content of 2:1 minerals and of 1:1 or 2:2 minerals in a clay sample. If an x-ray analysis and a chemical analysis are available also, it is possible to determine the amount of each mineral species present.

Although it is well established that the major portion of the crystal-lattice water for the various minerals is lost at different temperature ranges, particularly as indicated by differential thermal analysis, there is a sufficient overlap among them to preclude the use of thermogravimetric analysis in a particular temperature range for a quantitative determination of a single mineral.

The determination of water content of clay samples in a paste form at onset of gelation and in the air-dry state at about 50% relative humidity is also useful for differentiating among the different clay mineral species. At onset of gelation, depending on the exchangeable ion, the montmorillonites take up 5 to 30 g of water per gram of clay, whereas the kaolinite and illite minerals take up only 0.5 to 1.5 g of water per gram of clay (Barshad, 1955). Large differences exist also in the air-dry state, where montmorillonites adsorb 20 to 30 times more water per gram of clay than do kaolinites.

From the position and size of the endothermic peaks in DTA curves, it may be inferred that the adsorbed water and the crystal-lattice water

Table 7-2. Adsorbed and crystal-lattice water of clay minerals and the temperature at completion of desorption, and at start and completion of dehydroxylation (Barshad, 1965).

Clay mineral	Adsorbed water†		Crystal-lattice water	Temperature at completion of desorption		Temperature at start of dehydroxylation	Temperature at completion of dehydroxylation
	Layer water	Cavity water		Layer water	Cavity water		
	%‡			°C			
Ca Montmorillonite	20.16	3.37	5.08	250	370	370	1000
Na Montmorillonite§	14.00	0.00	5.05	150	--	150	1000
Ca Vermiculite§	20.00	3.08	4.80	250	700	250	1000
Na Vermiculite	10.14	4.66	4.76	150	700	150	1000
Ca Illite§	5.16	1.06	4.97	150	370	370	1000
Na Illite	3.45	0.66	5.07	150	370	370	1000
Muscovite§	1.00	0.45	4.74	250	350	370	1000
Phlogopite§	0.4–1.0		4.51	350		350	1000
Biotite§	0.4–1.5		4.17	350		350	1000
Talc§	0.7–1.0		4.99	350		350	1000
Pyrophylite§	0.5–1.5		5.26	350		350	1000
Kaolinite¶	0.2–1.2		16.20	350		350	1000
Halloysite (hydrated)	13.2		16.20	250		350	1000
Chrysotile and antigorite	1.0–2.0		14.95	250		350	1000
Clinochlore#	0.2–0.5		14.81	250		370	1000
Ripidolite	0.2–0.5		13.39	250		370	1000
Sheridinite	0.2–0.5		14.81	250		370	1000
Gibbsite	0.5–0.6		52.95	100		150	350
Goethite	0.0–0.1		33.80	100		200	370
Brucite	0.5–0.6		44.70	170		200	370

† See text for meaning of terms. ‡ On the ignited basis. § 2:1 minerals. ¶ 1:1 minerals. # 2:2 minerals.

require different amounts of energy for their separation from the solid phase and that several energy levels exist for both the adsorbed and the crystal-lattice water (Barshad, 1952, 1955, 1960).

If the calorie equivalent of a given endothermic reaction is known from DTA and the quantity of water released is known from TGA (when both analyses are carried out under comparable conditions), the activation energy for the release of water in a particular endothermic reaction may be determined. Measurements of the quantity of water released may be made by means of an automatic recording balance (Chevenard et al., 1944; Merveilli & Boureille, 1950; Mauer, 1954).

A careful comparison of DTA curves with TG and DTG curves of identical samples which were heated at approximately the same rate reveals that endothermic peaks in DTA curves correspond to relatively large water losses at a rapid rate and that small water losses at slow rates are not recorded at all on DTA curves. Therefore, in studying heats involved in desorption of water from clays by DTA, it is necessary to determine only the water losses between the initial and final temperatures of the desired endothermic peak in DTA curves. The water losses in the temperature region between the endothermic breaks should be ignored in such a study.

The rate of heating during TG analysis and the temperature at which weight losses are measured, when automatically recording balances are not available, depend on the purpose for which the analysis is made. For studying heats of reactions in conjunction with DTA, the rate of heating should be the same as during DTA, and the measurements of loss in weight should be made at the initial and final temperature of each endothermic break. For the purpose of differentiating between adsorbed and crystal-lattice water, however, the heating should be at a rate such that the adsorbed water is lost before the loss in crystal-lattice water begins. For this purpose, heating samples at known fixed temperatures for periods between 3 and 12 h is more practical than is heating them at a continuously rising temperature. The number of weight-loss measurements can thus be reduced to a small number. For Na-saturated samples, these measurements can be made at temperatures of 150, 370, and 1000°C.

The best condition of a sample to be analyzed depends also upon the purpose of the analysis. A Na-saturated sample freed of organic matter is most suitable for differentiating between adsorbed and crystal-lattice water, whereas a Mg- or Ca-saturated sample is more suitable for differentiating between the 2:1 and 1:1 minerals on the basis of adsorbed water.

7–4 REFERENCES

Arens, P. L. 1951. A study on the differential thermal analysis of clays and clay minerals. Proefschrift Landbouw Hogeschool, Wageningen. Excelsiors drukkerij, s'Gravenhage.

Barshad, I. 1952. Temperature and heat of reaction calibration of the differential thermal analysis apparatus. Am. Mineral. 37:667–694.

Barshad, I. 1955. Adsorptive and swelling properties of clay-water system. Proc. Natl. Conf. Clays Clay Technol. 1:70–77.

Barshad, I. 1960. Thermodynamics of water adsorption and desorption on montmorillonite. Clays Clay Miner. 8:84–101.

Barshad, I. 1965. Thermal analysis techniques for mineral identification and mineralogical composition. *In* C. A. Black (ed) Methods of soil analysis, part 1. Agronomy 9:699–742.

Brennan, W. P. 1974. Application of differential scanning calorimetry for the study of phase transitions. p. 103–117. *In* R. S. Porter and J. F. Johnson (ed.) Analytical calorimetry. Plenum Press, New York.

Chevenard, P., X. Wache, and R. de La Tullaye. 1944. Etude de la corrosion seche des metaux au moyen d'une thermobalance. Bull. Soc. Chim. France, Ser. 5, 11:41–47.

Flynn, J. H. 1974. Theory of differential scanning calorimetry.—Coupling of electronic and thermal steps. p. 17–32. *In* R. S. Porter and J. F. Johnson (ed.) Analytical calorimetry. Plenum Press, New York.

Hendricks, S. B., and L. T. Alexander. 1940. Semiquantitative estimation of montmorillonite and clays. Soil Sci. Soc. Am. Proc. 5:95–99.

Jackson, M. L. 1956. Soil chemical analysis—advanced course. Dept. Soils, Univ. Wisconsin, Madison, WI.

McAdie, H. G. 1969. Progress towards thermal analysis standards: a report from the Committee on Standardization, International Confederation for Thermal Analysis. *In* R. F. Schwenker and Gran (ed.) Thermal Analysis. Academic Press, London. Vol. 1:693–706.

Mackenzie, R. C. 1969. Nomenclature in thermal analysis. Talanta 16:1227–1230.

Mackenzie, R. C. (ed). 1970. Differential thermal analysis. Academic Press, London and New York.

Mackenzie, R. C. 1972. How is an acceptable nomenclature system achieved. J. Thermal Anal. 4:215–221.

Mackenzie, R. C., and S. Caillere. 1975. The thermal characteristics of soil minerals and the use of these characteristics in the qualitative and quantitative determination of clay minerals in soils. p. 529–571. *In* J. E. Gieseking (ed.) Soil components. Vol. 2. Inorganic components. Springer-Verlag, New York.

Mackenzie, R. C., C. J. Keattoh, D. Dollimore, J. A. Forrester, A. A. Hodgson, and J. P. Redfern 1972. Nomenclature in thermal analysis—II. Talanta 19:1079–1081.

Martin, R. T. 1955. Reference chlorite characterization for chlorite identification in soil clays. Clays Clay Miner. 3:117–145.

Matejka, J. 1922. Thermal analysis as a means of detecting kaolinite in soils. Chem. Listy 16:8–14.

Mauer, F. A. 1954. An analytical balance for recording rapid changes in weight. Rev. Sci. Inst. 25:598–602.

Merveilli, J., and A. Boureille. 1950. Identifications de argiles ceramique par la thermobalance. Bull. Soc. France Ceram. 7:18–27.

Murphy, C. B. 1962. Differential thermal analysis. A review of fundamental developments in analysis. Anal. Chem., Review Issue 34:298R–301R.

Redfern, J. P. 1970. Complementary methods. p. 123–158. *In* R. C. Mackenzie (ed.) Differential thermal analysis. Academic Press, New York.

Russell, M. B., and J. I. Haddock. 1940. The identification of clay minerals in five Iowa soils by the thermal method. Soil Sci. Soc. Am. Proc. 5:90–94.

Saito, H. 1969. Progress of thermobalance and thermobalance analysis in Japan. p. 11–24. *In* R. F. Schwenker, Jr. and P. D. Garn (ed.) Thermal analysis, Vol. 1. Instrumentation, organic materials and polymers. Academic Press, New York.

Schnitzer, M., and H. Kodama. 1972. Differential thermal analysis of metal-fulvic acid salts and complexes. Geoderma 7:93–103.

Schultze, D. 1969. Differential thermo-analyses. Verlag Chemie. GmbH. Weinheim/Bergstr.

Skinner, H. A. 1969. Theory, scope, and accuracy of calorimetric measurements. p. 1–32. *In* H. D. Brown (ed.) Biochemical microcalorimetry. Academic Press, New York.

Smothers, W. J., and Yao Chiang. 1966. Handbook of differential thermal analysis. Chem. Publ. Co., Inc., New York.

Tan, K. H. 1975. Infrared adsorption similarities between hymatomelanic acid and methylated humic acid. Soil Sci. Soc. Am. Proc. 39:70–73.

Tan, K. H. 1978. Formation of metal-humic acid complexes by titration and their characterization by differential thermal analysis and infrared spectroscopy. Soil Biol. Biochem. 10:123–129.

Tan, K. H., and F. E. Clark. 1969. Polysaccharide constituents in fulvic and humic acids extracted from soil. Geoderma 2:245–255.

Tan, K. H., and B. F. Hajek. 1977. Thermal analysis of soils. p. 865–884. *In* J. B. Dixon et al. (ed.) Minerals in soils environments. Soil Sci. Soc. Am., Madison, WI.

Wendlandt, W. W. 1974. Thermal methods of analysis. Part 1. Intersci. Publ., John Wiley & Sons, New York.

8 Petrographic Microscope Techniques

JOHN G. CADY

Johns Hopkins University
Baltimore, Maryland

L. P. WILDING AND L. R. DREES

Texas A&M University
College Station, Texas

8-1 GENERAL INTRODUCTION

The microscope was first used in soil studies to make an inventory of the minerals found in a variety of soils. Early examples of this approach are in publications by McCaughey and Fry (1913, 1914). Among the objectives of the early work were determination of source of parent materials and estimation of potential fertility reserves. Most of the work of this period was geological-mineralogical and was not applied to solution of soil development problems.

Kubiena's text, *Micropedology*, and work by his associates at Iowa State University introduced U.S. workers to micromorphology (Perterson, 1937; Johnston and Peterson, 1941). Shortly thereafter Frei and Cline (1949), Nikiforoff et al. (1948), and Cady (1950) applied microfabric analyses to pedogenesis and mineral transformation in soils and parent bedrocks. Micromorphology as applied in the USA is a tool—not a discipline. It is used as an ancillary technique to support other morphological, physical, chemical, mineralogical, and biological methods. In this country, micromorphology is in its infancy and hasn't reached the discipline status it has in Europe and Australia. Brewer (1976), Kubiena (1970), and FitzPatrick (1980) reviewed recent developments in applying micromorphology to pedogenic studies. Its applications in developing diagnostic criteria for *Soil Taxonomy* (Soil Survey Staff, 1975) and its historical evolution in the USA have recently been reviewed by Wilding and Flach (1985).

The microscope is a simple, direct observational tool; the user sees the features without intermediate calculations or inferences. However, the observations must be analyzed and their meanings synthesized and interpreted; hence, the results are affected by the bias, experience, skill, and aptitude of the observer. The petrographic microscope may be used in soils investigations to determine (i) identity, size, shape, and condition

 Methods of Soil Analysis, Part 1. Physical and Mineralogical Methods—Agronomy Monograph no. 9 (2nd Edition)

of single grains and mineral aggregates in the silt and sand size range, and (ii) the distribution and interrelations of constituents of the soil in thin sections. Microscopic observations are often valuable when used to interpret data obtained by other physical and mineralogical methods.

The petographic microscope differs in two ways from the basic or biological microscope: it has devices for polarizing light, one below the condenser and one above the objective; and it has a rotating stage, graduated in degrees, for measuring angles. Other special refinements exist, but the foregoing two are the essential ones. Descriptions of petrographic microscopes and instructions for their use are in many standard mineralogy texts, including those by Kerr (1977), Wahlstrom (1979), and Phillips (1971).

The purpose of this chapter is not to repeat information and instructions easily found in standard texts on optical mineralogy and petrography, particularly sedimentary petrography, but to describe methods, minerals, and structures that are peculiar to soils, based on the writers' experience in the application of microscopy to soil genesis and classification.

8–2 GRAINS

8–2.1 Introduction

Single grains of sand and silt size may be identified and described by examination under a petrographic microscope. Knowledge of the nature and condition of the minerals in these fractions provides information on the source of the parent material; on the presence of lithological discontinuities or overlays in the solum or between the solum and the underlying material; and on the degree of weathering in the soil as a key to its history, genetic processes, and possible fertility reserve. Examples of applications of such studies of sand or silt can be found in publications by Haseman and Marshall (1945), Marshall and Jeffries (1945), Ruhe (1956), Cady (1940), Willman et al. (1963), Hunter (1967), Khangarot, et al. (1971), and Brewer (1976).

Many of the procedures used for preparation and identification are identical with those used in sedimentary petrography (Milner, 1962; Krumbein and Pettijohn, 1938). Modifications of these procedures that apply to soils consider sampling and pretreatment prior to fractionation analysis.

The first step in the study of grains is the separation of the desired size fraction from the remaining material. Procedures for dispersion and separation are given in chapter 15. Combined with this step, or subsequent to it, is the cleaning of the mineral grains, for which procedures are given in chapter 5. It is sometimes desirable to examine soil separates before rigorous cleaning treatments are applied, because aggregates, concretions, partly weathered mineral grains, and other types of grains that are worth attention and study may be removed by the cleaning treatments.

All of the fractions from coarse sand to fine silt can be studied, but the most suitable ones are those in the middle part of this size range. There are usually too few grains of coarse sand on a slide to constitute a good sample, and it is difficult to observe optical properties on large grains; the grains in the fine silt range may be too small for observation of all their important optical properties. Smithson (1961) published some special instructions for studying the silt fraction.

Selection of the sample depends on the purpose of the analysis. For most work, such as checks on discontinuities or estimation of degree of weathering in different horizons, one or two of the size fractions that make up a relatively large weight percentage of the soil are selected. Examination of more than one size fraction is necessary in such problems as checking for an admixture of wind-blown material in dominantly coarse-textured residuum, and studying weathering where the process might cause minerals formerly dominant in one size class to shift to a smaller one.

The sample must be well mixed because the subsample on the slide is small. If a sample of sand is in a beaker or a vial, for example, shaking or jarring may cause heavy grains to settle and platy or prismatic grains to accumulate toward the top. Stirring with a small flat-bladed implement will usually mix the sample sufficiently. Steel needles or spatulas should be avoided because they will attract the magnetic minerals. Small sample-splitting devices are available that effectively subdivide sand and silt samples down to amounts suitable for single slides or replicates.

A sample of the whole fraction should be examined first; and then, if the nature of the material and requirements of the problem justify it, the heavy and light minerals can be separated. With some types of specimen it is advantageous to separate and weigh the magnetic fraction, either before or after the heavy liquid separation. Wrapping a thin sheet of flexible plastic around the magnet facilitates making this separation quantitative. Such separations can be done on either dry material or dispersed suspensions of silt and clay.

8–2.2 Heavy-liquid Separations

Heavy minerals are often indicative of provenance, weatheirng intensities, and parent material uniformity. To facilitate the study of this important suite of minerals, concentration by specific gravity separations using a suitable heavy liquid is required. Heavy minerals will sink in the heavy liquid while light minerals float. The heavy mineral suite is made up of those mineral grains with a specific gravity greater than about 2.90. Because of the differences in the specific gravities of separating liquids, no single value is universally accepted as defining the heavy minerals. The sand and silt fractions of most soils, however, are dominated by light minerals such as quartz and (or) feldspars. The heavy minerals exhibit a wide range in weatherability and mineralogy, but commonly comprise < 1% of the grains.

Common liquids for heavy mineral separation include bromoform (sp gr 2.89), tetrabromoethane (sp gr 2.95), and methylene iodide (sp gr 3.33). These liquids may be diluted with other solvents such as acetone, ethanol, dimethylsulphoxide (DMSO), or nitrobenzene to produce specific gravity ranges for concentration of other minerals such as mica, feldspars, or opal. A light liquid (sp gr <2.3) has been useful for concentrating plant opal and sponge spicules (Wilding et al., 1977). Such a light liquid can also be used to concentrate volcanic ash.

Separations in the sand fraction can usually be carried out by gravity alone in a separatory funnel or cylinder. A lower stopcock allows removal of the heavy minerals by gravity flow. Finer-size fractions require centrifugation. In a centrifuge tube, the lower portion may be frozen in a freezer (Matelski, 1951), with solid carbon dioxide (Fessenden, 1959), or liquid nitrogen (Scull, 1960), while the lighter portion is decanted. For quantitative separation of light and heavy minerals, the separation techniques should be repeated several times. For grain sizes < 5 μm, light-heavy mineral separations are difficult or impractical because of particle interaction and aggregation. A surfactant may be used to help prevent this problem (Henderson et al., 1972). More detailed descriptions of procedures and apparatus are given by Carver (1971) and Mitchell (1975) as well as by standard petrology texts.

Heavy-liquid separations are most effective on well-cleaned grains. Grain coatings or organic matter may act to bind smaller grains together to form larger aggregates. Grain coatings may also cause slight alterations in the specific gravity of some grains. This may be significant if these grains are near the specific gravity of the heavy liquid.

Precautions should be exercised in using heavy liquids and diluting solvents, many of which are toxic and flammable (Hauff & Airey, 1980). DMSO is especially hazardous, as it serves as a carrier for other toxic substances. Contact with skin should always be avoided and the work area should be well ventilated.

8–2.3 Slide Preparation

With a microspatula, enough sample is taken so that an area about 22 mm square on a slide can be covered uniformly without having individual grains touch each other. A few drops of water containing a little alcohol to reduce surface tension are added, and the grains are spread uniformly in the liquid with a pointed nonmagnetic instrument. After the slide has dried, the spacing of the grains can be checked with a microscope, and, if satisfactory, the mounting medium is applied. If this procedure is followed, the grains will lie in one plane and will not drift out when the cover glass is applied. They can be fixed more securely if a little gelatin or gum arabic is added to the water.

A number of media are available for permanent grain mounts. Canada balsam is the traditional material and still retains many advantages. Its refractive index (RI) of 1.54 is close to that of quartz, and this aids in distinguishing quartz from other colorless minerals, especially feldspars. Epoxy resins are now available in the refractive index range of 1.53 to 1.56. Piperine (RI 1.86) and Hydrax (RI 1.7) are often used in mounting heavy minerals; their refractive index is close to that of many of the common heavy minerals, and this facilitates their identification. Directions for slide preparation using some of the above resins are given by Swift (1971) as well as other standard reference works.

Permanent mounts are necessary where the same slide is needed for several purposes; they can be kept as records, and they are almost essential for percentage analysis by counting. There are, however, several advantages in the use of immersion liquids of known refractive index Swift, 1971; Fleischer et al., 1984). The refractive indexes of minerals can be determined exactly, and identification is aided in other ways as well. Permanent mounts may have all the prismatic or platy grains in a preferred-orientation position. In an oil a given grain can be moved into different orientations by moving the cover slip. The optical properties of anisotropic minerals vary with crystallographic directions, and so it is often a valuable aid in identification to see the same grain in different positions and to observe these variations.

8–2.4 Grain Analyses

8–2.4.1 GENERAL PROCEDURE

The first step is to survey the slide with a low-power objective to define the grain assemblage and to make a rough estimate of the relative abundance of the minerals and other grains present. By becoming familiar with the minerals first, whether they are identified or not, observers can categorize minerals as they are seen. This serves two purposes: it avoids wasting time going through identifying criteria, and it enables one to appraise mineral properties as seen in grains of different shapes and sizes in different positions. The most abundant minerals should be identified first. These will probably be the easiest to identify, and their elimination will decrease the number of possibilities to consider when the difficult ones are to be attacked. Furthermore, there are certain likely and unlikely assemblages of minerals, and awareness of the overall types present gives clues to the minor species that may be expected. Practical working procedures for identifying soil mineral grains were outlined by Fry (1933), Kerr (1977), FitzPatrick (1980), and Fleischer et al. (1984).

In actual practice, minerals are identified by a combination of familiarity with a few striking features and a process of elimination . If one sees a dog with very short legs, it is either a dachshund, a basset hound, or a Welsh corgi. If the short-legged dog is solid colored, it is a dachshund; if spotted, it is a basset hound; and if its ears stand up, a Welsh corgi. It is not necessary to make a number of observations on the length of the hair or the configuration of the teeth. At least 80% of the sand and silt grains in soil are identified like this, by a combination of a few distinctive features. Unfortunately, some nontypical specimens are usually present. Minerals and other grains not identifiable as specific minerals are common in many soils and may be important. They should be accurately described even if they can not be identified.

A mineralogical analysis of a sand or silt fraction may be entirely qualitative, or it may be quantitative to different degrees. For many purposes a list of minerals is sufficient information. It is easy to accompany such a list with an estimate of relative abundance. A crude scale, such as one based on numbers from 1 to 10, can be used to express the amounts. Presence, absence, scarcity, or abundance of certain minerals or mineral groups can sometimes confirm the source of the soil parent material, the presence of overlays, and the reserve of weatherable minerals.

To detect more subtle distinctions among samples, analysis is based on a count of grains, from which a volume percentage can be obtained. Weight percentages then can be calculated using specific gravity, and various useful ratios can be calculated from count percentages.

The counting procedure and number of grains counted depend on the requirements of the job, the number and proportions of minerals present, and the distribution on the slide. Uniform coverage of the whole area of the mount is important because grains may be segregated by shape, size or density in spite of care taken in slide preparation. Because of the natural variations in soils and because of the opportunity for sampling error by the time a heavy mineral concentrate is mounted on a slide, differences in amounts of minerals must be large and consistent to be interpreted with confidence. Mineral count percentages should usually be reported in whole numbers only, with no more than two significant figures.

Counting of numbers of grains of individual species can proceed on arbitrarily or regularly spaced traverses. If the grains are large or sparsely distributed, all grains can be counted. Various sampling methods may be used where only a portion of the grains is to be counted. In work on 20- to 50-μm fractions with well-populated slides, all grains in individual fields evenly spaced over the slide in a preset grid pattern may be counted. Another method is to count all the grains lying within an arbitrarily selected quadrant of the field of vision as the slide is moved past the objective. Still another method is to count all the grains touched by the cross-hair intersection in a continuous traverse.

If only a few species are present, identification of 100 to 300 grains will provide a good approximation of composition. As the number of species increases, the count should increase, within limits of practicability. It is rarely necessary to count more than 1000 grains, however; in most work, 500 to 600 is a more usual number. A multi-unit laboratory counter can be used to tally the most abundant species without having to take one's eyes from the microscope. Some of these counters sum the count and ring whenever 100 counts have been accumulated. Noting the composition of the first 100 will provide an idea of the number of grains that should be counted to give a good sample. Discussions of the statistics of analysis of mineral separates by counting may be found in the books by Krumbein and Pettijohn (1938), Milner (1962), and Brewer (1976).

The detail of the analysis can be adjusted to fit the need. In some instances only the amount or percentage of one mineral is of interest. In others, the ratio between two minerals is needed as an indicator of source. In still other cases, ratios between certain known weatherable minerals and known resistant minerals are indicative of weathering or age. For purposes other than a complete enumeration, several minor species of uncertain identify should not cause undue concern about the validity of the count on which interest is centered. In addition to identity and amount of the different grains in the sand and silt fractions, it is often important to record their morphology and condition. Evidence of wear or abrasion and evidence of chemical alteration or weathering is the most frequently sought information. Wear during transportation shows as rounding, especially in chemically resistant minerals that do not have good cleavage. Quartz, zircon, and rutile are good mineral species to examine for evidence of mechanical abrasion. Easily weathered minerals can be rounded by solution; apatite, for example, is often found in ovoid grains.

In connection with observations of rounding, it should be noted that a grain may have a round outline but still be a flat plate. If a truly rounded grain is observed in crossed polarized light, the interference colors, which can be read like the contour lines on a map, will rise smoothly without steps or interruption from low order at the periphery to high order in the thickest part. Weathering can have several manifestations ranging from slight bleaching of color, or slight lowering of refractive index, to replacement of one mineral by another or complete removal of a species. Effects depend upon the chemical composition, crystal structure, and habit of the mineral and upon the environment. Corrosion or solution results in etching and pitting of surfaces. Minerals with pronounced cleavage or a fibrous or columnar habit are usually attacked most along these planes of weakness. Hornblende, for example, appears to weather most readily at ends of the columnar grains and in a direction parallel to the long axis. The ends of the grains become forked and pinnacled, and pits in the sides are elongated with the length. Garnet is isometric and corrosion is random. Decomposition of feldspars follows cleavage and twinning planes.

Weathering can produce coatings of clay or mixed oxides, create open channels which may be filled with clay, iron oxide, or gibbsite, or completely alter the mineral to another mineral with little change in form. Observations on weathering in single grains are best made in two or more mounting media. A medium that closely matches the refractive index of the grain makes the interior of the grain visible and tends to expose contrasting coatings. A medium having a refractive index a few hundredths of a unit away from that of the grain will show the condition of the grain surface.

Resistance to both dissolution and alteration to secondary minerals varies greatly. Some minerals, magnetite for example, may be resistant in reducing environments and easily weathered in oxidizing environments. Such differences in weathering can exist between horizons in a soil profile. Lists of minerals arranged in order of resistance are given in many publications and are valuable guides; but, like all generalizations, they must be used with caution. Observations on weathering will be discussed further in section 8–3.5.

8–2.4.2 MINERALS

8– 2.4.2.1 Criteria Used in Identification. Properties important in grain identification are listed below in approximate order of ease and convenience of determination. Often estimates of several or even two or three of these properties will allow identification of a grain; therefore detailed or extremely accurate measurements are seldom necessary. In the finer soil separates, grains may be either too small or improperly positioned to permit measurement of some properties such as optic angle or optic sign. It is helpful to crush, sieve, and mount a set of known minerals for practice in estimating properties and for standards to compare with unknowns.

Refractive index can be estimated by relief or determined accurately by use of calibrated immersion liquids. When relief is used to estimate refractive index, allowance must be made for grain shape, color, and surface texture. Thin platy grains may be estimated low; colored grains and grains with rough, hackly surface texture may be estimated high. Estimation is aided by comparing unknown with nearby known minerals.

Birefringence, the difference between highest and lowest refractive index of the mineral, is estimated by interference color (see the chart in Phillips, 1971), taking into account grain thickness and orientation. Several grains of the same species must be observed because they may not all lie in positions that show the extremes of refractive index. Mica, for example, has high birefringence, but the refractive indexes of the two crystallographic directions in the plane of the plates are very close together, so that the birefringence appears low when the plate is perpendicular to the microscope axis. The carbonate minerals have extremely high birefringence (0.17 to 0.24), most of the ferromagnesian minerals are intermediate (0.015 to 0.08), orthoclase feldspar is low (0.008), and apatite is very low (0.005).

Color aids in discriminating among the heavy minerals. Pleochroism, the change in color or light absorption with stage rotation when one polarizer is in, is a good diagnostic characteristic for many colored minerals. Tourmaline, biotite, and hornblende are examples of pleochroic minerals.

Shape, cleavage, and *crystal form* are characteristic or unique for many minerals. Cleavage may be reflected in the external form of the grain, or may appear as cracks within it showing as regularly repeated straight parallel lines or sets of lines intersecting at definite repeated angles. The crystal shape may be quite different from the cleavage-fragment shape. Plagioclase feldspars, kyanite, and the pyroxenes have strong cleavage. Zircon and rutile usually appear in crystal forms.

Extinction angle and *character of extinction* observed in crossed polarized light are valuable criteria for some groups. The grain must show its cleavage or crystal form for extinction angles to be measured, and the angle may be different along different crystallographic axes. Some minerals have sharp, quick total extinction; in others extinction is more gradual; and in some minerals with high light dispersion, a dimming and change of interference color takes place at the extinction position.

Optic sign, optic angle and *sign of elongation* are useful, sometimes essential, determinations but are often difficult to make unless grains are large or in favorable orientation. To determine optic sign, grains that show dim, low-order interference colors or no extinction must be sought. Grains with bright colors and sharp, quick extinction will rarely give usable interference figures.

8–2.4.2.2 Useful Differentiating Criteria for Particular Species. The following are the outstanding diagnostic characteristics of the most common minerals and single-particle grains found in the sand and silt fractions of soils. The refractive indexes given are the intermediate values. If these minerals and the ones in section 8–2.4.3.2 can be learned, it is safe to say that one can identify over 80% of the grains in most soil mineral assemblages.

Quartz has irregular shapes. The refractive index, 1.54, is close to that of balsam. The interference colors are of low order but are bright and warm. There is sharp extinction within a small angle of rotation ("*blink* or *wavy* extinction"). Crystal forms are sometimes observed and usually indicate derivation from limestone, or other low-temperature secondary origin.

Feldspars: Orthoclase may resemble quartz, but the refractive index is about 1.52 (just below that of quartz), birefringence is lower, and the mineral may show cleavage. *Microcline* has a refractive index of 1.53, and twinning intergrowth produces a plaid or grid effect in crossed polarized light. The refractive indexes in the *plagioclase* group increase with increasing proportion of calcium. The refractive index of albite, 1.53, is below that of quartz; the refractive index of anorthite, 1.58, is noticeably above. Plagioclase feldspars almost always show a type of twinning that appears as alternating dark and light bands in crossed polarized light. Cleavage is good. Lath and prismatic shapes are common.

Mica occurs as platy grains that often are very thin. The plate view shows very low birefringence; the edge view, very high birefringence. Plates are commonly equidimensional and may appear as hexagons or may have some 60° angles. *Biotite* is green to dark brown. Paler colors, lowering of refractive index, and distortion of extinction and interference figure indicate weathering to *hydrobiotite* or *vermiculite. Muscovite* is colorless and has a moderate refractive index (about 1.59 in the plate view). If the identification is in doubt, it is desirable to use an oil mount, so that the grains can be seen from different angles.

Amphiboles are fibrous to platy or prismatic minerals with parallel to slightly inclined extinction. Color and refractive index increase as the iron content increases. They have good cleavage at angles of about 56 and 124°. Refractive index in the group ranges from 1.61 to 1.73. *Hornblende* is the most common; it is slightly pleochroic, has refractive index close to Piperine, and usually has a distinctive color close to olive-green. It is often used as an indicator of weathering.

Pyroxenes: Enstatite and *hypersthene* are prismatic and have parallel extinction; hypersthene has unique and striking green-pink pleochroism. *Augite* and *diopside* have good cleavage at angles close to 90° and large extinction angles; colors are usually shades of green. Refractive indexes of the pyroxenes are in a somewhat higher range than the amphiboles (1.65 to 1.79).

Olivine is colorless to very pale green, is usually irregular in shape (weak cleavage), has vivid, warm interference colors, and has a refractive index close to that of Piperine. It is an easily weathered mineral and may have cracks filled with serpentine or seams or crack fillings of goethite.

Staurolite is pleochroic yellow to pale brown; it sometimes contains holes, giving a "Swiss cheese" effect. Its refractive index is about 1.74. The grains usually have a foggy or milky appearance, possibly caused by colloidal inclusions.

Epidote is a common heavy mineral, but the forms occurring in soils may be hard to identify positively. Typical epidote, with its high refractive index (1.72 to 1.76), strong birefringence, and pleochroism that includes the pistachio-green color, is unmistakable. However, epidote is modified by weathering or metamorphism to colorless forms with lower birefringence and lower refractive index; and, furthermore, close relatives of epidote, *zoisite* and *clinozoisite*, are more abundant than some of the literature indicates. These minerals of the epidote group commonly appear as colorless, pale-green, or bluish-green, irregularly shaped or roughly platy grains with high (1.70 to 1.73) refractive index. Most show anomalous interference colors (bright pale blue) and no complete extinction. They can be confused with several other minerals such as kyanite and diopside. Identification usually depends on establishing properties on many grains.

Kyanite is common but seldom abundant. Its pale blue color, platy, angular cleavage flakes, large cleavage angles, and large extinction angles can usually be observed and make it easy to identify.

Sillimanite and *andalusite* are two fibrous to prismatic minerals with straight extinction that resemble each other; however, their sign of elongation is different; and sillimanite is colorless, but andalusite commonly has a pink color.

Garnet grains are irregularly shaped and equidimensional. They are isotropic and have high refractive index (1.77 and higher). Garnet of the size of fine sand and silt is often colorless; pale pink colors are diagnostic in larger grains.

Tourmaline has a refractive index close to Piperine. Prismatic shape and strong pleochroism are characteristic; some tourmaline is almost opaque when at right angles to the vibration plane of the polarizer.

Zircon occurs as tetragonal prisms with pyramidal ends, has very high refractive index (>1.9), straight extinction, and bright, strong interference colors. Broken and rounded crystals are found frequently. Zircon crystals and grains are almost always clear and fresh-appearing.

Sphene in some forms resembles zircon, but the crystal forms have oblique extinction; and the common form, a rounded or subrounded grain, has with crossed polarizers a color change through ultra blue instead of extinction because of its high dispersion. It is the only pale-colored or colorless high-index mineral that gives this effect. The refractive index is slightly lower than that of zircon, and the grains are often cloudy or rough-surfaced.

Rutile grains have prismatic shape. The refractive index and birefringence are extremely high (2.6 and 0.29). The interference colors are usually obscured by the brown, reddish-brown or yellow colors of the mineral. Other TiO_2 minerals, *anatase* and *brookite*, also have very high refractive indexes and brown colors and may be difficult to distinguish in small grains. The latter two usually occur as tabular or equidimensional grains.

Apatite is fairly common in youthful soil materials. It has a refractive index slightly below that of Piperine (1.63) and very low birefringence. Crystal shapes are common and may appear as prisms; rounding by solution produces ovoid forms. It is easily attacked by acid and may be lost in pretreatments.

Carbonates: Calcite, dolomite, and *siderite,* in their typical rhombohedral cleavage forms, are easy to identify by their extremely high birefringence. In soils they have other forms—scales and chips, cementing material in aggregates, microcrystalline coatings, and other fine-grained masses often mixed with clay and other minerals. The extreme birefringence always is the clue to identification; it is shown by the bright colors in crossed polarized light and by marked change in relief when the stage is rotated with one polarizer in. The three can be distinguished by refractive index measurements; siderite is the only one with both indexes above balsam.

Gypsum occurs in platy or prismatic flat grains with refractive index about the same as orthoclase.

Opaque minerals, of which *magnetite* and *ilmenite* are the most common, are difficult to identify, especially when they are worn by transportation or otherwise affected by weathering. Observations on color and luster by reflected light, aided by crystal form if visible, are the best procedures. Magnetic separations will help confirm the presence of magnetite and ilmenite.

Many grains that appear opaque in plane polarized light become translucent if viewed in strong crossed polarized light. Most such grains are altered or are aggregates, rather than opaque minerals.

8–2.4.3 MICROCRYSTALLINE AGGREGATES AND AMORPHOUS SUBSTANCES

8–2.4.3.1 Criteria Useful in Identification. Most microcrystalline aggregates have one striking characteristic feature: they show birefringence but do not have definite, sharp, complete extinction in crossed polarized light. Extinction may occur as dark bands that sweep through the grain or parts of the grain when the stage is turned, or it may occur in patches of irregular size and shape. In all positions, some part of the grain is bright except in a few types of grains such as well-oriented mineral pseudomorphs and certain clay-skin fragments. Aggregates and altered grains should be examined with a variety of combinations of illumination and magnification in both plane and crossed polarized light. The principal properties that can be used to identify or at least characterize aggregates are given below.

Color, if brown to bright red, is usually related to iron content and oxidation of the iron. Manganese and organic matter may contribute black and grayish-brown colors.

Refractive index is directly related to density. Elemental composition, atom packing, water content, and porosity all influence refractive index.

Strength of birefringence is a clue to identity of the minerals. Even though the individual units of the aggregate are small, birefringence can be estimated by interference color and brightness.

Morphology may provide clues to the composition or origin of the aggregate. Some aggregates are pseudomorphs after primary mineral grains, and characteristics of the original minerals such as cleavage traces, twinning, or crystal form can still be observed. Morphology can sometimes be observed in completely altered grains—even in volcanic ash shards and basalt fragments. Other morphological characteristics may be observable in the individual units or overall structure; for example, the units may be plates or needles, or there may be banding.

8–2.4.3.2 Useful Differentiating Criteria for Particular Species. For purposes of studies of soil genesis, the aggregates in sand and silt fractions are not of equal significance. Some are nuisances, but must be accounted for, and others are particles with important diagnostic value. Useful differentiating criteria for some of the commonly found types of aggregates are given below.

Rock fragments (Lithorelicts) include chips of shale, slate, schist, and fine-grained igneous rocks such as rhyolite. Identification depends on recognition of structure and individual components and on consideration of possible sources.

Clay aggregates occur in a wide variety of forms. Silt and sand bound together into larger grains by a nearly isotropic brownish material usually indicate faulty dispersion. Clay skins may resist dispersion and consequently may appear as fragments in grain mounts. Such fragments are usually brown or red and translucent, with wavy extinction bands. Care may be needed to distinguish them from weathered biotite. Clay aggregates may be mineral pseudomorphs. Kaolinite pseudomorphs after feldspar are common, and montmorillonite aggregates, pseudomorphic after basic rock minerals, have been observed. Montmorillonite in this form shows high birefringence, and its extinction is mottled or patchy on a small scale. Coarse kaolinite flakes, books, and vermicular aggregates resist dispersion and may be abundant in sand and silt; these particles may resemble muscovite, but they are cloudy, show no definite extinction, and have very low birefringence.

Volcanic glass is isotropic and has a low refractive index—lower than most of the silicate minerals, ranging from 1.48 in the colorless siliceous glasses to as high as 1.56 in the green or brown glasses of basalt composition. Shapes vary, but elongated, curved shard forms, often with bubbles, are fairly common. This glassy material may be observed sticking to other minerals, and particles may contain small crystals of feldspar or incipient crystals with needle and dendritic forms. The basic glasses weather easily, and so the colorless siliceous types are more common in soils.

Allophane is present in many soils derived from volcanic ash. It can seldom be identified directly, but its presence can be inferred when sand and silt are cemented into aggregates by isotropic material with low refractive index, especially if volcanic ash shards are also present.

Opal, an isotropic material, occurs as a cementing material and in separate grains, some of which are of organic origin (plant opal, sponge spicules, diatoms). Its refractive index is very low (<1.45, which is lower than the value for volcanic ash). Identification may depend in part on form and occurrence.

Iron oxides may occur separately or as coatings, cementing agents, and mixtures with other minerals. They impart brown and red colors and raise the refractive index in the mixtures. *Goethite* is yellow to bright red. The refractive index and birefringence are higher in the red varieties, which seem to be better crystallized, often having a prismatic or fibrous habit. Aggregates have parallel extinction. In oriented aggregates, the interference colors often have a greenish cast. *Hematite* has higher refractive index than goethite and is granular rather than prismatic. Large grains of hematite are nearly opaque.

Gibbsite often occurs as separate, pure, crystal aggregates, either alone or inside altered mineral grains. The grains may appear to be well-crystallized single crystals, but close inspection in crossed polarized light shows patchy, banded extinction, indicating intergrown aggregates. It is colorless, and the refractive index (1.56 to 1.58) and birefringence are higher than the values for quartz. The bright interference colors and aggregate extinction are characteristic.

Chert occurs as aggregate grains with patchy extinction. The refractive index is slightly lower than that of quartz, and the birefringence is lower than that of gibbsite. It sometimes occurs in pseudomorphs after fossils and sometimes in grains with the exterior form of quartz crystals.

Glauconite occurs in the form of an aggregate of small micaceous grains with high birefringence. When fresh, it is dark green and almost opaque, but it weathers to brown and more translucent forms. It is difficult to identify on optical evidence alone.

TiO_2 aggregates have been tentatively identified in the heavy mineral separates of many soils. These bodies have an extremely high refractive index and high birefringence like rutile, and the yellow to gray colors are similar to those of anatase. They are granular and rough-surfaced. This habit of growth, with its little spurs and projections, suggest that TiO_2 aggregates may be secondary.

8-3 THIN SECTIONS

8-3.1 Introduction

"A crushed or pulverized soil is related to the soil formed by nature like a pile of debris to a demolished building" (Kubiena, 1938). The architecture of a building can no more be determined from a pile of rubble than the structure or site-specific composition of a soil from a crushed bulk sample. Micropedology (micromorphology) may be defined as the science that studies microfabrics of soils in their natural undisturbed arrangement. Examination of thin sections with a polarizing light microscope can be considered an extension of field morphological studies. More recently this technique has been coupled with submicroscopy methods (Bisdom, 1981) and contact microradiography (Drees & Wilding, 1983). Originally, the primary thrust of micropedology was to further understanding of pedogenesis (Kubiena, 1938, 1970; Parfenova & Yarilova, 1965), but recent work also relates microfabrics to applied soil considerations such as soil strength and structural faulting (Morgenstern & Tchalenko, 1967; McCormack & Wilding, 1974; Crampton, 1974; Douglas, 1980; Low et al., 1982; Collins & McGown, 1983), influence of ped cutanic surfaces on root-soil interaction (Soileau et al., 1964; Khalifa & Buol, 1968; Miller & Wilding, 1972; Gerber et al., 1974; Blevins et al., 1970), and quantitative measurement of soil structure and porosity (Jongerius et al., 1972; Bullock & Murphy, 1980).

The results of micromorphological studies are most useful when they are combined with all other available field and laboratory information. Such information includes complete description of the profile sampled, including study of structure and special features with a hand lens. Levels of resolution should increase progressively from the visible range to field

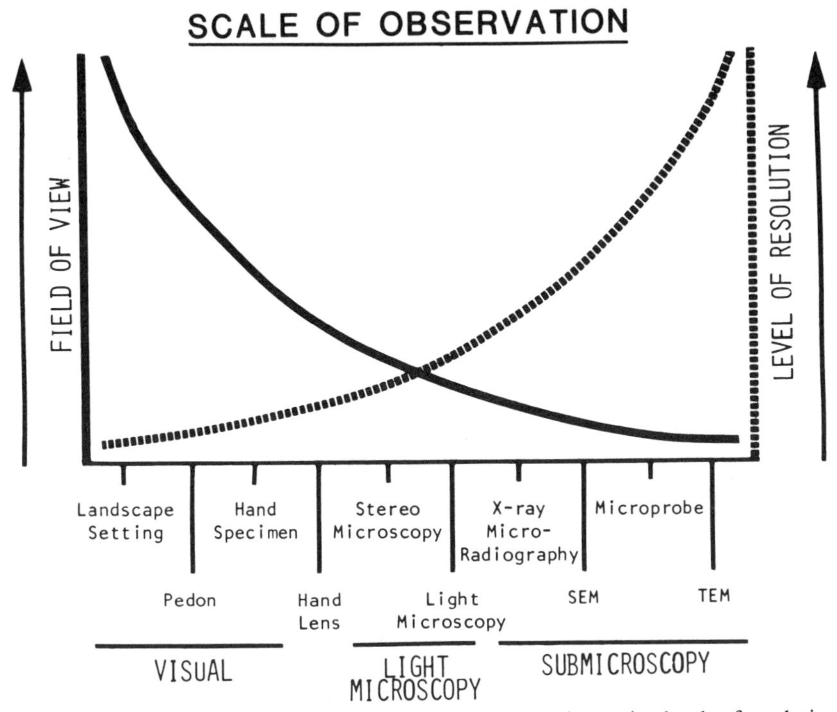

Fig. 8–1. Schematic illustration of the relationship between increasing levels of resolution and the area of the field under view.

hand lens, to steroscopic microscopy (Fig. 8–1). The microscopist should take the samples personally, so he or she will know how the detailed view relates to the whole object and will know what the problems and hypotheses are. One takes for granted that such field study includes notes on parent material, geomorphic position and history, macroclimate and microclimate, past vegetation, and present land use. Some of this information may not be available and may itself be the object of the investigation. Micromorphology has been used to look for effects of past vegetation and to identify layered or transported parent materials, for example.

For the purposes of communicating information about objects and features observed in thin section, a nomenclature specific to soils has been developed. The purpose of microfabric terminology is to provide a shorthand whereby a single term imparts a mental image of the microscopic field. It is an attempt to avoid long descriptions of microfabrics. A major problem is the difficulty in separating descriptive observation from genetic interpretation. Brewer's (1976) terminology is descriptively based without genetic inference. This clarifies communication because a number of microfabric features may form by different genetic processes (e.g., argillans). The terminology developed by Brewer was adapted from earlier terms of Kubiena (1938) and from the geologic literature. Most workers in the USA have adopted the terminology of Brewer (1976) for microfabric descriptions. It is more complete and systematic than terminology by Kubiena (1938). A brief listing of nomenclature terms commonly used in describing thin sections is given as a glossary at the end of this chapter and is taken from Brewer (1976).

Microfabric features can also be recorded by taking photomicrographs and/or sketches of characteristic elements of the fabric. Examples of this approach are given by Grossman et al. (1959), Alexander and Cady (1962), Flach et al. (1969), and Brewer (1976). A reference scale should be placed directly on the photomicrograph and the polarizing light mode should be noted in the caption.

8–3.2 Sampling

Samples may be natural clods, cores, or any other blocks or units of soil volume that can be collected without serious physical disturbance. Moist, fine-textured, cohesive structured units remain intact, but may part to smaller structural aggregates upon drying. Aluminum foil or plastic sheeting can be wrapped around the samples to avoid this problem in transporting samples to the laboratory. For loose, noncohesive, coarse-textured materials, a perforated can may be used to carry the samples. It may be gently pressed into the soil to be sampled; excess soil is trimmed off, and the can capped for transit. Upon reaching the laboratory, the sample may be dried and impregnated prior to removal from the container. Kubiena boxes have been widely used in Europe and Australia for sample collection (FitzPatrick, 1980). An alternative technique for soils that are organic and incoherent with an abundance of roots or a high percentage of coarse skeletal materials is to freeze a soil block with liquid N_2 and transport it to the laboratory in a frozen state (Blevins et al., 1968). Some investigators (Brasher et al., 1966; Murphy et al., 1981) coat delicate or fragile samples in the field with paraffin or synthetic resin (such as liquid Saran) before shipping to the laboratory. One hazard of this approach is possible incompatibility between the coat and the impregnating agent if the coat is not completely removed. Also, fabric disruption of the soil may occur during removal of the coat prior to impregnation. The vertical orientation should always be marked directly on the sample if possible; it may also be marked on the sample container or with tags as appropriate.

8–3.3 Impregnation

Most soil samples are not sufficiently indurated to permit thin section preparation without fabric alteration or disruption. Thus, soil samples must be impregnated with some suitable material that will harden, not alter the soil fabric due to volume changes, remain clear and isotropic in polarized light, and have an acceptable refractive index.

Several impregnating media meet the above requirements. Polyester resins have been, and continue to be, widely used. These resins may be diluted with monomeric styrene or acetone to lower their viscosity and improve impregnation. Styrene will mix with most polyester resins and not affect hardening. Acetone, however, must be allowed to volatilize before hardening can be completed. The hardening time can be controlled by adjusting the quantity of organic oxidant catalyst. Heating at 70 to 80°C completes the hardening process.

More recently, epoxy resins have been successfully used for soil impregnation. Materials such as Scotchcast No. 3 (Trade name for 3M epoxy) have low viscosity, refractive index of 1.53, short polymerization time, and high thermal stability. The high thermal stability is important for electron microprobe analysis.

For impregnating with any of these materials, samples must be dry. For best results, the impregnating solution is slowly added to the sample under vacuum. After impregnation the sample may be allowed to harden at atmospheric conditions or under a pressure of several atmospheres. The pressure aids in forcing the impregnating solution into unfilled voids. Procedures and media for impregnating with polyester resins are given by Bourbeau and Berger (1947), Buol and Fadness (1961), Jongerius and Heintzberger (1963), Cent and Brewer (1971) and Ashley (1973). Methods for epoxy resin impregnation are outlined by Hagni (1966), Sinkankas (1968), Innis and Pluth (1970), and Middleton and Kraus (1980).

Organic soil materials usually contain abundant amounts of water and shrink upon air- or oven-drying. Unless samples are freeze-dried, techniques which require a dry sample are not suitable because of possible natural fabric alteration. Moist organic soil materials have been successfully impregnated with Carbowax 6000 (Union Carbide Chemical Co.) (Mackenzie & Dawson, 1961; Langton & Lee, 1965). Carbowax 6000 melts at about 55°C and is soluble in water. Clods are submerged in Carbowax for several days to a week at 60 to 65°C to allow the Carbowax to replace the water. Carbowax has the disadvantage of being quite soft and slightly birefringent. Care must be exercised during cutting and grinding. An alternate method for moist samples is to slowly replace the water with acetone, then impregnate with a polyester resin (Miedema et al., 1974). These wet sample methods can also be used for soils that contain large amounts of allophane.

Each soil behaves somewhat differently in the various impregnating media; in actual practice each worker must develop his or her own satisfactory impregnating procedure. In general, soils of medium and coarse textures and those containing predominantly kaolinite clays are easy to impregnate. Medium- and fine-textured samples that contain swelling clays are more difficult to impregnate and require special care at all stages. No medium or procedure will work with all soil samples; in many respects, impregnation remains more of an art than a science.

8-3.4 Cutting and Finishing

If the sample can be successfully impregnated, the rest of the process of cutting, polishing, mounting, and finishing is essentially the same as that used by geologists for preparing rock sections. A complete set of general instructions for making sections of a variety of materials is given in articles by Reed and Mergner (1953) and Ireland (1971). Sometimes reimpregnation of surfaces with one of the epoxy or polyester cements is needed after cutting. Samples containing smectite often must be cut, ground, and polished in the absence of water—either dry, or in kerosene or mineral oil.

Making thin sections is an art that requires some practice and patience. Essentially it consists of the following steps after impregnation.

1. A rectangular block or slab is cut to the dimensions of a finished section.
2. One side of this block is ground and polished to produce a flat, smooth surface free of scratches or imperfections. The condition can be checked with a binocular microscope or a hand lens.
3. After polishing, the block is carefully cleaned and cemented to a clean microscope slide with epoxy or balsam.
4. Excess material is removed with a diamond cut-off wheel and the specimen is ground to the required thickness with successively finer grades of abrasive powder. A useful slide holder for this stage has been devised by Cochran and King (1957).
5. The slide is carefully cleaned and dried, and a cover glass is cemented in place.

Individual workers make adaptations to this general procedure to meet conditions and frequently change it to cope with new kinds of soil and new problems. Some additional tips are given in the following paragraphs.

Often impregnation will be incomplete, or the impregnating liquid will not reach the center of the specimen. After the specimen has been cut into rough blocks for finishing, the faces can be reimpregnated if soft spots are seen. Re-treatment with the same plastic may not be successful because the solvent may soften the previously set plastic. However, a low-viscosity epoxy such as used for mounting the cover glass may be spread over the surface and allowed to soak into the empty pores. An alternative method is to coat the surface with a layer of precooked balsam or other synthetic balsam-like resin and heat in an oven at about 100°C for several hours. Polishing the block for mounting must be done with care. The excess layer of balsam or epoxy must be removed without exceeding the penetration depth of these impregnants.

Polishing of the surface for mounting requires a lap wheel for abrasive powders or abrasive paper discs. Successively finer grit sizes are used (200—600 grit) until the surface is ground smooth and flat. The polished surface should have an almost mirror-like finish. For most soil samples, glycol or a light cutting oil is used as a lubricant during polishing. For highly cemented or indurated samples lacking swelling clays, water may be used as a lubricant.

Epoxy has generally replaced Canada balsam or Lakeside 70 as the preferred mounting medium. Several commercial epoxies are especially made for thin section work. They have low viscosity, a refractive index near 1.54, a long pot life, and they cure at low temperatures. Hardening at room temperature avoids possible warping and differential expansion–contraction effects that may occur when samples are heated using thermally cured resins. Because epoxy may take several hours to cure, a pressure clamp or bonding jig is important to hold the block in place, to apply a uniform pressure across the surface so that the film of epoxy between the slide and block will be of uniform thickness.

An accurate diamond cut-off wheel is used to remove excess material. Remaining excess material is removed by using successively finer abrasives, with final polishing done by hand on a piece of plate glass. Thickness is usually checked by interference colors of sand grains, but many finer-textured soils and high iron-oxide materials that are opaque or have very fine structural elements must be ground thinner than the standard 30 μm. Thickness control depends on repeated examination and grinding until exposure of as much fine structure as possible is accomplished without losing the section. When the thickness is being checked, the section should be coated with a thin film of immersion oil or index oil to avoid the strong contrast in refractive index between the air and the section. Without oil, it may look thicker and more featureless than it really is. Most sections are thinner at the edges than in the center because of the motions used in grinding, but this may be an advantage, enabling one to see morphology in a third dimension.

Cover glasses may be permanently mounted with epoxy or synthetic resins. If the slide is to be used for other purposes, such as staining or microprobe analysis, the surface can be protected with a viscous mounting oil and a cover glass while in storage. By this procedure, the thin section may be examined and photographed with a cover glass that can be easily removed for subsequent analyses.

Cleanliness in thin-section preparation is essential. When a shift from a coarser to a finer abrasive is made, the sample must be washed, brushed, or air-blown to remove coarser grit that will likely scratch the section. When the block is mounted on the slide, it should be as clean as possible so that the grinding compound is not occluded in the surface pores; likewise, the finished section should be well cleaned before the cover glass is applied. Lintless paper or lens tissue should be used to clean the section because entrapped lint is highly birefringent and may cause image confusion.

Abrasive powders and papers are obtained from many scientific supply houses or firms dealing in geologic, metallurgical, or lapidary equipment and supplies. A number of commercial firms manufacture specialized equipment for thin section preparation (Microtec, Logitech, Hillquist, Highland Park, Leco, Buehler). This equipment ranges from simple cut-off wheels to automatic apparatus that cut and grind to final thickness. The advantages in making one's own sections are that a large number of sections can be made from any sample at a variety of orientations, and the specimen can be studied at all stages of the work—as a polished section by reflected light, for example. Structural features of interest that are noted can be preserved in the sections. Special treatments can be used, such as staining or acid washes to remove carbonates, thus revealing other structures.

8–3.5 Observations and Interpretations

8–3.5.1 UNITS OF ORGANIZATION, DESCRIPTION, AND ANALYSIS

For structure and fabric analysis, soil materials can conveniently be placed in the following units of organization: *peds, pedological features.* and *s-matrices* (Brewer, 1976). Selected italicized terms are defined in the Glossary (section 8–5). The primary ped is the basic unit of description in *pedal soil* material, and the whole soil in *apedal* materials. The s-matrix is the material within the simplest peds or apedal material. It excludes pedological features other than *plasma separations.* The elements of the s-matrix include *skeleton grains, plasma*, and associated *voids.* Pedological features are recognizable units either within the s-matrix or forming surfaces of weakness between peds. They are distinguishable from the associated material because of origin, concentration of plasma, mineralogical composition, or differences in arrangement of the constituents. The following sections address identification, composition and micromorphology of these units of organization.

8–3.5.1.1 S-Matrix.

8–3.5.1.1.1 Skeleton Grains

Identification of sand and silt grains in thin sections is carried out by standard methods given in petrography texts (Kerr, 1977). The general approach is the same as that outlined in sections 8–2.4.2 and 8–2.4.3, except that refractive index can be used only roughly, and more weight is placed on the other optical and morphological properties. It is rarely necessary to be concerned with minerals that occur in small quantities or to attempt quantitative mineralogical analysis with a thin section. The usual soil thin section contains too few grains to be usable for such work. Rough information on particle-size distribution can be obtained in some materials with replicate sections, however.

The thickness of the section limits the size of grain that can be identified. If the section is 30 μm thick, grains smaller than this will be overlapping or buried in the matrix and can not be seen clearly enough to be identified, unless they have some outstanding property such as extremely high birefringence or refractive index (calcite and rutile, for example).

If identification and mineralogical analysis are important for the problem being studied, it is best to do this work on separated size fractions and to use the thin sections mainly for information about the arrangement of the components. The methods and types of samples supplement each other: grains in thin sections are seen as slices in random orientation; cleavage and interior structure can often be seen best in the sections; and the whole range of sizes is also visible.

Visible grains will be, in part, single minerals (usually quartz); but aggregates and compound grains are common, especially in transported material like loess and till. Recognition of aggregates, concretions, pseudomorphs, and weathered grains is more important in thin-section studies than in sand and silt petrography. It can be easier because interior structures are exposed. Grains of this type may be important in soil genesis studies but are often destroyed or eliminated by sample preparation procedures that separate sand, silt, and clay. Most sections are thinnest at the edge, and examination here with high power reveals patchy aggregate extinction, birefringence and refractive index data, and morphology of aggregates and small structural features.

When skeleton grains are in the finer silt separates, identification of individual grains is difficult or impossible optically. For this reason x-ray diffraction, thermal analysis, or some other analytical technique is needed for identifications and relative estimates of minerals in fractions too small for optical verification.

8.3.5.1.1.2 Plasma

Plasma in thin sections occurs as clay, free oxides, organic matter, and soluble fractions subject to mobilization, concentration, and/or reorganization. Clay may occur in the soil matrix (*s-matrix*) not only as individual domains (*plasma aggregates*) distributed randomly through the s-matrix, but also as preferred patterns of plasma aggregates due to stress (*plasma separations*) or illuviation (*plasma concentrations*). The latter commonly occur as clay infillings, coatings, and bridges (*argillans*). Even though the individual clay particles are submicroscopic, the clay can be described, characterized, and sometimes identified. Completely dispersed and randomly arranged clay will exhibit no birefringence and will appear isotropic in crossed polarized light. Seldom, however, does the clay in a soil occur as dispersed individual domains of that size. More commonly clay in the s-matrix consists of individual plasma aggregates of larger size, which are randomly oriented and yield a flecked birefringence pattern. Plasma assembles into birefringent plasma aggregates when plate-shaped particles are reoriented by pressure or by translocation and illuviation.

Except for halloysite, the silicate clay minerals in soils have a platy shape. The *a* and *b* crystallographic axes are within the plane of the plate, and the *c* axis is almost perpendicular to this plane. The crystals are monoclinic, but the distribution of atoms along the *a* and *b* axes is so nearly the same, and the *c* axis is so nearly perpendicular to the other axes, that the minerals are pseudohexagonal. The optical properties as well as the crystal structure and general habit of clay particles are analogous to those of the micas, and the micas can be used as a model when establishing and describing the properties of clays.

The speed of light traveling in the direction of the *c* axis and vibrating parallel to the *a* axis is almost the same as that vibrating parallel to the *b* axis; therefore, the refractive indexes are very close together, and interference effects seen in crossed polarized light will be small when observed along the *c* axis.

Light vibrating parallel to the *c* axis travels faster than in other directions, and hence the refractive index is lower. If the edges of crystals or plasma aggregates are viewed along the *a-b* plane between crossed polarizers, two straight extinction positions can be seen and interference colors will show in other positions.

Figures 8–2A and 8–2B illustrate opitcal properties for plasma concentrations. If the plasma aggregates are organized into patterns where the domains are predominantly parallel, highly birefringent, and continuous, optical effects will be observed with band extinction. Thus the degree of orientation of plates into aggregates, the pattern of aggregate orientation, and the process that oriented the aggregates can be deduced partly or in whole from optical phenomena.

Kaolinite has refractive indexes slightly higher than quartz and has low birefringence. In the average thin section, interference colors are gray to pale yellow. In residual soils derived from coarse-grained igneous rocks, it often occurs as book-like and accordion-like aggregates of silt and sand size called plasma aggregates.

Halloysite, because of tubular habit, should not show birefringence even though it could form oriented aggregates. It sometimes does have very faint patternless birefringence, possibly caused by impurities or refraction of light at interfaces between particles.

The 2:1 lattice minerals have high birefringence and show bright intermediate-order interference colors when the edges of the aggregates are viewed. It is seldom possible to distinguish between clay-size smectite, mica, vermiculite, and chlorite in thin section. These clay minerals usually occur as a mixture of the minerals, and are often stained by and mixed with iron oxides and organic matter.

Oxides, oxyhydroxides, and hydroxides of iron, manganese, and aluminum pose more problems in identification—especially iron and manganese, which are poorly crystalline, fine-grained, and often opaque to transmitted light. Reflected light may help in differentiating organic plasma fragments from iron components. Organic matter will appear pearly or milky in reflected light; iron and manganese oxides have a black or rusty-brown metallic luster.

Carbonates commonly form the plasmic element in crystic fabrics (disseminated carbonates) of soils in semiarid and arid environments (Mollisols, Aridisols, Vertisols, and Alfisols). Carbonates occur as aggregate structures whether of lithogenic or pedogenic origin. They can often be identified under high magnification by their high birefringence and rhombohedral forms. Further, they readily react to acids. Gypsum may be identified by its euhedral crystal form, but more soluble salts such as halite are rarely identified in thin sections.

Dispersed organic matter in soils gives thin sections a dusky, dull appearance and decreases significantly the light intensity passing through the section. Fragments of well-decomposed materials are nearly opaque; but organic matter with slight decomposition will exhibit red to strong brown colors in plane polarized or crossed polarized light, and may be confused with argillans if cell or fiber structure is not visible.

8–3.5.1.1.3 Voids

Voids associated with the packing volumes of primary minerals in the s-matrix are called simple packing voids (Brewer, 1976). In fine-grained s-matrices, the voids commonly are ultrafine (<5 μm) and beyond the resolution of a light microscope. If such voids are important to the objectives of the study they can be resolved and quantified by SEM backscatter techniques (Jongerius & Bisdom, 1981). In coarser-textured s-matrices, packing voids are easily identified. In void analysis the observer must remember that the two-dimensional thin-section slice can intersect voids of various geometry at any angle, that some of the features can be artifacts, and that the lower limit of the observable size range of voids is controlled by the thickness of the section.

Size, shape (wall conformation and smoothness), arrangement, and morphological classification of voids must be described. Brewer (1976) classifies voids as *packing voids, vughs, vesicles, channels, chambers,* and *planes.* He provides an excellent discussion of their morphology and possible mechanisms of formation. Smoothed walls (by coatings or pressure rearrangement) are called *meta* voids and indicate biological or physical origin, while those walls that appear to be physically and chemically unaltered are called *ortho* voids.

Packing voids are due to random packing of individual skeleton grains (simple) or nonaccomodating structural units (compound). Vughs are relatively large voids, other than packing voids, that are generally irregular in shape and not interconnected with other voids of similar size. Vesicles differ from vughs principally in their smooth regular wall conformation, generally circular in cross-section, that is believed to be formed by expansion and movement of gas bubbles. Channels are tubular voids that are significantly larger than those resulting from normal packing of skeleton grains. They commonly have smoothed walls, regular conformation, and relatively uniform cross-section diameters. Chambers differ from vughs in that their walls are regular and smoothed, and from vesicles and vughs in that they are interconntected through channels. Biological activity of fauna and flora is likely responsible for most channels and chambers observed. Planes are voids that are linear in shape and form surfaces of weakness along accomodating ped structural surfaces. They may be ortho, but more commonly are meta where the surfaces are smoothed by plasma concentration coatings or pressure rearrangement of the plasma. Brewer (1976) has suggested subclasses of joint, craze, and skew planes based on surface wall conformation, regularity, and distribution of planar voids.

Visible porosity can range from essentially none, as in some fragipans (Fig. 8–3C) and glacial tills, to situations where large voids make up a significant part of the thin section, as in A horizons of Mollisols with well-developed structure and in coarse, well-sorted sands.

Large-scale pore space can be described by its relation to shapes of the grains or aggregates and by estimates of percentage. Areas can be measured with a grid eyepiece or by various transect-measuring statistical methods like those used for modal analysis of rocks (Anderson & Binnie, 1961). Although only areas are measured, the volume of various constituents can be estimated in this way. More quantitative analysis can be made by application of a Quantimet (Jongerius & Bisdom, 1981; Bullock & Murphy, 1980). If pore area or volume and pore arrangement are important aspects of the study, a soluble dye may be added to the impregnating plastic (Lockwood, 1950) so that colorless sand grains can be distinguished from pores without using crossed polarized light.

Voids are distinguished from colorless skeleton grains by checking for anisotropism when the stage is rotated with polarizers crossed. Voids are isotropic unless an anisotropic impregnating agent is used. A grain being viewed down the optic axis will also exhibit an isotropic image; therefore a check for interference figure should be made.

For final verification, the section should be viewed in plane polarized light; voids should be white unless the impregnant is colored. Mineral grains are often clearly differentiated from voids by sharp, straight-edged boundaries, by fractures and inclusions, by relief differences, and by color and/or pleochroism.

8–3.5.1.2 Pedological Features.

8–3.5.1.2.1 Papules and Argillans

By far the greatest interest in micromorphology, at least in the USA, has centered on the arrangement of clay as an indication of the genetic processes that have operated or are currently operating, and on the relation of clay to soil structure. Examples of several common types of clay arrangement and associated features follow.

In soils developed from residual materials, or in saprolite, clay weathers from primary minerals and occurs as *papules* in a form roughly pseudomorphic after rock minerals (Figs. 8–3A and 8–3B). Clay may also occur as plasma aggregates such as the vermicular or accordian-like kaolinite books shown in Fig. 8–4B. Such clay pedological features often exhibit continuous patterns of plasma aggregates with band extinction. Regular, intact arrangement of these clay materials generally is diagnostic of in situ clay formation and may be diagnostic for residual materials. Clay aggregates can also become rearranged by pressure applied differentially to produce shear; these striated aggregates are optically identified as *plasma separations.* Stress-oriented argillans occur when platy clay particles become oriented by slip along a plane, as in slickensides and pressure faces of Vertisols or in soils derived from fine-textured glacial till (McCormack & Wilding, 1974). Plasma separations may also occur within the s-matrix (*masepic* plasmic fabric), along root channels or planar voids (*vosepic*), and around skeleton grains (*skelsepic*). Root pressure, differential wetting and drying, mass movement, and crystal ice growth can produce pressure orientation.

Pressure orientation can be inferred when smooth faces with no separate coating are seen on structural units. It cannot be seen in plane polarized light. In plane polarized light, the clay in the section may be homogeneous and rather featureless. In crossed polarized light, a reticulate pattern of striated orientation appears, consisting of bright lines showing aggregate birefringence often intersecting at regular angles. The effect is that of a network in a plaid pattern (*lattisepic fabric*). There may be numerous sets of micro-shear planes that will appear in different positions as the stage is turned. Pressure-oriented clay is illustrated in Fig. 8–2C.

Illuviated argillans (translocated clay) have several features that distinguish them from stress argillans and from residual clay as papules. Illuviated argillans occur as distinct zones of plasma concentration along conductive voids. They usually have a sharp boundary with the s-matrix and conform closely to convoluted or irregular void walls. Several common forms of translocated clay are shown in Fig. 8–2A, 8–2B, 8–2D, 8–3A, 8–3D, 8–4A, 8–4B, and 8–4C. It is more homogeneous than matrix clay and is usually finer (see Fig. 8–2, 8–3, 8–4). It is often of different composition from the matrix, especially if it came from another horizon. It shows lamination, indicating deposition in successive increments. And, finally, these bodies of translocated clay will show birefringence and extinction, indicating that they are continuously oriented plasma aggregates of parallel plates. If they are straight, they will have parallel extinction; if curved, a dark band will be present wherever the composite *c* axis and composite *a* and *b* axes are parallel to the vibration planes of the polarizers. These dark bands sweep through the argillan when the stage is rotated. Features of such curved plasma concentrations are shown in Fig. 8–3D.

Swelling, slump, and mass movement in soils may cause illuviated argillans to become distorted and broken. Pores may collapse or be plugged with translocated clay and the lining then becomes an elongated argillan. New ped faces develop, and the old argillans are found as embedded fragments in the s-matrix; ultimately they may be reincorporated into the matrix and disappear (see Fig. 8–4A).

Many of the translocated cutans are composed of more than one kind of material. Complex cutans include absorbed organic matter (*organo-argillans*) and iron oxides (*ferri-argillans*). Compound cutans consist of alternate layers of mineralogically and/or chemically different substances or of different fabrics (i.e. bands of argillans, ferrans, and mangans; Brewer, 1976).

Other kinds of plasma concentrations consisting of soluble salts, carbonates, sesquioxides, and silica occur as coatings along or subcutanic to conductive voids. When they impregnate the s-matrix subjacent to voids they are termed *neocutans*, but when they are not immediately subjacent to a void surface they are termed *quasicutans* (Brewer, 1976). The latter occur as mottles in the s-matrix. Criteria for optical identification of these substances are given in the early part of the chapter.

8–3.5.1.2.2 Glaebules

Glaebules are defined as approximately equant to prolate three-dimensional bodies within the s-matrix that differ from the host material because of a greater concentration of some constituent (often plasma) or a difference in fabric (Brewer, 1976). They are of variable composition, including clay minerals (papules as discussed under section 8–3.5.1.2.1), carbonates, sesquioxides, manganese oxides, silica, and pedorelect soil material of different source from the s-matrix (Brewer, 1976). The sharpness of the glaebule boundary may be helpful in differentiating allogenic versus authigenic origin. Further, size, shape, and mineralogy of skeleton grains and clay orientation within the glaebule should be the same as in the host s-matrix if it formed in situ.

Distribution of skeleton grains is commonly less dense in glaebules than in host s-matrix because of the greater concentration of plasma. Crystallization of the plasma impregnating a nodule can force skeleton grains further apart than in the original host s-matrix; such is the case with carbonate and some iron nodules.

Concretions have a concentric banded plasma distribution, while *nodules* have undifferentiated plasmic fabric. Concretions are commonly high in sesquioxides and/or MnO_2 and are nearly opaque, although they may exhibit light and dark bands. Others may exhibit banded carbonate fabrics of different grain size and iron impurities. Concretions in the form of tubules occur along channels, apparently in response to root activity that causes fluxes in redox, pH, and soil moisture content. These concentrations may be of calcite, lepidocrocite, or goethite, or ferro-manganiferous in composition (Brewer, 1976; Miller & Wilding, 1972; Chen et al., 1980).

Nodules commonly occur in soils when neo- and quasi-forms of mangans, ferrans, sesquans, and calcans are further impregnated with respective plasma. This represents a conversion of soft mottles and segregtions to indurated nodules. Upon weathering, mafic-rich lithorelicts, such as shale, may also be transformed to sesquioxidic nodules (Ritchie, et al., 1974).

A somewhat related type of dense body is illustrated by soil nodules of residual B-horizon material left isolated in the lower E horizon, as the E horizon tongues down into the B horizon. These have a higher clay content than the surrounding matrix, and boundaries are regular and usually rather sharp. Such pedorelic nodules may become centers of accumulation of iron and manganese. With development, they may become smaller and rounder and ultimately may become pellet-like.

Mottles, concretions, and nodules are cemented to varying degrees, and so they may move as sand or gravel when material is transported. Comparing their interior composition with that of the surrounding matrix and examining the continuity (or lack of it) at their boundaries will enable one to decide whether these aggregates are formed in place or not.

8-3.5.1.2.3 Organic Matter

Amorphous coatings of organic matter with or without admixed iron and aluminum are common, especially in Spodosols or soils influenced by the podzolization process. This material is dark brown to black, isotropic or faintly birefringent, and often flecked with minute opaque grains. It occurs as the bridging and coating material in the horizons of sandy Spodosols (Soil Survey Staff, 1975, Fig. 5A–B, p. 101).

Organic residues such as living and partly decomposed roots are usually recognizable by their cellular structure. The birefringence of many plant fibers often causes them to be confused with minerals. Chitinous remains of arthropods and egg capsules also may resemble an inorganic structure. Anything with unusually symmetrical shape or regular cellular form should be suspected to be of organic origin.

8-3.5.2 ARTIFACTS

Artifacts are caused by grains tearing out of the section, by scratches during grinding, and by splitting of the section when the cover glass is pressed down. These features usually have unnatural-looking boundaries with a ragged appearance. Splitting may follow natural structural lines; if it does, the face will show some evidence of a coating or of compression or alignment of grains. If it is a random split, it will cross natural features. In some soils, sand grains have compressed oriented clay coatings. Such grains may fall out during grinding and leave a smooth, coated hole. Recognition of such holes may depend on comparison of their shape and lining with the situation around grains of similar size. Other artifacts that may cause confusion are: grains of the abrasive grinding powder (carborundum, Al_2O_3, garnet); cellulose fibers, hairs and bristles from cleaning tissues, towels, and brushes (these often have high birefringence, but their fibrous structure is apparent); and air bubbles, which have high negative relief.

8-4 APPLICATIONS

Optical petrography may be useful in investigating the origin of soil parent material. Certain suites of minerals are associated with specific rock sources; hence, overlays and unconformities may be suspected from certain kinds of discontinuities in the mineralogical composition of samples taken at different depths in a given profile. For example, the finding of fresh feldspar and hornblende in the silt fraction in the A and B horizons and only resistant minerals in the C horizon may indicate an overlay of eolian deposits younger than the underlying material. Evidence of volcanic ash (glassy shards, plagioclase feldspar types) may be critical in explaining some soil properties.

Shape, size, and spacing of the primary mineral grains may be related to source of parent material, mode of deposition, and changes caused by weathering and soil development. Large, angular quartz grains suggest granite or related rocks, and material in place or transported only a short distance. Rounding indicates water transportation, and rounding plus frosting and pitting indicates movement by wind. These latter observations, however, can be more easily made on cleaned separates with a stereoscopic or SEM microscope than in thin section. Stratification of coarse and fine particles usually indicates alluvium. Alignment of platy particles can indicate depositional stratification but can also be caused by pressure, shear, or frost action. Spacing of primary or skeletal grains can be a valuable reference datum for volume changes in weathering and for interhorizonal and local losses and gains; leaching or removal of material leaves the resistant grains closer together; additions such as translocated clay or iron oxide may force them apart. Ratios between resistant minerals can serve as a check on the homogeneity of the original parent material in a soil development study and as a base for calculations of loss and gain (Barshad, 1964; Brewer, 1976).

Effects of weathering in formation of parent materials and in soil profiles can be investigated by optical petrography. Decrease or disappearance of minerals or groups when compared against resistant minerals provides an index of weathering. The condition of remaining minerals also answers some questions: for example, one can tell whether the minerals are altering to clay or other secondary products or whether the products are being carried away in solution. Thin sections make possible observations of minerals in the process of alteration with the products in place. Examples are alteration of feldspar to halloysite, augite to smectite, feldspars or mica to kaolinite, and micas to vermiculite. Stages of weathering can be followed, and the source of secondary minerals can be observed directly.

Clay illuviation can be observed in a sequence of thin sections from the clay-depleted E horizon, to the tongued transition zone at the top of the B/E horizon, to the Bt horizon with its variety of clay accumulation features, and to the C horizon with its indigenous clay. Clay distribution and arrangement vary greatly among different soil orders and much work remains to be done on origins of soil clay and reasons for its distribution and concentration. A review of this work is provided by Wilding and Flach (1985).

Fig. 8–2.

A. Longitudinal section of a channel argillan. Plane light (bar length 0.5 mm). Miami series, Typic Hapludalfs, Indiana.

B. The same field as (A) in crossed polarized light. The sides of the split channel are highly birefringent because the oriented plasma aggregates are seen from the edge or normal to the *c* axis. The center of the channel is dark because the view is parallel to the composite *c* axis of the aggregate. A large quartz grain (Q) occurs at the lower right. The white area in (A) at the lower left is the edge of the thin section. Soil same as in (A) above.

C. Stress-oriented plasma separations (S) occur along a planar void (V) in a Vertisol. When the stage is rotated, the other sets of plasma separations corresponding to other shear directions appear. Crossed polarized light. (Bar length 0.5 mm.) Houston series, Typic Chromuderts, Alabama.

D. An illuviated argillan (Ar) along a channel void (V). (Bar length 0.5 mm.) Gee series, Typic Glossudalfs, Washington.

Fig. 8–3.

A. Weathering effects and clay arrangement in a soil derived from gabbro. Plagioclase feldspar clay pseudomorph or papule (P), has weathered in place to clay, leaving the grain outlines and twinning planes still visible. The augite (Au) (upper left and right center) is weathering, but the clay produced has moved to form illuviated argillans (Ar) around adjacent voids (V). Plane light. (Bar Length 0.55.) Iredell series, Typic Hapludalfs, North Carolina.

B. Same field as (A) in crossed polarized light. The clay replacing the feldspars is almost isotropic, indicating that it is not organized into oriented plasma aggregates; however, some of the clay in these bodies is halloysite, which shows only weak birefringence, if any. The strongly birefringence areas are oriented clay concentrations (argillans, Ar) and unweathered augite. (Bar length 0.5 mm.) Iredell series, Typic Hapludalfs, North Carolina.

C. Plagioclase feldspar (F) weathering in place in A horizon of a Xeralf. Alteration follows the cleavage and twinning planes. The s-matrix is a mixture of silt, clay and organic matter; note the large voids (V) and inter-grain bridging. Plane light. (Bar length 0.5 mm). Ramona series, Typic Haploxeralf, California.

D. Large void (V) with kaolinitic ferriargillan (AR) along wall. The dark extinction bands through the clay deposit indicate where the composite *a* and *b* axes and the composite *c* axis are parallel to the planes of polarized light. The bands will sweep through the argillan when the stage is rotated. (Bar length 0.5 mm.) Soil is a Undult from Zaire.

Fig. 8–4.

A. Illuviated argillan (Ar) in ped surface (right). Embedded illuviated argillan (EAr) on former ped surface that is being incorporated into the matrix (upper left). Crossed polarized light. (Bar length 0.5 mm.)

B. Coarse vermicular, accordion-like or book-like aggregates (papules, P) of well-crystallized kaolinite. Large voids (V) lined with ferriargillan (Ar) composed of kaolinite. In crossed polarized light the pore is dark, and the complex pattern of extinction bands in the void argillan can be seen. The larger kaolinite aggregates resemble mica but have lower birefringence. (Bar length 0.5 mm.) Davison series, Rhodic Paleudults, North Carolina.

C. Clay-filled channel in a fragipan. The banded morphology suggests that the pore was filled from lower left to upper right in several successive stages. Note the close packing of the s-matrix (M). The white areas are quartz (Q) grains. At the top is an opaque concretion. Plane light. (Bar length 0.5mm.) Beltsville series, typic Fragiudults, Maryland.

D. Weathering and clay formation in soil derived from gabbro. Plagioclase feldspars (F) weather first, largely to halloysite (top), and the clay stays more or less in place. In the lower part of the section, the ferromagnesian mineral weathers later; montmorillonite is synthesized and accumulates as oriented argillans (Ar) in voids (V). Plane light. (Bar length 0.5mm.) Iredell series. Typic Hapludalfs, Virginia.

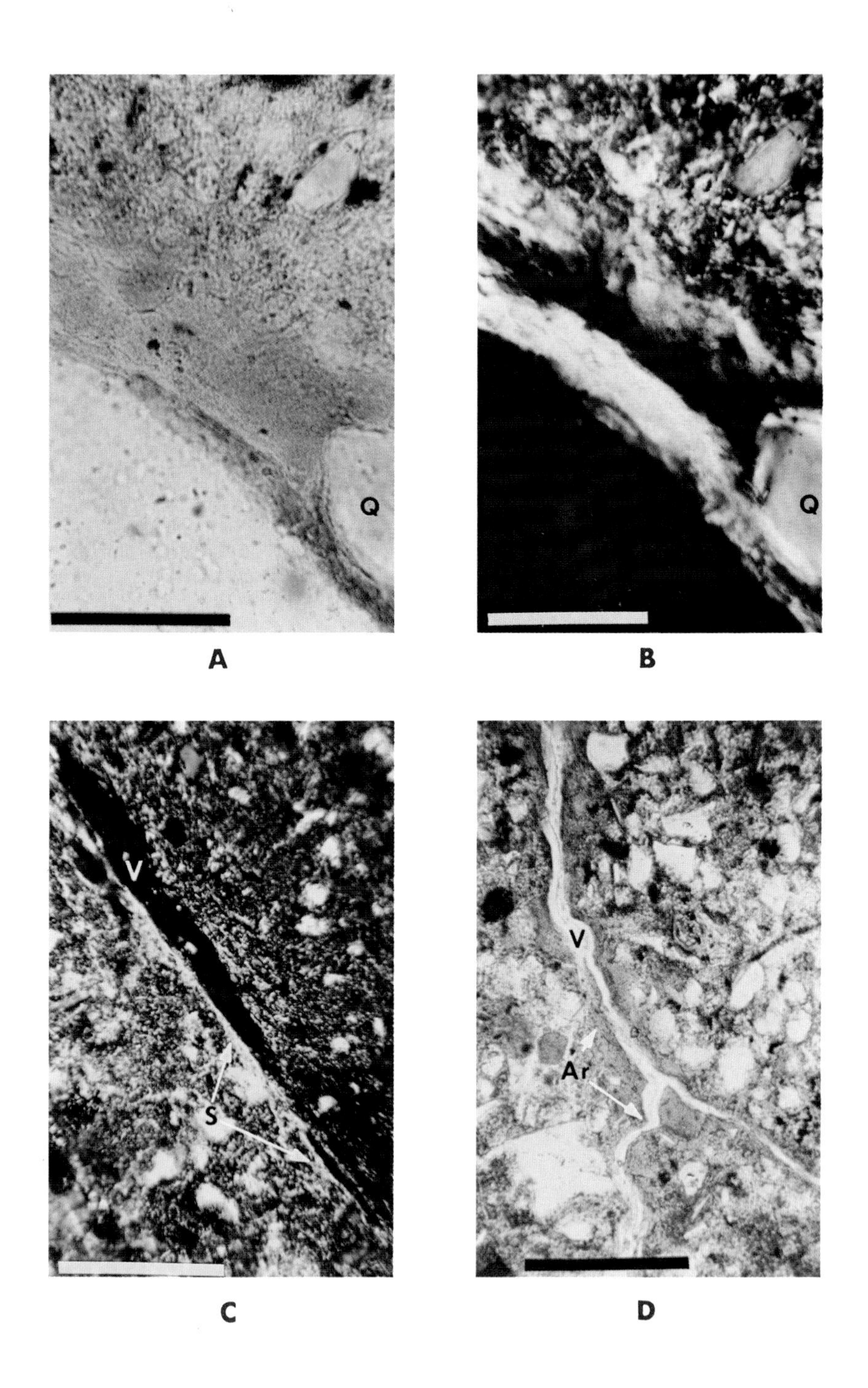

Fig. 8–2.

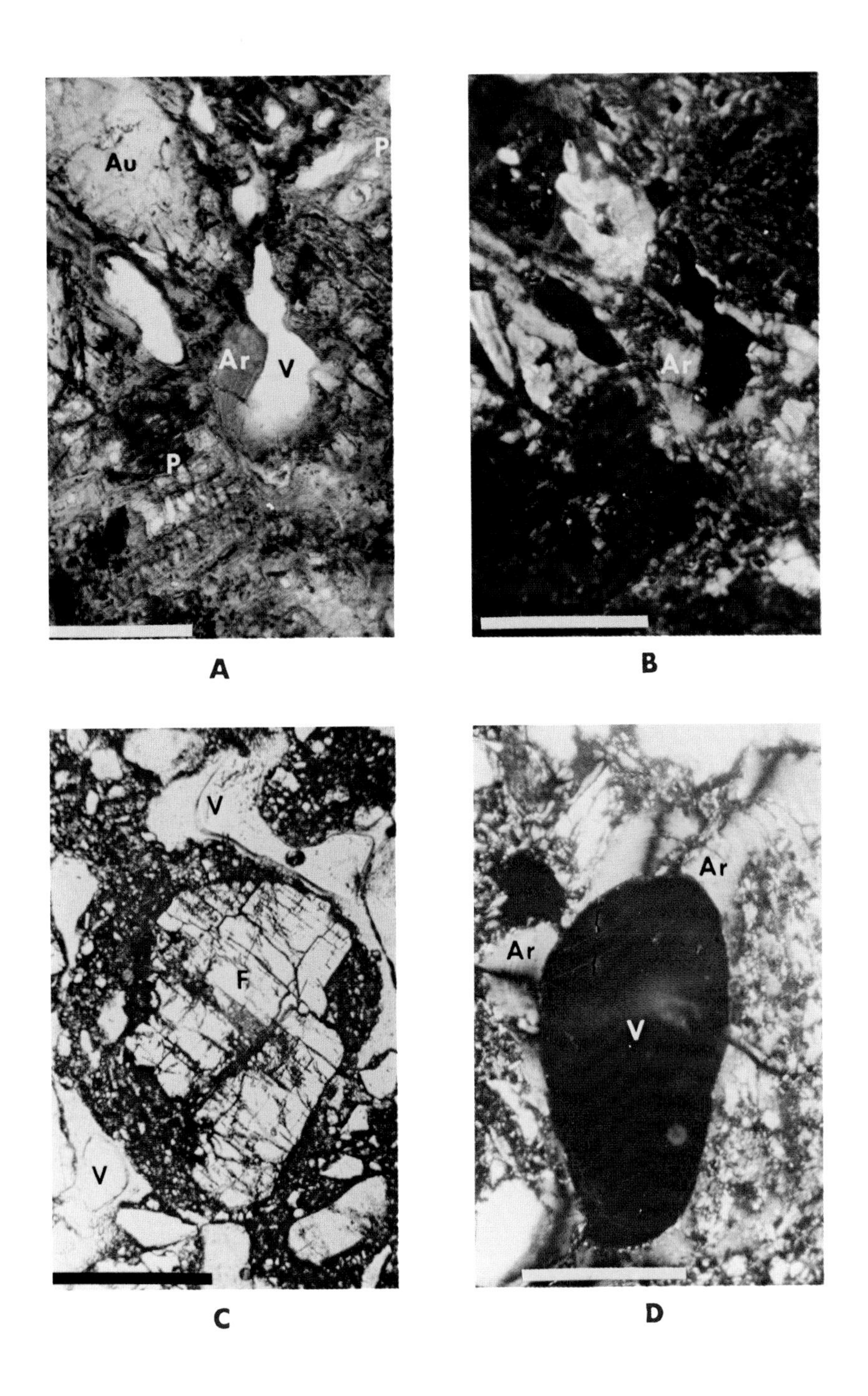

Fig. 8–3.

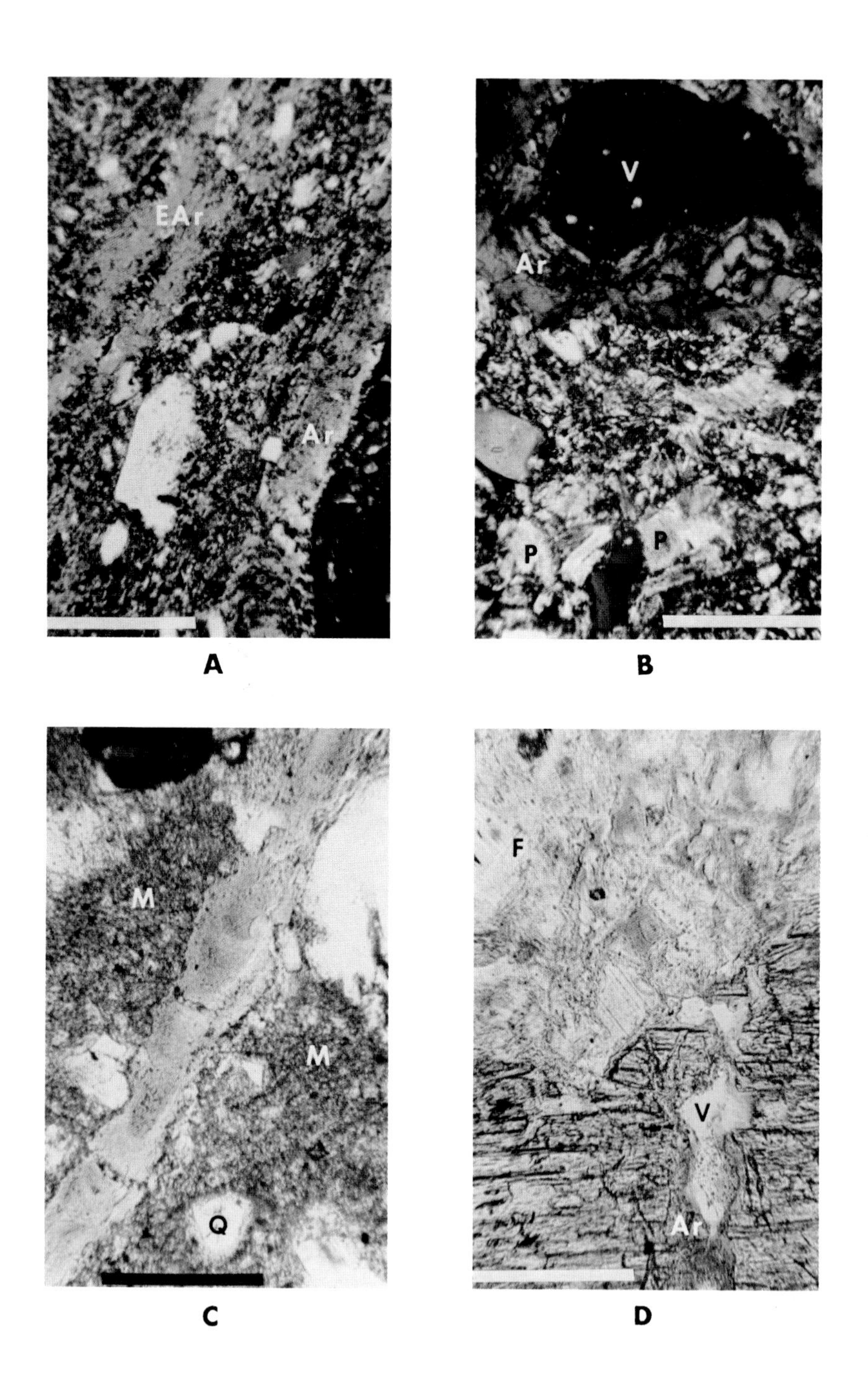

Fig. 8–4.

Petrographic studies may aid in understanding the development of certain types of concretions, laterite, and pans. Rearrangement and recrystallization of oxides to cause cementation and sequences of changes whereby mottles become hard plinthite nodules may be investigated (Alexander and Cady, 1962).

The microscopic view of soil approaches the view seen by the plant root and permits close, direct investigation of the physical and mineralogical environment of roots (Blevins, et al., 1970). Thus, optical petrography is useful also in investigations of tilth, seedling emergence, and penetration of air, water, and roots (Miller & Wilding, 1972; Jongerius et al. 1972). The shape and sometimes the binding agent of soil aggregates can be observed. Arrangement and continuity of pore space can be traced. Compaction and orientation of particles in "traffic" pans can be confirmed. Surface crusts can be observed directly (Evans & Buol, 1968).

Microscopy gives information about the form and location of chemical elements in the soil (Brewer, 1973). It provides a direct look at the interior arrangement of the soil—the location and condition of the sand, silt, and clay, and the distribution and character of the secondary minerals. Such information is a helpful adjunct to classification and mapping of soils and to the development of improved systems that will serve more accurately to extend the knowledge on specific areas to other areas having similar properties (Wilding & Flach, 1985).

8–5 GLOSSARY OF MICROMORPHOLOGY TERMS

Agglomeroplasmic fabric—The plasma occurs as discontinuous or incomplete fillings in the intergranular spaces between skeleton grains.

Alban—a cutan composed of materials that have been strongly reduced.

Apedal—Applied to soil materials without peds.

Argillan—A cutan composed dominantly of clay minerals.

Asepic fabric—Plasmic fabrics that have dominantly anisotropic plasma with anisotropic domains that are unoriented with regard to each other; they have a flecked extinction pattern and no plasma separations.

Calcan—A cutan composed of carbonates.

Chambers—Vesicles or vughs connected by a channel or channels.

Channel—A tubular-shaped void.

Concretion (micromorphological). A glaebule with a generally concentric fabric about a center which may be a point, line, or plane.

Craze plane—Planar voids with a highly complex conformation of the walls due to interconnection of numerous short planes.

Cutan—A modification of the texture, structure, or fabric at natural surfaces in soil materials due to concentration of particular soil constituents or in situ modification of the plasma.

Fecal pellets—The excreta of fauna.

Ferran—A cutan composed of a concentration of iron oxides.

Ferri-argillan—A cutan composed of intimately mixed clay minerals and iron oxides.

Granular fabric—There is no plasma, or all the plasma occurs as pedological features.

Glaebule—A three-dimensional pedogenic feature within the s-matrix of soil material that is approximately prolate to equant in shape.

Gypsan—A cutan composed of gypsum.

Joint Planes—Planar voids that traverse soil in a regular parallel or subparallel pattern.

Lithorelict—A pedological feature derived from the parent rocks; usually recognized by their rock structure and fabric.

Mangan—A cutan containing enough manganese to effervesce upon application of H_2O_2.

Matran—A cutan that contains s-matrix skeletal grains within the plasma concentration.

Nodules—Glaebules with an undifferentiated fabric; in this context undifferentiated fabric includes recognizable rock and soil fabrics.

Organan—A cutan composed of a concentration of organic matter.

Packing voids (simple). Voids formed by the random packing of single skeletal grains.

Packing voids (compound). Voids formed by the random packing of pods that do not accommodate each other.

Papules—Glaebules composed dominantly of clay minerals with continuous and/or lamellar fabric and sharp external boundaries.

Pedal—Applied to soil materials, most of which consist of peds.

Pedological features—Recognizable units within a soil material which are distinguishable from the enclosing material for any reason such as origin (deposition as an entity), differences in concentration of some fraction of the plasma, or differences in arrangement of the constituents (fabric).

Phytoliths—Inorganic bodies derived from replacement of plant cells; they are usually opaline.

Plasma—That part of the soil material that is capable of being or has been moved, reorganized, and/or concentrated by the processes of soil formation. It includes all the material, mineral or organic, of colloidal size and relatively soluble material that is not contained in the skeleton grains.

Plasma aggregate—Preferential alignment of individual plasma grains into larger anisotropic domains that can be recognized in thin sections.

Plasma concentration—Concentration of any of the fractions of the plasma in various parts of the soil material.

Plasma separation—Features characterized by significant change in arrangement of the constituents rather than change in concentration of some fraction of the plasma. For example, aligning of plasma aggregates by stress at or near the surface of slickensides.

Porphyroskelic fabric—The plasma occurs as a dense groundmass in which skeleton grains are set after the manner of phenocrysts in a porphyritic rock.

Primary fabric—The fabric within an apedal soil material or within the primary peds in a pedal soil material; it is an integration of the arrangement of all the pedological features enclosed in the s-matrix and the basic fabric, or fabric of the s-matrix.

Rock nodules—Nodules with recognizable rock fabrics.

Sepic fabric—Plasmic fabrics with recognizable anisotropic domains that have various patterns of preferred orientation; i.e., plasma separations with a striated extinction pattern are present.

Sesquan—A cutan composed of a concentration of sesquioxides.

Skeletan—A cutan composed of skeleton grains.

Skeleton grains—Individual grains that are relatively stable and not readily translocated, concentrated or reorganized by soil-forming processes; they include mineral grains and resistant siliceous and organic bodies larger than colloidal size.

Skew planes—Planar voids that traverse the soil material in an irregular manner and are formed mostly by soil desiccation.

S-matrix (of a soil material)—The material within the simplest peds, or composing apedal soil materials, in which the pedological features occur; it consists of the plasma, skeleton grains, and voids that do not occur as pedological features other than those expressed by specific extinction (orientation) patterns. Pedological features also have an internal s-matrix.

Unaccommodated—Applied to peds. Virtually none of the faces of adjoining peds are moulds of each other.

Vesicles—Relatively large, smooth-walled, regular vughs.

Vughs—Relatively large voids, usually irregular and not normally interconnected with other voids of comparable size; at the magnifications at which they are recognized they appear as discrete entities.

8–6 REFERENCES

Alexander, L. T., and J. G. Cady. 1962. The genesis and hardening of laterite in soils. U.S. Dep. Agric. Tech. Bull. 1282.

Anderson, D. M., and R. R. Binnie. 1961. Modal analysis of soils. Soil Sci. Soc. Am. Proc. 25:499–503.

Ashley, G. M. 1973. Impregnation of fine-grained sediments with a polyester resin: a modification of Altemueller's method. J. Sediment. Petrol. 43:298–301.

Barshad, I. 1964. Chemistry of soil development. p. 1–70. *In* F .F. Bear (ed.) Chemistry of the soil. Reinhold Publ. Corp., New York.

Bisdom, E. B. A. 1981. A review of the application of submicroscopic techniques in soil micromorphology: I. Transmission electron microscope (TEM) and scanning electron microscope (SEM). p. 67–116. *In* E. B. A. Bisdom (ed.) Submicroscopy of soils and weathered rocks. Centre for Agriculture Publishing and Documentation. Wageningen, Netherlands.

Blevins, R. L., G. M. Aubertin, and N. Holowaychuk. 1968. A technique for obtaining undisturbed soil samples by freezing in situ. Soil Sci. Soc. Am. Proc. 32:741–742.

Blevins, R. L., N. Holowaychuk, and L. P. Wilding. 1970. Micromorphology of soil fabric at tree root-soil interface. Soil Sci. Soc. Am. Proc. 34:460–465.

Bourbeau, G. A., and K. C. Berger. 1947. Thin sections of soils and friable materials prepared by impregnation with plastic "Castolite." Soil Sci. Soc. Am. Proc. 12:409–412.

Brasher, B. R., D. P. Franzmeier, V. Vallassis, and S. E. Davidson. 1966. Use of saran resin to coat natural soil clods for bulk-density and moisture- retention measurements. Soil Sci. 101:108.

Brewer, R. 1973. Micromorphology. A discipline at the chemistry mineralogy interface. Soil Sci. 115:261–267.

Brewer, R. 1976. Fabric and mineral analysis of soils. Robert E. Krieger Publishing Co., New York.

Bullock, P., and C. P. Murphy. 1980. Towards the quantification of soil structure. Microscopy. 120:317–328.

Buol, S. W., and D. M. Fadness. 1961. A new method of impregnating fragile material for thin sectioning. Soil Sci. Soc. Am. Proc. 25:253.

Cady, J. G. 1940. Soil analyses significant in forest soils investigations and methods of determination; 3. Some mineralogical characteristics of podzol and brown podzolic forest soil profiles. Soil Sci. Soc. Am. Proc. 5:352–354.

Cady, J. G. 1950. Rock weathering and soil formation in the North Carolina Piedmont Region. Soil Sci. Soc. Am. Proc. 15:337–342.

Carver, R. E. 1971. Heavy mineral separatons. p. 427–452. *In* R. E. Carver (ed.) Procedures in sedimentary petrology. Wiley-Interscience Publ., New York.

Cent, J., and R. Brewer. 1971. Preparation of thin sections of soil materials using synthetic resins. C.S.I.R.O. (Aust.). Div. of Soils, Tech. Pap. no. 7.

Chen, C. C., J. B. Dixon, and F. T. Turner. 1980. Iron coatings on rice roots: morphology and models of development. Soil Sci. Soc. Am. J. 5:1113–1119.

Cochran, M. and A. G. King. 1957. Two new types of holders used in grinding thin sections. Am. Mineral 42:422–425.

Collins, K., and A. McGown. 1983. Micromorphological studies in soil engineering. *In* P. Bullock and C. P. Murphy (ed.) Soil micromorphology. Vol. 1: Techniques and applications. AB Academic Publ. Berkhamsted, Herts, England. pp. 195–218.

Crampton, C.T. 1974. Micro shear-fabrics in soils of the Canadian North. p. 655–664. *In* G.K. Rutherford (ed.) Soil microscopy. The Limestone Press, Kingston, Ontario.

Douglas, L.A. 1980. The use of soils in estimating the time of last movement of faults. Soil Sci. 129:345–352.

Drees, L.R., and L.P. Wilding. 1983. Microradiography as a submicroscopic tool. Geoderma 30:65–76.

Evans, D. D., and S. W. Buol. 1968. Micromorphological study of soil crusts. Soil Sci. Soc. Am. Proc. 32:19–22.

Fessenden, F. W. 1959. Removal of heavy separates from glass centrifuge tubes. J. Sediment. Petrol. 29:621.

FitzPatrick, E. A. 1980. The micromorphology of soils. Dep. of Soil Science, Univ. of Aberdeen, Scotland.

Flach, K. W., W. D. Nettleton, L. H. Gile, and J. G. Cady. 1969. Pedocementation: induration by silica, carbonates and sesquioxides in the Quaternary. Soil Sci. 107:442–453.

Fleischer, M., R. E. Wilcox, and J. J. Matzko. 1984. Microscopic determination of nonopaque minerals. U. S. Geol. Surv. Bull. 1627. U. S. Dep. of Interior. U. S. Government Printing Office, Washington, DC.

Frei, E., and M. G. Cline. 1949. Profile studies of normal soils of New York: II. Micromorphological studies of the Gray-Brown Podzolic-Brown Podzolic soil sequence. Soil Sci. 68:333–344.

Fry, W. H. 1933. Petrographic methods for soil laboratories. U.S. Dep. Agric. Tech. Bull. 344.

Gerber, T. D., L. P. Wilding, and R. E. Franklin. 1974. Ion diffusion across cutans: a methodology study. p. 730–746. *In* G. K. Rutherford (ed.) Soil microscopy. The Limestone Press, Kingston, Ontario.

Grossman, R. B., L. Stephen, J. B. Fehrenbacher, and A. H. Beavers. 1959. Fragipan soils of Illinois. III. Micromorphological studies of Hosmer silt loam. Soil Sci. Soc. Am. Proc. 23:73–75.

Hagni, R. D. 1966. The preparation of thin sections of fragmental materials using epoxy resin. Am. Mineral. 51:1237–1242.

Haseman, J. F., and C. E. Marshall. 1945. The use of heavy minerals in studies of the origin and development of soils. Missouri Agric. Exp. Stn. Res. Bull. 387.

Hauff, P. L., and J. Airey. 1980. The handling, hazards, and maintenance of heavy liquids in the geologic laboratory. U.S. Geol. Surv. Circ. 827.

Henderson, J. H., R. N. Clayton, M. L. Jackson, J. K. Syers, R. W. Rex, J. L. Brown, and I. B. Sachs. 1972. Cristobalite and quartz isolation from soils and sediments by hydrofluosilicic acid treatment and heavy liquid separation. Soil Sci Soc. Am. Proc. 36:830–835.

Hunter, R. E. 1967. The petrography of some Illinois Pleistocene and recent sands. Sediment. Geol. 1:57–75.

Innis, R. P., and D. J. Pluth. 1970. Thin section preparation using an epoxy impregnation for petrographic and electron microprobe analysis. Soil Sci. Soc. Am. Proc. 34:483–485.

Ireland, H. A. 1971. Preparation of thin-sections. p. 367–383. *In* R. E. Carver (ed.) Procedures in sedimentary petrology. Wiley-Interscience Publ., New York.

Johnston, J. R., and J. B. Peterson. 1941. Microscopic study of soils from five great soil groups. Soil Sci. Soc. Am. Proc. 6:360–367.

Jongerius, A., and E. B. A. Bisdom. 1981. Porosity measurements using the Quantimet 720 on backscattered electron scanning images of thin sections of soils. p. 207–216. *In* E. B. A. Bisdom (ed.) Submicroscopy of soils and weathered rocks. Centre for Agriculture Publishing and Documentation. Wageningen, Netherlands.

Jongerius, A., and G. Heintzberger. 1963. The preparation of mammoth-sized thin sections. Soil Survey Papers No. 1., Netherlands Soil Surv. Inst., Wageningen, Netherlands.

Jongerius, A., D. Schoonderbeek, A. Jager, and St. Kowalinski. 1972. Electro-optical soil porosity investigation by means of Quantimet B equipment. Geoderma. 7:177–198.

Kerr, P. F. 1977. Optical mineralogy. McGraw-Hill Book Co., Inc., New York.

Khalifa, E. M., and S. W. Buol. 1968. Studies of clay skins in a Cecil (Typic Hapludult) soil: I. Composition and genesis. Soil Sci. Soc. Am. Proc. 32:857–861.

Khangarot, A. S., L. P. Wilding, and G. F. Hall. 1971. Composition and weathering of loess mantled Wisconsin- and Illinoian-age terraces in central Ohio. Soil Sci. Soc. Am. Proc. 35:621–626.

Krumbein, W. C. and F. J. Pettijohn. 1938. Manual of sedimentary petrography. Appleton-Century-Crofts, New York.

Kubiena, W. L. 1938. Micropedology. Collegiate Press, Ames, Iowa.

Kubiena, W. L. 1970. Micromorphological features of soil geography. Rutgers University Press, New Brunswick, NJ.

Langton, J. E., and G. B. Lee. 1965. Preparation of thin sections from moist organic soil materials. Soil Sci. Soc. Am. Proc. 29:221–223.

Lockwood, W. N. 1950. Impregnating sandstone specimens with thermosetting plastics for studies of oil-bearing formations. Bull. Am. Assoc. Petrol. Geol. 34:2061–2067.

Low, A. J., L. A. Douglas, and D. W. Platt. 1982. Soil-pore orientation and faults. Soil Sci. Soc. Am. J. 46:789–792.

Mackenzie, A. F., and J. E. Dawson. 1961. The preparation and study of thin sections of wet organic soil materials. J. Soil Sci. 12:142–144.

Marshall, C. E., and C. G. Jeffries. 1945. The correlation of soil types and parent materials, with supplementary information on weathering processes. Soil Sci. Soc. Am. Proc. 10:397–405.

Matelski, E. P. 1951. Separation of minerals by subdividing solidified bromoform after centrifugation. Soil Sci. 71:269–272.

McCaughey, W. J., and W. H. Fry. 1913. The microscopic determination of soil-forming minerals. U.S. Dep. Agric. Bur. Soils Bull. 91.

McCaughey, W. J., and W. H. Fry. 1914. *In* W. O. Robinson (ed.) The inorganic composition of some important American soils. U.S. Dep. Agric. bull. 122:16–27.

McCormack, D. E., and L. P. Wilding. 1974. Proposed origin of lattisepic fabric. p. 761–771. *In* G. E. Rutherford (ed.) Soil microscopy. The Limestone Press, Kingston, Ontario.

Middleton, L. T., and M. J. Kraus. 1980. Simple technique for thin-section preparation of unconsolidated materials. J. Sediment. Petrol. 50:622–623.

Miedema, R., Th. Paper, and G.J. van de Waal. 1974. A method to impregnate wet soil samples, producing high-quality thin sections. Neth. J. Agric. Sci. 22:37–39.

Miller, M. H., and L. P. Wilding. 1972. Microfabric studies in relation to the root-soil interface. pp. 75–110. *In* R.Protz (ed.) Proceedings of a symposium on microfabrics of soil and sedimentary deposit. Dep. of Land Resource Science. Univ. of Guelph, Ontario. C.R.D. Pub. No. 69.

Milner, H. B. 1962. Sedimentary Petrography, 4th ed. The Macmillan Co., New York.

Mitchell, W. A. 1975. Heavy minerals. p. 449–480. *In* J.E. Gieseking (ed.) Soil components. Vol. 2. Inorganic components. Springer-Verlag, New York.

Morgenstern, N. R., and J. S. Tchalenko. 1967. Microstructural observations on shear zones from slips in natural clays. Proc. Geotechemical Conf. Oslo, Norway. 1:147–152.

Murphey, J. B., E. H. Grissinger, and W. C. Little. 1981. Fiberglass encasement of large, undisturbed, weakly cohesive soil samples. Soil Sci. 131:130–134.

Nikiforoff, C. C., R. P. Humbert, and J. G. Cady. 1948. The hardpan in certain soils of the coastal plain. Soil Sci. 65:135–153.

Parfenova, E. I., and E. A. Yarilova. 1965. Mineralogical investigations in soil science. Israel Program for Scientific Translations Ltd., available from U.S. Dep. of Commerce, Springfield, VA.

Peterson, J. B. 1937. The micromorphology of some loessial soils of Iowa. Soil Sci. Soc. Am. Proc. 2:9–13.

Phillips, W. R. 1971. Mineral optics: principles and techniques. W. H. Freeman Co., San Francisco.

Reed, F. S., and J. L. Mergner. 1953. Preparation of rock thin sections. Am. Mineral. 38:1184–1203.

Ritchie, A. L., L. P. Wilding, G. F. Hall, and C. R. Stahnke. 1974. Genetic implications of B horizons in Aqualfs of northeastern Ohio. Soil Sci. Soc. Am. Proc. 38:351–358.

Ruhe, R. V. 1956. Geomorphic surfaces and the nature of soils. Soil Sci. 82:441– 455.

Scull, B. J. 1960. Removal of heavy liquid separates from glass centrifuge tubes-alternate method. J. Sediment. Petrol. 30:626.

Smithson, F. 1961. The microscopy of the silt fraction. J. Soil Sci. 12:145–157.

Sinkankas, J. 1968. High pressure epoxy impregnation of porous materials for thin section and microprobe analysis. Am. Mineral. 53:339–342.

Soil Survey Staff. 1975. Soil taxonomy: a basic system of soil classification for making and interpreting soil surveys. Agric. Handb. No. 436. U.S. Government Printing Office, Washington, DC.

Soileau, J. M., W. A. Jackson, and R. J. McCracken. 1964. Cutans (clay films) and potassium availability to plants. J. Soil Sci. 15:117–123.

Swift, D. J. P. 1971. Grain mounts. p. 499–510. *In* R.E. Carver (ed.) Procedures in sedimentary petrology. Wiley-Interscience Publ., New York.

Wahlstrom, E. E. 1979. Optical crystallography. 5th ed. John Wiley & Sons, Inc., New York.

Wilding, L. P., and K. Flach. 1985. Micropedology and *Soil Taxonomy*. p. 1–16. *In* L.A. Douglas and M. L. Thompson (ed.) Soil micromorphology and soil classification. Spec. Pub. 15. Soil Science Society of America, Madison, WI.

Wilding, L. P., N. E. Smeck, and L. R. Drees. 1977. Silica in soils: quartz, cristobalite, tridymite, and opal. p. 471–552. *In* J. B. Dixon and S. B. Weed (eds.) Minerals in soil environments. Soil Sci. Soc. Am., Madison, WI.

Willman, H. B., H. D. Glass, and J. C. Frye, 1963. Mineralogy of glacial tills and their weathering profiles in Illinois. Illinois State Geol. Surv. Circ. 347.

9 Magnetic Methods

M. B. Mc BRIDE

Cornell University
Ithaca, New York

9-1 INTRODUCTION

Magnetism is a property of solids which one would expect to hold a natural interest for soil scientists because of the presence of numerous magnetic clay and primary minerals in soils. However, it has only been fairly recently that the magnetic properties of soil components (mineral and organic) have been investigated in any detail. This late development had to await the modern understanding of magnetism, requiring quantum mechanics for an adequate explanation. In addition, magnetic resonance spectroscopy, including electron spin resonance (ESR) and nuclear magnetic resonance (NMR), have advanced remarkably over the past two decades, providing powerful methods in studies of magnetic atoms and nuclei.

Although strong magnetism arises from the magnetic moments associated with moving electrons in atoms, atomic nuclei may also possess magnetic moments. A nucleus is magnetic if the proton and neutron spins, with values $I = 1/2$, are unable to pair completely, leaving a nuclear spin, I. The existence of spin and charge confers a magnetic moment, μ_N, to the nucleus which is proportional to the magnitude of the spin

$$\mu_N = g_N \beta_N I . \quad [1]$$

Here, g_N is a dimensionless constant called the nuclear g-factor and β_N is the nuclear magneton, given by $e\hbar/2Mc$, where e, M, c, and $\hbar$ are the charge and mass of the proton, the speed of light, and Planck's constant divided by 2π, respectively. Nuclei with *even* atomic mass and *even* atomic number are not magnetic, since $I = 0$. However, all nuclei with *odd* atomic mass possess spin.

Using an analogous approach for electrons, an unpaired electron has spin $S = 1/2$, so that the magnetic moment, μ_e, is given by

$$\mu_e = -g\beta S . \quad [2]$$

Here, g is the dimensionless constant called the electron g-factor and β

 Methods of Soil Analysis, Part 1. Physical and Mineralogical Methods—Agronomy Monograph no. 9 (2nd Edition)

is the electronic Bohr magneton given by $e\hbar/2mc$, where $-e$ and m are the charge and mass of the electron. Since electron spins tend to pair in atomic and molecular orbitals, the net spin and electronic magnetic moment of many materials is zero, and they are nonmagnetic. However, the electrons in the d and f orbitals of transition and rare-earth metals are commonly unpaired, so that magnetic properties are generally associated with these elements. In addition, some molecules may have one or more unpaired electron(s) in molecular orbitals, although these are usually unstable species. A notable example of a relatively stable magnetic molecule is oxygen gas ($S = 1$).

In comparing the magnitude of the magnetic moments of nuclei and electrons, the ratio μ_e/μ_N is easily shown to be $-gM/g_N m$, or 658.2, when the nucleus of a hydrogen atom with spin $I = 1/2$ and a single unpaired electron ($S = 1/2$) are compared. This large electron-proton magnetic moment ratio indicates that magnetism associated with nuclei is very weak compared to that arising from electrons. Therefore, the phenomenon of magnetic susceptibility of minerals must be explained by theories of electronic structure. In addition to susceptibility measurement, this chapter introduces nuclear magnetic resonance (NMR), electron spin resonance (ESR), and the application of these methods in investigation of soil minerals and organics.

9–2 MAGNETIC SUSCEPTIBILITY

9–2.1 Principles

Different classes of magnetic solids can be defined based upon the type of ordering of the magnetic dipoles (i.e. electron spins) with respect to one another. However, most solids have no unpaired electrons and therefore no magnetic dipoles. These substances, when placed in an external magnetic field, H, have their electron orbitals perturbed to produce a negative change in the orbital dipole moment vector, μ (Martin, 1967):

$$\Delta\mu = (-e^2r^2/4mc^2)\mathbf{H}\,. \qquad [3]$$

Here, r is the radius for the simple case of an electron in a circular orbital. It is clear that μ is changed in a direction *opposite* to H, representing *diamagnetic* induction. For the general case of N randomly oriented ions or atoms per unit volume, each containing several electrons with radii r_i, the volume magnetic susceptibility, K is given as

$$K = N\Delta\mu/H = (-Ne^2/6mc^2)\Sigma_i\overline{r_i^2}\,. \qquad [4]$$

This is the classical equation for diamagnetism, using the mean value, $\overline{r_i^2}$ for elliptical and circular orbitals. Note that magnetic susceptibility of a material is defined as the ratio of the magnetization to the magnetic field inducing it. Most solids, with their electrons paired in closed shells, show only this diamagnetic response to an applied magnetic field. In actual fact, diamagnetism is a universal property of matter, but is a weak effect which is usually unobservable experimentally in substances containing incompletely paired electrons. As Eq. [4] shows, it is independent of the temperature of the material.

In contrast to diamagnetism, *paramagnetic* magnetization is in the *same* direction as the applied magnetic field and is due to partial alignment of magnetic dipole moments (associated with unpaired electrons) in the field. Generally, paramagnetism occurs in chemical compounds and metallic alloys of transition and rare-earth elements, pure metals containing conduction electrons (weakly paramagnetic), and organic molecules with unpaired electrons (free radicals). Of special interest to soil scientists is the class of paramagnetic substances including transition elements, since compounds of iron, manganese and other *iron-group* elements are part of this class. The 3d orbitals of a free transition element are not completely filled with electrons, and when the element forms a compound, these inner electrons are not fully paired up in chemical bonds. The outer 4s electrons are, however, generally involved in bond formation. There is a strong relationship between the volume susceptibility, K, of paramagnetic materials and temperature, T, given by the Curie-Weiss law

$$K = C/(T - \Theta) \qquad [5]$$

where C is the "Curie constant" and Θ is the "Weiss constant." Again, K is defined as the ratio of the magnetization, M (i.e. the total magnetic moment per unit volume) to the magnetic field, H, inducing it:

$$K = M/H \qquad [6]$$

where M and H have units of magnetic flux density or gauss (10^4 gauss $= 1$ tesla $= 1$ weber m^{-2}), but can also be expressed in magnetic field strength or oersted (ampere m^{-1}) units. Equation [5] breaks down for T comparable with or less than Θ, but at higher temperature typical "Curie-Weiss" behavior is observed for many iron-group compounds; that is, there is a linear relationship between $1/K$ and T (Martin, 1967). For example, a crystal of $Cu_2SO_4 \cdot 5H_2O$ demonstrates this behavior. The value of Θ is generally near zero for dilute compounds of transition metals or transition metals dispersed in a diamagnetic matrix, but may be on the order of 10K or greater for concentrated salts.

To understand the strong temperature dependence of paramagnetism, one can consider non-interacting magnetic ions (Fig. 9–1a) subjected to the orienting influence of an applied magnetic field. Each ion has a magnetic moment given by (Cotton & Wilkinson, 1966)

$$\mu_{S+L} = [4S(S+1) + L(L+1)]^{1/2}\, \mu_B \qquad [7]$$

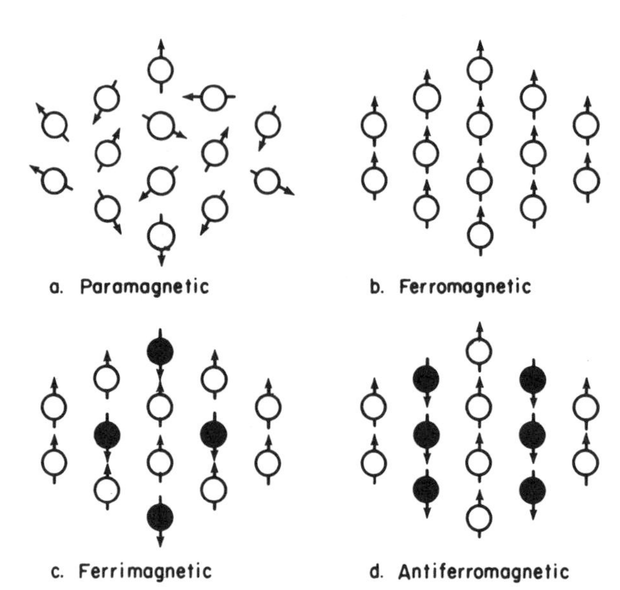

Figure 9–1. Diagrams of ionic magnetic moment vectors in the different classes of magnetic materials.

where S is the spin quantum number for the ion as a whole, L is the orbital angular momentum quantum number for the ion, and μ_B is the Bohr magneton. Taking the example of Mn^{2+}, with 5 unpaired 3d electrons, one in each d-orbital, $S = 5/2$ and $L = 0$. Therefore, there is no orbital angular momentum to contribute to the magnetic moment, and a spin-only moment of $\mu = 5.92\ \mu_B$ is expected. Each of these magnetic dipoles tends to line up parallel or antiparallel to the applied magnetic field, H. The energy of interaction between the dipoles and the field is given by $E_H = -\mu_s \cdot H$. This orienting potential is generally much smaller than kT except at very high external magnetic field and very low temperatures. Summing over all N ions per unit volume that contribute to the paramagnetic magnetization, M, it is found that (Martin, 1967)

$$K = (Ng^2\mu_B^2/3kT)J(J+1) \qquad [8]$$

where $J = L + S$ and k is the Boltzmann constant. For the case of Mn^{2+} and other ions with no orbital contributions to μ Eq. [8] becomes:

$$K = N\mu^2/3kT \qquad [9]$$

and the relationship to the Curie-Weiss law is shown. If Θ in Eq. [5] is assumed to be zero, then

$$K = C/T = N\mu^2/3kT \qquad [10]$$

and the Curie constant must have the value

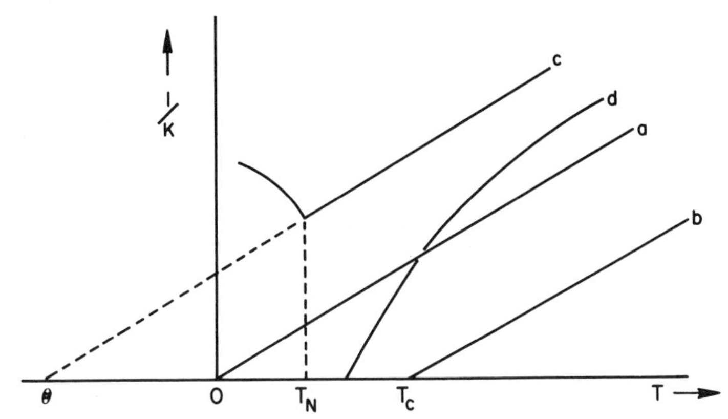

Figure 9–2. Relationship between the reciprocal of magnetic susceptibility ($1/K$) and temperature (T) (adapted from Craik, 1971) for: a. paramagnetic that obeys the Curie law (no interactions). b. paramagnetic that obeys the Curie-Weiss law above the Curie temperature, T_c, but has magnetic ordering below T_c due to a positive interaction. c. antiferromagnetic with negative interaction term, Θ, with the Néel temperature (T_N) indicating the breakpoint between ordering and paramagnetic behavior. d. ferrimagnetic, showing the approach to paramagnetic behavior at high temperature.

$$C = N\mu^2/3k\,. \qquad [11]$$

From Eq. [10], if one measures K for a solid at several temperatures and plots $1/K$ against T, a straight line of slope $1/C$ which intersects the origin should be obtained (Fig. 9–2). The substance would then obey Curie's law, $K = C/T$. However, it has been observed that the Curie-Weiss law (Eq. [5]) is often more accurate, where the "Weiss constant" is now seen to be the temperature at which the line intersects the T axis (Fig. 9–2).

A physical picture of the above mathematical description of paramagnetism is easily given. The N magnetic moments, μ, of the paramagnetic atoms or ions in a sample tend to become aligned when placed in a magnetic field. However, at the same time, thermal motion tends to randomize the orientations of these dipoles, resulting in the reduction of magnetic susceptibility at higher temperatures. According to the Curie law, as the temperature approaches 0K, K approaches infinity because of the near-perfect alignment of magnetic dipoles in the magnetic field. The Curie-Weiss law is a modification of the picture, as the Weiss constant takes into account interactions between neighboring magnetic dipoles, which tend to influence the alignment in the magnetic field.

From Eq. [10], at any given temperature

$$\mu = (3k/N)^{1/2}\,(KT)^{1/2} \qquad [12]$$

Experimentally, volume susceptibility, K, is often replaced by mass susceptibility, χ, where

$$\chi = K/d \qquad [13]$$

and d is the density of the substance in g cm^{-3}. For pure compounds, χ_M can be defined as the molar susceptibility, related to χ by the molecular weight, **M**:

$$\chi_M = \mathbf{M}\chi \, . \qquad [14]$$

The molar susceptibility should be corrected for diamagnetic effects (and temperature-independent paramagnetism), leaving a "corrected" molar susceptibility, χ_M^{corr}, which can be attributed to paramagnetic species in the sample. For molar quantities, N in Eq. [12] is given by Avogadro's number and this equation becomes (Cotton & Wilkinson, 1966)

$$\mu = 2.84 \, (\chi_M^{corr} \cdot T)^{1/2} \, . \qquad [15]$$

Clearly, it should be possible to calculate the magnetic moment of the ion, atom, or molecule responsible for paramagnetism from Eq. [15] if the magnetic susceptibility of the sample is measured.

In contrast to the weak positive magnetism of disordered ions in paramagnetic solids, magnetically ordered materials demonstrate much stronger magnetism. *Ferromagnetism* refers to the intense response of certain solids (e.g., pure iron) to an applied magnetic field. It arises from the *spontaneous* long-range ordering of the directions of magnetic moments of some electrons within domains of the sample (Fig. 9–1b). The *direction* of spontaneous magnetization of a domain can be changed by the application of a magnetic field, but its *magnitude* is very weakly dependent on the intensity of the applied field. However, by raising the temperature, the spontaneous magnetization can be decreased, approaching zero at a critical temperature known as the Curie point, T_c. Above T_c, the substance follows the Curie or Curie-Weiss law (with positive Θ), since thermal agitation breaks down long-range dipolar ordering and produces paramagnetic behavior (Fig. 9–2). For many substances, T_c is an extremely high temperature (e.g. 1393K for Co metal), meaning that the forces which cause spontaneous magnetization are exceedingly strong. Of course, one magnetic dipole will exert aligning forces on its neighbors, but these *classical* magnetic forces are insufficient to explain magnetic ordering (Martin, 1967). However, quantum mechanics predicts an "exchange" interaction between two closely spaced electrons which tends to order the spins in a parallel or antiparallel configuration. The exchange forces are electrostatic, with strengths many orders of magnitude greater than the classical magnetic forces between spin moments. The expected qualitative relationship between magnetic susceptibility and temperature is compared for paramagnetism and ferromagnetism in Fig. 9–3.

Strong magnetism is also observed in *ferrimagnetic* materials (e.g., magnetite). These substances have magnetic behavior qualitatively similar to that of ferromagnetics, exhibiting spontaneous magnetization of domains below a critical temperature, T_c. However, at temperatures well

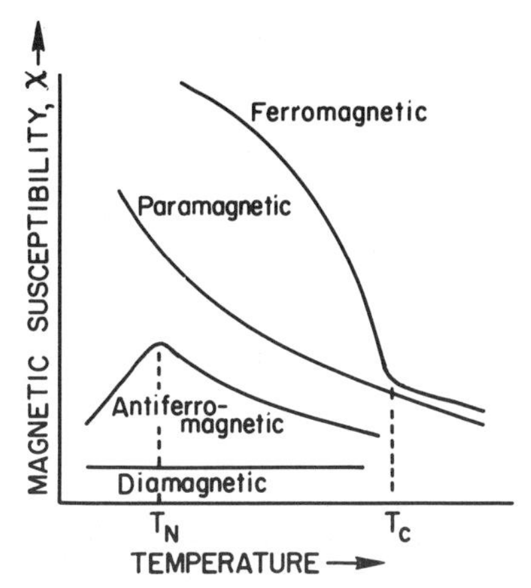

Figure 9–3. Relationship between susceptibility, χ, and temperature for the main classes of magnetic minerals.

above T_c, where magnetic ordering is destroyed, the paramagnetic susceptibility follows the Curie-Weiss law with negative Θ, (Fig. 9–2), rather than the positive Θ diagnostic of ferromagnetic metals. This difference is evidence that the strongest interactions between magnetic ions in ferrimagnetics are negative, favoring *antiparallel* alignment of neighboring magnetic dipoles. The ferrimagnetic crystal must then contain at least two coexisting sublattices which are spontaneously magnetized in different directions (Fig. 9–1c). The net spontaneous magnetization is the resultant of these two or more sublattice magnetizations. The type of crystal structure will generally determine the sublattice arrangement. For example, the spinel and garnet mineral structures are ferrimagnetic when they contain transition metals. For magnetite, Fe_3O_4, the Fe^{3+} ions occupy one type of site and form a sublattice with magnetization directed opposite to the sublattice composed of Fe^{2+} ions in the other type of site. Since the Fe^{3+} sites are more numerous, a net spontaneous magnetization or "ferrimagnetism" results.

In some cases, the two or more ordered sublattices are spontaneously magnetized in directions that produce a resultant magnetization of zero. Such a crystal is considered *antiferromagnetic* (see Fig. 9–1d), since it is highly ordered magnetically, but has a relatively weak magnetic susceptibility. For antiferromagnetic crystals, there is a characteristic temperature, T_N, called the Néel temperature (Fig. 9–3). Above T_N the solid behaves as a simple paramagnetic as magnetic order is lost, with reduced susceptibility at higher temperature. Below T_N, the antiparallel alignment forces dominate and the magnetic susceptibility again decreases. Thus, a maximum of susceptibility is observed very near T_N (Fig. 9–3). Well above T_N, a Curie-Weiss relation is followed with a negative value of Θ (Fig. 9–2). Below T_N, magnetic susceptibilities are usually slightly field-dependent since large applied magnetic fields may alter the direction of magnetic ordering.

Table 9–1. Main classes of magnetic behavior (adapted from Cotton and Wilkinson, 1966).

Class	Range of χ_M (cgs units)	Dependence of χ_M on H	Origin
Diamagnetism	-1 to -500×10^{-6}	Independent	Polarized electron orbitals
Paramagnetism	0 to $+10^{-2}$	Independent	Spin and oribital magnetic moment of unpaired electrons on individual atoms
Ferro (ferri) magnetism	$+10^{-2}$ to 10^{6}	Dependent	Exchange interactions between unpaired electrons in neighboring atoms
Antiferromagnetism	0 to $+10^{-2}$	May be dependent	

Table 9–2. Magnetic properties of soil minerals (data from Mullins, 1977 and Martin, 1967).

Mineral	Formula	Magnetism class	Critical temperature (K)
Magnetite	Fe_3O_4	Ferrimagnetic	$T_c = 858$
Hematite	α-Fe_2O_3	Antiferromagnetic	$T_N = 953$
Maghemite	γ-Fe_2O_3	Ferrimagnetic	--
Goethite	α-FaOOH	Antiferromagnetic	$T_N = 390$
Lepidocrocite	γ-FeOOH	Antiferromagnetic	$T_N = 70$
Ilmenite	$FeTiO_3$	Antiferromagnetic	--
Pyrrhotite	Fe_7S_8	Ferrimagnetic	--
Pyrolusite	MnO_2	Antiferromagnetic	$T_N = 84$

A summary of the main classes of magnetic behavior is given in Table 9–1. It is seen that any ferromagnetic or ferrimagnetic crystals in a sample would probably largely determine the value of χ in that sample, preventing the measurement of paramagnetism originating from isolated atoms of transition elements. In addition, the field dependence of strong magnetism requires that a measured magnetization be extrapolated to a zero-field value. This is often done by subjecting the sample to a very strong field and measuring the magnetization which equals the spontaneous magnetization plus a small increment produced by the applied field. Then, by repeating this measurement for several values of the strong field, a zero-field magnetization is determined by extrapolation.

The magnetic properties of a number of minerals are summarized in Table 9–2. All are either ferrimagnetic or antiferromagnetic. Although most clay minerals are expected to be paramagnetic due to the presence of Fe^{3+}, Mn^{2+}, and other transition metals in their structures, measured susceptibilities cannot be wholly attributed to these ions if highly magnetic mineral impurities are present.

9–2.2 Apparatus

The magnetic susceptibility of a sample is obtained by measuring the magnetic moment per unit volume or mass induced in a sample by an applied magnetic field, H. This magnetic moment can be determined by two general types of methods (Zijlstra, 1967): (i) measurement of the force exerted on the magnetic moment by an inhomogeneous field, and (ii) measurement of the voltage induced in an electrical circuit by a varying magnetic moment. The first type includes the Faraday and Gouy methods, while the second includes the weak alternating field bridge technique which some researchers have preferred for studying soils (Mullins, 1977).

One apparatus employing the latter method, described by Lukshin et al. (1968), is well adapted to soil studies, allowing large samples to be used (Mullins, 1977). In addition, the weak magnetic fields applied to the sample avoid the problem of field-dependent susceptibility that is observed at higher field strengths with strongly magnetic minerals. Thus, magnetic susceptibilities of ferrimagnetic minerals measured in different laboratories can be compared. The apparatus is sensitive, on the order of 10^{-6} cgs units, and can be made portable for in situ field measurements (Mooney, 1952). However the susceptibility in soil samples may be dependent on the frequency of the alternating field (Mullins, 1977), so that values must be compared at the same frequency.

The Faraday and Gouy methods are most commonly used to determine paramagnetic susceptibilities. Both utilize a balance to measure the force exerted on a sample placed in an inhomogeneous magnetic field. A sample having volume dV and mass dM in a horizontal magnetic field, H (i.e., H is in the x direction) is subjected to a force in the vertical direction (z-axis) given by (Lindoy et al., 1972)

$$dF = K \cdot dV \cdot H_x (\partial H_x/\partial z) \qquad [16]$$

where K is the volume susceptibility of the sample. In the Gouy method, the sample is packed into a relatively long cylindrical tube of length l, and uniform cross-sectional area A. The tube is suspended between the poles of a magnet so that one end is subjected to the magnetic field, H_1, while the other end is in a region of near zero field, H_0 (Fig. 9–4a). Integrating Eq. [16] over the full sample length, one obtains

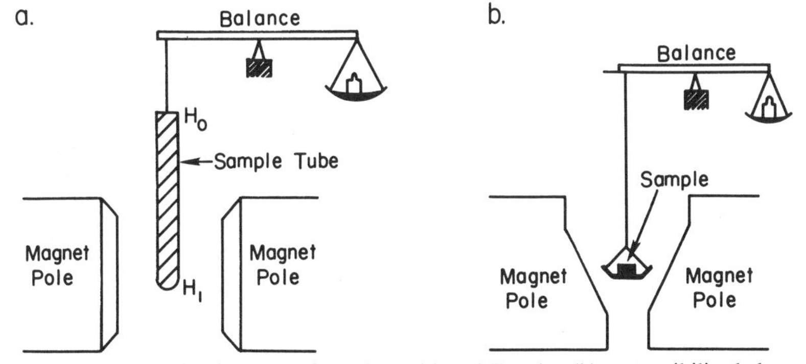

Figure 9–4. Schematic diagrams of the Gouy (a) and Faraday (b) susceptibility balances.

$$F = (KV/2l)(H_1^2 - H_0^2) \quad [17]$$

where V is the whole sample volume. Thus

$$F = (KA/2)(H_1^2 - H_0^2) . \quad [18]$$

In general, if H_1 is >10 times H_0, Eq. [18] shows that H_0 can be assumed to be zero with an error of $<1\%$. Therefore, a simple relationship between the force exerted on the sample and the magnetic field is obtained

$$F = (KA/2)H_1^2 . \quad [19]$$

This method has been popular because of the simplicity of the equipment. It has good sensitivity but fairly large samples are required (up to 1.0 g). Sample homogeneity and uniformity of packing in the tube are essential for reproducible measurements. Inhomogeneous compression of powder samples in the tube can be a problem.

The Faraday method is generally more popular for accurate susceptibility measurements. A very small sample (5–50 mg) is suspended in a region between the poles of a magnet that has been modified to provide a constant field gradient in the vertical direction [i.e., constant $H(\partial H/\partial z)$]. The gradient is usually obtained by fitting oblique pole faces on the magnet, so that for the apparatus diagrammed in Fig. 9–4b, the force on the sample will be downward if K is positive. From Eq. [16], the force on the sample is

$$F = K \cdot V \cdot H\,(\partial H/\partial z) . \quad [20]$$

While in the Gouy method, accuracy was dependent on the homogeneity of the sample within the tube, the Faraday method depends only upon the uniformity of the field gradient, $\partial H/\partial z$ in the region of the sample. With properly designed pole faces, a constant gradient can be maintained over about 1 cm (Craik, 1971). However, $\partial H/\partial z$ is not usually uniform over a wide range of H, generally decreasing for both very strong and very weak fields. For this reason, the Faraday apparatus is sometimes modified to use gradient coils rather than oblique pole faces to obtain a more constant gradient (Zijlstra, 1967). The Faraday method is considered more valid than the Gouy method for measuring ferromagnetism and ferrimagnetism (Lindoy et al., 1977). Reproducibility of sample positioning in the field gradient is critical to the method, a requirement that favors the use of an electrobalance rather than a standard balance because of the unchanging position of the sample holder. It is possible to suspend the sample within a cryostat for variable temperature measurements, a great advantage in interpreting magnetic behavior of samples (Lindoy et al., 1977).

9–2.3 Procedure

The method of magnetic susceptibility measurement for the generally preferred Faraday magnetic balance is as follows. The force, F, on the sample (see Eq. [20]) is calculated from the measured change in weight, ΔW, in the presence and absence of the magnetic field

$$F = g\Delta W \qquad [21]$$

where g is the gravitational constant. Since the density of the sample, d, is the weight (W) divided by the volume (V), and $K = \chi d = \chi(W/V)$, then Eq. [20] can be rearranged:

$$\chi = g\Delta W/[W \cdot H(\partial H/\partial z)] \qquad [22]$$

where χ is the mass susceptibility. With a given apparatus at constant H, χ is then simply related to the value of ΔW, and sample volume or geometry (for small samples) has no effect on the measured susceptibility

$$\chi = \Delta W \cdot B/W . \qquad [23]$$

B is a constant that could theoretically be calculated from the magnetic field properties, but it is usually measured using a standard compound of known susceptibility (e.g. $HgCo(SCN)_4$ or $[Ni(en)_3]S_2O_3$). A quartz sample holder is used to minimize paramagnetic impurities in the holder. Although pure quartz is diamagnetic, the change in sample weight contributed by the sample holder, ΔW_h, may still be temperature-dependent because of the possibility of paramagnetic impurities or defects. Thus, ΔW_h should first be measured at the temperature (or temperatures) to be used for the sample. Then a standard of known susceptibility is placed in the holder, and B is calculated from the following equation:

$$\chi(T) = B(\Delta W - \Delta W_h)/W \qquad [24]$$

where $\chi(T)$ is mass susceptibility at temperature T. Finally, a weighed sample with unknown $\chi(T)$ is placed in the sample holder, the value of ΔW is determined, and $\chi(T)$ is calculated from Eq. [24] using the known B and ΔW_h values.

The analogous procedure can be applied to the Gouy method, converting Eq. [19] to the following form:

$$\chi(T) = (2g \cdot \Delta W \cdot l)/(W \cdot H_1^2) . \qquad [25]$$

For a given magnetic field strength and sample geometry, this can be reduced to

$$\chi(T) = \Delta W \cdot B'/W \qquad [26]$$

which has exactly the same form as the Faraday equation. However, B' is sensitive to sample geometry and homogeneity, while B is not. Equation [24] can then be used for both the Faraday and Gouy methods by determining B for the specific apparatus.

The usual standards used to calibrate susceptibility measurements, $HgCo(SCN)_4$ and $[Ni(en)_3]S_2O_3$, are readily prepared and nonhygroscopic. The latter compound has a lower susceptibility, being more suitable for higher magnetic fields. The mass susceptibility of $HgCo(SCN)_4$ is 16.44×10^{-6} cgs units at 20°C, following the Curie-Weiss law (Lindoy et al., 1972)

$$\chi(T) = 4985 \times 10^{-6}/(T + 10). \qquad [27]$$

Thus, χ can be calculated for the standard at that temperature to be used for sample measurement. The susceptibility of $[Ni(en)_3]S_2O_3$ is 10.82×10^{-6} cgs units at 20°C, obeying the Curie law for a range of temperatures (Lindoy, et al., 1972)

$$\chi(T) = 3172 \times 10^{-6}/T. \qquad [28]$$

To obtain molar magnetic susceptibility, χ_M, $\chi(T)$ is multiplied by molecular weight. These molar susceptibilities can be corrected for diamagnetism of the substituent atoms of the compound by using Pascal's constants (Figgis & Lewis, 1960). In the case of $[Ni(en)_3]S_2O_3$ with a molecular weight of 351.5 *g*, $\chi_M = 3803 \times 10^{-6}$ at 20°C. The diamagnetic correction for this compound is $\chi_{\text{diam}} = -199 \times 10^{-6}$ (independent of T). Therefore the *corrected* χ_M is given by: χ_M^{corr} (20°C) $= \chi_M - \chi_{\text{diam}} = 4002 \times 10^{-6}$ cgs units.

The mass susceptibilities of a number of common soil minerals are listed in Table 9–3, showing the extreme range from diamagnetic minerals such as quartz to ferrimagnetics such as magnetite.

9–2.4 Comments

For certain soil or mineral samples containing strongly magnetic solids, the force exerted by the magnetic field of the Gouy apparatus may be great enough to cause the sample to sway toward one pole of the magnet. This effect should be avoided by diluting the sample with a diamagnetic material (e.g., $CaCO_3$). In addition, however, the observed susceptibility of these strongly magnetic minerals dispersed in the sample will depend on particle size, shape and concentration (Mullins, 1977). Only the cooperative phenomena of ferromagnetism and ferrimagnetism demonstrate these effects, resulting from the demagnetizing field induced at both ends of a magnetic particle upon the application of an external field.

Particle size has a fundamental influence on magnetic ordering, since

Table 9–3. Typical measured mass magnetic susceptibilities of common primary and secondary minerals (data from Vadyunina and Babanin, 1972 and Mullins (1977).

Mineral	$\times 10^6$ (cgs units)†
Corundum	−0.34
Orthoclase, calcite	−0.38
Quartz	−0.46
Dolomite	+0.9
Muscovite	+1.0–12
Biotite	+12–52
Amphiboles	+13–55
Pyroxenes	+3.0–75
Hematite	+20
Siderite	+93
Magnetite	up to +800 000
Lepidocrocite, goethite	+42
Maghemite	up to +30 000
Kaolinite	−1.5
Montmorillonite	+2.2
Nontronite	+69
Vermiculite	+12

† To convert from cgs units ($cm^3\ g^{-1}$) to rationalized SI units ($m^3\ kg^{-1}$), multiply values by 4×10^{-3}. See Quickenden and Marshall, 1972 for further details.

large particles contain a number of regions (domains) within which the electron spin magnetic moments are aligned, while the magnetization vectors of adjacent domains are not aligned. By the application of a very strong magnetic field, the domain structure is eliminated, as all magnetic moments become aligned in the field. The magnetization of the solid is then at its maximum value, M_s, and the solid is *magnetically saturated.* For an assembly of particles, the removal of the saturating field results in a decay of magnetization, M, with time (Craik, 1971)

$$M(t) = M_s \cdot e^{-t/\tau} \qquad [29]$$

where t is the time after removal of the field and τ is the relaxation time for decay of M. Then, $1/\tau$ can be considered the probability per second that thermal agitation will reverse the magnetization. For a particle of volume V, which has a stable orientation of magnetization in the absence of the field due to magnetic anisotropy, the activation energy of orientation is $CV/2$ (where C is a magnetic anisotropy constant), and the probability of reorientation is related to the required activation energy by the Boltzmann equation

$$(1/\tau) \propto \exp(-CV/2kT)\,. \qquad [30]$$

A *superparamagnetic* system can then be arbitrarily defined as one in which M_s decays at a fast rate (i.e., $\tau < 10^2$ s). For spherical particles, only crystal anisotropy plays a role in reorientation, and a critical particle diameter can be obtained from Eq. [30]. For diameters a little smaller than this critical value, the decay is very rapid, since $CV/2 \approx kT$ and there is a finite probability that the magnetization will be spontaneously reversed. Such continuously fluctuating magnetization in small particles creates the phenomenon of superparamagnetism. Particles slightly larger than this diameter have values of τ so large that the system appears magnetically stable. One may consider that a sufficiently small particle has only a single domain. For example, for nearly spherical magnetite, the critical diameter is near 30nm (Mullins, 1977). The magnetic moment of each small particle can flip from one easy axis of magnetization to another. With an applied magnetic field, these moments will spend a greater fraction of time oriented in the field direction, much like the behavior of individual electron spins of paramagnetic materials in a magnetic field. However, because the whole-particle magnetic moment is much larger than the moment of a single unpaired electron, the alignment in the field is much more efficient, explaining the very high values of magnetic susceptibility found in superparamagnetism. Within the region of superparamagnetism, the susceptibility of a particle is proportional to its volume. In the narrow range of particle size near the critical diameter, which separates superparamagnetic behavior from stable single domain behavior, magnetic properties are a complex function of time, and *magnetic viscosity* (time-dependent magnetization) is observed. For example, using the alternating field bridge to measure soil susceptibility, the frequency of the alternating field has been observed to change the apparent result (Mullins, 1977). In soils, antiferromagnetic clay minerals such as hematite and goethite have been shown by Mossbauer spectroscopy to possess superparamagnetic properties (on the atomic scale) at room temperature, indicating very small particle size (Bigham et al., 1978). However, particle geometry and ionic substitution can influence the critical particle diameter, the latter being evident in Al-substituted goethites which have large critical diameters. From Eq. [30], it is clear that a lowered temperature will reduce the probability of reorientation, $1/\tau$, and a value of T will be reached where superparamagnetism no longer occurs. In some soil goethites and hematites, temperatures as low as 4K may be necessary to restore domain ordering on the time scale of Mossbauer spectroscopy (Bigham et al., 1978; Longworth et al., 1979).

The paramagnetic species of most general interest to soil chemists and mineralogists are the iron-group (3d) transition metals. In strong crystal fields, the orbital and spin angular momenta of the 3d electrons can be treated separately to the first approximation. As a result, the orbital magnetic momentum of the electrons can be shown not to have a finite component in any given direction, making no contribution to the magnetic moment of the atom. Thus, for 3d ions in crystals, the magnetic moment arises largely from the electron spin, and the orbital momentum is said to be "quenched." This permits the L terms in Eq. [7] to be ignored and the moment is given by

$$\mu_s = 2[S(S+1)]^{1/2}\mu_B . \qquad [31]$$

Table 9–4. Theoretical and experimental magnetic moments for 3d transition metals (data from Cotton and Wilkinson, 1966 and Craik, 1971).

Ion	Ground state quantum numbers S	L	Magnetic moment μ_S (theoretical)	μ_{S+L} (theoretical)	μ (observed)
			Bohr magnetons		
Sc^{2+} ($3d^1$)	1/2	2	1.73	3.00	1.7
Ti^{2+} ($3d^2$)	1	3	2.83	4.47	2.8
Cr^{3+}, V^{2+} ($3d^3$)	3/2	3	3.87	5.20	3.8
Cr^{2+} ($3d^4$)	2	2	4.90	5.48	4.9
Fe^{3+}, Mn^{2+} ($3d^5$)	5/2	0	5.92	5.92	5.9
Fe^{2+} ($3d^6$)	2	2	4.90	5.48	5.1–5.5
Co^{2+} ($3d^7$)	3/2	3	3.87	5.20	4.1–5.2
Ni^{2+} ($3d^8$)	1	3	2.83	4.47	2.8–4.0
Cu^{2+} ($3d^9$)	1/2	2	1.73	3.00	1.7–1.8

This is generally found to be a good approximation, as shown in Table 9–4. For example, exchangeable Mn^{2+} on montmorillonite has μ between 6 and 7 Bohr magnetons, reasonably close to the predicted value of 5.9 for high spin Mn^{2+} (McBride, et al., 1975). In this case, iron oxide impurities may have biased the result. Biotite samples have measured magnetic moments per Fe^{3+} and Fe^{2+} ion in the structure of 6 to 7 and 5.1 to 5.4 Bohr magnetons, respectively (Anagnostopoulos & Calamiotou, 1973). This type of calculation can be made only when structural paramagnetic ions such as Fe^{2+}, Fe^{3+}, and Mn^{2+} are dispersed in the diamagnetic mineral structure (i.e., no cooperative magnetic phenomena occur). The molar magnetic susceptibility of the mineral is then equal to the sum of the susceptibilities contributed by each paramagnetic species

$$(\chi_M)_{\text{total}} = \Sigma(\chi_M)_i \,. \qquad [32]$$

Equations [5], [11], [13], and [14] are then used to obtain a "molar susceptibility"

$$(\chi_M)_i = (MV/W)\{\mathrm{N_i}\mu_i^2/[3k(T - \Theta)]\} \qquad [33]$$

where M for a mineral such as a mica can be taken as the formula weight, N_i and μ_i are the number and magnetic moment of the ith paramagnetic species in the structure, and V and W are the volume and weight of the mineral sample. Since W/M is the number of "moles" (formula weights) of the mineral, n, then N_i/n is the number of paramagnetic ions per "mole" of mineral, and Eq. [33] can be rewritten

$$(\chi_M)_i = V(N_i/n)\{\mu_i^2/[3k(T - \Theta)]\} \,. \qquad [34]$$

In the absence of exchange interactions between paramagnetic ions, $\Theta = 0$, and Eq. [32] and [34] can be used to assign magnetic contributions to the paramagnetic ions of minerals having known chemical composition. By measuring $(\chi_M)_{total}$, one can determine μ_i for each structural ion, although the presence of ferrimagnetic impurities in the mineral would invalidate these calculations.

Magnetic separation techniques have utilized the different magnetic properties of natural minerals. Thus, magnetite, being ferrimagnetic, is easily removed from soil samples using a small permanent magnet. Very strong magnetic fields allow strongly paramagnetic and ferrimagnetic minerals to be separated from diamagnetic and weakly paramagnetic minerals.

9–3 ELECTRON SPIN RESONANCE (ESR)

9–3.1 Principles

The magnetic moment of an electron, μ_e, is proportional to the spin angular momentum of the electron, $\hbar\mathbf{S}$, as was previously shown by Eq. [2]:

$$\mu_e = -g\beta\mathbf{S}\ .$$

For "pure" spin angular momentum, the g-factor has the value of 2, but g can vary for electrons in different atoms, since it determines the ratio of the total magnetic dipole moment to the total angular momentum of the electron in states where the angular momentum is partly spin and partly orbital. In the general case, where spin-orbit coupling results in an orbital contribution to the magnetic moment, the J quantum number must be defined with possible values ranging from $L + S$ to $|L - S|$ in integral steps, where L and S are the total orbital and spin quantum numbers for the atom. Then the g value is given by the Landé formula (Wertz & Bolton, 1972)

$$g_J = 1 + [J(J + 1) + S(S + 1) - L(L + 1)]/[2J(J + 1)] \quad [35]$$

and the magnetic moment of the atom is

$$\mu = -g_J\beta\mathbf{J}\ . \quad [36]$$

For example, the Ti^{3+} ion has a single unpaired 3d electron, so that $S = 1/2$, $L = 2$, and $J = 3/2$ or $5/2$. For transition metals in which the d shell is less than half full, the quantum state with the minimum J value is lowest in energy, while the reverse is true for metals with d shells more than half full. Thus, Ti^{3+} in the ground energy state has $J = 3/2$, and Eq. [35] can be solved to give $g_J = 0.8$. However, the first transition series metals are generally present in a medium-strength crystal field, whether they exist as ligand complexes in solution or ionic components of crystalline solids. The crystal field energy then exceeds the spin-orbit coupling energy, as the d-electrons of the atom interact strongly with the crystalline electric field. Some d-electrons will interact more strongly with the field than others, and the energies of the d-orbitals will be split apart. This *crystal field splitting* is usually great enough to allow only the lowest-lying states to be thermally populated. If the lowest state has no orbital degeneracy, the magnetic moment of the atom will be close to the contribution from the spin angular momentum, meaning that the orbital angular momentum has been "quenched" by the crystal field; that is, the average value for the angular momentum is zero. The quantum number J is no longer valid when the spin-orbit coupling is small relative to the crystal field energy, and if $L = 0$ for an orbitally nondegenerate ground state, then J is replaced by S in the Landé formula

$$g_S = 1 + [S(S + 1) + S(S + 1)]/[2S(S + 1)] = 2 . \qquad [37]$$

The g value of 2.00 is evidence of "spin-only" magnetic moment, and should be observed for transition metals in crystal fields, as well as electrons in s-orbitals ($L = 0$) of atoms or molecules. Thus, organic free radicals generally have g-values very near 2. The exact "free-electron" g-value is 2.0023 as a result of a small quantum electrodynamical correction (Abragam & Bleany, 1970). In real systems, although many first-row transition metals have g-values near 2, some orbital contributions are usually involved. An observed g-value for $Ti(H_2O)_6^{3+}$ is about 1.2 (Carrington & McLachlan, 1967), obviously between the predicted fully spin-orbit coupled value of 0.8 and the pure-spin value of 2.

The ESR experiment relies upon the interaction between the electron magnetic moment, μ_e, and an applied magnetic field, **H**. If the magnetic field is defined to be applied in the z-direction, the electron spin (S = 1/2) is quantized in that direction, having the two allowed values of M_s = 1/2 and $-1/2$. This can be visualized as two possible orientations of the electron spin in the magnetic field, parallel or antiparallel to H. This alignment of electron dipoles is depicted in Fig. 9–5, where a collection of randomly oriented electrons (i.e., a paramagnetic sample) is subjected to a field, creating an energy of interaction

$$E = -\mu_e H = g\beta H M_s . \qquad [38]$$

The allowed energies, E, of the unpaired electrons are then $+1/2\ g\beta H$ and $-1/2 g\beta H$, the difference being $\Delta E = g\beta H$ (see Fig. 9–5c). If electromagnetic radiation of proper frequency, ν, is applied perpendicular to **H**, transitions between these two Zeeman energy levels can be induced. The resonance condition

$$h\nu = \Delta E = g\beta H \qquad [39]$$

must be met for transitions to be possible. In the ESR experiment, the

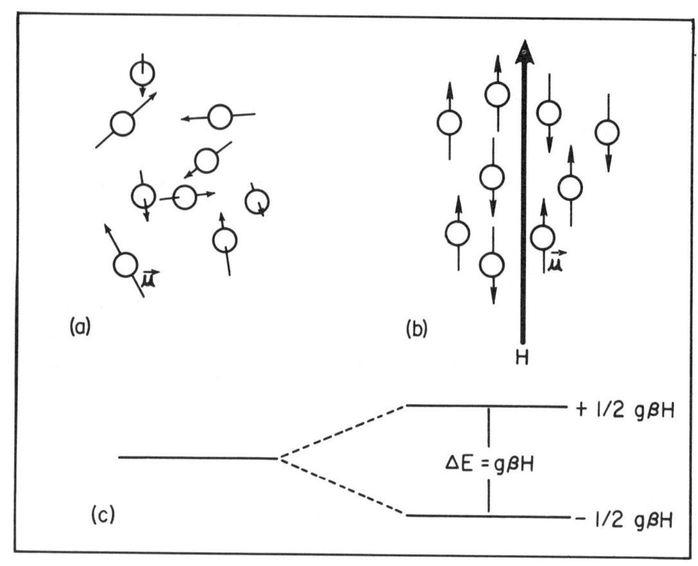

Figure 9–5. Orientation of electron magnetic moments in the absence (a) and presence (b) of an applied magnetic field, H. Relative energy levels of electrons oriented parallel and antiparallel to H are shown in (c).

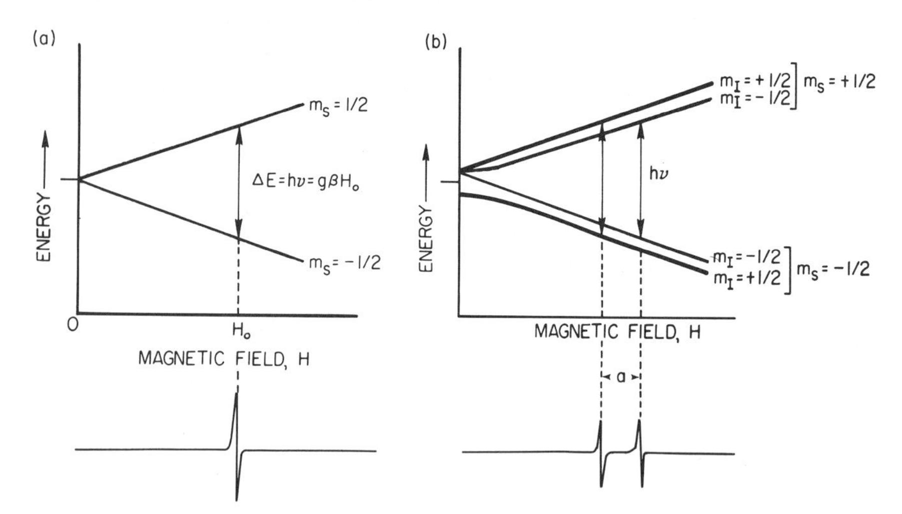

Figure 9–6. Effect of an applied magnetic field upon the energy levels of an electron ($S = 1/2$) for the cases of no nuclear interaction (a) and interaction with a nucleus having $I = 1/2$ (b). The electron transitions and observed first-derivative spectra are shown for the ESR experiment, with a representing the nuclear hyperfine coupling.

frequency of the radiation is usually fixed near $\nu = 9.5 \times 10^9$ Hz (*X*-band), corresponding to microwave radiation with a wavelength of about 3 cm. Spectra are obtained by scanning through a magnetic field until absorption of radiation is detected at a certain field strength, H_0, the value of H which satisfies Eq. [39]. In Fig. 9–6a, the resonance position,

H_0, is shown schematically. For a "free electron" (g = 2.0023), H_0 is about 0.34 tesla when ν is near 10^{10} Hz. Net absorption of electromagnetic radiation (and therefore signal detection) can occur only if there is a population difference between the two spin levels, since an electron has equal probability of undergoing a transition ("flipping") at the resonance condition regardless of its spin state. The Boltzmann equation shows that there will be a slight excess of spins in the lower energy level ($m_s = -1/2$) at room temperature and $\nu \approx 10^{10}$ Hz:

$$N_+/N_- = \exp\{-g\beta H/kT\} \cong 0.9984 \qquad [40]$$

where N_+/N_- is the ratio of number of electrons in the upper and lower energy states. Thus, only a very small fraction of the unpaired electrons in a sample actually contribute to the absorption signal. In addition, spins promoted by radiation to the higher energy level must be able to return rapidly to the lower level by relaxation processes. Otherwise, the spin system becomes "saturated," with equal numbers of spins in both levels and no net absorption of energy.

9–3.1.1 THE g-FACTOR

If all unpaired electrons in samples had spin-only g values of 2, the determination of g from the resonance position in ESR spectra would have little value. However, local magnetic fields induced in atoms by the external magnetic field may add to or subtract from the external field, H, thereby shifting the resonance position of the electron. The observed g-factor ($g = h\nu/\beta H_0$) varies depending on the characteristics of the molecule or atom in which the unpaired electrons are located. The main source of local magnetic fields causing g to deviate from the pure-spin value of 2.0023 is orbital magnetic moment generated from spin coupling to excited electronic states (spin-orbit coupling) as described earlier in this section for the case of transition metals. For most atoms or molecules, the spin-orbit coupling is anisotropic (orientation-dependent). Therefore, the magnitude of the local induced magnetic field depends on the orientation of the molecule relative to the external magnetic field, and g is orientation-dependent. The anisotropy of g, which is mathematically expressed as a second-rank tensor, can provide detailed information on the orientation of atoms and molecules in single crystals. For all g-tensors, there exists a principal axis system (X,Y,Z) which permits complete description of the tensor by only three principal values, g_{xx}, g_{yy}, and g_{zz}. The axes are generally coincident with molecular symmetry axes so that octahedral, tetrahedral, or cubic symmetries produce an isotropic g-factor ($g_{xx} = g_{yy} = g_{zz}$). Molecules with one threefold or higher axis of symmetry, labeled the z-axis by convention, have $g_{xx} = g_{yy} \neq g_{zz}$, since the x and y axes are equivalent. The g-factors in such molecules of axial symmetry [e.g., $Cu(H_2O)_4{}^{2+}$] are usually labelled $g_{\parallel}$ (the g-factor parallel to the symmetry axis, i.e. $g_{\parallel} = g_{zz}$) and $g_{\perp}$ (the g-factor perpendicular to the symmetry axis, i.e. $g_{\perp} = g_{xx} = g_{yy}$). If molecules have no threefold or higher axis of symmetry, the three principal g-factors are all different ($g_{xx} \neq g_{yy} \neq g_{zz}$).

For paramagnetic molecules in solution, anisotropy of the g-tensor is usually averaged and a single isotropic g-factor is exhibited, given by $g_0 = 1/3\,(g_{xx} + g_{yy} + g_{zz})$. However, averaging will occur only if molecular tumbling is fast relative to the ESR time-scale. The tumbling frequency must be much greater than $\Delta g \beta H/h$, where Δg is the difference between the lowest and highest values of g. Molecules of high molecular weight or molecules in solutions of high viscosity often cannot undergo rotational motion rapidly enough to average the g-factor anisotropy, and therefore demonstrate "rigid-limit" spectra with resolved principal g-factors. Such is the case for humic acid complexes of divalent copper and vanadyl ions, with rigid-limit spectra observed for aqueous suspensions (McBride, 1978).

Many ESR investigations of natural systems produce "powder" rather than "single-crystal" spectra, since the matrix containing the paramagnetic species is composed of randomly oriented particles (e.g., clays or organic colloids). The result is a loss of information regarding the alignment of the species relative to the crystal or molecular axes of the particles. In addition, anisotropy of the g-factor produces broadening of the resonance lines. However, if the principal g-factors are significantly different, the individual g values can still be resolved despite the random orientation. For example, if a paramagnetic ion with axial symmetry (e.g. Fe^{3+}) were contained in the structure of a clay particle, the "powder" spectrum of the clay would show resonance peaks at two values of H (Fig. 9–7)

$$H_{\parallel} = h\nu/g_{\parallel}\beta, \qquad H_{\perp} = h\nu/g_{\perp}\beta\,. \tag{41}$$

Resonance can occur only when H is between $H_{\parallel}$ and $H_{\perp}$. Therefore, as the spectrum is recorded with H approaching $H_{\parallel}$ from low field, absorption of energy suddenly begins at $H = H_{\parallel}$ (Fig. 9–7a). Absorption continues between $H_{\parallel}$ and $H_{\perp}$, becoming much more intense as H approaches $H_{\perp}$, since the probability that ions in the powder have their z-axes oriented nearly in the *plane* perpendicular to **H** is much higher than the probability that they have their z-axes almost parallel to **H**.

9–3.1.2 HYPERFINE SPLITTING

In addition to the local fields induced by the external field which result in different g-factors, other local fields exist which do *not* depend on the presence of an external field. These permanent local fields usually arise from magnetic nuclei (i.e., nuclei with spin) in the molecule, where the magnetic moment of the nucleus, μ_N, is proportional to the magnitude of the nuclear spin, I (see Eq. [1]). The different values of the nuclear g-factor (g_N) and I in Eq. [1] distinguish one nucleus from another. The

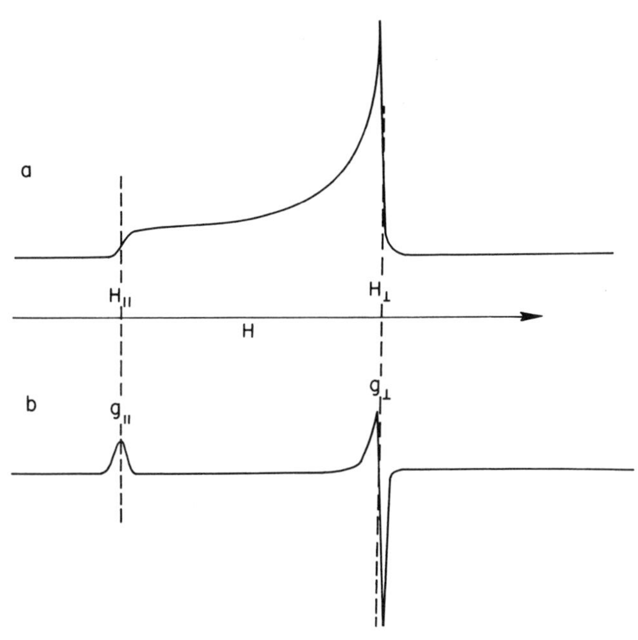

Figure 9–7. ESR spectrum of a randomly oriented paramagnetic species with axial symmetry and $g_{\parallel} > g_{\perp}$. The absorption (a) and first-derivative (b) powder spectra are shown with vertical broken lines indicating the positions of $g_{\parallel}$ and $g_{\perp}$.

interaction of the magnetic moment of an unpaired electron, μ_e, with μ_N is called the *nuclear hyperfine interaction*, producing *hyperfine splitting* of ESR spectra. Like the electron spin, the z-axis component of the nuclear spin is quantized in a magnetic field and allowed to have only certain values defined by the quantum number M_I. For a nucleus with $I = 1/2$, M_I can have only the values $+1/2$ or $-1/2$. Therefore, the local magnetic field generated by the two possible orientations of the nuclear spin vector I will add to or subtract from H, splitting each electron spin state into two possible energy levels. The result for an $S = 1/2$, $I = 1/2$ system (e.g., the hydrogen atom) is that two resonance lines appear, split by a tesla, where a is the hyperfine splitting constant (Fig. 9–6b). The selection rules for ESR are $\Delta M_S = \pm 1$ and $\Delta M_I = 0$, so that electron transitions involving simultaneous reorientation of the nuclear spin are forbidden. In general, for a nuclear spin I, there are $(2I + 1)$ possible nuclear spin states, and therefore $(2I + 1)$ resonance lines due to hyperfine splitting in the $S = 1/2$ system. Different isotopes of the same element *may* have different nuclear spins, so that hyperfine splitting may be affected by isotope ratios. For example, ^{12}C has $I = 0$ and does not produce hyperfine splitting in organic radicals, but the natural abundance of ^{13}C ($I = 1/2$) is about 1%, creating a very weak hyperfine doublet.

An electron in an s-orbital of an atom has equal probability of all orientations relative to the nuclear magnetic moment vector, μ_N, so that the average value of the local magnetic field at the electron produced by the nuclear spin is zero according to classical theory. However, the *Fermi contact interaction*, which is the quantum-mechanical effect of an electron having a finite probability of being found at the nucleus, produces an *isotropic* hyperfine interaction given by the coupling constant, *a*:

$$a = (8\pi/3)(g\beta g_N\beta_N|\psi(0)|^2) \quad [42]$$

where a has dimensions of energy and $|\psi(0)|^2$ is the squared amplitude of the electron orbital wave-function at the nucleus. The hydrogen atom, with an unpaired electron in the ls orbital, has a large isotropic hyperfine splitting ($I = 1/2$) as a result of the finite value of $|\psi(0)|^2$ for s-orbitals. In general, electrons in s-orbitals have isotropic hyperfine interactions, while electrons in p-, d-, and f-orbitals have orientation-dependent (anisotropic) hyperfine splitting constants arising from dipole-dipole interactions between the electron and nucleus. Since the p-, d-, and f-orbitals have nodes at the nucleus (i.e., $|\psi(0)|^2 = 0$), an unpaired electron should have zero probability of being found at the nucleus, and Eq. [42] predicts *no* isotropic hyperfine splitting. However, transition metals with unpaired electrons in d-orbitals *do* exhibit isotropic coupling, apparently by polarization of the s-orbitals through interaction with the unpaired d-electrons, producing a net electron magnetic moment at the nucleus. Therefore, for transition metals in solution, rapid thermal rotation averages the dipole-dipole (anisotropic) interaction to zero, but an isotropic hyperfine splitting always remains.

In addition to hyperfine interactions, other phenomena can give rise to splitting of spectral lines. *Fine structure* occurs in spin systems in which there is more than one unpaired electron per atom or molecule (i.e., $S>1/2$), an uncommon case for organic radicals but very common with transition metal ions. The magnetic field then splits the energy of the electrons into $2S + 1$ sublevels, each characterized by an M_s quantum number. Using the selection rule that $\Delta M_s = \pm 1$, $2S$ electron transitions are allowed. For example, Mn^{2+} and Fe^{3+} in clay mineral structures have five unpaired 3d-electrons ($S = 5/2$), so that six electronic sublevels are created in a magnetic field and five transitions are possible.

Superhyperfine splitting is occasionally observed in transition metal-ligand complexes if the ligands (e.g., ^{14}N) have nuclear spin. The unpaired d-electrons of the metal must be partially delocalized, spending a fraction of time at the ligand atoms. This phenomenon is differentiated from hyperfine splitting, since the electron spin interacts with nuclei of atoms other than that of its own.

9–3.1.3 FERROMAGNETIC RESONANCE

Early in this chapter, strong magnetism was shown to arise from cooperative interaction between unpaired electron spins in solids. Since the resonance experiment in ESR is applicable to all materials that contain unpaired electrons, ferromagnetic and ferrimagnetic materials should theoretically produce ESR spectra. The phenomenon of microwave radiation absorption by strongly magnetic materials is referred to as ferromagnetic resonance, and is significant in the study of soil materials because of the ferromagnetic resonance peaks commonly observed in ESR spectra of soils.

Unlike isolated paramagnetics, the usual subject of ESR investigation, ferromagnetics have uncompensated magnetic-spin moments of lattice ions oriented parallel to one another below their Curie temperatures, T_c. Although the electrostatic exchange interaction creating this magnetic ordering has no direct effect on the resonance, it produces a very large net magnetization as well as a large *internal* magnetic field. When the magnitude and orientation of the magnetization, M, change in a crystal, the resonance condition of ESR (Eq. [39]) can be significantly affected. The external field, H, is no longer the field experienced by the electrons, and an effective field, H_{eff}, must be defined so that (Martin, 1967)

$$h\nu = g\beta H_{eff} = g\beta(H + 2K_1/M_s) \quad [43]$$

where K_1 is the first anisotropy coefficient and M_s is the spontaneous magnetization of a crystal. (This equation must be modified if nonspherical particles are considered.) The angular dependence of the external magnetic field, H, at which resonance occurs then allows the K_1 contribution to the effective magnetic field to be separated. The "true" g-value for the electrons can then be determined, since the K_1 term can obviously produce apparent experimental g values that are orientation-dependent and very different from the spin-only value of 2. The "true" g value as defined in Eq. [43] is generally very nearly 2 for ferromagnetics, indicating that orbital magnetic moment makes only a small contribution to the magnetization of the solid. The crystal field of the lattice "quenches" the orbital magnetic moment almost completely, and g-values are generally only about 5% $>$2.00 (Vonsovskii, 1966). Ferromagnetics have high spontaneous magnetizations (about 0.1 tesla), and since magnetic resonance absorption is proportional to magnetic susceptibility, ESR absorption peaks due to ferromagnetics will be at least three orders of magnitude more intense than peaks due to paramagnetics.

Ferrimagnetics also demonstrate "ferromagnetic resonance," but the interpretation of measured g-factors (after correction for anisotropic internal magnetic fields as described for ferromagnetics) must take into account the presence of two or more magnetic sublattices. If all sublattice magnetic moments precess in phase throughout a crystal, the effective g-factor is a weighted mean over the sublattices (Martin, 1967):

$$g_{eff} = \Sigma_i \, g_i p_i / \Sigma_i \, p_i \quad [44]$$

where p_i is the spin momentum per ion in the ith lattice. For "spin-only" magnetism, $g_i = 2$ and $g_{eff} = 2$. In actual fact, g_{eff} is 2.15 for magnetite

(Fe_3O_4), but interpretation of this deviation from 2.00 as a contribution from orbital magnetic moment is not as straightforward as for ferromagnetics. If g_i for the different sublattices differs, a slightly temperature-dependent g_{eff} is expected because the p_i values vary differently with temperature.

Since ferromagnetic and ferrimagnetic materials lose long-range magnetic ordering above the Curie point, they behave as paramagnetics at high temperatures. The expected effect on their ESR spectra is a drastic reduction in resonance absorption intensity.

Antiferromagnetic crystals, since they have no resultant magnetization, should not in principle demonstrate an ESR signal. However, above the Néel temperature, the ordering is destroyed and paramagnetic resonance is induced. The g-factors for these compounds are near 2.00. As the Néel point is approached from higher temperatures, the paramagnetic absorption line broadens and decreases rapidly to zero intensity, while the g-factor remains constant. The line-broadening arises from the range of resonance conditions created by increasing short-range magnetic order. At temperatures well below the Néel point, there is no net dipole moment to couple to the applied electromagnetic field of the microwave radiation, but changes in the colinear alignment of the magnetizations of the sublattices can occur. This involves the turning of an electron spin moment out of the antiparallel alignment with its neighboring spins. Because of strong exchange interactions, the energy of this transition is large, and resonant frequencies occur in the extreme infrared rather than the microwave region. Thus, ESR spectroscopy does not produce resonances for purely antiferromagnetic crystals well below the Néel point unless the crystals are magnetically dilute (Martin, 1967).

A number of "antiferromagnetic" crystals have slight spontaneous magnetizations because of imperfect ordering by exchange forces. This phenomenon of *weak ferromagnetism* is observed in the similar structures of $MnCO_3$, $FeCO_3$, and α-Fe_2O_3 (hematite). The unit cell of the $MnCO_3$ crystal is a rhombohedron with Mn^{2+} ions in the corners and center, with the corner ions forming one magnetic sublattice and the center ions forming the second antiparallel sublattice. The antiferromagnetic sublattice magnetic moments are not strictly antiparallel because of a weak spontaneous magnetization that is generally less than 1% of the magnetizations of the individual sublattices (Turov, 1966a). ESR absorption and magnetic susceptibility should then be relatively weak, in the approximate range of paramagnetics (see Table 9–1). ESR absorption in weak ferromagnetics results from weak oscillations of the sublattice magnetizations, M_1 and M_2, about their equilibrium directions, which are determined by the angle between the applied magnetic field and the crystal axes. Only one of the two expected "spin wave" resonances of this type is in the centimeter wavelength range, so that a single ESR absorption for natural single crystals of hematite has been found, with the resonance field position (H_0) being dependent on the angle between **H** and the z-axis of the

Table 9–5. Magnetic properties of iron minerals (adapted from Angel and Vincent, 1978).

Mineral	Formula	Magnetic properties	Room temp. ESR signal intensity ($g \approx 2$)
Ferrihydrite	$\approx Fe(OH)_3$	Paramagnetic	Medium
Hematite	α-Fe_2O_3	Weakly ferromagnetic above −10°C	Medium
Goethite	α-FeOOH	Antiferromagnetic with Neel temperature of 120°C	Weak
Akaganeite	β-FeOOH	Antiferromagnetic with Neel temperature of 20°C	--
Maghemite	γ-Fe_2O_3	Ferrimagnetic	Strong
Lepidocrocite	γ-FeOOH	Antiferromagnetic with Neel temperature of −200°C	Medium
Magnetite	Fe_3O_4	Ferrimagnetic, ESR linewidth decreases with decreasing temperature to −143°C	Strong

crystal. Thus, "apparent" g-values are dependent on crystal orientation because of the internal field generated by the small imbalance of magnetic sublattices. In hematite, this internal magnetic field has been measured at about 2.3 tesla (Turov, 1966a). In summary, then, certain antiferromagnetics become weakly ferromagnetic by the application of a strong external magnetic field, a phenomenon observed for hematite above the temperature at which it converts to pure antiferromagnetism (about −10°C), as well as for $MnCO_3$ and $FeCO_3$. A summary of the magnetic properties of iron oxides of interest in soil science is given in Table 9–5. It should be stressed that iron oxides and hydroxides are not the only soil minerals capable of "ferromagnetic resonance," since manganese (e.g., MnO_2), titanium (e.g., $FeTiO_3$), and other transition metals may form minerals with ESR spectra.

ESR spectra of soils and clays commonly show a very broad, intense, ferromagnetic resonance centered at approximately $g = 2.00$. An explanation of these very wide resonances in ferromagnetic crystals is very difficult, although several mechanisms are known to create the broadening. For nonspherical particles, demagnetizing fields cause shifts in the effective magnetic fields, thereby creating a range of resonance field values for a collection of particles with different shapes. In addition, crystal magnetic anisotropy can shift the resonance field position because of the usual case of random orientation of the particles (and their anisotropy axes) relative to the applied magnetic field (Craik, 1971). Significant porosity or crystal defects give rise to internal demagnetizing fields which further increase absorption linewidth. Impurities in the crystals can produce substantial broadening as well. Some ferromagnetics have linewidths on the order of 0.1 tesla (e.g., magnetite). Polycrystalline solids produce very broad linewidths for the same reason that a collection of randomly oriented crystals shows this effect. The linewidths of ferromagnetic resonances are usually temperature-dependent, although the relationship between linewidth and temperature depends on the type of ferromagnetic, substituted impurities, and surface properties. The more ideal the crystal and the more perfect the state of the surface, the smaller is the linewidth (Turov, 1966b). The usual mechanisms given for the observed ESR linewidths of paramagnetics, spin-spin and spin-lattice relaxation, can account for only an insignificant effect in ferromagnetics. The most likely important contribution to linewidth in polycrystalline samples is the magnetic anisotropy of the individual crystallites, where the linewidth (ΔH) should be approximated by the anisotropy, H_A, if there is weak magnetic coupling between crystallites. A linewidth on the order of 0.01 to 0.1 tesla could result from this mechanism, and the resonance would not generally be symmetrical (Turov, 1966b). However, if the crystallite dipoles were strongly coupled, dipole-dipole exchange mechanisms would narrow the linewidth. Since powder samples are analogous to polycrystallites with weak magnetic coupling, large values of ΔH for ferromagnetic resonances in soil clays are expected.

9–3.2 Apparatus

A typical X-band electron spin resonance microwave spectrometer is composed of an electromagnet giving a continuous magnetic field that can be varied in strength (usually 0-1 tesla), a resonance cavity where the sample is positioned, and a microwave source (klystron) with a frequency of 9.5×10^9 Hz (Fig. 9–8). The microwave radiation is sent to the sample by means of a waveguide, setting up a standing wave in the cavity. As the magnetic field is continuously swept, the resonance condition is met at a specific value of H [$H_0 = (h\nu/g\beta)$], and microwave radiation is absorbed by the sample. This creates an imbalance in the

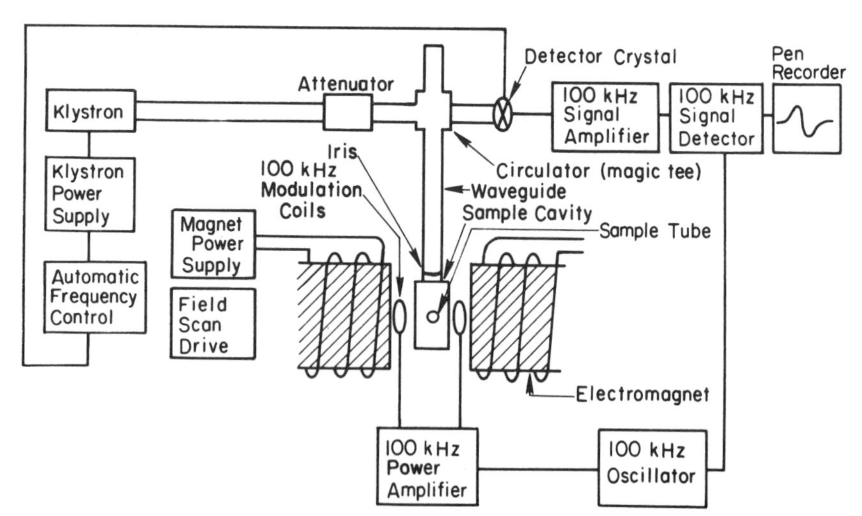

Figure 9–8. Diagram of a typical X-band ESR spectrometer (adapted from Wertz & Bolton, (1972)).

microwave circulator arms, and radiation is reflected from the cavity to a crystal detector (Fig. 9–8). The current variation thereby induced in the detector gives a weak absorption curve. The intensity of this absorption is greatly increased by modulating the magnetic field at low frequency, usually 100 kHz, but the *derivative* of the absorption peak is thereby recorded by an ESR spectrometer.

9–3.3 Procedure

There is little sample preparation required, since the ESR spectrometer can analyze samples in any form—solid, liquid, or gas. Solids may be single crystals or powders, and are generally placed into a quartz tube with inside diameter of several millimeters. The tube is placed in the cavity, which has a detection region about 2 cm high, with the greatest sensitivity being at the center of the cavity. Single crystals may be oriented relative to the magnetic field direction; this is sometimes done by gluing the crystal to the end of a glass rod and turning the rod in the cavity. Oriented clay films may be aligned in the cavity by mounting the films on flat quartz "tissue cells" and positioning the plane of the cell at various angles relative to the field (0–90°).

Aqueous solutions or samples with high water contents (e.g., clay suspensions) must be placed in capillary tubes or specially designed silica flat cells to prevent excessive dielectric absorption of microwave radiation by the water molecules. For organic solvents with dielectric constants less than 10, standard cylindrical sample tubes with an internal diameter of about 3 mm can be used. Most silica glasses have ESR signals due to defects and are not suitable as sample tubes. Highly pure quartz or fused silica is used in sample holder construction, having the advantages of no ESR signals and very low dielectric loss.

It is sometimes advisable to degass a sample before analysis, in order to remove oxygen. The O_2 molecules, with $S = 1$, can broaden ESR spectra by dipole-dipole interaction. Dipolar broadening can also occur if the concentration of paramagnetic species in a sample is too great. For solids, where intrinsic spectral linewidths are usually greater than 1×10^{-4} tesla, paramagnetic concentrations of 10^{-2} to 10^{-3} M are sufficient to prevent line-broadening. This can sometimes be attained by diluting the paramagnetic centers in a diamagnetic matrix. In liquids, intrinsic spectral linewidths are often very narrow, and a concentration of 10^{-4} M or less is recommended to avoid broadening (Wertz & Bolton, 1972).

After the sample has been placed in the ESR cavity, the resonance frequency of the cavity is tuned to the sample and the magnetic field is scanned. However, precautions must be taken to avoid signal distortion or loss by improper instrument settings. Although the signal output of the instrument is generally proportional to the square root of the microwave power level, $P^{1/2}$, the power cannot always be set at maximum. Paramagnetics with long relaxation times become "saturated" at high power, and signal intensity is reduced or lost. Higher modulation settings generally improve sensitivity, but modulation amplitudes greater than a fraction of the peak-to-peak linewidth of the resonance produce distortion of spectral line shape.

According to Curie's law for paramagnetics, maximum sensitivity will be attained at the lowest possible sample temperature. Therefore, a temperature-control apparatus is usually attached to the sample cavity, and spectra are commonly recorded at liquid nitrogen temperature. It should be realized that in many systems, drastic temperature changes may alter the environment of the paramagnetic species, thereby changing the nature of the spectrum.

The spin systems of transition metals are generally efficiently coupled to the lattice vibrations, producing short spin-lattice relaxation times, T_1, at room temperature. The short lifetime of a given spin state (Δt) introduces an uncertainty in the energy, ΔE, of the spin transition according to the Heisenberg uncertainty principle, thereby broadening resonance linewidths. Most transition-metal spectral linewidths are governed by this effect, with short T_1 and broad lines expected for ions with excited electronic states lying very close to ground state (i.e., ions with an efficient spin-orbit coupling mechanism). For this reason, ESR spectra of a number of transition metals such as Fe^{2+} and Ni^{2+} are too broad to be detected at room temperature.

Quantitative ESR is possible by using either a "concentration" or "absolute spin number" standard. To obtain the concentration of a paramagnetic species in a liquid or solid, a concentration standard is used; but the following conditions must apply to both standard and unknown (Wertz and Bolton, 1972): (i) the same matrix and same sample geometry should be used, (ii) the amplitude of the ESR signals should be proportional to $P^{1/2}$ (no saturation of sample), and (iii) the unknown and standard should be at the same temperature. With these conditions, the concentration of the unknown paramagnetic species, X, with spin, S, is given by:

$$[X = (\text{standard}) \times \left(\frac{A \cdot [\text{scale}]^2}{[\text{gain}][\text{mod}]g^2 \cdot S(S+1)} \right)_{\text{unknown}} \times \left(\frac{[\text{gain}][\text{mod}]g^2 \cdot S(S+1)}{A \cdot [\text{scale}]^2} \right)_{\text{standard}} \quad [45]$$

where

A = measured area under the ESR absorption curve (obtained by double integration of the first-derivative curve);
scale = horizontal scale on the chart paper of the spectrum (tesla/unit length);
gain = relative signal amplifier gain;
mod = modulation amplitude (tesla); and
(standard) = concentration of paramagnetic species in standard (M).

If hyperfine splitting of the standard or unknown occurs, the area under the hyperfine lines must be summed to obtain A. Commonly used concentration standards are DPPH (α,α'-diphenyl-β-picrylhydrazl), $K_2NO(SO_3)_2$ (potassium peroxylamine disulfonate), and $MnSO_4 \cdot H_2O$ or $CuSO_4 \cdot 5H_2O$ salts. One of the main problems with quantitative ESR is error involved in obtaining A, since broad signals have considerable intensity in the shoulder regions which is especially difficult to quantify from first-derivative ESR spectra.

Absolute numbers of spins can theoretically be determined in a sample, but the sample volume must be small and weighed quantities of sample and standard must be placed in equivalent positions in the cavity.

The spectral parameters [g factor and hyperfine splitting (a)] can be measured by taping a small amount of standard with known g-value (e.g., DPPH, g = 2.0037) to the sample tube, and using the signal as a field "marker." For the standard, $h\nu = g_s\beta H_0^s$, and for the sample, $h\nu = g\beta H_0$. Since ν is identical for both, $H_0/H_0^s = g_s/g$, and g is readily calculated from the ratio of field positions (the value of H_0^s is calculated from ν and g_s).

The hyperfine splitting is measured as the separation (in tesla) between hyperfine lines of a spectrum on calibrated chart paper. The accuracy of the magnetic field calibration can be checked by using a standard with widely separated hyperfine lines and accurately known hyperfine splittings (e.g., Mn^{2+} in $CaCO_3$).

9–3.4 Comments

An interesting illustration of the concepts presented above is the Cu^{2+} ion adsorbed on an expanding 2:1 clay mineral, hectorite. The ESR spectrum, arising from the single unpaired electron of the d^9 ion, appears to be a single, broad, nearly symmetrical line for Cu^{2+}–hectorite equilibrated at 100% relative humidity (Fig. 9–9a). A similar spectrum is obtained with excess water added to the clay, although the resonance line is then fully symmetrical. The g-value of the resonance is 2.17. When this clay is cooled to $-100°C$ (or air-dried), an orientation-dependent rigid-limit spectrum of Cu^{2+} is obtained (Fig. 9–9b), and the $g_{\parallel}$ and $g_{\perp}$ components of the g-tensor are resolved. The Cu^{2+} nuclei have $I = 3/2$, so that $g_{\parallel}$ is split into four hyperfine lines ($M_I = -3/2, -1/2, +1/2, +3/2$) with a hyperfine splitting constant, $a_{\parallel} = 0.0123$ tesla. The splitting of the $g_{\perp}$ component is too small to be resolved in this spectrum, a result of the pronounced anisotropy of the hyperfine splitting. It should be noted that Cu is an isotopic mixture of ^{63}Cu(69.2%) and ^{65}Cu(30.8%); both nuclei have $I = 3/2$, but the value of a is slightly different for the two nuclei.

The greater intensity of the $g_{\parallel}$ component and reduced intensity of the $g_{\perp}$ component of the Cu^{2+} spectrum observed when the plane of the clay film is oriented perpendicular to **H** is evidence that the z-axis of the Cu^{2+} ion is aligned normal to the plane of the clay platelets. Evidently, planar $Cu(H_2O)_4{}^{2+}$ ions become aligned in the interlamellar regions of clay platelets due to the removal of interlayer water by freezing (McBride, 1976). Since the $g_{\parallel}$ and $g_{\perp}$ factors are 2.35 and 2.08, respectively, a fre-

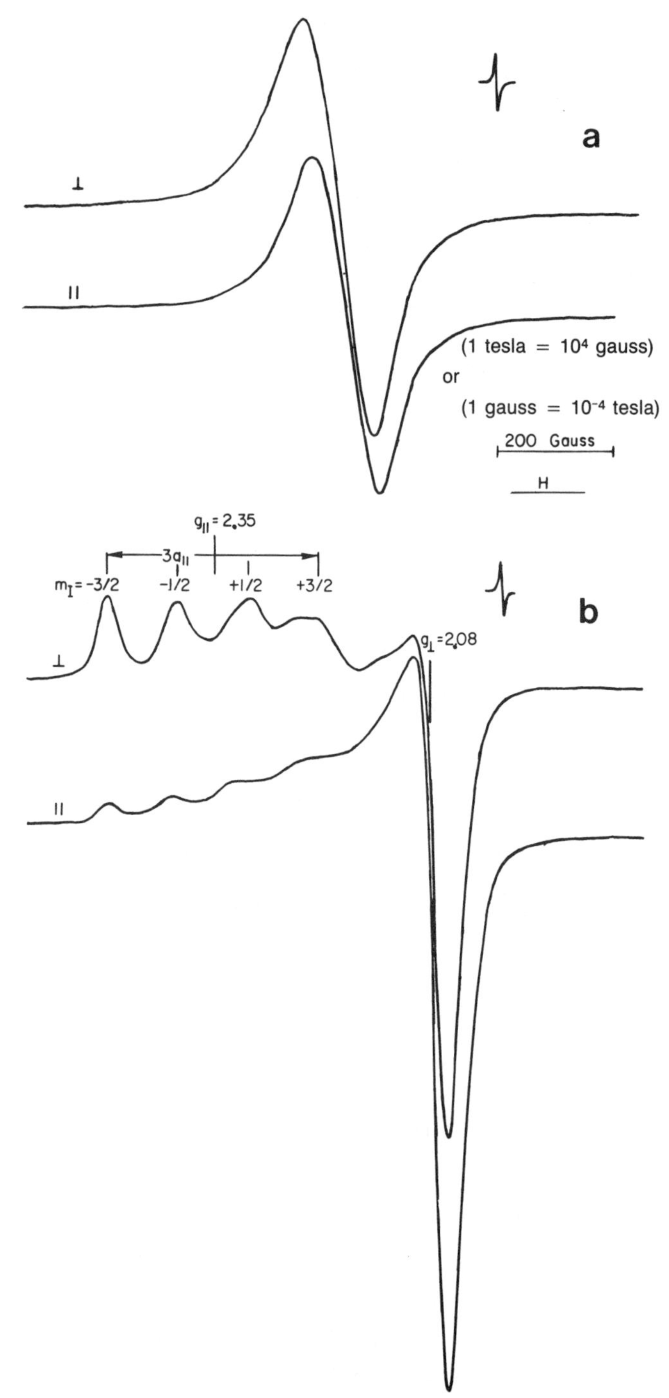

Figure 9–9. ESR spectra of Cu^{2+} adsorbed on hectorite equilibrated at 100% relative humidity and (a) room temperature (b) −100°C. Clay films were oriented perpendicular (⊥) and parallel ($_{\parallel}$) to *H*. The small strong-pitch signals are centered at g = 2.0027.

quency of ionic rotation greater than this anisotropy (i.e. > $(g_{\parallel} - g_{\perp})\beta H/h$), or about 10^9 s^{-1}, would begin to average $g_{\parallel}$ and $g_{\perp}$, producing the symmetrical spectrum (Fig. 9–9a) with $g_0 = 1/3\ (g_{\parallel} + 2g_{\perp})$. Addition of excess water to the Cu^{2+}-hectorite allows full expansion to a ~2 nm d-spacing, and $Cu(H_2O)_6{}^{2+}$ then tumbles rapidly enough to completely average the g-tensor and hyperfine splitting anisotropy. However, the expected four-line hyperfine structure of rapidly tumbling Cu^{2+} is not observed at room temperature because of relaxation mechanisms that broaden the lines (Poupko & Luz, 1972).

Example ESR spectra of layer silicates are shown in Fig. 9–10. A signal attributable to structural Fe^{3+} in micas appears at low-field position ($g = 4.3$), apparently due to rather large crystal field splittings (Abragam & Bleaney, 1970). This signal is generally composed of an anisotropic and an isotropic g-factor and has been observed in kaolinites, vermiculites, micas, and smectites (Angel & Hall, 1972; McBride et al., 1975b). The isotropic and anisotropic signals have been suggested to arise from octahedral $FeO_4(OH)_2$ with OH groups in *cis* and *trans* configurations, respectively (Olivier et al., 1975). Upon weathering of a phlogopite, there appears to be a loss in intensity of these Fe^{3+} signals, but a very broad resonance near $g = 2$ appears that might be attributed to ferrimagnetism arising from Fe^{2+} oxidation to Fe^{3+} (Fig. 9–10a). The broad resonance may represent clusters of structural Fe^{3+} created upon oxidation, or a small amount of ferric oxide created by expulsion of iron from the structure during weathering. Vermiculites (i.e., naturally weathered micas) often demonstrate a strong ferromagnetic resonance (Fig. 9–10b).

Other signals that are common in layer silicates are six-line spectra (Fig. 9–10a, 9–10c) arising from Mn^{2+} in octahedral sites, and signals near $g = 2$ that are attributed to structure defects.

Soil clays often demonstrate a strong ferromagnetic resonance which can be largely eliminated by a single citrate-dithionite treatment to remove iron oxides. Other spectral details tend to be obscured by this broad signal, but exchangeable Mn^{2+} and organic radical spectra are commonly observed. The latter signal has a g-value very near the "free electron" position (2.0030–2.0040) and little or no observable hyperfine structure, and has been attributed to quinone-type structures largely concentrated in the humic acid fraction. The intensity of the radical signal increases markedly with increasing pH (Wilson and Weber, 1977).

ESR can often provide information about the paramagnetic ions present in clays and soil colloids, their environments (including ligand type and symmetry) and degree of mobility in the adsorbed state, and their orientations within mineral structures or on mineral surfaces. In many cases it is useful to add a paramagnetic "probe" to natural clay or organic systems in order to study surface properties by ESR. Ions commonly used for this purpose are Mn^{2+}, Cu^{2+}, and VO^{2+} (McBride, 1978). Spectra of precipitated complexes of fulvic acid with VO^{2+} and Cu^{2+} are shown in Fig. 9–11. The VO^{2+} spectrum is complex, having a highly anisotropic

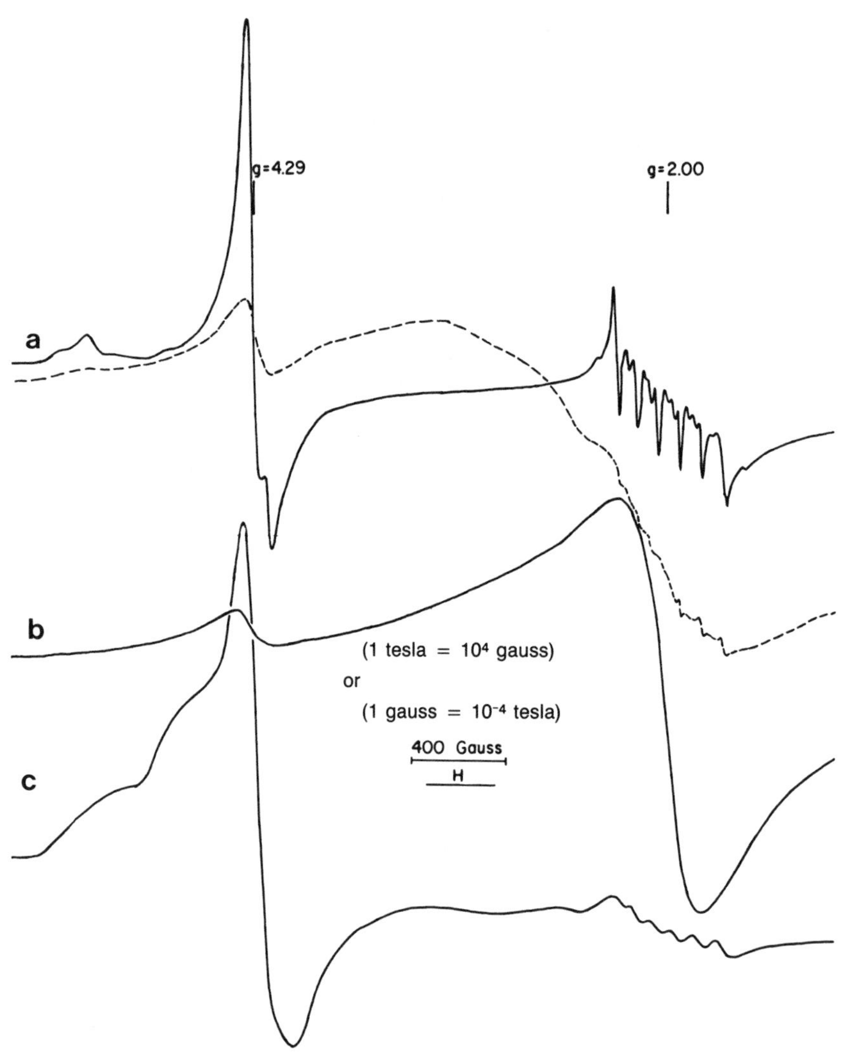

Figure 9–10. ESR spectra of layer silicate minerals: a. phlogopite before (solid line) and after (broken line) several months' weathering by Na^+-tetraphenylboron–NaCl solution. b. vermiculite (Transvaal). c. muscovite.

hyperfine splitting, with eight hyperfine lines due to the 7/2 nuclear spin of ^{51}V. Since the VO^{2+} ion is rigidly bound in soil organic matter, a rigid-limit spectrum of 16 partially overlapping lines results, composed of two sets of eight hyperfine lines for the $g_{\parallel}$ and $g_{\perp}$ components (Fig. 9–11a). For VO^{2+}, the separation between adjacent hyperfine lines changes significantly as a function of field position, as second-order effects shift some hyperfine lines more than others. As a result, accurate determination of

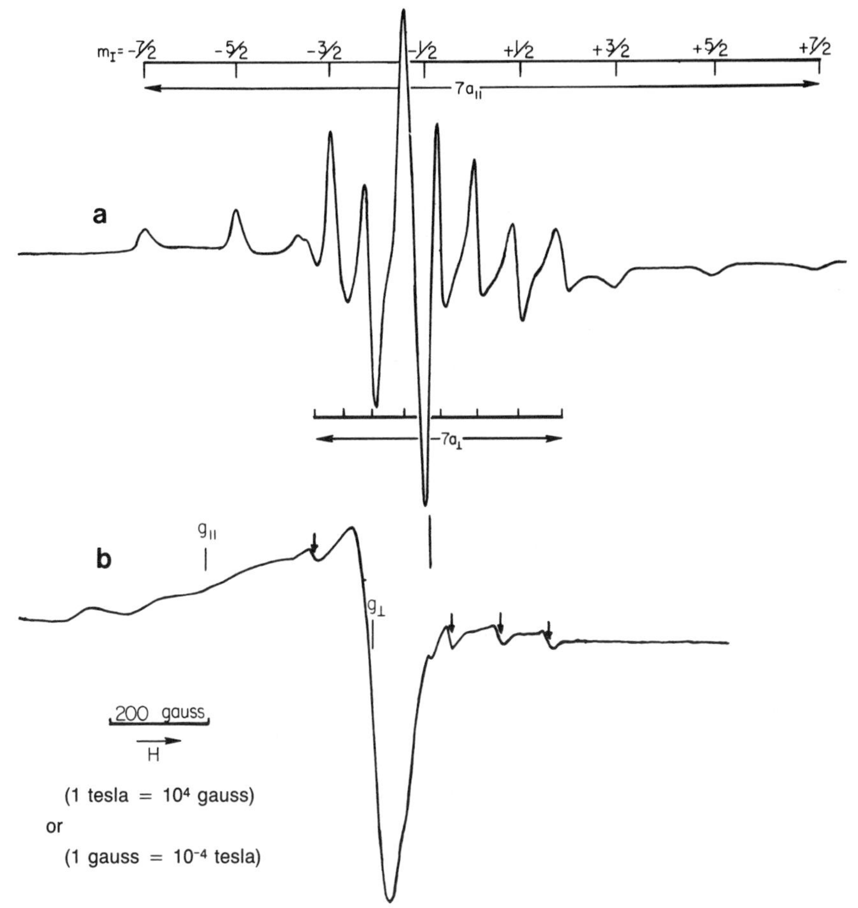

Figure 9–11. Rigid-limit ESR spectra of VO^{2+}-fulvic acid (a) and Cu^{2+}-fulvic acid (b) precipitates in the moist state. Vertical arrows denote natural Mn^{2+} resonances superimposed on the Cu^{2+} signal.

a and g may require computer simulation of spectra. The rigid-limit Cu^{2+} spectrum is simpler (Fig. 9–11b), but not all of the four hyperfine lines of the $g_{\parallel}$ component can be resolved at room temperature. This broadening suggests the presence of Cu^{2+} complexes with a range of *a* and g values.

It should be stressed that the absence of an ESR signal in a sample does *not* prove the absence of paramagnetic species, since relaxation effects can prevent the observation of a spectrum. In contrast, magnetic susceptibility measurements will detect paramagnetics unequivocally. The detection limit at room temperature of an ESR spectrometer operating at X-band (9.5 GHz) is on the order of 10^{-9} *M* for samples with low dielectric constants. Aqueous solutions have a limit of about $10^{-7}M$.

9–4 NUCLEAR MAGNETIC RESONANCE (NMR)

9–4.1 Principles

The NMR experiment is based on the same principle as the ESR experiment, with the nuclear magnetic moment, $\mu_N = g_N\beta_N I$ (see Eq. [1] for symbol definitions) replacing the electron magnetic moment, μ_e. The values of g_N and I are different for different nuclei, distinguishing one nucleus from another in a magnetic resonance experiment. Some common nuclei and their spins and g-values are listed in Table 9–6. The application of a magnetic field, H, to the nuclear magnetic moment, μ_N, produces an energy of interaction

$$E = -\mu_N \cdot H = -g_N\beta_N H M_I \qquad [46]$$

where M_I is the nuclear spin quantum number. For the most common NMR experiment, the proton ($I = 1/2$) is the nucleus of interest, and M_I can have only the values of $+1/2$ or $-1/2$. Since β_N is positive and g_N is positive for the proton (Table 9–6), the lower level with $M_I = +1/2$ corresponds to the parallel alignment of H and the nuclear moment, μ_N, while the upper level is the antiparallel alignment. The difference in energy of the two levels is $g_N\beta_N H$, so that a transition between the two nuclear spin levels can be induced by an electromagnetic field applied perpendicular to H if the frequency, ν, of this radiation satisfies the resonance condition, $h\nu = g_N\beta_N H$ (see Fig. 9–6a for analogous ESR transition). An assembly of protons will produce a detectable absorption of energy only if there are more protons in the lower energy state. The Boltzmann equation describes this distribution, but the much weaker magnetic moment of a nucleus compared with an electron results in a

Table 9–6. Nuclear spins and g-values (from Carrington and McLachlan, 1967).

Nucleus†	Natural isotopic abundance	I	g_N
	%		
^{1}H	99.985	1/2	5.585
^{2}H	0.015	1	0.857
^{7}Li	92.44	3/2	2.171
^{13}C	1.11	1/2	1.405
^{14}N	99.63	1	0.403
^{15}N	0.37	1/2	−0.567
^{17}O	0.037	5/2	−0.757
^{19}F	100	1/2	5.257
^{23}Na	100	3/2	1.478
^{29}Si	4.70	1/2	−1.111
^{31}P	100	1/2	2.263
^{33}S	0.76	3/2	0.429
^{35}Cl	75.77	3/2	0.548
^{37}Cl	24.23	3/2	0.456
^{39}K	93.08	3/2	0.261

† The following common nuclei have no spin ($I = 0$): ^{12}C, ^{16}O, ^{28}Si, ^{32}S, ^{40}Ca.

very small difference in energy between the two states, and at room temperature there is an excess of only about seven protons per one million in the lower energy state if the applied magnetic field is 1 tesla! This tiny difference in populations accounts for the NMR absorption peak, resulting in a much lower inherent sensitivity of NMR compared to ESR. In addition, the frequency of radiation, ν, required for resonance to occur at a given magnetic field strength, H, is widely different for different nuclei depending on the value of g_N and I (Table 9–6), but is 42×10^6 Hz for the proton if $H = 1$ tesla. Obviously, proton resonance occurs at a frequency almost a thousand times lower than electron resonance, in the radio frequency range. Experimentally, as in ESR, ν is normally fixed, and the spectrum is obtained by sweeping the magnetic field through the resonance value.

Unlike ESR g-values, NMR g_N values are taken by convention to be inherent properties of the atom, remaining constant. The observed NMR frequency of a given nucleus (e.g., proton) in different molecules differs slightly from the theoretical value of $g_N\beta_N H/h$ for the "bare" nucleus, but this deviation is considered to be a *chemical shift* that arises from the magnetic effect of electronic currents induced in atomic orbitals by the external magnetic field. The currents produce a small local field at the nucleus which opposes H and has a strength proportional to H. The *effective* magnetic field which the nucleus "feels" becomes:

$$H_{\text{eff}} = (1 - \sigma)H \qquad [47]$$

where σ is the *screening constant* with a value on the order of 10^{-6}. The nuclear resonance energy then becomes:

$$h\nu = g_N\beta_N(1 - \sigma)H \qquad [48]$$

and magnetic resonance at fixed ν will appear at slightly higher field than for the free nucleus. The value of σ, calculated from the field position of resonance, depends on electronic structure (i.e., bonding environment) and is very useful in the identification of compounds containing magnetic nuclei. Using the example of proton NMR, σ is not measured relative to the "bare" proton but relative to a reference compound. The relative chemical shift of a sample proton and reference proton is easily shown to be:

$$\sigma_1 - \sigma_{\text{ref}} = (H_1 - H_{\text{ref}})/H_0 \qquad [49]$$

where H_1, H_{ref}, and H_0 are the resonance positions of the sample, reference, and free proton, respectively. This relative shift can also be given in resonance frequency units, symbolized as δ:

$$\delta_1 - \delta_{\text{ref}} = (\sigma_1 - \sigma_{\text{ref}})\nu_0 \qquad [50]$$

where ν_0 is the resonance frequency of the free proton. The σ and δ shifts are generally expressed in parts per million (ppm)

$$\sigma(\text{ppm}) = (\sigma_1 - \sigma_{\text{ref}}) \times 10^6 \quad [51]$$

$$\delta(\text{ppm}) = (\delta_1 - \delta_{\text{ref}}) \times 10^6 . \quad [52]$$

An internal standard such as tetramethylsilane (TMS) is commonly used in liquid samples, and chemical shifts are measured relative to the proton resonance position of TMS, which is at higher field than almost all organic protons. Thus, letting δ for TMS equal zero, chemical shift can be expressed for protons on a scale relative to $\delta = 0$, with larger negative values of δ corresponding to lower field position. The value of δ for the proton is characteristic of the type of organic compound, as shown by Fig. 9–12. Protons attached to electronegative atoms or groups tend to resonate at lower field as these atoms draw electrons away from the hydrogen nucleus, reducing the screening effect. For example, carboxylic acid protons have large values of δ (10–13 ppm), indicating weak screening.

Although the value of δ may be useful in identification of compounds containing protons, *spin-spin coupling* may provide even more detailed information. The proton spin moments couple through the intervening bonding electrons because of the tendency of bonding electrons to pair spins with the nearest proton. Considering the example compound CH_3CHCl_2, there are two chemically and magnetically nonequivalent types of protons which have very different chemical shifts. In addition, spin-spin coupling splits the two main proton resonances into several components. The local magnetic field at the CH_3 group is slightly increased or decreased by a value, H', depending on whether the proton of

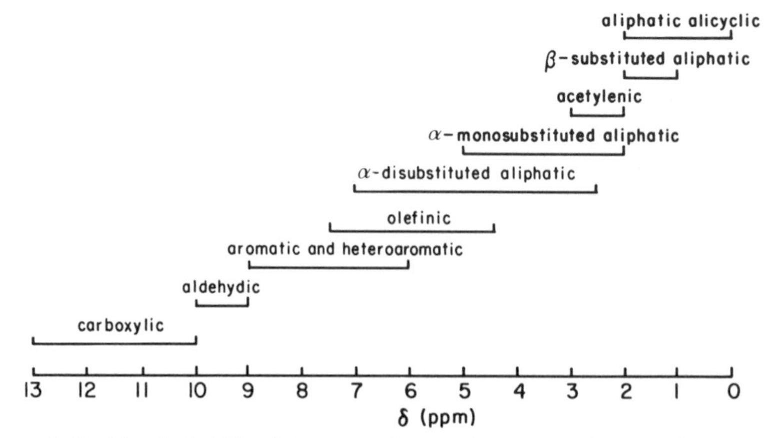

Figure 9–12. Chemical shifts of the proton in organic compounds (after Silverstein and Bassler (1967)).

the $CHCl_2$ group has spin $+1/2$ or $-1/2$. As a result, the three CH_3 protons produce a doublet (Fig. 9–13). The proton of the $CHCl_2$ group experiences *four* possible values of the local field, depending upon the orientations of the three CH_3 proton spins relative to the magnetic field. These are shown in Fig. 9–14 to produce four resonances separated by J_{HH}, the spin-spin coupling parameter, with relative intensities of 1:3:3:1 arising from the possible combinations of spin orientation. In general, the splitting of proton NMR spectra of organic compounds in solution by spin-spin coupling are interpreted as follows:

1. Splitting of a proton absorption is effected by neighboring protons, with the multiplicity of splitting given by the value of $n + 1$, where n is the number of magnetically equivalent neighboring protons. For nuclei other than protons with spin (I) not equal to 1/2, the multiplicity is $2nI + 1$.
2. Relative intensities of the peaks of a multiplet depend on n, with the coefficients of the expanded formula $(a + b)^n$ giving the relative intensities (e.g., for $n = 2$, $(a + b)^n = a^2 + 2ab + b^2$, and a triplet with intensities 1:2:1 is observed).
3. The integrated intensities of multiplets are proportional to the relative number of protons causing the multiplet in the compound.

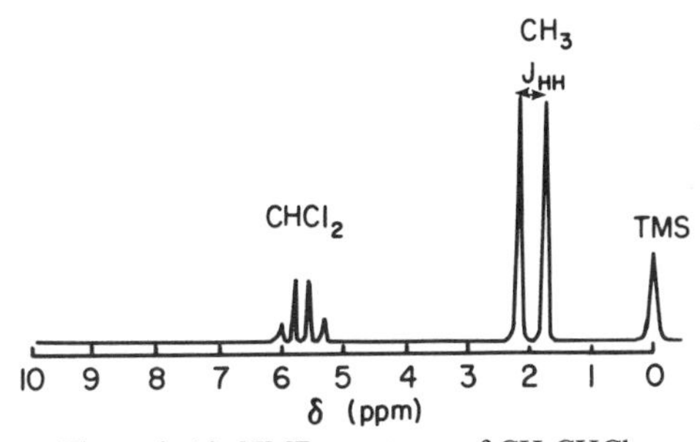

Figure 9–13. NMR spectrum of CH_3CHCl_2.

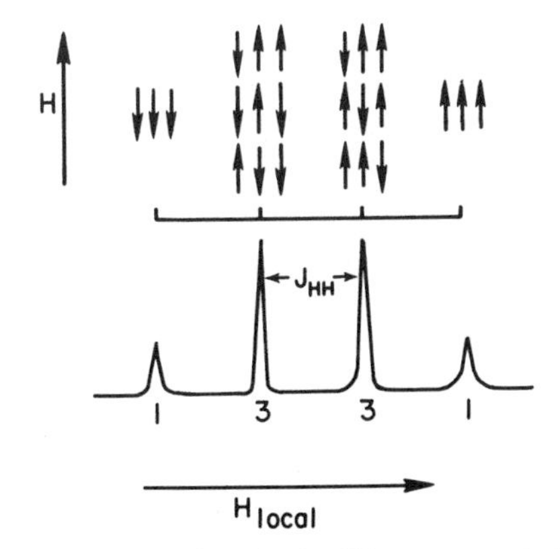

Figure 9–14. Diagram of spin orientations in the CH_3 group producing the 1:3:3:1 splitting in the NMR resonance of a neighboring proton.

Characteristics of the *spin-spin splitting constant*, J, useful in identifying NMR spectra are:

1. The spin interaction is transmitted through the chemical bonds, and J values depend on electronic structure, being independent of H, the external field strength. Although direct through-space dipole-dipole interaction between magnetic nuclei is important in solids, it averages to zero in tumbling molecules of liquids (standard NMR spectroscopy requires that the substance be in solution.)
2. J is usually insensitive to a change of temperature or solvent.
3. J tends to decrease as the number of intervening bonds between magnetically coupled nuclei increases, being generally too small to be observed for four or more bonds.
4. J_{AB} increases rapidly as a function of $Z_A Z_B$, the product of the atomic numbers of the two coupled nuclei, A and B.
5. Magnetically equivalent protons (or other atoms) do not split one another, so that similar protons are treated as a group.
6. Chlorine and most other nuclei with quadrupole moments (nuclei with $I \geq 1$) do not produce observable splittings. These nuclei rapidly reorient through their allowed spin orientations, so that the average field generated at a neighboring proton (or other atom) is zero (NH_4^+ is an exception, with ^{14}N ($I = 1$) splitting the proton resonance into a 1:1:1 triplet).

9–4.1.1 PULSED NMR

With conventional NMR equipment, the magnetic field is slowly swept through the resonance position to obtain an absorption peak. It is often possible to obtain relaxation times of spin systems with these continuous wave spectrometers, where the spin-spin relaxation time (T_2) is derived from the width of the spectrum at radiofrequency (rf) power below saturation, and the spin-lattice relaxation time (T_1) is calculated from data in the region of signal saturation (Poole & Farach, 1971). However, pulsed NMR methods are better suited to the measurement of T_1 and T_2, requiring specialized equipment capable of producing short calibrated bursts of rf energy. These methods will be discussed after a brief description of T_1 and T_2, the characteristic relaxation times of a spin system (T_1 and T_2 are defined in an analogous manner for ESR).

If a collection of nuclei ($I = 1/2$) is suddenly subjected to a magnetic field, H_0, the spin system will initially at zero-time ($t = 0$) have $n_+ = n_-$, where n_- and n_+ are the number of spins in the higher ($M_I = -1/2$) and lower ($M_I = +1/2$) energy states, respectively. This spin system is not at equilibrium at $t = 0$, since the Boltzmann equation requires that a slight excess of spins reside in the lower energy state. Defining this excess as $n_{ex} = n_+ - n_-$, the rate of change of spin excess after applying the magnetic field defines the spin-lattice relaxation time, T_1 as

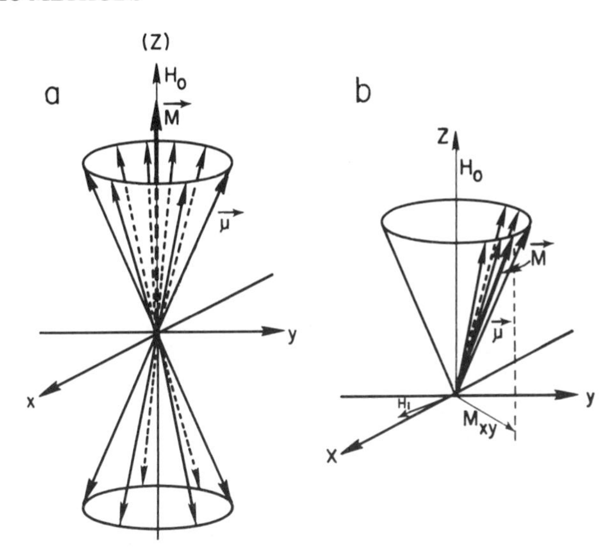

Figure 9–15. The precession of spin magnetic moments, μ, in a magnetic field, H_0, in the absence (a) and presence (b) of an applied rf field, H_1, at right angles to H_0. Only one of the two spin envelopes is shown in (b) to simplify the diagram.

$$dn_{ex}/dt = [(n_{ex})_{equil} - n_{ex}]/T_1 \qquad [53]$$

where $(n_{ex})_{equil}$ is the equilibrium excess. Therefore, for any excess of spins not equal to the equilibrium value, adjustment with time, t, occurs as follows:

$$n_{ex} = (n_{ex})_{equil}\,[1 - \exp(-t/T_1)]\,. \qquad [54]$$

During this process of spin-lattice (or longitudinal) relaxation, energy must be transferred between the spin system and the "lattice." In addition, the longitudinal component of magnetization, $\mathbf{M}(\mathbf{M} = n_{ex}\boldsymbol{\mu}$, where $\boldsymbol{\mu}$ is the magnetic moment of one spin), changes during this relaxation process (Fig. 9–15a).

Considering the same collection of spins precessing in phase in a magnetic field with $\mathbf{M} = n_{ex}\boldsymbol{\mu}$, the random positions of the $\boldsymbol{\mu}$ vectors shown in Fig. 9–15a cause the components of $\boldsymbol{\mu}$ along the x and y axes to sum to zero. When an rf electromagnetic field, H_1, is applied at the precession (Larmor) frequency, ω_0, the $\boldsymbol{\mu}$ vectors acquire phase coherence, merging into a single resultant vector, **M**. This vector is now tilted with respect to the z-axis as shown in Fig. 9–15b, and an xy magnetization (M_{xy}) now exists. The receiver coil of an NMR spectrometer can detect this M_{xy} component, so that the resonance condition shown in Fig. 9–15b produces a signal as the frequency, ω, is slowly swept through ω_0. Upon removing the applied rf field, H_1, the $\boldsymbol{\mu}$ vectors again randomize and M_{xy} approaches zero as the spin system reverts to the condition diagrammed in Fig. 9–15a. The rate of phase randomization is determined by the *spin-spin* (or transverse) relaxation time, T_2, defined by the equations:

$$dM_x/dt = -M_x/T_2, \qquad dM_y/dt = -M_y/T_2 . \qquad [55]$$

T_2 relaxation does *not* involve energy transfer, unlike T_1 relaxation. In T_2 relaxation, spin-spin exchange occurs between spins of the *same* spin state with no change in energy of the system, but a shortening of the lifetime of each spin state. The result is an NMR linebroadening of $\Delta\omega = 2/T_2$ where $\Delta\omega$ is the resonance linewidth in frequency units.

In an NMR pulse experiment, an rf pulse is applied to a sample *at resonance*, causing the magnetization to tip away from the H_0 direction (Fig. 9–15b). This magnetization, **M**, precesses around H_0, decaying exponentially with time as described by Eq. [55] (see Fig. 9–16a). Thus, the observed signal intensity decays exponentially, as a function of $\exp(-2t/T_2)$. Theoretically, the shape of the decay curve observed at the end of the pulse ($t = 0$) could be used to measure T_2. However, magnetic field inhomogeneity of the instrument introduces a further apparent relaxation time, T'_2, and the *observed* time constant of signal decay, T^*_2, is given by

$$1/T^*_2 = 1/T_2 + 1/T'_2 . \qquad [56]$$

Pulses of rf power can be applied for definite periods of time, t_p, so that **M** will move through a definite angle during the pulse, given by $\Theta = \gamma_N H_1 t_p$, where γ_N is the nuclear magnetogyric ratio and H_1 is the magnitude of the rf magnetic field. The angular precession of **M** at resonance is occurring about the effective magnetic field, given by the vector sum of $\mathbf{H}_0$ and $\mathbf{H}_1$, and is *not* the same as the Larmor precession of **M** about $\mathbf{H}_0$. A 90° pulse involves applying rf power to the sample for time, t_p, so that $\pi/2 = \gamma_N H_1 t_p$. This pulse turns **M** completely into the *xy* plane, a nonequilibrium situation which reverts to equilibrium after the pulse by spin-spin and spin-lattice relaxation. A 180° pulse "flips" **M** into the negative *z*-axis direction.

The pulse method can be used to measure T_1 relaxation times of liquids, T_1 being on the order of about one second for most liquids. This is done by applying a 180° pulse, waiting a time τ for decay of signal to occur, and applying a 90° pulse. This sequence, shown in Fig. 9–16b, first flips the **M** vector 180° to the negative *z*-axis direction so that no signal is observed after the pulse ($M_x = M_y = 0$). However, during the time τ, decay of the M_z component of **M** occurs at the rate

$$dM_z/dt = -(M_z - M)/T_1 \qquad [57]$$

by the relaxation of spins back to the positive *z*-axis (spin-lattice relaxation). No signal can be detected during τ because no M_{xy} component of

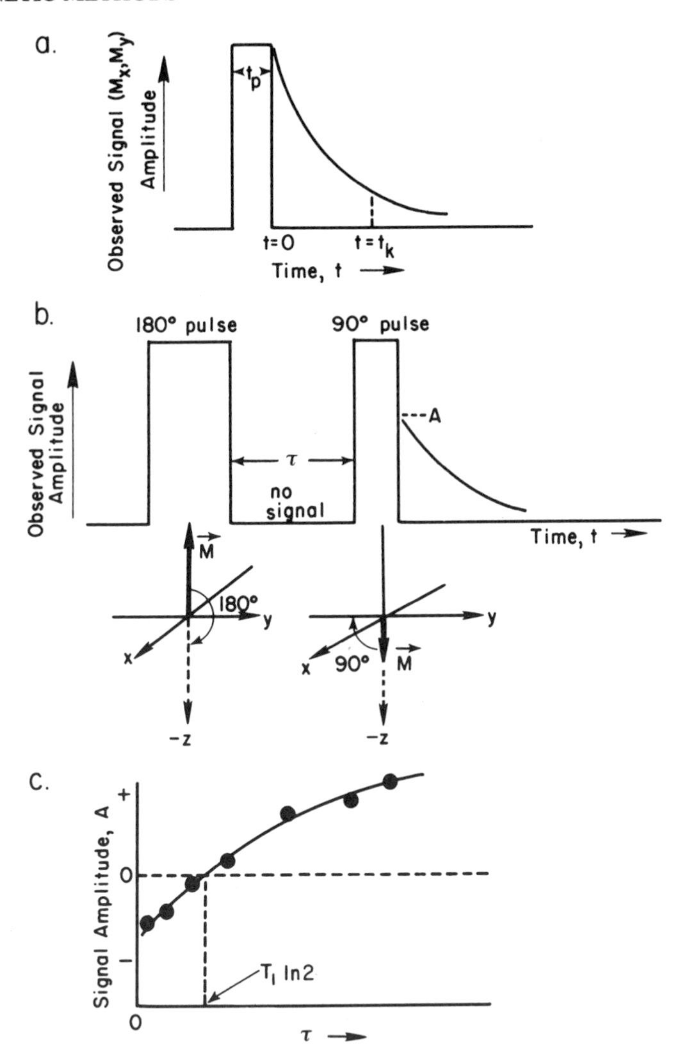

Figure 9–16. The pulse NMR experiment, showing the decay of signal strength after application of a pulse for t_p seconds (a), the pulse sequence used to determine T_1 and the effect on M (b), the graphical method of measuring T_1 from a sequence of 180°, τ, 90° pulse experiments (c).

M exists. The 90° pulse then "samples" the decay process by flipping **M** from the $-z$ direction into the xy plane, allowing **M** to be detected. For example, if the 90° pulse had been applied when half of the spins had returned to the $+z$ direction, no net magnetization would exist ($M = 0$) and no signal would be observed after the pulse. Experimentally, a series of 180° pulse, τ, 90° pulse experiments is done, varying τ for each experiment and measuring the signal amplitude (A) detected after application of the 90° pulse. For very short τ, A will be negative, since **M** has not had enough time to relax back to the $+z$ axis direction. For long τ, A will become positive and approach a limiting value, A_∞ (Fig. 9–16c). The equation for signal amplitude, A_τ, as a function of τ is given by:

$$\ln (A_\infty - A_\tau) = \ln 2A_\infty - \tau/T_1 . \qquad [58]$$

At the zero amplitude point, $\tau = T_1 \ln 2$, so that T_1 is readily calculated.

Although T_2^* can be measured from the shape of the free induction decay (FID) curve observed after a pulse, Eq. [56] shows that it includes magnet inhomogeneity, and the true T_2 of the spin system is not determined in this way. The *spin-echo* pulse experiment can circumvent this problem, measuring T_2 only. The experiment requires that the sequence: 90° pulse, τ, 180° pulse be applied (Fig. 9–17a). After a time of 2τ following the 90° pulse, a signal (spin echo) is observed. This experiment can be visualized in the xy plane of a rotating frame of reference, where the 90° pulse initially flips the **M** vector into the xy plane (Fig. 9–17b). The time period τ then permits the individual moment vectors to fan out slowly in the xy plane, caused by field inhomogeneity which creates slightly different precession frequencies for the spins in the sample. After this time period, τ, a 180° rf pulse is applied, turning the vectors through 180°. Since the vectors will precess at the same frequency before and after the 180° pulse, the inverted positions created by the pulse results in a perfect reclustering of the pulses at time 2τ, regardless of the degree of field inhomogeneity (Fig. 9–17b). This spontaneously creates a signal

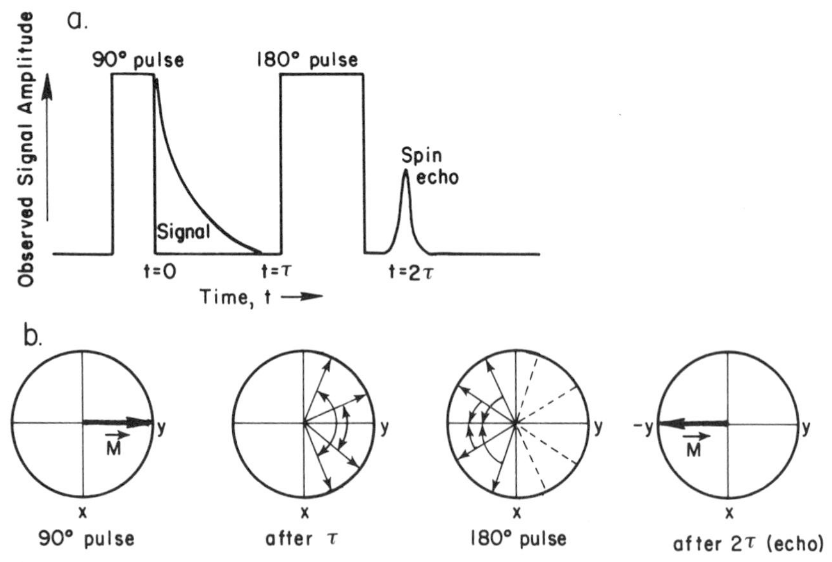

Figure 9–17. The pulse NMR spin echo experiment showing the 90°, τ, 180° sequence of events (a) and a schematic representation of spin moments in the xy plane of the rotating frame of reference (b) (axis system rotates at the Larmor frequency).

or spin echo at $t = 2\tau$. Following the echo, the signal decays as the vectors again fan out. To measure T_2, a series of 90°, τ, 180° spin-echo experiments is done at various intervals of τ. The echo intensity follows the exponential decay of $\exp(-2t/T_2)$, so that T_2 is determined without the contribution from field inhomogeneity.

T_1 and T_2 relaxation times are important in understanding continuous wave proton NMR spectral linewidths. Most organic liquids have equal T_1 and T_2 relaxation times which are on the order of seconds. As a result, linewidths (which are proportional to $1/T_2$) become very narrow. Aqueous solutions may have shorter $T_1 = T_2$ values and broader lines. For solids, T_1 is generally long (e.g., 10 min for ice) but T_2 is very short ($\leq 10^{-4}$ s). Lines are very broad as a result of the nuclear dipole-dipole interactions. Paramagnetic impurities in liquids cause relaxation because of the strong magnetic field of unpaired electrons, T_1 and T_2 are shortened, and broader lines result (which are less readily saturated). In precise NMR work, dissolved O_2 must be removed from liquid samples to prevent line broadening. Also, nuclei with quadrupole moments (e.g., ^{14}N) can shorten the relaxation time of a neighboring nucleus (e.g., proton), producing a broad NMR resonance. However, as noted above, magnetic field inhomogeneity also produces line broadening by spreading out the frequency range of resonance in a sample. This produces an *apparently* shortened T_2. Sample spinning in the NMR instrument can partly average out field inhomogeneity by moving a given molecule so rapidly through a range of fields that the NMR signal responds to an average field. Samples are usually spun at 1000 to 2000 rpm for high-resolution work to obtain narrow lines.

Since the random magnetic fields in a liquid provide a mechanism for nuclear relaxation, molecular motions can be studied by measuring T_1 and T_2. The rate of rotational and translational motion of a water molecule, for example, affects the proton spin-lattice relaxation time. Chemical exchange of protons between two sites in a molecule, or between two different molecules, changes T_1 and T_2 so that line shapes can be used to study such exchanges. Proton exchange rates in aqueous solutions have been measured in this manner. Highly viscous liquids do not allow protons to rotate quickly enough to completely average out dipole-dipole interactions, and protons experience local magnetic fields over a range. This creates line broadening.

9–4.1.2. BROAD-LINE SOLID NMR

Although most NMR work is done in liquids in order to produce very narrow linewidths (high resolution), allowing the spin-spin interactions and chemical shifts to be measured very accurately, NMR spectra can also be obtained on solids. Rapid molecular tumbling in liquids averages various anisotropic interactions to zero, so that information on these interactions is lost in liquid spectra. Solids produce broad absorption lines, so that their study is often referred to as *broad-line NMR*. The most important interaction between two nuclear spins is the dipole-dipole coupling between their magnetic moments. In liquids this interaction is motionally averaged out, but in solids the magnitude of the dipole coupling varies according to the orientation of the magnetic field relative to the molecules or crystal. A detailed study of this anisotropy permits the separation distance and relative orientation of the nuclei to be obtained. NMR spectra of water adsorbed on clays may produce orientation-dependent spectra, allowing the calculation of orientation of H_2O molecules, as will be discussed later.

A simple example of a broad-line NMR spectrum is that of a gypsum crystal, $CaSO_4 \cdot 2H_2O$. Since the water molecules are well separated, intermolecular proton-proton interactions can be neglected. However, the two protons within a water molecule are coupled by a dipolar interaction. If the water molecules in the crystal were not distinguishable, dipolar interaction would produce a pair of resonance lines at the field values given by the equation (Carrington & McLachlan, 1967):

$$H = H^* \pm 3/4\ g_N\beta_N(1 - 3\cos^2\Theta)/r^3 \quad [59]$$

where Θ is the angle between the interproton vector and the applied magnetic field, r is the interproton distance for the water molecules, and H^* is the resonance position of a single proton without dipolar interactions. In fact, $CaSO_4 \cdot 2H_2O$ contains two types of water molecules with nonparallel proton-proton vectors. Thus, a four-line spectrum is observed for certain orientations of the crystal in the magnetic field. From Eq. [59], it is clear that when $3\cos^2\Theta = 1$ (i.e., $\Theta = 54.7°$) for both proton pairs, the dipolar interaction term becomes zero and one proton resonance line is observed at H^*. This "magic angle" of 54.7° can be used to advantage to narrow the resonance linewidth and improve the resolution of NMR spectra of solids; the important field of magic-angle spinning (MAS) NMR has thereby developed and will be discussed in more detail below.

When three or more nuclei in a crystal or rigid molecule have strong dipolar interactions, very complex NMR spectra are expected. However, in most cases the lines are not resolved, since more distant nuclei contribute to the linewidth, and a broad resonance is observed. The *second moment* of the absorption line (i.e., the mean-square width measured from the center of the resonance) can be evaluated graphically from the observed line shape. The value of the second moment can sometimes be used to determine crystal structures or degrees of molecular motion in the solid state (Carrington & McLachlan, 1967).

An interesting case of solid NMR is that involving magnetically ordered solids. For example, NMR has been used to determine the internal fields generated by nearby electrons at the ^{57}Fe ($I = 1/2$) nucleus of ferrimagnetic minerals (Martin, 1967). Also, although antiferromagnetic minerals have no *net* magnetic moment, the nucleus of an atom in a sublattice is subjected to a "hyperfine field" created by near electrons, and its resonance position is shifted by this internal field. The immense fields at nuclei of magnetically ordered solids (about 40 tesla at the Fe nucleus in goethite) can shift resonances of nuclei very far from their normal field positions. Paramagnetics, because of the random orientation of electron dipoles, do not create a hyperfine field at the nucleus, but do broaden the NMR line by dipole-dipole interactions.

9–4.1.3 HIGH-RESOLUTION NMR AND MAGIC-ANGLE SPINNING

A rapidly developing field of NMR spectroscopy, which shows great promise for applications in soil chemistry and mineralogy, is the technique of high-resolution NMR. The utilization of (i) very strong magnetic fields (and therefore, high resonance frequencies), (ii) signal accumulation by repeatedly pulsing the sample with rf radiation and Fourier-transforming the data from time domain into the conventional frequency domain, (iii) double resonance procedures to decouple the strong dipolar interactions of nuclei with protons, and (iv) signal enhancement of low-abundance magnetic nuclei (e.g., ^{13}C) by a cross-polarization or transfer of magnetization from an abundant nucleus (i.e., ^{1}H), has converted NMR into a sensitive and powerful method for the study of magnetic nuclei in solutions. One additional technique, that of magic-angle spinning (MAS), is required before high-resolution NMR spectra can be obtained from solid samples. The principles and procedure of magic-angle spinning are discussed by Andrew (1981) and will not be covered in detail here. Briefly, the solid sample is rapidly rotated about an axis which is inclined at 54.7° to the applied magnetic field direction. The effect of the rotation is to eliminate nuclear magnetic dipolar interactions (as described in the section on broad-line NMR) by forcing the orientation-dependent dipole-dipole interaction term to zero (see Eq. [59]). The rotation also averages to zero the anisotropic chemical shifts and interactions of nuclei having $I \geqslant 1$ with their own electric quadrupoles. As a result, NMR linewidths of solids are dramatically reduced, and resolution of isotropic nuclear spin-spin interactions can often be achieved. These interactions split the resonance into several lines as a result of the presence of neighboring magnetic nuclei in the solid, and are classified as J couplings, generally following the same rules in the splitting of NMR spectra as those outlined earlier for spin-spin coupling in NMR spectra of liquids. The isotropic spin-spin interactions arise from the indirect coupling of nuclear spins via electron orbitals, since it has already been noted that direct through-space spin-spin interactions are averaged to zero by rapid rotation of the sample.

The typical high-resolution MAS NMR spectrum of a magnetic nucleus in a solid will, then, reveal a resonance line (which may be split into several lines by nearby magnetic nuclei) at a specific chemical shift position for each chemically nonequivalent nucleus. For example, ^{29}Si spectra of aluminosilicates can be used to determine whether the SiO_4 tetrahedron in framework silicates is connected to 0, 1, 2, 3, or 4 AlO_4 tetrahedra (Lippmaa et al., 1981), based upon the different chemical shift of the ^{29}Si nucleus in each environment (Fig. 9–18). Some minerals have diagnostic ^{29}Si chemical shifts; it has been suggested that a shift of -78

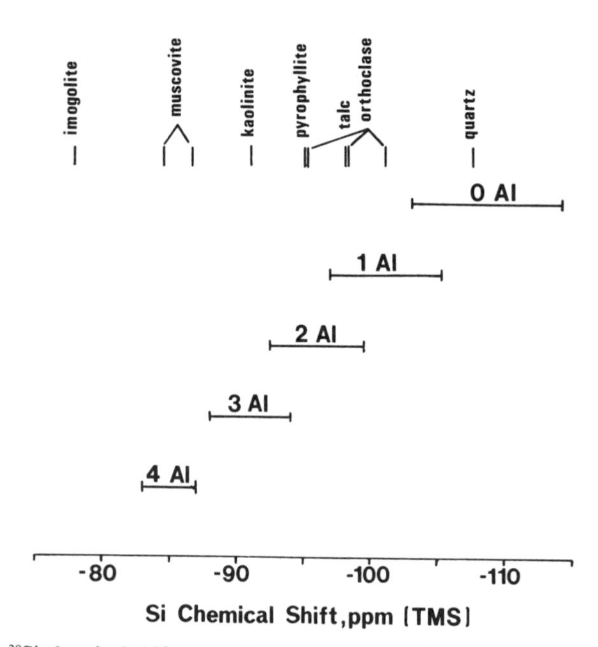

Figure 9–18. ^{29}Si chemical shift ranges for framework aluminosilicates, with the number of Al atoms sharing oxygen atoms with Si equal to 0, 1, 2, 3, and 4. Measured chemical shifts of common minerals are also indicated. (Data from Lippmaa & co-workers, 1981).

ppm for ^{29}Si can be used to identify imogolite in soil clays (Barron et al., 1982). Measured chemical shifts of ^{29}Si in a few common minerals are indicated in Fig. 9–18. In general, increasing Si-O-Si bonding in silicates produces increasing diamagnetic shielding and a greater chemical shift, so that framework silicates have greater chemical shifts than layer or chain silicates, and Al substitution for Si reduces the chemical shift (as Fig. 9–18 shows). Some aluminosilicate minerals (e.g., orthoclase) reveal several ^{29}Si lines because of the presence of nonequivalent Si environments arising from different degrees of neighboring tetrahedral Al substitution.

In addition to ^{29}Si NMR, high-resolution ^{27}Al NMR has been applied to mineral samples, easily distinguishing between tetrahedral and octahedral Al based upon chemical shifts. The ^{31}P spectra of solids can also be used to identify chemical forms of phosphates in minerals and soils based upon chemical shifts (Williams et al., 1981).

The application of high-resolution ^{1}H and ^{13}C NMR to studies of soil organic matter has been reviewed by Wilson (1981). Sensitivity has been improved through Fourier transform techniques and the use of high resonance frequencies. In addition, the low inherent sensitivity of ^{13}C NMR, a result of the low natural abundance of ^{13}C, can be improved by cross-polarization methods. The result is that ^{1}H and ^{13}C NMR spectroscopy have attained sufficient sensitivity to be useful in the study of soil organic matter extracts. Magic angle spinning has been applied to solid organic materials to obtain highly resolved ^{13}C NMR spectra (Wilson et al., 1983). The NMR spectra allow complex organic materials such as humic acids to be characterized, based upon chemical shift positions of the ^{1}H and ^{13}C resonances. (Fig. 9–12 indicates typical shifts for protons in different chemical environments.) In particular, ^{13}C and ^{1}H NMR spectra are useful in estimating the relative aromatic and aliphatic character of complex organics (Hayes & Swift, 1978; Wilson, 1981).

The main requirement for obtaining narrow, well-resolved spectra of solids is that the frequency of sample spinning be about the same order of magnitude as the breadth of the broad-line spectrum of the static sample (expressed in frequency units). This may require spinning speeds of several kHz. Rotation sidebands are thereby generated on either side of the resonance line, shifted from the line by multiples of the spinning frequency (Andrew, 1981).

9–4.2 Apparatus

A high-resolution continuous-wave NMR spectrometer (Fig. 9–19) commonly has a permanent magnet with field strength of about 1.4 tesla. The field strength must be stable to about one part in 10^8 and homogeneous. Proton spectra are recorded at about 60 MHz (megahertz) with a magnet of this strength. A sweep generator and sweep coils allow the field to be varied continuously over a narrow range. However, electromagnets are more flexible, allowing proton spectra to be obtained at frequencies as high as 100 MHz, as well as permitting spectra of ^{19}F, ^{11}B, ^{13}C, and ^{31}P to be recorded at appropriate combinations of frequency and magnetic field strength. In addition to the magnet, the spectrometer includes a radio-frequency transmitter and receiver as well as a recorder. The sample holder positions the sample in the magnetic field adjacent to the transmitter coil and receiver coil, and spins the sample to average

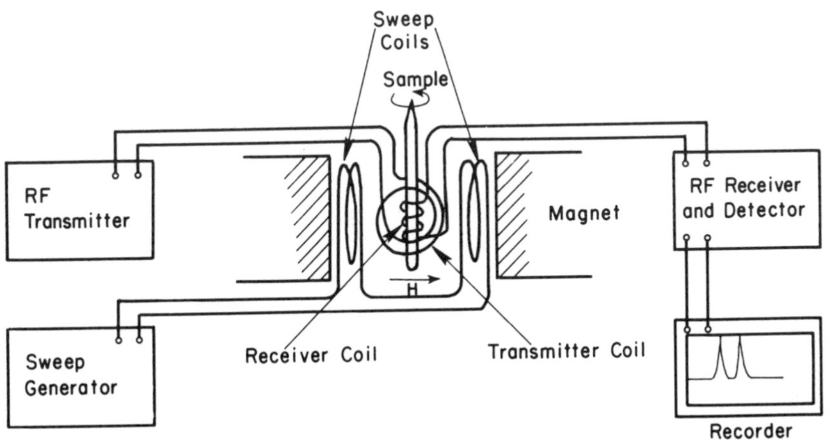

Figure 9–19. Diagram of an NMR spectrometer (from Silverstein & Bassler, (1967)).

magnetic field inhomogeneities. Frequency shifts of resonance peaks from a reference marker (commonly TMS) can be measured with an accuracy of ±1 Hz.

Instrumentation required for pulse NMR experiments will not be discussed here, although it relies on the basic magnetic resonance apparatus shown in Fig. 9–19.

9–4.3 Procedure

The continuous-wave NMR spectrum is recorded as a series of absorption peaks by scanning the magnetic field, with the integrated peak areas being proportional to the number of protons (or nuclei) they represent.

The sample for high resolution NMR (a liquid or solute in a suitable solvent) is placed in a glass tube. Usually, about 0.4 mL of liquid or 10 to 50 mg of sample dissolved in 0.4 mL of solvent is needed. Inherent sensitivity of NMR is much lower than that of ESR, so that more sample is required. However, this problem can be minimized by repeated scans, accumulating signal and averaging out noise. The signal-to-noise ratio improves in proportion to $N^{1/2}$, where N is the number of repeated scans. Since pulse NMR FID curves can be recorded much more quickly than continuous wave spectra, the usual procedure for obtaining high-resolution spectra is to conduct many pulse experiments in rapid succession, and perform a Fourier transform on the accumulated signal to obtain the conventional NMR spectrum. This procedure vastly improves NMR sensitivity for both liquid and solid samples.

The solvent used in proton NMR should contain no protons, have a low boiling point, and be nonpolar and inert. A common solvent with these properties is carbon tetrachloride (CCl_4); however, many samples are not soluble in it. Deuterated chloroform ($CDCl_3$) is more expensive, but most commonly used. Deuterated solvents normally demonstrate a weak peak due to proton impurities, but this rarely causes a serious interference. Care must be taken to avoid paramagnetic or ferromagnetic impurities in samples, since they can broaden NMR absorption peaks.

Identification of compounds from high-resolution liquid NMR spectra will not be discussed in detail here. By measurement of chemical shift, spin-spin coupling, and relative peak intensities, it is often possible to identify pure organic compounds. The important details of compound identification are given by Silverstein and Bassler (1967). The Sadtler index of NMR spectra can also be useful in this regard.

9–4.4 Comments

Proton NMR has been used with success in clay-water systems. Considering that a water molecule has two protons, each of which can have one of two orientations ($M_I = \pm 1/2$) relative to the applied magnetic field, H, a given proton is subjected to an effective field that is slightly greater or less than H. As a result, resonance of a collection of water

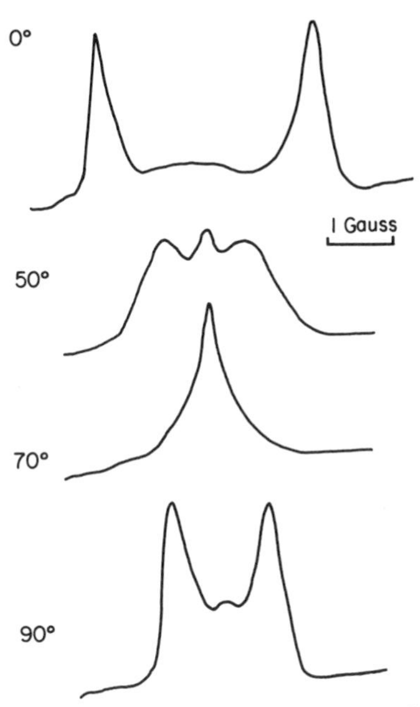

Fig. 9–20. Proton NMR spectra of the one-layer hydrate of Li^+-hectorite (from Fripiat, (1980).

molecules produces an NMR doublet *if* the proton-proton vector maintains a certain orientation with respect to H. This orientation can be accomplished if the water molecule is rigid on the NMR time scale, or if the molecule undergoes anisotropic motion in such a way that the proton-proton vector reorients rapidly about a rotation axis with a fixed orientation relative to H (Hougardy et al., 1975). If, however, molecular rotation of water molecules is random (as in liquid water) or if protons undergo rapid exchange between water molecules (as in acid aqueous systems), an NMR singlet is observed. In the case of random motion, dipolar interactions are averaged to zero, while rapid proton exchange causes a given proton to "see" an average of spin orientations of neighboring protons as it moves quickly from molecule to molecule.

Oriented, air-dry films of Li^+-hectorite have proton NMR doublets with splittings that are dependent upon the angle between the external magnetic field and the normal to the plane of the film (Fig. 9–20). When the angle between the proton-proton vector and H is 54.7°, the doublet merges into a singlet as described by Eq. [59]. This is in fact observed at an angle of about 70° between H and the normal to the plane of the film (Fig. 9–20). A weaker central line observed in the proton spectra of clays becomes stronger at higher temperature and has been attributed to H^+ exchange between molecules of adsorbed water. Clearly, NMR can distinguish between adsorbed and free water and provide information on orientation of surface-bound water molecules. Clays with low paramagnetic ion contents must be used in these NMR studies to avoid excessive spectral broadening; hectorite and Llano vermiculite have been favored minerals for this reason. It is also possible to obtain $^{23}Na^+$ and $^{7}Li^+$ NMR spectra of Na^+- and Li^+-exchanged clays (Conard, 1976), allowing the symmetry and motion of surface-adsorbed metal ions to be studied.

Pulse proton NMR experiments on clays have also been done (Fripiat, 1980), providing information on the dynamics of proton motion. Since liquid water has $T_1 \cong 1$ s, while adsorbed water on clays has been found to have T_1 on the order of 10^{-3} s (Hougardy et al., 1975), pulse NMR can easily distinguish bound water from free water. Although solids (e.g., ice) generally have long T_1 values, the short T_1 observed for adsorbed water results from the fact that paramagnetic impurities in clays dominate the relaxation process of protons in the vicinity of the surfaces. Interlamellar water diffusion coefficients of two water layer Ca^{2+}-montmorillonite has been calculated to be about 10^{-6} cm^2sec^{-1} using pulse NMR to obtain proton T_1 values (Hougardy et al., 1975).

Despite the complete loss of information about orientation of chemical species in solids, high-resolution MAS NMR is likely to have a great impact on soil mineralogy and surface chemistry. This powerful method, adaptable to a wide range of soil chemical problems because of the large number of suitable magnetic nuclei (for a partial list, see Table 9–6), overcomes the past limitation of NMR in soils research: low sensitivity. In addition, paramagnetic and ferromagnetic "impurities," ubiquitous in soil clays and organics, are evidently less of a problem in the broadening of high-resolution solid-state NMR spectra than they are in solution NMR spectra. For example, several percent Fe in soil clay samples do not normally prevent the determination of high-resolution ^{29}Si spectra, probably because most of the Fe is concentrated in mineral phases physically separated from the silicate minerals.

9–5 REFERENCES

Abragam, A., and B. Bleaney. 1970. Electron paramagnetic resonance of transition ions. Oxford University Press, London.

Anagnostopoulos, T., and M. Calamiotou. 1973. Magnetic behavior of some biotite samples from West Thrace, N.E. Greece. Clays Clay Miner. 21:459–464.

Andrew, E. R. 1981. Magic angle spinning. Int. Rev. Phys. Chem. 1:195–224.

Angel, B. R., and P. L. Hall. 1972. Electron spin resonance studies of kaolins. p. 47–60. *In* J. M. Serratosa (ed). Proc. Int. Clay Conf. Div. of Ciencias, C.S.I.C., Madrid.

Angel, B.R., and W.E.J. Vincent. 1978. Electron spin resonance studies of iron oxides associated with the surface of kaolins. Clays Clay Miner. 26:263–272.

Barron, P.F., M.A. Wilson, A.S. Campbell, and R.L. Frost. 1982. Detection of imogolite in soils using solid state ^{29}Si NMR. Nature (London) 299:616–618.

Bigham, J.M., D.C. Golden, L.H. Bowen, S.W. Buol, and S.B. Weed. 1978. Iron oxide mineralogy of well-drained ultisols and oxisols: I. Characterization of iron oxides in soil clays by Mossbauer spectroscopy, X-ray diffractometry, and selected chemical techniques. Soil Sci. Soc. Am. J. 42:816–825.

Carrington, A., and A.D. McLachlan. 1967. Introduction to magnetic resonance (with applications to chemistry and chemical physics). Harper and Row, New York.

Conard, J. 1976. Structure of water and hydrogen bonding on clays studied by ^{7}Li and ^{1}H

NMR. p. 85–93. *In* H. A. Resing and C. G. Wade (ed.) Magnetic resonance in colloid and interface science. ACS Symp. Ser. 34, Am. Chem. Soc., Washington, DC.

Cotton, F. A., and G. Wilkinson. 1966. Advanced inorganic chemistry. A comprehensive text. 2nd ed. Interscience Publishers, New York.

Craik, D. J. 1971. Structure and properties of magnetic materials. Applied Physics Series (no. 1), H.J. Goldsmid (ed.) Pion Ltd., London.

Figgis, B. N., and J. Lewis. 1960. Modern coordination chemistry. Interscience Publishers, New York.

Fripiat, J. J. 1980. The application of NMR to the study of clay minerals. p. 245–315. *In* J. W. Stucki and W. L. Banwart (eds.) Advanced chemical methods for soil and clay minerals research. D. Reidel, Boston.

Hayes, M. H. B., and R. S. Swift. 1978. The chemistry of soil organic colloids. p. 179–320. *In* D.J. Greenland and M.H.B. Hayes (ed.) The chemistry of soil constituents. Wiley-Interscience, New York.

Hougardy, J., W. Stone, and J. J. Fripiat. 1975. Structure and motion of water molecules in the two layer hydrate of Na^{+}-vermiculite. p. 191–199. *In* S. W. Bailey (ed). Proc. Int. Clay Conf., Mexico City. Applied Publishing Ltd., Wilmette, IL.

Lindoy, L. F., V. Katovic, and D. H. Busch. 1972. A variable temperature Faraday magnetic balance. J. Chem. Ed. 49:117–120.

Lippmaa, E., M. Magi, A. Samoson, M. Tarmak, and G. Engelhardt. 1981. Investigation of the structure of zeolites by solid-state high-resolution ^{29}Si NMR spectroscopy. J. Am. Chem. Soc. 103:4992–4996.

Longworth, G., L. W. Becker, R. Thompson, F. Oldfield, J. A. Dearing, and T. A. Rummery. 1979. Mossbauer effect and magnetic studies of secondary iron oxides in soils. J. Soil Sci. 30:93–110.

Lukshin, A. A., T. I. Rumyantseva, and V. P. Kovrigo. 1968. Magnetic susceptibility of the principal soil in the Udmurt A.S.S.R. Soviet Soil Sci. 88–93.

Martin, D. H. 1967. Magnetism in solids. M.I.T. Press, Cambridge, Mass.

McBride, M.B. 1976. Nitroxide spin probes on smectite surfaces. Temperature and solvation effects on the mobility of exchange cations. J. Phys. Chem. 80:196–203.

McBride, M. B. 1978. Transition metal bonding in humic acid: an ESR study. Soil Sci. 126:200–209.

McBride, M. B., T. J. Pinnavaia, and M. M. Mortland. 1975a. Electron spin relaxation and the mobility of manganese (II) exchange ions in smectites. Am. Mineral. 60:66–72.

McBride, M. B., T. J. Pinnavaia, and M. M. Mortland. 1975b. Perturbation of structural Fe^{3+} in smectites by exchange ions. Clays Clay Miner. 23:103–107.

Mooney, H. M. 1952. Magnetic susceptibility measurements in Minnesota. Part I: Technique of measurement. Geophysics 17:531–543.

Mullins, C. E. 1977. Magnetic susceptibility of the soil and its significance in soil science—a review. J. Soil Sci. 28:223–246.

Olivier, D., J. C. Vedrine, and H. Pezerat. 1975. Resonance paramagnetique electronique du Fe^{3+} dans les argiles alteres artificiellement et dans le milieu naturel. Proc. Int. Clay Conf., Mexico City. p. 231–238.

Poole, C. P., and H. A. Farach. 1971. Relaxation in magnetic resonance. Academic Press, New York.

Poupko, R., and Z. Luz. 1972. ESR and NMR in aqueous and methanol solutions of copper (II) solvates. Temperature and magnetic field dependence of electron and nuclear spin relaxation. J. Chem. Phys. 57:3311–3318.

Quickenden, T. I., and R. C. Marshall. 1972. Magnetochemistry in SI units. J. Chem. Ed. 49:114–116.

Silverstein, R. M., and G. C. Bassler. 1967. Spectrometric identification of organic compounds. John Wiley and Sons, Inc., New York.

Turov, E. A. 1966a. Magnetic resonance in ferromagnetics and antiferromagnetics as excitation of spin waves. p. 78–126. *In* S.V. Vonsovskii (ed.) Ferromagnetic resonance. Pergamon Press, London.

Turov, E. A. 1966b. Line width of ferromagnetic resonance absorption. p. 184–230. *In* S.V. Vonsovskii (ed.) Ferromagnetic resonance. Pergamon Press, London.

Vadyunina, A. F., and Babanin, V. F. 1972. Magnetic susceptibility of some soils in the U.S.S.R. Soviet Soil Sci. 4:588–599.

Vonsovskii, S. V. 1966. Magnetic resonance in ferromagnetics. p. 1–11. *In* S.V. Vonsovskii (ed.) Ferromagnetic resonance. Pergamon Press, London.

Wertz, J. E., and J. R. Bolton. 1972. Electron spin resonance. Elementary theory and practical applications. McGraw-Hill, New York.

Williams, R. J. P., R. G. F. Giles, and A. M. Posner. 1981. Solid state phosphorus N.M.R. spectroscopy of minerals and soils. J.C.S. Chem. Comm. 1051–1052.

Wilson, M. A. 1981. Applications of nuclear magnetic resonance spectroscopy to the study of the structure of soil organic matter. J. Soil Sci. 32:167–186.

Wilson, M. A., S. Heng, K. M. Goh, R. J. Pugmire, and D. M. Grant. 1983. Studies of litter and acid-insoluble soil organic matter fractions using ^{13}C-cross polarization nuclear magnetic resonance spectroscopy with magic angle spinning. J. Soil Sci. 34:83–97.

Wilson, S. A., and J. H. Weber. 1977. Electron spin resonance analysis of semiquinone free radicals of aquatic and soil fulvic and humic acids. Anal. Lett. 10:75–84.

Zijlstra, H. 1967. Experimental methods in magnetism 2. Measurement of magnetic quantities. *In* E.P. Wohlfarth (ed.) Selected topics in solid state physics. Interscience Publishers, New York.

10 Electron Microprobe Analysis

B. L. SAWHNEY
The Connecticut Agricultural Experiment Station
New Haven, Connecticut

10-1 INTRODUCTION

Electron microprobe analysis or electron probe microanalysis offers a powerful tool for in-situ determination of the chemical composition of microscopic regions on specimen surfaces. The technique permits quantitative analysis of elements of atomic number above 11 (Na) and semi-quantitative analysis of elements down to atomic number 4 (Be). Although determinations of most elements by an electron microprobe are limited to concentrations of 50 to 100 ppm, its ability to analyze microscopic volumes permits nondestructive measurement of elements in amounts as low as 10^{-16} g within a volume of 1 μm^3. In this respect, the electron microprobe surpasses other techniques, as these amounts may be too small for analysis by those methods. Electron microprobe analysis, developed and perfected in the last 30 years, combines instrumentation and techniques of electron microscopy and x-ray spectrochemical analysis. The first successful use of the microprobe was described in a doctoral thesis by Castaing (1951) at the University of Paris. Now the equipment incorporates various modifications and sophisticated devices for automation of data analysis and is manufactured by several scientific companies. It is used in numerous investigations in metallurgy, geology, mineralogy, soil science, electronics, biology, and medicine.

In soils, the electron microprobe has been used in analyses of silicate minerals and metal oxides and in studies of chemical and biological weathering of minerals, cation exchange, structure and composition of Fe and Mn concretions, phosphate reactions, trace element diffusion, and geochemical processes.

In this chapter, various aspects of electron microprobe analysis are discussed briefly, while details of these and related phases of the technique are given in books listed in section 10-8.

10-2 PRINCIPLES

Electron microprobe analysis is based upon measurements of wavelength and intensities of the characteristic lines in the x-ray spectra emitted by elements bombarded by an electron beam. When an electron beam of sufficient energy strikes a chemical element, electrons from the inner shells of the atom are removed. Consequently, electrons from the outer shell fall into the vacancy created in the inner shell, emitting x-ray photons. If removal of electrons or ionization occurs in the K shell, K radiation is emitted. As the energy (E_1) of an electron in the outer shell of an element is higher than the energy (E_2) of an electron in the inner shell, the x-ray photon of energy E_1-E_2 is emitted. This energy (E) is related to the wavelength (λ, expressed in nanometers) of the emitted x-ray as follows:

$$E = E_1 - E_2 = 1240 \text{ eV}/\lambda .$$

The wavelength of the radiation is uniquely characteristic of the element, while the intensity of the radiation depends upon its concentration. Qualitative analysis thus requires identification of the emitted radiations, while quantitative analysis involves comparisons of their intensities from the specimens to those from standards of known composition.

10–3 INSTRUMENT

The electron microprobe consists of a source of electrons, an electron optics system to focus the electron beam on the sample, a specimen stage, an optical microscope to select the specimen area for analysis and an X-ray detector system. Fig. 10–1 is a schematic diagram showing components of an electron microprobe.

10–3.1 Electron Source and Optics

Just as in electron microscopes, an electron beam is produced from a 100- to 150-μm heated tungsten filament in an electron gun operating under a low accelerating potential (5-50 kV). The electron beam is focused on the specimen by means of a set of electromagnetic lenses; current in the condenser lens coil is regulated to control the focal length of the lens and thus the size of the beam, while the current in the objective lens coil is adjusted to focus the beam on the microscopic region with a final probe diameter of 5 nm to 1 μm on the specimen to be analyzed. The power supplies for both the electron gun and the magnetic lenses must be stable in order to maintain constant beam size, current, and focus.

10–3.2 Specimen Stage and Optical Microscope

Most instruments include a mechanical stage equipped with fine micrometers, which permit movement of the stage along X and Y directions with a precision of 1 μm. The translation distance varies between 20 and 80 mm in different probes. In some instruments, however, the stage can

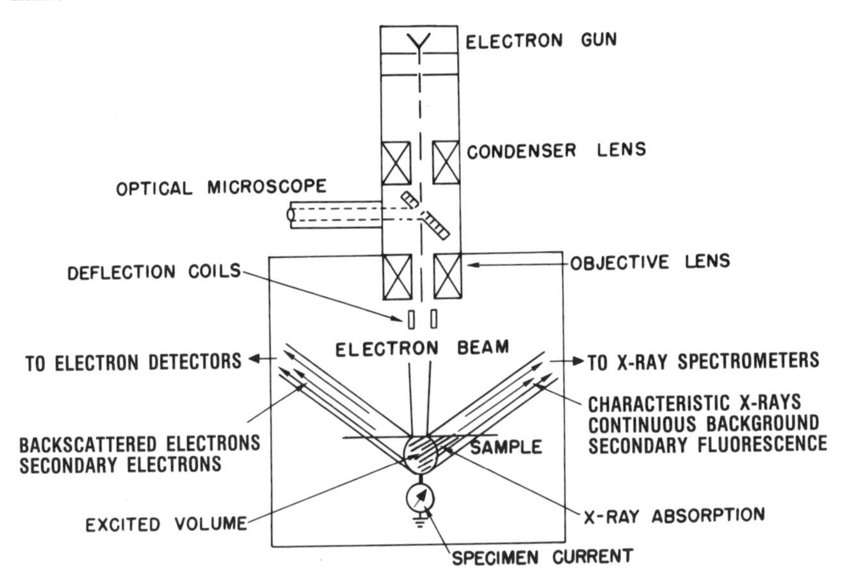

Fig. 10–1. Schematic diagram showing components of an electron microprobe and the signals produced from a specimen surface irradiated with an electron beam.

be moved in the *Z* direction also. For example, the CAMECA (Courbevoie, Seine, France) instrument uses a scanning microscope stage which can be translated over 20 to 50 mm along the *X* and *Y* directions and moved 40 mm along the *Z* direction. For quantitative analyses across varying composition regions, automatic translations at low speed (1 μm/min) and high speed (20 μm/min) are attainable. Large specimens or a number of small specimens and standards can be mounted on the stage simultaneously. The specimen stage is enclosed in a chamber which can be connected to the vacuum system and includes suitable windows for emergent X-rays and provision for specimen observation with a light microscope.

A light microscope with good resolution is included for visual observation of the specimen and for selection of an area to be analyzed. In some instruments, the light optical and electron optical systems are coaxial, and the specimen can be viewed constantly during analysis. In others, the specimen is moved out of the electron beam to allow high quality viewing, but the specimen cannot be observed during analysis. Still other arrangements of positioning the magnetic lens, microscope, and specimen have been used.

10–3.3 Electron-beam Scanning Devices

When an electron beam strikes the surface of a solid specimen, a number of interactions occur. Some of these produce signals that can be used for investigation of topography and surface chemistry of the sample as described below.

10–3.3.1 BACK-SCATTERED ELECTRON AND SPECIMEN CURRENT IMAGES

On entering the specimen, beam electrons are deflected in different directions. A few electrons scatter through large angles and travel back to the surface and escape. This process is called back-scattering and is strongly dependent on the average atomic number of the sample matrix. For samples of low atomic number, most electrons penetrate and are absorbed; while for samples of high atomic number, a considerable number of incoming electrons may be back-scattered. To display an image of the back-scattered electrons, the electron beam is deflected to sweep across an area of the specimen. Beam deflection is generally attained by a set of electromagnetic deflection coils or electrostatic deflection plates placed within the electron optical column so that the beam traverses the specimen surface in a raster pattern over an area of about 400 μm^2 or more. The back-scattered electrons can be used to modulate the brightness of a cathode-ray tube scanned in synchronism with the electron probe, but over a larger area of about 10 cm^2. Thus, an enlarged "image" of the back-scattered electrons showing contrasts due to the differences in atomic numbers of the elements on the specimen surface is obtained. Similarly, topographic contrasts on specimen surfaces are obtained from secondary electron images. These images stem from low-energy electrons which are produced within a few tens of nanometers of the surface, providing a high spatial resolution and topographic contrast. As the area of secondary electron generation is limited by the electron probe spot size, these electrons are often used for studying micromorphology of surfaces.

As stated above, the smaller the fraction of the beam electrons back-scattered, the larger the fraction absorbed, and hence the greater is the current flowing through the specimen. A specimen current image produced by the current flowing through the specimen can be displayed on the cathode-ray tube in a similar manner as the back-scattered electrons and secondary electron images. The specimen current image is essentially a compliment of the back-scattered electron image.

10–3.3.2 X-RAY IMAGES

In contrast to the above, when brightness of the cathode-ray tube is controlled by the electrical output from the x-ray spectrometers, x-ray images are produced. As the brightness of any spot on the cathode-ray tube is governed by the intensity of the characteristic x-rays from the element being analyzed, the x-ray image shows the distribution of the element in the area being scanned.

Distribution of an element across a specimen can also be measured by attaching a motor to the X or Y drive of the specimen stage and recording the x-ray output from the desired element using a focused electron beam. A line scan showing relative concentrations of the element across the specimen is obtained. Alternatively, a single raster line can be used to produce the line scan.

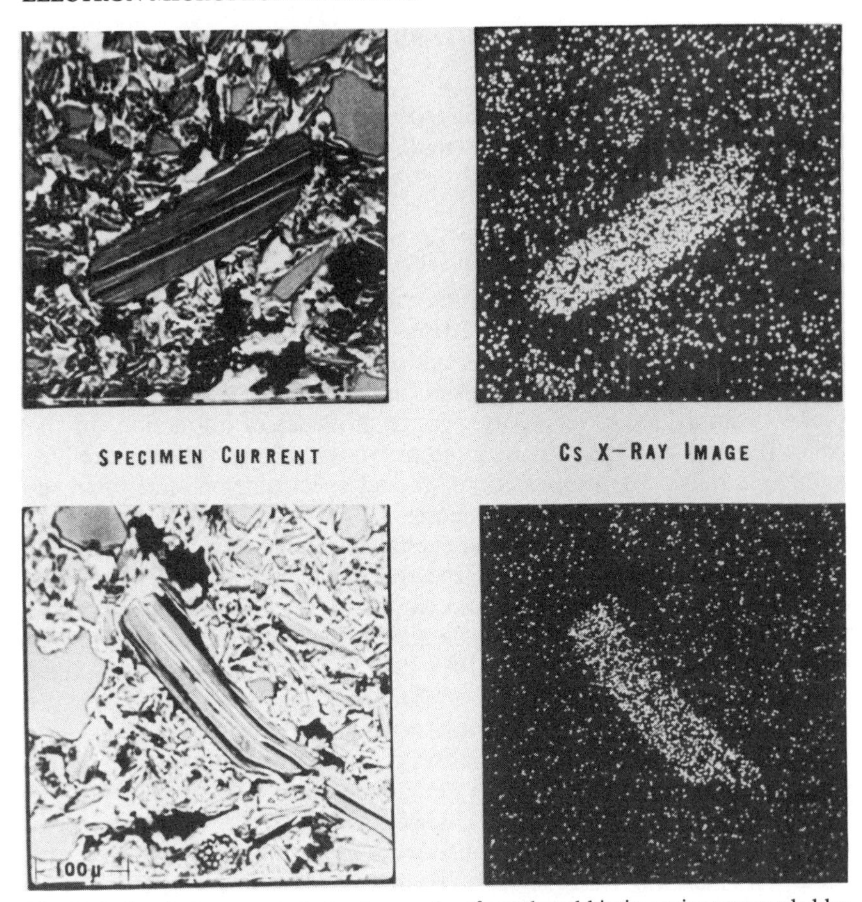

Fig. 10–2. Specimen current photomicrographs of weathered biotite grains surrounded by silty matrix (left) and congruent Cs x-ray images following Cs-saturation (right). Cs distribution illustrates loci of cation exchange in the specimen (from Hill & Sawhney, 1969).

Fig. 10–2 shows specimen current pictures of two weathered biotite grains surrounded by silty matrix (left), illustrating wide variation in topography and electron absorption and the corresponding Cs x-ray images following sorption of Cs by the soil (right). Large sorption of Cs by the weathered biotite grain is illustrated by the high density of bright spots corresponding to the high intensity of the characteristic x-ray from Cs (Hill & Sawhney, 1969).

10–3.4 X-ray Detectors

Analyses of the x-rays emitted by different elements in the specimen involves dispersion of the characteristic wavelengths and measurement of their intensity. Two types of x-ray detection systems used in electron microprobe are discussed below.

10–3.4.1 CRYSTAL SPECTROMETERS OR WAVELENGTH DISPERSIVE SYSTEMS

In this system, x-rays from the specimen are collimated onto a diffracting crystal, where the wavelengths are dispersed according to the Bragg equation

$$n\lambda = 2d \sin\Theta$$

where n represents the order of diffraction, λ is the wavelength of x-rays diffracted, Θ is the angle of diffraction, and d is the crystal interplanar spacing. The spectrometer may be equipped with either a flat or a curved diffracting crystal. The curved crystals are commonly used in the microprobe because they cover a larger range of angles of diffraction for the collection of x-rays and provide an improved wavelength resolution. Different geometrical arrangements of crystal spectrometers and their advantages have been discussed by Long (1977).

A large number of diffracting crystals covering a wide range of interplanar spacings with different spectral dispersion and reflection efficiencies are available. As a crystal cannot diffract wavelengths greater than its $2d$ spacing, crystals with small spacing, such as LiF with 2d = 0.28 and 0.4 nm, cannot be used for long wavelengths emitted by light elements with atomic number < 9. For these analyses, crystals such as Pb-stearate, with $2d = 10$ nm, are available. Choice of a crystal also requires that any fluorescent radiation generated within the crystal does not interfere with the wavelength being measured.

As the angle of diffraction is changed from 0 to 90°, the wavelength diffracted will change in accordance with the Bragg equation, covering the whole spectrum in sequence. In practice, servo-systems are commonly used to control spectrometer settings for Bragg angles predetermined for maximum intensity of x-ray peaks from different elements. Also, programmed automatic systems can be used for a sequence of determinations at predetermined diffraction angles.

X-rays diffracted by the crystal are passed through a slit system before their intensities are measured by the detector. The crystal and the detector are moved together by a goniometer so that for each angular setting of the crystal and the x-ray beam Θ, the detector is at 2Θ. Most commercial instruments are equipped with several spectrometers so that more than one diffracting crystal can be used simultaneously. The detectors commonly used in crystal spectrometer systems for converting x-ray photons into electrical energy are proportional counter type. Proportional counters are filled with a gas such as argon, often mixed with methane or carbon dioxide, that is readily ionized by the x-ray photons of long wavelengths. For transmitting long-wavelength x-rays, the proportional counters are fitted with thin windows of Mylar, and for short and medium wavelengths beryllium or aluminum windows are used. Electrons of the ionizeed gas produce electrical impulses as they reach the counter wire, which is at a high positive potential. Pulses from these counters have to be magnified before they are measured. After amplification, the pulses are passed through a pulse height selector where pulses from natural radiation, produced by fluorescence in the crystal, from scattering, and from higher orders (n = 2 and above) in the Bragg equation are rejected. This results in a higher peak/background ratio. The pulses can then be counted by a ratemeter or recorded on a chart as a spectrum of the sample over a selected range of angles.

10–3.4.2 ENERGY DISPERSIVE SPECTROMETERS

In these systems, characteristic x-rays emitted from the specimen are separated according to their energy, rather than their wavelength as in crystal spectrometers. Each characteristic x-ray has a specific quantum energy which is related to wavelength by

$$E(\text{keV}) = 1.2396/\lambda(\text{nm})\,.$$

Thus, when x-rays from a specimen fall on a detector in which response to each x-ray photon is proportional to its quantum energy, the whole spectrum from a specimen can be recorded. Lithium-drifted silicon, which consists of a single-crystal slice of silicon anodized with lithium in an applied electric field, provides such a detector. Absorption of x-rays on one side of the silicon plate causes ionization and production of electron-hole pairs. The process is similar to ionization in a gas-filled proportional counter. It differs from the gas ionization in that no amplification occurs in the silicon detector. The electron-hole pairs form charge pulses, which are collected at the other side of the plate. These pulses are amplified and passed to a multichannel analyzer where they are sorted by voltage and can be displayed on a cathode-ray tube or recorded, as illustrated in Fig. 10–3. The horizontal energy scale of the spectrum consists of as many as 1024 channels, permitting simultaneous identification of a number of elements in the specimen. The vertical axis represents the intensity scale that allows quantitative analyses of the elements present. The entire spectrum can be stored in a computer for subsequent elemental analysis.

10–3.4.3 COMPARISON BETWEEN WAVELENGTH-DISPERSIVE AND ENERGY-DISPERSIVE SYSTEMS

Wavelength-dispersive systems generally are more sensitive and have higher resolution, especially for low-energy x-ray photons. In addition, since the x-rays measured at a given wavelength setting come primarily from the element being analyzed, they are not count-rate limited. Energy-dispersive systems, on the other hand, have higher detector efficiency. As the whole spectrum is obtained in one measurement, the analysis is much faster, generally 2 min for a typical spectrum. However, since all characteristic and back-scattered energy photons from the specimen are measured, some unexpected elements are also detected. Also, because of

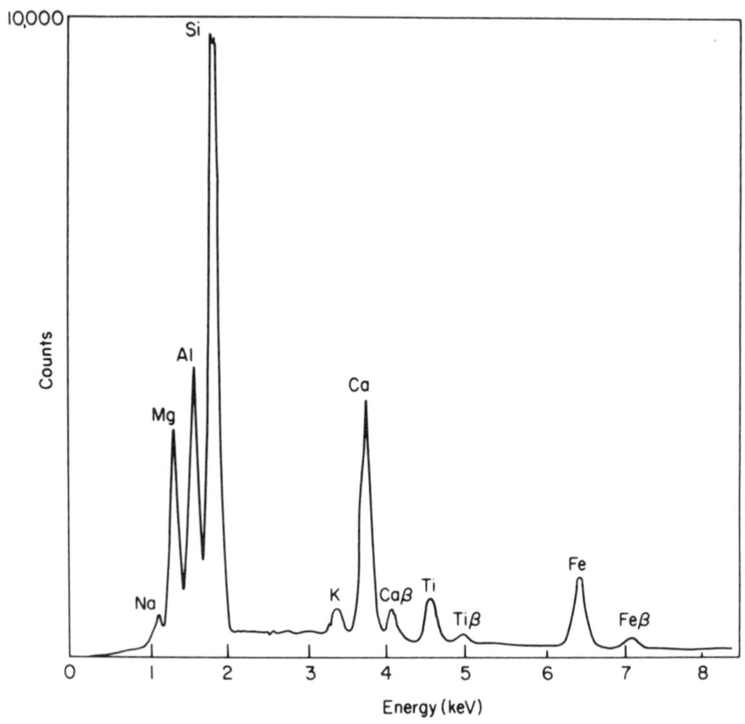

Fig. 10–3. X-ray energy spectrum from kaersutite, showing K_α radiations from Na, Mg, Al, Si, K, Ca, Ti, and Fe and K_β radiations from Ca, Ti, and Fe in the specimen (from Long, 1977).

the much longer dead-time (duration for processing each individual pulse) in comparison to the proportional counter, the maximum counting rate is limited. At high counting rates, pulse overlap may occur. However, "pile-up rejection circuits" capable of detecting two separate pulses are included to reduce this effect. Because of lower resolution, the peak-to-background ratio is much lower in energy-dispersive systems. Despite these limitations, energy-dispersive systems are widely used in elemental analysis, primarily because of the speed in analysis. Nevertheless, for analyses of trace elements with high precision and for complex rare earth mixtures and minerals with heavy metals in which severe peak overlap may occur, crystal spectrometers offer a more appropriate method.

10–4 SPECIMEN PREPARATION

10–4.1 Specimen Mounting

Samples of soils to be examined in the electron microprobe can be mounted, using epoxy resin, in short brass tubes (Long, 1977) or in holes drilled in brass rods (Cescas et al., 1968). Various kinds of resins have been used successfully, as for sample preparation in electron microscopy. Alternatively, whole soil sections may be impregnated with plastic and thin sections affixed to glass slides (Hill and Sawhney, 1971). Plastic materials should be hard enough to withstand a high degree of polish and should have low vapor pressure so that they are thermally stable under the electron beam. Some materials such as mica flakes can be affixed on a glass slide with a high-vacuum grease (Sawhney & Voigt, 1969) or on silica discs with a thin collodion film (LeRoux et al., 1970), especially if the samples are repeatedly examined for a continuation of specific reactions. Similar techniques can be employed for examination of plant material (Sawhney & Zelitch, 1969).

10–4.2 Specimen Polishing

For quantitative analysis, the embedded soil samples must be polished to produce a relief-free flat surface. Since the region of x-ray production consists of a specific volume (typically 1–4 μm^3) underneath the point of impact of the incident electron beam, a scratch at the surface may create an additional absorption path (Fig. 10–1) for characteristic x-rays before they leave the specimen surface. Consequently, the decreased intensity would give lower estimates of the elements determined. The error is larger for elements of low atomic number and for instruments employing low takeoff angles. The magnitude of increased absorption of the emergent x-rays by a surface scratch or pit and different takeoff angles are discussed by Long (1977).

A number of polishing techniques have been described in the literature, each material presenting different problems. Initial grinding can be done with wheels and abrasive paper. For final polishing to produce a relief-free surface, successively finer grades (8, 3, 1, and 0.25 μm) of diamond pastes or alumina are used.

10–4.3 Surface Coating

In nonconducting materials such as soils, the electrons can produce a negative charge near the irradiated surface and cause deflection of the incident beam. In addition, a voltage drop between the specimen and the ground, due to large resistance of the sample, causes a decrease in the intensity of the emitted x-rays. Therefore, polished specimens of soils and silicates are coated with a thin layer (10–20 nm) of a conducting material such as carbon, to carry away the current in the specimen. For heavy metals analysis, a conducting surface on the specimen can be produced by evaporating a thin metallic layer of Al, Ag, Cu, Cr, or Mn. The metal used should not be one of the elements to be analyzed. The thickness of the film can be controlled by observing the color of the film deposited on polished brass or on porcelain, or can be measured by an interferometer. To compensate for attenuation of electrons or x-rays by the conducting film, specimens and standards are generally coated simultaneously. The standards used should be homogeneous on a submicron scale and stable under electron bombardment, and their composition should be accurately known. For analysis of soils and silicate minerals, single oxides or metal standards for each element have been successfully used by Sweatman and Long (1969).

10–5 QUANTITATIVE ANALYSIS

10–5.1 Instrumental Factors

For accurate quantitative analysis, it is essential that certain instrumental parameters are carefully controlled.

10–5.1.1 SPECIMEN ALIGNMENT

The standards and specimens should be maintained in a fixed plane so that the takeoff angle of x-rays from both is the same as used for the absorption correction. Error introduced in the determination of an element increases with the tilt of the specimen's surface, especially in instruments with low takeoff angles.

10–5.1.2 ELECTRON OPTICAL SYSTEM ALIGNMENT

The current incident on the specimen should be maintained constant by varying the condenser lens and not by altering the electron-gun current. Altering the electron-gun current produces a change in beam accelerating potential. Constant value of the accelerating voltage, corresponding to the excitation potential of an element, must be maintained because this value is used later in matrix correction.

10–5.1.3 X-RAY DETECTORS

The operating conditions of detectors for x-ray measurements should be adjusted to give a stable and linear response over the counting rate used. Furthermore, counting rates should be restricted for minimum error arising from resolving time of the instrument, although most computer programs now contain corrections for the dead time.

10–5.1.4 BACKGROUND MEASUREMENT

Estimates of background intensity, from x-ray continuum, are usually obtained by making one measurement on each side of the peak. Care should be taken to avoid characteristic lines of other elements in selecting positions for background measurements. When the peak is only slightly above background, a profile of the peak may be drawn and mean background obtained by interpolation.

10–5.2 Matrix Corrections

Quantitative analysis requires relating the intensities of the characteristic x-rays from an element to its concentration in the specimen. This can be achieved by comparing intensities of the characteristic x-rays from the specimen and standard under the same conditions. Thus,

$$C_u/C_s = I_u/I_s$$

where C_u and C_s are the concentrations and I_u and I_s are the intensities of the characteristic x-rays from the element in the unknown and the standard, respectively. However, because of the various interactions of the incident electrons and the emitted x-rays within the specimen, the relationship between the intensities and the concentrations of the element are not linear and depend upon the specimen matrix. Therefore, the observed intensities have to be corrected for the matrix factors: x-ray generation in specimen (atomic number factor), absorption of the emerging x-rays (absorption factor), and enhancement of the characteristic x-rays by secondary excitation in the specimen (fluorescence factor).

10–5.2.1 ATOMIC NUMBER CORRECTION

When an electron beam enters a target, the electrons lose energy by interaction with orbital electrons of the target atoms. The distribution of the incident energy between desired and other matrix elements is determined by their relative absorption or "stopping power." Only a portion of the incident energy interacts with the desired element to produce the characteristic x-rays of that element. Lighter elements with more orbital electrons absorb more energy. Consequently, the intensity of x-rays generated by a given concentration of an element in a matrix with light elements (low Z) would be less than in a matrix of heavy elements (high Z).

Also, when an electron beam enters a target, some electrons undergo large-angle deflection and are back-scattered. As mentioned earlier, the higher the average atomic number of the sample matrix, the greater is the back-scattering of electrons. Consequently, a smaller fraction of the incident beam is available for producing x-rays from the desired element in a matrix of heavy elements than in a matrix of light elements. The absorption of the incident energy and the back-scattering of electrons in a matrix thus produce opposite effects on the measured intensity and may partially cancel each other.

There are several methods for applying corrections for the above two components of the atomic number factor, stopping power of elements and back-scattering losses. A simplified treatment adopted by Sweatman and Long (1969) is described below.

The corrected concentration C' of an element is given by

$$C' = C_0(R_0/R_1)(S_1/S_0)$$

where C_0 is the measured concentration, R_0 and S_0 are the back-scattering losses and stopping powers of the standard, and R_1 and S_1 are the corresponding quantities for the specimen.

10–5.2.1.1 Calculation of S_1/S_0. The stopping power (S_i) of an element A is given by

$$S_i = K(1/E)(Z_i/A_i)\ln[1.66(E/J_i)]$$

where K is a constant; E is the mean energy of the electrons; Z_i and A_i are the atomic number and atomic weight of element A; and J_i is the mean ionization potential of the element. Mean stopping-power (S) of a compound can then be calculated from

$$S = K \cdot (\overline{Z/A})\,(1/E)(4.54 + \ln E - \ln \overline{Z})$$

where $(\overline{Z/A})$ and $\overline{Z}$ are the weighted means and $J = 11.5Z$. Stopping power of the standard is similarly calculated and the ratio S_1/S_0 determined.

10–5.2.1.2 Calculation of R_0/R_1. The electron back-scattering coefficient of a compound has been shown to be approximately equal to the weighted mean of the coefficients of individual elements and is given by

$$\overline{R} = \sum_{i=1}^{n} C_i R_i$$

where $\overline{R}$ is the back-scattering loss from a compound and C_i and R_i are the concentration and the back-scattering loss from an individual element. Curves giving back-scattering losses from different elements as a function of Z and accelerating potential have been prepared and can be used for calculations of R_0/R_1.

10–5.2.2 ABSORPTION CORRECTION

The absorption factor $f(x)$ by which the intensity of the measured x-rays from the desired element is attenuated depends upon the mass absorption coefficient (μ/ρ) of the specimen and the cosec of the takeoff angle and is given by:

$$f(x) = \frac{1 + h}{[1 + (x/\sigma)]\{1 + h\,[1 + (x/\sigma)]\}}$$

where $x = (\mu/\rho)\text{cosec}\ \Theta$, $h = 1.2\ A/Z^2$ (and for a compound $\overline{h} = \Sigma c_i h_i$) and linear absorption coefficient for electrons $\sigma = 4.5 \times 10^5/E_0^{1.65}$ –

$E_c^{1.65}$. The concentration of an element corrected for absorption (C′) is given by

$$C' = C_0\,[f(x)_0/f(x)_1]$$

where $f(x)_0$ and $f(x)_1$ are the absorption factors for the standard and the specimen respectively.

10–5.2.3 FLUORESCENCE CORRECTION

When the interaction of the electron beam with certain elements produces characteristic x-rays with energies greater than the critical excitation potential of the analyzed element, additional atoms of the element may be excited. Consequently, the measured intensity would include this characteristic fluorescence. In soils and silicates, the fluorescence enhancement is usually small because of the diluting effect of oxygen and the particular combination of elements in the minerals. Increase in measured intensity can also be caused by continuous fluorescence where excitation by a range of wavelengths in continuous spectra occurs. The continuous fluorescent effect is relatively small, however.

In order to obtain the true concentrations of the elements in a specimen, the measured concentrations have to be corrected for the atomic number factor, the absorption factor, and the fluorescence factor. The calculations are generally carried out by the aid of digital computers because of the successive use of improved factors and concentrations during the iterative procedure. A number of programs for calculating matrix corrections, differing mainly in format, have been used.

10–6 MONTE CARLO METHOD

The Monte Carlo method for determination of matrix correction involves calculations of the total effect of a large number of electrons in the target, rather than the average behavior of the electrons as described in the above procedures. The method calculates the trajectories and the energy loss of electrons along their path until the electrons can no longer cause ionization. The number of trajectories for the various elements and the number of steps along each trajectory in some cases are extremely large and are thus very costly in terms of computer time. The calculations are, however, convenient for determining the spatial distribution of x-rays, the back-scattering, and the absorption factors in electron microprobe analysis.

10–7 APPLICATIONS IN SOIL ANALYSIS

10–7.1 Chemical and Biological Weathering

The electron microprobe has been used effectively in a number of investigations of the weathering in micaceous minerals. One of the first

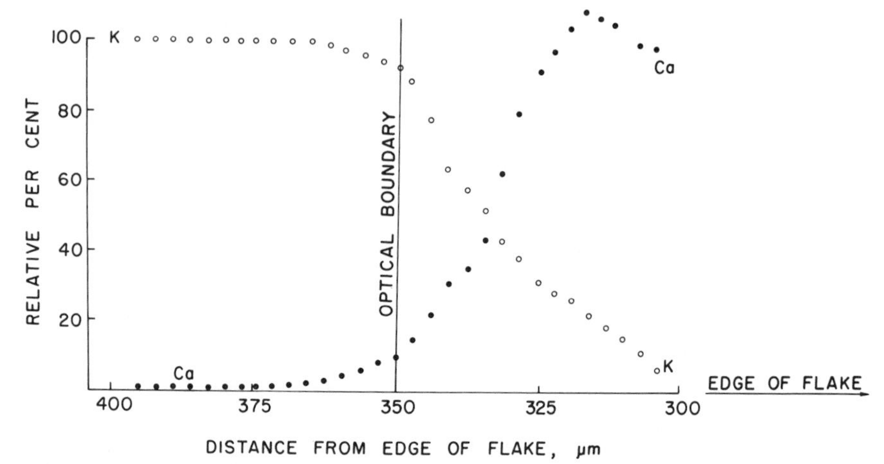

Fig. 10–4a. Electron beam traverses (in 3-μm steps) across a mica flake, showing replacement of K by Ca in the weathered edge (from Rausell-Colom et al., 1965).

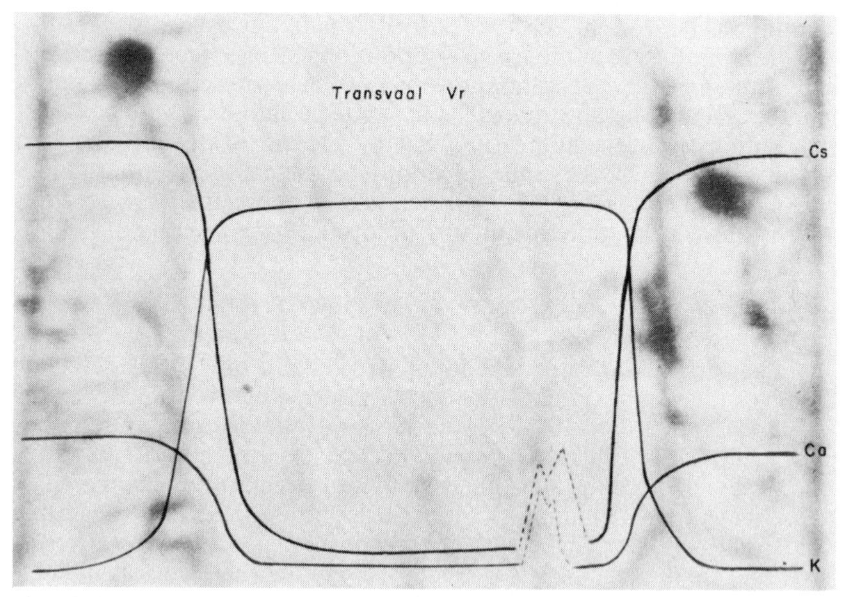

Fig. 10–4b. Smoothed traces of electron microprobe traverses for Cs, Ca, and K x-rays showing relative concentrations of these elements across a 1-mm Transvaal vermiculite flake; treatment with CsCl solution produced a mixed layer Cs-Ca system in the weathered edge, while no diffusion of Cs or Ca ions occurred in the unweathered portion of the flake.

such investigations was by Rausell-Colom et al. (1965), who treated different mica minerals with various salt solutions. Mica flakes were sealed in ampules containing salt solutions. The ampules were opened at intervals, solutions and the flakes were analyzed, and the solutions renewed. Hydrated cations of the salt solutions diffused through the edges of the

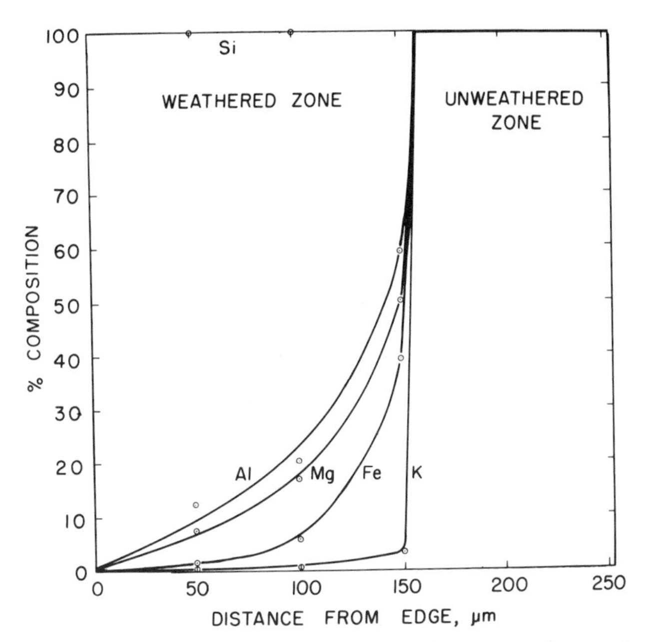

Fig. 10–5. Chemical composition of the weathered zone (as percent of unweathered zone) in a vermiculite flake weathered in a tulip poplar system, showing progressive increase in cation concentration as the unweathered zone is approached (from Sawhney & Voigt, 1969).

flakes, producing weathered boundaries surrounding the central unweathered core. Changes in the composition of the flakes were measured using the electron microprobe. An electron microprobe traverse across a mica flake treated with $CaCl_2$ solution (Fig. 10–4a) illustrates the replacement of K^+ ions by the Ca^{2+} ions in the altered boundary. Similarly, Fig. 10–4b (Sawhney, unpublished) shows a photomicrograph of a mica flake with weathered edges produced by $CaCl_2$ treatment. Superimposed on the photomicrograph are smoothed traces of electron microprobe line scans, illustrating the distribution of K^+ and Ca^{2+}, and of Cs^{2+} ions which were used to study diffusion of Cs into weathered micas. Clearly, replacement of K^+ by Ca^{2+} in the weathered edge occurred almost completely. In the subsequent Cs^{2+} treatment however, Ca^{2+} was only partially replaced by Cs^{2+}. X-ray diffraction analysis of the weathered edge (removed with a razor blade) revealed that it consisted of a randomly interstratified mixture of Cs^{2+}-saturated and collapsed interlayers and Ca-saturated and expanded interlayers.

Experiments on biological weathering of Transvaal vermiculite in a tulip poplar system (Sawhney & Voigt, 1969) (Fig. 10–5) show a progressive decrease in concentration of Al, Mg, and Fe in the weathered edges proceeding away from the unweathered central core. These observations of in-situ weathering have been possible only with an electron microprobe.

10–7.2 Phosphates in Soils

Using an electron microprobe, Cescas et al. (1970) discovered inclusions of apatite in certain primary minerals in a rhyolite pumice ash and beach sands from New Zealand. The occurrence of apatite in these minerals reflects the composition of the magma and environmental conditions during mineral crystallization.

Norrish (1968) confirmed the presence of complex phosphate minerals, plumbogummites, in soils on the basis of their chemical composition determined by electron microprobe. Phosphorus in these minerals was found to be associated with several alkaline and rare earth metals. Similarly, electron microprobe analysis of discrete phosphate grains in soils and lake sediments (Sawhney, 1973) revealed that phosphates may occur as complex compounds containing Al, Fe, Ca, Si, and P (Fig. 10–6a) or as individual apatite grains (Fig. 10–6b). These observations support the results of Koritnig (1965), indicating the replacement of Si by P in silicate minerals.

10–7.3 Ferromanganiferous Concretions

Microprobe analysis of the zoned concretions from sand fractions of soils of the Morrow plots at the University of Illinois (Cescas et al., 1968) showed that the core of zoned concretions was high in Fe, while the outer shell was dominated by Mn. Nonzoned concretions, on the other hand, contained Fe and Mn uniformly distributed throughout. Zoned concretions were thought to be formed under a slowly oxidizing environment, and nonzoned concretions under a rapidly oxidizing environment.

10–7.4 Silicate Minerals

Sweatman and Long (1969) performed quantitative analyses of a number of silicate minerals, using simple oxides and minerals as standards. They concluded that an accuracy of 1% or better can be obtained when matrix corrections are applied. Smith (1965) suggested the use of calibration curves prepared from standard minerals for each silicate mineral group. However, large number of standards of homogeneous composition are required for preparing the curves. Besides, analyses of the unknown specimens must be carried out under conditions identical to those used for establishing the calibration curves.

10–7.5 Geochemical Investigations

In a recent study, Krug (1981) used the electron microprobe to analyze profiles of soils occupying different toppographic positions in the New Jersey Pine Barrens. Based on the micromorphological characteristics of soils and the elemental composition of ironstones in the Bx

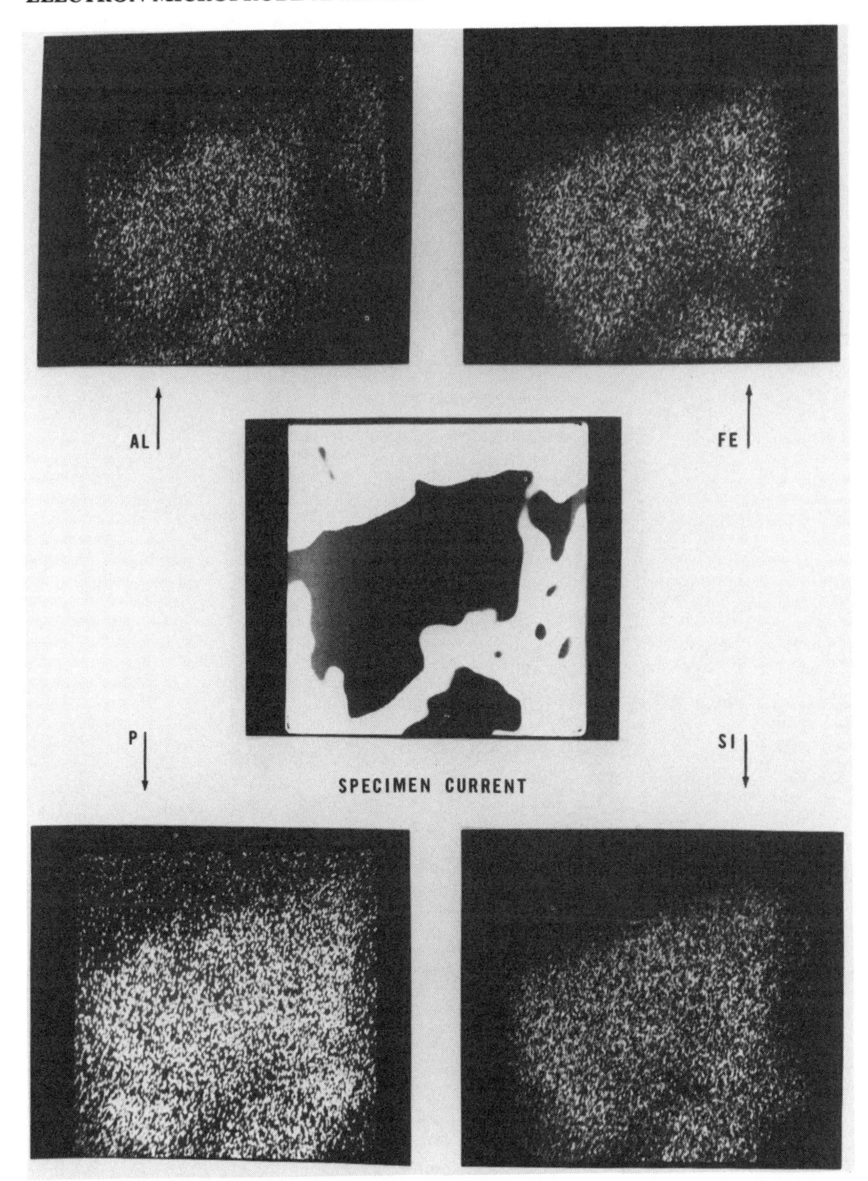

Fig. 10–6a. Specimen current image of a phosphate grain (center) and x-ray images for Al, Fe, P, and Si, showing the distribution of these elements in the grain (from Sawhney, 1973).

horizons (Fig. 10–7), the mechanism of ironstone formation and existence of geochemical soil catena were established.

Electron microprobe analysis continues to provide useful information where other analytical techniques cannot be readily applied. However, because of the high cost, the electron microprobe has not been extensively

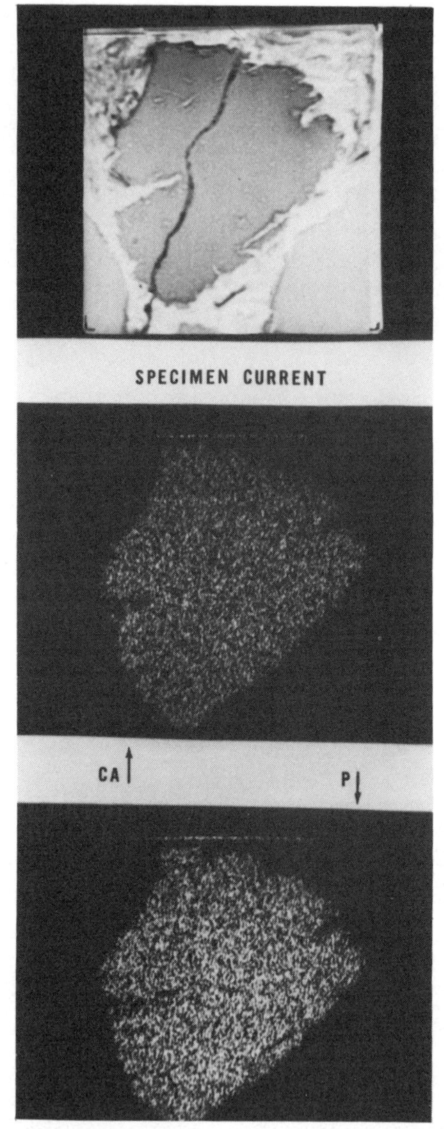

Fig. 10–6b. Specimen current image (top) and x-ray images for Ca and P in a calcium phosphate crystal (from Sawhney, 1973).

used in soil analysis. In recent years, the scanning electron microscope has been increasingly used in studies of the micromorphology of soils and soil minerals surfaces. Instruments combining the electron microprobe and the scanning electron microscope, providing capability of determining detailed micromorphology and elemental analysis, are now available and are in use in several laboratories. This combined instru-

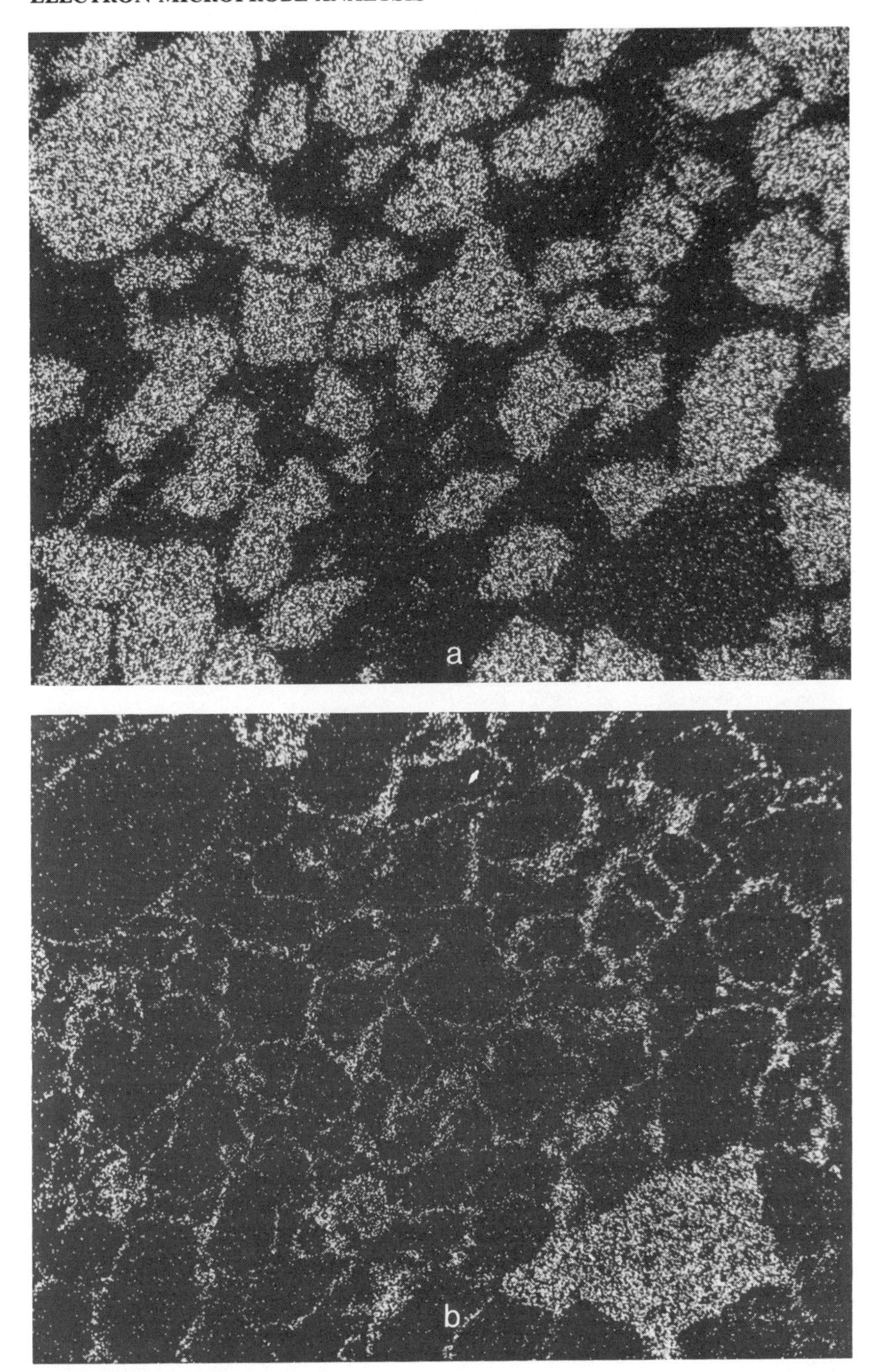

Fig. 10–7. Silicon and Fe x-ray images from pedogenic bog iron sandstone, showing the distribution of quartz sand grains (a) and a sesquioxide body in the lower left hand corner and ferruginous cutans (or coatings) on the grains (b).

mentation is certain to produce rapid expansion in the application of the electron microprobe in soils investigations.

10–8 REFERENCE BOOKS

Birks, L. S. 1963. Electron probe microanalysis. Interscience Publishers, New York.

Goldstein, J. I., and H. Yakowitz (ed.) 1977. Practical scanning electron microscopy. Plenum Press, New York.

McKinley, T. D., K. F. J. Heinrich, and D. B. Wittry (ed.) 1966. The electron microprobe. John Wiley and Sons, Inc., New York.

Reed, S. J. B. 1975. Electron microprobe analysis. Cambridge University Press, Cambridge.

Zussman, J. (ed.) 1977. Physical methods in determinative mineralogy. Academic Press, New York.

10–9 REFERENCES

Castaing, R. 1951. Application of electron probes to local, chemical and crytallographic analyses. Ph.D. Thesis, University of Paris, Paris, France.

Cescas, M. P., E. H. Tyner, and L. J. Gray. 1968. The electron microprobe X-ray analyzer and its use in soil investigations. Adv. Agron. 20:153–198.

Cescas, M. P., E. H. Tyner, and J. K. Syers. 1970. Distribution of apatite and other mineral inclusions in a rhyolitic pumice ash and beach sands from New Zealand: an electron-microprobe study. J. Soil Sci. 21:78–84.

Hill, D. E., and B. L. Sawhney. 1969. Electron microprobe analysis of thin sections of soil to observe loci of cation exchange. Soil Sci. Am. Proc. 33:531–534.

Hill, D. E., and B. L. Sawhney. 1971. Electron microprobe analysis of soils. Soil Sci. 112:32–38.

Koritnig, S. 1965. Geochemistry and phosphorus. I. The replacement of Si^{4+} by P^{5+} in rock-forming silicate minerals. Geochim. Cosmochim. Acta 29:361–371.

Krug, E. C. 1981. Geochemistry of pedogenic bog iron and concretion formation. Ph.D. Thesis, Rutgers, The State University of New Jersey. Diss. abstr. No. 81-15220.

LeRoux, J., C. I. Rich, and P. H. Ribbe. 1970. Ion selectivity by weathered micas as determined by electron microprobe analysis. Clays Clay Min. 18:333–338.

Long, J. V. P. 1977. Electron probe microanalysis, p. 273–341. *In* J. Zussman (ed.) Physical methods in determinative mineralogy. Academic Press, New York.

Norrish, K. 1968. Some phosphate minerals in soils. Trans. Int. Congr. Soil Sci., 9th 2:713–723.

Rausell-Colom, J. A., T. R. Sweatman, C. B. Wells, and K. Norrish. 1965. Studies in the artificial weathering of mica. p. 40–72. *In* E. G. Hallsworth, and D. V. Crawford (eds.) Experimental pedology. Butterworths, London.

Sawhney, B. L. 1973. Electron microprobe analysis of phosphates in soils and sediments. Soil Sci. Soc. Am. Proc. 37:658–659.

Sawhney, B. L., and G. K. Voigt. 1969. Chemical and biological weathering in vermiculite from Transvaal. Soil Sci. Soc. Am. Proc. 33:625–629.

Sawhney, B. L., and I. Zelitch. 1969. Direct determination of potassium ion accumulation in guard cells in relation to stomatal opening in light. Plant Physiol. 44:1350–1354.

Smith, J. V. 1965. X-ray emission microanalysis of rock-forming minerals. I. Experimental techniques. J. Geol. 73:830–864.

Sweatman, T. R., and J. V. P. Long. 1969. Quantitative electron-probe microanalysis of rock forming minerals. J. Petrol. 10:332–379.

11 Infrared Spectrometry

JOE L. WHITE AND CHARLES B. ROTH
Purdue University
West Lafayette, Indiana

11-1 GENERAL INTRODUCTION

At the time of preparation of this chapter for the first edition of *Methods of Soil Analysis* (Black et al., 1965), the application of infrared spectrometry to the study of inorganic and organic phases had been made primarily by mineralogists and organic chemists. It was remarked that infrared techniques had not been used extensively in the study of soils. In spite of the many spectacular advances in instrumentation, preparative techniques, and data-processing and handling capabilities that have occurred in the past two decades, this statement is, unfortunately, still valid. Lack of easy access to infrared instruments and/or sympathetic colleagues with some first-hand experience in application of infrared techniques to soil components are largely responsible for this situation.

Many of the principles and potential uses of infrared techniques in soil studies have been pointed out previously (Fripiat, 1960; Lyon, 1964; Mortensen et al., 1965; Chaussidon, 1972a, 1972b; Fieldes et al., 1972; Mitchell et al., 1964). Farmer & Russell (1967), Farmer et al. (1968), Farmer (1968, 1976, 1979), and White (1971, 1977) have reviewed some of the earlier advances in this area. Tuddenham and Stephens (1971) have thoroughly discussed the application of infrared techniques to the solution of geochemical and mineralogical problems.

The total energy of a molecule consists of the translational, rotational, vibrational, and electronic energies. Transitions among the different types of energy levels occur in different regions of the electromagnetic spectrum. Infrared radiation is associated with transitions between rotational and vibrational energy levels of the ground state. Organic molecules may be considered as weakly interacting entities possessing bonds and groups that vibrate independently of each other. Thus, the identification of functional groups in organic analysis was the major application of infrared spectrometry in the earlier stages of its development. In contrast to organic molecules, minerals tend to have few isolated vibrating groups and solid-state interactions must be taken into account. As a result of the lack of theoretical developments for the treatment of solid-state spectra, much of the early progress in understanding and interpreting the spectra of

 Methods of Soil Analysis, Part 1. Physical and Mineralogical Methods—Agronomy Monograph no. 9 (2nd Edition)

minerals in which isomorphous substitutions occurred was achieved through the study of mineral phases produced in hydrothermal syntheses (Stubican & Roy, 1961a, 1961b; Tarte, 1962).

Because of the complexity and diversity of the soil organic matter components, there is considerable ambiguity and uncertainty in the interpretation of infrared spectra of organic matter preparations. When used with care, infrared spectrometry can (i) provide specific and unique information about the nature, reactivity, and structural arrangement of oxygen-containing functional groups, (ii) establish the occurrence of protein and carbohydrate constituents, (iii) demonstrate the presence or absence of inorganic impurities (metal ions, aluminosilicate clays, etc.) in isolated humic fractions, and (iv) be used for quantitative analysis of components. Other possible applications include the study of interaction of pesticides and other organic molecules with humic acids, and characterization of metal-organic complexes.

The application of infrared spectrometry to the study of organic compounds that naturally occur in the soil, or that may interact with soil components when added as pesticides, herbicides, fertilizers, sewage sludge, etc., has progressed significantly since the preparation of the material in chapter 51 of *Methods of Soil Analysis* (Black et al., 1965). Schnitzer (1971), Schnitzer and Khan (1978), Stevenson and Butler (1969), and Stevenson and Goh (1971), among others, have made important contributions in the development of techniques of sample preparation and in interpretation of infrared data for humic substances. The recent book, *Humus Chemistry: Genesis, Composition, Reactions*, by Stevenson (1982), provides an extended treatment of the application of infrared spectrometry in humus research and should be consulted by those contemplating research in this area. Recent examples of the fruitful application of infrared techniques to the study of soil organic matter include reports by Boyd et al. (1979, 1980) and Schaumberg et al. (1980). These workers used infrared techniques to monitor the changes occurring in the humic acid fraction of soils produced by the application of large amounts of sewage sludge.

The most comprehensive references on infrared studies of minerals include the monograph *The Infrared Spectra of Minerals*, edited by Farmer (1974); a chapter on "Characterization of Soil Minerals by Infrared Spectroscopy," by Farmer and Palmieri (1975); the book *Infrared Spectra of Minerals and Related Inorganic Compounds*, by Gadsden (1975); an *Atlas of Infrared Spectroscopy of Clay Minerals and Their Admixtures*, by van der Marel and Beutelspacher (1976); sections in *Data Handbook for Clay Materials and other Non-metallic Minerals*, edited by van Olphen and Fripiat (1979); and the *Sadtler Catalog of Infrared Spectra of Clay Minerals*, edited by Ferraro (1982).

Infrared spectrometry is most useful in mineralogical studies when used in conjunction with x-ray diffraction and other similar techniques. The method can be used in identifying inorganic compounds and minerals which have well-defined absorption bands (e.g., 1:1 vs. 2:1 layer silicates), in determining whether a layer silicate is dioctahedral or trioctahedral in composition, in studying isomorphous substitutions, in investigating the hydration of minerals, and in quantitative analysis. Infrared techniques provide information about the nature and identity of compounds that may be amorphous to x-rays.

11–2 PRINCIPLES

11–2.1 Origin of Spectra

No attempt will be made to consider the details of the theoretical principles of infrared spectrometry. The reader should consult texts such as those by Conley (1972), Miller and Stace (1972), and Smith (1979). The monograph edited by Farmer (1974) also provides an up-to-date treatment of the theory of the interaction of infrared radiation with crystals and the application of this in infrared studies of minerals.

The region of the electromagnetic spectrum of interest in the application of infrared spectrometry to the study of soil components is that region consisting of radiant energies of slightly greater wavelengths than those associated with visible light. The terms used to specify a position in the infrared range are the *wavelength* (λ), in units of micrometers (μm) per wavelength, and a so-called frequency or wavenumber (ν), in units of waves per centimeter (cm^{-1}). The relationship of wavenumber to wavelength may be expressed as follows:

$$\nu\ (\mathrm{cm}^{-1}) = 1/\lambda\ (\mathrm{cm}^{-1}) = 10^4/\lambda\ (\mu\mathrm{m})\,.$$

True frequency ($\bar{\nu}$) is related to so-called frequency or wavenumber (ν) in the following manner: $\bar{\nu}(\mathrm{s}^{-1}) = c\nu$, where c is the velocity of light (3×10^{10} cm/s). The so-called frequency unit ν (cm^{-1} or waves/cm) is used rather than the more fundamental unit of true frequency $\bar{\nu}$ (cycles or waves per second) as a matter of convenience. *Wavenumber* is the preferred term for the frequency designation.

The overall infrared region may be subdivided as follows: near-infrared (overtone), 13 300 to 4000 cm^{-1} (0.75–2.5 μm); mid-infrared or fundamental (rotation-vibration), 4000 to 400 cm^{-1} (2.5–25 μm); and far-infrared (sketetal vibration), 400 to 20 cm^{-1} (25–500 μm). The region of primary interest in the study of soil components is between wavenumbers of 4000 and 200 cm^{-1} (2.5–50 μm).

The frequency of oscillation of atoms and molecules about their equilibrium positions is 10^{13} to 10^{14} cyles/s. Infrared radiation has frequencies in this range and promotes transitions in a molecule between rotational and vibrational energy levels of the ground electronic energy state.

There are two types of bond vibration modes in simple molecules: stretching (sometimes designated ν) and bending or deformation (sometimes designated δ). The former refers to the periodic stretching of the bond A–B along the bond axis. Bending modes for the A–B bond involve displacements that occur at right angles to the bond axis. These vibrations produce periodic displacements of atoms with respect to one another, causing a change in interatomic distance. When the vibrations are accompanied by a change in dipole moment, they give rise to absorption of radiation in the infrared region.

Spectral information from infrared spectrometry can be supplemented with information from Raman spectrometry. Infrared and Raman spectra are not exact duplicates of each other because of the different manner in which the photon energy is transferred to the molecule. Whereas, as pointed out earlier, the infrared absorption intensity depends on the change in the permanent dipole moment of the molecule, Raman emission intensity is dependent upon a change in the polarizability of the molecule. Therefore vibrations that are active in the infrared may be inactive in the Raman and vice versa, and the two techniques complement one another in providing information related to molecular structure. Since the Raman effect is very inefficient, the use of Raman spectrometry is normally limited to liquids and solids. Raman spectrometry is especially useful in the study of molecules in aqueous solution, a medium that transmits infrared radiation very poorly.

11–2.2 Infrared Spectrometers

Smith (1979) has classified commercially available infrared spectrometers into two groups: (i) the sequential dispersive spectrometers and (ii) multiplex nondispersive interferometer spectrometers. The most commonly used sequential dispersive instruments are double-beam instruments that employ gratings and band-pass filters to cover a complete spectrum. In such instruments, a source of infrared radiation, such as the Globar, a bonded silicon carbide rod, or the Nernst glower (a mixture of zirconium, thorium, and yttrium oxides), provides radiation over the region 4000 to 100 cm^{-1}. This radiation passes through the sample cell and is dispersed by the system of gratings and filters that are a part of the monochromator. The slits of the monochromator select a narrow frequency range, the energy of which is measured by a detector. Thus, information is collected sequentially in time, each spectral element being scanned in turn. Detectors used to measure the radiant energy include thermocouples, bolometers, Golay or pneumatic detectors, and photon detectors. The detector transforms the energy received into an electrical signal, which is then amplified and recorded in synchronization with the monochromator. Spectra are plotted as percent transmittance or absorbance as a function of wavenumber (frequency). The development of microprocessor computerized infrared systems and accompanying software has dramatically improved instrument performance, computer compatibility, and data-handling capability. Hannah and Coates (1980) have remarked that the problem-solving capacity of infrared spectroscopy has increased by at least tenfold and the sensitivity of the technique has been improved by two to three orders of magnitude. Sample preparation is currently one of the largest sources of error in an infrared measurement.

In contrast to the sequential dispersive spectrometer discussed in the preceding paragraphs, the multiplex nondispersive interferometer spectrometer, typified by the Fourier transform infrared spectrometer (FT-IR), is able to utilize infrared energy much more efficiently than the more common sequential dispersive instrument because it makes use of all frequencies from the source simultaneously rather than individually (Smith, 1979). An FT-IR spectrometer basically consists of two parts: an optical system that uses an interferometer, and a dedicated computer (Griffiths, 1975). The computer controls optical components, collects and stores data, performs computations on data, and displays spectra. Most interferometers have been designed to obtain 1 cm^{-1} resolution while making approximately 10 scans/s. This capability of rapid scans of the complete spectral region makes the FT-IR well-adapted to the study of transient systems (Durana & Mantz, 1979). Elegant software and data acquisition and analysis capabilities have revolutionized the potential usefulness of FT-IR as well as the previously discussed sequential dispersive instrument. The interfacing of gas chromatographs, HPLC, and other instruments with FT-IR has enhanced even further the power of this method of analysis. Operations such as signal enhancement, signal averaging, digital smoothing, spectra subtraction or addition, on-line spectral search (comparison of sample spectrum standards in data bases such as the Sadtler libraries or user-created libraries), and quantitative analysis of multiple component systems can now be performed with virtually all the instruments and software currently being marketed. Application of these recent developments will bring a new dimension to the indentification and characterization of the soil and soil components.

Depending on the degree of flexibility desired and cost constraints, available "dispersive" infrared spectrometers that operate in a sequential mode range from low-resolution, high-ordinate-accuracy "IR colorimeters" with built-in computational capabilities at relatively low cost, to a double-beam spectrometer of a fairly high degree of sophistication in terms of microprocessor and computational developments, at a moderate to high cost. Griffiths (1975) suggests that FT-IR is superior to sequential dispersive spectroscopy in (i) applications requiring high resolution over a wide spectral range, (ii) rapid scan applications, and (iii) applications where the signal is so weak that an unacceptably long time is needed to measure the spectrum conveniently. FT-IR instruments that provide such flexibility are in the moderate to rather high cost range.

For details of spectrometer operations, optimization of recording parameters, calibration of the spectrometer, etc., the reader should consult manuals provided by the instrument manufacturer as well as practical guides to infrared instrumentation and techniques (Potts, 1963; Miller & Stace, 1972; Smith, 1979; Stewart, 1970; Griffiths 1975; Brame & Grasselli, 1976, 1977a, 1977b; Ferraro & Basile, 1978, 1979, 1982).

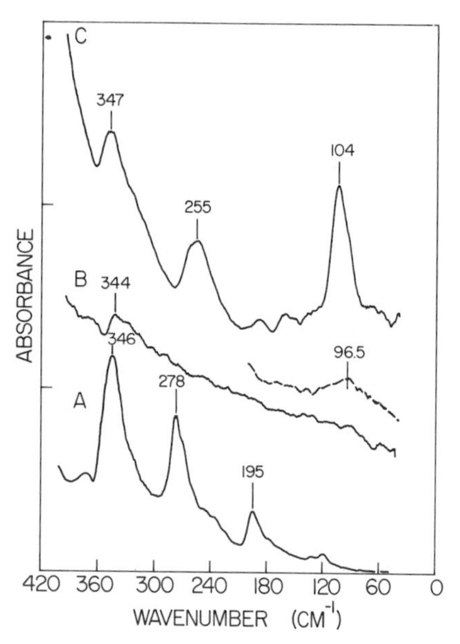

Fig. 11–1 Far-infrared spectra of (A) kaolinite, (B) montmorillonite, untreated (solid line) and after K^+-saturation (broken line), and (C) muscovite (C. B. Roth, unpublished).

As mentioned previously, the infrared spectrum is comprised of the near-infrared, the mid-infrared, and the far-infrared regions. Although this chapter is concerned primarily with applications in the mid-infrared range (4000–400 cm^{-1}), it is useful to briefly mention applications and give references for the near-infrared and far-infrared regions.

Near-infrared measurements may be made by spectrometers similar to those used for the visible region. The usefulness of visible and near-infrared spectra of minerals and rocks has been thoroughly documented by Hunt and Salisbury (1970, 1971a, 1971b) and Hunt et al. (1971a, 1971b, 1973a, 1973b, 1973c, 1974). Diffuse reflectance spectrometry in the near-infrared has proven useful for quantitative analysis of protein, oil, and moisture in grains and other agricultural products (Watson, 1977).

Far-infrared spectrometry deals with internal stretching, bending, torsional, rotational, and lattice vibrations (Griffiths, 1975). The relatively low energy of sources generally used for far-infrared spectrometry severely limited the performance of grating spectrometers and led to the commercial development of far-infrared interferometers. The use of FT-IR for far-infrared spectrometry will undoubtedly increase. Applications of far-infrared spectrometry to layer silicates have been described by Ishii. Shimanouchi, and Nakahira (1967) and Ishii, Nakahira, and Takeda (1969). Absorption bands in the region 1200 to 250 cm^{-1} have been tentatively assigned to O–Si–O deformation and OH libration frequencies in the tetrahedral and octahedral layers. Bands in the 250 to 60 cm^{-1} region have been assigned to interlayer vibrations. The far-infrared spectra of kaolinite, muscovite, and a montmorillonite are shown in Fig. 11–

Table 11-1. Characteristic infrared absorption frequencies of humic substances (from Stevenson, 1982).

Frequency	Assignment
cm^{-1}	
3400–3300	O–H stretching, N–H stretching (trace)
2940–2900	Aliphatic C–H stretching
1725–1720	C=O stretching of COOH and ketones (trace)
1660–1630	C=O stretching of amide groups (amide I band), quinone C=O and/or C=O of H-bonded conjugated ketones
1620–1600	Aromatic C=C, strongly H-bonded C=O of conjugated ketones (?)
1590–1517	COO^- symmetric stretching, N–H deformation + C=N stretching (amide II band)
1460–1450	Aliphatic C–H
1400–1390	OH deformation and C–O stretching of phenolic OH, C–H deformation of CH_2 and CH_3 groups, COO^- antisymmetric stretching
1280–1200	C–O stretching and OH deformation of COOH, C–O stretching of aryl ethers
1170–950	C–O stretching of polysaccharide or polysaccharide-like substances, Si–O of silicate impurities

1. The band near 100 cm^{-1} in the muscovite and K-saturated montmorillonite has been attributed to vibrations of the interlayer cation.

11–2.3 Organic Spectra

The absorption spectrum of most organic molecules is simplified due to the fact that certain groups of atoms vibrate with the same frequency irrespective of the molecule to which they are attached. The relative constancy of group frequencies is due to the uniformity of bond force constants from molecule to molecule. These absorption bands can be used to characterize and identify functional groups in organic molecules (Bellamy, 1975; Colthup et al., 1975; Rao, 1963). Comparison of many compounds having known structures with infrared absorption spectra has resulted in the assignment of absorption frequency bands to functional groups (Bellamy, 1975; Colthup et al., 1975; Rao, 1963) (see Fig. 11–2). The correlations are based on both frequency and intensity measurements. Correlations between frequency and functional groups on soil organic matter preparation are fairly extensive but not without some controversy (Stevenson, 1982). Table 11–1 summarizes the major infrared group frequencies that have been reported for organic matter preparations.

Renewed interest in the application of infrared spectrometry to the study of organic matter has been rekindled by the recent emphasis on the influence of sewage sludge on soil properties and the environment (Boyd et al., 1979, 1980; Schaumberg et al., 1980).

11–2.4 Inorganic Spectra

Although considerable progress has been made in the application of solid-state physics and group theory to an understanding of crystal vi-

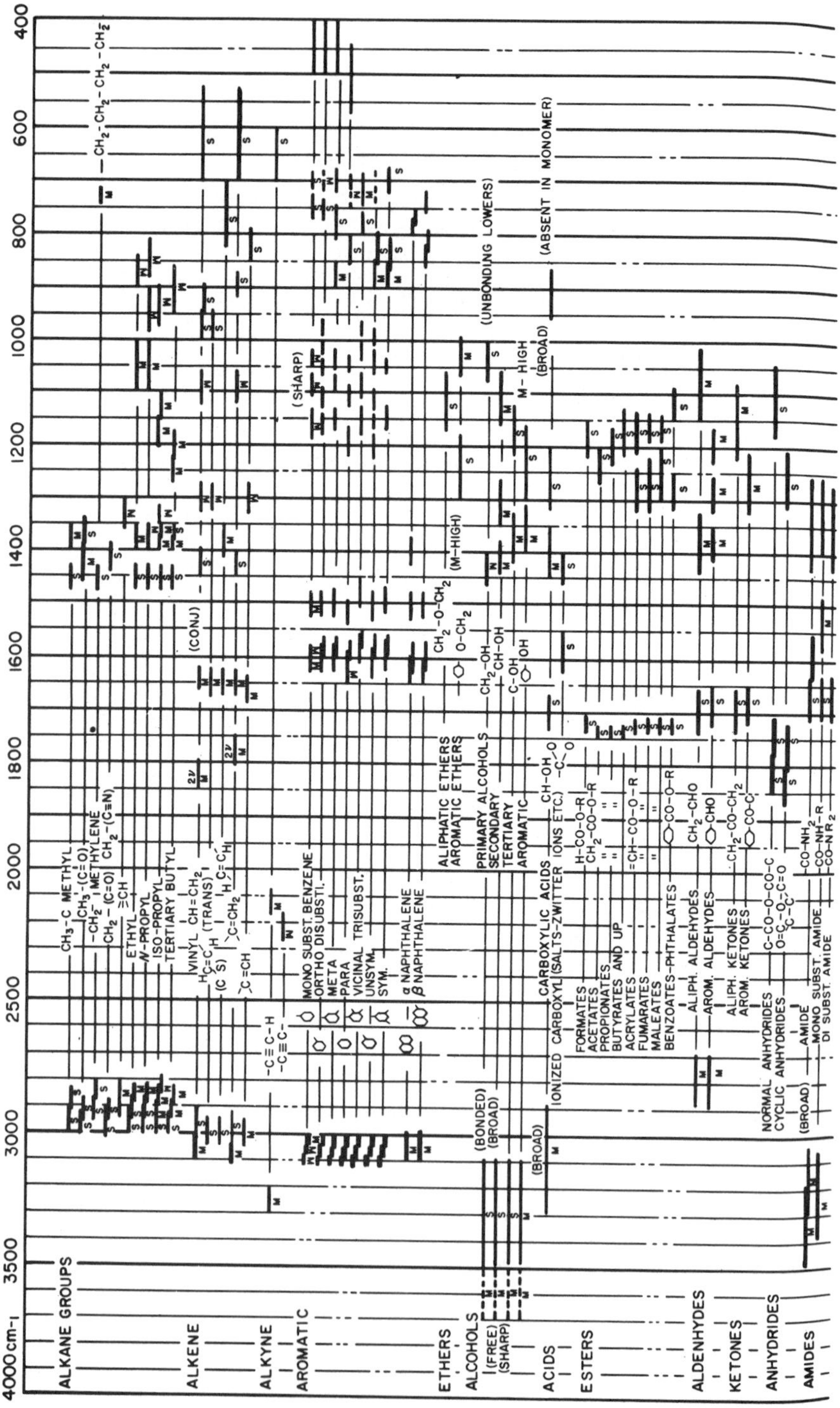

4000 cm-1 3500 3000 2500 2000 1800 1600 1400 1200 1000 800 600 400

ALKANE GROUPS
CH3-C METHYL
CH3-(C=O)
-CH2- METHYLENE
CH2-(C=O) CH2-(C≡N)
≡CH
ETHYL
N-PROPYL
ISO-PROPYL
TERTIARY BUTYL
-CH2-CH2-CH2-CH2-

ALKENE
VINYL CH=CH2
(TRANS)
(C S)
C=CH2
C=CH
2ν
(CONJ)

ALKYNE
-C≡C-H
-C≡C-

AROMATIC
MONO SUBST. BENZENE
ORTHO DISUBSTI.
META
PARA
VICINAL TRISUBST.
UNSYM.
SYM.
α NAPHTHALENE
β NAPHTHALENE
(SHARP)

ETHERS
ALIPHATIC ETHERS
AROMATIC ETHERS
CH2-O-CH2
O-CH2
(M-HIGH)

ALCOHOLS
(FREE)
(SHARP)
(BONDED)
(BROAD)
PRIMARY ALCOHOLS
SECONDARY
TERTIARY
AROMATIC
CH2-OH
CH-OH
C-OH
OH
M-HIGH
(UNBONDING LOWERS)

ACIDS
(BROAD)
CARBOXYLIC ACIDS
IONIZED CARBOXYL (SALTS-ZWITTER IONS ETC.)
CH-OH
-C-O O
(BROAD)
(ABSENT IN MONOMER)

ESTERS
FORMATES H-CO-O-R
ACETATES CH2-CO-O-R
PROPIONATES " "
BUTYRATES AND UP " "
ACRYLATES =CH-CO-O-R
FUMARATES " "
MALEATES "
BENZOATES-PHTHALATES CO-O-R

ALDENHYDES
ALIPH. ALDEHYDES CH2-CHO
AROM. ALDEHYDES CHO

KETONES
ALIPH. KETONES CH2-CO-CH2
AROM. KETONES CO-C

ANHYDRIDES
NORMAL ANHYDRIDES C-CO-O-CO-C
CYCLIC ANHYDRIDES O=C-O-C=O C-C

AMIDES
(BROAD)
AMIDE -CO-NH2
MONO SUBST. AMIDE -CO-NH-R
DI SUBST. AMIDE -CO-NR2

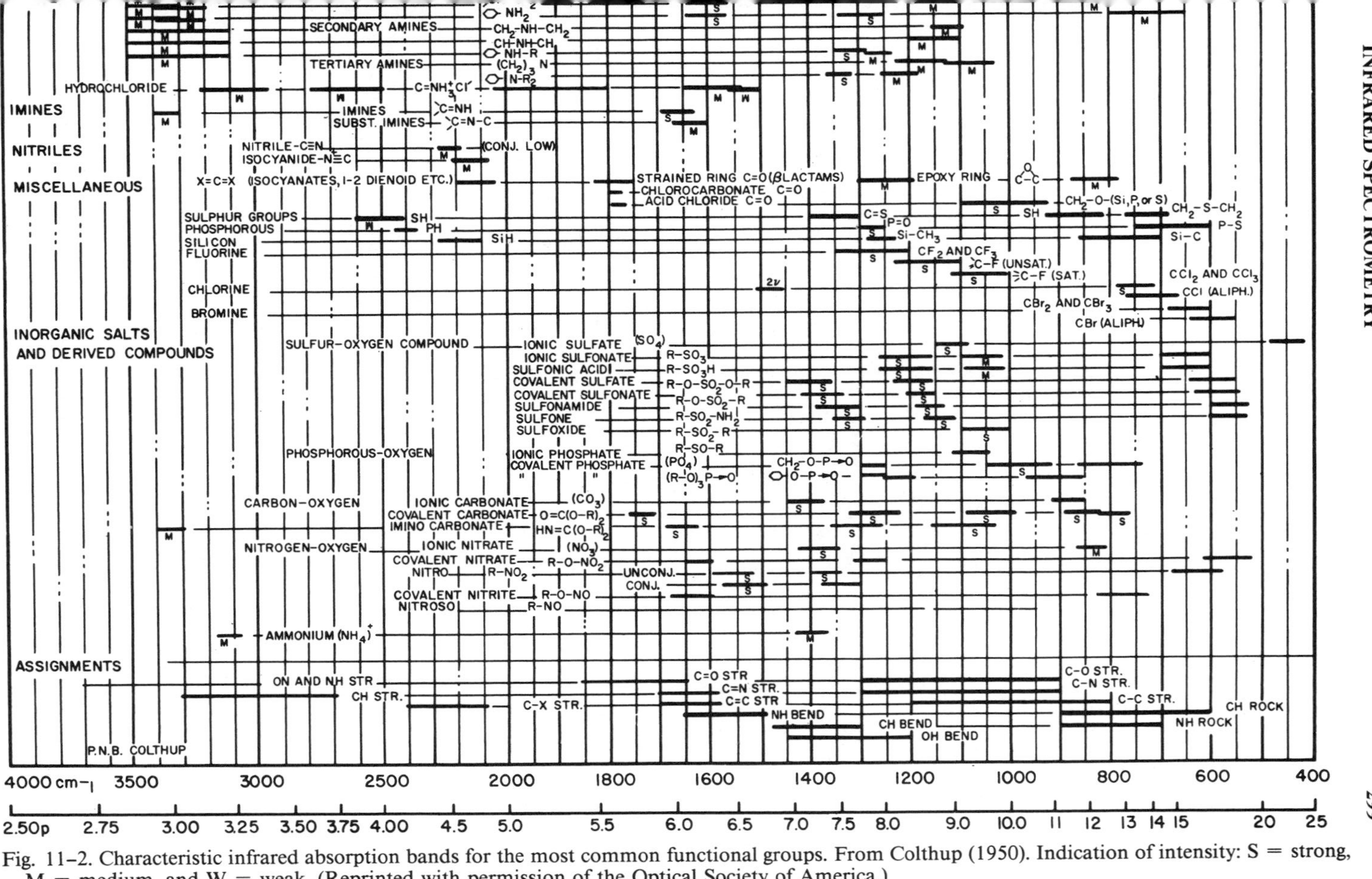

Fig. 11-2. Characteristic infrared absorption bands for the most common functional groups. From Colthup (1950). Indication of intensity: S = strong, M = medium, and W = weak. (Reprinted with permission of the Optical Society of America.)

brations in minerals, the uncertainties introduced by, for example, order-disorder effects, and compositional variations in these rather complex systems have somewhat restricted the usefulness of this approach. Careful studies of well-characterized families of natural and synthetic minerals have provided a sound basis for relating structural and compositional variations to spectral features. For discussions of assignments of infrared absorption bands to structural elements and groups, symmetry considerations, etc. the reader is referred to the monograph *The Infrared Spectra of Minerals* (Farmer 1974). Published studies of spectrum-composition relationships for specific groups of minerals include those for micas and smectites (Stubican & Roy, 1961a, 1961b; Vedder, 1964; Farmer & Russell, 1964; Wilson, 1966, 1970; Farmer et al., 1967; Wilkins, 1967; Ishii et al., 1969; Russell et al., 1970; Akhmanova & Alekhina, 1971; Russell & Fraser, 1971; Farmer et al., 1971; Velde, 1978, 1983), chlorites (Hayashi & Oinuma, 1965, 1967), zeolites (Oinuma & Hayashi, 1967; Flanigen et al., 1971), manganese oxides (Potter & Rossman, 1979a, 1979b), iron oxides (Fysh & Fredericks, 1983), and carbonates, sulfates, nitrates, and phosphates (Moenke, 1962, 1966; Lehr et al., 1967). A generalized as-

Table 11-2. Characteristic infrared group absorption frequencies found in minerals and other inorganic compounds.

Wavenumber	Assignment
cm^{-1}	
	O-H Vibrations
3700	free O-H stretch
3675-3540	O-H stretch
3390-2500	bonded O-H stretch
1700-1610	H-O-H bending
	O-H Librations (dioctahedral)
950-915	Al_2OH
~890	$Fe^{3+}AlOH$
~840	MgAlOH
~800	$MgFe^{3+}OH$
~800	$Fe^{2+}Fe^{3+}OH$
	Si-O Vibrations
1100-970	Si-O-Si antisymmetric stretch
800-600	Si-O-Si symmetric stretch
540-400	Si-O + miscellaneous vibrations
	NH_4^+ Vibrations
3330-3030	NH-stretching
1485-1390	NH-deformation
	CO_3^{2-} Vibrations
1490-1410	asymmetric stretch
1085-1050	symmetric stretch
875-860	out-of-plane bend
750-680	in-plane bend
	SO_4^{2-} Vibrations
1180-1100	stretching
680-580	bending
	PO_4^{3-} Vibrations
1100-1000	antisymmetric stretch
500-635	bending

signment of absorption frequency bands to functional groups common in soil minerals and selected inorganic compounds is summarized in Table 11–2.

Because of the range of wavelengths involved, particle size has a pronounced effect on the absorption spectra. Prerequisites for obtaining satisfactory infrared spectra include having mineral samples whose particle size is less than the wavelength of the infrared radiation to be used. Since the usual wavelength range is from 2.5 to 50 μm (wavenumbers 4000–200 cm^{-1}) the clay fraction (< 2 μm) is of appropriate size and requires no further reduction in particle size. Silt and sand fractions need to be reduced in particle size by careful grinding.

The percentage content of a mineral that can be detected qualitatively in a sample depends largely on the intensity of the diagnostic absorption bands of mineral. Minerals having strong diagnostic bands, such as gibbsite, kaolinite, quartz, and calcite, can be detected at concentrations of 1 to 2%; scale expansion and computer processing of spectral data can lower detection levels to 0.1% and below under favorable conditions. Minerals whose structures accommodate isomorphous substitutions, such as micas, smectites, and chlorites, may be difficult to identify in polycomponent samples because distinctive features of their spectra vary with chemical composition and may overlap in key regions of the spectrum. Spectral subtraction capabilities of instruments having microprocessor and appropriate software components may overcome this limitation to a large degree.

Criteria for classification of clay minerals include the ratio of silica tetrahedral sheets to alumina octahedral sheets; thus micas and smectites are examples of 2:1 layer silicates, while kaolinites and serpentines are examples of 1:1 layer silicates. Differences in these two groups of minerals are reflected in the environment of the hydroxyls; for example, 2:1 minerals like micas and smectites have hydroxyls in one main environment associated with octahedral cations located at the base of the hole in the tetrahedral surface of the clay (designated the *inner hydroxyl*). Since all are in nearly the same environment, all have nearly the same frequency and give one strong absorption band in the 3600 cm^{-1} region. Nonuniformity of octahedral cation composition may give rise to slight shifts in hydroxyl stretching frequencies. The 1:1 group of layer silicates has hydroxyl groups on a crystal surface in addition to those common to the 2:1 minerals. These two different hydroxyl environments give rise to four distinct hydroxyl absorption frequencies for kaolinite. Variations of this pattern occur for other kaolin-group minerals.

Proton environments in hydroxides and oxyhydroxides differ considerably from those of the aluminosilicates; consequently, the OH-stretching region of the infrared spectra of certain of the minerals in this group, (e.g., gibbsite) may be essentially unique.

Mid-infrared spectra for kaolinite, montmorillonite, and gibbsite are shown in Fig. 11–3

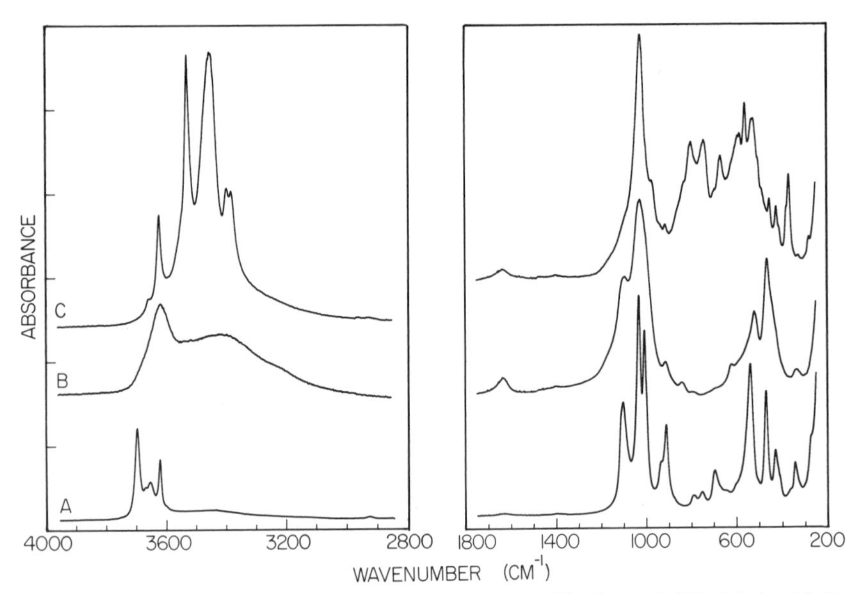

Fig. 11–3. Infrared spectra of (A) kaolinite, (B) montmorillonite, and (C) gibbsite (C. B. Roth, unpublished).

Another criterion for classification of clay minerals is the level of occupancy of octahedral cation sites; minerals in which 3 out of 3 sites are occupied are "trioctahedral" and those in which only 2 out of 3 sites are occupied are "dioctahedral." The trioctahedral composition usually results in the orientation of the O–H bond axis of the inner hydroxyl normal to the OOl plane of the clay layer; as the angle of an oriented sample is decreased with respect to the incident infrared beam, there is an increase in hydroxyl absorption. This orientation property is referred to as *pleochroism*. In the case of dioctahedral minerals, the hydrogen of the inner hydroxyl is repelled toward the empty octahedral site. The O–H bond axis is at a low angle to the OOl surface and the diagonal pattern of orientations results essentially in the cancellation of any changes in absorption with change in angle. Thus, dioctahedral minerals show no significant pleochroism of the inner hydroxyl band at about 3620 cm^{-1}.

11–3 SAMPLE PREPARATION

11–3.1 Gases

Infrared analysis of gaseous samples involves the normal procedures encountered in gas analysis. A variety of sample cells for containing the gas sample while exposing it to infrared radiation is available from the manufacturers of infrared instruments and accessories. The cells usually consist of a glass or metal cylinder with a gas inlet near one end and an outlet near the other, gas flow being controlled by stopcocks. The end windows through which the infrared radiation passes are made of materials such as NaCl, ZnS (Irtran-2), and AgBr, which are mechanically suitable and sufficiently transparent to infrared radiation. Accessories for temperature control of the gas cell can be obtained. The usual path length of a gas cell is 10 cm. When this is too short to measure the spectra of minor components or substances encountered in trace analysis, a variable-path cell may be used; this provides path lengths in steps of 1.5 m for 20-, 40-, and 120-m cells. The light path is folded using internal gold-surfaced mirrors and gold-plated or stainless steel metal components. Further gains in sensitivity can be realized by increasing the pressure of the gas sample in the cell to 1.0MPa. The long-path gas cells are intended for measurements in the range for a few parts per million and lower, concentration ranges encountered in air pollution and air monitoring determinations. If the spectral region of interest includes that in which water vapor or carbon dioxide absorption occurs, it may be necessary to use a dual cell system for compensation. The use of FT-IR and other computer-assisted instruments may eliminate this difficulty.

A related application of gas cells is the development of accessories for analysis of gases generated as the result of pyrolysis of less volatile materials.

11–3.2 Liquids

In most chemical studies, the majority of samples are analyzed as liquids. Usually, a solvent or diluent must be used in conjunction with the substance to be analyzed. The solvent or diluent should be chosen after considering the following requirements:

1. A solvent must dissolve the substance to be analyzed. The solvent or diluent used must maintain the substance in the proper chemical form.
2. A solvent or diluent must not injure the sample cell. Usually, there is a possibility of damaging the crystalline windows in the cell; adequate precautions must be taken to prevent such damage.
3. The infrared absorption spectrum of the solvent or diluent must not interfere with that of the substance being investigated.

Willard et al. (1981) gives a list of infrared transmitting materials (p. 202) and a figure (Fig. 7–19, p. 203) illustrating the transmission characteristics of selected solvents.

Since the desirable optical path length is much shorter for liquids than for gases, sample cells for liquids usually consist of two cell windows sealed and separated by thin gaskets. The gaskets may be of copper and lead that have been wetted with mercury. The assembly is securely clamped together and permanently mounted in a stainless steel holder. The cell is provided with tapered fittings to accept hypodermic syringe needles for filling and emptying. Each cell should be labeled with its precise path length as measured by the interference fringe or standard absorber technique. A demountable cell involves two window pieces and a Teflon fitting, which forms the leak-proof seal when the cell is slipped into a mount and knurled nuts are fastened until finger-tight.

Variable path-length cells offer great flexibility in the analysis of liquid samples. This type of cell consists of a cylindrical, Teflon-lined, stainless steel chamber with parallel windows. The path length may be continuously adjusted from 0.005 to 5 mm and reproduced to within 0.001 mm. A vernier and scale provided on the cylinder permit the cell thickness to be read within ± 0.0005 mm. Accuracy of the thickness settings are ± 0.001 mm or 1%, whichever is larger. The variable path-length cell can eliminate the requirement for a large collection of different thickness, fixed path-length cells. The windows in variable path-length cells are usually made of NaCl and cannot be used with liquid samples containing water; ZnS windows are now available on special order.

Minicells formed of AgCl and having fixed path-lengths of from 0.025 to 0.100 mm are available as low-cost disposable units.

Water would be the solvent of choice for many of the soluble organic and inorganic phases in soils, but two factors restrict the use of water as an infrared solvent for routine analysis: (i) its spectrum has intense absorption that covers the major part of the normal infrared region, 4000 cm^{-1} to 200 cm^{-1}, and (ii) most of the common cell window materials, such as sodium chloride (NaCl) and potassium bromide (KBr), are extremely soluble in water and hence readily attacked by aqueous solutions. The use of ZnS or barium fluoride (BaF_2) windows will overcome the problem of susceptibility to attack by aqueous solution. Coates (1978) has recently shown that the problem of intense absorption in aqueous solutions can be largely overcome by carefully following specific experimental guidelines and taking full advantage of computer facilities, such as automatic spectral difference, large ordinate expansion and digital smoothing. This approach has the potential of differentiating solvating water molecules from free water molecules.

A second means of overcoming the problem of intense absorption in aqueous systems is the use of the technique of multiple internal reflectance (also known as attenuated total reflectance or ATR). The design of the cell is such that radiation penetrates the sample only a few micrometers and is independent of sample thickness. Mulla et al. (1985) and Low and Roth (unpublished data) have successfully applied this technique to the measurement of the surface area of clays and soils through measurement of spectral features of water associated with the mineral surfaces. The recent development of a cylindrical internal reflection (CIR) cell designed to match the optics of FT-IR instruments (Spectra-Tech, Inc., 652 Glenbrook Road, Stanford, CT 06906) has made possible the routine analysis of aqueous solutions and other "infrared-opaque" liquids. With 10 reflections at a 45° angle of incidence, path lengths of 8 to 12 μm are routinely achieved.

The above-mentioned techniques will be useful in infrared studies of aqueous extracts of soil organic matter, such as humic and fulvic acids.

MacCarthy et al. (1975) have reported the use of digital subtraction of water bands in infrared studies of humic substances in aqueous solutions, using an absorbance mode of operation.

11–3.3 Solids

11–3.3.1 SAMPLE PRETREATMENT

11–3.3.1.1 Humic Substances. Humic substances such as humic acid, fulvic acid, and humin are the major components of soil organic matter (Schnitzer, 1982). Humic and fulvic acids are soluble in dilute base and thus dilute NaOH is a logical extractant. See Schnitzer (1982), part 2 (chapter 30) of this monograph, for principles and details of separation, fractionation, and purification of humic materials. Because of the aforementioned problems in infrared measurements on aqueous solutions, the humic materials are usually freeze-dried after separation and purification. Thus, they are available as both aqueous solutions and solids; but as a matter of convenience, samples for infrared analysis are commonly prepared from the solid forms.

11–3.3.1.2 Mineral and Inorganic Phases. For the examination of the mineral and inorganic phases of soils, the soils are usually subjected to various pretreatments designed to remove organic matter, iron oxides, and certain other materials, and to enhance dispersion. Most of these pretreatments were designed for the purpose of improving the appearance and resolution of x-ray diffraction patterns. It should be emphasized that these treatments may produce artifacts (Farmer & Mitchell, 1963) as well as change the properties of the minerals so that the infrared spectra are not representative of the mineral components in their original condition. These pretreatments may also thwart the capabilities of infrared spectrometry for detecting and characterizing amorphous and poorly crystalline phases. For example, high surface-area amorphous components may be selectively dissolved by extraction procedures designed to enhance the diffraction peaks of crystalline phases; obviously, the physicochemical properties of the soil cannot be completely accounted for by the lower surface-area, highly ordered core of the crystalline phases.

In the case of soils whose general composition is unknown, it may be worthwhile to make a preliminary examination of the infrared spectra of the clay fraction from some representative horizons of the soil profile before application of routine treatments for dispersion, organic matter removal, deferration, etc. A sample of the whole soil, after appropriate grinding treatment, may also be used in the preliminary examination (Fieldes et al., 1972). Identification of the predominant minerals can provide a basis for application of the most appropriate treatments for more detailed studies.

One of the major problems encountered in the application of infrared spectrometry to the study of soil minerals is the distortion of absorption bands due to light scattering by the mineral particles. As mentioned previously, the clay fraction (<2 μm) is of appropriate size, but larger size fractions need to be reduced in size.

The light scattering due to particle size is referred to as the Christiansen effect; the effect diminishes as particle size is reduced (Duyckaerts, 1959). Polycomponent mineral systems whose components differ in particle size cannot be quantitatively analyzed in a straight-forward manner by infrared techniques because of the Christiansen effect. This difficulty cannot be overcome by grinding the sample to reduce the particle size of the components having the larger particle size, since grinding differentially affects the other components because of variations in hardness of the minerals (Hlavay et al., 1977, 1978, 1979).

Particle-size reduction may be accomplished by filing, grinding, or sonification. Filing is most appropriate for preparing samples from layer silicate minerals of hand-specimen dimensions, such as chlorite and mica. Filing may be accomplished by using a small glass file and making a filing motion perpendicular to the planar surface of the mineral. Sedimentation is used to separate the various particle size fractions. Filing tends to minimize structural modifications in layer silicates, whereas the pressure and heat generated in grinding, especially dry grinding, can result in drastic changes in structure and properties.

There are two techniques for grinding: dry grinding and wet grinding. Dry grinding, whether by mortar and pestle or by ball-milling, causes destruction of the mineral structure and modification of the chemical and physical properties and should not be used for particle size reduction of samples to be examined by infrared spectroscopy. Wet grinding, in which a liquid such as acetone or ethanol is used to reduce the friction and pressure of the grinding operation, minimizes structural changes in minerals. Wet grinding may be carried out by placing 10 to 15 mg of the sample in a mullite or agate mortar, then adding 10 to 15 drops of ethanol to the mortar with an eye dropper. The sample is then ground with a vigorous rotary motion, confining the sample to ⅓ to ½ of the mortar surface until the ethanol evaporates completely. Do not continue grinding after the sample becomes dry. The wet grinding step may need to be repeated several times in order to reduce the particle size to <2 μm. Levitt and Condrate (1970) have described a wet-grinding procedure in which the mineral particles are impacted with themselves in a test tube containing acetone in a Wig-L-Bug vibrator.

Particle-size reduction may also be achieved by sonification—treatment of a suspension of the mineral particles with ultrasonic frequencies. The frequency and time period of treatment need to be adjusted through experimentation in order to minimize any structural modification.

Farmer (1974, p. 354) suggests that clays should be saturated with K to prevent the retention of cation hydration water to high temperatures which may occur with smaller exchangeable cations such as Ca^{2+} and Mg^{2+}. It is recommended that clay fractions to be used in preparation of KBr disks be freeze-dried or dried from benzene after washing in ethanol.

11–3.4 Techniques of Sample Presentation

11–3.4.1 ABSORPTION METHODS

11–3.4.1.1 Dispersions. The most commonly employed technique for the infrared study of organic and mineral and inorganic components of soils involves absorption of infrared radiation by a sample uniformly dispersed in a continuous optical medium placed directly in the infrared beam. The use of a medium such as an oil or KBr (which becomes plastic at high pressure) minimizes the problem of light-scattering at the particle/air interfaces.

One of the more common methods for solids is the mull technique, in which the sample is mixed or ground in some weakly absorbing, nonvolatile liquid (Bradley & Potts, 1958). The resulting paste is then uniformly spread as a capillary film between two transparent windows (e.g., NaCl or ZnS) and analyzed. Light-scattering is reduced to a minimum by using a liquid having an index of refraction near that of the solid being analyzed. Nujol and Fluorolube have been used successfully for soil and mineral samples. Nujol absorbs strongly in the 3000 to 2800 cm^{-1} region (C–H stretching) as well as at 1460 and 1375 cm^{-1}; these interferences must be considered when organic components are analyzed in the form of Nujol mulls. Fluorolube has essentially no absorption from 4000 to 1400 cm^{-1}.

Probably the most commonly used technique is the KBr pellet technique. This involves mixing the finely divided sample and finely powdered, spectroscopically pure KBr and pressing the mixture in an evacuable die at sufficient pressure to produce a transparent disk.

Most manufacturers of infrared spectrometers supply dies and pellet sample holders suitable for use with their instruments. In addition, a number of firms specializing in accessories for spectrometers have developed dies and holders adapted to most commercially available instruments. The specific directions in section 11–3.4.1.4 are for the analysis of samples using a Beckman 0.5-in. KBr die, but in all major details it will apply to other similar dies.

11–3.4.1.2 Films. Several methods of presenting mineral samples to the infrared beam as a thin film have been described (Fripiat, 1960; Fripiat et al., 1960). In cases for which it is desirable to make successive observations on a single sample after adsorption or desorption reactions (e.g., heat treatment to remove adsorbed water), the film technique is preferable to the mull or pressed-pellet technique. When circumstances make it advisable to examine the sample in the form of a thin film, one of the following methods may be useful:

11–3.4.1.2.1 Sedimented Films on Infrared Transparent Windows

Material which is <2 μm in size is dispersed in water. An amount of suspension sufficient to give a concentration of 1 to 2 mg/cm^2 is placed on infrared-transparent windows, such as AgCl or ZnS, and allowed to dry. Glass microscope slides and cover glasses may be useful window materials for exploratory scans in the region in which most OH-stretching frequencies occur (wavenumbers 4000 to 3000 cm^{-1}). Use of a duplicate glass slide or cover glass from the same box or lot will approximately compensate the OH absorption bands of the glass; the degree of compensation should be established by placing the pair of matched slides in the sample and reference beams and recording the spectrum. Care should be taken to insure that the thermal treatment of the reference slide is the same as that of the sample slide. Light-scattering effects arising from air/particle interfaces may be lessened by placing a drop of Nujol on the irradiated sample. Volatile organic liquids (e.g., isopropanol) may also be used as a suspending medium to hasten drying.

11–3.4.1.2.2 Self-supporting Films

Self-supporting films of many smectites, vermiculites, and fibrous clays may be formed by evaporating aqueous suspensions on foil or plastic films and separating the mineral film by pulling the foil or plastic sheet over a sharp edge (Farmer, 1968). The mineral film is more rigid than the plastic or foil and will separate in a manner similar to that of certain self-adhesive labels. The mineral films should have a density of 1 to 2 mg/cm^2. Total absorption of the infrared beam may occur in the 1100 to 970 cm^{-1} region due to the strong Si–O absorption bands.

Smooth polythylene film, such as used in storage of frozen foods, may be used. A piece of film 15 to 20 cm in width and of a convenient length is placed on a level sheet of plate glass on which has been placed several milliliters of distilled water. A straight-edge ruler covered with a piece of the plastic is used as a squeegee to smooth the plastic film so that the capillary film of water serves to hold the plastic film flat until the mineral suspension dries. A convenient volume (10–25 mL) of a 2% suspension of the mineral colloid is placed on the plastic film using a pipette held vertically at a fixed position. This will give a film having a circular shape, with a uniform thickness everywhere except at the edge of the circle.

Mylar film may also be used for preparation of self-supporting films. It is much stronger and smoother than polyethylene and may be fastened to the sheet of plate glass by use of cellophane tape, pulling it taut to make it as smooth as possible. The suspension is added in the manner described in the preceding paragraph.

After the mineral film is air-dry, the plastic with the attached film is carefully separated from the sheet of plate glass. The mineral film is then detached from the plastic film by slowly pulling the plastic film over a sharp edge at an angle of 90° or greater.

The clay film may then be cut into flakes of the desired size and shape with a sharp razor blade or cork borer. Because of the tendency of the clay films to curl, they should be stored between sheets of stiff paper until they are ready to be used.

The clay films may be conveniently mounted in the infrared spectrophotometer by placing them between two pieces of blotter paper (5.0 by 7.6 cm) in which openings of the desired size and in the correct position to intercept the infrared beam have been cut. When the clay film has been positioned in the proper place, a staple may then be placed in the blotter paper just outside each edge of the film. The clay film will be held firmly in position and can be removed for further treatment and studies if desired.

If the clay film is to be used in studies involving dehydration or heating in a vacuum infrared cell (Russell, 1974), it may be mounted in a holder fashioned of aluminum foil or stainless steel. The clay film in the holder can be easily moved from one part of the cell to another by tilting the cell or gently tapping the cell until the holder is in the proper position for recording the spectra.

Self-supporting films of amorphous silica–alumina gels and synthetic zeolites can be prepared by pressing the powders between polished steel faces at a pressure of about 10 000 psi (700 kg/cm^2; 68.9 MPa) (McDonald, 1958). Such films are very fragile and must be handled with great care.

11–3.4.1.3 Mull Method.

11–3.4.1.3.1 Materials

1. Nujol, Fluorolube LG 160 (Hooker Chemical), or other mulling medium.
2. Carbon tetrachloride (CCl_4).
3. Agate or mullite mortar (50 or 65 mm diameter) and pestle.

11–3.4.1.3.2 Procedure

Place about 10 to 15 mg of the powered sample in a 50- or 65-mm diameter agate or mullite mortar. Add a small drop of the mulling agent to the mortar by using the end of a partially unfolded paper clip or length of glass fiber. Grind the sample with a vigorous rotary motion until all the material is suspended in the mulling agent. It may be necessary to add another small drop of the mulling agent as this grinding is continued. The consistency of the final mixture is about that of vaseline.

This procedure can be made more quantitative by use of weighed samples, addition of precise amounts of mulling agent by means of a positive displacement micropipetting method (Absoluter® Micro-Pipetters, Tri-Continent Scientific, Inc., 12541 Loma Rica Drive, Grass Valley, CA 95945), and use of an internal standard or reference compound (Hesterberg, 1984).

The mull may be removed from the mortar and transferred to an infrared-transparent window (ZnS, AgCl, NaCl, etc.) by use of a rubber policeman that has been carefully cleaned to remove all traces of talc remaining from its manufacture. The mull may also be collected by means of a microspatula with the convex side up, collecting the mull at the leading edge on the convex side. After the mull is transferred to one window, a second window is placed on top of the first and the sample is evenly distributed between the plates. The concentration of the mull can be adjusted by gently squeezing out the excess mull or by separating the windows to give approximately one-half the initial concentration on each window. The film between the windows should appear slightly translucent and should be free of air bubbles when examined in front of a light. Carbon tetrachloride is used to clean the windows and the mortar and pestle.

It should be noted that mineral specimens prepared as films and mulls may have some degree of preferred orientation, whereas specimens prepared in KBr disks will tend to show random orientation of particles. This difference is of importance in samples that show pleochroic behavior because of the orientation of OH dipoles or other bonds.

11–3.4.1.4 Alkali Halide Pressed-Pellet Method.

11–3.4.1.4.1 Materials

1. Potassium bromide (KBr), spectroscopic grade.
2. Acetone or carbon tetrachloride (CCl_4)
3. Die for making KBr pellets: The die consists of a ram and an anvil enclosed in such a way that when they are pressed together by means of a hydraulic press, the sample is compressed into a pellet or disk by the confining space. Evacuation of the die by means of a vacuum pump removes entrapped air and moisture and results in a better pellet. Farmer (1957) has described a die design in which the finished pellet is mounted in a metal ring; this facilitates heat treatment and re-pressing of the pellet.
4. Laboratory hydraulic press for a stroke of at least 10 cm and capable of creating pressure on the confined sample equivalent to 9000 to 11 000 kg (20 000–24 000 lb) total force on the ram of a 13-mm diameter pellet-pressing die.
5. A small hand mortar and pestle.
6. Mechanical vacuum pump.

11–3.4.1.4.2 Procedure

To prepare a pellet 1 mm in thickness and 13 mm in diameter, mix approximately 0.5 to 3 mg of sample with 300 mg of spectroscopic-grade KBr. Satisfactory mixing can be carried out by hand, using a mortar and pestle. The appropriate amount of the sample (0.5–3 mg) is weighed on a microbalance and placed in a clean 50- or 65-mm (o.d.) agate or mullite mortar. Next, 200 mg ±5 mg of dry KBr powder (infrared quality such as Harshaw or Isomet), which has been dried at 105°C (378 K) for 12 h and stored in a desiccator over P_2O_5 or a zeolite desiccant, is weighed and 5 to 10 mg of the KBr is added to the mortar containing the sample. The pestle is used to mix the KBr and the sample with a gentle rubbing motion. The purpose of this step is only to mix the sample and the KBr; *do not grind the KBr* during this mixing step, as it will lead to adsorption of water on the KBr. Approximately 15 to 30 mg of KBr is added and mixed as before. Further addition and mixing of KBr in the above manner are continued until all the KBr has been added to the mortar. This procedure will produce a homogeneous mixture of the sample and the KBr and minimize the adsorption of water by the KBr. The mixture may be dried in an oven or vacuum oven for 60 min at 105 to 110°C (378–383 K) before pressing the pellet. A small camel's-hair brush is used to quantitatively transfer the mixture to the die for pressing the pellet.

The pellet is pressed as follows: the small lower plunger of the die is placed in position at the bottom of the die body. Sufficient mixed sample-KBr powder (~ 300 mg) to give a pellet 1 mm thick is poured into the bore of the die and the powder distributed as evenly as possible by lightly shaking the die. To complete the distribution, slowly insert the large upper plunger or ram and rotate it a few times using light finger pressure to hold it against the powder. The plunger is then withdrawn very slowly. The small upper plunger is then placed on top of the compacted powder, the large upper plunger inserted, and the die placed in a hydraulic press. Very slight pressure is placed on the die so as to produce a good air seal. A vacuum pump is attached and the air is exhausted from the die for 5 to 10 min. Following this, a pressure of 9000 to 11 000 kg total force is applied to the press and maintained for 10 to 15 min. The vacuum is first released; then the pressure on the hydraulic press is slowly released and the pellet removed from the die by forcing the plunger to extrude the pellet from the bottom of the die. The pellet should be transparent; care should be taken not to touch the pellet with the hands. Using tweezers, place the pellet in a suitable sample holder and introduce it in the infrared beam for analysis, or place the pellet in a desiccator over a zeolite desiccant, if it is necessary to store it. Finally, clean the die assembly immediately with water and acetone or CCl_4, to remove corrosive KBr. The die, KBr and other materials can be conveniently stored in a 50°C oven, which tends to minimize problems with moisture.

11–3.4.1.4.3 Comments

Potassium bromide pellets are not suitable for studies involving absorbed water or surface cations. The possibility of cation exchange reactions must be considered in cases where this may alter the initial state of the mineral. The hygroscopic nature of KBr can be minimized by drying the pellet in a vacuum oven at 50°C (323 K); the pellet may be dried at higher temperatures, but will need to be re-pressed in the die to restore the transparency of the pellet (Farmer, 1957).

A rapid procedure that does not require an expensive die is as follows: the KBr-sample mixture is placed in a 13-mm circular cavity in 100-lb blotting paper with a layer of aluminim foil above and below the sample in the cavity, and this sandwich is pressed between two stainless steel dies (20-mm diameter) in a hydraulic press at a total pressure of 9000 to 11 000 kg (20 000 to 24 000 lb.).

In cases necessitating an inert matrix, and particularly for use in the far infrared, materials such as polyethylene, polytetrafluoroethylene (Teflon), or paraffin wax can be used in the preparation of pressed pellets.

11–3.4.2 REFLECTANCE METHODS

In addition to the absorption methods previously discussed, techniques utilizing emission and reflection of infrared radiation are also available. Reflection methods include specular reflectance, diffuse reflectance, and internal reflectance. Because of the requirement of a highly polished surface, specular reflectance techniques are of limited interest in soil studies.

Diffuse reflectance in the visible and near infrared (0.35–2.5 μm) has been extensively used by Hunt and Salisbury (1970, 1971a, 1971b), Hunt et al., (1971a, 1971b, 1973a, 1973b, 1973c, 1974), Kahle and Goetz (1983), Kosmas (1984), and Kosmas et al. (1984) for the detection and identification of clay minerals as well as many other groups of minerals. Near-infrared spectrometry has also proven useful for quantitative analysis of protein, oil, and moisture in agricultural products (Watson, 1977); advances in microprocessor developments have made it possible for instruments to be calibrated for quantitative analysis of many specific compounds.

The third reflectance technique is based on internal reflection or attenuated total reflection. Attenuated total reflection (ATR), also referred to as frustrated multiple internal reflection (FMIR), represents a different approach to sampling and was designed primarily for materials that are opaque to infrared radiation. ATR spectroscopy is based on the fact that although complete internal reflection occurs at the interface between two transparent materials of different refractive indices, radiation does in fact penetrate a short distance (approximately 1 wavelength) into the rarer medium. This penetrating radiation can be partially absorbed by placing a sample in optical contact with the dense medium (prism), at which point-reflectance occurs. The reflected radiation can yield an absorption spectrum that closely resembles a transmission spectrum of the sample. Harrick (1967) has described this technique in his book *Internal Relfection Spectroscopy.* This technique has been applied to the study of water in clay-water systems by Mulla et al. (1985).

11–4 FUNCTIONAL-GROUP AND QUALITATIVE ANALYSIS OF ORGANIC COMPOUNDS

11–4.1 Materials

See section 11–3.4.1.4.

11–4.2 Procedure

Extract, fractionate, and purify the desired organic components following the procedures described by Schnitzer (1982) in sections 30.2, 30.3, and 30.4 of *Methods of Soil Analysis*, Part 2, 2nd edition (Page et al., 1982). Freeze-dry and store in a vacuum desiccator over activated Na-A type zeolite.

Prepare a pellet of the sample with KBr as described in section 11–3.4.1.4.2, using 1.5 to 2 mg of sample and 300 mg of KBr. Prepare other pellets containing a higher or lower proportion of sample if the initial pellet does not yield absorption bands of the desired intensity. Interference caused by adsorbed moisture may be minimized by heating the KBr pellet under vacuum at 100°C for 2 h (Stevenson and Goh, 1974). Pellets formed within a metal sleeve in the die described by Farmer (1957) and Russell (1974) may be re-pressed easily to restore the transparency of the heated pellet.

The sample may also be prepared as a mull (see section 11–3.4.1.3). Use of a "split mull" technique (Smith, 1979, p. 78), in which a fluorinated oil is used for the 4000 to 1340 cm^{-1} region and mineral oil (Nujol) for frequencies lower than 1340 cm^{-1}, can produce a combined spectrum without interference from the mulling liquids.

Dispersive spectrometers with computer processing capabilities and FT-IR instruments make it possible to record the spectra of aqueous solutions of humic substances through computer subtraction of the water bands (MacCarthy et al., 1975). Direct measurement of the infrared spectra of solute molecules in aqueous solutions as well as subtle perturbations in the vibrations of the water molecules is now possible (Coates, 1978) and has many potential applications in studies of organic and inorganic soil colloids.

11–4.3 Interpretation of Spectra

Even though the spectra of most organic matter preparations are generally relatively simple (consisting of 6 to 12 bands in the 4000 to 600 cm^{-1} region), there is considerable ambiguity and uncertainty in the interpretation of the spectra. The following general considerations are suggested:

1. As evidence for the occurrence or nonoccurrence of a functional group,

the presence of the absorption band in the proper locations is not as convincing as is the absence of an absorption band from the proper location.
2. All the bands in a spectrum cannot be interpreted.
3. Polymers in general have fewer, broader, and often less intense bands than the monomers from which they are derived.
4. Some unknowns cannot be identified.

McDonald (1984) has recently reviewed some of the dramatic advances in computer-assisted comparison and interpretation of infrared spectra; such techniques will assume an increasingly important role in the study of organic compounds.

Divide the spectrum into the characteristic-functional-group region (4000–1500 cm^{-1}) and the fingerprint region (below 1500 cm^{-1}). Examine the strongest absorptions first and then the medium ones. Refer to Table 11–1 and Fig. 11–2 for information on major groups. If a band lines up for a functional group, determine whether all absorptions of the group are present.

The main absorption bands occur in the regions of 3300 cm^{-1} (O–H and N–H stretching), 2900 cm^{-1} (aliphatic C–H stretching), 1720 cm^{-1} (C=O stretching of COOH and ketonic groups), 1610 cm^{-1} (aromatic C=C and/or H-bonded C=O), 1400 cm^{-1} (OH deformation and C–O stretching of phenolic OH and C–H deformation of CH_2 and CH_3), and 1250 cm^{-1} (C–O stretching and OH deformation of COOH). Infrared spectra of humic and fulvic acids from a till plain depression soil developed under forest are shown in Fig. 11–4.

Stevenson and Goh (1971) surveyed the infrared spectra of humic substances and classified them into three general types.

Type I. Humic acids: strong bands are present at 3400, 2900, 1720, 1600, and 1200 cm^{-1}. Intensities of 1600 and 1720 cm^{-1} are approximately equal.

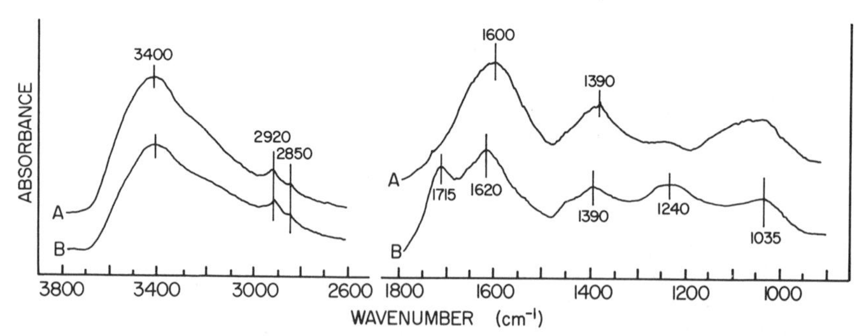

Fig. 11–4. Infrared spectra of (A) humic acid and (B) fulvic acid, from the A-horizon of a Kokomo silty clay soil (typic argiaquoll, fine, mixed mesic), formed in a till plain depression under forest (T. E. Moody, unpublished).

Type II. Low-molecular weight fulvic acids: the 1720 cm^{-1} absorption feature is much stronger than the 1600 cm^{-1} band; the latter is centered near 1640 cm^{-1}.

Type III. Bands indicative of protein and carbohydrate components are present; strong bands near 1540 cm^{-1} are present in addition to major absorption features typical of Types I and II. Absorption near 2900 cm^{-1} (aliphatic C–H absorption) is also more pronounced.

Using the above criteria, examine the locations and relative intensities of the absorption bands in the spectrum of the sample and determine the types of humic substances present.

Infrared spectra of humic and fulvic acids may be useful in characterizing these and related components from soil organic matter. More definitive information can be obtained through chemical treatments such as deuteration, acid hydrolysis, complexation, and chemical modifications such as methylation, acetylation, and saponification (Stevenson, 1982). The reader is referred to Stevenson (1982) for a more detailed discussion of the interpretation of infrared spectra of humus.

In addition to identification and characterization of isolated organic compounds, functional groups involved in adsorption of organic compounds on clays can be determined. Infrared absorption bands of both the clay adsorbent and organic adsorbate may be perturbed by adsorption reactions (Little, 1966). Provided sufficient adsorption of the organic compound occurs, difference spectra are particularly revealing (White, 1976). The spectrum of the adsorbed organic molecules can be obtained by placing in the reference beam an amount of clay equal to that in the clay-organic system; the clay-organic mixture is placed in the sample beam. Sample preparation may be by a quantitative adaptation of the KBr pellet technique or by use of self-supporting films. With the introduction of infrared spectrometers having data acquisition and analysis capabilities, the preparation of difference spectra by spectral subtraction is much more accurate and convenient.

11–5 IDENTIFICATION AND CHARACTERIZATION OF AMORPHOUS AND CRYSTALLINE INORGANIC OR MINERAL PHASES

11–5.1 Materials

See sections 11–3.4.1.1, 11–3.4.1.2, 11–3.4.1.3.1, and 11–3.4.1.4.1.

11–5.2 Procedure

See sections 11–3.4.1.1, 11–3.4.1.2, 11–3.4.1.3.2, and 11–3.4.1.4.2.

11–5.3 Interpretation of Spectra

11–5.3.1 AMORPHOUS INORGANIC PHASES

Since infrared spectra are sensitive to assemblages of groups in which atoms are in specific environments (e.g., the protons in the various hydroxyl groups of gibbsite), the presence of these environments in structures may be detected even when long-range order is absent. Thus, infrared spectrometry makes a unique contribution in the areas of variable structural order and composition. Examples of applications of infrared techniques to the study of amorphous materials or structures with short-range order include studies by Russell et al. (1969), Mitchell et al. (1964), White (1971) and White et al. (1976). Infrared data can provide compositional and structural information for materials and components for which x-ray diffraction techniques are inapplicable.

The presence of a broad band in the 3400 cm^{-1} region together with a relatively strong band at about 1640 cm^{-1} (H_2O deformation or bending) in the spectrum of a sample amorphous to x-rays may be attributed, in part, to adsorbed water and suggests the presence of an amorphous phase or phases having a high surface area. The use of ATR techniques for studying clay-water systems developed by Mulla et al. (1985) and P. F. Low and C. B. Roth (unpublished data), and particularly for measuring the surface area of both amorphous and crystalline inorganic materials, suggests this approach would be ideal for characterizing amorphous gels in their natural, moist state.

Selective dissolution and differential infrared spectroscopy have been used by Wada and Greenland (1970) to characterize the "amorphous constituents" in soil clays. With spectral subtraction and data-handling capabilities of current FT-IR and dispersive instruments, this can be done easily, and should become a routine part of soil mineralogy and genesis studies.

11–5.3.2 CRYSTALLINE PHASES

Farmer (1974) suggests that the vibrations of layer silicates can be separated into those of the hydroxyl groups, the silicate anion, the octahedral cations, and the interlayer cations. The high-frequency OH-stretching vibrations occur in the 3400 to 3750 cm^{-1} region, and the OH-bending (librational) frequencies occur in the 600 to 950 cm^{-1} region. The Si-O stretching vibrations lying in the 700 to 1200 cm^{-1} region are only weakly coupled with other vibrations of the structure; however, Si-O bending vibrations between 150 and 600 cm^{-1} are strongly coupled with vibrations of the octahedral cations and with translatory vibrations of hydroxyl groups. The interlayer cations have vibrations between 70 and 150 cm^{-1}, which may be localized.

From a comparison of the absorption bands in the spectrum of the sample with the assignments in Table 11–1 and Fig. 11–2, determine the major anionic components in the sample (e.g., silicates, hydroxides, carbonates, phosphates). Consult Farmer (1974) for more specific information on the spectral features of layer silicates, hydroxides, carbonates, and phosphates.

Identification of the mineral phases in the sample may be undertaken by referring to Table 11–3 to check for the presence of specific minerals. The spectrum should next be compared to spectra of reference clay minerals for components thought to be present. (Infrared spectra of kaolinite, montmorillonite, and gibbsite are shown in Fig. 11–2.) A number of compilations of clay mineral spectra are available and should be consulted. These include the following: *The Infrared Spectra of Minerals*, edited by Farmer (1974); *Atlas of Infrared Spectroscopy of Clay Minerals and Their Admixtures*, by van der Marel and Beutelspacher (1976); *Data Handbook for Clay Materials and Other Non-metallic Minerals*, edited by van Olphen and Fripiat (1979); *The Sadtler Infrared Spectra Handbook of Minerals and Clays*, edited by Ferraro (1982); and a chapter on "Characterization of Soil Minerals by Infrared Spectroscopy" by Farmer and Palmieri (1975). Libraries of fully digitized spectra of clay minerals are being developed and will greatly facilitate the routine analysis of clay materials by infrared spectrometry (J. M. Brown & J. J. Elliott, unpublished data).

To determine the orientation of structural hydroxyl groups in crystalline clay minerals, prepare a sample in the form of a film (see sections 11–3.4.1.2.1 and 11–3.4.1.2.2). Record a spectrum of the OH-stretching region with the film in the normal orientation (perpendicular to the incident infrared beam), then rotate the film on its vertical axis so that the acute angle between the plane of the film and the infrared beam is about 45° and record the spectrum again. The use of a protractor to establish the angle and tape to hold the film in position will give qualitative results; a sample holder with a goniometer is necessary for more quantitative measurements. An increase in the intensities of OH-stretching frequencies upon this change in angle with respect to the incident beam signifies that the orientation of the OH-bond axis is essentially normal to the 001 plane of the clay layer. For clay minerals of the 2:1 groups, this variation in intensity of the OH-stretching frequency with increasing angle of incidence suggests the presence of a component having trioctahedral character. In kaolinite, a 1:1 mineral, the 3620-cm^{-1} band intensity is independent of orientation, indicating that the hydroxyls are directed towards the vacant octahedral positions (as in dioctahedral 2:1 minerals). However, the hydroxyls on the hydroxylic surface of kaolinite are normal to the plane of the silicate sheet, and thus the 3695-cm^{-1} band intensity increases with increasing angle of incidence.

For samples in which the major component has a 2:1 structure, inferences concerning the octahedral composition may be drawn by comparison of the observed OH-stretching and OH-bending frequencies with those given in Farmer (1974, pages 332–343), Vedder (1964), Wilkins (1967), and White (1971, Table 1 and Table 2).

Table 11-3. Infrared absorption bands for selected soil minerals (White, 1971).

Mineral	Wavenumber												
	cm^{-1}												
Kaolinite	3695	3670	3650	3620							1108		1038
Halloysite	3695			3620		3400†		1640‡			1100		1040
Montmorillonite				3620		3400†		1640‡			1100		1040
Nontronite					3560	3400†		1640‡			1130		1050
Muscovite				3628							1120		
Biotite			3658		3550								
Vermiculite					3550		3380†	1640‡					
Chlorite Al-rich			3620	3520		3340							1004
Gibbsite				3610	3525	3445	3395				1102		1030
Quartz										1172		1084	
Microcline											1110		1030
Calcite									1435				
Kaolinite	1012		940	915			700		540		472		
Halloysite	1020			918			695		545		474		
Montmorillonite	1020			915					520		470		
Nontronite					827						490	430	
Muscovite	1020		928		828	750			535		480		
Biotite	1000					760	690				465	445	
Vermiculite		985			812			670			480		
Chlorite Al-rich		940		825		692		528		475			
Gibbsite		975			800	745		670	560	540			
Quartz					800	780	697		512		462		
Microcline	1000					769	727	647					
Calcite				877			712						

† OH-stretching frequency for water molecules.
‡ OH-bending frequency for water molecules.

11-5.3.3 SURFACE INTERACTIONS

In addition to identification of crystalline components, infrared spectrometry provides means for characterizing the surfaces of both crystalline and amorphous components. The extent and nature of the surfaces may be measured in many cases. The ATR technique for measuring surface area through changes in the stretching and bending vibrations of water, recently developed by Mulla et al. (1985), has been mentioned previously.

Another useful technique for measuring the extent and accessibility of surfaces is that of deuteration and other types of exchange studies. Nail et al., (1976) and White et al. (1976), using a vacuum infrared cell and D_2O treatments, followed the aging process in aluminum hydroxide gels through the extent of the OD—OH exchange and were able to distinguish between amorphous and crystalline aluminum hydroxide phases. This technique is illustrated in Figures 11-5 and 11-6.

Crystalline clay minerals such as kaolinite and montmorillonite react with a wide variety of organic compounds (Little, 1966, p. 334–351; Mortland, 1970; Theng, 1974, 1979; and Solomon & Hawthorne, 1983). Clay minerals in soils may detoxify adsorbed pesticides by catalyzing their decomposition (Mortland & Raman, 1967; Cruz et al., 1968). These catalytic properties are ascribed to the ability of clay minerals to either donate protons and so act as Brønsted acids, or accept electrons and behave as Lewis acids. Brønsted acidity derives essentially from the dissociation of water molecules coordinated to the exchangeable cations, giving rise to active protons. Mortland et al. (1963), Farmer and Mortland (1966), and Mortland (1966) have shown by infrared absorption studies that this type of acidity is strongly influenced by the hydration status of the clay and by the kind of cation occupying exchange sites on the silicate surface. Low water contents and highly polarizing cations promote Brønsted acidity. Lewis acidity has usually been associated with specific ions in the silicate structure, such as aluminum ions exposed at the edges

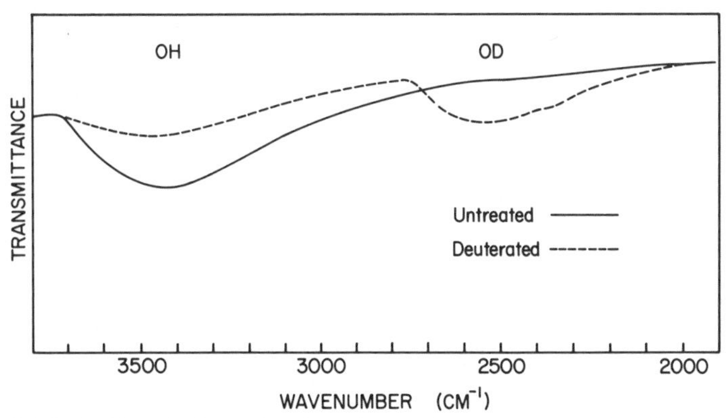

Fig. 11-5. The 3700 to 2000 cm^{-1} region of the infrared spectrum of a slightly aged aluminum hydroxide gel before and after deuteration (from White et al., 1976).

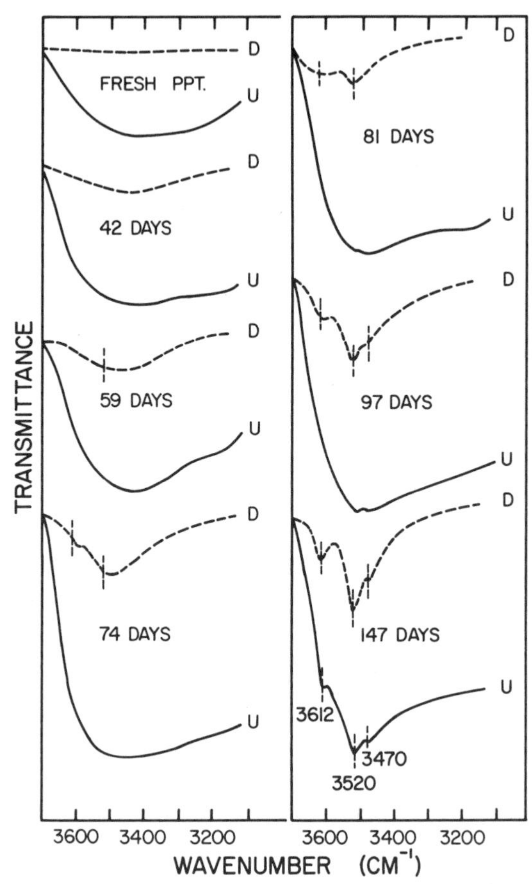

Fig. 11–6. Changes in the hydroxyl stretching frequency region of the infrared spectra of aluminum hydroxide gel for untreated (U) and deuterated (D) samples during aging at 25°C for time periods from 0 to 147 days (from White et al., 1976).

of the clay crystals. However, exchangeable cations that are electron acceptors (e.g., Cu^{2+}, Fe^{2+}, Fe^{3+}, and Mn^{2+}) can also be Lewis sites.

To study interactions of organic compounds with clay minerals, prepare a sample in the form of a sedimented or self-supporting film (see sections 11–3.4.1.2.1 and 11–3.4.1.2.2). Consult the references in the preceding paragraph for details of the preparation of specific clay-organic complexes. Preparation of the sample may involve interacting the clay with an organic compound in an aqueous suspension prior to evaporation of water to form the film, or the pre-formed film may be exposed to a non-aqueous solution of the organic compound. Vapor phase adsorption may be appropriate for more volatile compounds. The concentration of adsorbed organic species which can be detected by infrared spectrometry is limited by the amount of surface area of the clay absorbent, thus the method is most useful for systems in which the organic compound is adsorbed into the interlamellar space of the mineral (e.g., smectites, vermiculites and certain kaolinite intercalation complexes).

11-6 QUANTITATIVE ANALYSIS

11-6.1 Introduction

The laws of absorption that form the basis of any system of quantitative analysis consist primarily of relationships enunciated independently by P. Bouguer and A. Beer and their ultimate combination into the Bouguer-Beer Law (Smith, 1979). The combined relationship is written

$$A = -\log (I/I_0) = \log (I_0/I) = abc$$

where A = absorbance, a = absorptivity, b = path length, c = concentration, I_0 = intensity of the incident radiation, and I = intensity of the transmitted radiation.

This equation states that a straight-line relationship should be obtained from a plot of absorbance against concentration or thickness of a given species in a homogeneous mixture or solution; it also predicts that at the same wavelength, absorbances of bands from different species will be additive.

Since gases and liquids can readily be made homogeneous, they are suitable candidates for quantitative analysis of a high degree of accuracy. For details of the practice of quantitative analysis of gases and liquids, the reader should consult standard references such as Smith (1979), Miller and Stace (1972) and Alpert et al. (1973).

In contrast to the easily achieved homogeneity for gas and liquid samples, solids, and particularly minerals, usually exhibit inhomogeneities that preclude highly accurate quantiative analysis. Minerals are particulate in nature and light scattering phenomena (the Christiansen effect) impose inherent limitations on the relationship between the concentration of a given mineral component and the infrared radiation (Duyckaerts, 1959). Problems associated with particle-size reduction in mineral mixtures as well as the lack of "standard" reference minerals having the same structure, particle-size distribution, composition, and spectral features as the component in the sample further limit the cases in which quantitative determinations can be successfully made (Hlavay et al., 1977, 1978, 1979) with infrared spectrometers having little or no computing capabilities.

Both organic and inorganic fractions of soils are multicomponent systems; consequently, there is often considerable overlapping of bands in the infrared spectra. These interferences limit the number of isolated infrared bands that can be used to uniquely identify and quantify the presence of components in the mixture. Techniques for physical and chemical fractionation and separation may reduce the interferences, but the results obtained may not justify the immense effort expended.

The basis of quantitative infrared analysis is the direct or indirect comparison of the absorbance of the unknown at a given wavelength (usually the peak of a strong absorption band) with the absorbance of the same material in a standard of known concentration. Peak absorbance is easily measured and is directly related to concentration. For compounds and minerals whose composition is relatively invariant, e.g., kaolinite and gibbsite, it is relatively easy to find reference minerals similar in composition and properties to those in the sample being analyzed. The procedures and discussion of the calibration curve that follow are directed primarily to this most favorable case for minerals of invariant composition. Brief remarks concerning multicomponent analysis of samples containing compounds and minerals with variable composition (e.g., isomorphous substitution) by use of FT-IR and dispersive instruments with computing capabilities are given in the comments (section 11–6.5) at the end of this topic.

11–6.2 Materials

See sections 11–3.4.1.3.1 and 11–3.4.1.4.1.

11–6.3 Procedure

If the material is >2 μm, reduce it to that size by appropriate grinding procedure (see section 11–3.3.1.2).

For the KBr pellet technique, use a quantity of sample such that the analytical absorption band shows a transmittance near 37% (absorbance near 0.43); in practice, a range of 20 to 60% transmittance (0.2–0.7 absorbance) is quite satisfactory. Intimately mix 2.0 to 5.0 mg or other appropriate amount of dried sample with 3.000 g of spectroscopic-grade KBr in a small mortar for 5 min and dry the mixture overnight in vacuo over activated Na-A zeolite, Weigh 300.0 mg of the mixture and prepare a pellet as described in section 11–3.4.1.4.2. Determine the absorbance or peak height and obtain the percentage composition from the standard curve prepared in section 11–6.4.

The mull technique may also be used for the preparation of samples for quantitative analysis. Because of the problem of reproducibility in path length inherent in the mull technique, use of an internal standardization procedure may be necessary. In this procedure, a known amount of noninterfering material having an isolated and easily measured absorption band is mixed with the weighed unknown, and the ratio of band absorbances compared. Materials that have been used as internal standards include lead thiocyanate, potassium thiocyanate, calcium carbonate, and sodium azide. The internal standardization procedure may also be used in the KBr pellet technique. Hesterberg (1984) has recently described the use of deuterated brucite [$Mg(OD)_2$] as an internal standard for quantitative analysis of clay minerals using a mull technique. The procedure which follows is that developed by Hesterberg (1984).

1. Weigh 20.0 mg of the unknown sample and 5.0 mg of synthetic $Mg(OD)_2$

to the nearest 0.1 mg on a single powder weighing paper, using a sensitive balance set to read to 0.00001 g.

2. Transfer the sample to an agate mortar (50–65 mm diameter) and accurately add 80 μl of Fluorolube with an adjustable 30- to 100-μL positive-displacement Absoluter® micropipette (Tri-Continental Scientific, Inc.)
3. Mix the sample, internal standard, and mulling agent moderately but thoroughly for 10 min. Frequently scrape the mull to the center of the mortar with the pestle.
4. Remove a portion of the sample mull with a spatula and place between two polycrystalline ZnS windows. Cover the windows with a piece of clean tissue paper; then, holding the windows slightly off center between the thumb and middle finger, rub gently in a circular fashion to evenly distribute the mull and produce nonparallel window surfaces.
5. With the infrared spectrometer set to monitor the OD-stretching frequency of $Mg(OD)_2$ at 2728 cm^{-1}, adjust the sample pathlength until the relative intensity of this band shows a transmittance near 37% (absorbance of 0.43). If the pathlength is too long, squeeze along one edge of the ZnS windows to remove some of the excess mull and shorten the pathlength. If the absorbance is too low, separate the ZnS windows, add more sample and adjust to appropriate pathlength.

Record the spectrum of the sample and determine the absorbances of the analytical band(s) of the component(s) being measured, as well as that of the internal standard—in this case, $Mg(OD)_2$ at 2728 cm^{-1}. Compare the ratio of band absorbances of the unknown vs that of the internal standard with the calibration curve prepared as described in section 11–6.4.

11–6.4 Calibration Curve

From a spectrum of the compound or material under study, choose an analytical absorption band outside regions of extremely low transmission percentages. Ideally, the band should be free from background interference and should be of the shape on which a good baseline can be drawn. Draw the baseline as nearly as possible where the pen-tracing would go if the band were not present. Determine the absorbance or percentage transmission in the spectral region of the analytical absorption band on a number of samples containing varying amounts of the pure component. Select concentrations which give band intensities within the range of 20 to 60% transmittance (0.2–0.7 absorbance).

Construct a calibration curve by plotting sample weight against peak absorbance [log (I_0/I)]; in the case of internal standardization, plot the ratio of the absorbance of the pure component vs the absorbance of the internal standard against sample weight of the pure component.

11–6.5 Comments

The application of computer analysis to infrared spectrometry makes possible collection and processing of spectral data to give results, which, prior to the advent of computer-assisted infrared spectrometers, could not be duplicated by any combination of conventional instrumentation and experimental measurements. The sheer magnitude of effort required to carry out the equivalent experimental work would be prohibitive.

In spite of the recent advances, spectral analysis of multicomponent systems is still one of the greatest challenges facing infrared spectroscopists. The procedures that have been proposed for doing this require adherence to the Bouguer-Beer law (i.e., the mixture spectrum may be expressed as a linear combination of its constituent pure components). Unfortunately, many samples fail to fulfill this criterion due to instrumental factors and inherent properties of the sample. Hirschfeld (1979) has written a thorough review of problems of quantitative analysis using FT-IR.

Depending on how many of the pure components in a mixture are known, computer analysis may be carried out by least-squares programs, factor analysis, spectral subtraction, and curvefitting, often in combination with interactive real-time graphics. Gillette and Koenig (1984) have proposed an iterative linear least-squares algorithm which makes it possible to selectively subtract known spectral components from a mixture in an automated procedure. This approach overcomes the subjective decisions encountered with interactive real-time graphics programs; it also make possible simultaneous removal and/or optimization of scaling coefficients of multiple known pure components in order to produce the difference spectrum.

J. M. Brown and J. J. Elliott (Exxon Research and Engineering Co., Linden, NJ 07036) reported on "The Quantitative Analysis of Minerals by FT-IR Spectroscopy" in a workshop on "Applications of IR Methods to the Study of Clay Minerals," sponsored by the Clay Minerals Society, 1 Oct. 1983 in Buffalo, NY. They developed a methodology for spectral analysis of minerals by FT-IR which involves spectral subtraction and a modification of the curvefitting program developed by Koenig and coworkers. They concluded that FT-IR mineral analysis can provide a reasonably rapid means of measuring mineral compositions to ± 10% on a relative basis. They feel that a major limitation in the further development of procedures for quantitative analysis by infrared techniques is the lack of satisfactory reference minerals and reference sets of minerals.

As one of the authors wrote earlier (White, 1977), "The compilation of a catalog of infrared spectra of well-characterized minerals by a cooperative effort within or among the several mineralogical organizations would provide considerable impetus for more extensive use of infrared spectroscopy in mineralogical studies." This statement is still valid. The great advances in spectral data collection and manipulation resulting from the development and incorporation of microprocessors and computer capabilities into the design of infrared spectrometers only serve to emphasize the great need for the establishment of an extensive library of digitally recorded infrared spectra of carefully selected and well-characterized reference minerals. The establishment of an authoritative library of reference mineral infrared spectra through collaboration among interested mineralogical organizations would be similar to the work of the Joint Committee on Powder Diffraction Standards in publishing the *Powder Diffraction File*, but on a much more modest scale. Access to such a library of reference mineral spectra, together with less expensive infrared spectrometers having computer capabilities, and increasing availability of easy-to-use software programs, could provide the impetus for soil scientists, mineralogists, ceramicists, civil engineers, and others to more fully exploit the potential usefulness of infrared spectrometry as both a routine and research analytical tool.

11–7 SPECTRAL DATA COLLECTION AND MANIPULATION

The recent advances in infrared spectrometry have resulted primarily from the development of microprocessor-based computerized infrared systems and accompanying sophisticated software for the collection and evaluation of spectroscopic data. The problem-solving capacity and sensitivity of the technique have been increased in terms of orders of magnitude. This has resulted in significant improvements in the performance of dispersive infrared instruments and the development of competitively priced Fourier transform infrared (FT-IR) instruments with their inherent advantages.

The use of computers in infrared spectrometry has led to advances in spectra interpretation via (i) spectral addition to improve signal-to-noise ratio, (ii) spectral subtraction or difference spectroscopy to allow investigation of low-concentration or trace components, and (iii) derivative spectra to evaluate small changes in infrared absorption peaks. Improvements in microcomputer hardware and software now allow the spectroscopist to use these techniques with instruments having interactive graphics capabilities.

McDonald (1984) has reviewed recent advances in all phases of infrared spectrometry. Among the advances mentioned are easier-to-use computer system software and hardware, interactive graphics, large disk memories, factor analysis, and multiple regression techniques. The availability of large disks makes feasible libraries of fully digitized spectra and provides a search and identification capability matching or surpassing that of a spectroscopist. The mathematical routines involving factor analysis and multiple regression are being used effectively for determination of the spectra of pure components of multicomponent mixtures, which can only be partially resolved by physical separation techniques. Improved microcomputer on-line software provides for various methods of resolution enhancement via derivative spectra and Fourier deconvolution.

11-8 REFERENCES

Akhmanova, M. V., and L. G. Alekhina. 1971. Infrared spectroscopic study of isomorphism in minerals. p. 243–267. *In* A. P. Vinogradov (ed.) Probl. isomorinykh zameschchenii at. kirst. (Russ) "Nauka", Moscow, USSR.

Alpert, N. L., W. E. Keiser, and H. A. Szymanski. 1973. IR—Theory and practice of infrared spectroscopy. 2nd ed. Plenum/Rosetta ed., Plenum Publishing Co., New York.

Bellamy, L. J. 1975. The infrared spectra of complex molecules, Vol. 1, 3rd ed. John Wiley and Sons, Inc., New York.

Black, C. A., D. D. Evans, J. L. White, L. E. Ensminger, and F. E. Clark (ed.). 1965. Methods of soil analysis, Part 1. Agronomy 9. Am. Soc. Agron. Inc., Madison, WI.

Boyd, S. A., L. E. Sommers, and D. W. Nelson. 1979. Infrared spectra of sewage sludge fractions: evidence for an amide metal binding site. Soil Sci. Soc. Am. J. 43:893–899.

Boyd, S. A., L. E. Sommers, and D. W. Nelson. 1980. Changes in the humic acid fraction of soil resulting from sludge application. Soil Sci. Soc. Am. J. 44:1179–1186.

Bradley, K. B., and W. J. Potts, Jr. 1958. The internally standardized Nujol mull as a method of quantitative infrared spectroscopy. Appl. Spectrosc. 12:77–80.

Brame, E. G., Jr., and J. G. Grasselli (ed.). 1976. Infrared and Raman spectroscopy, Vol. 1, Part A, Series on practical spectroscopy. Marcel Dekker, Inc. New York.

Brame, E. G., Jr., and J. G. Grasselli (ed.) 1977a. Infrared and Raman spectroscopy, Vol. 1, Part B, Series on practical spectroscopy. Marcel Dekker, Inc., New York.

Brame, E. G., Jr., and J. G. Grasselli (ed.) 1977b. Infrared and Raman spectroscopy, Vol. 1, Part C, Series on practical spectroscopy. Marcel Dekker, Inc., New York.

Chaussidon, J. 1972a. Application of infrared spectroscopy in the study of clay minerals. Ceramurgia 3:210–213 (Ital.).

Chaussidon, J. 1972b. Application of infrared spectroscopy to the study of mineral weathering. p. 53–56. *In* Potassium—soil, Proc. Colloq. Int. Potash Inst., Bern, Switzerland.

Coates, J. P. 1978. The analysis of aqueous solutions by infrared spectroscopy. Eur. Spect. News 16:25–33.

Colthup, N. B. 1950. Spectra-structure correlations in the infrared region. J. Opt. Soc. Am. 40:397–400.

Colthup, N. B., L. H. Daly, and S. E. Wiberley. 1975. Introduction to infrared and Raman spectroscopy. 2nd ed. Academic Press Inc., New York.

Conley, R. T. 1972. Infrared spectroscopy. 2nd ed. Allyn and Bacon, Inc., Boston.

Cruz, M. I., J. L. White, and J. D. Russell. 1968. Montmorillonite—s-triazine interactions. Israel J. Chem. 6:315–323.

Durana, J. F., and A. W. Mantz. 1979. Laboratory studies of reacting and transient systems. p. 1–72. *In* J. R. Ferraro and L. J. Basile (ed.) Fourier transform infrared spectroscopy (Vol.2); applications to chemical systems. Academic Press, Inc., New York.

Duyckaerts, G. 1959. The infrared analysis of solid substances. A review. Analyst 84:201–214.

Farmer, V. C. 1957. Effects of grinding during the preparation of alkali halide disks. Spectrochim. Acta 8:374—389.

Farmer, V. C. 1968. Infrared spectroscopy in clay mineral studies. Clay Miner. 7:373—387.

Farmer, V. C. (ed.). 1974. The infrared spectra of minerals. Mineral. Soc., London.

Farmer, V. C. 1976. The role of infrared spectroscopy in a soil research institute. Eur. Spectrosc. News 7:13–15.

Farmer, V. C. 1979. The role of infrared spectroscopy in a soil research institute: characterization of inorganic materials. Eur. Spectrosc. News 25:25–27.

Farmer, V. C., and B. D. Mitchell. 1963. Occurrence of oxalates in soil clays following hydrogen peroxide treatment. Soil Sci. 96:221–229.

Farmer, V. C., and M. M. Mortland. 1966. An infrared study of the coordination of pyridine and water to exchangeable cations in montmorillonite and saponite. J. Chem. Soc. (London) A, 344–351.

Farmer, V. C., and F. Palmieri. 1975. The characterization of soil minerals by infrared spectroscopy. p. 573–670. *In* J. E. Gieseking (ed.) Soil components. Vol. 2, Inorganic components. Springer-Verlag, Berlin.

Farmer, V. C., and J. D. Russell. 1964. The infrared spectra of layer silicates. Spectrochim. Acta 20:1149–1173.

Farmer, V. C., and J. D. Russell. 1967. Infrared absorption spectrometry in clay studies. Clays Clay Miner. 15:121–142.

Farmer, V. C., J. D. Russell, and J. L. Ahlrichs. 1968. Characterization of clay minerals by infrared spectroscopy. Trans. Int. Congr. Soil Sci., 9th (Adelaide, Aust.) 3:101–110.

Farmer, V. C., J. D. Russell, J. L. Ahlrichs, and B. Velde. 1967. Vibration du groupe hydroxyle dans les silicates en couches. Bull. Groupe Fr. Argiles 19:5–10.

Farmer, V. C., J. D. Russell, W. J. McHardy, A. C. D. Newman, J. L. Ahlrichs, and J. Y. H. Rimsaite. 1971. Evidence of loss of protons and octahedral iron from oxidized biotites and vermiculites. Mineral. Mag. 38:121–137.

Ferraro, J. R. (ed.) 1982. The Sadtler infrared spectra handbook of minerals and clays. Sadtler Research Lab., Philadelphia.

Ferraro, J. R., and L. J. Basile (ed.) 1978. Fourier transform infrared spectroscopy (Vol. 1): applications to chemical systems. Academic Press, Inc., New York.

Ferraro, J. R., and L. J. Basile (ed.) 1979. Fourier transform infrared spectroscopy (Vol. 2): applications to chemical systems. Academic Press, Inc., New York.

Ferraro, J. R., and L. J. Basile (ed.) 1982. Fourier transform infrared spectroscopy (Vol. 3): techniques using Fourier transform interferometry. Academic Press, Inc., New York.

Fieldes, M. R., J. Furkert, and N. Wells. 1972. Rapid determination of constituents of whole soils using infrared absorption. N. Z. J. Sci. 15:615–627.

Flanigen, E. M., H. Khatami, and H. Szymanski. 1971. Infrared structural studies of zeolite frameworks. Adv. Chem. Ser. 101:201–229.

Fripiat, J. J. 1960. Application de la spectroscopie infrarouge à l'étude des mineraux argileux. Bull. Groupe Fr. Argiles 12(7):25–41.

Fripiat, J. J., J. Chaussidon, and R. Touillaux. 1960. Study of dehydration of montmorillonite and vermiculite by infrared spectroscopy. J. Phys. Chem. 64:1234–1241.

Fysh, S. A., and P. M. Fredericks. 1983. Fourier transform infrared studies of aluminous goethites and hematites. Clays Clay Miner. 31:377–382.

Gadsden, J. A. 1975. Infrared spectra of minerals and related inorganic compounds. Butterworth and Co., London.

Gillette, P. C., and J. L. Koenig. 1984. Objective criteria for absorbance subtraction. Appl. Spectrosc. 38:334–337.

Griffiths, P. R. 1975. Chemical infrared Fourier transform spectroscopy. Vol. 43, Chemical analysis. John Wiley and Sons, Inc., New York.

Hannah, R. W., and J. P. Coates. 1980. Microprocessors for infrared spectroscopy. Eur. Spectrosc. News 32:30–35.

Harrick, N. J. 1967. Internal reflection spectroscopy. Interscience Publishers, Inc., New York.

Hayashi, H., and K. Oinuma. 1965. Relationship between infrared absorption spectra in the region of 450–900 cm^{-1} and chemical composition of chlorite. Am. Mineral. 50:476–483.

Hayashi, H., and K. Oinuma. 1967. Si-O absorption band near 1000 cm^{-1} and OH absorption bands of chlorite. Am. Mineral. 52:1206–1210.

Hesterberg, D. L. R. 1984. Preferential erosion of clay minerals from reclaimed mine soils as detected by infrared spectroscopic techniques. M. S. Thesis, Purdue University, West Lafayette, IN.

Hirschfeld, T. 1979. Quantitative FT-IR: a detailed look at the problems involved. p. 193–242. *In* J. R. Ferraro and L. J. Basile (ed.) Fourier transform infrared spectroscopy (Vol. 2): applications to chemical systems. Academic Press, Inc., New York.

Hlavay, J., and J. Inczedy. 1979. Sources of error of quantitative determination of the solid crystalline minerals by infrared spectroscopy. Acta Chim, Budapest 102:11–18.

Hlavay, J., K. Jonas, S. Elek, and J. Inczedy. 1977. Characterization of the particle size and the crystallinity of certain minerals by infrared spectrophotometry and other instrumental methods: I. Investigations on clay minerals. Clays Clay Miner. 25:451–456.

Hlavay, J., K. Jonas, S. Elek, and J. Inczedy. 1978. Characterization of the particle size and the crystallinity of certain minerals by infrared spectrophotometry and other instrumental methods: II. Investigations on quartz and feldspar. Clays Clay Miner. 26:139–143.

Hunt, G. R., and J. W. Salisbury. 1970. Visible and near infrared spectra of minerals and rocks: I. Silicate minerals. Modern Geol. 1:283–300.

Hunt, G. R., and J. W. Salisbury. 1971a. Visible and near-infrared spectra of minerals and rocks: II. Carbonates. Modern Geol. 2:23–30.

Hunt, G. R., and J. W. Salisbury. 1971b. Visible and near-infrared spectra of minerals and rocks: III. Oxides and hydroxides. Modern Geol. 2:195–205.

Hunt, G. R., J. W. Salisbury, and C. J. Lenhoff. 1971a. Visible and near-infrared spectra of minerals and rocks: IV. Sulphides and sulphates. Modern Geol. 3:1–14.

Hunt, G. R., J. W. Salisbury, and C. J. Lenhoff. 1971b. Visible and near-infrared spectra of minerals and rocks: V. Halides, phosphates, arsenates, vanadates, and borates. Modern Geol. 3:121–132.

Hunt, G. R., J. W. Salisbury, and C. J. Lenhoff. 1973a. Visible and near-infrared spectra of minerals and rocks. VI. Additional silicates. Modern Geol. 4:85–105.

Hunt, G. R., J. W. Salisbury, and C. J. Lenhoff. 1973b. Visible and near-infrared spectra of minerals and rocks: VII. Acidic igneous rocks. Modern Geol. 4:217–224.

Hunt, G. R., J. W. Salisbury, and C. J. Lenhoff. 1973c. Visible and near-infrared spectra of minerals and rocks: VIII. Intermediate igneous rocks. Modern Geol. 4:237–244.

Hunt, G. R., J. W. Salisbury, and C. J. Lenhoff. 1974. Visible and near-infrared spectra of minerals and rocks: IX. Basic and ultrabasic igneous rocks. Modern Geol. 5:15–22.

Ishii, M., M. Nakahira, and H. Takeda. 1969. Far-infrared absorption spectra of micas. Proc. Int. Clay Conf., 1969, Tokyo (L. Heller, ed.), Vol. 1, p. 247–259. Israel University Press, Jerusalem.

Ishii, M., T. Shimanouchi, and M. Nakahira. 1967. Far-infrared absorption spectra of layer silicates. Inorg. Chim. Acta 1:387–392.

Kahle, A. B., and A. F. H. Goetz. 1983. Mineralogic information from a new airborne thermal infrared multispectral scanner. Science 222:24–27.

Kosmas, C. S. 1984. Visible spectra and color of synthetic Al-substituted goethites and hematites. Ph.D. Thesis, Purdue University, West Lafayette, IN. Dissert. Abstr. Int. 45/07 p. 1969-B.

Kosmas, C. S., N. Curi, R. B. Bryant, and D. P. Franzmeier. 1984. Characterization of iron oxide minerals by second-derivative visible spectroscopy. Soil Sci. Soc. Am. J. 48:401–405.

Lehr, J. R., E. H. Brown, and A. W. Frazier. 1967. Crystallographic properties of fertilizer compounds. TVA, Chem, Eng. Bull. No. 6, Muscle Shoals, AL.

Levitt, S. R., and R. A. Condrate, Sr. 1970. The preparation of fine mineral powders for infrared spectroscopy. Am. Mineral. 55:522–524.

Little, L. H. 1966. Infrared spectra of adsorbed species. Academic Press, Inc., London.

Lyon, R. J. P. 1964. Infrared analysis of soil minerals. p. 170–199. *In* C. I. Rich and G. W. Kunze (ed.) Soil clay mineralogy. Univ. North Carolina Press, Chapel Hill, NC.

MacCarthy, P., H. B. Mark, Jr., and P. R. Griffiths. 1975. Direct measurement of the infrared spectra of humic substances in water by Fourier transform infrared spectroscopy. J. Agric. Food Chem, 23:600–602.

Marel, H. W. van der., and H. Beutelspacher. 1976. Atlas of infrared spectroscopy of clay minerals and their admixtures. Elsevier Science Publishers, Amsterdam.

McDonald, R. S. 1958. Surface functionality of amorphous silica by infrared spectroscopy. J. Phys. Chem. 62:1168–1178.

McDonald, R. S. 1984. Infrared spectrometry. Anal. Chem. 56:349R–372R.

Miller, R. G. T., and B. C. Stace. 1972. Laboratory methods in infrared spectroscopy. 2nd ed. Heyden and Son, Ltd. London.

Mitchell, B. D., V.C. Farmer, and W. J. McHardy. 1964. Amorphous inorganic materials in soils. Adv. Agron. 16:327–383. Academic Press, Inc., New York.

Moenke, H. 1962. Mineralspektren, I. Akademie Verlag, Berlin.

Moenke, H. 1966. Mineralspektren, II. Akademie Verlag, Berlin.

Mortensen, J. L., D. M. Anderson, and J. L. White. 1965. Infrared spectrometry. *In* C. A. Black, et al., (ed.) Methods of soil analysis. Part I. Agronomy 9:743–770. Am. Soc. Agron., Inc., Madison, WI.

Mortland, M. M. 1966. Urea complexes with montmorillonite: an infrared absorption study. Clay Miner. 6:143–156.

Mortland, M. M. 1970. Clay-organic complexes and interactions. Adv. Agron. 22:75–117.

Mortland, M. M., and K. V. Raman. 1967. Catalytic hydrolysis of some organic phosphate pesticides by copper (II.) J. Agric. Food Chem. 15:163–167.

Mortland, M. M., J. J. Fripiat, J. Chaussidon, and J. B. Uytterhoeven. 1963. Interaction between ammonia and the expanding lattices of montmorillonite and vermiculite. J. Phys. Chem. 67:248–258.

Mulla, D. J., P. F. Low, and C. B. Roth. 1985. Measurement of the specific surface area of clays by internal reflectance, spectroscopy. Clays Clay Miner. 33:391-396.

Nail, S. L., J. L. White, and S. L. Hem. 1976. IR studies of development of order in aluminum hydroxide gels. J. Pharm. Sci. 65:231–234.

Oinuma, K., and H. Hayashi. 1967. Infrared absorption spectra of some zeolites from Japan. J. Tokyo Univ. General Educ. (Nat. Sci.) 8:1–12, Tokyo.

Olphen, H. van, and J. J. Fripiat. 1979. Data handbook for clay materials and other non-metallic minerals. Pergamon Press, Oxford.

Potter, R. M., and G. R. Rossman. 1979a. The tetravalent manganese oxides: identification, hydration, and structural relationships by infrared spectroscopy. Am. Mineral. 64:1199–1218.

Potter, R. M., and G. R. Rossman. 1979b. Mineralogy of manganese dendrites and coatings. Am. Mineral. 64:1219–1226.

Potts, W. J., Jr. 1963. Chemical infrared spectroscopy. Vol. 1. Techniques. John Wiley and Sons, Inc., New York.

Rao, C. N. R. 1963. Chemical applications of infrared spectroscopy. Academic Press, Inc., New York.

Russell, J. D. 1974. Instrumentation and techniques. p. 11–25. *In* V. C. Farmer (ed.) The infrared spectra of minerals. Mineral. Soc., London.

Russell, J. D., V. C. Farmer, and B. Velde. 1970. Replacement of OH by OD in layer silicates, and identification of the vibration of these groups in infrared spectra. Mineral. Mag. 37:869–879.

Russell, J. D., and A. R. Fraser. 1971. Infrared spectroscopic evidence for interaction between hydronium ions and lattice OH groups in montmorillonite. Clays Clay Miner. 19:55–59.

Russell, J. D., W. J. McHardy, and A. R. Fraser. 1969. Imogolite: a unique aluminosilicate. Clay Miner. 8:87–99.

Schaumberg, G. D., C. S. LeVesque-Madore, G. Sposito, and L. J. Lund. 1980. Infrared spectroscopic study of the water-soluble fraction of sewage sludge-soil mixtures during incubation. J. Environ. Qual. 9:297–303.

Schnitzer, M. 1971. Characterization of humic constituents by spectroscopy. p. 60–95. *In* A. D. McLaren and J. Skujins (ed.) Soil biochemistry. Vol. 2. Marcel Dekker, Inc., New York.

Schnitzer, M. 1982. Organic matter characterization. *In* A. L. Page et al. (ed.) Methods of soil analysis, part 2. 2nd ed. Agronomy 9:581–594.

Schnitzer, M., and S. U. Khan. 1978. Soil organic matter. Elsevier North-Holland, Inc., New York.

Smith, A. L. 1979. Applied infrared spectroscopy: fundamentals, techniques, and analytical problem-solving. Chemical analysis Vol. 54. John Wiley and Sons, Inc., New York.

Solomon, D. H., and D. G. Hawthorne. 1983. Chemistry of pigments and fillers. John Wiley and Sons., Inc., New York.

Stevenson, F. J. 1982. Humus chemistry: genesis, composition, reactions. John Wiley and Sons, Inc., New York.

Stevenson, F. J., and J. H. A. Butler. 1969. Chemistry of humic acids and related pigments. p. 534–557. *In* G. Eglinton and M. T. J. Murphy (ed.) Organic geochemistry. Springer-Verlag, New York.

Stevenson, F. J., and K. M. Goh. 1971. Infrared spectra of humic acids and related substances. Geochim. Cosmochim. Acta 35:471–483.

Stevenson, F. J., and K. M. Goh. 1974. Infrared spectra of humic acids: elimination of interference due to hygroscopic moisture and structural changes accompanying heating with KBr. Soil Sci. 117:34–41.

Stewart, J. E. 1970. Infrared spectroscopy: Experimental methods and techniques. Marcel Dekker, Inc., New York.

Stubican, V., and R. Roy. 1961a. Isomorphous substitution and infrared spectra of the layer lattice silicates. Am. Mineral. 46:32–51.

Stubican, V., and R. Roy. 1961b. Infrared spectra of layer-structure silicates. J. Am. Ceram. Soc. 44:625–627.

Tarte, P. 1962. Infrared study of orthosilicates and orthogermanates. New method for interpretation of spectra. Spectrochim. Acta 18:467–483.

Theng, B. K. G. 1974. The chemistry of clay-organic reactions. Adam Hilger, Ltd., London.

Theng, B. K. G. 1979. Formation and properties of clay-polymer complexes. Developments in Soil Science 9, Elsevier Science Publishing Co., Amsterdam.

Tuddenham, W. M., and J. D. Stephens. 1971. Infrared spectrophotometry. p. 127–168. *In* R. E. Wainerdi (ed.) Modern methods of geochemical analysis. Plenum Publishing Corp., New York.

Vedder, W. 1964. Correlations between infrared spectrum and chemical composition of mica. Am. Mineral. 49:736–768.

Velde, B. 1978. Infrared spectra of synthetic micas in the series muscovite-magnesium-aluminum celadonite. Am. Mineral. 63:343–349.

Velde, B. 1983. Infrared OH-stretch bands in potassic micas, talcs and saponites; influence of electronic configuration and site of charge compensation. Am. Mineral. 68:1169–1173.

Wada, K., and D. J. Greenland. 1970. Selective dissolution and differential infrared spectroscopy for characterization of amorphous constituents in soil clays. Clay Miner. 8:241–254.

Watson, C. A. 1977. Near infrared reflectance spectrophotometric analysis of agricultural products. Anal. Chem. 49:835A–840A.

White, J. L. 1971. Interpretation of infrared spectra of soil minerals. Soil Sci. 112:22–31.

White, J. L. 1976. Determination of susceptibility of s-triazine herbicides to protonation and hydrolysis by mineral surfaces. Archives Environ. Contamin. Toxicol. 3:461–469.

White, J. L. 1977. Preparation of specimens for infrared analysis. p. 847–863. *In* J. B. Dixon and S. B. Weed (ed.) Minerals in soil environments. Soil Sci. Soc. Am., Inc. Madison, WI.

White, J. L., S. L. Nail, and S. L. Hem. 1976. Infrared technique for distinguishing between amorphous and crystalline aluminum hydroxide phases. p. 51–59. Proc. 7th Conf. Clay Mineral. Petrology, Karlovy Vary, Czechoslovakia.

Wilkins, R. W. T. 1967. The hydroxyl-stretching region of the biotite mica spectrum. Mineral. Mag. 36:325–333.

Willard, H. H., L. L. Merritt, Jr., J. A. Dean, and F. A. Settle, Jr. 1981. Instrumental methods of analysis. 6th ed. D. Van Nostrand Co., New York.

Wilson, M. J. 1966. The weathering of biotite in some Aberdeenshire soils. Mineral. Mag. 35:1080–1093.

Wilson, M. J. 1970. A study of weathering in a soil derived from a biotite-hornblende rock: I. Clay Miner. 8:291–303.

12 X-Ray Diffraction Techniques

L. D. WHITTIG AND W. R. ALLARDICE

University of California
Davis, California

12-1 GENERAL INTRODUCTION

Both the physical and chemical properties of any soil are controlled to a very large degree by the minerals of the soil, and especially by those constituting the clay fraction. Identification, characterization, and an understanding of properties of the different minerals materially aid in evaluation of soils in relation to classification, agronomic practices, and engineering properties.

Clay fractions of soils are commonly composed of mixtures of one or more secondary phyllosilicate minerals together with primary minerals inherited directly from the parent material. Positive identification of mineral species and quantitative estimation of their proportions in such polycomponent systems usually require the application of several complementary qualitative and quantitative analyses. One of the most useful methods is x-ray diffraction analysis. Hadding (1923) and Rinne (1924) were the first to apply x-rays to the study of clay minerals, and Hendricks and Fry (1930) and Kelley et al. (1931) were the first to demonstrate that soil clays contain crystalline mineral components that yield x-ray diffraction patterns. Investigations of the structure, properties, and occurrence of soil clay minerals by x-ray diffraction methods have become major efforts in soil science.

Continued improvements in x-ray instrumentation, techniques of sample preparation, and definition of criteria for identification and characterization of clay mineral species in recent years have advanced the field of clay mineralogy to a point where mineralogical analyses yield a wealth of information relative to the properties and genesis of soils. X-ray diffraction has contributed more to mineralogical characterization of clay fractions of soils than has any other single method of analysis.

12-2 PRINCIPLES OF X-RAY DIFFRACTION

The nature and properties of x-rays and principles of x-ray diffraction are discussed thoroughly in a number of texts (Buerger, 1942; Clark, 1955;

Auleytner, 1967; Wormald, 1973; Klug & Alexander, 1974; Cullity, 1978; and others). A brief treatment of these subjects is presented here.

M. von Laue is credited with characterization of the nature of x rays (after Cullity, 1978), following their discovery by Roentgen in 1895. Laue, in 1912, reasoned that if mineral crystals were composed of regularly spaced atoms which might act as centers of scattering of x-rays, and if x-rays were electromagnetic waves of wavelength about equal to interatomic distances in crystals, then it should be possible to diffract x-rays with crystals. The classical experiments directed by Laue to test these theories established the wave nature of x-rays and the periodicity of arrangement of atoms within crystals.

Crystalline structures are characterized by a systematic and periodic arrangement of atoms (or ions) in a three-dimensional array. Because crystals are composed of regularly spaced atoms, each crystal contains planes of atoms which are separated by a constant distance. The distances between planes are characteristic of the crystalline species.

X-rays consist of electrostatic and electromagnetic fields that oscillate in periodic cycles in planes perpendicular to one another and to their direction of propagation through space. X-rays, with wavelengths of the order of 10^{-3} to 10^{1} nm, are generated within an evacuated x-ray tube by bombardment of a metal target (anode) with high-velocity electrons. Monochromatic x-rays utilized for x-ray diffraction have finite energy and wavelength characteristic for the particular target metal. Transfer of energy from high-velocity electrons to target atom electrons elevates the latter to higher energy levels, thus momentarily creating electron orbital vacancies. These electron vacancies in excited target atoms are filled by transfer of electrons from higher energy levels. Each electron transfer from a higher to a lower energy state results in emission of a quantum of energy (x-ray photon) equivalent to the difference in energy between the two levels involved in the transition. Since the electron transfers involve quantum changes, the energies and wavelengths of emitted x-ray photons are finite and characteristic for the particular atoms of the target metal. Photons generated by electron transfer to the ls level (K shell) from 2s and 2p levels (L shell) are utilized for most x-ray diffraction applications. For most practical purposes, differences in energy and wavelength between $K_{\alpha 1}$ ($2p \rightarrow 1s$) and $K_{\alpha 2}$ ($2s \rightarrow 1s$) photons are relatively insignificant.

X-ray photons may be considered as "packets" of monochromatic electromagnetic waves, generated as random events, emanating outward in a spherical array from their point source. A narrow, directed beam of randomly displaced x-ray photons is obtained by allowing a segment of the spherical waves to pass through a window-collimator system. (Fig. 12-1.

The phenomenon of diffraction involves the scattering of x-rays by atoms of a crystal and the reinforcement of scattered rays in definite directions away from the crystal. Reinforcement of the scattered rays is quantitatively related to the distance of separation of atomic planes as

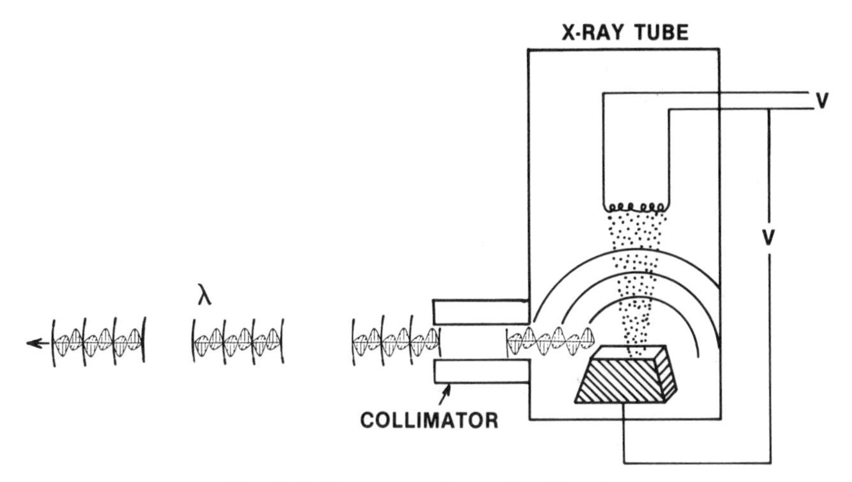

Fig. 12–1. X-ray generation and collimation.

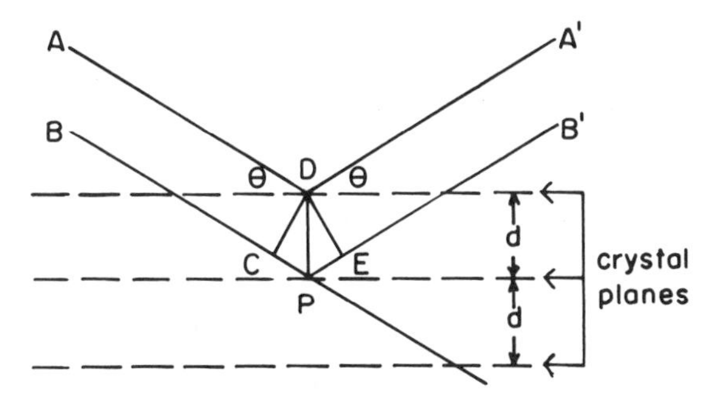

Fig. 12–2. Diffraction from crystal planes according to Bragg's law, $n\lambda = 2d \sin \theta$.

defined by Bragg's law,

$$n\lambda = 2d \sin \theta \, .$$

When a collimated beam of monochromatic x-rays of wavelength λ strikes a crystal, the rays penetrate and are partially scattered from many successive planes within the crystal (Fig. 12–2). For a given interplanar spacing, d, there will be a critical angle, θ, at which rays scattered from successive planes will be in phase along a front as they leave the crystal. A ray following the path BPB′, for example, will have traveled some whole number of wavelengths, $n\lambda$, farther than a ray traveling along the path ADA′. The angle between the normal to the emerging wave front and the atomic planes will equal the angle between the normal to the primary wave front and the atomic planes. Diffraction from a succession of equally spaced lattice planes results in a diffraction maximum which has sufficient intensity to be recorded.

Since no two minerals have exactly the same interatomic distances in three dimensions, the angles at which diffraction occurs will be distinctive for a particular mineral. The interatomic distances within a mineral crystal then result in a unique array of diffraction maxima, which serves to identify that mineral.

Diffraction can occur whenever the Bragg law, $n\lambda = 2d \sin \theta$, is satisfied. The wavelength of radiation is characteristic and constant for the particular x-ray tube used. The angle of incidence, θ, of the primary radiation with the crystal planes can be varied, however.

When n is equal to 1, diffraction is of first order. At other angles, where n is equal to 2, 3, or a greater number, diffraction is again possible, giving rise to second, third, and higher orders of diffraction (Fig. 12–3). Although d remains the same, d/n values will be different depending upon the value of n.

With an x-ray spectrometer, the angle of incidence is varied by rotating the sample in the path of the primary x-ray beam. A suitable detector (Geiger, proportional, or scintillation counter), used to intercept and measure the diffracted rays, also moves in such a way as to maintain an angle with the sample which is equal at all positions to the angle of incidence of the primary beam. From the chart of a direct recording x-ray spectrometer, the value of 2θ (with reference to the primary beam) is available directly.

The angle θ may be effectively varied by analysis with stationary sample and recorder, the recorder in this case being a photographic film. Crystals to be analyzed are reduced to a very fine powder and placed in the path of a beam of monochromatic x-rays. The particles of the powder are tiny crystals oriented at random with respect to the primary beam.

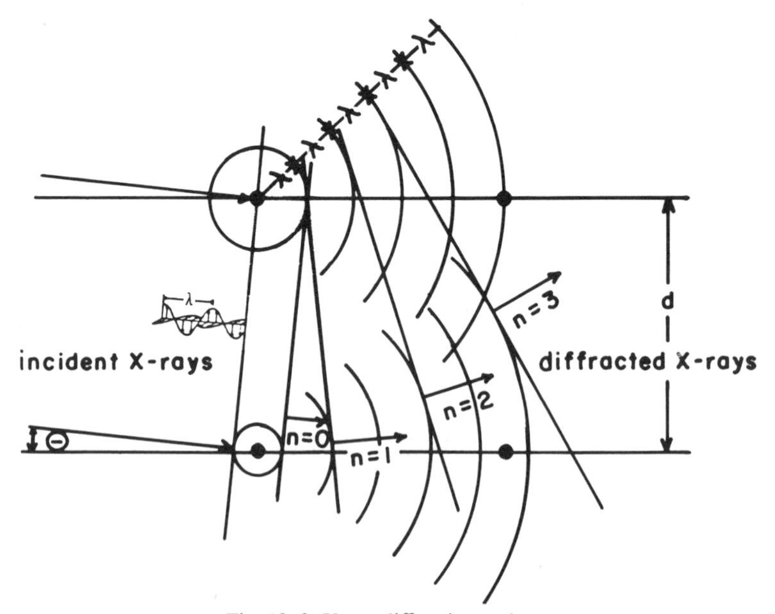

Fig. 12–3. X-ray diffraction orders.

Because of the large number of crystals in the mass of fine powder, there will be sufficient crystals properly oriented so that every set of lattice planes will be capable of diffraction. The mass of powder is equivalent to a single crystal rotated, not about one axis, but about all possible axes. In the event that several crystal species are present in the powder mixture, each component species registers its own diffraction maxima independently. The diffracted rays from a powder mixture may be registered on a photographic film or plate that is geometrically placed in relation to the sample, so as to allow determination of the angle of diffraction for each maximum and subsequent calculation of interatomic spacing.

A cylindrical film pattern registers diffraction-cone intercepts at angles, θ, from the diffracting crystal planes (2θ as measured from the primary beam) as shown in Fig. 12–4. The geometry of cylindrical camera systems requires that θ be related to the camera radius, R, and the distance between the primary beam position and each line position, L, as measured on the film in a flat position, by the equation

$$L/2\pi R = 2\theta/360°$$

Knowing both the radius of the camera and the measured value of L in centimeters, one may readily calculate θ for each diffraction line.

After determining θ, or more usually 2θ, for each diffraction line on either a chart-recorded or film-recorded pattern, d may be calculated from Bragg's equation. Operation manuals of the various commercial diffraction instruments usually contain tables for ready conversion of θ or 2θ to d.

Efficiency of diffraction from a mineral sample is dependent upon the wavelength of radiation used and the physical and chemical nature of the sample. Selection of the proper wavelength of radiation and careful sample preparation are factors to be considered by the analyst to ensure that maximum diffraction intensity and quality are obtained in any particular analysis.

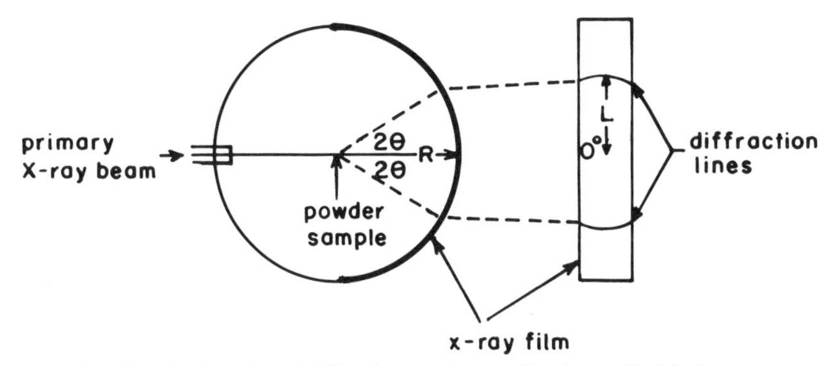

Fig. 12–4. Registration of diffraction maxima on film in a cylindrical camera.

12-3 PREPARATION OF SAMPLES

12-3.1 Introduction

Because of the necessity for application of several diagnostic tests for positive identification of certain phyllosilicate species, a sample may have to be analyzed two or more times. The number of analyses, preparatory treatments, and manner of presentation of a sample to the x-ray diffractometer will depend upon the particular mineral assemblage in a sample.

To ensure detection of mineral species in a soil or clay sample by x-ray diffraction analysis, it is distinctly advantageous to concentrate the individual species as much as possible. Although it is impossible to physically separate mineral species from polycomponent systems, it is possible in many cases to concentrate species by fractionation of samples according to particle size.

Segregation of size-fractions in polycomponent mineral samples can be accomplished only if the samples are thoroughly dispersed. Dispersion is generally aided by chemical removal of flocculating and aggregate-cementing agents. In some cases, dispersion may be effected by mechanical and electronic means (section 15-2).

Great dependence is placed on diffraction from (00*l*) planes of phyllosilicate species, since diffraction intensities from these planes are normally highest per unit of clay. Both the distances between crystal layers and the intensity of diffraction from expansible phyllosilicates vary with the nature of the liquid present between layers. Preparation of unknown clay samples for x-ray analysis should always be designed to allow detection of expansible phyllosilicates that may be present.

Both film-recorded and chart-recorded x-ray patterns of a sample may be desirable if both types of equipment are available. Chart-recorded x-ray spectrometer analyses are generally sufficient for identification of species. A film-recorded x-ray pattern of a random powder sample is especially advantageous, however, for analyis of very small samples and for identification of species other than phyllosilicates. One or more patterns of oriented aggregates are taken with the direct-recording spectrometer for most efficient identification of phyllosilicate species. Although this sequence is most convenient, all analyses can be performed with either type of equipment. The procedures described are designed to allow the necessary sequence of analyses with either or both methods.

12-3.2 Sample Dispersion for Particle-Size Segregation

Soil samples are customarily subjected to specific chemical treatments prior to particle-size fractionation. Treatments to remove flocculating and aggregate-cementing agents are designed to enhance mineral particle dispersion and to improve the quality of x-ray diffraction from segregated samples. Soluble salts and polyvalent cations adsorbed on charged surfaces promote flocculation of soil colloids. Common soil aggregate-cementing agents include alkaline-earth carbonates, organic matter, oxides and hydroxy oxides of iron, and amorphous silica and alumina. Specific treatments for removal of these components are described in sections 5–2 to 5–4 and 15–2.

Although the referenced chemical pretreatments for soil mineral dispersion are accepted as standard in most laboratories, including those of the present authors, care must be exercised to avoid or minimize, in certain cases, deleterious effects of chemical treatments on soil mineral components. Douglas and Fiessinger (1971) have shown that the use of H_2O_2 for oxidation of soil organic matter in poorly buffered systems can cause the systems to become very acidic and thus promote degradation of particular clay mineral species. This result is consistent with the conclusion of Grim (1968) that several 2:1 clays are solubilized at low pH. Brewster (1980), in a study of the effect of chemical pretreatment on x-ray diffraction characteristics of clay minerals derived from volcanic ash, concluded that certain of the clay minerals in the soils examined were chemically altered by pretreatments. He felt that enhanced x-ray diffraction intensities and apparent increase in the amount of expandable clay minerals within a volcanic soil Ae horizon, for example, were probably due to hydrolysis and removal of interlayer aluminim when H_2O_2 and sodium citrate-bicarbonate-dithionite were used to remove organic matter and iron oxides, respectively. Harward and Theisen (1962) and Harward et al. (1962), in studies of Oregon soil clays, observed that clay minerals identified in a given sample may be dependent upon a number of factors, including method of iron removal, dispersion reagents, and cation saturation of clay.

Reports such as those cited suggest the need for (i) continued investigation of possible modifying effects of chemical pretreatments on mineral properties and mineralogical interpretation from x-ray diffraction analyses and (ii) minimizing to the extent possible the number and severity of chemical pretreatments.

The effectiveness of ultrasonic vibration of soil suspensions, with minimal or no chemical pretreatments, has been demonstrated for specific cases by a number of investigators in recent years (Edwards & Bremner, 1964; Carroll, 1970; Genrich & Bremner, 1972; Busacca et al., 1984; and others). Further testing may validate this method as effective for general applications.

Aside from their influence on mineral particle dispersion, soil aggregate-cementing agents may affect the quality of x-ray diffraction from segregated samples. Unless they are removed, these cementing agents may (i) act as diluents, thus reducing diffraction intensity of crystalline species present, (ii) prevent efficient orientation of phyllosilicate species during preparation of an oriented aggregate specimen, (iii) cause attenuation of the primary x-ray beam, and (iv) cause an increase in the general level of scatter of x-rays from the sample being analyzed.

Items (iii) and (iv) deserve special comment. It is a common observation that materials amorphous to x-rays cause nondirectional scatter of x-radiation from a sample, resulting in a general increase in background radiation. Such an effect lowers the ratio of diffraction maxima to background and thus decreases the sensitivity of an analysis. Organic matter, amorphous alumina, and amorphous silica produce this effect.

Removal of oxides of iron can improve diffraction intensity if copper radiation is employed. Iron will absorb copper radiation, the amount of absorption more or less depending upon the concentration of iron. This will reduce the intensity of both primary and diffracted radiation. As a result of the absorption, the iron in turn will fluoresce and emit its own characteristic x-radiation, which will be nondirectional and will add to the general background level. Crystal monochromators, however, available as standard accessories for recent x-ray diffractometers, effectively eliminate fluorescent radiation and general nondirectional scatter by allowing selective passage of diffracted K_α radiation to the x-ray detector.

12–3.3 Separation of Particle-Size Fractions

12–3.3.1 PRINCIPLES

Soil clay samples are normally separated into two or more subfractions before x-ray diffraction analysis. The customary subfractions are coarse (2–0.2 μm), medium (0.2–0.08 μm), and fine (<0.08 μm) clay. Separation of particles into two subfractions (2–0.2 μm and <0.2 μm) is generally sufficient for most purposes. Separation of fine clay (< 0.08 μm) from medium clay (2–0.2 μm) is sometimes advantageous for detection of relatively small quantities of smectites among other mineral species in clay, since smectites tend to concentrate in the smaller particle-size range.

The tendency of particular soil mineral species to occur within rather well-defined size limits was demonstrated by a number of investigators in early clay mineral studies (Marshall, 1930; Truog et al. 1937a, 1937b; Nagelschmidt, 1939; Jackson & Hellman, 1942; Jackson et al., 1948). According to Jackson and Sherman (1953), the tendency for specific colloidal mineral species to concentrate within specific size-fractions is a function of their resistance to weathering and of intensity of weathering. Minerals that are more resistant to chemical weathering, such as members of the smectite series, tend to persist in greater quantities in the finer clay fractions. Weathering susceptibility of less resistant minerals, such as feldspars and micas, results in their extinction before they reach fine-clay dimensions.

The intensity of diffraction from a particular mineral species is affected by a number of factors in addition to the concentration of the mineral. Included among these factors are crystal size and crystal perfection. Since finer particles within the clay-size fractions yield weaker diffraction intensities than larger particles of the same species, and since finer particles might be expected to exhibit crystal imperfections to a greater extent, it is especially important that each fraction be analyzed separately. It has been a common practice of many investigators to analyze, without further separation, the entire clay fraction (particles <2 μm diameter). Such practice may lead to serious error in identification of species present and in estimation of relative proportions of the various species. Analysis of various size fractions of a sample, rather than analysis of bulk samples, will often enable detection of small amounts of a mineral or minerals which may otherwise be obscured by the more abundant or more easily detected species present.

The procedures for particle dispersion and separation of sand fractions are given in section 15–4. Clay fractions may be physically separated from dispersed silt plus clay following the prescribed particle size analysis. Alternatively, separation of small quantities of clay fractions specifically for x-ray analysis may be accomplished using small quantities of soil and with elimination of several steps involved in complete fractionation.

Subdivisions within the clay fraction are accomplished by differential sedimentation of particles under centrifugal force. The reader is referred to Jackson (1969) for principles involved in separation of fractions with a tube centrifuge and with a supercentrifuge.

An integrated form of Stokes' law, as proposed by Svedberg and Nichols (1923), allows calculation of the time required for sedimentation of particles of particular diameter and specific gravity under centrifugal acceleration. The formula takes the following form for separation of particles in a tube centrifuge:

$$t=[n \log (R/S)]/[3.81\ N^2\ r^2\ (\Delta S)] \qquad [1]$$

where t is the time in seconds, R is the radius in centimeters of rotation of the top of the sediment in the tube, S is the radius in centimeters of rotation of the surface of the suspension in the tube, N is revolutions per second, r is the particle radius in centimeters, n is the viscosity of the liquid in poises at the existing temperature, and ΔS is the difference in specific gravity between the particles and the suspension liquid.

The proper flow rate, F, for separation of fractions with a continuous flow supercentrifuge can be calculated by the following formula (derivation of this formula is given in detail by Jackson, 1969):

$$F\ (\text{mL/min}) = [60A_wC_w\Delta S(D_L)^2]/18n \qquad [2]$$

where A_w is the area of the inner wall of the supercentrifuge bowl in square centimeters, C_w is centrifugal force in dynes, ΔS is the difference in density between the particles and the suspension liquid (g/cm^3), D_L is the particle limiting-diameter in micrometers, and n is the viscosity of the liquid at the existing temperature.

12-3.3.2 METHOD

12-3.3.2.1 Special Apparatus.

1. International no. 2 centrifuge (International Equipment Co., Boston, MA) equipped with a no. 240 head and 100-mL centrifuge tubes.
2. Centrifuge speed indicator.
3. Supercentrifuge (The Sharples Corp., Philadelphia, PA) equipped with a cellulose acetate liner (made by Eastman Kodak Co. and sold by Transilwrap Co., 281 W. Fullerton Ave., Chicago, IL).
4. Sample blender.
5. Time clock.

12-3.3.2.2 Reagents.

1. Sodium carbonate (Na_2CO_3) solution, pH 9.5.
2. Sodium chloride (NaCl).
3. Methanol, 95%.

12-3.3.2.3 Procedure.

Start with enough soil to ensure separation of an adequate amount of each clay fraction (150 mg of each is sufficient for all analyses usually performed). Remove the soluble salts, carbonates, organic matter, and free iron oxides as described in chapter 5. Separate the sand and coarse silt from the sample (as outlined below) to remove the bulk of the coarse fractions and facilitate subsequent separation of clay.

Transfer the pretreated soil to a 250-mL tall-form beaker (if the quantity of soil taken is large, distribute the soil between two or more beakers to limit the sediment volume and hence facilitate subsequent separation by sedimentation and decantation). Stir the suspension and allow it to stand for at least 2 min for each 5-cm depth. Make a mark on the side of the beaker with a wax pencil at a point exactly 5 cm above the top of the sediment in the bottom of the beaker. Decant the supernatant liquid (containing particles < 20 μm in diameter) into an appropriate vessel. Add Na_2CO_3 solution to the sediment in the beaker to just below the 5-cm mark, and stir the suspension thoroughly with a stirring rod equipped with a rubber policeman. When the suspension is thoroughly mixed, remove the stirring rod, wash the adhering mineral particles into the suspension with water from a wash bottle, and make the suspension to the 5-cm mark with a jet of water from the wash bottle. Allow the suspension to stand for about 2 min. (Determine the exact time at the existing temperature by calculation from Stokes' law or from the nomograph given by Jackson, 1969.) After the appropriate sedimentation time, decant the supernatant suspension into the vessel reserved for the $<$20-μm particles. Resuspend the sediment in Na_2CO_3 solution, stir the suspension thoroughly, and repeat the sedimentation and decantation until the supernatant liquid is practically free of suspended material. As many as 5 to 8 repetitions of the procedure may be required to effect nearly complete separation. (If only the clay fractions are desired, complete separation of particles at 20 μm is not necessary.)

Particles < 20 μm in diameter may be separated directly be centrifugation and decantation as described below. It is usually advantageous, however, first to reduce the suspension volume by allowing particles > 2 μm to sediment by gravity.

If the latter procedure is followed, allow the suspension containing <20-μm particles to stand for at least 8 h for each 10-cm depth. Decant the supernatant liquid containing suspended <2-μm particles into an appropriate vessel. The sediment will contain all >2-μm and some <2-μm particles. Resuspend the sediment in Na_2CO_3 solution and transfer the suspension to 100-mL centrifuge tubes, keeping the suspension depth exactly 10 cm. To facilitate subsequent separation, limit the depth of sediment in each tube (as determined after the first centrifugation) to approximately 1 cm. Centrifuge the suspension in an International no. 2 centrifuge at the appropriate speed and for the time required to allow all >2-μm particles to settle to the bottom of the tube. Calculate the speed and time with the aid of Eq. [1], or use the nomograph given by Jackson (1969). Decant the supernatant liquid into the vessel reserved for <2-μm particles. Resuspend the sediment in Na_2CO_3 solution, mix the suspension thoroughly, centrifuge it for the appropriate time, and decant the supernatant liquid. Repeat the procedure until separation is essentially complete, as indicated by constant turbidity of the supernatant liquid after successive centrifugations. From 5 to 8 repetitions of the resuspension, centrifugation, and decantation procedures are usually required. Gradual physical breakdown of silt particles during the process of resuspension prevents complete clearing of the supernatant liquid.

A preliminary separation of <0.08-μm particles may be desirable as a means of reducing suspension volume. If the suspension volume is less than about 2 L, however, it is usually more efficient to proceed with separation of >0.2-μm particles.

To start the separation at 0.2 μm, transfer the suspension of <2-μm clay to a group of 100-mL centrifuge tubes, keeping the depth of suspension in the tubes exactly 10 cm. Centrifuge the suspensions in the International no. 2 centrifuge for the appropriate time and speed to allow >0.2-μm particles to sediment to the bottom. Determine the time and speed by calculation from Eq. [1] or from the nomograph given by Jackson (1969). Decant the supernatant liquid containing <0.2-μm particles into an appropriate vessel. Continue the procedure until all the clay suspension has been collected in the tubes. Resuspend the sediment in the tubes in Na_2CO_3 solution, make the suspension to volume, and repeat the centrifugation and decantation until separation is essentially complete. From 5 to 8 repetitions are usually sufficient.

After the last decantation, transfer the sediment in the bottom of the tubes (coarse clay, 2–0.2 μm) to a small flask, and save it for x-ray analysis.

Because of the fine size of the particles remaining in suspension and the force required to effect sedimentation, the separation of <0.08-μm particles is accomplished with a supercentrifuge. Operation of the supercentrifuge and adjustment of flow rate are outlined in the Instruction Manual supplied by the manufacturer.

The suspension of medium and fine clay decanted in the previous separation is then placed in the feed reservoir of the supercentrifuge. Fit a wetted cellulose acetate liner around the interior of the clarifier bowl and install the bowl in the supercentrifuge. Place a 6-L flask under the upper delivery spout of the separator assembly and start the supercentrifuge. When the desired speed is attained, start the flow of suspension through the bowl and catch the effluent containing <0.08-μm particles in the collection flask. When nearly all the suspension has passed from the reservoir, wash down the sides of the reservoir with Na_2CO_3 solution from a wash bottle. When the wash solution has nearly drained away, add about 300 mL of Na_2CO_3 solution to the reservoir and allow it to flush out the suspension remaining in the bowl. When the effluent flow ceases, allow the centrifuge to coast freely to a stop, remove and open the bowl, and extract the cellulose acetate liner holding the sedimented particles. Carefully wash the sediment from the liner and from the interior of the bowl into a blender. Add Na_2CO_3 solution and mix the suspension in the blender for 1 to 2 min. Dilute the suspension to a concentration of 1% or less, and repeat the entire supercentrifugation. A total of 4 to 5 repetitions of the resuspension and centrifugation procedure allows essentially complete separation of particles at the 0.08-μm diameter limit. After the last centrifugation, collect the sedimented particles (medium clay, 0.2—0.08 μm), and save this fraction for x-ray analysis.

Add sufficient NaCl to the effluent from the supercentrifugation to render it approximately 1 *M* with respect to NaCl. Let the suspension stand until the clay flocculates and settles; then remove the clear supernatant liquid with a siphon or by careful decantation. Transfer the flocculated clay to 100-mL centrifuge tubes, and centrifuge them at 1500 rpm ($628 \times g$) for 5 min. Decant the supernatant liquid and wash the clay five times with 95% methanol to remove most of the NaCl. Save the clay for x-ray diffraction analysis.

12–3.4 Saturation of Exchange Complex

12– 3.4.1 PRINCIPLES

Since expansible phyllosilicates may retain different amounts of interlayer water, depending upon the nature of interlayer exchangeable cations (Barshad, 1950; Norrish, 1954; Mielenz et al., 1955), it is imperative that a clay sample prepared for diffraction analysis be homoionic to ensure that expansion as a result of hydration will be uniform within all crystals of a species. Also, since clay samples are commonly analyzed after drying in air, it is advisable to exchange-saturate the clay with a cation that will minimize changes in interlayer water adsorption due to fluctuations in relative humidity. Magnesium (Mg), which allows relatively uniform interlayer adsorption of water by expansible phyllosilicates, and potassium (K), which specifically restricts interlayer adsorption of water by vermiculite, are most commonly used for exchange-saturation (Hellman et al., 1943; MacEwan, 1946; Walker, 1957). Walker (1957, 1958) further suggested that differentiation of vermiculites from smectites is aided by pretreating clays with Mg.

Formation of a stable, two-layer water complex between layers of air-dried, Mg-saturated members of the smectite and vermiculite series results in an interatomic spacing of approximately 1.4 nm between (00*l*) planes. This spacing, which results from expansion of the layers, allows the differentiation of these species from the nonexpansible 2:1 phyllosilicates, the interatomic spacing of which is approximately 1.0 nm.

Detection of nonexpansible 1.4-nm chlorites (2:2 phyllosilicates) that may be present requires further diagnostic tests for positive differentiation from 1.4-nm vermiculite. Vermiculites will collapse to essentially non-expanded structures (1.0 nm) when saturated with K (Barshad, 1948), whereas most chlorites will be unaffected. The difference in behavior of these species when saturated with K is utilized as a means for their differentiation.

Following the early reports of Rich and Obenshain (1955), Tamura (1955), and Klages and White (1957) on the the occurrence of nonexchangeable interlayer aluminum (as hydroxy complexes) in some expansible soil clays, it has become generally accepted that hydroxy interlayers (Al, Fe, or Mg) are common in soil vermiculite and smectite species, depending upon the chemical environment in which they occur. The hydroxy interlayers, as islands between phyllosilicate layers, are generally randomly spaced as compared to continuous, fully polymerized hydroxy interlayers in chlorites. The reader is referred to Barnhisel (1977) for a comprehensive treatment of the subject of hydroxy interlayered minerals.

The presence of interlayer hydroxy complexes in vermiculites, in particular, complicates the differentiation of smectites from chlorites. Heating a K-saturated sample at 550°C for 2 h, however, will collapse discontinuous hydroxy complexes. The heat treatment will not affect most chlorites. The effects of heating of samples will be discussed in greater detail in section 12–5.

As many as two Mg-saturated samples and two K-saturated samples may be required for the complete succession of analyses, depending upon the mineral species present and the manner of presentation of the sample to the x-ray equipment. The procedure for exchange-saturation with Mg or K is the same for all preparations regardless of subsequent treatments. It is usually convenient to prepare sufficient Mg- and K-saturated samples to allow the maximum number of analyses that may be necessary.

12–3.4.2 METHOD

The method is essentially that given by Jackson (1969). The method described allows preparation of both random-power and oriented-aggregate specimens.

12–3.4.2.1 Reagents.

1. Hydrochloric acid (HCl), 0.1 *M*.
2. Magnesium chloride ($MgCl_2$), 0.5 *M* and 5 *M*.
3. Magnesium acetate [$Mg(OAc)_2$], 0.5 *M*.
4. Potassium chloride (KCl), 1 *M*
5. Silver nitrate ($AgNO_3$), 0.1 *M*.
6. Methanol, 50 and 95%.
7. Acetone, 95 and 100%.
8. Brom phenol blue indicator, 0.04% in ethanol.

12–3.4.2.2 Procedure. To saturate the clay with Mg, transfer an aliquot of dispersed clay suspension containing approximately 25 mg of clay to a beaker. Add 0.1 *M* HCl dropwise with stirring until the pH of the suspension is between 3.5 and 4.0, as determined by a spot-plate test with brom phenol blue indicator. [The suspension is acidified before addition of Mg solutions to prevent precipitation of $Mg(OH)_2$ if clay has been dispersed in highly alkaline dispersant.] Add sufficient 5 *M* $Mg(OAc)_2$ to render the suspension approximately 0.5 *M* with respect to Mg. Transfer the clay suspension into a 15-mL centrifuge tube, centrifuge the tube for 5 min at 1500 rpm ($628 \times g$), and decant and discard the clear supernatant liquid. Complete the exchange-saturation with Mg by washing the sample two times with 0.5 *M* $Mg(OAc)_2$, which is effective in removing H^+ ions from the acidified suspension, and then two times with 0.5 *M* $MgCl_2$. To effect each washing, add approximately 10 mL of the salt solution, mix the suspension thoroughly, centrifuge, and decant the supernatant liquid. After exchange-saturation is complete, remove excess salts from the sample by washing (centrifugation and decantation) once with 50% methanol, once with 95% methanol, and finally with 95% acetone until the clear decantate gives a negative test for chloride with 0.1 *M* $AgNO_3$. If the sample is to be analyzed as a random powder, transfer it to a large watchglass with acetone, and allow it to dry. Do not dry samples which are to be prepared as oriented aggregates.

To saturate a sample of clay with K, transfer an aliquot of dispersed clay suspension containing approximately 25 mg of clay to a 15-mL centrifuge tube. Add enough 1 *M* KCl to the suspension to flocculate the clay. Centrifuge the suspension and discard the supernatant solution. Wash the sample four times (centrifugation and decantation) with 1 *M* KCl to complete saturation of the clay with K (the suspension need not be acidified before treatment with KCl, since there is no danger of precipitation of K during the treatments).

Remove excess salts from the sample by washing once with 50% methanol, once with 95% methanol, and finally with 95% acetone until the decantate gives a negative test for chloride with $AgNO_3$.

If the sample is to be analyzed as a random powder, transfer it to a watchglass with acetone and allow it to dry. Do not dry the sample if it is to be analyzed as an oriented aggregate.

12–3.5 Solvation with Glycerol

12-3.5.1 PRINCIPLES

The similarity in basal spacing of Mg-saturated vermiculite and smectite species necessitates further differentiation to distinguish between the two groups of minerals. The ability of smectite members to adsorb double sheets of glycerol [$C_3H_5(OH)_3$] molecules between adjacent layers to yield a basal spacing of approximately 1.77 nm (MacEwan, 1944: Bradley, 1945; White and Jackson, 1947) is utilized as a means of differentiating the two groups. Solvation of vermiculite with glycerol does not materially change its interlayer expansion (Walker, 1950; Barshad, 1950; Brindley, 1966; Suquet et al., 1975)

Smectite readily forms complexes with many other polar molecules of moderate size, including monohydric alcohols (MacEwan, 1946); polyhydric alcohols, benzene, and ethers (Bradley, 1945); and amines and polyamines (Gieseking, 1939; Bradley, 1945). MacEwan and Wilson (1980) provide a thorough discussion of organic complexes with expansible phyllosilicates.

Among the many organic compounds which may be used for differentiation of expansible phyllosilicates, glycerol offers the maximum number of advantages. As summarized by MacEwan (1946), the main advantages of glycerol are given below.

1. Resolution of diffraction maxima:
   a. The intense, first-order, basal-diffraction maximum of glycerol-smectite (1.77 nm) is particularly well separated from other maxima likely to occur in clay diffraction patterns.
   b. None of the other glycerol-smectite basal-diffraction maxima interfere seriously with maxima from other likely minerals.
2. Stability of the complex:
   a. The spacings and intensities of the glycerol-smectite basal-diffraction maxima are unaffected by wide variations in water content.
   b. The extremely low volatility of glycerol causes the complex which it forms to be extremely stable.

12–3.5.2 METHOD (Jackson, 1969)

12–3.5.2.1 Reagents.

1. Benzene.
2. Glycerol.
3. Benzene–ethanol, 10.1 (vol/vol) and 200:1 (vol/vol).
4. Benzene–ethanol–glycerol, 1000:100:4.5 (vol/vol/vol)

12–3.5.2.2 Procedure. To prepare a glycerol-solvated, random-powder sample, add 10 mL of 20:1 benzene–ethanol to a Mg-saturated sample as prepared in section 12–3.4.2. Mix the suspension thoroughly, centrifuge it for 5 min at 1500 rpm (628 $\times$ *g*) and decant the supernatant liquid. Repeat the benzene–ethanol washing once. Wash the sample three times with 10-mL portions of the ternary mixture of benzene, ethanol, and glycerol to effect solvation of the clay with glycerol. Wash the sample once with 200:1 benzene–ethanol solution to remove most of the excess glycerol. Resuspend the sample in benzene and transfer the mixture to a large watchglass. After evaporation of the benzene, the sample is ready for analysis as a Mg-saturated, glycerol-solvated, random-powder sample (section 12–3.6).

To prepare a glycerol-solvated, oriented-aggregate sample, add approximately 2 mL of distilled water and 2 drops of glycerol to a Mg-saturated sample as prepared in section 12–3.4.2. Thoroughly mix the suspension to ensure complete dispersion. The sample is ready for mounting as an oriented aggregate (section 12–3.6).

Alternatively, a Mg-saturated sample may be mounted on a porous ceramic plate as an oriented aggregate (section 12–3.6) prior to solvation with glycerol. The mounted sample may then be glycerol-solvated by passing a 1:1 glycerol-water solution (two times) through the sample by suction.

12–3.5.2.3 Comment. The correct amount of glycerol to add to a sample being prepared as an oriented aggregate by the centrifuge method will depend upon the proportion of expansible phyllosilicates present and the particle-size range of the mineral particles. Two drops of glycerol/25-mg sample is sufficient without excess in most cases. The correct amount may have to be determined by trial and error, however. Direct solvation of a sample by suction through a porous ceramic plate essentially eliminates the uncertainty.

12–3.6 Mounting

12–3.6.1 PRINCIPLES

Two different methods of sample mounting are commonly used in diffraction analysis of clay samples: the sample is mounted either as a random powder or as an oriented aggregate. There are several ways in which a random powder may be mounted and at least two ways in which an oriented aggregate may be mounted. Both the random-powder and oriented-aggregate mounts have advantages for particular purposes.

12–3.6.2 PREPARATION OF SAMPLES

In a random-powder sample, in which crystals lie in all possible positions, there will be a sufficient number of crystals properly oriented to yield diffraction maxima from all atomic planes within the crystals. Different sets of crystals simultaneously register the diffraction maxima corresponding to all interatomic plane distances within each crystal species. The random sample, therefore, enables one to obtain all possible diffraction spacings from minerals present in a sample. This often assists in identification of species, since a family of diffraction maxima is obtained for each species. Relative intensities of diffraction maxima obtained from a random powder are more nearly proportional to the number of crystals present than are the maxima obtained from an oriented aggregate.

Preferential orientation of layer-silicate species, as attained in an oriented aggregate, results in enhancement of basal (00*l*) diffraction maxima. This often allows detection of small quantities of phyllosilicate species which might otherwise be obscured among more crystalline or more perfect diffracting species in the sample. MacEwan (1946) has reported detection of as little as 1% smectite in polycomponent mixtures using such a mount. The oriented mount has the further advantage that variation in basal spacings may be examined more critically and minor variations more easily detected. Preferential orientation of species, on the other hand, decreases the number of *hkl* planes in position to diffract x-rays.

The oriented-aggregate mount is especially advantageous when a direct-recording x-ray spectrometer is employed. If both film-recording and direct-recording equipment are available, the two types of mounts are sometimes used to advantage in conjunction with each other.

An oriented aggregate mounted on a porous ceramic plate offers some distinct advantages. Clay on a porous plate may be saturated with Mg or K or solvated with glycerol by applying successive portions of the appropriate solution to the surface and pulling the solution through the plate by suction. A single sample mounted on a porous plate can be Mg-saturated, glycerol-solvated, K-saturated, and heated by successive treatments and x-rayed after each treatment, thus eliminating the need for preparation of a separate mount for each treatment.

12–3.6.2.1 Rod Method for Random Orientation. Random powder samples are commonly formed into a rod 0.3 to 0.5 mm in diameter for analysis. Dried glycerol-solvated samples are slightly sticky and may be readily formed into a thin rod without addition of a binding substance. A binder usually must be added to nonsolvated samples before they can be formed into a rod, however. Gum tragacanth, which offers little interference to x-rays, is a suitable binding substance.

To form such a rod, mix a small amount of the dry clay sample with one-fifth to one-tenth its volume of powdered gum tragacanth, and add enough water to form a thick paste. After a few moments, shape the mixture into a rough rod with the fingers. Perfect the rod to the desired diameter by rolling it back and forth between two glass slides.

Although the above-described rod is the most commonly used form of cylindrical mount, alternative mounts may be employed. For example, the sample may be enclosed in a borosilicate glass capillary or coated on a gummed glass fiber (Henry et al., 1951).

12–3.6.2.2 Wedge Method for Random Orientation(Jackson, 1969). Powder samples are frequently mounted in a specially designed metal wedge (Jeffries and Jackson, 1949). To form the mount, place a small quantity of sample on a clean glass slide and gently push it into the recess of the wedge with a spatula or with the edge of another glass slide. Carefully smooth the surface of the mounted sample, so that it forms a sharp edge flush with the tip of the wedge.

12–3.6.2.3 Glass Plate Method for Oriented Aggregates (Jackson, 1969). Oriented-aggregate specimens may be formed directly from Mg-saturated, K-saturated, or Mg-saturated glycerol-solvated samples as prepared in section 12–3.4 or 12–3.5.

Add sufficient water to a sample to make a suspension of approximately 2-mL volume. Thoroughly mix the suspension to ensure complete dispersion. Extract the suspension with a pipette, and carefully transfer it to a glass microscope slide (2.6 by 4.6 cm) resting on a level surface. Add as much suspension to the slide as can be held by film tension. The total amount of clay per slide should be between about 15 and 25 mg. Allow the suspension to dry completely on the slide before the sample is analyzed.

12–3.6.2.4 Porous Ceramic Plate Method for Oriented Aggregates. An alternative method for preparation of oriented aggregates employs a porous ceramic plate as a mounting surface. Unglazed tile is commonly used for this purpose. The tile may be cut with a diamond or carborundum saw or scored with a diamond pencil and broken to appropriate dimensions (approximately 2.6 by 4.6 by 0.5 cm). The surface should be smoothed by wet-grinding on a glass plate with fine-grade corundum abrasive.

The reader is referred to Kinter and Diamond (1956) for specific details of the procedure and special apparatus needed for preparation of the porous-ceramic-plate-mounted aggregate. In general, the procedure involves the following:

A dilute suspension (containing between 15 and 25 mg of clay) is added to the surface of a porous ceramic plate. The liquid is separated from the clay by applied suction or by centrifugal force. After the liquid is removed through the plate, the clay remains oriented on the surface. Preferential orientation of phyllosilicates can often be increased by smoothing the surface of a moist mounted sample with a spatula (Theisen & Harward, 1962) or a glass slide held at an angle to the sample. Theisen and Harward and the present authors have found that the smoothing technique markedly increases the basal (00*l*) x-ray diffraction maxima from phyllosilicate species in many cases.

12–3.6.2.5 Preparation of Clay Films and Clay Rods from Salted Clay Pastes. An alternative procedure for analysis of clay samples involves preparation of clay films and clay rods from glycerol-ethanol-treated, salted clay pastes. The reader is referred to Barshad (1960) for specific details of the procedure for preparation of the pastes, clay films, and clay rods.

In general, the procedure for preparation of rods for analysis by film techniques involves extrusion of the salted paste from a hypodermic needle. For analysis by the diffractometer method, the clay paste may be spread on the surface of a microscope slide with a smooth-edged instrument.

The rods formed by extrusion from a hypodermic needle present both oriented and unoriented clay particles to the x-ray beam. Particles closest to the surface of the rods are preferentially oriented in the direction parallel to the axis of the rods, whereas those in the interior of the rods are unoriented.

12–4 X-RAY EXAMINATION OF SAMPLES

12–4.1 Special Apparatus

Commercial x-ray diffraction equipment includes a high-voltage generator, an x-ray source, an x-ray beam collimating system, and a detecting and recording system. Crystal monochrometers for selective discrimination of diffracted x-rays from nondirectional x-ray scatter are available for use with recent x-ray spectrometers. Film-recording systems require suitable x-ray film (no-screen medical x-ray film is most commonly used) and film-developing reagents and equipment.

12–4.2 Procedure

Clay samples prepared for analysis should be examined in the following order for most efficient identification and differentiation of species: (i) Mg-saturated, air-dried sample; (ii) Mg-saturated, glycerol-solvated sample; (iii) K-saturated, air-dried sample; and (iv) K-saturated, heated sample (550°C).

12–4.2.1 EXAMINATION WITH CYLINDRICAL CAMERA

Cylindrical cameras are equipped with a bracket for mounting either a cylindrical or a wedge-mounted sample. Film is loaded after mounting the sample in some cameras. In others, film is loaded before mounting the samples.

If a cylindrical sample is analyzed, carefully position the sample squarely in the path of the x-ray beam. If a wedge mount is employed, position the wedge so that the sample intercepts approximately two-thirds of the primary beam. Expose the sample to x-rays for sufficient time to yield a sharp, contrasting pattern of diffraction maxima. (The proper time of exposure will depend upon the radius of the camera used, the wavelength of radiation employed, the instrument voltage and current settings, and the chemical and mineralogical composition of the sample. Appropriate exposure times must be determined experimentally with each type of equipment.) After exposure, develop the x-ray film according to the procedure recommended by the film manufacturer.

12–4.2.2 EXAMINATION WITH DIRECT-RECORDING SPECTROMETER

Place a sample in the sample holder provided on the instrument, and position the goniometer to start its angular scan. The practical lower limit for 2θ is 2° for most analyses. The upper limit for 2θ is dictated by the geometry of the goniometer system (maximum allowable is 180°) and the mineral assemblage in the sample. Start the goniometer in synchronism with the chart recorder, and scan through the desired angular range.

12–4.3 Comments

The principal components which the analyst must consider for particular applications are the x-ray tube and the detecting and recording system.

The characteristic wavelength produced by an x-ray tube will depend on the metal used in forming the anode target. Metals commonly used for the purpose include molybdenum (Mo), copper (Cu), nickel (Ni), cobalt (Co), iron (Fe), and chromium (Cr). The wavelength of the characteristic radiation produced is different for each metal and is inversely related to atomic number of the metal. The choice of wavelength for diffraction analysis will be dictated by the minerals present in samples to be analyzed, the elemental composition of the samples, the resolution required, and the amount of absorption of the particular wavelength by air.

Chromium radiation (K_α, 0.228 nm) suffers relatively high absorption by the air that it must pass through between the source and the detector. Elements within a sample such as Al, Fe, and calcium (Ca), all present in relatively high concentrations in most soil or clay samples, also have relatively high absorption coefficients for Cr radiation. Absorption of the radiation in the air path or within the sample decreases the intensity of diffraction maxima obtained and when a monochrometer is not used, increases the general background level. The relatively long wavelength of Cr radiation, on the other hand, provides a high degree of resolution in a diffraction pattern. For a given interatomic spacing, the angle at which diffraction occurs will increase with the wavelength used, thus allowing a greater distance between recorded diffraction maxima.

Molybdenum radiation (K_α, 0.071 nm), in contrast to that of Cr, is little affected by absorption by air or by elements within a soil or clay sample. Offsetting these advantages, however, is the relatively poor resolution attainable with Mo radiation. This is of particular significance in detection and separation of maxima at low angles of diffraction.

Copper radiation (K_α, 0.154 nm), is almost universally used in diffraction work because it is not greatly affected in most cases by air or sample absorption, and it provides adequate resolution for most mineral analyses. Absorption of Cu radiation is significant, however, in samples with a high concentration of Fe. This problem can often be alleviated by selective removal of free iron oxides from samples (section 12–3.2) before x-ray analysis. Fluorescent radiation from Fe in a sample can be largely eliminated by passing diffracted x-rays through a monochrometer.

The most common detecting-recording systems used for mineral analysis are cylindrical film-loaded cameras and Geiger- or proportional-counter detectors with chart recorders. As mentioned in section 12–3.6, these two systems have their particular advantages for specific applications. If both systems are available, they may both be used to advantage in analysis of a sample.

12–5 CRITERIA FOR DIFFERENTIATION OF PHYLLOSILICATE SPECIES

Soils nearly always contain a number of mineral species. Many of the species can be identified easily by their distinctive diffraction maxima from a single x-ray pattern. Tables of diffraction spacings for crystalline substances (Anon., 1980) are adequate for identification of species other than phyllosilicates. Phyllosilicate species, however, have many similar or identical structural features which make their differentiation and identification more difficult. Differentiating analyses and criteria for identification of layer silicate species are discussed in detail by Warshaw and Roy (1961) and Brown and Brindley (1980). The compilation of mineral diffraction maxima presented by Brown and Brindley is sufficient for identification of mineral species likely to occur in clays.

X-ray diffraction criteria for differentiation of a few phyllosilicates common in soils are presented in Tables 12–1 and 12–2 for the purpose of illustrating effects of differentiating sample treatments (section 12–3.4 and 12–3.5). Approximate d/n values for diagnostic basal ($00l$) diffraction maxima are presented for ready reference. The effects reflected by treatments for diffraction of common individual phyllosilicates are further illustrated in Fig. 12–5.

As seen in Table 12–1, a diffraction spacing of approximately 1.4 nm obtained from a Mg-saturated, air-dried sample may be contributed by smectite, vermiculite, or chlorite or by a mixture of these species. Solvation with glycerol allows separation and positive identification of smectite. Saturation with K similarly allows separation of vermiculite from chlorite, which does not collapse.

Heating of a sample to 550°C serves two important functions. It effects collapse of vermiculite, which contains nonexchangeable interlayer hydroxy complexes, and it destroys the kaolin minerals. When chlorite is present in a sample, it normally yields a second-order maximum at nearly the same position as the first-order maximum of kaolinite (0.715 nm) or serpentine (0.71–0.73 nm). If a 0.715-nm maximum, obtained

Table 12-1. Diagnostic X-ray diffraction maxima obtained from (00*l*) planes of a few common phyllosilicates as related to sample treatment.

Diffraction spacing	Mineral (or minerals) indicated
nm	
	Mg-saturated, air-dried
1.4 -1.5	Smectite, vermiculite, chlorite
0.99–1.01	Mica (illite), halloysite
0.72–0.75	Metahalloysite
0.71–0.73	Serpentine
0.715	Kaolinite, chlorite (2nd-order maximum)
	Mg-saturated, glycerol-solvated
1.77–1.80	Smectite
1.4 -1.5	Vermiculite, chlorite
1.08	Halloysite
0.99–1.01	Mica (illite)
0.72–0.75	Metahalloysite
0.71–0.73	Serpentine
0.715	Kaolinite, chlorite (2nd-order maximum)
	K-saturated, air-dried
1.4 -1.5	Chlorite, vermiculite (with hydroxy interlayer)
1.24–1.28	Smectite
0.99–1.01	Mica (illite), halloysite, vermiculite (contracted)
0.72–0.75	Metahalloysite
0.71–0.73	Serpentine
0.715	Kaolinite, chlorite (2nd-order maximum)
	K-saturated, heated (550 °C)
1.4	Chlorite
0.99–1.01	Mica, vermiculite (contracted), smectite (contracted)
0.71–0.73	Serpentine
0.715	Chlorite (2nd-order maximum)

Table 12-2. X-ray diffraction maxima obtained from (00*l*) planes of binary, regularly alternating phyllosilicates as related to sample treatment.

Interstratified mixture	Mg-saturated, air-dried	Mg-saturated, glycerol-solvated	K-saturated, heated (550 °C)
	diffraction spacings (nm)		
Mica-vermiculite	2.4	2.4	1.0
Mica-chlorite	2.4	2.4	2.4
Mica-smectite	2.4	2.8	1.0
Vermiculite-chlorite	2.8	2.8	2.4
Vermiculite-smectite	2.8	3.2	1.0
Smectite-chlorite	2.8	3.2	2.4

from an unheated sample, disappears or decreases in intensity after heating at 550°C, the presence of kaolinite is confirmed.

A technique for differentiation of kaolinite from chlorite has been described by Andrew et al. (1960). Kaolinite may be expanded to 1.4 nm by intersalation with potassium acetate. Replacement of the interlayer potassium acetate with NH_4NO_3 results in a spacing of 1.16 nm for ka-

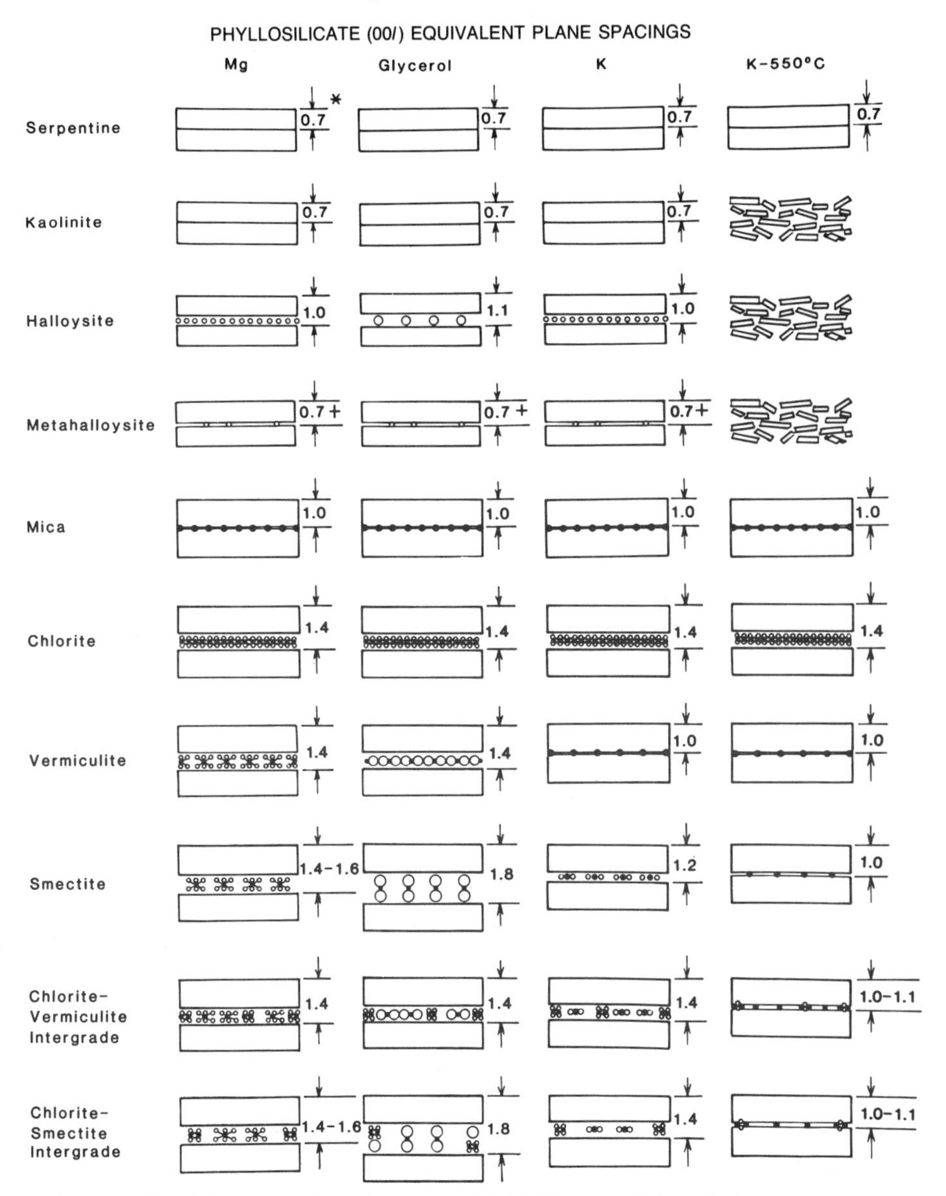

Fig. 12-5. Pictorial representation of response of phyllosilicates to differentiating treatments. *Approximate spacings (nm).

olinite. This spacing does not coincide with spacings of other common clay minerals.

Soil clays often contain interstratified mixtures of phyllosilicate species. Two or more species may be interstratified within a single crystal in either a regular or random manner. Diffraction effects from mixed crystals are quite different from those obtained from crystals containing only a single species. Diffraction from a few of the more common interstratified mixtures will be considered here.

Regular alternation of two species within a crystal produces repeating diffraction planes at a distance equal to the sum of the (00*l*) distances of the two species. Thus, a regular alternation of Mg-saturated smectite and mica, chlorite and mica, or vermiculite and mica yields a diffraction spacing of 2.4 nm (1.4 nm + 1.0 nm in each case). Regular alternation of smectite with either chlorite or vermiculite, or chlorite with vermiculite, yields a spacing of approximately 2.8 nm (1.4 nm + 1.4 nm in each case).

Individual species in regularly alternating binary mixtures can be identified by the succession of analyses of samples treated as prescribed in section 12–3. Species within each possible mixture will give the normal response to glycerol solvation, K-saturation, and heat treatment.

Two species randomly interstratified yield a well-defined spacing intermediate between the normal (00*l*) spacing of the two individual members (Hendricks and Teller, 1942; Reynolds, 1980). Randomly interstratified mica and vermiculite, for example, yield an "average" spacing between 1.0 and 1.4 nm. The exact spacing distance will depend upon the relative proportions of the two species in the mixture. A randomly interstratified mixture of vermiculite and chlorite yields a single spacing of 1.4 nm. Potassium saturation of such a sample will collapse the vermiculite to 1.0 nm and will leave the chlorite at 1.4 nm, resulting in an average spacing somewhere between 1.0 and 1.4 nm.

Other random binary combinations can be differentiated by the succession of analyses of samples treated as prescribed in section 12–3. The reader is referred to Reynolds (1980) for a detailed discussion of diffraction effects from randomly interstratified systems.

More complicated interstratified mixtures involving three or more species do occur in soil, but they are practically impossible to differentiate into individual component species. Such mixtures usually yield only broad, indistinct diffraction effects.

Diffraction maxima obtained from common regularly alternating interstratified binary mixtures are given in Table 12–2.

12–6 QUALITATIVE INTERPRETATION OF DIFFRACTION PATTERNS

Qualitative interpretation of diffraction patterns involves identification of crystalline species from the array of diffraction maxima obtained from a sample. Identification may be accomplished by either (i) direct comparison of diffraction patterns of unknown samples with patterns obtained from known minerals or (ii) measurement of diffraction spacings and comparison of these spacings with known spacings of standard minerals.

12–6.1 Direct Comparison

In the direct-comparison method, a film or chart-recorded pattern is compared, edge to edge, with patterns obtained from standard minerals. Lines on the pattern of the unknown that match lines on the known pattern, both in position and relative intensities, can be identified in this way. When more than one mineral is present, comparison with patterns of several known minerals may be necessary to identify all species present.

12–6.2 Measurement of Diffraction Spacings

When several minerals are present in a sample, identification of species is usually accomplished most easily and positively by determining the interatomic spacings giving rise to the various maxima and by comparing these with known spacings of minerals. Each maximum will represent either a specific interatomic distance, d, or a diffraction order, d/n, of a given distance.

The position of diffraction maxima on a pattern is determined by the geometry of the analyzing system, the wavelength of radiation used, and the distance between diffracting planes within crystals in the sample. In any particular analysis, the geometery of the system and the wavelength are fixed. Therefore, the angle of diffraction from diffracting planes will be directly dependent upon the interplanar distances. Interception of diffraction cones at a known distance from the sample provides a means for determination of diffraction angles and, hence, interplanar spacings.

Diffraction angles may be determined directly in terms of 2θ from a direct-recording spectrometer pattern. By reference to standard conversion tables, one may obtain the corresponding diffraction spacings.

Determination of diffraction angles from film patterns requires physical measurement of the position of each maximum on the film with respect to zero degrees θ (position of undiffracted primary beam) and conversion of these measured values to angular displacements.

Measurement of the distance of a diffraction maximum from the position of the undiffracted beam is done on a film placed in a flat position on an illuminated viewing device fitted with a ruler and a vernier cross-hair indicator. The exact center of the pattern is found by determining the midpoint between two corresponding maxima (representing arcs of a single diffraction cone) equidistant on both sides of the center. The distance between each maximum and the focus of the pattern is measured in centimeters. The diffraction angle, θ, corresponding to each maximum can then be calculated directly by the equation $\theta = 360°L/2\pi R$, where R represents the radius of the camera in centimeters. Interatomic spacings, d/n, can then be obtained by reference to standard conversion tables or by calculation from Bragg's equation.

12–6.3 Identification of Mineral Species

Both the diffraction spacings and the intensity of each maximum are recorded to enable one to identify species. Intensities of maxima obtained with an x-ray spectrometer may be recorded directly as counts per second or measured on a chart-recording as height above background. Maxima from photographically recorded patterns are usually measured visually and assigned relative intensities as very strong, strong, medium, weak, or very weak.

Identification of some of the more common phyllosilicates can be accomplished by comparing diffraction maxima obtained from Mg-saturated, glycerol-solvated, K-saturated and heated samples with diagnostic maxima tabulated in section 12–5. Identification of other mineral species may require reference to the tables of x-ray maxima compiled by Brown and Brindley (1980) or to the Hanawalt or Fink indexes in the *Mineral Powder Diffraction File* (Anon. 1980). Use of the Hanawalt and Fink indexes are explained in the latter publication.

12–7 QUANTITATIVE INTERPRETATION OF DIFFRACTION PATTERNS

The intensities of diffraction maxima are related to the number of corresponding diffraction planes in a sample. Thus, the relative intensities of maxima theoretically provide a basis for estimation of concentrations of mineral species present. There are a number of factors relative to the physical and chemical nature of samples, however, which can also greatly influence diffraction intensities and hence the validity of quantitative estimation of mineral species. These factors will be considered briefly before discussion of methods for quantitative estimation of diffraction patterns, to acquaint the reader with some of the difficulties inherent in the methods.

According to Jackson (1969), the principal factors that influence diffraction intensities, in addition to concentration of species, are particle size, crystal perfection, chemical composition, variations in sample packing, crystal orientation, and presence of amorphous substances. Hydroxy interlayering, a common factor in vermiculites and smecitites in particular environments (Barnhisel, 1977), will also influence diffraction intensities. Klages and Hopper (1982) group sources of errors encountered in quantitative clay mineral estimations into three categories: those related to clay minerals themselves, those related to sample material and those related to procedures.

When particles are greater than about 10 μm in diameter, x-ray patterns become spotty in character because there are not sufficient crystals in all possible orientations to yield continuous diffraction-cones. If particles are less than about 0.02 μm in diameter, the diffraction maxima become broad and diffuse. Diffraction intensity decreases simply as a function of very fine particle size in this range. Soil clays can be expected to have a relatively wide range in particle size, and the size of a particular species may be different in different samples.

The degree of crystal imperfection can vary among members of a species, and it depends upon conditions of formation of the mineral, conditions of weathering of its mineral grains, or both. Any discontinuities due to crystal imperfections will reduce x-ray intensities. Hydroxy interlayers in some smectities and vermiculites may be considered, in a broad context, as special examples of crystal imperfections.

Phyllosilicate clays frequently occur as interstratified mixtures. In such mixtures there may be too few planes of one species in any one zone to allow detection or to be represented by proportional diffraction intensity (Jackson et al., 1952).

The chemical composition of a sample can affect diffraction intensities in two principal ways. The presence in a sample of an element or elements capable of absorbing the wavelength of radiation employed for the analysis will reduce diffraction intensities. Diffraction intensities from a particular species will be affected also by the constituent elements of the species, since atoms vary in their ability to scatter x-rays.

The density of packing of a powder sample will also affect the intensity of diffracted rays. In general, the more dense (tightly packed) a sample is, the more planes will contribute to diffraction. If a sample is too tightly packed, however, the aborption of both primary and diffracted rays increases.

The degree of orientation or, on the other hand, the degree of randomness of the crystals of a powder will cause a variation in diffraction intensity independent of the quantity of a mineral species present.

The presence of amorphous substances, acting as nondiffracting diluents, will also lead to an erroneous estimation of the mineral composition of a sample. If the amorphous substances cannot be detected, one is apt to conclude that the sample is composed entirely of the detectable crystalline components.

If it were possible to hold each of these variables constant, or if one were able to properly evaluate the influence of these variables, a precise quantitative estimation of species would be possible. Unfortunately, difficulties in controlling all ameliorating factors in many cases preclude acceptance at this time of a standard, generally applicable method for quantitative estimation of minerals from x-ray diffraction analysis. Quantitative x-ray mineralogy is presently in a state of flux. Improvements in analytical techniques in recent years have significantly enhanced estimations of mineral contents in specific instances. The reader is referred to Brindley (1980) for a comprehensive review of quantitative x-ray mineral analysis of clays.

A few examples of techniques designed to compensate for factors which complicate mineral quantification are presented here.

Thomson et al. (1972) described a reproducible method of specimen preparation to avoid preferential orientation of clay minerals with platy morphology. The method for preparing random powder mounts involves mixing clay with polyester foam in a vibrating ball mill. The clay coats the honeycomb-like structure of the polyester foam in a random manner.

Other techniques for preparation of randomly oriented specimens for x-ray diffraction analysis include imbedding clay in thermoplastic cement (Bridley & Kurtossy, 1961), end-loading specimens in a sample holder (Niskanen, 1964), spray-drying specimens on a sample mount (Jonas & Kuykendall, 1966; Hughes & Bohor, 1970), and mechanical vibration during filling of a sample holder (Brindley & Wardle, 1970).

The use of an internal standard is an effective means of compensating for diffraction intensity variations caused by x-ray absorption within a sample. The internal standard method is particularly amenable for quantitative estimates of mineral species that have little or no variability in elemental composition and crystallinity, that are not components of interstratified systems, and that do not contain hydroxy interlayers.

A number of different internal standards have been employed for this purpose. Corundum is commonly accepted as a suitable internal standard for most applications (Brindley, 1980). In use of an internal standard, a known quantity of some mineral not already present in a sample is added and thoroughly mixed with each standard mixture. The mixtures are analyzed, and the intensities of diffraction maxima of the added internal standard mineral and of each component mineral in each standard mixture are measured. The ratio of the diffraction intensity of each mineral of the standard mixture to the diffraction intensity of the internal standard is plotted against the weight ratio of standard mineral to internal standard for each case. The internal standard is then added to each test sample in the same weight ratio as in the standard mixtures, and the test mixtures are analyzed. The diffraction-intensity ratios (ratios of intensity of mineral being analyzed to intensity of internal standard) are determined and compared to the plot obtained for the standard-mineral internal-standard mixtures to obtain the weight ratio of mineral to internal standard. Finally, by multiplying the weight of internal standard added to the test sample by the weight ratio of mineral to internal standard, the weight of the mineral in the sample is obtained. Knowing the total weight of the test sample, the percentage of the determined mineral (or minerals) can be calculated.

The internal standard chosen should have absorption characteristics similar to the sample. If the standard is properly chosen, absorption effects will be the same for the internal standard and the component minerals in the sample.

In specific cases, where variations in diffraction intensities from the cited causes can be considered negligible, quantitative estimations of clay mineral percentages may be made by comparison of intensities of diffraction maxima contributed by minerals in a sample to diffraction intensities obtained from the same species in artificial standard mixtures (Willis et al., 1948; Talvenheimo & White, 1952). By this method, standard mineral mixtures composed of two or more minerals in various proportions are prepared and analyzed. Calibration curves equating diffraction intensities obtained from component minerals in the mixtures to known concentrations of the minerals are then prepared. Quantitative estimates of minerals in test samples are made by measuring the intensities of their diffraction maxima and by comparing these intensities to those obtained from known concentrations in the prepared standards.

Ruhe and Olson (1979) have described a variation of the method of additions for estimating mineral composition of soil clay samples. The Ruhe and Olson method involves additions of increasing proportions of soil and clay to a standard composed of equal weights of ternary mixtures of kaolinite, illite, and montmorillonite. Intensity ratios for species in the standard mixture, in an unknown sample and in a prepared mixture containing a proportion of unknown, allow calculation of species content in the unknown.

Measurements of x-ray attenuation coefficients for samples and appropriate corrections of x-ray maxima intensities have also been employed to improve mineral quantification by x-ray diffraction analysis (Brindley, 1980).

In summary, quantitative analysis of minerals by x-ray diffraction can be reliable in some cases, but generally speaking, variations in chemical composition, crystal perfection, amorphous substances, and particle size are very difficult to evaluate and compensate for in analysis of soil or clay samples. In most cases, estimation of mineral percentages from x-ray diffraction patterns is only semiquantiative at best. For the most reliable and accurate estimation, the use of x-ray diffraction analysis in conjunction with other methods such as differential-thermal, integral-thermal, infrared, surface-area, elemental analysis, and other species-specific chemical methods (Alexiades & Jackson, 1966) is advisable.

12–8 REFERENCES

Alexiades, C. A., and M. L. Jackson. 1965. Quantitative clay mineralogical analysis of soils and sediments. Clays Clay Miner. 14:35–52.

Andrew, R. W., M. L. Jackson, and K. Wada. 1960. Intersalation as a technique for differentiation of kaolinite from chloritic minerals by x- ray diffraction. Soil Sci. Soc. Am. Proc. 24:422–423.

Anonymous. 1980. Mineral Powder Diffraction File Search Manual. JCPDS International Centre for Diffraction Data, Swarthmore, PA.

Auleytner, J. 1967. X- ray methods in the study of defects in single crystals. Pergamon Press, Oxford.

Barnhisel, R. I. 1977. Chlorites and hydroxy interlayered vermiculite and smectite. p. 331–356. *In* J. B. Dixon and S. B. Weed (ed.). Minerals in soil environments. Soil Sci. Soc. Am. Monogr. Soil Science Society of America, Madison, WI.

Barshad, I. 1948. Vermiculite and its relation to biotite as revealed by base exchange reactions, x-ray analysis, differential thermal curves, and water content. Am. Mineral. 33:655–678.

Barshad, I. 1950. The effect of the interlayer cations on the expansion of the mica-type crystal lattice. Am. Mineral. 35:225–238.

Barshad, I. 1960. X-ray analysis of soil colloids by a modified salted paste method. Clays Clay Miner. 5:350–364.

Bradley, W. F. 1945. Molecular associations between montmorillonite and some polyfunctional organic liquids. J. Am. Chem. Soc. 67:975–981.

Brewster, G. R. 1980. Effects of chemical pretreatment on x-ray powder diffraction characteristics of clay mienrals derived from volcanic ash. Clays Clay Miner. 28:303–310.

Brindley, G. W. 1966. Ethylene glycol and glycerol complexes of smectites and vermiculites. Clay Miner. 6:237–260.

Brindley, G. W. 1980. Quantitative x-ray mineral analysis of clays. p. 411–438. *In* G. W. Brindley and G. Brown (ed.). Crystal structures of clay minerals and their X-ray identification. Mineral. Soc. Monogr. 5. Mineralogical Society, London.

Brindley, G. W., and S. S. Kurtossy. 1961. Quantitative determination of kaolinite by x-ray diffraction. Am. Mineral. 46:1205–1215.

Brindley, G. W., and R. Wardle. 1970. Monoclinic and triclinic forms of pyrophyllite and pyrophyllite anhydride. Am. Mineral. 55:1259–1272.

Brown, G., and G. W. Brindley. 1980. X-ray diffraction procedures for clay mineral identification. p. 305–359. *In* G. W. Brindley and G. Brown (ed.). Crystal structures of clay minerals and their X-ray identification. Mineral. Soc. Monogr. 5. Mineralogical Society, London.

Buerger, M.J. 1942. X-Ray crystallography. John Wiley and Sons, Inc., New York.

Busacca, A. J., J. R. Aniku, and M. J. Singer. 1984. Dispersion of soils by an ultrasonic method that eliminates probe contact. Soil Sci. Soc. Am. J. 48:1125–1129.

Carroll, D. 1970. Clay minerals: a guide to their x-ray identification. Geol. Soc. Am. Spec. Paper 126.

Clark, G. L. 1955. Applied x-rays. McGraw-Hill Book Co., Inc., New York.

Cullity, B.D. 1978. Elements of x-ray diffraction. Addison-Wesley Publ. Co., Reading, MA.

Douglas, L. A., and F. Fiessinger. 1971. Degradation of clay minerals by H_2O_2 treatments to oxidize organic matter. Clays Clay Miner. 19:67–68.

Edwards, A. P., and J. M. Bremner. 1964. Use of sonic vibration for separation of soil particles. Can. J. Soil Sci. 44:366.

Genrich, D. A., and J. M. Bremner. 1972. A reevaluation of the ultrasonic vibration method of dispersing soils. Soil Sci. Soc. Am. Proc. 36:944–947.

Gieseking, J. E. 1939. The mechanism of cation exchange in the montmorillonite-beidellite-nontronite type of clay minerals. Soil Sci. 47:1–13.

Grim, R. E. 1968. Clay mineralogy, 2nd ed. McGraw-Hill, New York.

Hadding, A. 1923. Eine röntgenographische Methode kristalline und kryptokristalline Substanzen zu identifizieren. Z. Kristallogr. 58:108– 112.

Harward, M. E., and A. A. Theisen. 1962. Problems in clay mineral identification by x-ray diffraction. Soil Sci. Soc. Am. Proc. 26:335–341.

Harward, M. E., A. A. Theisen, and D. D. Evans. 1962. Effect of iron removal and dispersion methods on clay mineral identification by x-ray diffraction. Soil Sci. Soc. Am. Proc. 26:535–541.

Hellman, N. N., D. G. Aldrich, and M. L. Jackson. 1943. Further note on an x-ray diffraction procedure for the positive differentiation of montmorillonite from hydrous mica. Soil Sci. Soc. Am. Proc. 7:194–200.

Hendricks, S. B., and W. H. Fry. 1930. The results of x-ray and microscopical examinations of soil colloids. Soil Sci. 29:547–580.

Hendricks, S. B., and E. Teller. 1942. X-ray interference in partially ordered layer lattices. J. Chem. Phys. 10:147–167.

Henry, N. F. M., H. Lipson, and W. A. Wooster. 1951. The interpretation of x-ray diffraction photographs. Macmillan and Co., Ltd., London.

Hughes, R., and B. Bohor. 1970. Random clay powders prepared by spray drying. Am. Mineral. 55:1780–1786.

Jackson, M. L. 1969. Soil chemical analysis—advanced course. Publ. by author, Dep. of Soils, Univ. of Wisconsin, Madison, WI.

Jackson, M. L., and N. N. Hellman. 1942. X-ray diffraction procedure for positive differentiation of montmorillonite from hydrous mica. Soil Sci. Soc. Am. Proc. 6:133–145.

Jackson, M. L., Y. Hseung, R. B. Corey, E. J. Evans, and R. C. Vanden Heuvel. 1952. Weathering sequence of clay-size minerals in soils and sediments: II. Chemical weathering of layer silicates. Soil Sci. Soc. Am. Proc. 16:3–6.

Jackson, M. L., and G. D. Sherman. 1953. Chemical weathering of minerals in soils. Adv. Agron. 5:219–318.

Jackson, M. L., S. A. Tyler, A. L. Willis, G. A. Bourbeau, and R. P. Pennington. 1948. Weathering sequence of clay size minerals in soils and sediments. I. Fundamental generalizations. J. Phys. Colloid Chem. 52:1237–1260.

Jeffries, C. D., and M. L. Jackson. 1949. Mineralogical analysis of soils. Soil Sci. 68:57–73.

Jonas, E. C., and J. R. Kuykendall. 1966. Preparation of montmorillonites for random powder diffraction. Clay Mineral. 6:232–236.

Kelley, W. P., W. H. Dore, and S. M. Brown. 1931. The nature of the base exchange material of bentonite, soils, and zeolites, as revealed by chemical investigations and x-ray analysis. Soil Sci. 31:25–55.

Kinter, E. B., and S. Diamond. 1966. A new method for preparation and treatment of oriented-aggregate specimens of soil clays for x-ray diffraction analysis. Soil Sci. 81:111–120.

Klages, M. G., and R. W. Hopper. 1982. Clay minerals in northern plains coal overburden as measured by x-ray diffraction. Soil Sci. Soc. Am. J. 46:415–418.

Klages, M. G., and J. L. White. 1957. A chlorite-like mineral in Indiana soils. Soil Sci. Soc. Am. Proc. 21:16–20.

Klug, H. P., and L. E. Alexander. 1974. X-ray diffraction procedures for polycrystalline and amorphous materials, 2nd ed. John Wiley & Sons, Inc., New York.

MacEwan, D. M. C. 1944. Identification of the montmorillonite group of minerals by x-rays. Nature (London) 154:577–578.

MacEwan, D. M. C. 1946. The identification and estimation of the montmorillonite group of minerals, with special reference to soil clays. J. Soc. Chem. Ind. 65:298–305.

MacEwan, D. M. C., and M. J. Wilson. 1980. Interlayer and intercalation complexes of clay minerals. p. 197–248. *In* G. W. Brindley & G. Brown (ed.). Crystal structures of clay minerals and their X-ray identification. Mineral. Soc. Monogr. 5. Mineralogical Society, London.

Marshall, C. E. 1930. The orientation of anisotropic particles in an electric field. Trans. Faraday Soc. 26:173–189.

Mielenz, R. C., N. C. Schieltz, and G. E. King. 1955. Effects of exchangeable cation on x-ray diffraction patterns and thermal behavior of a montmorillonite clay. Clays Clay Miner. 3:146–173.

Nagelschmidt, G. 1939. The identification of minerals in soil colloids. J. Agric. Sci. 29:477–501.

Niskanen, E. 1964. Reduction of orientation effects in the quantitative X-ray diffraction analysis of kaolin minerals. Am. Mineral. 49:705–714.

Norrish, K. 1954. The swelling of montmorillonite. Disc. Faraday Soc. 18:120–134.

Reynolds, R. C. 1980. Interstratified clay minerals. p. 249–303. *In* G. W. Brindley and G. Brown (ed.). Crystal structures of clay minerals and their x-ray identification. Mineral. Soc. Monogr. 5. Mineralogical Society, London.

Rich, C. I., and S. S. Obenshain. 1955. Chemical and mineral properties of a red-yellow podzolic soil derived from sericite schist. Soil Sci. Soc. Am. Proc. 19:334–339.

Rinne, F. 1924. Röntgenographische Untersuchungen an einigen feinzerteilten Mineralien. Kunstprodukten and Dichten gesteinen. Z. Kristallogu. 60:55–69.

Ruhe, R. V., and C. G. Olson. 1979. Estimate of clay-mineral content: additions of proportions of soil clay to constant standards. Clays Clay Miner. 27:322–326.

Suquet, H., C. de la Calle, and H. Pezerat. 1975. Swelling and structural organization of saponite. Clays Clay Miner. 23:1–9.

Svedberg, T., and J. B. Nichols. 1923. Determination of size and distribution of size by centrifugal methods. J. Am. Chem. Soc. 45:2910–2917.

Talvenheimo, G., and J. L. White. 1952. Quantitative analysis of clay minerals with the x-ray spectrometer. Anal. Chem. 24:1748–1789.

Tamura, T. 1955. Weathering of mixed-layer clays in soils. Clays Clay Miner. 4:413–422.

Theisen, A. A., and M. E. Harward. 1962. A paste method for preparation of slides for clay mineral identification by x-ray diffraction. Soil Sci. Soc. Am. Proc. 26:90–91.

Thomson, A. P., D. M. L. Duthie, and M. J. Wilson. 1972. Randomly oriented powders for quantitative x-ray determination of clay minerals. Clay Miner. 9:345–348.

Truog, E., J. R. Taylor, R. W. Pearson, M. E. Weeks, and R. W. Simonson. 1937a. Procedure for special type of mechanical and mineralogical soil analysis. Soil Sci. Soc. Am. Proc. 1:101–112.

Truog, E., J. R. Taylor, R. W. Simonson, and M. E. Weeks. 1937b. Mechanical and mineralogical subdivision of the clay separate of soils. Soil Sci. Soc. Am. Proc. 1:175–179.

Walker, G. F. 1950. Vermiculite-organic complexes. Nature (London) 166:695–697.

Walker, G. F. 1957. On the differentiation of vermiculites and smectites in clays. Clay Miner. Bull. 3:154–163.

Walker, G. F. 1958. Reactions of expanding-lattice clay minerals with glycerol and ethylene glycol. Clay Miner. Bull. 3:302–313.

Warshaw, C. M., and R. Roy. 1961. Classification and a scheme for the identification of layer silicates. Bull. Geol. Soc. Am. 72:1455–1492.

White, J. L., and M. L. Jackson. (1947) Glycerol solvation of soil clays for X-ray diffraction analysis. Soil Sci. Soc. Am. Proc. 11:150–154.

Willis, A. L., R. P. Pennington, and M. L. Jackson. 1948. Mineral standards for quantitative x-ray diffraction analysis of soil clays. Soil Sci. Soc. Am. Proc. 12:400–406.

Wormald, J. 1973. Diffraction methods. Oxford University Press, London.

13 Bulk Density[1]

G. R. BLAKE

University of Minnesota
St. Paul, Minnesota

K. H. HARTGE

University of Hanover
Hanover, Federal Republic of Germany

13–1 GENERAL INTRODUCTION

Soil *bulk density,* ρ_b, is the ratio of the mass of dry solids to the bulk volume of the soil. The bulk volume includes the volume of the solids and of the pore space. The mass is determined after drying to constant weight at 105 °C, and the volume is that of the sample as taken in the field.

Bulk density is a widely used value. It is needed for converting water percentage by weight to content by volume, for calculating porosity and void ratio when the particle density is known, and for estimating the weight of a volume of soil too large to weigh conveniently, such as the weight of a furrow slice or an acre-foot.

Bulk density is not an invariant quantity for a given soil. It varies with structural condition of the soil, particularly that related to packing. For this reason it is often used as a measure of soil structure. In swelling soils it varies with the water content (Hartge, 1965, 1968). In such soils, the bulk density obtained should be accompanied by the water content of the soil at the time of sampling.

The determination usually consists of weighing and drying a soil sample, the volume of which is known (core method) or must be determined (clod method and excavation method). These methods differ in the way the soil sample is obtained and its volume determined. A different principle is employed with the radiation method. Transmitted or scattered gamma radiation is measured; and with suitable calibration, the density of the combined gaseous-liquid-solid components of a soil mass is determined. Correction is then necessary to remove the components of density attributable to liquid and gas that are present. The radiation method is an in situ method.

---

[1]Paper no. 11718 of the Scientific Journal Series, Minnesota Agricultural Experiment Station, St. Paul, MN.

Clod and core methods have been used for many years. Excavation methods were developed in recent years, chiefly by soil engineers for bituminous and gravelly material. More recently the excavation method has found use in tillage research, or where surface soil is often too loose to allow core sampling, or where abundant stones preclude the use of core samplers. Radiation methods have been used since the 1950s, particularly in soil engineering.

Bulk density is expressed in SI units or units derived from them. The most straightforward would be kg m^{-3}. However, derived units such as tons m^{-3}, g cm^{-3}, or Mg m^{-3}, which are numerically equal to each other, may be more convenient, as they give values for soils which vary from about 1.2 to 1.7 (rather than from 1200 to 1700, as when units of kg m^{-3} are used). Obsolete terms such as "volume weight" (weight · volume^{-1}) and "bulk specific gravity" or "apparent specific gravity" are sometimes found in the older literature and in some foreign language literature. Specific gravity terms are relative densities, i.e. density of a substance with respect to water at 4°C, and are nearly equal numerically to bulk density. At standard gravitation (g = 9.8 m s^{-2}), kilogram weight and kilogram mass are equal, and under this condition "volume weight" is numerically equal to bulk density. In many engineering and commercial applications, bulk density is expressed in lb ft^{-3}, which one may convert to g cm^{-3} by dividing by 62.4 (which is the mass, in pounds, of a cubic foot of a substance whose density is unity, i.e., water at 4 °C).

13–2 CORE METHOD

13–2.1 Introduction

With this method, a cylindrical metal sampler is pressed or driven into the soil to the desired depth and is carefully removed to preserve a known volume of sample as it existed in situ. The sample is dried to 105 °C and weighed. The core method is usually unsatisfactory if more than an occasional stone is present in the soil.

13–2.2 Method

Core samplers vary in design from a thin-walled metal cylinder to a cylindrical sleeve with removable sample cylinders that fit inside. Samplers are usually designed not only to remove a relatively undisturbed sample of soil from a profile, but also to hold the sample during transport and eventually during further measurements in the laboratory, such as pore-size distribution or hydraulic conductivity. For the latter measurement it is desirable to have core diameters not less than 75 mm and preferably 100 mm to minimize the effect of disturbed soil interfacing the cylinder wall. For the same reason it is desirable that the height of the cylinder not exceed the diameter.

A widely used and very satisfactory sampler consists of two cylinders fitted one inside the other. The outer one extends above and below the inner to accept a hammer or press at the upper end and to form a cutting edge at the lower. The inside cylinder is the sample holder. The inside diameters of the two cylinders when nested are essentially the same at the lower end, the inner being fitted against a shoulder cut on the inner surface of the outside cylinder. Figure 13–1 shows such a sampler (available in slightly different design from the Utah State University Technical Services, UMC 12, Logan, UT 84322).

Where densities at various depths in a soil profile are to be determined, one can obtain samples with a hydraulically driven probe mounted on a pickup truck, tractor, or other vehicle. The probe is forced into the soil and removed hydraulically. The probe tube has a 2- to 3-cm wide slit running most of the length of the tube, through which one can insert a rounded knife or spatula to slice off segments of the soil as desired. Segments typically 10 cm in length are cut and removed from the tube and placed in containers for transport to the laboratory. Depending on the probe model, samples can be taken to about 1-m depth, though extensions for greater depths are available for many models. Probe samplers are available from Giddings Machine Co., P.O. Drawer 2024, 401 Pine St., Fort Collins, CO 80522; A. D. Bull Enterprizes, 1904 South 21st Street, Chickasha, OK 73018; or Soiltest Inc., 2205 Lee St., Evanston, IL 60202.

Fig. 13–1. Typical double-cylinder, hammer-driven core sampler, for obtaining soil samples for bulk density.

Numerous hand-driven samplers have been described in the literature. Some of the more accessible ones are described by Lutz (1947), Jamison et al. (1950), and U.S. Department of Agriculture (1954, p. 159). McIntyre (1974) describes types of core samples and their properties and gives additional references.

13–2.2.1 PROCEDURE

The exact procedure for obtaining the samples depends on the kind of sampler used. The following steps apply when the widely known double-cylinder sampler is used.

Drive or press the sampler into either a vertical or horizontal soil surface far enough to fill the sampler, but not so far as to compress the soil in the confined space of the sampler. Carefully remove the sampler and its contents so as to preserve the natural structure and packing of the soil as nearly as possible. A shovel, alongside and under the sampler, may be needed in some soils to remove the sample without disturbance. Separate the two cylinders, retaining the undisturbed soil in the inner cylinder. Trim the soil extending beyond each end of the sample holder (inner cylinder) flush with each end with a straight-edged knife or sharp spatula. The soil sample volume is thus established to be the same as the volume of the sample holder. In some sampler designs, the cutting edge of the sampler has an inside diameter slightly less than the sample holder, so as to reduce friction as the soil enters the holder. In these cases, determine the diameter of the cutting head and use this to calculate the sample holder volume. Transfer the soil to a container, place it in an oven at 105 °C until constant weight is reached, and weigh it. The bulk density is the oven-dry mass of the sample divided by the sample volume.

13–2.2.2 COMMENTS

It is not necessary that soil be kept undisturbed during transport to the laboratory and drying. A single sample cylinder can be reused if each sample is transferred to another container. It is often desired, however, to make other measurements such as pore-size distribution, conductivity, or water retention in addition to bulk density on the same samples. These require that they be kept undisturbed, each sample being transported in the sample cylinder in which it was taken. Thus one must provide for sufficient cylinders. Frequently other measurements to be made in the laboratory require that samples be kept at field water-content. In that case cylinders must be placed in containers that do not permit loss of water during transport. Waxed paper or plastic containers with lids are satisfactory for this purpose.

Core samples should be taken in soils of medium water content. In wet soils, friction along the sides of the sampler and vibrations due to hammering are likely to result in viscous flow of the soil and thus in compression of the sample. When this occurs the sample obtained is unrepresentative, being more dense than the body of the soil. Compression may occur even in dry soils if they are very loose. Whenever a sample is taken, one should carefully observe whether the soil elevation inside the sampler is the same as the undisturbed surface outside the sampler. One can only roughly estimate in this manner whether the density of the sample is changing because of sampling.

In dry or hard soils hammering the sampler into the soil often shatters the sample, and an actual loosening during sampling may occur. Pressing the sampler into the soil usually avoids the vibration that causes this shattering. Close examination of the soil sample usually allows one to estimate whether serious shattering occurs. And, as in the case of wet soils, soil level inside and outside the sampler must remain the same if the sample is to be considered satisfactory. (see also McIntyre, 1974.)

13–3 EXCAVATION METHOD

13–3.1 Introduction

Bulk density is determined in this method by excavating a quantity of soil, drying and weighing it, and determining the volume of the excavation. In the sand-funnel method, the volume is determined by filling the hole with sand, of which the volume per unit mass is known. In the rubber-balloon method, the volume is determined by inserting a balloon into the excavation and filling it with water or other fluid until the excavation is just full. The volume of the excavated soil sample is then equal to the volume of the fluid dispensed. If the excavation is carefully done it is possible simply to measure its dimensions and calculate the volume. Mensuration apparatus is described that enables one to determine the volume of an irregular excavation.

13–3.2 Method (ASTM, 1958, p. 422–441)

13–3.2.1 SPECIAL APPARATUS

13–3.2.1.1. Sand-Funnel Apparatus (see Fig. 13–2) (Soiltest, Inc., 2205 Lee Street, Evanston, IL 60202).

1. A metal funnel 15 to 18 cm at its largest diameter, fitted with a valve on the stem. Attached to the stem when the funnel is inverted is a sand container.
2. A standard sand that is clean, dry, and free-flowing. Particle size should be fairly uniform to avoid possible separation in the dispenser with consequent error in calibration. Sand particles passing a no. 20 sieve and retained on a no. 60 sieve are recommended (0.841–0.25 mm).
3. A template consisting of a thin, flat, metal plate approximately 30 cm square, with a hole 10 to 12 cm in diameter in its center.
4. Scales to weigh to 5 g.

Fig. 13–2. Apparatus for sand-funnel technique of determining soil bulk density in place.

13–3.2.1.2. Rubber-Balloon Apparatus.

1. A thin-walled rubber balloon (may be purchased from Barr Inc., 1531 First Street, Sandusky, OH 44870, and the Anderson Rubber Co., 310-T N. Howard Street, P.O. Box 170 Akron, OH 44309).
2. A 1000-cm^3 graduated cylinder and a water container.
3. A template, described in section 13–3.2.1.1 (3).

Rubber-balloon density apparatus is available from several manufacturers supplying soil testing equipment. (One supplier is Soiltest, Inc., 2205 Lee Street, Evanston, IL 60202.) The apparatus made commercially has the convenience of a volumetrically calibrated water container-dispenser, with suction facilities for returning the water to the container for re-use (Fig. 13–3).

13–3.2.1.3 Mensuration Apparatus.

1. Tape measure marked in millimeters.
2. Flat metal plate approximately 50 cm square, with 30 to 40 evenly spaced holes forming a grid through which the tape measure can be inserted. Alternatively, 3 wooden beams 3 by 3 by 80 cm with 5-cm markings may be used.
3. Four wood stakes approximately 10 cm long, one end sharpened.
4. Scales to weigh to approximately 10 g.

13–3.2.2. PROCEDURE

13–3.2.2.1. Sand-Funnel Procedure. Level the soil surface and remove loose soil at the test site. Place the template on the soil. Excavate a soil sample through the center hole of the template, leaving a hole with a diameter of approximately 12 cm and a depth of approximately 12 cm,

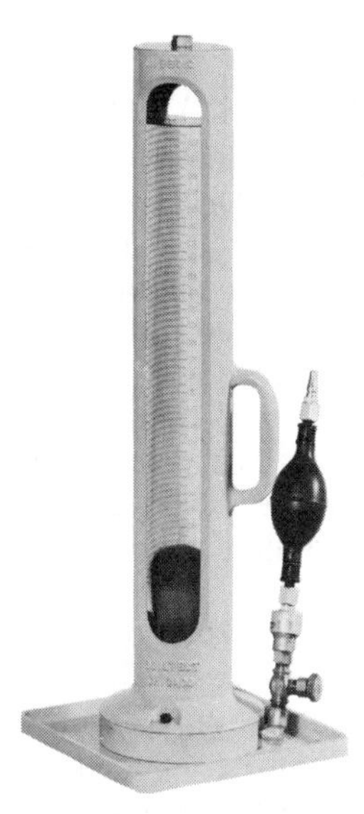

Fig. 13–3. Apparatus for determining soil bulk density in place by the rubber-balloon technique.

or other value as desired. A large spoon is convenient for excavating. Recover all evacuated soil in a container, being careful to include any loose soil that has fallen in from the sides of the excavation. Determine the oven-dry soil mass including stones by drying the soil to 105 °C and weighing it.

Determine the volume of the test hole by filling it with sand to the level of the bottom of the template. Level the sand with a spatula if necessary, but disturb it as little as possible to avoid packing the free-flowing sand. (Dispensing the sand through a funnel placed on the template, as is done with commercially available equipment, avoids the problem of leveling the sand. The excavation as well as the funnel is filled by free flow of sand, the predetermined weight required to fill the funnel being subtracted as a tare.)

Determine the weight of sand required to fill the test excavation by weighing it to the nearest 5 g. Precalibrate the mass-to-volume ratio of sand by letting sand fall from a similar height and at a similar rate of flow as in the test procedure. Using the calibration curve or values derived from it, determine the volume of the excavation from the measured mass of sand dispensed.

13–3.2.2.2. Rubber-Balloon Procedure. Level the soil surface, place the template on the surface, and excavate a soil sample as described in the preceding section. Place the rubber balloon in the test hole and fill the balloon with water to the bottom of the template. Determine the volume of water required to the nearest 2 cm^3. (A 1000-cm^3 graduate has markings to 10 cm^3, but one can estimate to 2 cm^3 if the graduate is placed on a horizontal surface.)

Calculate bulk density from the oven-dry mass of the excavated sample and the volume of the test excavation.

13–3.2.2.3. Mensuration Procedure. Prepare the soil surface as described in previous procedures. Drive wooden stakes into the soil at four corners of a square about 40 cm on a side, allowing them to project 1 to 3 cm above the soil surface. Place the metal plate on the stakes. Measure the distance from the upper surface of the plate to the soil surface through each of the holes. Remove plates and excavate soil from an area about 30 by 30 cm to a depth of 10 cm. Weigh the soil including stones. Remove an aliquot of reasonable size for determination of water content and discard the remainder. Replace the plate on the stakes and measure the distance from the top of the plate to the excavated surface through each of the holes as before.

If wooden beams are used in place of a metal plate, place two beams parallel, each resting on two stakes. With the third beam, bridge across the other two and from this datum, measure distance to the soil surface on a 5-cm grid using care that the tape is perpendicular to the beam. As above, excavate the soil sample and again measure distance from the third beam to excavated soil surface on a 5-cm grid. Weigh the soil including stones. Remove an aliquot for water-content measurement and discard the remainder of the excavated soil.

Calculate bulk density, ρ_b, from the weight of the soil corrected to oven dryness and the volume of the excavated soil. The volume is determined by summing the volumes around each depth measurement as follows:

$$V = (\Sigma d - \Sigma d_0)A$$

where

d = depth after excavation,

d_0 = depth before excavation, and

A = area covered by each measurement of the ruler. If holes are on 5-cm centers in the plate or measurements are made at 5-cm centers with the wooden beams, $A = 25$ cm^2.

By subsequent excavation of deeper layers in the same holes, one can determine bulk density deeper in the profile by measurements from the same datum.

13–3.2.3 COMMENTS

Bulk density can be estimated accurately by excavation methods carried out carefully. Holes should have smooth, rounded walls. Protruding stones should be included in the sample, care being used to round and smooth the area from which stones are taken. A heavy pair of scissors can be used to cut roots at the wall surface so the surrounding soil is undisturbed.

The relatively large sample (a cylinder of 12-cm diameter and of 12-cm depth has a volume of 1357 cm^3) has the advantage that small errors

Table 13-1. Comparison of surface nuclear gauge and sand-cone method for determining soil bulk density (Mintzer, 1961).

Soil material	Number of comparisons	Mean difference	Extreme difference
		%†	
	Wet bulk density, ρ_{bw}		
Brown sand, trace silt	9	2.03	−0.25–4.28
Brown silt and clay	4	0.89	−0.64–1.87
Brown sand and gravel, some silt, trace clay	6	1.01	−2.27–7.75
Brown till	4	−1.19	−3.97–1.60
	Dry bulk density, ρ_b		
Brown sand, trace silt	9	2.05	−1.21–4.98
Brown silt and clay	4	7.21	5.51–9.27
Brown sand and gravel, some silt, trace clay	6	1.48	−1.71–8.32
Brown till	4	−0.38	−4.53–3.21

† A positive value indicates nuclear method gave higher bulk density value.

in measuring water volumes or sand weights are insignificant. The disadvantage is the lack of discrimination to a localized horizon. An error of 5 cm^3 in liquid volume will give an error of 0.005 in a sample having a bulk density of 1.36 g cm^{-3}. Though one determines the water dispensed to perhaps 2 cm^3, a much greater source of error in water measurement arises in determining when the water in the excavation is level with the bottom of the template. Extreme care and judgment are required in this. The volume of the balloon itself, being of the order of 2 cm^3, will give a considerably smaller error than the volume measurement, and can be neglected.

An error of 7 g in weighing the sand gives an error of 0.005 if the bulk density is 1.36 g cm^{-3}. As in the balloon technique, greater error is likely to result in the precision with which one can determine the sand level at the bottom of the template. An error of 1 mm in this level will result in an error of 0.01 in the bulk density. Extreme care is therefore required to assure that the sand level is at the template bottom. The need for the dispensing funnel in reducing this error is obvious.

If one assumes an excavation of 30 by 30 by 10 cm in the mensuration method and measures depth 36 times on a 5-cm grid, assuming a cumulative error of 1 mm in each of the depth measurements the error in volume measurement is 1%.

A comparison of the sand-funnel and radiation methods was made by Mintzer (1961), and the results are summarized in Table 13-1.

13-4 CLOD METHOD

13-4.1 Introduction

The bulk density of clods, or coarse peds, can be calculated from their mass and volume. The volume may be determined by coating a clod of known weight with a water-repellent substance and by weighing it first in air, then again while immersed in a liquid of known density, making use of Archimedes' principle. The clod or ped must be sufficiently stable to cohere during coating, weighing and handling.

13–4.2 Method

13–4.2.1 SPECIAL APPARATUS

1. A balance, modified to accept the clod suspended below the balance arm by means of a nylon thread or thin wire, to allow weighing the clod when it is suspended in a container of liquid.
2. A fine nylon thread or 28 to 30 gauge wire, to attach the clod to the balance. A fine nylon hairnet makes a good container for the clod.
3. Saran solution. Dissolve 1 part by weight of Saran resin (Dow Saran F-310; Dow Chemical Co., Suite 500/ Tower No. 2, 1701 West Golf Road, Rolling Meadows, IL 60008) in 7 parts by weight of methylethyl ketone in a 1-gallon container in sufficient quantities to fill the container about three-quarters full. Manufacturer's instructions for safe handling of solvents should be carefully followed. Dissolution requires about an hour, with vigorous stirring. Since the solvent is flammable and explosive when its vapors are mixed with air, either hand-stirring or use of an air-driven stirrer should be employed in a well ventilated hood. The solution can be stored for long periods if kept in a tightly closed container to prevent evaporation of the solvent.

13–4.2.2 PROCEDURE

Secure the clod with two loops of the thread or wire, loops being at right angles to one another, leaving sufficient thread or wire to connect to the balance arm. Weigh the clod and thread. Holding it by the thread, dip the clod into the saran solution. Suspend it in air under a hood for 15 to 30 min to allow the solvent to evaporate. Repeat dipping and drying one or more times as needed, to waterproof the clod. Weigh the clod, with its coating and the thread. Weigh it again when it is suspended in water and note the water temperature. Determine the tare weight of the thread or wire. To obtain a correction for water content of the soil, break open the clod, remove an aliquot of soil, and weigh the aliquot before and after oven-drying it at 105 °C.

Calculate the oven-dry mass of the soil sample W_{ods} as follows, from the water content of the aliquot removed from the clod after other weights are taken:

$$W_{ods} = W_{sa}/(1 + \theta_w)$$

where θ = water content of the subsample in g/g and W_{sa} = net weight of clod or ped in air at its original water content.

Calculate bulk density as follows:

$$\rho_b = \rho_w W_{ods}/[W_{sa} - W_{spw} + W_{pa} - (W_{pa}\rho_w/\rho_p)]$$

where

ρ_w = density of water at temperature of determination,
W_{ods} = oven-dry weight of soil sample (clod or ped),
W_{sa} = net weight of clod or ped in air,
W_{spw} = net weight of soil sample plus saran in water,
W_{pa} = weight of saran coating in air, and
ρ_p = density of saran.

13–4.2.3 COMMENTS

The clod method usually gives higher bulk-density values than do other methods (Tisdall, 1951). One reason is that the clod method does not take the interclod spaces into account.

Extreme care should be exercised to get naturally occurring masses of soil. Clods on or near the soil surface are likely to be unrepresentative, for these are often formed by packing with tillage implements. Natural soil masses, or coarse peds, that are more representative should be sought.

If bubbles appear on the saran when the sample is weighed in water or if the weight in water increases with time, water is penetrating the clod, and the sample must be discarded.

Brasher et al. (1966) first proposed use of saran coating. They suggested that for clods with large pores, a more viscous saran solution of 1 part resin to as little as 4 parts methylethyl ketone could be used. They gave the density of saran to be 1.3 g cm^{-3}.

Precision in calculating the bulk density would require a correction for the difference of the weight of the wire in air and in water. However, the error is negligible with thread or a 28-gauge wire.

Using clods as small as 40 g oven-dry weight and weighing to 10 mg gives a standard deviation in the bulk density with 25 replications of a single measurement of 0.07 g cm^{-3} (Hartge, 1965). This can be reduced by using larger clods or by weighing to 1 mg or both. Obviously, using a greater number of samples for a determination would also reduce the standard deviation.

Several other substances have been used to seal the clod against water, including paraffin, rubber, wax mixtures, and oils.

13–5 RADIATION METHODS

13–5.1 Introduction

The transmission of gamma radiation through soil or scattering within soil varies with soil properties, including bulk density. By suitable calibration, measurements of either transmission or scattering of gamma radiation can be used to estimate bulk density.

In the transmission technique, two probes at a fixed spacing are lowered into previously prepared openings in the soil. One probe contains a Geiger tube, which detects the radiation transmitted through the soil from the gamma source located in the second probe. The scattering technique employs a single probe containing both gamma source and detector separated by shielding in the probe. It can be used either at the soil surface or placed in a hole, depending on design of the equipment.

Radiation methods have several advantages, among which are minimum disturbance of the soil, short time required for sampling, accessibility to subsoil measurement with minimum excavation, and the possibility of continuous or repeated measurements at the same point.

Both transmission and scattering techniques measure the bulk density of all components combined. The densities of gaseous components are insignificant in comparison to those of the solid or liquid components, and can therefore be ignored. It is necessary, however, to determine the water content of the soil at sampling time and to apply a correction to obtain bulk density on a dry soil basis.

13–5.2 Methods

13–5.2.1 SPECIAL APPARATUS

Transmission apparatus is supplied by Troxler Electronics Laboratories, P.O. Box 12057, Cornwallis Road, Research Triangle Park, NC 27709, following a design by Vomocil (1954). A design, including a discussion of the theory of the method, calibration, and methods of making measurements was included in the first edition of *Methods of Soil Analysis,* Part 1 (Blake, 1965).

Scattering apparatus is supplied by Troxler Electronic Laboratories, P.O. Box 12057, Cornwallis Road, Research Triangle Park, NC 27709 and by Soiltest Inc., 2205 Lee Street, Evanston, IL 60202.

13–5.2.2 PROCEDURE

It is recommended that the instructions supplied with the commercial apparatus be followed. One may also wish to refer to the first edition of *Methods of Soil Analysis,* Part 1 (Blake, 1965).

13–5.2.3 COMMENTS

There is some radiation hazard with these methods. Gamma photons are high-energy radiation. Some will pass through several centimeters of lead shielding. Commercially available equipment, as well as designs described in the literature, reduce the hazard to safe levels. But it is important to adhere strictly to time limits, distances, and other conditions described by the manufacturers. One should be equipped for and knowledgeable in means of checking the equipment for radiation levels according to the way it is handled in actual sampling. If there is doubt, the equipment should be checked for safety by a competent testing laboratory.

Since radiation transmitted from a source to a detector is dependent on probe spacing or sample thickness, care must be exercised with the two-probe sampler to assure that access holes are parallel and spaced exactly as in the calibration.

Mintzer (1961) reported comparisons of the surface-density probe and the sand-cone method on four engineering projects. He reported his comparisons on both the wet and dry bulk-density bases. He used a surface neutron meter for water content where the surface-density probe was used. His results are summarized in Table 13–1.

13–6 REFERENCES

Am. Soc. Test. Mater. 1958. Procedures for testing soils. American Society for Testing and Materials, Philadelphia.

Blake, G. R. 1965. Bulk density. *In* C. A. Black et al. (ed). Methods of soil analysis, Part 1. Agronomy 9:383–390.

Brasher, B. R., D. P. Franzmeier, V. Valassis, and S. E. Davidson. 1966. Use of Saran Resin to coat natural soil clods for bulk density and moisture retention measurements. Soil Sci. 101:108.

Hartge, K. H. 1965. Vergleich der Schrumpfung ungestörter Böden und gekneteter Pasten. Z. Friedr. Wilh. Univ. Jena. (math-nat. Reihe) 14:53–57.

Hartge, K. H. 1968. Heterogenität des Bodens oder Quellung? Trans. Int. Congr. Soil Sci., 9th 3:591–597.

Jamison, V. C., H. H. Weaver, and I. F. Reed. 1959. A hammer-driven soil core sampler. Soil Sci. 69:487–496.

Lutz, J. F. 1947. Apparatus for collecting undisturbed soil samples. Soil Sci. 64:399–401.

McIntyre, D. S. 1974. Soil sampling techniques for physical measurements, chapter 3; Bulk density, chapter 5; and Appendix 1. *In* J. Loveday (ed.) Methods of analysis of irrigated soils. Technical Communication no. 54, Comw. Bur. Soils, Comw. Agric. Bureaux. Farnham Royal, Bucks, England.

Mintzer, S. 1961. Comparison of nuclear and sand-cone methods of density and moisture determinations for four New York State soils. *In* Symposium on nuclear methods for measuring soil density and moisture. Am. Soc. Testing Mater., Spec. Tech. Pub. 293:45–54.

Tisdall, A. L. 1951. Comparison of methods of determining apparent density of soils. Aust. J. Agric. Res. 2:349–354.

U. S. Department of Agriculture. 1954. Diagnosis and improvement of saline and alkali soils. USDA Handb. 60.

Vomocil, J. A. 1954. In situ measurement of soil bulk density. Agric. Eng. 35:651–654.

14 Particle Density[1]

G. R. BLAKE

University of Minnesota
St. Paul, Minnesota

K. H. HARTGE

University of Hanover
Hanover, Federal Republic of Germany

14-1 INTRODUCTION

Particle density of soils refers to the density of the solid particles collectively. It is expressed as the ratio of the total mass of the solid particles to their total volume, excluding pore spaces between particles. Convenient units for particle density are megagrams per cubic meter (Mg m^{-3}), or the numerically equal grams per cubic centimeter (g cm^3).

Particle density is used in most mathematical expressions where volume or weight of a soil sample is being considered. Thus interrelationships of porosity, bulk density, air space, and rates of sedimentation of particles in fluids depend on particle density. Particle-size analyses that employ sedimentation rate, as well as calculations involving particle movement by wind and water, require information on particle density.

14-2 PRINCIPLES

Particle density of a soil sample is calculated from two measured quantities, namely, the mass and volume of the sample. The mass is determined by weighing; the volume, by calculation from the mass and density of water (or other fluid) displaced by the sample. The pycnometer and the submersion methods are based on the same principle. Both have long been in use. They are simple, direct, and accurate if done carefully.

---

[1]Paper no. 11121 of the Scientific Journal Series, Minnesota Agricultural Experiment Station, St. Paul, MN.

14–3 PYCNOMETER METHOD (ASTM, 1958, p. 80; U.S. Dep. Agric., 1954, p. 122)

14–3.1 Special Apparatus

A pycnometer (specific-gravity flask) is employed. A pycnometer is a glass flask fitted with a ground-glass stopper that is pierced lengthwise by a capillary opening. A thermometer is sometimes an integral part of the stopper, the glass-enclosed mercury reservoir being in contact with the fluid in the flask, with the stem extending above the ground joint. A 10-mL pycnometer has sufficient capacity.

A small volumetric flask (25, 50, or 100 mL) may be used in place of a pycnometer when the sample is large enough to compensate for the decrease in precision of measuring fluid volume.

14–3.2 Procedure

Weigh a clean, dry pycnometer in air. Add about 10 g of air-dry soil sieved through a 2-mm sieve. If a 100-mL volumetric flask is used, add 50 g of soil. Clean the outside and neck of the pycnometer of any soil that may have spilled during transfer. Weigh the pycnometer (including stopper) and its contents. Determine the water content of a duplicate soil sample by drying it at 105 °C.

Fill the pycnometer about one-half full with distilled water, washing into the flask any soil adhering to the inside of the neck. Remove entrapped air by gentle boiling of the water for several minutes, with frequent gentle agitation of the contents to prevent loss of soil by foaming.

Cool the pycnometer and its contents to room temperature, and then add enough boiled, cooled, distilled water at room temperature to fill the pycnometer. Insert the stopper and seat it carefully. Thoroughly dry and clean the outside of the flask with a dry cloth, using care to avoid drawing water out of the capillary. Weigh the pycnometer and its contents, and determine the temperature of the contents after they have cooled to room temperature.

Finally, remove the soil from the pycnometer and thoroughly wash it. Fill the pycnometer with boiled, cooled distilled water at the same temperature as before, insert the stopper, thoroughly dry the outside with a cloth, and weigh the pycnometer and contents, being careful that the temperature remains the same as before.

Calculate the particle density as follows:

$$\rho_p = \rho_w (W_s - W_a)/[(W_s - W_a) - (W_{sw} - W_w)] \qquad [1]$$

where

ρ_w = density of water in grams per cubic centimeter at temperature observed,

W_s = weight of pycnometer plus soil sample corrected to oven-dry water content,
W_a = weight of pycnometer filled with air,
W_{sw} = weight of pycnometer filled with soil and water, and
W_w = weight of pycnometer filled with water at temperature observed.

14–4 SUBMERSION METHOD (Capek, 1933)

14–4.1 Special Apparatus

1. A laboratory balance with a thin wire attached to the weighing beam, to which a light frame can be suspended. The frame serves as a platform for placing a weighing dish so that both frame and dish can be immersed in a container of liquid during weighing (see also section 13–4.2.1).
2. Sample containers. Aluminum weighing dishes of about 5-cm diameter and 3-cm height are suitable.
3. A container for water or a nonpolar liquid such as xylene or toluene, into which the weighing dish and sample can be immersed. Surface diameter should be about three times that of the weighing dishes.

14–4.2 Procedure

Moisten about 25 g of soil to a plastic consistency and force it by hand through a 2-mm sieve to form spaghetti-like threads. Dry the soil in a tared weighing dish to 105 °C, cool it to room temperature in a desiccator with a drying agent, and weigh it.

Add water to the dish to cover the soil, place weighing dish and soil in a vacuum desiccator, and evacuate for about 10 min to eliminate entrapped air from between the threads. Transfer weighing dish and sample to the weighing frame attached with a wire to the balance. Submerge weighing dish, frame, and soil sample into container of water and carefully reweigh while they are suspended in the water. Remove and discard the sample, clean the weighing dish, and weigh it while it is submerged in water. Determine the temperature of the water, and from handbook tables, determine its density.

When a series of samples is analyzed using the same organic liquid, it is convenient at this point to submerge and weigh a small piece of metal such as 30 to 50 g of brass in the same container of liquid. Constancy of its submerged weight after each soil sample weighing assures the analyst that the organic liquid is not contaminated and allows re-use of the same liquid.

Calculate particle density as follows:

$$\rho_p = \rho_l (W_{sd} - W_d)/[(W_{sd} - W_d) - (W_{sdl} - W_{dl})] \qquad [2]$$

where

ρ_l = density of water or organic liquid used, g cm^{-3},
W_{sd} = oven-dried weight of soil with weighing dish,
W_d = weight of weighing dish,
W_{sdl} = weight of sample and dish submerged in liquid, and
W_{dl} = weight of dish alone, submerged in liquid.

14–5 COMMENTS

The pycnometer method has the advantage of giving very precise densities if volumes and weights are carefully measured. The submersion method sacrifices some precision but offers ease of measurement, especially when measurements are made on a series of samples. It does not require a calibrated pycnometer, it avoids accurate drying and cleaning of containers during repeated measurements, and it is less laborious, since the care needed to obtain reproducible accuracy in filling the pycnometer or flask is unnecessary. These advantages of the submersion method are best realized when a nonpolar organic liquid is used. A disadvantage of the submersion method is that it cannot be used on sandy soils where coherence may be too small to allow one to make the spaghetti-like threads.

With the pycnometer or a flask, a weighing error of 1 mg on a 10-g soil sample gives an error in particle density of only 0.0003 g cm^{-3}. A weighing error of 10 mg on a 30-g sample gives a particle density error of 0.001 g cm^{-3}. Greater errors can result from lack of precision in the volume measurement. If W_{sw} in Eq. [1] is based on a volume that exceeds the volumetric flask marking by 0.2 mL, and W_w on a volume 0.2 mL deficient of the marking, the compounded particle density error is 0.05 g cm^{-3} on a 40-g sample. The analyst should check the calibration marking on the flask as well as his or her ability to measure a reproducible volume, by making a number of preliminary weighings of water in the flask to be used for the analysis. The submersion method, if performed as described with 25 samples each between 20 and 30 g, gives a standard error of 0.005 g cm^{-3} for homogenized material. If unmixed replicate samples are used from surface soils, standard error tends to be several times greater. In addition to weight and volume errors, one must assume some error due to nonrepresentative sampling in either method.

Particle density values for finely divided active soil obtained by weighing in water are greater than those obtained with nonpolar organic liquids. There appears to be little difference between organic liquids. Anderson and Mattson (1926) found the average specific gravity of the clay fractions of six soils to be greater in water than in toluene by 0.13, while Capek (1933), using xylol, benzol, petroleum ether, and benzene, found an increase averaging 0.001 for quartz and 0.01 to 0.1 or even more for loam and chernozem soils. Smith (1943) found that water gave higher values for five soils than xylene, tetralin, or dichloroethyl ether by 0.01 to 0.03; and Gradwell (1955) found that the value for the specific gravity increased as the content of minerals with expanding lattices increased, or as the presence of finely divided, amorphous minerals increased. As the internal surface of non-allophane minerals increased from 33 to 306 m^2 g^{-1}, the increase in specific gravity determined in water over that determined in toluene varied from 0.014 to 0.094. Allophane values were greater in water by 0.05 to 0.3.

Water density is known to be affected by surfaces of finely divided particles. Though interactions of nonpolar organic molecules with clay surfaces are incompletely understood, it seems evident that the more accurate particle densities of clays would be obtained by use of nonpolar inorganic liquids in a pycnometer. Nevertheless, as Gradwell (1955) pointed out, where finely divided amorphous minerals or minerals with expanding lattices are present, it may be undesirable to substitute other liquids for water in determinations of specific gravity if the measurements are to be applied in computing the volume of solids in a soil in contact with water. For many applications, however, densities inaccurate by 0.05 g cm^{-3} will suffice. Whether to use water or organic liquids is thus largely a question of how the data are to be used.

An advantage of nonpolar organic liquids is that soil samples, especially those high in organic matter, are wetted more easily than they are with water. Boiling is unnecessary when the pycnometer or flask is used; gentle shaking or stirring lightly with a glass rod is sufficient. It is desirable, however, when using nonpolar liquids, to evacuate the half-filled container in a vacuum desiccator for 10 min to facilitate removal of air. Another advantage in using organic liquids, especially for organic soils and peat, is that the soil or organic particles sediment faster after stirring than they would in water. In the submersion method this is important in reducing buoyancy when one weighs the sample in a weighing dish submerged in the fluid. Disadvantages of using organic liquids are their high vapor pressure and their low heat capacities. Because of the former, work in a well-ventilated hood is necessary. Since the low heat capacity presents the hazard of thermal dilation, it is essential to use only tongs for handling the containers.

Both the pycnometer and the submersion methods give the weighted mean density of all particles in the sample. This is the value needed for calculations mentioned in the introduction. Densities of individual soil grains may vary widely from the weighted mean. For example, handbook densities of silt and sand-sized particles are 2.65 for quartz, 2.5 to 2.8 for feldspars, 2.7 to 3.3 for micas, and 3.1 to 3.3 for apatite. The density of humus is usually < 1.5 Mg m^{-3}.

14–6 REFERENCES

Am. Soc. Test. Mater. 1958. Procedures for testing soils. American Society for Testing and Materials, Philadelphia.

Anderson, M. S., and S. Mattson. 1926. Properties of the colloidal soil material. U.S. Dep. Agric. Bull. 1452.

Capek, M. 1933. Cited by DiGleria, J., A. Klimes-Szmik, and M. Dvoracsek. 1962. Bodenphysik und Bodenkolloidik. German edition jointly by Akademiai Kiado, Budapest, and VEB Gustav Fischer Verlag, Jena.

Gradwell, M. W. 1955. The determination of specific gravities of soils as influenced by clay-mineral composition. N.Z.J. Sci. Technol. 37B:283–289.

Smith, W. O. 1943. The density of soil colloids and their genetic relations. Soil Sci. 56:263.

U.S. Department of Agriculture. 1954. Diagnosis and improvement of saline and alkali soils. USDA Handb. 60.

15 Particle-size Analysis[1]

G. W. GEE

Battelle, Pactific Northwest Laboratories
Richland, Washington

J. W. BAUDER

Montana State University
Bozeman, Montana

15-1 INTRODUCTION

Particle-size analysis (PSA) is a measurement of the size distribution of individual particles in a soil sample. The major features of PSA are the destruction or dispersion of soil aggregates into discrete units by chemical, mechanical, or ultrasonic means and the separation of particles according to size limits by sieving and sedimentation.

Soil particles cover an extreme size range, varying from stones and rocks (exceeding 0.25 m in size) down to submicron clays (< 1 μm). Various systems of size classification have been used to define arbitrary limits and ranges of soil particle size. Soil particles smaller than 2000 μm are generally divided into three major size groups: sands, silts and clays. These groups are sometimes called soil separates and can be subdivided into smaller size classes. Figure 15-1 shows the particle size, sieve dimension, and defined size class for the system of classification used by the U. S. Department of Agriculture (USDA), the Canadian Soil Survey Committee (CSSC), the International Soil Science Society (ISSS) and the American Society for Testing and Materials (ASTM). The American Society of Agronomy has adopted the USDA classification [i.e., sands ($<$2000–50 μm), silts ($<$50–2 μm), and clays ($<$2 μm)]. Although the USDA classification scheme will be emphasized in the following methods, it should be recognized that other systems are frequently cited, particularly in engineering literature, hence, care should be taken to specify clearly which system is being used when reporting results.

Particle-size analysis data can be presented and used in several ways, the most common being a particle-size distribution curve. An example of this type of curve is shown in Figure 15-2. The percentage of particles

---

[1] Prepared for the U.S. Department of Energy and the U.S. Nuclear Regulatory Commission under Contract DE-AC06-76RLO 1830.

 Methods of Soil Analysis, Part 1. Physical and Mineralogical Methods—Agronomy Monograph no. 9 (2nd Edition)

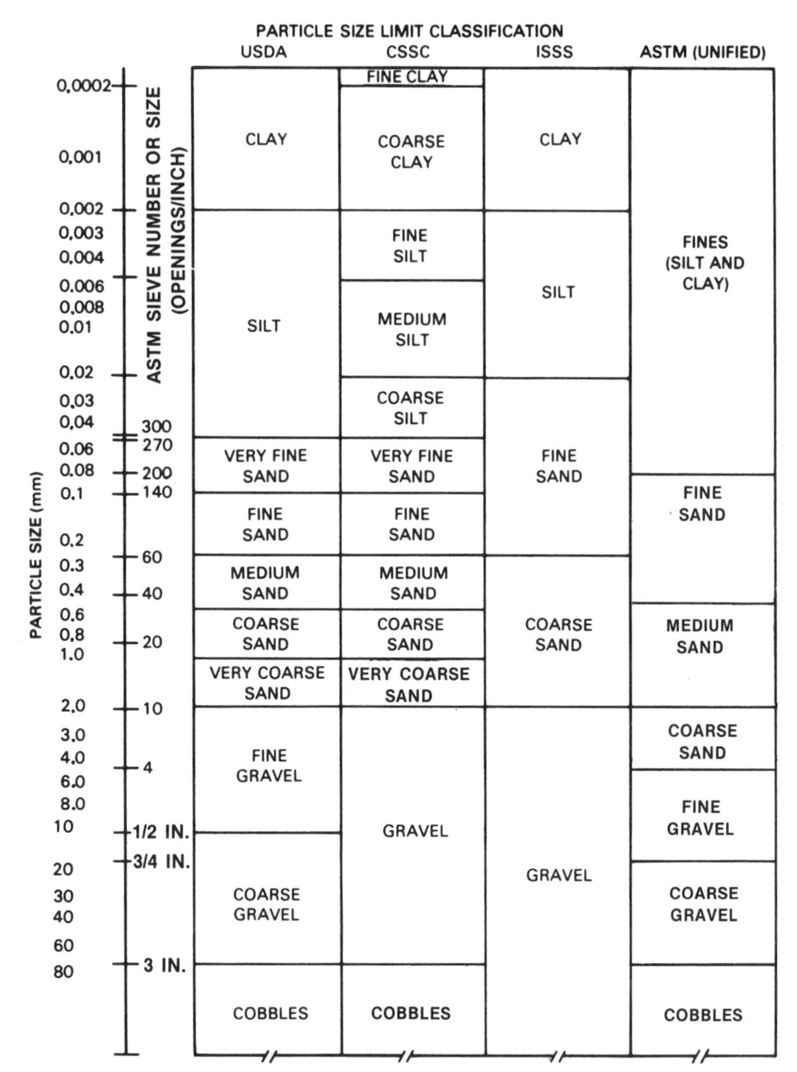

Fig. 15–1. Particle-size limits according to several current classification schemes.

less than a given particle size is plotted against the logarithm of the "effective" particle diameter. Particle-size distribution curves, when differentiated graphically, produce frequency distribution curves for various particle sizes. Frequency curves usually exhibit a peak or peaks representing the most prevalent particle sizes.

Particle-size distribution curves are used extensively by geologists in geomorphological studies to evaluate sedimentation and alluvial pro-

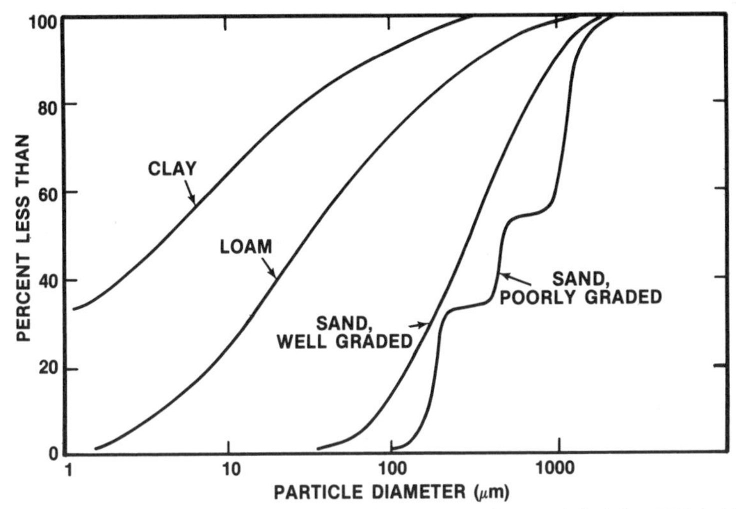

Fig. 15-2. Particle-size distribution curves for several soil materials (after Hillel, 1982).

cesses, and by civil engineers to evaluate materials used for foundations, road fills, and other construction purposes. Details of the use of these curves are given by Krumbein and Pettijohn (1938) and Irani and Callas (1963).

Particle-size analysis is often used in soil science to evaluate soil texture. Soils rarely consist entirely of one size range. Soil texture is based on different combinations of sand, silt, and clay separates that make up the particle-size distribution of a soil sample. Figure 15-3 shows the USDA defined limits for the basic soil textural classes. Details for interpretation of the textural triangle for soil classification purposes are given by the Soil Survey Staff (1975). The ASTM (Unified) engineering classification system is used widely for delineating soil types for construction purposes (Fig. 15-4). In this system, liquid limits and plasticity indexes must be known in order to properly classify the soil type (ASTM, 1985a,b).[2]

Hydrologists often use PSA as a means of predicting hydraulic properties, particularly for sands (Todd, 1964). Recently, Bloemen (1980) and Arya and Paris (1981) have used PSA as a means to predict water retention and unsaturated hydraulic conductivity of soils. These predictive methods appear to work best on sands or structureless soil materials.

15-2 PRINCIPLES

15-2.1 Pretreatment and Dispersion Techniques

Pretreatment of samples to enhance separation or dispersion of aggregates is a key step in PSA and is generally recommended, since many

[2]Stevens (1982) has published a BASIC program for computing the Unified (ASTM) classification for a tested soil. A BASIC program for computing the USDA textural classes is available upon request from the authors.

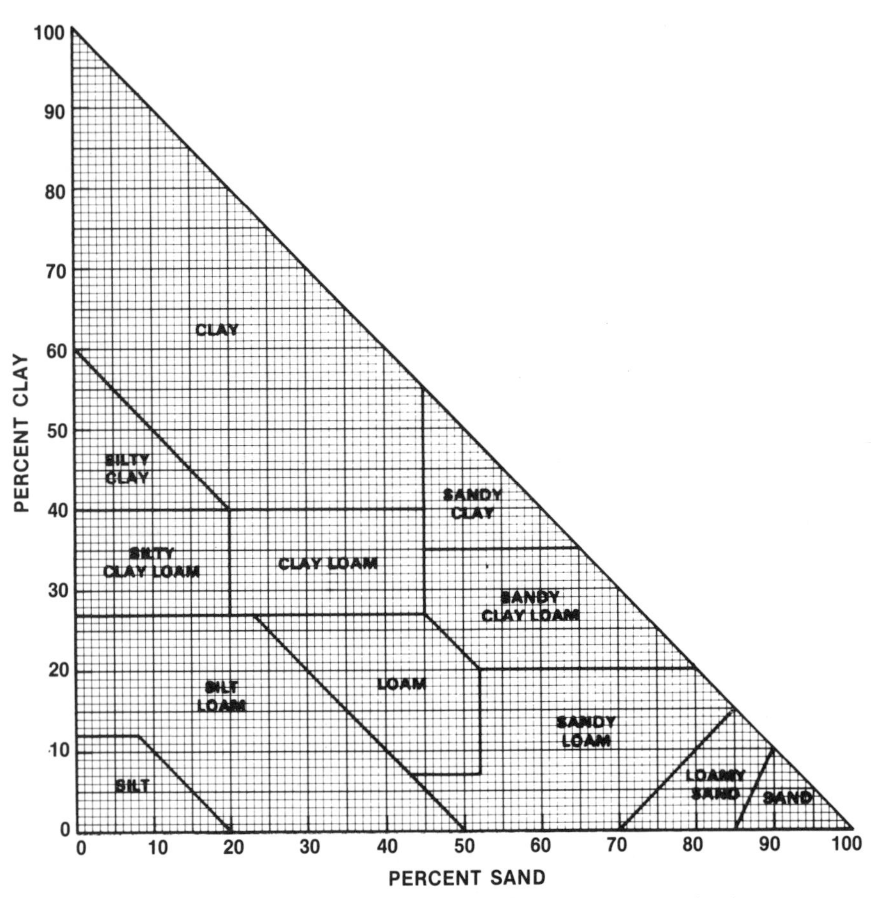

Fig. 15–3. Textural triangle for soil textural analysis using the USDA classification scheme.

soils contain aggregates that are not readily dispersed. Soils generally contain organic matter and often contain iron oxides and carbonate coatings that bind particles together. Chemical pretreatments are used for removal of these coatings; however, chemical treatment can result in destruction and dissolution of some soil minerals. Physical treatments are also used, but standardization of treatment and adequate testing of specific methods are needed, since the very process of separation by mechanical or ultrasonic means can fragment the individual particles into further subunits. Procedures should clearly specify the sample pretreatment, the separation method, and the purpose for which the size analysis is intended for a particular soil.

Standard PSA methods require that soil particles be dispersed in an aqueous solution by both chemical and physical means. After pretreatment, chemical dispersion is often accomplished using a dilute alkaline solution of sodium polyphosphate. The effectiveness of the chemical dispersing agent depends on its ability to create and maintain repulsive forces between soil particles. Some soils (e.g., those of volcanic ash origin) that have been highly weathered disperse more readily in acid media; hence, some pretesting may be required to determine effects of soil mineralogy and other factors on soil dispersibility and to select an appropriate method to achieve complete dispersion. Physical dispersion of particles is accomplished by shearing action or turbulent mixing, using mechanical shakers, electrical mixers, or ultrasonic probes.

Dispersibility of soils low in organic matter depends primarily on soil mineralogy. Highly oxidized soils are particularly difficult to disperse. Examples include the "subplastic" soils of Australia (McIntyre, 1976; Brewer & Blackmore, 1976; Walker & Hutka, 1976; Blackmore, 1976; Norrish & Tiller, 1976). Depending on the method of chemical treatment and physical dispersion used, measured clay content for an individual soil sample can vary by factors of two to four or more.

Volcanic ash soils are high in amorphous (noncrystalline) clay-sized materials and have great resistance to dispersion, particularly after air or oven drying (Kubota, 1972; Schalscha et al., 1965; Espinoza et al., 1975; Maeda et al., 1977). Kubota (1972) reported clay contents ranging from 1 to 56 wt% for one volcanic ash soil, depending on pretreatment. Maximum clay content was obtained when the soil was retained at field moisture prior to ultrasonic dispersion. Warkentin and Maeda (1980) recommend that volcanic ash soils be left at field moisture and dispersed at either pH 3 or above pH 9. Tama and El-Swaify (1978) and El-Swaify (1980) have observed that soils with variable charge are particularly difficult to disperse unless the dispersant solution is well below or above the zero-point of charge.

Highly aggregated, stable clay soils may behave like coarse sands in terms of water infiltration; hence they may be identified in the field as sands or coarse loams. These same soils, having significant microporosity and high exchange capacities, retain water and nutrients much better than sands. For agricultural purposes, these soils should be texturally classed in a much finer category than they appear in the field. For soils where these uncertainties are known to exist, measurements such as a simple dispersive index (Sherard et al., 1976), ASTM dispersion test (ASTM, 1985c), or the water-stability of aggregates (see chapter 17) would be necessary and useful information. Also, a calculated clay content, determined from a ratio of the cation exchange capacity (CEC) of the total soil to the CEC of the clay-size material (Norrish & Tiller, 1976), can be used to estimate the theoretical maximum clay fraction of the soil material.

The method that produces the most complete dispersion of a soil sample is generally the more acceptable method. However, the chemical treatment and mechanical work done on the soil are dictated by somewhat arbitrary decisions, so there is no "absolute" size-distribution for a given sample. Intense mechanical or ultrasonic dispersion, coupled with appropriate chemical treatment, should yield a sample with most of the

Criteria for Assigning Group Symbols and Group Names Using Laboratory Tests[A]					Soil Classification	
					Group Symbol	Group Name[B]
Coarse-Grained Soils More than 50% retained on No. 200 sieve	Gravels More than 50% of coarse fraction retained on No. 4 sieve	Clean Gravels Less than 5% fines[C]	$Cu \geq 4$ and $1 \leq Cc \leq 3$[E]		GW	Well-graded gravel[F]
			$Cu < 4$ and/or $1 > Cc > 3$[E]		GP	Poorly graded gravel[F]
		Gravels with Fines More than 12% fines[C]	Fines classify as ML or MH		GM	Silty gravel[F,G,H]
			Fines classify as CL or CH		GC	Clayey gravel[F,G,H]
	Sands 50% or more of coarse fraction passes No. 4 sieve	Clean Sands Less than 5% fines[D]	$Cu \geq 6$ and $1 \leq Cc \leq 3$[E]		SW	Well-graded sand[I]
			$Cu < 6$ and/or $1 > Cc > 3$[E]		SP	Poorly graded sand[I]
		Sands with Fines More than 12% fines[D]	Fines classify as ML or MH		SM	Silty sand[G,H,I]
			Fines classify as CL or CH		SC	Clayey sand[G,H,I]
Fine-Grained Soils 50% or more passes the No. 200 sieve	Silts and Clays Liquid limit less than 50	inorganic	$PI > 7$ and plots on or above "A" line[J]		CL	Lean clay[K,L,M]
			$PI < 4$ or plots below "A" line[J]		ML	Silt[K,L,M]
		organic	Liquid limit - oven dried / Liquid limit - not dried	<0.75	OL	Organic clay[K,L,M,N]
						Organic silt[K,L,M,O]
	Silts and Clays Liquid limit 50 or more	inorganic	PI plots on or above "A" line		CH	Fat clay[K,L,M]
			PI plots below "A" line		MH	Elastic silt[K,L,M]
		organic	Liquid limit - oven dried / Liquid limit - not dried	<0.75	OH	Organic clay[K,L,M,P]
						Organic silt[K,L,M,Q]
Highly organic soils		Primarily organic matter, dark in color, and orgnic odor			PT	Peat

Fig. 15-4. Unified soil classification system including plasticity chart (ASTM, 1985a). Continued on p. 389.

[A]Based on the material passing the 3-in. (75-mm) sieve.
[B]If field sample contained cobbles or boulders, or both, add "with cobbles or boulders, or both" to group name.
[C]Gravels with 5 to 12% fines require dual symbols:
GW-GM well-graded gravel with silt
GW-GC well-graded gravel with clay
GP-GM poorly graded gravel with silt
GP-GC poorly graded gravel with clay
[D]Sands with 5 to 12% fines require dual symbols:
SW-SM well-graded sand with silt
SW-SC well-graded sand with clay
SP-SM poorly graded sand with silt
SP-SC poorly graded sand with clay
[E]$Cu = D_{60}/D_{10}$ and $Cc = \frac{(D_{30})^2}{D_{10} \times D_{60}}$
[F]If soil contains ≥ 15% sand, add "with sand" to group name.
[G]If fines classify as CL-ML, use dual symbol GC-GM, or SC-SM.
[H]If fines are organic, add "with organic fines" to group name.
[I]If soil contains ≥ 15% gravel, add "with gravel" to group name.
[J]If Atterberg limits plot in hatched area, soil is a CL-ML, silty clay.
[K]If soil contains 15 to 29% plus No. 200, add "with sand" or "with gravel," whichever is predominant.
[L]If soil contains ≥ 30% plus No. 200, predominantly sand, add "sandy" to group name.
[M]If soil contains ≤ 30% plus No. 200, predominantly gravel, add "gravelly" to group name.
[N]PI ≥ 4 and plots on or above "A" line.
[O]PI < 4 or plots below "A" line.
[P]PI plots on or above "A" line.
[Q]PI plots below "A" line.

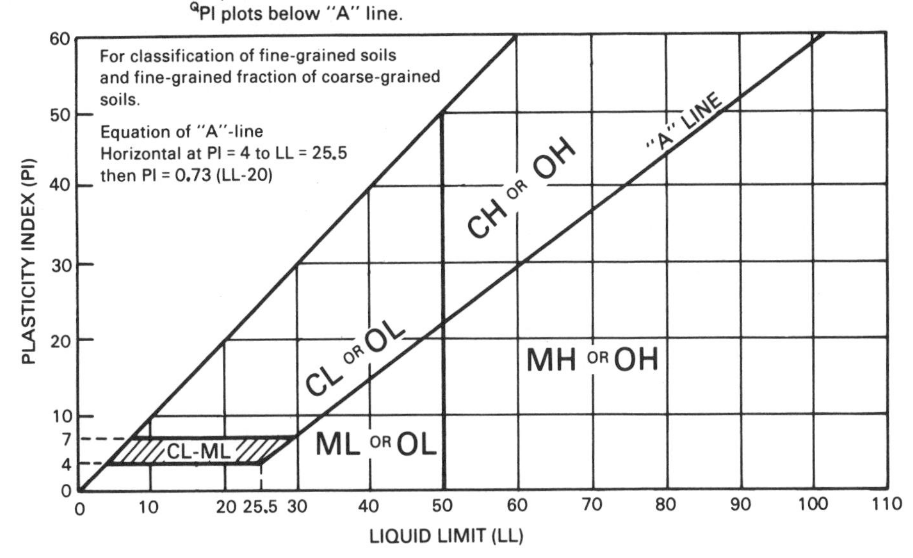

Fig. 15–4. Continued.

clay minerals in the measured clay fraction. In contrast, a less drastic chemical treatment and/or little mechanical dispersion may reflect the more "natural" particle-size distribution of the soil. Comparisons of PSA results should always include comparisons of the pretreatment and dispersion methods used.

15–2.1.1 ORGANIC MATTER REMOVAL

Removal of organic matter is often a first step in the chemical pretreatment of many soils. The necessity and difficulty of organic matter removal depends on the intended use of the analytical results of the PSA, the nature and concentration of organic matter in the sample to be analyzed, the pH of the soil, and the associated presence in the soil of free carbonates, gypsum, oxides, and soluble salts. A variety of reagents have been used in the past to successfully remove organic matter. Notable among these are hydrogen peroxide (H_2O_2), sodium hypochlorite, sodium hypobromite, and potassium permanganate. Hydrogen peroxide has been recommended as the standard oxidant for most soils (Day, 1965).

15–2.1.2 REMOVAL OF IRON OXIDE

Coatings and crystals of various iron oxides, such as hematite and goethite, often act as cementing and binding agents in soils. Removal of these cementing agents aids in dispersion of the silicate portion of the soil and is often necessary for accurate mineralogical analysis. Mehra and Jackson (1960) recommend the use of a bicarbonate-buffered, sodium dithionite–citrate system for iron oxide removal. This method, compared with several other methods for removal of free iron oxides from latosolic soils, was found to be the most effective. In addition, this method was the least destructive of iron silicate clays, as indicated by least loss of cation exchange capacity. Mehra and Jackson (1960) indicated that the optimum pH for maximum iron oxide removal was approximately 7.3. Since considerable OH^- is expended in the sodium dithionite–citrate reaction with iron oxide, a buffer is needed to hold the pH at the optimum level. Sodium bicarbonate has proven to be an effective buffer. This procedure minimizes the formation of sulfide, iron sulfide, zinc oxalate or other unwanted precipitates during iron oxide removal.

In soils where iron oxides are part of the dominant mineralogy, it is not recommended that iron oxides be removed, since many of the primary mineral grains in the clay fraction could be destroyed (El-Swaify, 1980).

15–2.1.3 REMOVAL OF CARBONATES

Removal of carbonate from soils prior to dispersion and sedimentation can be accomplished relatively easily by acidification of the sample. Heating accelerates the reaction. Samples that are acidified before organic matter removal with H_2O_2 will usually be free of carbonates. Hydrogen chloride (HCl) treatment can cause destruction of crystalline lattice of clay minerals; therefore, acid treatment with 1 *M* NaOAC at pH 5 is preferred.

15–2.1.4 REMOVAL OF SOLUBLE SALTS

A variety of soluble salts including sodium, calcium, and magnesium chlorides and carbonates are commonly found in alkaline soils. High concentrations of soluble salts can cause flocculation of soil suspensions. Alkaline salts can cause decomposition of H_2O_2, decreasing its effectiveness as an oxidizing agent for soil organic matter. In addition, many soluble salts interfere with saturation of the exchange complex. Calcium and magnesium salts, commonly occurring as carbonates, are relatively unstable and are often measured as part of the clay and silt fractions.

The most common procedure for removal of soluble salts is to leach the salts with distilled water. Sample washing with distilled water can be accomplished by use of a filter candle or by centrifuging. The procedure should be repeated until the leachate salt concentration drops below 10 m*M*. The washing treatment is then followed by chemical and physical dispersion.

15–2.1.5 SAMPLE DISPERSION

Dispersion of soils is accomplished by a combination of methods. The methods for dispersion can be classified as either chemical or physical. Numerous methods of chemical dispersion have been investigated and reported (Theisen et al., 1968; Norrish & Tiller, 1976). Soils are chemically dispersed after oxidation of organic matter and removal of carbonates and iron oxides. Chemical dispersion is based primarily on the concept of particle repulsion, as a result of elevation of the particle zeta potential. This process is usually accomplished by saturating the exchange complex with sodium. Physical or mechanical methods of dispersion involve separation of the individual particles by means of some mechanical or physical process, such as rubbing, rolling, shaking, or vibrating. During the past 20 years, electronic dispersion, primarily by the use of ultrasonics, has become increasingly popular. Most researchers have found that a combination of chemical and physical or electronic methods provides the most complete and stable dispersion (Maeda et al., 1977; Mikhail & Briner, 1978).

15-2.1.5.1 Chemical Dispersion. Following removal of cementing and flocculating agents, samples must be dispersed and maintained in a dispersed state until sedimentation measurements are completed. A number of dispersing chemicals have been used. These include Na-hexametaphosphate (HMP), Na_2PO_7, NaOH, Na_2CO_3, and NaOBr. Of these, HMP appears to be the most commonly used dispersant. Commercial detergents contain quantities of HMP and other soluble phosphates, but uncertainty exists as to the exact amounts (Yaalon, 1976; Veneman, 1977).

For this reason, reagent grade HMP, which is commercially available, is the recommended chemical dispersant for the pipet and hydrometer tests described later in this chapter.

The exact amount of chemical dispersant needed to prevent flocculation is dependent on soil type (mineralogy, etc.). Flocculation often can be prevented by increasing the concentration of the dispersant solution. It should be noted that the pipet analysis requires only 0.5 g/L HMP, compared to a 5 g/L HMP solution for the hydrometer analysis. The lower amount needed for pipet analysis is likely due to pretreatment (organic matter, iron oxide, and soluble salt removal). Specific amounts used in these analyses have been established by empirical methods.

15–2.1.5.2 Physical Dispersion. Several methods of physical dispersion have been used in conjunction with pretreatment and chemical dispersion. The ASTM (1985d) recommends either an electric mixer with specially designed stirring paddles or an air-jet stirrer (Chu & Davidson, 1953; Theisen et al., 1968). For the hydrometer method, Day (1965) recommends a 5 min mixing with a standard electrical mixer (malted milk style), but cautions that the mixer blades deteriorate rapidly by abrasion and should be replaced after 1 or 2 h of use or when showing signs of wear. Reciprocating shakers have also been used. Overnight shaking is prescribed in the pipet procedure and can be used in the hydrometer method. However, the larger sample (40 g) used in the hydrometer method will pack to the bottom of 250 mL bottles; hence, larger (>500 mL) shaking bottles are recommended for the larger samples to avoid this problem. High-speed reciprocating shakers have been used effectively on small samples of 10 g or less (El-Swaify, 1980). These high-speed shakers optimize dispersion when the liquid-to-solid ratio is about 5:1.

15–2.1.5.3 Ultrasonic Dispersion. The principle behind ultrasonic dispersion is the transmission of vibrating sound waves in the soil solution. The sound waves produce microscopic bubbles, which collapse, producing *cavitation.* The release of intense energy of cavitation literally blasts the soil aggregates apart, causing dispersion even in highly aggregated soils.

Much work has been done in testing the use of ultrasonic dispersion of soils, but no standard procedures have been adopted (Edwards & Bremner, 1964, 1967; Saly, 1967; Bourget, 1968; Watson, 1971; Kubota, 1972; Mikhail & Briner, 1978). An initial concern with this method of dispersion was the possible destruction of primary particles, but Saly (1967) reported that ultrasonic vibration did not cause destruction of the crystalline lattice or breakdown of primary grains. Edwards and Bremner (1964, 1967) investigated the use of ultrasonic dispersion in the absence of a dispersing agent. For mineralogical analysis, ultrasonic dispersion was preferred, since dispersion was achieved without soil pretreatment or addition of a dispersing agent. Edwards and Bremner summarized the following advantages of ultrasonic dispersion: (i) the resultant suspension is stable, hence flocculation does not occur during sedimentation; (ii) the method works well for dispersing calcareous soils, organic soils, and soils with high clay content; (iii) ultrasonic dispersion does not cause destruction of organic matter; and (iv) ultrasonic dispersion does not alter the soil pH, electrical conductivity, or cation exchange capacity. In contrast to the work of Edwards and Bremner, Mikhail and Briner (1978) reported that the most satisfactory method of pretreatment and dispersion involved the following steps; oxidation of organic matter, removal of carbonates and acid washing, and sodium saturation followed by ultrasonic dispersion. The results indicated that the highest degree of dispersion was achieved by this technique. Kubota (1972) reported that a sonic dispersion at low pH was effective in dispersing peroxide-treated volcanic ash soils. Each of the above authors used a different ultrasonic power and dispersion time, indicating that effective dispersion with ultrasonics is soil dependent.

For routine PSA, there is no standard method for ultrasonic mixing proposed at this time. Much additional research is needed to determine the effectiveness or limitations of ultrasonic dispersion for a wide range of soil materials.

15–2.2 Sieving

The typical particle size range for sieving is 2000 to 50 μm. Several limitations of sieving have been noted in the past. Day (1965) indicated that the probability of a particle passing through a sieve in a given time of shaking depends on the nature of the particle, the number of particles of that size, and the properties of the sieve. Particle shape and sieve opening shape affect probability of passage. For example, a particle whose shape permits its passage only in one orientation has a limited chance of getting through, except after prolonged shaking. Sieve openings are generally unequal in size, and extensive shaking is required before all particles have had the opportunity of approaching the largest openings. In fact, it is rare that complete sorting of a given size range can be achieved. Good reproducibility requires careful standardization of procedure.

15–2.3 Sedimentation

Sedimentation analysis relies on the relationship that exists between settling velocity and particle diameter. Settling velocity is related to the diameter of a spherical particle in the following way. The force acting downward on each particle due to its weight in water is

$$F_{\text{down}} = 4/3\ \pi\ (X^3/8)\ (\rho_s - \rho_l)g \qquad [1]$$

where X = particle diameter, ρ_s = particle density, ρ_l = liquid density, and g = acceleration due to gravity. Because of the viscous resistance of the water, the opposing upward force is

$$F_{up} = 3\ \pi\ X\eta v \qquad [2]$$

where η = fluid viscosity and v = velocity of fall. The resisting force is zero where velocity, v, is zero at time $t = 0$, and it increases with increasing v until it is equal to the downward force. For sedimenting particles in a dilute dispersent solution, it can be shown that the terminal velocity for silt- and clay-size particles is reached in a relatively short time (a few seconds).

Equating F_{down} and F_{up} relates the terminal velocity to the particle diameter as follows:

$$v = g\ (\rho_s - \rho_l)X^2/(18\ \eta)\,. \qquad [3]$$

A form of this relationship was first developed by Stokes (1851) and is now known as Stokes' Law. Basic assumptions used in applying Stokes' Law to sedimenting soil suspensions are:

1. Terminal velocity is attained as soon as settling begins.
2. Settling and resistance are entirely due to the viscosity of the fluid.
3. Particles are smooth and spherical.
4. There is no interaction between individual particles in the solution.

Gibbs et al. (1971) have shown that assumptions (1) and (2) are met by soil particles $< 80\ \mu m$ in diameter. Since soil particles are not smooth and spherical, X must be regarded as an "equivalent" rather than actual diameter. The assumptions of Stokes' Law as applied to soils are discussed fully by Krumbein and Pettijohn (1938).

In mineralogical analysis there is often a need to separate various clay fractions for specific analysis. The removal of the clay fraction by sedimentation can be accomplished by homogenizing a soil suspension and decanting all that remains above the plane $z = -h$ after time, t, where

$$t = 18\ \eta h/[g\ (\rho_s - \rho_l)X^2]\,. \qquad [4]$$

Quantitative separation by decantation requires that the residue be resuspended and decanted repeatedly to salvage those particles that were not previously at the top of the suspension at the start of the sedimentation period.

15–2.3.1 PRINCIPLE OF THE PIPET METHOD

The pipet method is a direct sampling procedure. It depends on taking a small subsample by a pipet at a depth h, at time t, in which all particles coarser than X have been eliminated. Using Stokes' Law in the form of Eq. [4], settling times for the clay fraction ($<2\ \mu m$) can be calculated for sampling at a given depth for a given temperature. Table 15–1 lists sampling times for the clay fraction for a 10-cm sampling depth at selected temperatures for the pipet technique. Tables 15–2 and 15–3 list sampling depths and times for various selected size fractions and specified settling times.

Experimental measurements with HMP solutions (Gee, unpublished data) show the following relationships for solution viscosity and density:

$$\rho_l = \rho^\circ(1 + 0.630\ C_s) \qquad [5]$$

where

ρ_l = solution density at temperature t, g/mL,
ρ° = water density at temperature t, g/mL,
C_s = concentration of HMP, g/mL,

and

$$\eta = \eta^\circ(1 + 4.25\ C_s) \qquad [6]$$

Table 15–1. Settling times for 2-μm clay at various temperatures. Calculated for a 10-cm sampling depth in distilled water, 0.5 g/L, and 5 g/L HMP solutions; with a particle density equal to 2.60 Mg/m^3.

	Viscosity			Settling time		
Temperature	Distilled H_2O	0.5 g/L HMP	5.0 g/L HMP	Distilled H_2O	0.5 g/L HMP	5.0 g/L HMP
°C	10^{-3} kg m^{-1} s^{-1}			h		
18	1.0530	1.0553	1.0759	8.39	8.41	8.58
20	1.0020	1.0042	1.0238	7.99	8.00	8.16
22	0.9548	0.9569	0.9756	7.61	7.63	7.78
24	0.9111	0.9131	0.9310	7.26	7.28	7.42
26	0.8705	0.8724	0.8895	6.94	6.95	7.09
28	0.8327	0.8345	0.8508	6.64	6.65	6.78
30	0.7975	0.7992	0.8149	6.36	6.37	6.50

Table 15–2. Selected depths for 2-μm clay at specified times and temperatures, assuming a particle density of 2.60 Mg/m^3 and dispersion of 0.5 g/L HMP solution.

		Sampling depth			
Temperature	Viscosity	4.5 h	5.0 h	5.5 h	6.0 h
°C	10^{-3} kg m^{-1} s^{-1}	cm			
20	1.0042	5.6	6.2	6.9	7.5
21	0.9800	5.8	6.4	7.0	7.7
22	0.9569	5.9	6.5	7.2	7.9
23	0.9345	6.0	6.7	7.4	8.1
24	0.9131	6.2	6.9	7.6	8.2
25	0.8923	6.3	7.0	7.7	8.4
26	0.8724	6.5	7.2	7.9	8.6
27	0.8532	6.6	7.4	8.1	8.8
28	0.8345	6.8	7.5	8.3	9.0
29	0.8166	6.9	7.7	8.4	9.2
30	0.7992	7.1	7.8	8.6	9.4

Table 15-3. Sampling times for 5-μm and 20-μm size fractions at a 10-cm sampling depth for pipet in 0.5 g/L HMP solution, over the temperature range 20 to 30 °C for selected particle densities.

Temperature	5-μm Particle size, Particle density (Mg/m^3)			20-μm Particle size, Particle density (Mg/m^3)		
	2.4	2.6	2.8	2.4	2.6	2.8
°C	time (min)					
20	87.7	76.8	68.3	5.5	4.8	4.3
21	85.7	75.0	66.7	5.4	4.7	4.2
22	83.7	73.2	65.1	5.2	4.6	4.1
23	81.7	71.5	63.6	5.1	4.5	4.0
24	79.9	69.9	62.1	5.0	4.4	3.9
25	78.0	68.3	60.7	4.9	4.3	3.8
26	76.3	66.8	59.3	4.8	4.2	3.7
27	74.6	65.3	58.0	4.7	4.1	3.6
28	73.0	63.9	56.8	4.6	4.0	3.5
29	71.4	62.5	55.6	4.5	3.9	3.5
30	69.9	61.2	54.4	4.4	3.8	3.4

where

η = solution viscosity at temperature t, 10^{-3} kg $m^{-1}s^{-1}$ (cpoise), and
η° = water viscosity at temperature t, 10^{-3} kg $m^{-1}s^{-1}$ (cpoise).

Equations [5] and [6] apply to HMP solutions in the range of 0 to 50 g/L. For tests with HMP solution concentrations in the range 0 to 5 g/L, $<$ 0.3% error in settling time results when the solution density is assumed to be that of pure water. Similarly low error results when 0.5 g/L HMP solutions are assumed to have the viscosity of pure water. However, settling-time errors as great as 2% result from not correcting for increased viscosity when using 5 g/L HMP solutions. Water densities and viscosities at various temperatures are available from Weast (1983).[3]

Particle densities should be known with a precision of at least $\pm$ 0.05 Mg/m^3. Settling-time errors in excess of 2% occur if particle densities are not known with at least this precision (see Table 15-3).

15-2.3.2 THEORY OF THE HYDROMETER METHOD

The hydrometer method, like the pipet method, depends fundamentally upon Stokes' Law, which for the hydrometer may be written as

$$X = \theta t^{-1/2} \qquad [7]$$

where θ is the sedimentation parameter and is a function of the hydrometer settling depth, solution viscosity, and particle and solution density. This relationship follows from Eq. [4] by rearranging terms such that

$$X = (18\eta h'/[g(\rho_s - \rho_l)])^{1/2}\, t^{-1/2}\,. \qquad [8]$$

[3]Note that Weast (1983) reports viscosity in centipoise (cpoise). For conversion to SI units, 1 cpoise = 10^{-3} kg $m^{-1}s^{-1}$.

Hence

$$\theta = (18\eta h'/[g(\rho_s - \rho_l)])^{1/2} \quad [9]$$

where h' = hydrometer settling depth, cm.

The hydrometer settling depth, h', is a measure of the effective depth of settlement for particles with diameter X. It can be related to the hydrometer stem reading, R, by considering the specific design and shape of the hydrometer (Kaddah, 1974; ASTM, 1983d). The relationship of the settling depth to the hydrometer dimensions can be approximated by

$$h' = L_1 + 1/2\,(L_2 - V_B/A) \quad [10]$$

where

L_1 = distance along the stem of the hydrometer from the top of the bulb to the mark for a hydrometer reading, cm,

L_2 = overall length of the hydrometer bulb, cm,

V_B = volume of hydrometer bulb, cm^3, and

A = cross sectional area of the sedimentation cylinder, cm^2.

For the ASTM 152H hydrometer and a standard sedimentation cylinder: L_1 = 10.5 cm for a reading, R, of 0 g/L and 2.3 cm for a reading, R, of 50 g/L; L_2 = 14.0 cm; V_B = 67.0 cm^3; and A = 27.8 cm^2. Substitution of these values into Eq. [10] and solving in terms of R yields

$$h' = -0.164\,R + 16.3 \quad [11]$$

where R is the uncorrected hydrometer reading. The use of Eq. [11] and [8] to calculate particle diameter is detailed in section 15–5.2.5.

Sedimentation parameter values, θ, as a function of hydrometer readings, R, have been tabulated for the ASTM 152H hydrometer for temperatures of 30 °C by Day (1965) and for 20 to 25 °C by Green (1981). Correction factors for other temperatures and for particle densities other than 2.65 g/cm^3 are given by Day (1965). However, the use of Eq. [9] and [11] provides a straight-forward method to determine θ for any given temperature and particle density; hence tabulated θ values are not reported here.

ASTM 152H hydrometers are calibrated at 20 °C directly in terms of soil solution concentration, expressed as grams of soil per liter of solution (ASTM, 1985d). Correction of hydrometer readings for other temperatures and for solution viscosity and density effects is made by taking a hydrometer reading, R_L, in a blank (no soil) solution. This reading should be taken immediately after the uncorrected reading, R, is taken. The corrected concentration of soil in suspension at any given time is $C = R - R_L$, where C is expressed in g/L.

Differences in particle density for different soils affect particle settlement time, hence requires the correction of hydrometer readings and sedimentation parameter values. However, Gee and Bauder (1979) and

ASTM (1985d) show that moderate changes in particle density have only small effects on a given size determination. For example, errors in particle density of ± 0.1 g/cm^3 result in errors of $< \pm 0.5$ wt% clay for soils with clay contents up to 50 wt%.

15–3 SAMPLE PREPARATION

15–3.1 Apparatus

1. Drying trays
2. Wooden rolling pin
3. Sodium hexametaphosphate (HMP) solution (50 g/L)
4. Sieves. Large 20.5 cm (8 in.) diameter, with a 2 mm (2000 μm) square hole screen.
   Other screen sizes needed include: 5, 20, and 75 mm (USDA 1982); 5 mm (#4), 13 mm (1/2 in.), 20 mm (3/4 in.), 25 mm (1 in.), 50 mm (2 in.), and 75 mm (3 in.) (ASTM, 1985d).
5. Ruler or caliper capable of measuring to 250 mm (10 in.).

15–3.2 Method

Spread the bulk sample thinly (in 2 to 3 cm thick layers, maximum) on trays and air-dry. Thoroughly mix and roll the sample with a wooden rolling pin to break up clods to pass a 2-mm sieve. Sieve out the >2-mm size fractions. Continue rolling and sieving until only coarse fragments that do not slake in water or HMP solution remain on the 2-mm screen. Use a rubber roller for samples with easily crushed coarse fragments. Sieve larger size fractions, record weights, and use total sample weight to calculate the percentage of total sample < 2 mm.

15–3.3 Comments

Sometimes it is desirable to keep the sample at field moist conditions. If this is determined appropriate, force the field moist sample through the 2-mm screen by hand, using a large rubber stopper, double bag the sample in plastic, and store for further use. From a separate subsample determine the water content, so that a check can be made on possible drying effects during storage.

Whether material over 2 mm in diameter is sieved depends on the purpose for the data set. For soil survey purposes, methods specified by the USDA (1982) may be used. For engineering purposes, the material >2 mm can be sieved according to requirements specified by ASTM method D-2487 (ASTM, 1985a).

Sample size depends upon the maximum size fragments present. Suggested sample sizes are:

1. Particles up to 20 mm diameter—use 5 kg or more

2. Particles up to 75 mm diameter—use 20 kg or more
3. Particles up to 250 mm diameter—use 100 kg or more.

Because of the large samples required, the volume percent of particles coarser than about 20 mm is usually estimated. A suggested procedure for handling coarser fragments follows.

Weigh and sieve the entire sample through 75- and 20-mm screens. Weigh the >75-mm and the 75- to 20-mm fractions. Take a subsample of the <20-mm fraction for laboratory processing. Weigh the <20-mm sample before and after air-drying and correct the total sample weight for the loss of water from field conditions. Separate and weigh the 2- to 5-mm and the 5- to 20-mm fractions. If fine earth adheres to the coarse fraction, wash the coarse material, dry, reweigh, and apply the appropriate corrections. Calculate the coarse fractions as a percentage of the <20 mm material (or the <75 mm or the <250 mm depending upon the size limit involved in sampling). Note that for taxonomic (classification) purpose, stones or rock fragments >250 mm (10 in.) are separated and used to estimate the volume of coarse fragments for family placement of soils. A large caliper or ruler can be used to check the dimensions of the >250-mm material. In addition, weight measurements and volume displacement techniques can be used to evaluate coarse fragment volume.

15–4 PIPET METHOD

The pipet method is often used as a standard method from which other PSA methods are compared. This procedure has been adapted from Day (1965) and Green (1981).

15–4.1 Apparatus and Reagents

1. Beakers—100 mL to 1000 mL; centrifuge bottles, both glass and plastic—250 mL.
2. Centrifuges—low speed, about 1500 rpm, and high speed, about 12 000 rpm, with 250-mL bottles.
3. Filter candle—Porus ceramic tube, 0.05 MPa (0.5 bar) pressure rated.
4. Shakers—horizontal reciprocating shaker, sieve shaker, wrist action shaker, holders for 250-mL centrifuge bottles on paint shaker.
5. Cylinders—1000 mL (height of 1000-mL mark, 36 ± 2 cm).
6. Large (no. 13) rubber stoppers for 1000-mL cylinder.
7. Stirrers—electric stirrers for mechanical mixing (available from Soil Test, Inc., Evanston, IL, or other source),[4] hand stirrer made by joining a brass rod about 50 cm long to the center of a thin circular piece of perforated brass or plastic sheeting. The circular plate should

[4]Trade names are used in this chapter soley for the purpose of providing specific information. Mention of a trade name does not constitute a guarantee of the product, nor does it imply an endorsement over other products not mentioned.

be cut to fit easily into the sedimentation cylinder. A 6-cm-diameter plate is normally adequate. If brass is used, place a wide rubber band around the edge of the brass sheeting to prevent scratching of the cylinder.

8. pH meter.
9. Pipet rack—device to permit sliding the pipet laterally and lowering the pipet to a percise depth in the sedimentation cylinder (Clark, 1962; Day, 1965; see also Fig. 15–4).
10. Lowy pipets—25 mL capacity (available from Sargent-Welch Co. or other source).
11. Weighing bottles—(beakers can be used).
12. Set of sieves—square mesh with bronze wire cloth, 7.6 cm (3 in.) diameter with the following openings: 1000, 500, 250, 106, 53, or 47 μm.
13. Reagents—hydrogen peroxide (~30%); 1 *M* NaOAc (adjusted to pH 5); citrate-bicarbonate buffer: prepare 0.3 *M* sodium citrate (88.4 g/L) and add 125 mL of 1 *M* sodium bicarbonate (84 g/L) to each liter of citrate solution; sodium dithionite (hydrosulphite); saturated NaCl solution; 10% NaCl solution; 1 *M* $AgNO_3$; 1 *M* $BaCl_2$; acetone; Na-hexametaphosphate (HMP), 50 g/L stock solution; 1 *M* $CaCl_2$; 1 *M* HCl.

15–4.2 Procedures

15–4.2.1 PRETREATMENT

15–4.2.1.1 Removal of Carbonates and Soluble Salts. Weigh a small portion of the <2-mm fraction of air-dry soil into a 250 mL centrifuge bottle (10 g for clays, 20 g for loams, 40 g for sandy loams and loamy sands, and 80 g for sands). Weights are optional, but these are generally suitable if clay samples are required for mineralogy. Add approximately 100 mL of water, mix, and add 10 mL 1 *M* NaOAc (adjusted to pH 5). Centrifuge (about 10 min at 1500 rpm) until the supernatant is clear, then pour it off. Wash the soil twice by shaking with 50 mL of water, centrifuging and discarding the centrifugate if it is clear. If the centrifugate is not clear (as is often the case for soils containing high amounts of soluble salts and soils containing gypsum), further washing may be necessary. Washing through a filter candle to remove salts is a permissible substitute for centrifugation, but this procedure takes considerably longer than centrifugation. Check for salts by testing with $AgNO_3$ for Cl^- and $BaCl_2$ for SO_4^{2-}.

15–4.2.1.2 Removal of Organic Matter. After carbonate removal, add 25 mL of water to the soil in the centrifuge bottle, and shake on a wrist action shaker. Transfer samples containing high amounts of organic matter (>5%) to 1000 mL beakers. Add 5 mL of (H_2O_2) to the soil suspension, stir, cover, and observe closely for several minutes. If excessive frothing occurs, cool the container in cold water. Add more H_2O_2 when the reaction subsides. Note that MnO_2 decomposes H_2O_2, so if present in measurable amounts, steps should be taken to complex or remove before peroxide treatment. Heat to 90 °C when frothing has ceased, remove cover, and evaporate excess water (do not take to dryness). Continue peroxide and heat treatment until most of the organic matter has been destroyed (as judged by the rate of reaction and the bleached color of the sample). Rinse down the sides of the reaction vessel occasionally. Heat for about an hour after the final addition of peroxide to destroy excess peroxide. Transfer the sample to a 250-mL glass centrifuge bottle.

15–4.2.1.3 Removal of Iron Oxides. Add citrate-bicarbonate buffer to the peroxide treated sample in the centrifuge bottle to bring the total volume of solution to approximately 150 mL. Shake to disperse the soil. Add 3 g of sodium dithionite ($Na_2S_2O_4$) gradually, as the sample may froth. Put the bottle into a water bath at 80° C and stir the suspension intermittently for 20 min. Remove the sample from the bath, add 10 mL of saturated NaCl, mix, centrifuge, and decant off the centrifugate. It may be combined with subsequent centrifugates, if any, and analyzed for dithionite-extractable Fe, Al, Mn, etc. If the sample is completely gray (gleyed), proceed to the next step. If brownish color remains, repeat the previous step. Wash the sample once with 50 mL of citrate–bicarbonate buffer plus 20 mL of saturated NaCl (shake, centrifuge, and decant). Wash the sample twice with 50 mL of 10% NaCl, then twice with 50 mL of distilled water. If the wash solution is not clear, transfer the sample to a plastic centrifuge bottle and centrifuge at high speed. If this fails to yield clear centrifugate, add acetone, warm the sample, and re-centrifuge. Add 150 mL of water, shake the sample, and check the pH. It should be above pH 8 if the soil is Na-saturated. Transfer the suspension to a 1-L shaker bottle, add 400 mL of distilled water and 10 mL of HMP (dispersant) stock solution, and shake overnight on a horizontal shaker.

15–4.2.2 SEPARATION OF THE SAND FRACTIONS

Pour the suspension through a 270-mesh (53 μm) sieve into a 1-L sedimentation cylinder. A 20-cm-diameter (8-in.) sieve is placed in a large funnel held by a stand above the cylinder. Tap the funnel gently and wash the sand thoroughly on the sieve. A soap solution placed on the sieve will aid in wetting the fine screen. Collect the washings in the cylinder. Transfer the sand to a tared beaker or aluminum weighing dish, dry (105 °C), and weigh.

Transfer the dried sand to the nest of sieves arranged from top to bottom with decreasing size in the following order: 1000-, 500-, 250-, 106-, 53-μm, and pan. Shake the sieves on a sieve shaker. A 3-min shaking time is usually adequate. Weigh each sand fraction and the residual silt and clay that passed through the 53-μm (270-mesh) sieve. Weighing precision of 0.01 g is adequate.

15–4.2.3 DETERMINATION OF SILT FRACTIONS

The 20 and 5 μm fractions can be determined by pipet by following the procedure outlined in the next section for clay and using Eq. [4] or Table 15–3 for determining the required settling times.

15–4.2.4 DETERMINATION OF CLAY (< 2 μm)

Place the cylinder containing the silt and clay suspension in a water bath; add 10 mL of HMP solution and make up to 1 L volume with distilled water; cover with a watch glass. Let the suspension stand at least several hours to equilibrate.

After equilibration, stir the suspension thoroughly with a hand stirrer for at least 30 s using an up-and-down motion. Note the time at completion of stirring and the temperature of the water bath. It is convenient to complete stirring of adjacent suspensions at intervals of about 3 min. An alternative to hand stirring is stoppering the sedimentation cylinder and shaking end-over-end for 1 min.

After the appropriate time interval (see Tables 15–1 through 15–3), lower the closed Lowy pipet carefully to the appropriate depth, turn on the vacuum, and withdraw a 25-mL sample in about 12 s (see Fig. 15–5). A device for controlling the vacuum is required.

Discharge the sample into a tared and numbered weighing bottle, beaker, or aluminum dish. Rinse the pipet with distilled water and add the rinse water to the clay suspension in the weighing bottle. Evaporate the water, dry the clay at 105 °C, cool in a desiccator, and weigh.

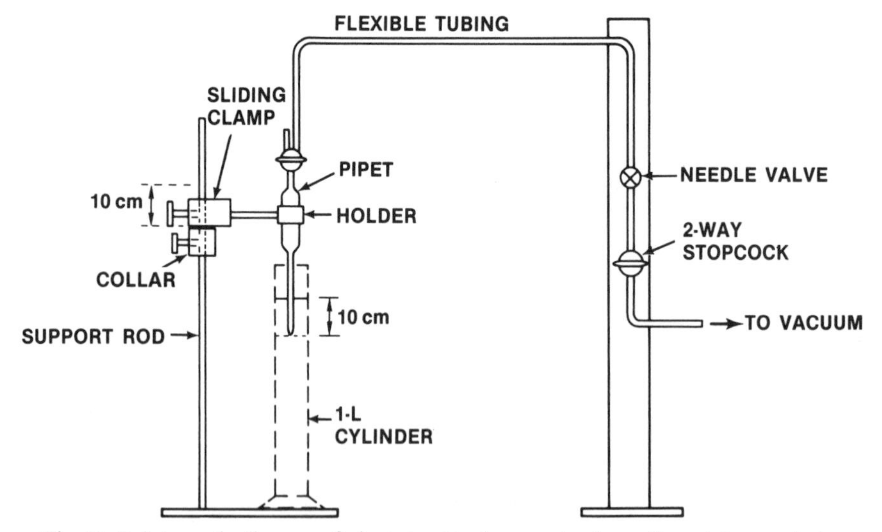

Fig. 15–5. Schematic diagram of pipet stand and apparatus for sedimentation analysis.

15–4.2.5 DETERMINING THE WEIGHT OF TREATED SOIL

Add 10 mL of 1 *M* $CaCl_2$ and 1 mL of 1 *M* HCl to the suspension remaining in the cylinder to prevent $CaCO_3$ formation. Siphon off the clear solution after flocculation has occurred. Transfer the soil from the cylinder to a tared beaker, evaporate, dry at 105 °C, and weigh.

Differences between original soil weight and weight found in the cylinder are attributed to pretreatment soil loss, solution loss, sieving loss, and sample removal for pipet sieving analysis. The total oven-dry weight of the treated sample is used as the basis for calculating the size fraction. The total oven-dry weight can be expressed as:

$$W_s + W_p + W_r = W_t \qquad [12]$$

where

W_s = oven dry weight of sand fraction,
W_p = corrected oven dry weights of pipet samples,
W_r = corrected oven dry weight of residual silt and clay, and
W_t = total weight of treated sample.

W_r and W_p are corrected by subtracting the weight of the dispersing agent (Table 15–4).

15–4.3.6 Calculations

Table 15–4 shows how the pipet method is used to determine size-fraction percentages using a 25-mL pipet.

15–4.4 Comments

Flocculation of clay from suspension has been observed in soils containing large amounts of gypsum (Kaddah, 1975; Hesse, 1976; Rivers et al., 1982). Flocculation is recognized by a distinct separation of clear liquid and suspended clay (flocculated clay often has the appearance of a cloudy gel-like precipitate). Removal of soluble salts (Section 15–4.2.1.1) helps prevent flocculation in most soils. However gypsum, having a low but measurable solubility, can cause flocculation by replacement of Na

Table 15–4. Example calculations of three particle-size percentages using a 25-mL pipet.

Particle size	Sample weight	Concentration	Corrected concentration†	Percent less than‡
mm	g	g/L		%
0.020	0.114	4.56	4.06	39.8
0.005	0.073	2.92	2.42	23.7
0.002	0.057	2.28	1.78	17.4

† Dispersing agent concentration = 0.5 g/L.
‡ Based on oven-dry weight of treated sample, W_t = 10.21 g.

with Ca. Procedures for removal of gypsum are available (Rivers et al., 1982). Options for removal of gypsum include adding barium (Hesse, 1976) or increasing the concentration of HMP dispersant (Kaddah, 1975). Flocculation must be prevented for sedimentation analysis (pipet, hydrometer, etc.) to provide meaningful results.

Errors in PSA values using the pipet analysis are mainly associated with sampling and weighing. With care, clay fractions can be determined with a precision of ±1 wt% using pipet procedures.

15–5 HYDROMETER METHOD

Particle-size analysis can be done conveniently with a hydrometer which allows for nondestructive sampling of suspensions undergoing settling. The hydrometer method provides for multiple measurements on the same suspension so that detailed particle-size distributions can be obtained with minimum effort. The hydrometer method outlined is that modified from Day (1965) and ASTM (1985d).

15–5.1 Apparatus and Reagents

1. Standard hydrometer, ASTM no. 152 H, with Bouyoucos scale in g/L (Fig. 15–6).
2. Electric stirrer (malted-milk-mixer type, with 10 000-rpm motor).
3. Plunger or rubber stoppers for 1000-mL sedimentation cylinders.
4. Sedimentation cylinders with 1-L mark 36 ±2 cm from the bottom of the inside.
5. Metal dispersing cups and 600-mL beakers.
6. Amyl alcohol.
7. Sodium-hexametaphosphate (HMP) solution (50 g/L).
8. Set of sieves—7.6-cm (3 in.) diameter square mesh woven bronze wire cloth, with the following openings: 1000, 500, 250, 106, 75, and 53 μm.
9. Electric oven and weighing jars.

15–5.2 Procedure

15–5.2.1 CALIBRATION OF HYDROMETER

Add 100 mL of the HMP solution to a cylinder and make the volume to 1 L with room temperature distilled water. Mix thoroughly with plunger

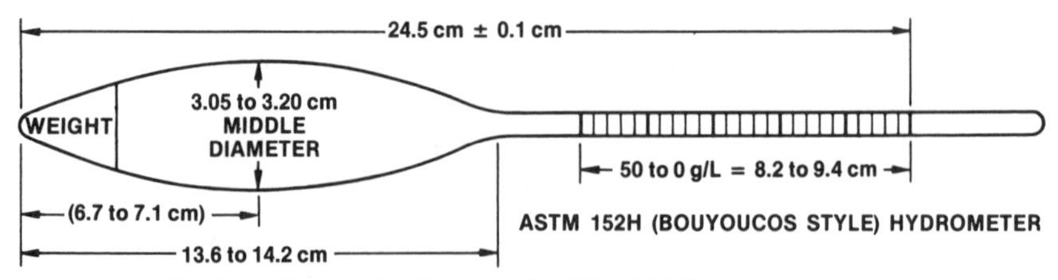

Fig. 15–6. Schematic diagram of ASTM 152 H-type hydrometer.

and record temperature. Lower the hydrometer into the solution and determine R_L, the hydrometer-scale reading of the (blank) solution. Read the upper edge of the meniscus surrounding the stem. Periodically recheck R_L during the course of the hydrometer tests (section 15–5.2.3). The calibration value R_L is used in the analysis to correct for solution viscosity and to correct the soil solution concentration, C.

15–5.2.2 DISPERSION OF SOIL

Weigh 40.0 g of soil into a 600-mL beaker, add 250 mL of distilled water and 100 mL of HMP solution, and allow the sample to soak overnight. The exact sample size depends upon soil texture. For fine-textured soils—silts or clays—10 to 20 g may be adequate. For coarse sands, 60 to 100 g will be needed in order to obtain reproducible results. Most temperate zone soils can be air dried prior to testing. However, for many tropical soils and soils of volcanic origin, samples must be stored at field moisture. Weigh another sample of the soil (about 10 g) for determination of oven-dry weight. Dry overnight at 105 °C, cool, and weigh.

Transfer the HMP-treated sample to a dispersing cup and mix for 5 min with the electric mixer, or transfer the suspension to shaker bottles and shake overnight on a horizontal shaker. Transfer the suspension to a sedimentation cylinder and add distilled water to bring the volume to 1 L.

15–5.2.3 HYDROMETER MEASUREMENTS

Allow time for the suspension to equilibrate thermally and record temperature. Insert plunger into cylinder and mix the contents thoroughly. Hold bottom of cylinder to prevent tipping. Dislodge sediment from the bottom using strong upward strokes of plunger. Finish stirring with two or three slow, smooth strokes. An alternative mixing procedure is to stopper the cylinder and use end-over-end shaking for 1 min. Add a drop of amyl alcohol if the surface of the suspension is covered with foam. As soon as mixing is completed, lower the hydrometer into the suspension and take readings after 30 s and again at the end of 1 min. Remove the hydrometer, rinse, and wipe it dry. Reinsert the hydrometer carefully about 10 s before each reading and take readings at 3, 10, 30, 60, 90, 120, and 1440 min. Times of reading can be modified according to need. Remove and clean the hydrometer after each reading. Record the reading R at each time. Read the hydrometer after placing it in the blank solution (containing no soil), and record the blank reading as R_L and the temperature at each time.

15–5.2.4 SEPARATION OF SAND FRACTIONS

Quantitatively transfer the sediment and suspension from the 1-L sedimentation cylinder through a 270-mesh (53-μm) sieve. A 20-cm-diameter (8 in.) sieve is placed over a sink. The sediment is washed onto the 53-μm screen using a wash bottle or gentle stream of water. The 53-μm screen can be dipped in a soap solution to improve the wettability of the screen and speed the flow. Transfer the sand to a tared beaker or aluminum weighing dish, dry (105 °C), and weigh.

Transfer the dried sand to the nest of sieves arranged from top to bottom in the following order: 1000, 500, 250, 106, and 53 μm. Shake on a sieve shaker for 3 min. Weigh each sand fraction and the residual silt and clay that has passed through the 53-μm sieve.

15–5.2.5 CALCULATION OF PARTICLE SIZE

Determine C, the concentration of soil in suspension in g/L, where $C = R - R_L$, with R, the uncorrected hydrometer reading in g/L, and R_L, the hydrometer reading of a blank solution. R and R_L are taken at each time interval. Determine P, the summation percentage for the given time interval, where $P = C/C_o \times 100$ and C_o = oven-dry weight of the soil sample.

Determine X, the mean particle diameter in suspension in μm at time t, using Eq. [7], [9], and [11]:

$$X = \theta t^{-1/2} . \qquad [13]$$

For the special case that X and t are reported in μm and min, respectively, and all other terms expressed in cgs units, the sedimentation parameter is commonly written as

$$\theta = 1000(Bh')^{1/2} , \qquad [14]$$

where $B = 30\eta/[g\,(\rho_s - \rho_l)]$, and $h' = -0.164R + 16.3$ (Eq. [11]), and with each term expressed in the following units:[5]

θ = sedimentation parameter, μm min$^{1/2}$,
h' = effective hydrometer depth, cm,
η = fluid viscosity in poise, g cm^{-1}s^{-1},
g = gravitational constant, cm/s^2,
ρ_s = soil particle density, g/cm^3, and
ρ_l = solution density, g/cm^3.

Equations [5] and [6] can be used to provide approximate corrections for density and viscosity variations for HMP solutions.

Plot a summation percentage curve (P vs. log X) using hydrometer readings taken over a time period from 0.5 min to 24 h coupled with sieve data. From this curve determine sand, silt, and clay percentages.

For routine textural analysis a summation percentage curve has more detail than is required; hence, the following procedure may be used.

[5]The sedimentation parameter and associated terms have not been expressed in S.I. units in order to maintain consistency with reported tables (Day, 1965; Weast, 1984).

15–5.2.5.1 Simplified Clay Fraction Procedure.

1. Take hydrometer readings at 1.5 and 24 h only (record both R and R_L values).
2. Determine effective particle diameter X and summation percentage P for 1.5- and 24-h readings using Eq. [7] and [13].
3. Compute $P_{2\mu m}$ (summation percentage at 2 μm) as follows:

$$P_{2\mu m} = m \ln (2/X_{24}) + P_{24} \quad [15]$$

where

X_{24} = mean particle diameter in suspension at 24 h (from Eq. [7]),

P_{24} = summation percentage at 24 h,

$m = (P_{1.5} - P_{24})/\ln (X_{1.5}/X_{24})$ = slope of the summation percentage curve between X at 1.5 h and X at 24 h,

$X_{1.5}$ = Mean particle diameter in suspension at 1.5 h, and

$P_{1.5}$ = summation percentage at 1.5 h.

15–5.2.5.2 Sand Fraction Calculation. Compute the 50-μm summation percentage, using the same procedure as for $P_{2\mu m}$, but use the 30- and 60-s hydrometer readings rather than the 1.5- and 24-h readings, respectively, and subtract the computed $P_{50\mu m}$ value from 100 to obtain the sand percentages. A standard sieve analysis can be run for comparison, using a 53- or 47-μm screen (section 15-5.2.4).

15–5.2.5.3 Silt Fraction Calculation. Determine the percent silt by difference as

$$\% \text{ silt} = 100 - (\% \text{ sand} + \% \text{ clay}) . \quad [16]$$

Calculations for sand, silt, and clay are conveniently made with a programmable desk calculator or microcomputer. BASIC and FORTRAN programs for clay fraction and textural determinations are available from the authors upon request.

15–5.2.6 COMMENTS

Flocculation of clay by soluble salts or gypsum during sedimentation may cause significant errors in the hydrometer method, since no pretreatment is used. Kaddah (1975) recommends increasing the concentration of HMP to levels high enough to maintain dispersion. If higher concentrations are used, the blank solution must contain the same concentration of HMP as that used in the soil solution so that the blank reading, R_L, corrects for the increased solution viscosity and density. If soil is high in soluble salts or gypsum, pretreatment procedures (section

15–4.2.1.1), removal techniques (Rivers et al., 1982), or chemical treatment (Hesse, 1976) may be needed.

The Bouyoucos procedure (Bouyoucos, 1962) has been used by a number of laboratories to estimate sand, silt, and clay from hydrometer measurements. Readings at 40 s and 2 h are used to estimate sand and clay percentages, respectively. From basic sedimentation theory, the 2-hr reading cannot yield correct estimates of the 2-μm clay fraction. Based on theoretical considerations, the 2-h hydrometer reading is a closer estimate of the 5-μm silt fraction than it is of the 2-μm clay fraction, and errors in clay contents using the 2-h reading often exceed 10 wt% for clay soils (Gee & Bauder, 1979). Similar problems arise when using the 40-s hydrometer reading to estimate the sand fraction. Differences between sieve and 40-s hydrometer measurement often exceed 5 wt%. The correlations between silt and clay and the 40-s and 2-h readings are empirical. In some cases, they seem adequate for textural class identification, but cannot be used to accurately define the particle size, hence, the Bouyoucos procedure is not recommended.

Walter et al. (1978) compared pipet and hydrometer measurements of 2μm size fraction in glacial till soils and found agreement well within 5%. Liu et al. (1966) also found generally good agreement between pipet and hydrometer analysis. Calculated correlation coefficients (r values) varied between 0.90 and 0.99 for 155 samples of soils from eleven states. These and other results suggest that pipet and hydrometer can give comparable results, with major differences arising largely from differences in pretreatment techniques.

A detailed error analysis for the hydrometer has been made by Gee and Bauder (1979). They indicate that the major source of error is in the hydrometer reading. An error of ± 1 g/L hydrometer reading results in an error of about ± 2 wt% for clay-size particles.

15–6 OTHER METHODS

In addition to sieving and sedimentation procedures, there are numerous techniques for measurement of particle-size distribution that have been developed for powder technology and other applications. These techniques include optical microscopy, transmission electron microscopy (TEM), scanning electron microscopy (SEM), electrical sensory zone (Coulter counter) methods, and light-scattering methods such as laser-light scattering, turbidimeters, holography, and x-ray centrifuges. An excellent discussion of these and other methods for particle-size distribution is given by Allen (1981).

Pennington and Lewis (1979) and Lewis et al. (1984) describe a procedure for using Coulter counters for particle-size distribution and textural analysis. Tama and El-Swaify (1978) have used turbidimeters to qualitatively assess clay contents in tropical soils. Weiss and Frock (1976) and Cooper et al. (1984) detail the use of laser light scattering methods for PSA and textural analysis. Laser-light instruments normally do not operate into the clay range; hence, a correction factor is used to estimate clay-size materials (Cooper et al., 1984). Soil mineralogy, particle shape, and density all affect this correction factor.

Standard procedures for PSA using Coulter counters, turbidimeters, or laser-light techniques are not proposed at this time. High cost of instrumentation coupled with uncertainties in correction factors make these methods less attractive than the pipet or hydrometer methods for most routine applications. However, in such applications as the analysis of runoff sediments, where great numbers of tests are required, the speed of these methods has encouraged their use, particularly when only relative values of particle size are considered adequate.

15–7 REFERENCES

Allen, T. 1981. Particle size measurement. 3rd ed. Chapman and Hall, New York.

American Society for Testing and Materials. 1985a. Standard test method for classification of soils for engineering purposes. D 2487-83. 1985 Annual Book of ASTM Standards 04.08:395–408. American Society for Testing and Materials, Philadelphia.

American Society for Testing and Materials. 1985b. Standard test method for liquid limit, plastic limit, and plasticity index of soils. D4318–84. 1985 Annual Book of ASTM Standards 04.08:767—782. American Society for Testing and Materials, Philadelphia.

American Society for Testing and Materials. 1985c. Standard test method for dispersive characteristics of clay soil by double hydrometer. D 4221-83a. 1985 Annual Book of ASTM Standards 04.08:733-735. American Society for Testing and Materials, Philadelphia.

American Society for Testing and Materials. 1985d. Standard test method for particle-size analysis of soils. D 422-63 (1972). 1985 Annual Book of ASTM Standards 04.08:117-127. American Society for Testing and Materials, Philadelphia.

Arya, L. M., and J. F. Paris. 1981. A physicoempirical model to predict the soil moisture characteristic from particle-size distribution and bulk density data. Soil Sci. Soc. Am. J. 45:1023–1030.

Blackmore, A. V. 1976. Subplasticity in Austrialian soils. IV. Plasticity and structure related to clay cementation. Aust. J. Soil Res. 14:261–272.

Bloemen, G. W. 1980. Calculation of hydraulic conductivities of soils from texture and organic matter content. Z. Pflanzenernaehr Bodenkd. 143:581–605.

Bourget, S. J. 1968. Ultrasonic vibration for particle size analyses. Can. J. Soil Sci. 48:372–373.

Bouyoucos, G. J. 1962. Hydrometer method improved for making particle size analysis of soils. Agron. J. 54:464–465.

Brewer, R., and A. V. Blackmore. 1976. Subplasticity in Austrialian soils. II. Relationship between subplasticity rating, optically oriented clay, cementation and aggregate stability. Aust. J. Soil. Res. 14:237–248.

Chu, T. Y., and D. T. Davidson. 1953. Simplified airjet apparatus for mechanical analysis of soils. Proc. Highway Res. Board 33:541–547.

Clark, J. S. 1962. Note on pipetting assembly for the mechanical analysis of soils. Can. J. Soil Sci. 41:316.

Cooper, L. R., R. L. Haverland, D. M. Hendricks, and W. G. Knisel. 1984. Microtrac particle size analyser: an alternative particle-size determination method for sediment and soils. Soil Sci. 138(2):138–146.

Day, P. R. 1965. Particle fractionation and particle-size analysis. p. 545–567. *In* C. A. Black et al. (ed.) Methods of soil analysis, Part I. Agronomy 9:545–567.

Edwards, A. P., and J. M. Bremner. 1964. Use of sonic vibration for separation of soil particles. Can. J. Soil Sci. 44:366.

Edwards, A. P. , and J. M. Bremner. 1967. Dispersion of soil particles by sonic vibration. J. Soil Sci. 18:47–63.

El-Swaify, S. A. 1980. Physical and mechanical properties of oxisols. p. 303–324. *In* B. K. G. Theng (ed.) Soils with variable charge. New Zealand Society of Soil Science, Lower Hutt, New Zealand.

Espinoza, W., R. H. Rust, and R. S. Adams, Jr. 1975. Characterization of mineral forms in andepts from Chile. Soil Sci. Soc. Am. Proc. 39:556–561.

Gee, G. W., and J. W. Bauder. 1979. Particle size analysis by hydrometer: a simplified method for routine textural analysis and a sen sitivity test of measurement parameters. Soil Sci. Soc. Am. J. 43:1004–1007.

Green, A. J. 1981. Particle-size analysis. p. 4–29. *In* J. A. McKeague (ed.) Manual on soil sampling and methods of analysis. Canadian Society of Soil Science, Ottawa.

Gibbs, R. J., M. D. Matthews, and D. A. Link. 1971. The relationship between sphere size and settling velocity. J. Sed. Petrol. 41:7–18.

Hesse, P. R. 1976. Particle-size distribution in gypsic soils. Plant Soil 44:241–247.

Hillel, D. 1982. Introduction to soil physics. Acadamic Press, New York.

Irani, R. R., and C. F. Callis. 1963. Particle size. Measurement, interpretation and application. John Wiley and Son, New York.

Jackson, M. L. 1969. Soil chemical analysis—advanced course. 2nd ed. University of Wisconsin, Madison, WI.

Kaddah, M. T. 1974. The hydrometer method for detailed particle size analysis. I. Graphical interpretation of hydrometer reading and test of method. Soil Sci. 118:102–108.

Kaddah, M. T. 1975. The hydrometer method for particle size analysis. 2. Factors affecting the dispersive properties of glossy Na-polyphosphate in calcareous saline soil suspensions. Soil Sci. 120:412–420.

Krumbein, W. C., and F. J. Pettijohn. 1938. Manual of sedimentary petrography. D. Appleton-Century Co., New York.

Kubota, T. 1972. Aggregate-formation of allophanic soils: effects of drying on the dispersion of the soils. Soil Sci. Plant Nutr. 18:79–87.

Lewis, G. C., M. A. Fosberg, and A. L. Falen. 1984. Identification of Loess by particle size distribution using the Coulter Counter TA II. Soil Sci. 137:172–176.

Liu, T. K., R. T. Odell, W. C. Etter and T. H. Thornburn. 1966. Comparison of clay contents determined by hydrometer and pipette methods using reduced major axis analysis. Soil Sci. Soc. Am. Proc. 30:665–669.

Maeda, T., H. Takenaka, and B. P. Warkentin. 1977. Physical properties of allophane soils. Adv. Agron. 29:229–263.

McIntyre, D. S. 1976. Subplasticity in Australian soils. I. Description, occurrence and some properties. Aust. J. Soil Res. 14:227–236.

McKeague, J. A. (ed.) 1978. Manual on soil sampling and methods of analysis. Canadian Society of Soil Science, Ottawa, Canada.

Mehra, O. P., and M. L. Jackson. 1960. Iron oxide removal from soils and clays by a dithionite–citrate system buffered with sodium bicarbonate. p. 237–317. *In* Clays and clay minerals. Proc. 7th Conf. Natl. Acad. Sci. Natl. Res. Counc. Pub., Washington, DC.

Mikhail, E. H., and G. P. Briner. 1978. Routine particle size analysis of soils using sodium hypochlorite and ultrasonic dispersion. Aust. J. Soil Res. 16:241–244.

Norrish, K., and K. G. Tiller. 1976. Subplasticity in Australian soils. V. Factors involved and techniques of dispersion. Aust. J. Soil. Res. 14:273–289.

Pennington, K. L., and G. C. Lewis. 1979. A comparison of electronic and pipet method for mechanical analysis of soils. Soil Sci. 128:280–284.

Rivers, E. D., C. T. Hallmark, L. T. West, and L. R. Drees. 1982. A technique for rapid removal of gypsum from soil samples. Soil Sci. Soc. Am. J. 46:1338–1340.

Saly, R. 1967. Use of ultrasonic vibration for dispersing of soil samples. Sov. Soil Sci. 1967:1547–1559.

Schalscha, E. B., C. Gonzales, I. Vergara, G. Galindo, and A. Schatz. 1965. Effect of drying on volcanic ash soils in Chile. Soil Sci. Soc. Am. Proc. 29:481–482.

Sherard, J. L., L. P. Dunnigan, and R. S. Decker. 1976. Identification and nature of dispersive soils. Am. Soc. Civ. Eng. J. Geotech. Eng. 101(11846):69–85.

Soil Survey Staff. 1975. Soil taxonomy: A basic system of soil classification for making and interpreting soil surveys. USDA-SCS Agric. Handb. 436. U.S. Government Printing Office, Washington, DC.

Stevens, J. 1982. Unified soil classification system. Civil Engineering. December. p. 61–62.

Stokes, G. G. 1851. On the effect of the lateral friction of fluids on the motion of pendulums. Trans. Cambridge Phil. Soc. 9:8–106.

Tama, K., and S. A. El-Swaify. 1978. Charge, colloidal, and structural stability interrelationships for oxidic soils. p. 41–52. *In* W. W. Emerson, R. D. Bond, and A. R. Dexter (eds.) Modification of soil structure. John Wiley and Sons, New York.

Theisen, A. A., D. D. Evans, and M. E. Harward. 1968. Effect of dispersion techniques on mechanical analysis of Oregon soils. Oregon Agric. Exp. Stn. Tech. Bull. 104.

Todd, D. K. 1964. Groundwater. p. 13–8, 13–9, 13–10. *In* V. T. Chow (ed.) Handbook of applied hydrology. McGraw-Hill, New York.

U.S. Department of Agriculture. 1982. Procedures for collecting soil samples and methods of analysis for soil survey. Soil Survey Investigations Report no. 1. Soil Conservation Service, Washington, DC.

Veneman, P. L. M. 1977. "Calgon" still suitable. Soil Sci. Soc. Am. J. 41:456.

Walker, P. H., and J. Hutka. 1976. Subplasticity in Austrialian soils. III. Disaggregation and particle-size characteristics. Aust. J. Soil Res. 14:249–260.

Walter, N. F., G. R. Hallberg, and T. S. Fenton. 1978. Particle-size analysis by the Iowa State University Soil Survey Laboratory. p. 61–74. *In* G. R. Hallberg (ed.) Standard procedures for evaluation of quaternary materials in Iowa. Iowa Geological Survey, Iowa City, IA. TIS 8.

Warkentin, B. P., and T. Maeda. 1980. Physical and mechanical characteristics of andisols. p. 281–301. *In* B. K. G. Theng (ed.) Soils with variable charge. New Zealand Society of Soil Science, Lower Hutt, New Zealand.

Watson, J. R. 1971. Ultrasonic vibration as a method of soil dispersion. Soil Fertil. 34:127–134.

Weast, R. C. (ed.) 1983. CRC handbook of chemistry and physics. 64th ed. CRC Press, Boca Raton, FL.

Weiss, E. L., and N. H. Frock. 1976. Rapid analysis of particle-size distribution by laser light scattering. Powder Technol. 14:287–293.

Yaalon, D. H. 1976. "Calgon" no longer suitable. Soil Sci. Soc. Am. J. 40:333.

Yong, R. N., and B. P. Warkentin. 1966. Introduction to soil behavior. Macmillan Co., New York.

16 Specific Surface

D. L. CARTER
Agricultural Research Service, USDA
Kimberly, Idaho

M. M. MORTLAND
Michigan State University
East Lansing, Michigan

W. D. KEMPER
Agricultural Research Service, USDA
Kimberly, Idaho

16–1 INTRODUCTION

Surface area largely determines many physical and chemical properties of materials. Physical adsorption of molecules, heat loss or gain resulting from that adsorption, swelling and shrinking, and many other physical and chemical processes are closely related to surface area. Surface or exposed area is also closely related to and often the controlling factor in many biological processes. Soils vary widely in their reactive surface because of differences in mineralogical and organic composition and in their particle-size distribution. Water retention and movement, cation exchange capacity, and pesticide adsorption are closely related to the *specific surface* (defined as the surface area per unit mass of soil). Specific surface is usually expressed in square meters per gram (m^2/g).

Clay-size particles, and particularly some layer silicate minerals, contribute most of the inorganic surface area to soils. Nonexpanding layer silicates such as kaolinite and some micas have only external surfaces. The specific surface of these minerals ranges from 10 to 70 m^2/g. Expanding layer silicates, such as montmorillonites, other smectites, and vermiculites, have extensive internal as well as external surface, giving specific surfaces up to 810 m^2/g, depending upon the amount of internal surface exposed by expansion. Consequently, the types of minerals present in soil largely determine soil specific surface and related properties.

Many methods and approaches have been employed to measure the specific surface of soils and minerals. Some methods are thermodynamically sound but too time-consuming, or they require too much spe-

 Methods of Soil Analysis, Part 1. Physical and Mineralogical Methods—Agronomy Monograph no. 9 (2nd Edition)

cialized apparatus to be used on a routine basis. Therefore, other more rapid but less accurate methods have been developed. Cihacek and Bremner (1979) suggested that the assumptions required and the several factors influencing results from most of these latter proceedures give us an assessment but not a real measure of surface area. Absolute measurements are difficult to attain and interpret because of the interaction of factors such as the adsorbed cation, the orientation of the adsorbed molecules, mono- or duo-layer coverage in the interlayer spacing, and sample water-content.

The purpose of surface-area measurements is commonly to determine the accessibility of internal surfaces of the clay mineral complex to molecules or ions which can be adsorbed thereon. This accessibility is generally determined by the relative adsorption forces of the introduced molecules or ions to the clay mineral surfaces as compared to the attraction forces between adjacent clay mineral platelets. The systems of most general interest are hydrated systems. Consequently, a measurement procedure which estimates surface area that would be accessible under hydrated conditions is commonly desired. Both ethylene glycol and ethylene glycol monoethyl ether appear to be adsorbed on clay mineral surfaces and to adsorbed cations to essentially the same extent, and probably by the same mechanisms, as is water in hydrated systems.

16–2 PRINCIPLES

16–2.1 Adsorption Isotherms

Gas molecules close to the solid surface are attracted by forces arising from solid-phase surface atoms. Essentially, all gases tend to be adsorbed on solid surfaces in response to the force field at or very near the surface. The quantity of gas adsorbed can provide a measure of surface area.

Langmuir (1918) provided the equation

$$P/V = 1/k_2 V_m + P/V_m$$

where V is the volume of gas adsorbed per gram of adsorbent at pressure P, k_2 is a constant, and V_m is the gas volume adsorbed per gram when a complete monomolecular layer has been formed. The surface area is obtained by plotting P/V vs. P at constant temperature. The slope of the plot is equal to $1/V_m$. Knowing this value, the specific surface of the adsorbent may be calculated by determining the number of molecules in V_m and multiplying by the cross-sectional area of the adsorbate. The Langmuir equation is based upon assumptions that (i) only one layer of molecules is adsorbed and (ii) the heat of adsorption is uniform during adsorption of the monomolecular layer.

Brunauer, Emmett, and Teller (1938) derived an equation, now commonly called the BET equation, from multimolecular adsorption theory that provided for calculating the number of adsorbate molecules in a monolayer. The BET equation is

$$P/V(P_0 - P) = (1/V_mC) + [(C - 1)P/V_mCP_0]$$

where V is the gas volume adsorbed at pressure P, V_m is the volume of gas required for a single molecular layer over the entire adsorbent surface, P_0 is the gas pressure required for saturation at the temperature of the experiment, and $C = \exp [E_1 - E_2]/RT$, where E_2 is the heat of liquification of the gas, E_1 is the heat of adsorption of the first layer of adsorbate, R is the gas constant, and T is the absolute temperature. The BET equation involves assumptions that (i) the heat of adsorption of all molecular layers after the first is equal to the heat of liquification and (ii) at equilibrium, the condensation rate on the surface is equal to the evaporation rate from the first or subsequent layers. The BET equation is most useful between relative pressures (P/P_0) of 0.05 to 0.45. By plotting $P/V(P_0 - P)$ vs. P/P_0, V_m can be calculated from the intercept and slope of the linear portion of the curve. The surface area can then be calculated by multiplying V_m by the adsorbate molecule cross-sectional area. The density of the adsorbate is usually assumed to be either that of the liquified or solidified gas. Since the calculated area per molecule depends upon the density used, there is a degree of uncertainty in the absolute surface areas measured.

The BET equation has been applied by many researchers utilizing nitrogen, ethane, water, ammonia, and other gases on soils and minerals. The apparent surface area of montmorillonites has been shown to depend upon the nature of the adsorbate used. Weakly adsorbed nitrogen, for instance, does not penetrate the interlayer surfaces, so that the measurement obtained is only for external surfaces—whereas polar molecules such as water and ammonia are strongly adsorbed and penetrate into the interlayer surfaces, giving more nearly total surface-area measurements. The use of the BET equation in determining surface area of clay minerals from water adsorption has been criticized because water molecules tend to cluster around cation sites (Quirk, 1955).

16–2.2 Retention of Polar Liquids

Dyal and Hendricks (1950) introduced a method for estimating surface area of clays based upon adsorption of ethylene glycol to form a monomolecular layer over the entire surface. Their work suggested that other polar liquids could also be used for measuring surface area. The method involves adding ethylene glycol to properly pretreated soil or clay mineral samples and evaporating the excess in an evacuated system. The evaporation rate decreases when all free ethylene glycol is gone and only that adsorbed in a monomolecular layer is left. The quantity of glycol retained at the moment the evaporation rate decreases should be proportional to the surface area. Utilizing a theoretically specific surface for bentonite of 810 m^2/g and the amount of ethylene glycol retained when the evaporation rate decreased, Dyal and Hendricks (1950) found that 0.00031 g of ethylene glycol was needed to form a monomolecular layer on each square meter of bentonite surface.

Bower and Gschwend (1952) utilized ethylene glycol adsorption as a measure of surface area and interlayer swelling of layer silicates. Samples were weighed at consecutive time intervals to determine the transition from a high rate of free glycol loss to a lower loss rate for that bound to the clay surface. The vapor pressure of ethylene glycol was considered of critical importance in gravimetric methods for estimating surface area. Where ethylene-glycol-saturated samples were maintained in an evacuated chamber over dry $CaCl_2$, the vapor pressure was low and losses beyond the transition from high to low rate loss occurred. Martin (1955) included a free surface of ethylene glycol with the samples in order to maintain a higher ethylene glycol vapor pressure. As long as both free ethylene glycol and $CaCl_2$ are present, the amount of ethylene glycol retained by a sample should reach and maintain a constant value.

Bower and Goertzen (1959) introduced the use of a $CaCl_2$–ethylene glycol complex ($CaCl_2$-monoglycolate) to maintain an ethylene glycol vapor pressure slightly lower than required for monolayer formation. Mixtures of an anhydrous and a solvated form, or mixtures of two solvated forms of salts, have a definite vapor pressure at a given temperature. This pressure is independent of the relative proportions of the two forms of the salt present. (For an explanation of this phenomenon, see Prutton & Maron, 1951, p. 372–375.)

Another approach was to include a large quantity of bentonite, containing just enough ethylene glycol to form a monomolecular on all surfaces, to maintain an ethylene glycol vapor pressure in the evacuated desiccator equal to that necessary for forming a monomolecular layer on the surfaces of small samples. The environment provided by this system allows a true equilibrium value to be reached (Sor & Kemper, 1959). The ethylene glycol retained by samples in this latter system is slightly higher than the amount retained in the environment provided by the presence of both the $CaCl_2$–monoglycolate and free $CaCl_2$. Using a $CaCl_2$–monoglycolate with $CaCl_2$ to stabilize the ethylene glycol vapor pressure in an equilibrium procedure has the advantage of using pure, commercially available materials.

Glycerol has also been used as a polar molecule to measure the surface area of layer silicate minerals (Diamond & Kinter, 1956). They found that montmorillonite forms a stable complex with glycerol in the presence of glycerol vapor at high temperatures, and that a monomolecular layer of glycerol molecules occurs in the interlayer spacings of most layer silicates.

Carter et al. (1965) introduced the use of ethylene glycol monoethyl ether (EGME) as the polar molecule for determining the surface area of layer silicate minerals and soils. This polar liquid has a higher vapor pressure at room temperature than does ethylene glycol. Hence, it evaporates more rapidly from treated samples, requiring a shorter time to evaporate free liquid and to attain equilibrium with a monomolecular layer. In addition to the advantages of being more rapid, the EGME method generally has greater precision than the ethylene glycol method because samples are handled fewer times and the opportunities for introducing water vapor and other sources of errors are decreased.

Molecular coverage for EGME was calculated with the same assumption applied earlier by Dyal and Hendricks (1950) for ethylene glycol. Using 810 m^2/g as the theoretical specific surface for montmorillonite, and a measured value of 23.7 mg EGME retained per gram of clay, the calculated quantity of EGME to cover 1 m^2 of clay surface with one molecular layer is 0.000286 g, with a molecular coverage of 5.2×10^{-15} cm^2/molecule. Applying these coverage values for EGME and assuming 0.00031 g of ethylene glycol per square meter as reported by Dyal and Hendricks (1950), extensive comparisons for layer silicates, layer silicate mixtures of known composition, and soils were made between the EGME and ethylene glycol methods. Excellent agreement was found in all comparisons, indicating that the two polar molecules cover the same surface (Carter et al., 1965; Heilman et al., 1965).

The EGME forms a solvate with $CaCl_2$ that is stable at 70 °C. The EGME/$CaCl_2$ ratio is 1.5:1, indicating that three molecules of EGME solvate two molecules of $CaCl_2$. This solvate may be used in the desiccator with $CaCl_2$ to assure an EGME vapor pressure near that of an adsorbed monomolecular layer. Maintaining the adsorbate vapor pressure in the sample environment is generally easier with the EGME method than with the ethylene glycol method because of the shorter time required for attaining a monomolecular layer and the fewer sample-handling events. Samples must be weighed within a reasonable time after evacuating the desiccator. It is advisable to include the solvate along with $CaCl_2$ when time periods such as overnight or over a weekend are involved, in order to prevent some EGME from being lost from the monolayer. Eltanawy and Arnold (1973) suggested that such losses occur even in the presence of EGME–$CaCl_2$ solvates and contended that a free EGME surface is required in the desiccator to provide sufficient vapor pressure to assure a monomolecular layer. Their own data, however, indicate that EGME in excess of that required to form a monomolecular layer is retained under conditions involving a free EGME surface.

The convenient and rapid EGME method for estimating surface area of soils and clays has been widely accepted. Nevertheless, it involves assumptions and has limitations. One assumption involves the method used to calculate molecular coverage. Similar to the assumption made for ethylene glycol, it was assumed that EGME covered all interlayer and external surfaces. This assumption is difficult to prove. It is possible that the 0.000286 g of EGME required for a monomolecular layer on 1 m^2

of surface is not absolutely correct. It is probable that EGME molecules are associated with exchangeable cations in thicknesses greater than required for a monomolecular layer, as has been reported for ethylene glycol (McNeal, 1964). It is also possible that some voids occur on certain minerals so that coverage is incomplete in the interlayer spaces (Dowdy & Mortland, 1967). There are also questions regarding the density of adsorbed EGME in comparison to the density of bulk liquid EGME. The latter has been used in calculations of molecular coverage.

As is the case with ethylene glycol, vermiculite likely adsorbs only one layer of EGME molecules between adjacent clay platelets. Hence, the assumption that all surfaces are covered with a monomolecular layer of EGME probably does not hold for vermiculite, hydrated halloysite, and possibly some intergrade layer silicates, as is the case for ethylene glycol.

The wide acceptance of the EGME method is related to the simplicity and ease with which it can be used in the laboratory. This, in turn, has led to more widespread relation of surface-area values of soils and layer silicates to various physical, chemical, and biological processes. Ross (1978) and Low (1980) have shown that shrink-swell properties are highly correlated with the surface areas of soils and clays. Supak et al. (1978) related the adsorption of aldicarb by clays to their specific surface values. Van der Staay and Focht (1977) investigated the effects of soil surface area upon bacterial denitrification rates. Moreale and Van Bladel (1979) related specific surface to herbicide-derived aniline residue. Also, because surface-area estimates are easily obtained by the EGME method, the property is commonly used in characterizing soils and in relating specific surface to other soil properties (Bingham et al., 1978; DeKimpe & Laverdiere, 1980; DeKimpe et al., 1979; Farrar & Coleman, 1967; Galindo & Bingham, 1977; Gallez et al., 1976; Singer & Navrot, 1977).

Some other organic chemicals may have equal or greater potential for use in estimating surface area of soils and clay minerals. Porter (1971) reported that 2-h desorption curves for dioxane, 2-methoxy-ethanol, and tetrahydrofluran from Na-saturated bentonite were quite similar to the desorption curve for EGME. He found excellent relationships between the EGME retained by various ratios of Na-kaolinite to Na-bentonite and the 2-methoxyethanol, dioxane, and tetrafluran retained by the same mixtures. He concluded that these materials should be as effective and rapid as EGME in measuring surface areas of Na-saturated systems.

A method for measuring surface area of clays involving methylene blue as the adsorbate has been published by Pham and Brindley (1970). An advantage of this method is that adsorption is carried out in fully hydrated systems. Clays are exposed to various concentrations of methylene blue and the amount adsorbed is measured spectrophotometrically by measuring the decrease in concentration of the supernatent solution. The surface area is calculated by taking the area of the adsorbed molecule as 1.30 nm^2, which corresponds to flat orientation of the molecule on the mineral surface. A good relationship was found between surface area by this method and by use of a BET method.

16–3 METHOD

16–3.1 Special Apparatus

1. Vacuum desiccator 25 cm or larger in diameter with a plate to hold samples above the desiccant.
2. Vacuum pump capable of reducing pressure to 0.250 mm of Hg.
3. Aluminum cans having a diameter of 6 to 7 cm and a height not exceeding 2 cm, with lids.
4. Culture chambers consisting of a glass dish with cover, having a diameter of about 20 cm and a height of about 7.5 cm.
5. Support for holding aluminum cans in each culture chamber approximately 2 cm above the bottom of the chamber. The support may consist of a circular piece of hardware cloth with openings of 0.5 to 1.0 cm, with brass machine screws or similar devices for legs attached near the perimeter of the hardware cloth by means of nuts and washers. Alternatively a square piece of hardware cloth with the corners bent down to serve as legs may be used.

16–3.2 Chemicals

1. Ethylene glycol monoethyl ether (2-ethoxyethanol) (EGME), reagent grade.
2. Phosphorus pentoxide (P_2O_5).
3. Calcium chloride ($CaCl_2$), 40-mesh (0.425 mm opening) anhydrous, reagent grade.

16–3.3 Procedure

16–3.3.1 PREPARATION OF $CaCl_2$–EGME SOLVATE

Weigh approximately 120 g of 40-mesh $CaCl_2$ into a 1-L beaker and dry in an oven at 210 °C for 1 h or more to remove all traces of water. Weigh 20 g of EGME into a 400-mL beaker. Remove the $CaCl_2$ from the oven, weigh out 100 g without cooling, and add it to the beaker containing the EGME. Mix immediately and thoroughly with a spatula. The heat of the $CaCl_2$ facilitates solvation. After the solvate has cooled, transfer it to a culture chamber and spread it uniformly over the bottom. Store the chamber and contents in a sealed desiccator.

16–3.3.2 SAMPLE PRETREATMENT

Treat the sample with H_2O_2 as described in chapter 5, to remove organic matter. Saturate the sample with Ca by leaching or repeated shaking and centrifuging with an excess of 1.0 *M* $CaCl_2$. Remove the excess $CaCl_2$ with three successive water washings. Air dry the sample and pass it through a 60-mesh sieve, grinding if necessary. If a measure of only the external surface is desired, heat the sample at 600 °C for 2 h to suppress interlayer swelling.

16–3.3.3 SORPTION TECHNIQUE

Weigh approximately 1.1 g of soil or clay into a tared aluminum can, including a lid, and spread the sample evenly over the bottom of the can. Place the can, with lid beneath, in a vacuum desiccator over about 250 g of P_2O_5, evacuate the desiccator by applying a vacuum pump for one hour, close the stopcock, and dry to constant weight. Constant weight is usually attained in about 6 to 7 h for groups of four to six samples. The drying is most conveniently accomplished overnight. Weigh the dried sample, using care to minimize adsorption of atmospheric water. Wet the sample with approximately 3 mL of reagent-grade ethylene glycol monoethyl ether (EGME) to form a soil- or clay-adsorbate slurry. Place the can containing the sample-adsorbate slurry, with lid beneath, in a culture chamber on the hardware cloth support over the $CaCl_2$–EGME solvate. Place the lid on the culture chamber. (Elevation of the lid with a small block to leave a space approximately 2 mm wide between lid and chamber will better allow gases to escape.) Place the entire culture chamber in a vacuum desiccator containing $CaCl_2$. Allow 30 min or more for the sample-solvate slurry to equilibrate. Evacuate the desiccator with a vacuum pump for about 45 min. Allow the desiccator to stand at room temperature for 4 to 6 h, release the vacuum, open the desiccator and culture chamber, and place the lid on the aluminum can to prevent the sample from adsorbing atmospheric water. Weigh the can, lid, and sample. Return the can, with lid beneath, to the culture chamber and the culture chamber to the desiccator. Evacuate the desiccator by applying a vacuum pump for 45 min. Weigh the samples at 2- to 4-h intervals, evacuating between weighings, until constant weight is attained. Generally constant weight will be indicated by the second or third weighing. If surface area determinations are made on several samples concurrently, a point is often reached when some samples appear to gain and some to lose a fraction of a milligram between two successive weighings. This is good indication that equilibrium has been attained. Use the mean of two successive weights that agree within a few tenths of a milligram to calculate the quantity of ethylene glycol monoethyl ether retained by the sample.

16–3.3.4 CALCULATING SPECIFIC SURFACE

Calculate the specific surface by the equation

$$A = W_a/(W_s \times 0.000286)$$

where A = specific surface in m^2/g, W_a = weight of ethylene glycol monoethyl ether (EGME) retained by the sample in g, W_s = weight of P_2O_5-dried sample in g, and 0.000286 is the weight of EGME required to form a monomolecular layer on a square meter of surface.

16–3.4 Comments

Smaller samples, 0.3 to 1.1 g, may be used for samples having high specific surface.

The vacuum pump and desiccators should be connected with tight-fitting, vacuum-type rubber tubing. A glass tube filled with the 8-mesh anhydrous $CaCl_2$ should be inserted in the line to prevent undesirable vapors from entering the pump. High-vacuum stopcock lubricant should be used to seal glass joints.

Adsorption of atmospheric water by the sample during weighing operations is controlled by allowing air to flow back through the $CaCl_2$ tube into each desiccator when releasing the vacuum, by placing the lid on each can promptly after releasing the vacuum, and by weighing rapidly. Determinations may be made on as many as six samples concurrently.

Use of the $CaCl_2$–EGME solvate and the associated culture chamber is optional for this method. Inclusion of the solvate insures equilibrium with approximately a monomolecular layer. However, reasonable results are also attainable without the solvate (Heilman et al., 1965; Cihacek & Bremner, 1979). The migration of excess EGME vapor to the $CaCl_2$ spontaneously forms a small amount of $CaCl_2$–EGME solvate.

It is desirable to prepare fresh $CaCl_2$–EGME solvate, where used, for each set of determinations, as some absorption of atmospheric water with continuing use is unavoidable. The P_2O_5 employed for drying may be used until it absorbs sufficient water to develop a syrupy consistency.

According to Diamond and Kinter (1956), some montmorillonites re-expand after being heated at 600 °C. Therefore, measurements on samples which have received this pretreatment do not always provide an unbiased estimate of external surface. Prior saturation with a less readily hydrated cation, such as K^+, should help in this regard, though recalibration of the procedure with pure montmorillonite will also be required.

Specific surface determinations for large numbers of samples can be completed in a 2-day routine (Heilman et al., 1965). Drying over P_2O_5 is accomplished the 1st day or overnight. The equilibration to constant weight, indicative of a monomolecular layer, can then be accomplished the 2nd day or overnight.

Specific surface determinations for many samples can be made with reasonable accuracy by this method without H_2O_2 pretreatment to destroy organic matter (Cihacek & Bremner, 1979).

Samples may be oven-dried at 110 °C for 24 h instead of drying over P_2O_5 in most situations. Eltanawy and Arnold (1973) claim more complete drying in the oven at 110 °C.

16-4 REFERENCES

Bingham, J. M., D. C. Golden, S. W. Buol, S. B. Weed, and L. H. Bowen. 1978. Iron oxide mineralogy of well-drained Ultosols and Oxisols: II. Influence on color, surface area and phosphate retention. Soil Sci. Soc. Am. J. 42:825–830.

Bower, C. A., and J. O. Goertzen. 1959. Surface area of soils and clays by an equilibrium ethylene glycol method. Soil Sci. 87:289–292.

Bower, C. A., and F. B. Gschwend. 1952. Ethylene glycol retention by soils as a measure of surface area and interlayer swelling. Soil Sci. Soc. Am. Proc. 16:342–345.

Brunauer, S., P. H. Emmett, and E. Teller. 1938. Adsorption of gases in multi-molecular layers. J. Am. Chem. Soc. 60:309–319.

Carter, D. L., M. D. Heilman, and C. L. Gonzalez. 1965. Ethylene glycol monoethyl ether for determining surface area of silicate minerals. Soil Sci. 100:356–360.

Cihacek, L. J., and J. M. Bremner. 1979. A simplified ethylene glycol monoethyl ether procedure for assessment of soil surface area. Soil Sci. Soc. Am. J. 43:821–822.

DeKimpe, C.R., and M.R. Laverdiere. 1980. Amorphous material and aluminum interlayers in Quebec Spodosols. Soil Sci. Soc. Am. J. 44:639–642.

DeKimpe, C.R., M.R. Laverdiere, and Y.A. Martel. 1979. Surface area and exchange capacity of clay in relation to the mineralogical composition of gleysolic soils. Can. J. Soil Sci. 59:341–347.

Diamond, Sidney, and E.B. Kinter. 1956. Surface area of clay minerals as derived from measurements of glycerol retention. Clays Clay Miner. 5:334–347.

Dowdy, R.H., and M.M. Mortland. 1967. Alcohol-water interactions on montmorillonite surfaces. I. Ethanol. Clays Clay Miner. 15:259–271.

Dyal, R.S., and S.B. Hendricks. 1950. Total surface of clays in polar liquids as a characteristic index. Soil Sci. 69:421–432.

Eltanawy, I.M., and P.W. Arnold. 1973. Reappraisal of ethylene glycol mono-ethyl ether (EGME) method for surface area estimations of clays. J. Soil Sci. 24:232–238.

Farrar, D.M., and J.D. Coleman. 1967. The correlation of surface area with other properties of nineteen British clay soils. J. Soil Sci. 18:118–124.

Galindo, G. G., and F. T. Bingham. 1977. Homovalent and heterovalent cation exchange equilibria in soils with variable surface charge. Soil Sci. Soc. Am. J. 41:883–886.

Gallez, A., A. S. R. Juo, and A. J. Herbillon. 1976. Surface and charge characteristics of selected soils in the tropics. Soil Sci. Soc. Am. J. 40:601–608.

Heilman, M. D., D. L. Carter, and C. L. Gonzalez. 1965. The ethylene glycol monoethyl ether (EGME) technique for determining soil-surface area. Soil Sci. 100:409–413.

Langmuir, I. 1918. The adsorption of gases on plane surfaces of glass, mica, and platinum. J. Am. Chem. Soc. 40:1361–1402.

Low, P. F. 1980. The swelling of clay: II. Montmorillonites. Soil Sci. Soc. Am. J. 44:667–676.

Martin, R. T. 1955. Ethylene glycol retention by clays. Soil Sci. Soc. Am. Proc. 19:160–164.

McNeal, B.L. 1964. Effect of exchangeable cations on glycol retention by clay minerals. Soil Sci. 97:96–102.

Moreale, A., and R. Van Bladel. 1979. Soil interactions of herbicide-derived aniline residues: a thermodynamic approach. Soil Sci. 127:1–9.

Pham, T.H., and G.W. Brindley. 1970. Methylene blue adsorption by clay minerals. Determination of surface areas and cation exchange capacities. Clays Clay Miner. 18:203–212.

Porter, L. 1971. Relationship between the retention of 2-ethoxyethanol by sodium-saturated clays and the retention of dioxane, tetrahydrofuran, and 2-methoxyethanol. Soil Sci. 112:156–160.

Prutton, C.F., and S.H. Maron. 1951. Fundamental principles of physical chemistry. MacMillan Publishing Co., New York.

Quirk, J.P. 1955. Significance of surface areas calculated from water vapor sorption isotherms by use of the B.E.T. equation. Soil Sci. 80:423–430.

Ross, G. J. 1978. Relationships of specific surface area and clay content to shrink-swell potential of soils having different clay mineralogical compositions. Can. J. Soil Sci. 58:159–166.

Singer, A., and J. Navrot. 1977. Clay formation from basic volcanic rocks in a humid Mediterranean climate. Soil Sci. Soc. Am. J. 41:645–650.

Sor, K., and W. D. Kemper. 1959. Estimation of hydrateable surface area of soils and clays from the amount of adsorption and retention of ethylene glycol. Soil Sci. Soc. Am. Proc. 23:105–110.

Supak, J. R., A. R. Swoboda, and J. B. Dixon. 1978. Adsorption of Aldicarb by clays and soil organo-clay complexes. Soil Sci. Soc. Am. J. 42:244–248.

Van der Staay, R. L., and D. D. Focht. 1977. Effects of surface area upon bacterial denitrification rates. Soil Sci. 123:18–24.

17 Aggregate Stability and Size Distribution[1]

W. D. KEMPER and R. C. ROSENAU

Snake River Conservation Research Center
Agricultural Research Service, USDA
Kimberly, Idaho

17-1 INTRODUCTION

17-1.1 Definition and General Approaches

An aggregate is a group of primary particles that cohere to each other more strongly than to other surrounding soil particles. Most adjacent particles adhere to some degree. Therefore, disintegration of the soil mass into aggregates requires imposition of a disrupting force. Stability of aggregates is a function of whether the cohesive forces between particles withstand the applied disruptive force.

When comparing size distribution or stability of soil aggregates, it is necessary that the disruptive force(s) be standardized. If the measurements are to have practical significance, forces causing disintegration should be related to forces expected in the field. The better methods to determine aggregate-size distribution have evolved as a result of efforts to standardize the disruptive forces and make them comparable to those in field phenomena.

Forces involved in aggregate size and stability studies include (i) impact and shearing forces administered while taking and preparing the samples, (ii) abrasive and impact forces during sieving, and/or (iii) forces involved in the entry of water into the aggregate. These forces are generally related to cultivation, erosion (wind and water), and wetting of soils, respectively. However, the disintegrating forces occurring during sample taking, preparation, and analysis do not duplicate the field phenomena. Consequently, the relationship between aggregate-size distribution obtained in the laboratory and that existing in the field is somewhat empirical. Recognizing the empiricism involved in relating aggregate-

---

[1]Contribution from the U.S. Department of Agriculture, Agricultural Research Service, Snake River Conservation Research Center, Kimberly, ID 83341.

size measurements to field phenomena, most investigators have decided to use stability of the aggregates rather than aggregate-size distribution as an index of soil structure in the field. The bases of these decisions have generally been (i) that a simpler procedure involving only one size fraction may be used for stability analysis, (ii) that results of stability analysis are highly correlated with aggregate-size distributions and field phenomena, and (iii) that the ability of the aggregates to resist breakdown by continuing or increasing disruptive forces is often an important factor in the phenomenon being studied.

Large pores in the soil generally favor high infiltration rates, good tilth, and adequate aeration for plant growth. Immediately after cultivation, most soils contain an abundance of these large pores. Their continued existence in the soil depends on the stability of the aggregates. Erodibility of soils decreases as aggregate stability increases.

In determining aggregate stability, known amounts of some size fraction of aggregates are commonly subjected to a disintegrating force designed to simulate some important field phenomenon. The amount of disintegration is measured by determining the portion (by weight) of the aggregates that is broken down into aggregates and primary particles smaller than some selected size. This determination is usually made by sieving or sedimentation. Since the disintegrating forces of interest range from wind erosion to slaking by water, no single type of disintegrating force can be recommended for all objectives. In the following sections, the principles underlying the development of methods for measuring aggregate stability and size distribution will be considered. Detailed descriptions of, or references to, methods that are reproducible and sensitive to factors important to field phenomena conclude this section.

17–1.2 Forces Involved in Aggregation

Two of the primary forces holding particles together in aggregates in moist soils are the surface tension of the air and water interface and the cohesive tension (negative pressure) in the liquid phase. Briggs (1950) has shown that the cohesive tension of water can have values up to 26 MPa. As a soil dries, the water phase recedes into capillary wedges surrounding particle-to-particle contacts and films between closely adjacent platelets. The interfacial tension and internal cohesive tension pull adjacent particles together with great force as soil dries. Soluble compounds such as silica, carbonates, and organic molecules are concentrated in the liquid phase as the soil dries. As the capillary wedges of supersaturated soil solution retreat toward the particle-to-particle contact point, the potential adsorption spots which offer the lowest free energy for the solute molecules or ions are at the junctions of the adjacent particles. Many of these solute molecules and ions thus precipitate as inorganic semicrystalline compounds or amorphous organic compounds around these particle-to-particle contacts, cementing them together.

As the highly adhesive liquid phase pulls adjacent mineral particles into closer proximity, there are also more opportunities for hydrogen bonding between adjacent oxygen and hydroxyl groups and for other intermolecular, interionic, and crystalline bonds to develop. When relative humidity in the soil goes down below 75%, the last molecular layer of water adsorbed on mineral surfaces begins to leave. If the relative humidity goes below 30%, the water molecules hydrated to adsorbed Ca^{2+} are drawn off into the gaseous phase. With the water gone, most soils still generally have great strength due to the cementation discussed above. However, this cementation is generally crystalline and brittle and once broken by mechanical force (e.g., wheel traffic on a dry earth road) does not reform unless the wetting and drying process is repeated.

17-2 PRINCIPLES

17-2.1 Dry Aggregates

17-2.1.1 DISINTEGRATION BY IMPACT AND ABRASION

Wind is a primary natural vector disintegrating dry soil. The wind itself is a relatively mild and broad-based force, which can detach only those particles that are extremely loosely held by underlying particles. However, once these loose particles are moving and can be accelerated for finite time by the wind, their kinetic energy becomes substantial and when they hit the soil surface the force per unit area of impact can be large. These impacts frequently break other particles loose from the surface, which in turn become missles bombarding downwind soil.

One method of controlling the geometrically intensified erosion of this type is to create a rough surface composed of large nonerodible aggregates or clods (Chepil, 1951). Then the smaller wind-moved particles are trapped in the lee of, and in the crevasses between, the clods before they knock additional particles loose from the larger units. One practical method of achieving large numbers of big clods is to cultivate dry soils with implements which gently lift, fracture, and disorient a thick dry surface layer, but do not crush and pulverize the soil.

While the resulting aggregate-size distribution is definitely a function of the cultivation method, it is also a function of soil properties. To measure differences in size distribution of dry aggregates, Chepil (Chepil & Bisal, 1943; Chepil, 1962) developed a rotary sieve that could separate samples of broken soil into dry aggregates of several size ranges. The sieve itself abrades the aggregates to some extent, so that the average aggregate size is reduced somewhat below what it was in the field. An indication of how much this abrasion is reducing the aggregate size can be obtained by running the soil through this sieve a second time. Comparison of the amounts of soil in the various aggregate sizes after the first and second sieving and extrapolation provides an estimate of how much abrasion changed the aggregate sizes during the first sieving. Chepil (1951) found that the abrasion was reproducible and that this abradability was closely related to susceptibility of the soil to wind erosion.

17-2.1.2 EQUIPMENT DESIGNED TO APPLY DISINTEGRATING FORCES AND MEASURE RESULTS

Chepil constructed three versions of his rotary sieve (Chepil & Bisal, 1943; Chepil, 1950, 1951, 1952, 1962). Each successive version improved the performance. Lyles et al. (1970) further improved the rotary sieve and substantially improved the accuracy with which the sample is separated into specific sized fractions.

Aggregate-size distribution is affected strongly by the tillage treatment(s) to which the soil has been subjected and the procedure by which the sample was taken and prepared for sieving. Consequently, by standardizing the sampling procedure, the rotary sieve can be used to obtain good measures of the effects of tillage implements on the aggregate-size distribution, which Chepil (1941) found to be a primary factor controlling wind erosion.

When tillage factors and sampling procedure are kept constant, dry sieving is a sensitive measure of soil structure differences due to amendment, fertilization, and cropping of soils (Chepil, 1962).

17-2.2 Wet Aggregates

17-2.2.1 WEAKENING AND DISINTEGRATION BY WETTING

The wetting process can be highly disruptive. Ion hydration and osmotic swelling forces pull water in between clay platelets, pushing them apart and causing swelling of the aggregates in which they are incorporated. If an aggregate is wetted quickly, the wetted portion can swell appreciably compared to the dry portion, causing a shear plane to accompany the wetting front which can break many of the bonds.

Some of the bonding materials are soluble and dissolve as water enters the soil. Others are hydratable and may become weaker, but more flexible when hydrated. If soil aggregates are wetted slowly at atmospheric pressure or quickly under a vacuum, the bonding is still sufficiently strong to hold most of the primary particles together in aggregates.

If dry aggregates are wetted quickly at atmospheric pressure, much more disintegration and slaking occur (Panabokke & Quirk, 1957).

Lyles et al. (1974) found that more than twice as much soil was detached from large dry aggregates by raindrop impact as from aggregates which had been moistened prior to the rainfall event.

These facts, plus the observable emergence of air bubbles from quick-wetted slaking aggregates, indicate that entrapped air plays a role in their distintegration. It has been found (Kemper et al., 1985a) that when a soil sample is stored at relative humidities $< 50\%$, substantial amounts of O_2 and N_2 are absorbed on the mineral surfaces. When the sample is wetted by immersion, these O_2 and N_2 molecules are displaced by the more tightly adsorbed water molecules. This O_2 and N_2 from the adsorbed phase join the air entrapped inside the pores and the pressure increases as capillarity pulls water into the aggregate. Finally the aggregate ruptures and the air bubble emerges.

Kemper et al. (1985b) also observed that the amount of soil eroded from a furrow was related to how quickly the soil in the furrow was wetted. Quick wetting allows the water to encompass more of the aggregates before the O_2 and N_2 have escaped from their interior.

17–2.2.2 EQUIPMENT TO APPLY DISINTEGRATING FORCES AND MEASURE RESULTS

Much of the earliest work on the stability of hydrated aggregates utilized elutriation through various sized tubes to separate the aggregates into several size groupings, the effective sizes of which were calculated according to Stokes' Law. Yoder (1936) pointed out the deficiencies of this method and mechanized and modified a wet-sieving procedure. In Yoder's procedure, a nest of six sieves was placed in a holder and suspended in a container of water. Fifty grams of air-dry soil was placed on the top sieve of each nest, and the nest was lowered to the point where the soil sample on the top screen was just covered with water. A motor and a mechanical arrangement lowered and raised the nest of sieves through a distance of 3.18 cm at a rate of 30 cycles/min for 30 min. The amounts of soil retained on each sieve were determined by drying and weighing. Size fractions smaller than 0.1 mm in diameter were determined by a procedure utilizing sedimentation, decanting, drying, and weighing. Much of the work on wet aggregate stability has utilized Yoder's method or some variation of it.

Many investigators (e.g. Kemper & Koch, 1966) have concluded that aggregate stability can be determined using a single sieve with much less investment and is essentially as well correlated with practically important field phenomena. The single sieve method also avoids the calculations (described in the next section) necessary to convert size distributions to single numbers that can be used in ranking soils and in statistical correlations.

Kemper and Koch (1966) constructed a sieving machine similar to that developed by Yoder, but modified so that only one screen was used and stroke length and frequency could be varied. They also varied sieve size, sample size, aggregate size, length of sieving time, method of wetting, period of soaking prior to sieving, and other factors to optimize the reproducibility and sensitivity of the analyses.

The sieving machine diagrammed in Fig. 17–1 (Five Star Cablegation and Scientific Supply, 303 Lake St., Kimberly, ID 83341) incorporates the characteristics which Kemper and Koch (1966) found to be optimum, plus some modifications found by Kemper et al. (1985b) to facilitate the measurements. The sieves are stainless steel screens, 24 mesh/cm, wire diameter 0.165 mm, hole size 0.26 mm. The machine provides a stroke

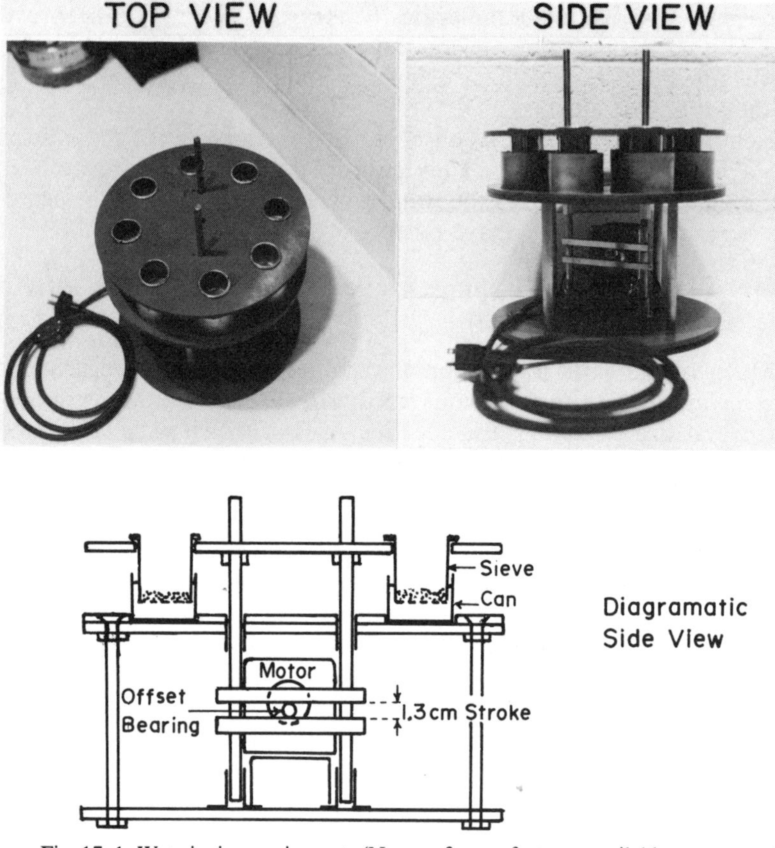

Fig. 17–1. Wet sieving equipment. (Name of manufacturer available on request.)

length of 1.3 cm to the sieves at a frequency of 35 cycles/min. It is designed for use with aggregates in the 1- and 2-mm diameter size range. The primary disintegrating factor applied to these aggregates is the forces involved during wetting. Much less disintegration takes place when samples are wet under vacuum than when they are wet under atmospheric pressure (Kemper & Koch, 1966).

Mubarak et al. (1978) used an ultrasonic nebulization technique to produce an aerosol, which was passed over the aggregates. This wetted the aggregates slowly, resulting in little disintegration, and the stability measurement was more reproducible than when vacuum wetting was used. Their success with the procedure lead Kemper et al. (1985b) to try commercially available humidifiers to produce aerosols in the chamber shown in Fig. 17–2. Vapor moves up through the aggregate samples on the screens. Samples wetted in this manner had high and reproducible stabilities when subjected to the wet-sieving process.

Soils from humid regions tend to have higher aggregate stabilities (Kemper & Koch, 1966). The gentle vacuum or aerosol wetting methods

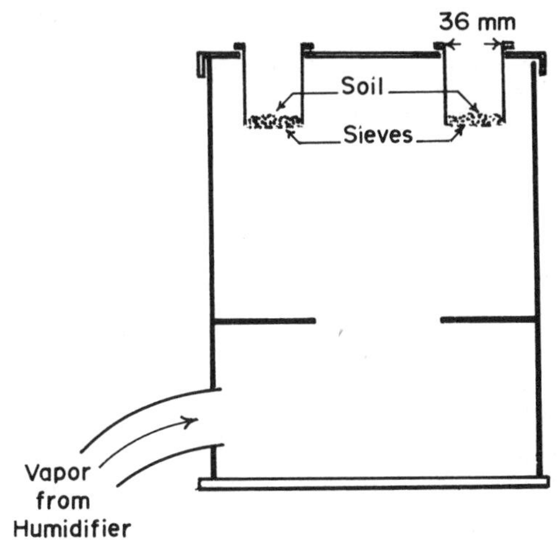

Fig. 17–2. Equipment for vapor wetting aggregates.

tend to leave those soils nearly 100% stable. Differences between their stabilities are generally more readily detectable if they are subjected to the greater disruptive forces involved in direct immersion of dry aggregates at atmospheric pressure.

The sieving frequency of 35 cycles/min and stroke length of 1.3 cm provide a water velocity with respect to the screen on the downstroke which just lifts the 1- to 2-mm aggregates clear of the screen. This provides opportunity for disintegration products that are smaller than the screen holes to pass through the screen within three minutes, as is indicated in Fig. 17–3. Continued sieving causes abrasion, and the resulting slow rate of breakdown is also indicated in that figure. Greater frequency and/or stroke length increases the slopes of these breakdown curves. This is reasonably analogous to the increased furrow erosion which occurs when flow rate of the water is increased. The initial rapid breakdown due to wetting and sustained abrasive breakdown apparent in Fig. 17–3 are closely

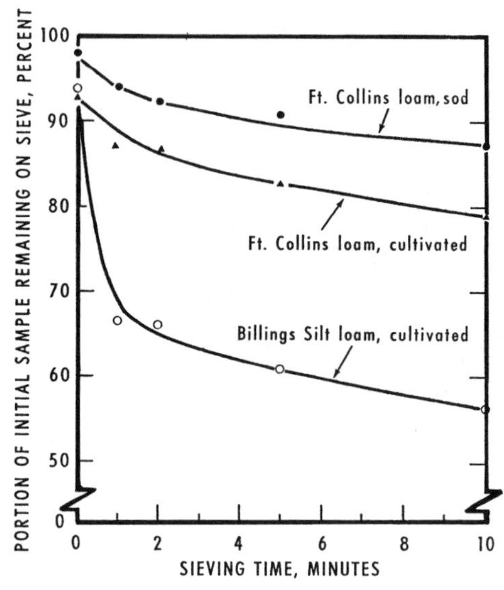

Fig. 17–3. Effect of wet sieving time on the portion of the initial sample remaining on the sieve.

related to the high sediment load in initial furrow runoff, followed by lower sustained sediment loads when runoff rate is constant.

17–2.3 Methods of Expressing the Size Distribution of Aggregates

Multiple-sieve techniques give data on the amount of aggregates in each of several size groupings. To evaluate treatments or to rank soils, a single parameter to represent a soil sample is necessary. The problem then is to decide on the importance of each group to the final parameter. If equal weights of the different size groups of aggregates are of equal importance, the amounts in each of the different sizes should be added, so that making a number of separations would provide no advantage over a single separation obtained using the smallest size screen and determining the weight of all the aggregates retained. It is generally considered, however, that a specific weight of large aggregates is more indicative of good structure for most agricultural purposes than is an equal weight of small aggregates. This concept is incorporated in certain mathematical techniques for expressing aggregation data in the form of a single parameter.

17–2.3.1 MEANWEIGHT DIAMETER

Van Bavel (1949) proposed that aggregates be assigned an importance or weighting factor that is proportional to their size. The parameter that he called the *meanweight diameter* (MWD) is equal to the sum of products of (i) the mean diameter, $\bar{x}_i$, of each size fraction and (ii) the proportion of the total sample weight, w_i, occurring in the corresponding size fraction, where the summation is carried out over all n size fractions, including the one that passes through the finest sieve:

$$\mathrm{MWD} = \sum_{i=1}^{n} \bar{x}_i \, w_i \, . \qquad [1]$$

The entire sample is passed through an 8-mm sieve prior to analysis. In his original definition, Van Bavel used an integration that is equivalent to a summation made over a very large number of very small increments.

Van Bavel's (1949) concept of the MWD has been used widely. However, its calculation in the integral form involves plotting points on a graph and determining the area enclosed. This is a time-consuming process. Youker and McGuinness (1956) suggested that the summation type calculation in the above equation be used in place of the graphical approach. The summation equation shown above generally overestimates the original MWD when only five fairly broad size fractions are used. However, the correlation using five size-fractions is excellent. Because of the general shape of the curve, overestimation of the MWD from use of the summation equation is generally somewhat less if sieves with openings of 4.76, 2.00, 1.00, and 0.21 mm are used rather than sieves with openings of 2.00, 1.00, 0.50, and 0.20 mm.

17-2.3.2 GEOMETRIC MEAN DIAMETER

Mazurak (1950) suggested that the *geometric mean diameter* (GMD) be used as an index of the aggregate size distribution. The geometric mean diameter is calculated approximately by the equation:

$$\mathrm{GMD} = \exp\,[\Sigma_{i=1}^{n} \, w_i \log \bar{x}_i / \Sigma_{i=1}^{n} \, w_i] \qquad [2]$$

where w_i is the weight of aggregates in a size class with an average diameter x_i and $\Sigma_{i=1}^{n} w_i$ is the total weight of the sample. The use of the GMD is supported by Gardner's (1956) finding that the aggregate size distribution in most soils is approximately log-normal rather than normal. This log-normal distribution provides the opportunity to describe the actual aggregate size distribution of most soils with two parameters, the geometric mean diameter and the log standard deviation (Gardner, 1956). If data are summarized in the form of these two parameters, the MWD and other aggregate size indexes may be calculated from these parameters. The main disadvantage of expressing data in terms of GMD and log standard deviation is the extensive work involved in obtaining them. The log standard deviation must be obtained by either graphical or differential interpolation from the data.

17-2.3.3 OTHER PROPOSALS

Other parameters that have been developed to express aggregate size distribution in terms of one or two numbers are the weighted mean diameter and standard deviation (Puri & Puri, 1939) and the coefficient of aggregation (Retzer & Russell, 1941). De Boodt et al. (1961) found that change in mean weight diameter when dry soil samples are wetted was correlated with crop yields. They obtained correlation coefficients that averaged about 0.9.

17-2.3.4 COMPARISON AND EVALUATION

Schaller and Stockinger (1953) compared five methods for expressing aggregation data on several soils. The best methods for presenting aggregate size distribution data seemed to be MWD or GMD and the log standard deviation. The correlation coefficient between MWD and GWD was about 0.9. The GMD and the log standard deviation give a more complete description of the size distribution than the MWD. However, the MWD is easier to calculate and easier for most individuals to visualize. Both the MWD and GMD can be used to represent aggregate size distribution for statistical analysis. Stirk (1958) maintains that the reliability of a mean is questionable if the aggregate distributions are extremely skewed. Consequently, he favors the use of a GMD and a log standard deviation on basic considerations. However, he points out that use of these parameters may complicate the calculations to a point beyond that which is most efficient for interpreting individual experiments.

17-3 METHODS

17-3.1 Dry Aggregates: Equipment and Procedure for Determining Size Distribution and Resistance to Abrasion

The rotary sieve, developed by Chepil (1962) and improved by Lyles et al. (1970), is recommended. Some advantages of the rotary sieve are (i) it is the most consistent dry-sieving method thus far devised, (ii) it is not subject to a personal factor, (iii) it gives fairly consistent results irrespective of the size of soil sample used, (iv) it causes less breakdown of clods than the mechanical flat-sieve method, (v) it virtually eliminates clogging, and (vi) it is well suited to resieving soil any number of times to determine the relative resistance of the soil to breakdown by mechanical forces.

Limitations are that the cost of a rotary sieve is appreciable and experienced manufacturers are difficult to find. (One manufacturer which has experience and is willing to produce these machines on order is "Five Star Cablegation and Scientific Supply," 303 Lake Street, Kimberly, ID 83341.) The number and size of the sieves used influences the amount of aggregate breakdown. Therefore, it is necessary to have the same sizes and number of sieves in all comparable tests.

Samples for dry sieve analysis should be taken when the soil is reasonably dry, to avoid breakdown or change in structure. A flat, preferably square-cornered spade is pushed under the sample to lift it and place it in a suitable tray. Air-drying at humidities in the general range expected in the field is recommended. Since clod structure varies with depth, the same depth of sampling must be used for all results expected to be comparable.

Since the objective of determining size distribution is generally to characterize an existing condition, crushing the soil by bagging it or by other means should be avoided. The sample is gently slid from the drying tray into the feed bin of the rotary sieve, the sieve is turned on and the various sizes of dry aggregates are caught in the pans and weighed.

Resistance of the aggregates to abrasion is determined by pouring the weighed aggregates back into the drying pan, sliding them off this pan back into the feed bin of the rotary sieve, sieving and weighing again, and determining the changes in aggregate size distribution.

For more details and cautions concerning the procedure, the reader is encouraged to consult Chepil (1962) and Lyles et al. (1970).

17–3.2 Wet Aggregates

17–3.2.1 STABILITY

17–3.2.1.1 Wetting the Sample. The primary factor governing the wet stability of aggregates is generally the method of wetting. Procedures for achieving ranges of initial water contents and rates of wetting and stabilities obtained therefrom are presented by Kemper et al. (1985a). When studying effects of other factors the method of wetting must be standardized. Vacuum wetting as proposed by Kemper and Koch (1966) or slow wetting with an aerosol produced by the equipment indicated in Fig. 17–2 provided minimum disruption and the aggregate stabilities resulting therefrom are reproducible within a few percent. If the samples are wetted slowly with vapor (taking about 30 min) to water content near their field capacities before immersion, the stability of the aggregates will be slightly higher than when they are wetted quickly under vacuum. Water content of the wetting aggregates can be monitored by obtaining the initial weight of the sample and sieve and weighing them again at appropriate intervals during the vapor wetting process. Monitoring water content in this manner and observing the time required and color changes involved allow an operator to achieve the desired range of water content by time and/or visual inspection.

17–3.2.1.2 Determining the Stability.

1. Weigh 4 g of 1- to 2-mm air-dried aggregates into sieves of the type indicated in Fig. 17–1.
2. Bring the aggregates to the desired water content as described above.
3. Place sufficient distilled water in the weighed and numbered cans to cover the soil when the sieve is at the bottom of its stroke, and place these cans in the recessed holders in the sieving apparatus.

4. Place the numbered sieves in the sieve holder and lower the assembly so that each sieve and aggregates contained therein enter the corresponding can.
5. Start the motor and allow it to raise and lower the sieves 1.3 cm, 35 times/min, for 3 min ± 5 s.
6. At the end of this time, raise the sieves out of the water and place the numbered cans (containing the particles and aggregate fragments that have broken loose from the aggregates and come through the sieves) on a tray.
7. Replace these cans with another set of numbered and weighed cans containing 100 cm^3 of dispersing solution (containing 2 g sodium hexametaphosphate/L) for soils with pH >7 or 2 g NaOH/L for soils with pH <7.
8. Resume and continue sieving until only sand particles are left on the sieve. If some aggregates remain stable after 5 min of sieving in the dispersing solution, stop the sieve and rub them across the screen with a rubber tipped rod until they are disintegrated.
9. Continue sieving until materials smaller than the screen openings have gone through.
10. Raise the sieves and place the numbered cans on a separate tray. These cans contain the materials from the aggregates which were stable, except for sand particles too large to get through the screen.
11. Both sets of cans are placed in a convection oven at 110 °C until the water has evaporated.
12. The weight of materials in each can is then determined by weighing the can, plus contents, and subtracting the weight of the can. In the cans which were filled with dispersing solution, there will be 0.2 g of the dispersing solute along with the soil. Consequently, 0.2 g should be subtracted from the weight of the contents to obtain the soil weight.
13. The fraction stable is equal to the weight of soil obtained in the dispersing solution dishes divided by the sum of the weights obtained in the two dishes.

Dispersion of the aggregates may also be achieved using an ultrasonic probe, in which case the dispersing solution (Step 7) can be distilled water rather than the sodium salt solution, which eliminates the need for the 0.2-g correction indicated in Step 12. To facilitate insertion of the probe, the sieve is removed from the holder and inserted into its respective numbered can. Holding the probe in the water inside the sieve for 30 s at medium frequency generally disintegrates the aggregates into primary particles. The sieve and can are then replaced in the holder. After dispersing each sample in this manner Steps 8 to 13 are followed with the appropriate modification.

17–3.2.1.3 Precautions and Special Considerations. Compression of samples when they are taken from the soil causes some variation in aggregate stability; consequently, sampling instruments such as shovels and tubes that cause little compression are better than augers. Compression is more likely to affect the results obtained with samples taken from soil that is relatively moist than from soil that is dry. Where proper care is exercised in sampling, similar results may be obtained with samples taken at any water content between the field capacity and the permanent wilting point (Kemper & Koch, 1966).

Drying the soil at high temperatures sometimes causes irreversible or slowly reversible dehydration of the bonding materials and clay particles. This dehydration is particularly important in soils containing considerable exchangeable sodium. In such soils, oven drying may cause a transient stability of otherwise unstable aggregates. This transient stability decreases with increasing time allowed for rehydration before sieving (Kemper & Koch, 1966). The stability of samples dried at room temperatures does not change appreciably with time allowed for rehydration (Russell & Feng, 1947; Emerson & Grundy, 1954) as long as hydration takes more than 20 min. Apparently dehydration at room temperature is not too severe, or else it is rapidly reversible. Consequently, drying at room temperatures is suggested.

Aggregate stability increases slowly with time of storage (Kemper & Koch, 1966). Miller and Kemper (1962) found that increases in aggregate stability occurring following the incorporation of fresh organic material into soil were detected more easily if the analyses were made within 2 weeks after the sample was dried. For these reasons it is recommended that analyses be made as soon as feasible after air-drying is completed. Since soils dried slowly develop more strength, it is recommended that for comparison purposes, all the soils be spread out in a layer < 3 mm thick on an area where the air circulation is good, so they will dry within 24 h.

The larger an aggregate is when dry, the smaller the probability that its disintegrated components will pass through a sieve of given size when it is wet. Consequently, results are more reproducible if aggregates of a limited size range are selected for the determination of stability. Aggregates of 1 to 2 mm in diameter have been found to be best suited to the screen size and sieving speed employed in this procedure. Observations of Bryant et al. (1948) and data by Strickling (1951), Schaller and Stockinger (1953), and Panabokke and Quirk (1957) all indicate that results from simple one- and two-sieve methods of determining aggregate stability are closely correlated with results using several size ranges of aggregates and sieves, when expressed in terms of mean weight diameters. Consequently, a single size range of aggregates is employed in preference to more time-consuming methods such that of De Boodt et al. (1961), using several size ranges of aggregates and sieves.

Kemper and Koch (1966) found a slight tendency for aggregate stability to decrease when the temperature of water used to wet and sieve the samples was increased from 20 to 30 °C. Consequently, the temperature of the water should be maintained within the range of 22 to 25 °C.

Appreciable salt in the water can cause changes in the ionic status and stability of the soils; therefore, it is suggested that the salt content of the water be low enough so that the electrical conductivity is <0.01 dS m^{-1} (<0.01 mho cm^{-1}). On the other hand, low electrolyte in the water increases swelling pressures between particles which can lower stabilities of sodic soils (e.g., Shainberg et al., 1981). The electrolyte content of the water should be standardized and specified when reporting aggregate stabilities of sodic soils (Shainberg et al., 1981).

Some soils (e.g., soils from humid areas containing large amounts of free iron oxides) may be nearly 100% stable when wet by the vacuum or slow wetting techniques and sieved gently. If differences in aggregate stability among such stable soils are to be determined, the disrupting force must be increased. Wetting dry soil by immersion at atmospheric pressure accomplishes this purpose by allowing entrapment of air within the aggregates. Sieving more violent and prolonged than described in this procedure may also provide sufficient disruption of aggregates to detect differences among highly stable soils.

Aggregates that do not break down under a rubber-tipped rod or a jet of water from a wash bottle are referred to as concretions. They are commonly held together by $CaCO_3$, iron oxides, and so forth. They may be considered and treated in two ways:

1. Since concretions are extremely stable and will not break down under normal cultivation practices, they may be considered as sand >0.26 mm mesh. If this is done, no further treatments should be used to break them down and wash them through the sieve.
2. On the other hand, since concretions usually have some porosity and appreciable internal surface area and exchange capacity, they may be treated as stable aggregates. If they are considered in this way, they must be broken down when separating out the sand so the fine particles can pass through the sieve or be dissolved. Soaking $CaCO_3$-bonded concretions in 1 *M* HCl causes complete disintegration. Concretions bonded with organic matter and iron usually disintegrate when soaked in 0.5 *M* NaOH.

It is suggested that concretions be treated as stable aggregates unless the investigator has some specific reason for considering them as sand.

Reproducibility of the measurement varies with the texture of the soil. Aggregate stability measurements carefully carried out on coarse-textured soils on successive days were found to have coefficients of variation of 4.0% (Kemper & Koch, 1966). Under the same conditions, aggregate stability measurements on fine-textured soils had an average coefficient of variation of 1.2%. Where large numbers of samples have been handled in routine fashion, somewhat higher coefficients of variation have been obtained for duplicate runs on different dates. It is recommended that duplicate subsamples be analyzed on different dates for all analyses.

17–3.2.2 SIZE DISTRIBUTION

According to Stokes' Law, the velocity with which a small sphere settles through a viscous fluid is proportional to the square of its radius, proportional to the difference between its density and the density of the fluid, and inversely proportional to the viscosityy of the fluid.

The dependency of settling velocity on particle radius can be used to separate aggregates of various equivalent sizes in the same manner that it is used to separate primary particles into various size fractions (see chapter 15 on particle-size distribution). Determination of aggregate size distribution by this method would require that aggregates have the same density. Indications are that this is about as true of aggregates as of primary particles (e.g. Chepil, 1950). Although aggregates are not spherical, as assumed in use of Stokes' Law, they approximate spherical shape more closely than do most clay particles.

The major limitation of sedimentation procedures is that aggregates >1 mm in diameter settle too rapidly to be measured precisely. Yoder (1936), Kolodny & Joffe (1940), and others have used sedimentation techniques to determine the amounts of small aggregates in soils. Electronic equipment is now available to make measurements more quickly (Keren, 1980).

Davidson and Evans (1960) extended the sedimentation technique to measure amounts of aggregates in the larger size ranges by using a liquid (1 part water to 9 parts glycerol) that had a viscosity about 140 times that of water. Settling rate was measured photometrically. The method yields reproducible results and is easily adapted to automatic recording of the type used for smaller aggregates. Stability of aggregates in glycerol-water mixtures is correlated (Davidson & Evans, 1960) with stability of aggregates in water.

17–3.2.2.1 Elutriation. Water flowing upward tends to carry aggregates with it. However, if the aggregates are large enough and dense enough that their velocity of settling with respect to the water is as fast as or faster than the upward movement of the water, they will not be carried upward.

Kopecky (1914) connected a series of successively larger tubes, placed the sample in the smallest tube, and passed water upward through the series of tubes at a set volume per unit time. The upward velocity of the water was smaller in the larger tubes; hence, successively smaller equivalent sizes of particles remained in the successively larger tubes. Baver and Rhoades (1932) and others used elutriation to determine aggregate size distributions.

Elutriation methods have been criticized because it is difficult to avoid some turbulence in water flowing in wide columns, and some aggregates that would stay in the tube if only laminar flow occurred are carried out of the tube as a result of turbulence. Aggregate density should be determined if actual sizes of aggregates are to be calculated. Water flowing upward in a tube normally has a parabolic velocity distribution, with water adjacent to the sides moving more slowly than water at the center of the tube. While there is a sound basis for all of these criticisms, elutriation may still be used to make a separation of aggregates into several approximate size fractions. If care is used and good equipment is available, the separation of aggregates < 2 mm in diameter will be fairly accurate.

17–3.2.2.2 Sieving. This technique involves wetting the sample and then separating aggregates into various sizes by sieving the sample through a nest of sieves under water. Care is taken to cause as little mechanical disruption of the aggregates as possible. Tiulin (1928) and Yoder (1936) were among the early workers to use this procedure extensively.

As discussed in section 17–2.2.1, direct immersion of dry soil in water at atmospheric air pressure causes a great disruption of aggregates into smaller aggregates and primary particles. A much smaller degree of disruption is caused when soils are wetted slowly under tension by passing vapor through them or by adding water under a vacuum.

The immersion wetting process is fairly comparable to wetting the soil surface by irrigation. If the purpose of the aggregate analysis is related to infiltration rates of flooded soils or the formation of soil crusts, immersion wetting is probably the preferred procedure. On the other hand, soils below the surfaçe are wet more slowly under tension, and therefore are disrupted much less. Wetting the dry soils under a tension is preferred when geometry of soil particles and voids below the immediate surface is being studied. Kemper and Koch (1966) found that soils wet under tension had about the same aggregate stabilities whether the tension in the water was 50 or 1500 Pa. Aggregate stability of soils wet under a vacuum was found to be closely correlated with aggregate stability of soils wet under a tension. Since the vacuum wetting procedure produced more reproducible results and was faster, it was used to approximate disruptive forces involved in wetting soil below the soil surface.

A study concerning the reproducibility of aggregate size distribution values was reported by the Soil Science Society of America Committee on Physical Analysis (Van Bavel, 1953). Analyses were made at several laboratories using a Yoder-type procedure and using both immersion at atmospheric air pressure and vacuum wetting techniques. They found extreme variability of size distribution of a given soil when it was measured by different laboratories using their vacuum wetting technique. This technique involved transferring wet soils from nonstandardized containers, in which they were wet under vacuum, to the sieves. The committee believed that the large variation using the vacuum wetting technique probably resulted from nonuniform handling procedures. Some of the variation probably was caused by wetting the samples when only a partial vacuum had been achieved. It was assumed that the pressure in the vacuum desiccator had reached the vapor pressure of water when water left in the desiccator began to boil. It is probable that some of the technicians mistook the formation of air bubbles and their escape from the water for the onset of boiling. This mistaken assumption that the proper degree of vacuum had been reached would cause them to wet the samples when appreciable air was left in the desiccator.

If water brought to equilibrium with atmospheric air pressure is used to wet the samples in the evacuated desiccators, air begins to come out of the water as it enters the desiccators. When the water contacts the soil, the soil appears to catalyze the release of the remaining air from the water. Formation of a bubble in water in or near an aggregate causes some disruption of the particle. The degree of disruption depends on how fast the water enters the desiccator and other factors. Kemper and Koch (1966) found that the average aggregate stability increased and the reproducibility of the measurement was improved if the samples were wet by de-aerated water under vacuum.

Aggregate size distribution often is measured to gain information on the size of the aggregates as they exist in the mass of soil. This requires a wetting treatment that will cause as little disintegration as possible. Consequently, wetting by an aerosol, or under vacuum, is proposed for this purpose.

During the past 20 years, as simpler and quicker methods of wet aggregate stability (the single sieve method) and size distribution sedimentation techniques (e.g., as used by Keren, 1980) have come into use, the multiple sieve technique has been used less often for determining stability and size distribution of aggregates. Consequently, it will not be outlined in detail here. Investigators who wish to utilize the multiple sieve technique are referred to pages 506 to 509 of the 1965 Methods of Analysis Monograph (Kemper & Chepil, 1965) and to the original studies by Yoder (1936).

17-4 REFERENCES

Baver, L. D., and H. F. Rhoades. 1932. Aggregate analysis as an aid in the study of soil structure relationships. Agron. J. 24:920–930.

Briggs, L. J. 1950. Limiting negative pressure of water. J. Appl. Phys. 21:721–722.

Bryant, J. C., T. W. Bendixen, and C. S. Slater. 1948. Measurement of the water-stability of soils. Soil Sci. 65:341–345.

Chepil, W. S. 1941. Relation of wind erosion to the dry aggregate structure of soil. Sci. Agric. 21:488–507.

Chepil, W. S., and F. Bisal. 1943. A rotary sieve method for determining the size distribution of soil clods. Soil Sci. 56:95–100.

Chepil, W. S. 1950. Methods of estimating apparent density of discrete soil grains and aggregates. Soil Sci. 70:351–362.

Chepil, W. S. 1951. Properties of soil which influence wind erosion: III. Effect of apparent density on erodibility. Soil Sci. 71:141–153. IV. State of dry aggregate structure. Soil Sci. 72:387–401. V. Mechanical stability of structure. Soil Sci. 72:465–478.

Chepil, W. S. 1952. Improved rotary sieve for measuring state and stability of dry soil structure. Soil Sci. Soc. Am. Proc. 16:113–117.

Chepil, W. S. 1962. A compact rotary sieve and the importance of dry sieving in physical soil analysis. Soil Sci. Soc. Am. Proc. 26:4–6.

Davidson, J. M., and D. D. Evans. 1960. Turbidimeter technique for measuring the stability of soil aggregates in a water-glycerol mixture. Soil Sci. Soc. Am. Proc. 24:75–79.

De Boodt, M., L. De Leenheer, and D. Kirkham. 1961. Soil aggregate stability indexes and crop yields. Soil Sci. 91:138–146.

Emmerson, W., and G. M. F. Grundy. 1954. The effect of rate of wetting on water uptake and cohesion of soil crumbs. J. Agric. Sci. 44:249–253.

Gardner, W. R. 1956. Representation of soil aggregate-size distribution by a logarithmic-normal distribution. Soil Sci. Soc. Am. Proc. 20:151–153.

Kemper, W. D., and W. S. Chepil. 1965. Size distribution of aggregates. *In* C. A. Black et al. (ed.) Methods of soil analysis, Part 1. Agronomy 9:499–510.

Kemper, W. D., and Koch, E. J. 1966. Aggregate stability of soils from the western portions of the United States and Canada. U. S. Dep. Agric. Tech. Bull. 1355.

Kemper, W. D., R. C. Rosenau, and S. Nelson. 1985a. Gas displacement and aggregate stability of soils. Soil Sci. Soc. Am. J. 49:25–28.

Kemper, W. D., T. J. Trout, M. J. Brown, and R. C. Rosenau. 1985b. Furrow erosion and water and soil management. Trans. Am. Soc. Agric. Eng. 28:1564–1572.

Keren, R. 1980. Effects of titration rate, pH and drying process on cation exchange capacity reduction and aggregate size distribution of montmorillonite hydroxy-aluminum complexes. Soil Sci. Am. J. 44:1209–1212.

Kolodny, L., and J. S. Joffe. 1940. The relation between moisture content and micro-aggregation or the degree of dispersion in soils. Soil Sci. Soc. Am. Proc. (1939) 4:7–12.

Kopecky, J. 1914. Ein Beitrag zur Frage der neuen Einteilung der Kornungsprodukte bei der mechanischen Analyse. Int. Mitt. Bodenk. 4:199–202.

Lyles, L., J. D. Dickerson, and L. A. Disrud. 1970. Modified rotary sieve for improved accuracy. Soil Sci. 109:207–210.

Lyles, L., J. D. Dickerson, and N. F. Schmeidler. 1974. Soil detachment from clods by rainfall: effects of wind, mulch cover and essential soil moisture. Trans. ASAE 17:697–700.

Mazurak, A. P. 1950. Effect of gaseous phase on water-stable synthetic aggregates. Soil Sci. 69:135–148.

Miller, D. E., and W. D. Kemper. 1962. Water stability of aggregates of two soils as influenced by incorporation of alfalfa. Agron. J. 54:494–496.

Mubarak, A., M. G. Klages, and R. A. Olsen. 1978. An improved moistening technique for aggregate stability measurement. Soil Sci. Soc. Am. J. 42:173–174.

Panabokke, C. R., and J. P. Quirk. 1957. Effect of water content on stability of soil aggregates in water. Soil Sci. 83:185–195.

Puri, A. N., and B. R. Puri. 1939. Physical characteristics of soils: II. Expressing mechanical analysis and state of aggregation of soils by single values. Soil Sci. 47:77–86.

Retzer, J. L., and M. B. Russell. 1941. Differences in the aggregation of a Prairie and a Gray-Brown Podzolic soil. Soil Sci. 52:47–58.

Russell, M. B., and C. L. Feng. 1947. Characterization of the stability of soil aggregates. Soil Sci. 63:299–304.

Schaller, F. W., and K. R. Stockinger. 1953. A comparison of five methods for expressing aggregation data. Soil Sci. Soc. Am. Proc. 17:310–313.

Shainberg, I., J. D. Rhoades, and R. J. Prather. 1981. Effect of low electrolyte concentration on clay dispersion and hydraulic conductivity of a sodic soil. Soil Sci. Soc. Am J. 45:273–277.

Stirk, G. B. 1958. Expression of soil aggregate distributions. Soil Sci. 86:133–135.

Strickling, E. 1951. The effect of soybeans on volume weight and water stability of soil aggregates, soil organic matter content and crop yield. Soil Sci. Soc. Am. Proc. 15:30–34.

Van Bavel, C. H. M. 1949. Mean weight diameter of soil aggregates as a statistical index of aggregation. Soil Sci. Soc. Am. Proc. 14:20–23.

Van Bavel, C. H. M. 1953. Report of the committee on physical analyses 1951–1953, Soil Science Society of America. Soil Sci. Soc. Am. Proc. 17:416–418.

Yoder, R. E. 1936. A direct method of aggregate analysis of soils and a study of the physical nature of erosion losses. J. Am. Soc. Agron. 28:337–351.

Youker, R. E., and J. L. McGuiness. 1956. A short method of obtaining mean weight-diameter values of aggregate analyses of soils. Soil Sci. 83:291–294.

18 Porosity

R. E. DANIELSON AND P. L. SUTHERLAND

Colorado State University
Fort Collins, Colorado

18-1 INTRODUCTION

The structure of the soil is related to many important soil physical properties, especially those pertaining to the retention and transport of solutions, gases, and heat. Soil structure can be measured in various ways, but perhaps it is most meaningfully evaluated through some knowledge of the amount, size, configuration, or distribution of soil pores. It is often the information concerning these pore spaces, rather than the soil particles, that is most useful in characterizing the soil as a medium for plant growth or other uses. Within the soil matrix there exists an array of complex interaggregate and intraaggregate cavities which vary in amount, size, shape, tortuosity and continuity. Precise quantification of these characteristics of soil pores is essentially impossible, due to their extremely complicated nature. However, the total pore space can be determined with relatively high precision, and by making certain assumptions, the size distribution of the larger pores can be made with at least useful accuracy for both laboratory and field purposes.

An important problem associated with the characterization of soil pores is the lack of standard terminology related to their classification into distinct size ranges. The need for a standard classification scheme has been identified by various researchers, and suggestions for such a classification have been made. Typical examples are the proposed index for soil pore size distribution of Cary and Hayden (1973), the suggested classification of micro-, meso- and macroporosity by Luxmoore (1981), and the responses of Bouma (1981), Beven (1981) and Skopp (1981). A second problem relates to the need for identifying pore size in terms of an equivalent cylindrical diameter. This results from the complex and variable pore shapes and their interconnected nature. Commonly these equivalent diameters are estimated from liquid retention and capillary pressure considerations, and the computed size may be somewhat dependent upon the method used.

A thorough discussion relating to the general subject of pore structure may be found in the text of Dullien (1979), and a review of techniques that have been used for measuring pore sizes in soil is provided by Law-

 Methods of Soil Analysis, Part 1. Physical and Mineralogical Methods—Agronomy Monograph no. 9 (2nd Edition)

rence (1977). Nitrogen sorption is an estabished practice for determining pore-size distribution, but although it has been used, the method appears to have limited potential for evaluating a wide range of pore sizes in soils. Wilkins et al. (1977) have described a procedure for direct measurement of soil pores larger than 20 μm by vacuum impregnation with a fluorescent polyester resin followed by sectioning and photographing under black light.

Although the porosity and pore size distribution of many porous materials have been satisfactorily measured by a variety of techniques, only a few have been adequately evaluated and commonly accepted for use with soils. In this chapter, two methods, calculation from density measurements and direct evaluation by the gas pycnometer, are provided for determining total pore space. Also, two methods, water retention and mercury intrusion, are described for identifying the pore size distribution.

18–2 TOTAL POROSITY

18–2.1 Calculation from Particle and Bulk Densities

18–2.1.1 PRINCIPLES

Total soil porosity may be calculated if the particle density, ρ_p, and the dry bulk density, ρ_b, are known. The ratio ρ_b/ρ_p is the fraction of the total volume occupied by solids, and this value subtracted from unity gives the fraction of the total volume occupied by pores. Total porosity, S_t, is therefore calculated by:

$$S_t = (1 - \rho_b/\rho_p) \,. \qquad [1]$$

18–2.1.2 APPARATUS

The apparatus required for the various methods used in measuring bulk and particle densities are described in chapters 13 and 14, respectively.

18–2.1.3 PROCEDURE

Measure the particle density and the bulk density of the soil using one of the procedures described in chapters 13 and 14. The procedure to be chosen depends upon the accuracy desired in calculating the total porosity. The number of soil samples to be obtained must be based on the spatial variability of the soil. Using the density values, or averages of values, calculate total porosity using Eq. [1].

18–2.1.4 COMMENTS

Particle density of soil is essentially constant over a reasonable time period. Thus, the value obtained for total porosity will vary with bulk

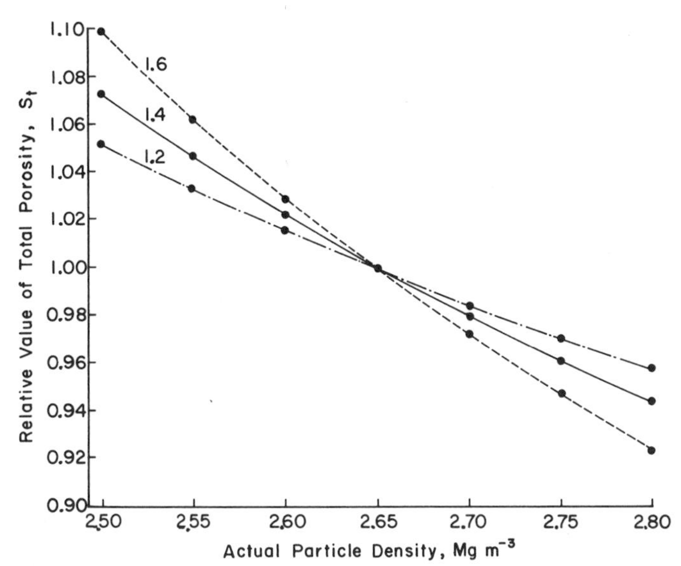

Fig. 18–1. Relative error in calculation of total porosity of soil when particle density is assumed to be 2.65 Mg m^{-3}. The three curves represent bulk density values of 1.2, 1.4, and 1.6 Mg m^{-3}.

density variations. If an aggregate (clod) is used for the evaluation of ρ_b, the calculated porosity will reflect only the aggregate and not the inter-aggregate pore volume. The influence of shrinking and swelling of the soil must be taken into consideration when bulk density is measured. The bulk volume of the sample must be determined at the water content of interest to allow calculation of S_t at that degree of wetness.

When a highly accurate determination of total porosity is not required, an estimate of soil particle density may be satisfactory. For many mineral soils ρ_p is approximately 2.65 Mg m^{-3}. If this value is used instead of the actual measurement, the error in calculating S_t by Eq. [1] is shown in Fig. 18–1. As an example, if the true values for ρ_p and ρ_b are 2.55 and 1.40 Mg m^{-3}, respectively, the error resulting by estimating the particle density as 2.65 will be $< 5\%$.

18–2.2 Gas Pycnometer Method

18–2.2.1 PRINCIPLES

The pycnometer method for measuring gas-filled volume is based on Boyle's Law of volume-pressure relationships. Assume a given amount of gas (moles) confined in a given volume at a specified temperature and pressure. Boyle's Law states that the product of temperature (T) and pressure (P) will remain constant if the gas is compressed or allowed to expand and again brought to the previous temperature. Thus, $P_1V_1 = P_2V_2$, where the subscripts refer to the initial and final pressure and volume.

In practice a sample chamber is used in conjunction with a reservoir chamber. Either of two methods may be utilized. The volume of the reservoir may be changed after sealing the system and the resulting pressure change measured, or the reservoir and sample chamber may initially contain gas at different pressures and the resulting pressure change in the sample chamber measured after they are pneumatically connected. Sketches of systems utilizing these two methods are shown in Fig. 18–2.

When the variable volume pycnometer is used, the volume of gas in the complete system after compression, V_2, may be calculated using Eq. [2]. Since $P_1V_1 = P_2V_2$ and $V_1 = V_2 + \Delta V$ (where ΔV is the change in volume after the system is sealed), it follows that $P_1 (V_2 + \Delta V) = P_2V_2$. Thus, $V_2 = (P_1 \Delta V)/(P_2 - P_1)$ or

$$V_2 = P_1\Delta V/\Delta P \qquad [2]$$

where ΔP is the change in gas pressure resulting from ΔV. It must be remembered that Eq. [2] holds only when the amount of gas is constant (no leaks) and the temperature is constant (all materials are at ambient temperature and heat resulting from gas compression has dissipated). If the sample chamber is sealed at atmospheric pressure, P_1 may be obtained from a barometer reading, ΔP is read from the pressure gage, and ΔV must be known.

The volume of gas in the system after compression, V_2, may be determined with and without a soil sample in the chamber. The difference, [V_2 (without sample) $-$ V_2 (with sample)], is the volume of solids and liquids in the sample. If this value is subtracted from the sample bulk volume, the result is the volume of gas-filled pores in the soil sample.

When the constant volume pycnometer is used, the volume of gas in the sample chamber, V_c, may be calculated by equating the sum of the

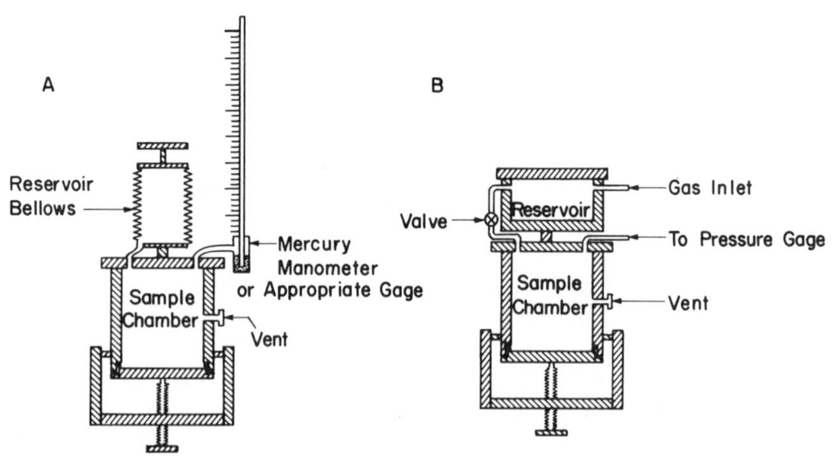

Fig. 18–2. Diagrammatic sketches of two types of gas pycnometer, the variable volume system (A) and constant volume system (B).

pressure-volume products of the chamber and the reservoir to that of the combined system after the valve is opened (Fig. 18–2). This is expressed as $V_cP_c + V_rP_r = (V_c + V_r)P$, where subscripts c and r refer to the sample chamber and the reservoir, respectively, and P is the pressure in the combined system after equilibration. Solving for V_c:

$$V_c = (P_r - P)\, V_r/(P - P_c)\,. \quad [3]$$

As with the previous case, the volume of liquid and solid in a soil sample may be obtained from the difference in V_c determined with and without the sample in the chamber. The gas-filled porosity is then calculated from the bulk volume of the soil sample.

18–2.2.2 APPARATUS

Construct the gas pycnometer as indicated in Fig. 18–2. The aid of a professional machine shop will usually be required. The sample chamber must be designed for easy closing after a soil sample of known bulk volume is inserted. It is expedient to have the chamber just large enough to contain a sampling cylinder used to obtain bulk samples from the field. A volume of 80 to 100 cm^3 should be useful for most purposes. A leakproof vent must be available on the sample chamber to establish atmospheric pressure when desired. A bellows, with reproducible open and closed stops, is recommended for the variable volume pycnometer. The change in volume, ΔV, when the bellows of the variable volume pycnometer is collapsed, and the volume of the reservoir of the constant-volume instrument may be approximately one-third the volume of the sample chambers for effective use over a range of soil porosity values. An automobile tire valve may be used for a gas inlet device on the reservoir. Tubing required for connections on the pycnometer should be rigid and of small bore. Any suitable pressure-sensing device may be installed, providing it has a sensitivity suitable to the results desired. A mercury manometer is ideally suitable, providing the pycnometer is adequately supported in an upright position to prevent spillage. The device should be capable of measuring gage pressure to approximately 40 kPa.

Construct a set of identical sampling cylinders with outside diameter and length to easily fit into the sample chamber of the pycnometer. If the calibration method is to be used, prepare solid calibration discs which fit closely into the sampling cylinders and which occupy specific fractions of the total cylinder volume. Discs of 5, 10, 20, 25, and 50% are suggested.

Various commercial gas pycnometers are available. They are relatively expensive and are capable of higher precision than is usually warranted for soil porosity measurements, because sampling errors are usually significantly larger than measurement errors.

18–2.2.3 PROCEDURE

18–2.2.3.1 Variable Volume Pycnometer. Obtain a soil core sample in a sampling cylinder designed for the pycnometer. Oven-dry the sample, cool to ambient temperature, insert the cylinder with the soil into the sample chamber of the pycnometer, and seal. Open the vent to establish atmospheric pressure in the chamber, open the reservoir bellows to the maximum, close the vent, and collapse the bellows to the stop. Read and record the pressure gage, ΔP, as soon as a constant pressure is attained. If the pressure decreases, a leak exists in the system and the measurement is invalid. Open the vent to return the pressure to atmospheric. A second measurement may be obtained, if desired, before removing the sample from the chamber. Calculate the volume of air in the pycnometer after compression, V_2, using Eq. [2]. Repeat the procedure when the sample chamber contains a clean sampling cylinder without soil.

The volume of solid phase in the soil sample is obtained by subtracting V_2 measured with the soil in the chamber from V_2 measured without soil. The total porosity, S_t, of the soil sample may be calculated from Eq. [4].

$$S_t = \frac{[(V_2 \text{ without soil}) - (V_2 \text{ with soil})}{\text{volume of sampling cylinder}}. \quad [4]$$

In order to use Eq. [2], the atmospheric pressure, P_1, may be measured with a barometer. The volume change due to collapse of the bellows, ΔV, is usually difficult to measure directly; however, the value may be calculated from the values of Eq. [2] obtained when the sample cylinder is filled with air and when it is filled with solid calibration discs. Let the symbols of Eq. [2] represent the conditions when the cylinder is air-filled and let ΔP^* be the change in pressure resulting from ΔV when the sample cylinder of volume V_s is completely filled with calibration discs. Then

$$V_2 = P_1\Delta V/\Delta P \text{ and } V_2 - V_s = P_1\Delta V/\Delta P^* .$$

Thus

$$\Delta V = V_s/[(P_1/\Delta P) - (P_1/\Delta P^*)] . \quad [5]$$

18–2.2.3.2 Constant Volume Pycnometer. Obtain a soil sample in a sampling cylinder designed for the pycnometer. Oven-dry, cool to ambient temperature, insert into the sample chamber of the pycnometer, and seal. Close the vent on the chamber and open the valve between the chamber and reservoir. Using an appropriate pump, inflate the system to give a gage pressure of approximately 35 kPa. The desired gage pressure may be obtained by over inflation followed by careful release of air through the vent. Test for leaks by observing the gage for a few minutes. Measure atmospheric pressure with a barometer and record the absolute air pressure, P_r, in the reservoir. Close the valve between sample chamber and reservoir. Open the vent to establish atmospheric pressure in the chamber, P_c, and then close it again. Open the valve to allow equilibration between chamber and reservoir, allow time for thermal equilibrium and record the absolute pressure, P, of the combined system. The volume of air, V_c, in the sample chamber may be calculated by Eq. [3]. Repeat the procedure using a clean sampling cylinder without soil. Calculate the total porosity of the soil, S_t, using Eq. [4] when V_2 is replaced by V_c.

18–2.2.3.3 Use of Calibration Curves. The above procedures require that the barometric pressure and certain volume values of the pycnometers are known so that Eq. [2] and [3] can be solved. It may be expedient to use calibration curves rather than the equations for determining the air-filled pore space of the soil sample. This is accomplished by measuring the value ΔP or P, in case of the variable or constant volume pycnometers, respectively, when the sampling cylinder contains known fractions of air-filled volume. The appropriate procedure described above for soil samples is used repeatedly to obtain gage pressure readings when a clean sampling cylinder, containing various combinations of calibration discs, is inserted into the sample chamber. Prepare a graph relating gage reading to air-filled pore space, S_t. Use this curve to determine S_t when soil samples are used in the pycnometer. When the calibration method is used with the constant volume pycnometer, the system must be inflated to precisely the same gage pressure when measuring porosity of soil samples and when using the calibration discs.

A calibration curve prepared under a specific value of atmospheric pressure is not applicable if the pycnometer is used at a different elevation. Thus, separate calibrations must be obtained for each value of atmospheric pressure at which the instrument is to be used. Usually, the variation in barometric pressure at a given elevation is not of sufficient magnitude to be of concern; however, it is advisable to check the calibration curve periodically by measuring the gage pressure at one or two known values of S_t.

When the barometric pressure is measured and the characteristics of the pycnometer are known, so that Eq. [2] or [3] together with [4] are used, the instruments may be used at any elevation.

18–2.2.4 COMMENTS

Further details of the construction and operation of gas pycnometers for use in the field are given by Russell (1950) for the variable-volume type and by Page (1948) for the constant-volume system. Commercial instruments are usually designed to use helium as the gas instead of air, since helium does not absorb on most solids and more closely obeys the perfect gas law. Identical temperature is required before and after the pressure change in the system but may vary from one measurement to another. The soil in the pycnometer sampling cylinder may shrink to some degree when oven-dried. As a result, the "pore space" within the cylinder may not all be within the soil core; however, the total porosity measured represents the soil at sampling time.

Since the gas pycnometer may be used to determine the volume of solid phase in an oven-dried soil sample, the particle density may be determined in this manner. Accuracy will not equal that of the method described in chapter 14 for particle density measurement unless commercial pycnometers using helium gas are utilized.

If the calibration method is to be the only method used, the pressure-sensing system on the pycnometer does not have to accurately measure gage pressure. The gage reading may be used for the calibration curve even though it is not a true pressure measurement. This is useful when a mercury manometer serves as the pressure sensor, because the scale does not have to be set for a true zero reading.

Although the procedure is written to use an oven-dried soil sample in order to measure total porosity, a wet sample may also be used in the pycnometer. The method then measures the air-filled porosity at the water content of the soil. This might be useful for evaluating soil aeration potential at a given time; for example when the soil is at the "field capacity." Total soil porosity, S_t, can be calculated by adding the volumetric water content to the air-filled porosity measured by the pycnometer.

18–3 PORE SIZE DISTRIBUTION

18–3.1 Water Desorption Method

18–3.1.1 PRINCIPLES

Evaluation of the size, configuration, and distribution of the soil pores is essentially impossible due to their extremely complicated nature. However, by making certain assumptions, the size distribution of the larger pores can be measured with at least useful accuracy. If we start with a saturated soil sample, drain it by steps, and measure the volume of water removed between consecutive steps, we can equate the volume of water removed to the soil pore-volume drained. Then, if the size-range of pores drained during each step can be calculated, the pore-size distribution can be determined. In theory, the largest pores should drain first, followed by successively smaller and smaller pores. Actually the drainage of the "pore" will be determined by the size of the largest "opening" of that pore to a larger one.

The pressure difference (ΔP) across an air-water meniscus, as in a capillary tube, is expressed as

$$\Delta P = 2\sigma\ r_c^{-1} = 2\sigma\ \cos\theta\ r_p^{-1} \qquad [6]$$

where σ is the surface tension of water (J m^{-2}), r_c is the radius of curvature of the meniscus (m), θ is the contact angle (degrees) of water to the solid, and r_p is the radius of the tube (m). ΔP is then expressed as pascals. If the contact angle is assumed to be zero, the cosine of θ becomes unity and the radius of meniscus curvature is equal to the radius of a circular capillary tube of size equivalent to that of the pore. Thus, Eq. [6] can be altered to

$$\Delta P = 2\sigma \, r_p^{-1} \qquad [7]$$

where r_p is considered to be the equivalent cylindrical pore radius.

As water is removed by any means from the porous medium, the radius of curvature of air-water interfaces will decrease. When r_c decreases to the effective radius of a given pore, that pore will drain as further water is removed. All pores with effective radii of a lesser value will still be water filled. When a given water content in a soil sample is attained by bringing the water matric potential to a desired value using a porous plate, the pore size dividing water-filled pores and drained pores may be calculated from Eq. [7]. When equilibrium has been reached, the pressure difference across the meniscus (ΔP) will be equal to the pressure difference across the porous plate.

18–3.1.2 APPARATUS

A device is required whereby a saturated soil sample can be drained stepwise to known matric potentials and a volume measurement obtained of the water removed during each step. A filter funnel modified as shown in Fig. 18–3 is ideal. For less precise measurements the funnel may be simply attached with flexible, thick-walled, transparent tubing to a suitable buret fitted with a stopcock. In this case the pressure differential established across the porous plate of the funnel is limited by the extent to which the buret can be lowered below the level of the plate. When a water column of 2 m is established, the smallest soil pores drained at equilibrium would have an equivalent diameter of approximately 15 μm.

The filter funnel should be of convenient size to contain the soil sample in an appropriate sampling cylinder. Funnels of various diameters and with a wide range of air-entry values are commercially available. Plates with maximum pore diameter of approximately 5 μm are appropriate. Those with smaller pores have reduced permeability values and require longer times for sample equilibration during drainage steps.

The funnel system diagrammed in Fig. 18–3 may be easily constructed by an experienced glass-blower. The welds attaching the inner and outer tubes to the funnel should be made as close to the porous plate as possible. A battery of funnels of this construction is shown in the photograph of Fig. 18–4. The plate diameter and overall length of these units are 65 mm and 60 cm, respectively. A good quality stopcock is required at the bottom to avoid inadvertent loosening under the pressure of mercury in the outer tube. The inner tube should be of uniform inside diameter along its length and carefully centered within the outer tube.

A cathetometer for precision measurements of elevation is essential to determine the volume of water removed from the soil sample during

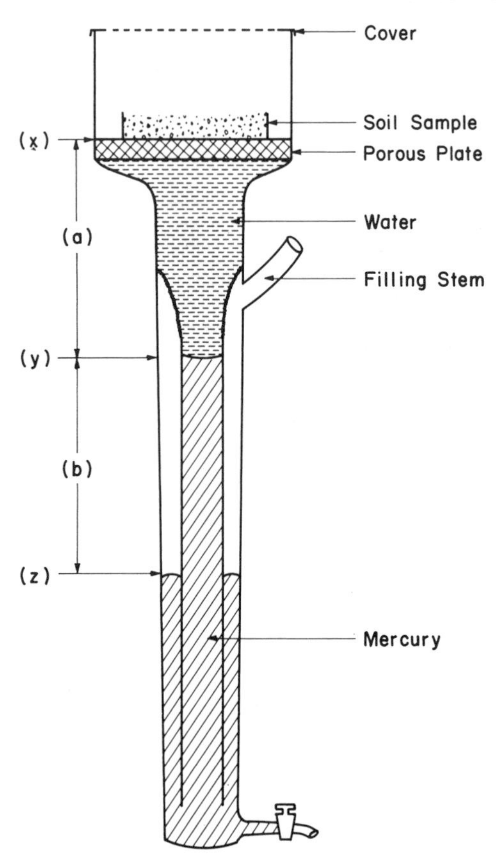

Fig. 18–3. Porous funnel apparatus for determining pore-size distribution in soil.

each drainage step. This instrument, shown in Fig. 18–4, should allow easy readings to 0.1 mm.

18–3.1.3 PROCEDURE

Specific instructions for measuring pore-size distribution of a soil sample using water desorption data will depend on the specific equipment used. The procedure provided in this section relates to the use of porous funnel systems comparable to the design shown in Fig. 18–3 and 18–4. The general methodology is comparable for use of equipment of other design.

Prepare the apparatus, as shown in Fig. 18–3, by inverting the funnel, opening or removing the stopcock, and slowly adding previously boiled water at room temperature through the filling stem (boiling under vacuum is an ideal way to remove air from the water). When the entire system is full of water, close the stopcock, return the funnel to the upright position, and clamp it to a ring stand or other support. Check to be sure

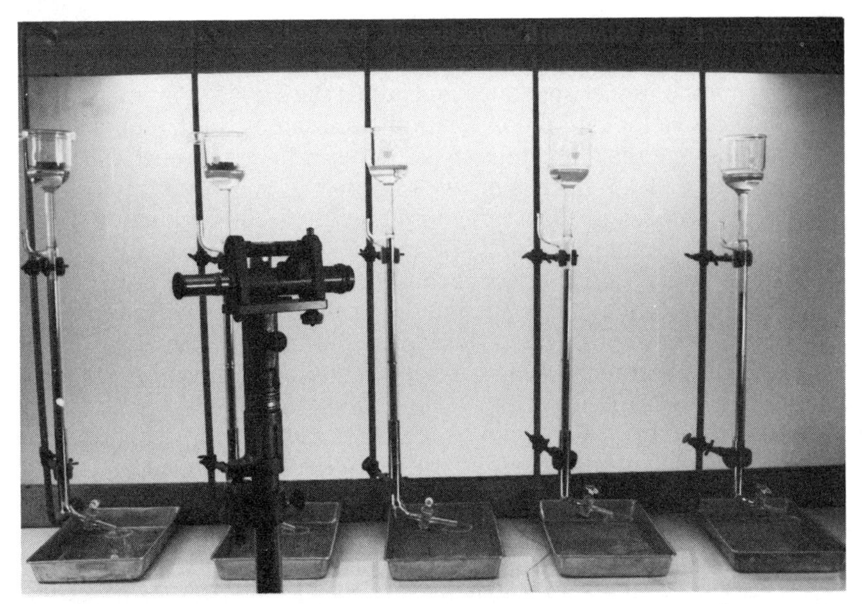

Fig. 18–4. Filter funnels used for pore-size distribution and a cathetometer for measuring elevations.

there are no air bubbles under the porous plate or anywhere within the inner glass tube.

Add water through the filling tube until it overflows; then slowly add mercury to force water out of the filling tube and upward through the porous plate. Add mercury until the Hg–H_2O interface (y) is a short distance below the weld of the inner tube and the Hg–air interface (z) is approximately level with y. Make certain there are no air bubbles below the porous plate. Remove the water from above the porous plate and from above the mercury in the outer tube. Cover the funnel to eliminate evaporation.

Calibrate the funnel as follows. Lower the level of z about 5 cm by removing mercury through the stopcock. Measure the level of y with the cathetometer and record to the nearest 0.1 mm. Lift cover to pipette 1.00 cm^3 of water onto the porous plate. Establish equilibrium, measure, and record the resulting level of y. Repeat as needed to obtained a series of values, lowering z when necessary. Calculate the changes in the elevation of y for each addition of 1.00 cm^3 of water, and record the average value as the calibration factor, k.

The soil sample to be used for pore-size distribution analysis should be contained in a metal or plastic cylinder about 1 cm high and of such diameter that it can be placed in, and removed from, the funnel with a large tweezers or forceps. Refer to chapter 13 section 13–2.2.1 for field sampling techniques. Secure high-density filter paper to the bottom of the sample cylinder by forcing the cylinder into a larger ring to hold the paper in place, or secure paper with an elastic band. Trim away excess filter paper. If a disturbed sample is used, secure the paper across the sample cylinders, add the soil, and pack as desired.

Prepare the funnel apparatus and adjust the Hg–H_2O interface, y, as described above for calibration. Place the soil sample in the funnel. Add mercury through the filling tube so that water is forced through the plate. Continue to add as much mercury as possible, add water around the sample as needed, and allow time for the soil to become saturated. Refer to soil sample wetting technique (chapter 26, section 26–6.2.2).

Lower the Hg–air interface, z, and continue adjusting until y is in the calibrated portion of the tube and the distance b is as small as possible. The objective is to bring the system to equilibrium at as small a suction as possible, so that the soil sample remains saturated. It is possible, with careful adjustment, to have the value of b approximately 1 mm. Water may be removed or added above the porous plate to adjust the level of y. When the system is at equilibrium, with the funnel sealed to prevent evaporation, measure and record the elevations of x, y, and z to the nearest 0.1 mm. A convenient format for recording measured and calculated data is shown in Table 18–1.

Withdraw mercury to lower the level of y, to establish the second equilibrium at a value for b of approximately 1 cm. Again measure and record elevation of y and z. Continue to establish equilibrium steps with values for b of approximately 2.5, 5, 10, 15, and 20 cm of mercury. Longer and longer equilibrium times will be required as the matric potential of the soil water decreases. Periodic readings of y with the cathetometer will allow detection of equilibrium. A day or so may be required at the greatest suction, depending on soil type and depth of sample.

Table 18–1. Data for determination of pore-size distribution using the water desorption method. An example.

Funnel no. 1   Calibration factor, k 1.19   Elevation of x 94.33 cm
Temperature 22°C   Surface tension 0.07244 J m^{-2}

Eq. no.	(Y)	(Z)	(a)	(b)	h	θ_w	θ_v	θ_v/S_t	Max diameter saturated pore
	cm				cm H_2O				µm
1	80.82	80.41	13.51	0.41	19.1	0.243	0.388	1.00	155
2	80.46	79.55	13.87	0.91	26.2	0.229	0.366	0.94	113
3	79.70	77.94	14.63	1.76	38.5	0.199	0.318	0.82	77
4	78.21	73.99	16.12	4.22	73.3	0.142	0.227	0.58	40
5	77.10	67.42	17.23	9.68	148.3	0.098	0.157	0.41	20
6	76.87	61.23	17.46	15.64	229.2	0.089	0.142	0.37	13
7	76.65	56.05	17.68	20.60	296.6	0.081	0.129	0.33	10
8...									

Gross wet wt. soil sample 36.56 g   Calculated bulk density, ϱ_b 1.60 Mg m^{-3}
Gross dry wt. soil sample 34.81 g   Calculated total porosity, S_t 0.3879
Tare wt. 13.17 g   Particle density 2.61 Mg m^{-3}
θ_w at final equilibrium 0.0809

When values for the final equilibrium have been recorded, remove the soil sample from the funnel, transfer the soil and filter paper to a weighing dish, obtain wet and oven-dry weights, and calculate the final water content on the weight basis (θ_w). Record the laboratory temperature and the surface tension of the water.

Calculate the soil water suction, h, in cm of water at each equilibrium by

$$h = a + \rho_{Hg}\, b \qquad [8]$$

where ρ_{Hg} is the density of mercury in Mg m^{-3} at the laboratory temperature. Calculate the water content, θ_w, at each of the other equilibrium values by

$$\theta_w = \text{final } \theta_w + \rho_{H_2O}\,(a_f - a)/w\,k \qquad [9]$$

where ρ_{H_2O} is the density of water (1.00 Mg m^{-3} may be used), a_f is the value of a at the final equilibrium, w is the oven-dry mass of the soil sample, and k is the calibration factor for the funnel.

Multiply θ_w at each equilibrium by the sample bulk density to obtain the volume fraction of water, θ_v. The bulk density, Mg m^{-3}, may be calculated from the sample dry mass and bulk volume or determined by assuming the soil sample is saturated at the first equilibrium value. The latter case will be appropriate for many soils, since the matric potential is only negative 1 or 2 kPa. In that case, the bulk density may be calculated if the particle density is known. The total porosity, S_t, may be expressed as:

$$S_t = (\rho_b)\,(\theta_w \text{ at equilibrium no. 1})\,. \qquad [10]$$

Combining with Eq. [1] results in:

$$\rho_b = 1/[(\theta_w \text{ at eq. no. 1}) + (1/\rho_p)]\,. \qquad [11]$$

Calculate S_t using Eq. [1] or [10]. Divide the value of θ_v at each equilibrium by S_t to obtain the fraction of total pore space occupied by water. Calculate the diameter in μm of the largest water-filled pore at each equilibrium by:

$$d_p = 4\,\sigma \times 10^5/\rho_w g\, h \qquad [12]$$

where σ and ρ_w are surface tension (J m^{-2}) and density (Mg m^{-3}) of water, g is the gravitational acceleration (m s^{-2}), and h is the matric suction (centimeters of water).

Table 18–1 provides an example of a convenient manner for recording data and demonstrates the results for a typical soil core. The fraction

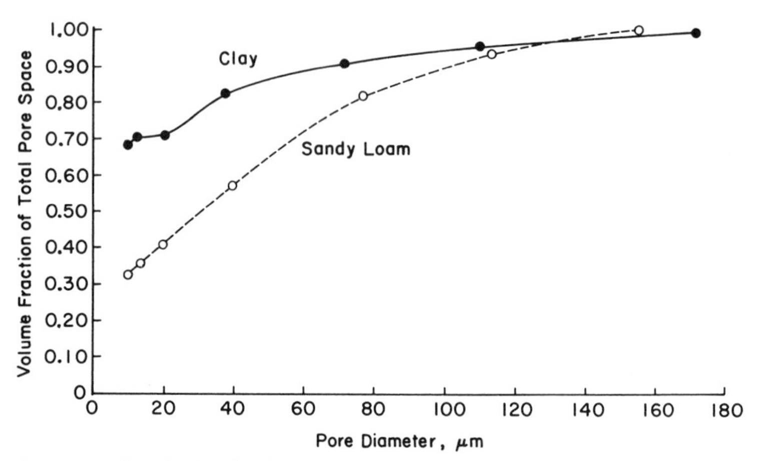

Fig. 18–5. Pore-size distribution for two soils. Data for the sandy loam are given in Table 18–1.

of total pore space filled with water, θ_v/S_t, may be plotted as a function of pore diameter to express the pore-size distribution (Fig. 18–5).

18–3.1.4 COMMENTS

Certain errors are inherent to the water desorption method of pore-size distribution. Some plate drainage could occur as suction is increased. This may be checked during calibration of the funnel by measuring the change in elevation of y when the elevation of z is lowered. With good quality equipment this drainage is negligible. The calculation of pore size is based on the assumption that the pores are of capillary-tube shape and that the contact angle of water to soil (Eq. [6]) is zero. It is customary to ignore these obvious problems by expressing pore size in terms of "equivalent" radius or diameter. The use of filter paper to contain the sample, and thus allow complete removal of all soil from the plate, enters into the determined pore-size determination. This error is minimized by having a small ratio of paper to soil sample. Evaporation losses during establishment of equilibrium should be avoided as far as possible. A cover of waxed paper or plastic sheet may be secured over the funnel with tape or elastic bands, so that it is water-vapor-tight but flexible to maintain the air around the sample at atmospheric pressure. Errors due to normal temperature changes are minimal, the greatest influence being on the calculation of pore size from Eq. [12]. Water density (ρ_w) changes are negligible, but the value for surface tension (σ) changes about 0.2% for each degree change in temperature. A constant-temperature room is preferable, but ordinary laboratory conditions may be satisfactory if caution is used to prevent temperature change due to air currents or direct sun radiation on the equipment.

The greatest chances for error undoubtedly relate to determination of sample volume or bulk density and to assurance of sample saturation.

When soil is wetted its bulk volume will increase to an extent depending on the particular soil. Thus, the use of cylinder size to represent sample volume may not be satisfactory for calculating bulk density and total pore space. Since it is very difficult to measure the soil bulk volume when it has swelled, it is probably most precise to use Eq. [10] and [11] if care is taken to have minimum matric suction at equilibrium no. 1. The effort expended to achieve complete saturation and to prevent errors due to gas production within the sample by microorganisms depend upon how the data are to be used. For some soils, the assumption that the contact angle is zero may be the dominant error of the determination.

The matric suction values established at each equilibrium will depend on pore-size values desired. The values for b indicated in the described procedure will provide points to plot a satisfactory pore-size distribution curve for most soils. Intermediate steps may be utilized if desired.

As pointed out by Russell (1949), bimodal pore size distribution curves are commonly found for well aggregated soils as a consequence of the intraaggregate and interaggregate pores. Voorhees et al. (1966) have used a unique method for measuring the pore volume between aggregates. They utilized small glass beads as a non-imbibing "fluid" to fill the intercrumb space. The volume of these macropores was calculated from the mass of the beads and their predetermined bulk density.

Regardless of the limitations to the water desorption method for measuring pore size distribution, it serves as the most commonly used technique.

18–3.2 Mercury Intrusion Method

18–3.2.1 PRINCIPLES

Mercury intrusion porosimetry has been utilized as a reliable method for determining pore-size distribution of a wide variety of porous solids (Ritter & Drake, 1945a, 1945b; Winslow & Diamond, 1970). The technique has been further expanded for use with clay systems (Aylmore & Quirk, 1967; Diamond, 1970; Sills et al., 1973a, 1973b) and with soil (Klock et al., 1969; Nagpal et al., 1972; Sills et al., 1974; Lawrence, 1977; Ragab et al., 1982).

The theoretical basis for using mercury intrusion to determine pore-size distribution is similar to that for the water desorption method, except that pressure is required to force the mercury into the pores instead of suction to remove water from a saturated system. Because mercury is a nonwetting liquid for most materials, the contact angle, θ, has a value greater than 90°. Thus, it will enter the pores of a solid only if an external pressure is applied. The pressure required is dependent on the value of the contact angle, the size and geometry of the pore, and the surface tension. If the porous material is initially evacuated, Eq. [6] may be restated as:

$$r_p = -2\sigma \cos\theta \, P^{-1} \quad [13]$$

where r_p is the calculated pore radius (m), σ the surface tension of mercury (J m^{-2}), θ the contact angle of mercury on the porous solid, and P is the absolute pressure applied (N m^{-2}). Since $\cos\theta$ is negative for the obtuse angle of contact, a negative sign is required in the equation to obtain a positive value for the radius of the pore, r_p.

The porous sample is dried, evacuated, and inundated in mercury, and pressure is applied hydraulically in discrete steps. The diminution of the bathing mercury is measured and equated to the volume of pores invaded at each pressure step. Equation [13] is used to calculate the equivalent radii of the smallest pores filled with mercury, assuming a model of cylindrical capillary tubes of differing sizes. Because the pore geometry is not that simple, it may be necessary to use a correction factor in order to obtain "effective" pore size-volume relationships that are useful in predicting certain soil properties.

18–3.2.2 APPARATUS

Two types of mercury intrusion porosimeters are commercially available. One type is designed to obtain data by applying pressure incrementally; the second type is the scanning porosimeter which provides for a constant increase or decrease in pressure and the recording of intruded or extruded mercury volume on a continuous basis. The mercury volume is measured using a capacitance bridge.

Although the commercial porosimeters differ in certain aspects, each instrument consists of two components. One is used for mercury intrusion into the porous material by pressures less than atmospheric. This requires allowance for evacuation of the sample and subsequent mercury-filling by desired fractions of atmospheric air pressure. The second component consists of a system for establishing pressures greater than atmospheric. The most advanced instruments have an upper pressure limit of over 400 MPa (4000 bars), which allows evaluation of equivalent cylindrical pore radii to a minimum of approximately 1.5 nm.

18–3.2.3 PROCEDURE

Specific step-by-step procedures are dependent on the porosimeter chosen for the analysis. The instructions for the equipment should be followed. An oven-dried soil sample is placed in the penetrometer assembly. The penetrometer is then inserted into the filling device, sealed, and evacuated. Pressure on the mercury in the penetrometer is increased by small increments and readings of pressure and mercury volume are recorded for each step. When a continuous scan porosimeter is used, pressure and mercury volume are recorded continuously.

The equivalent pore radius at each applied pressure value is calculated from Eq. [13]. It has been observed that pore radii are consistently underestimated using the mercury intrusion technique. Thus, an empirical correction factor may be necessary. Klock et al. (1969) suggested that the correction factor required for their measurements was 1.31, to be multipled by the calculated pore size. It has been observed by Nagpal et al. (1972) that the correction factor may not be the same for all soils. They suggested that it varies linearly with the clay content of the soil. Ragab et al. (1982) compared the water desorption and mercury intrusion methods of determining pore size distribution for use in evaluating soil-water characteristic curves and hydraulic conductivity relationships. A correction factor of 2.60 for the mercury method was found to be necessary for their data. Once the "effective" pore size is calculated by multiplying the equivalent value obtained from Eq. [13] by the desired correction factor, the pore volume obtained from the amount of mercury intruded into the sample can be related to the size of pores.

Pore-size distribution may be determined in a manner similar to that described for the water desorption method. The volume of pores not intruded with mercury at the maximum applied pressure may be determined by subtracting the volume of intruded mercury from the total pore volume, S_t, as calculated from Eq. [1]. A check on the volume of intruded mercury may be made by weighing the sample assembly containing the soil and mercury after the final pressure step. Subtracting from this the mass of the sample holder and of the oven-dry soil provides the mass of mercury in the pores, and when divided by ρ_{Hg} the volume is obtained (Nagpal et al., 1972).

18–3.2.4 COMMENTS

The mercury intrusion porosimetry method for determining pore-size distribution in soils is a very convenient and fast method. However, several sources of error have been identified. The value of the mercury-particle contact angle is uncertain. Reported values range from 112 to 150° for various types of porous materials (Lawrence, 1977). The value for clay-mercury systems has been reported to range from 139° for montmorillonite to 147° for illite and kaolinite (Diamond, 1970). A value of 130° has been used by Sills et al. (1973a) and Nagpal et al. (1972). Contact angle hysteresis may also influence the accuracy of the mercury porosimetry measurements when a continuous scan instrument is used. The effect has been observed with several porous media (Lowell & Shields, 1981a, 1981b). The result is a greater mercury volume at a given pressure during extrusion than during intrusion.

Uncertainty also exists concerning the surface tension of mercury. Values have been reported to range from 0.43 to 0.52 J m^{-2}. The most probable range is from 0.472 to 0.487; a value of 0.473 has commonly been used for soil porosimetry measurements. Lawrence (1977) points out that the error associated with surface tension variation is not very important when compared to those related to contact angles. A rise in temperature with a consequent decrease in surface tension is expected when the pressure is increased. The error associated with this effect can be alleviated by waiting for temperature equilibrium between pressure steps.

Further sources of error may be associated with mercury compression, nonuniform pressure distribution within the soil sample, impurities in the mercury, and entrapped air in the sample after initial evacuation. The collapse of the pores and subsequent compression of the soil sample during the measurement period may also be a source of error. There appears, however, to be little information available concerning the effect of mercury intrusion on pore distortion.

18-4 REFERENCES

Aylmore, L. A. G., and J. P. Quirk. 1967. The micropore size distribution of clay mineral systems. J. Soil Sci. 18:1–17.

Beven, K. 1981. Micro-, meso-, macroporosity and channeling flow phenomena in soils. Soil Sci. Soc. Am. J. 45:1245.

Bouma, J. 1981. Comment on "micro-, meso-, and macroporosity of soil." Soil Sci. Soc. Am. J. 45:1244–1245.

Cary, J.W., and C.W. Hayden. 1973. An index for soil pore size distribution. Geoderma 9:249–256.

Diamond, S., 1970. Pore size distributions in clays. Clays Clay Miner. 18:7–23.

Dullien, F. A. L. 1979. Pore structure. p. 75–155. *In* Porous media. Academic Press, New York.

Klock, G. O., L. Boersma, and L. W. DeBacker. 1969. Pore size distributions as measured by the mercury intrusion method and their use in predicting permeability. Soil Sci. Soc. Am. Proc. 33:12–15.

Lawrence, G. P. 1977. Measurement of pore sizes in fine-textured soils: a review of existing techniques. J. Soil Sci. 28:527–540.

Lowell, S., and J. E. Shields. 1981a. Influence of contact angle on hysteresis in mercury porosimetry. J. Colloid Interface Sci. 80:192–196.

Lowell, S., and J. E. Shields. 1981b. Hysteresis, entrapment, and wetting angle in mercury porosimetry. J. Colloid Interface Sci. 83:273–278.

Luxmoore, R. J. 1981. Micro-, meso-, and macroporosity of soil. Letter to the editor. Soil Sci. Soc. Am. J. 45:671–672.

Nagpal, N. K., L. Boersma, and L. W. DeBacker. 1972. Pore size distributions of soils from mercury intrusion porosimeter data. Soil Sci. Soc. Am. Proc. 36:264–267.

Page, J. B. 1948. Advantages of the pressure pycnometer for measuring the pore space in soils. Soil Sci. Soc. Am. Proc. 12:81–84.

Ragab, R., J. Feyen, and D. Hillel. 1982. Effect of the method for determining pore size distribution on prediction of the hydraulic conductivity function and of infiltration. Soil Sci. 134:141–145.

Ritter, H. L., and L. C. Drake. 1945a. Pore-size distribution in porous materials: pressure porosimeter and determination of complete macropore-size distributions. Ind. Eng. Chem. Anal. Ed. 17:782–786.

Ritter, H. L., and L. C. Drake. 1945b. Pore-size distribution in porous materials: macropore-size distributions in some typical porous substances. Ind. Eng. Chem. Anal. Ed. 17:787–791.

Russell, M. B. 1949. Methods of measuring soil structure and aeration. Soil Sci. 68:25–35.

Russell, M. B. 1950. A simplified air picnometer for field use. Soil Sci. Soc. Am. Proc. 14:73–76.

Sills, I. D., L. A. G. Aylmore, and J. P. Quirk. 1973a. An analysis of pore-size in illite-kaolinite mixtures. J. Soil Sci. 24:480–490.

Sills, I. D., L. A. G. Aylmore, and J. P. Quirk. 1973b. A comparison between mercury injection and nitrogen sorption as methods of determining pore size distributions. Soil Sci. Soc. Am. Proc. 37:535–537.

Sills, I. D., L. A. G. Aylmore, and J. P. Quirk. 1974. Relationship between pore size distributions and physical properties of clay soils. Aust. J. Soil Res. 12:107–117.

Skopp, J. 1981. Comment on "micro-, meso-, and macroporosity of soil." Soil Sci. Soc. Am. J. 45:1246.

Voorhees, W. B., R. R. Allmaras, and W. E. Larson. 1966. Porosity of surface soil aggregates at various moisture contents. Soil Sci. Soc. Am. Proc. 30:163–167.

Wilkins, D. E., W. F. Buchele, and W. G. Lovely. 1977. A technique to index soil pores and aggregates larger than 20 micrometers. Soil Sci. Soc. Am. J. 41:139–140.

Winslow, D. N., and S. Diamond. 1970. A mercury porosimetry study of the evolution of porosity in Portland cement. J. Mater. 5:564–585.

19 Penetrability

J. M. BRADFORD

National Soil Erosion Laboratory, Agricultural Research Service, USDA, and Purdue University
West Lafayette, Indiana

19-1 INTRODUCTION

Soil penetrability is a measure of the ease with which an object can be pushed or driven into the soil. Any device designed to measure resistance to penetration may be called a penetrometer. There are two principal types of penetrometers and ways of measuring penetration: (i) the *dynamic* (penetration is accomplished by driving the tool into the soil with a hammer or falling weight) and (ii) the *static* (the probe is pushed steadily into the soil without impact). The dynamic method has found limited application in soil science and will not be discussed.

Penetrometer technology has been advanced by civil engineers in search of methods to survey such subsoil conditions as relative density, shear strength, bearing capacity, and settlement. Comprehensive engineering studies of penetrometers and penetration techniques are in the Proceedings of the European Symposium on Penetration Testing (Swedish Geotechnical Society, 1974) and Sanglerat's (1972) *The Penetrometer and Soil Exploration.* The soil scientist's concern, however, is to relate penetrometer resistance to root growth, crop yields, and soil physical properties descriptive of tilth. For these purposes, the penetrometer guidelines provided by the engineer, e.g., the high rates of penetration and the large-diameter drive-rods, do not readily apply.

The purpose of the particular investigation and the soil conditions encountered dictate the penetration method chosen. Penetration testing can be simple or highly mechanized and very sophisticated. In this chapter three general methods are presented; they range from the inexpensive and simple (the pocket penetrometer) to the sophisticated (the motorized friction-sleeve cone penetrometer).

19-2 PRINCIPLES

Static penetrometers are designed to measure either (i) the *cone* or *point resistance* (resistance to penetration developed by the cone); (ii)

 Methods of Soil Analysis, Part 1. Physical and Mineralogical Methods—Agronomy Monograph no. 9 (2nd Edition)

both cone resistance and *friction-sleeve resistance* (the resistance to penetration developed by the moveable sleeve—friction sleeve—located above the cone and surrounding a central rod), separately; or (iii) total resistance (the sum of cone and friction-sleeve resistance). Of the penetrometers described here, the cone penetrometer is of the first type; the friction-sleeve cone penetrometer is of the second type; and the pocket penetrometer, the third type. Cone resistance is calculated as the vertical force applied to the cone divided by its basal area; friction-sleeve resistance is equal to the vertical force applied to the sleeve divided by its surface area.

Penetration resistance is influenced by soil and probe characteristics and by mode of soil failure. Penetrometer factors affecting penetration resistance are cone angle, diameter, roughness, and rate of penetration. The rougher the penetrometer surface, the greater the resistance to penetration. Resistance to penetration is comprised of two principal forces: (i) force to deform the soil by wedge-action of the conical point and (ii) soil-to-metal friction against the surface. For a given probe diameter, as the cone angle decreases the cone length increases, greatly increasing the surface area. Therefore, as the cone angle decreases, the theoretically increasing mechanical advantage of the wedge is overcome by the friction from the increased surface area. This "cross-over" commonly occurs at a cone angle of about 30°. Consequently, for cone angles less than 30° cone resistance normally decreases as the angle increases (Greacen et al., 1968; Voorhees et al., 1975) and for cone angles exceeding about 30° cone resistance increases as the angle increases (Voorhees et al., 1975; Durgunoglu & Mitchell, 1975b). Soil-to-soil friction becomes the dominant resistance component. The magnitude of the effect of cone angle depends upon the cohesive and adhesive properties of the soil-probe system.

In agricultural soils investigations, penetrometer diameter becomes an important design consideration because of the soil macrostructure. For structureless or homogeneous soils, penetration resistance below a certain critical depth (Fig. 19–1) is independent of probe diameter. In strongly structured soils, if the diameter of the cone relative to structural unit size is large, the variability in cone resistance is low and the resistance is primarily a function of interaggregate strength. If the cone diameter relative to structural size is small, the cone resistance variability is large because of the lower resistances in the cracks between the structural units of the soil; cone resistance is mainly a function of intraaggregate strength.

Establishment of guidelines for penetration rate is difficult because an increase in rate can increase, decrease, or not influence probe resistance depending upon soil properties and water conditions. The relationship of penetration resistance to penetration rate is influenced to a large degree by the pore water pressures generated by the advancing cone, the rate of dissipation of these pressures, and the dilatency properties of the soil under shear.

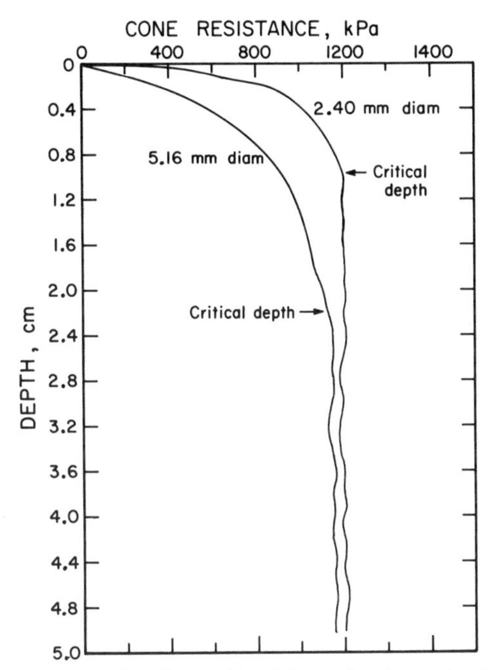

Fig. 19–1. Cone resistance as a functionn of depth in confined, remolded cores of Dickinson loam, $\rho = 1.43$ g cm^{-3}, $\psi = -33$ kPa.

Soil factors influencing penetration resistance are matric potential (or water content), bulk density, soil compressibility, soil strength parameters, soil structure, and others. The dependence of cone resistance on matric water potential and soil bulk density is shown in Fig. 19–2 from the studies of Taylor and Gardner (1962). Numerous papers have been published on these factors, both in the engineering and agronomic literature.

Both penetrometer geometry and soil conditions influence soil failure mode. Durgunoglu and Mitchell (1975a) reviewed the failure mechanisms proposed for static penetration. Penetration resistance increases with depth to a certain depth (as shown in Fig. 19–1), which is dependent on soil hardness and probe diameter. Below this critical depth, the failure mechanism changes from shear to one of shear and compression, and the rate of increase of resistance with depth is less or the resistance is constant with depth. Cavity expansion theory has been used to describe such failure (Landanyi, 1963; Farrell & Greacen, 1966; Vesic, 1972). Compression is assumed to occur in two main zones: a zone of compression with plastic failure surrounding the probe, and outside this, a zone of elastic compression (Farrell & Greacen, 1966). Cone resistance is assumed to be made up of a pressure component to expand the cavity and a cone frictional component determined by the properties of the probe. The total cone or point resistance (q_c) is defined as

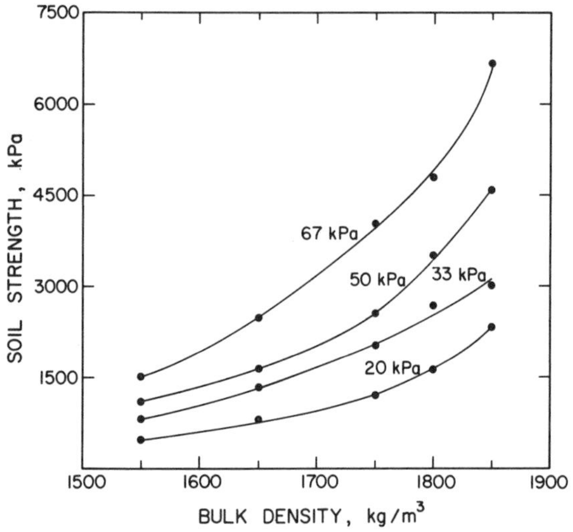

Fig. 19–2. Effect of bulk density and water tension of Amarillo fine sandy loam on penetration resistance (Taylor and Gardner, 1962).

$$q_c = \sigma_n (1 + \tan \phi' \cot \alpha) \quad [1]$$

where α is the included semiangle of the cone, ϕ' is the coefficient of soil-metal friction, and σ_n is the normal point resistance, i.e., the normal stress on the basal surface of the probe. The coefficient of soil-metal friction can be obtained with a direct shear machine by filling the upper half of the shear box with a natural or remolded soil specimen and the lower half with a metal block of the penetrometer cone material (Durgunoglu & Mitchell, 1975b; Bradford, 1980). Since plant roots have a low skin-friction with the soil, the normal point resistance more closely correlates with root elongation than does the total point resistance (Greacen et al., 1968; Cockcoft et al., 1969; Voorhees et al., 1975).

The tests described herein can be used either in the laboratory or in the field. For laboratory determinations on either natural or remolded core samples, the ratio of core diameter to penetrometer diameter should not fall below a certain critical value, which for loam soil material at less than −100 kPa water potential is normally between 20 and 30 (Greacen et al., 1968). At lower ratios, the confining effect of the container or sampling tube increases the resistance. As soils become drier or more brittle (less compressible), the critical ratio increases.

19–3 POCKET PENETROMETER

19–3.1 Introduction

The pocket penetrometer is a hand-operated, calibrated-spring penetrometer, originally developed as an improvement on the thumb-fin-

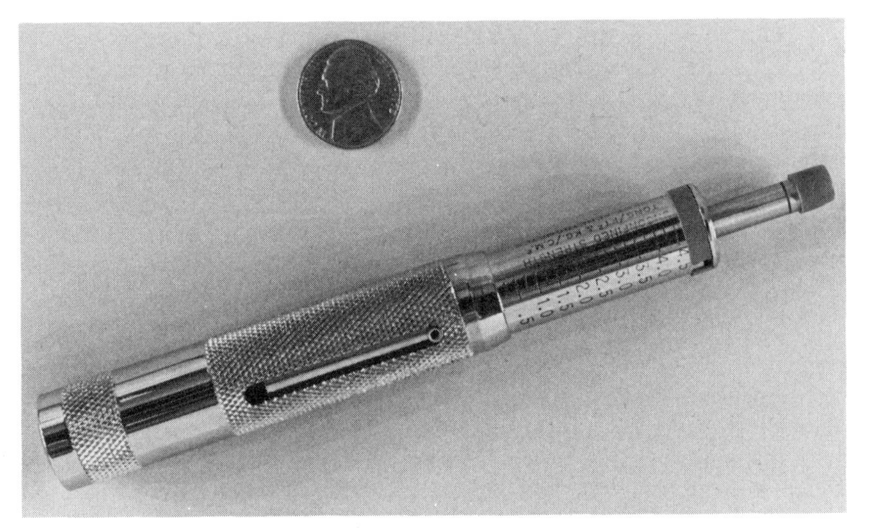

Fig. 19–3. A pocket penetrometer.

gernail technique for estimating the engineering consistency of cohesive, fine-grained soils (Fig. 19–3). The maximum deformation of the spring as the piston needle is pushed into silty clay or clay soil has been correlated with unconfined compressive strength of soil in tons per square foot or kilograms per square centimeter. The latter values are the scale on the piston barrel. Since correlations have been developed between root growth and point resistance, Bradford (1980) suggested converting the unconfined compressive strength scale to read in units of total probe resistance by calibrating the penetrometer scale against a load cell or set of known weights.

The pocket penetrometer is simple to operate. It can be pushed into the surface in agricultural fields, into soil in a sampling tube, into an undisturbed soil block, into soil in an open pit excavation, or into molded soil specimens.

The method is useful in comparing relative strengths among similar soil types or in determining hardpans, zones of compaction, or dense soil layers, if matric potential is measured jointly. The pocket penetrometer test is primarily supplemental, and usually does not eliminate the need for more precise field or laboratory testing.

19–3.2 Method

19–3.2.1 SPECIAL APPARATUS

Direct-reading pocket penetrometers in several different models and sizes are commercially available (Fig. 19–3). They weigh from 170 to 200 g and are from 160 to 180 mm in length. All have a diameter of 19.1 mm and a piston needle diameter of 6.4 mm.

Commercial sources[1] of penetrometers include the following: Soiltest, Inc., 2205 Lee St., Evanston, IL 60202; Wykeham Farrance, Inc., 8000 Glenwood Ave., P.O. Drawer 30967, Raleigh, NC 27622; Humboldt Manufacturing Co., 7300 W. Agatite Ave., Chicago, IL 60656; Engineering Laboratory Equipment, Inc., 2205 Lee St., Evanston, IL 60202.

19–3.2.2 PROCEDURE

Select the test location and determine the associated soil properties by referring to section 19–4.2.2.

Move the indicator sleeve to the lowest reading (zero) on the penetrometer scale. Grip the handle and push the piston needle, with a steady rate of penetration, into the soil until the engraved line 6 mm from the blunt tip is flush with the surface of the soil. Remove the penetrometer from the soil and read the scale. Clean the piston and return the sliding indicator to its zero position. Repeat the test several times in different areas to find an average value for unconfined compressive strength.

19–3.2.3 COMMENTS

Readings can be converted to total probe resistance by calibrating the penetrometer scale with a set of known weights. Plot the added weight (kg) per tip area (0.317 cm^2) against the sleeve reading. Convert each unconfined compressive strength value to total point resistance from the calibration plot.

19–4 CONE PENETROMETER

19–4.1 Introduction

The cone penetrometer was developed by the U.S. Army Corps of Engineers at the Waterways Experiment Station for predicting the carrying capacity of cohesive, fine-grained soils for army vehicles in off-road military operations. It has been used extensively in agricultural soils research to locate hardpans or traffic compaction areas, to correlate soil strength parameters with root growth and crop yield, and to quantify the physical state of soil. The applied force required to press the cone penetrometer into a soil is an index of the shear resistance of the soil and was called the "cone index" (Department of Army Staff, 1960).

The cone penetrometer was designed for hand operation; however, vehicle-mounted systems have been used by Williford et al. (1972), Cassel et al. (1978), and Smith and Dumas (1978). The main advantage of these units is the controlled rate of penetration and the reduction in man-hours

[1]Trade names and company names, included for the benefit of the reader, do not imply endorsement or preferential treatment of the product listed by the USDA.

required for data collection. Regardless of the driving system, the penetrometer dimensions should conform to the requirements given below.

The commercial cone penetrometer model is supplied with a proving-ring dial-gauge. For hand-operated penetrometers, the test is best performed by two persons: an operator, who presses the cone penetrometer into the soil and calls out times for readings, and a recorder, who reads the dial readings. If the cone penetrometer is used frequently, electronic load-measuring recording devices are recommended (Prather et al., 1970; Anderson et al., 1980). The dial gauge in the proving ring is replaced with a linear displacement transducer, or the proving ring with a strain-gauge load cell. Connecting the electronic transducer to a recorder produces a continuous recording of penetration resistance with depth.

The guidelines below conform to the cone penetration standards recommended by the American Society of Agricultural Engineers (1982) and by the Department of Army Staff (1960).

19–4.2 Method

19–4.2.1 SPECIAL APPARATUS

The Corps of Engineers cone penetrometer (Fig. 19–4) is commercially available. It consists of a handle, a 68-kg proving ring and dial gauge, a cone, and a graduated driving shaft. Two cone base sizes and corresponding shafts are recommended: (i) a 9.5-mm diameter (3/8 in.) shaft with a 12.8-mm diameter, 1.3-cm^2 (0.2 $in.^2$) cone for "hard" soils, and (ii) a 15.9-mm diameter (5/8 in.) rod with a 20.3 mm-diameter, 3.2 cm^2 (0.5 $in.^2$) cone for "soft" soils. The driving shafts are 46 cm long, and for the 3.2-cm^2 cone an extension 46 mm long is used. The cones are smoothly machined stainless steel and have a 30° included angle.

The load measuring device for electronic recording hand-operated units should have a penetration resistance capacity of about 50 kg/cm^2 for the large cone and 100 kg/cm^2 for the small cone.

19–4.2.2 PROCEDURE

Select the test location. For cultivated areas, the location must be defined in relation to tillage relief, wheel tracks, plant rows, and other horizontal nonuniformities. Suggested positions are (i) the row axes, (ii) the upper interrow shoulders about 15 to 20 cm from the row axes, and (iii) midway between the rows in both wheel-tracked and nontracked interrows. The specific position selected depends upon the study objectives—e.g., characterization of effects of tillage compaction or soil erosion. Penetrometer test results are most usefully interpreted if matric water potential or water content, soil bulk density, and soil structure are determined under testing conditions. Determine matric potential (or water content) and bulk density by following the procedures given in chapters 13 and 23. Penetrability testing at the in situ field water-capacity is recommended because this water content is repeatable from season to season

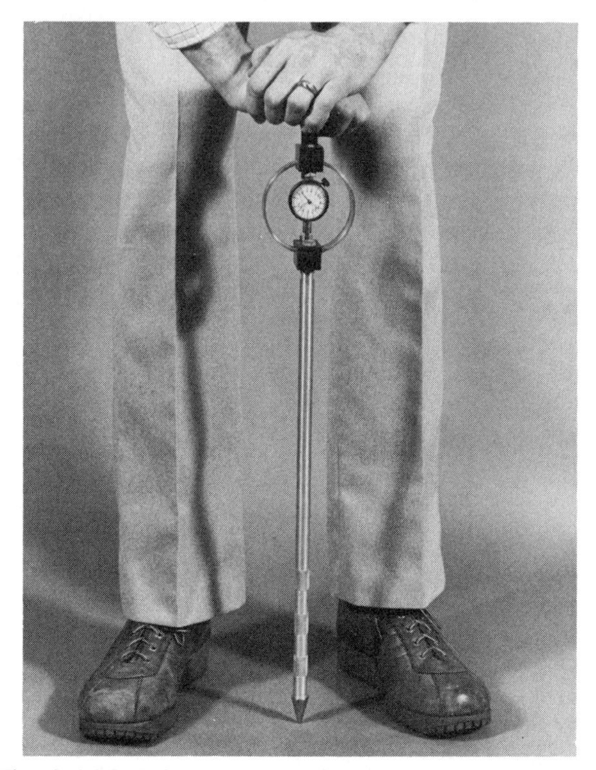

Fig. 19–4. The U.S. Army Corps of Engineers cone penetrometer.

and from time to time within a given season. Morphologically describe each site and classify soil structure for each horizon according to the Soil Conservation Service's system, described in Agricultural Handbook 436 (Soil Survey Staff, 1975).

Allow the penetrometer to hang vertically from its handle and set the dial gauge to the zero position. If an extension shaft is added or removed during testing, reset the instrument to zero. Push the shaft vertically downward into the soil at a constant rate of about 3 cm/s. Take dial readings just as the base of the cone is flush with the ground surface and at depth intervals predetermined by soil conditions and experimental objectives (the commercial models normally mark graduations on the shaft at 2.54-cm intervals). By using one extension shaft on the 3.2 cm^2 cone, the operator can take readings to a depth of 91 cm.

Withdraw the cone from the soil and wipe it clean. Repeat the test at least five times at locations at least 50 to 60 cm apart to prevent erroneous readings because of soil disturbance. If the soil is extremely nonuniform, additional replications should be made based on a presampling survey of the area (Cassel, 1981). From this presampling, the sample variance and mean are determined, and the number of samples required to attain the desired confidence level is calculated. The minimum number of samples, n, required to obtain an acceptable estimate of the mean of a population is given by (Steel and Torrie, 1960; Cassel, 1981)

$$n = t_{\alpha}^{2}\, s^{2}/d^{2} \quad [2]$$

where t_{α} is the Student's t with $(n - 1)$ degrees of freedom at the α probability level, s^2 is the sample variance, and d is the specified acceptable error. The number of replications, however, is very often determined largely by the funds and time available for the experiment.

Using the proving-ring calibration chart, convert each dial reading to penetrometer force in kilograms. Calculate the penetration resistance in kg/cm^2 by dividing the force by the cone's projected cross-sectional area (cm^2). Convert the resistance from kg/cm^2 to kN/m^2 (kPa) by multiplying by 98.07. Average the resistance values (kPa) obtained at each depth increment and calculate the standard deviation and coefficient of variation. Plot the average penetration resistance against depth for each test location (ordinate scale, depth of penetration; abscissa scale, penetration resistance).

For vehicle-mounted drive systems, the penetrometer dimension guidelines given above should be followed. Vehicle-mounted systems should use electronic load cell readout systems for recording data. The penetrometer rate can be reduced from 180 cm/min to about 10 cm/min in order to minimize damage to the thrust rod and load cell if small stones or rocks are encountered and to approach more of a drained failure condition.

19–4.2.3 COMMENTS

The cone penetrometer has a small resistance component in addition to the cone resistance. Even though the cone diameter is larger than the push-shaft diameter, a slight frictional resistance develops between the shaft and the soil. This resistance depends on soil volume change properties and the drive unit, and it increases with depth.

In dry or hard soils or in soils containing pebbles or stones, it is difficult to obtain consistent penetrometer measurements. In stony soils care should be taken not to damage the cone or overstress the proving ring.

For extremely soft soils (e.g., near-saturated plow layers), another commerically available cone penetrometer can be substituted for the Corps of Engineers penetrometer described above. The design is similar to the Corps penetrometer except that a 19.1-mm diameter push rod, a 28.7-mm diameter 30° cone, and a 113-kg proving ring are used.

Penetrometer cones should be regularly checked for changes in surface roughness, cone angles, and base diameter.

Since plant roots are flexible and grow through zones of least resistance, minimum cone resistance values possibly have more meaning than maximum or average horizon resistance values. The minimum resistance value might be taken as an average of the interaggregate resistance. To determine the minimum resistance, several replicates of a continuous trace of the resistance with depth is required.

19–5 SMALL-DIAMETER FRICTION-SLEEVE CONE PENETROMETER

19–5.1 Introduction

Cone penetrometers with the additional capacity of measuring the rod- or sleeve-friction component of the total resistance are called friction-sleeve cone penetrometers or simply friction cone penetrometers. Many types of friction cone penetrometers have been designed for engineering purposes (Sanglerat, 1972). Soil scientists have not adapted these penetrometers to their studies, because the cone base diameters normally exceed 35 mm and are often as large as 80 mm. This large-diameter penetrometer increases the depth at which point resistance becomes a constant or a function mainly of soil properties and not of probe size. It also requires a larger soil core for laboratory studies.

Soil scientists have designed smaller-diameter friction cone-type penetrometers to study root elongation, soil structure, and rate of penetration. A 3.0-mm diameter penetrometer was designed by Barley et al. (1965) and slightly modified by Bradford et al. (1971), Voorhees et al. (1975), and Bradford (1980), as shown in Fig. 19–5. The penetrometer consists of a shaft or central push-rod used for advancing the cone into the soil. The cone is attached to the end of the push rod. Surrounding the central rod is moveable sleeve with outside diameter equal to the cone base diameter. Barley et al. (1965) used two proving rings to record both the point resistance and the sleeve friction, whereas Bradford et al. (1971) and Voorhees et al. (1975) recorded only the point resistance.

The penetrometer is easily adapted to either laboratory or field determinations. For laboratory studies, the penetrometer can be pushed into the soil with any rate-controlled laboratory load machine. For field tests with penetrometers less than about 5 mm in diameter, a drive machine such as shown in Fig. 19–6 (Bradford, 1980) is used. The weight of the machine is used to drive the penetrometer.

The push rods must be strong enough to sustain the force required to advance the penetrometer cone without buckling. Rods therefore must be relatively short. For 5-mm diameter penetrometers, this length is about 150 mm. Penetrometer length determines test depth; hence, deeper in situ determinations must be conducted in a soil pit, or core samples may be tested using guidelines for size in section 19–5.2.2.

Penetrometer design and procedures to follow are intended for soils research. They are used to study the effect of rate of penetration, spatial variability, structural features, physical and chemical properties, etc. on

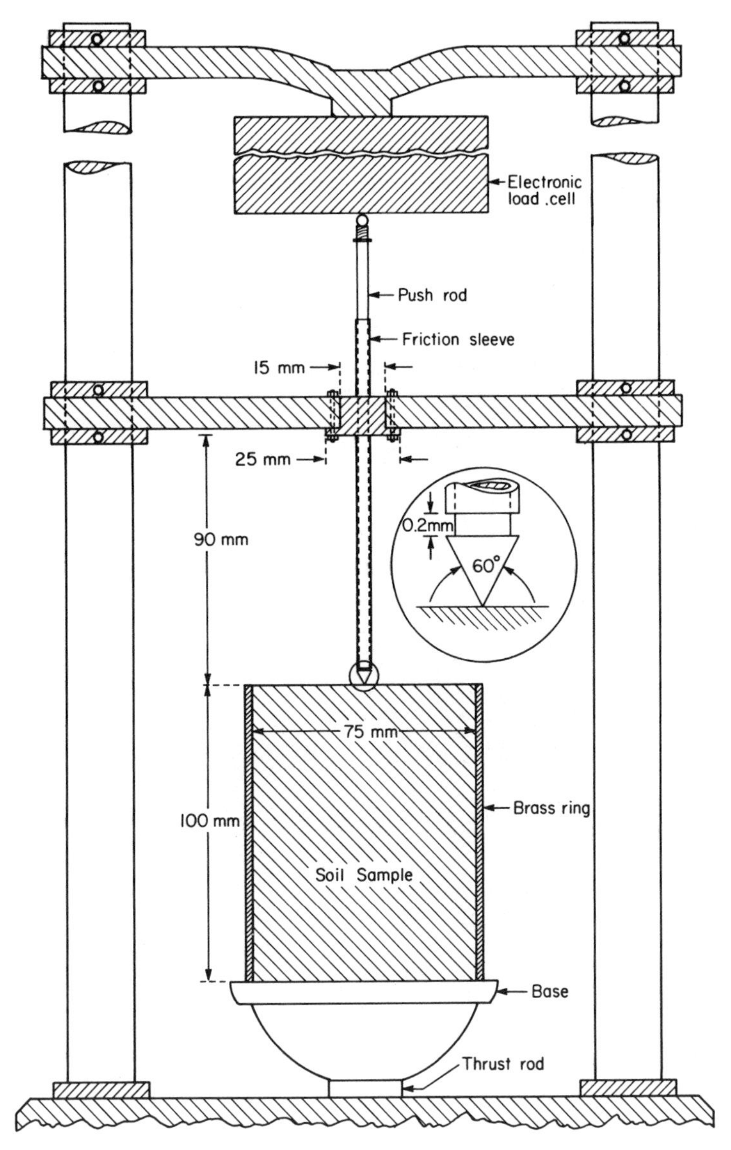

Fig. 19–5. A friction-sleeve cone penetrometer.

cone resistance in soils. They can be used in tillage studies if small pits can be dug or core samples taken at the study site.

19–5.2 Method

19–5.2.1 SPECIAL APPARATUS

1. Friction-sleeve cone penetrometer (Barley et al., 1965; Bradford et al., 1971): The cone has a 60° angle and a base diameter of 3.74 ± 0.02

Fig. 19–6. Friction-sleeve cone penetrometer drive machine.

mm; the cone is made from stainless steel or brass and has a polished surface. The push rod has an outside diameter of 2.76 ± 0.02 mm and is made from stainless hypodermic needle tubing (12-gauge). The length is about 150 mm. Inner rods can be inserted into the push rod (the hypodermic tubing) for added strength. The friction sleeve has an outside diameter of 3.74 ± 0.02 mm and an inside diameter of 3.00 ± 0.02 mm, and is made from stainless steel hypodermic needle tubing (9-gauge). The length is about 140 mm.

Different sizes of friction-sleeve cone penetrometers can be constructed from other tubing sizes. Recommended cone diameters are: 5.16 ± 0.02 mm (sleeve, 6-gauge; rod, 8-gauge); 2.40 ± 0.02 mm (sleeve, 13-gauge; rod, 16-gauge); and 1.08 ± 0.02 mm (sleeve, 19-gauge; rod, 25-gauge).

2. Load weighing system: The system consists of a 10- to 50-kg capacity

strain gauge load cell, a signal conditioning unit, and a recorder. This apparatus is available commercially.

3. Penetrometer drive machine: For laboratory tests, any motorized compression- or strength-testing machine with variable speed drive can be used.

   For in situ tests, a portable lightweight drive machine is used. The machine must advance the penetrometer at a constant rate and have a stroke of at least 150 mm. To minimize cost, the design in Fig. 19–6 can be used in the field and the laboratory.

19–5.2.2 PROCEDURE

For laboratory determinations: Attach the friction-sleeve penetrometer to the penetrometer drive machine as shown in Fig. 19–5. The friction sleeve of the penetrometer is secured to a metal cross-bar which is then fastened to the compression machine guide rods. A 4- to 8-mm diameter steel ball is positioned between the top end of the load shaft and the base of the load cell. While holding the push rod and steel ball against the load cell, move the crossbar upwards or downwards until the distance between the cone base and the end of the friction sleeve is about 0.2 mm. If a high-deflection proving-ring is used in place of a load cell, this distance should be increased so that at maximum cone resistance the distance is about 0.2 mm.

Position the soil core on the compression machine platen. The core diameter should be at least 20 times the probe diameter and should be radially confined in metal rings; for brittle soil materials the core-to-probe diameter ratio should exceed 30. Advance the core upward so that the penetrometer enters the soil at a constant rate of 0.05 cm/min. For penetration rate studies, select additional rates of 0.0005, 0.005, and 0.5 cm/min. For re-formed soil cores, only two replications are needed; for natural soil cores, at least three to five replications are recommended (sampling cost and time usually limit replications).

Convert the load cell reading (kg) into cone resistance (kPa) by dividing the load (kg) by the cone basal area (0.1099 cm^2) and multiplying by 98.07. Plot cone resistance in kPa on the abscissa against penetration depth on the ordinate (see Fig. 19–1).

Determine the core bulk density and matric potential. Classify the type and class of soil structure.

For in situ determinations: Attach the friction-sleeve cone penetrometer to the field drive unit by following the procedures given for laboratory determinations. Position the penetrometer above the test area and advance the penetrometer into the soil at a constant rate of 5.0 cm/min to a depth of about 10 cm. Repeat the test 5 to 10 times or using the guidelines in section 19–4.2.2. Remove the soil overburden to the next desired depth and repeat the procedure. Continue the sequence until the desired depth is reached.

The layout of the test position and the depth increment is determined by the objectives of the experiment; however, adjacent penetration readings should be no closer than 20 cm for the 3.74-mm penetrometer.

Plot cone resistance (kPa) against penetration depth and report soil bulk density, matric potential, and soil structural type and class.

19–5.2.3 COMMENTS

To record sleeve friction, the above design must be modified by attaching a friction sleeve 5 mm long (Barley et al., 1965) to the outer sleeve immediately above the cone. This requires that both the push rod and outer sleeve diameters be reduced enough that the diameter of the 5-mm-long friction sleeve equals the cone base diameter. The sleeve friction is recorded either: (i) by using the two-proving-ring design of Barley et al. (1965) (the push rod is secured to a smaller, inner proving ring and the friction sleeve is attached to a larger, outer ring); or (ii) by attaching the friction sleeve to a miniature donut-shaped load cell (commercially available) and by screwing the push rod into the base of an upper load cell. Since the friction-sleeve resistance is partly that of a remolded soil, it is less than that of the undisturbed soil friction.

Fine particles entering the space between the outer sleeve and the inner rod cause excessive friction and errors in the cone resistance readings. Experience has shown that a 0.2- to 0.3-mm clearance between the cone base and the end of the friction sleeve is enough to prevent entry of soil particles. However, if this becomes a problem, a silicone rubber sleeve can be placed in the void between the cone base and the sleeve end. Erroneous readings also result when the push rod is overloaded and bends, which creates excessive friction on the sleeve. After each test, check for binding caused by soil particles or bent rods or sleeves.

The recommended penetration rates for the laboratory (0.05 cm/min) and the field (5.0 cm/min) are established arbitrarily as a practical matter and not from theory. Some soils will have different cone resistance for the two rates; some will not. In situ tests on several horizons in a soil profile at a rate less than 5.0 cm/min would be too time consuming. However, the lower rate is desirable to create failure under drained conditions. At the faster penetration rates, the cone resistance is influenced more by pore water pressure and hydraulic conductivity and less by effective soil stresses; e.g., whether failure aproaches the undrained condition depends on the degree of saturation. For root and crop yield investigations, undrained failure does not simulate the mechanics of root elongation.

For correlation studies between root elongation and cone resistance, the normal point resistance, σ_n, should be determined from Eq. [1]. The coefficient of soil-metal friction is obtained with a direct shear machine using methods of Durgunoglu and Mitchell (1975b) or Bradford (1980).

One of the major problems with the small friction-sleeve penetrometer is that removal of soil overburden in test pits or on natural soil samples reduces cone resistance determinations. The magnitude of this decrease in resistance depends upon soil compressibility, pore water pressure, and the depth of the overburden removed. Studies that evaluate cone resistance before and after excavation are needed.

19-6 REFERENCES

American Society of Agricultural Engineers. 1982. Soil penetrometer. Agricultural Engineers Yearbook. ASAE Standard: ASAE S313.1. p. 246. American Society of Agricultural Engineers, St. Joseph, MI.

Anderson, G., J. D. Pidgeon, H. B. Spencer, and R. Parks. 1980. A new hand-held recording penetrometer for soil studies. J. Soil Sci. 31:279–296.

Barley, K. P., D. A. Farrell, and E. L. Greacen. 1965. The influence of soil strength on the penetration of a loam by plant roots. Aust. J. Soil Res. 3:69–79.

Bradford, J. M. 1980. The penetration resistance in a soil with well-defined structural units. Soil Sci. Soc. Am. J. 44:601–606.

Bradford, J. M., D. A. Farrell, and W. E. Larson. 1971. Effect of soil overburden pressure on penetration of fine metal probes. Soil Sci. Soc. Am. Proc. 35:12–15.

Cassel, D. K. 1981. Tillage effects on soil bulk density and mechanical impedance. p. 45–67. *In* P. W. Unger and D. M. Van Doren, Jr. (ed.) Predicting tillage effects on soil physical properties. Spec. Pub. 44. American Society of Agronomy and Soil Science Society of America, Madison, WI.

Cassel, D. K., H. D. Bowen, and L. A. Nelson. 1978. An evaluation of mechanical impedance for three tillage treatments on Norfolk sandy loam. Soil Sci. Soc. Am. J. 42:116–120.

Cockroft, B., K. P. Barley, and E. L. Greacen. 1969. The penetration of clays by fine probes and root tips. Aust. J. Soil Res. 7:333–348.

Department of Army Staff. 1960. Soils trafficability. Dep. of Army Tech. Bull. TB ENG 37. U.S. Government Printing Office, Washington, DC.

Durgunoglu, H. T., and J. K. Mitchell. 1975a. Static penetration resistance of soils: I - Analysis. p. 151–171. *In* Proc. of the Conf. on In Situ Measurement of Soil Properties, Vol. 1. American Society of Civil Engineers, New York.

Durgunoglu, H. T., and J. K. Mitchell. 1975b. Static penetration resistance of soils: II - Evaluation of theory and implications for practice. p. 172–189. *In* Proc. of the Conf. on In Situ Measurement of Soil Properties, Vol. 1. American Society of Civil Engineers, New York.

Farrell, D. A., and E. L. Greacen. 1966. Resistance to penetration of fine probes in compressible soil. Aust. J. Soil Res. 4:1–17.

Greacen, E. L., D. A. Farrell, and B. Cockroft. 1968. Soil resistance to metal probes and plants roots. Int. Cong. Soil Sci. Trans. 9th 1:769–779.

Ladanyi, B. 1963. Expansion of a cavity in a saturated clay medium. J. Soil Mech. Found. Div., ASCE 89:127–161.

Prather, O. C., J. G. Hendrick, and R. L. Shafer. 1970. An electronic hand-operated recording penetrometer. Trans. ASAE 13:385–386, 390.

Sanglerat, G. 1972. The penetrometer and soil exploration. Elsevier Publishing Co., New York.

Smith, L. A., and W. T. Dumas. 1978. A recording soil penetrometer. Trans. ASAE 21:12–14, 19.

Soil Survey Staff. 1975. Soil taxonomy: A basic system of soil classification for making and interpreting soil surveys. USDA-SCS Agric. Handb. 436. U.S. Government Printing Office, Washington, DC.

Steel, R. G. D., and J. H. Torrie. 1960. Principles and procedures of statistics. 1st ed. McGraw-Hill Book Co., New York.

Swedish Geotechnical Society. 1974. Proceedings of the European Symposium on penetration testing. Swedish Geotechnical Society, Stockholm, Sweden.

Taylor, H. M., and H. B. Gardner. 1962. Penetration of cotton seedling taproots as influenced by bulk density, moisture content, and strength of soil. Soil Sci. 96:153–156.

Vesic, A. S. 1972. Expansion of cavities in infinite soil mass. J. Soil Mech. Found. Div., ASCE 98:265–290.

Voorhees, W. B., D. A. Farrell, and W. E. Larson. 1975. Soil strength and aeration effects on root elongation. Soil Sci. Soc. Am. Proc. 39:948–953.

Williford, J. R., O. B. Wooten, and F. E. Fulgham. 1972. Tractor mounted field penetrometer. Trans. ASAE 15:226–227.

20 Compressibility

J. M. BRADFORD

National Soil Erosion Laboratory, Agricultural Research Service, USDA, and Purdue University
West Lafayette, Indiana

S. C. GUPTA

University of Minnesota
St. Paul, Minnesota

20-1 INTRODUCTION

Soil compressibility is the ease with which a soil decreases in volume when subjected to a mechanical load. The process that describes the decrease in soil volume (soil densification) under an externally applied load is defined as compression. Compression of soils is due to (i) exclusion of air or water from the void spaces, (ii) rearrangement of soil particles, (iii) compression and deformation of solid particles, and (iv) compression of the liquid and gas within the voids. If the voids are filled mainly with air, the addition of a load on the soil mass will result in volume change without appreciable water drainage. On the other hand, if the voids are nearly or completely filled with water, very little or no volume change will take place immediately upon the application of a load. Only as the water drains from the soil mass can volume change take place. The mechanism by which compression occurs and during which the applied loads are gradually transferred from the pore water to the soil matrix is called consolidation. Complete consolidation is reached at a state of zero pore water pressure. If the water can readily drain from a highly permeable soil mass, consolidation takes place within a short period of time; however, if the soil permeability is low, complete consolidation under an applied load may require several years.

Literature in structural and foundation engineering deals with the consolidation of soils in relation to the construction of dams, buildings, roadbeds, and embankments. This information is of limited use to agronomists and soil scientists for three reasons: (i) the loads due to agricultural implements and equipments are normally smaller than the loads due to superstructures such as buildings and dams, (ii) most compression tests reported in foundation engineering literature are conducted on saturated soils (consolidation), while agricultural soils are mostly unsaturated when

 Methods of Soil Analysis, Part 1. Physical and Mineralogical Methods—Agronomy Monograph no. 9 (2nd Edition)

implements and equipment are used in the field, and (iii) the compressibility of unsaturated soils with various degrees of structural development differs considerably from that of dense, massive-structured subsurface layers.

Soil compressibility and consolidation should not be confused with the terms compactibility and soil compaction when describing volume change phenomena. Compactibility is the maximum density to which a soil can be packed by a given amount of energy. The American Society for Testing and Materials (1979) gives a standard method for determining soil compactibility. In this procedure, sometimes called the Proctor test, the soil is compacted in layers in a metal mold by a set number of impacts from a free-falling hammer. The relationship between the density to which the soil can be compacted by a set number of blows and water content is determined, and the peak of the curve is defined as "the optimum moisture content" of the soil. In contrast to compactability, soil compaction is frequently defined as the volume change produced by momentary load application caused by rolling, tamping, or vibration. Soil compaction involves an expulsion of air without significant change in the amount of water in the soil mass. The most common causes of agricultural soil compaction are trampling by livestock and pressures imposed by vehicles or tillage equipment.

The practical significance of the soil compressibility determination lies in its relation to the mechanics of root growth or probe penetration and to soil compaction from vehicular traffic in unsaturated soils. Penetration resistance decreases and compaction by field machines increases as the compressibility of soil increases. The procedures described in this chapter do not necessarily simulate the dynamics or stress conditions of field compaction, but may be used to better understand soil properties affecting soil compaction and to correlate compressibility with potential compaction.

Data in the literature on the compressibility of agricultural soils are limited. One of the earlier investigations on this subject was done by Scott Blair (1937) and Scott Blair and Cashen (1938), who used compressibility curves to quantitatively describe tilth of agricultural soils. In recent investigations of Larson et al. (1980), Stone and Larson (1980), Larson and Gupta (1980), and Gupta and Larson (1981), the traditional consolidation test was modified to measure the effects of unsaturated conditions on the compressibility of agricultural soils.

20–2 PRINCIPLES

Laboratory tests for measuring soil compressibility are normally conducted either in a consolidometer (also called an oedometer) or in a triaxial compression cell. We shall confine our attention to uniaxial compression in a consolidometer, as shown in Fig. 20–1. In the uniaxial compression test, a soil sample confined in a metal ring is placed between

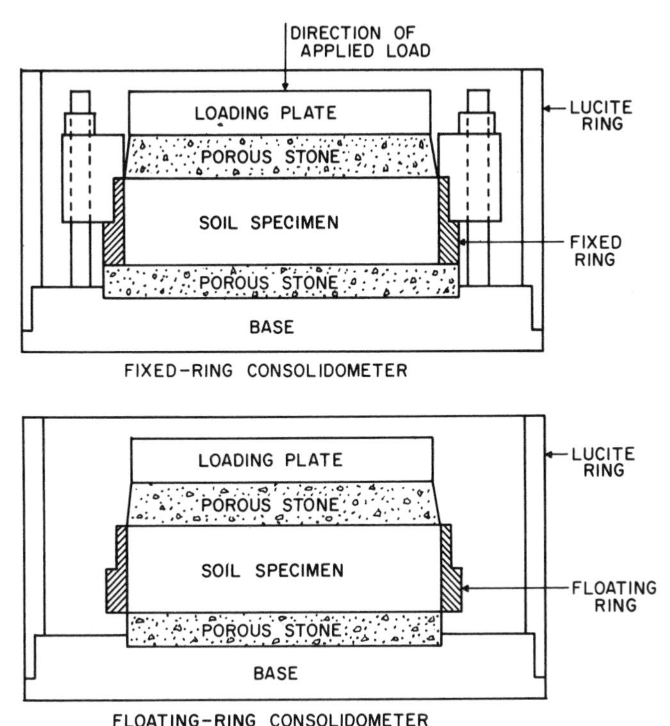

Fig. 20–1. Fixed-ring and floating-ring consolidometers.

two porous stones, a static load is applied to the specimen, and readings of the compression are taken at appropriate time intervals after the application of the load. The stresses on the soil created by the confining metal ring are not measured. The decrease in the void ratio (defined as the ratio of the volume of voids to the total volume of solid particles) due to the application of an external stress (load per unit area) is time-dependent. The general shape of the soil deformation vs. time curve at a given load increment is similar to the plot in Fig. 20–2. The curve for a saturated soil consists of: (i) initial compression—possibly because of sample disturbance, incomplete saturation, and particle readjustment, (ii) primary compression—resulting from the drainage of pore water because of the hydrostatic excess pressure, and (iii) secondary compression—caused by deformation of individual particles and relative movements of individual particles with respect to each other. For organic and highly compressible inorganic soils, the amount of secondary compression can be larger than that of the primary compression. For unsaturated soils the time required for complete compression by a given load increment depends upon the magnitude of the load and the soil and water properties, especially initial soil density, water content, and particle-size distribution.

The void ratio (e) at the end of compression plotted against the logarithm of effective stress gives a straight line relationship over much of the stress range, as in Fig. 20–3. For saturated soils the stress is plotted

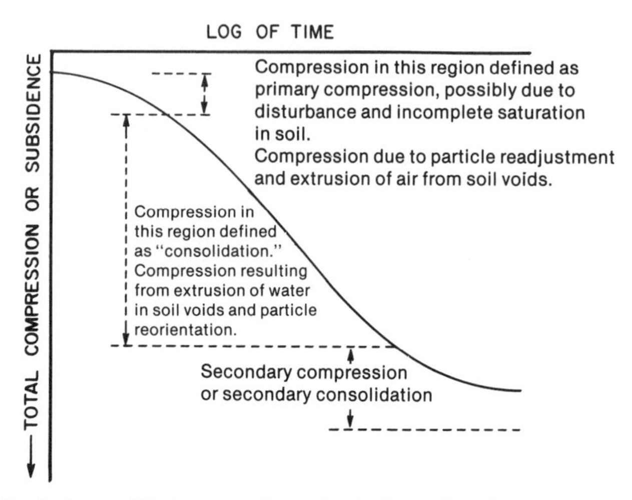

Fig. 20–2. Typical consolidation curve for a clay (redrawn from Yong & Warkentin, 1966).

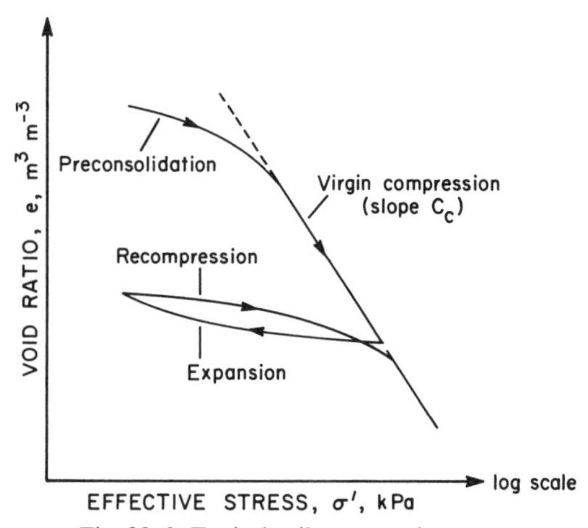

Fig. 20–3. Typical soil compression curve.

in terms of effective stress, σ'(the difference between the total applied stress, σ_a, and the pore water pressure, u_w). This definition of effective stress also holds for soils having pore water pressure less than the air entry value; i.e.,

$$\sigma' = \sigma_a + T \quad [1]$$

where T is the pore water suction. For unsaturated soils the determination is much more complicated (Lee & Donald, 1968; Towner & Childs, 1972) and is presently unsolved; thus Larson et al. (1980), Stone and Larson (1980), and Larson and Gupta (1980) chose to plot stress in terms of σ_a.

The straight-line portion of the curve in Fig. 20–3 is called the *virgin compression line* (VCC) and is described by the equation

$$e = e_1 - C_c \log (\sigma'/\sigma_1) \quad [2]$$

where e is the computed void ratio corresponding to σ', e_1 is the void ratio at the known effective stress, σ_1, and the slope of the line

$$C_c = - de/(d \log \sigma') \quad [3]$$

is called the compression index. Normally, σ' is expressed in kPa; C_c is dimensionless.

Two other indices are commonly used to represent soil compressibility. The first is the *coefficient of compressibility*, a_v, defined as the slope of the linear portion of a plot of e vs. σ', i.e.,

$$a_v = - de/d\sigma' . \quad [4]$$

The second is the *coefficient of volume change*, m_v, defined as the volume change per unit volume per unit increase in effective stress. If for an increase in effective stress from σ_1' to σ_2', the void ratio decreases from e_1 to e_2, then

$$m_v = - [1/(1 + e_1)] [(e_1 - e_2)/(\sigma_1' - \sigma_2')] = - a_v/(1 + e_1) . \quad [5]$$

The units for both a_v and m_v are kPa^{-1}.

To be consistent with agronomy and soils literature, Larson et al. (1980) substituted bulk density, ρ, for void ratio and applied stress for effective stress in defining a compression index, C_ρ, for unsaturated soils as:

$$C_\rho = d\rho/d \log \sigma_a \quad [6]$$

If the soil has been previously subjected to an overburden pressure or a desiccation stress (i.e., it is preconsolidated), the void ratio vs. stress relationship initially increases along a line with a slight slope until it joins the virgin compression line. The compression curve in Fig. 20–3 is typical of a preconsolidated material.

If the applied stress at some point on the virgin consolidation curve is released to zero, a certain rebound (an increase in void ratio) will occur (Fig. 20–3). The slope of the rebound curve relates directly to the swelling properties of the soil. In high-swelling soils, this rebound slope approaches the slope of the loading curve (Yong & Warkentin, 1966). Stone and Larson (1980) found rebound to be usually < 50 kg m^{-3} when measured at stresses between 100 and 1000 kPa and pore water pressures of 0 to -100 kPa.

Some of the soil factors affecting soil compressibility are soil fabric and structure, soil mineralogy, particle surface forces, pore water chemistry, stress history, and temperature. Soil compression index values (C_c

from Eq. [3]) for natural soil deposits normally range between 0.2 and 2. Compression indices (C_c) > 9 have been reported for undisturbed samples of Mexico City clay (highly organic, montmorillonite, thixotropic, and very sensitive); values for remolded samples were about 5.8 (Mesri et al., 1975). Rahman (1973) reported a compression index decrease from 0.54 to 0.26 as the negative pore water potential decreased from −70 to −20 700 kPa for an eastern Iowa loessial soil. Larson et al. (1980) showed that the VCC shifted to the right with a decrease in the water content of the soil (Fig. 20–4). Also, VCC's of a soil were nearly parallel and had the same compression index at water contents corresponding to initial pore water pressure of −5 to −60 kPa.

Larson et al. (1980) grouped compression curves of unsaturated agricultural soils at approximate pore water potentials of −30 kPa into four categories, as shown in Fig. 20–5. Curve A represents the amorphous materials such as allophanes; curves B and C represent the medium- and fine-textured soils from the tropics and the temperate regions of the world, respectively; and curve D represents a sandy soil. Differences in the slope of VCC of curve B and C result from differences in the type of clay minerals and the amount of organic matter. Temperate region soils are high in swelling clay minerals and in organic matter, while tropical region soils contain small amounts of organic matter and are dominated by nonswelling clay minerals. Figures 20–6 and 20–7 show the relationships between C_ρ and clay content for temperate and tropical region soils, respectively. Soils described in Fig. 20–6 contained montmorillonite, allophane, vermiculite, and hydrous mica clays, whereas soils in Fig. 20–7 were highly weathered soils and were dominated by kaolinite or iron oxide in the clay fraction. In both figures, C_ρ increases up to clay contents

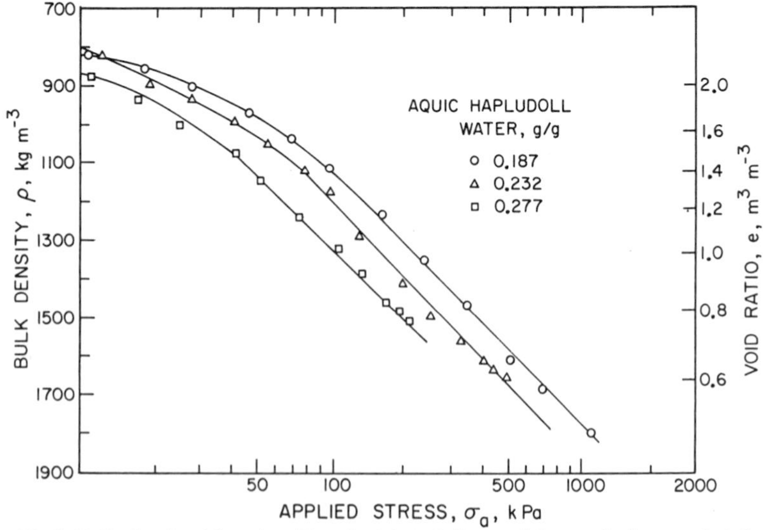

Fig. 20–4. Bulk density (ρ) and void ratio (e) as influenced by applied stress (σ_a) for an Aquic Hapludoll at three water contents (Larson et al., 1980).

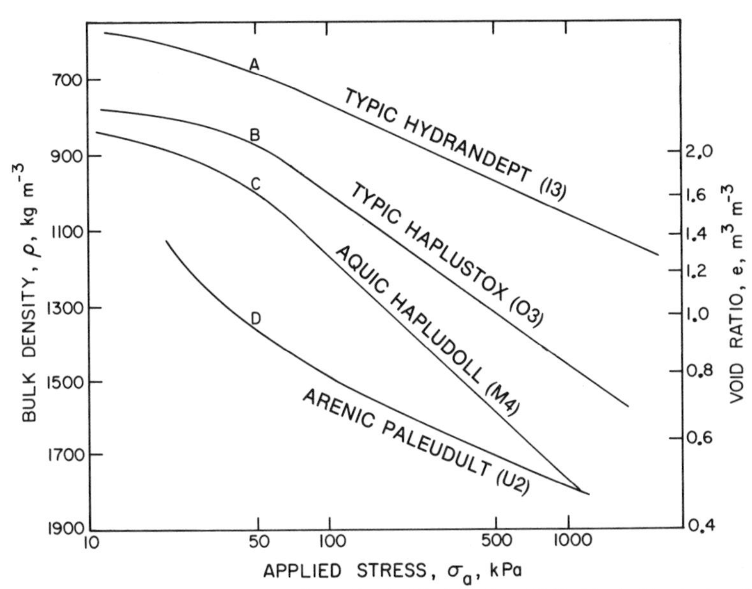

Fig. 20–5. Soil compression curves for a Typic Hydrandept (I 3), Typic Haplustox (O 3), Aquic Hapludoll (M 4), and Arenic Paleudult (U 2) at pore water potentials of approximately −30 kPa (Larson et al., 1980).

of about 33% and then levels off. Larson et al. (1980) reasoned that this trend resulted because soils with clay contents >33% behave essentially as a clay matrix with coarse material embedded in the clay. The maximum value of C_ρ for soils having expanding-type clay was 0.59, as compared with 0.56 for highly weathered soils.

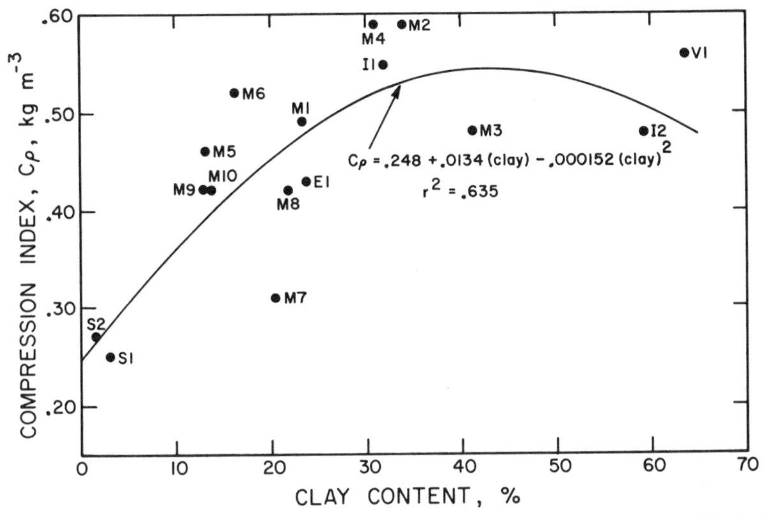

Fig. 20–6. Relationship between percent clay and compression index, C_ρ, for Mollisols (M), Spodosols (S), Entisols (E), Inceptisols (I), and Vertisols (V) (Larson et al., 1980).

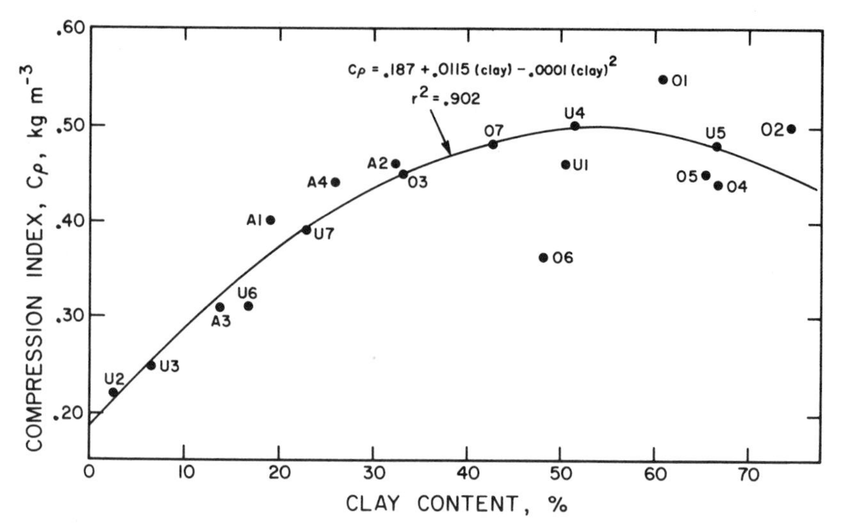

Fig. 20–7. Relationship between percent clay and compression index, C_p, for Alfisols (A), Utisols (U), and Oxisols (O) (Larson et al., 1980).

In addition to Eq. [2], several other methods and models have been suggested in the literature (Harris, 1971) to describe the compression behavior of soils. However, most are more complex, and according to the literature, few have been used to account for the effect of water content on compression behavior of soils. Larson et al. (1980) modified Eq. [2] to account for the effect of water content on compression curves:

$$\rho = [\rho_k + \Delta_T (S_1 - S_k)] + C_p \log \sigma_a/\sigma_k) \quad [7]$$

where ρ is the computed bulk density corresponding to σ_a and at a desired degree of water saturation, S_1; ρ_k is the bulk density at a known applied stress, σ_k, and at a known degree of water saturation, S_k; and Δ_T is the slope of ρ_k vs. the degree of water saturation.

Gupta and Larson (1981) gave statistical relationships that describe the changes in ρ_k, Δ_T and C_p with particle-size distribution. These relationships along with Eq. [7] could be used to approximately describe the compression behavior of soil. Gupta and Larson (1981) indicated that predicted compression curves were sensitive to the errors in the estimate of ρ_k and thus recommended laboratory measurement of ρ_k at one water content for the desired soil.

20–3 METHODS

20–3.1 Special Apparatus

20–3.1.1 CONSOLIDOMETER

Most consolidometers used in the USA are of the fixed-ring and floating-ring types, illustrated in Fig. 20–1. Both types are available commercially, and either is suitable for compressibility testing. In the floating-ring container, compression movement occurs from the top and bottom toward the center; thus it has the advantage of having less friction between the ring and the soil. In the fixed-ring container, all specimen movement relative to the container is downward during compression. The fixed-ring container can be adapted for permeability tests and for negative pore water pressure measurements in unsaturated samples. The consolidation rings are stainless steel or aluminum and range from 50 to 112 mm in diameter and 20 to 38 mm in depth. The diameter of the ring is determined by the size of the undisturbed tube samples which, in turn, is determined by the maximum size of the particles or the soil structural class. However, we recommend that the ring diameter and loading weights be selected to give unit stresses in kPa.

Porous stones of suitable porosity and compressibility are required at the top and bottom of the specimen to transmit the vertical axial stress to the soil specimen and at the same time to permit free drainage of the specimen at all stages of the consolidation test. For consolidation tests on soft fine-grain soils, a fine-grade stone is used; for all other soils, a medium grade is used. The top porous stone for fixed-ring consolidometers and both the top and bottom stones for the floating-ring consolidometer should have clearances of about 0.2 to 0.5 mm all around in order to permit free compression without binding. For compression tests on unsaturated soils where negative pore water pressure measurements are required, the bottom porous stone should have an air-entry value between −33 and −100 kPa and should be cemented around the outer diameter to the base plate.

Vertical deformation may be measured with a dial micrometer with 0.002-mm divisions or an electronic transducer with similar sensitivity.

20–3.1.2 LOADING DEVICE

Several types of load apparatus for applying vertical compression stresses are commercially available. Three common types are a hydraulic load-applying machine, a dead-weight lever arm oedometer, and a stress-controlled strength-testing machine. The loading device should be capable of producing compression stresses up to 2500 kPa and of maintaining a constant load of ± 1% of the applied load for at least 24 h. It should permit application of a given load increment within a period of 2 s without impact.

20–3.1.3 EQUIPMENT FOR PREPARING SPECIMENS

Remolded soil specimens of a particular void ratio are prepared (i) by compressing moist soil into a metal cylinder with metal plungers or (ii) by pouring soil from a constant height into a consolidation ring having an extension collar about 5 cm in depth and the same diameter as the container and, if necessary, vibrating the sample in the assembled cylinder to obtain the predetermined density. Natural soil specimens are either handcarved from a test pit or sampled with metal tubes. The apparatus for trimming specimens consists of a cutting stand, wire saws, knives, and a metal straightedge.

20–3.2 Procedure

20–3.2.1 PREPARATION OF UNDISTURBED SPECIMENS

Using a wire saw or knife, trim the specimen into an approximately cylindrical shape with a diameter about 10 mm greater than the inside diameter of the specimen ring. Place the consolidation ring on the soil specimen. By carefully trimming the specimen, gently force the ring down over the specimen. Cut off the portion of the sample remaining above the ring with a wire saw or knife. For many soils, extreme care must be taken in cutting off this portion to avoid disturbing the sample. Place a glass plate over the ring and turn the specimen over. Cut off the soil extending beyond the bottom of the ring in the same manner. Place another glass plate on this surface to control evaporation until the specimen is placed in the loading device.

20–3.2.2 PREPARATION OF REMOLDED SPECIMENS

Remolded samples can be obtained by trimming compressed cores to fit into the consolidation ring or by pouring or vibrating soil in a consolidation ring. For compressed specimens, compress a known weight of moist soil at a known water content into a metal cylinder to give a specific bulk density or void ratio. Extrude the specimen from the metal cylinder and trim to fit the consolidation ring using the procedures described above. Another method of packing soil is to pour or vibrate a known weight of soil into a consolidation ring with a 5-cm long extension collar. Remove the extension collar and trim the excess material flush with the ring surface with a straight-edged cutting tool. Weigh the specimen and determine the dry density. If the dry density is not within 20 kg m^{-3} of the desired density, repeat the preparation procedures until the required density is obtained.

20–3.2.3 CALIBRATION OF EQUIPMENT

In the compressibility test using a consolidometer, only measurements of volume change of the soil specimen are desired; therefore, corrections must be applied for any significant deformation caused by the compressibility of the apparatus itself. This is done by placing the consolidometer with saturated porous stones and filter papers in the loading device and applying the load increments to be used in the compressibility test. Read and record the dial indicator for each load. Apply each correction to each dial reading during the soil compression test.

20–3.2.4 DETERMINATION OF SOIL COMPRESSIBILITY

Record all identifying information for the specimen, such as horizon designation, depth, soil series, date of sampling, and location. Measure and record the height and cross-sectional area of the specimen. Record the weight of the specimen ring and glass plates. Record the weight of the specimen plus tare (ring and glass plates), and from the soil trimmings determine the soil particle density and water content.

20–3.2.4.1 Saturated Specimens. Fill the base of the consolidometer with water. Next place the porous stone, which already has been saturated with water, on the base of the consolidometer. Add enough water so that the water level is at the top of the porous stone. Place a moist filter paper (Whatman no. 1 or equivalent) over the porous stone. Place the ring containing the specimen on the porous stone. Secure the ring to the base by means of the consolidometer clamp and screws. Place a moist filter paper on the top of the specimen, and then place the previously saturated top porous stone and loading plate in position (see Fig. 20–1).

Place the consolidometer containing the specimen in the loading device. Adjust the loading device until it just makes contact with the top loading-plate. The seating load should not exceed 1 kPa. Adjust the dial indicator to its zero reading. Record this reading on the data sheet.

With the specimen assembled in the loading device, apply a load of 25 kPa. For saturated samples, immediately fill the chamber surrounding the ring and the soil specimen with water. When the vertical deflection readings indicate no further volume change, record the dial indicator reading.

Continue compression of the specimen by applying the next load increment. Use loads of 25, 50, 100, 200, 400, 600, 800, 1200, and 1600 kPa. The maximum load should be great enough to establish the straight-line portion of the void ratio vs. log stress curve. Allow each load increment to remain on the specimen for a minimum of 24 h, or at least until the slope of the secondary consolidation is apparent. Record on the data sheet the dial reading for each load increment corresponding to complete compression.

After completing the test, remove the entire sample from the steel ring, weigh, oven-dry, and reweigh to recheck the weight of the solids.

20–3.2.4.2 Unsaturated Specimens. A special apparatus has been described by Larson et al. (1980) to measure one-dimensional compression of unsaturated soils while recording changes in the pore water pressure during the compression test. Figure 20–8 shows the experimental setup used in this test. The apparatus consisted of a porous ceramic plate embedded in a brass press plate and connected to an automatic pore water pressure readout system. The following procedure should be followed when using this setup.

Flush the space below the porous plate of air bubbles with deaerated water. Force water through the press plate by creating a pressure gradient as in a siphon. Gently stroke the back of the press plate by hand to remove any entrapped air bubbles in the space below the porous plate. Fill the wet port of a pressure transducer with water using a squeeze bottle, and

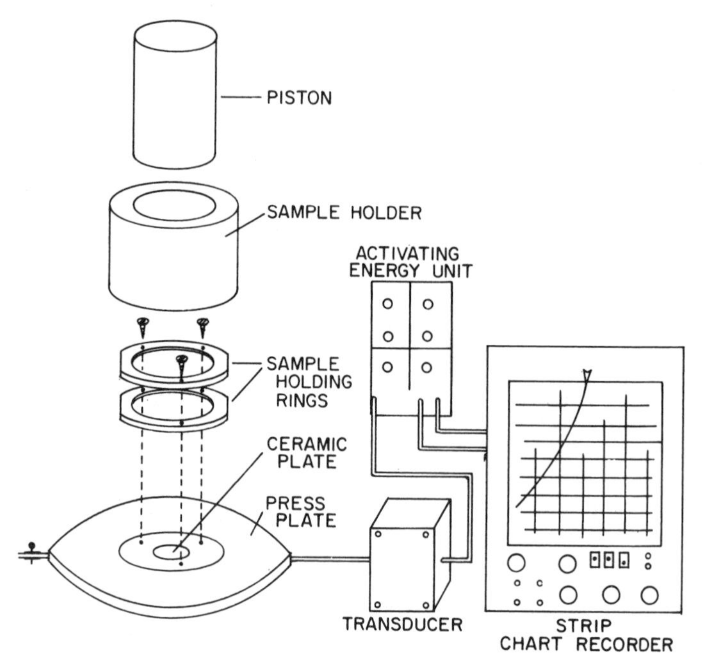

Fig. 20–8. Apparatus used for determination of uniaxial compression in unsaturated soils.

connect the pressure transducer to the press plate with a compression nut attached at the end of rubber tubing. After slightly tightening the compression nut, pinch off the rubber tubing on the other end of the press plate with a hosecock clamp. The pore water pressure may be continuously monitored without appreciable change in the amount of water in the soil sample during the compression test using this setup. Connect the transducer to appropriate recording instrumentation. Fix the dry port of the transducer at the same elevation as the porous plate. Establish the zero base output from the transducer by taking an initial reading while keeping the porous plate under a thin layer of water. This reading represents the zero pore water pressure and is subtracted from readings during the compression test to obtain the pore water pressure corresponding to each applied load. Select transducers that require a negligible transfer of water across the porous ceramic plate to record changes in the pore water pressure. If no monitoring of changes in pore water pressure is needed, a compression test on unsaturated soil could be performed by pinching off the rubber tubing connected to the transducer with a hosecock clamp.

As with the consolidation test, a dead-weight load system and a microdial gauge could be used, respectively, to load the sample and monitor the changes in its length. Larson et al. (1980), however, used an Instron Universal Testing Machine (IUTM)[1] to measure the load and thickness

[1]Trade names and company names, which are included for the benefit of the reader, do not imply endorsement or preferential treatment of the product listed by the USDA.

of the sample during compression tests. The loading schedule may be modified to meet the objectives of the study (Larson et al., 1980). When the pore water pressure becomes zero, the test can be stopped or continued using the procedures for saturated soils.

20–3.2.5 CALCULATIONS AND PRESENTATION OF RESULTS

From the recorded data, compute and record the initial and final water contents, heights of solids, void ratio before and after compression, initial and final degree of saturation, and initial and final dry density. From the final dial reading for each load increment, calculate the corresponding void ratio.

For saturated specimens, plot the void ratio, e, on the arithmetic scale (ordinate) and the corresponding effective stress, σ', in kPa on the logarithmic scale (abscissa) as shown in Fig. 20–3. Calculate the coefficients of compressibility using Eq. [3] through [5].

For unsaturated specimens, from the plot of bulk density vs. applied stress, calculate the coefficient of compressibility using Eq. 20–6.

20–3.3 COMMENTS

For unsaturated specimens, care should be taken that the air-entry pressure of the porous plate is less than pore water pressure of the sample and that the transducer range is large enough to record pore water pressure changes in the desired sample.

Equilibrium deformation value of the sample can be obtained by extrapolation from the deformation vs. time curves. Similarly, the equilibrium pore water pressure value can be estimated by extrapolation from the pore water pressure vs. time curve. Although large time-intervals are desirable for each increment load applied, care should be taken to prevent loss of soil water during these long time-intervals. Using the IUTM, Larson et al. (1980) obtained a nearly constant value of soil deformation and pore water pressure after about 30 min. During the compression test in their study, the applied stress was kept constant, since the machine crosshead freely moved to maintain this stress.

Compression curve data may be useful in establishing ranges of water content and applied stresses that are not conducive to detrimental levels of soil compaction, and in predicting bulk density profiles after passage of agricultural equipment (Gupta & Larson, 1981). One must recognize the limitations and possible ramifications of estimates and predictions developed using compression curve data obtained from a consolidometer. The total stress conditions of the application may be quite different than those imposed on the soil in the consolidometer test.

20–4 REFERENCES

American Society for Testing and Materials. 1979. Standard test methods for moisture-density relations of soils and soil-aggregate mixtures using 5.5-lb (2.49-kg) rammer and

12 in. (305 mm) drop. *In* 1979 Annual Book of ASTM Standards. Part 19:201–207. American Society for Testing and Materials, Philadelphia.

Gupta, S. C., and W. E. Larson. 1981. Modeling soil mechanical behavior during tillage. p. 151–178. *In* P. W. Unger and D. M. Van Doren, Jr. (ed.) Predicting tillage effects on soil physical properties and processes. Spec. Pub. 44. American Society of Agronomy and Soil Science Society of America, Madison, WI.

Harris, W. L. 1971. Methods of measuring soil compaction. p. 9–44. *In* Compaction of agricultural soils. American Society of Agricultural Engineering, St. Joseph, MI.

Larson, W. E., and S. C. Gupta. 1980. Estimating critical stress in unsaturated soils from changes in pore water pressure during confined compression. Soil Sci. Soc. Am. J. 44:1127–1132.

Larson, W. E., S. C. Gupta, and R. A. Useche. 1980. Compression of agricultural soils from eight soil orders. Soil Sci. Soc. Am. J. 44:450–457.

Lee, I. K., and I. B. Donald. 1968. Pore pressures in soils and rocks. p. 58–81. *In* I. K. Lee (ed.) Soil mechanics—selected topics. American Elsevier Publishing Co., New York.

Mesri, G., A. Rokhsar, and B. F. Bohor. 1975. Composition and compressibility of typical samples of Mexico City clay. Geotechnique 25:527–554.

Rahman, A. 1973. The behavior of loess in confined compression. Ph.D. diss. Univ. of Iowa, Iowa City. MI (Diss. Abstr. 73–30, 973).

Scott Blair, G. W. 1937. Compressibility curves as a quantitative measure of soil tilth. J. Agric. Sci. 27:541–556.

Scott Blair, G. W., and G. H. Cashen. 1938. Compressibility curves as a quantitative measure of soil tilth, II. J. Agric. Sci. 28:367–378.

Stone, J. A., and W. E. Larson. 1980. Rebound of five one-dimensionally compressed unsaturated granular soils. Soil Sci. Soc. Am. J. 44:819–822.

Towner, G. D., and E. C. Childs. 1972. The mechanical strength of unsaturated porous granular material. J. Soil Sci. 23:481–498.

Yong, R. N., and B. P. Warkentin. 1966. Introduction to soil behavior. The Macmillan Co., New York.

21 Water Content

WALTER H. GARDNER
Washington State University
Pullman, Washington

21-1 GENERAL INFORMATION

Direct or indirect measures of soil water content are needed in practically every type of soil study. In the field, knowledge of the water available for plant growth requires a direct measure of water content or a measure of some index of water content. In the laboratory, determining and reporting many physical and chemical properties of soil necessitates knowledge of water content. In soils work, water content traditionally has been expressed as the ratio of the mass of water present in a sample to the mass of the sample after it has beeen dried to constant weight, or as the volume of water present in a unit volume of the sample. In either case the amount of water in the sample is needed. To determine this the water must be removed and measured, or the mass of the sample must be determined before and after removal of water. And criteria must exist for deciding at what point the sample is "dry." For a great many purposes where high precision and reproducibility are not required, a precise definition of the term "dry" is not required. Where high precision does become important it should be recognized that "dry" is a subjective term and must be defined. Where high precision is implied in water content figures, such figures must be accompanied by the definition and a description of the procedure used to determine the water content. Traditionally, the most frequently used definition for a dry soil is the mass of a soil sample after it has come to constant weight in an oven at a temperature between 100 and 110°C. The choice of this particular temperature range appears not to have been based upon scientific consideration of the drying characteristics of soil; and in practice, samples are not always dried to constant weight.

Water content as usually used in soils work is either a dimensionless ratio of two masses or two volumes or is given as a mass per unit volume. When either of the dimensionless ratios is multiplied by 100, such values become percentages, and the basis (mass or volume) should be stated. Where no indication is given, the figure generally may be assumed to be on a mass basis because the determination usually involves getting mass-basis figures first and then converting them to volume-basis figures.

Occasionally the ratio of mass of water to that of the wet soil is used. Conversions from one basis to the other easily are made by use of the formulas,

$$\theta_{ww} = \theta_{dw}/(1 + \theta_{dw}), \quad \theta_{dw} = \theta_{ww}/(1 - \theta_{ww}) \qquad [1]$$

where θ is the water content, and the subscripts *dw* and *ww* refer to dry-weight and wet-weight basis.

If water content is desired on a volume basis, the volume from which the sample was taken must be known. A sampling device which takes a known volume of soil (Richards & Stumpf, 1960) may be used; or the bulk density of the soil (mass of soil per unit volume), as obtained from independent measurements in the same area, may be used. Where a bulk density is known or assumed, volume-basis water contents may be obtained from the mass-basis figures by use of the formula

$$\theta_{vb} = (\rho_b/\rho_w)\theta_{dw} \qquad [2]$$

where ρ_b is the bulk density of the soil, ρ_w is the density of water (this usually may be taken as unity in g/cm^3 or Mg/m^3 units, but must be present for dimensional consistency), and the subscripts *vb* and *dw* refer to volume basis and dry-weight basis. The accuracy of the inferred volume-basis water content depends upon the accuracy of the bulk density figure used as well as upon the accuracy of the dry-weight water-content figure. Both depend upon the accuracy of the figure for the "dry" mass of the soil. In some types of work water content also is expressed as the ratio of the volumetric water content to that which would exist at saturation, or the *saturation ratio.* On this basis the water content of a saturated soil is unity. Further information on measurement of bulk density may be found in chapter 13.

Computations of water content on a volume basis require a correct measure of bulk density. Considering the variability of soil, even in an area as small as 1 m^2, some error nearly always is involved. However, for many purposes where volume-basis water contents are needed, the error probably is no more important than the error involved in representing the water content at a particular depth in the field by a single number. Taylor (1955), working with Millville loam (which is a relatively uniform soil insofar as observable physical and chemical properties are concerned), reported coefficients of variation of 17 and 20% for gravimetric measurements of the water content of samples of soil from field plots under irrigation by furrow and sprinkler methods, respectively. Similar variation has been observed on apparently uniform areas of Palouse silt loam in the dryland areas of southeastern Washington. For precise work, statistically sound methods must be used for sampling and for reporting such measurements (see chapters 1, 2 and 3; Cline, 1944).

When water is applied to soil by irrigation or rainfall, the quantity applied is reported as the depth of the water if it were accumulated in a layer. To obtain the volume of water applied to a given area would require multiplication of the depth by the area, measured in the same length units. In like manner the quantity of water in a soil profile may be reported as the depth of the water present if it were to be accumulated in a layer. In this case, however, the number is not quantitatively meaningful without an association being made with some particular profile depth defined by a number or by a description of the part of the profile under consideration, such as the rooting zone, plow depth, or depth to bedrock. In the case of uniform rainfall the concept of "depth of water" matches the usual way of visualizing the amount of water involved, as well as the way in which it is measured in an ordinary catchment device. However, when the amount of water in the soil is to be measured or computed the number used is volume of water in unit volume of soil (measured on a volume basis or inferred using Eq. [2]) multiplied by the incremental volume used (unit area × length of depth increment) and summed for the entire profile under consideration. To reduce this to the same basis as for rainfall, the area (unit, in this case) is divided out:

$$(\text{vol/vol})(\text{unit area} \times \text{length})/(\text{unit area}) = \text{length}\,. \qquad [3]$$

Specification of a "depth" of accumulated water in such fashion, unless the context makes its meaning unequivocal, should be accompanied by a modifier such as "in the rooting zone." The length or "depth" may be specified in millimeters, centimeters or meters. If specified as a depth per some depth increment, as has been common in the past, this dimensionless number would be multiplied by the number of increments of concern to give the total depth of water in a defined soil depth, providing that the water content over the total depth is uniform and that units are properly taken into account.

Common expressions for water in soil in the USA have been "inches per foot," "acre-inches," "acre-feet," and "acre-feet per acre." The term "acre-feet" (and similarly, "acre-inch") often has been improperly used to mean "acre-feet per acre" or "feet" of water rather than a volume of water, which it really represents. Similar problems have existed with other units, such as the hectare and metric units of length. Eq. [3] may be seen to be independent of units except for the length of the depth increment. This can be in any convenient system of units. To obtain the volume of water contained in soil on a given size of farm unit, as may be needed in farm water accounting, the depth or length computed by Eq. [3] would be multiplied by the area of the field in question, measured in appropriate units.

21–2 DIRECT METHODS

21–2.1 General Principles

Determination of water content of soils may be accomplished by direct and indirect methods. Direct methods may be regarded as those methods wherein water is removed from a sample by evaporation, leaching, or chemical reaction, with the amount removed being determined. Determination of the amount removed may be by one or more of the following methods: (i) measurement of loss of weight of the sample, (ii) collection by distillation or absorption in a desiccant and measurement of the amount of water removed, (iii) extraction of the water with substances which will replace it in the sample and measurement of some physical or chemical property of the extracting material that is quantitatively affected by water content, or (iv) quantitative measurement of reaction products displaced from a sample. In each of these methods the water and soil are somehow separated, and the amount of water removed is measured or inferred.

The key problem in water content determination in porous materials has to do with the definition of the dry state. Soil is made up of colloidal and noncolloidal mineral particles, organic materials, volatile liquids, water, and chemical substances dissolved in water. Of the mineral fraction of the soil, the noncolloidal particles probably are the easiest with which to deal. At ordinary room temperatures and room humidities, such materials will have a small quantity of adsorbed water that is removed easily from their surfaces by heating. Only at elevated temperatures does water of crystallization associated with these minerals come off. The status of water in the colloidal particles of the mineral fraction is more complicated. Water present in the colloidal fraction may be considered in two categories. *Structural water*, is water derived from components of the mineral lattice itself, whereas *adsorbed water* is water that is attached to the mineral lattice but is not a structural component of the lattice. It often is difficult to distinguish between the two categories. Some adsorbed water may be located with respect to the lattice in such a way that the difficulty of removal is comparable to that for structural water. (Difficulty here refers generally to the temperature required for removal and possibly to the fineness of the particles.) The exact situation depends upon the nature of the particular mineral.

Some representative thermal dehydration curves (after Nutting, 1943) are presented in Fig. 21-1. (It should be recognized that the character of the thermal dehydration curve depends upon the purity of the sample, so that it is unlikely that curves for the same type of minerals constructed by various workers would be identical.) The curves in Fig. 21-1 show the weight loss of samples that were initially at room temperature and in equilibrium with room humidity of approximately 60%, as the samples were heated to constant weight at various temperatures. The weight at a temperature of 800°C or slightly higher is taken to be unity. It is not clear from the curves what part of the water is adsorbed water and what part is structural water. But it should be obvious from these curves that for colloidal minerals it is important to specify the exact temperature of drying if precision is required. Considering that the temperature control of ordinary laboratory ovens is not very good, precision water content

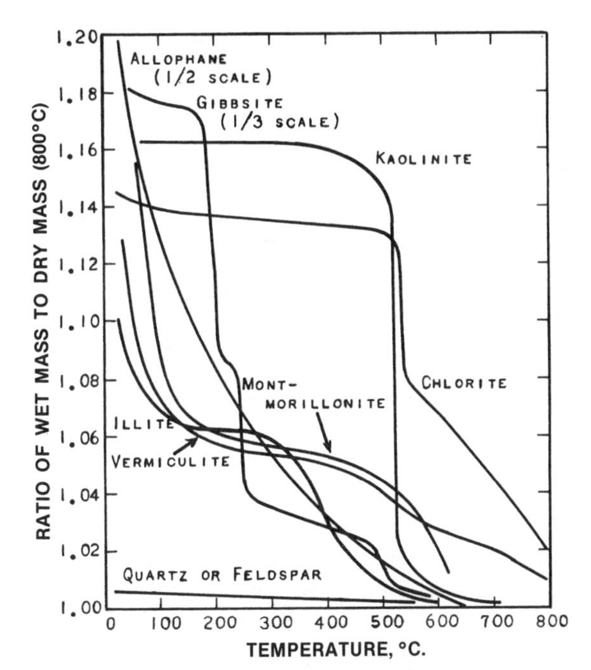

Fig. 21–1. Wet mass of a unit dry mass of a clay mineral as a function of drying temperature. The sample is presumed to be dry at a temperature of 800°C or greater. Water content percentages on a mass basis for this drying temperature may be obtained by subtracting 1 from the ordinate figure and multiplying by 100. (The numbers to the right of the decimal thus may be regarded as percentages.) (Nutting, 1943)

measurements should not be expected except where extraordinary attention is given to temperature measurement and control.

Although it is not possible to generalize on the precision of oven temperature control, it can be noted that unless the temperature control of an oven has been checked, it is not safe to assume that control in the specified 100 to 110°C range is achieved in ordinary laboratory ovens. Several ovens in Washington State University soils laboratories showed variations as large as +15°C over a period of a few weeks (without a sample load) and temperature differences within the ovens of as much as 40°C. This was true of forced-draft ovens as well as convection ovens. Temperature variation with elevation in the oven was found to be as great at 40 to 100°C for several convection-type ovens at the Massachusetts Institute of Technology (Lambe, 1949). Since radiation heating could be a factor in some ovens, precision work requires that temperature measurements be made within the soil sample rather than in the oven atmosphere, as is done with conventional oven thermometers.

Insofar as most of the colloidal minerals are concerned, the temperature sensitivity of water content measurements would be reduced materially by selection of a temperature range where water loss changes with increasing temperature are smaller than they are from 100 to 110°C.

Judging from the curves of Fig. 21–1, the range between 165 and 175°C would be better than the traditional range of 100 to 110°C. However, excessive oxidation and decomposition of soil organic matter in this high temperature range would prevent its general adoption, although in special cases it might be appropriate.

The problem of defining "dry conditions" in the organic fraction of the soil is even more difficult than defining such conditions for the colloidal mineral fraction. The organic fraction of soils consists of undecomposed plant fragments, such as roots, and resistant decomposition products such as polysaccharides or polyuronides as well as intermediate products. Volatile liquids other than water also may be present. If subjected to too high temperatures, organic materials are oxidized or decomposed and lost from the sample. The temperature at which excessive oxidation occurs would be difficult to specify, but 50 °C seems to be a common temperature used to dry organic materials and is thought by many investigators to be sufficiently low to avoid loss of organic matter by oxidation.

Little is said in the soils literature about weight changes due to oxidation and decomposition of the organic fraction in soil samples undergoing drying. However, many investigators readily admit that they often have observed weight changes in soils being dried over periods of many days to many weeks. In our own laboratory we have observed weight changes in samples during drying over periods as long as 15 days. A silt loam soil, containing about 3% organic matter and with illite predominating in the clay fraction, lost weight which corresponded to 0.3% water content from the 2nd to the 10th day in a forced-draft oven at 100 to 110°C. Whether this represents water loss or oxidation and decomposition is difficult to tell.

Extensive studies have been made on the drying of food products by food technologists interested in dehydration. Here efforts seem to be more in the direction of finding a drying technique that will yield reproducible results rather than in finding a method wherein the dry status is unequivocal. A reference method used by food technologists (Makower, 1950) involves drying ground vegetable material at room temperature in a vacuum over a desiccant that permits practically no water vapor pressure (magnesium perchlorate). Drying time is long (6 to 9 months). The assumption is made that with no air present decomposition and oxidation are negligible. Even as a reference standard, this method is painfully slow and other methods have been sought.

One modification that has reduced the drying time from months to days (Makower and Nielsen, 1948) involves lyophilization or freeze-drying. The fresh weight of the organic matter is obtained first and then water is added to the sample to cause it to swell. In the swollen state, the sample then is frozen, and most of the water is removed by sublimation. In this state the mechanical structure of the sample essentially is fixed, and evaporation can proceed much more rapidly than under conditions where the porosity of the sample is reduced drastically through shrinking. After most of the water has been removed, drying can be continued at higher temperatures. A redrying procedure involving drying at temperatures between 50 and 70°C also has been developed by Makower et al. (1946) as a reference standard. They reported that considerable decomposition of vegetables takes place at temperatures above 70°C.

Since many soils contain only small amounts of organic material, much of which is fairly stable, inaccuracies introduced by uncertainties in drying of organic materials often may be negligible. However, since drying temperatures normally used for soils are well above those considered safe for organic materials, decomposition and oxidation should be expected and considered in the development of drying techniques and taken into account in reporting data. Where precise water content values are required, precautions must be taken to assure that minimum losses due to oxidation or decomposition occur and that all samples to be compared are dried to constant weight or to some arbitrary state of dryness which, by careful timing, can be reproduced. The measured water content of a soil which contains more than negligible quantities of organic material must be regarded as an arbitrary value, depending upon a state of dryness defined by the method used rather than upon a dry state which could be regarded as unequivocal.

In many salt-affected soils, considerable material exists in the soil solution. If water is leached out of the sample by some liquid such as alcohol, most of the dissolved materials would be removed along with the water. On the other hand, if the soil is dried in an oven only the volatile materials are lost. Depending upon the quantities of dissolved substances present, the method of water removal could make considerable difference. It is possible to have water content differences as great as several tenths of a percent, depending upon whether water was removed by leaching or by evaporation. It is a moot question whether or not the dissolved substances should be considered to be part of the "dry" soil. Most procedures in common use involve oven-drying or calibration against oven-drying data, so that dissolved salts are counted as part of the "dry" soil.

It appears from the literature that oven-drying methods often may have been accepted too readily as reference standards for water content measurement. The situation is complicated by lack of detail in the literature concerning oven-drying procedures used for the water content measurements that are reported. Drying time rarely is mentioned, and the specification of oven temperature appears to be perfunctory if mentioned at all. There usually is no evidence to suggest that an investigator has questioned oven-drying procedures. By implication, scatter in many kinds of data based upon or related to water content measurements fails to be associated with the water content measurement. In some cases, other methods for obtaining water content are compared to oven-drying methods, and any lack of reproducibility observed is ascribed automatically to the other method rather than to the oven-drying procedure. Accuracy itself is measured against oven-drying methods. As a consequence of this misplaced trust in oven-drying procedures, there may be good reason to re-evaluate some of the other methods for water content determination and possibly a few other types of analytical procedures when water content determinations are critical.

Gravimetric water content measurements usually involve three independent measurements, the wet (x) and dry (y) weights of the sample and the tare weight (t). These may be combined in this way

$$[(x - t)/(y - t)] - 1 = \theta \quad [4]$$

to provide water content θ. With each weighing, there is associated an error e, which is the balance error. The balance error is made up of two parts, a random error having to do with the reproducibility of a particular balance reading, and a bias which is the difference between an average reading and the true value. The bias is likely to be different at different weight values, but since a balance may be calibrated such bias may be eliminated to any desired degree of accuracy. A constant bias, such as might result from an improper adjustment of the zero on an automatic balance, introduces no error in the water content ratio provided it is the same for each weighing and the tare weight is not balanced out. This may be seen when such an error e is applied uniformly to each measurement appearing in Eq. [4]

$$[(x + e) - (t + e)]/[(y + e) - (t + e)] - 1 = \theta\,. \quad [5]$$

The error term may be seen to cancel out.

The effect of random error may be evaluated by determining the variance σ^2 of the ratio, $(x - t)/(y - t)$, in Eq. [4]. The variance of this ratio with reasonable assumptions may be approximated by

$$\sigma^2_{x-t/y-t} = [1/(y - t)^2]\{\sigma^2_{x-t} + [(x - t)^2/(y - t)^2]\sigma^2_{y-t}\}\,. \quad [6]$$

The variances of the numerator and denominator are given as

$$\sigma^2_{x-t} = \sigma^2_x + \sigma^2_t \quad [7]$$

$$\sigma^2_{y-t} = \sigma^2_y + \sigma^2_t\,. \quad [8]$$

Substituting Eq. [7] and [8] into [6] gives

$$\sigma^2 x - t/y - t = \frac{1}{(y-t)^2}\left[\sigma^2_x + \sigma^2_t + \frac{(x-t)^2}{(y-t)^2}\left(\sigma^2_y + \sigma^2_t\right)\right]. \quad [9]$$

Since the variances due to the random error of all three weighings are likely to be nearly the same, Eq. [9] is approximated by

$$\sigma^2_{x\text{-}t/y-t} = 2\sigma^2_{x,y,t}\left[1 + \frac{(x - t)^2}{(y - t)^2}\right] \Big/ (y - t)^2 \qquad [10]$$

Combining Eq. [4] with Eq. [10] and replacing $(y - t)$ by the dry mass of the soil z gives

$$\begin{aligned}\sigma^2_\theta = \sigma^2_{x-t/y-t} &= (2\sigma^2_{x,y,t}/z^2)[1 + (\theta + 1)^2] \\ &= (2\sigma^2_{x,y,t}/z^2)(\theta^2 + 2\theta + 2)\,. \end{aligned} \qquad [11]$$

For water contents and sample sizes ordinarily encountered in soils work, θ^2 is negligible so that Eq. [11] further reduces to

$$\sigma^2_\theta = (4\sigma^2_{x,y,t}/z^2)(\theta + 1) \qquad [12]$$

where the subscript, $(x - t)/(y - t)$, is replaced by θ for water content. Written in terms of standard deviations $(V{=}\sigma^2)$, this becomes

$$\sigma_\theta = (2\sigma_{x,y,t}/z)\,(\theta + 1)^{1/2}\,. \qquad [13]$$

If a tare compensation device is used so that each water content determination involves only two weighings, Eq. [6] becomes

$$\sigma^2_{x/y} = (1/y^2)[\sigma^2_x + (x/y)^2\,\sigma^2_y]\,, \qquad [14]$$

and Eq. [9] is not needed, since the tare weight is not obtained independently. However, if the tare compensation involves a constant bias, this does affect the accuracy of the resulting water content figure. The effect on water content values of a constant bias due to a tare error or other causes, which may be either positive or negative, may be obtained by solving explicity for E in the equation

$$[(x + e)/(y + e)] - 1 = \theta + E \qquad [15]$$

which, when combined with $\theta = (x - y)/y$, reduces to

$$E = (-e\theta)/(e + y) \qquad [16]$$

or when e is small compared to y, it reduces to

$$E = -\,\theta(e/y)\,. \qquad [17]$$

If the ratio e/y does not exceed 0.001 (e.g., 100 mg/100 g, 10 mg/10 g, or 1 mg/1 g), the error will not exceed 0.1% water content up to water contents of 100%. If the ratio e/y does not exceed 0.0001, the error will not exceed 0.01% water content over a similar range. This applies to any constant bias of a balance as well. The random error when tare compensation is used may be obtained from Eq. [14] using a similar development and with the same assumptions as used for Eq. [13]. The standard deviation is

$$\sigma_\theta = \sigma_{x,y} [1 + (\theta + 1)^2]^{1/2}/y \qquad [18]$$

which, when θ^2 is regarded as negligible, reduces to

$$\sigma_\theta = (2)^{1/2} \sigma_{x,y} (\theta + 1)^{1/2}/y \qquad [19]$$

when tare compensation is used.

Balance precision and accuracy are reported in various ways by manufacturers, and the meaning of the terminology used is not always clear. Hence, it would be difficult to provide general instructions for using balance ratings as provided by the manufacturer to compute the variance or standard deviation in a water content measurement. However, the standard deviation for a particular balance at a specified load may be obtained experimentally with small effort. Hence, water content measurement errors are given here in terms of standard deviations. (For statistical methods for computing standard deviations, consult an elementary book on statistical analyses or books on chemical analyses such as *Chemical Computations and Errors* by Crumpler and Yoe (1946).)

If the limit of precision in water content measurement E due only to random weighing errors is taken to be 3σ, then from Eq. [13],

$$E = 3\sigma_\theta = 6\sigma_{x,y,t} (\theta + 1)^{1/2}/z \qquad [20]$$

where a tare weighing is involved. At 3σ, 99.7% of the measurements would be within $\pm E$; at 2σ, 95%; and at σ, 68%. If the dry weight z in Eq. [11] is replaced by the wet weight, $z' = z/(\theta + 1)$, then Eq. [20] (again neglecting θ^2 and higher powers) becomes:

$$E = 6\sigma_{x,y,t} (\theta + 1)^{3/2}/z' = 6\sigma_{x,y,t} (3\theta + 1)^{1/2}/z' . \qquad [21]$$

Neglect of θ^2 and higher powers leads to E values that are slightly too small for large values of θ in Eq. [20] and [21]. However, the magnitude is correct, and the practical value of the expressions is enhanced by their simplicity. In cases where tare compensation is used, the equation corresponding to Eq. [20] is

$$E = 3\sigma_\theta = 3(2)^{1/2}\sigma_{x,y}(\theta + 1)^{1/2}/y = 4.24\sigma_{x,y}(\theta + 1)^{1/2}/y . \qquad [22]$$

A balance should be read to $\pm\ \sigma$. (The precision of some balances is

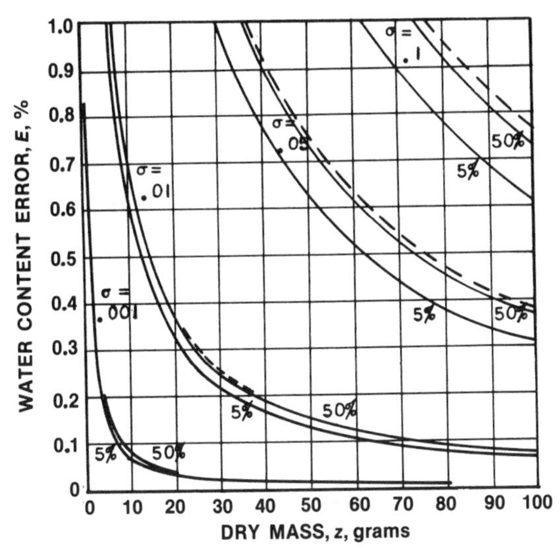

Fig. 21–2. Error in water content measurements as a function of the dry mass of the sample at various weighing precisions and for 5 and 50% water contents on a dry-mass basis (Eq. [20]). The value of E is 3 σ_θ, and the weighing precision is the standard deviation σ for the balance in grams. When tare compensation is used, E should be multiplied by $2^{1/2}/2$, and where n replicate determinations are made, by a factor of $1/n^{1/2}$. If z is taken as the wet mass of the sample, the error E from the graph should be multiplied by (θ + 1), as can be seen from Eq. [21]. The dashed curves, for 50% water content, result when the θ^2 term in Eqs. [11] and [18] is not ignored.

reported as $\pm e$, where e is taken to be 3σ. The Metler K-7 is of this type, the precision being given as $\pm$ 0.03 g. The standard deviation at loads of approximately 100 g was found experimentally to be about 0.01 g on four different balances of this type. The bias of each balance was $< \pm 0.01$ g at the same loading.)

The precision measurement may be improved by a factor of $1/n^{1/2}$ by making n replicate determinations. Values of σ for various balance standard deviations $\sigma_{x,y,z}$, sample dry masses E, and water contents are shown in Fig. 21–2. When tare compensation is used the value of E should be multiplied by $2^{1/2}/2$, and where n replicate determinations are to be made, by a factor of $1/n^{1/2}$.

21–2.2 Gravimetry With Oven Drying

21–2.2.1 INTRODUCTION

Water content measurements by gravimetric methods involve weighing the wet sample, removing the water, and reweighing the sample to determine the amount of water removed. Water content then is obtained by dividing the difference between wet and dry masses by the mass of the dry sample to obtain the ratio of the mass of water to the mass of the dry soil. When multiplied by 100, this becomes the percentage of water in the sample on a dry-mass (or, as often expressed, on a dry-weight) basis. Water content may be described in other ways as indicated in section 21–1. Water may be removed from the sample in any of a number of ways, the principal method in common use being the oven-drying method described here. Accuracy and reproducibility of water content measurements, assuming that weighing precision is consistent with desired precision of water content measurement, depend upon the drying technique and the care with which it is used. (See discussion in section 21–2.1).

21–2.2.2 METHOD

21–2.2.2.1 Special apparatus. Apparatus required for gravimetric determination of water content may be used in many different forms, and so exact specifications are not needed. Requirements include an auger or sampling tube or some other suitable device to take a soil sample, soil containers with tight-fitting lids, an oven with means for controlling the temperature to 100 to 110°C, a desiccator with active desiccant, and a balance for weighing the samples. In the field, if soil samples are taken under conditions where evaporation losses may be of sufficient magnitude to affect the desired accuracy of measurement, special equipment for weighing the samples immediately or reducing evaporative loss must be used. Both convective and forced-draft ovens are used, and for precise work a vacuum oven is of particular value. Balances used range all the way from analytical balances to rough platform scales, depending upon the size of the sample to be taken and the precision of measurement desired.

21–2.2.2.2 Procedure The procedure to be used must vary with the circumstances of measurement and the equipment. Since these vary widely it is impossible to specify a detailed standard procedure that will fit all of the many uses made of water content measurements. The procedure given here is intended for use in routine work where moderate precision (say, measurements having a precision of $\pm$ 0.5% water content) is desired. Replication must depend upon the nature of the sample and soil system for which water content is desired, but it is suggested that samples be run in duplicate as a minimum.

Place samples of 1 to 100 g of soil in weighing bottles or metal cans with tight-fitting lids. Weigh the samples immediately, or store them in such a way that evaporation is negligible. Refer to Fig. 21–2 to find the required weighing precision. (The balance need not be read to a precision greatly exceeding the standard deviation for the balance.) Place the sample in a drying oven with the lid off, and dry it to constant weight. Remove the sample from the oven, replace the cover, and place it in a desiccator containing active desiccant (e.g., magnesium perchlorate or calcium sulfate) until cool. Weigh it again, and also determine the tare weight of the sample container. Compute the water content by one of the following formulas:

$$\theta_{dw} = \frac{(\text{weight of wet soil + tare}) - (\text{weight of dry soil + tare})}{(\text{weight of dry soil + tare}) - (\text{tare})} \quad [23]$$

$$= \frac{(\text{weight of wet soil + tare}) - (\text{tare})}{(\text{weight of dry soil + tare}) - (\text{tare})} - 1 \quad [24]$$

$$= \frac{\text{weight of wet soil}}{\text{weight of dry soil}} - 1. \quad [25]$$

The third of these equations is useful where standardized cans are used and the tare weight is balanced out in the weighing process so that the sample weight is obtained directly. Multiplication by 100 gives the percentage of water in the sample on a dry-mass basis.

21–2.2.2.3 Comments. The time necessary to reach constant weight (the term being loosely used here, since constant weight rarely is obtained except for very sandy soils containing little or no organic matter) will depend upon the type of oven used, the size or depth of the sample, and the nature of soil. If a forced-draft oven is used, 10 h usually is considered sufficient. If a convection oven is used, samples should be dried for at least 24 h, and precautions should be taken to avoid adding wet samples during the last half of the drying period. Also, additional time should be added if the oven is loaded heavily. Water contents for samples that are to be compared should be determined using precisely the same method for each measurement. For more precise work, other considerations are involved, which are discussed below.

An alternative method may be used for drying soil. Radiation-drying using an infrared or ordinary heat lamp, often in association with a built-in balance, can be used for soil water content measurements where low precision is adequate. Several such instruments containing a built-in infrared heat lamp, a torsion or analytic balance, and a scale for direct reading of wet-mass basis water contents are available from scientific supply houses under the name of "moisture determination balance."

The uncertainty of the drying temperature makes radiation-drying methods less accurate than those using closely controlled constant-temperature drying ovens. However, the method is rapid, requiring only a few minutes to dry the soil; and when the built-in balance is used, wet-mass basis water content values ordinarily may be read directly from a scale. These may be converted to dry-mass figures by Eq. [1]. When using radiation-drying, care should be taken to avoid excessive heating of the sample.

Water content values for stony or gravelly soils, both on a mass and volume basis, can be grossly misleading. The problem arises from the fact that a large rock can occupy appreciable volume in a soil sample and contribute appreciably to the mass without making a commensurate contribution to the porosity or water capacity of the soil. Mass-basis water content figures are lower than corresponding values for a soil on a rock-free basis because of the excessive contribution to the dry mass made by a rock which may have a bulk density of about 2.6 g cm^{-3} compared to that of the finer soil material, which will usually range from 1.0 to 1.6 g cm^{-3}. A mass-basis water content figure of 10% based upon the dry mass of a gravelly or stony soil having a bulk density of 2.0 g cm^{-3} would represent a water content of 20% if based upon the dry mass of the finer fraction with a bulk density of 1.6 g cm^{-3}.

It is important when presenting water content data on gravelly or stony soils to specify the basis of measurement—particularly the size fraction on which it is made. Water retention in stony soils also is discussed in chapter 26.

The two types of water content figures of greatest interest are water content per unit bulk volume and water content per unit mass of the fine fraction. The volume-basis figure makes it possible to compute the volume of water per unit area in, say, a root zone. The mass-basis figure for the fine fraction is usually the figure obtained from gravimetric analyses or from a wilting-point water-content matric potential determination, and is the figure which would ordinarily be used to compare water conditions from place to place in a soil. The relationship between these two types of water content values is

$$\theta_{vb} = (\theta_{dmf}\rho_b/\rho_w)/(1 + M_{\text{stone}}/M_{\text{fines}}) \qquad [26]$$

where θ_{vb} is the volume of water in a unit volume of the whole soil, θ_{dmf} is the water content on a dry-mass basis for the fine fraction, ρ_b and ρ_w are the bulk densities of the whole soil (including stones) and of water, and M_{stone} and M_{fines} are the dry masses of the stone and fine fractions.

Water content on a volume basis or bulk density may be determined by taking a sample of known volume, oven-drying it according to procedures already described, and dividing the difference between wet and oven-dry mass by the volume to give water content, or dividing the oven-dry mass by the volume to give bulk density. The heterogeneity of gravelly and stony soils and the variability that usually exists from point to point in the soil make for low precision. Because of this low precision, it is possible to discard large stones prior to oven-drying without greatly affecting the precision. Rocks and stones to be discarded are carefully and quickly brushed to avoid soil loss and to reduce evaporation loss, and then are weighed.

Conventional cylindrical tube samplers may be used in some gravelly and stony soils. However, as the number or size of stones increases, the utility of such sampling devices diminishes. In these kinds of soils it is important to determine the volume sampled each time a water content determination is to be made. Where core-type samplers can be used this is not difficult. However, where large rocks and stones interfere seriously, other methods must be used. One useful method involves sampling with a spade or shovel and determining the volume of the hole which is dug.

The volume may be measured by placing a rubber or plastic membrane in the hole and filling it with water from a container filled to a known volume. The quantity of water used in filling the hole is determined easily. Grain millet or dry sand, which flow easily and pack easily to constant bulk density, also may be used to fill the hole; and if the material is to be discarded, the rubber or plastic membrane is not needed. A description of the method is given in chapter 13. No simple, inexpensive methods have as yet been developed for sampling beyond shallow depths without digging an access hole.

Certain general requirements must be met in the development of a procedure for obtaining accurate and reproducible water content measurements. Foremost of these is the requirement that the sample be dried at a specified temperature to constant weight with nothing being lost but water. This rarely is possible with a colloidal material like soil, particularly if it contains any appreciable organic matter, as is discussed earlier in the section on general principles (21–2.1). Weight losses during drying at 100 to 110°C for periods as long as 15 days have been observed in soils ranging from fine sands to silty clay loams. Because of this it is important to specify the details of the drying procedure used in reporting water contents where precise values are needed.

Since accuracy in water content determinations hinges upon existence of a definable dry condition which, with soil, can only be based upon subjective judgment, it is more appropriate to refer to reproducibility than to accuracy. Reproducibility in water content measurements can be achieved in two ways: (i) treating every sample of a set to be compared exactly the same way in terms of such things as sample size and depth in the container, drying temperature, and drying time; or (ii) following techniques that lead to equilibria which are as nearly independent of such variables as is possible. From the latter point of view, the nature of the thermal dehydration curves (Fig. 21–1) suggests the desirability of choosing a drying temperature in a region where the weight change of the colloidal constituents with temperature is at a minimum, as is discussed in section 21–2.1. Also, as is discussed in the same section, vacuum drying at relatively low temperature, with the temperature being controlled carefully and specified when reporting, probably provides the most reproducible drying data.

Reproducibility in water content measurements also may depend upon the technique used to avoid absorption of water from the air during cooling and prior to weighing. While absorption may be negligible in dry ambient conditions or for low-precision work, in certain mineralogical studies or studies of other types, where small weight changes may be confounding, an unacceptable error may be introduced. A method involving flame-sealing of small vials containing the soil prior to cooling has been described by Kittrick and Hope (1970).

21–2.3 Gravimetry With Microwave Oven Drying

21–2.3.1 INTRODUCTION

An alternative to conventional oven-drying of soil (and of plant materials, as well) involves heating by means of microwave energy, which will penetrate into and vaporize water throughout a soil sample. Although not designed specifically for highest efficiency in drying soil samples, commercial microwave ovens, using 0.9- to 300-GHz frequencies in the electromagnetic spectrum, may be used. Polished metals reflect microwave energy, but heat is generated to some degree in most materials by absorption of microwave energy, and to a considerably greater degree in polar water than in most soil materials. Thus, soil water temperature quickly is raised to the boiling point where, because of consumption of heat in vaporizing the water, it remains until absorption of microwave energy by the soil and water exceeds the rate of energy consumption by vaporization. At this time the temperature rapidly rises. If heating is discontinued after the absorbed water is gone and before soil temperature rises to a point where organic matter is appreciably oxidized or structural water is driven off, the dryness achieved should match that achieved using conventional oven-drying. However, this is not easily accomplished and it constitutes the most serious obstacle to achieving accuracy and reproducibility in microwave drying.

The amount of microwave energy absorbed, which determines drying time needed, depends upon the quantity and absorptive characteristics of the material being dried, and hence upon the nature and quantity of energy-absorbing materials present and their water content. And since a flat plateau does not exist in the water loss vs. time curve for microwave drying (Fig. 21–3) (as is assumed and is approximately true for most practical measurements using conventional ovens where temperature is held constant), strictly speaking only a single sample—or duplicate samples of the same soil and same size and assumed to have nearly the same

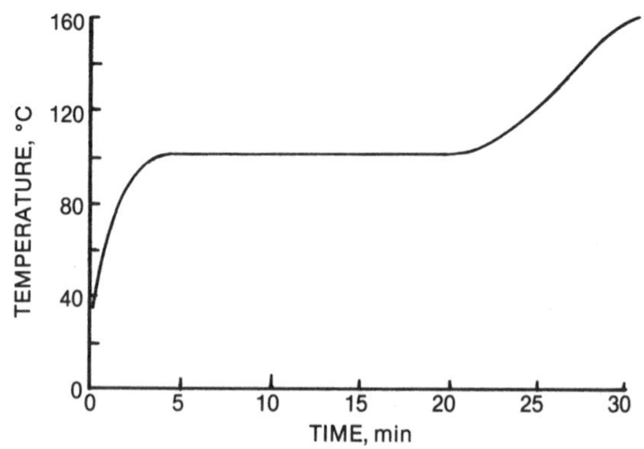

Fig. 21–3. Sample temperature as a function of time in oven. The length of the plateau varies with sample size, water content, and soil type.

water content—may be dried during a single oven run. However, for many purposes errors associated with temperature rise above the standard drying temperature may not always reduce the accuracy and reproducibility of a moisture content determination beyond acceptable limits (Gee and Dodson, 1981; Gilbert, 1974; Hankin and Sawhney, 1978; Miller et al., 1974). Curves shown in Fig. 21–4 illustrate something of the nature of the problem. Here, samples of soils at several different initial water contents have been removed from the microwave oven at the end of different drying periods and the water content computed under the assumption that the sample is dry. After drying periods ranging from 6 to 20 min the computed water content tends to approach a constant value, which may be taken as the correct initial water content. It is evident that selection of an unequivocal value is not always possible.

21–2.3.2 METHOD

21–2.3.2.1 Special Apparatus. A household microwave oven may be used. Preferably the oven will have a uniform microwave field over the load surface, which is achieved by means of a rotating microwave reflector or by a rotating loading platform, or possibly by both. It also will have an automatic shut-off switch, which operates when the oven load, as a consequence of drying, is reduced and excessive microwave energy is reflected back to an absorbing heat sensor.

Glass or paper cups holding about 25 g of soil are required. Some plastic materials can be used, but many plastics melt at temperatures reached by the container or its sample. Where saturated or near-saturated samples are to be dried, a lid with a small vent may be used to prevent loss of sample from boiling.

21–2.3.2.2 Procedure. The procedure to be used must vary with type

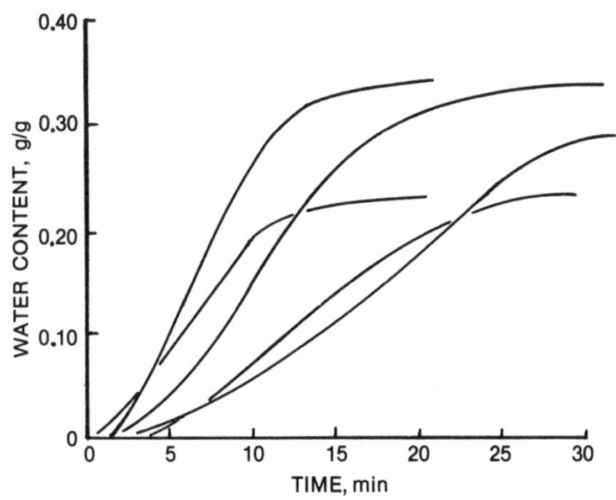

Fig. 21–4. Computed water content of several different soil samples, based upon sample weight reached at various times in the oven. These curves vary widely with sample size, water content, and soil type.

of soil, accuracy desired, and the performance of the microwave oven to be used. Where accuracy of ± 1% is satisfactory (under some conditions this may be ± 0.5%), if the oven has previously been used for soil water content determination and found to have a fairly uniform distribution of microwave energy over the area where samples are placed, and if temperatures reached by samples are not more than about 300°C, the technique to be used is approximately the same as that for conventional oven-drying. Drying times of the order of 20 min have been used on samples of approximately 20 g, but some experience with a particular oven, sample size, and water content may be needed to determine this. Where greater accuracy is required, additional precautions must be taken. These precautions depend upon expected sources of errors, but could involve making drying tests on uniform samples to evaluate uniformity of the microwave energy field, making temperature measurements on samples of differing size and water content at different drying times, and determination of water loss vs. temperature relationships for the particular soils involved. In situ temperature measurements are difficult to make because of the absorption of microwave energy by the temperature sensing device and its possible destruction. However, satisfactory measurements may be made by inserting a low thermal mass thermocouple into the soil sample immediately upon opening the oven door.

21–2.3.2.3 Comments. Since drying time and temperature reached by the sample depend upon its size, water content, and composition (including organic matter), standardization of a drying technique for all materials and water contents is not possible. Hence, where high accuracy is important the standard of reference must be a reliable oven-drying technique, preferably with a vacuum oven having accurate and precise temperature control (see section 21–2.2.2.3). Under many conditions where drying times are chosen so that temperature always rises above the boiling point of water, the computed water contents always will be equal to or higher than the standard value for 105 to 110°C drying. Experience with a given oven, soil, sample size, and range of water content sometimes may permit subtraction of a small corrective term to reduce such bias.

Because the microwave field in a microwave oven is not completely uniform, temperatures reached by samples distributed over the oven floor also may not be uniform. As indicated earlier, small errors associated with nonuniformity in heating often may be neglected where high accuracy is not required. However, where accuracy is important some guidance is required as to the size of possible errors which may result from unequal heating. Two major sources of such errors are organic matter oxidation and removal of structural water. Both of these would result in bias so as to yield water content values that are too large. The organic matter problem exists in conventional oven-drying and is not easily resolved, except that an upper limit is defined by the proportion of the sample which is organic and subject to loss through oxidation.

As discussed in section 21–2.1, General Principles, and shown in Fig.

21–1, loss of structural water upon drying of clay minerals depends upon drying temperature, and for accurate work, drying temperature must be controlled closely. The size of errors, $\Delta\theta$, to be expected from excessive heating—which always results in a value too high—may be obtained as follows. The mass basis water content, θ_T of a wet sample, M_{ws}, dried to constant weight, M_T at a temperature, T, is $(M_{ws}/M_T - 1)$, so that

$$\Delta\theta = \theta_T - \theta_{105} = M_{ws}/M_T - 1 - (M_{ws}/M_{105} - 1)$$
$$= [(M_{ws}/M_T) - M_{ws})/M_{105})] \,. \qquad [27]$$

An illustration of the magnitude of possible errors which might occur with different clay minerals may be obtained using data from the temperature–water loss curves shown in Fig. 21–1. The ordinate in this figure gives the mass of a sample which would be reduced to a constant mass of unity after drying at a temperature of 800°C; thus the ordinate number minus unity and divided by unity would give the water content on a g/g/ basis, e.g., $(1.04 - 1)/1 = 0.04$, or the ordinate value is $(\theta_{800} + 1)$. Multiplication of Eq. [27] by M_{800}/M_{800} permits rewriting this equation as

$$\Delta\theta = [(M_{800}/M_T) - (M_{800}/M_{105})]M_{ws}/M_{800}$$
$$= [1/(\theta_{T,800} + 1) - 1/(\theta_{105,800} + 1)](\theta_{ws,800} + 1) \qquad [28]$$

where $\theta_{T,800} = M_T/M_{800} - 1$; $\theta_{105,800} = M_{105}/M_{800} - 1$ and $\theta_{s,800} = M_{ws}/M_{800} - 1$. Quantities $\theta_T + 1$ and $\theta_{105} + 1$ may be obtained from the ordinate in Fig. 21–1 for different drying temperatures and for several different clay minerals. The quantity $\theta_{s,800}$ is the sample water content based upon drying to constant weight at 800°C. The range of water contents of interest on this basis will not differ appreciably from what the range would be for 105°C drying. Assuming this range to be from about 0.05 to 0.60 g/g for illite clay the approximate error $\Delta\theta$ for drying temperatures different from the standard 105°C drying temperature may be obtained. The bracketed portion of Eq. [28] for 200 and 105°C becomes $(1/1.062) - 1/1.067) = 0.0044$. Multiplying this by $0.05 + 1$ and by $0.60 + 1$ gives 0.0046 and 0.0070 or an error range of approximately 0.46 to 0.70% water content. Similarly, computations for 300 and 105°C over the same range of water contents result in an error range of approximately 0.75 to 1.14%, and for 400 and 105°C the range is approximately 3.5 to 5.4%. For illite the water loss curve is nearly flat in the 150 to 300° range and steepens appreciably in the 300 to 500°C range.

Most common soils contain appreciable soil particles larger than clay size and which are less affected by drying at higher than usual drying temperatures. Assuming that only 1/3 of a sample is clay and 2/3 is unaffected by over-heating the error need be associated only with 1/3 of the sample so that the error range for 200°C becomes 0.15 to 0.23%; for

300°, 0.25 to 0.38%; and for 400°, 1.2 to 1.8%. Larger errors would be found for some clay soils as may be inferred from the curves given for other clay types, particularly if higher temperatures than those considered here are reached in a part of the sample.

Further error may occur due to nonuniform microwave heating in the various mineral components of a nonuniform sample. Sizes of errors anticipated by the above computations have been found by the author and by Gee and Dodson (1981) with soils of several different types.

Special microwave equipment for automatic drying, weighing, and computation of water content is commercially available. However, most of the same limitations as discussed above apply. Furthermore, the automatic weighing feature limits measurements to single samples and a measurement time of 10 to 20 min each. Thus, these have limited usefulness in much soils work where multiple samples are required.

21-3 INDIRECT METHODS

21-3.1 Introduction

Certain physical and physical-chemical properties of soil vary with water content. However, the relationship between such properties and water content usually is complicated. Both the pore structure and constituents of the soil solution are involved in these relationships. Also, some of these properties, even under conditions where all other factors are held constant, are not determined uniquely by water content. The wetting history of the soil is a factor in many instances, particularly where water must flow into and out of a sensing device such as a porous block. Despite many limitations, some of these properties, with appropriate calibration, can be useful in characterizing soil with respect to its water content.

Wetting history often must be considered because water content of soil for a given energy status (temperature being constant) will be greater if the soil has reached a given water content by drying than by wetting. This phenomenon, known as hysteresis, is discussed in more detail in Chapter 26. For present purposes however, it is sufficient to point out that whether or not a pore becomes water-filled during a wetting process depends upon the size of the pore itself, whereas in a drying process the emptying of the pore depends upon the size of the channels connecting it with other pores in the system. It is possible, therefore, to have two soil samples with identical porosities but different water contents at equilibrium with each other if their wetting histories have been different.

Indirect methods involve measurement of some property of the soil that is affected by soil water content or measurement of a property of some object placed in the soil, usually a porous absorber, which comes to water equilibrium with the soil. The water content of a porous absorber at equilibrium depends upon the energy status of the water rather than upon the water content of the soil with which it is in contact. For example, a soil with fine pores will contain more water than a soil with coarse pores at equal matric potential. Hence, if the properties of the porous absorber are to provide an indication of water content, calibration against the water content of a sample of the soil in which the absorber is to be used is required, together with some indication of the wetting history of the soil. Electrical or thermal properties of the absorber or weight changes in the absorber are indications of its water content. Methods involving weighing the absorber (Richards and Weaver, 1943) have not become widely used, probably because of technical complications associated with the weighing process.

Neutron scattering and neutron and gamma ray absorption are affected by water content of a porous material and may be adapted to water content measurement. Although generally requiring calibration and therefore considered to be indirect methods, under some conditions radiation measurements can be converted directly to water content on the basis of theoretical considerations. Hence, such methods might be considered to be direct methods. There are conditions where radiation water content measurements are subject to less error than gravimetric measurements. For example, where spatial variation is high and is a sizeable confounding factor in repeated gravimetric sampling, radiation measurements, which can be made repeatedly in the same soil volume, have a definite advantage.

The need for indirect methods for obtaining water content or indices of water content is evident when the time and labor involved in gravimetric sampling are considered. In addition to requiring a waiting time for oven-drying, such determinations are destructive, and therefore each sample must be taken at a different place in the soil system under study. Destructive sampling may disturb an experiment and may increase the possibility that a change in water content with position in a sampling area may be interpreted falsely as a change in water content with time at a particular location. Many of the indirect methods permit frequent or continuous measurements in the same place and, after equipment is installed, with only small expenditure of time. Thus, if a suitable calibration curve is available, changes in water content with time can be approximated.

Although this chapter is concerned primarily with measurement of water content or inferences of water content from other measurements, it should be pointed out that for many purposes water content is less useful than certain other properties of the soil-water system which depend upon water content. For example, in studies involving plant growth, matric potential in the soil has greater meaning provided that, in soils containing more than small quantities of soluble salts, a term taking into account osmotic potential is added. It is common practice to calibrate some of the indirect methods for evaluating water conditions in soil in terms of matric potential rather than water content. The subject of water potential is discussed in greater detail and methods for its measurement are given in chapters 23 through 26.

21-3.2 Electrical Conductivity and Capacitance

21-3.2.1 PRINCIPLES

Electrical and thermal conductivity and electrical capacitance of porous materials vary with water content. Such properties of materials can ordinarily be measured with great precision; and if a reliable correlation with water content existed, methods based upon measurement of these properties would have considerable utility. Unfortunately, such measurements made directly in soil rather than in a porous body inserted in soil have not resulted in unique correlations with water content and have not come into general use. The most thoroughly tested of the methods has involved measurement of electrical resistance (e.g., Edlefsen & Anderson, 1941; Kirkham & Taylor, 1950). Soil heterogeneity, which prevents uniform flow of current in the soil mass, and uncertain electrical contact between electrodes and soil seem to be the major obstacles to successful use of direct electrical resistance methods. Electromagnetic properties of soil measured by means of buried coaxial cables have been used for water content measurement (Topp, 1980).

Many of the problems involved in measurement of electrical and thermal conductivity and electrical capacitance in soil are avoided by use of porous blocks containing suitable electrodes and imbedded in the soil. When these blocks reach equilibrium, i.e., when water ceases to flow into or out of the blocks, their electrical or thermal properties often are regarded as an index of soil water content. However, the associated soil water content must be obtained from a calibration curve made using soil from the site where the block is used, because the equilibrium between a block and soil is a matric potential equilibrium and not a water-content equilibrium. Different soils have different water content vs. matric potential curves, variations among soils often being as much as several hundred percent. For example, a fine sandy loam may have a water content of 5% at −1.5 MPa matric potential, whereas a clay loam may have a water content of, say, 13% at the same potential. As a consequence, calibration of a porous block against matric potential often may be considered more reasonable and more useful than calibration against water content (see chapter 25).

Hysteresis enters into the problem of inferring water content from measurements made on porous blocks even though a calibration curve for a particular soil is available. The water content of both soil and block depends in part upon wetting history. Ideally, two calibration curves are needed: one for drying, extending from very wet to very dry, and one for wetting, where the starting point is in the very dry range. These two curves are considered to close at the endpoints and to provide an envelope which would contain all possible intermediate curves. However, because it is difficult to wet a soil only part way, the wetting curve is usually not made. And in many practical situations, the starting point is unknown. Hence, the curve traced out as the soil wets or dries is unknown, and water content can be known only to lie at some point between the limiting wetting and drying curves. Resulting errors in water content inferences depend upon the nature of the soil in question and its wetting or drying history, but easily can be 20% or more (Taylor et al., 1961). Nevertheless, blocks often are used to indicate water content of soil even though the precision in such use is rather low. However, the popular use of porous blocks likely stems from their utility as indicators of water conditions favorable or unfavorable to plant growth (matric potential as opposed to water content), rather than from their ability to indicate soil water content.

Thermal conductivity and electrical capacitance measurements in porous blocks, although favorably reported on from time to time in the literature, have not come into general use (see Fletcher, 1939; Shaw & Baver, 1940; Anderson & Edlefsen, 1942; de Plater, 1955; Bloodworth & Page, 1957; Phene, et al., 1971a, 1971b). This probably results from the fact that electrical conductivity is measured easily and porous blocks for such use are easy to construct. Therefore, the most common porous-block technique involves measurement of electrical conductivity. The porous-block method which follows is for such measurements.

21–3.2.2 METHOD

21–3.2.2.1 Special Apparatus.

1. Wheatstone Bridge for measuring resistance: Bridges in common use are of the alternating current type (usually 1000 cycle) to avoid polarization at the electrodes in the porous block. Both null-point and deflection-type instruments are used. Digital read-out instruments which can be used in data acquisition systems are useful. Resistances to be measured range from a few hundred ohms to 200 000 or more ohms, a single calibration curve often covering as much as 100 000 ohms.
2. Porous blocks: Blocks now available are made of a variety of porous materials ranging from nylon cloth (Bouyoucos, 1949) and fiberglass (Coleman & Hendrix, 1949; Cummings & Chandler, 1949; England, 1965) to casting plasters (Bouyoucos & Mick, 1940; Perrier & Marsh, 1958; Cannell & Asbell, 1964; and numerous others), the most common being some form of gypsum. Various grades and kinds of casting plaster are mixed with different amounts of water and in a variety of ways, including in some instances pouring the mix into the mold in a partial vacuum. In some instances resin is added to the mix, which changes the electrical characteristics and decomposition rate in the soil (Bouyoucos, 1953). The method of preparation as well as the mix itself governs the porosity of the block and the resulting response

curve. Some types of gypsum last longer in soil than others.

Several different electrode systems are in common use. The simplest consists of two tinned wires about 35 mm long (made using ordinary twin-conductor, rubber-insulated lamp cord). These are imbedded 1 or 2 cm apart in a rectangular porous block roughly 1 by 3 by 5 cm in size. Cylindrical blocks also are in use. These consist of a cylindrical screen (usually made of stainless steel) surrounding a central post or in some instances a second, smaller cylindrical screen. Such blocks are 2 to 3 cm in diameter and about 3 cm long. A parallel-screen system, using rectangular blocks, has been used by several investigators and is reported to have less lag in coming to equilibrium than other types of blocks.

3. Calibration container: Prepare a small screen box (window screen or hardware cloth soldered together), open at one end and of suitable dimensions to contain the block and a layer of soil at least 2 cm thick around the block (Kelley, 1944).
4. Equipment for determining a reference water content of the soil used in the calibration (section 21–2.2.2.1 or 21–2.4.2.1).

21–3.2.2.2 Procedure. Calibrate each block in soil typical of the site in which it is to be used, and with packing to about the same bulk density. To carry out the calibration, saturate the block with water, preferably with vacuum soaking, and weigh the wet block, its attached leads, and the screen box together to obtain a tare weight. Moisten the soil to be used to a state where it can be packed around the block to approximately its field density, mix the soil thoroughly, and take a sample for water content determination. Then pack the soil around the block in the screen box and weigh the entire apparatus. Using the water content determined independently by gravimetric methods (see section 21–2.2 or 21–2.3), compute the dry mass of the soil in the box. This will be

$$\text{soil dry mass} = \frac{(\text{wet mass} + \text{tare}) - (\text{tare})}{\%\ \text{water content}/100 + 1} \qquad [29]$$

where the water content figure is on a dry-mass basis. At all subsequent calibration points the water content percentage will be

$$\begin{array}{c}\text{water content, \%,}\\ \text{dry mass basis}\end{array} = \frac{100\left[(\text{tare} + \text{wet soil}) - (\text{tare} + \text{dry soil})\right]}{\text{dry soil}}. \qquad [30]$$

Wet the soil in the screen box to near saturation, weigh the entire assembly, and then measure the block resistance to determine the first calibration point. Allow water to evaporate from the apparatus in the air until the desired weight for the next calibration point is reached. After the desired weight is reached, place the entire apparatus in a closed container (such as a desiccator without desiccant) in the dark at uniform temperature. Leave it overnight or longer, to permit the water to equilibrate in the block and soil. Alternatively, embed the block in soil on the porous plate used in obtaining the water content vs. matric potential curve (see chapter 26). Water content then may be varied by changing the gas pressure used. After the soil comes to equilibrium water content, measure the resistance. Plot the resistance as a function of water content. (Three-cycle semilogarithmic paper is convenient.)

To install a block in the field, form a hole vertically from the surface or horizontally in the side of a trench. Wet the block thoroughly, place it in the hole, and then pack soil around it to assure good contact with the surrounding soil. Bring the leads to the surface, running them horizontally for a short distance beneath the surface to assure that no continuous channel exists along the leads for passage of free water. After equilibrium is reached, usually overnight, make resistance measurements as desired and convert them to water content values with a calibration curve.

21–3.2.2.3 Comments. In calibration as well as under field conditions, true water content equilibria rarely are reached, particularly in the dry range. However, the uncertainty of the water content inference, at best, does not justify an elaborate and time-consuming calibration. Under practical conditions, when the resistance reading approaches a constant value, equilibrium may be assumed to be close. In the wet range, for most porous blocks, the resistance change with changing water content is small and the precision is low. Precision also is affected by changes in the calibration curve over successive wetting and drying cycles (Cannell, 1958).

Restricted water flow at the interface between the smooth face of a porous block and coarse soil materials creates some problems in the use of blocks in sandy soil. However, coating a block with a porous material such as diatomaceous earth can reduce this problem appreciably. An envelope of fine-grained material placed around tensiometer cups (A. Cass and G.S. Campbell. Tensiometer response in coarse grained soils, Western Society of Soil Science, Eugene, OR, June 1981 unpublished), has produced a significant improvement in interface conductivity. Nonetheless, some problems do exist in use of blocks in coarse soils where flow of water is restricted by low unsaturated conductivities.

It is common practice to place blocks into uniform groups according to their resistance at saturation and to calibrate only selected blocks from each group (Tanner et al., 1949). This practice does not completely ensure obtaining groups of blocks with like calibration curves; but considering the low precision of the method when it is used for water content determination, it probably is an adequate procedure.

The calibration procedure described is for desorption. The procedure cannot be reversed easily to provide a sorption curve because of difficulties associated with partially wetting a soil mass. However, two tedious processes have been used in the author's laboratory to produce sorption curves in the dryer part of the water content range where wetting is difficult. One involves condensing water into soil from the air under carefully controlled temperature and humidity at the water potential required to provide the desired water content. The second involves thorough mixing of finely pulverized ice in appropriate quantities into dry soil at low temperatures and then slowly warming to melt the ice. Control of pore size distribution affecting wet range water content is difficult in the latter method.

Calibration often is carried out in the field by obtaining water content values gravimetrically and plotting them against resistance readings. Or, blocks are placed in pots containing growing plants and the water content determined by weighing the pot, with an estimate of plant weight being subtracted (Cannell, 1958).

Electrical conductivity of a porous block depends upon the electrolyte concentration of the conducting fluid as well as upon the cross section of this fluid or water content. In a porous block made from an inert material, the electrolytes that carry the current come from the soil solution. Even a small change in electrolyte concentration will influence the resistance. In blocks made from gypsum the electrolyte concentration corresponds primarily to that of a saturated solution of calcium sulfate. Variations in the soil solution due to fertilization have relatively little influence upon the electrolyte concentration in such blocks and therefore relatively little influence on resistance. Such blocks also may be used without serious difficulty in slightly saline soils (where soil extract conductivities are less than approximately 2 mmho/cm^2 (Taylor, 1955).

Blocks made from gypsum compounds gradually deteriorate in soil, particularly in sodic soils and in soil where the water table frequently is at high levels. However, those made from hydrocal have lasted for upwards of 6 years in some soils. Blocks made from ordinary plaster of paris have been known to deteriorate in a single season beyond the point where they can be used.

It is difficult to specify an expected precision for water content measurements using electrical conductivity blocks because of the many sources of error involved. The precision depends not only upon the care used in manufacture, selection, and calibration of blocks but also upon factors of hysteresis which are out of the control of the operator. However, it appears that precision better than +2% water content should not be expected and that errors as great as 100% easily are possible. On-the-site checking as the blocks are used appears necessary if confidence is to be developed in water content inferences to be made. On the other hand, where porous blocks are used as a measure of matric potential rather than water content, considerably better performance is possible. Calibration against matric potential may be carried out using porous-plate and pressure-membrance equipment (chapter 25) with special pass-through electrical contacts.

21-3.3 Neutron Thermalization

21-3.3.1 PRINCIPLES

Hydrogen nuclei have a marked property for scattering and slowing neutrons. This property is exploited in the neutron method for measuring water content. High-energy neutrons (5.05 MeV) emitted from a radioactive substance such as radium–beryllium or americium–beryllium [$^{9}Be(\alpha,n)^{12}C$] are slowed and changed in direction by elastic collisions with atomic nuclei. This process is called *thermalization*, the neutrons being reduced in energy to about the thermal energy of atoms in a substance at room temperature.

Neutrons interact with matter in two general ways: by elastic and inelastic scattering, and by interactions leading to capture with a consequent emission of energy or of other nuclear particles. The probability of any particular interaction depends upon neutron energy and characteristics of the nuclei encountered. These characteristics are described generally by the nuclear cross section, measured using a unit of area called the "barn," which is 10^{-24} cm^2. The larger the probability of a particular interaction, the larger is the nuclear cross section.

The two major factors involved in scattering and slowing of neutrons are the transfer of energy at each collision and statistical probability of collision.

The average energy transfer at collision of a neutron with other nuclei depends largely upon the mass number of the nuclei encountered. The average number of collisions required to slow a neutron from 2 MeV to thermal energies is 18 for hydrogen, 67 for lithium, 86 for beryllium, 114 for carbon, 150 for oxygen, and $9A + 6$ for nuclei with large mass numbers A (Weinberg & Wigner, 1958, Table 10.1).

The statistical probability of collision is dealt with using the concept of "scattering cross section," which is a statistically derived, cross-sectional area measured in barns which is proportional to the probability of collision—in this case, between neutrons and other nuclei. This scattering cross section depends upon the nature of the nuclei encountered and the energy of the neutron. The scattering cross section for hydrogen varies from about 1 barn at 10 MeV to about 13 barns at 0.1 MeV. The cross section varies considerably as the neutron continues to lose energy, but is somewhat higher in the thermal energy range. Other elements found in soil with appreciable scattering cross sections (2–5 barns) are beryllium, carbon, nitrogen, oxygen, and fluorine.

Considering both energy transfer and scattering cross-section, it is evident that hydrogen, having a nucleus of about the same size and mass as the neutron, has a much greater thermalizing effect on fast neutrons than any other element. In addition, when both hydrogen and oxygen are considered, water has a marked effect on slowing or thermalizing neutrons. This is particularly true in the thermal range.

The quantity of hydrogen in the soil ranges from near zero for dry coarse sand to as much as 8% of the mass of a fine-textured soil with 50% water content when structural water (see section 21–2.1) is included in the computation. Most of the hydrogen in soil is associated with water, and lesser amounts with organic matter.

As fast neutrons lose energy and become thermalized, another nuclear-matter interaction becomes increasingly important—neutron capture with the release of other nuclear particles or energy. Of the elements usually present in soil in quantities of 1% or greater, the capture cross-section for thermal neutrons is greatest for iron (2.53 barns) and potassium (2.07 barns). The other most common elements in the soil, silicon, aluminum, hydrogen, carbon, and oxygen, have capture cross-sections of 0.16, 0.23, 0.33, 0.003, and 0.0002 barns. Several elements present in soil in small (or even minute) quantities, such as cadmium with an absorption cross section of 2450 barns, boron with 755 barns, lithium with 71 barns, or chlorine with 34 barns, can have an appreciable effect on neutron capture.

When a fast neutron source is placed in moist soil it immediately becomes surrounded by a cloud of thermal neutrons. The density of this cloud represents an equilibrium between the rate of emission of fast neutrons, their thermalization by nuclei such as those of hydrogen, and their capture by absorbing nuclei, as determined by their concentration and capture cross-section.

The scattering cross-section and the concentration of hydrogen nuclei determine the distance from the source a fast neutron must travel before making a sufficient number of collisions to become thermalized. The farther a neutron travels from the source the larger the volume which will be occupied by thermal neutrons and the lower their density. With the number of slow neutrons involved, the absorption capacity of the soil for neutron capture is essentially infinite, and the rate of capture depends only upon thermal neutron concentration and the combined capture cross-section of the elements in the soil. If the capture cross-section, except for that due to water, remains constant (i.e., chemical composition constant), then the thermal neutron density may be calibrated against water concentration on a volume basis. Thermal neutron density is easily measured with a detector, insensitive to fast neutrons, which is placed in the vicinity of the fast neutron source. Thermal neutrons interact with a boron trifluoride gas in the detector, releasing an alpha particle, which is attracted to a negative high-voltage electrode within the detector. This creates a short electrical pulse, which registers as a count in an associated electronic scaling unit. The source usually is placed at the bottom of the detector tube or against the side or as an annular ring about the detector. This probe can be lowered through an access hole into the soil and measurements obtained for conversion to water content (Belcher et al., 1950; Gardner & Kirkham, 1952).

The nature of the neutron-scattering and thermalization process imposes an important restriction on the resolution of water content measurements. The volume of soil involved in the measurement will depend upon the concentration of scattering nuclei and thus largely upon water content, and upon the energies of the emitted fast neutrons. The strength of the neutron source affects the thermal neutron density and is involved in the counting statistics, but does not affect the range of the fast neutrons. Experimental work with neutron sources indicates that the practical resolution ranges from about a 16-cm radius at saturation to 70 cm at near zero water content (Van Bavel et al., 1956). The radius of the sphere of influence accounting for 95% of the neutron flux which would be obtained in an infinite medium is given by Olgaard (1965) in an empirical equation as

$$R = 100 \text{ cm}/(1.4 + 10\ m) \qquad [31]$$

where m is the water content in g/cm^3. However, the parameters in the equation may be expected to change somewhat with differences in chemical composition of the soil. Lack of high resolution makes it impossible to detect accurately any discontinuity or sharp change in water content gradient in a soil profile (McHenry, 1963). In particular, measurements close to the soil surface are unreliable because of the discontinuity at the interface between soil and air; and measurements usually are not made with well-type equipment any closer than about 18 cm from the surface. Therefore, water-content distribution curves for soil profiles containing steep water-content gradients will be rounded and inaccurate in detail; however, they are likely to be of sufficient accuracy for many practical uses. In particular, water content *changes* within a profile can be obtained with considerably greater accuracy than is possible in the determination of the profile water content itself.

Surface probes, in which the slow-neutron detector is laid horizontally on the surface of the soil with the fast-neutron source beside it, make it possible to obtain water content measurements in surface soil where the well-type unit is inadequate, but with considerably less accuracy. Most surface probes involve use of a moderator rich in hydrogen (such as paraffin or polyethylene) over the top of source and detector to compensate partially for the discontinuity at the interface between soil and air. Experimental work with surface units indicates a sensitive depth of from 15 to 35 cm (Van Bavel et al., 1961; de Vries & King, 1961; Phillips et al., 1960). Where water content of the surface soil is not uniform, say under conditions of rapid surface evaporation or superficial wetting by low rainfall, or where the surface of the soil is rough, precision falls off materially (Van Bavel et al., 1961).

In theory it should be possible to determine the density of thermalized neutrons in the vicinity of a fast-neutron source from the chemical composition of the soil in the absence of water, and then to relate changing water content to increases in this thermalized neutron density as measured with an appropriate detector. However, the several types of neutron-matter interactions, together with the many chemical constituents of soil and variations in bulk density (Holmes, 1966; Lal, 1974; Greacen & Hignett, 1979), make this impractical and some form of calibration becomes necessary. Some success has been achieved through laboratory calibration of soil materials (Vachaud et al., 1977). Although still difficult, calibration is somewhat simplified by the fact that for most commercial neutron source–detector units the calibration curve relating thermal neutron density to water content is essentially linear over the range of interest, so that the calibration equation or curve may be defined by as few as two points if they are available with sufficient accuracy. Further simplification, although the calibration process remains difficult, is associated with the fortuitous coincidence that some differences in composition and texture among soils are compensated partially by differences in the neutron-matter interactive processes, so that a single calibration curve may approximately fit a small group of soils of similar chemical composition or a soil which varies with depth or spatial distribution. However, for high accuracy or under conditions where a soil is suspected to deviate markedly from the normal, extensive calibration procedures may be required. See Van Bavel et al. (1961), Holmes and Jenkinson (1959), Stolzy and Cahoon (1957), Holmes (1956), Olgaard (1965), Sinclair and Williams (1979), Zuber and Cameron (1966), Vachaud et al., (1977), and Lal (1974) for these procedures.

The quantity of hydrogen in soil, apart from absorbed water, depends upon organic matter content and the nature of the mineralogical components. Certain clay minerals contain appreciable structural water (see section 21–2.1), which is not removed by oven-drying at 100 to 110°C. For several minerals (Fig. 22–1) as much as 20% additional water can be removed by heating to about 800°C. On the other hand, coarse sands composed of quartz or feldspars have almost no water associated with them at 100 to 110°C. It is significant that the construction of many calibration curves for neutron water-content equipment has involved use of sand to provide some water contents and loams and clays for other water contents, often without apparent difficulty. The explanation for this apparent anomaly possibly involves a nearly perfect balance between increased scattering due to hydrogen in organic matter or in the structural hydrogen of clays, which tends to increase thermal neutron density, and increased neutron capture associated with a different chemical composition, which tends to decrease thermal neutron density. Clay materials that retain large amounts of structural water also are known to contain higher concentrations of such elements as boron, lithium, chlorine, and iron, which are good neutron absorbers compared to the elements composing sands. Evidence that small quantities of good neutron absorbers can affect a calibration curve has been obtained by Holmes and Jenkinson (1959). They added boron to a soil at rates of 65, 156, and 245 mg/kg and noted that increasing boron concentration decreased the slope of the counting-rate vs. water-content curve. It is evident that the right amounts of neutron absorbers could compensate for increased neutron-scattering due to organic or structural hydrogen in a soil. It also is evident that a soil containing uncommon excesses of such elements as boron, chlorine, or iron could have a different calibration curve than that for a more normal soil. An additional fact which must be acknowledged in the calibration problem is that tightly bound hydrogen nuclei interact differently with neutrons than do hydrogen nuclei of water.

21–3.3.2 METHOD

21–3.3.2.1. Special Apparatus.

1. Neutron moisture depth probe and meter consisting of a source of fast neutrons (usually americium-241/beryllium), a detector for thermalized neutrons, a protective shield composed of lead (for gamma ray absorption) and polyethylene or paraffin (for neutron absorption) which serves also as a reference standard, and a scaler for registering counts or, in conjunction with a built-in computer, a meter for direct display of water content. The neutron source usually is either a small capsule located on the side of the detector cylinder or an annular ring placed around the detector. Such units are commercially available with several different neutron-source strengths and arrangements. Separate gamma ray density units for back-scatter bulk density measurements or a density probe combined with a moisture probe also are available commercially. A separate neutron moisture meter for use on surface soil also is available.
2. Soil auger for installing access tubing, slightly smaller than the tubing to assure a tight fit.
3. Thin-wall aluminum, steel, or plastic access tubing. Several sizes of tubing are used, but the size should be consistent with the probe size to reduce errors associated with an air gap, which may be created between the probe and tubing wall.
4. Calibration curves, or calibration parameters for units involving an integral computer, and a moisture content read-out.
5. Film badges, leak test-kits.
6. State or federal license, as may be required.

21–3.3.2.2 Procedure.

1. Use the soil auger to form the hole for installation of the access tubing. In many soils, particularly where loose materials are present, it is advisable to drill the access hole through the access tubing, advancing the tubing little by little as the hole is drilled. This is important because of the influence that air space in the vicinity of the access hole has upon the thermalized neutron distribution and consequently the neutron count. The access tubing usually is left so as to protrude about 10 cm above the soil surface and is covered with an empty can or

stopper between readings to keep water and debris out. In situations where water might enter, a tight cap or rubber stopper may be necessary at the lower end of the tube.

2. To make a measurement, place the probe unit over the access tube preparatory to lowering it into the hole. Select an appropriate counting time and make several standard counts while the probe is in the shield. The measurement used then will be the ratio of the count taken at a particular position of the probe to the average of counts taken in the standard. This will correct for any electronic changes in the counting circuits which otherwise might confound the measurement.
3. Make one or more counts at each selected depth in the profile. Since the zone of influence for 5-MeV neutrons is roughly spherical with a radius of about 15 cm in wet soil and 70 cm in dry soil, the depth increment should be no greater than 15 cm.
4. Use the calibration curve to reduce count ratios (count in soil/count in standard shield) to volumetric water content, or read water content directly if the equipment has the required built-in computer.
5. The procedure for use of the surface probe is comparable, except that precautions must be taken to assure a smooth surface so that air gaps are not present between the surface probe and the soil.

21–3.3.2.3. Comments. With reasonable attention to safety rules supplied by the manufacturer, the health hazard involved in using the equipment is small. The important precautions are the following: (i) Keep the probe in its shield at all times except when it is lowered into the soil for measurement. (ii) Reduce exposure to the small amount of radiation escaping the shield by keeping several feet away, except when changing the position of the probe, and by keeping the open end of the probe and shield pointed away from personnel. (iii) Carry the probe in the field on a cart or on a sling between two persons if more than a few minutes is involved in getting the equipment to a position for use. (iv) Transport the probe in the back of a truck, a car trunk, or for short periods in the unoccupied rear seat. (v) Have operators wear a film badge at waist level. (vi) When the probe is not in use, lock it in a storage room. Label the container plainly to indicate radioactive content. (vii) Have a leak test performed on the source by a competent safety officer semiannually, or as may be prescribed by the radiation license (the manufacturer can advise on this). (viii) See that probe maintenance is performed only by personnel trained in servicing radioactive equipment.

For many practical uses of the neutron equipment, the manufacturer's calibration curve is adequate. However, calibration may be required if a soil has unusual neutron-absorption characteristics or if the access tubing is of different size or material from that in which the probe was calibrated. For maximum reproducibility of water content measurements, the probe should fit the access tubing as closely as practical.

A field calibration check may be used for detecting large differences in calibration due to unusual soil conditions. However, field calibration should not be relied upon for precision work because of the inaccuracies associated with the determination of the volume-basis water content values required. These inaccuracies arise from errors in measurement of both the mass-basis water content (see section 21–2.2) and the bulk density. Field variation in both water content and bulk density (which can occur over a relatively short distance, particularly in a heterogeneous soil) introduces an additional confounding factor. For these reasons, unless the calibration curve is considerably in error, water content inferences from the neutron measurements may well be better than measurements obtained by direct sampling.

Accurate calibration curves require the use of large homogeneous bodies of soil with carefully regulated water content. Some useful checking can be done with substitute neutron absorbers (such as $CdCl_2$–water solutions) in small containers. However, the techniques are sufficiently complicated that persons attempting calibration should consult original literature sources such as Van Bavel et al. (1961), Holmes (1956), Holmes and Jenkinson (1959), Stone et al. (1960), Hewlett et al. (1964), Sinclair and Williams (1979), and Zuber and Cameron (1966), or a review by Visvalingham and Tandy (1972).

An empirical equation for water content in terms of the count ratio, used over the range of water contents of usual interest, is

$$\theta = a + bf \qquad [32]$$

where f is the count rate ratio, I/I_{std}, and a and b are parameters which depend upon soil characteristics and the standard count when the neutron source and detector are in the shield. The count ratio usually has a range of from near zero to about 1.7 at saturation, depending upon the characteristics of the soil and composition of the standard absorber. At low water contents the equation departs from linearity because in real soil the count rate, I, never is zero. The parameter a depends in part upon the bulk density and is closest to zero in soils which have a low bulk density and, when dry, contain the smallest quantities of moderating substances. The parameter b is essentially independent of bulk density and depends upon the presence of substances in the soil, such as structural water (as opposed to absorbed water) and chemical materials, which are effective in the neutron thermalization and capture process. Such substances as iron, boron, molybdenum, and cadmium significantly affect the value of b. Inhomogeneities, such as stratification in mineral density and composition, or even sharp changes in water content at interfaces, where the porosity may change abruptly, can have a marked effect upon the calibration curve where it is constructed from field measurements. Sharp changes in water content distort the shape of the volume-of-influence boundary, which is given approximately by Eq. [31].

Laboratory calibration usually can be done with greater precision than field calibration, providing that the volume of soil used exceeds that of the volume of influence as given by Eq. [31]. However, when the laboratory calibration curve is used on field measurements it must be recognized that it will deviate from a true calibration to the degree that a nonhomogeneous field soil deviates from the calibration soil both spatially and in its neutron thermalizing and capture constituents. Thus, the accuracy of a calibration curve, whether done in the field or in the laboratory, often is highly subjective.

Bearing in mind the many difficulties associated with calibration, high precision and accuracy require painstaking effort and (usually) separate curves covering the area where the method is to be used. However, for precision and accuracy comparable to that usually achieved with gravimetric methods (roughly + 0.005 g/cm^3), considering field variability, reasonable calibration curves easily may be produced. At this level of accuracy it often is adequate to determine, with considerable care, two points on the calibration curve, one very wet and one dry, and to draw a straight line between. Alternatively, calibration data may be obtained over a range of water contents and a calibration equation determined statistically.

Both water content and change in water content with time at a fixed position are of interest. Ignoring spatial variation in water content and assuming the validity of the empirical equation relating water content and neutron count ratio, f, given by Eq. [32], error equations may be written, one for water content and one for change in water content at a fixed position,

$$\theta = a + bf \quad [32]$$

$$\Delta\theta = a + bf_1 - (a + bf_2) = b(f_1 - f_2) = \mathrm{b}\Delta f. \quad [33]$$

The corresponding variances are

$$\sigma_\theta^2 = \sigma_a^2 + b^2\sigma_f^2 + f^2\sigma_b^2 \quad [34]$$

$$\sigma_{\Delta\theta}^2 = b^2\sigma_{\Delta f}^2 + \Delta f^2\sigma_b^2 . \quad [35]$$

Variances σ_a^2 and σ_b^2 for water content depend upon spatial variations in soil constituents and profile structure and the degree to which calibration conditions resemble particular field conditions. Such variations can involve bias as well as unidentified error associated with site variation. Detailed determination of these variances would involve site studies, but reasonable estimates of the magnitude of the standard deviations often may be made. Assuming that accurate placement of the source-detector probe is possible, σ_a^2 and σ_b^2 become essentially zero for water content change at a fixed location in the soil. An exception to this may occur if sharp interfaces or wetting-fronts exist in a manner so as to seriously alter the shape or size of the volume of influence as the water content changes.

An additional factor of some importance where high accuracy in measurements is desired involves perturbations in water content caused by temperature changes in soil surrounding a neutron-probe access-tube induced by heat conduction along the tube. However, Hanks and Bowers (1960) measured changes in water content around access tubes and concluded that the influence usually would be minor compared with other sources of error.

Errors in water content measurements associated only with random neutron emission may be computed with Eq. [34], where variances involving the calibration parameters are taken as zero. The variance in water content, σ_θ^2, is $b^2\sigma_f^2$ or $b^2\sigma_{It/I_{std}t}^2$, where t is the time counted at rate I and I_{std}. The error may be reduced by counting in the standard for a longer time, t', which usually is practical because of the infrequent need for making this count during a series of measurements. Inclusion of this in the error analysis requires replacing $I_{std}t$ with $I_{std}t'/n$, with $t'/t = n$. Thus,

$$\sigma_\theta^2 = \sigma_{nIt/(I_{std}t')}^2 = b^2[\sigma_{nIt}^2 + (nIt)^2/(I_{std}t')^2\,\sigma_{I_{std}t'}^2]/(I_{std}t')^2 \qquad [36]$$

which, with $\sigma_{nIt}^2 = n^2\sigma_{It}^2 + (It)^2\sigma_n^2$ and with $\sigma_n^2 = 0$, reduces to

$$\sigma_\theta^2 = n^2b^2[It/(I_{std}t') + (It)^2/(I_{std}t')^2]/(I_{std}t') \qquad [37]$$

where, for a Poisson distribution in neutron emission from the source, $\sigma_{It}^2 = It$ and $\sigma_{I_{std}t'}^2 = I_{std}t'$. Recalling that $I_{std}t' = nI_{std}t$ and that $f = It/I_{std}t$, Eq. [37] becomes

$$\sigma_\theta^2 = b^2(f + f^2/n)/(I_{std}t) = fb^2(1 + f/n)/I_{std}t;$$

$$\sigma_\theta = b[f(1 + f/n)/I_{std}t]^{1/2} \qquad [38]$$

From Eq. [38] it may be seen that increasing the value of n by making long counts in the standard will reduce the standard deviation in water content caused by random neutron emission. The variance is largest in wet soil, so that for a large value of f, say 1.6, $b = 0.5$ and for the counting time in the standard (e.g., $I_{std}t = 4 \times 10^4$) being the same as for other counts ($n = 1$) the standard deviation in water content is 0.005 g/cm^3. Increasing counting time in the standard by a factor of 5 reduces this to 0.0036 g/cm^3. For 95% certainty ($2\sigma_\theta$) water content may be known to about 0.007 g/cm^3 or to about 0.7%, neglecting errors in a and b of the calibrating equation.

Variance in the difference in two counts, $\Delta f = f_1 - f_2$, is the sum of the two variances, σ_{f1}^2 and σ_{f2}^2, so that the variance and standard deviation in water content due solely to variation in the count ratio, f, are

$$\sigma^2_{\Delta\theta} = 2\bar{f}b^2(1 + \bar{f}/2)/I_{std}t; \qquad \sigma_{\Delta\theta} = b[2\bar{f}(1 + \bar{f}/2)/(I_{std}t)]^{1/2} \quad [39]$$

where $(f_1 + f_2)/2 = \bar{f}$, the average of the counts made at two different times. In the wet range where the variance is greatest, say $\bar{f} = 1.6$, *and for* $\mathbf{t} = 5$, $\mathbf{b} = 0.5$ *and* $\mathbf{I}_{std}t = 40\,000$, $\sigma_{\Delta\theta} = 0.005$ g/cm³. Thus, at a certainty of 95%, water content change may be known to the nearest 1% water content (2 × 0.005 × 100). Little improvement is achieved by going beyond $t = 5$. The accuracy in measurements of water content *change* associated with random neutron emission is slightly less than that for water content itself, but errors due to variations in parameters a and b usually are not involved as in water content measurements, so that the overall accuracy generally is better. Neither variance in water content nor water content change given in Eq. [38] and [39] have involved variance due to spatial variation in water content, except as parameters a and b may be involved. Spatial variation can be large, exceeding 0.001 (standard deviation of $1 \times 10^{-3/2} = 0.03$ g/cm³), thus somewhat reducing the importance of high accuracy in measurements of a and b and of f.

The parameters, a and b in Eq. [32] and subsequent equations, for a given soil sample of large enough volume to encompass the effective volume of influence, may be determined to almost any desired precision by carefully adjusting bulk density and water content to known values and making long neutron counts. At a single location in the field, errors in determination of these parameters depend upon the accuracy of the independent measure of water content in the soil volume that is seen by the source-detector unit. Such errors would impart a fixed bias to subsequent measurements. Larger errors, however, usually are associated with field variation of soil properties that determine the size of parameters a and b. The sizes of both a fixed bias and variation in spatial distribution of soil properties affecting a and b can only be determined by studies made in the field or upon samples taken from the field. However, experience often will permit estimates of the probable size of such errors. The effect of such errors upon water content inference may be determined by converting them to variances and applying Eq. [34].

21–3.4 Gamma Ray or Neutron Attenuation

21–3.4.1. PRINCIPLES

Principles of absorption by matter of both gamma rays and neutrons are well known. The degree to which a beam of monoenergetic gamma rays is attenuated or reduced in intensity in passing through a soil column depends upon its constituent elements and the overall density of the column. If the constituents and bulk density of the soil without its water remain constant, then changes in attenuation represent changes in water content. Or, if measurements are made at two different gamma ray energies two attenuation equations may be solved simultaneously to provide both water content and soil bulk density. Since bulk density often changes somewhat with wetting and drying, use of the dual gamma technique (Gardner & Calissendorff, 1967; Soane, 1967; Corey et al., 1971; Gardner et al., 1972) improves the accuracy of water content measurements over what is possible where bulk density must be assumed to remain constant.

The attenuation equation, neglecting air, is

$$I_m/I_o = \exp[-S(\mu_s\rho_s + \mu_w\theta) - 2S'\mu_c\rho_c] \quad [40]$$

where I_m/I_o is the ratio of the transmitted to incident flux for the moist soil, μ_c, μ_s, and μ_w are the mass attenuation coefficients for the container material, soil, and water respectively, θ is mass of water per unit bulk volume of soil, ρ_c is the density of the container, S' is the thickness of its wall, ρ_s is the bulk density of the soil and S is the thickness of the soil column. The intensity of the incident monoenergetic gamma beam is proportional to the intensity at the source and inversely proportional to the square of the distance from the source. Since gamma ray sources are of various sizes and dimensions, some deviation from the inverse square of distance rule is possible. This rule holds strictly only for uniform radiation from an infinitely small source. However, for a ^{137}Cs source, believed to have an active projected area of at least several square millimeters, the falloff was closely proportional to the inverse square of the distance in the range of from 30 to 90 cm as measured using a 12-cm long by 0.33 cm by 1.2 cm collimator against the source and a 10-cm long by 0.1 cm collimator against the scintillation crystal. A correction for air adsorption was made.

The corresponding equation for a dry soil is

$$I_d/I_o = \exp(-S\mu_s\rho_s - 2S'\mu_c\rho_c)\,. \quad [41]$$

Division of Eq. [39] by Eq. [40] yields

$$I_m/I_d = \exp(-\mu_w\theta S) \quad [42]$$

or

$$\theta = \ln (I_m/I_d)/-\mu_w S \quad [43]$$

which is useful under conditions where bulk density may be presumed to be constant. If a perfectly collimated gamma beam of uniform energy were used, and if all scattered and secondary radiation were eliminated, the established value of μ_w, available from independent work, could be used. However, satisfactory calibration curves can be obtained experimentally, usually with greater practicality.

The single gamma ray attenuation method has been successfully used to follow water content change in laboratory columns by Gurr and Marshall (1960), Ferguson and Gardner (1962), Gurr (1962) Rawlins and

Gardner (1963), Davidson et al., (1963), Reginato (1974) Reginato and Van Bavel (1964), and by more recent investigators. It was used to monitor water content change in the root zone in growing plants by Ashton (1956) and by Hsieh et al., (1972).

Two techniques are available for dual gamma scanning. The first involves independent measurements of gamma ray attenuation closely spaced in time, usually using ^{241}Am at 0.060 MeV and ^{137}Cs at 0.662 MeV, where either the soil column or the scanning gamma ray equipment is moved with precision from one set-up to another for the separate measurements. Use of the attenuation measurements and available parameters permits solution of the two simultaneous equations, for energy *a*

$$I_a = I_{oa}\exp\{-S(\mu_{sa}\rho + \mu_{wa}\theta) - S'\mu_{ca}\rho_c\} \quad [44]$$

and for energy *b*

$$I_b = I_{ob}\exp\{-S(\mu_{sb}\rho + \mu_{wb}\theta) - S'\mu_{cb}\rho_c\} \quad [45]$$

to yield

$$\theta = [\mu_{sb}\ln(I_a/I_{ca}) - \mu_{sa}\ln(I_b/I_{cb})]/Sk \quad [46]$$

$$\rho = [\mu_{wa}\ln(I_b/I_{cb}) - \mu_{wb}\ln(I_a/I_{ca})]/Sk \quad [47]$$

for water content, θ, and bulk density, ρ, where the attenuation coefficients for dry soil and water are μ_s and μ_w with additional subscripts, *a* and *b* to designate the gamma energy used, S is the column thickness, $k = \mu_{sa}\,\mu_{wb} - \mu_{sb}\,\mu_{wa}$, and the container counts, I_{ca} and I_{cb}, replace I_{oa} and I_{ob} to eliminate the container terms from Eq. [44] and [45].

The second dual gamma scanning technique involves simultaneous measurement, using two single channel analyzers or a multichannel analyzer to make simultaneous gamma counts at the two energy levels involved, and placing the ^{137}Cs gamma source behind the ^{241}Am source so that the higher-energy ^{137}Cs gamma rays pass through the ^{241}Am source. With appropriate geometry so that the soil traversed by both the collimated cesium and americium beams is close to the same length, the only additional problem, beyond those involved where two separate measurements are taken at close to the same time, is that of correcting the americium count for down-scattered gammas coming from the cesium source. Nofziger and Swartzendruber (1974) have shown that the probability for downscatter of 0.662 MeV gammas for cesium into the 0.060 MeV energy range of americium in the soil traversed by a narrowly collimated gamma beam is negligible and that the major downscatter occurs in the scintillation crystal. They have shown that it is possible to obtain a curve relating the intensity of low-energy gammas in the 0.060 MeV range to the 0.662 MeV gamma intensity, thus producing a correction term to be subtracted from the measured low-energy americium intensity. The empirical correction, *e*, in counts/second to be subtracted from the low-energy count rate is of the form of a cubic polynomial

$$e = A + Bi_c + Ci_c^2 + Di_c^3 \quad [48]$$

where i_c is the measured cesium intensity in counts/sec. For the gamma ray system of Nofziger and Swartsendruber (1974), the coefficients are given by $A = 11.343$ counts/s, $B = 0.10056$, $C = 9.5979 \times 10^{-8}$ s/count and $D = -2.1916 \times 10^{-11}$ $s^2/count^2$ over the range of cesium intensities from 3000 to 14000 counts/s. This correction equation was obtained from measurements made with the americium removed and replaced with a brass plug. The correction was shown to be independent of the material in the gamma-ray beam.

Water-content measurement techniques using neutron attentuation rather than gamma radiation are similar, but more specific to water. Although neutrons are scattered and absorbed in some degree by all kinds of nuclei in the soil, hydrogen is by far the most effective. Hence, attentuation change is relatively sensitive to water-content change (see section 21–3.3). However, high neutron fluxes, not readily available outside of a reactor facility, are required and field application would be difficult. Because of the current impracticality of the neutron attentuation method for general use, the method will not be described here in detail, and the reader is referred to an AEC report (Stewart & Gardner, 1969) for further information. The method has been successfully used at the 100-kW nuclear reactor at Washington State University, where fast and thermal neutron fluxes available at beam ports are 4×10^{12} and 1.2×10^{12} neutrons/cm^2 per s.

21–3.4.2. METHOD

21–3.4.2.1. Special Apparatus.

21–3.4.2.1.1 Sources of Gamma Radiation

Cesium-137, which emits gamma rays at 0.662 MeV and has a half-life of 30 years, and ^{241}Am at 0.060 MeV and a half-life of 470 years, are well suited for water-content measurements. The size of sources required depends upon the use to be made of the equipment. Sources of 20 or 25 mCi have been satisfactorily used. However, where rapidly changing water content is to be followed, so that counting times of only a few seconds are required, or where resolution of the order of a millimeter or less is required, much larger sources are desirable. Sources from 100 to 500 mCi have proved satisfactory under these more stringent conditions. However, self-absorption limits the practical size of ^{241}Am. Lead shielding required for safe operation increases, of course, with increasing source size. Either of the sources, or both for concurrent measurements, may be used.

21–3.4.2.1.2 Lead Shields and Collimators

Sources are housed in lead shields with suitable collimating holes or slits. With protective plugs for collimating slits, shields serve also as storage containers. Ideally, collimating holes should be drilled appropriately to serve for gross measurements and to accept lead plugs with smaller collimating holes or slits for measurements where greater resolution is desired. Holes about 3 cm in diameter will accommodate a collimating plug containing a 1- by 20-mm slit or any number of other slits or cylindrical openings that might be desired. Collimating plugs should have the form of bolts with large heads to cover the space between bolt and lead block.

For good spatial resolution, the gamma rays, emitted from the source in a solid angle, should be passed through a collimator which is as long as possible, consistent with source size and desired count rate, so as to be close to parallel as they enter the detecting crystal. For errors associated with collimation of the order of 0.001 g/cm^3 in a soil column 10 cm thick, the collimator for ^{137}Cs should be about 8 cm long on both source and detector side. For a tapered collimator slit 0.45 $\times$ 45 mm at the detector crystal face, the spatial resolution for cesium has been found to be only slightly greater than slit size, or about 0.5 mm. For dual gamma measurements the ^{241}Am system should have similar collimators, so that the shape of the gamma ray beam as it passes through the soil is comparable to that for the cesium. However, the practical size of an americium source is limited by self-absorption to roughly 22 mCi/mm^2, so that path and collimator length suited to cesium may severely reduce the count rate possible with americium and make compromise necessary. The length of collimator material needed for the 0.060-MeV gamma photons from ^{241}Am is roughly 1/50 of that needed for 0.62 gamma photons from ^{137}Cs. Although lead is used ordinarily for collimators and protective shielding, tungston absorbs gamma energy from two times as well (at a thickness of 1.2 cm) as lead to many times as well, and may be machined to closer tolerances. Hence, it often is used in place of lead.

The dimensions of protective shielding depend upon source strength; but, for greatest convenience where weight is not an important factor, blocks should be thick to reduce radiation in the vicinity to values only slightly above natural background, so that special precautions required in the vicinity of such radiation hazards may be minimized. Normal background varies from 0.01 to 0.03 mR (milliroentgen)/hour. One milliroentgen in air is about 5% less than 1 mrad in tissue (the rad is the unit of physical radiation dose), so that for practical considerations they may be regarded as equivalent.

In the design of shielding and collimators, it is necessary to determine the dose rate in terms of source strength, shielding thickness, and distance from the source. This may be determined by the following formula. The dose rate in mrad per hour at a distance R in cm from a point-source of gamma radiation at strength A in mCi and energy E_o in MeV, and with shielding of thickness t in cm of a material with a linear attenuation coefficient in cm^{-1}, is (as inferred from Eq. [2], [3], and [4] of Blizard (1958))

$$D(R,t) = 2.134 \times 10^6 B(\mu,t)\, (\mu_a/\rho)\, E_o[\mathrm{A} \exp(-\mu t)/(4\pi R^2)] \quad [49]$$

where $B(\mu,t)$ is the build-up factor, μ_a/ρ in cm^2/g is the energy-absorption mass-attenuation coefficient for the source energy and the material in which the dose is to be calculated (ρ is its density) and where the numerical constant has units of g $\times$ rad/MeV $\times$ mCi $\times$ hr). For gamma sources with more than one peak in an energy range that would contribute significantly to the dose rate, Eq. [49] must be applied to each peak and the dose rates summed.

For 0.662-MeV gamma radiation from a ^{137}Cs source and for 0.060-MeV gamma radiation from a ^{241}Am source, μ_a/ρ, measured in tissue, is about 0.0323 cm^2/g and 0.033 cm^2/g (Evans, 1968). For the same two gamma sources, μ is 0.717/cm and 46.4/cm, and the build-up factor $B(\mu,t)$ for lead thicknesses from 2 to 15 cm is approximated by $1.4 + 0.21t$ (a dimensionless factor) for the ^{137}Cs. The build-up factor for ^{241}Am is close to unity. The dose rate equations for ^{137}Cs and ^{241}Am inferred from data in Tables 1 and 4 (Blizard, 1958) and Evans (1968), are:

$$^{137}\mathrm{Cs}: D(R,t) = (5.1 + 0.76t)10^3 A \exp(-0.717t)/R^2 \text{ mrad/hour} \quad [50]$$

$$^{241}\mathrm{Am}: D(R,t) = 336.24 A \exp(-46.4t)/R^2 \text{ mrad/hour} \quad [51]$$

where R is the sum of the shield thickness, t, and the distance from the shield surface.

For a 500-mCi ^{137}Cs source with 13.6 cm of lead shielding, the radiation at the surface of the lead is about 2.5 mrad/hour, and at 1 m from the surface it is about 0.03 mrad/hour, or the equivalent of background radiation. Shielding required for a ^{241}Am source of comparable strength is approximately 3 mm, with radiation at 1 m being negligible compared to natural background.

The maximum exposure limitation given by the U. S. Nuclear Regulatory Commission is 1.25 rem/calendar quarter (the rem is the "roentgen equivalent, man") for persons 18 years of age or older. For gamma radiation, 1 rem is the equivalent of 1 rad, so that for a 7-hour exposure/day for 65 days (ordinary working days/quarter year), 0.03 to 2.5 mrad/hour results in an accumulation of 0.014 to 1.14 rad/quarter, well within U. S. regulations. For a more complete description of applicable regulations in the USA see part 20 of the Federal Code (U.S. Nuclear Regulatory Commission, 1985).

21–3.4.2.1.3 Gamma-sensitive Probe

Scintillation-type gamma-sensitive probes are the most satisfactory. A 5-cm phototube probe containing a preamplifier and equipped with a 2.5-cm thick thallium-activated sodium iodide crystal is adequate for many measurements. However, the crystal size may be increased for greater sensitivity. A second lead collimator about 5 cm long for ^{137}Cs (or much less for ^{241}Am), containing a thin slit or hole, is aligned with the source collimator, with an intervening space large enough to accommodate the soil container. The scintillation crystal, with a surrounding lead shield of approximately the same thickness as used around the source, is placed against the collimator. Where sequential measurements at two gamma energies are to be made, two probes may be desired.

21–3.4.2.1.4. Soil Container and Mechanism for Orienting Soil in Beam

Specifications for the soil container depend upon the nature of an experiment. However, two factors should be considered for optimum water content measurement. First, the container walls through which the beam will pass should be as thin as practical and of low density, and should not absorb water. Where a small source is to be used so that counting rate is limiting, it often is desirable to arrange holes in the container at desired counting positions, covering them with Mylar film. Second, the thickness of the soil through which the beam is to pass should be about 10 to 35 cm for ordinary soil density when ^{137}Cs is used as a gamma source. For ^{241}Am, thickness should be 2 to 8 cm. For dual gamma measurements the curves in Figure 21–5 should be used. Equation [61] in section 21–3.4.2.3 can be used to compute optimum soil column thickness for single gamma. The mechanism for positioning the soil container in the beam can take many forms, depending upon the nature of an experiment. In some cases, because of the weight of the lead shielding, it is easiest to move the soil container in the beam. A rack-and-pinion-operated sliding table or elevator works well.

21–3.4.2.1.5 Scaler or Rate Meter

A number of different types of scalers or rate meters can be used. A single- or multiple-channel analyzer with a built-in adjustable amplifier-discriminator and with a resolution time of 1 μs or less and a preset timer is desirable. Data acquisition systems, incorporating computer programs to accept and refine data, may be used for direct output of water contents and bulk densities.

21–3.4.2.2 Procedure. Before it is possible to infer water content

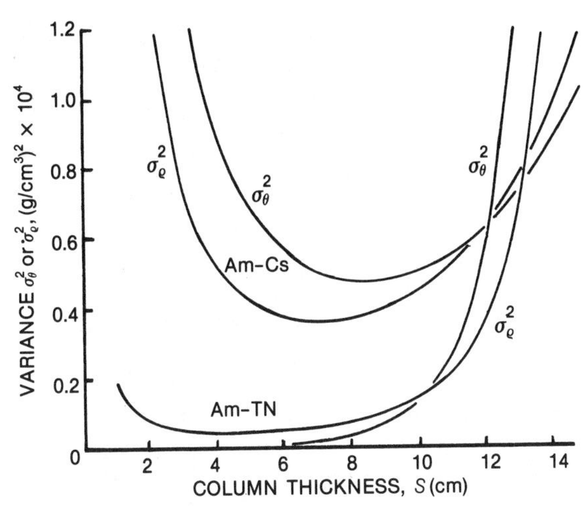

Fig. 21-5. Contribution to variance in bulk density or water content from random emission of radiation sources for double gamma measurements. Mass attenuation coefficients have the following values: for ^{241}Am μ_{sa} = 0.305 cm^2/g, μ_{wa} = 0.2036; for ^{137}Cs μ_{sb} = 0.0769, μ_{wb} = 0.0858; for thermal neutrons (TN) μ_{sn} = 0.4130 and μ_{wn} = 2.632.

from gamma ray measurements, it is necessary to evaluate the attenuation coefficient, μ_w, and the count through dry soil and the container, I_d, in Eq. [43], when a single gamma energy is to be used. Or, for concurrent measurement of water content and bulk density using two gamma energies with Eqs. [46] and [47], it is necessary to evaluate the attenuation coefficients for dry soil and for water, μ_s and μ_w, at the two different energies. Also, the gamma count for the container walls, I_c, must be obtained at the two energies and the column thickness, S, must be measured.

Measurements of I_d and I_c are similar, I_d being obtained by packing the soil column to be used with oven-dry soil and scanning the column over its length, using sufficiently long counting times to reduce the error due to random emission of gamma photons to any desired level (see section 21-3.4.2.3 for a discussion of errors). Air-dry soil may be used, providing that the count is corrected using Eq. [42], where θ is the water content of the soil determined gravimetrically, g/g, and converted to g/cm^3 through multiplication by the density of water, usually taken as unity. In single-energy systems values of I_d and I_c must be uniform throughout the length of a soil column under study, or their values at measuring positions along the column must be obtained and tabulated for use in water-content computations to be made from gamma ray counts at each position. Also, I_d may change with swelling or shrinking of the soil in wetting and drying situations, which introduces error in single-energy systems. Using container materials of uniform dimensions, I_c likely will be uniform over a column so that a single value may be obtained using the empty container. This value may then be used throughout a series of measurements. However, it is the ratio of the count in moist soil, I, to the count in the empty container, I_c, that is used in the attenuation equations for the dual-energy systems. This fact makes it possible to reduce the effect of any instrument drift on measurements by making periodic measurements of I_c and using them to obtain the ratio, I/I_c. This may be done using an absorber made from two pieces of the container wall material, so as to simulate the wall. Other absorbers may be used as well. Attenuation in air usually may be ignored, as it is negligible in most measurements involved with soil.

For dual gamma measurements, the values of the attenuation coefficients for soil and water must be obtained with considerable precision (see section 21–3.4.2.3). These measurements are made by placing soil in small containers having parallel walls at precisely known spacing, with the same geometry as that for the soil columns to be used. The bulk density is computed from mass and volume measurements. Careful counts are made on the containers without soil to determine I_c, and air-dry soil then is packed into the containers as uniformily as possible with a correction later made for the air-dry water content. (Oven-dry soil may be used to eliminate need for a water-content correction. However, absorption of moisture from the atmosphere during measurement may introduce some error.) Gamma counts then are made at several positions of the gamma beam in the soil, and the average value computed. Counting times may be set at values to produce any desired precision in the attenuation coefficients (the larger the count, the greater the precision). A similar measurement is made for water, with the values obtained usually being close to the theoretical value but usually not identical because of dependence upon the geometry of the attenuation system. (Mass attenuation coefficients for water generally are reported to be 0.197 and 0.088 cm^2/g for 0.060- and 0.662-MeV gamma rays. Experimental values for gamma beams used on soil columns often are from 1 to 6% less.) Eq. [43], with $\theta = 1$ g/cm^3 (usually) and I_d replaced by the container count, I_c, is used to compute the attenuation coefficients for water. The equation for computing the values for soil is

$$\mu_s = [-\mu_w\theta_{ad} - \ln(I_{ad}/I_c)/S]/\rho_s \quad [52]$$

where ρ_s is the dry bulk density in g/cm^3 computed for the container soil, θ_{ad} is the air-dry water content in g/cm^3, S is the thickness of the soil column in centimeters, and I_{ad}/I_c is the ratio of count in air-dry soil to count through the empty container.

21–3.4.2.3. Comments. The precision of gamma ray methods for measuring water content and bulk density varies with the thickness and density of the soil column, the adsorption characteristics of the soil, and the size of the gamma count in a moist and dry soil column. The error analysis of the single-gamma technique which follows provides useful design criteria for systems, including optimum column thickness. Since the variance in gamma-ray emission is the gamma-ray count itself, count intensity, I, must be replaced by the number of counts, It, where t is the counting time in Eq. [43]; thus,

$$\theta = -\ln (I_m t/I_d t)/(\mu_w S) \tag{53}$$

and the variance, with reasonable assumptions, becomes

$$\sigma_\theta^2 = \sigma^2_{\ln(Imt/Idt)}/(\mu_w^2 S^2) \tag{54}$$

so that the standard deviation is

$$\sigma_\theta = \sigma_{\ln(Imt/Idt)}/(\mu_w S) \tag{55}$$

To reduce the error, the counting time for $I_d t$, which is measured infrequently and usually under conditions where time is not pressing, is increased by a factor of 3 or 4 over that of $I_m t$. However, the computation requires that t be the same for both $I_m t$ and $I_d t$. Hence the longer count $I_d t'$ is reduced to comparable size through division by n, where $n = t'/t$. Thus $I_d t'/n$ is the time-equivalent count to $I_d t$ and may replace it in Eqs. [54] and [55].

The standard deviation of $\ln(I_m t/I_d t'/n)$ is

$$\sigma_{\ln}(I_m t/I_d t'/n) = \frac{I_d t'/n}{I_m t}\,\sigma(I_m t/I_d t'/n)$$

$$= \frac{I_d t'/n}{I_m t}\,\frac{1}{I_d t'/n}\left[\sigma^2_{I_m t} + \frac{(I_m t)^2}{(I_d t'/n)^2}\,\sigma^2_{(I_d t'/n)}\right]^{1/2}. \tag{56}$$

Using Eq. [56] with

$$\sigma_{Imt} = (I_m t)^{1/2} \text{ and } \sigma_{(Idt'/n)} = (I_d t')^{1/2}/n$$

Eq. [55] becomes

$$\sigma_\theta = 1/[\mu_w S\,(I_m t)^{1/2}]\,[1 + I_m t/(I_d t')]^{1/2}$$
$$= 1/(\mu_w S\,(I_m t)^{1/2})\,[1 + I_m t/(n I_d t)]^{1/2}. \tag{57}$$

Where n is 3 or 4, the second term under the radical may be neglected with only small effect on σ_θ (ca 10%), so that Eq. [57] becomes

$$\sigma_\theta = 1/[\mu_w S\,(I_m t)^{1/2}] \tag{58}$$

or, multiplying Eq. [40] by t/t and substituting into Eq. [58] yields

$$\sigma_\theta = \exp[S(\mu_s \rho_s + \mu_w \theta)/2 + S' \mu_c \rho_c]/[\mu_w S (I_0 t)^{1/2}]. \tag{59}$$

The optimum value of soil-column thickness, S, may be obtained by equating $d\sigma_\theta/dS$, obtained by differentiation of Eq. [59], to zero:

$$\frac{d\sigma_\theta}{ds} = \frac{\exp\left[S(\mu_s\rho_s + \mu_w\theta)/2\right]\exp(S'\mu_c\rho_c)}{\mu_w S(I_0 t)^{1/2}} \times [1/2(\mu_s\rho_s + \mu_w\theta) - (1/S)] = 0 \quad [60]$$

and

$$S = 2/(\mu_s\rho_s + \mu_w\theta) \quad [61]$$

For ^{241}Am with $\mu_s = 0.307$ and $\mu_w = 0.195$ cm^2/g and for a bulk density of 1.2 g/cm^3, the optimum soil-column thickness ranges between 5 and 4 cm for low to high water-contents. Similarly, for ^{137}Cs with $\mu_s = 0.076$ and $\mu_w = 0.082$ cm^2/g, the optimum soil-column thickness ranges from 20 to 15 cm.

A complete error analysis for use with Eq. [46] and [47] for dual gamma measurements is given in Gardner et al., (1972). However, a brief discussion of the two most important factors involved in the design of experimental soil columns—errors associated with random gamma ray emission which determine the optimum column thickness, and errors in measurement of column thickness—is given here. The error equation may be written

$$\sigma_\theta^2 = \left(\frac{\partial\theta}{\partial I_a t}\right)^2\sigma_{I_a}^2 + \left(\frac{\partial\theta}{\partial I_{ca} t}\right)^2\sigma_{I_{ca}}^2 + \left(\frac{\partial\theta}{\partial I_b t}\right)^2\sigma_{I_b}^2 + \left(\frac{\partial\theta}{\partial I_{cb} t}\right)^2\sigma_{I_{cb}}^2 + \left(\frac{\partial\theta}{\partial S}\right)^2\sigma_S^2 + \left(\frac{\partial\theta}{\partial\mu_{sa}}\right)^2\sigma_{\mu_{sa}}^2 + \left(\frac{\partial\theta}{\partial\mu_{sb}}\right)^2\sigma_{\mu_{sb}}^2 + \left(\frac{\partial\theta}{\partial\mu_{wa}}\right)^2\sigma_{\mu_{wa}}^2 + \left(\frac{\partial\theta}{\partial\mu_{wb}}\right)^2\sigma_{\mu_{wb}}^2 . \quad [62]$$

For bulk density, θ is replaced by ρ. Carrying out the partial differentiation indicated in this equation yields appropriate equations giving the contribution to variance in water content or bulk density of each of the parameters in terms of a variance which may be assumed for measurement of that particular parameter. Gamma ray emission follows a Poisson distribution for which the variance, σ_{It}^2, is the number of counts, It. Hence, it is possible, for a particular count, to obtain relationships between variances in water content or bulk density and variances associated with (i) the container in the absence of soil (this count to be made over appropriate times so as to be the same for both gamma sources), (ii) the various absorption coefficients, (iii) the water content and bulk density, and (iv) the thickness of the soil column. For 10^6 counts through the empty container and mid-range values of water content and bulk density (0.15 and

1.2 g/cm³), the curves for variance in water content and in bulk density are shown in Fig. 21–5 in terms of soil-column thickness. Curves also are given for the combination of ^{241}Am and thermal neutrons. From these curves it may be observed that optimum column thickness for bulk density is about 7 cm (a range of 5–10 cm) and for water content is about 8.5 cm (average of about 7–10 cm). A similar analysis involving variation in column thickness from a designated value (a bias rather than a variance for repeated measurements at the same position) shows that for an accuracy contribution in water content of 0.001 g/cm³ the error in column thickness must be less than 0.67%, or for 0.01 g/cm³ less than 6.7%. Similarly, for an accuracy contribution in bulk density of 0.001 g/cm³, the error in column thickness must be less than 0.083%, or for 0.01 g/cm³ less than 0.83%. For water and homogeneous soil the attenuation coefficients may be determined by repeated measurements to any desired accuracy, using a particular gamma-scanning set-up. Hence these factors do not enter into design, but the relevant accuracy evaluation may be carried out similarly using the appropriate term in Eq. [62].

The above analyses assume a rectangular soil column. If a cylindrical column is used it is obvious that the attenuation coefficients must be carefully made on cylindrical samples, with the column positioned precisely in the gamma beam to avoid error. (The difference in column thickness over a gamma-beam cross-section would be absorbed in the attenuation coefficients in this case.)

For all gamma counting, a correction for the resolving time of the gamma-ray counting system must be used. This is the time between successive counts during which an additional count cannot be recorded (see Fritten, 1969; Gardner et al., 1972). A corrected count rate, N, for the counting period used may be computed using the equation

$$N = R/(1 - \tau R) \qquad [63]$$

where R is the observed count rate over the prescribed counting time and τ is the dead time per count. The value of τ may be determined experimentally using particular equipment in the range of settings to be involved and by making counts through air, R_0, separately for each of two iron blocks of equal thickness (thickness such as to give a counting rate at least as high as that expected in measurements to be made) and averaging them, R_1; and making counts for both blocks together, R_2. The equation for τ (Gardner et al., 1972) is

$$\tau = (R_1^2 - R_0R_2)/[R_1^2(R_0 + R_2) - 2R_0R_1R_2]\,. \qquad [64]$$

The basic gamma count rate I_0 with only air between the source and scintillation crystal, and hence the precision and resolution possible in water-content measurement, depends upon the strength of the gamma source, the range of gamma energies counted, the geometry of the equipment, and the efficiency with which gamma radiation is measured. The source emits gamma rays in all directions, and the intensity of the gamma field decreases with the square of the distance from the source (neglecting the absorption of air in the path and assuming a point source) (section 21–3.4.1). Under these conditions, the fraction of the radiation reaching the face of the scintillation crystal depends upon the degree of collimation and the distance from the source. With an absorber in the beam, some scattered and secondary radiation reaches the crystal so that the inverse square law no longer holds. However, if a single-channel or multichannel analyzer or a discriminator is used so that only gamma rays of the maximum energy emitted from the source are counted, the effect of scattered and secondary radiation is eliminated.

With the gamma-ray detector set to count only a narrow range of gamma energies, the resolution possible for given precision, source strength, and experimental arrangement can be computed. If the area of the collimator at the face of the crystal is a (assuming this to be the minimum collimator restriction) and the distance from the crystal face is r, then the fraction of the total radiation reaching the crystal, neglecting the absorption of air in the path, is $a/(4\pi r^2)$. The count rate I_0 is proportional to the product of this fraction and the strength of the source Z:

$$I_0 = k(a/4\pi r^2)Z \qquad [65]$$

with

$$k = 3.70 \times 10^7 f[1 - \exp(-\mu_p s)] \qquad [66]$$

where the constant, 3.70×10^7, is the number of disintegrations per second associated with 1 mCi of any radioactive nuclide; f is the number of photoelectrons produced per disintegration (0.851 for ^{137}Cs and 0.359 for ^{241}Am), μ_p is the linear attenuation coefficient for the thallium-activated sodium iodide crystal, which is the fractional decrease in intensity per unit length traversed, s. The value of μ_p is 0.0352 cm^{-1} for 0.62-MeV photons of ^{137}Cs and 22.12 cm^{-1} for 0.060-MeV photons of ^{241}Am. For a scintillation crystal approximately 5 cm in diameter and 5 cm deep, k has the value of 5.08×10^6 photoelectrons s^{-1} mCi^{-1} for ^{137}Cs and 13.18×10^6 photoelectrons s^{-1} mCi^{-1} for ^{241}Am. Eq. [65] gives the maximum count rate obtainable in air for a particular equipment geometry, collimation, and source strength. Using narrow collimation, values considerably less than this may be expected where perfect alignment of collimators is difficult to achieve.

Replacing the count N_0 in Eq. [59] by its equivalent $I_0 t$, substituting this into Eq. [65], and solving explicitly for the source strength leads to

$$Z = 4\pi r^2 \exp[S(\mu_s\rho_s + \mu_w\theta) + 2S'\mu_c\rho_c]/(ka\mu_w^2 S^2\sigma_\theta^2 t). \qquad [67]$$

For a typical situation at high water content (0.4 g/cm^3) and bulk density (1.3 g/cm^3) where the error is largest, a 10-cm wide soil column contained in a plastic box with 0.62-cm thick walls having a mass attenuation coefficient of 0.08747 cm^2/g and a density of 1.19 g/cm^3 would require a 200-mCi ^{137}Cs source for 60-s counting to yield a precision of ± 0.5% water content ($3\sigma_\theta = 0.005$ g/cm^3), with tapering collimation at the source and on the detector side having a slit cross-section of 1 by 40 mm at the face of a 5-cm deep scintillation crystal. With imperfect alignment of a narrow slit, the count rate would be significantly reduced, and a somewhat larger source strength would be required. The attenuation coefficients used in the above computation are 0.07785 and 0.08559 cm^2/g for soil and water, and the value of k is 5.081×10^6 photoelectrons/s.

If long counting-times are practicable in an experiment, high precision is possible with relatively small sources and at high resolution. However, if rapid measurement is important, then either the resolution must be low or the source strength high. In such cases, since k increases with increasing crystal size, large scintillation crystals are desirable. If resolution and counting time are not important factors, as in an experiment involving measurement of gross water content in a container of soil with growing plants, source strengths of the order of 20 mCi may be adequate. Also, small crystals are considerably less expensive than large ones, and crystals of the order of 2 to 3 cm may often be adequate under such conditions.

In the foregoing discussion it has been assumed that the soil traversed by the gamma beam had spatially uniform physical and chemical properties so that the density and attenuation coefficients could be represented by single values. In stratified soil this is not always the case, particularly in systems that change with time, and the interpretation of calculated bulk densities and water contents becomes complicated. Special problems to be encountered in measurements made in stratified soil and some methods for dealing with them are discussed by Goit et al. (1978) and by Nofziger (1978).

21-4 REFERENCES

Anderson, A. B. C., and N. E. Edlefsen. 1942. Electrical capacity of the 2-electrode plaster of Paris block as an indicator of soil moisture content. Soil Sci. 54:35–46.

Ashton, F. M. 1956. Effects of a series of cycles of alternating low and high soil water contents on the rate of apparent photosynthesis in sugar cane. Plant Physiol. 31:266–274.

Belcher, D. J., T. R. Cuykendall, and H. S. Sack. 1950. The measurement of soil moisture and density by neutron and gamma ray scattering. Civil Aeron. Admin. Tech. Develop. Rep. no. 127. Washington, DC.

Blizard, E. P. 1958. Nuclear radiation shielding. *In* H. Etherington (ed.) Nuclear engineering handbook. McGraw-Hill, New York.

Bloodworth, M. E., and J. B. Page. 1957. Use of thermistors for the measurement of soil moisture and temperature. Soil Sci. Soc. Am. Proc. 21:11–15.

Bouyoucos, G. J. 1949. Nylon electrical resistance unit for continuous measurement of soil moisture in the field. Soil Sci. 67:319–330.

Bouyoucos, G. J. 1953. More durable plaster of Paris moisture blocks. Soil Sci. 76:447–451.

Bouyoucos, G. J., and A. H. Mick. 1940. An electrical resistance method for the continuous measurement of soil moisture under field conditions. Michigan Agric. Exp. Stn. Tech. Bull. 172. East Lansing, MI.

Cannell, G. H. 1958. Effect of drying cycles on changes in resistance of soil moisture units. Soil Sci. Soc. Am. Proc. 22:379–382.

Cannell, G. H., and C. W. Asbell. 1964. Prefabrication of mold and construction of cylindrical electrode-type resistance units. Soil Sci. 97:108–112.

Cline, M. G. 1944. Principles of soil sampling. Soil Sci. 58:275–288.

Colman, E. A., and T. M. Hendrix. 1949. Fiberglass electrical soil-moisture instrument. Soil Sci. 67:425–438.

Corey, J. C., S. F. Peterson, and M. A. Wakat. 1971. Measurement of attenuation of ^{137}Cs and ^{241}Am gamma rays for soil density and water content determinations. Soil Sci. Soc. Am. Proc. 35:215–219.

Crumpler, T. B., and J. H. Yoe. 1946. Chemical computations and errors. John Wiley & Sons, Inc., New York.

Cummings, R. W., and R. F. Chandler, Jr. 1940. A field comparison of the electrothermal and gypsum block electrical resistance methods with the tensiometer method for estimating soil moisture in situ. Soil Sci. Soc. Am. Proc. 5:80–85.

Davidson, J. M., J. W. Biggar, and D. R. Nielsen. 1963. Gamma-radiation attenuation for measuring bulk density and transient water flow in porous materials. J. Geophys. Res. 68:4777–4783.

de Plater, C. V. 1955. A portable capacitance-type soil moisture meter. Soil Sci. 80:391–395.

de Vries, J., and K. M. King. 1961. Note on the volume of influence of a neutron surface moisture probe. Can. J. Soil Sci. 41:253–257.

Edlefsen, N. E., and A. B. C. Anderson. 1941. The four-electrode resistance method for measuring soil moisture content under field conditions. Soil Sci. 51:367–376.

England, C. B. 1965. Changes in fiber-glass soil moisture-electrical resistance elements in long-term installations. Soil Sci. Soc. Am. Proc. 29:229–231.

Evans, R. D. 1968. X-ray and γ-ray interactions. *In* F. H. Attix and W. C. Roesch (ed). Radiation dosimetry, Vol. 1, Fundamentals. 2nd ed. Academic Press, New York.

Ferguson, A. H., and W. H. Gardner. 1962. Water content measurement in soil columns by gamma ray absorption. Soil Sci. Am. Proc. 26:11–14.

Fletcher, J. E. 1939. A dielectric method for determining soil moisture. Soil Sci. Soc. Am. Proc. 4:84–88.

Fritton, D. D. 1969. Resolving time, mass absorption coefficient, and water content with gamma ray attenuation. Soil Sci. Soc. Amer. Proc. 33:651–655.

Gardner, W. H., and C. Calissendorff, 1967. Gamma ray and neutron attenuation measurement of soil bulk density and water content. p. 101–113. *In* Isotope and radiation techniques. Proc. of Symp. Techniques in Soil Physics and Irrigation Studies, Instanbul, IAEA Vienna.

Gardner, W. H., G. S. Campbell, and C. Calissendorff. 1972. Systematic and random errors in dual gamma energy soil bulk density and water content measurements. Soil Sci. Soc. Am. Proc. 36:393–398.

Gardner, Wilford, and D. Kirkham. 1952. Determination of soil moisture by neutron scattering. Soil Sci. 73:391–401.

Gee, G. W., and M. E. Dodson. 1981. Soil water content by microwave drying: a routine procedure. Soil Sci. Soc. Am. J. 45:1234–1237.

Gilbert, P. A. 1974. Microwave oven used for rapid determination of soil water contents. Rep. No. 13, Misc. paper No. 3-478, U. S. Army Engineer Waterways Exp. Stn., Vicksburg, MS.

Goit, J. B., P. H. Groenevelt, B. D. Kay, and J. G. P. Loch. 1978. The applicability of dual gamma scanning to freezing soils and the problem of stratification. Soil Sci. Soc. Am. J. 42:858–863.

Greacen, E. L., and C. T. Hignett. 1979. Sources of bias in the field calibration of a neutron meter. Aust. J. Soil Res. 17:405–415.

Gurr, C. G. 1962. Use of gamma rays in measuring water content and permeability in unsaturated columns of soil. Soil Sci. 94:224–229.

Gurr, C. G., and T. J. Marshall. 1960. Unsaturated permeability—its measurement and its estimation from other properties of the material. Trans. Int. Congr. Soil Sci., 7th. 1:306–310.

Hankin, L., and B. L. Sawhney. 1978. Soil moisture determination using microwave radiation. Soil Sci. 126:313–315.

Hanks, R. J., and S. A. Bowers. 1960. Neutron meter access tube influences soil temperature. Soil Sci. Soc. Am. Proc. 24:62–63.

Hewlett, J. D., J. E. Douglass, and J. L. Clutter. 1964. Instrumental and soil moisture variance using the neutron-scattering method. Soil Sci. 97:19–24.

Holmes, J. W. 1956. Calibration and field use of the neutron scattering method of measuring soil water content. Aust. J. Appl. Sci. 7:45–58.

Holmes, J. W. 1966. Influence of bulk density of the soil on neutron moisture meter calibration. Soil Sci. 102:355–360.

Holmes, J. W., and A. F. Jenkinson. 1959. Techniques for using the neutron moisture meter. J. Agric. Eng. Res. 4:100–109.

Hsieh, J. J. C., W. H. Gardner, and G. S. Campbell. 1972. Experimental control of soil water content in the vicinity of root hairs. Soil Sci. Soc. Am. Proc. 36:418–421.

Kelley, O. J. 1944. A rapid method for calibration of various instruments for measuring soil moisture in situ. Soil Sci. 58:433–440.

Kirkham, D., and G. S. Taylor. 1950. Some tests of a four-electrode probe for soil moisture measurement. Soil Sci. Soc. Am. Proc. 14:42–46.

Kittrick, J. A., and E. W. Hope. 1970. Preventing water resorption in weight loss determinations. Soil Sci. Soc. Am. Proc. 34:536–537.

Lal, R. 1974. The effect of soil texture and density on the neutron and density probe calibration for some tropical soils. Soil Sci. 117:183–190.

Lambe, T. W. 1949. How dry is a dry soil? Proc. 29th Annual Meeting, Highway Research Board 29:491–496.

Makower, B. 1950. Determination of water in some dehydrated foods. Adv. Chem. 3:37–54.

Makower, B. and E. Nielsen. 1948. Use of lyophilization in determination of moisture content of dehydrated vegetables. Anal. Chem. 20:856–858.

Makower, B., S. M. Chastain, and E. Nielsen. 1946. Moisture determination in dehydrated vegetables. Vacuum-oven method. Ind. Eng. Chem. 38:725–731.

McHenry, J. R. 1963. Theory and application of neutron scattering in the measurement of soil moisture. Soil Sci. 95:294–307.

Miller, R. J., R. B. Smith, and J. W. Biggar. 1974. Soil water content: microwave oven method. Soil Sci. Soc. Am. Proc. 38:535–537.

Nofziger, D. L. 1978. Errors in gamma-ray measurement of water content and bulk density in nonuniform soils. Soil Sci. Soc. Am. J. 42:845–850.

Nofziger, D. L., and D. Swartzendruber. 1974. Material content of binary physical mixtures as measured with a dual-energy beam of gamma rays. J. Appl. Phys. 45:5443–5449.

Nutting, P. G. 1943. Some standard thermal dehydration curves of minerals. U. S. Geol. Survey, Prof. Paper 197-E.

Olgaard, P. L. 1965. On the theory of the neutronic method for measuring the water content of soil. Danish Atomic Energy Comm., Riso Rep. 97.

Perrier, E. R., and A. W. Marsh. 1958. Performance characteristics of various electrical resistance units and gypsum materials. Soil Sci. 86:140–147.

Phene, C. J., G. J. Hoffman, and S. L. Rawlins. 1971a. Measuring soil matric potential in situ by sensing heat dissipation within a porous body: I. Theory and sensor construction. Soil Sci. Soc. Am. Proc. 35:27–32.

Phene, C. J., G. J. Hoffman, and S. L. Rawlins. 1971b. Measuring soil matric potential in situ by sensing heat dissipation within a porous body: II. Experimental results. Soil Sci. Soc. Am. Proc. 35:225–229.

Phillips, R. E., C. R. Jensen, and D. Kirkham. 1960. Use of radiation equipment for plow-layer density and moisture. Soil Sci. 89:2–7.

Rawlins, S. L., and W. H. Gardner. 1963. A test of the validity of the diffusion equation for unsaturated flow of soil water. Soil Sci. Soc. Am. Proc. 27:507–511.

Reginato, R. J. 1974. Gamma radiation measurements of bulk density changes in a soil pedon following irrigation. Soil Sci. Soc. Am. Proc. 38:24–29.

Reginato, R. J., and C. H. M. van Bavel. 1964. Soil-water measurement with gamma attentuation. Soil Sci. Soc. Am. Proc. 28:721–724.

Richards, L. A., and H. T. Stumpf. 1960. Volumetric soil sampler. Soil Sci. 89:108–110.

Richards, L. A., and L. R. Weaver. 1943. The sorption-block soil moisture meter and hysteresis effects related to its operation. J. Am. Soc. Agron. 35:1002–1011.

Shaw, B. T., and L. D. Baver. 1940. An electrothermal method for following moisture changes in soil in situ. Soil Sci. Soc. Am. Proc. (1939) 4:78–83.

Sinclair, D. F., and J. Williams. 1979. Components of variance involved in estimating soil water content and water content change using a neutron moisture meter. Aust. J. Soil Res. 17:237–47.

Soane, B. D. 1967. Dual energy gamma-ray transmission for coincident measurement of water content and dry bulk density of soil. Nature (London) 214:1273–1274.

Stewart, G. L., and W. H. Gardner. 1969. Water content measurement by neutron attenuation for application to study of unsaturated flow of soil water. AEC Rep. No. RLO-1543-5. Washington, DC.

Stolzy, L. H., and G. A. Cahoon. 1957. A field-calibrated portable neutron rate meter for measuring soil moisture in citrus orchards. Soil Sci. Soc. Am. Proc. 21:571–575.

Stone, J. F., R. H. Shaw, and D. Kirkham. 1960. Statistical parameters and reproducibility of the neutron method of measuring soil moisture. Soil Sci. Soc. Am. Proc. 24:435–438.

Tanner, C. B., E. Abrams, and J. C. Zubriski. 1949. Gypsum moisture-block calibration based on electrical conductivity in distilled water. Soil Sci. Soc. Am. Proc. 13:62–65.

Taylor, S. A. 1955. Field determinations of soil moisture. Agric. Eng. 36:654–659.

Taylor, S. A., D. D. Evans, and W. D. Kemper. 1961. Evaluating soil water. Utah Agric. Exp. Stn. Bull. 426.

Topp, G. C. 1980. Electromagnetic determination of soil water content: measurements in coaxial transmission lines. Water Resources 16:574–582.

U. S. Nuclear Regulatory Commission. 1985. Rules and regulations, Title 10, Chap. 1, Code of Federal Regulations, Part 20, Standards for protection against radiation. U.S. Supt. of Public. Doc. Washington, D.C.

Vachaud, G., J. M. Royer, and J. D. Cooper. 1977. Comparison of methods of calibration of a neutron probe by gravimetry or neutron-capture model. J. Hydrol. 34:343–356.

Van Bavel, C. H. M., D. R. Nielsen, and J. M. Davidson. 1961. Calibration and characteristics of two neutron moisture probes. Soil Sci. Soc. Am. Proc. 25:329–334.

Van Bavel, C. H. M., N. Underwood, and R. W. Swanson. 1956. Soil moisture measurement by neutron moderation. Soil Sci. 82:29–41.

Visvalingam, M., and J. D. Tandy. 1972. The neutron method for measuring soil moisture content—a review. J. Soil Sci., 23:499–511.

Weinberg, A. M., and E. P. Wigner. 1958. The physical theory of neutron chain reactors. Univ. of Chicago Press, Chicago.

Zuber, A. and J. F. Cameron. 1966. Neutron soil moisture gauges. Atomic Energy Rev. 4:143–167.

22 Water Potential: Piezometry

R. C. REEVE
Agricultural Research Service, USDA
Columbus, Ohio

22–1 INTRODUCTION

Peizometry is a useful technique for measuring important parameters of hydraulic flow systems. Of particular interest is the measurement of "hydraulic head," a concept of fluid mechanics that pertains to the energy status of water in water flow systems. It is useful for describing flow not only in conduits and other hydraulic structures, but also in soils and other porous media. It is a concept that simplifies the description of some liquid flow phenomena and often is relatively easy to measure. "Hydraulic head" in water flow systems is analogous to "potential" or "voltage" in electrical flow problems and to "temperature" where heat flow is involved. Hydraulic-head measurements are particularly useful in determining the direction of flow of water in soils. Evaluating the space rate of change of hydraulic head is required for applying the Darcy equation to solve various flow problems (see chapters 29 and 30). Both the Darcy flow equation and hydraulic head measurements are used extensively in hydrology, hydraulics, irrigation, drainage, and other fields of science where flow of water in the liquid state is involved.

The piezometric methods described herein pertain particularly to the measurement of hydraulic head in soils. Methods are described for measuring hydraulic head both above and below a water table. While the interpretation of hydraulic-head readings is much the same for both cases, the equipment and procedures are quite different. In general, the measurement of hydraulic head above a water table, where an equivalent pressure that is less than atmospheric pressure is measured, is more difficult than below a water table where hydrostatic pressure in bulk water is usually involved. Two procedures are described for installing piezometers.

Chapter 23 should be consulted for directions and information on the use of tensiometers. Soil-suction and hydraulic-head measurements are discussed therein; however, hydraulic-head measurements are treated here in somewhat greater detail. An attempt is made to relate hydraulic-head measurements made with piezometers to those made with tensiometers in order that water flow in a given system involving both saturated and unsaturated flow can be evaluated. The method of installing tensiometers is the same whether suction values are desired or whether hydraulic head is to be measured. The principal difference is in the setting of the scales.

22-2 PRINCIPLES

The defining equation for hydraulic head comes from a consideration of the law of conservation of energy as applied to a liquid system by Bernoulli in 1738. This subject is treated in standard textbooks on fluid mechanics such as Dodge and Thompson (1937).

The Bernoulli equation describes the energy status of a flowing liquid in terms of kinetic, potential, and pressure energies. When the energy is expressed as energy per unit weight of water, it has the physical dimensions of length (L). This length, which is a vertical distance, i.e., parallel to the gravity force field, is termed "head."

For unit weight of water located at a point where the pressure is p, the velocity is v, and the elevation above a reference level is z, the *hydraulic head* h at the point in question in a steady flow system is as follows:

$$h = (v^2/2g) + (p/w) + z \quad [1]$$

where g = the acceleration due to gravity, and w = specific weight of water ($w = \rho g$, where ρ = density of water). The individual components of the equation are *velocity head* ($v^2/2g$), *pressure head* (p/w) and *position head* (z), representing the kinetic, pressure-potential, and position-potential energies, respectively. For flow of water in soils or other porous media, flow velocities are usually very low; and, for all practical purposes, the velocity head can be neglected. The hydraulic-head equation then becomes.

$$h = p/w + z\,. \quad [2]$$

The quantities of Eq. [2] are illustrated in Fig. 22-1 for (A) saturated conditions and (B) unsaturated conditions. Piezometers are used for measuring hydraulic head for the saturated case. The piezometer pipe makes connection with the soil water through the open end of the pipe as shown at point "A", Fig. 22-1A. The pressure head is the length of the water column in the pipe above point "A" and in this case is positive. In accordance with Eq. [2] the hydraulic head at point "A" is equal to the sum of the pressure head p_A/w and the position head z_A; or, in other words, it is the height the water level stands in an open pipe above the reference elevation.

For the unsaturated case, a porous membrane, which is usually a ceramic body (Richards, 1942), is required for hydraulic connection be-

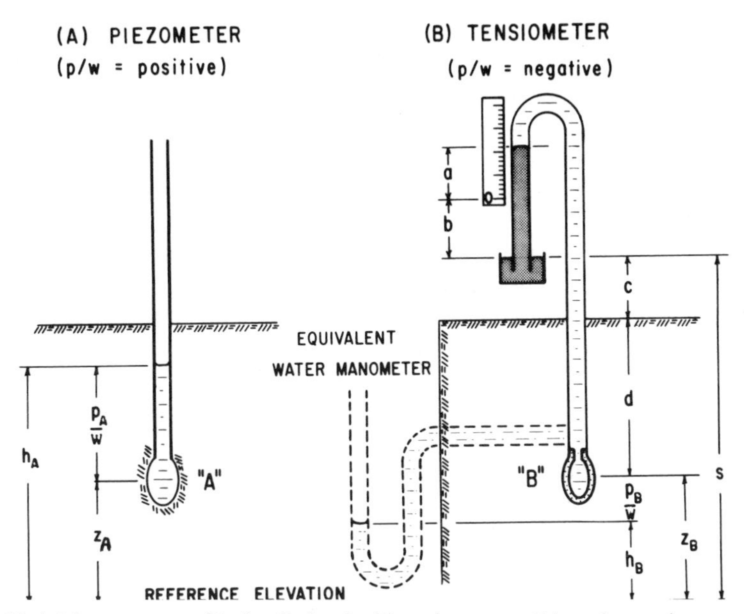

Fig. 22–1. Measurement of hydraulic head with a piezometer (A), and a tensiometer (B).

tween the liquid water in the soil and that in the tensiometer. The subatmospheric pressure in the tensiometer is measured with a mercury manometer, or a Bourdon vacuum gauge, or other vacuum-sensing device. The ceramic body or cup is required to restrict the entry of air into the tensiometer, where the pressure in the water is less than atmospheric, and at the same time allow free movement of water to and from the tensiometer to attain hydraulic equilibrium with the soil water. Specifications for ceramic cups and other tensiometer parts, together with procedures for the proper use of tensiometers, are given in chapter 23.

A reference elevation is required as a base for hydraulic-head measurements in any given flow system. It is convenient to select a reference elevation at some depth below the lowest hydraulic-head value in the system. Mean sea level is often used for this purpose, but any arbitrary reference elevation that is convenient will do. Hydraulic head increases in a positive direction upward from this reference elevation.

Tensiometers have been used for many years to read soil suction. To avoid negative numbers, it has become customary to consider the soil suction as a positive value. This convention is quite satisfactory for many purposes; but for evaluating the energy status of water in a flow system for the purpose of determining direction of flow and evaluating driving forces or hydraulic gradients in a gravity field, it is necessary to consider this value in relation to the position-potential energy component z_B, which by convention increases in an upward direction opposite to the direction of the gravity force. With z_B positive in an upward direction, the pressure head in unsaturated soil is a negative value as is illustrated in Fig. 22–

1B by the phantom water manometer shown in dashed lines. Note that the water surface in the phantom manometer is below the point of measurement, thus indicating that the equivalent pressure in the soil water is less than atmospheric; i.e., the pressure-head value is negative. The hydraulic head h_B is the algebraic sum of the pressure head p_B/w and the position head z_B in accordance with Eq. [2].

It is impractical to use a water manometer, such as illustrated in Fig. 22–1B, to measure this negative pressure head because to do so would require that the measurement be made below the soil surface. In practice, a manometer with a fluid of density greater than that of water, such as mercury, is used to measure remotely either the hydraulic head or the negative pressure head at some depth in the soil from a position above the soil surface.

Whether the tensiometer reads the negative pressure-head component or the hydraulic head depends upon the initial positioning of the tensiometer scale. The scale may be positioned to read zero opposite the mercury level of the manometer when any desired reference pressure or hydraulic head occurs at the cup. To read only the negative pressure-head component, the reference elevation is taken at the center of the cup. For hydraulic-head readings, the scale zero is set to correspond to zero hydraulic head at some convenient elevation, such as the soil surface, and all tensiometers in a given system are zeroed to the same elevation.

If s is the distance from the reference elevation to the surface of the mercury in the mercury reservoir (Fig. 22–1B), ρ_m = density of mercury, and ρ_w = density of water, then the height of the scale zero above the mercury surface b is given by the equation

$$b = s(\rho_w)/(\rho_m - \rho_w)\,. \qquad [3]$$

In the derivation of this equation, it is assumed that the mercury level in the mercury pot remains constant. If the surface area of the mercury pot is large compared to the area of the mercury manometer, the correction required because of the change in level of the mercury in the mercury reservoir is small and for many purposes can be neglected. There is also a depression of the mercury column as a result of surface tension forces in the manometer tube. A correction for this effect may also be made, but it also is negligible for most purposes and is not included in Eq. [3]. For additional information on this subject, the paper by Richards (1949) may be consulted.

If the reference elevation is to be set at the center of the tensiometer cup, s is taken equal to $(c + d)$, and Eq. [3] becomes

$$b = (c + d)(\rho_w)/(\rho_m - \rho_w)\,. \qquad [4]$$

When a mercury tensiometer is used to measure hydraulic head, a scale calibrated to read in length units should be specified. The centimeter is the unit most commonly used; however, any desired length unit may be selected. For reading soil suction values, manometer scales are commonly calibrated in millibars (1 bar = 1×10^6 dynes per cm^2 = 10^3 millibars = 10^2 kPa). For all practical purposes, the millibar scale may be used as a centimeter scale (1 millibar = 1.0227 cm of water at 21°C). When properly zeroed, such a scale may be used to read hydraulic head directly. The ratio of scale length to head is

$$a/h_B = (\rho_w)/(\rho_m - \rho_w) \quad [5]$$

where a = height of the mercury column above the scale zero (Fig. 22–1B). Thus for a mercury manometer, $\rho_m = 13.5 \text{ g cm}^{-3}$, $\rho_w = 1 \text{ g cm}^{-3}$ and $a/h_B = 1/12.5$.

The working range for tensiometers is from zero to about two-thirds of the negative pressure required for complete vacuum, or from zero to about 700 cm of water. Since the zero setting of the tensiometer scale (length b, Fig. 22–1B) utilizes a part of the working range of the tensiometer scale, it is desirable to keep b relatively small to maintain the usable scale range as large as possible. For this reason, it is helpful to select a reference elevation that is not far distant from the elevation of the mercury reservoir. The ground surface is often used for this purpose. All tensiometers in a given system should be referenced to the same elevation. Hydraulic-head readings referred to one elevation can be transferred to another reference elevation by adding or subtracting the elevation difference as the case may be.

Tensiometers and piezometers have a response time that is not zero because of the volume of water that must move into or out of the piezometer or tensiometer to register a pressure change that occurs in the soil water. In the case of the tensiometer, this response time depends primarily upon the sensitivity of the manometer, the conductance of the porous cup, and the hydraulic conductivity of the soil in which the tensiometer is placed. For piezometers, the diameter of the pipe, the size and shape of the cavity at the base of the pipe, and the hydraulic conductivity of the soil are the major factors. For static hydraulic-head conditions in the soil, the time required for the instrument to register the hydraulic pressure of the soil, following a head displacement within the instrument, depends upon this response rate. Where hydraulic head in the soil changes with time, there will be a time-lag in instrument readings that also is related to this response rate, and for which proper correction should be made. It is beyond the scope of this paper to provide the details required for this correction. The aim here is merely to call this source of error to the attention of the reader. Additional information can be obtained on this subject by consulting the papers by Klute and Gardner (1962) for tensiometers and Hvorslev (1951) for piezometers. Null-type tensiometers that eliminate the response-time problem have been designed and used in special research studies (Miller, 1951; Leonard & Low,

1962). Chapter 23 should also be consulted for details of the use of pressure transducers in tensiometry.

Piezometers are installed both by driving and jetting. In general, the driving method is limited to relatively shallow depths (up to 8 to 10 m), whereas piezometers may be jetted to depths of 30 to 50 m or greater. A qualitative log of subsoil materials penetrated is obtained in the process of jetting piezometers, which is not possible by the driving method. The choice of method of installation will, therefore, depend primarily upon the nature of the problem under study and the information that is desired. The equipment needed and the procedures required are different for these two methods. Therefore both methods are described in the following sections, and the choice of method is left to the operator. For additional information the papers by Christiansen (1943), Reeve and Jensen (1949), Reger et al. (1950), Donnan and Bradshaw (1952), U. S. Salinity Laboratory Staff (1954), and Mickelson et al. (1961) should be consulted.

22–3 METHOD OF INSTALLING PIEZOMETERS BY DRIVING

22–3.1 Special Apparatus (Fig. 22–2)

1. Driving hammer, fence-post type. Nominal pipe sizes: (a) ¾ inch; (b) ¾ inch, (c) 1½ inch.

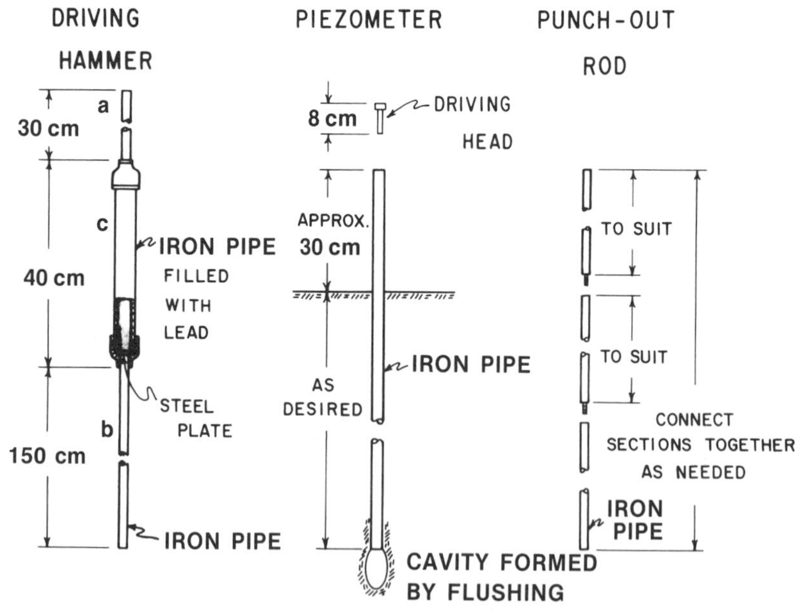

Fig. 22–2. Special equipment for installing piezometers by driving (Christiansen, 1943). Select iron pipe sizes to accommodate piezometer tube. Driving hammer (approx. 7–10 kg) to fit loosely over piezometer. Punch-out rod to fit loosely inside piezometer.

2. Driving head.
3. Punch-out rod. Nominal pipe size: 1/8 inch.
4. Rivets to fit freely in piezometer pipe (2 needed per piezometer).
5. Piezometer pipe, black or galvanized iron, 1.3 cm (3/8-inch nominal) inside diameter, lengths as needed. (The actual inside diameter of 3/8-inch pipe is slightly greater than 1.3 cm.) Other sizes may be used.

22–3.2 Procedure

Cut the pipe into desired lengths (3–4 meter lengths are convenient), and mark the pipe in 30-cm intervals. Place a rivet in the lower end of the pipe and insert the driving head in the upper end. Using the drive hammer first with the long end of the hammer (end b, Fig. 22–2) over the piezometer pipe, and lastly with the short end of the hammer over the piezometer pipe (end a, Fig. 22–2), drive the pipe to the desired depth in the soil. With punch-out rod, push rivet about 5 cm beyond the lower end of the piezometer pipe.

22–3.3 Comments

Pipe lengths up to about 4 m can be started easily and driven in the manner described in the procedure. Greater pipe-lengths can be started by use of a stepladder. Additional lengths can be added with standard pipe couplings as driving progresses, until the pipe reaches the desired depth in the soil.

The pipe should be left extending approximately 30 cm above ground surface. A stake driven alongside the piezometer will serve as a protection against damage and as a location marker. If hydraulic-head readings are desired at several depths at a given location, drive pipes of different lengths into the soil, spacing the pipes laterally with a separation of at least 30 cm. It is convenient for recording and interpreting hydraulic-head readings to set the tops of all piezometers at a given location to the same elevation. This can be done with a carpenter's level or with a transit or surveyor's level. With a transit or surveyor's level, an adhesive marker placed on the upper end of the driving hammer at such a position as to give the desired piezometer elevation is an easy way to set the tops of several piezometers to the same elevation, especially where the piezometers are separated by more than a few feet.

In some soils, the rivet in the end may not be necessary. When a piezometer is driven without a rivet, a soil plug of about 7 to 15 cm in length may lodge in the lower end of the pipe. In many soils, this plug can be flushed out in the flushing operation in less time than is required to punch the rivet out.

22–4 METHOD OF INSTALLING PIEZOMETERS BY JETTING

22–4.1 Special Apparatus

1. Commercial spray rig with tank and high-pressure pump: Pressure $\geq$ 3 MPa (400 psi).
   Tank capacity = 750 to 1000 L (200–300 gallons)
   Flow capacity = 0.3 to 1.3 L s^{-1} (5–20 gal min^{-1}).
2. High-pressure hose 2-cm diameter (¾-inch), 8 m (25 feet) or more, as needed.
3. Swivel joint, diameter to fit piezometer (swivel joint, Style No. 20, Chiksan Company, Brea, CA 92621)

22–4.2 Procedure

Connect the upper end of the piezometer pipe to the swivel joint, which in turn is connected to the hose leading from the pump, as shown in Fig. 22–3. If a jetting rig with a frame for hoisting and handling the pipe is used (Reger et al., 1950), read and record the depth of penetration as jetting progresses directly from the adjustable measuring tape that is suspended from the frame alongside the piezometer pipe. If the pipe is installed by hand, mark the pipe at some convenient intervals before starting, so that depth of penetration can be read directly from the pipe.

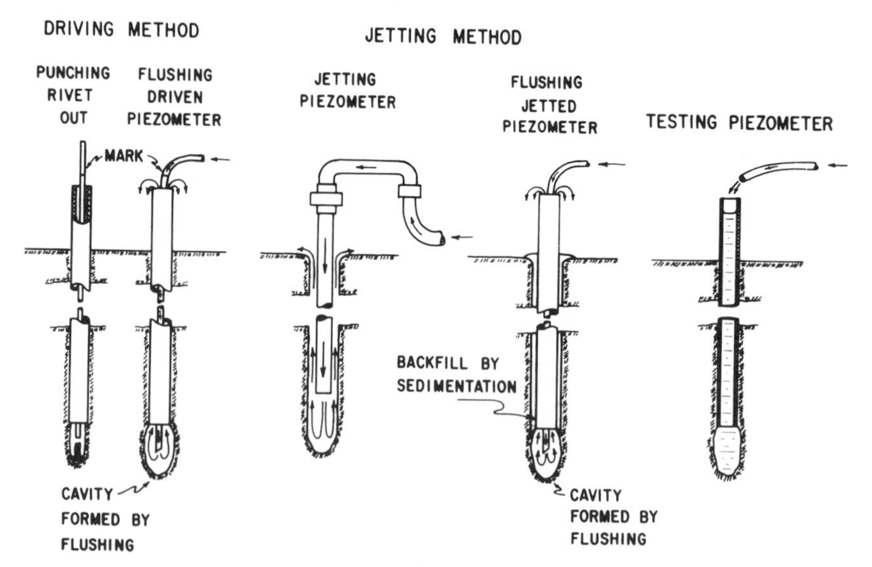

Fig. 22–3. Installing, flushing, and testing piezometers. Piezometers may be installed by driving or by jetting. A cavity is formed at the base of the piezometer by flushing, and the piezometer is tested for responsiveness by filling it with water and observing the rate of recession of the water level.

Start the pump and direct the jet of water issuing from the lower end of the pipe into the soil. As the pipe penetrates the soil behind the jetting stream, oscillate the pipe up and down to assist the pipe in penetrating the soil. Estimate and record the texture and consolidation of the material penetrated as installation progresses.

22–4.3 Comments

The jetting method makes use of the eroding and lubricating properties of a stream of water issuing from the end of the pipe for opening a passage into the soil. The movement of the pipe up and down helps to penetrate resistant materials. Logging of the soil materials penetrated provides helpful information for better understanding of ground-water movement and interpreting hydraulic-head readings. It also helps in the selection of subsequent termination points for additional piezometers.

If the pressure developed by the pump is less than about 1.5 MPa the end of the pipe can become plugged with sediment from the driving action, especially when penetrating clays.

Logging subsurface layers by jetting requires experience that can be gained and checked by jetting in soils for which data on stratigraphy are available from independent logging procedures. An estimate of texture and consolidation of the material is made from (i) the nature of the vibrations transmitted from the material penetrated through the pipe to the hands of the operator, (ii) the rate of downward progress, (iii) examination of sediments carried by the effluent, and (iv) observation of color changes of the effluent.

Return flow may be lost and penetration may stop in premeable sands and gravels. A driller's mud added to the water is effective for maintaining return flow in coarse materials. Approximately 6 kg of driller's mud per 100 L of water has been found suitable for most conditions (10 lb/100 gal). It is necessary to add this preparation to the water supply slowly and to agitate it thoroughly as it is added.

Where several hydraulic-head measurements are desired at different depths, the deepest pipe is usually installed first. The log from the first pipe serves for selecting depths at which additional pipes are terminated. It is often desirable to terminate piezometers in sandy lenses to increase the rate at which they respond to hydraulic-head changes in the soil.

The sealing of the soil around the pipe at the base of the piezometer requires careful attention. Jetting should be stopped immediately as each pipe reaches the desired depth, so that excessive washing of material from around the pipe will not occur. The material in suspension settles back around the pipe and usually provides a satisfactory seal. When this precaution is not taken, leakage may occur along the pipe or from one pipe to another, causing invalid readings.

The size of the jetting stream is important. In general, the smallest stream-size that will maintain a return flow of water to the ground surface should be used to conserve water. A flow of 0.5 L/s (8 gal/min) has been found to be satisfactory in many soils. Where deep sandy soils are involved, flows up to 1.5 L/s have been required. Because the flow capacity must be determined at the outset when the pump is selected, it is important that advance consideration be given to the kind of soils likely to be encountered.

Other pipe sizes may be used quite satisfactorily for piezometers; however, the flow capacities suggested herein are for 1.3-cm pipe (3/8 in.). If pipe size is greatly increased, water flow rates must also be increased to give equal jetting velocities. High flow rates also mean that a greater water storage capacity will be required. Pipe sizes smaller than 1.3 cm in diameter are not recommended because of inadequate rigidity for driving or jetting and because the bore is too small for the convenient measurement of water elevations.

22–5 METHOD OF FLUSHING AND TESTING PIEZOMETERS

22–5.1 Special Apparatus

1. Plastic tubing (Saran or nylon), size to fit inside piezometer, 7 meters or more as needed.
2. Bucket pump, hand operated, or other suitable water source.

22–5.2 Procedure

After the piezometer is driven or jetted into place and the rivet is punched from the end of the pipe (where a rivet is used), flush the soil material out from the base of the pipe to form a cavity 7 to 10 cm long, as shown in Fig. 22–3. To do this, push the plastic tubing (which has been previously marked with paint or tape to indicate the pipe length) to the bottom of the pipe, and pump water through the tube. As the flushing proceeds, move the tube up and down to help loosen the soil. Soil material and water will return and overflow the top of the pipe through the annular space between the tubing and pipe. After the cavity at the base of the pipe is formed, test the piezometer for response rate by filling it with water and observing the rate at which the water level drops (see Fig. 22–3). In sands and gravels, the rate of entry of water may be so great that no overflow can be obtained during flushing, whereas in clays, the rate of drop may be so slow that it is hardly noticeable. If the level of the water in the pipe does not drop, repeat the flushing operation (without unduly extending the plastic tube below the end of the pipe) until the rate of change of the water level in the pipe, after filling, is perceptible. Allow the water level in the piezometer to come to equilibrium with the ground water.

Cap the piezometer to prevent the entry of insects and/or the filling of the piezometer by children or vandals. Attach a standard pipe-coupling to the top of the piezometer and insert a rivet, as shown in Fig. 22–4. Remove the rivet with a magnet to make water-level readings.

22–5.3 Comments

Because of the time lag involved in piezometer measurements, the above test of responsiveness of the piezometer is important. The time required for the water level in the piezometer to come to equilibrium with the ground water, after being displaced, is inversely proportional to the permeability of the soil.

From a practical standpoint, the time-lag difficulty can best be met by terminating piezometers in sandy lenses whenever this is possible, as described under procedure for the jetting method. When this is not practical, the difficulty can sometimes be overcome by enlarging the cavity by flushing. In any event, a measurement of the responsiveness of the piezometer should be made as a basis for interpretation of water-level readings. Because of the sealing and plugging of piezometers that may occur with time, it is good practice to reflush and retest piezometers periodically.

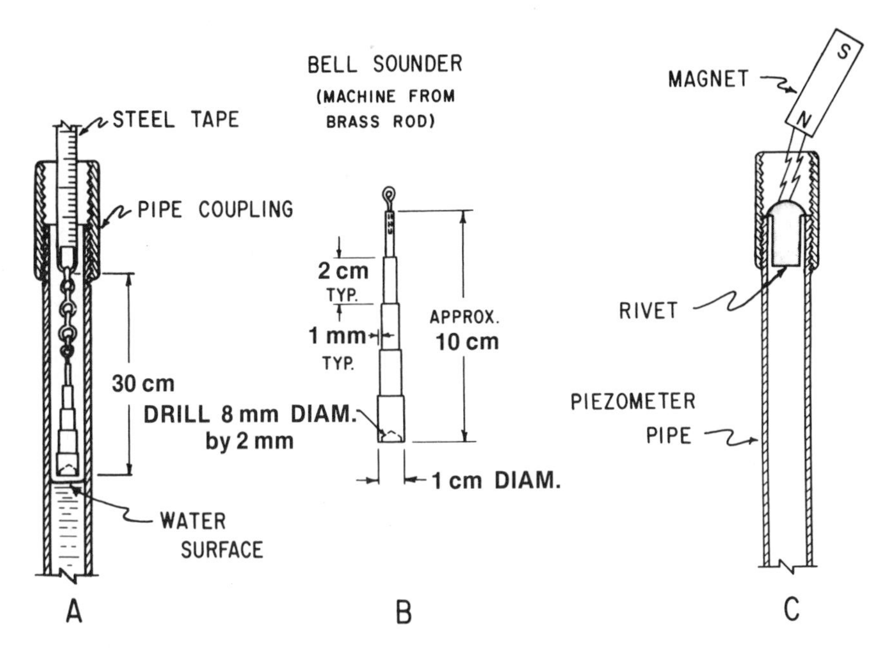

Fig. 22–4. Method for measuring depth to water level and for capping piezometers. (A) The "bell sounder," lowered on a steel tape into the piezometer, makes a sound when contact is made with the water. (B) The bell sounder is made by machining and drilling a solid rod. (C) A rivet placed in the top of the piezometer within a standard pipe coupling serves as a vented cap. The rivet is removed with a magnet for readings.

22-6 METHOD OF MEASURING WATER LEVELS IN PIEZOMETERS

22-6.1 Special Apparatus (Fig. 22-4)

1. Bell sounder.
2. Steel measuring tape, graduated in metric units, of width to fit inside piezometer.
3. Magnet.

22-6.2 Procedure

Remove the rivet from the top end of the piezometer with the magnet, as shown in Fig. 22-4. Attach the bell sounder to the end of the measuring tape, and lower it into the piezometer pipe. Ascertain the reading on the tape at the top of the piezometer pipe when the bell sounder impacts the water level. The impact is denoted by the sound produced as the "bell" strikes the water. First obtain an approximate reading for the impact point, and then refine the reading by subsequent short vertical strokes of the tape and sounder. Begin the vertical strokes with the bell a short distance above the water surface; and while stroking, gradually lower the bell until contact is made.

22-6.3 Comments

Inasmuch as the commercial steel tape is graduated to begin at zero at the end of the tape, it is necessary to make a correction for the additional length of chain used to attach the bell. It is convenient to use an even 30-cm (1-foot) length, as shown in Fig. 22-4.

Other types of sounders are also satisfactory. An electrical sounder may work well, but it is more complex and subject to difficulties such as loss of power, shorting, improper contact, and other problems associated with electrical circuits.

The air-tube method is a simple alternative. A length of plastic tubing (Saran or nylon), with length graduations marked on the outside, is inserted and lowered into the piezometer pipe until it contacts the water surface. The contact of the end of the tube with the water surface is discerned by blowing through the tube and listening for the bubbling sound of air through the water.

Other caps can be used for capping piezometers, such as a standard pipe cap; but, whatever cap is used, it must be vented to admit air.

22-7 METHOD OF INSTALLING TENSIOMETERS

22-7.1 Procedure

This subject is covered in detail in chapter 23, where special apparatus and materials are listed. The following statement of procedure gives only the general requirements. The reader is referred to the above section for detailed directions and information for this test. Make a hole to the desired depth in the soil, using a soil tube or an auger of such size as to give a close fit for the ceramic cup of the tensiometer. Insert the soil tube into the hole, and fill the tube and the hole with water. Withdraw the soil tube and place the tensiometer by pushing it to the bottom of the hole. Fill the tensiometer with air-free water by injecting water into the manometer system through the air-trap vent. Force water into the system until it overflows the mercury cup to remove the free air in the manometer system. Stopper the air-trap vent, and allow the tensiometer to come to equilibrium with the water in the soil.

22–7.2 Comments

In some soils, the walls of the hole are not stable enough to permit filling the hole with water, and the soil tube is used as a liner in the hole while filling. It is important that the ceramic cup be in contact with the soil. For this reason, oversize holes should be avoided.

Water can be freed of air by boiling it. The water should be allowed to cool before filling the tensiometer. Tensiometer readings are unreliable if the units are not substantially filled with water. Tensiometers with transparent manometer lines are preferable to nontransparent ones for the reason that air bubbles can be easily detected.

Tensiometers are also subject to errors from temperature changes, especially those from heat conduction through the main stem of the tensiometer from the above-ground portion to the cup in the soil. This problem is satisfactorily overcome by use of tensiometers made of materials such as plastics that are poor conductors of heat.

Before a tensiometer is installed, the scale should be positioned to give the desired reference-level reading. With the manometer system filled with air-free water, the tensiometer should be stood on end in water so that the water level is at the desired reference-level position with respect to the center of the cup. The zero mark of the tensiometer scale then should be positioned opposite the mercury level of the manometer. When the reference level is set at the center of the cup, tensiometer readings will equal the negative-pressure-head component or soil suction. If the reference level is set at the ground surface line (or other desired reference level), the reading will include both the negative-pressure and the position-head components. Where several tensiometers are installed, each to a different depth for the purpose of measuring hydraulic head, it is advantageous to select a common reference level such as the soil surface. Moreover, it is helpful to mount the manometer tubes of the several tensiometers so as to be read from a single scale and to insert them into a common mercury reservoir. An unusually large scale-displacement (value b, Eq. [3]) is avoided by selecting a reference level not far distant from the mercury level in the reservoir. The scale displacement b can be calculated by use of Eq. [3]; however, for greater precision it is advisable to set the scale as outlined above. The calculated value is most useful for detecting gross errors in scale position.

22–8 INTERPRETATION OF HYDRAULIC-HEAD READINGS

Figure 22–5 (Richards, 1952) shows hydraulic-head readings for both tensiometers and piezometers for A the static case, B downward flow from ponded water on the soil surface, and C upward flow from a pressure aquifer below. Points of measurement for the three cases shown in the figure are indicated by open circles for tensiometers and by shaded circles for piezometers. The measurement points are positioned in a vertical line downward through the profile. The line that connects from each circle to an equivalent (or hypothetical) water manometer in the case of tensiometers (open circles) and to the open end of a pipe for piezometers (shaded circles) represents a water line which provides hydraulic attachment from the manometer or piezometer to the point in question. A tensiometer is sketched for the first depth, Case A, to illustrate the fact that tensiometer hydraulic-head readings in all cases are actually made

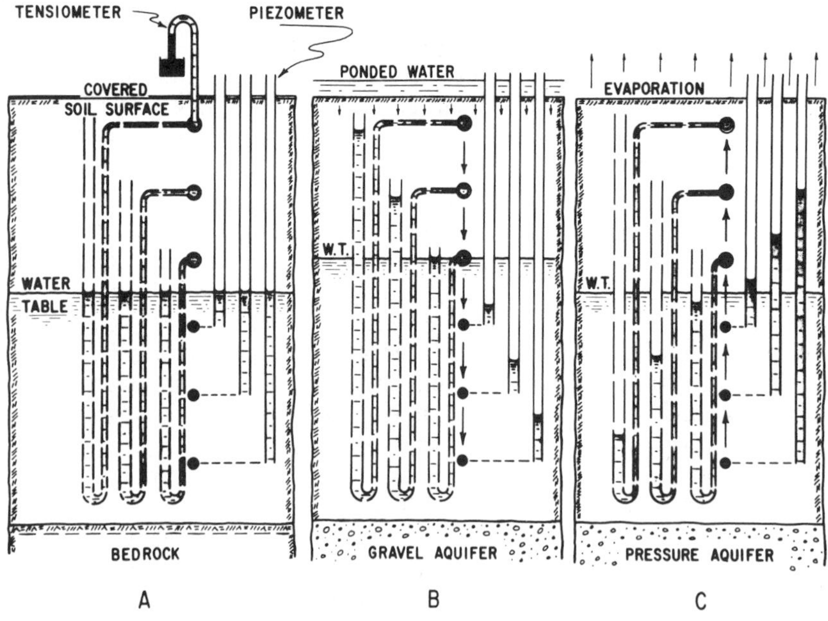

Fig. 22–5. Distribution of hydraulic head for several flow systems. (Case A) Static equilibrium under gravity—no flow. (Case B) Downward flow to a gravel aquifer from water ponded on the soil surface. (Case C) Upward flow from an aquifer under pressure to the soil surface where water is evaporated. Lines connecting from the point of measurement for tensiometers (open circles) and for piezometers (shaded circles) represent water lines for hydraulic attachment to equivalent or hypothetical manometers in the case of tensiometers, or open-end pipes in the case of piezometers (Richards, 1952).

with the tensiometer scale above the soil surface. The equivalent or hypothetical manometers are diagrammatic only as an aid in interpretation of hydraulic-head readings.

From thermodynamics, it is well known that flow is in the direction of decreasing energy level. Since hydraulic head is a measure of energy per unit weight of water, flow is also in the direction of decreasing hydraulic head. For Case A, the hydraulic head is the same at all depths in the soil as indicated by equal water levels in all tubes. In this case there is no flow. For Case B, the hydraulic head at the soil surface is equal to the elevation of the ponded water; and as can be seen by the levels in the open tubes, it decreases downward in the soil at each succeeding point of measurement. Flow in this case is downward. The reverse is true for Case C where the hydraulic head decreases from the pressure aquifer to the soil surface, indicating that in this case flow is upward.

Where there is no vertical flow component (Case A), the equal hydraulic-head values read in all tubes represents the position of the water table. The water table is defined as the loci of points in the soil where the soil water is at atmospheric pressure. Under conditions of vertical flow of water (Cases B and C), the hydraulic-head value, for either a tensiometer or piezometer, coincides with the water table only if the end of the pipe or the center of the cup is located at the water-table elevation (see Case B). That is to say, when the piezometer terminates or the tensiometer cup is located at the water table elevation, the pressure head p/w equals zero.

Hydraulic head is a point function in three-dimensional space. Procedures and methods from standard textbooks that are used for the analysis of other physical flow processes, such as flow of heat and electricity, are applicable to the analysis of water flow in premeable media. An example of the use of equal-hydraulic-head lines to represent the hydraulic-head distribution in a plane or profile for two-dimensional flow of water into an open drain is shown in Fig. 22–6. Here the hydraulic-head values and distances are plotted as ratios to the drain depth on a cross-sectional drawing; and, by the procedures of interpolation and extrapolation, equal-hydraulic-head lines are constructed. Convenient hydraulic-head intervals may be selected, extending over the range of measured values for hydraulic head. Usually an interval is selected that allows a number of equal-hydraulic-head lines to be sketched on the same profile. The component of flow in the plane of the profile is normal to lines of equal-hydraulic-head, if the profile is plotted to a 1:1 scale. For profiles where the scales are equal, flow lines can be sketched in at right angles to the equal-hydraulic-head lines, with arrows to show the direction of flow. If the vertical scale is exaggerated, the relation between steam lines and equal-hydraulic-head lines on the plotted profile is no longer orthogonal. Where the vertical and horizontal scales are not equal, the hydraulic-head distribution may be properly plotted, but flow lines should not be indicated. While Fig. 22–6 shows only hydraulic-head values obtained

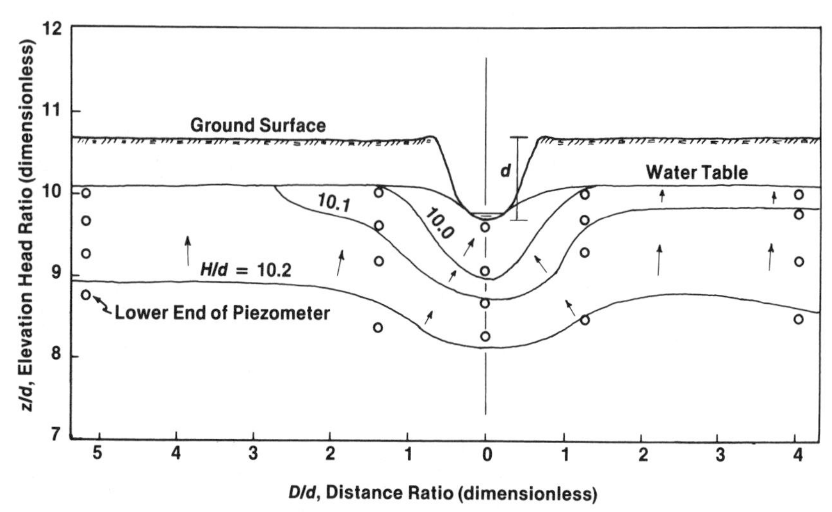

Fig. 22–6. Equal hydraulic-head lines below a water table on a profile section in the vicinity of an open drain. Example from Delta area, Utah. The direction of the hydraulic gradient is represented by arrows and indicates upward movement from an underlying source. d = drain depth, Z = elevation head, P/w = pressure head, $v^2/2g$ = velocity head $\equiv 0$, H= hydraulic head = $Z + P/w + v^2/2g$)

by the use of piezometers below a water table, the lines of equal hydraulic head in the unsaturated region above the water table can be constructed in a similar manner by use of tensiometer readings. In this case, they would be essentially horizontal lines above the water table with a downward dip near the open drain, showing flow toward the soil surfaces where evaporation takes place.

Hydraulic-gradient values for use in the Darcy flow equation are obtained by determining the change in hydraulic head per unit flow distance in the soil. The usual practice is to use average gradient values, which are obtained by dividing the difference in hydraulic head at two points by the distance between points. If the two points in question happen to lie along a stream path, the result will represent the true average hydraulic gradient. Frequently, hydraulic-head measurements are made in a selected plane in three-dimensional space which may or may not coincide with the true flow direction. Where such measurements are used, the results represent only some space component of the true hydraulic gradient. Vector analysis is applicable for determining true flow direction or true hydraulic gradients in three-dimensional space. For flow of water in isotropic soils, the gradient vector coincides in direction with the velocity vector.

22–9 REFERENCES

Christiansen, J.E. 1943. Ground water studies in relation to drainage. Agric. Eng. 24:339–342.

Dodge, R. A., and M. J. Thompson. 1937. Fluid mechanics. McGraw-Hill Book Company, Inc., New York and London.

Donnan, W. W., and G.B. Bradshaw. 1952. Drainage investigation methods for irrigated areas in Western United States. U.S. Dep. Agric. Tech. Bull. 1065.

Hvorslev, M.J. 1951. Time lag and soil permeability in ground-water observations. Waterways Experiment Station, Corps of Engineers, U.S. Army, Vicksburg, MS. Bull. 36.

Klute, A., and W.R. Gardner. 1962. Tensiometer response time. Soil Sci. 93:204–207.

Leonard, R.A., and P.F. Low. 1962. A self-adjusting null-point tensiometer. Soil Sci. Soc. Am. Proc. 26:123–125.

Mickelson, R.H., L.C. Benz, C.W. Carlson, and F.M. Sandoval. 1961. Jetting equipment and techniques in a drainage and salinity study. Trans. Am. Soc. Agric. Eng. 4:222–223, 225, 228.

Miller, R.D. 1951. A technique for measuring tensions in rapidly changing systems. Soil Sci. 72:291–301.

Reeve, R.C., and M.C. Jensen. 1949. Piezometers for ground-water flow studies and measurement of subsoil permeability. Agric. Eng. 30:435–438.

Reger, J.S., A.F. Pillsbury, R.C. Reeve, and R.K. Petersen. 1950. Techniques for drainage investigations in the Coachella Valley, California. Agric. Eng. 31:559–564.

Richards, L. A. 1942. Soil moisture tensiometer materials and construction. Soil Sci. 53:241–248.

Richards, L.A. 1949. Methods of measuring soil moisture tension. Soil Sci. 68:95–112.

Richards, L.A. 1952. Water conducting and retaining properties of soils in relation to irrigation p. 523–546. *In* Miriam Balaban (ed). Proc. Internat. Symp. Desert Res. Research Council of Isreal in cooperation with UNESCO. Jerusalem Post Press, Jerusalem.

U.S. Salinity Laboratory Staff. 1954. Diagnosis and improvement of saline and alkali soils. U.S. Dep. Agric. Handb. 60.

23 Water Potential: Tensiometry

D. K. CASSELL

North Carolina State University
Raleigh, North Carolin

A. KLUTE

Agricultural Research Service, USDA, and
Colorado State University
Fort Collins, Colorado

23–1 INTRODUCTION

Soil water studies have long been conducted using soil water content measurements. The values are reported on either a weight basis or a volume basis (See chapter 21, Soil Water Content). For many studies, soil water content information is of primary interest. However, for studies involving water transport and storage in soils and soil-water-plant relationships, the energy status of the soil solution phase (soil water) and/or the energy status of the chemical species *water* in the soil are required.

The retention of water by soil and its relationship to the soil water energy level was discussed by Buckingham (1907), and has become known as "the potential concept of soil water". Gardner et al. (1922) proposed that porous, ceramic-walled equipment could be used to measure the relation between soil water content and the energy status of the soil water. Porous ceramic cups connected to vacuum gauges or manometers were proposed for measuring the capillary potential of the water in a soil (Richards, 1928; Heck, 1934; Rogers, 1935). These three investigators independently reasoned that the water inside the tensiometer would equilibrate with water in the water films in the soil.

23–2 PRINCIPLES

The essential parts of a tensiometer for field use are shown in Fig. 23–1. Many modifications of this design exist. A few will be discussed in detail in section 23–3.1. Tensiometer designs for laboratory use are described in section 23–4. The porous tip or cup of the tensiometer is sealed to the barrel or connecting tube. A removable air-tight cap, which facilitates filling the tensiometer with water, is used to seal the barrel at the top. A device to measure the pressure in the water in the tensiometer

 Methods of Soil Analysis, Part 1. Physical and Mineralogical Methods–Agronomy Monograph no. 9 (2nd Edition)

cup (a Bourdon-vacuum gauge, a water manometer, a mercury-water manometer, or an electrical pressure transducer) is attached near the upper end of the barrel. The connecting tube and all pores in the porous cup are filled with deaerated water.

Tensiometers are used to measure the energy status of the soil solution phase (commonly called the soil water). As the water content of the soil surrounding the water-filled porous tensiometer cup decreases, the energy level of the soil water decreases relative to that of the water in the tensiometer cup, and water moves out of the tensiometer through the pores in the tensiometer cup and into the soil. The pressure in the water in the tensiometer cup is reduced. If the soil surrounding the porous cup receives additional water, the soil water pressure is increased, and soil water flows through the walls of the porous cup into the tensiometer, thereby increasing the pressure of the water in the tensiometer cup. The energy status of the soil water is obtained from that of the tensiometer water, assuming that the latter is in equilibrium with the soil water.

The water in a tensiometer system is probably never in thermodynamic equilibrium (i.e., thermally, mechanically, and chemically) with the soil water. The solutes in the soil solution phase generally will not be in equilibrium with those in the water in the tensiometer cup, so that chemical equilibrium will not exist. At best, the equilibrium attained in a tensiometer-soil water system is a mechanical or hydraulic equilibrium and a thermal equilibrium. If the cup material were sufficiently impermeable to some of the solutes, the pressure observed in the tensiometer cup could be affected by the lack of chemical equilibrium, i.e., the instrument could function as a "leaky" osmometer. However, the porous cups and barriers commonly used in tensiometers are quite permeable to solutes, and the pressure in the water in the tensiometer cup is not affected significantly by the lack of thermodynamic equilibrium.

The condition of mechanical or hydraulic equilibrium in a tensiometer may be discussed using a potential function, which is a measure of the energy per quantity of *solution phase*. The unit of quantity of solution phase may be *mass, volume,* or *weight*. At mechanical equilibrium the pressure in the tensiometer water, P_T, is that pressure necessary to mechanically or hydraulically equilibrate the tensiometer solution with the soil solution phase. The mechanical potential of the tensiometer solution phase has two components; a gravitational component and a pressure component. Table 23-1 lists the definitions of these potentials on a mass, volume and weight basis, assuming that the density of the solution in the tensiometer, d_T, is constant. It is customary to assume that the density of the tensiometer solution and that of the soil solution are the same.

The gravitational component of the mechanical or hydraulic potential of the tensiometer solution is determined from the elevation of the cup, z, relative to a reference level. The pressure component is usually calculated using the ambient atmospheric pressure, P_a, as the reference. The pressure potential can be split into pneumatic and matric components. The pneumatic potential accounts for the fact that the gas phase pressure, P_G, in the soil may be different than atmospheric and is zero if $P_G = P_a$.

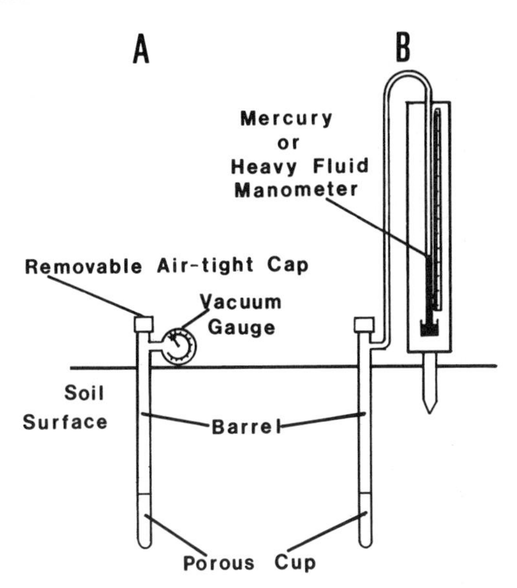

Fig. 23–1. Diagram of two common types of tensiometers. (A) Vacuum gauge. (B) Mercury-water manometer. A heavy fluid immiscible with water may be used in place of mercury.

The matric component of pressure potential is related to the water content of the soil and is calculated from the difference between the pressure in the tensiometer water and the gas phase pressure in the soil.

Within the soil pores, the mechanical forces acting on the solution phase are the gravitational force, a pressure gradient force, and various solid-liquid interaction forces (adsorptive forces). If we assume that potentials can be defined for each of these force fields, the mechanical potential of the soil-solution phase would consist of three components; gravitational, pressure and adsorptive. Since the gravitational potential of the solution in the soil just outside the cup is equal to that of the solution in the tensiometer cup, the pressure of the tensiometer solution, P_T, is an equivalent pressure, which reflects the influence of the adsorptive forces, the capillary effects on the pore water pressure, and in some cases, the fact that the soil gas phase pressure is not atmospheric. The pressure P_T cannot be identified with the pressure in the solution phase of the soil unless the influence of adsorptive forces on the solution phase in the pores is negligible. In rather wet, coarse-textured materials (sands), it is appropriate to identify P_T with the soil water pressure. In sands at low water content and in finer textured materials with high specific surface, a greater fraction of the soil solution phase is under the influence of the adsorptive forces, and P_T must be interpreted as an equivalent pressure and not as the solution phase pressure in the pores of the soil.

In swelling soils there is a contribution to the pressure potential due to the overburden load that may be present on the soil at the location of the tensiometer cup (Philip, 1970; Groenevelt & Bolt, 1972; Mahony, 1975). In that case, the matric component of the pressure potential is a function of both water content and the external load, in contrast to the

Table 23–1. Definitions of the components of the mechanical potential of the soil solution phase. The overburden component is not included.

Basis	Units	Gravitational potential	Pressure potential†		
			Total	Pneumatic	Matric
Mass	Energy/mass of solution. Dimensions: length squared per time squared	$\Phi_g = gz$	$\Phi_P = (P_T - P_a)/d_T$	$\Phi_a = (P_G - P_a)/d_T$	$\Phi_m = (P_T - P_G)/d_T$
Volume	Energy/volume of solution. Dimensions: mass per unit length per unit time	$\Psi_g = d_T gz$	$\Psi_P = P_T - P_a$	$\Psi_a = P_G - P_a$	$\Psi_m = P_T - P_G$
Weight	Energy/weight of solution. Dimensions: length	$Z = z$	$h = (P_T - P_a)/d_T g$	$h_a = (P_G - P_a)/d_T g$	$h_m = (P_T - P_G)/d_T g$

† P_T = pressure in tensiometer cup water; P_a = atmospheric pressure; and P_G = soil gas phase pressure.

situation in nonswelling soils where the matric component of pressure potential is related to the water content only. In either case, the tensiometer measures the potential of the solution phase, including the effects of gas phase pressure, adsorptive forces, and overburden load. The potential so obtained is the proper one to be used in analyses of direction and rate of flow based on the Darcy equation. However, if the pressure potential obtained from the tensiometer readings is to be related to soil water content, then the effects of gas phase pressure and overburden load must be considered. If the gas phase pressure at the region of the tensiometer cup is known, the pneumatic component may be calculated and subtracted from the total pressure head. If it is desired to relate the remaining matric component of pressure head to the water content of the soil, an assessment of the contribution of the overburden load to the pressure head must be made. Techniques for this assessment have been described by Talsma (1977), Talsma and van der Lelij (1976), and Rose et al. (1965).

Numerous terms and units are in use for the potential and its components. The matric component of the pressure potential, often called the *matric potential* or *capillary potential*, is negative in unsaturated soil. In much work with soil water it is convenient to express the potential in energy/weight units or length units. Then it is called the *hydraulic head*, H, with components *gravitational head*, Z, and *pressure head*, h. The matric component of the pressure head, h_m, is negative in unsaturated soil. *Suction* or *tension* is the negative of the matric pressure head. Hence, suction and tension values for water in unsaturated soil are positive.

The energy status of the chemical species *water* in the tensiometer may be expressed in terms of the *partial specific Gibbs free energy*, g_w, or *thermodynamic potential* of the water. In the tensiometer water it is defined by

$$g_w = v_w(P_T - P_a) + g_{wo} + g_z \qquad [1]$$

where the terms on the right are the pressure, osmotic, and gravitational components, respectively. The thermodynamic potential and its components in Eq. [1] are expressed in units of energy per mass of the chemical species *water*. The thermodynamic potentials are often expressed on an energy per mole basis, and sometimes on an energy per weight basis. If we assume that the lack of chemical equilibrium between the tensiometer and soil solution phases does not significantly affect P_T, the pressure component of the thermodynamic potential of the tensiometer water may be calculated from the pressure P_T and the partial specific volume of the water, v_w. The pressure potential of the water obtained in this way will reflect the influence of capillary forces, adsorptive forces, and the gas phase pressure upon the equivalent pressure P_T.

Two characteristics of the porous cup are of interest, viz., the *bubbling pressure* and the *cup conductance*. The *bubbling pressure* is the pressure difference required to force a gas phase through a wetted porous cup. The hydraulic radius of the largest continuous pore through the tensiometer cup material will determine the bubbling pressure of the cup. The bubbling pressure of the cup material must be selected to be at least as high as the highest suction (atmospheric pressure minus the pressure in the water in the cup of the tensiometer) to be measured.

The *cup conductance*, k, is defined by

$$k = V/(t\Delta H) \qquad [2]$$

where V is the volume of water that flows through the cup in time t when the hydraulic head difference ΔH is applied. The dimensions of k are length squared per time. The cup conductance depends on the cup geometry and the hydraulic conductivity of the cup material.

The *gauge sensitivity* of the tensiometer system, and the cup conductance determine the *response time* of the instrument. The *gauge sensitivity*, S_G, of the tensiometer system is the change of pressure head per unit volume of water in the tensiometer system, and is defined by:

$$S_G = dh_T/dV \qquad [3]$$

where V is the volume of water in the tensiometer system, including the pressure measuring device. The gauge sensitivity has dimensions of reciprocal length squared. The term *gauge* is used here in a generic sense, to include any pressure measuring device and the tubing and valves connecting it to the porous cup. The pressure measuring device requires a displacment of fluid to register a pressure change. In addition, both the flexibility of the materials used to construct the tensiometer and the compressibility of the fluids in the instrument will affect the gauge sensitivity of the system. Table 23–2 lists the gauge sensitivity of some pressure measuring devices.

The *response time* of a tensiometer is a measure of its responsiveness to changes in soil water pressure head at the external surface of the cup. The response time is the time required for a fraction $(1-1/e = 0.632)$ of a step change in pressure head, which has been applied to the cup, to be registered by the pressure measuring device of the tensiometer. If the water in the tensiometer is initially at a pressure head h_1 and the cup is suddenly exposed to water at a constant pressure head h_2, it can easily be shown that the time dependence of the pressure head of the water in the tensiometer is given by (Klute and Gardner, 1962)

$$h_T = (\mathrm{h}_1 - \mathrm{h}_2)\exp(-t/T_r) + h_2 \qquad [4]$$

where t is the time and T_r is the response time. The response time is related to the gauge sensitivity of the system and the cup conductance by

$$T_r = 1/(kS_G)\,. \qquad [5]$$

From Eq. [5] it can be seen that increasing the cup conductance and the

Table 23-2. Volume change with pressure head for various components of tensiometers.

Item	dV/dh	dh/dV Gauge sensitivity
	cm^2	cm^{-2}
Water manometer, 1.5-mm diameter	1.8×10^{-2}	5.7×10^{1}
Mercury manometer, 1.5-mm diameter	1.3×10^{-3}	7.6×10^{2}
Vacuum gauge†	5×10^{-5}	2×10^{4}
Pressure transducer‡	1.5×10^{-5}	3×10^{4}
Pressure transducer§	1×10^{-6}	
Plug valve, O-ring seals¶	1×10^{-5}	
Ball value, TFE ball#	2.4×10^{-6}	
Tubing, nylon, 0.060 in. i.d.††	1×10^{-7} per cm of tube length	
Tubing, vinyl, 1/16 in. i.d.‡‡	1.5×10^{-7} per cm of tube length	
Water, 20 °C	0.5×10^{-7} cm^{-3}	
Air, @ 100 kPa	1×10^{-3} cm^{-3}	
@ 50 kPa	2×10^{-3} cm^{-3}	
@ 20 kPa	5×10^{-3} cm^{-3}	

† From Model 2700 Tensiometer, Soilmoisture Equipment Corp., P.O. Box 30025, Santa Barbara, CA 93105.
‡ Model DP-15, ± 5 p.s.i., Validyne Engineering Corp., 8626 Wilbur Ave., Northridge, CA 91324.
§ Model 4-321, ± 25 p.s.i., CEC Division, Bell and Howell, 360 Sierra Madre Villa Ave., Pasadena, CA 91109.
¶ 9500 Series, 1/4 in. NPT, Circle Seal Controls, 1111 N. Brookhurst St., P.O. Box 3666, Anaheim, CA 92803.
# Model B4F2, 1/8 in. NPT, Whitey Co., 318 Bishop Rd., Highland Heights, OH 44143.
†† Scanivalve Corp., 10222 San Diego Mission Rd., San Diego, CA 92120.
‡‡ Clippard Instrument Co., 7350 Colerain Ave., Cincinnati, OH 45239.

gauge sensitivity will decrease the response time of the tensiometer. A time equal to about five to six response times will be required for a tensiometer to display essentially all of a step change in the suction applied to the cup. The response time defined by Eq. [5] is a function of the instrument only, and assumes that the suction gradient in the soil just outside the cup, which produces the flow of water into or out of the cup, is negligible. If the hydraulic conductivity of the soil surrounding the cup is sufficiently low, the response of the instrument may become limited by the conductivity of the soil, and the response time will be greater than that calculated from Eq. [5] (Klute & Gardner, 1962). Towner (1980) has analyzed in more detail the response of a tensiometer in the situation where the properties of the soil are limiting.

One or more of the following phenomena may occur as the pressure of the water in the tensiometer system decreases:

1. Dissolved gases will come out of solution as the pressure in the tensiometer cup water is reduced.
2. If the pressure in the water in the tensiometer system is reduced to the level of the vapor pressure of water at the ambient temperature of the system, the liquid water will spontaneously convert to water vapor; i.e., the water will boil at the ambient temperature.
3. If the difference between the gas phase pressure and the pressure in the tensiometer cup water equals or exceeds the bubbling pressure, air will be drawn into the cup.

Any of the above phenomena will introduce a gas phase into the tensiometer system and seriously interfere with its operation. The presence of gases, which can expand and contract with changes in pressure, will cause the instrument to be sluggish in its response to changes in the pressure of the soil water. The dissolution of dissolved gases and the spontaneous vaporization of the water cause the measurement range of a tensiometer to be restricted to suctions less than about 85 kPa when the ambient atmospheric pressure is approximately 76 cm of mercury. Because of the reduction in atmospheric pressure with elevation, the operating range of tensiometers used at higher elevations is correspondingly reduced.

23–3 FIELD TENSIOMETRY

23–3.1 Apparatus

23–3.1.1 SELECTION OF A TENSIOMETER

It is important to select an appropriate tensiometer design for a given tensiometer application. Some factors to be considered in selection of a tensiometer are:

1. The desired accuracy, sensitivity, and method of recording the data from the tensiometer. This has a bearing on the choice of pressure measurement device.
2. The rapidity of the anticipated changes in soil water suction that are to be measured. The cup conductance and the gauge sensitivity of the tensiometer system determine the response time of the tensiometer.
3. The number of tensiometers required. Compromises will probably have to be made between the requirements of the design of the experiment and the resources (funds and tensiometer equipment) available to the researcher.
4. The size of the tensiometer. The length of the barrel will obviously depend on the depths in the soil profile at which measurements are to be made. The choice of the tensiometer cup dimensions will depend on the required degree of spatial resolution in the soil water suction measurements. The size of the cup relative to soil structural units may need to be considered (see section 23–3.4).
5. The range of soil water suction to be measured. Most tensiometers for field use are designed for the 0- to 100-kPa suction range, i.e., with cups that have bubbling pressures of at least 100 kPa. In some cases, it may be possible to use tensiometer cups with a bubbling pressure < 100 kPa , if it is known that the suction range to be encountered in an experiment is < 100 kPa.
6. The availability of commercial tensiometers to suit the application. The size and configuration of the tensiometer cup desirable for a given experiment may not be available from commercial sources.
7. The durability of the materials used for tensiometer construction un-

der field environmental conditions. Exposure to sun, rain, and fluctuating temperature may cause failure of some types of plastic and tubing materials.

A cup conductance of 1×10^{-5} cm^2/s is sufficient for most field studies. Standard ceramic tensiometer cups for field use have a conductance on the order of 3×10^{-5} cm^2/s. A tensiometer constructed with such a cup and a mercury-water manometer will have a response time of 40 to 60 seconds (estimated from Eq. [5] using gauge sensitivity data from Table 23–2). If a Bourdon vacuum gauge were to be used instead of the mercury-water manometer, the response time would be reduced to about 5 s. A response time on the order of 1 min is adequate for most field applications. If, for some reason, a shorter response time appears to be required, the tensiometer may be constructed with special high conductance cups which are available with conductances on the order of 1×10^{-3} cm^2/s.

The sensitivity of measurement of the pressure in the tensiometer with a Bourdon vacuum gauge is about 10 to 20 cm of water. The pressure may be determined with greater precision and accuracy than this with a mercury-water manometer—perhaps to 1 or 2 cm of water. In applications where the hydraulic gradient is to be measured, mercury-water manometers are generally preferred. However, Bourdon gauges have a higher gauge sensitivity than a mercury-water manometer (see Table 23–2), and consequently the response time of mercury-water manometer tensiometers will be longer. This is not usually a significant limitation.

Tensiometers with Bourdon gauges or mercury-water manometers are not suitable for automated data collection systems. Electrical pressure transducers provide an electrical output that may be used in a data collection system. Field tensiometers, with a transducer for each tensiometer cup, have been constructed (Strebel et al., 1973). However, the high cost of transducers discourages such use. Long (1982) describes the use of a lower cost semiconductor pressure transducer that appears to be suitable for field use.

A hydraulic switching system may be used to connect one transducer to more than one tensiometer cup to reduce the cost of an installation. Several such systems have been described (Rice, 1969; Watson, 1965; Williams, 1978; Daian, 1971). Leakage in the hydraulic valve switching system may occur, which can be minimized or prevented by proper design and use of the hydraulic switching valve. A hydraulic switch that has been used is Scanivalve Model No. W0601/1P.24T, connencted to a rotary solenoid stepping motor.[1] Fluctuations of temperature may induce transient pressure fluctuations in the transducer-tensiometer system (Watson & Jackson, 1967), which can be reduced by insulating the transducer and the connecting tubing.

---

[1]Scanivalve Corp., P. O. Box 20005, 10222 San Diego Mission Rd., San Diego, CA 92120. Reference to trade names is purely for the convenience of the reader and does not imply endorsement by either USDA-ARS, N. Carolina State University, or Colorado State University, nor does it imply that other similar products obtainable from other suppliers may not be satisfactory.

Marthaler et al. (1983) describe a method for the use of one transducer to read (manually) many tensiometers. The transducer is connected to a syringe needle, which can be inserted through a septum into the top of the tensiometer barrel, to obtain a reading of the suction. This instrument is commercially available.[2]

It is useful to provide a space in the tensiometer barrel above the point of connection of the vacuum sensing element to serve as a gas trap (Fig. 23–1). Dissolved gases tend to accumulate in the water in the tensiometer by diffusion through the wetted cup from the soil water to the water in the tensiometer cup. Water vapor may accumulate due to vaporization of the water in the tensiometer. To facilitate visual determination of the presence and quantity of air and water vapor in the tensiometer, the upper portion of the barrel should be transparent.

23–3.1.1.1 Commercial Tensiometers. A wide range of commercial tensiometers and components is available. Standard commercial tensiometers have barrels approximately 20 mm in diameter and are available in lengths appropriate for installation of the porous cup at depths of 15, 30, 45, 61, 76, 91, 106, 122, or 152 cm below the soil surface. Other lengths are available on special order. A vacuum gauge is standard on commercial instruments. Both adjustable (zeroing capability) and non-adjustable gauges are available. Some off-the-shelf, commercially available tensiometer innovations include portable nullpoint tensiometers, porous ceramic cups of various sizes, fast response porous cups, small tensiometers for greenhouse use, replaceable cups, and multipoint tensiometers. Kits for complete servicing of tensiometers are also available.[3]

23–3.1.1.2 Custom-made Tensiometers. The tensiometer described below is similar in size and design to the commercially available tensiometers discussed above. Researchers working on a limited budget may find it worthwhile to construct their own tensiometers, particularly if the labor resources are available to do so. The porous cups for building this tensiometer may be obtained from the suppliers mentioned.[3] Other parts can be obtained from plumbing and tubing supply sources.

The construction details for a tensiometer, modified slightly from those published by Henderson and Rogers (1963), are shown in Fig. 23–2. Because air will diffuse through many plastics, it is important to use a material that will minimize this problem. A PVC pipe, which is available at many pipe suppliers and hardware stores, is satisfactory. Porous ceramic cups that fit snugly into nominal size 1/2-inch Schedule 80 PVC pipe are available.[1] It may be necessary to ream the inside of the pipe to obtain a snug fit between the cup and the PVC pipe. The cup is sealed to the PVC pipe with epoxy cement. A PVC slip-coupling is permanently attached to the upper end of the PVC pipe with PVC cement. A length

---

[2]Soil Measurement Systems, 1906 S. Espina, Suite Six, Las Cruces, NM 88001.

[3]The only U. S. manufacturers of tensiometers, components and supplies known to the authors are (i) Irrometer Co., P O. Box 2424, Riverside, CA 92516, (ii) Soilmoisture Equipment Corp., P. O. Box 30025, Santa Barbara, CA 93105, and (iii) Perma Rain Irrigation, Inc., P. O. Box 880, Lindsay, CA 93247.

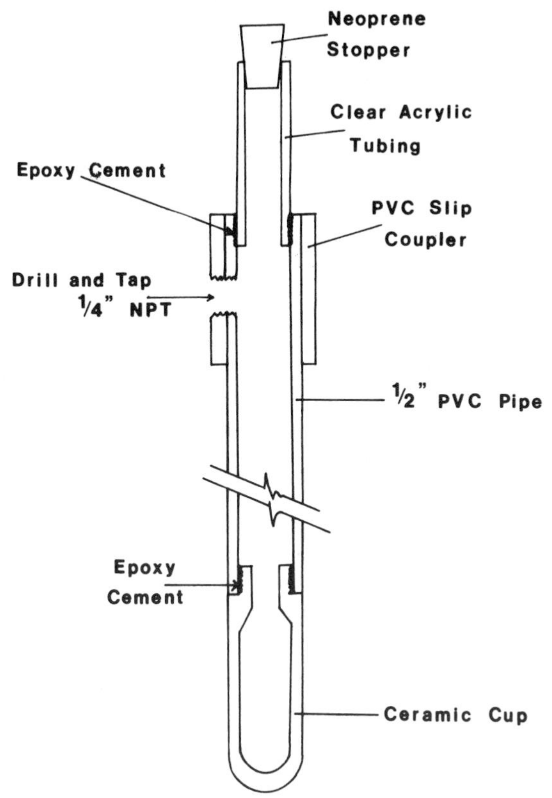

Fig. 23–2. Construction details for a custom-built tensiometer. Adapted from Henderson and Rogers (1963).

(about 10 cm) of 5/8-inch (16 mm) o.d. clear acrylic tubing is affixed to the inside of the PVC pipe with epoxy cement. The acrylic tubing serves as a sight-glass for detecting air accumulation in the tensiometer. A vacuum gauge is attached to the barrel of the tensiometer by drilling and tapping a 1/4 inch pipe thread in the coupling. If a manometer is to be used in place of the vacuum gauge, the slip coupling need not be used. Instead, a hole to accept nylon high-pressure tubing (0.096 inch o.d., × 0.066 inch i.d.) is drilled directly into the PVC pipe. Suitable nylon tubing is available from the Soilmoisture Equipment Co., Santa Barbara, CA, or from many plastic product suppliers. One end of a length of the tubing (sufficient to form the mercury-water manometer and reach from the manometer to the tensiometer barrel) is attached to the PVC pipe with epoxy cement.

In some field and greenhouse research applications it is desirable to use tensiometers with cups of smaller diameter than that of the commercially available tensiometers or the tensiometer just described. The "miniature" tensiometer described below and shown in Fig. 23–3 is modified from the design published by Arya et al. (1975a). These tensiometers are designed for use with mercury manometers, although they could be

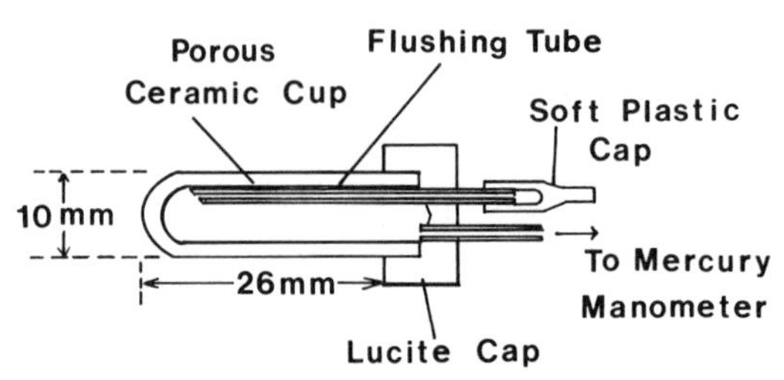

Fig. 23–3. Construction details for "miniature" tensiometer. Adapted from Arya et al. (1975a).

adapted for use with vacuum gauges. Ceramic cups, 2.9 cm long by 0.95 cm o.d. with 0.16-cm thick walls (Soilmoisture Equipment Corp., Santa Barbara, CA), are used. Two holes to accept high-pressure nylon tubing (0.096-inch o.d. by 0.066-inch i.d.) are drilled in a 10-mm length of 13-mm diameter acrylic rod. The open end of the porous cup fits into a recess in the disk made with a 3/8-inch drill and is fastened with epoxy cement. One end of a length of pressure tubing (labeled A in Fig. 23–3) is cut at an angle with a sharp knife, inserted into one of the holes in the acrylic disc, and pushed to the bottom of the tensiometer cup. A second length of nylon pressure tubing (B) is cut perpendicular to its longitudinal axis and inserted through the second hole in the acrylic disk so that the end of the tubing is flush with the bottom of the disk. Both pieces of nylon tubing are sealed to the disk with epoxy cement. Tube B must be cut long enough to be inserted into a pool of mercury and serve as a manometer. The tensiometer and both lengths of nylon pressure-tubing are filled with water by flushing water from a wash bottle into the cup through tube A and out through tube B. The air is carried out of the cup and through the mercury pool by the flow of water. Tube A is then closed by slipping a snug-fitting plastic cap over it. The cap should be filled with water, and care should be taken to avoid trapping an air bubble as the cap is put in place. The cap may be made by heat sealing one end of a 1-cm length of 1/16-inch i.d. by 1/8-inch o.d. vinyl tubing.

If several tensiometers are installed in the same vicinity in the field, it is advantageous to connect them to a common mercury well to reduce the equipment requirements and facilitate the interpretation of the readings of the tensiometers when hydraulic head measurements are made. The multiple manometer well shown in Fig. 23–4 is adapted from a design given by Doering and Harms (1972). This mercury well can serve as many as 20 or 30 tensiometers. The pressure tubing that extends into the well must be cut at an angle so that mercury can freely flow into and out of the tubing. If the set of tensiometers is being used to measure hydraulic head, the mercury level should be maintained above the dividers. If the set of tensiometers is being used to obtain only pressure head data, it is not necessary to keep the mercury level above the dividers. The multiple manometer well is particularly advantageous when several of the tensiometers undergo extreme fluctuations, (e.g., those at shallow depths) while the other tensiometers at greater depths do not. The dividers

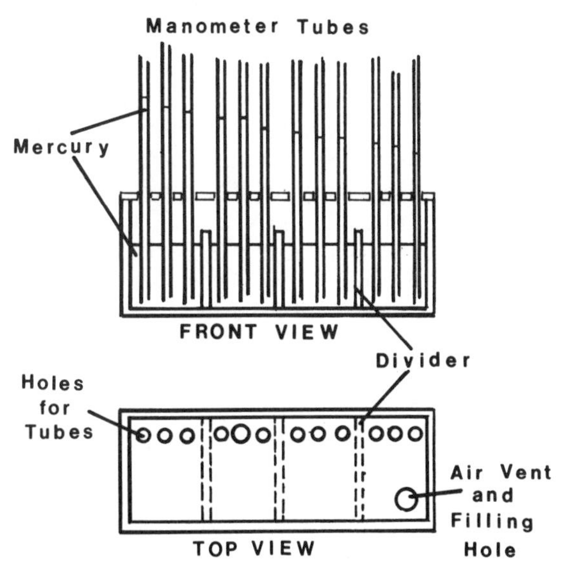

Fig. 23–4. Diagram of a mercury manometer well for a multiple tensiometer system. Adapted from Doering and Harms (1972).

prevent the loss of operation of all the tensiometers in the occasional case when mercury is sucked out of the mercury well by one of the tensiometers. The manometer tubing should extend at least 76 cm above the mercury level to prevent this from happening.

23–3.2 Procedures

23–3.2.1 PREPARATION AND TESTING

The tensiometer is a simple instrument to use. Yet many needless frustrating hours have been spent trying to get tensiometers to work properly. These frustrations usually arise because of (i) improper testing and preparation of the tensiometers before installation, (ii) improper installation of the tensiometers, (iii) improper servicing of the tensiometers, or (iv) improper storage of the tensiometers when they were removed from the soil after a previous use. Tensiometers should last indefinitely if provided proper care.

The following section deals with the use of commercial tensiometers or custom-made tensiometers with vacuum gauges, which are used in field situations. The concepts are applicable to all tensiometers.

Each tensiometer should be tested prior to field installation. Remove the cap from the tensiometer, stand the tensiometer upright in a pail of water, and allow the cup to soak overnight. During this period the water will wet the cup, move through it, and accumulate in the barrel. If water does not move into the cup, the walls of the cup are plugged and the cup conductance is too low. It may be possible to clean the cup by a combination of sanding the outer surface and treating the cup with warm

HCl solution (one part acid to three parts water). If such treatment is unsucessful or if one does not wish to attempt it, the cup should be replaced.

If water does collect in the tensiometer after the soaking period, pour it out and fill the tensiometer with deaerated water. The use of deaerated water is imperative. Water that has not been deaerated contains dissolved gases which will come out of solution when the water in the tensiometer is subjected to a pressure less than atmospheric. Deaerated water may be prepared by boiling and cooling the water. During the cooling the air-water surface area should be minimized to reduce the uptake of air by the water.

With a hand-operated vacuum pump and attached vacuum gauge (Fig. 23–5), apply a vacuum of 60 to 80 kPa to the tensiometer. This process will tend to remove air bubbles from the porous cup and the Bourdon gauge, and will also remove the minute bubbles adsorbed to the imperfections in the walls of the material of the barrel. Hold the vacuum for 30 to 60 s to allow time for the bubbles to rise into the air trap. Tilting and tapping the tensiometer will expedite the movement of bubbles to the air trap. Release the vacuum and refill the air trap with deaerated water. Replace the cap tightly and develop a suction of 60 to 80 kPa in the tensiometer by allowing water to evaporate from the ceramic cup. Additional air may accumulate in the air trap. If it does, refill the

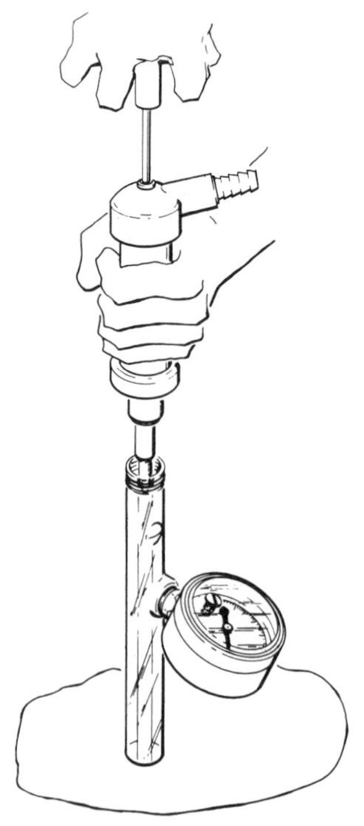

Fig. 23–5. De-airing tensiometer with a hand-held vacuum pump. Drawing supplied by Soilmoisture Equipment Corp. Santa Barbara, CA 92105.

air trap with deaerated water and again develop the suction in the tensiometer by evaporation of water from the cup. Repeat the cycle of refilling the air trap and development of suction by evaporation until the amount of air that accumulates in the trap is negligible.

If the above process is not successful, there is a leak in the instrument, which must be found. Leaks are most likely at the point of connection of the vacuum gauge to the tensiometer barrel, but the cup may be defective (cracked), or there may be a crack at the junction of the ceramic cup and the barrel. Close observation of the instrument while the suction is increasing due to evaporation will usually provide clues as to where the leak is occurring. To test the bubbling pressure of the cup and the seal at the cup-barrel juncture, soak the tensiometer cup in water to wet it. Remove the vacuum gauge and insert a pipe plug in its place. Remove any accumulated water from the inside of the tensiometer cup, and apply air pressure inside the tensiometer while the cup and the cup-barrel seal are submerged in water. It should be possible to apply an air pressure somewhat in excess of 100 kPa without forcing air through the cup. Air flow through the cup can be detected as a stream of bubbles from the cup when it is submerged in water.

The suction may also be developed in the tensiometer with an applied vacuum. Puckett and Dane (1981) describe such a procedure for testing tensiometers. The method appears to be faster than the evaporation method.

The response time of the tensiometer should be checked by developing a suction of 60 to 80 kPa in the instrument by evaporation from the cup, and then submerging the cup in water. The gauge reading should drop to the 0 to 5 kPa range within 5 min. If the gauge reading does not approach zero, the gauge may be faulty, it may require rezeroing, or there may still be some entrapped air in the system, perhaps in the gauge. Appropriate steps must be taken to correct such problems. If the cup conductance is very low the response time will be too long, and the cup must be cleaned or replaced (see above).

The porous cups should be kept wet until the tensiometers are installed in the field.

23–3.2.2 INSTALLATION

Tensiometers may be installed in moist loose soil at shallow depths by pushing them into the soil to the desired depth. In most situations the tensiometer can be installed in an access hole made by a thin walled metal tube with the same diameter as the barrel of the tensiometer. In soils that are severely compacted or stony or that contain other foreign objects, a pointed steel rod may be driven into the soil with a sledge hammer to make the access hole. A soil auger with a diameter of about 5 cm may also be used to make the access hole. The cup should be imbedded in a slurry of soil at the bottom of the access hole. Pour about 50 mL of water or a similar quantity of a soil slurry into the hole, insert the tensiometer, and seat it with a slight twisting motion. If a tube or auger with a diameter larger than that of the tensiometer barrel has been used to make the access hole, backfill the hole around the barrel and tamp the soil firmly into place with a metal rod to prevent water from running into the hole along the tensiometer barrel. The surface of the soil around the tensiometer should be 5 to 8 cm higher than the prevailing soil surface.

After installation, fill the tensiometer with deaerated water. Allow time for dissipation of the hydraulic disturbance introduced by the installation (perhaps 3 to 6 h) before taking the first readings.

23–3.2.3 MAINTENANCE AND SERVICING

The tensiometers must be serviced regularly after they are installed. Each tensiometer should be inspected for air accumulation at least twice a week and preferably more often. In general, more frequent maintenance is required under hot dry soil conditions than under cool wet soil conditions. Even though there may be a small amount of air in the trap, the reading of the tensiometer may be acceptable and should be recorded before purging the air from the system.

To purge the instrument, remove the cap and refill the barrel with deaerated water. Purge air bubbles from the tensiometer by creating a suction in the tensiometer with the hand-operated vacuum pump. Maintain the suction for 15 to 30 s, or long enough to allow the bubbles to rise to the top of the water in the tensiometer barrel. Add deaerated water to fill the instrument and replace the cap. Time should be allowed for the dissipation of the hydraulic disturbance introduced into the soil water by the purging operation (perhaps 1 or 2 h) before taking readings.

If mercury-water manometers are being used, the purging operation must be slightly modified. In addition to the application of vacuum to the tensiometer system as described above, the manometer tubing must be flushed free of air. This can be done by forcing water from the region of the air trap through the manometer tubing and out through the pool of mercury. A 50-cm^3 plastic syringe fitted with a large needle inserted in a rubber stopper is convenient for this purpose. The rubber stopper should be chosen to provide a tight seal to the top of the air trap.

When the suction approaches the limit of operation of the tensiometer, the accumulation of air in the tensiometer will become so excessive that the readings are unreliable. Whether or not this is the case can generally be evaluated from the previous history of readings of the tensiometer. If it is known that the instrument has reached the operating limit, it might as well be "retired" from use, until the soil around the cup is rewetted by rain or irrigation. At that time the tensiometer can be placed back in service by filling it with deaerated water and using the vacuum pump to purge air bubbles from the system as described above.

23–3.2.4 INTERPRETATION OF READINGS

Tensiometers are used to measure the hydraulic head and/or the pressure head of the soil water. To illustrate the definition of the hydraulic

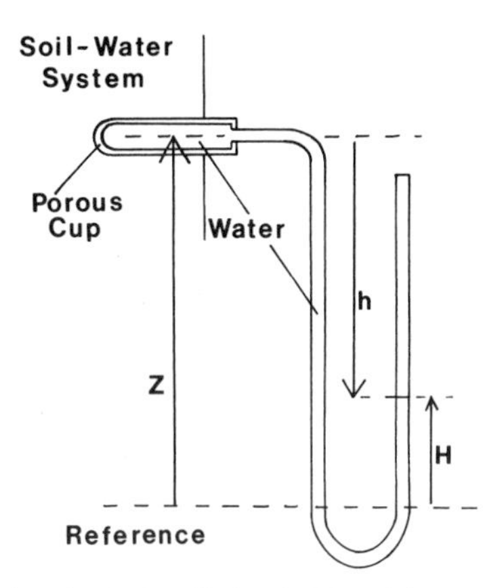

Fig. 23–6. Diagram showing the hydraulic head, H, and its components, pressure head h and gravitational head Z, as they are observed on a water-manometer tensiometer.

head and its components, the simple tensiometer system with a water manometer (shown in Fig. 23–6) may be used. A reference elevation and pressure must be chosen for the measurement of hydraulic head . The choice of reference elevation is arbitrary and is usually based on convenience for the experimenter. The ambient atmospheric pressure is commonly used as the reference pressure. The hydraulic head, H, of the water in the tensiometer cup is defined as the elevation of the water surface (subjected to atmospheric pressure) in the open arm of the water manometer connected to the tensiometer cup, and may be positive or negative. The pressure head, h, is the distance from the tensiometer cup to the water surface in the manometer, and will be negative if the water surface is below the cup. The gravitational head, Z, is the elevation of the cup above the chosen reference. If the cup is located below the reference level the gravitational head is negative. The hydraulic head is the sum of the pressure and gravitational heads. The head components are measured in length units , but they may be converted to other hydraulic potential units with the relationships shown in Table 23-1. It is assumed that the water in the tensiometer and that in the soil just outside the cup are in hydraulic equilibrium, so that the hydraulic head of the soil water is the same as that of the tensiometer water.

If the pressure head of the soil water is negative, as it is in unsaturated soil, a pit would have to be dug beside the installation site of the tensiometer cup in order to use a water manometer. Furthermore, a water manometer much more than a meter in height, equivalent to about 10 kPa, is impractical, and the suction range that can be measured with such a manometer is severely limited.

To obviate this difficulty, tensiometers for field use commonly make use of the Bourdon gauge, or a combination mercury-water manometer. An additional advantage of the Bourdon gauge tensiometer is that the gauge is less likely to be damaged during cultivation and other field operations.

Consider first the mercury-water system, shown in Fig. 23–7A. The

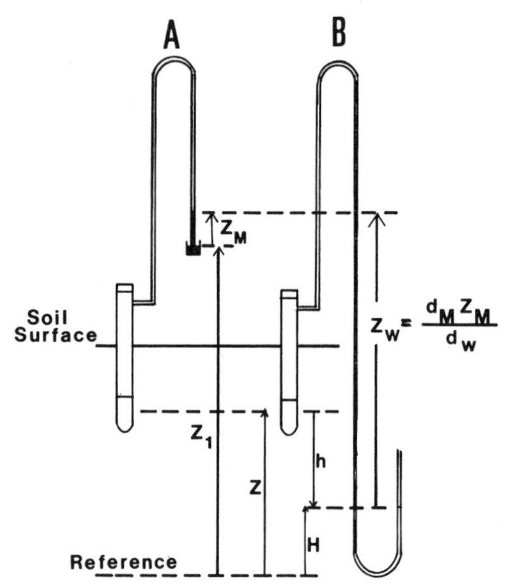

Fig. 23–7. (A) Diagram of a mercury-water manometer tensiometer and (B) its equivalent water manometer system. The symbols are defined in the text.

mercury column of height z_M may be replaced with an equivalent water column of height z_w as shown in Fig. 23–7B. The hydraulic head of the water in the tensiometer cup, and hence that of the soil water just outside the cup, is given by

$$H = z_1 + z_M(1 - d_M/d_w) \quad [6]$$

where d_M is the density of mercury, and d_w the density of water.

The pressure head is given by

$$h = H - Z = z_1 - Z + z_M(1 - d_M/d_w)\,. \quad [7]$$

If the pressure head at the cup is zero, the mercury will rise above the level in the mercury reservoir a distance z_{Mo} given by

$$z_{Mo} = (z_1 - Z)/(d_M/d_w - 1)\,. \quad [8]$$

An alternative expression for the pressure head is

$$h = (z_M - z_{Mo})(1 - d_M/d_w)\,. \quad [9]$$

Commercial mercury-water tensiometers are fitted with a scale whose calibration is based on Eq. [9], with a zero located a distance z_{Mo} above the mercury level in the reservoir.

A diagram of an installed gauge type tensiometer is shown in Fig. 23–8A. The gauge is often calibrated in centibars (kilo-Pascals) of suction.

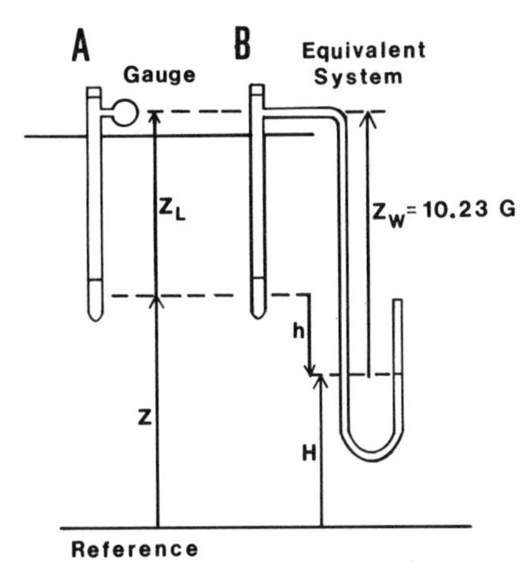

Fig. 23–8. (A) Diagram of a gauge tensiometer and (B) its equivalent water manometer system. The symbols are defined in the text.

The suction indicated is that in the water at the level of the gauge on the tensiometer. We may replace the gauge with an equivalent water manometer as shown in Fig. 23–8B. If G is the gauge reading in kilo-Pascals, the height of the equivalent water column z_w is given by

$$z_w = 10.23\ G$$

(10.23 cm of water at 25 °C are equivalent to 1 kPa). The hydraulic head of the water in the tensiometer cup, in centimeters of water, is given by

$$H = Z + z_L - 10.23\ G \qquad [10]$$

and the pressure head by

$$h = z_L - 10.23\ G\,. \qquad [11]$$

In these equations, z_L is positive, i. e., it is measured from the tensiometer cup to the gauge.

23–3.2.5 REMOVAL AND STORAGE

When readings of the tensiometers are discontinued, e.g., at the end of a growing season, the tensiometers should be removed from the soil. Suppliers of commercial tensiometers recommend scouring the porous cup of the tensiometer with soil to prevent plugging of the cup pores by precipitation of salts accumulated at the cup surface. Flushing the cup pores by water flow from within the cup will also reduce the possibility of plugging. If the tensiometer is to be reinstalled within 1 or 2 weeks, fill the tensiometer with water and store it with the cup immersed in water. If the instrument is to be stored for longer periods, clean the cup by washing and flushing, and remove the water. Store the instrument where it will not be damaged by physical abuse, and where residual water in the Bourdon gauges will not be frozen.

23–3.3 Applications

23–3.3.1 IRRIGATION SCHEDULING

Tensiometers are widely used to schedule the application of water for a large variety of tree and field crops. Irrigations for container-grown plants and plants grown in greenhouse beds are also often scheduled using tensiometers. When the suction indicated by the tensiometer(s) installed at appropriate depths in the root zone reaches a prescribed value, irrigation water is applied. The following discussion is directed toward use of tensiometers for scheduling in field situations.

It is essential to make an inventory of the soil resources in the field where the irrigations are to be scheduled by tensiometers. Information concerning the kinds of soils, their area, slope, and water-holding properties should be collected. The amount of water available to a plant in a given soil is dependent on the texture and structure of the soil. A greater percentage of the water available to a plant is retained by coarse textured soils at suctions less than 80 kPa than is the case for fine-textured soils (see Fig. 23–9). Hence the use of tensiometers for scheduling irrigations will be more successful in coarse textured soils than it will in fine textured soils. For the latter an alternative method of scheduling should be sought.

Tensiometers should be installed at sites or stations that are representative of the major areas of kinds of soil and of water application practices in a field. No general recommendation can be given regarding the number of stations required for scheduling irrigations, because fields differ with regard to size, uniformity of water applications, and uniformity

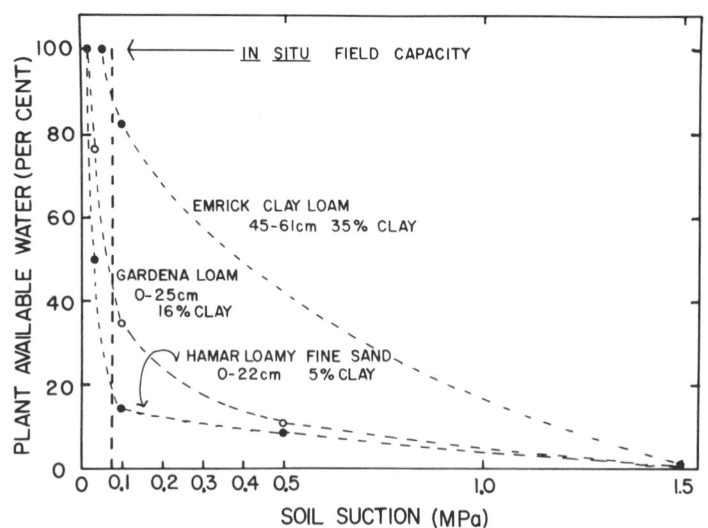

Fig. 23–9. Percent plant-available water versus suction for several soils of various texture. Adapted from Cassel and Sweeney (1975).

of soils. It is desirable to have more than one station for each kind of soil. The number of stations for a given field is determined in part by the flexibility and uniformity of the irrigation system for applying water. If the system is such that water can be applied at different rates and/or different amounts on various parts of the field, it is desirable to have stations on each of the major kinds of soil in the field that can be irrigated separately. If on the other hand the irrigation system can only apply water to the whole field, tensiometer stations should be established on the major kind of soil in the field, and the timing of the application of water determined in a manner that will do the largest area of the field the most good.

Each tensiometer station must be located so that it will not be damaged by machinery or laborers. For row crops, the tensiometers are usually installed in the row. The soil in the vicinity of the tensiometer should not be compacted by foot or vehicle traffic, which may reduce the infiltration of water. If it is considered desirable to protect the gauge or manometer of the tensiometer with a cover, care must be taken that the cover does not interfere with the application of water to the soil where the tensiometer cup is located.

The number of tensiometers installed at each station is dependent upon the kind of crop and its stage of growth. Two tensiometers per station are often used, with the porous cup of one at a depth equal to one-fourth of the active rooting depth and the cup of the other at the bottom of the rooting zone. The upper tensiometer is used to schedule the irrigations and the lower one is used as an indicator of leaching. If the rooting depth of the crop is less than about 50 cm, e.g., strawberries (*Fragaria* × *ananassa*) or potatoes (*Solanum tuberosum* L.), one tensiometer per station is often used at a depth equal to about two-thirds of the active rooting depth. Knowledge of the rooting pattern of the crop to be irrigated as it might be modified by the presence of tillage pans, levels of toxic aluminum, or other structural and textural features of the soil profile will be useful in selecting the depths at which the tensiometers are to be placed.

Scheduling irrigation for deeper rooted crops using only one tensiometer per station has been proposed for soybeans [*Glycine max* (L.) Merr.] and sugar beets (*Beta vulgaris* L.) (Cassel & Bauer, 1976, Cassel et al., 1978). The quantity of water applied at each irrigation is based on knowledge of the soil water characteristic and the changes in the rooting depth with time. Further details concerning depths of tensiometer installation, the number of stations per field, etc., are discussed for selected crops by Hagan et al. (1967).

The tensiometers should be read often enough to detect trends in the soil water suction. Three to four or more readings per week are required for this purpose in coarse textured soils under conditions of high evaporative demand. It is useful to plot the suction values for each tensiometer versus time on a graph (Fig. 23–10). Updating this chart each day or two allows an experienced operator to predict the readings for the next few

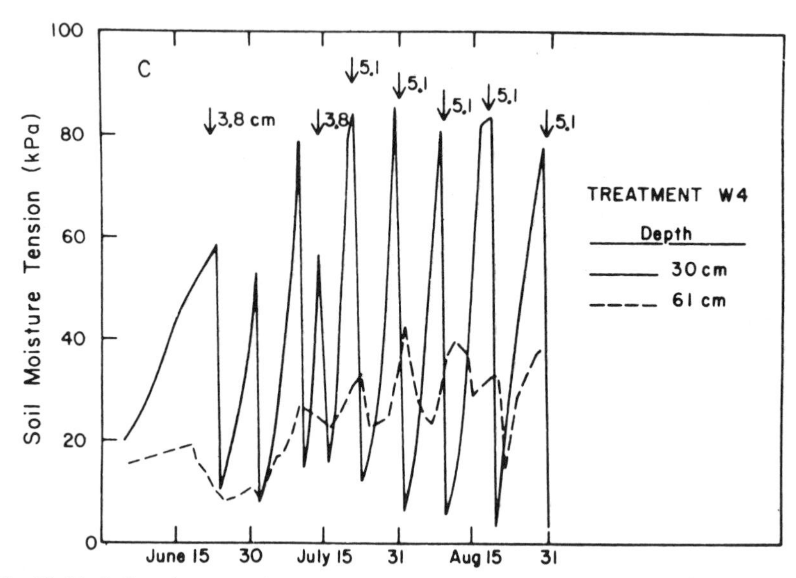

Fig. 23-10. Soil moisture tension at the 30 and 61-cm depths for sugar beets on Maddock sandy loam. After an irrigation of 3.8 or 5.1 cm the suction decreased and then increased, gradually at first and then at a faster and faster rate, until irrigation water was applied again (Cassel & Bauer, 1976).

days and therefore to anticipate the date when the next irrigation will be required.

If tensiometers are to be used to schedule irrigations for plants grown in containers and greenhouse beds, one tensiometer per container is usually sufficient, because the depth of soil in such containers does not usually exceed 20 to 30 cm. Irrigation is initiated when the suction indicated by the tensiometer reaches a prescribed level, and is terminated when water begins to flow from the drainage holes in the base of the container, or when the computed amount of water, based on the water retention characteristic of the soil in the container, has been applied.

23-3.3.2 ROOT ZONE DELINEATION

In some research on soil-plant relationships, it may be desirable to determine the rooting zone of the plants. The temporal and spatial pattern of soil water suction allows the active portion of the root zone to be defined. Information on rooting volumes is useful for selecting installation depths of tensiometers, for evaluating the effects of crop management practices (such as tillage) on rooting development (Cary & Rasmussen, 1979; Weatherly & Dane, 1979), and to acertain the presence of root-restricting barriers in soil (Campbell et al., 1974, Doty et al., 1975, Campbell & Phene, 1977). In the absence of a shallow water table in a relatively uniform soil profile (i.e., the soil water characteristic is uniform with depth), the rate of increase in suction at each position is related to the rate of water taken up by the active roots. For soils with a wider range in texture and/or structure in the soil profile, the rooting depth can still be defined with tensiometers, but because the water retention characteristics vary with texture and structure, the relationship of rooting activity and the rate of suction change is more complex than in the former situation.

Arya et al. (1975a, 1975b) used a set of the miniature tensiometers described in section 23–3.1.1.2 to measure the suction and water loss patterns of soybeans following several irrigations. Daily tensiometer readings were made on a 10 cm square grid to a depth of 70 cm between rows spaced 80 cm apart. Each vertical tier of tensiometers was installed in an access hole with a diameter of 7.8 cm and a depth of 70 cm. The tensiometers were installed at 10 cm depth intervals as the hole was backfilled with soil and packed to its original bulk density.

Tensiometers have also been used to evaluate the effects of tillage or irrigation upon the rooting patterns of soybeans grown on soils having tillage-induced pans (Martin et al., 1979; Kamprath et al., 1979; Weatherly & Dane, 1979).

23–3.3.3 HYDRAULIC GRADIENT MEASUREMENTS

In studies of soil water flow in unsaturated soil profiles, the distribution of hydraulic gradient in the profile is often needed. Tensiometers can be used to measure the hydraulic head (section 23–3.2.4). The hydraulic gradient distribution in depth and time can be developed from the tensiometer data. Because the hydraulic head data are being differentiated to estimate the hydraulic gradient, it is necessary to determine the hydraulic head with sufficient precision and sensitivity to allow a reasonable estimate of the gradient. For this purpose, tensiometers with mercury-water manometers are generally better than those with vacuum gauges. The interpretation of the tensiometer readings in terms of hydraulic head and hydraulic gradient is greatly facilitated if a common mercury reservoir (see Fig. 23–4) is used for all the manometers that are connected to the tensiometers in a given profile "stack."

The instantaneous profile method for measuring in situ unsaturated hydraulic conductivity of field soils employs replicated tensiometers installed at a series of soil depths (Nielsen et al., 1964; Rose et al., 1965; see also chapter 30 on field measurement of conductivity and diffusivity). Water is applied to wet the soil profile to a desired depth. Then the water source is removed, and the hydraulic head and suction distribution in the soil profile are monitored with tensiometers while evaporation is prevented at the soil surface. Often, tensiometers have been installed at arbitrary depths, but more recently there is an awareness that the tensiometers should be installed at depths based on soil morphology. The soil water flux is calculated from the changes in the distribution of soil water content using the continuity equation. The water content is either measured (often with a neutron meter) or inferred from the suction values with the aid of water retention data for the soil.

Another in situ method for the measurement of hydraulic conductivity is the "plane of zero flux" method (Arya et al., 1975a; Cooper,

1980). Tensiometers are used to determine the depth in the profile at which the hydraulic gradient is zero, and hence the depth at which the flux is zero.

23–3.4 Comments

The size of the tensiometer cup relative to the size of structural inhomogeneities in the soil should be considered, when measurements are to be made in soils with highly developed structure and well developed macropores, such as root and worm holes. An example of the phenomena and problems encountered in such a case was described by Bouma et al. (1982). The soil in that study had prominent worm and root holes. Water was infiltrated into the soil and allowed to redistribute. Tensiometers with large cups relative to the spacing between the macropores quickly showed low suction during the infiltration. Tensiometers with small cups which were embedded in the soil between the macropores did not show as rapid a decrease of suction during the infiltration. The suction displayed by a tensiometer is some average value of the range of suctions to which the cup is exposed. Even though only a small fraction of the surface of the large tensiometer cup was exposed to the water in the macropores, which was at a low suction, the reading of the tensiometer was dominated by that water. The water in the soil between the macropores was at a higher suction, and at a given depth in the profile, was not at equilibrium with the water in the macropores. Consequently, the water in the small tensiometer cups in the intermacropore soil was at a higher suction. The question to be considered is the scale of averaging of the suction that is needed at a given depth. The answer will depend in part upon the scale of averaging employed in the model of the soil in which the tensiometer data are to be used.

Soil and air temperatures below freezing present a hazard in the use of tensiometers in field situations in late fall, winter, and early spring. Damage to the cup, gauge, or manometer, and in some cases even the barrel of the tensiometer may occur.

When it is known that tensiometers will be used under freezing conditions, they are often buried below ground level or installed with the gauges or manometers in heated enclosures. Wendt et al. (1978) filled tensiometers with a methanol-water solution (30% methanol by volume) and found that the tensiometer functioned properly at temperatures as low as −18 °C. The maximum error reported as a result of using the methanol solution instead of water was 3 kPa. The long term effect of the alcohol solution on the life of the tensiometer is unknown.

Fluctuations of air and soil temperature cause fluctuations of the tensiometer readings. The effects of temperature fluctuations are greatest for tensiometers at shallow depths. Temperature changes cause changes in the density and surface tension of the fluid in the tensiometer (and in the soil). Hence the suction of the soil water is a function of temperature. Conduction of heat from the air to the soil around the tensiometer cup via the barrel of the tensiometer can occur if the barrel is made of metal

(Haise & Kelly, 1950). Construction of the barrel from plastic materials reduces this problem.

23-4 LABORATORY TENSIOMETRY

23-4.1 Apparatus

23-4.1. SELECTION OF POROUS BARRIER

There are many variations of form and kind of porous barrier for laboratory tensiometers. In the following discussion the term "cup" will refer to the porous barrier used to establish hydraulic contact between soil water and tensiometer water, even though the actual form may be a flat plate or strip. Tensiometer cups may be constructed from porous ceramic, sintered metal, or fritted glass beads (Chow, 1977; Nielsen & Phillips, 1958), or various types of porous plastic membranes. In some cases a layer of fine soil material can serve as a tensiometer cup. The porous barrier may be a small tube, which extends across a flow column; a cup, which protrudes into a flow column; a flat plate or strip in the wall of a rectangular flow cell; or a strip in the wall of a flow cell of cylindrical cross section. Spring loading may be used to maintain contact between the tensiometer cup and the soil. Narrow strips of porous barrier material may be cemented in place over a narrow groove to provide a channel for flushing the air from the tensiometer. Figure 23-11 shows two strip tensiometers of this kind. Inlet and outlet tubes may be placed in a tensiometer cup to provide flushing capability, as was shown in the miniature tensiometer described in section 23-3.1.1.2.

The flow-through tensiometer shown in Fig. 23-12 is useful in laboratory soil columns. The porous cup can be a tube fabricated from porous plastic or porous ceramic, or it may be a fritted glass tube (obtainable on special order from Corning Glass Works, Corning, NY). The

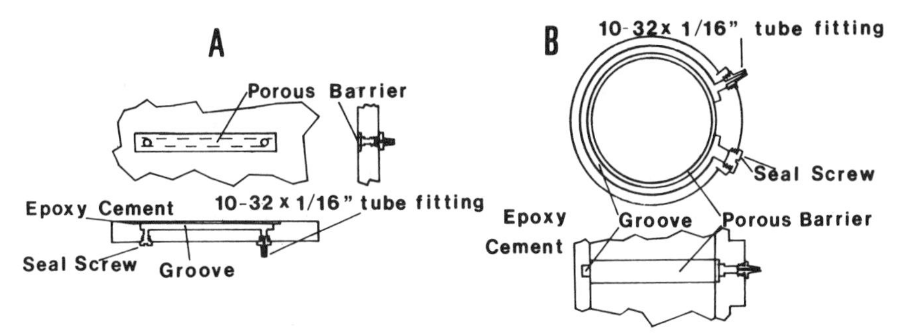

Fig. 23-11. Two strip-tensiometer designs for use in laboratory flow systems. (A) Rectangular flow cell. (B) Cylindrical flow cell. The 10-32machine screw to 1/16-inch hose fitting for hydraulic connection to the transducer via 1/16-inch i.d. vinyl tubing is available from the Clippard Instrument Co., 7390 Colerain Ave., Cincinnati, Oh 45239. Alternatively, a 1/16-inch NPT to 1/8-inch tube compression fitting may be used if metal tubing is used.

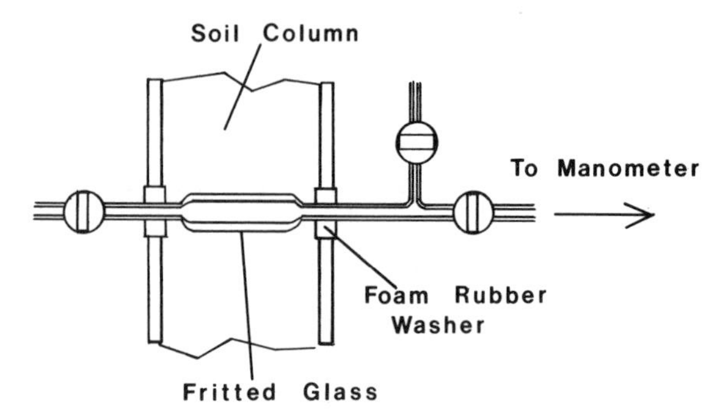

Fig. 23–12. Diagram of a flow-through tensiometer for installation in a laboratory flow column.

stopcocks facilitate the flushing of air from the tensiometer. The glass tee serves as an air trap. The foam rubber washers to support the connecting tubing minimize the danger of breakage of the fritted glass and connecting tubes.

Table 23–3 lists some porous materials for tensiometer construction and their hydraulic characteristics. The choice of cup material is dictated to a high degree by the required bubbling pressure and by whether or not the material can be fabricated into the desired physical form for use in the laboratory flow system. A material with a bubbling pressure no higher than necessary should be chosen in order to obtain the highest possible conductance for the cup, and hence a more responsive tensiometer. Many tensiometer cups in laboratory flow systems will inherently have a small area of contact with the soil in the flow system, and will consequently have low conductance. Porous ceramics and fritted glass in the form of cups or plates usually have sufficient strength and rigidity to provide their own physical support. Porous plastic materials are sufficiently rigid to be self supporting if the shape of the barrier is a narrow strip, no more than about 6 mm wide. Because of the variability of the bubbling pressure of many of the porous materials used for tensiometer cup construction and the possibility of leaks in the fabricated tensiometer, it is desirable to test each cup before it is placed in service.

The cup conductance defined in Eq. [2] is an appropriate conductance parameter for porous bodies of irregular geometry, whose thickness and cross sectional area are difficult or impossible to define. This conductance parameter may be called the "per piece" conductance. If the area through which the flow is taking place can be defined, another useful conductance parameter is the conductance "per unit area," k, defined by

$$k = V/(At\Delta H) \qquad [12]$$

where k has dimensions of reciprocal time.

23–4.2. THE PRESSURE MEASUREMENT SYSTEM

Most laboratory tensiometers use a manometer or a pressure transducer for the measurement of the pressure. Water and combination mer-

Table 23-3. Hydraulic characteristics of some porous materials for tensiometer construction.

Material	Bubbling pressure	Con- ductivity	Conductance per piece	Conductance per area
	cm water	cm s^{-1}	cm^2 s^{-1}	s^{-1}
Ceramic plates†				
1/4 bar, 0.64 cm thick	240–320	7.0×10^{-5}		1.1×10^{-4}
1/1 bar, 0.64 cm thick	490–630	1.7×10^{-6}		2.7×10^{-5}
1/2 bar, Hi-Flow, 0.64 cm thick	490–630	3.2×10^{-5}		5.0×10^{-5}
1 bar, 0.64 cm thick	1400–2100	3.5×10^{-7}		5.5×10^{-7}
1 bar, Hi-Flow, 0.64 cm thick	1400–2100	8.7×10^{-6}		1.4×10^{-5}
3 bar, 0.64 cm thick	3200–4900	1.7×10^{-5}		2.7×10^{-7}
Ceramic cups				
1 bar, 0.7 cm o.d., 0.3 cm i.d., 3.0 cm long, Cat. no. 2105-1†	1400–2100		3.0×10^{-5}	
1 bar, 1.9 cm o.d., 6.7 cm long, Standard Tensiometer cup, Cat. no. 2325†	1400–2100		5.0×10^{-5}	
1 bar, 2.2 cm o.d., 6.7 cm long, Hi-Flow Tensiometer cup†	1400–2100		1.0×10^{-3}	
P-16-C, 1.1 cm o.d., 5.5 cm long‡	175–260		5.0×10^{-3}	
P-10-C, 1.1 cm o.d., 5.5 cm long‡	280–420		1.2×10^{-3}	
P-3-C, 1.1 cm o.d., 5.5 cm long‡	1300		5.0×10^{-5}	
Fritted glass bead plates				
28-μm beads§	250–270	4.0×10^{-5}		
44–100-μm beads¶	110–120	2.7×10^{-3}		
Fritted glass				
Immersion filter, 0.9-cm diameter flat plate, porosity grade M††	240–250		4.8×10^{-4}	7.7×10^{-4}
Porous plastic				
MicroPorous Plastic, Grade 10‡‡	250–500			7.0×10^{-5}

† Soilmoisture Equipment Corp., P.O. Box 30025, Santa Barbara, CA. Ceramic cups and plates may also be obtained from Selas Corp. of Am., Flotronics Div., P.O. Box 220TR, Dresher, PA 19025.
‡ Coors Porcelain Co., Golden, CO 80401.
§ Nielsen and Phillips (1958).
¶ Chow (1977).
†† Corning Glass Works, Corning, NY 14831.
‡‡ Amerace Corp., 555-T 5th Ave., New York, NY 10017.

cury-water manometers may be used. However, because of the small gauge sensitivity of such manometers, pressure transducers of the flat diaphragm type, which have relatively large gauge sensitivities, are much to be preferred (Klute & Peters, 1968; Watson, 1965). This is particularly the case when measurements are to be made in unsteady state flow systems, in which rapid changes of the pressure head may be occurring. The generally smaller area of contact between the tensiometer barrier and the soil leads to lower values of plate conductance, which tends to increase the response time. However, if the tensiometer system is made as rigid as possible, and if a diaphragm type transducer with a large gauge sensitivity is used, the response time of the instrument can be made as short as a few seconds or even less. For example, the tensiometer system shown in Fig. 23–13, composed of a flat diaphragm type pressure transducer,[4] two plug valves,[5] about 50 cm of 0.060 in. (1.5 mm) i.d. nylon tubing,[6] and a 100 kPa bubbling pressure porous ceramic strip 10 cm long, 0.32 cm wide and 0.32 cm thick,[7] will have a response time of about 0.3 s.

Since there are many possible variations in design, form, and choice of components for a laboratory tensiometer system, it is not possible to describe a system that will serve all needs. The system shown in Fig. 23–13 has been found useful for the measurement of pressure and hydraulic head in laboratory flow systems. The tubing which connects the "cup" to the transducer should be metal for rigidity, if the shortest response time system is desired, but a very satisfactory system can be made using nylon tubing or even small diameter tygon tubing. The valves used in the system should be ball valves or "stopcock-like" plug valves, which introduce nearly zero volume change in the system when they are opened or closed. Ball valves, which have almost no dead volume for entrapment of air and which are consequently easier to flush, are to be preferred over plug valves. Plug valves are usually constructed with O-ring seals, which are somewhat flexible and thus reduce the gauge sensitivity of the tensiometer system. Glass stopcocks may be used in some cases, where their fragility can be tolerated. Fairly rigid connections may be made between the glass stopcock and the pressure transducer by cementing appropriate tube or pipe fittings to the glass with epoxy cement.

[4]Model DP-15, ± 20 psi, Validyne Engineering Corp., 8626 Wilbur Ave., Northridge, CA 91324.

[5]9500 Series, Circle Seal Controls, 1111 N. Brookhurst St., P. O. Box 3666, Anaheim, CA 92803.

[6]Scanivalve Co., P. O. Box 20005, 10222 San Diego Mission Rd., San Diego, CA 92120.

[7]Hi-Flow ceramic, Soilmoisture Equipment. Corp., P. O. Box 30025, Santa Barbara, CA 93105.

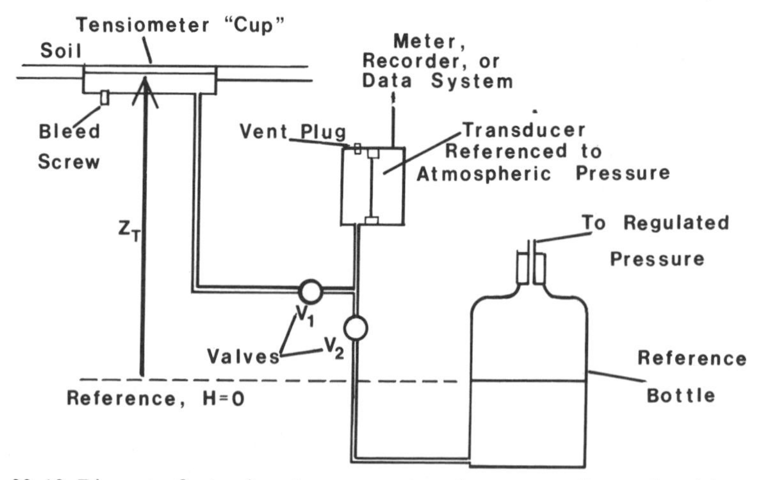

Fig. 23–13. Diagram of a tensiometer-pressure transducer system for use in a laboratory flow cell.

The volume change per unit pressure head change of a tensiometer system may be estimated from the sum of the volume changes with pressure of each of the components of the sytem. Table 23–2 lists the volume change with pressure head for various components that may be used in a tensiometer system. The values for the two manometers were calculated from the diameter of the manometer tubes. The value for water was obtained from its compressibility, and that for air from the gas law. The volume change per unit pressure head change for the other components was measured by connecting each water-filled item to a calibrated capillary tube. Air pressure was applied to the other end of the capillary tube, and the change in position of an air-water meniscus with change in the applied air pressure was used to calculate the volume change per pressure change. It may be seen from these data that one of the important contributions to the gauge sensitivity of an assembled tensiometer system is from the valves. It may also be seen that the presence of even a small amount of air in the tensiometer system will significantly increase the response time of the instrument. For example, a gas bubble with a volume of 0.1 cm^3 in the system described above and shown in Fig. 23–13 will increase the response time from 0.3 s to approximately 3 s.

When measurements are to be made in steady-state or nearly steady-state systems, the response time of the instrument is of somewhat less concern. However, even here, the time for the flow system to attain steady state may in some cases be determined by the quantity of water that must be transferred through the tensiometer cup to bring it to equilibrium with the flow system, and an instrument with a reasonably short response time will facilitate measurements on the flow system.

The details of the read-out system for the transducer will vary with the type of transducer. Flat diaphragm transducers, which use bonded resistance strain gauges, usually incorporate a Wheatstone bridge circuit. Direct-current power for the bridge may be supplied by mercury cells or a well regulated power supply. For reasonably small deviations from the balance condition of the bridge, the voltage vs. pressure relation of the transducer is linear. The output may be observed on a meter from which data may be recorded by hand at intervals. Alternatively, a strip chart recorder or some type of data collection system may be used to capture the data from the tensiometer system. Thus, an additional benefit of the use of an electrical pressure transducer is the possibility of automated data collection.

Variable reluctance pressure transducers require AC excitation and a demodulation circuit to produce a DC output, which may then be displayed on a meter or recorded.

There are many types of electrical pressure transducers available in addition to those mentioned here, some of which may have desirable features. For example, piezoelectric transducers have a very small volume change per unit of pressure change, and in association with appropriate electronic readout equipment provide a robust output signal.

23–4.2. Procedures

The procedure to be given will be appropriate for the tensiometer-transducer system shown in Fig. 23–13 and the tensiometer cups (strips) shown in Fig. 23–11. The procedure may easily be adapted to other tensiometer systems.

When the tensiometer system is first placed in service, the air must be removed from the system and the cup must be wetted. Removal of the air may be accomplished by flushing the system with deaerated water. The reference bottle (Fig. 23–13) is a convenient source of water for flushing the system. Apply sufficient pressure to the water in the bottle to cause water to flow from the bottle and out of the open vent screw on the tensiometer (Fig. 23–11) until all air bubbles are displaced. The cavity of the transducer must also be flushed by forcing water from the reference bottle and out through the vent screw on the transducer.

If air accumulates in the transducer system during its use, the above procedure may be used to remove it.

The transducer is calibrated as follows:

1. Place the free water surface in the reference bottle at a level which is to be designated as zero hydraulic head.
2. Close valve V_1 and open valve V_2.
3. Record the output (voltage, meter scale reading, or recorder chart scale reading) of the transducer readout system.
4. Apply a second known hydraulic head to the transducer by either raising or lowering the reference bottle or by applying a known air pressure to the surface of the water in the reference bottle.
5. Record the output of the transducer system. The slope and intercept of the (assumed) linear relation between transducer output and hydraulic head can be calculated from the readings taken in steps 2 and 5. It is quite often possible to adjust the readout apparatus to produce an output such that the slope of this relation is a power of ten and the intercept is zero. If this can be done the conversion of the output to hydraulic head values is greatly facilitated.

To obtain the hydraulic head from the transducer output, use the linear calibration relation obtained above. To obtain the pressure head of the water in the tensiometer cup, measure the elevation of the cup z_T relative to the level designated as zero hydraulic head. Calculate the pressure head from

$$h = H - z_T . \qquad [13]$$

If the cup is above the zero head reference level, z_T is positive.

23–4.3 Applications and Comments

Flow systems are often used in laboratory studies of the transport of water, solutes, gases, or heat in porous media. Tensiometers may be used in such flow systems to measure the hydraulic and/or pressure head of the water. In some experiments the pressure head at one or more points in the flow system is required to relate the energy state of the soil water to some other property of the medium (e.g., water content), or to the response of plants that may be growing in the medium. In other cases, it may be necessary to measure the hydraulic gradient in the flow system, as in a permeability cell (see chapter 28 on the measurement of hydraulic conductivity and diffusivity). The precision of measurement of the hydraulic gradient can be increased if a differential transducer arrangement is used. In this application, the cavity on each side of the diaphragm of a pressure transducer is connected to a tensiometer to measure the head difference between the tensiometers. The pressure transducer must be capable of withstanding corrosive fluids on each side of the diaphragm. Water is regarded as a corrosive fluid in this context.

If a tensiometer is to be used in a flow cell in which the gas-phase pressure is greater than atmospheric, the porous material used for the tensiometer cup must have a bubbling pressure somewhat higher than the highest gas phase gauge pressure to be used in the cell.

If the flushing procedure given above is not successful one or more of the following techniques may be found useful for the removal of gases from the system:

1. Assemble the tubing system under water. Plug valves may be taken apart and reassembled under water to insure complete filling of the annular space in and around the O-ring seals in the valves.
2. Saturate the tensiometer cups before insertion in the flow system and keep them wet until the tensiometer is assembled.
3. Flush the tensiometer system with CO_2 followed by flushing with deaerated water. Carbon dioxide is much more soluble in water than the predominant gaseous components of air.
4. Disconnect the tubing connection to the tensiometer cup at the tensiometer cup. Flush the transducer and valve system with deaerated water and temporarily close the tubing with a suitable fitting. Leave the valve leading to the reference-flushing bottle open and apply gas pressure to the bottle. This pressure must be within the range of the transducer. Apply the pressure for several hours or even overnight. The increased pressure in the water in the valves and transducer tends to dissolve any minute gas bubbles. At the end of the period of pressurization, connect the tubing to the tensiometer and flush the system with freshly deaerated water.

If water is flushed through the tensiometer cup, the soil water suction in the soil next to the cup will be affected. Consequently, the readings from the tensiometer system will not reflect the "true" soil suction until the disturbance has had time to dissipate.

The assumption of a linear relation between the pressure applied to the diaphragm and the output of the transducer may not be valid if the range of pressure includes zero pressure differential across the diaphragm (i.e., the "slack" diaphragm condition). In some cases, the slope of the calibration curve of the variable reluctance transducer identified in Fig. 23–13 will be found to change as the pressure applied passes through zero pressure differential. It is good practice to check the linearity of the calibration of a given transducer using the calibration procedure described above.

Because of the relatively high cost of the transducer and readout system, it may be necessary to use one transducer with more than one tensiometer cup. A system for hydraulic switching of the transducer from one cup to another is then needed. The valves used in a hydraulic switching system must operate with minimum volume change of the system as they are opened and closed, to minimize the occurrence of pressure transients, which may disturb the flow system and which must be dissipated before valid measurements of the pressure and hydraulic head can be obtained.

A valve manifold system for manual selection of a given tensiometer cup may be constructed using plug or ball valves. Additional valves in the system will increase the response time of the tensiometer. Electrically operated hydraulic switching systems may be constructed, and several models of at least one are available commercially (Scanivalve Corp., San Diego, CA).

If plastic tubing, such as nylon, is used in the tensiometer, it may be found that air will accumulate in the system by diffusion through the tubing, especially if the pressure of the water in the tubing is significantly less than atmospheric. For this reason, metal tubing is recommended instead of nylon or other plastic tubing if the suction levels in the tensiometer are expected to be more than approximately 150 cm of water for extended periods of time.

Temperature fluctuations will cause a change in the output of the tensiometer-transducer system (Watson & Jackson, 1967). Some of this effect is due to the effects of changing temperature on the electrical characteristics of the transducer. Also, stresses may be induced in the transducer by the changing temperature. Temperature changes will also cause expansion or contraction of the material of the tensiometer system and of the water in it, and these will be translated into a change of pressure in the system. When the exchange of water between the tensiometer cup and the soil is inhibited, as by poor contact or by the low hydraulic conductivity of the soil, the effects of temperature change on the output of the tensiometer system will be greatly magnified. Tensiometers that are installed in sands which are at or near the residual water content of the sand will, in general, be very sensitive to the effects of temperature change. The effects of temperature change can be minimized by operating the system in a constant temperature environment, which can be attained by a constant temperature room or enclosure or by insulating the tensiometer-transducer system.

23–5 REFERENCES

Arya, L. M., D. A. Farrell, and G. R. Blake. 1975a. A field study of soil water depletion patterns in presence of growing soybean roots: I. Determination of hydraulic properties of the soil. Soil Sci. Soc. Am. Proc. 39:424–430.

Arya, L. M., G. R. Blake, and D. A. Farrell. 1975b. A field study of soil water depletion patterns in presence of growing soybean roots: II. Effect of plant growth on soil water pressure and water loss patterns. Soil Sci.Soc. Am. Proc. 39:430–436.

Bouma, J., C. F. M. Belmans, and L. W. Dekker. 1982. Water infiltration and redistribution in a silt loam subsoil with vertical worm channels. Soil Sci. Soc. Am. J. 46:917–921.

Buckingham, E. 1907. Studies on the movement of soil moisture. U. S. Dep. Agric. Bur. Soils Bull. 38.

Campbell, R. B., and C. J. Phene. 1977. Tillage, matric potential, oxygen and millet yield relationships in a layered soil. Trans. Am. Soc. Agric. Eng. 20:271–275.

Campbell, R. B., D. C. Reicosky, and C. W. Doty. 1974. Physical properties and tillage of Paleudults in the southeastern Coastal Plains. J. Soil Water Conserv. 29:220–224.

Cary, J. W., and W. W. Rasmussen. 1979. Response of three irrigated crops to deep tillage of a semiarid silt loam. Soil Sci. Soc. Am. J. 43:574–577.

Cassel, D. K., and Armand Bauer. 1976. Irrigation schedules for sugarbeets on medium and coarse textured soils in the northern Great Plains. Agron. J. 68:45–48.

Cassel, D. K., A. Bauer, and D. A. Whited. 1978. Management of irrigated soybeans on a moderately coarse-textured soil in the Upper Midwest. Agron. J. 70:100–104.

Cassel, D. K., and M. D. Sweeney. 1975. In situ soil water holding capacities of selected North Dakota Soils. Bull. 495, Agricultural Exp. Stn., North Dakota State University, Fargo, ND.

Chow, T. L. 1977. Fritted glass bead materials as tensiometers and tension plates. Soil Sci. Soc. Am. J. 41:19–22.

Cooper, J. D. 1980. Measurement of moisture fluxes in unsaturated soil in Thetford Forest. Rep. no. 66 , Institute of Hydrology, Wallingford, Oxon, England.

Daian, J. F. 1971. On a system of automatic scanning and recording of fluid pressures with particular application to the study of water balance made in situ. C. R. Acad. Sci. 272:348–350.

Doering, E. J., and J. P. Harms. 1972. Improved mercury well for multiple tensiometer systems. Soil Sci. Soc. Am. Proc. 36:849–850.

Doty, C. W., R. B. Campbell, and D. C. Reicosky. 1975. Crop response to chiseling and irrigation in soils with a compact A2 horizon. Trans. Am. Soc. Agric. Eng. 18:668–672.

Gardner, W., O. W. Israelsen, N. E. Edlesfsen, and D. Clyde. 1922. The capillary potential function and its relation to irrigation practice. (Abstract). Phys. Rev. 20:196.

Groenevelt, P. H., and G. H. Bolt. 1972. Water retention in soil. Soil Sci. 113:238–245.

Hagan, R. M., H. R. Haise, and T. W. Edminster (ed.) 1967. Irrigation of agricultural lands. Agronomy 11.

Haise, H. R., and O. J. Kelley. 1950. Cause of diurnal fluctuations in tensiometers. Soil Sci. 70:301–313.

Heck, A. F. 1934. A soul hygrometer for irrigated cane lands of Hawaii. J. Am. Soc. Agron. 26:274–278.

Henderson, D. W., and E. P. Rogers. 1963. Tensiometer construction with plastic materials. Soil Sci. Soc. Am. Proc. 27:239–240.

Kamprath, E. J., D. K. Cassel, H. D. Gross, and D. W. Dibb. 1979. Tillage effects on biomass production and moisture utilization by soybeans on Coastal Plain soils. Agron. J. 71:1001–1005.

Klute, A., and W. R. Gardner. 1962. Tensiometer response time. Soil Sci. 93:204–207.

Klute, A., and D. B. Peters. 1968. Hydraulic and pressure head measurement with strain gauge pressure transducers. p. 156–165. *In* P. E. Rijtema and H. Wassink (ed.) Water in the unsaturated zone. Int. Assoc. Sci. Hydrology. Proc. of the Wageningen Symposium, Wageningen, The Netherlands

Long, F. L. 1982. A new solid state device for reading tensiometers. Soil Sci. 133:131–132.

Mahony, J. J. 1975. Tensiometer measurements in anisotropically loaded swelling soils. Soil Sci. 120:421–427.

Marthaler, H. P., W. Vogelsanger, F. Richard, and P. J. Wierenga. 1983. A pressure transducer for field tensiometers. Soil Sci. Soc. Am. J. 47:624–627.

Martin, C. K., D. K. Cassel, and E. J. Kamprath. 1979. Irrigation and tillage effects on soybean yield in a Coastal Plain soil. Agron. J. 71:592–594.

Nielsen, D. R., J. M. Davidson, J. W. Biggar, and R. J. Miller. 1964. Water movement through Panoche clay loam soil. Hilgardia 35:491–506.

Nielsen, D. R., and R. E. Phillips. 1958. Small fritted glass bead plates for determination of moisture retention. Soil Sci. Soc. Am. Proc. 22:574–575.

Philip, J. R. 1970. Reply to comments by E. G. Youngs and G. D. Towner on hydrostatics and hydrodynamics in swelling soils'. Water Resour. Res. 6:1248–1251.

Puckett, W. E., and J. H. Dane. 1981. Testing tensiometers by a vacuum method. Soil Sci. 132:44–445.

Rice, Robert. 1969. A fast response field tensiometer system. Trans. Am. Soc. Agric. Eng. 12:48–50.

Richards, L. A. 1928. The usefulness of capillary potential to soil moisture and plant investigators. J. Agric. Res. 37:719–742.

Rogers, W. S. 1935. A soil moisture meter depending on the capillary pull of the soil. J. Agric. Sci. 25:326–343.

Rose, C. W., W. R. Stern, and J. E. Drummond. 1965. Determination of hydraulic conductivity as a function of depth and water content for soil in situ. Aust. J. Soil Res. 3:1–9.

Strebel, O., M. Renger, and W. Geisel. 1973. Soil suction measurements for evaluation of vertical flow at greater depths with a pressure transducer tensiometer. J. Hydrol. 18:367–370.

Talsma, T. 1977. Measurement of the overburden component of total potential in swelling field soils. Aust. J. Soil Res. 15:95–102.

Talsma, T., and A. van der Lelij. 1976. Infiltration and water movement in an in situ swelling soil during prolonged ponding. Aust. J. Soil Res. 14:337–349.

Towner, G. D. 1980. Theory of time response of tensiometers. J. Soil Sci. 31:607–621.

Watson, K. K. 1965. Some operating characteristics of a rapid response tensiometer system. Water Resources Res. 1:577–586.

Watson, K. K., and R. D. Jackson. 1967. Temperature effects in a tensiometer-pressure transducer system. Soil Sci. Soc. Am. Proc. 31:156–160.

Weatherly, A. B., and J. H. Dane. 1979. Effect of tillage on soil water movement during corn growth. Soil Sci. Soc. Am. J. 43:1222–1225.

Wendt, C. W., O. C. Wilke, and L. L. New. 1978. Use of methanol-water solutions for freeze protection of tensiometers. Agron. J. 70:890–891.

Williams, T. H. L. 1978. An automatic scanning and recording tensiometer system. J. Hydrol. 39:175–183.

24 Water Potential: Thermocouple Psychrometry

STEPHEN L. RAWLINS
Agricultural Research Service, USDA
Beltsville, Maryland

GAYLON S. CAMPBELL
Washington State University
Pullman, Washington

24–1 INTRODUCTION

Thermocouple psychrometers infer the water potential of the liquid phase of a soil sample from measurements within the vapor phase in equilibrium with it. The water potential, ψ, is related to the relative humidity, p/p_o, by

$$\psi = \text{energy/mass} = (RT/M) \ln (p/p_o) \quad [1]$$

where M is the molecular weight of water (0.018 kg mol^{-1}), R is the ideal gas constant (8.31 J K^{-1} mol^{-1}), T is the Kelvin temperature of the liquid phase, p is the water vapor pressure in equilibrium with the liquid phase, and p_o is the saturated water vapor pressure of the liquid phase. Within the range of water potentials commonly encountered in agricultural soils, $\ln(p/p_o)$ can be approximated by $(p/p_o) - 1$. Therefore the water potential, in J kg^{-1}, is

$$\psi \simeq 461\ T\,(p/p_o - 1)\,. \quad [2]$$

The major difficulty in making this measurement stems from the fact that the relative humidity in the soil gas phase changes only slightly within the growth range of plants. For example, at a water potential of -1500 J kg^{-1}, a value often associated with the permanent wilting of plants, the relative humidity calculated from the above equation at 25°C, is still about 0.99. Thus, practically all measurements of interest to soil scientists lie in the narrow relative humidity range between 0.99 and 1.00.

The development of an instrument to measure relative humidity in equilibrium with a plant or soil sample within this range began with Spanner (1951). Major developments since then have been primarily

 Methods of Soil Analysis, Part 1. Physical and Mineralogical Methods–Agronomy Monograph no. 9 (2nd Edition)

concerned with improving the accuracy and reliability of measurements, as well as simplifying psychrometer construction and measurement techniques. Spanner's psychrometer, as well as most of those now used, consists of a miniature thermocouple junction, placed within a sample chamber, that can be cooled by the Peltier effect to condense water on it. It is then connected to a voltmeter to measure its temperature depression as the water evaporates. An alternative method first suggested by Richards and Ogata (1958) requires that a droplet of water be placed on a silver loop that is at the thermocouple measuring junction. With either method, the temperature depression of the measuring junction relative to a dry junction at the same location varies with the rate of evaporation, which, in turn, increases as the relative humidity of the atmosphere decreases.

Recently, Neumann and Thurtell (1972) introduced an improved technique that controls the temperature of the measuring junction at the dew point by a feedback control loop that controls the Peltier cooling rate. Measuring the dew point rather than the wet-bulb temperature depression to estimate the relative humidity has some distinct advantages that will be discussed below.

Because relative humidity measurements are made within such a narrow range, the major source of error arises from a difference in temperature between the reference junction and the liquid phase within the sample whose water potential is being measured. At 20°C a difference of 1 °C between these two points introduces an error of > 13 kJ kg^{-1}. This means that if water potential is to be inferred to within 10 J kg^{-1}, temperature differences must be controlled to within $< 10^{-3}$ °C.

Thermocouple psychrometers are normally calibrated empirically, with solutions of known water potential replacing the sample within the psychrometer chamber. The accuracy with which the water potential of a sample can be measured is, therefore, limited by how closely conditions during calibration of the instrument match those that exist during the measurement. Because it is seldom possible to match these conditions exactly, it is important to understand the operating characteristics of the psychrometer in sufficient detail that the magnitude of error that can be expected by specific deviations from the calibration conditions can be evaluated. Rawlins (1966) and Peck (1968) developed the basic heat and mass transport theory for the operation of typical thermocouple psychrometers that has been used to make designs to increase sensitivity and decrease dependence on changes in environmental conditions.

24-2 PRINCIPLES OF OPERATION

Consider a psychrometer consisting of a spherical chamber of radius r_c, with a spherical wet junction of radius r_j positioned at its center by thermocouple wires of radius r_w. Sample material at the temperature of the chamber wall is assumed to cover the entire interior surface of the psychrometer chamber, and vapor flows from it to the wet junction by simple diffusion through the air. The thermal emissivity of the sample is considered to be the same as the wet junction.

Heat flows to the wet junction from the sample by radiation, by conduction through the wires, and by conduction through the air in the chamber. If r_c does not exceed a few centimeters, heat and vapor transport by convection are negligible compared to conduction for the temperature depressions that normally occur in soil psychrometers. This heat flow to the wet junction is balanced by the cooling effect of water evaporating from it. The actual temperature the wet junction attains depends on factors that influence heat flow to it and vapor flow from it.

For a given wet bulb depression, heat flow to the wet junction is controlled by the dimensions r_c, r_j, and r_w, as well as by the thermal conductivity of the thermocouple wires and the air separating the junction from the chamber wall. Cooling at the wet junction is proportional to the product of the evaporation rate and the latent heat of vaporization. The rate of evaporation is primarily a function of the relative humidity of the chamber, but also varies with the diffusivity of water in air. The latent heat of vaporization, thermal conductivity, and water vapor diffusivity of air increase with increasing temperature. In addition, the diffusivity of water in air decreases with atmospheric pressure.

By inserting the known temperature and pressure dependencies of these parameters into the heat and mass flow equations applicable to thermocouple psychrometers, Rawlins (1966) explained the temperature dependence reported by Klute and Richards (1962) and the pressure dependence reported by Richards et al. (1964) of thermocouple psychrometers.

Pressure and temperature effects are conveniently summarized using the conventional psychrometer equation, rewritten to give relative humidity (Campbell, 1979):

$$p/p_o = 1 - [(s + \gamma^*)/p_o]\Delta T . \qquad [3]$$

Here s is the slope of the saturation vapor pressure curve ΔT is the wet bulb depression, and γ^* is the apparent psychrometer constant, the product of the thermodynamic psychrometer constant and the ratio of vapor to heat transfer resistance. The term in square brackets determines the psychrometer sensitivity. Temperature dependence comes mainly from the temperature dependence of s and p_o. Pressure dependence comes mainly from the effect of pressure on γ^*. When pressure doubles, γ^* doubles. Equation [3] is useful to correct for temperature and pressure dependence of psychrometer sensitivity.

Rawlins (1972) showed that by reducing r_j relative to r_c and minimizing r_w, the psychrometer sensitivity could be made independent of its dimensions. This he termed an ideal psychrometer. Effectively, choosing these ideal dimensions reduces heat flow to the wet bulb by radiation and thermal conduction down the wires to values that are negligible compared to conduction through the air. Because heat flow to the junction and vapor flow away from the junction occur through identical pathways through air only, any variation of dimensions that affects one will cause an equal and offsetting effect in the other.

Maintaining the wet junction at the dew point by controlling the cooling current significantly improves the instrument's performance. Heat flow to the wet junction from its surroundings is exactly offset by adjustments in the cooling current, so no water evaporates or condenses. Unlike the case for the wet-bulb depression of a psychrometer, the dew point is maintained independent of the rate of heat flow to the wet junction. This eliminates most of the temperature-dependent parameters that influence the wet-bulb temperature, in addition, because no water vapor diffusion occurs, the pressure dependence of the measurement is also eliminated. The fact that the dew-point temperature depression is greater than the wet-bulb temperature depression increases the sensitivity of an ideal constantan–chromel thermocouple psychrometer from about 5.6 nV J^{-1} kg^{-1} to about 7.8 nV J^{-1} kg^{-1}.

Any deviation from the ideal psychrometer design discussed above or from the ideal environmental conditions can cause significant decreases in the psychrometer sensitivity. But this does not always lead to systematic error in the measurement of water potential. Psychrometers are usually calibrated with salt solutions of known water potential supported on filter paper lining the chamber. The only conditions that lead to systematic error are those that cause the psychrometer sensitivity to be different during a measurement than it was during its calibration. The discussion which follows treats four problem areas that can cause such systematic error, and design considerations to minimize these. The problem areas are (i) temperature fluctuations with time, (ii) temperature gradients, (iii) vapor pressure gradients, and (iv) variation in wet junction size and shape.

24–2.1 Temperature Fluctuations with Time

Temperature fluctuations with time can induce temperature gradients in the psychrometer as the result of differing heat capacities and conductivities of its parts, as will be discussed below, but in addition they can cause another problem. Rawlins and Dalton (1967) pointed out that as air is heated, its water-holding capacity increases. This causes a decrease in relative humidity unless sufficient water enters the air during heating to compensate for it.

If the sample area is small compared to the chamber wall area, heat can enter and leave more readily than water vapor, causing the relative humidity to vary. In the extreme case of a sealed chamber where water vapor can neither enter nor leave, the error in water potential resulting from changes in temperature of the chamber would be of the order of 10 kJ kg^{-1}°C^{-1}. The error will be less as the sample area increases relative to the chamber wall area, and will be least for the ideal psychrometer where the junction is completely surrounded by the sample, and heat flow to and from the junction is restricted to conduction through air. In this case the chamber is completely shielded by the sample and cannot change temperature except as the sample changes temperature. Heat and vapor flow to and from the chamber in phase, maintaining the relative humidity invariant with fluctuating temperature.

If the sample consists of soil or solution imbibed in porous ceramic of the type used in tensiometers, the required hydraulic conductivity to permit unrestrained transport of water to and from the chamber would generally be several orders of magnitude greater than that required for hourly temperature changes of the order of 1 °C. (For plant leaf samples with intact epidermal layers, on the other hand, vapor flow can be seriously restrained, requiring smaller temperature fluctuation rates if errors are to be avoided.)

Because the error induced by fluctuating temperature is the consequence of an actual change in the relative humidity within the chamber, both wet-bulb and dew-point instruments are subject to it.

24–2.2. Temperature Gradients.

Temperature gradients within the psychrometer can arise from such obvious sources as temperature gradients in the environment, but they can also arise from the heat produced by respiration within the sample, heating of the reference junctions and thermocouple wires of Peltier cooled psychrometers, and by radiation absorbed by the psychrometer. Temperature gradients can introduce systematic error either by causing the relative humidity of the chamber to be different from that which would be in equilibrium with the sample under isothermal conditions, or they can simply induce instrumental errors in the measurement of the wet junction temperature. Because the latter are easiest to cope with, they will be dealt with first.

24–2.2.1 INSTRUMENTAL ERRORS

Any difference in the temperature between the reference junction and the wet junction of the psychrometer is included in the electromotive force (emf) measured during evaporation of water. As long as the sample temperature is the same as the chamber temperature (that is, the reference junction is isolated from the sample), a correction for this temperature difference can be made by measuring it independently. For a Peltier psychrometer, if the temperature difference is reasonably stable with time, it can be measured independently immediately prior to cooling the wet junction and corrected for either by direct subtraction or, if it is drifting with time, by extrapolating the temperature curve to the time the wet junction temperature is read and then subtracting. Psychrometers with permanently wet junctions, such as the wet-loop psychrometer (Richards & Ogata, 1958) do not offer this possibility. However, a technique described by Kreeb (1965), whereby a permanently wet thermistor junction is covered between measurements with a remotely-actuated vapor-tight cup, does offer some potential for permitting the dry junction temperature to be estimated.

An obvious solution to the problem of temperature difference between junctions is to place the measuring junction and the reference junction close enough together so that they are always in the same thermal environment. This has been impractical for psychrometers using copper leads from the reference junction to the measuring circuitry because copper is such a good thermal conductor. The use of wires between the reference junction and the measuring circuit that have the same thermoelectric properties as cooper but have considerably lower thermal conductivity, as proposed by Millar et al. (1970), may eliminate this difficulty for permanently wet psychrometers. The psychrometer they proposed is effectively a wet-loop psychrometer that includes an auxiliary Peltier junction to wet the measuring junction remotely.

It is even less practical to place the reference junction of a Peltier psychrometer close to the measuring junction, because it dissipates heat. Hsieh and Hungate (1970) and Calissendorff (1970) modified the Peltier psychrometer by placing two constantan–chromel junctions in the chamber, using one of them simply to measure the temperature of the chamber.

Heat generated at the reference junctions of a Peltier psychrometer can, if the junctions do not provide for its dissipation, cause the reference junctions to differ in temperature from the dry measuring junction temperature. But this leads to systematic error only if the heat dissipation is different during the measurement than during the calibration. This could be a particularly serious problem if in situ psychrometers were calibrated in the laboratory exposed only to air, but were later imbedded in soil.

Because the dew-point hygrometer also requires a reference junction to measure the wet junction temperature, it suffers from identical instrumental errors as the psychrometer.

24–2.2.2 ERRORS CAUSED BY DIFFERENCES BETWEEN CHAMBER AND SAMPLE TEMPERATURE

Because the psychrometer measures the relative humidity of the air in equilibrium with the sample, any difference in temperature between the sample and the chamber air will introduce a systematic error unless this difference is measured and corrected for. Large temperature differences can permit vapor to condense or move in response to thermally induced convection, causing errors that cannot be corrected for (see Campbell, 1979; Wiebe & Brown, 1979).

The temperature-difference error results from the fact that equilibrium occurs at equal vapor pressures between the sample and the chamber, not at equal relative humidities. The error in water potential is of the order of 10 kJ $kg^{-1}{}^{\circ}C^{-1}$. Sample temperature could be raised above that of the chamber by respiration within the sample, but this has not been reported for soil. (It can be a problem for plant samples.) Shielding the chamber with the sample, as called for by the ideal psychrometer design, goes a long way toward eliminating this error. Nonuniform absorption of radiant energy is easily avoided with soil samples, although it can be very difficult to avoid in making in situ water potential measurements of intact leaves.

Thermal gradients within in situ soil psychrometers are caused most often by naturally occurring soil temperature gradients, or by rapidly fluctuating temperatures that induce thermal gradients in nonsymmetrically designed psychrometers. Both of these problems can be eliminated for laboratory measurements on discrete soil samples with sufficient thermal ballast and insulation. In situ, however, they can only be eliminated by symmetrical, thermally shielded psychrometer design. One can envision the ideal in situ psychrometer as consisting of a spherical chamber with walls constructed from a porous wettable material with high thermal conductivity, such as sintered silver powder. The measuring junction should be centrally located and the reference junction thermally tied to the chamber wall.

24–2.3 VAPOR PRESSURE GRADIENTS

Excluding temperature-induced gradients, vapor pressure gradients in a psychrometer chamber can be caused by extraneous sources or sinks for water vapor, or by barriers to water vapor diffusion between the sample and the chamber. The latter, although a serious problem with plant samples, should rarely cause problems with soil samples.

Extraneous sources of water within the psychrometer chamber will rarely occur with Peltier psychrometers, but separation of the sample water into droplets by condensation resulting from inadequate temperature control can cause serious errors that persist for a long time. Care needs to be taken never to allow condensation to occur after the sample is placed in the psychrometer chamber. Even though the temperature is well controlled during a measurement, if condensation has occurred by insufficient temperature control before the measurement, for example by placing field samples directly into psychrometer chambers for transfer to the laboratory, serious problems can occur. The magnitude of the error induced by condensed water within the psychrometer chamber will depend on its proximity to the measuring junction and to the exposed surface area of the droplets compared to that of the sample.

Extraneous sinks for water vapor can result from adsorption of water on exposed surfaces within the chamber. Like extraneous sources for water vapor, sinks are particularly serious when they represent an area that is large compared to the sample area. The effect of water adsorption on exposed wall surface area on the rate of approach to equilibrium within a psychrometer chamber is clearly shown in Fig. 24–1. Figure 24–2 shows the results of an experiment to determine the magnitude of the vapor sink represented by different psychrometer chamber wall materials. Although a coating of resolidified grease was the least absorptive, its inconvenience would preclude its use in many instances. Millar (1971b) suggested the use of stainless steel as a convenient compromise. The more nearly the psychrometer design approaches that of the ideal psychrometer, with the sample lining the interior of the chamber, the less likelihood there is that vapor sinks or sources will cause a problem.

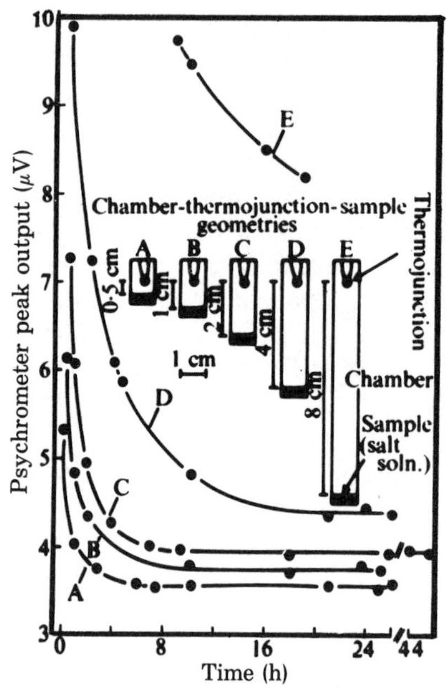

Fig. 24–1. The effect of varying sample chamber lengths on equilibrium time. The water potential of the salt solution at the bottom of each chamber was -900 J kg^{-1} (after Millar, 1971b).

24–2.4 Wet Junction Characteristics

The model for an ideal psychrometer discussed above assumed the wet junction to be spherical, with free evaporation from its entire surface. Rawlins (1966) and Peck (1968) pointed out that water condensed by Peltier cooling on a thermocouple junction does not constitute a spherical wet bulb, but rather the wet area of the junction varies in size and shape depending upon its wetting characteristics, as well as with the duration of cooling and the time elapsed following the cooling cycle. Systematic error resulting from variations in the cooling period and time of reading following cooling can be minimized by matching the precise conditions that existed during psychrometer calibration. But changes in the wettability of the psychrometer junction due to contamination can only be eliminated by cleaning and/or recalibration. Specific precautions and cleaning procedures are discussed below.

24–3 METHODS

Specific details concerning the use of thermocouple psychrometers for measurement of water potential in both the laboratory and the field are given in Wiebe et al. (1971) and Brown and van Haveren (1972). Much of what follows is abstracted from articles contained in these publications.

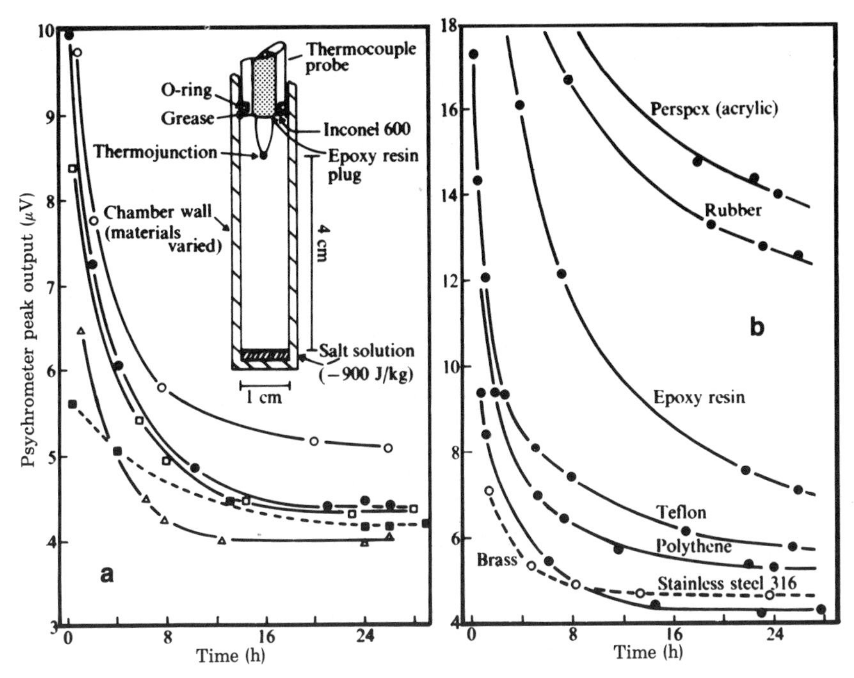

Fig. 24–2. The effect of different psychrometer chamber wall materials on the rate of approach to equilibrium. Wall materials in the left-hand figure (a) are: Rhodium-plated brass; Inconel 600; Stainless steel 316; Pyrex glass; Inconel 600, but with the exposed face of the thermocouple probe coated with resolidified Apiezon grease "M." The insert in the left-hand figure (b) shows the geometry of the psychrometer and the usually exposed materials of the thermocouple probe for obtaining the data in both the left and right-hand subfigures (after Millar, 1971b).

24–3.1 Apparatus

24-3.1.1 LABORATORY PSYCHROMETERS

Several different types of psychrometers have been used for routine measurement of soil water potential in the laboratory. Three typical examples are illustrated below.

The psychrometer of Millar (1971a) shown in Fig. 24–3 permits water potential to be measured within an intact soil core. This geometry approaches that of the ideal psychrometer by nearly surrounding the wet junction with the sample. It is essentially the same design as that proposed by Richards and Ogata (1958), except that the latter used a wet loop measuring junction, by including a small silver ring to hold a droplet of water. The entire apparatus in Fig. 24–3 is immersed in a water bath controlled so that temperature fluctuations are $< 10^{-3}$ °C.

The sample changer psychrometer illustrated in Fig. 24–4 is a modification of that described by Campbell et al. (1966). Samples are contained within stainless steel cups and are packed with a device that produces a conical depression in the sample for the thermocouple. Samples are rotated to the single thermocouple for measurement. A lever and

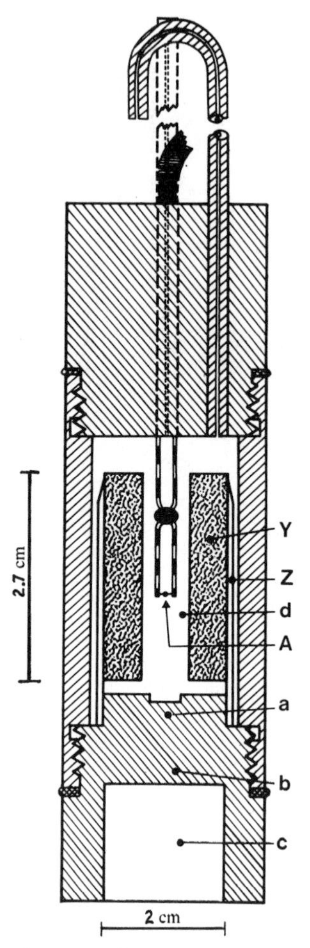

Fig. 24–3. Thermocouple psychrometer for measuring the water potential of intact soil cores. The measuring junction (A) is made from chromel and constantan wire 25-μm in diameter and is suspended between two 254-μm diameter wires of the same metals in a central hole (d) bored through the intact soil core (Y) (after Millar, 1971b).

piston arrangement raises the sample and seals the chamber for each measurement. The ideal configuration of sample surrounding the thermocouple is not as closely approximated here as with the design in Fig. 24–3, but the sample changer permits rapid measurements and direct comparison of samples with calibration standards. The sample changer can be fitted with either a Peltier-type thermocouple or with a modified Richards junction on which the silver loop used by Richards has been replaced by a small ceramic bead. The bead is wetted by dipping it into water placed in one of the sample cups. The massive aluminum housing provides sufficient thermal stability for most applications, though some insulation is required if room temperature changes are rapid or extremely precise measurements are needed.

The third example of a laboratory soil psychrometer, illustrated in

Fig. 24–4. Thermocouple psychrometer sample changer capable of measuring water potential successively on 9 or 10 samples (depending on mode of operation). Samples are placed in stainless steel cups (foreground). The cups are placed in the massive aluminum body of the changer to provide thermal stability and are rotated to the thermocouple by turning the knob. The measuring junction may be either the Richards or Peltier type (Model SC–10A, Decagon Devices, Inc., Pullman, WA 99163).

Fig. 24–5, allows small samples contained in metal sample holders to be inserted into a position under a thermocouple transducer without disassembling the apparatus. It does not permit the measuring junction to protrude into a cavity within the soil. The metal thermocouple holder is pressed against the sample holder and sealed with an O-ring when the screw handle is turned. The metal-to-metal contact and thermal shielding provided by the aluminum housing provide adequate thermal stability for measurements to be made in the laboratory without a controlled-temperature water-bath. Experience has shown, however, that additional insulation is required if ambient temperature is not relatively constant. An older version of the psychrometer in Fig. 24–5 (Model C–51) was made from nylon, and does not provide adequate temperature stability under most conditions.

24–3.1.2 FIELD PSYCHROMETERS

For in situ measurements of soil water potential, the thermocouple is protected by a cup-shaped device that maintains a void in the soil. The thermocouple shield should permit rapid vapor equilibration between the soil and the chamber. Ceramic cups and bulbs of various shapes and sizes have been used, as well as fine-mesh screen cages. In the absence of temperature gradients or rapid changes in temperature, their shape or size is not particularly critical, as long as the psychrometer is calibrated with the same geometry as exists during a measurement. But nonsym-

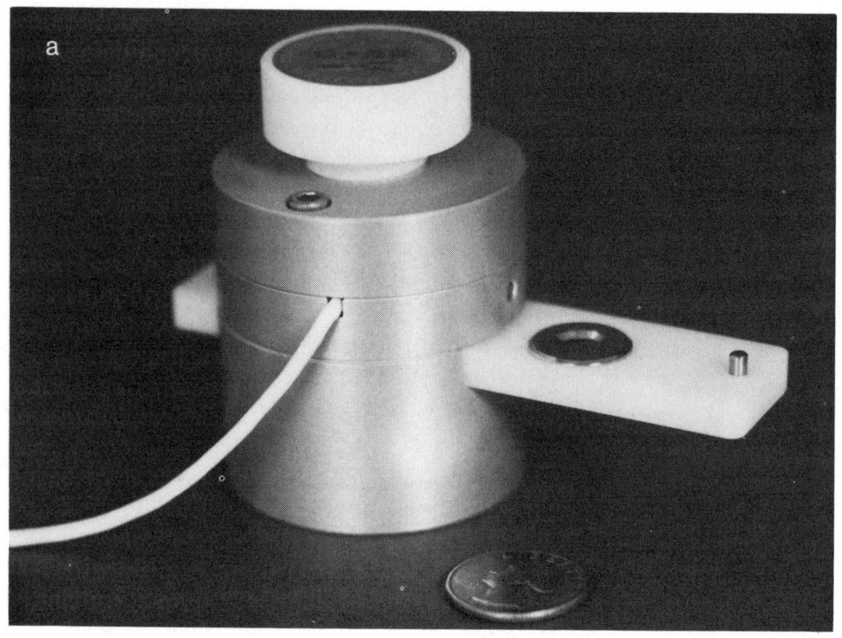

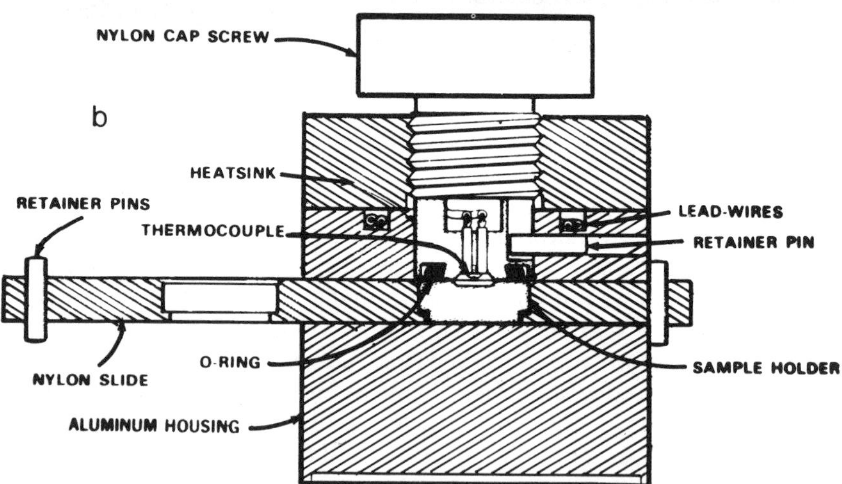

Fig. 24–5. Sample chamber psychrometer (a) and cut-away view (b) of a thermocouple psychrometer for single samples. A sample is placed in a shallow sample holder which is positioned under the thermocouple with the nylon slide. The metal thermocouple mounting unit, which serves as a heat sink, is pressed against the sample head with the screw, and is sealed by the O-ring. Readings are made either in the psychrometric or dew-point mode after temperature and vapor equilibration (Model C–52, Wescor, Inc., Logan, UT 84321).

metrical geometry can lead to serious errors in measurement in nonisothermal media. Figure 24–6 shows the effect of a temperature gradient in the soil on apparent water potential for commercially available psychrometers from Wescor, Inc., Logan, UT (see Appendix). As is shown,

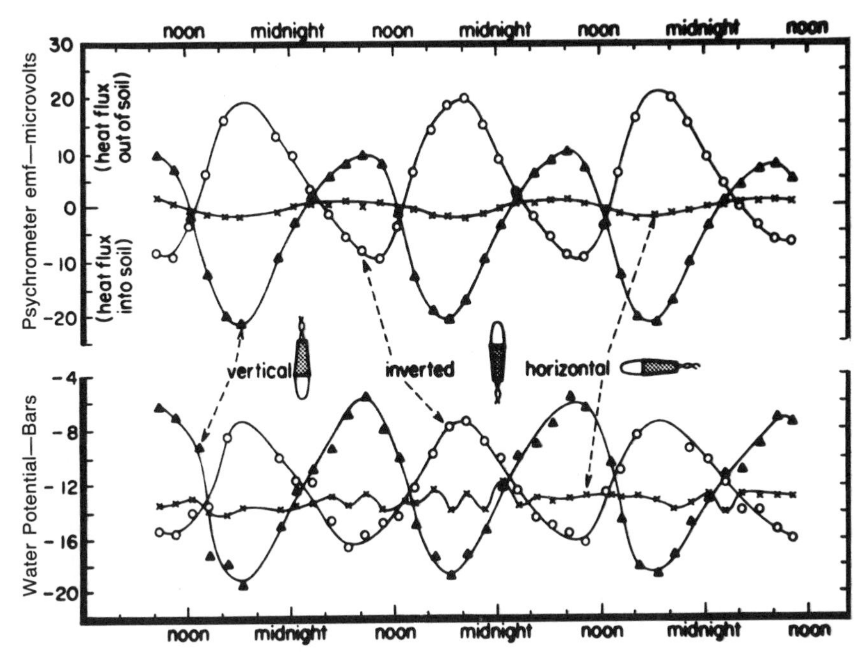

Fig. 24–6. The emf (upper portion) and apparent water potential (lower portion) from sealed-cup psychrometers installed in different orientations at a depth of 25 cm in the soil. The emf values were recorded with the measuring junctions of the psychrometers unwetted. Data shown were collected in the summer with no crop canopy present. The emf multiplied by 60 yields the temperature difference in °C between the measuring and reference junctions of the thermocouples, and hence is a measure of the thermal gradient along the axis of the psychrometer. (from Merrill & Rawlins, 1972)

this error is the reverse direction if the psychrometer is inverted, or can be nearly eliminated if the psychrometer is oriented so that its axis of symmetry is perpendicular to the temperature gradient. Campbell (1972) found that inserting a nonwettable plug into the end of the ceramic cup of this psychrometer to increase the symmetry of the design (Fig. 24–7) reduced errors resulting from temperature changes.

Wiebe et al. (1977) compared performance in a temperature gradient of several modifications of the Wescor design (Fig. 24–7) and other commercial psychrometers having a screen window over the end of a small stainless steel tube for their cup. The end-window design represents almost the ultimate in departure from the ideal psychrometer described earlier. They found that the temperature-gradient sensitivity of the Wescor design was closely correlated with thermocouple length, and were able to suggest modifications that substantially reduced the temperature-gradient sensitivity of the Wescor units. Predictably, they found that the end-window psychrometers were so sensitive to temperature gradients that they are practically useless for any field measurement.

Several new in situ psychrometer and hygrometer designs have recently been introduced. Campbell (1979) constructed soil psychrometers from materials with high thermal conductivity to minimize reference

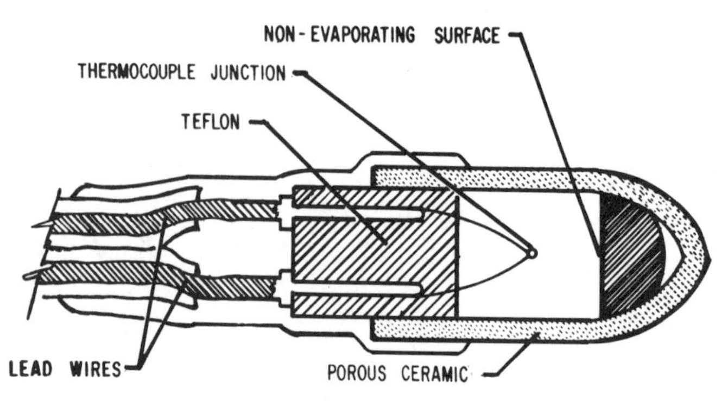

Fig. 24–7. Soil psychrometer modified to reduce error due to thermal gradients by using an additional nonevaporating surface in the chamber (Campbell, 1972).

junction temperature drift. A symmetric design minimized temperature-gradient errors. Brunini and Thurtell (1982) constructed a soil hygrometer from solid copper, with windows made from a porous silver membrane. Both designs offer substantial improvement in performance over older designs when temperature gradients are present. They do not, however, eliminate the errors introduced by psychrometer interference with thermally induced water flow (Campbell, 1979; Wiebe & Brown, 1979).

24–3.1.3 MEASURING EQUIPMENT

Readings are made with Peltier psychrometers by first measuring the voltage with both junctions dry. Current is then passed through the thermocouple in the direction to cause cooling of the measuring junction to condense water on it. The change in voltage caused by the temperature depression of this junction as this water evaporates is then measured. Requirements for the amplifiers and switching circuitry to accomplish this measurement are outlined in detail in Wiebe et al. (1971). Commercial sources of measuring equipment are given in the Appendix.

With a dew-point hygrometer, the measuring junction is cooled to and maintained at the dew point. In the original design (Neumann & Thurtell, 1972), a four-wire thermocouple junction is required. Cooling current is passed through one pair of wires, and the temperature depression is measured with the other pair. The procedure consists of first plotting a curve of temperature depression vs. cooling current with an x-y recorder, with the thermocouple pair exposed to dry air. This procedure is then repeated on the same piece of paper with the thermocouple exposed to a calibration solution or to an unknown sample. The point at which the curve for the moist sample deviates from that of the dry sample is the dew point. In practice it has been found necessary to proceed a short distance past the dew point on this new curve to identify the dew point clearly, and then reduce the cooling current until the temperature is at the dew point.

Campbell et al. (1973) designed circuitry that uses a single measuring junction for a dew-point measurement, intermittently cools and measures, and then electronically searches for and remains at the dew point. This equipment is commercially available from Wescor. The improved accuracy that one would expect from the dew-point measurement because of its higher sensitivity is partially offset by its greater sensitivity to zero errors (Campbell et al. 1973). In practice, accuracy is about comparable with hygrometric and psychrometric techniques using commercially available equipment. The dew-point system of Brunini and Thurtell (1982), however, appears to give accuracies which are two to five times better than have been attained previously.

24–3.2 Calibration

Calibration of laboratory psychrometers (or hygrometers) is accomplished by suspending KCl or NaCl solutions on filter paper in the chamber in the same geometrical position the sample will occupy and determining psychrometer output. This is done for a range of solution concentrations for each psychrometer, to construct a calibration curve. If water potentials are to be measured at different temperatures, calibration curves should also be made at different temperatures that cover the expected range.

Water potentials for NaCl solutions for a range of concentrations and temperatures are given by Lang (1967). Similar data for KCl, given by Campbell and Gardner (1971), are shown in Table 24–1.

The psychrometer sensitivity and range of measurement of water potential are strongly dependent on the Peltier cooling current and the

Table 24–1. Water potentials of KCl solutions at temperatures between 0 and 40 °C (after Campbell & Gardner, 1971).

	Temperature								
Molality	0 °C	5 °C	10 °C	15 °C	20 °C	25 °C	30 °C	35 °C	40 °C
					J kg^{-1}				
0.05	− 214	− 218	− 222	− 225	− 229	− 233	− 237	− 241	− 245
0.10	− 421	− 429	− 436	− 444	− 452	− 459	− 467	− 474	− 482
0.15	− 625	− 637	− 648	− 660	− 671	− 683	− 694	− 705	− 716
0.20	− 827	− 843	− 859	− 874	− 890	− 905	− 920	− 935	− 949
0.25	−1029	−1049	−1068	−1087	−1107	−1126	−1145	−1164	−1182
0.30	−1229	−1253	−1277	−1300	−1324	−1347	−1370	−1392	−1414
0.35	−1429	−1457	−1485	−1512	−1540	−1568	−1594	−1620	−1646
0.40	−1628	−1661	−1693	−1724	−1757	−1788	−1819	−1849	−1879
0.45	−1826	−1864	−1901	−1936	−1973	−2009	−2043	−2077	−2111
0.50	−2025	−2067	−2108	−2148	−2190	−2230	−2268	−2306	−2344
0.55	−2222	−2270	−2316	−2360	−2406	−2451	−2493	−2536	−2578
0.60	−2420	−2473	−2523	−2572	−2632	−2672	−2719	−2765	−2811
0.65	−2617	−2675	−2731	−2784	−2840	−2894	−2945	−2995	−3046
0.70	−2814	−2878	−2938	−2996	−3057	−3116	−3171	−3226	−3280
0.75	−3011	−2080	−3146	−3208	−3274	−3338	−3398	−3457	−3516
0.80	−3208	−3283	−3353	−3421	−3492	−3561	−3625	−3688	−3752
0.85	−3405	−3485	−3561	−3633	−3710	−3784	−3852	−3920	−3988
0.90	−3601	−3687	−3769	−3846	−3928	−4007	−4080	−4153	−4225
0.95	−3797	−3889	−3977	−4059	−4147	−4231	−4309	−4386	−4463
1.00	−3993	−4092	−4185	−4272	−4366	−4455	−4538	−4620	−4702

cooling period. Millar (1971b) pointed out that the cooling current for maximum psychrometer sensitivity varies with water potential. Although a cooling current of 3 mA may be optimum for high water potentials (as was found by Dalton and Rawlins, 1966), measurements can be extended to far lower water potentials if it is increased. Millar found that by increasing the cooling current to 8 mA for a period of 60 s, with the psychrometer illustrated in Fig. 24–3 he could extend the range of measurement to -13kJ kg^{-1}. However, at the higher water potentials, a current of this magnitude caused undesirable transients resulting from Joule heating of the wires. Millar suggested that if maximum range of measurement is desired, it may be necessary to make calibrations at more than one cooling current. In any case, the cooling current and period, as well as the time period following cooling during which the emf is read, must be the same for measurement of the water potential of an unknown as they were for calibration.

The sample changer (Fig. 24–4) permits frequent calibration of psychrometers by routinely including a calibrating solution in one or two of the positions. This permits the operator to detect irregularities calling either for a change in the calibration curve or for thermocouple cleaning or repair.

In situ soil psychrometers (or dew-point hygrometers) enclosed in ceramic cups can be immersed directly into a flask containing a standard solution for calibration. The flask must be maintained at a constant temperature, which is most easily accomplished in a controlled temperature bath. Psychrometer outputs are measured after temperature equilibrium is attained and readings are constant with time. The bath temperature can then be changed to a new temperature, and another set of readings taken. This procedures is repeated for different calibrating solutions, making certain the ceramic is in equilibrium with the new solution before readings are recorded. After calibration, the psychrometers must be washed for several hours in changes of distilled water to remove traces of solute.

Screen-cage psychrometers are calibrated by mounting the complete psychrometer in a calibration chamber similar to that described above for laboratory psychrometers.

Merrill and Rawlins (1972) buried 33 Wescor sealed ceramic-cup psychrometers in the soil for 8 months and found that the calibration of 45% of them shifted $< 5\%$, 33% shifted between 5 and 10%, 21% shifted between 10 and 20%, and only one instrument shifted $> 25\%$. The median change was 5.3%. Calibrations of psychrometers that were not sealed shifted considerably more. A number of these psychrometers displayed very fine, funguslike threads on the thermocouple wires, suggesting microbial contamination. The frequency with which psychrometers need to be cleaned or recalibrated will depend on how they are used, and must be found by testing for specific situations.

24–3.3 Temperature Correction.

Except for the sample changer, where the measurements and standards may be run simultaneously, measurements will seldom be made at the calibration temperature. Adjustments must therefore be made, knowing the calibration temperature and the measurement temperature. With dew-point hygrometer methods the measurement itself is relatively insensitive to temperature, but the cooling coefficient required for proper dew-point control is temperature-dependent. This must be set to the value appropriate for the psychrometer temperature before the measurement is made.

Corrections to psychrometer measurements can be made by calibrating at several temperatures and then interpolating between calibration curves to determine sample potential for a particular temperature. It is often more convenient, however, to use Eq [3]. Combining Eq. [1] and [3], and substituting V/α for ΔT we obtain

$$\psi = (RT/M) \ln \{1 - [(s + \gamma^*)/p_o]V/\alpha\} \quad [4]$$

where α is the thermocouple sensitivity (60 μV K^{-1}), and V is the psychrometer output (microvolts). Rearranging, we obtain an equation for determining the psychrometer constant for a given psychrometer from calibration data:

$$\gamma^* = \{\alpha p_o [1 - \exp(M\psi/RT]/V\} - s\,. \quad [5]$$

The values for p_o can be looked up in a table, or calculated from the approximate equations

$$p_o = \exp(a - b/T)$$

$$s = [b \exp(a - b/T)]/T^2 \quad [6]$$

with $a = 19.017$ and $b = 5327$. Pressure will be in kPa. Temperature must be in Kelvin. Once γ^* is known for a particular psychrometer, Eq. [4] can be used with Eq. [6] to find water potential for any microvolt reading and temperature. Some psychrometers show a nonzero output when the water potential is zero. This offset must be subtracted from the reading before the psychrometer equations are applied.

24–3.4 Sample Preparation

Accurate measurements of soil water potentials require that serious attention be given to sample preparation and loading. Water loss from samples during loading can result in large measurement errors. For this reason it is highly desirable to transfer samples to psychrometers in a humid transfer box or glove box. The box should be lined with wet blotter paper, have a viewing window at the front or top, and have access holes for the operator's hands. If the blotter paper is kept wet, the humidity in the box will remain near saturation. Because the samples are also near saturation, water vapor exchange with the surroundings is minimized.

Low intensity, indirect illumination of the box is also desirable to minimize radiant heating and drying of the samples.

Water loss is generally more serious for dry soil samples, which show a greater drop in water potential for a given quantity of evaporation than wetter samples.

If the sample chamber and sample are warmer than the constant-temperature bath in which measurements are to be made, condensation of water may occur in the chamber when the psychrometer is placed in the bath. Condensation problems may be avoided if the chamber is covered and cooled on ice for a few minutes before the psychrometer is assembled and placed in the bath. As long as the sample remains cooler than the chamber surfaces, no condensation will occur on the chamber walls.

One of the most important factors in making good psychrometer readings is clean chamber surfaces and thermocouples. Good results have been obtained by rinsing all surfaces that are exposed inside the chamber with boiling distilled water after each used and removing the excess water by shaking. Thermocouples have also been cleaned by boiling water or steam, as well as dipping in a 20% ammonia solution or acetone if grease is a problem. Contaminated thermocouple junctions are the greatest cause for calibration shift.

24–3.5 Measuring Extremely Low Water Potentials

Psychrometers in sample changers have been used to measure water potentials as low as -80 kJ kg^{-1} (Wilson & Harris, 1968) by condensing water on the thermocouple over a -1 kJ kg^{-1} KCl solution for 10 min and then quickly rotating the cylinder to the appropriate chamber and making a reading as the water evaporates. The Decagon sample changer in Fig. 24–4, when equipped with the ceramic bead thermocouple, makes this measurement even simpler because the wet bulb is obtained by dipping the thermocouple in water. Measurements to -300 kJ kg^{-1} are possible with it. This range of water potentials requires a more soluble calibrating solute than NaCl or KCl. Campbell and Wilson (1972) provide lithium chloride concentrations for solutions having water potentials ranging from -5 to -100 kJ kg^{-1}. Humidities of saturated salt solutions, which can be used for calibration in this range, are given by Greenspan (1977).

24–3.6 Separation of Osmotic and Matric Potential Components

To this point we have discussed only the measurement of total water potential. In some cases it is important to know which components make up the water potential.

Richards and Ogata (1961) separated the matric and osmotic components by bringing soil samples to a given matric potential on a pressure membrane and then measuring the total water potential with the thermocouple psychrometer. Oster et al. (1969) combined these two measurements into a single apparatus to more precisely measure the separate components. Figure 24–8 is a simplified drawing of the apparatus which consists of a pressure vessel filled with the soil sample and contains a thermocouple psychrometer inside an enclosure constructed of ceramic with a bubbling pressure of 1500 kPa. The total soil water potential is determined with the thermocouple psychrometer as with the in situ ther-

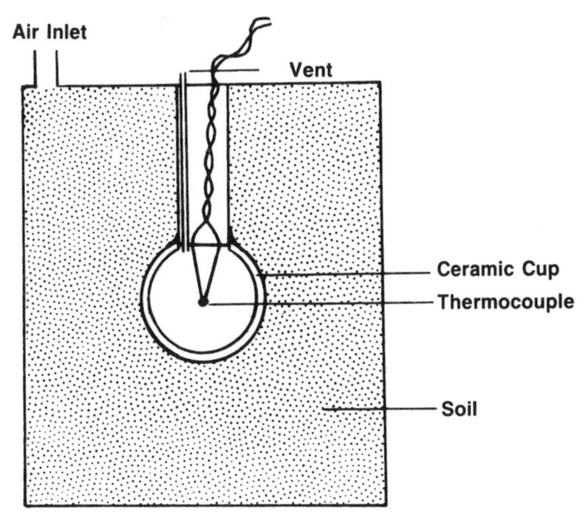

Fig. 24–8. Combination pressure membrane and soil psychrometer apparatus for determining the osmotic and matric components of the soil water potential (after Oster et al., 1969).

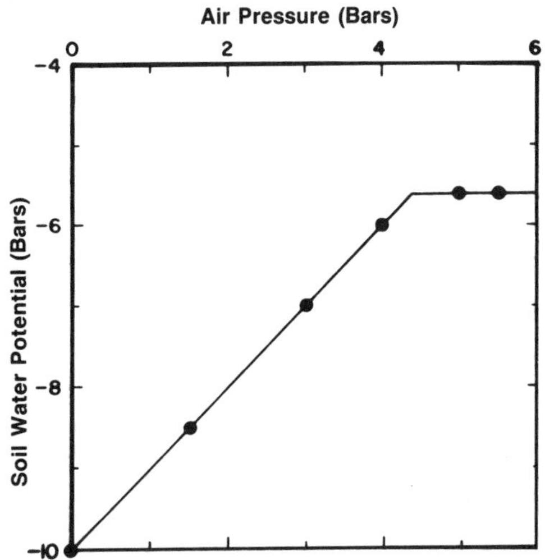

Fig. 24–9. Typical data set for determining the osmotic and matric components of the soil water potential using the apparatus of Oster et al. (1969). The soil water potential is given by the psychrometer reading at 0 bars (0 J/kg) air pressure (soil water potential of −10 bars, 1 J/kg). The osmotic potential is obtained from the psychrometer reading when further increases in air pressure do not change the reading (−5.6 bars, −560 J/kg). The matric potential is the difference (−4.4 bars, −440 J/kg).

mocouple psychrometer designed for field use, with the pressure in the vessel at atmospheric pressure. The pressure in the vessel is then increased in increments and the water potential measured by the psychrometer is recorded. Because the inside of the psychrometer chamber is vented to the atmosphere, each increment of pressure increase raises the matric potential of the water in the ceramic wall by an equal increment, as is shown in Fig. 24–9. When the pressure potential exactly counterbalances the original matric potential of the soil sample, free solution at zero matric potential exists on the inside wall of the chamber. Any increases of pressure beyond this point merely cause more water to flow, but the potential does not increase further. At the inflection point the psychrometer measures the osmotic potential of the soil solution in the ceramic, the original matric potential is given by the pressure required to counterbalance it, and the original total potential was measured initially by the thermocouple psychrometer. The standard error of measurement varied between 3 and 4 J kg^{-1} for these experiments. This represents the practical maximum precision that can be obtained with Peltier psychrometers under ideal laboratory conditions.

For less precise work, an estimate of the osmotic potential can be obtained from the electrical conductivity (EC) of the saturation extract. The osmotic potential of the saturation extract is calculated from

$$\psi_{os} = -36 \text{ EC}$$

where ψ_{os} is in J kg^{-1} and EC in dS m^{-1} (mmho cm^{-1}). The osmotic potential of the soil solution is then calculated assuming that it is an ideal solution, and ignoring anion exclusion and precipitation of solutes with low solubility. For an ideal solution

$$\psi_o = \psi_{os} (\theta_s/\theta)$$

where θ and θ_s are the water content and the water content at saturation.

Once ψ_o is known, ψ_m is obtained as the difference between total and osmotic potentials.

24–4 APPENDIX

Commercial Sources of Psychrometer Equipment

Laboratory psychrometers
Wescor, Inc.
459 South Main
Logan, UT 84321
Decagon Devices, Inc.
P. O. Box 835
Pullman, WA 99163

In situ soil psychrometers
Wescor, Inc.
459 South Main
Logan, UT 84321
J. R. D. Merrill Equipment
RFD Box 140A
Logan, UT 84321

Electronic readout equipment
Wescor, Inc.
459 South Main
Logan, UT 84321
Decagon Devices, Inc.
P. O. Box 835
Pullman, WA 99163
J. R. D. Merrill Equipment
RFD Box 140A
Logan, UT 84321
Campbell Scientific, Inc.
Box 551
Logan, UT 84321

24-5 REFERENCES

Brown, R. W., and B. P. van Haveren. (ed.) 1972. Psychrometry in water relations research. Utah Agric. Exp. Stn., Utah State Univ. Logan, UT.

Brunini, O., and G. W. Thurtell. 1982. An improved thermocouple hygrometer for in situ measurements of soil water potential. Soil Sci. Soc. Am. J. 46:900–904.

Callissendorff, C. 1970. An in situ leaf and soil water psychrometer having temperature sensitivity. M. S. thesis, Washington State Univ. Pullman, WA.

Campbell, E. C. 1972. Vapor sink and thermal gradient effects on psychrometer calibration. p. 84–97. *In* R. W. Brown and B. P. van Haveren (ed.) Psychrometry in water relations research. Utah Agric. Exp. Stn., Utah State Univ., Logan.

Campbell, E. C., G. S. Campbell, and W. K. Barlow. 1973. A dewpoint hygrometer for water potential measurement. Agric. Meteorol. 12:113–121.

Campbell, G. S. 1979. Improved thermocouple psychrometers for measurement of soil water potential in a temperature gradient. J. Phys. E: Sci. Instrument. 12:1–5.

Campbell, G. S., and W. H. Gardner. 1971. Psychrometric measurement of soil water potential: temperature and bulk density effects. Soil Sci. Soc. Am. Proc. 35:8–12.

Campbell, G. S., and A. M. Wilson. 1972. Water potential measurements of soil samples. p. 142–149. *In* R. W. Brown and B. P. van Haveren (ed.) Psychrometry in water relations research. Utah Agric. Exp. Stn., Utah State Univ., Logan.

Campbell, G. S., W. D. Zollinger, and S. A. Taylor. 1966. Sample changer for the thermocouple psychrometers: construction and some applications. Agron. J. 58:315–318.

Dalton, F. N., and S. L. Rawlins. 1966. Design criteria for Peltier-effect thermocouple psychrometers. Soil Sci. 113:102–109.

Greenspan, L. 1977. Humidity fixed points of binary saturated aqueous solutions. J. Res. Nat. Bureau Stds. A: Phys. Chem. 81A:89–96.

Hsieh, J. C., and F. P. Hungate. 1970. Temperature compensated Peltier psychrometer for measuring plant and soil water potentials. Soil Sci. 110:253–257.

Klute, A., and L. A. Richards. Effect of temperature on relative vapor pressure of water in soil. Apparatus and preliminary measurements. Soil Sci. 93:391–396.

Kreeb, K. 1965. Untersuchungen zu den osmotischen Zustandsgrossen. II. Mitteilung: eine electronische Methode zur Messgung der Saugspannung (NTC-Methode). Planta 66:156–164.

Lang, A. R. G. 1967. Osmotic coefficients and water potentials of sodium chloride solutions from 0 to 40°C. Aust. J. Chem. 20:2017–2023.

Merrill, S. D., and S. L. Rawlins. 1972. Field measurement of soil water potential with thermocouple psychrometers. Soil Sci. 113:102–109.

Millar, A. A., A. R. G. Lang, and W. R. Gardner. 1970. Four-terminal Peltier type thermocouple psychrometer for measuring water potential in nonisothermal systems. Agron. J. 62:705–708.

Millar, B. D. 1971a. Improved thermocouple psychrometer for measurement of plant and soil water potential. I Thermocouple psychrometry and an improved instrument design. J. Exp. Bot. 22:875–890.

Millar, B. D. 1971b. Improved thermocouple psychrometer for measurement of plant and soil water potential. II. Operation and calibration. J. Exp. Bot. 22:891–905.

Neumann, H. H., and G. W. Thurtell. 1972. A Peltier cooled thermocouple dewpoint hygrometer for in situ measurement of water potentials. p. 103–112. *In* R. W. Brown and B. P. van Haveren (ed.) Psychrometry in water relations research. Utah Agric. Exp. Stn., Utah State Univ., Logan.

Oster, J. D., S. L. Rawlins, and R. D. Ingvalson. 1969. Independent measurement of matric and osmotic potential of soil water. Soil Sci. Soc. Am. Proc. 33:188–192.

Peck, A. J. 1968. Theory of the Spanner psychrometer. I. The thermocouple. Agric. Meteorol. 5:433–447.

Rawlins, S. L. 1966. Theory for thermocouple psychrometers used to measure water potential in soil and plant samples. Agric. Meteorol. 3:293–310.

Rawlins, S. L. 1972. Theory of thermocouple psychrometers for measuring plant and soil water potential. *In* R. W. Brown and B. P. van Haveren (ed.) Psychrometry in water relations research. Utah Agric. Exp. Stn., Utah State Univ., Logan.

Rawlins, S. L., and F. N. Dalton. 1967. Psychrometric measurement of soil water potential without precise temperature control. Soil Sci. Soc. Am. Proc. 31:297–301.

Richards, L. A., P. F. Low, and D. L. Decker. 1964. Pressure dependence of the relative vapor pressure of water in soil. Soil Sci. Soc. Am. Proc., 28:5–8.

Richards, L. A., and G. Ogata. 1958. Thermocouple for vapor-pressure measurement in biological and soil systems at high humidity. Science 128:1089–1090.

Richards, L. A., and G. Ogata, 1961. Psychrometric measurements of soil samples equilibrated on pressure membranes. Soil Sci. Soc. Amer. Proc. 25:456–459.

Spanner, D. C. 1951. The Peltier effect and its use in the measurement of suction pressure. J. Exp. Bot. 11:145–168.

Wiebe, H. H., and R. W. Brown. 1979. Temperature gradient effects on in situ hygrometer measurements of water potential. II. Water movement. Agron. J. 71:397–401.

Wiebe, H. H., R. W. Brown, and J. Barker. 1977. Temperature gradient effects on in situ hygrometer measurements of water potential. Agron. J. 69:933–939.

Wiebe, H. H., G. S. Campbell, W. H. Gardner, S. L. Rawlins, J. W. Cary, and R. W. Brown. 1971. Measurement of plant and soil water status. Utah State Univ., Exp. Stn. Bull. 484.

Wilson, A. M., and G. A. Harris. 1968. Phosphorylation in crested wheatgrass seeds at low water potentials. Plant Physiol. 43:61–65.

25 Water Potential: Miscellaneous Methods

GAYLON S. CAMPBELL

Washington State University
Pullman, Washington

GLENDON W. GEE

Battelle Pacific Northwest Laboratory
Richland, Washington

25-1 INTRODUCTION

A knowledge of the energy status of water in soil is fundamental to most studies involving transport, plant growth, or microbial activity. While no ideal method exists at present for measuring water potential in soil, a number of techniques are available. Several are discussed in this volume. All of the methods require that some medium, whose water potential can be measured or inferred, be equilibrated with soil water. The soil water potential is then found from the known water potential of the medium. Four methods for measuring water potential will be discussed in this chapter. In the first three methods, soil water is equilibrated with water in a standard matrix. At equilibrium, the water potential of the soil is determined by measuring the water content of the standard matrix using electrical resistance, heat dissipation, or weighing. In the last method, the soil is brought to the water potential of the standard medium through vapor exchange. The water potential of the soil is then that of the standard, which is determined by the concentration of salts in the standard.

In the first three methods, salts are usually free to diffuse into the porous material in which the measurement is made. Equilibration is therefore primarily between matric forces acting on the water, and the methods measure matric potential. In the last method, equilibration is across a vapor gap, so at equilibrium, the sum of matric and osmotic potentials in the soil should equal the osmotic potential of the standard. If the standard matrix were equilibrated across a vapor gap, then the first three methods would also measure the sum of matric and osmotic potentials, but in most cases equilibration through a vapor gap is so slow that these methods would not be used in this manner.

25-2 WATER POTENTIAL MEASUREMENTS WITH ELECTRICAL RESISTANCE SENSORS

25-2.1 Principles

A standard matrix is equilibrated with the soil solution, generally under conditions such that both water and solutes are exchanged. The measurement is made when the matric potential of the standard matrix in the sensor equals the potential of the soil solution. The matric potential of the sensor is inferred from a measurement of electrical resistance of the sensor and a previously determined relationship between electrical resistance and water potential of the matrix. The standard matrix can be of any material that desaturates over the water potential range of interest to the investigator. Materials in common use are gypsum, nylon, and fiberglass. Ions from the soil solution provide the conducting medium in the fiberglass and nylon units. In the gypsum sensors, the gypsum dissolves to provide Ca^{2+} and SO_4^{2-} in solution. The electrical resistance of the sensor will be determined primarily by the water content of the sensor. If the soil solution electrical conductivity (EC) approaches or exceeds 2 mmho/cm, the resistance of the sensor will be determined by both the EC of the soil and the water content of the sensor. The electrical resistance of nylon and fiberglass sensors will always be functions of both variables.

25-2.2 Apparatus

A description of the equipment necessary for these measurements is given in chapter 21 (Soil Water Content) of this volume. Installation and reading procedures are also given there. The measurement requires a meter that measures EC using alternating current. The meter and the sensors are available commercially, but can also be constructed by the researcher. A good, modern meter design is given by Goltz et al. (1981). Strangeways (1983) describes simple circuitry for automatic recording of moisture blocks. Construction techniques for gypsum blocks are given by Taylor et al. (1961). The sensors, or blocks, as they are often called, are small (generally < 3 cm in diameter by 3 cm high) and perhaps the least expensive of all water potential monitoring devices ($<$ \$5/unit).

25-2.3 Calibration

A calibration for electrical resistance sensors (blocks) is obtained by equilibrating the blocks with soil at various water potentials and measuring the electrical resistance of the blocks at these potentials. The calibration is most easily made using a pressure plate apparatus (Tanner et al., 1948; Haise & Kelley, 1946).

A modification of the pressure chamber is commercially available (Soilmoisture Equipment Corp., P.O. Box 30025, Santa Barbara, CA 93105) such that electrical connections can be made through the wall or lid of the chamber, allowing measurements of block resistance to be made without releasing the pressure and removing the lid. When several blocks are calibrated at once in a single chamber, it is important to run separate leads for each block rather than running a single common lead for all blocks, since the soil between the blocks provides a conductive path if the blocks have a common lead.

The calibration is performed by placing the blocks in saturated soil on the pressure plate. Soil should be from the location where the blocks will be used so that the EC of the calibrating soil solution is as similar as possible to that of the measurement. This is particularly true for nylon and fiberglass units where soil solution concentration determines conductance. Pressure is increased in the desired increments, and readings are made after outflow ceases. See chapter 26 (Water Retention: Laboratory Methods) for correct pressure plate procedures.

An indirect calibration procedure has been used by Cannell and Asbell (1964). Blocks are placed randomly in a large soil container which is subsequently planted with wheat or a rapid-growing grass species. After complete root penetration of the soil mass, the electrical resistance of the blocks and water content of the soil are determined periodically as the soil dries to wilting. Soil water content is obtained from a knowledge of the soil mass, the starting mass of water, and the loss of water over time. By relating the water content to matric potential from a previously determined water release curve for the test soil, the matric potential vs. electrical conductivity calibration for the sensors can be determined. Although not as precise a method as the pressure-plate method, one advantage of this technique is that a relatively large number of units can be calibrated at once in an environment that reflects field conditions more closely than those found in the pressure-plate method.

The data obtained from the calibration can be transformed in various ways to obtain useful calibration curves. Taylor et al. (1961) and Goltz et al. (1981) suggested plotting the log resistance as a function of log potential to give linear, or near-linear relationships between the variables. The meter described by Goltz et al. (1981) does the log conversion electronically.

For nylon and fiberglass resistance blocks the linear range (of the log-transformed calibration) extends from near saturation to approximately -1 bar (-100 J/kg) (Fig. 25–1). Rapid desaturation of the fiberglass, due to limited fine porosity, results in these units displaying their greatest sensitivity in the 0 to -1 bar (-100 J/kg) range. Upon further drying, these units have resistance values that increase rapidly; hence they exhibit only limited sensitivity below -1 bar (-100 J/kg). In contrast, gypsum blocks have more fine porosity and are most sensitive when the soil matric potential is less than -0.3 bar (-30 J/kg) (Fig. 25–1) (Bourget et al., 1958).

A much simpler calibration procedure for gypsum sensors, which may be adequate for some studies, is outlined by Tanner et al. (1948).

They saturated blocks under vacuum with distilled water, and grouped them according to their saturated resistance. Representative calibration curves for each group were then obtained by the pressure plate technique and these were used for all blocks in that group.

It is important to note the temperature at which resistance sensors are calibrated, and to correct field readings for the difference between their temperature in the field and the laboratory calibration conditions. Temperature correction data are given by Bouyoucos and Mick (1940), Colman and Hendrix (1949), and Cary (1981). Some manufacturers supply temperature correction data for their units. The temperature correction is typically around 3% per degree, so a temperature correction adequate for many purposes is

$$R_c = R[1 + 0.03\ (T - T_c)]$$

where R_c is the resistance at calibration temperature, T_c, and R is the resistance measured in the field at temperature, T, when the temperature is above freezing.

25–2.3.1 TEMPERATURES BELOW FREEZING

High electrical resistance (low conductivity) has been observed for sensors in frozen soils, hence electrical resistance measurements have been used to measure or infer frozen soil conditions (Colman & Hendrix, 1949; Wilin et al., 1972). The low electrical conductivity (high resistance) of ice is due to the lack of mobile ions to assist in the transport of an electrical current. There is therefore a sharp rise in electrical resistance

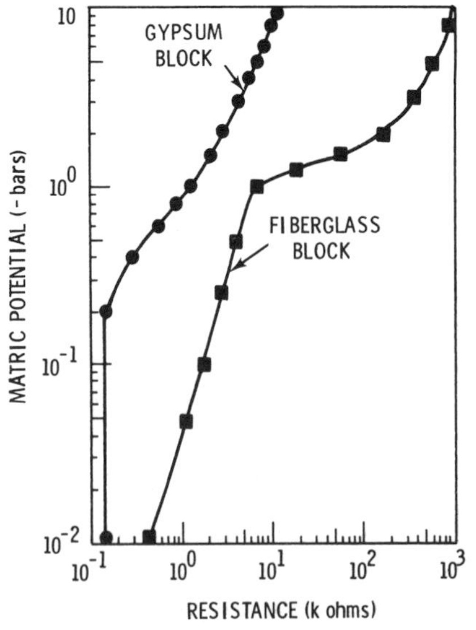

Fig. 25–1. Typical calibration curves for gypsum and fiberglass block soil moisture sensors.

as the water in the electrical resistance block freezes. Since soil solution contains varying amounts of salts, which can lower the freezing point below that of pure water, frozen soils generally occur at temperatures somewhat below 0 °C.

25–2.4 Precautions and Errors

There are a number of errors and uncertainties in using electrical resistance methods to infer soil water potential. The most serious are (i) improper selection of measurement range over which the sensor is to perform, (ii) effects of changing soil solution concentration on sensor resistance, (iii) hysteresis in the water potential-sensor resistance relationship (the sensor resistance is a function of sensor water content; the relationship of water content to potential depends on the wetting history of the sensor), (iv) improper contact between sensor and surrounding soil causing sensor isolation and sluggish response, (v) improper measurement of sensor resistance because of polarization of the sensor with improperly designed readout equipment, and (vi) changes in calibration over time due to changes in sensor matrix characteristics.

Figure 25–1 illustrates the importance of selecting the right sensor for a particular application. It is apparent from the figure that fiberglass block units work best for water potential estimates in the wet range, and gypsum block units are best for drier materials. The uncertainties under items (ii) and (iii) are part of the system, and there is little hope of eliminating errors arising from these sources. Figure 25–2 illustrates the effect of changing salt concentration on a fiberglass block calibration curve. Errors resulting from soil solution effects are minimized by using gypsum

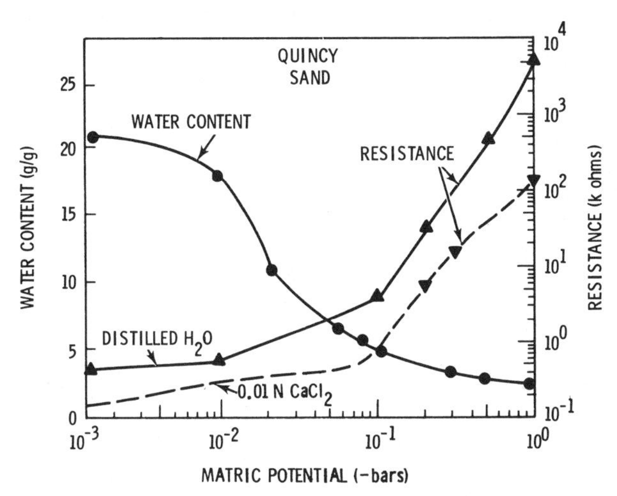

Fig. 25–2. Comparison of water retention of Quincy sand with fiberglass resistance cell calibration curves. Soil brought to saturation by adding distilled water (H_2O) or calcium chloride (0.002 *M* $CaCl_2$). Resistance data points represent mean values of 10 blocks for each calibration pressure.

resistance sensors and calibrating the sensors using the soil for which they are intended.

Where highly saline soil conditions occur, special precautions must be taken. Attempts have been made to correct electrical resistance readings for soil salt effects by making simultaneous salinity sensor measurements. Scholl (1978) described a two-element sensor used to measure matric potential in saline soils. The electrical resistance is monitored both in a desorbing ceramic and in a companion salinity sensor (a ceramic whose bubbling pressure is high enough to maintain saturation over the matric potential range of interest). The ceramic matric potential sensor is calibrated using soil solutions of known salt composition and the field readings are corrected using the companion salinity sensor data.

Sensor hysteresis seldom, if ever, matches the hysteresis found in the natural soil materials. Hence, the drying and wetting response of the sensor as well as that of the tested soil material must be considered for accurate estimation of matric potential with electrical sensors. To minimize hysteresis effects, field drying-cycle data should be matched with drying-curve calibration data. Errors in using calibrations based on drying curves to estimate matric potential during wetting may exceed several bars in the plant growth range (−0.3 to −1.5 bars, −30 to −1500 J/kg) (Tanner & Hanks, 1952; Bourget et al., 1958). Figure 25–3 shows the hysteresis in a wetting and drying gypsum block calibration curve.

Instrument errors are minimized by using conductivity bridges that operate at frequencies at or above 1 kHz and calibrating the sensors using the same meter that is to be used in the field. Frequencies < than 1 kHz are not recommended because they polarize the sensor electrodes.

Errors due to changes in the properties of the matrix with time can be minimized by frequent calibration. England (1965) observed deterioration of fiberglass units after 2 to 3 years and complete failure of some units after 10 years in wet soils in North Carolina. England (1965) reported significant changes in sensor calibration (decreased resistance for

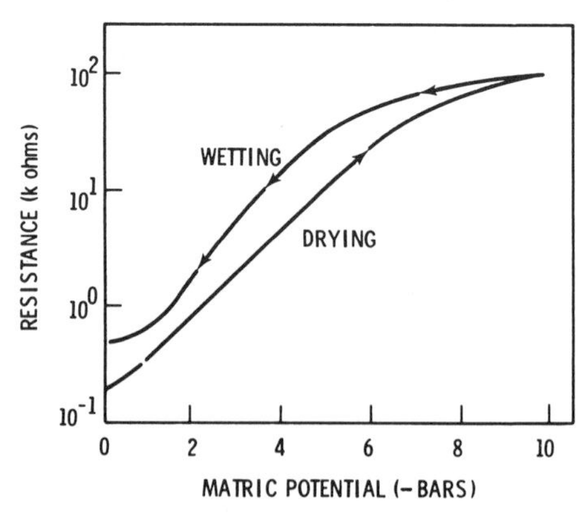

Fig. 25–3. Hysteresis in gypsum block resistance vs. matric potential calibration curve (after Tanner & Hanks, 1952).

the same water content) after several years. He attributed these effects to observed penetration of the fiberglass fabric by fine soil material and to gradual decomposition of the fiberglass which possibly shortened the interelectrode spacing of the sensor. Complete failure after long-term soil exposure was observed on units where the monel case of the sensor had corroded at the spot welds and had separated. Based on these observations and the small cost of the sensors, periodic replacement of units may be a more suitable procedure than recalibration when fiberglass units are used in long-term studies.

Taylor et al. (1961) provided statistical data on various types of electrical resistance sensors that are in common use. A meaningful measure of precision for those intending to use electrical resistance sensors in the field would be the number of sensors required to estimate a given potential with a given certainty. Taylor et al. (1961) made measurements under greenhouse conditions on four soils using four types of sensors. The number of gypsum sensors required to estimate a soil water potential of -5 bars (-500 J/kg) to within ± 0.5 bars ($\pm$ 50J/kg), 95% of the time, ranged from 5 to 10. The number of fiberglass sensors needed for the same precision ranged from 9 to 60, indicating the strong effect of changing soil solution concentration on these sensors. Cary (1981) found coefficients of variation for gypsum blocks in soil wetter than -1.5 bars (150 J/kg) to be around 12%.

Even though the uncertainties in measurements using electrical resistance sensors are relatively large, it is important to keep them in perspective. Cary (1981) investigated the use of gypsum sensors for scheduling irrigation. He found that, even though uncertainties with the sensors are large, field variability is so much larger that the overall measurement error results mainly from spatial variability of the soil matric potential. This fact, along with the low cost and simplicity of operation make electrical resistance sensors attractive as a field method for monitoring water potential, particularly when the purpose of the measurement is to assess soil water status for plant growth. The uncertainties are probably too great to make these units useful for measuring gradients to determine water fluxes in soil.

25–3 WATER POTENTIAL MEASUREMENT WITH HEAT DISSIPATION SENSORS

25–3.1 Principles

A standard matrix is equilibrated with the soil solution. Generally, water and solutes are exchanged. At equilibrium, the water content of the standard matrix is determined by measuring the heat dissipation characteristics of the matrix. The standard matrix is typically a porous ceramic (Phene et al., 1971a), but other porous materials have been used (Shaw & Baver, 1939; Willis & Hadley, 1959; Phene et al., 1971a). The thermal conductivity of a porous material is a complicated function of water content, pore size distribution, and makeup of the porous material. The selection of a porous material that has its greatest sensitivity in the water potential range of interest has therefore largely been by trial and error (Phene et al., 1971a).

Heat dissipation (or thermal diffusivity) is determined by applying a heat pulse to a heater within the ceramic and monitoring the temperature at the center of the ceramic before and after heating. The temperature difference is a function of the thermal diffusivity, and therefore of the water content of the ceramic (chapter 39). Because the thermal conductivity of the surrounding soil may differ substantially from that of the reference matrix, it is important that the reference matrix be large enough to contain the entire heat pulse over the period of measurement.

25–3.2 Sensor Construction

Basic sensor design and construction procedures are described in detail by Phene et al. (1971a). A heater and thermal sensor are placed in the center of a ceramic or other porous reference material. The thermal sensor used by Phene was a P-N junction diode around which insulated copper wire was wrapped to serve as an external heater. Figure 25–4 shows a cross-sectional sketch of the heat-dissipation sensor. Perhaps the most difficult part of the construction is the fabrication of the reference porous material. Requirements for this material are (i) that it have a pore-size distribution or drainage characteristic that will provide maximum sensitivity over the water potential range of interest and (ii) that it will remain stable with time (i.e., will not dissolve or deform with time, requiring sensor replacement or repeated calibration). Because of these considerations, specially constructed porous ceramics have been used most successfully as the reference matrix material. A procedure for making a high-porosity, stable ceramic matrix has been described by Scholl (1978).

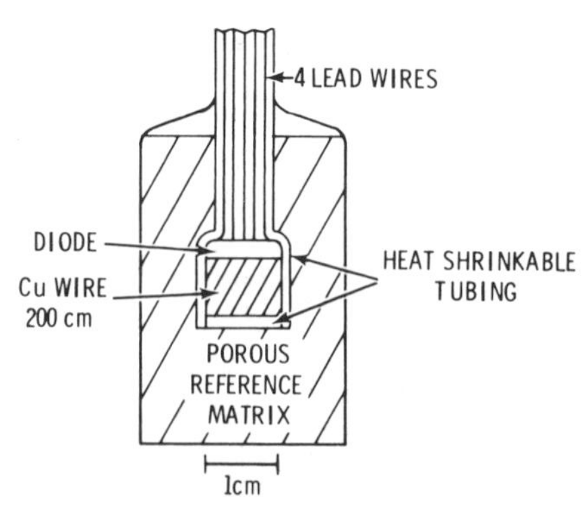

Fig. 25–4. Cross-sectional sketch of the Phene heat dissipation sensor (after Phene et al., 1971a, 1971b).

25–3.3 Calibration Procedures

Calibration of heat-dissipation sensors is similar to that outlined previously for resistance blocks, except that the choice of calibration medium is less critical. Electrical conductivity of the soil solution should have little effect on heat-dissipation sensors. Figure 25–5 shows a typical calibration curve for a commercial heat-dissipation unit.

25–3.4 Installation

Installation procedures are similar to those for electrical resistance sensors, and are described in chapter 21 (Soil Water Content).

25–3.5 Measurement Techniques and Interpretation

The measurement system used for heat dissipation is a diode bridge circuit that measures the electromotive force (emf) generated from the change in diode temperature in the sensing element as a heater current is applied to the sensor for a fixed period of time. The heat dissipation away from the sensor is determined by the thermal conductivity and diffusivity of the reference matrix material in which the thermal sensor is imbedded. The higher the water content, the greater the thermal conductivity and diffusivity of the material; hence, the lower the measured emf. As the soil drains, causing the reference material to desorb, the thermal conductivity and diffusivity decrease causing an increase in temperature in the reference matrix material. Measured emf data are related directly to matric potential through appropriate calibration curves (Fig. 25– 5). The emf data can be recorded by automatic data-logging equipment that periodically scans the sensor after appropriate heat pulsing. It should be noted that the scan frequency is limited because several minutes

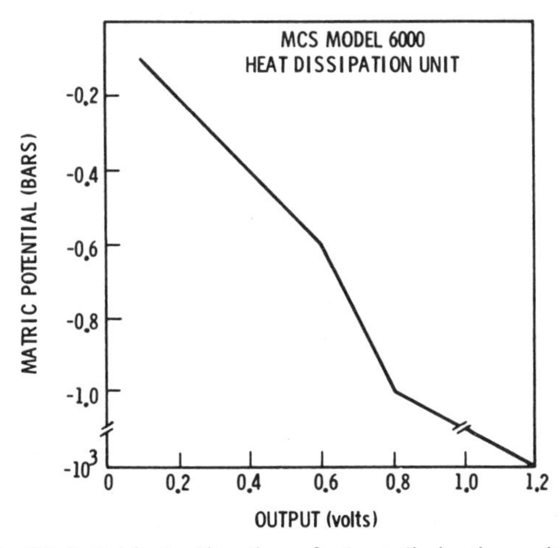

Fig. 25–5. Typical calibration of a heat dissipation unit.

are required for the heat pulse to dissipate in the porous reference material. Generally, sensors are not interrogated more often than on an hourly basis, to assure that the heat pulse has completely dissipated. With appropriate calibration, these heat-dissipation units have been used successfully in measuring and controlling soil matric potential under irrigated soil conditions (Phene & Beale 1976). Precision of measurement in the range 0 to −1 bar (−100 J/kg) matric potential is estimated at ±0.1 bar (±10 J/kg) or better. Less sensitivity exists at lower water potentials. Salinity effects are minimal, since the thermal diffusivity and conductivity are not directly influenced by salts in the soil solution. Phene et al. (1971a) indicated that the measurement is completely independent of soil solution concentration, but supplied no data on which to base that statement.

25–3.6 Precautions and Errors

Several errors associated with the heat-dissipation unit are similar to those found in the electrical resistance block units. Hysteresis, improper selection of the measurement range over which the sensor is expected to operate for a particular soil, improper contact between sensor and surrounding soil, and inadequate calibration under frozen soil conditions are all precautions discussed previously for resistance blocks, which present similar problems for heat-dissipation units. Additional precautions necessary in using heat-dissipation units include providing an adequate time interval for measurement, and determining that the influence of the heat pulse on the soil water content is small by confining the heat pulse to the sensor.

Because the method is not a direct method for matric potential measurement and empirical calibrations are necessary, it is unlikely that these units will be useful in measuring precise matric potential gradients. Nevertheless, for irrigation monitoring and control these units have some advantage over tensiometers, for example, because of the ease with which the units can be automated.

25–4 WATER POTENTIAL MEASUREMENT USING THE FILTER PAPER TECHNIQUE

25–4.1 Principle

Filter paper is used as the standard matrix. The paper is equilibrated with soil water, either through the vapor phase or through combined liquid and vapor phase flow. The soil water potential is determined by measuring the water content of the filter paper and using a moisture release curve to determine the water potential of the paper. If moisture exchange is entirely in the vapor phase, then the potential measured by the paper is the sum of soil matric and osmotic potentials. If exchange is through liquid phase flow, then it is the matric potential.

25–4.2 Measurement Procedure

Details of the measurement procedure are outlined by McQueen and Miller (1968) and Al-Khafaf and Hanks (1974). Filter paper (Schleicher and Schuell no. 589 White Ribbon) is treated to inhibit microbial decomposition by dipping in a 3% solution of pentachlorophenol (Dowcide - 7) in ethanol. Following dipping, the papers are air-dried and stored for use. Soil cans, slightly larger in diameter than the filter paper and large enough to hold 100 g of sample, and disks from paper towel and polyethylene, cut to the diameter of the cans, are also needed. If equilibration through the vapor phase is desired, then a disk of stainless steel screen, cut to the size of the can, is also needed.

To make the measurement, a filter paper disk is sandwiched between the polyethylene disk and the paper towel so that soil will not directly contact the filter paper, and these are placed in the bottom of sample cans. The sample is collected and placed on top of the towel material so that there is good contact between the soil and the towel. If equilibration through the vapor phase is desired, the stainless steel screen is placed on top of the soil sample, and a filter paper disk is placed on top of the screen. The container is closed, sealed with plastic electricians tape, and placed in an insulated box in a constant- temperature environment to equilibrate. McQueen and Miller (1968) suggest a 7-day equilibration time, while Al-Khafaf and Hanks (1974) say 2 days is sufficient. Some experimentation may be required to determine the exact equilibration time necessary for a particular set of conditions. After the paper has equilibrated with the soil moisture, it is removed from the sample can (under humid conditions, if possible) and quickly placed in a smaller weighing container to determine its water content by oven drying. Once the water content of the filter paper is known, the water potential is determined from moisture-release curves for the filter paper. McQueen & Miller (1968) showed that the moisture release curves for the filter paper can be fitted by two exponential functions:

$$\psi = -\exp(7.46 - 16.7\,w), \text{ for } w < 0.54$$

and

$$\psi = -\exp(-0.23 - 2.4\,w), \text{ for } w \geq 0.54$$

where w is the mass-basis water content of the paper (grams of water per gram of dry paper) and ψ is in bars (multiply results by 100 for J/kg).

For precise work, or if filter paper of a different make than that used by McQueen and Miller is used, the moisture-release function for the filter paper should be determined. This may be done either by equilibrating filter paper with salt solutions of known water potential (section 5 of this chapter), by measuring water potential of filter paper at various water contents using the thermocouple psychrometer (chapter 24), or by equilibrating soil in a pressure plate or tension table (chapter 26) to known potentials and using these samples to equilibrate filter paper. A combination of several of these methods may be necessary to obtain a calibration over the entire range of water potentials.

25–4.3 Precautions and Errors

Possible sources of error in the filter paper method are (i) errors in determining the water content of the paper, (ii) errors in the water content vs. water potential relationship for the filter paper, and (iii) errors due to failure of the paper to equilibrate with the soil water. Sources of error in water content measurements are discussed in detail in chapter 21. Errors from uncertainty in the water content vs. water potential relationship result from variation in paper properties, hysteresis, and uncertainty in the original calibration. Statistical datea were not given in the original papers, but from the scatter in the data, it appears that these uncertainties are less than 10% of the reading at any potential. Hysteresis errors are minimized because the paper always starts dry and remains on a wetting curve. See Hamblin (1981) for new and useful information on the method.

Equilibration errors may be of at least two types. Obvious errors result if insufficient time is allowed for the paper to come to potential equilibrium with the soil. Liquid exchange at high potential is rapid, and equilibrium should be reached within one to two days. Equilibrium through the vapor phase is much slower, and at high potentials, where large quantities of water are transferred, equilibrium may take very long times. The second type of equilibrium error results from temperature gradients within the soil sample and container. For the vapor exchange method, a 1 °C temperature difference between the soil sample and the filter paper will result in an error of around 80 bars (8 kJ/kg in the potential reading. Temperature differences must therefore be <0.001 °C for acceptable accuracy, especially at high (wet) water potential. Errors can also result from thermally induced liquid flow along temperature gradients in the sample. The magnitude of these errors is hard to assess, but can be substantial. To minimize temperature-gradient errors, samples should be equilibrated in styrofoam containers in rooms that have minimal diurnal temperature fluctuations. The boxes should not be set directly on a laboratory bench because of temperature gradients caused by conduction of heat through the bottom of the box to the bench. Boxes should be elevated a few centimeters above the bench surface using narrow strips of styrofoam as supports. The actual temperature of the room apparently is much less critical than the extent to which it remains constant during equilibration (Al-Khafaf & Hanks, 1974).

25–5 WATER POTENTIAL MEASUREMENT USING VAPOR EQUILIBRATION

25–5.1 Principle

The water potential in the vapor phase surrounding a soil sample is maintained at some constant predetermined value by osmotic agents. The soil sorbs or desorbs water until equilibrium with the vapor phase is reached. At equilibrium, the sample is at the water potential of the osmotic solution. The water content of the sample at several water potentials can be determined to produce a moisture-release curve which can then be used with water-content measurements of the soil to infer water potential. It would also be possible to determine the water potential of a soil sample directly by equilibrating subsamples with a set of graded osmotic solutions and determining (by interpolation) the osmotic potential that results in no uptake or loss of water from the sample.

25–5.2 Procedure

Many experimental arrangements are possible for vapor equilibration. The points to remember in choosing an experimental arrangement are (i) a temperature difference of 1 °C between the osmotic medium and the sample will result in a water potential difference of around 80 bars (8 kJ/kg), and (ii) vapor flow is relatively slow when gradients are small, and quantities of water to be transferred between the controlling medium and the soil are often relatively large. These factors argue for a minimum of separation between the sample and the osmotic medium, and a maximum surface area exposed to exchange. The method of Harris et al. (1970) meets these requirements, and is therefore outlined here.

Harris et al. (1970) used solute-amended agar gels of known osmotic potential to control water potential in the vapor phase of petri dishes. The gels are made from 2% water agar using salt solutions of the desired concentration. Concentrations of KCl required to give potentials to −45 bars (−4.5 kJ/kg) are given in Chapter 24. Table 25–1 shows concentrations of LiCl for lower potentials. Potential can also be controlled using saturated salt solutions. Water activities for a number of saturated solutions are given by Robinson and Stokes (1965). The salt-amended agar is allowed to solidify in 88-mm diameter petri dishes and then is transferred to lids of 97-mm diameter dishes. These agar gel dishes can be prepared in bulk and stored in moisture-tight bags until used. The sample is placed in a small flat container in the bottom of the petri dish and the lid with agar is put in place. The agar adhers to the lid so that it is suspended immediately above the sample. The dishes are then placed in an insulated container to await equilibrium. It is important that the samples remain isothermal during equilibration, so they are placed in a styrofoam chamber, insulated from the laboratory bench, in an environment that minimizes diurnal temperature fluctuations. Time for equilibration

Table 25–1. Osmotic potential of LiCl solutions at 25 °C.†

Molality	Osmotic potential
	bar or J/kg × 10^{-2}
1.0	− 50
1.5	− 80
2.0	−113
2.5	−150
3.0	−191
3.5	−237
4.0	−287
4.5	−342
5.0	−401

† Values were computed using osmotic coefficient data from Robinson and Stokes (1965). To find values for other temperatures, multiply the values in the table by $T/298$), where T is the Kelvin temperature.

will vary with water potential and characteristics of the medium. Saturated salt solutions should keep water potential constant even when some water is lost, but with the other solutions it may be necessary to change agar lids during equilibration to prevent errors resulting from concentration of the salts through loss of water to the sample. Equilibrium may be checked by removing the samples periodically and weighing them. It may be possible to speed equilibrium by partially evacuating a container containing the dishes. The reduced pressure increases the diffusion coefficient for vapor in the system.

25–5.3 Precautions and Errors

The most serious errors in this method result from temperature differences within the sample container and failure to reach equilibrium. Weighing errors in determining water content and desiccation of the controlling osmoticum can also result in error. With careful technique, Harris et al. (1970) reported an accuracy of ± 1 to 2 bars. (± 100 to 200 J/kg).

25–6 REFERENCES

Al-Khafaf, S., and R. J. Hanks. 1974. Evaluation of the filter paper method for estimating soil water potential. Soil Sci. 117:194–199.

Bourget, S. J., D. E. Elrick, and C. B. Tanner. 1958. Electrical resistance units for moisture measurements: their moisture hysteresis, uniformity and sensitivity. Soil Sci. 86:298–304.

Bouyoucos, G. J., and A. H. Mick. 1940. An electrical resistance method for the continuous measurement of soil moisture under field conditions. Michigan Agric. Exp. Stn. Tech. Bull. 172, East Lansing.

Cannell, G. H., and C. W. Asbell. 1964. Prefabrication of mold and construction of cylindrical electrode-type resistance units. Soil Sci. 97:108–112.

Cary, J. W. 1981. Projecting irrigation with soil instruments: error levels and microprocessing design criteria. p. 81–90. *In* Irrigation scheduling for water and energy conservation in the 80's. Pub. 23-81. American Society of Agricultural Engineers, St. Joseph, MI.

Colman, E. A., and T. M. Hendrix. 1949. The fiberglass, electrical soil-moisture instrument. Soil Sci. 67:425-438.

England, C. B. 1965. Changes in fiberglass soil moisture-electrical resistance elements in long-term installations. Soil Sci. Soc. Am. Proc. 29:229-231.

Goltz, S. M., G. Benoit, and H. Schimmelpfennig. 1981. New circuitry for measuring soil water matric potential with moisture blocks. Agric. Meteorol. 24:75-82.

Haise, H. R., and O. J. Kelley. 1946. Relation of moisture tension to heat transfer and electrical resistance in plaster of Paris blocks. Soil Sci. 61:411-422.

Hamblin, A. P. 1981. Filter paper method for routine measurement of field water potential. J. Hydrol. 53:355-360.

Harris, R. F., W. R. Gardner, A. A. Adebayo, and L. E. Sommers. 1970. Agar dish isopiestic equilibration method for controlling the water potential of solid substrates. Appl. Microbiol. 19:536-537.

McQueen, I. S., and R. R. Miller. 1968. Calibration and evaluation of a wide-range gravimeteric method for measuring moisture stress. Soil Sci. 106:225-231.

Phene, C. J., and D. W. Beale. 1976. High-frequency irrigation for water nutrient management in humid regions. Soil Sci. Soc. Am. J. 40:430-436.

Phene, C. J., G. J. Hoffman, and S. L. Rawlins. 1971a. Measuring soil matric potential in situ by sensing heat dissipation within a porous body: I. Theory and sensor construction. Soil Sci. Soc. Am. Proc. 35:27-32.

Phene, C. J., G. J. Hoffman, and S. L. Rawlins. 1971b. Measuring soil matric potential in situ by sensing heat dissipation within a porous body: II. Experimental results. Soil Sci. Soc. Am. Proc. 35:225-229.

Robinson, R. A., and R. H. Stokes. 1965. Electrolytic solutions. Butterworths, London.

Scholl, D. G. 1978. A two-element ceramic sensor for matric potential and salinity measurements. Soil Sci. Soc. Am. J. 42:429-432.

Shaw, B., and L. D. Baver. 1939. An electrothermal method for following moisture changes of the soil in situ. Soil Sci. Soc. Am. Proc. 4:78-83.

Strangeways, I. C. 1983. Interfacing soil moisture gypsum blocks with a modern data-logging system using a simple, low-cost, DC method. Soil Sci. 136:322-324.

Tanner, C. B., and R. J. Hanks. 1952. Moisture hysteresis in gypsum moisture blocks. Soil Sci. Soc. Am. Proc. 16:48-51.

Tanner, C.B., E. Abrams, and J. C. Zubriski. 1948. Gypsum moisture-block calibration based on electrical conductivity in distilled water. Soil Sci. Soc. Am. Proc. 13:62-65.

Taylor, S. A., D. D. Evans, and W. D. Kemper. 1961. Evaluating soil water. Utah Agric. Exp. Stn., Bull. 426. Logan.

Wilin, B. O., W. P. MacConnell, and L. F. Michelson. 1972. Reliable and inexpensive soil frost gage. Agron. J. 64:804-841.

Willis, W. O., and J. R. Hadley. 1959. Electrothermal unit for measuring soil suction. U. S. Salinity Lab. Res. Rep. 91.

26 Water Retention: Laboratory Methods

A. KLUTE

Agricultural Research Service, USDA, and
Colorado State University
Fort Collins, Colorado

26-1 GENERAL PRINCIPLES

The relation between the soil water content and the soil water suction is a fundamental part of the characterization of the hydraulic properties of a soil. The relationship is identified in the literature by various names, including *water retention* function, *moisture characteristic*, and the *capillary pressure-saturation* curve. The function relates a *capacity factor*, the water content, to an *intensity factor*, the energy state of the soil water. The term *soil water* is commonly used for the solution or liquid phase of the soil. One of the various terms synonymous to suction, such as capillary potential, capillary pressure head, matric pressure head, tension, matric potential, or pressure potential, may be used instead of suction. Matric pressure head values for unsaturated soils are negative. Suction and tension are the negative of the matric pressure head, and hence are positive.

The potential of the soil water may be expressed in units of energy per unit mass, energy per unit volume, or energy per unit weight of the soil water. Energy per unit volume is dimensionally equivalent to force per unit area or pressure. Energy per unit weight has dimensions of length. The pressure head is thus expressed as the length of a fluid column of a given density. The fluid is usually water at the ambient temperature of the soil water system.

The water content, solution phase content, or wetness of the soil may be expressed on a weight, volume, or degree of saturation basis. For analysis of water flow in soil profiles, the volume basis is most useful.

The water retention function is primarily dependent upon the *texture* or particle-size distribution of the soil, and the *structure* or arrangement of the particles (Salter & Williams, 1965; Richards & Weaver, 1944; Reeve et al., 1973; Sharma & Uehara, 1968; Croney & Coleman, 1954). The organic matter content and the composition of the solution phase also play a role in determining the retention function. Organic matter has a *direct* effect on the retention function because of its hydrophilic nature, and an *indirect* effect because of the modification of the soil structure

 Methods of Soil Analysis, Part 1. Physical and Mineralogical Methods—Agronomy Monograph no. 9 (2nd Edition)

that may be effected due to the presence of the organic matter. In soils that contain swelling clays, the composition and concentration of the soil solutes affects the amount of water retained at a given pressure head (Bolt, 1956; El-Swaify & Henderson, 1967; Thomas & Moodie, 1962; Warkentin et al., 1957).

The traditional method of determining the water retention function involves establishing a series of equilibria between water in the soil sample and a body of water at known potential. The soil-water system is in hydraulic contact with the body of water via a water-wetted porous plate or membrane. At each equilibrium, the volumetric water content, θ, of the soil is determined and paired with a value of the matric pressure head, h_m, determined from the pressure in the body of water and the gas phase pressure in the soil. The data pair (θ, h_m) is one point on a retention function. A *drainage* curve is mapped by establishing a series of equilibria by drainage from zero pressure head. A *wetting* curve is obtained by equilibrating samples wetted from a low water content or low pressure head (large negative value).

The retention function is *hysteretic*; i.e., the water content at a given pressure head for a wetting soil is less than that for a draining soil (Topp, 1969; Haines, 1930; Pavlakis & Barden, 1972). The principal features of the hysteresis of the retention function are shown in Fig. 26–1. The drainage curve that starts at complete saturation of the porous medium is the initial drainage curve (IDC). In many porous media, as water is removed through a porous plate, the matric pressure head decreases (i.e., becomes more negative), and the water content approaches a limit called the residual water content, θ_r. The main wetting curve (MWC) is obtained by wetting the soil from a low water content. The initial water content of the sample that is used to determine the MWC will affect the relationship

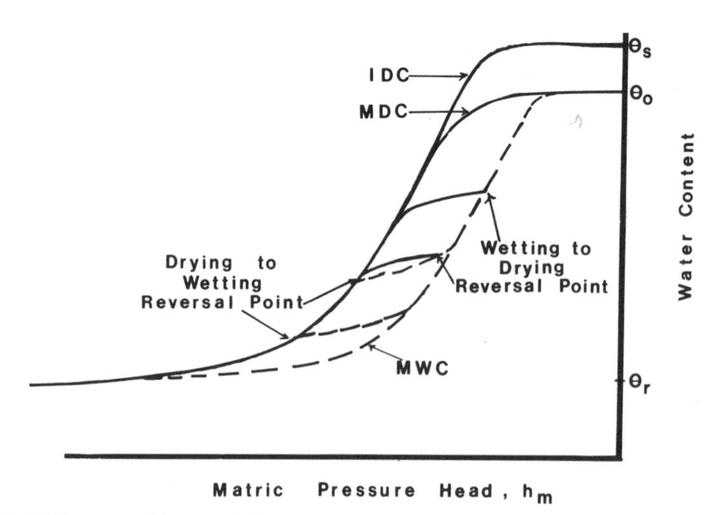

Fig. 26–1. Diagram of hysteresis in the water retention function of a coarse-textured soil of stable structure. Drying curves are shown as solid lines; wetting curves as dashed lines.

found between θ and h_m. In many instances, the MWC is determined starting at the residual water content, θ_r. As the soil is wetted along the MWC and the pressure head approaches zero, the water content approaches a value, θ_o, that is less than the total porosity, θ_s, due to the presence of entrapped air. Usually θ_o is about $0.8\theta_s$ to $0.9\theta_s$. The water content, θ_o is called the *natural saturation* or the *satiated* water content. The drainage curve obtained beginning at θ_o is called the main drainage curve (MDC). The MDC merges asymptotically with the IDC as the pressure head decreases. An infinite set of scanning curves lies inside the envelope of the IDC and the MDC. A few are shown in Fig. 26–1. The particular scanning curve applicable in a given situation depends on the sequence of reversal points that has occurred in the drying and wetting of the soil.

The retention curves of soils with a relatively rigid structure display the capillary fringe region, where the water content is nearly constant as the matric pressure head is decreased from zero. In these soils, a decrease of water content requires the entry of air. If the matrix of the soil can be deformed as the suction in the soil water increases (shrinking and swelling soils), the volumetric water content can decrease as the matric pressure head decreases from zero, without the entry of air into the pores. The retention curves of such materials do not have a region near zero suction where $d\theta/dh_m$ is essentially zero. If the soil has developed structure, i.e., aggregates, etc., its water retention function will display some of the attributes of the retention function of a rigid soil matrix, as shown in Fig. 26–1, and some of the characteristics of the retention function of a shrinking and swelling soil.

In some cases, only the drainage curve (either the IDC or the MDC) is required for the analysis of the water flow. In other cases, only a wetting curve (e.g., the MWC) is required. Most "natural" time-dependent boundary conditions on water flow in soils will insure that the flow is hysteretic. Recent numerical methods of solution of the flow equation for soil water are beginning to incorporate hysteresis information. Mualem (1974) and Mualem and Dagan (1975) have developed methods for calculating the scanning curves from the MDC and the IDC. The methods are easily applicable to the numerical solution process. The possibility of using the envelope curves (MDC and MWC) in analyses of soil water flow increases the need for such information.

The equipment for determining the retention function is one of two types—*suction* cell apparatus or *pressure* cell apparatus. In the suction apparatus (Haines, 1930), the wet soil sample is in hydraulic contact with bulk water through a porous plate. Atmospheric pressure is applied to the soil and the pressure in the bulk water is reduced to subatmospheric levels, thereby reducing its hydraulic head. Water flows out of the sample until hydraulic equilibrium is reached. The water content and the matric pressure head at equilibrium are then determined. The absolute pressure in the bulk water cannot be reduced below its vapor pressure at the ambient temperature, because it then spontaneously vaporizes (boils). Consequently, the theoretical lowest pressure head that can be established in the suction apparatus is given by

$$h_{\min} = (P_{\text{vap}} - P_{\text{atm}})/(dg)$$

where P_{vap} is the vapor pressure. In practice, because of dissolution of gases from the bulk water, the suction apparatus is limited to less than about 850 cm of water suction at low elevations. The suction range is reduced at higher elevations because of the decrease of P_{atm} with elevation. Pressure cell apparatus (Richards, 1941, 1947; Richards & Fireman, 1943) avoids this limitation by keeping the body of water under the porous plate at about atmospheric pressure and raising the gas phase pressure applied to the soil sample, so that no water in the system is actually subjected to pressures greatly less than atmospheric.

The analysis of hydraulic equilibrium when the ambient gas phase pressure on the soil sample is not atmospheric requires recognition of a *pneumatic* component of the pressure head, h_a, which is given by $(P_g - P_{\text{atm}})/dg$, where d is the density of water, g is the acceleration of gravity, and P_g is the gas phase pressure in the soil. Thus, when the gas phase pressure is not atmospheric, the *total* pressure head, h is given by

$$h = h_a + h_m \,. \qquad [1]$$

It is the matric pressure head that is related to the water content, not the total pressure head. However, water flow occurs in response to gradients of total pressure head and gravitational head.

A diagram of a combined pressure and suction apparatus is shown in Fig. 26–2A. When a wet soil sample is placed on the wetted porous plate in a pressure cell and the gas phase pressure in the cell is raised above atmospheric, the pneumatic pressure head and hence the total

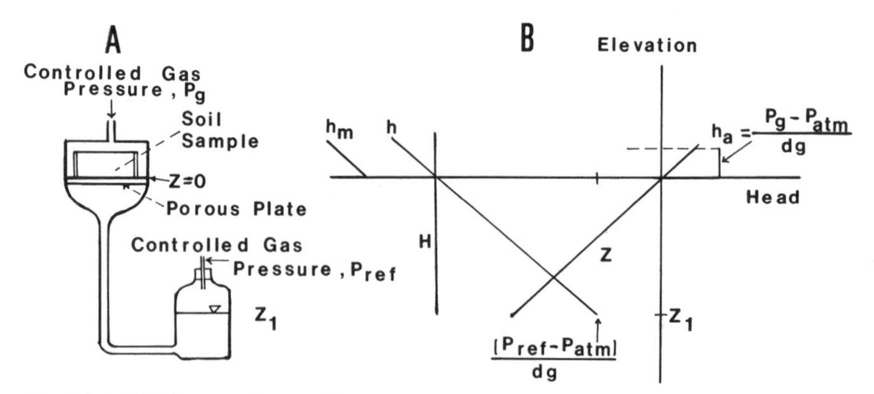

Fig. 26–2. (A) Diagram of a combined pressure and suction apparatus for measurement of water retention. (B) Equilibrium distribution of hydraulic head, total pressure head, matric pressure head, and pneumatic pressure head.

pressure head and the hydraulic head of the soil water are increased, and water flows out of the sample through the porous plate. As the soil drains, the matric component, h_m, of the pressure head, and the hydraulic head of the water in the sample decrease. Eventually, hydraulic equilibrium is established between the soil water and the water in the reference system. The equilibrium distributions of the hydraulic head and its components are shown in Fig. 26–2B.

The hydraulic head of the water in the reference bottle can be raised or lowered by changing the elevation of the water surface in the bottle or by changing the pressure, P_{ref}, on it. It is usually not convenient to change the physical position of the reference bottle more than about a meter, and more convenient to control the hydraulic head by changing P_{ref}. At the surface of the water in the reference bottle, the hydraulic head is given by

$$H_{ref} = z_1 + (P_{ref} - P_{atm})/dg \qquad [2]$$

where d is the density of water and g is the acceleration of gravity. The elevation reference ($z = 0$) has been chosen at the soil-plate contact. At equilibrium, the hydraulic head of the water in the system, including the soil, will be constant and equal to H_{ref}. The total pressure head in the soil sample varies linearly with elevation. If the height of the soil sample is small (e.g., 1 or 2 cm) the variation in matric pressure head from top to bottom of the sample may be neglected in comparison to the magnitude of the matric pressure head. The matric pressure head of the soil water at the bottom of the sample is given by

$$h_m = -(P_g - P_{atm})/dg + z_1 + (P_{ref} - P_{atm})/dg \qquad [3]$$

where $(P_g - P_{atm})$ is the cell air-gauge pressure, d is the density of water, $(P_{ref} - P_{atm})$ is the gauge pressure applied in the reference bottle, and z_1 is the elevation of the water in the reference bottle. If the latter is below the soil sample, z_1 is negative.

It is emphasized that as long as the soil sample is in the pressure cell at equilibrium, the pneumatic pressure head is positive, and the matric pressure head has a certain negative value. When the gas pressure is released and the soil sample is removed from the porous plate, it is usually assumed that the matric pressure head is unchanged. There are at least two reasons why this assumption may not be valid. First, if there are isolated bodies of entrapped air in the solution phase of the sample, they will tend to expand when the reduction of the gas phase pressure is transferred to the liquid phase. The expansion of the entrapped air will affect the curvature of the interface between the solution phase and the continuous gas phase, and hence will affect the matric pressure head. Usually, the matric pressure head will be increased, i.e., become less negative. Chahal and Yong (1965) and De Backer and Klute (1967) compared the matric pressure head of a sample equilibrated in a pressure cell with that of the same sample after the apparatus was converted to a suction cell, with no change of water content. Of necessity, the comparison could only be made within the working range of the suction system. When the gas pressure on the sample was released, the matric pressure head of the water in the sample increased. The change of pressure head upon release of gas pressure was largest at pressure heads corresponding to the onset of drainage of the sample. At more negative values of pressure head there was less shift of pressure head. Since the probability of entrapped air becomes smaller as the matric pressure head and water content decrease, it is likely that this phenomenon is of less significance as the matric pressure head becomes < -1000 cm of water, where pressure apparatus must be used.

A second reason why the matric pressure head of the water in a sample equilibrated in a pressure cell and then removed from the cell may not be as expected is that the sample may have been disturbed when it was removed. If the soil matrix is disturbed when the soil is removed from the chamber, the matric pressure head will be increased (cf. Jensen & Klute, 1967). Of course, this effect applies equally well to samples equilibrated in suction apparatus.

Thermodynamic equilibrium is not generally attained in the apparatus; i.e., mechanical and thermal equilibrium are achieved, but chemical equilibrium is not usually achieved. The soil solutes are free to move through the porous barriers that are usually used for water retention measurements, and no significant pressure difference develops across the porous plate because of "osmotic" effects. As a consequence the operation of the apparatus can be analyzed using the concept of the mechanical potential of the solution phase of the soil, e.g., the hydraulic head of the soil solution phase.

There are two basic measurement options available for the determination of the water retention function. If a nondestructive method for the determination of the soil water content is available, one can map a retention curve by carrying one or a few samples through a series of equilibria. Alternatively, if the determination of the water content of the sample requires destruction of the sample, the curve can be mapped by establishing a series of equilibria in a limited number of suction or pressure cells, using different samples for each equilibrium. In this case, the different samples must be replicates with regard to their water-holding properties.

The establishment of each equilibrium usually requires from one to seven days, depending on the nature of the soil and the height of the samples. In the above options, the equilibria are established *sequentially*, and a great deal of elapsed time can be required to map a curve. The basic difficulties with the above options are (i) the use of a limited number of samples and/or (ii) the use of a limited number of suction or pressure cells. Both of these require the sequential establishment of the equilibria.

The elapsed time for mapping a curve can be very much reduced by using a larger number of replicate samples and a larger number of suction or pressure cells, so that the equilibria at each of the points on the curve can be established *simultaneously*. In the logical limit of this approach, a set of cells is used, one at each level of matric pressure head desired, with three to six replicate samples of the soil in each cell, and all cells are in operation at the same time. Equipment is substituted for time.

The approach that is used will be determined by the number of samples to be run, the number of data points desired, and the equipment available.

In many instances detailed mapping of the water retention function is not required. For example, estimates of the available water-holding capacity of soils are often made by determining two points on the water retention curve, which are considered to represent the "limits" of available water. This approach is described in chapter 36.

26–2 SAMPLES

The soil samples for determining the retention curve may be either repacked samples or samples of "natural" structure. Since the structure of the sample affects the water retention, especially in the low suction range, it is generally best to use samples of natural structure. A core sampler may be used to obtain soil samples of relatively undisturbed structure, which are held in a cylindrical sleeve. Thin-walled metal cylinders with a sharpened edge may be pressed into the soil, or soil cores may be obtained in metal cylinders that fit into a sleeve which has a sharpened edge[1] (see chapter 13 on measurement of bulk density). McIntyre (1974) discussed methods of procuring undisturbed samples for physical measurements.

The sample dimensions will generally be in the range of 5 to 15 cm diameter and 1 to 5 cm high.. The time for reaching equilibrium is proportional to the square of the height of the sample, and the height should be small to reduce the equilibration time. On the other hand, core samples with lengths of 1 cm or less are difficult to handle. A practical length is 2 to 3 cm. The diameter of the sample should be large (perhaps larger by a factor of 10) relative to the size of the structural units in the sample (peds, cracks, worm and root holes, aggregates, etc.) over which the retention data are to be averaged. This may dictate an impractically large size for the core sample. Also, the larger the diameter of the samples the fewer the number of samples that can be placed on each porous plate, so that either more equipment must be used, or more sequential equilibria

---

[1]For example, the model 200A core sampler from the Soilmoisture Equipment Corp., P. O. Box 30025, Santa Barbara, CA 93105. Mention of trade names is for the convenience of the reader and does not constitute official endorsement by the USDA–ARS or Colorado State University.

must be established. The latter may require an unreasonable increase in the elapsed time to obtain the data. Generally, samples with a diameter of 5 to 8 cm and a height of 2 to 3 cm are used. However, the considerations mentioned above with respect to sample size in relation to the structural elements of the soil should be kept in mind when making and interpreting these measurements.

Water retention measurements on subsoils, especially those in which there is a significant degree of structural development, should be made on "undisturbed" core samples. The use of fragmented and repacked samples for such materials can give results that are not representative of the soil in situ, even if the fragmented soil is repacked to the same bulk density as the soil in situ. Crushing, drying, and sieving of the soil destroys or severely modifies any structural units the soil may have had in situ, and repacking, even to the original bulk density, cannot reproduce the field structure.

Soil in a profile in situ is subjected to an external load due to the weight of the overlying soil. In swelling soils, the external load on the sample affects the water retention curve (Collis-George & Bridge, 1973; Philip, 1969). When such soils are to be studied, consideration should be given to a method of applying, measuring, and controlling the external load on the soil sample. Techniques for pore pressure measurements in consolidating clay samples have been available for some time (Bishop, 1961; Rowe & Barden 1966), but apparatus for determining water retention data of swelling soils is not well developed for routine use on large numbers of samples.

It some cases, "undisturbed" cores may not be available, either because it is not expedient to invest the effort required to obtain them, or because it is desired to characterize a given sample that has been collected, and its field structure has been lost. It may also be desired to obtain water retention data on soil that is to be used in studies of various transport processes in a laboratory setting. In such cases, the use of repacked samples is the best that can be done, and may be quite appropriate. A procedure is given below for packing samples to a known bulk density. Repacked samples are probably adequate for many coarse-textured materials that have little structural development. However, even coarse-textured soils may have structural features, such as root and worm holes, that will affect the retention characteristics of the soil and that will be destroyed by crushing, sieving, and repacking of the soil.

The practice of placing fragmented samples on the porous plate with no control or knowledge of the bulk density of the packing is not recommended. Valid volumetric water content data at a given suction cannot, in general, be obtained by multiplying the weight-percentage water content obtained on such fragmented samples by a bulk density value that is considered appropriate for the in situ soil (cf. Young & Dixon, 1966). The volumetric water content obtained by such a procedure is greatly in error at low suctions. However, at higher suctions, where the retention of water is primarily by adsorption forces and is proportional to the specific surface of the soil, this practice may yield more valid results (cf. Elrick & Tanner, 1955).

26-3 WETTING FLUID

The chemical composition of the wetting fluid can affect the water retention of the samples, particularly in fine-textured soils that contain significant amounts of swelling clays. In coarse-textured soils, little attention need be paid to the chemical make-up of the wetting fluid. Generally a fluid with a chemical composition similar to that of the indigent soil water should be used. However, the composition of the in situ soil water is usually not known.

A deaerated 0.005*M* $CaSO_4$ solution is suggested as a general test fluid, unless other reasons are known that would dictate the use of another solution. The preparation of such a solution is described in chapter 28, sect. 28-3. Wetting with distilled or deionized water or freshly drawn tap water is not generally recommended. The former promotes dispersion of the clays in the sample, and the latter is often supersaturated with dissolved air that may come out of solution after wetting the samples, causing the water content at a given pressure head to be affected. Of course, if water retention measurements are made in a study in which the chemical nature of the wetting fluid is a factor, the objectives of the study will determine the specific nature of the wetting fluid.

26-4 WETTING OF SAMPLES

The wetting procedure will be determined by whether one wants to determine the IDC or the MDC. For the IDC a wetting procedure must be used that attains complete saturation. Two methods are commonly used: wetting under vacuum, or flushing with CO_2 followed by wetting and flushing with deaerated water. When core samples are used in water retention measurements, the vacuum-wetting procedure is generally more convenient. The procedure is described below.

If the MDC is to be determined, the cores must be wetted in a manner that produces the "natural saturation." Soaking the cores in water that is at a level just below the top of the core is a convenient method. The entrapped air bubbles in the pores are not stable if the water pressure is atmospheric (Peck, 1969). However, the removal of the gas bubbles by solution and diffusion of the dissolved gas to the exterior of the sample is very slow, and on the time scale of measurement of the retention function the entrapped gas bubbles are effectively stable.

Sudden immersion of the dry soil in water can cause slaking of the soil aggregates, which is a significant change of soil structure. When a dry soil aggregate is suddenly surrounded by liquid water, capillary and adsorptive forces drive the water into the aggregate, compressing the entrapped gas in the intraaggregate pores. The gas pressure increase is often sufficient to disrupt the aggregates, a process known as slaking. Slaking will not occur if the vacuum wetting procedure is followed. Wetting by imbibition from a shallow layer of water minimizes the slaking action.

26–5 TEMPERATURE EFFECTS

The intensity of the forces retaining the water in the soil at a given matric pressure head is temperature-dependent. The surface tension of the soil water decreases with increasing temperature, which leads to a reduction of water content at a given pressure head. The effects of temperature on the adsorptive forces are somewhat obscure, but again it appears that an increase in temperature leads to a reduction of the water content of the soil at a given matric pressure head. The effects are not extremely large, but may be significant in some studies.

Fluctuations of temperature in the cell tend to cause the distillation of water from the wetted samples and the porous plate to the walls of the chamber enclosing the samples and plate. If volume outflow is being monitored to determine the condition of the samples, this distillation process makes it difficult to infer the water content of the samples from the outflow volume. For this reason and because of gas diffusion through the wetted plate (discussed below), volume outflow measurement is not generally recommended as a criterion of the status of the water content of the sample.

26–6 METHOD

26–6.1 Apparatus

The nature of the apparatus required will depend upon the range of matric pressure head in which the retention measurements are to be made. In general, the lower the pressure head, the higher the bubbling pressure requirement of the porous plate, and the greater the strength requirements of the pressure chamber. Three systems, each suited to a given range of measurement, will be described.

26–6.1.1 LOW-RANGE SYSTEM

This system (shown in Fig. 26–3) is especially suited to measurements in the matric pressure head range 0 to approximately −200 cm of water. The major components of the system are (i) the sample chamber, (ii) the cell pressure-control system, and (iii) the suction control system. The cell may be operated in the suction mode, in the pressure mode, or in the combined pressure-suction mode.

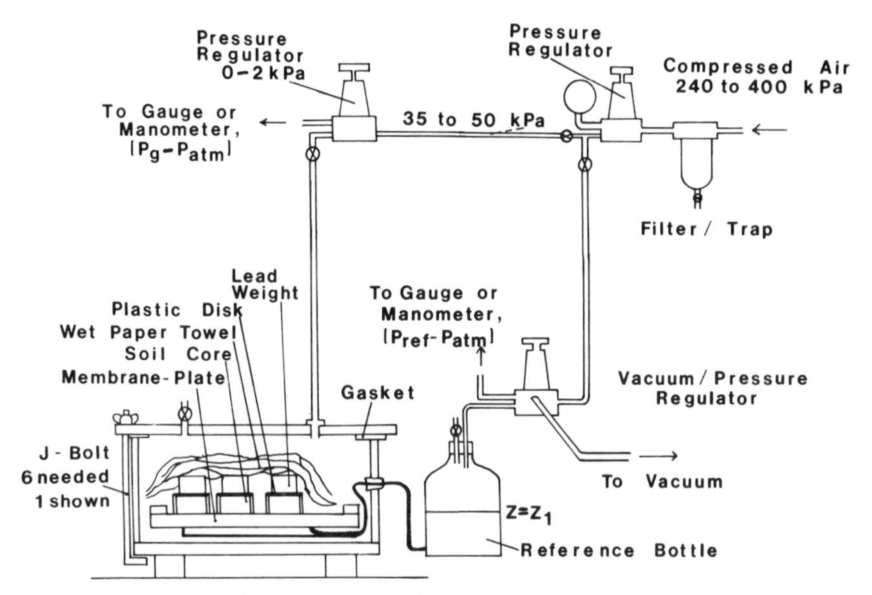

Fig. 26–3. Low-range system for water retention measurement.

The sample chamber is made from a section of PVC pipe. The inside diameter of the chamber is 30 cm; its depth is 15 cm; the wall thickness is 7 mm. The lid and bottom of the chamber are made from 0.95 cm- (3/8 inch) thick acrylic plastic sheet. If the cell is to be operated as a pressure chamber, J-bolts are used to hold the lid. The gas pressure in the chamber should not exceed about 500 cm of water.

Figure 26–4 shows a membrane "plate" suited to measurements in this range. A number of available membranes may be used. Acropore or Versapore (porosity grade 200 or 450, Gelman Sciences, Inc. 600 S. Wagner Rd., Ann Arbor, MI 48106) membrane has been found suitable. The membrane is clamped and sealed at the edges by an O-ring or a flat rubber gasket, and one or two layers of fiberglass window screen are used under the membrane to facilitate lateral movement of water to and from the outlet connections. The bubbling pressure of the assembled membrane-plate system should be checked before putting it into service.

The chamber and membrane plate described here can easily be fabricated, are inexpensive, and provide a convenient way of increasing the amount of equipment available for retention measurements. It is, of course, possible to use the chamber and plates described below for mid-range measurements for determinations in the low range. However, the membrane plates have proven to be more satisfactory than ceramic plates for measurements in the low suction range.

The cell gas-phase pressure regulation system utilizes a pressure regulator (bleed type), which is capable of regulation in the output pressure range 0 to 15 kPa (model 15–30, Moore Products Co., Spring House, PA 19477; various regulated pressure output ranges are available.). The pres-

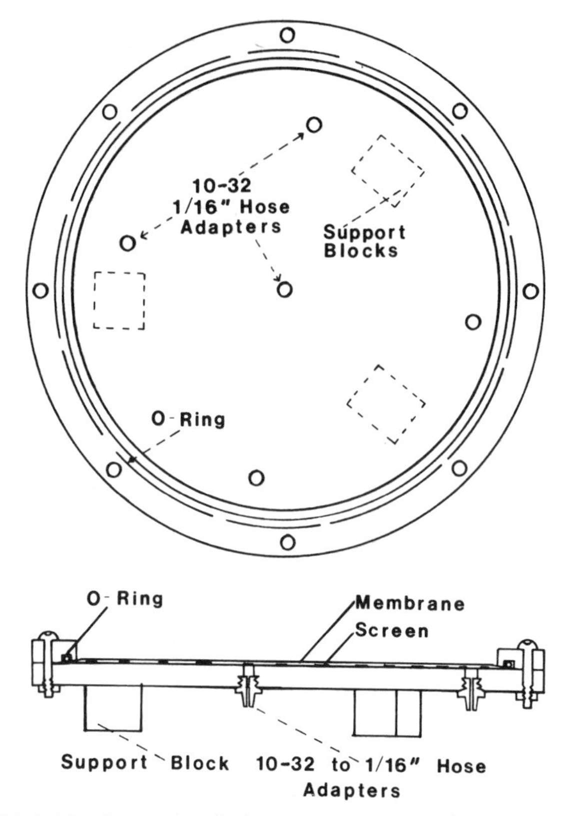

Fig. 26–4. Membrane-plate for low range water retention measurement.

sure in the chamber is conveniently measured with a water manometer up to about 80 or 100 cm of cell pressure, and beyond that with a mercury manometer. The cell pressure regulator requires a drive pressure at about 35 to 50 kPa, which can be obtained from another pressure regulator, operating from a compressed-air source at a higher pressure.

The suction control system consists of the reservoir bottle (a 2- or 4-L aspirator bottle), a subatmospheric pressure regulator (model 44–20, Moore Products Co., Spring House, PA 19477), and a suitable manometer or pressure gauge. For very low suctions (less than approximately 25 cm of water), control may be achieved by adjustment of the free water level in the reservoir bottle relative to the plate, without the pressure regulator—i.e, with P_{ref} equal to atmospheric pressure. If the subatmospheric pressure regulator is used, a vacuum source will be needed. If a laboratory vacuum line is not available, a system may be constructed from standard components. The main components are (i) a tank such as that used for air compressors, (ii) a vacuum pump, and (iii) a vacuum-operated switch and relay system to control the pump.

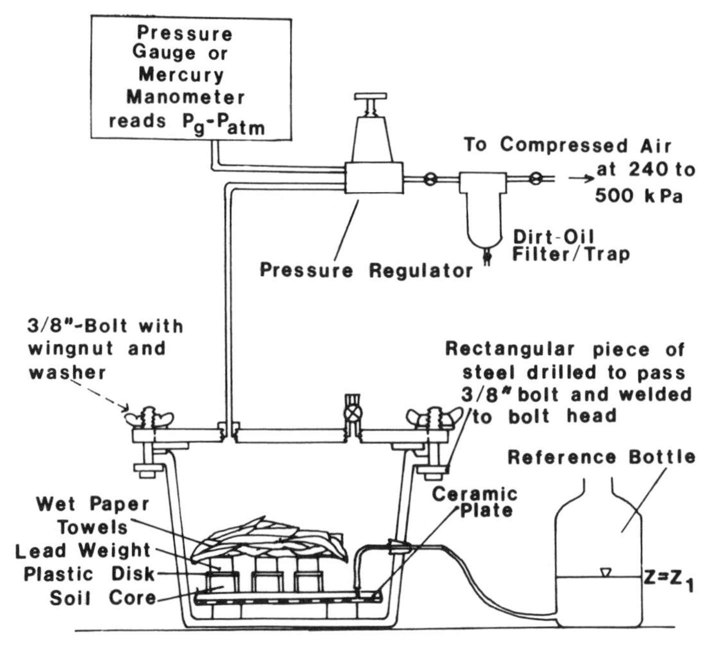

Fig. 26–5. Mid-range system for water retention measurement.

26–6.1.2 MID-RANGE SYSTEM

This system is suited to measurements in the matric pressure head range −200 to −1000 cm of water. A diagram of the system is shown in Fig. 26–5.

The chamber is a standard 16-qt. (14.5 L) pressure cooker, which is available at outlets for kitchen utensils. The author has found considerable difficulty with air leaks with the standard quick-lock lid that comes with the chamber. It is suggested that the standard lid be discarded, and replaced with the flat lid shown in the diagram.

This system is most conveniently operated as a pressure cell. The gas-phase pressure regulation system is basically the same as that for the low-range system, except that a higher-range pressure regulator is used (model 15–30, Moore Products Co., Spring House, PA 19477).

The outflow system consists of the connection from the outlet from the plate, and the reference bottle. The latter may be a 1-L aspirator bottle or just a beaker of water with the outflow tube dipped into the water.

The porous plate is ceramic, with a bubbling pressure of at least 10 m of water (1 bar). A rubber backing sheet is attached to the plate at the edges, with cement and wire. An outlet from the space between the plate and the backing sheet leads through the ceramic. A layer of screen between the plate and the rubber expedites lateral movement of water to the outlet.

The range of measurement of the mid-range system may be extended to matric pressure heads of −30 m of water (−3 bars) by using a ceramic plate with a bubbling pressure of 30 m of water (300 kPa), a suitable higher output range pressure regulator, and a 500-kPa (5 bar) chamber.[2]

26–6.1.3 HIGH-RANGE SYSTEM

The measurement range of this system is from 100 to 1500 kPa (1–15 bars) suction. The essential components, shown in Fig. 26–6, are (i) a pressure chamber, (ii) a ceramic plate with a bubbling pressure of at least 1500 kPa, and (iii) a gas pressure supply and regulation system capable of pressure regulation to 1500 kPa. If the amount of work of this kind is limited, a tank of compressed gas (nitrogen or air) with a nonbleed type pressure regulator may be used. If frequent measurements are to be made, it is better to use a special high-pressure, low-capacity compressor, which can supply compressed air at approximately 2 MPa (300 psi). The pressure regulator must be capable of control in the range from 0.1 to 1.6 MPa. (All components are available from the Soilmoisture Equipment Co., Santa Barbara, CA 93105)

26–6.1.4 OTHER EQUIPMENT

Other items of needed equipment are:

1. Core sampler to collect soil samples of undisturbed structure.
2. A set of numbered sample retainer rings to fit the core sampler, one for each individual soil sample to be placed on a plate.
3. Spatula with a wide blade, for transferring core samples to and from the porous plate.
4. Apparatus for packing cores to known bulk density, if cores of natural

[2]The 100- and 300-kPa ceramic plates, the 500-kPa chamber, and many other items of equipment for water retention measurements are available from the Soilmoisture Equipment Corp., P. O. Box 30025, Santa Barbara, CA 93105.

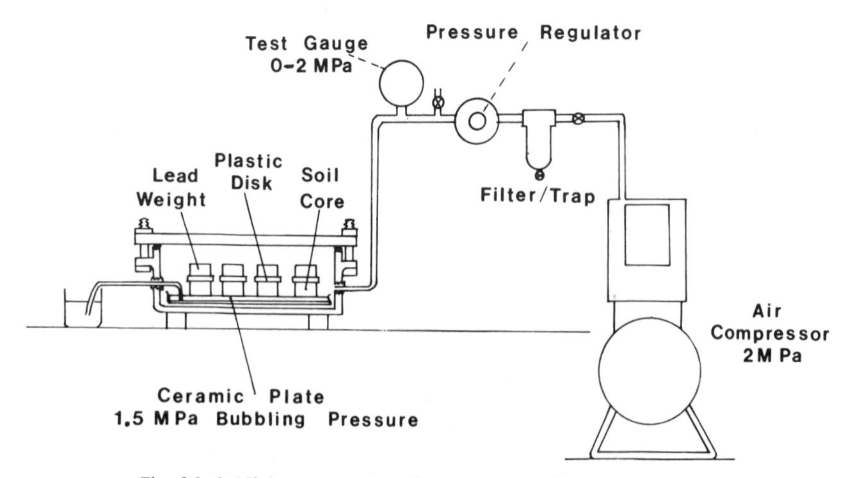

Fig. 26–6. High range system for water retention measurement.

structure are not to be used. This includes a piston that fits a ring of the same inside diameter as the soil retainer rings (see Fig. 26–7) and a hydraulic press.

5. A set of 0.32-cm (1/8 inch) thick plastic disks of a diameter to cover the top of the core samples.
6. A set of lead weights (approximately 700 g for a 5-cm diameter core). These may be made by casting lead in 2-oz metal moisture boxes.

The weights and plastic disks are used to impose an external load on the soil cores. The load helps maintain hydraulic contact between the soil and the plate, prevents free swelling of the soil, and also simulates, to some extent, the external load that is present on the soil in situ.

26–6.2 Procedure

26–6.2.1 SAMPLE PREPARATION

If cores of natural structure have been collected, make sure that the ends of the cores are trimmed flat.

If cores are to be repacked, the following procedure is suggested:

Place enough soil to pack the desired number of cores for each soil in a plastic bag, such as a garbage bag. Add sufficient water to the soil to make it slightly cohesive. A water content of about 5% (weight basis) is generally suitable for sands; 8 to 10% is suitable for soils of finer texture. Close the bag to minimize evaporative loss of water, and mix the soil thoroughly three to five times over a period of 2 to 3 days to allow the water to distribute throughout the soil.

Determine the water content (weight basis) of the soil in the bag. Three to five replicate samples should be used.

From the contained volume of the sample ring, V_s, the desired bulk density, ρ_b, and the water content of the soil, p_w, calculate the weight of moist soil, M, that will give the desired bulk density in the sample ring:

$$M = \rho_b (1 + p_w) V_s . \qquad [4]$$

Fasten a piece of cheesecloth to the lower end of the sample ring with

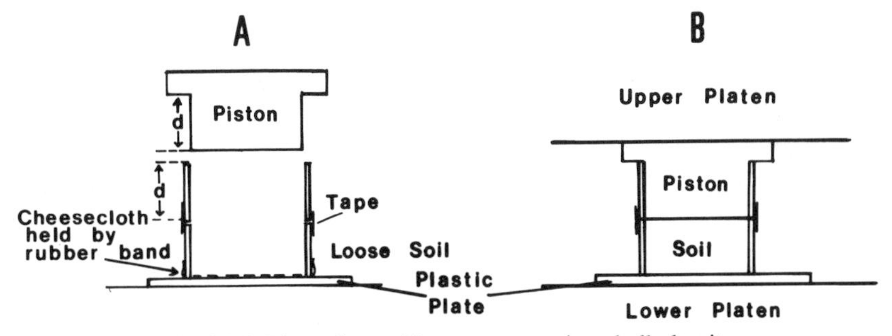

Fig. 26–7. Piston for packing cores to a given bulk density.

a rubber band. Using masking tape, fasten the piston ring to the top of a sample ring (see Fig. 26–7). Place the sample ring and attached piston ring on a 0.32-cm (1/8 inch) plastic sheet.

Weigh out the required amount of moist soil and pour it into the rings. Insert the packing piston into the piston ring and compress the soil into the sample ring with the hydraulic press.

A sufficient number of samples of each soil should be prepared so that three to five replicates will be available at each suction at which the water content is to be determined.

26–6.2.2 SAMPLE WETTING

The method of wetting depends on whether points on the IDC or the MDC are to be determined. See the discussion above in section 26–4 on the method of wetting.

If points on the MDC are to be determined, place the replicate samples for a given suction on a prewetted plate or membrane that has an appropriate bubbling pressure. Place a lead weight on a plastic disk on top of each sample. Wet the samples on the plate by immersing the plate and the samples in water to a level just below the top of the samples. It is best to do this in the chamber in which the samples are to be drained to equilibrium.

If points on the IDC are desired, a vacuum wetting procedure to attain complete saturation must be used. For this purpose, place the samples on the dry porous plate on which they will later be equilibrated. Place the samples and plate in a chamber that can be evacuated. A pressure cooker with a flat gasketed transparent (3/4 inch Lucite) lid is convenient. Evacuate the chamber to the maximum vacuum possible with a water aspirator for at least 15 min, preferably longer. Then allow deaerated water to flow into the chamber until the samples are covered, while maintaining the vacuum in the chamber. When the samples are covered with water, release the vacuum and transfer the plate and samples to the chamber in which they are to be equilibrated. Add the lead weights and plastic disks to the top of each sample.

26–6.2.3 EQUILIBRATION

Remove any excess water from the porous plate with a syringe or siphon. Connect the outflow tube to the pass-through connector in the wall of the chamber, and connect it in turn to the outflow reservoir bottle. Wet some paper toweling with water and wring out the excess water. Cover the samples with several layers of wetted toweling.

Close the chamber. Apply the desired cell gas pressure and/or the desired suction to the outflow system. Allow the samples to come to equilibrium. For core samples that are about 2 to 3 cm high, an equilibration time of 2 to 3 days has been found sufficient. (See section 26–7 below for additional discussion on this point.) While the samples are equilibrating, monitor the regulated pressures and suctions applied to the apparatus. Suggestions are given below (section 26–7) on actions to be taken if the regulated pressures fluctuate during a run.

26–6.2.4 SAMPLE REMOVAL

Weigh and number a set of moisture cans, each of sufficient capacity to hold the soil from a core. When the samples are judged to have come to equilibrium, record the cell air pressure, $P_g - P_{atm}$, and the elevation difference, $z_0 - z_1$ (see Fig. 26–2). Place a clamp on the outflow tube to prevent backflow of water, and release the cell air pressure. Quickly transfer the soil from each sample ring to a moisture can and replace the lid of the can. Determine the wet weight of the soil plus can, W_w. Dry the samples at 105°C and determine the oven dry weight of the soil plus can, W_d.

26–6.3 Calculations

Calculate the volumetric water content, θ, and bulk density, ρ_b, of each sample from

$$\theta = (W_w - W_d)/(dV_s) \quad [5]$$

$$\rho_b = (W_d - W_c)/V_s \quad [6]$$

where d is the density of water.

Calculate the matric pressure head using Eq. [3].

26–7 COMMENTS

Soil loss that could occur during a transfer of the wet samples is avoided by wetting the samples on the plate on which they will be equilibrated, rather than in a separate container.

In many cases, there will be some fluctuation of the regulated cell pressure during the time that the samples are equilibrating. If these fluctuations are small enough to be within the specified tolerances of the regulator, it is best not to readjust the regulator. The action to be taken if the pressure fluctuations exceed regulator specifications should be chosen to prevent the introduction of hysteresis into the results, and consequently depends on whether drainage or wetting data (see section 26–7.1) are being determined. if drainage data are being collected and the pressure in the cell decreases more than one would expect from the regulator specifications, it is appropriate to increase the regulated output pressure to the desired value. If the pressure in the cell increases, the regulator should not be readjusted to a lower cell pressure. If wetting retention data are being collected and an increase of cell air pressure occurs, the regulator should be readjusted to the desired output pressure. If the pressure in the cell decreases, the regulator should not be readjusted.

If repacked cores are used, a bulk density must be chosen. It may be chosen to match the in situ bulk density of the soil, or it may be arbitrarily selected. In the later case, preliminary packing tests should be made to check the suitability of the selected density. The density may be too high and cannot be attained with the packing procedure, or the density may be so low that the packing is unstable, and the sample settles to a higher density when it is wetted.

When the cores are placed in a vacuum chamber for evacuation and subsequent wetting, they must be dry enough so that they do not contain entrapped air in the pores. Entrapped air will expand under vacuum and cause disruption of the soil matrix.

The soil penetrates the open weave of the cheesecloth and makes direct contact with the plate, and the fibers of the cheesecloth are wetted by water. In addition, the cloth assists in obtaining a relatively clean separation of the soil core from the porous plate. If the soil has sufficient cohesion, the cheesecloth can be omitted. A porous material other than cheesecloth may be used, but care should be taken that it does not desaturate at the matric pressure head in the sample. If the barrier drains, its conductivity is reduced, and the drainage of the soil core is hindered. A layer of fine material, such as silt, is sometimes used to improve hydraulic contact between the soil and the plate. This procedure is quite effective if the material is fine enough to retain water at the prevailing suction. However, it may introduce an ambiguity into the determination of the water content of the core, because it is sometimes difficult to clearly separate the core and the fine material in which it is imbedded.

The wet paper towel placed in the chamber while the samples are equilibrating raises the humidity of the air in the chamber and prevents significant loss of water from the samples by vapor transfer.

The elevated gas pressure in the pressure cell apparatus increases the concentration of dissolved gas in the water in the pores of the porous plate, because the solubility of a gas is proportional to the pressure of the gas in contact with the water. Diffusion of the dissolved gas occurs toward the water under the plate in response to a concentration gradient of dissolved gas. Eventually the dissolved gas in the water beneath the plate begins to come out of solution. If the pressure in the outflow system is subatmospheric, the dissolution process is enhanced. The gas accumulates in the space beneath the plate, and displaces water. If volumetric outflow measurements are being used to detect equilibrium or to infer the water content of the sample, the accumulating gas confuses the measurements. The process of gas diffusion is much faster through a thin membrane than it is through the usual ceramic plates, probably because of the shorter diffusion path and the increased concentration gradient in the membrane. For this reason, and because of uncertainties associated with the distillation of water vapor to the walls of the pressure chamber, volume outflow measurements are not recommended for detection of equilibrium, except at low cell air pressure levels and/or low soil water suctions.

An equilibration-time study was conducted by the author on soil cores 2.5 cm high, at several levels of suction. Repacked cores of two soils, a sand and a silty clay loam, were used, and tests were made for both drying and wetting. The cores came to constant water content in 2 to 3 days, regardless of texture, level of suction, or whether they were drying or wetting. Based on these results, the practice of allowing a minimum of two and preferably three, days for equilibration was adopted. If the height of the cores were increased, the equilibration time would be increased. Tests should be run to determine the appropriate equilibrium time if there is reason to suspect that longer equilibration times may be required.

If the amount of soil or number of soil cores is limited, it may be necessary to determine the water content at more than one matric pressure head on a given soil core. This requires the sequential equilibration of the core at two or more pressure heads. When the core has equilibrated at the first pressure head, it is removed from the plate, its gross wet weight (moist soil, ring, and cloth retainer) is determined, and it is replaced on the plate and equilibrated at the next lower pressure head. At the final pressure head, the gross wet weight of the core and the water content of the soil in the core are determined. The water content at each of the previous pressure heads can be calculated from the change in gross weight of the core between equilibria. A correction must be made for the change of water retained in the cheesecloth. This correction can be made with sufficient accuracy by running blank determinations of the weight of a standard size of wetted cheesecloth retainer as a function of pressure head.

When the core samples are replaced on the plate to be equilibrated at the next pressure head, the surface of the plate should be moistened with a fine spray or mist of water and the core placed in it. This practice has been found to successfully establish hydraulic contact of the soil with the plate. The rewetting of the sample that occurs because it is placed in the film of water on the plate is localized, and of no consequence by the time the next value of pressure head is attained.

Because of the increased elapsed time required for determining retention data when a limited number of cores is carried sequentially through a series of equilibria, and because of the possibility that there *may* be difficulty with re-establishing hydraulic contact of the soil with the porous plate, the practice of replacing cores for equilibration at a series of pressure heads should be used only when there is no alternative.

26–7.1 Determination of Wetting Curves

The procedures described above are suitable for determining drainage curves. If a wetting curve, such as the MWC, is to be determined, several modifications must be made in the apparatus and procedure.

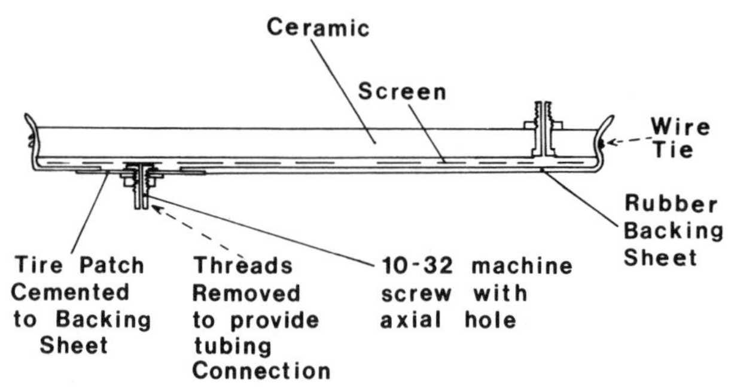

Fig. 26–8. Modification of rubber-backed ceramic plate for determining wetting retention data.

A wetting curve can be determined by wetting the core samples to the natural saturation, draining them to the desired pressure head for the start of the wetting curve, equilibrating by rewetting them at a series of increasing pressure heads, and determining the water content at each pressure head. The accumulation of gas by diffusion through the wetted porous plate that was described above can block the supply of water, so that the sample does not wet to equilibrium at the next higher pressure head. An apparatus for the removal of accumulated air from the outflow system by a flushing process was described by Tanner and Elrick (1958) and is available commercially from Soilmoisture Equipment Corp., Santa Barbara, CA. This apparatus may be used to determine wetting curves in the range of matric pressure heads > -1000 cm of water.

Wetting curves starting at matric pressure heads of -20 to -150 m of water may be determined with rubber-backed ceramic plates modified to permit crossflow of water through the space between the plate and the backing sheet. An additional tublature is placed in the rubber backing sheet as shown in Fig. 26–8. A low flow rate (0.5–1 L h^{-1}) of deaerated water is established through the screen under the plate by connecting a constant-head supply-bottle to one of the tublatures of the plate. The head difference driving the flow is kept symmetrical with respect to the plate and as small as possible. The objective is not to prevent all accumulation of air, but just to prevent the hydraulic blockage of the supply system by air. If the head difference driving the crossflow can be kept small, perhaps less than 5% of the level of suction at which the sample is to be equilibrated, little error will be introduced in the determination of the equilibrium pressure head.

Alternate mounting arrangements for ceramic plates that provide for crossflow under the plates are shown in Fig. 26–9. The ceramic plate can be mounted in epoxy adhesive in an acrylic plastic backing. Support rings or posts must be provided to prevent breakage of the plastic backing or the ceramic plate under the applied cell air pressure. The epoxy should be somewhat flexible, and the plate should not be subjected to excessive temperature fluctuations to avoid breaking the epoxy-to-plastic seal by differential thermal expansion. The plate can also be mounted in an

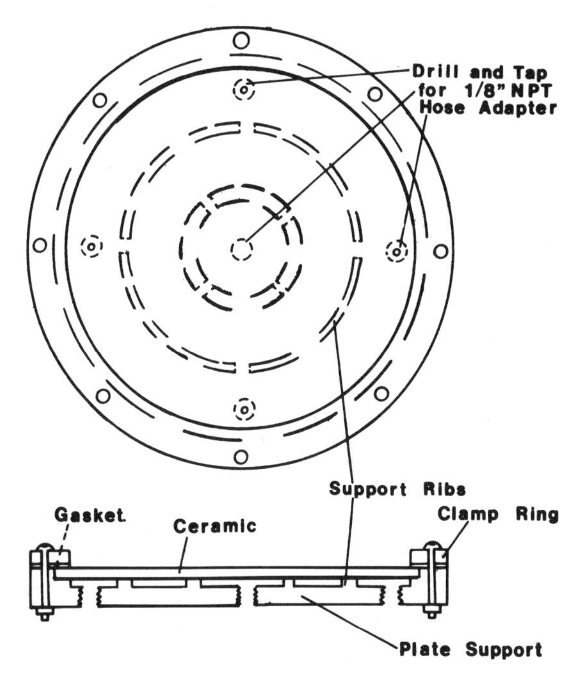

Fig. 26–9. A method for mounting ceramic plates to permit crossflow beneath the plate. The clamp ring and support plate are made of acrylic plastic.

acrylic plastic backing and sealed with a flat gasket coated with a waterproof gasket-sealing compound. This design seems to be more reliable than the epoxy-cemented plate. With these designs, crossflow beneath the plate is much easier to establish and maintain with a small head difference. The membrane plates shown in Fig. 26–4 may also be used with a crossflow system to obtain wetting data in the low suction range.

The procedure for determining the MWC is as follows:

1. Pack a set of replicate cores by the procedure described in section 26–6.2.1. The number of cores depends on the number of data points to be collected on the MWC and the number of replicate cores desired at each pressure head. For example, if three replicates at each pressure head and eight data points on the MWC are desired, 24 cores are required.
2. Place all the cores on ceramic plates with a bubbling pressure at least as high as the suction of the reversal point selected for the MWC. Wet the plates and cores by soaking in water, and drain them to equilibrium at the pressure head of the reversal point.
3. Prepare a set of pressure and/or suction chambers, one for each point on the wetting curve. Provide a crossflow system for each plate. Wet the porous plates in these cells and pre-stress them to the pressure head at which the soil cores will be equilibrated. Clamp the outflow tube to prevent backflow of water to the plate while the chamber is open for loading the samples.

4. Weigh the cores, W_4. This weight includes the ring, cheesecloth and rubber band. Distribute the desired number of replicate cores to each chamber for equilibration. When the cores are transferred to a given plate, moisten the surface of the plate with a mist of water as described above and place the cores in the film of water on the plate.
5. After equilibration, remove the cores as described in section 26–6.2.4. In addition to the weights indicated in that section, determine the weight, W_5, of the soil, ring, cheesecloth, and rubber band. Calculate the bulk density and water content of the sample from Eq. [5] and [6]. Calculate the water content at the reversal-point suction for the MWC from the final water content and the increase in weight, $W_5 - W_4$, corrected for the change in weight of the cheesecloth (see the discussion above concerning this correction).

The standard size plate (Soilmoisture Equipment Corp., Santa Barbara, CA 93105) holds 12 to 15 cores with a diameter of 5 cm. Since three to five replicate cores are usually sufficient, cores from several soils can be loaded into the same chamber, thus making effective use of the equipment.

26–7.2 Water Retention of Stony Soils

Soils containing stones pose special problems for the determination of water retention relationships (Coile, 1953; Reinhart, 1961; Hanson & Blevins, 1979). The term "stones" in this context refers to materials $>$ 2 mm in diameter. Obtaining representative samples of the bulk soil and of the proportions of soil and stones in it may be quite difficult. A volumetric core sampling of the entire soil mass is usually out of the question if a significant amount of stones is present in the soil. The sampling of stony soils is easily subject to bias due to the tendency of the sampler to reject some stones and not others depending on the size of the stones, or the ease with which they may be obtained as part of the soil sample.

In some cases, it may be possible to obtain core samples of "undisturbed" structure from the soil material between the larger fragments in the in situ soil. Such cores could then be used for water retention measurements as described in this chapter. The data obtained must then be corrected for the presence of stones. Berger (1976) gives equations for estimating the water content of stony soils from the proportion of stones in the soil and the water content of the stone and the soil material. The water-retention properties of the entire soil (including the stones) may be estimated with these equations if the retention properties of the soil and the stony materials are known. The stony material can hold considerable water, which may be important in estimating the water available to plants growing in such soils (Coile, 1953; Hanson & Blevens, 1979). These references should be consulted for the details of various approaches to the estimation of the water content and water retention properties of stony soils.

26-8 OTHER METHODS

Numerous variations of procedure for the determination of the water retention function exist in the literature. Some of these are briefly identified below. One of these methods may be found useful. The cited literature should be consulted for details on each method.

26-8.1 Weighable Cells

In this procedure, first described by Reginato and van Bavel (1962), a soil sample in a cylindrical sleeve is held between two end caps. One of the caps contains a porous plate; the other may be connected to a source of regulated gas pressure. The cell may be operated either as a pressure cell or as a suction cell.[3] The soil cores may be repacked or may be of natural structure. The range of measurement is restricted to matric pressure heads greater than about -1000 cm of water.

Each cell is carried through a sequence of equilibria, and its weight is determined at each. At the final equilibrium, the water content of the sample is determined, and the water contents at all the previous pressure heads are obtained by back-calculation using the changes in weight.

The cells are useful for determining the hysteresis of the water retention function. Excellent retention data can be obtained on each core. However, the procedure has the disadvantage that since each core is carried sequentially through a series of equilibria, the elapsed time to obtain the data is rather great. Each soil core requires a set of end caps, and thus the method is not well suited to measurements on a large number of cores.

26-8.2 Suction Tables

Various simple "suction table" types of apparatus have been described in the literature (e.g., Leamer & Shaw, 1941; Jamison & Reed, 1949; Jamison, 1958). These systems are designed to operate in the suction mode and can handle a relatively large number of cores. The measurement range is restricted to suctions less than the bubbling pressure of the material used for the porous barrier of the system (usually < 200 cm of water). Desk blotters and sand-silt packings are among the materials that have been used. The method has primarily been used for the determination of water retention, by drainage, at low suctions.

26-8.3 High-Range Membrane Method

The measurement range of the pressure cell method can be extended to much higher suctions with membranes that have very high bubbling

[3]The cells are available from the Soilmoisture Equipment Corp., P. O. Box 30025, Santa Barbara, CA. 93105.

pressure (Richards, 1941, 1947; Coleman & Marsh, 1961). Measurements at suctions as large as 10 MPa have been made.

26–8.4 Psychrometer Method

Thermocouple psychrometers have been employed to determine water retention as a function of the chemical potential of the chemical species *water* in the soil (Williams, 1968; Riggle & Slack, 1980). The potential function measured by the psychrometer is not the same as that measured by porous plate apparatus. The latter measures the mechanical energy state of the *solution phase*, while the psychrometer measures the energy state of the *water*, which includes an osmotic component. If the osmotic component can be neglected, the potential measured by the psychrometer becomes equivalent to that measured by the porous plate apparatus. To determine the retention function, a thermocouple psychrometer is imbedded in a soil sample. The soil sample is wetted, then dried by evaporation or by suction-removal of water (usually the former). The chemical potential of the water in the soil sample and the water content of the sample are measured periodically. The method is best suited to measurements in the range of potential covered by the thermocouple psychrometer, which is generally -0.1 to -7.0 MPa. Chapter 24 on thermocouple psychrometry should be consulted for details of their use.

26–8.5 Vapor Equilibrium Method

This method involves equilibrating soil samples with an atmosphere of known relative humidity. At equilibrium, the water content of the sample is determined and paired with the thermodynamic potential of the water calculated from the relative humidity. Control of relative humidity is achieved with water-salt solutions of known potential. Saturated salt solutions are convenient for this purpose. The soil samples are usually small (5–50 g) and of disturbed structure. Both drying and wetting data may be obtained. The range of measurement is in the very dry range, at soil water potentials less than about -1.5 MPa. Careful control of the temperature of the system is necessary. The time for equilibration can be very much shortened by evacuating the air from the chamber, because the diffusion coefficient of the water vapor is inversely proportional to the absolute pressure in the chamber. The method is described in chapter 25.

26–8.6 Osmotic Method

Polyethylene glycol (PEG) solutions, separated from a soil water system by a membrane that is impermeable to the PEG, have been used to control the soil water matric potential (Zur, 1966; Painter, 1966; Graham-Bryce, 1967; Williams & Shaykewich, 1969; Pritchard, 1969). Since the membrane is permeable to the normal soil solutes, the method controls the matric pressure head of the soil solution phase. By equilibrating a series of soil samples with PEG solutions of varying concentration (and hence osmotic pressure) and determining the resulting water content of the soil, a water-retention function can be determined. Membrane breakdown by biological action is a problem, and the control of the sample structure in the procedures that have been described in the literature is not good. Because a membrane of rather low conductance is used, it would appear that the time for the sample to attain equilibrium would tend to be rather long, compared to that required by conventional porous ceramic plate apparatus, unless the dimensions of the sample are kept small.

26–8.7 Null Method

In the methods described above, the soil water was brought to equilibrium with a body of water in contact with the soil via a porous plate. Alternatively, a rapid-response tensiometer system may be used to measure the existing soil water matric pressure head. Then the potential of the water in the tensiometer can be adjusted until it matches that of the soil. Null methods of this type have been described by Miller (1951), Leonard and Low (1962), Croney and Coleman (1954), and Dumbleton and West (1968). Most measurements have been restricted to suctions < 0.1MPa, but Dumbleton and West extended the range beyond 0.1MPa by combining pressure and suction cell systems. Water retention data can be obtained by drying samples of soil to various water contents and determining the matric pressure head of the soil water at the resulting water content. Recently, Su and Brooks (1980) described a variation of the null method which was used to determine both drainage and wetting retention data. The method is quite rapid and appears to be advantageous for measurements in the low suction range.

26–8.8 Dynamic Methods

In all the above procedures, a soil sample is equilibrated at a given matric pressure head. Thus they might be called "static" methods. In principle, if one can measure the water content and matric pressure head as a function of time at a given point in a soil water flow system, a "dynamic" water retention function can be obtained by pairing the values of water content and pressure head at a given time. The flow may be either steady or unsteady. The approach requires a nondestructive measurement of the water content of a soil sample, which can be done with gamma attenuation measurements. Rapid-response tensiometry can be used to measure the matric pressure head. Measurements of this kind have been conducted by Topp et al. (1967), Sedgely (1967), Watson (1965), Wana-Etyem (1982), and Perroux et al. (1982). The measurements have been limited to the tensiometer range of suction. Comparison of drainage retention data obtained by the dynamic method and that obtained from the conventional static method has indicated that there is often a greater amount of water held at a given suction in the dynamic case. The reasons for and implications of this observation continue to be a subject of investigation.

Other examples of dynamic approaches to the measurement of water retention properties of soil include the parameter-identification approach described by Zachmann et al (1982), the infiltration method of Ahuja and El Swaify (1976), and the instantaneous profile methods as used by Rogers and Klute (1971), Watson (1965), and Vachaud et al (1972).

26–9 REFERENCES

Ahuja, L. R., and S. A. El-Swaify. 1976. Determining both water characteristics and hydraulic conductivity of a soil core at high water contents from a transient flow experiment. Soil Sci. 121:198–204.

Berger, E. 1976. Partitioning the parameters of stony soils, important in moisture determinations, into their constituents. Plant Soil 44:201–207.

Bishop, A. W. 1961. The measurement of pore pressure in the triaxial test. p. 38–46. *In* Pore pressure and suction in soils, Butterworths, London.

Bolt, G. H. 1956. Physico-chemical analysis of the compressibility of pure clays. Geotechnique 8:86–90.

Chahal, R. S., and R. N. Yong. 1965. Validity of the soil water characteristics determined with the pressurized apparatus. Soil Sci. 99:98–103.

Coile, T. S. 1953. Moisture content of small stone in soil. Soil Sci. 75:203–207.

Coleman, J. D., and A. D. Marsh. 1961. An investigation of the pressure membrane method for measuring the suction properties of soil. J. Soil Sci. 12:343–362.

Collis-George, N., and B. J. Bridge. 1973. The effect of height of sample and confinement on the moisture characteristic of an aggregated swelling clay soil. Aust. J. Soil Res. 11:107–120.

Croney, D., and J. D., Coleman. 1954. Soil structure in relation to soil suction (pF). J. Soil Sci. 5:75–84.

De Backer, L. S., and A. Klute. 1967. Comparison of pressure and suction methods for soil-water content-pressure head determinations. Soil Sci. 104:46–55.

Dumbleton, M. J., and G. West. 1968. Soil suction by the rapid method: An apparatus with extended range. J. Soil Sci. 19:40–46.

Elrick. D. E., and C. B. Tanner. 1955. Influence of sample pretreatment on soil moisture retention. Soil Sci. Soc. Am. Proc. 19:279–282.

El-Swaify, S. A., and D. W. Henderson. 1967. Water retention by osmotic swelling of certain colloidal clays with varying ionic composition. J. Soil Sci. 18:223–232.

Graham-Bryce, I. J. 1967. Method of supplying water to soil at osmotically controlled potentials. Chem Ind. 9:353–354.

Haines, W. B. 1930. The hysteresis effect in capillary properties and the modes of moisture distribution associated therewith. J. Agric. Sci. 20:96–105.

Hanson, C. T., and R. L. Blevins. 1979. Soil water in coarse fragments. Soil Sci. Soc. Am. J. 43:819–820.

Jamison, V. C. 1958. Sand-silt suction column for determination of moisture retention. Soil Sci. Soc. Am. Proc. 22:82–83.

Jamison, V. C., and I. F. Reed. 1949. Durable asbestos tension tables. Soil Sci. 67:311–318.

Jensen, R. D., and A. Klute. 1967. Water flow in an unsaturated soil with a step-type initial water content distribution. Soil Sci. Soc. Am. Proc. 31:289–296.

Leamer, R. W., and B. T. Shaw. 1941. A simple apparatus for measuring noncapillary porosity on an extensive scale. J. Am. Soc. Agron. 33:1003–1008.

Leonard, R. A., and P. F. Low. 1962. A self adjusting null point tensiometer. Soil Sci. Soc. Am. Proc. 26:123–125.

McIntyre, D. S., 1974. Procuring undisturbed cores for soil physical measurements p. 154–165. *In* J. Loveday (ed.) Methods for analysis of irrigated soils. Tech. Commun. no. 54, Commonwealth Bureau of Soils. Farnham Royal, Bucks, England.

Miller, R. D. 1951. A technique for measuring soil moisture tensions in rapidly changing systems. Soil Sci. 72:291–301.

Mualem, Y. 1974. A conceptual model of hysteresis. Water Resour. Res. 10:514–520.

Mualem, Y., and G. Dagan. 1975. A dependent domain model of capillary hysteresis. Water Resour. Res. 11:452–460.

Painter, L. I. 1966. Method of subjecting growing plants to a continuous soil moisture stress. Agron. J. 58:459–460.

Pavlakis, George, and L. Barden. 1972. Hysteresis in the moisture characteristics of clay soil. J. Soil Sci. 23:350–361.

Peck. A. J. 1969. Entrapment, stability and persistence of air bubbles in soil water. Aust. J. Soil Res. 7:79–90.

Perroux, K. M., P. A. C. Raats, and D. E. Smiles. 1982. Wetting moisture characteristic curves derived from constant-rate infiltration into thin samples. Soil Sci. Soc. Am. J. 46:231–234.

Philip, J. R. 1969. Moisture equilibrium in the vertical in swelling soils. I. Basic theory. Aust. J. Soil Res. 7:99–120.

Pritchard, D. T. 1969. An osmotic method for studying the suction/moisture content relationships of porous materials. J. Soil Sci. 20:374–383.

Reeve, M. J., P. D. Smith, and A. J. Thomasson. 1973. The effect of density on water retention properties of field soils. J. Soil Sci. 24:355–367.

Reginato, R. J., and C. H. M. van Bavel. 1962. Pressure cell for soil cores. Soil Sci. Soc. Am. Proc. 26:1–3.

Reinhart, K. G. 1961. The problem of stones in soil-moisture measurement. Soil Sci. Soc. Am. Proc. 25:268–270.

Richards, L. A. 1941. A pressure-membrane extraction apparatus for soil solution. Soil Sci. 51:377–386.

Richards, L. A. 1947. Pressure-membrane apparatus—construction and use. Agric. Eng. 28:451–454.

Richards, L. A., and M. Fireman. 1943. Pressure plate apparatus for measuring moisture sorption and transmission by soils. Soil Sci. 56:395–404.

Richards, L. A., and L. R. Weaver. 1944. Moisture retention by some irrigated soils related to soil moisture tension. J. Agric. Res. 69:215–235.

Riggle, F. R., and D. C. Slack. 1980. Rapid determination of soil water characteristic by thermocouple psychrometry. Trans. Am. Soc. Agric. Engr. 23:99–103.

Rogers, J. S., and A. Klute. 1971. The hydraulic conductivity-water content relationship during nonsteady flow through a sand column. Soil Sci. Soc. Am. Proc. 35:695–700.

Rowe, P. W., and L. Barden. 1966. A new consolidation cell. Geotechnique 16:162–170.

Salter, P. J., and J. B. Williams. 1965. The influence of texture on the moisture characteristics of soils. Part I: A critical comparison of techniques for determining the available water capacity and moisture characteristic curve of a soil. J. Soil Sci. 16:1–15.

Sedgely, R. H. 1967. Water content pressure head relationships of a porous medium. Ph. D. diss. Univ. of Illinois, Urbana. (Diss. Abstr. 68–1852).

Sharma, M. L., and G. Uehara. 1968. Influence of soil structure on water relations in low humic latosols: I. Water retention. Soil Sci. Soc. Am. Proc. 32:765–770.

Su, C., and R. H. Brooks. 1980. Water retention measurement for soils. J. Irrig. Drain, Div., Proc. Am. Soc. Civ. Engr. 106:105–112.

Tanner, C. B., and D. E. Elrick. 1958. Volumetric porous plate apparatus for moisture hysteresis measurements. Soil Sci. Soc. Am. Proc. 22:575–576.

Thomas, G. W., and J. E. Moodie. 1962. Chemical relationships affecting the water-holding capacities of clays. Soil Sci. Soc. Am. Proc. 26:153–155.

Topp, G. C. 1969. Soil water hysteresis measured in a sandy loam compared with the hysteretic domain model. Soil Sci. Soc. Am. Proc. 33:645–651.

Topp, G. C., A. Klute, and D. B. Peters. 1967. Comparison of water content-pressure head data obtained by equilibrium, steady state, and unsteady state methods. Soil Sci. Soc. Am. Proc. 31:312–314.

Vachaud, G., M. Vauclin, and M. Wakil. 1972. A study of the uniqueness of the soil moisture characteristic during desorption by vertical drainage. Soil Sci. Soc. Am. Proc. 36:531–532.

Wana-Etyem, C. 1982. Static and dynamic water content pressure head relations of porous media. Ph. D. diss., Colorado State Univ., Ft. Collins, CO. (Diss. Abstr. 8306597).

Warkentin, B. P., G. H. Bolt, and R. D. Miller. 1957. Swelling pressure of montmorillonite. Soil Sci. Soc. Am. Proc. 21:495–497.

Watson, K. K. 1965. Non-continuous porous media flow. Rep. 84, Water Res. Lab., Univ. of New South Wales, Manly Vale, N. S. W., Australia.

Williams, J. B. 1968. Measurement of total and matric suctions of soil water using thermocouple psychrometer and pressure membrane apparatus. J. Appl. Ecol. 5:263–272.

Williams, John, and C. F., Shaykewich. 1969. An evaluation of polyethylene glycol (P.E.G.) 6000 and P.E.G. 20000 in the osmotic control of soil water matric potential. Can. J. Soil Sci. 49:397–401.

Young, K. K., and J. D. Dixon. 1966. Overestimation of water content at field capacity from sieved sample data. Soil Sci. 101:104–107.

Zachmann, D. W., P. C. Duchateau, and A. Klute. 1982. Simultaneous approximation of water capacity and hydraulic conductivity by parameter identification. Soil Sci. 134:157–163.

Zur, B. 1966. Osmotic control of the matric soil water potential I. Soil water system. Soil Sci. 102:394–398.

27 Water Retention: Field Methods[1]

R. R. BRUCE

Southern Piedmont Conservation Research Center
Agricultural Research Service, USDA
Watkinsville, Georgia

R. J. LUXMOORE

Oak Ridge National Laboratory
Oak Ridge, Tennessee

27-1 INTRODUCTION

The evaluation of the water content–matric potential relationship, θ (ψ), in the field requires considerable time and effort as well as much equipment. Before initiating such an activity certain questions should be answered to ensure that particular data needs will be satisfied. These questions relate primarily to site characteristics and intended use of the data.

The water content-potential relationship generally has the form of the curves in Fig. 27–1. Since instrumentation, equipment, time, and effort requirements depend upon whether data are required over the range from zero to -15 or -20×10^2 kPa or only from 0 to -50 kPa, it is well to determine range of interest. As Fig. 27–1 indicates, this relationship is hysteretic (see chapter 26); therefore, the need for both wetting and drying curves must be considered. For certain uses and for certain soils a drying or draining curve may be satisfactory (Watson et al., 1975). In spite of the effort and expense involved in defining a reliable water content–matric potential relationship, this relationship is fundamental to the characterization of water capacity, water retention, and flow of water in soil. The nature of the relationship is more critical in those potential ranges where the slope is large or is changing, for example 0 to -30 kPa (Fig. 27–1). If the retention capacity over a wide potential range is the primary need—for example, retention volume between -10^3 and -10^1

[1]Contribution from Southern Piedmont Conservation Research Center, ARS-USDA, Watkinsville, GA 30677 in cooperation with University of Georgia Agricultural Experiment Stations and Environmental Sciences Division, Oak Ridge National Laboratory, Oak Ridge, TN 37830 as publication no. 1674. Research at ORNL sponsored by Office of Health and Environmental Research, U. S. Dep. of Energy under contract DE-AC05-840R21400 with Martin Marietta Energy Systems, Inc.

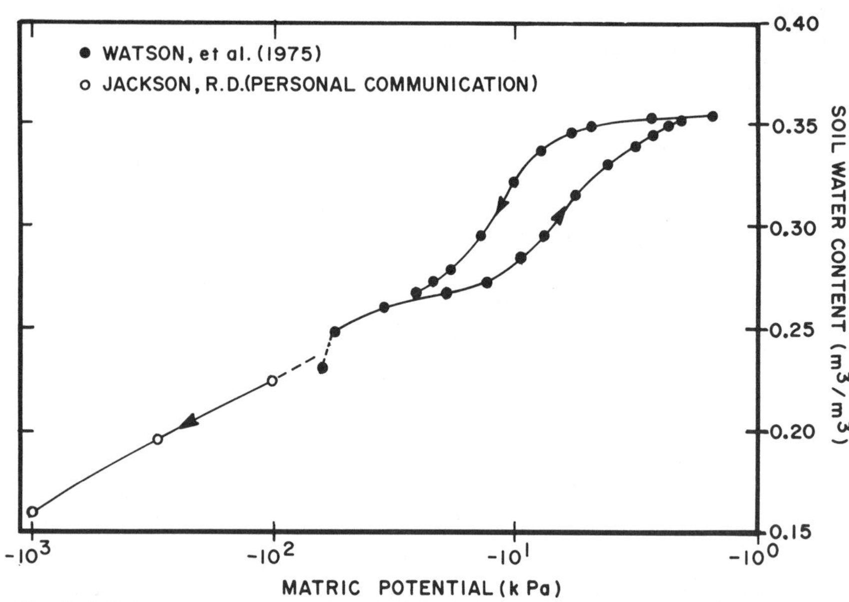

Fig. 27–1. Soil water content-matric potential relationship at 8- to 9-cm depth of Avondale clay loam (Torrifluventic Haplustolls) during wetting and drying (Watson et al., 1975).

kPa—and definition of the relationship at many intermediate points is of no interest, approximate methods of evaluation might be considered.

The soil area and depth for which the water content-matric potential relation is needed must be carefully defined. Any field soil characterization must face the reality of variability and include procedures for dealing with variability in order that objectives can be met. Because field determination of the water content–potential relation is done on relatively small areas, the selection of this small area is important and should depend upon its relation to existing variability in the larger field site that may be the ultimate objective of the characterization. In some situations the soil water characteristics vary more with depth than with area and a decision upon the depth of characterization affects methodology. Certainly the various applications of such data by agronomists, soil scientists, engineers, hydrologists, or others should affect the measurement site selection if the effort in obtaining data is to produce satisfactory results.

27–2 PRINCIPLES

27–2.1 Site-related Considerations

Some soil information is generally available for all situations. Current soil classification programs in the USA have provided maps where the classification unit is the polypedon, which involves variability within the named soil series at the particular scale of mapping, plus some inclusions

(Soil Survey Staff, 1975). Data application may make it desirable to attempt a further classification of the variability found within the polypedon as mapped. It may seem feasible to draw a map that further defines the variation in erosion, drainage, slope, or other readily determined features that are related to water properties. In any case, a pedon is selected as the basic sampling or measurement unit (Soil Survey Staff, 1975). The water content-matric potential relations obtained by measurement on this soil volume are assumed to represent the smallest unit of site classification within the existing variability. Measurements should be made on more than one pedon in order to quantify the variability.

To get measurements over the desired range in all soil layers, each field measurement unit must be wetted, dried, or both. Soils with horizon characteristics that restrict water flow will impede wetting and drainage, thereby reducing the range of water potential observation. For example, when horizon characteristics and sequences restrict drainage, such as is the case of a clay layer over sand, evaporation or extraction of water by plant roots may be used to obtain the desired water potential range.

27–2.2 Instrument-related Considerations

All instrumentation used in these field measurements is described elsewhere (chapters 21, 23, 24 and 25), where their characteristics and operation can be reviewed. Particular attention should be given to the sampling volume of the instrument or technique. The pedon or soil volume being characterized has both vertical and horizontal heterogeneity. Ideally, the instrument or method selected to measure water content or potential should sample a pore configuration that includes the range of large and small pores in each horizon of the pedon being characterized. Since it is unlikely that the selected method or instruments will sample both pore configuration and range of pore size at one position, measurements should be made in enough positions in the soil horizon to allow a statement of variability. Unfortunately, the volumes sampled in the measurement of soil water content and of water potential are never coincident, although partial sample volumes may be common to both. Not only is the magnitude of sample volume usually different, but the sample volumes are necessarily separated to avoid measurement interference. Physical separation of sampling volumes of water content and water potential should be minimal or matched (Van Bavel et al., 1968; Hewlett & Douglass, 1961). Since measurements will be made on transient state systems with a variety of pore sizes and configurations, instruments—such as tensiometers—that depend upon water flux across porous cups or membranes should present sufficient interface with soil to reduce lag (Klute and Gardner, 1962; see also chapters 23, 24, and 25). On the other hand, the instrument sample volume can be excessively large, and thus thin soil layers cannot be sampled independently from adjacent layers (e.g., water content by the thermal neutron method, chapter 21).

Measurements will be made on transient state systems that are frequently subject to sizeable temperature change. Therefore, instrument time and temperature response need careful consideration (chapters 21, 23, 24 and 25). By insulation and appropriate housing of instruments the field temperature can be moderated and changes minimized. Time-response requirement of instrumentation depends upon the particular soil system. Examples of matric potential changes at four positions at the 10-cm depth in an Ap soil horizon are shown in Fig. 27–2. Watson et al. (1975) have presented a picture of simultaneous change in water content and matric potential during infiltration and drainage that may be instructive.

27–2.3 Nature and State of Soil Volume

To describe the water content–potential relation it is necessary to make measurements over a large soil-water range. To change the soil water, a period of water application is involved until thorough wetting to the desired depth in the soil volume is achieved. During this period swelling will occur to varying degrees, depending upon quantity and type of clay present. Subsequently, during drying by drainage, evaporation, or plant extraction, shrinking may occur. In nonswelling, unsaturated soils, water moves in response to a gradient of total potential, ϕ, made up of matric potential, ψ, and gravitational potential, $-z$, so that $\phi = \psi - z$. The ψ is readily measured in situ with a tensiometer over its functional

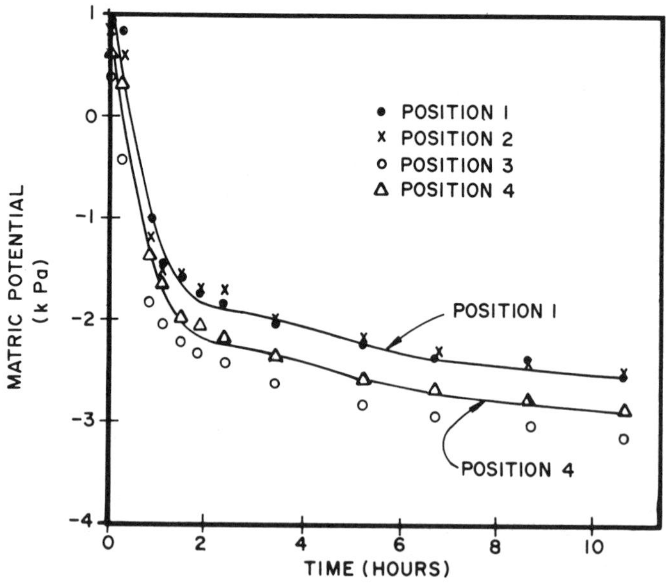

Fig. 27–2. Matric potential during first 10 h of drainage at 0.1-m depth in four positions of 2.5- by 2.5-m plot of Cecil sandy loam (Typic Hapludults). Curves represent sliding polynomial smoothing of data at positions 1 and 4. (Bruce et al., 1983).

range. If one-dimensional volume change occurs as in swelling soils, overburden potential, Ω, is added to the total potential as follows: $\phi = \psi + \Omega - z$. The tensiometer measures $\psi + \Omega$ in such cases, not simply ψ. Measurement of Ω is required before ψ may be evaluated. For measurement of overburden potential refer to Talsma (1977), Sposito et al. (1976), and associated literature citations.

Air entrapment by the system during wetting and draining is unavoidable, although it may be responsible for high temperature-dependence of matric potential (Cary, 1967, 1975) and thus contribute to hysteresis. Amount of air entrapment may be affected by water application procedures, soil characteristics, and environmental conditions. However, the need for measurements during wetting may largely determine the water application procedure. There is little reported evidence to guide the choice of water application procedure to ensure less air entrapment. Wetting from the bottom of the soil profile with a slowly rising watertable, where possible, will entrap less air (Royer & Vachaud, 1975).

27–3 METHOD

Using the principles stated in section 27–2, select site(s) for water content and potential measurements. The area for measurement should be large enough to accommodate the sampling and instrument installation and to meet the pedon definition for the particular soil series. Plot areas have ranged from < 1 to > 600 m^2 (Table 27–1) depending upon extent of investigations. Install an earthen dike around the perimeter of the plot sufficient to allow ponding of water over the surface. Metal, plastic, or wood may be driven or dug into the soil to a depth of 10 or 12 cm to serve as a dike.

Accurately assess layer or horizon thickness for each plot before installing any instruments. To provide an adequate log, a competent soil morphologist should examine several cores or augered material to the required depth of characterization on all sides of the plot and within a meter of the plot perimeter.

27–3.1 Method for 0 to −50 kPa Range

27–3.1.1 APPARATUS AND EQUIPMENT

1. Water supply.
2. Shovel.
3. Bucket soil auger (approximately 6-cm diameter).
4. Soil sample cans.
5. Balance.
6. Oven for drying at 105 °C.
7. Three or four tensiometers per horizon.
8. Neutron-depth soil water probe.

Table 27–1. Characteristics of some field $\theta(\psi)$ determinations.†

Instrument		Plot		Liner		Measure-ment depth, m	Soil	Wetting (+) Drying (−)	Comment (pressure range; other measurements)	Citation
θ	ψ	Shape	Size, m	Depth, m	Type					
Gravimetric	Hg manometer	Square	2.63 by 2.63	--	Wood frame	0.8	Pachappa sandy loam (Mollic Haploxeralfs)	−	Bare soil	Richards et al. (1956)
Gravimetric	Hg manometer	Square	2.63 by 2.63	--	Wood frame	0.8	Pachappa sandy loam (Mollic Haploxeralfs)	−	Bare soil, no evaporation	Ogata & Richards (1957)
Neutron method	Hg manometer	Rectangular	5 by 10	0.08	Plywood	1.6	Adelanto clay loam (Xeralfic Haplargids)	−	Bare soil followed by sorghum crop	Van Bavel et al. (1968)
Gamma-ray attenuation	Hg manometer	Rectangular	0.3 by 0.19	0.3	Acrylic plastic and marine plywood	0.01	Sunshine sandy loam (Boralfic cryoborolls)	±	Millipore-membrane tensiometer controlled water table data for 1 soil layer	de Vries (1969)
Neutron method	Hg manometer	Square	2.4 by 2.4	0.8	Plastic	0.76	Ramona sandy loam (Typic Haploxeralfs)	−	Bare soil	Roulier et al. (1972)
Neutron method	Hg manometer	Rectangular	61.1 by 1.4	2.13	Plastic	1.78	Halewood sandy loam (not classified inactive series)	−	Soil removed and repacked in field following lining of the sloping trench	Scholl and Hibbert (1973)
Neutron method	Transducer	Undefined	Undefined	2.5	None	2.5	Fine sand	±		Royer & Vachaud (1974, 1975)

(continued on next page)

Table 27–1. Characteristics of some field θ (ψ) determinations.†

Instrument		Plot		Liner		Measurement depth, m	Soil	Wetting (+) Drying (−)	Comment (pressure range; other measurements)	Citation
θ	ψ	Shape	Size, m	Depth, m	Type					
Gravimetric	Hg manometer	Square	3 by 3	0.3	Plastic	0.9	Waukegan loam (Typic Hapeudolls)	−	Bare plot	Arya et al. (1975)
Neutron method	Transducer	Circular	0.55 diameter	0.87	metal drum	0.75	*Capilano* and *Strachan* gravelly sandy loam	−	Laterally inserted tensiometers	Cheng et al. (1975)
Gamma-ray attenuation	Transducer	Equilateral Hexagon	0.76 (side)	1.27	Steel plate	1.1	Avondale clay loam (Typic Torrifluvents)	±	Soil monolith	Watson et al. (1975)
Neutron method	Hg manometer	Square	2 by 2	0.16	Wood planks	1.4	Troup loamy sand (Grossarenic Paleudults)	−		Dane (1980)
Neutron method	Hg manometer	Rectangular	18.3 by 30.5	--	None	3	Yolo silt loam (Typic Xerochrepts)	±	Irrigated crop study	Simmons et al. (1979)

† θ = soil water content; ψ = matric potential.

9. Bulk density sampler.
10. Rainfall shelter for plot.
11. Plastic sheet (of at least plot dimensions).
12. Insulation material.

27–3.1.2 PROCEDURE

When the site has been selected and before the dike is constructed, as suggested above, kill vegetation and prepare the soil surface to suit measurement objectives: e.g., tilled, nontilled, or with plant residue. Provision should be made to keep most traffic off the plot during instrument installation. Assuming the validity of principles stated in section 27–2, tensiometers installed in clusters, with all depths sampled in each cluster, are recommended for matric potential measurement. A cluster, in effect, samples one segment of the plot at all depths. Water-content sampling is then arranged at each depth in the vicinity of the cluster. Therefore, install three or four clusters of tensiometers with a corresponding arrangement for measuring soil water at each depth in the vicinity of each tensiometer cluster. Locate the tensiometer cup in the middle of each layer or horizon to be characterized; or if more than one is located in thick horizons, avoid proximity to horizon boundaries. Installation of instruments and soil sampling at a selected depth interval, without regard to soil profile horizonation, ignores the opportunity to begin classification of variability and is not recommended.

Tensiometers may be installed vertically or horizontally. For horizontal installation, trenching along at least two plot boundaries will likely be necessary to accommodate the number of instruments suggested above. These soil faces of the plot must be satisfactorily retained to prevent sloughing during wetting and drying. Vertically installed tensiometers should not allow free water to flow along the vertical shank or connection between cup and pressure-measuring device. Procedures for vertical installation of tensiometers must ensure that water applied to the plot surface does not reach the porous cup along a pathway resulting from the installation before it reaches the cup through the undisturbed soil body. For dependable installation, make a hole at the appointed plot position to a depth corresponding to slightly above the proposed elevation of the upper edge of the porous cup. The hole diameter, D, should be large enough to permit ready tamping of backfill around the tensiometer shank or connection between porous cup and soil surface (Fig. 27–3). A bucket auger with the appropriate diameter is a recommended hand tool if inaccessibility or unavailability prevents the use of power sampling-devices. Create a cylindrical cavity in the bottom of this hole, with a common center axis, of such length and diameter to provide good soil contact when the porous cup of the tensiometer is inserted (Fig. 27–3). If rocks or other difficult soil conditions are present, an alternative to creating a cavity for the tensiometer cup is to extend the auger hole of diameter D to the proposed elevation for the lower end of porous cup and to backfill

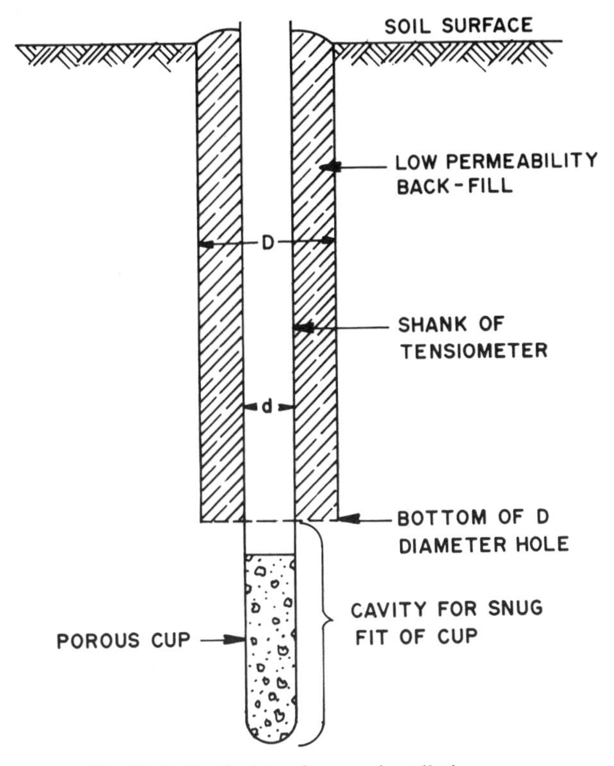

Fig. 27–3. Vertical tensiometer installation.

around the cup with dry, sieved soil similar to that excavated from that depth. In each case make sure to have the cup in intimate contact with a material having hydraulic characteristics similar to those of the soil volume being sampled. After inserting the tensiometer cup into the cavity or backfilling around the cup, backfill around the tensiometer shank to attain a low permeability cylinder in intimate contact with soil and tensiometer shank. A 50:50 by volume mixture of dry kaolin and building sand tamped in incremental amounts has proven satisfactory. A. W. Thomas and R. R. Bruce, unpublished data; Southern Piedmont Conservation Research Center, USDA, ARS, Watkinsville, GA 30677.

Read pressures from manometers using water (horizontal installation) or mercury (vertical or horizontal installation), or from pressure transducers. Response time is a primary criterion depending upon soil characteristics, as discussed in section 27–2.2.

To measure soil water content, θ, over the entire range from 0 to -50 kPa potential or less, use either thermal neutron or dual-source gamma transmission methods. In each plot, install access tubes in the vicinity of each tensiometer cluster. The single access tube for a neutron probe has been commonly located at the center of a circular tensiometer arrangement. Access tube installation procedures are discussed in chapter

21. Obtain calibrations applicable to the soil situation to ensure reliable water content evaluation.

When tensiometers and access tubes have been installed, a rainfall shelter may be needed. Although a rainfall shelter may be unnecessary in more arid climates, it is essential in many areas if uninterrupted periods of drying are to be maintained. However, rainfall shelters do not remove the possibility of water moving in from surrounding areas during extended periods of rainfall. Protection against rising water tables and water from surrounding areas may not normally be economically justified when compared with a restart.

With installation complete, the plot surface may need shallow hand-tillage and leveling to overcome effects of installation activity, even if the effects have been slight. Water may then be applied until a near steady-state condition is achieved or no further change in tensiometer readings is noted. The tensiometers may read negative, zero, or positive pressures, depending upon pedon characteristics. After all tensiometers are reading properly and the water-content measuring device is in place, cease water application and allow drainage. Read tensiometers and make neutron or gamma transmission counts frequently during early drainage; then appropriately reduce frequency as drainage proceeds (Fig. 27–2 and 27–4). As soon as free water leaves the soil surface, install a barrier to evaporative water loss, such as a polyethylene sheet. To moderate soil temperature, place abundant insulation over the soil surface. After soil water pressures become somewhat negative, soil sampling for water content determination can replace or supplement the neutron or gamma system. Soil samples should correspond with tensiometer positions. To convert water content from mass to volume basis, the bulk density for each position must be determined. Measure bulk density in a soil volume as close as possible to the volume in which water content was measured (Hewlett & Douglass, 1961).

When tensiometer readings do not change measurably during 7 to 10 days, the drainage is about complete. If you need water content-potential data during wetting, remove the insulation and polyethylene sheet from the plot surface and quickly establish and maintain a constant low head of water on the plot surface while recording ensuing changes in water content and matric potential with time. Estimate and record head of water on surface. Cease water application and measurement when steady state is achieved.

27–3.2. Method for Evapotranspiration Range of Potentials

Whereas the range of potential under consideration in section 27–3.1 was limited to that attainable by drainage, in this section the methodology will extend the range to about -1.5×10^3 kPa. Much of the previous discussion of methods in section 27–3 applies here.

27–3.2.1 APPARATUS AND EQUIPMENT

1. Water supply.
2. Shovel.
3. Bucket soil auger (approximately 6-cm diameter).
4. Soil sample cans.
5. Balance.
6. Oven for drying at 105°C.
7. Neutron-depth soil water probe.
8. Bulk density sampler.
9. Three or four tensiometers per horizon.
10. Rainfall shelter for plot.
11. Thermocouple psychrometers or electrical resistance sensors.

27–3.2.2 PROCEDURE

When the site-selection process has been completed as discussed earlier and soil pedon morphology described, plant an appropriate grass or other relatively short species to develop a uniform, deep-rooting system for profile water extraction. As soon as vegetation is well established, begin installing instruments to measure soil water content and water potential over the desired range, applying principles in section 27–2. Instrumentation is required for soil-water content and potential measurements in each soil horizon at three or four positions in the plot. If completely in situ measurements are not selected, provision must be made to obtain soil samples to measure water content or water potential in each horizon or selected depth at each of the positions in the plot.

If the water-potential range from near saturation or early drainage to less than -1.5×10^3 kPa requires characterization, and if completely in situ measurements are selected to measure down to -1.5 or -2.0×10^3 kPa, tensiometers can be installed as discussed in section 27–3.1.2 along with electrical resistance sensors or soil thermocouple psychrometers, to measure down to -70 or -80 kPa. The electrical resistance sensor is discussed in chapters 21 and 25 and the thermocouple psychrometer in chapter 24. Summary considerations are presented in Table 27–2. Where experience with the thermocouple psychrometer is unavailable and salinity is not a problem, electrical resistance sensors having reliable calibrations offer a satisfactory alternative in many situations. If thermocouple psychrometer equipment and experience are available for laboratory use, soil samples may be transported to the laboratory for determination of water potential (Papendick et al., 1971). In cases of significant soil salinity, the osmotic potential will need to be determined if matric potential is obtained from psychrometer data. Water content may be determined by the thermal neutron method in each cluster of tensiometers, electrical-resistance sensors, or psychrometers by locating access tubes at least 0.2 m from each sensor. As the soil dries, the sampling volume of the neutron probe may more than double and include adjacent

Table 27-2. Measurement of matric potential.

Method	Range	Operational considerations
	In situ	
Tensiometer	> -80 kPa	Direct and accurate determination Servicing required Temperature-sensitive
Electrical resistance sensor	-50 kPa to -1.5×10^3 kPa	Calibration-dependent Low maintenance Inexpensive Time-consuming calibration Affected by salinity
Thermocouple psychrometer	-0.1×10^3 to -5×10^3 kPa	Calibration-dependent Measures total water potential Temperature-sensitive Expensive and complex
	Samples taken to laboratory	
Thermocouple psychrometer	-0.1×10^3 to -5×10^3 kPa	Calibration-dependent Measures total water potential Expensive Accurate Relatively rapid

soil layers (chapter 21). If soil horizons < 0.15 m in depth must be characterized, other methods will be needed. Physical removal of soil samples from each designated level in the pedon for gravimetric determination of soil water may ultimately be necessary. If the plot will subsequently be flooded or exposed to rainfall, holes left after sampling must be filled with material having lower hydraulic conductivity than the soil profile. The 50:50 kaolin and sand mixture mentioned in section 27–3.1.2 is quite satisfactory. Since the mixture is white, it is also a good marker to identify disturbed soil.

When instruments have been installed and plans made for necessary sampling, apply water uniformly over the plot surface by sprinkling or ponding until the soil profile is wet to at least the required depth of measurement, as determined by tensiometers or other sensors. When steady state has been achieved, measure the potential and water content and cease water application. Measure water content and potential at intervals as drainage and drying proceed. To ensure that drying is not interrupted, a rainfall shelter may be installed with provision for removal to supply adequate light for vegetation. The alternative is to plan for long-term observation, with soil water patterns determined by the weather. The several wetting and drying periods that may be experienced can yield information about hysteresis in the water content-potential relation. Measurement of water content and potential can continue until characterization of the desired soil water range in each soil layer is achieved.

27–4 DATA HANDLING

The data derived from the series of field measurements obtained during drying or wetting procedures will require at least some manipulation, as follows: (i) conversion into appropriate units of volumetric water content, θ, or matric potential, ψ, (ii) smoothing into suitable $\theta(t)$ and $\psi(t)$ functions, where t = time, (iii) derivation of contemporary θ and ψ values from the smoothed functions, (iv) plotting of the pedon retention relationships θ (ψ), and (v) development of mathematical representation of the retention characteristic for each pedon (if desired). Most of these operations can be conveniently conducted with a computer; however, data may also be plotted manually. If computer data processing is anticipated, the field data collection can be made directly onto 80-column coding forms in a format that appropriately identifies the site, year, month, day, hour, minute, instrument type, instrument readings, and calibration data (e.g., standard count with the neutron probe).

27–4.1 Data Conversion

The first step in data handling is the conversion of the primary field records into volumetric water content and matric potential for each measurement time, using the appropriate calibration data. Upon conversion, the data for each soil depth should consist of water content and potential values at specified times over the desired range, at each of three or four positions in the plot. Neutron probe calibration should relate count ratio to volumetric water content (chapter 21) and tensiometer data must be converted to potential (chapter 23). Data obtained by other procedures should be appropriately handled to obtain volumetric water content and potential for each recorded item.

27–4.2 Smoothing $\theta(t)$, $\psi(t)$ Data

The temporal water content and matric potential data for each measurement position at each soil depth should be plotted and any spurious values checked for errors in data handling or instrument failure. Keypunching errors, weak battery in the neutron probe scaler, and gas formation in tensiometer lines are sources of error that can be identified from plotted data. The latter two errors are normally identified and corrected during field measurement; however, some data may have been collected before remedial action was taken. If justified rationally (see chapter 4), other widely deviating data should be deleted from the data set before data smoothing is attempted.

There are two main approaches to smoothing $\theta(t)$ and $\psi(t)$ data; these are the arbitrary or visual, and curve-fitting by mathematical and statistical procedures. In both cases it is usually advantageous to conduct the smoothing with the time values transformed logarithmically.

27–4.2.1 ARBITRARY SMOOTHING

This is a well known procedure of drawing a line through the data, alternatively called the "eyeball" method. Data are graphed and a line drawn where the operator thinks it ought to go. Straight line segments

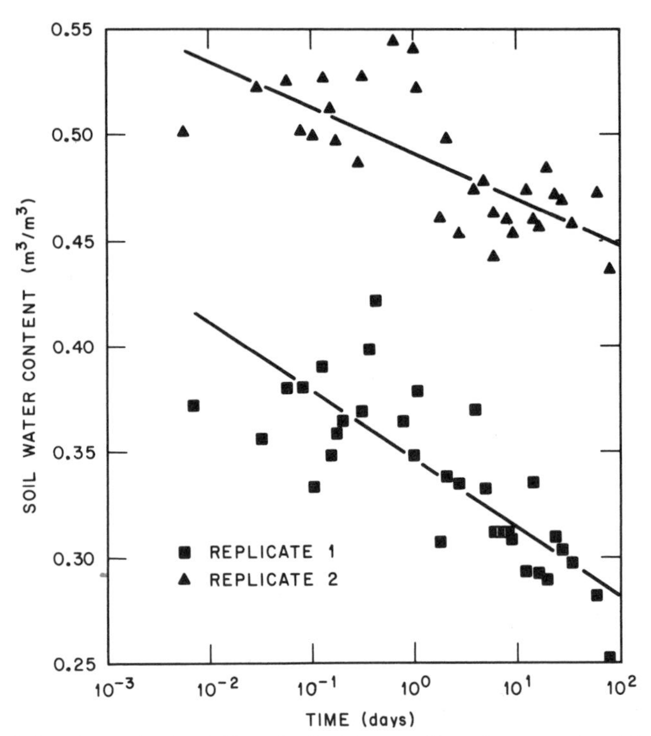

Fig. 27–4. Soil water content at 1.22-m depth at two positions 1 m apart in a 2-by 2-m plot of Sequoia silt loam (Typic Hapludults) as a function of time after beginning of drainage. The curves are cubic spline fits of data. (Luxmoore, 1982.)

can be fitted in some cases. Data that show considerable scatter, for example water content data such as those in Fig. 27–4, provide freedom for the operator to express his intuition. The bias that is included in this approach may greatly influence the outcome of subsequent calculations. Some data exhibit a well defined form (for example, matric potential data such as those in Fig. 27–5) and may readily be fitted by a hand-drawn line. A statement of statistical confidence may be derived for the arbitrary curve-fitting method by estimation of the differences between the field data and the curve. This statement can then be used to calculate a correlation coefficient.

27–4.2.2 FITTING SMOOTH FUNCTIONS TO DATA

There are several numerical procedures by which, to a large extent, the data determine the fitted function. These include smoothing procedures commonly involving spline (Erh, 1972), sliding polynomial (Snyder, 1976, 1980; Snyder et al., 1984) algorithms, or moving averages.

Parabolic spline (DuChateau et al., 1972) or cubic spline (Kimball, 1976) procedures have been advocated in recent years. Spline smoothing of unequally spaced data requires that the user choose the placement of

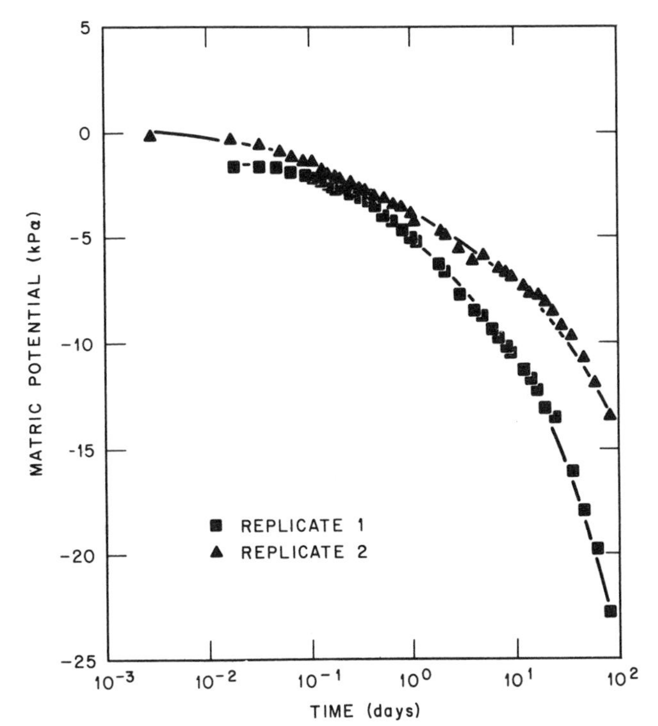

Fig. 27–5. Matric potential at 1.22-m depth at two positions 1 m apart in a 2- by 2-m plot of Sequoia silt loam (Typic Hapludults) as a function of time after beginning of drainage. The curves are cubic spline fits of data. (Luxmoore, 1982.)

"knots" which join together the segments of the polynomial. The requirement that the polynomial and the first and second derivatives be continuous at the knots make these methods sensitive to the placement of knots. Instruction from mathematicians or experienced scientists is helpful for those applying this procedure for the first time. Kimball (1976) has offered to provide his program to interested users. Many computer systems have subroutines available for processing data by polynomial spline smoothing. Figures 27–4 and 27–5 show a cubic spline fit made to some field data after spurious data were deleted from the analysis.

Sliding polynomial (Snyder, 1976, 1980; Thomas et al., 1977) algorithms are also useful for data smoothing and may provide mathematical advantages for some applications. An essential difference between splines and sliding polynomial is the form of mathematical continuity for joining piecewise arcs of functions. Splines are point-continuous, whereas sliding polynomials have functional continuity. Sliding polynomials also have reversibility, which means that they give consistent values of function, integral, and derivative in both the forward and backward calculation through a data set. As with spline and other procedures, there are some arbitrary aspects in using sliding polynomials, for example specifications of "base points." Smoothing is restricted to the range of observations unless special procedures are applied for limited extrapolation. Snyder (1976, 1980) provides some guidance on the choice of base points even for cases with continuously extending data, such as retention relationships. Program documentation is available to smooth data and determine gradients and standard deviation using sliding polynomials (Snyder, 1978). The data in Fig. 27–6 are fitted by sliding polynomial with accompanying standard deviation.

27–4.3 Derivation of Simultaneous θ and ψ Values

Hand-fitted $\theta(t)$ and $\psi(t)$ curves may be interpolated to find (θ,ψ) in the plot for a range of simultaneous times for each of the three or four positions at each depth. These (θ,ψ) pairs may then be plotted as θ (ψ). Sections of $\theta(t)$ and $\psi(t)$ curves with steep slopes may need to be plotted on an expanded time scale or on a log time scale to yield reasonable estimates of θ or ψ. For example, in drainage studies both θ and ψ change rapidly in the first 3 to 8 h, and it is difficult to estimate simultaneous (θ, ψ) pairs during this period if $\theta(t)$ and $\psi(t)$ are plotted for the whole study period (usually 20 to 50 days). This problem is avoided if polynomial smoothing functions have been fitted to the data. The respective

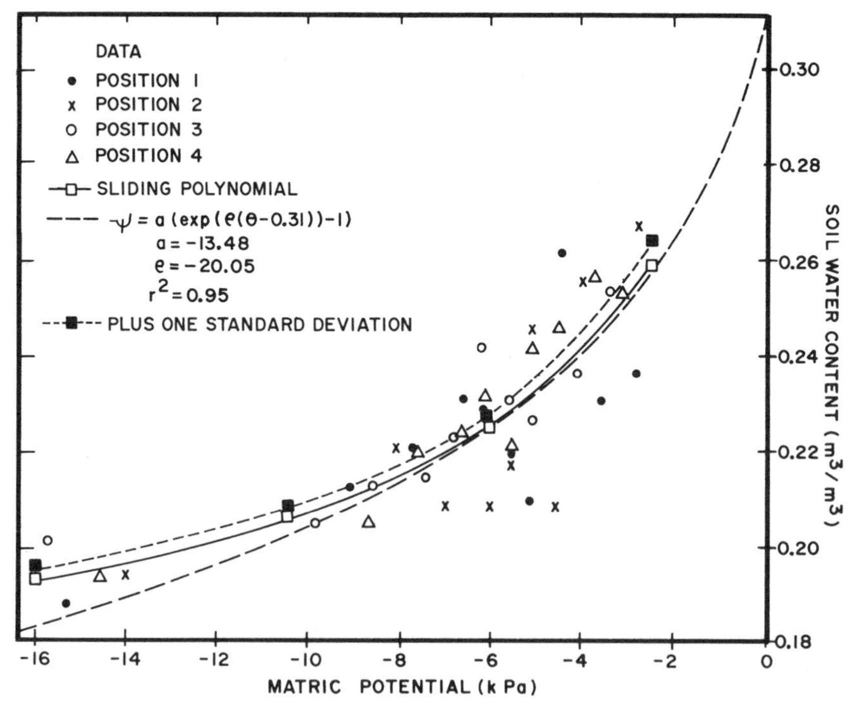

Fig. 27–6. Soil water content–matric potential at 0.1-m depth measured at four positions of 2.5- by 2.5-m plot of Cecil sandy loam; sliding polynomial smoothed curved, indicated standard deviation, and a fitted exponential by nonlinear regression. (Bruce et al., 1983).

$\theta(t)$ and $\psi(t)$ polynomials may be evaluated at a series of times to generate the (θ, ψ) pairs.

27–4.4 Determining $\theta(\psi)$ for the Pedon

Using the simultaneous θ and ψ values over the range of time for each of the three or four measurement positions at a given soil depth, a single $\theta(\psi)$ curve may be determined by mathematical smoothing procedures or by handfitting. The spatial variability that is exhibited may be expressed by the standard deviation or other statistic associated with the fitted curve. In Fig. 27–6, soil water content determined gravimetrically and matric potential determined at four positions in a 6.25-m^2 plot at 0.1-m depth are plotted with a sliding-polynomial fitted-curve and accompanying standard deviation. Mathematical options for expressing the pedon $\theta(\psi)$ relation and associated variability are discussed below. Expression of the $\theta(\psi)$ variability with depth, as well as area, in the profile may also be justified.

27–4.5 Mathematical Retention Functions

In this approach some preconceived functional form is selected for the relationship between water content and matric potential. Nonlinear regression procedures can then be applied to the data for selecting the most appropriate values for parameters required by the function. The following functions are examples of $\theta(\psi)$ or $\psi(\theta)$ relationships:

$$\psi(\theta) = -a \exp(-b\theta) \qquad \text{exponential}$$

$$\psi(\theta) = -a\, \theta^{-b} \qquad \text{power}$$

$$\psi(\theta) = a(f-\theta)^b/\theta^c \qquad \text{(Visser, 1966)}$$

$$\psi(\theta) = a\{\exp[\rho(\theta - \phi)] - 1\} \qquad \text{(Simmons et al., 1979)}$$

$$\theta(\psi) = [(\theta - \theta_r)/(1 - \theta_r)][1/(\psi/\psi_b)\lambda] \qquad \text{(Brooks \& Corey, 1964)}$$

$$\theta(\psi) = \theta_e + \frac{\theta_{15} - \theta_e}{\ln(\psi_e - \psi_{15} + 1)} \ln(\psi_e - \psi + 1) \qquad \text{(Rogowski, 1971)}$$

$$\theta(\psi) = [1/(1 + \alpha\psi n)]^m \qquad \text{(Van Genuchten, 1979)}$$

In the first two functions, a and b are estimated parameter values. The specified citations define the parameters in the other functions. White et al. (1970) recognized four zones in the drying characteristic and derived rather complex equations for each zone. McQueen and Miller (1974) identified three linear segments for the retention characteristic when expressed as a semilogarithmic plot (pF). These linear segments can be drawn for limited data according to simple guidelines.

In all these cases, estimates of the goodness of fit of the data to the function should be obtained both by visual inspection of the plotted data with the function results and by statistical use of correlation coefficients. The equation of Simmons et al. (1979) is fitted to data in Fig. 27–6. Sometimes the mathematical function can be a good predictor of a retention characteristic for part of the range, with poor agreement obtained in other parts of the range.

27–5 ERROR ANALYSIS

The principal sources of error in the field determination of retention characteristics are due to operator performance, instrument performance, calibration, and data manipulation. Several factors contribute to these error sources (Table 27–3); they are fully discussed in the respective chapters on the specific instrumental methods. Data manipulation introduces error that may be arbitrary, systematic, or both. Mathematical smoothing introduces both forms of bias. The arbitrary selection of knots for cubic spline or base points for sliding polynomial procedures, for example, introduces an arbitrary bias; the difference between using a spline or sliding polynomial fitting function represents a systematic bias. This systematic bias is probably less important than other sources of error.

The most significant source of error in retention relationships probably relates to the estimation of volumetric water content. The neutron probe calibration is usually a linear function in which volumetric water content (θ) is related to count ratio (R) as follows:

$$\theta = aR + b$$

where a and b are fitted parameters. Accurate values for both a and b are required for accurate estimation of θ. The standard deviation of the slope parameter, a, is often in the range of 10 to 20% of a (Fluhler et al., 1976). The 95% confidence intervals for a are thus about $\pm$30% of a, assuming a normal frequency distribution for a. The corresponding range

Table 27–3. Sources of error in tensiometer, neutron probe, and gravimetric water content determinations.

Instrument or method	Operator	Instrument performance	Instrument calibration
Tensiometer	Reading mercury or water manometer	Gas formation in manometer lines	Pressure transducer calibration
Neutron probe	Placing probe at the intended depth	Weak battery in scaler	Count ratio calibrated with gravimetric water content and bulk density
Gravimetric	Soil handling procedures to reduce vapor loss; Reading balance	Negligible	Balance calibration

in water content may be considerable. Errors in the *a* and *b* parameters introduce a systematic bias in the retention curve that will not change the shape of the curve, but will influence the overall position in the (θ,ψ) domain. Any change in probe calibration during the experiment will change the shape of the retention curve if a constant calibration is assumed. It is usually not feasible to check for drift in probe calibration when rapid drainage or infiltration is occurring during the critical phase of measurement.

The sources and propagation of errors in the experimental measurements can be estimated if the precision of the instrumental methods is determined. For example, (i) if tensiometers with mercury manometers can be read with an accuracy of ± 0.5 mm of Hg, then the reading lies within a range of 0.12 kPa, and (ii) if water content estimation with a calibrated neutron probe can be made within a range of $\pm$ 0.05 m^3/m^3, then any estimated retention curve would lie within a domain that can be plotted. Given these estimates, the error range is relatively dominated by errors in measurement of soil water, which is less precisely estimated.

27-6 SPATIAL HETEROGENEITY

Currently, there are two main approaches for characterization of variability in soil properties. These are statistical methods (Beckett & Webster, 1971) and scaling by the similar media concept (Miller & Miller, 1956). The techniques for determination of spatial isopleths for soil properties have been advanced with the development of the semivariogram and kriging techniques. These two latter procedures will not be discussed further here; the reader may refer to two papers by Burgess and Webster (1980a, 1980b) for an introduction to the methods (see also chapter 3).

27-6.1 Statistical Methods

The field measurement of retention characteristics, $\theta(\psi)$, at one location represents one determination from some population of retention curves for the polypedon. Additional field determinations at surrounding locations provide the necessary data for characterizing the population of retention characteristics. Results from 20 or more sites within the same polypedon can be plotted on one figure to show the range of heterogeneity (see Fig. 27-7A) (Warrick et al., 1977). The mean retention curve and an estimate of variance can be obtained by regression analysis.

27-6.2 Scaling

There is evidence that by a scaling procedure, one set of retention characteristics for a selected reference profile along with a frequency distribution of some measure of variability for the polypedon of interest can be used to derive the heterogeneous retention characteristics of a poly-

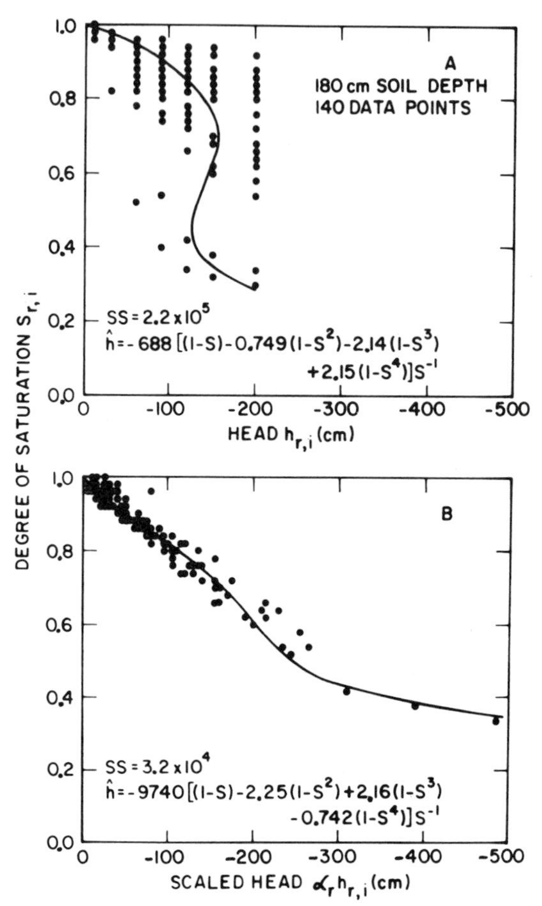

Fig. 27–7. Soil water characteristic data for 180-cm depth of Panoche soil: (A) unscaled and (B) scaled (Warrick et al., 1977)

pedon. Thus field characterization is reduced to two components: (i) measuring water content and matric potential at a site (reference soil), and (ii) determining a frequency distribution of site heterogeneity. The first component has been discussed at some length in this chapter and the second will be briefly discussed in this section. The scaling approach assumes that the similar media criterion can be applied to soils. This means that all soils under consideration have the same total porosity but differing mean particle sizes (characteristic lengths). The ratio of the characteristic lengths of a particular soil to that of the reference soil is the scaling factor, α. Soils with a finer pore-size distribution than the reference have $\alpha < 1$; those with a coarser pore-size distribution have $\alpha > 1$. The pressure head, h_i, of a particular soil at the same water content as the reference soil ($\theta_i = \theta_r$) may be obtained by scaling the reference pressure head, h_r, with the scaling factor, α_i, for the particular soil as follows:

$$h_i = h_r/\alpha_i .$$

This relationship may be applied to a range of (θ,h) retention data. Since the constant total porosity criterion is usually difficult to achieve in field soils, Warrick et al. (1977) introduced degree of saturation, s, (water content/total porosity) for scaling retention characteristics. This approach does not satisfy the constant total porosity requirement; however, scaling is still assumed to be meaningful.

In the Warrick et al. (1977) analysis, retention data (laboratory core data) from 20 locations and 6 depths in a 150-ha field (data from Nielsen et al., 1973) displayed considerable variability even when presented on a degree of saturation basis (Fig. 27–7A). The next step was the determination of scaling factors from the retention data. Warrick et al. (1977) discussed a method of deriving scaling factors. Applying the scaling factors to the retention data resulted in a set of scaled retention data (Fig. 27–7B) in which pressure heads less than the mean moved horizontally to lower pressures and those greater than the mean moved horizontally to higher pressures, providing a narrow band of scaled retention data. The frequency distribution of scaling factors represents a summary of site heterogeneity characteristics and was shown in the Warrick et al. (1977) study to be better represented by a lognormal function than by a normal distribution.

The frequency distribution of the scaling factor has also been determined from statistical analysis of field retention data (Simmons et al., 1979), infiltration data (Sharma et al., 1980), and in situ field determination of soil physical characteristics (Russo & Bresler, 1980). The frequency distribution of the scaling factor seemed to follow a lognormal model in the first two cases and a normal function in the third.

The studies of Warrick et al. (1977), Sharma et al. (1980), Simmons et al. (1979), and Russo and Bresler (1980) provided supporting evidence for the efficacy of scaling soil hydraulic properties. A major advantage of the procedure is the orderly treatment of field data while at the same time preserving some of the features of heterogeneity. In any application of the technique it behooves the investigator to demonstrate the appropriateness of scaling. Scaling is outlined here as a promising approach; we do not advocate it for routine use.

27–7 COMMENTS

The water content-potential relationships that have been applied to field situations have commonly been derived from some type of samples upon which laboratory measurements have been made. Substantial agreement between the soil water content-potential relation determined from soil cores in the laboratory and that determined in the field from tensiometers and soil samples for water content has been shown at greater than −50 kPa matric potential (Larue et al., 1968). In a field study of Brust et al. (1968), water content was determined in the field by the thermal neutron method instead of gravimetric methods on samples, and agreement with a laboratory-determined water content-potential relation was reported. In all cases, laboratory determinations were made on core samples procured from the field site and represented in situ soil structure and pore configuration as well as possible. The water content-potential relation of disturbed samples in the greater than −100 kPa range differs markedly from that determined from undisturbed core samples or in situ measurements (Brust et al., 1968; Bruce, 1972). Recently, a significant difference has been observed between water content-potential relationships determined from core samples and those obtained from in situ measurements (Cassel, 1985). These data, obtained on soils of the southeastern USA, generally indicate that core samples retain more water at a given matric potential than is measured in situ, signifying the need for caution in the general application of core sample data to the field situation.

The variability in the field situation must be adequately sampled both when taking samples for laboratory determination and when making in situ measurements. However, the controlled conditions of the laboratory allow a definition of a sample's water retention that is impossible in most field situations. For example, wetting of soil volumes in the laboratory can be readily controlled and described, as compared with the lack of control and certainty in wetting field soil volumes. In fact, the observed differences between water content potential relationships determined in situ and those determined in the laboratory may be due to the incomplete wetting of field soil volumes because of air-entrapment or macropore flow, and to our inability to describe the situation with existing field measurement techniques.

27-8 REFERENCES

Arya, L. M., D. A. Farrell, and G. R. Blake. 1975. A field study of soil water depletion patterns in presence of growing soybean roots: I. Determination of hydraulic properties of the soil. Soil Sci. Soc. Am. Proc. 39:424–430.

Beckett, P. H. T., and R. Webster. 1971. Soil variability: a review. Soils Fert. 34:1–15.

Brooks, R. H., and A. T. Corey. 1964. Hydraulic properties of porous media. Colorado State Univ., Fort Collins, Hydrology Paper no. 3.

Bruce, R. R. 1972. Hydraulic conductivity evaluation of the soil profile from soil water retention values. Soil Sci. Soc. Am. Proc. 36:555–561.

Bruce, R. R., J. H. Dane, V. L. Quisenberry, N. L. Powell, and A. W. Thomas. 1983. Physical characteristics of soils in the Southern Region: Cecil. Georgia Agric. Expt. Stn. and Univ. of Georgia Southern Coop. Ser. Bull. 267.

Brust, K. J., C. H. M. van Bavel, and G. B. Stirk. 1968. Hydraulic properties of a clay loam soil and the field measurement of water uptake by roots: III. Comparison of field and laboratory data on retention and of measured and calculated conductivities. Soil Sci. Soc. Am. Proc. 32:322–326.

Burgess, T. M., and R. Webster. 1980a. Optimal interpolation and isarithmic mapping of soil properties. I. The semi-variogram and punctual kriging. J. Soil Sci. 31:315–331.

Burgess, T. M., and R. Webster. 1980b. Optimal interpolation and isarithmic mapping of soil properties. II. Block kriging. J. Soil Sci. 31:333–341.

Cary, J. W. 1967. Experimental measurements of soil-moisture hysteresis and entrapped air. Soil Sci. 104:174–180.

Cary, J. W. 1975. Soil water hysteresis: temperature and pressure effects. Soil Sci. 120:308–311.

Cassel, D. K. (ed.) 1985. Physical characteristics of soils of the Southern Region—Summary of in situ unsaturated hydraulic conductivity. North Carolina State Univ. Southern Coop. Ser. Bull. 303.

Cheng, J. D., T. A. Black, and R. P. Willington. 1975. A technique for the field determination of the hydraulic conductivity of forest soils. Can. J. Soil Sci. 55:79–82.

Dane, J. H. 1980. Comparison of field and laboratory determined hydraulic conductivity values. Soil Sci. Soc. Am. J. 44:228–231.

de Vries, J. 1969. In situ determination of physical properties of the surface layer of field soil. Soil Sci. Soc. Am. Proc. 33:349–353.

DuChateau, P. C., D. L. Nofziger, L. R. Ahuja, and D. Swartzendruber. 1972. Experimental curves and rates of change from piecewise parabolic fits. Agron. J. 64:538–542.

Erh, K. T. 1972. Application of spline function to soil science. Soil Sci. 114:333–338.

Fluhler, H., M. S. Ardakani, and L. H. Stolzy. 1976. Error propagation in determining hydraulic conductivities from successive water content and pressure head profiles. Soil Sci. Soc. Am. J. 40:830–836.

Hewlett, John D., and James E. Douglass. 1961. Method for calculating error of soil moisture volumes in gravimetric sampling. For. Sci. 7:265–272.

Kimball, B. A. 1976. Smoothing data with cubic splines. Agron. J. 68:126–129.

Klute, A., and W. R. Gardner. 1962. Tensiometer response time. Soil Sci. 93:204–207.

Larue, M. E., D. R. Nielson, and R. M. Hagan. 1968. Soil water flux below a ryegrass root zone. Agron. J. 60:625–629.

Luxmoore, R. J. 1982. Physical characteristics of soils of the Southern Region: Fullerton and Sequoia Series. North Carolina State Univ. Southern Coop. Ser. Bull. 268.

McQueen, I. S., and R. F. Miller. 1974. Approximating soil moisture characteristics from limited data: empirical evidence and tentative model. Water Resour. Res. 10:521–527.

Miller, E. E., and R. D. Miller. 1956. Physical theory for capillary flow phenomena. J. Appl. Phys. 27:324–332.

Nielsen, D. R., J. W. Biggar, and K. T. Erh. 1973. Spatial variability of field measured soil-water properties. Hilgardia 42:215–260.

Ogata, G., and L. A. Richards. 1957. Water content changes following irrigation of bare-field soil that is protected from evaporation. Soil Sci. Soc. Am. Proc. 21:355–356.

Papendick, R. I., V. I. Cochran, and W. M. Woody. 1971. Soil water potential and water content profiles with wheat under low spring and summer rainfall. Agron. J. 63:731–734.

Richards, L. A., W. R. Gardner, and G. Ogata. 1956. Physical processes determining water loss from soil. Soil Sci. Soc. Am. Proc. 20:310–314.

Rogowski, A. S. 1971. Watershed physics: Model of soil moisture characteristics. Water Resour. Res. 7:1575–1582.

Roulier, M. H., L. H. Stolzy, J. Letey, and L. V. Weeks. 1972. Approximation of field hydraulic conductivity by laboratory procedures on intact cores. Soil Sci. Soc. Am. Proc. 36:387–393.

Royer, J. M., and G. Vachaud. 1974. Determinations directe de l'evapotranspiration et de l'infiltration par mesure des teneurs en eau et des succions. Hydrol. Sci. Bull. 19:319–336.

Royer, J. M., and G. Vachaud. 1975. Field determination of hysteresis in soil-water characteristics. Soil Sci. Am. Proc. 39:221–223.

Russo, D., and E. Bresler. 1980. Scaling soil hydraulic properties of a heterogeneous field. Soil Sci. Soc. Am. J. 44:681–684.

Scholl, D. G., and A. R. Hibbert. 1973. Unsaturated flow properties used to predict outflow and evapotranspiration from a sloping lysimeter. Water Resour. Res. 9:1645–1655.

Sharma, M. L., G. A. Bander, and C. G. Hunt. 1980. Spatial variability of infiltration in a watershed. J. Hydrol. 45:101–122.

Simmons, C. S., D. R. Nielsen, and J. W. Biggar. 1979. Scaling of field-measured soil water properties. Hilgardia 47:77–173.

Snyder, W. M. 1976. Interpolation and smoothing of experimental data with sliding polynomials. USDA-U.S. Government Printing Office, Washington, DC.

Snyder, W. M. 1978. Running details for fully computerized smoothed data and gradients using sliding polynomials. Lab Note SEWRP 097803, Southeast Watershed Research Lab, ARS-USDA, Tifton, GA 31793.

Snyder, W. M. 1980. Smoothed data and gradients using sliding polynomials with optional controls. Water Resour. Bull. 16(1):22–30.

Snyder, W. M., R. R. Bruce, L. A. Harper, and A. W. Thomas. 1984. Two-dimensional sliding polynomials. Georgia Agric. Exp. Stn. Res. Bull. 320.

Soil Survey Staff. 1975. Soil taxonomy: A basic system of soil classification for making and interpreting soil surveys. USDA-SCS Agric. Handbook 436. U.S. Government Printing Office, Washington, D.C.

Sposito, Garrison, Juan V. Giraldez, and Robert J. Reginato. 1976. The theoretical interpretation of field observations of soil swelling through a material coordinate transformation. Soil Sci. Soc. Am. J. 40-208-211.

Talsma, T. 1977. Measurement of the overburden component of total potential in swelling field soils. Aust. J. Soil Res. 15:95–102.

Thomas, A. W., W. M. Snyder, and R. R. Bruce. 1977. Smoothing, interpolation, and gradients from limited data. Agron. J. 69:747–750.

Van Bavel, C. H. M., G. D. Stirk, and K. J. Brust. 1968. Hydraulic properties of a clay loam soil and the field measurement of water uptake by roots: I. Interpolation of water content and pressure profiles. Soil Sci. Soc. Am. Proc. 32:310–317.

Van Genuchten, M. Th. 1979. Calculating the unsaturated hydraulic conductivity with a new closed-form analytical model. Res. Rep. 78-WR-08, Princeton University, Princeton, NJ.

Visser, W. C. 1966. Progress in the knowledge about the effect of soil moisture content on plant production. Tech. Bull. 45. Institute for Land and Water Management Research, Wageningen, The Netherlands.

Warrick, A. W., G. J. Mullen, and D. R. Nielsen. 1977. Scaling field-measured soil hydraulic properties using a similar media concept. Water Resour. Res. 13:355–362.

Watson, K. K., R. J. Reginato, and R. D. Jackson. 1975. Soil water hysteresis in a field soil. Soil Sci. Soc. Am. Proc. 39:242–246.

White, N. F., H. R. Duke, D. K. Sunada, and A. T. Corey. 1970. Physics of desaturation in porous materials. J. Irrig. Drain. Div. Am. Soc. Civ. Eng. 96(IR2):165–191.

28 Hydraulic Conductivity and Diffusivity: Laboratory Methods

A. KLUTE

Agricultural Research Service, USDA, and
Colorado State University
Fort Collins, Colorado

C. DIRKSEN

The Agricultural University, Wageningen
The Netherlands

28-1 INTRODUCTION

The rate of movement of water through soil is of considerable importance in many aspects of agricultural and urban life. The entry of water into soil, the movement of water to plant roots, the flow of water to drains and wells, and the evaporation of water from the soil surface are but a few of the obvious situations in which the rate of movement plays an important role. The soil properties that determine the behavior of soil water flow systems are the hydraulic conductivity and water-retention characteristics. The *hydraulic conductivity* of a soil is a measure of its ability to transmit water; the *water-retention characteristics* are an expression of its ability to store water. These properties determine the response of a soil water system to imposed boundary conditions. In some cases, the hydraulic or soil water *diffusivity*, which is the ratio of the hydraulic conductivity to the differential water capacity, may be used to analyze the behavior of a soil water system. These properties are often called the *hydraulic properties* of the soil.

In this chapter several laboratory methods of determining the hydraulic conductivity and hydraulic diffusivity are described. Many methods for determining these properties have been given in the literature. The choice of method depends upon such factors as (i) the available equipment, (ii) the nature of the soil, (iii) the kind of samples available, (iv) the skills and knowledge of the experimenter, (v) the soil-water suction range to be covered, and (vi) the purpose for which the measurements are being made. The methods given here are, in the opinion of the authors, methods that should be useful and applicable in many cases. However, in specific instances one or more of the many other methods presented

 Methods of Soil Analysis, Part 1. Physical and Mineralogical Methods—Agronomy Monograph no. 9 (2nd Edition)

in the literature may be more suitable than any of those described here. If a method is not described here, it should not be taken to mean that the method is unsatisfactory. In the closing section of the chapter, references are given to other methods that may be suitable for a given purpose.

28–2 GENERAL PRINCIPLES

Water moves through soil in response to various forces acting upon it. The chemical species *water* may be transported due to bulk movement of the liquid phase or soil solution, or it may be transported by diffusion relative to the mean motion of the liquid phase. In this chapter we shall be primarily concerned with bulk movement, under isothermal conditions, of the liquid phase in response to mechanical driving forces. However, the transport of water in the gas phase by vapor diffusion will be included in the measured hydraulic conductivity and diffusivity, especially at low water contents.

The *hydraulic conductivity* is defined by Darcy's law, which for one-dimensional, vertical flow may be written as:

$$q = -K(\theta)\ \partial H/\partial z \qquad [1]$$

where q is the volume flux density, Darcy velocity, or apparent velocity (i.e., the volume of liquid phase passing through unit cross-sectional area of soil in unit time), $\partial H/\partial z$ is the gradient of the hydraulic head H, and $K(\theta)$ is the hydraulic conductivity. The driving force is expressed as the negative gradient of the hydraulic head composed of the gravitational head, z, and the pressure head, h; i.e.,

$$H = h + z\,. \qquad [2]$$

The pressure head has two components, a matric pressure head, h_m, and a pneumatic pressure head, h_a. The latter represents the (possible) difference between the soil gas-phase pressure and atmospheric pressure.

The conductivity of a soil depends on the geometry of the pores and the properties of the fluid in them. The two fluid properties that directly affect the hydraulic conductivity are viscosity and density. The texture and structure of the soil are the principal determinants of the geometry of the water in the soil pores. In soils with appreciable clay content, the composition of the soil solution can significantly affect the hydraulic conductivity because of interactions between the soil solution and the solid matrix.

As the water content decreases from saturation, the large pores, which are most effective in conducting water, are the first to drain. Also, the tortuosity of the flow paths may increase, and the average properties (density and viscosity) of the soil solution may change. These factors contribute to a rapid decrease in conductivity with decreasing water content. For these reasons, as defined in Eq. [1], the hydraulic conductivity is regarded as a function of the soil water content. Because of the relationship between the soil water content and the matric pressure head of the soil solution phase, the hydraulic conductivity may be regarded as a function of the latter as well. Some typical $K(\theta)$ relations are shown in Fig. 28–1. Representative hydraulic conductivity–suction data are shown

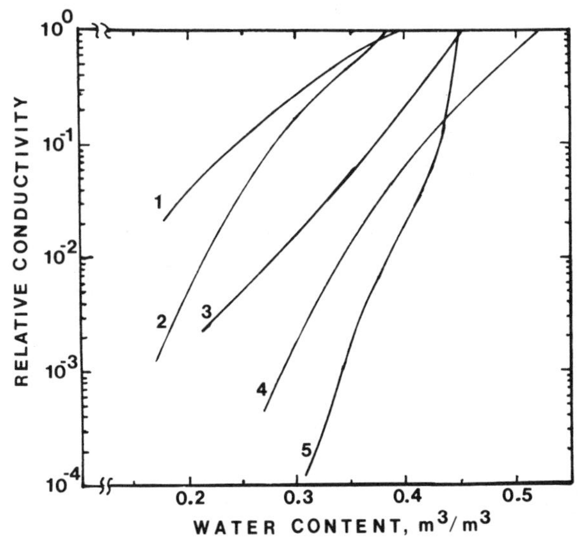

Fig. 28–1. Examples of relative conductivity–water content relations for various soils: (1) 500 μm sand, (2) Rubicon sandy loam, (3) Columbia silt loam, (4) Guelph loam, and (5) Rideau clay loam. Data plotted from Mualem (1976).

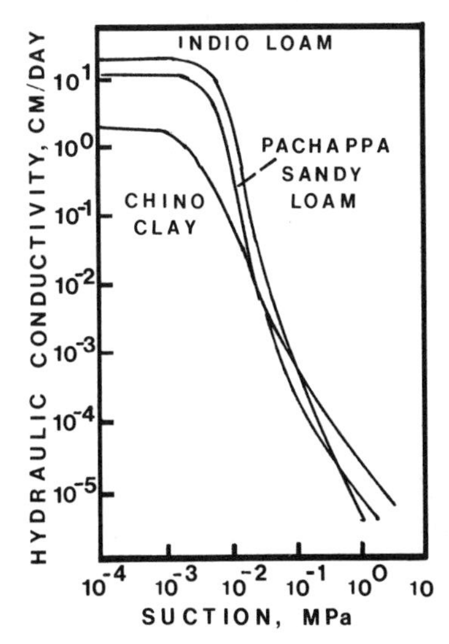

Fig. 28–2. Conductivity–suction relations for three soils. Redrawn from Gardner (1960).

in Fig. 28–2, and some examples of hydraulic diffusivity–water content data are shown in Fig. 28–3. Additional discussion of the relationships between K, θ, and h_m may be found in chapter 31.

The *intrinsic permeability* is related to the hydraulic conductivity by the relation:

$$k = K\eta/\rho g \quad [3]$$

where g is the acceleration of gravity, η is the viscosity, and ρ is the density of the percolating fluid. The dimensions of the intrinsic permeability are length squared. A practical unit for many applications is square micrometers. In developing Eq. [3], it has been assumed that the fluid properties are not affected by the nature of the solid matrix, that the matrix structure is not affected by the fluid, and that the intrinsic permeability is a function of only the pore-space geometry. According to these assumptions, the same intrinsic permeability will be obtained with different fluids. This ideal is seldom realized in soils, and changes in the nature of the percolating fluid will generally affect the matrix. Changes in the concentration and kinds of cationic species in the soil solution can cause large changes in the hydraulic conductivity of soils containing significant amounts of clay, especially swelling clays.

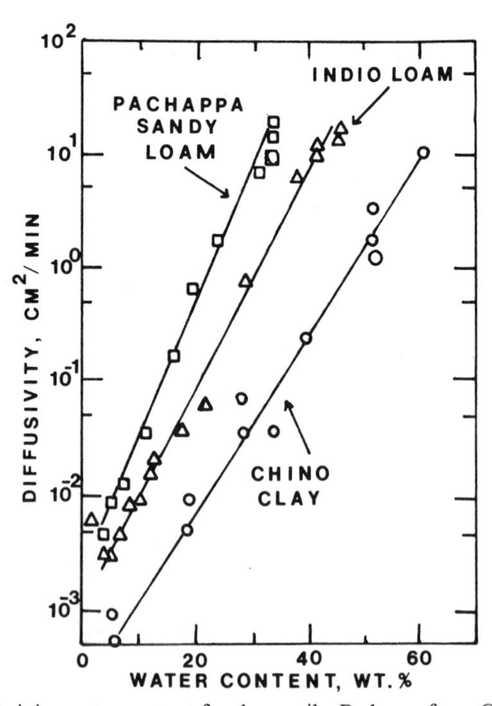

Fig. 28–3. Diffusivity–water content for three soils. Redrawn from Gardner (1958).

28-3 SAMPLES AND TEST FLUID

28-3.1 Samples

To characterize a field or plot of soil, in situ measurement of the hydraulic properties is preferred. Methods of conducting such measurements are described in chapters 27, 29, and 30. If in situ measurements are not possible, or are not desired, recourse must be made to laboratory measurements on samples of the soil material to be characterized. These samples should be as representative as possible of the structure of the structure of the soil in the field.

In some studies, such as those intended to examine principles of flow in porous media, repacked samples of disturbed structure may be appropriate. In most such studies uniformity of packing is desired, and it is not necessary that the samples represent a field or plot.

Some methods for determining the hydraulic properties lend themselves better than others to the use of soil cores with natural structure. The steady state and falling head methods described below for saturated soils can be used with either repacked or natural cores. Methods that require a long soil column generally are not well suited to the use of undisturbed core samples.

Soil samples, with either disturbed or undisturbed structure, for determining the hydraulic conductivity by the constant-head or falling-head method are usually held in metal or plastic cylinders to obtain one-dimensional flow. Soil samples of relatively undisturbed structure can be obtained in a number of ways (Uhland, 1950; Smith & Stallman, 1954). Thin-walled cylinders may be pressed into the soil, or soil cores may be obtained in metal cylinders that fit into a sampling tube.[1]. After the samples are taken, the cylinders served as retainers for the soil during the measurements. Because of the great number of possible variations in techniques to obtain undisturbed soil samples, no specific directions for sampling will be given. McIntyre (1974) gives a thorough discussion of methods for procuring undisturbed cores for physical measurements.

If repacked soil samples are to be used, we suggest the procedure for packing soil cores to known bulk density described in chapter 26.

The dimensions of samples may vary. Ideally, they should be large relative to the largest structural units in the soil. Such a sample may, however, be impractically large. Sample cylinders with diameters on the order of 2 to 10 cm and lengths of 5 to 25 cm are reasonably practical for measurements in the laboratory.

---

[1]For example, the Model 200A core sampler from the Soilmoisture Equipment Corp., P.O. Box 30025, Santa Barbara, CA 93105. Sleeve-type core samplers can also be used with a hydraulically powered soil sampling rig. Mention of trade names for apparatus and supplies is for the convenience of the reader and does not constitute endorsement by the contributing agencies.

28–3.2 Test Fluid

Careful consideration should be given to the choice of test fluid used for the measurements. When a fluid flows through soil, various biological, chemical, and physical processes may occur in the system that change the hydraulic conductivity. If the incoming water contains a high concentration of dissolved gases, as is the case in most tap-water supplies, the gas will come out of solution, accumulate in the pores, and reduce the hydraulic conductivity. For this reason, freshly drawn tap-water is not recommended as a test fluid. If there are swelling clay minerals in the soil samples, a change in the chemical composition of the soil solution may have large effects on the hydraulic conductivity (Reeve et al., 1954; Quirk et al., 1955; Dane & Klute, 1977; Rolfe & Aylmore, 1977). The conductivity tends to increase with the total electrolyte concentration. An increase in the Na content of the percolating solution tends to decrease the conductivity. If the percolating solution causes dispersion of the clays in the sample, the fine particles may migrate within the sample, and lodge in the pores of the sample in such a way as to reduce the conductivity. Microbial activity within the sample may also affect the conductivity, particularly when water is passed through the sample for long periods of time or when the samples have been submerged for long periods.

In general, deionized or distilled water should not be used to determine the conductivity, because the soil structure may change when the ambient soil solution is replaced by distilled water. However, if the resistance of the soil to structural breakdown by rainfall is being examined, the use of deionized or distilled water may be appropriate.

It is sometimes remarked that the best fluid to use for measuring the hydraulic conductivity is the ambient fluid in the soil pores. However, it is usually impractical to obtain sufficient fluid to conduct the measurements. Also, the composition of the soil solution may not be known, making duplication of the soil solution impossible. In spite of these difficulties, some consideration should be given to the use of a test solution that is not drastically different from the in situ soil solution.

Unless there are reasons to choose some other solution, we suggest the use of a deaerated, 0.005 M $CaSO_4$ solution, saturated with thymol. The calcium ion will reduce the dispersion of the clays in the samples, and the thymol will inhibit biological activity. Other biological activity inhibitors may be used, such as phenol (0.1%) or mercuric chloride (20–500 mg/L). No single inhibitor is completely effective and satisfactory in all respects. Organisms that are resistant to the inhibitor may develop, or the inhibitor may be rendered ineffective by reactions in the soil. The senior author has found thymol to be satisfactory and convenient to use. Sterilization of the soil samples with propylene oxide may be useful for inhibiting biological activity (Poulovassilis, 1972). Performing the measurements at a temperature just above the freezing point of water may also be useful.

The question may be raised whether one should inhibit the development of microorganisms while measuring the hydraulic conductivity. The answer depends on the purpose of the measurements. If they are made to analyze the response of a given soil to applied water, one could argue that that development of organisms should not be suppressed, because they are part of the naturally occurring soil. However, if the purpose of the measurements is to study the physical process of soil water flow, it seems logical to remove the confusing, time-dependent effects of the development of organisms upon the hydraulic conductivity.

For some purposes, as in studies of the principles of fluid flow in porous media, it may be appropriate or desirable to use a test fluid other than water or a water solution—for instance, the hydrocarbon core test-fluid, Soltrol Phillips Petroleum Co., Bartlesville, OK 74003. Users of this fluid should take care to work in well-ventilated laboratory space, as some users have experienced an allergic reaction to the vapors. This fluid has the advantage that its properties, especially surface tension, are less subject to change due to contamination than are those of water (Brooks & Corey, 1964). However, the translation of the properties measured with the core test-fluid into the corresponding properties for water is difficult, if not impossible, if the porous matrix interacts with the core test-fluid differently than it does with water (e.g., see Van Schaik, 1970).

28–3.3. Preparation of Deaerated Solution

The recommended calcium sulfate ($CaSO_4$)–thymol solution is prepared as follows:

Prepare a saturated solution of $CaSO_4$ and thymol by adding an excess of $CaSO_4$ and about 3 to 5 g of thymol to an 18-L carboy filled with deionized or distilled water. Shake or stir the solution at intervals until it is saturated with the $CaSO_4$ and thymol. An apparatus to deaerate the solution is shown in Fig. 28–4. Fill the carboy shown in Fig. 28–4 with a solution made by mixing one part of the saturated $CaSO_4$–thymol solution with two parts of deionized or distilled water. With valves V_1 and V_2 closed, turn on the aspirator and apply suction to the vacuum chamber. Open valve V_1 and allow solution to flow into the beaker to fill it, then close value V_1. Heat the solution in the beaker to 50 to 60°C. When the solution in the beaker is warmed, open valve V_2 and adjust needle valve NV_2 so that solution sprays into the vacuum chamber at about 200 mL/min (not critical), filling the vacuum chamber in 15 to 30 min. Open valve V_1 and adjust needle valve NV_1 to maintain the level in the beaker. When the vacuum chamber is filled, close valves V_1 and V_2 and open V_3 to relieve the vacuum. Remove any suspended matter in the solution by filtering it into a suitable container for storage. A 4-L pyrex glass bottom-feed bottle makes a convenient storage container, and the solution may be filtered directly into it with a Buchner funnel-filter paper-suction system. The deaerated solution may be held in storage for an indefinite period if the storage container is completely filled. The deaer-

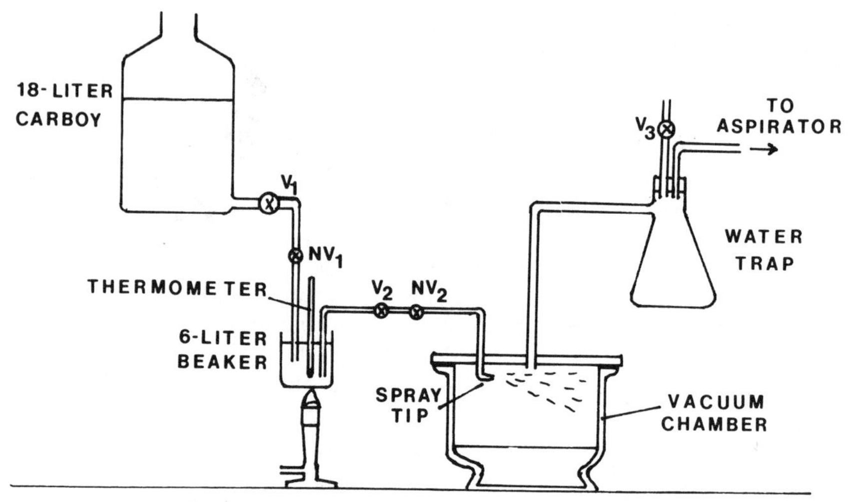

Fig. 28–4. Diagram of a system for deaerating water.

ated solution may be kept for a day or two in an open container if the air-water interfacial area is kept to a minimum and the solution is not agitated.

In the remainder of this chapter we will use the term water for the test fluid, with the understanding that it might also be another fluid for reasons outlined in section 28–3.2.

28–4 HYDRAULIC CONDUCTIVITY OF SATURATED SOILS

28–4.1 Constant Head Method

28–4.1.1 PRINCIPLES

Measurements of the hydraulic conductivity of saturated soils in the laboratory are based on the direct application of the Darcy equation to a saturated soil column of uniform cross-sectional area. A hydraulic head difference is imposed on the soil column, and the resulting flux of water is measured. The conductivity is given by

$$K_s = VL/[At(H_2 - H_1)] \qquad [4]$$

where V is the volume of water that flows through the sample of cross-sectional area A in time t, and $(H_2 - H_1)$ is the hydraulic head difference imposed across the sample of length L.

28–4.1.2 APPARATUS

A simple apparatus for measuring conductivity of saturated samples by the constant-head method is shown in Fig. 28–5. A rack may be built

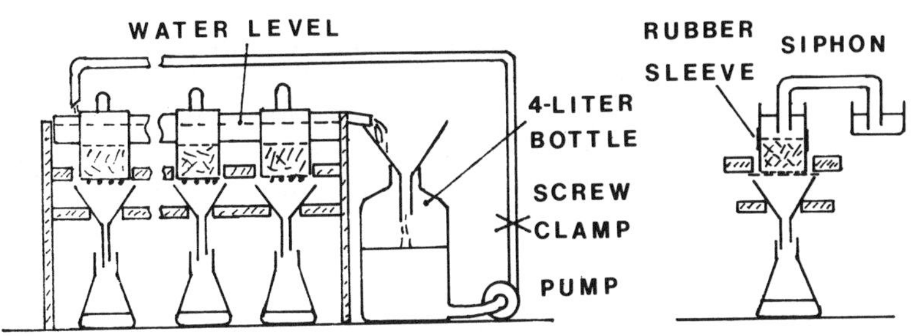

Fig. 28–5. Simple, multiple-core system for measuring hydraulic conductivity at saturation.

to hold 6 to 12 soil cores in a row. Water is siphoned from the common supply trough to the individual soil cores.

The recirculating water-supply system consists of a reservoir bottle to collect the overflow from the upper trough, a centrifugal pump to deliver water to the upper trough, and an overflow from the upper trough to the reservoir bottle. A magnetic drive pump such as Cole-Parmer Cat. no. K-7004 is suitable (Cole Parmer Instrument Co., 7425 N. Oak Park Ave, Chicago, IL 60621). Deaerated 0.005 *M* $CaSO_4$ solution with thymol (see section 28–3.2) may be used in the system.

28–4.1.3 PROCEDURE

Cover one end of each sample with a barrier to retain the soil in the core. In many cases, a double layer of cheesecloth held with a rubber band will be satisfactory. If the sample is fine textured, it may be necessary to use a finer-weave cloth or screen. The conductance of the retaining barrier should be as high as possible, so that the head loss across it will be negligible compared to that across the soil core.

Place the samples, cloth-covered end down, in a tray filled with water to a depth just below the top of the samples. Allow them to soak at least 12 h, or until the samples appear to be wetted.

Start the recirculating water-supply system. Connect an empty cylinder to the top of each sample using a wide rubber band or waterproof tape. Leave the lower part of the sample in the water while this is being done. Place a piece of blotting paper or filter paper on top of the sample. Slowly pour water into the upper cylinder until it is two-thirds to three-fourths filled. Slide a wide-blade spatula under the sample and quickly, but carefully, transfer the sample to the rack and start one of the siphons to maintain a constant head of water on the sample. Do not allow the water to drain from the top of the sample.

After the water level on top of the sample has become stabilized, collect the percolate in a beaker or flask. Measure the volume of water, V, that passes through the sample in time, t, and the hydraulic head difference, $(H_2 - H_1)$. The volume of water may conveniently be measured by collecting the water in a flask and weighing it. Sufficient water should be collected to give at least three significant figures in the measured volume.

Remove the sample in its retaining cylinder from the rack and remove the upper cylinder. Wipe excess water from the exterior of the sample cylinder. Extrude the soil, with the water it contains, into a sample can for moisture determination. Determine the wet sample weight, the oven-dry weight of the sample, and the tare weight of the sample container.

28–4.1.4 CALCULATIONS

Calculate the hydraulic conductivity using Eq. [4]. The conductivity has dimensions of velocity.

Calculate the volumetric water content of the sample from:

$$\theta = (W_w - W_d)/(d_w V_s) \qquad [5]$$

where W_w is the wet weight of the sample, W_d is the oven-dry weight, d_w is the density of water, and V_s is the volume of the sample.

Calculate the bulk density from:

$$d_b = W_d/V_s \,. \qquad [6]$$

28–4.1.5 COMMENTS

The soaking procedure given above to wet the samples will not completely saturate them. Air will be trapped in the pores, and will tend to disappear slowly as deaerated water is passed through the sample. The degree of saturation obtained in the sample can be estimated by comparing the volumetric water content with the total porosity calculated from the bulk and particle densities.

The degree of saturation obtained by soaking may be representative of that obtained in situ when a soil is flooded with water, and may be called the *natural saturation*. The term *satiated* has also been used for this state of wetting. If the conductivity at total saturation is desired, a vacuum wetting procedure may be employed (see chapter 26 for details). The samples may also be wetted to complete saturation by flushing them with carbon dioxide, followed by wetting with deaerated water. After a few pore volumes of water have passed through the sample, the CO_2 will be dissolved in the water and the pore space will be filled with liquid. This procedure works well with coarse-textured samples. Precautions must be taken to prevent the introduction of air into the pores of the sample after flushing with CO_2 and before wetting with water. In some cases, the CO_2 deaerated water flushing procedure may be undesirable, because of the acidic solution that is formed when the deaerated water is introduced into the sample.

A wide range of conductivity values will be encountered in soils. Figure 28–6 shows the magnitudes of conductivity to be expected for a range of materials.

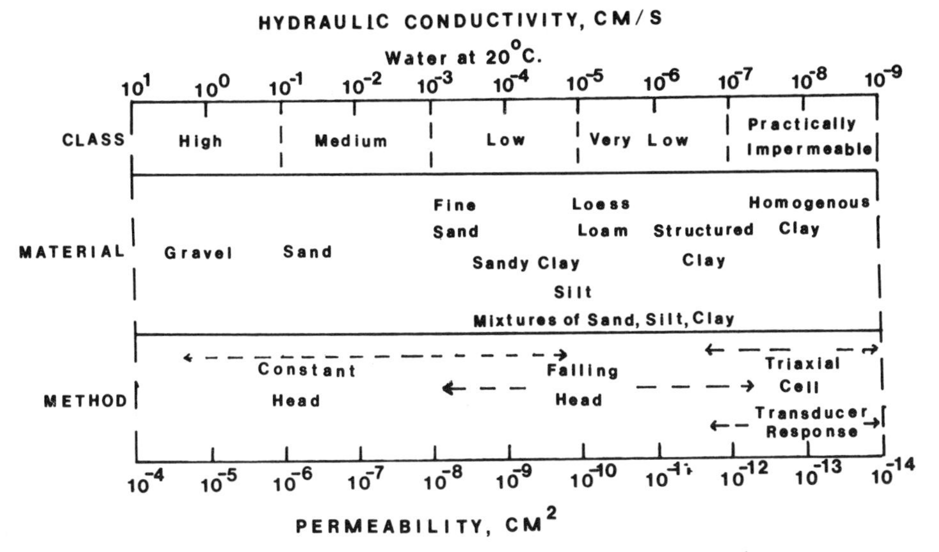

Fig. 28–6. Hydraulic conductivity of various materials at saturation.

The Darcy equation is not valid for all flow in porous media. Analysis of the various forces acting on the water passing through a porous medium shows that the Darcy equation should be valid when the inertial forces on the fluid are negligible compared to the viscous forces (Hubbert, 1957). As a practical matter, such a condition will prevail in silts and finer materials for any commonly occurring hydraulic gradient found in nature. In sands, especially the coarser sands, it will be necessary to restrict the hydraulic gradient to values less than about 0.5 to 1 to apply the Darcy equation. The range of validity of the Darcy equation can be demonstrated by measuring the flux density resulting from a series of applied gradients. The result should be a linear relation between the flux density and the hydraulic gradient. If the applied gradient is too large, the resulting flux density will be less than that predicted by the Darcy equation, using a conductivity calculated at low hydraulic gradients.

Swartzendruber (1969) discusses a number of reasons for deviations from the linear relation between flux density and gradient. Deviations have been observed, particularly in finer textured materials at low gradients. Some of these deviations have been traced to experimental problems, but in other cases no such explanation could be found. Non-Newtonian behavior of the fluid phase, changes in the soil matrix under flow, and electro-osmotic effects are some of the possible reasons for nonproportional behavior.

In many practical problems, nonproportional behavior is of little consequence, especially if the experimental error is large enough to mask the nonproportionality. If there is reason to doubt the applicability of the Darcy equation, tests should be made to determine its validity. Extremely large hydraulic gradients should be avoided, if possible, in con-

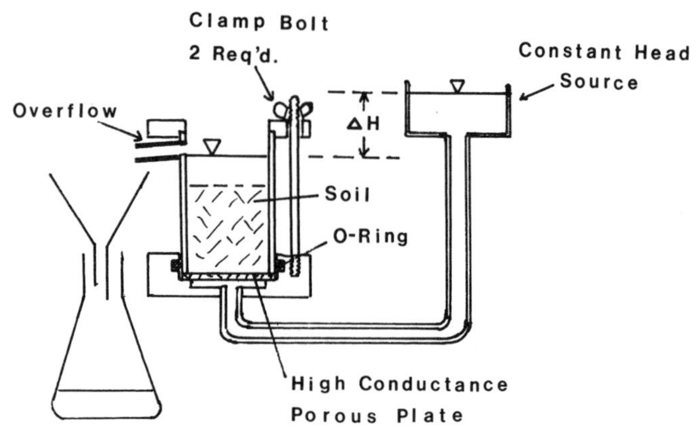

Fig. 28–7. Experimental arrangement for measuring hydraulic conductivity at low hydraulic gradient.

ductivity measurements if such gradients are unlikely to be encountered in the situations where the data are to be applied.

With the simple constant-head apparatus described above, the measurement error of the volumetric flow rate will become appreciable at 5 mL/h or less. If the sample has a diameter of 7.5 cm and the hydraulic gradient is about 1.5, this corresponds to a conductivity of approximately 2×10^{-5} cm/s. This indicates that a more sensitive method of measuring the volume flow rate is required for samples with low conductivity, or that another method must be used, such as the falling-head method described in section 28–4.2. The simple constant-head apparatus described above is not suitable for samples with very high conductivities, because the siphon tubes cannot deliver water fast enough to maintain a constant head of water on the sample. In such cases, an arrangement with hydraulic gradient less than unity, as shown in Fig. 28–7, may be used.

A more elaborate system for measuring conductivity may be warranted for studies of the physical principles of flow. The system for the steady-state measurement of the conductivity of unsaturated soil, described in section 28–5.1, may be used for the measurement of the conductivity of saturated soil. An alternative system is shown in Fig. 28–8 If a porous plate for the lower end of the sample cannot be found that has a high conductance relative to that of the soil sample, it is better to install piezometers at two points along the axis of flow in the sample and calculate the hydraulic gradient for the sample section defined by the position of the tensiometers. The hydraulic head of the water in the piezometers may be measured with water manometers, but a more accurate measurement of the hydraulic gradient can be obtained with a differential pressure transducer connected between the two piezometers. The time required for the system to attain steady state is shorter for a transducer system than for a manometer system.

An alternative to the use of piezometers in the soil is to measure the head loss across the soil-plate system and correct for the conductance of the plate. The plate conductance, k_b, is defined by

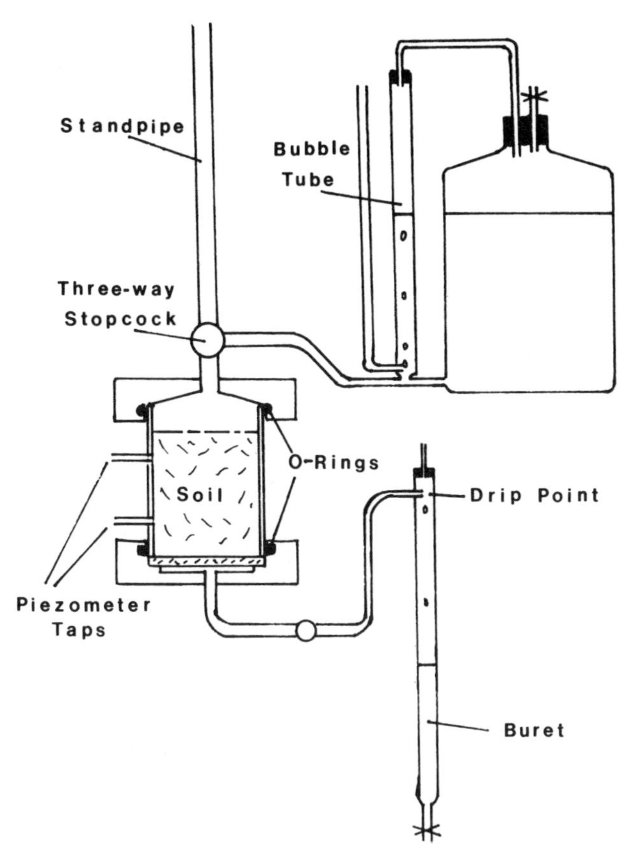

Fig. 28–8. A more elaborate apparatus for measuring hydraulic conductivity of saturated soils, which may be operated at constant head or falling head. The "side-arm" bubble tube minimizes aeration of the water supply.

$$k_b = (V/At)/\Delta H \qquad [7]$$

where ΔH is the hydraulic head difference across the plate, and V is the volume of water passing through the plate of cross sectional area A in time t.

A separate measurement of the plate conductance is made. The conductivity of the soil-plate system, $\langle K \rangle$, is then measured, and the conductivity of the soil, K, is calculated from

$$K = L \,/\, (L_t/\langle K \rangle - 1/k_b) \qquad [8]$$

where L_t is the thickness of the combined soil-plate system, and L is the thickness of the soil sample. In this procedure, it is assumed that the conductance of the plate does not change from the time that it is measured separately until it is used in the composite measurement, and that no significant contact resistance between the soil and the porous plate is present. The factor $1/k_b$, which is the resistance of the plate, should be less than about 10% of $L_t/\langle K \rangle$.

An appreciable error may be made in the measurement of low flow rates due to variations in the degree of "holdup" of water on the walls of a measuring buret. Evaporation losses also become relatively more significant. Precautions should be taken to minimize these errors when the flow rate is low. The movement of a meniscus in a *clean* horizontal calibrated tube may be used for low flow rate measurements with the flow cell shown in Fig. 28–8. The use of a short bubble (<1 cm) in a calibrated tube is subject to possible error due to water bypassing the bubble, and is not generally recommended.

28–4.2 Falling-Head Method

28–4.2.1 PRINCIPLES

A diagram of the system used for measuring the hydraulic conductivity by the falling-head method is shown in Fig. 28–9. With a hydraulic head difference, H, across the sample, the volume of water dV that passes through the sample in time dt is given by

$$dV/dt = -K(H/L) . \qquad [9]$$

The differential volume of water, dV, may be replaced by adH, where a is the cross-sectional area of the standpipe. Integrating between limits t_1, H_1 and t_2, H_2 and solving for the conductivity yields the result:

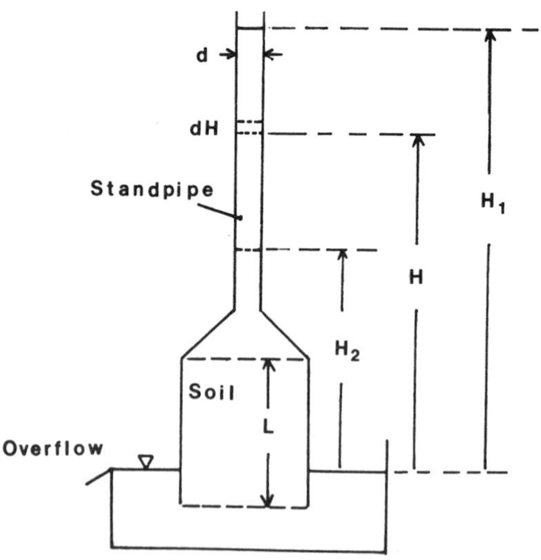

Fig. 28–9. Falling-head system for measuring hydraulic conductivity. L is the length of the soil sample, d is the diameter of the standpipe, dH is the change of hydraulic head that occurs in time dt, H is the hydraulic head difference across the sample at time t, H_1 is the initial and H_2 is the final hydraulic head difference.

$$K = (aL/At) \log_e (H_1/H_2) \qquad [10]$$

where A is the cross-sectional area of the sample.

28–4.2.2 APPARATUS

The actual form of the apparatus may be quite varied. One possible arrangement is shown in Fig. 28–10. The support for the sample should have a high conductivity relative to that of the soil. A suitable screen, gauze, or cloth barrier may be fastened to the bottom of the sample cylinder, or a very high conductance porous stone such as those used in consolidation apparatus, may be used. The diameter of the standpipe should be chosen so that an easily measured change in head will occur in a reasonable time, say between 1 and 100 min. The required tube diameter may be estimated from

$$d = [K\, t\, D^2/L\, (\log_e H_r)]^{1/2} \qquad [11]$$

using appropriate values for the soil sample length, L, the diameter of the sample, D, the time, t, the hydraulic head ratio, $H_r = H_1/H_2$, and the conductivity, K. With standpipe 2 cm in diameter and a soil sample with a diameter of 7.5 cm, a length of 5 cm, and a head ratio of 1.1, the largest conductivity that can be measured, using a fall time of 1 min, is

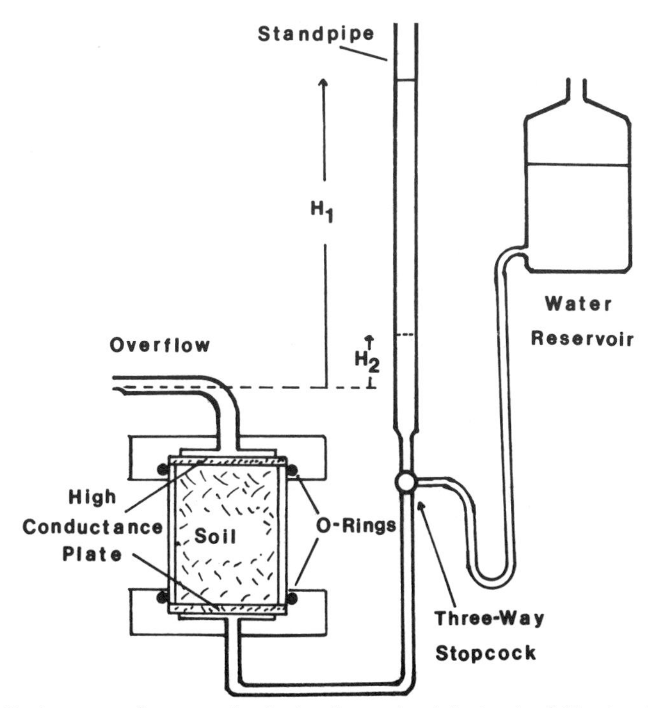

Fig. 28–10. Apparatus for measuring hydraulic conductivity by the falling-head method.

about 1×10^{-3} cm/s. The practical range of diameter for the standpipe is 0.2 to 2 cm. If the fall time is limited to 100 min or less and a standpipe with a diameter of 0.2 cm is used, the smallest K that can be measured is about 1×10^{-7} cm/s. This corresponds approximately to the lower limit of conductivity of silts and coarse clays.

28–4.2.3 PROCEDURE

The procedure given here is based on the apparatus shown in Fig. 28–10. If other falling-head apparatus is used, appropriate modifications in procedure will have to be made.

1. Install the end caps on the sample in its retaining cylinder, and wet the samples by supplying water to the bottom of the sample through the three-way stopcock and the lower porous plate. If complete rather than "natural" saturation of the sample is desired, the samples must be wetted under vacuum or by CO_2-deaerated water flushing (see comments on method of wetting in section 28–4.1.5).
2. Fill the space above the sample up to the overflow with water. This may be done by flow upward through the sample, or by introduction of water with a pipet or syringe at the top of the sample. A 50-mL hypodermic syringe fitted with a large needle may conveniently be used for this purpose.
3. Establish a water level in the standpipe somewhat above the level chosen for H_1, by introducing water through the three-way stopcock. The stopcock may be closed to hold the water in the standpipe until everything is ready to make the flow measurements.
4. Connect the standpipe to the sample by opening the stopcock, and measure the time for the water level to fall from H_1 to H_2.
5. Additional measurements may be taken by repeating steps 3 and 4.
6. When the flow-rate measurements are finished, drain the excess water from the top of the sample and adjust the level of the water in the standpipe to the top of the sample.
7. Remove the end caps from the soil core, and transfer the soil into a moisture can. Determine the wet weight of the soil sample, W_w. Obtain the oven-dry weight of the sample, W_d.

28–4.2.4. CALCULATIONS

Calculate the hydraulic conductivity from Eq. [10].

Calculate the volumetric water content from

$$\theta = (W_w - W_d)/(d_w V_s) .$$

Calculate the bulk density from

$$d_b = W_d/V_s .$$

28–4.2.5 COMMENTS

Most of the comments made in section 28–4.1.5 on the constant-head method of determining saturated hydraulic conductivity also apply to the falling head method.

The experimental arrangement shown in Fig. 28–8 may be used for either falling-head or constant-head measurements.

Conductivity measurements in relatively impervious materials ($K < 1 \times 10^{-7}$ cm/s) require excessively long measurement times with a standpipe of reasonable diameter. The conductivity of such materials may be measured by replacing the standpipe with a pressure transducer (Overman et al., 1968; Nightingale & Bianchi, 1970; Remy, 1973). The volume change per unit pressure change of many transducers is reasonably constant over a certain range of deformation of the diaphragm of the transducer. The volume change per unit pressure head change of the standpipe of a conventional falling-head system, i.e., its cross-sectional area, may be replaced by the reciprocal of the gauge sensitivity of the transducer, S, in the formula for calculating conductivity by the falling-head method

$$K = (L/AtS) \log_e (H_1/H_2) . \quad [12]$$

Because the volume change per unit pressure change of many transducers, especially the diaphragm type, is generally smaller than that for a standpipe, the range of measurement of K may be extended to much lower values. However, certain precautions must be observed. Deformation of the apparatus and sample with changing pressure must be kept to a minimum. Unconfined samples of materials that have conductivities low enough that this method could be used will generally exhibit a change in consolidation with change in water pressure, thereby complicating the interpretation of the transducer response.

28–5 CONDUCTIVITY AND DIFFUSIVITY OF UNSATURATED SOILS

28–5.1 Steady-state Head Control Method

28–5.1.1 PRINCIPLES

The Darcy equation is extended for use in unsaturated soils by assuming that the hydraulic conductivity is a function of the degree of saturation of the medium, i. e., the volumetric water content (see Eq. [1]). Characterization of the water-conducting properties of partially saturated media requires measurement of the conductivity–water content relationship, $K(\theta)$, or the conductivity–capillary pressure head function, $K(h_m)$. Alternatively, the difffusivity–water content relationship may be measured.

In the *steady-state* method of determination of $K(\theta)$, a time-invariant, one-dimensional flow of the liquid phase is established in a soil sample at a given water content. The volumetric flux density and the hydraulic gradient are measured, and the conductivity is calculated from the flux density/gradient ratio. The conductivity obtained is associated with the matric pressure head and water content of the region of the sample in which the gradient was measured. The drainage conductivity function is mapped by proceeding through a series of steady-state flows with progressively decreasing values of h_m, beginning with h_m approximately equal to 0. The wetting conductivity function is obtained by starting at a given negative pressure head and preceeding through a series of steady-state flows with increasing values of matric pressure head.

In the short-column version of this method (sample length less than 10 to 15 cm, Fig. 28–11 to 28–13), the soil sample is held between two porous plates or membranes that are preferentially wetted by water and serve to establish hydraulic contact with the water supply and removal systems connected to the inflow and outflow ends of the sample, respectively. The bubbling pressure of the plates must be at least as large as the magnitude of the most negative pressure head to be used in the measurements.

Because of the uncertain head losses across the porous plates at the ends of the soil sample and the possibility of variable contact conductance at the soil-plate interfaces, it is necessary to measure the hydraulic gradient in the sample with tensiometers. The bubbling pressure of the tensiometer cups should be compatible with that of the plates.

A flow cell operated with vertically downward flow at unit hydraulic gradient is preferred for measuring the conductivity. If the sample is uniform with respect to the $K(h_c)$ function, the pressure-head gradient within the soil will be zero, and the water content and pressure head will be uniform. The conductivity at the water content and pressure head in the sample is then equal to the absolute value of the volumetric flux density. Operation with downward flow at unit gradient makes the association of the conductivity with the water content or the pressure head straightforward.

The hydraulic conductivity is calculated using a finite-difference form of the Darcy equation:

$$q = -\langle K \rangle (\Delta H/\Delta z) \quad [13a]$$

$$= -\langle K \rangle (\Delta h/\Delta z + 1) \quad [13b]$$

where $\Delta z = z_2 - z_1$, is the elevation difference of the two points between

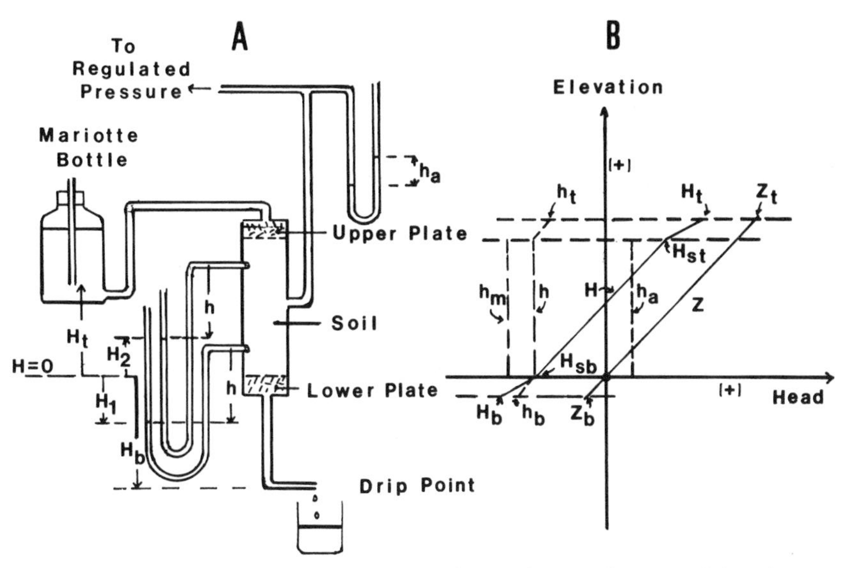

Fig. 28–11. (A) System for steady-state measurement of hydraulic conductivity of unsaturated soils. The cell air pressure is atmospheric and the applied hydraulic heads are controlled by "hanging water" columns. (B). Hydraulic, pressure, and gravitational head distributions in the cell. The symbols are defined in the text in section 28–5.1.1.

which the pressure head difference Δh is measured, and $\langle K \rangle$ is the average conductivity of the region of the sample between the points at which the head difference is measured. The average conductivity will depend on the range of pressure head, h_2 to h_1, involved in the flow, the magnitudes of q and Δz, and the conductivity function $K(h_m)$.

Constant hydraulic heads are applied at the top of the upper plate and at the bottom of the lower plate. If the hydraulic conductivities of the soil and the plates remain constant in time, a steady flow through the sample will be obtained. The soil sample is brought to an unsaturated condition by elevating the gas-phase pressure in the sample above atmospheric and (or) by reducing the pressure head and hydraulic head of the water at the top of the upper plate and at the bottom of the lower plate with hanging water columns.

Figure 28–11A is a diagram of a conductivity cell with hanging water columns to control the applied heads. The hydraulic head of the water supply to the upper end of the system, H_t, is controlled by the Mariotte bottle. The hydraulic head of the water at the bottom of the flow system, H_b, is kept constant by the fixed location of the drip point in the outflow system. For downward flow, H_t must be greater than H_b. The pressure head in the sample is controlled by the elevations of the supply-bottle bubble tube and the drip point relative to the sample, the gas-phase pressure in the cell, and the conductances of the upper and lower porous plates.

Figure 28–11B shows the distributions of the hydraulic head, pressure

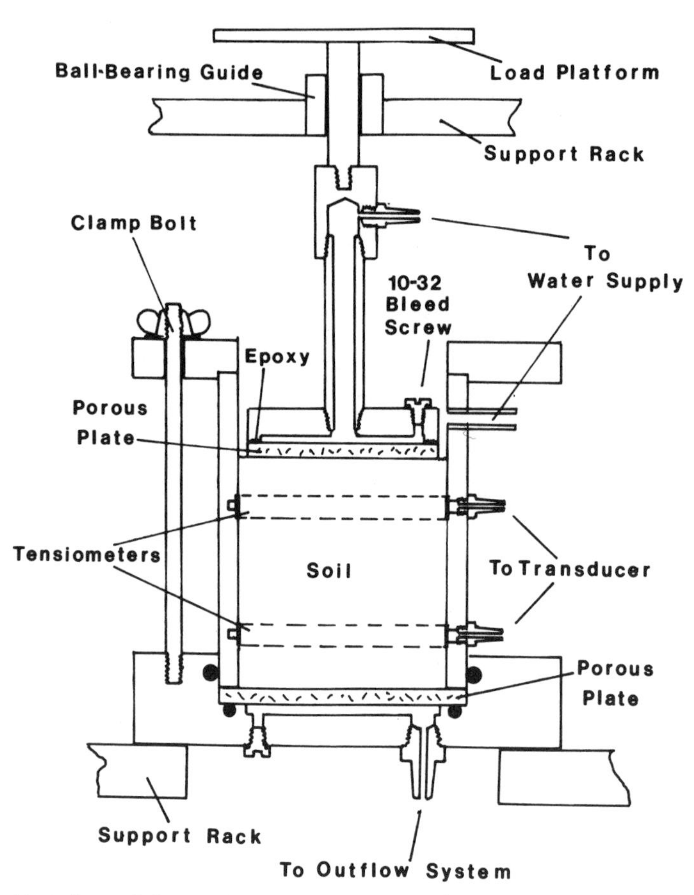

Fig. 28–12. A flow cell for steady-state measurement of the hydraulic conductivity of unsaturated soil.

head, and gravitational head versus elevation in the system of Fig. 28–11A, for steady flow at unit gradient, in which the pressure head is negative. The pressure head, h, in the soil is composed of a matric component, h_m, and pneumatic component, h_a. The latter is given by $(P_g - P_{\text{atm}})/d_w g$, where P_g is the absolute gas-phase pressure in the cell. In the case shown, h_a is positive, and h_m is less than h. Whether or not the sample will be partially desaturated at the matric pressure head in the soil depends on the relationship between h_m and the water content of the soil.

Under unit-gradient flow conditions at pressure head H in the soil, the required hydraulic head at the top of the upper barrier is given by

$$H_t = h + L + K/k_t \qquad [14]$$

where L is the length of the sample, k_t is the conductance of the upper barrier, and K is the hydraulic conductivity of the soil. Equation [14] is

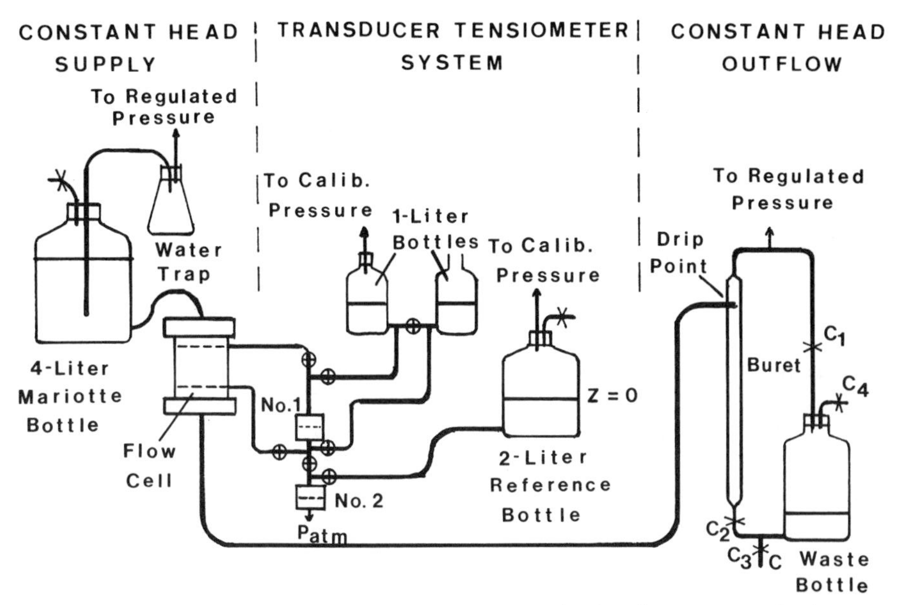

Fig. 28–13. Constant head water supply system, constant head outflow system and volume measurement system, and pressure transducer system to be used with the flow cell shown in Fig. 28–12. The two 1-L bottles are used for calibrating the differential transducer, no. 1. The valves are ball or plug type valves.

derived on the assumption that the reference for hydraulic head is at the bottom of the soil sample. The required hydraulic head at the bottom of the lower barrier is given by

$$H_b = h - K/k_b \qquad [15]$$

where k_b is the conductance of the lower plate. The plate conductance is defined by Eq. [7].

28–5.1.2 APPARATUS

A design for the conductivity cell is shown in Fig. 28–12. This cell has tensiometer rings built into the cylinder wall and is best suited for measurement on repacked samples. The porous barriers at the ends of the sample may be either membranes or ceramic plates. The upper barrier is mounted on a piston that is free to move in a vertical direction to maintain contact with the soil sample. A load may be placed on the platform connected to the piston. The porous barrier is cemented to the piston with epoxy cement. The lower porous barrier is sealed in place with O-rings. The details of the tensiometer rings in the wall are shown in Fig. 23–12, chapter 23.

A diagram of the measuring system for hydraulic conductivity is shown in Fig. 28–13. The system consists of a constant-head water supply, a flow cell, a transducer–tensiometer system, and a constant-head outflow system. The lower end of the bubble tube in the Mariotte bottle is placed slightly above the top of the soil in the flow cell. A pressure regulator capable of control at subatmospheric pressure as well as above atmospheric pressure (Model 44-20, Moore Products Co., Springhouse, PA, 19477) is connected to the bubble tube, to provide convenient control of the hydraulic head of the supply water without physical movement of the bottle. The water trap prevents accidental entry of water into the subatmospheric pressure regulator.

The drip point of the outflow system is located at an elevation slightly above the top of the soil in the flow cell. The air pressure at the drip point and in the waste bottle is controlled by another subatmospheric pressure regulator. The waste bottle is used to collect the outflow when flow rate measurements are not being made. The buret can be drained into the waste bottle without disturbance of the lower boundary condition by opening clamp C_2. When the waste bottle is full of water, it may be drained by isolating the bottle from the outflow system with tubing clamps, C_1 and C_2, and opening C_3 and C_4. During this period, the outflow from the cell can be collected in the buret.

Transducer no. 1 measures the hydraulic head difference between the two tensiometers. It is calibrated by establishing zero head difference between the two 1-L bottles and by applying a known increase (on the order of 5 cm of water) of air pressure to one of the bottles.

The hydraulic head at the lower tensiometer is measured by transducer no. 2. The water level in the 2-L reference bottle is set at zero elevation at atmospheric pressure, to provide a calibration point for the transducer at zero hydraulic head. A second calibration point is obtained by applying a known air pressure to the reference bottle. Chapter 23, section 23–4, on laboratory tensiometry should be consulted for additional information on the tensiometer system.

28–5.1.3 PROCEDURE

Measure the conductances of the upper and lower barriers by either a constant-head or falling-head method.

Pack the soil into the flow cell. The method of packing will vary with the nature of the soil and the objectives of the experiment. Air-dry soil can be placed in the cell through a funnel and tube, followed by tapping and vibration to settle the sample, or the soil may be packed to a known bulk density by the procedure described in chapter 26. Measure the bulk volume, water content, and mass of the air-dry soil in the cell, so that the bulk density of the packing may be determined.

Wet the sample with the test fluid (see section 28–3), using (i) vacuum saturation, (ii) CO_2-deaerated water flushing, if total saturation is desired, or (iii) soaking, if natural saturation is desired.

If vacuum saturation is to be used, place the entire flow cell including the top piston barrier in a vacuum chamber and evacuate the chamber with a vacuum pump or aspirator for 15 or 30 min. While the vacuum is maintained in the chamber, flow deaerated water into the chamber until the flow cell is covered with water. Then release the vacuum, close all tubing connections to the cell, remove it from the vacuum chamber, and place it in position for the conductivity measurements.

If the CO_2-deaerated water flushing technique is to be used, connect a tank of CO_2 with a pressure regulator to the bottom of the flow cell. With the upper piston removed, close the top of the flow cell with a large stopper with a tube dipping into a small beaker of water. Close the connections to the tensiometer rings. Connect a bottle of deaerated water to a tee in the tubing leading from the CO_2 tank to the flow cell. Fill the tube between the water bottle and the tee with water and close it with a clamp close to the tee. Flow CO_2 through the cell at the rate of approximately 50 to 70 bubbles/min for 15 to 30 min. Stop the flow of gas and force deaerated water into the cell until it ponds on the surface of the soil. Remove the stopper at the top of the flow cell. Vacuum-saturate the porous plate on the piston, and place the piston on the soil in the cell.

If soaking is to be used to wet the sample, connect a bottle of deaerated water to the bottom of the flow cell and force water up through the lower plate into the soil. Set the head of the water at or slightly below the top of the soil, and let the sample imbibe water for 12 to 24 h. Then pond water on top of the sample. Vacuum-saturate the porous plate in the upper piston, and place it on top of the soil.

Connect the water supply system to the upper piston-barrier and to the side inlet tube at the top of the flow cell. Connect the outflow system to the flow cell. Flush all connecting tubing free of air bubbles and clamp the outflow tube at the cell. Set the hydraulic head of the water in the inflow system so that a depth of 1 to 2 cm of ponded water will be applied to the top of the soil through the side inlet tube. Adjust the hydraulic head in the outflow system so that when the outflow tube is opened and flow is started, the soil will remain saturated at a slightly positive pressure head.

Connect the tensiometer rings to the transducer system, and flush the entire tensiometer-transducer system. This can conveniently be done by forcing water from the 4-L reference bottle through the various parts of the system.

Calibrate the transducers and readout system. Establish the zero-head reference water level in the 4-L bottle. A convenient choice of reference level is the lower soil-plate contact, or the lower tensiometer. The differential transducer system must be calibrated to read head differences on the order of the distance of separation of the upper and lower tensiometers. In the cell shown, this distance is 3 cm. The other transducer is calibrated to read hydraulic heads in the range h_{min} to 0, where h_{min} is the lowest pressure head to be encountered in the flow cell during the measurements.

Establish downward saturated flow in the cell. This need not be a unit-gradient flow. The hydraulic head difference applied to the flow cell should be increased until the hydraulic gradient in the soil is sufficiently large to permit a reasonably precise measurement. The pressure head in the soil should be slightly positive. Measure the flow rate and hydraulic head difference between the tensiometers. These data will be used to calculate the conductivity of the soil at saturation and the conductances of the end barriers (see Eq. [7], [16], and [17].

Using Eq. [14] and [15], estimate the hydraulic heads, H_t and H_b, that must be applied to the flow cell to maintain a unit gradient flow in the soil at a slightly negative pressure head (-1 to -3 cm). Use the previously measured conductances of the barriers and the conductivity of the saturated soil for this calculation.

Close the connection from the water supply to the side inlet at the top of the cell. Apply the estimated heads H_t and H_b, and measure the head difference between the tensiometers and the hydraulic head at the lower tensiometer, H_1. If the hydraulic gradient in the soil is not unity, adjust the heads H_t and H_b to produce unit gradient. This must be done in a manner to prevent introducing hysteresis into the conductivity data. See section 28–5.1.5 for a discussion of how to manage the boundary conditions.

Establish a series of unit-gradient flows at progressively decreasing pressure head values in the soil. At each flow, measure the pressure head at the lower tensiometer, the hydraulic head difference between the tensiometers, and the volumetric flow rate through the sample. Proceed in this manner until the most negative value of pressure head that is desired is reached, or the limit of the bubbling pressure of the plates or the tensiometers is reached.

28–5.1.4 CALCULATIONS

At each steady flow, calculate the hydraulic conductivity from Eq. [13]. Calculate the pressure heads at each of the tensiometers, and the average pressure head in the test section. Construct the $K(h)$ function as a table of conductivity values and average pressure heads for each flow rate.

The conductances of the upper and lower porous plates may be estimated from

$$k_t = (V/At)/(H_t - H_{ts}) \qquad [16]$$

$$k_b = (V/At)/(H_{bs} - H_b) \qquad [17]$$

where the hydraulic head at the upper soil-plate contact, H_{ts}, and the hydraulic head at the lower soil-plate contact, H_{bs}, are estimated by linear extrapolation of the two measured hydraulic heads at the two tensiometers in the flow cell.

28-5.1.5 COMMENTS

The procedure described in chapter 26 for packing samples to known bulk density may be adapted to this flow cell. However, care must be taken that the tensiometer rings are not damaged during compaction of the soil.

Porous barriers have been constructed of porous plastic, fritted glass beads (Nielsen & Phillips, 1958), various types of membrane filters (Elrick & Bowman, 1964), sintered metal, and porous ceramics (Richards & Moore, 1952). Table 24–3 in chapter 23 gives a list of barrier materials and their hydraulic characteristics. Barriers with the highest conductance for the required bubbling pressure should be chosen. This ordinarily requires more than one kind of barrier if the pressure-head range extends from zero to less than −500 cm or water. Membrane filters are highly conductive for a given bubbling pressure, but require physical support. The rate of gas diffusion (in solution) through the thin membranes is quite high, which causes problems with loss of suction control in the flow system and with the volumetric flow rate measurements. Fritted glass-bead plates provide a moderately high conductance and are quite satisfactory for use at pressure heads greater than about −200 cm of water.

It is not practical to raise or lower the elevation of the bubble tube system or the drip point more than about 100 to 150 cm. The hydraulic head of the water in the supply and removal systems can be varied without movement of the bubble tube and drip point by controlling the air pressure applied to the bubble tube and to the drip point.

Figure 28–14 shows the variation of H_t and H_b with the pressure head in the soil between the porous plates for a sequence of steady flows, starting at or near zero pressure head and proceeding toward more negative pressure heads. In developing these curves it was assumed that the cell gas phase was at atmospheric pressure, that the conductances of the porous plates remained constant, that no contact resistance developed at the planes of contact of the soil and the end plates, that the conductivity–pressure head relation was a Brooks-Corey function (Brooks & Corey, 1966), and that unit gradient was established in the sample at each successive steady flow. The head loss across the porous barriers will decrease as the pressure head becomes more negative and the flow rate through the sample decreases. The fraction of the total head difference that is dissipated across the porous plates will thus decrease as the pressure head in the sample becomes more negative. Depending on the relative conductances of the plates and that of the soil sample, H_b may have to be increased, decreased, or kept the same to proceed from a steady state such as that at h_1 to a steady state at a more negative pressure head such as h_2. If the parameter $(nK_o)/(k_b h_b)$ is greater than unity, the variation of H_b is like curve (1), Fig. 28–14; if $(nK_o)/(k_b h_b)$ is less than unity, H_b varies with h in the manner shown in curve (3).

As the steady flows at progressively more negative pressure heads in the soil are established, the boundary conditions on the cell should be

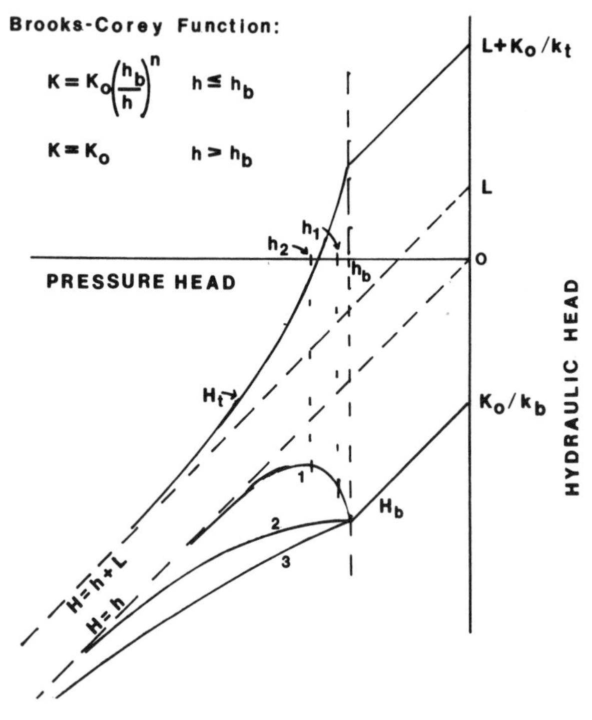

Fig. 28–14. Plot of the upper boundary head, H_t, and lower boundary head, H_b, vs. matric pressure head, at unit-gradient flow in the cell of Fig. 28–12. The reference level for the hydraulic head is the lower soil-plate contact.

managed in a manner to prevent introducing hysteresis into the measured relation between the conductivity and the pressure head. If hysteresis has been introduced, the same $h_m - \theta$ relation will not apply to all points in the soil. As one proceeds from one steady state flow to the next, the boundary conditions H_t and H_b required to produce the desired pressure head at unit hydraulic gradient in the soil must be guessed. As shown in Fig. 28–14, it may be necessary to increase, decrease, or make no change in the lower boundary condition, H_b, when changing from one flow to the next. Since the conductivity of the soil is unknown, it is difficult to know a priori what to do.

To avoid this boundary-condition management problem, the following procedure is suggested. First, measure the conductivity of the saturated soil and calculate the conductance of the soil core at saturation, where $k_s = K_{sat}/L$. Then select a porous plate or membrane which has a conductance, k_b, equal to or greater than k_s. Use this plate for measuring the conductivity to the limit of the bubbling pressure of the plate. Then choose another plate with a higher bubbling pressure, and use it to continue the measurements to higher suctions on another soil sample. An alternative to the above procedure is described by Mualem and Klute (1984), where a predictor-corrector approach is used to manage the

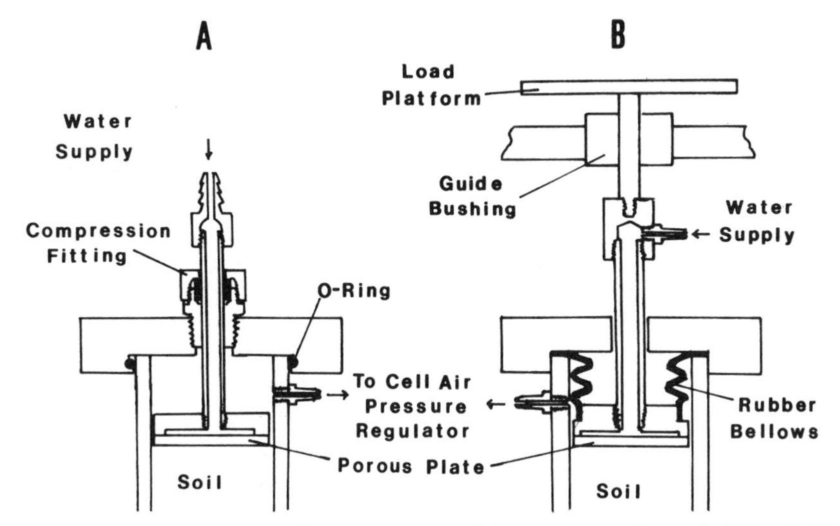

Fig. 28–15. Two arrangements for the upper cap of the steady-state flow cell of Fig. 28–12, to permit operation at controlled gas pressure. Upper piston held in place by a compression fitting (A), or by a dead weight load (B).

boundary conditions. That paper should be consulted for details of the procedure.

While the unit-gradient condition is ideal from the point of view of producing data without ambiguity in the relation between K and h_m, it is not always necessary to operate with the unit-gradient condition. Generally, an effort should be made to obtain unit gradient when measurements are made in the portion of the water retention relationship where θ changes rapidly with h_m. When measurements are made at pressure heads in the capillary fringe region or at lower pressure heads where θ is approaching the residual value, it is not as critical to use the unit-gradient condition.

If a gamma attenuation apparatus for the measurement of water content in soil columns is available, it may be used to measure the water content of the sample in the test region and thereby obtain $K(\theta)$ as well as $K(h_m)$. The ability to monitor the water content of the sample as well as the pressure and hydraulic heads also allows better control of the conditions in the sample.

The cell shown in Fig. 28–12 is designed to operate with the gas phase at atmospheric pressure. Alternate designs of the top cap, which permit the cell gas-phase pressure to be regulated, are shown in Fig. 28–15.

If samples of natural structure are to be used for the conductivity measurements, a sleeve-type core sampler may be used to collect the samples. The cylinders that hold the soil may be provided with appropriate openings through which small tensiometers can be inserted. Laliberte and Corey (1967) describe an experimental arrangement of this type.

28–5.2 Steady-state Flux Control Method

28–5.2.1 PRINCIPLES

The *long-column* version of this steady-state method uses columns of 50 to 200 cm length interposed between a constant flux supply of water at the upper end and a water table at the bottom. If the column is initially wetted and the above boundary conditions are applied, the column will drain to a condition of steady-state downward flow. In a uniform column, the upper part tends to drain to a constant pressure-head and water-content (Childs, 1969). Unit hydraulic gradient prevails in this region, and the absolute value of the flux density is equal to the conductivity associated with the water content and pressure head in the region. The pressure head can be measured with one appropriately placed tensiometer. By starting at saturation and proceeding through a series of progressively decreasing flow rates, one can determine a series of points on the drainage $K(h_m)$ function.

In the *short-column* (5–15 cm) version of this method a porous plate is used at the bottom of the column, with suction control at the lower side of the plate. A controlled flow rate is applied to the top of the column. The suction at the lower plate is adjusted to produce a unit gradient condition in the soil.

28–5.2.2 APPARATUS

The essential elements of the system are the soil column, a method for applying water at a constant flow rate to the top of the column, a tensiometer, and a system to measure the flow rate at the bottom of the column.

Ring tensiometers built into the wall of the column are suggested. Details of these are shown in chapter 23, Fig. 23–12. Alternatively, small "probe"-type tensiometers may be inserted in the wall of the soil column. Manometers (water or mercury–water) may be used on the tensiometers to measure the hydraulic and pressure head. If available, a transducer–tensiometer system, such as that shown in Fig. 23–14, chapter 23, is better. The column should be made in sections to facilitate soil sampling at the end of the experiments and to provide flexibility in the length of the column. The base of the column contains a porous plate of high conductance, or a suitable screen, to retain the soil.

A constant flow rate pump or a Mariotte bottle–capillary tube system may be used to provide satisfactory flow rate control. At low flow rates, temperature fluctuations in the laboratory may affect the gas pressure in the Mariotte bottle and cause the system to lose control of the head applied to the capillary tube. Peristaltic tubing pumps are available that will provide a constant flow rate. For small volume flow rates and for small total volumes of flow, syringe pumps can be used for flow control.

The volume outflow system shown in Fig. 28–13 may be used for measuring the flow rate. A meniscus in a horizontal calibrated tube may be used for measurements at low flow rates.

The apparatus for the short-column version of this method consists of a constant flow-rate control system the soil column with two tensiometers, and a suction control system at the lower end of the column.

28–5.2.3 PROCEDURE

28–5.2.3.1 Long-Column Version. Pack the soil in the column, and wet it. Total saturation may be obtained with the CO_2-deareated water flushing method, or "natural" saturation may be obtained by flooding the column. Connect the tensiometer(s) to the transducer(s) and flush the air from the tensiometer system. Connect the outflow system to the column and flush the air from it.

Use a constant-head supply system to pond water on the surface of the column. Measure the flow rate through the column and the head difference across the column.

Set up the constant-flow control system to provide a flow rate about 0.9 of the saturated flow rate. Maintain the controlled flow rate until the tensiometer readings are stabilized and the volumetric outflow rate is constant. Record the flow rate and the pressure head shown by the tensiometer.

Repeat this process at a series of decreasing flow rates, each time recording the flow rate and the pressure head in the upper part of the column as shown by the tensiometer. Continue until the smallest flow rate that is desired (or that can be controlled) is reached.

28–5.2.3.2 Short-Column Version. The sequence of measurements is similar to that described above. Starting at saturation, the flow rate is measured to obtain the conductivity at the degree of wetting of the column. The applied flow is controlled as above. The tensiometers are monitored, and the suction at the bottom is controlled to produce unit gradient in the column at each flow rate.

28–5.2.4 CALCULATIONS

Calculate the flux density at each flow rate from V/At, where V is the volume of water passing through the column of cross-sectional area A in time t. The hydraulic conductivity is equal to the flux density if unit gradient was attained. Otherwise, the conductivity is calculated from the flux density-to-gradient ratio. The pressure head to be associated with the conductivity is obtained from the tensiometer readings. If unit gradient was not obtained and the pressure heads indicated by the two tensiometers are unequal, use the arithmetic mean of the pressure heads.

28–5.2.5 COMMENTS

The long-column method, with a water table at the lower end, will require relatively long times for establishing steady conditions in the upper part of the column, and is generally limited to higher water contents and to coarser materials. The method is not well suited for undisturbed samples.

The short-column version requires less time to reach steady state, and can be used with materials of medium to fine texture and samples of "natural" structure.

The porous barrier at the bottom of the short column system should be chosen to have the highest conductance possible for the required bubbling pressure. The problem of managing the lower boundary condition described in section 28–5.1.5 may also be encountered here. The suggestions made there for selection of the lower porous plate are also applicable in this method.

If a gamma system for measurement of water content is available, the water content as well as the pressure head and conductivity may be obtained at each steady flow, thus providing $K(h_m)$, $K(\theta)$, and $\theta(h_m)$.

The conductivity function for drainage is obtained from the procedure described above. By starting at the lowest flow rate and establishing a series of steady flows at a series of increasing flow rates, the wetting function may be obtained.

28–5.3 Nonsteady-state Boltzmann Transform Methods

28–5.3.1 PRINCIPLES

These methods are based on the use of an effectively semi-infinite ($x > 0$), uniform, one-dimensional flow system. Either the effect of gravity is neglected, or the flow system is horizontal. The diffusivity form of the flow equation, viz.

$$\partial\theta/\partial t = \partial/\partial x\,(D(\theta)\partial\theta/\partial x) \quad [18]$$

is assumed to be applicable. The initial and boundary conditions on Eq. [18] are

$$\theta\,(x,0) = \theta_0,\ x > 0$$

$$\theta(0,t) = \theta_1,\ t > 0$$

$$\theta(x,t) \rightarrow \theta_0,\ x \rightarrow \infty,\ t > 0$$

The initial and boundary water contents, θ_0 and θ_1, are assumed constant. If θ_0 is $< \theta_1$, infiltration of water will occur. If θ_0 is $> \theta_1$, outflow of water will occur.

The Boltzmann variable, $\lambda = xt^{-1/2}$, is used to transform Eq. [18] to an ordinary differential equation. Integration, using the given boundary conditions, yields the following for the soil-water diffusivity:

$$D(\theta') = -1/2\ (d\lambda/d\theta)_{\theta=\theta'} \int_{\theta 1}^{\theta'} \lambda(\theta) d\theta \qquad [19]$$

where D and $d\lambda/d\theta$ are evaluated at the water content θ'.

Bruce and Klute (1956) described a method in which the spatial distribution of water content, determined by destructive gravimetric sampling at a fixed time in a horizontal infiltration flow system, was used to calculate the diffusivity function.

Whisler et al. (1968) described a method in which the water content as a function of time at a fixed position was measured. This procedure requires a nondestructive method of determining the water content in the soil column, such as gamma attenuation.

The water content vs. position at a series of fixed times, or the water content vs. time at a series of fixed positions can be used to construct a plot of λ vs. θ. If the flow is described by the nonlinear diffusion equation and the boundary and initial water contents are constant, the transformed water content–distance–time data should give a unique λ (θ) function. The derivative and integral in Eq. [19] can then be evaluated from this plot.

28–5.3.2 METHOD BASED ON WATER CONTENT DISTRIBUTION AT FIXED TIME

28–5.3.2.1 Apparatus. The apparatus is shown in Fig. 28–16.. The soil column is made of short sections of acrylic plastic tubing with a diameter of 2 to 3 cm. Most of the sections are 1 cm long. The section at the inflow end is 2 cm long, and there are about 10 sections with a length of 0.5 cm. The sections are held in a split acrylic plastic sleeve. The section at the inflow end of the column has a high-conductance, low bubbling-pressure, fritted glass-bead plate cemented on it. The water supply system is a Mariotte flask. The hydraulic head of the water supply is controlled by the bubble tube at the level of the midpoint of the soil column. The funnel and the two stopcocks S_1 and S_2 are used to start the flow.

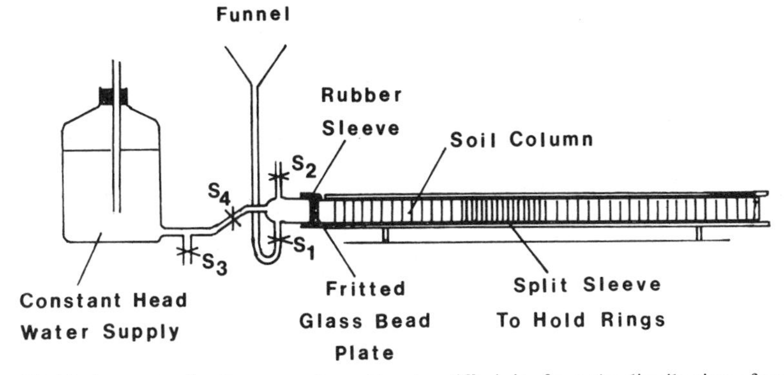

Fig. 28–16. Apparatus for the measuring soil water diffusivity from the distribution of water content at a fixed time in a horizontal infiltration system.

28–5.3.2.2 Procedure. Determine the water content (weight basis) of the air-dry soil to be used for the measurements.

Number and weigh a set of moisture cans, one for each of the sections in the soil column. Because of the small size of the samples, all weighings should be made to about 1 mg.

Assemble the column. Place the 0.5-cm sections adjacent to each other, at about 12 to 17 cm from the inflow end. Place the column in a vertical position with the inflow end down. Pack the soil by flowing air-dry soil into the column from a large funnel and tube system. After placing the soil in the column, it may be compacted by vibration or tapping of the column. When this is done, an attempt should be made to tap the column in a "uniform" manner by distributing the blows along the soil column. Collect data for the calculation of the average bulk density in the column. For this purpose, the tare weight of the empty column, its weight when filled with soil, the water content of the air-dry soil, and the length and cross-sectional area of the soil column must be measured.

Moisten the porous plate at the inflow end of the column with a few drops of water applied with a dropper. Flow water from the supply bottle to place the meniscus just at the entrance to the end compartment. Fill the funnel with water, and with stopcock S_2 open, open stopcock S_1 and fill the tube leading to the end compartment with water. Attach the end compartment to the column. With stopcock S_4 closed, open S_3 and drain water from the supply bottle until a bubble is poised ready to be released from the bubble tube.

As nearly as possible at the same time, open stopcocks S_1 and S_4, start a timer or stopwatch, and close stopcock S_2. While the inflow is occurring, lower the funnel to a position below that of the column. When the front has penetrated to the part of the column composed of 0.5-cm sections, open S_1 and S_2 to drain the end compartment, and close S_4. Record the time.

Remove the upper half of the split sleeve holding the column sections, and quickly transfer the soil in each of the column sections to the previously weighed moisture boxes. Close them and determine the water contents by the gravimetric procedure.

Clean the soil from the column and reassemble the apparatus. Repeat the process as described above for wetting front penetration distances of about 20 cm and about 30 cm. In each case, place the 0.5-cm sections in the column at the anticipated penetration distance of the wetting front.

28–5.3.2.3 Calculations. Use the data from all three runs to construct a similarity plot with $\lambda = x\,t^{-1/2}$ vs. θ, in the form shown in Fig. 28–17. Fit a curve to these data, and evaluate the derivative and integral from it. This procedure may be done graphically, numerically, or by analytical means if an equation for the fitted curve is available.

Calculate the diffusivity as a function of the water content from Eq. [19].

The bulk density of each of the samples from the column may also

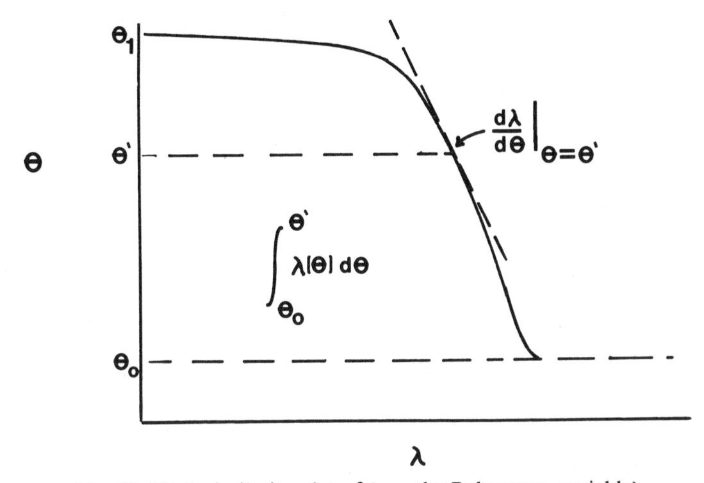

Fig. 28–17. A similarity plot of θ vs. the Boltzmann variableλ.

be calculated, if care has been taken to obtain all the soil in each ring during the sampling for water content.

28–5.3.3 METHOD BASED ON WATER CONTENT VS. TIME AT FIXED POSITION

28–5.3.3.1 Apparatus. The experimental arrangement for this method is similar to that shown in Fig. 28–16.

A nondestructive method for measuring the water content versus time at a fixed position in the column is required for the method. The gamma attenuation method provides this capability. A description of such systems and their use to measure water content is given in chapter 21.

The soil container is an acrylic plastic tray of rectangular cross-section, 50 to 100 cm long. To eliminate the effect of gravity, the height of the tray should be 1 to 2 cm. The optimal width depends on the gamma radiation source used. If the source is ^{137}Cs, the width should be in the range 7 to 20 cm; if it is ^{241}Am, the width should be from 3 to 6 cm. Vents are provided at intervals (about 10 cm) to keep the gas phase at atmospheric pressure. The cover of the tray should be easily and quickly removable so that the soil column may be sampled for water content immediately at the end of the infiltration.

A porous barrier of very low bubbling pressure and high conductance, such as a glass bead plate, is used at the inflow end of the column. The system for starting the flow at a definite time is the same as that described in section 28–5.3.2.1.

28–5.3.3.2 Procedure. Determine the water content of the air-dry soil to be packed into the column. Weigh the empty column. Place the column in a vertical position with the inflow end down. Introduce the soil into the column with a funnel and tube system. Settle the soil by tapping and vibrating the column. Distribute the blows along the column as evenly as possible.

Place the column in the horizontal gamma scan apparatus. Attach the end compartment to the column and make the constant-head supply ready as described in section 28–5.3.2.2.

Position the gamma beam on the column about 5 to 7 cm from the inflow end, and start obtaining count data. The counting time should be as short as possible while still getting acceptable statistical accuracy for the counting. The gamma system should be capable of obtaining approximately 100 thousand counts in 10 to 15 s when the gamma beam is on the soil column.

As nearly as possible at the same time, open stopcocks S_1 and S_4 and start a timer. Make the necessary notes and observations to correlate the elapsed time with the output of the gamma counting system.

Collect gamma count data as the front passes the position of the beam and until the gamma count rate appears to stabilize. After the rapid decrease of counting rate that corresponds to the passage of the wetting front, a position 10 to 15 cm from the inflow end may be selected for measurement. After the front passes this position, the gamma beam may be moved to other measurement positions further along the column. The water content versus time should be observed for at least three positions in the column.

When the gamma count rate at the measurement positions has stabilized, determine the water content of the soil at those positions.

28–5.3.3.3 Calculations. Use the initial water content of the soil and the final water contents measured at each observation point along the column to calibrate the gamma system. Let θ_o θ_f be the initial and final water contents, and I_o, I_f the corresponding initial and final count rates at a given position. Calculate the water content corresponding to a given time at that position from

$$\theta = \theta_f - (\theta_f - \theta_o) \log_e (I/I_f)/\log_e (I_o/I_f)$$

where I is the count rate corresponding to θ.

Plot the water content data vs. time from all positions as a similarity plot, i.e., as λ vs. θ. Obtain the diffusivity function as described in section 28–5.3.2.3.

28–5.3.4 COMMENTS

The methods have usually been applied to disturbed samples in relatively long columns (20 cm or more). Inflow or infiltration is usually employed, and thus the wetting $D(\theta)$ function is obtained. The initial water content is usually that of the air-dry soil material, and it is most convenient to maintain the boundary $x = \theta$ at or near saturation. The latter can be done with a porous barrier of very low bubbling pressure and high conductance.

If the hydraulic conductivity is required, the water capacity function must be obtained from a water retention curve measured on other samples. If infiltration is employed to measure $D(\theta)$, the water retention data must be for a corresponding wetting process.

One of the shortcomings of the Bruce-Klute method is that the diffusivity values obtained at high water contents are the most inaccurate. This was evident in the original paper by Bruce and Klute, and has been observed by others (e.g, Jackson, 1963) since that time. Recently, Clothier et al. (1983) have developed a method of fitting the primary $\lambda(\theta)$ data with a mathematical function that improves the diffusivity values obtained in the high water content range and which gives an analytical expression for the $D(\theta)$ function. Their paper should be consulted for the details of their analysis.

Rose (1968) has applied the principles of this method to measure the drying soil water diffusivity at low water contents in a soil column that was losing water by evaporation. The effects of gravity were neglected, and the evaporation was treated as an isothermal process.

Jackson (1964) used the method based on water content distribution measurements at a fixed time to measure the soil water diffusivity for absorption of water at low water contents.

28–5.4 Sorptivity Method

28–5.4.1 PRINCIPLES

The sorptivity method (Dirksen 1975a, 1979) entails the determination of the soil water diffusivity function, $D(\theta)$, by means of a series of one-dimensional absorption experiments. One-dimensional absorption, initiated at time zero by a step increase of the water content from θ_0 to θ_1 at the absorption interface $x = 0$, is described by

$$\partial\theta/\partial t = \partial/\partial x\,(D(\theta)\partial\theta/\partial x) \qquad [20]$$

subject to the conditions:

$$\theta = \theta_0 \text{ for } t < 0,\ x > 0 \qquad [21a]$$

$$\theta = \theta_1 \text{ for } t \geq 0,\ x = 0\,. \qquad [21b]$$

Using the Boltzmann transform, $\partial = xt^{-1/2}$, the nonlinear solution for the cumulative volume, i, of absorption per area of absorption interface is (Philip, 1969)

$$i = S\,[\theta_1,\theta_0]t^{1/2}\,. \qquad [22]$$

The sorptivity S defined by

$$S[\theta_1,\theta_0] = \int_{\theta_0}^{\theta_1} \lambda d\theta \qquad [23]$$

is a soil property which is a function of the initial water content and the water content at the absorption interface.

If the diffusivity is assumed to be a constant, D^*, over the moisture range θ_0 to θ_1, the resulting linearized solution is:

$$i = 2(\theta_1 - \theta_0)(D^* t/\pi)^{1/2}. \qquad [24]$$

Comparison of Eq. [22] and [24] shows that

$$D^* = \pi S^2/4(\theta_1 - \theta_0)^2. \qquad [25]$$

This solution would at first seem to be of little use, since soil water diffusivities are far from constant. However, according to Crank (1956, p. 256), for certain nonconstant diffusivities varying up to 200-fold, the initial rates of absorption are predicted by the linearized solution within 99% accuracy, when the diffusivity is set equal to the weighted mean diffusivity

$$\langle D_\gamma \rangle = (1 + \gamma)/(\theta_1 - \theta_0)^{1+\gamma} \int_{\theta_0}^{\theta_1} (\theta - \theta_0)\gamma D(\theta) d\theta \qquad [26]$$

where $\gamma = 0.67$. For soil water flow problems, the linearized solution for the weighted mean diffusivity appears to give more than adequate results. Combining Eq. [25] and [26] gives

$$\int_{\theta_0}^{\theta_1} (\theta - \theta_0)\gamma D(\theta)\, d\theta = \frac{1}{1+\gamma} \times \frac{\pi S^2}{4(\theta_1 - \theta_0)^{1-\gamma}}. \qquad [27]$$

Differentiation with respect to θ_1, keeping, θ_0 constant, yields

$$D(\theta_1) = \frac{1}{1+\gamma} \times \frac{1}{(\theta_1 - \theta_0)\,\gamma} \frac{d}{d\theta_1} \left[\frac{\pi S^2(\theta_1, \theta_0 = \text{constant})}{4(\theta_1 - \theta_0)^{1-\gamma}} \right]. \qquad [28]$$

Thus, the diffusivity function $D(\theta)$ can be calculated from Eq. [28], once the function $S(\theta_1, \theta_0 = \text{constant})$ is known. Measuring the latter has distinct advantages over measuring the former. The weighting parameter γ can be varied between 0.5 and 0.67 without significant effect. Calculations for three different soils yielded the most accurate results for $\gamma = 0.62$ (Dirksen, 1975b).

Equation [28] can be rewritten in a more convenient form:

$$D(\theta_1) = \frac{\pi S^2}{4(\theta_1 - \theta_0)^2} \left[\frac{(\theta_1 - \theta_0)}{(1+\gamma)\log e} \times \frac{d}{d\theta_1} (\log S^2) - \frac{1-\gamma}{1+\gamma} \right]. \qquad [29]$$

With this equation, $D(\theta_1)$ can be obtained by differentiating a polynomial regression of log S^2 in terms of θ_1, rather than obtaining the differentiation by visually determining slopes of appropriate graphs. This makes best use of the original data, and can be performed easily with a programmable calculator.

Since the soil water retention function θ (h_m) is hysteretic, so is the diffusivity function $D(\theta)$. From the nature of the sorptivity measurements, it follows that the diffusivity function obtained is only valid for wetting. If the wetting retentivity function is known, the hydraulic conductivity function $K(\theta)$ can be calculated from

$$K(\theta) = D(\theta)\,(dh/d\theta) \qquad [30]$$

where the functions on the right hand side of Eq. [31] are to be evaluated at the water content at which the conductivity is desired.

The sorptivity method thus involves determining the sorptivity function $S(\theta_1, \theta_0 =$ constant) by means of a series of one-dimensional absorption measurements, each initiated by a step increase of the water content from θ_0 to θ_1 at the absorption interface. Each measurement must start from the same uniform initial value θ_0 and needs to be continued only long enough to establish the familiar linear relationship of Eq. [22]. Theoretically, this relationship is only valid for one-dimensional absorption into a semi-infinite horizontal medium, and then for all time $t >$ 0. Practically, S can often be identified after only a few minutes, whether the flow is horizontal or vertical (Dirksen, 1975a, Table 2).

Sorptivity measurements require only one controlled boundary, while no measurements need to be made within the soil. Originally, Dirksen (1975a) used a Mariotte-type buret to supply water at constant negative pressure head through a porous ceramic plate to the soil surface. Unfortunately, the Mariotte buret regulated the pressure head applied to the ceramic plate, not that applied to the soil surface. Unless the conductance of the ceramic plate was negligible, which conflicted with the requirement of an appropriate bubbling pressure, the pressure head at the soil surface increased with time, as the flux decreased with time. Contact resistances between the plate and soil were even more serious. They can be very large, and variable with time and position on the plate-soil interface. As a result, wetting fronts were often uneven and the rates of absorption did not follow the $t^{1/2}$ relationship, preventing proper identification of sorptivities.

The solution to these problems with plate and contact resistances was found in shifting control of the flow rate from the soil to the apparatus; i.e., changing from a potential-controlled to a flux-controlled boundary condition (Dirksen, 1979). The experimental problem is then changed from determining S for a known regulated value of θ_1 to determining θ_1 for an imposed value of S. Theoretically, θ should be determined at the interface. Practically, the wetting front initially advances fast enough, and θ changes little in time and distance, so that it can be determined in a shallow surface layer.

For each imposed sorptivity, the wetting front needs to penetrate only a short distance, on the order of 1 to 2 cm. The area over which water can be applied is limited by practical considerations. Therefore, there is not much point in carrying out these measurements in situ, since the influence of larger structural units, macropores, cracks, etc. cannot be measured. Instead, the measurements are best made on undisturbed soil cores taken to the laboratory. These can then first be dried to uniform water content. This makes it possible to determine D over the entire water content range. For the same reason there seems to be little point in making these cores larger than the customary 5-cm diameter. Greater lengths (about 10 cm), however, have a distinct advantage. The cores should be sufficiently uniform that local variations do not obscure the relationship between S and θ to be derived from a number of measurements, each made on a different soil sample.

28–5.4.2 SPECIAL APPARATUS

A simple method of imposing sorptivities on soil columns is driving syringe pumps with rotating cams such that at any time the cumulative volume of water delivered is proportional to the square root of time. Plate and contact resistances are then automatically accommodated by the force the cam exerts on the syringe. If the cams are rotated at constant speed ω rad s^{-1}, they can be made according to the polar coordinates:

$$(\alpha, r) = [\alpha, R_O + (C\alpha)^{1/2}] . \qquad [31]$$

The displacement of the syringe plunger is then equal to the change in radial distance r, which starts with R_O at $\alpha = \omega t = 0$ and increases according to $(C\omega t)^{1/2}$. If the diameters of syringe and soil core are d and D, respectively, then

$$i = (d/D)^2(C\omega t)^{1/2} . \qquad [32]$$

Comparison of Eq. [32] with Eq. [22] shows that

$$S = (d/D)^2 \, (C\omega)^{1/2} . \qquad [33]$$

For a fixed diameter of the soil columns, the sorptivity can be varied by changing d, ω and C (Table 28–1). Figure 28–18 shows a "square root of time" syringe pump, together with a soil column apparatus designed specifically for making the flux-controlled sorptivity measurements. (The apparatus is available from Technische en Fysische Dienst voor de Landbouw, P. O. Box 356, 6700 AJ Wageningen, The Netherlands.)

Figure 28–19 gives a cross-section of the soil column apparatus for 10-cm diameter columns. The one shown in Fig. 28–19 is a 5-cm diameter

Table 28-1. Practical combinations of d, ω, and C covering the normal sorptivity range for soil columns 5 cm in diameter.

d	ω	C	$\log (S^2/m^2\ s^{-1})$
mm	rev min^{-1}	mm^2 rad^{-1}	
32	0.20	285	−6.00
32	0.05	285	−6.60
32	0.05	71.2	−7.20
16	0.05	285	−7.81
16	0.05	71.2	−8.41
8	0.05	285	−9.01
8	0.05	71.2	−9.61
4	0.05	285	−10.21
4	0.05	71.2	−10.82
4	0.05	17.8	−11.42
4	0.02	17.8	−11.82

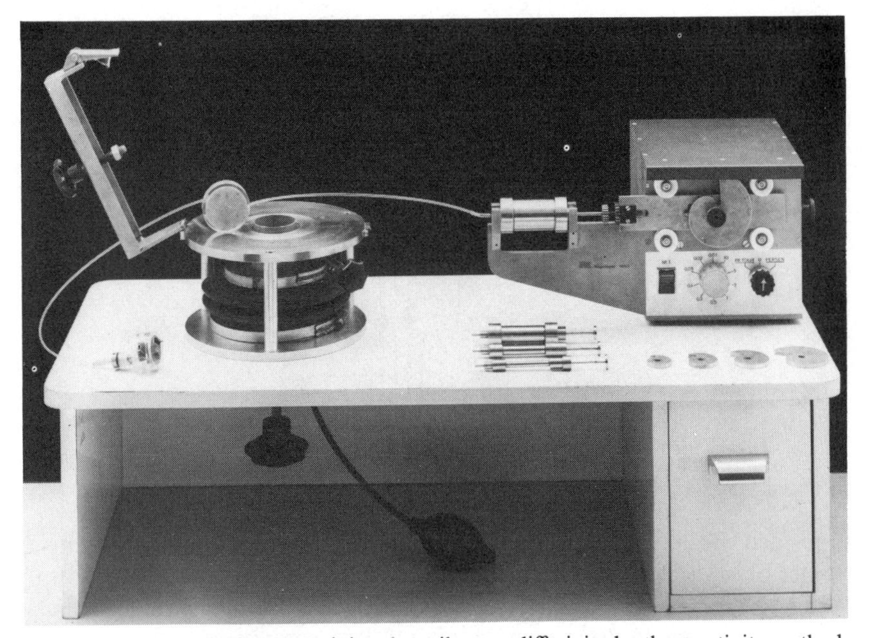

Fig. 28–18. Apparatus for determining the soil water diffusivity by the sorptivity method.

version. The major improvements compared to an earlier version (Dirksen, 1979) are continuous adjustment capability while preparing new soil surfaces and easy variation of ω.

For the cam mounted on the pump, R_O = 10 mm and C = 285 mm^2 rad^{-1}. The other cams have C values of 142, 71.2, 35.6, and 17.8 mm^2 rad^{-1}, respectively. The number of revolutions per minute can be set at 10, 5, 2, 1, 0.5, 0.2, 0.1, 0.05, 0.02 and 0.01. The inside diameter of the syringe on the pump is 32 mm; those of the others are 16, 8, and 4 mm, respectively. There is much overlap in the S values obtainable with these variables. Normally only a limited number of combinations

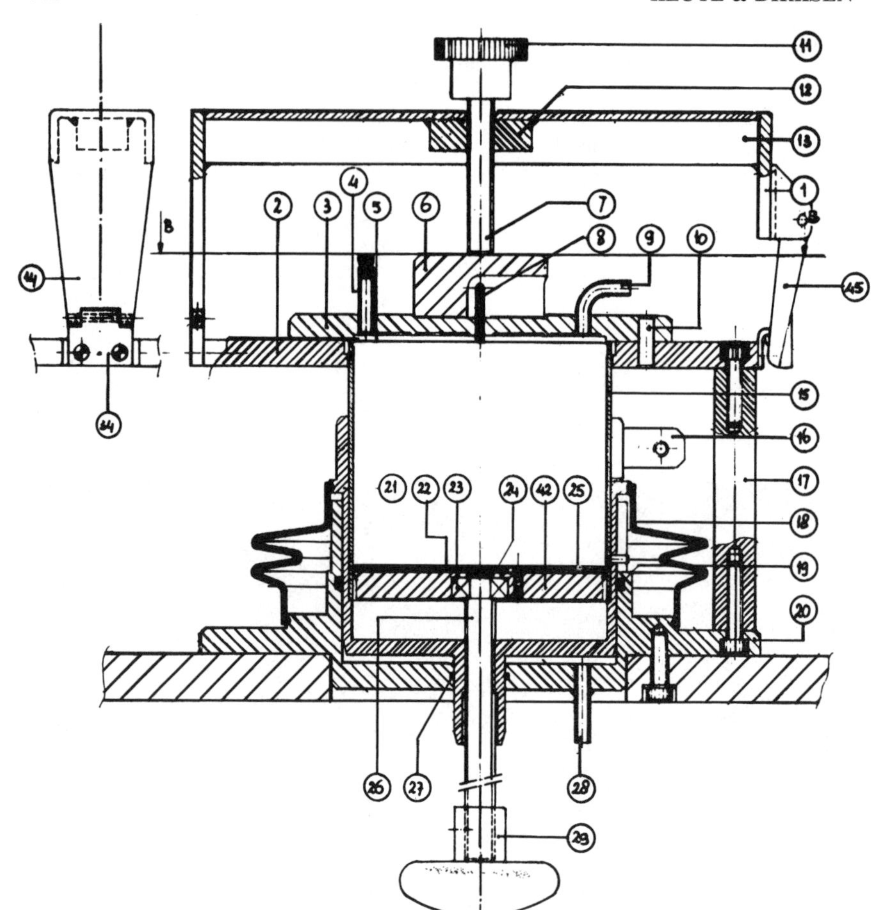

Fig. 28–19. Diagram of the sorptivity apparatus.

is needed. Table 28–1 shows 11 combinations that yield equally spaced values of log S_2 for 5-cm diameter soil columns covering the normal soil water content range. The apparatus is unsuitable for measuring S at saturation; this can be done easily with standard equipment. A small roller bearing attached to the drive shaft rolls over the perimeter of the cam and fits inside the slot near $\alpha = 0$. The beginning of this slot is kept somewhat less steep than its theoretical shape to prevent shearing. The extremities of the syringes are identical so they all fit in the same holder. With the cam in the $\alpha = 0$ position, the drive shaft can be brought snugly against the syringe by means of a threaded rod with lock-nut.

For the larger sorptivities, the pressure head at the absorption interface can be measured during a run with a small tensiometer in the center of the ceramic plate (see Fig. 28–19) and a pressure transducer. In this way the wetting soil water retentivity function, which is usually not available, can be determined simultaneously, allowing the calculation of the hydraulic conductivity from Eq. [30]. The tensiometer is an impermeable spot for the delivery of water, and depends for its proper functioning on lateral flow. It works best when its diameter is kept small (about 1.5 mm).

28–5.4.3 PROCEDURE

In the following step-by-step procedure the numbers in parenthesis refer to those in Fig. 28–19.

1. Pack soil columns in the cylinders (15) and proceed to step 4, or obtain undisturbed soil cores in the field in cylinders which fit the column apparatus.
2. Bring the soil cores to uniform water content, preferably air-dry. Excessive shrinkage should be avoided. Weigh cores for determination of the bulk densities of the cores.
3. Bring the plunger (42) down by turning knob (29), place the soil core in its cylinder in the soil column holder (21) and clamp the cylinder (16) (see horizontal knob in Fig. 28–18).
4. Move the soil column holder upward by pumping air into pipe (28) (see squeeze bulb in Fig. 28–18) until the top of the cylinder reaches well above the flat top (2).
5. Move the soil column upward by turning knob (29) until a thin dry layer of soil reaches above the top of the cylinder (15).
6. With knife, sandpaper, etc., and finally with a straight-edge, prepare a flat, smooth soil surface even with the top of the cylinder. This is a critical phase of the measurement, and should be done with the utmost care. If necessary, repeat steps 5 and 6. Some scraped-off soil may be sieved on the soil surface, pressed down, and the excess removed to fill incidental irregularities.
7. Remove the squeeze-bulb and push or pull the soil column holder assembly down [without turning knob (29)] such that the prepared soil surface is located a few millimeters below the flat top. Moving the column up and down can be made much easier by attaching a three-way valve to pipe (28) and switching to pressurized-air line and a vacuum line, respectively.
8. Select the proper combination of syringe size, cam size, and rotational speed for the desired imposed sorptivity (see Table 28–1).
9. Saturate the ceramic plate (5) and connect the filled syringe to tube (9) without enclosing air. Bring the cam and drive shaft of the syringe pump into starting position (use motor in reverse if supplied with slip-coupling). Place the syringe in holder and take out slack with threaded rod and lock-nut.
10. If the pressure head is to be measured, saturate the small tensiometer in the center of the ceramic plate (8) and connect it with the pressure transducer.
11. Remove excess water from the ceramic plate. Thoroughly clean the seat for the ceramic plate assembly in the flat top and place the ceramic plate assembly on the flat top. Place pressing block (6) over

the tubing. After careful adjustment and locking of the press rod (7), close the quick-coupling (45) to press the ceramic plate assembly snugly against the flat top.

12. Reconnect the squeeze bulb to the pipe (28). At the *same* time, start the syringe pump, and squeeze the bulb a few times. This pushes the soil column up against the ceramic plate, compressing the soil only to the extent that the ceramic plate (5) clears the housing (3). After that, the compressive force is carried by the soil cylinder (15).
13. If applicable, record the pressure transducer output.
14. At the end of the run, execute the following operations as quickly as possible: (i) disconnect the quick-coupling (45) and swing the arm (13) away: (ii) remove the ceramic plate assembly: (iii) turn the knob (29) until the desired height of soil sample emerges above the flat top: (iv) slice the emerging soil layer off the soil column and slide it in a soil moisture can for gravimetric water content determination: and (v) stop the syringe pump.
15. Inspect the wetting front for uniformity by turning the knob (29) and slicing successively more layers off the soil column. If the wetting front is extremely irregular, discard the soil sample and repeat the measurement on another sample.
16. If enough dry soil is left in the cylinder for the following measurement, repeat the above procedure starting at step 4; otherwise start at step 3 after the soil column holder has been cleaned from any soil that may hinder the proper seating of a new soil column. If the same syringe and ceramic plate are to be used, the syringe can be refilled through the ceramic plate under water without disconnecting it (step 9).
17. Make measurements over the desired range of sorptivities.

28–5.4.4 CALCULATIONS

Convert the gravimetric water contents to volumetric water contents using the average bulk density determined in step 2. Plot log S^2 vs. θ_1. Select and make runs with additional sorptivities until the relationship is established satisfactorily.

Obtain least-square polynomials of log S^2 in θ_1 of several degrees, n:

$$\log S^2 = a_0 + a_1\theta_1 + \ldots a_n(\theta_1)^n \ldots . \quad [34]$$

If another differentiable expression fits the data better it should be used instead.

Calculate diffusivities at regular θ intervals according to Eq. [29], using the various polynomials obtained above to perform the differentiation. Equation [34] must also be used to obtain the values of S^2 needed in Eq. [29]. The diffusivities that still increase monotonically with water content and are obtained with the highest degree polynomial are likely to be most accurate.

Calculate hydraulic conductivities at regular θ intervals from Eq. [30], if the wetting soil water retentivity function is known.

28–5.4.5 COMMENTS

There are no standard soils with known diffusivities with which the absolute accuracy of the sorptivity method, or any other method, can be established. The evaluation of a method must be based on its theoretical foundation and its inherent features that promote accuracy. The theoretical basis of the sorptivity method appears solid, although not entirely rigorous. The use of the weighted mean diffusivity is empirical, supported by theoretical calculations and experiments.

The main attraction of the sorptivity method is that it is a transient method which takes relatively little time, while it has the features and apparent accuracy of a steady-state method. The water content near the absorbing soil surface quickly attains its final value. As a result, the time of sampling, except for very low sorptivities, is not important, and the water content can be determined without concern for spatial resolution. The one sample, usually a few millimeters thick, can be taken very quickly, virtually eliminating errors due to evaporation and redistribution during sampling. The price that one pays for this accuracy is that each run yields only one data point. The measurements need to be repeated as many times as the number of data points required to establish the $S(\theta_1)$ relationship to the desired accuracy.

The apparatus is somewhat complicated, the expense depending upon the degree of sophistication. Recent attempts to drive the syringe with a microprocessor would seem to go in the wrong direction, and are not necessary. Once the equipment is assembled, the measurements are simple and relatively fast. For more accurate work, at least three soil samples of 10-cm length should be available. For soil survey type of work, e.g., to study spatial variability, measurements on one 10-cm core should give enough information in less than 12 hours. In the wet range, the sorptivity run should last only a few minutes; otherwise too much soil core is used. At the very low water contents, where steady-state methods take a prohibitive amount of time, a sorptivity run need not last more than an hour. Because of this short duration, and probably because water is constantly being pushed out of the ceramic plate into the soil, seemingly good measurements were obtained far beyond the tensiometer range.

28–6 ALTERNATIVE METHODS

In this section, some alternatives to the methods described above are given. The general principles of each method, and some comments on the apparent advantages, and disadvantages will be given. The original literature should be consulted for additional information about any given method.

28–6.1 Instantaneous Profile Method

The application of this method to field determination of the hydraulic conductivity of unsaturated soil is described in chapter 30. The method can also be applied to flow systems in a laboratory setting.

Integration of the continuity equation for soil water applied to one-dimensional flow in the absence of sources or sinks yields:

$$q(z, \langle t \rangle) = q(z_0, \langle t \rangle) - 1/\Delta t \int_{z_0}^{z} \Delta\theta \, dz \qquad [35]$$

where $q(z, \langle t \rangle)$ and $q(z_0, t)$ are flux densities at positions z, and z_0 at time $\langle t \rangle$, $\Delta\theta = \theta(z, t_2) - \theta(z, t_1)$, $\Delta t = t_2 - t_1$, and $\langle t \rangle$ is $(t_2 + t_1)/2$. If the water content distribution in space and time, $\theta(z, t)$, is measured, the integral can be evaluated, and if the flux density at some position designated as z_0 is known at the time $\langle t \rangle$, the other flux density can be calculated. The known flux density that is required at one position may be obtained by (i) closing one end of the flow system (flux density zero), (ii) finding a position in the flow column at which the hydraulic gradient is zero (flux density zero), or (iii) measuring the flux density at a boundary. Uniformity of the hydraulic properties of the flow system is not assumed, and the boundary conditions do not need to be constant or known in detail.

If the hydraulic head distribution $H(z, t)$ is also measured, the hydraulic gradient at a given position and time can be evaluated. The ratio of the flux density and gradient at the given time and position is the hydraulic conductivity at the water content and pressure head found at that position.

The instantaneous profile method based on the above concepts has been applied by a number of workers to laboratory flow columns (e.g., Watson, 1966; Flocker et al., 1968; Vachaud, 1967; Rogers & Klute, 1971; Weeks & Richards 1967; Gillham et al., 1976; Hamilton et al., 1981; Daniel, 1982).

There are three options with respect to the parameters to be measured: (i) the water content and pressure head distributions may both be measured, (ii) the water content distribution may be measured and the pressure head inferred from water retention data, and (iii) the pressure head distribution may be measured and the water content inferred from water retention data. The first option is inherently the best choice, since no question arises as to the applicability of water retention data obtained on separate samples. Options (i) and (iii) are limited to the tensiometer range, unless one is willing to assume that the osmotic component of the thermodynamic potential of the soil water is negligible, in which case thermocouple psychrometers (Hamilton et al., 1981) may be used. Option (ii) can be extended beyond the tensiometer range.

Drainage of a horizontal column, drainage of a vertical column, horizontal infiltration, vertical infiltration, and redistribution of infiltrated water are some of the flow situations that have been and may be employed. The initial condition need not be one of uniform water content or pressure head, as long as it does not inject an unknown element of hysteresis into the flow system behavior, which can confuse the interpretation of the data.

The boundary and initial conditions to be chosen will depend on whether wetting or drying conductivity hydraulic properties are desired. There is no need to apply step changes of potential or constant flux densities as boundary conditions, as is often required in many analyses of unsaturated flow. Time-dependent boundary conditions at an inflow or outflow end of the soil column can be used to prevent the occurrence of excessively steep gradients in the flow system. This has the advantage that a better measurement of the water content and hydraulic head profiles can be obtained with fewer measurement points in the flow system.

A method of measuring water content as a function of position and time in a nondestructive manner is required. Gamma attenuation apparatus (see chapter 21) is an obvious choice for this purpose. Rapid response tensiometry, such as a pressure transducer–tensiometer system is, in general, needed (see chapter 23) for the measuring hydraulic heads and pressure heads. Because of the large amount of data to be processed, it will be advantageous to use a data acquisition system for the water content and hydraulic head measurements.

The soil column can be either a repacked soil column or an "undisturbed" soil column collected with an appropriate core sampler. The method is suited for long columns, and has no advantage over the methods described in the previous sections when the column is short. Because of the column length the method tends to require considerable time.

28–6.2 Inflow-outflow Through a High-resistance Porous Plate

Ahuja (1974) described a method of determining the wetting hydraulic conductivity function from cumulative inflow into a uniform soil core through a porous plate of high hydraulic resistance. The method is based on the piecewise application of the Green and Ampt (1911) approach to infiltration. The measurements necessary to obtain the conductivity function are the cumulative inflow volume vs. time, the hydraulic resistance of the plate, the initial water content in the soil, and the soil water content–suction relationship of the soil. The method is applicable to short cores.

A one-step inflow-outflow procedure for determining both the hydraulic conductivity and water retention data for a soil core has been described by Ahuja (1975) and Ahuja and El-Swaify (1976). The water is introduced or removed from the soil column through a high-resistance porous plate. The cumulative volume of inflow or outflow and the soil water suction at the opposite end of the core are measured and used to calculate the hydraulic properties. The method appears to be worthy of consideration, especially for measurements in the wet range.

28–6.3 Unit-gradient Drainage Method

This method, which has been described for use in a field situation in chapter 30, may also be used in the laboratory. A soil column is wetted at the top and redistribution of water without evaporation is allowed to occur. In a uniform draining profile, the hydraulic gradient in the upper wetted part of the column will be approximately unity. An analysis can be developed which allows the calculation of the hydraulic conductivity from

$$K = L\,(d\langle\theta\rangle/dt) \qquad [36]$$

where $\langle\theta\rangle$ is the average water content above the depth L, and the conductivity is associated with the water content at depth L. Measurement of the water content in the upper part of the column can be accomplished with gamma attenuation equipment.

An alternate analysis leads to the following equation for the soil water diffusivity

$$D = L\,(dh/dt) \qquad [37]$$

which indicates that a measurement of the pressure head versus time in the unit-gradient draining region of the column will permit the determination of the soil water diffusivity as a function of pressure head. If an applicable soil water retention curve is available from which the water capacity can be calculated, the conductivity can then be obtained.

28–6.4 Parameter Identification Method

The concept of parameter identification has been applied to the determination of the parameters in the hydraulic conductivity and water retention functions (Zachmann et al., 1981a 1981b; 1982). The method involves the measurement of some aspect of the response of a soil water flow system to a set of applied boundary conditions. The response is also simulated with a numerical solution of the soil water flow equation, using a selected functional form for the hydraulic conductivity and water capacity. A "bestfit" between the measured and simulated response of the flow system is found by an optimization procedure which adjusts the parameters of the hydraulic conductivity and water capicity functions. To date, the method has been applied to the drainage of an initially saturated soil column and to the vertical infiltration process. The method has the advantage of rapid simultaneous determination of the hydraulic conductivity and the water retention properties of a soil.

28–7 REFERENCES

Ahuja, L. R. 1974. Unsaturated hydraulic conductivity from cumulative inflow data. Soil Sci. Soc. Am. Proc. 38:695–698.

Ahuja, L. R. 1975. One-step wetting procedure for determining both water characteristic and hydraulic conductivity of a soil core. Soil Sci. Soc. Am. Proc. 39:418–423.

Ahuja, L. R., and S. A. El-Swaify. 1976. Determining both water characteristics and hydraulic conductivity of a soil core at high water contents from a transient flow experiment. Soil Sci. 121:198–204.

Brooks R. H., and A. T. Corey. 1966. Properties of porous media affecting fluid flow. Proc. Am. Soc. Civ. Eng. 92:61–68.

Bruce, R. R., and A. Klute. 1956. The measurement of soil moisture diffusivity. Soil Sci. Soc. Am. Proc. 20:458–462.

Childs, E. C. 1969. The physics of soil water phenomena. John Wiley and Sons, New York.

Clothier, B. E., D. R. Scotter, and A. E. Green. 1983. Diffusivity and one-dimensional absorption experiments. Soil Sci. Soc. Am. J. 47:641–644.

Crank, J. 1956. The mathematics of diffusion. Oxford University Press, London.

Dane, J. H., and A. Klute. 1977. Salt effects on the hydraulic properties of a swelling soil. Soil Sci. Soc. Am. J. 41:1043–1049.

Daniel, D. E. 1982. Measurement of hydraulic conductivity of unsaturated soils with thermocouple psychrometers. Soil Sci. Soc. Am. J. 46:1125–1129.

Dirksen, C. 1975a. Determination of soil water diffusivity by sorptivity measurements. Soil Sci. Soc. Am. Proc. 39:22–27.

Dirksen, C. 1975b. Determination of soil water diffusivity by sorptivity measurements. Reply to Dr. Parlange's letter. Soil Sci. Soc. Am. Proc. 39:1012–1013.

Dirksen, C. 1979. Flux-controlled sorptivity measurements to determine soil hydraulic property functions. Soil Sci. Soc. Am. J. 43:827–834.

Elrick, D. E., and D. H. Bowman. 1964. Note on an improved apparatus for soil moisture flow measurements. Soil Sci. Soc. Am. Proc. 28:450–451.

Flocker, W. J., M. Yamaguchi, and D. R. Nielsen. 1968. Capillary conductivity and soil water diffusivity values from vertical soil columns. Agron. J. 60:605–610.

Gardner, W. R. 1958. Mathematics of isothermal water conduction in unsaturated soil. Highway Res. Board, Spec. Rep. 40, p. 78–87.

Gardner, W. R. 1960. Dynamic aspects of water availability to plants. Soil Sci. 89:63–73.

Gillham, R. W., A. Klute, and D. F. Heermann. 1976. Hydraulic properties of a porous medium: measurement and empirical representation. Soil Sci. Soc. Am. J. 40:203–207.

Green, W. H., and G. A. Ampt. 1911. Studies in soil physics. I. The flow of water and air through soils. J. Agric. Sci. 4:1–24.

Hamilton, J. M., D. E. Daniel, and R. E. Olson. 1981. Measurement of hydraulic conductivity of partially saturated soils. p. 182–196. *In* T. F. Zimmie and C. O. Riggs (ed.) Permeability and groundwater contaminant transport. American Society for Testing Materials Spec. Tech. Pub. 746.

Hubbert, M. K. 1957. Darcy's law and the field equations of the flow of underground fluids. Bull. Assoc. Int. Hydrol. Sci. 5:24–59.

Jackson, R. D. 1963. Porosity and soil water diffusivity relations. Soil Sci. Soc. Am. Proc. 27:123–126.

Jackson, R. D. 1964. Water vapor diffusion in relatively dry soil. I. Theoretical considerations and sorption experiments. Soil Sci. Soc. Am. Proc. 28:172–175.

Laliberte, G. E., and A. T. Corey. 1967. Hydraulic properties of disturbed and undisturbed soils. Permeability and capillarity of soils. American Society for Testing Materials Spec. Tech. Pub. 417, p. 57–71.

McIntyre, D. S. 1974. Procuring undisturbed cores for soil physical measurements. p. 154–165. *In* J. Loveday (ed.) Methods for analysis of irrigated soils. Tech. Commun. 54, Commonwealth Bureau Soils, Appendix I.

Maulem, Y. 1976. A catalogue of the properties of unsaturated soils. Israel Institute of Technology, Haifa, Israel.

Mualem, Y., and A. Klute. 1984. A predictor-corrector method for measurement of hydraulic conductivity and membrane conductance. Soil Sci. Soc. Am. J. 48:993–1000.

Nielsen, D. R., and R. E. Philips. 1958. Small fritted glass bead plates for determination of moisture retention. Soil Sci. Soc. Am. Proc. 22:574–575.

Nightingale, H. I., and W. C. Bianchi. 1970. Rapid measurement of hydraulic conductivity changes in slowly permeable soils. Soil Sci. 110:221–228.

Overman, A. R., J. H. Peverly, and R. J. Miller. 1968. Hydraulic conductivity measurements with a pressure transducer. Soil Sci. Soc. Am. Proc. 32:884–886.

Philip. J. R. 1969. Theory of infiltration. Hydrosci. 5:215–296.

Poulovassilis, A. 1972. The changeability of the hydraulic conductivity of saturated soil samples. Soil Sci. 113:81–87.

Quirk, J. P., and R. K. Schofield. 1955. The effect of electrolyte concentration on soil permeability. J. Soil Sci. 6:163–178.

Reeve, R. C., C. A. Bower, R. H. Brooks, and F. B. Gschwend. 1954. A comparison of the effects of exchangeable sodium and potassium upon the physical condition of soils. Soil Sci. Soc. Am. Proc. 18:130–132.

Remy, J. P. 1973. The measurement of small permeabilities in the laboratory. Geotechnique 23:454–458.

Richards, L. A., and D. C. Moore. 1952. Influence of capillary conductivity and depth of wetting on moisture retention in soil. Trans. Am. Geophys. Union 33:531–540.

Rogers, J. S., and A. Klute. 1971. The hydraulic conductivity–water content relationship during nonsteady flow through a sand column. Soil Sci. Soc. Am. Proc. 35:695–700.

Rolfe, P. F., and L. A. G. Aylmore. 1977. Water and salt movement through compacted clays: 1. Permeability of compacted illite and montmorillonite. Soil Sci. Soc. Am. J. 41:489–495.

Rose, D. A. 1968. Water movement in porous materials. III. Evaporation of water from soil. Br. J. Appl. Physics 1:1779–1791.

Smith, W. O., and R. W. Stallman. 1954. Measurement of permeabilities in ground water investigations. Symposium on permeability of Soils. American Society for Testing Materials Tech. Pub. 163, p. 98–114.

Swartzendruber, D. 1969. The flow of water in unsaturated soils. p. 215–292. *In* R. J. M. de Wiest (ed.) Flow through porous media. Academic Press, New York.

Uhland, R. E. 1950. Physical properties of soils as modified by crops and management. Soil Sci. Soc. Am. Proc. 14:361–366.

Vachaud, G. 1967. Determination of the hydraulic conductivity of unsaturated soils from an analysis of transient flow data. Water Resour. Res. 3:697–705.

Van Schaik, J. C. 1970. Soil hydraulic properties determined with water and with a hydrocarbon liquid. Can. J. Soil Sci. 50:79–84.

Watson, K. K. 1966. An instantaneous profile method for determining the hydraulic conductivity of unsaturated porous materials. Water Resour. Res. 1:577–586.

Weeks, L. V., and S. J. Richards. 1967. Soil water properties computed from transient flow data. Soil Sci. Soc. Am. Proc. 31:721–725.

Whisler, F. D., A. Klute, and D. B. Peters. 1968. Soil water diffusivity from horizontal infiltration. Soil Sci. Soc. Am. Proc. 32:6–11.

Zachmann, D. W., P. C. DuChateau, and A. Klute. 1981a. The calibration of the Richards flow equation for a draining column by parameter identification. Soil Sci. Soc. Am. J. 45:1012–1015.

Zachmann, D. W., P. C. DuChateau, and A. Klute. 1981b. The estimation of soil hydraulic properties from inflow data. p. 173–180. *In* V. Singh (ed.) Proc. of the Symposium on Rainfall-Runoff Modeling. May 1981. Mississippi State Univ. Water Resources Pub., Littleton, CO.

Zachmann, D. W., P. C. DuChateau, and A. Klute. 1982. Simultaneous approximation of water capacity and soil hydraulic conductivity by parameter identification. Soil Sci. 134:157–163.

29 Hydraulic Conductivity of Saturated Soils: Field Methods

A. AMOOZEGAR

North Carolina State University
Raleigh, North Carolina

A. W. WARRICK

University of Arizona
Tucson, Arizona

29–1 INTRODUCTION

The hydraulic conductivity is a measure of a soil's ability to transmit water. Water movement, whether under saturated or unsaturated conditions, is highly dependent on the hydraulic conductivity. The basic relationship describing soil water flow is Darcy's law, which relates flux density $\bar{v}$ to the hydraulic conductivity K and the gradient of the soil water potential H:

$$\bar{v} = -K \text{ grad } H . \quad [1]$$

The conductivity is a spatially variable characteristic, but is constant under saturated conditions for any given position in the field, at any given time. It is a key parameter for all aspects of water and solute movement.

Several methods have been developed to determine the saturated conductivity in the field. They include methods applied to areas with shallow water tables as well as those with deep water tables. In principle, K is calculated from Darcy's law after measuring soil water flux and hydraulic gradient.

29–2 SHALLOW WATER TABLE METHODS

The most common methods to determine the saturated hydraulic conductivity of the soil in the presence of a water table are the auger-hole method and the piezometer method. Other methods, such as the two well, four well, multiple well, well-point, pit bailing, and field mon-

 Methods of Soil Analysis, Part 1. Physical and Mineralogical Methods—Agronomy Monograph no. 9 (2nd Edition)

oliths are also available, but will be discussed in less detail. For methods primarily developed for groundwater systems, such as pumping or slug tests, see, for example, Bouwer (1978) or Freeze and Cherry (1979).

29–2.1 Auger-hole Method

29–2.1 PRINCIPLES

The auger-hole method introduced by Diserens (see Boersma, 1965a), is the procedure most widely used to measure the saturated hydraulic conductivity. Modifications and improvements have been introduced by Kirkham and van Bavel (1948), van Bavel and Kirkham (1948), Johnson et al. (1952), Kirkham (1958), van Beers (1958), Boast and Kirkham (1971), and others (see Bouwer & Jackson, 1974). The auger-hole method, in principle, involves preparation of a cavity extending below the water table, with minimum disturbance of the soil. To decrease the puddling effects, water is pumped out of the hole and then the cavity is allowed to refill several times (similar to priming water wells). After priming the hole, the water is allowed to equilibrate with the groundwater. At equilibrium, the level in the hole will be at the water table. The depth of water in the hole, H, the diameter of the hole, $2r$, and the distance between the bottom of the hole and the underlying impermeable layers, s, must be determined (Fig. 29–1). Then the water is pumped out of the hole and the rate of the rise of the water level measured, allowing calculation of the saturated conductivity of the surrounding soil.

Unlike laboratory methods, where simple equations can describe the saturated hydraulic conductivity as a function of the flux and hydraulic gradient, no simple equation is available for the accurate determination of the conductivity. This is because the flow of water into the auger hole is three-dimensional. In addition, the flow properties could be different in each direction and the cavity might extend through strata with different hydraulic characteristics. To overcome these problems, several variations of the auger-hole method have been developed. Maasland (1955, 1957), for example, presented the theory of the water flow in anisotropic soils

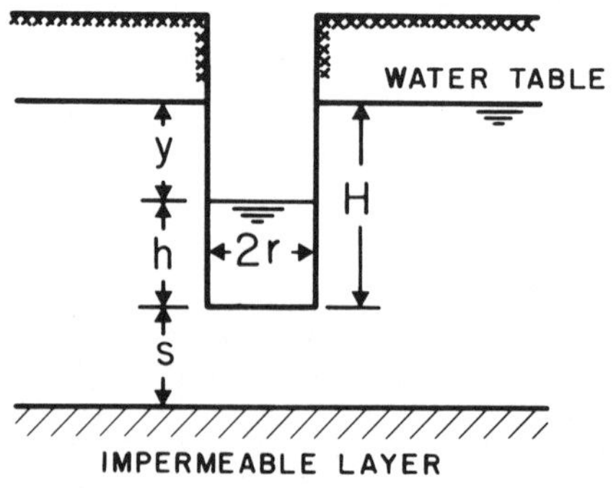

Fig. 29–1. Geometry of an auger hole.

as applicable to the auger-hole method. Luthin (1957) reviewed methods proposed for layered soils. Recently, Topp and Sattlecker (1983) introduced an inflatable assembly and hole liner which allows measurement of either horizontal or vertical components of saturated hydraulic conductivity.

Luthin (1957) as well as Bouwer and Jackson (1974) reviewed various relationships for calculating saturated hydraulic conductivity using the auger-hole method. Johnson et al. (1952) presented still another equation and nomographs to facilitate calculations. Ernst presented an approximate equation for K based on his numerical solution for an auger-hole sufficiently above an impermeable layer

$$K = \{4.63\ r^2/[y(H + 20r)(2 - y/H)]\}(\Delta y/\Delta t) \quad [2]$$

where r (units of length, L) is the radius of the hole, H (units L) is the depth of the groundwater in the hole, y (units L) is the difference between the depth of groundwater and the depth of the water in the hole, and $\Delta y/\Delta t$ is the rate of change of y with respect to time t (units of time, T) (see Bouwer & Jackson, 1974; Boast & Kirkham, 1971). For the case where the auger hole is extended to the impermeable layer, $s = 0$, the equation is

$$K = \{4.17\ r^2/[y(H + 10r)(2 - y/H)]\}(\Delta y/\Delta t)\,. \quad [3]$$

For both equations, K has the same units as $\Delta y/\Delta t$ (e.g., cm/day).

Boast and Kirkham (1971) presented the simple equation

$$K = (\Delta y/\Delta t)C/864 \quad [4]$$

where K is the saturated hydraulic conductivity, $\Delta y/\Delta t$ is the rate of rise of water in the auger hole, and C is a shape factor. Values of C for a variety of cases are presented in Table 29–1. In Eq. [4], C/864 is dimensionless; thus, K has the same units as $\Delta y/\Delta t$.

29–2.1.2 SPECIAL APPARATUS

1. Auger. Use any type of bucket auger to make a hole with a diameter in the range for which Table 29–1 can be used. A post-hole closed-end auger is recommended for coarse-textured soils.
2. Water pump or bailer. To remove the water from the hole as fast as possible use a pump with adequate capacity or a bailer. For small-diameter holes, a bailer constructed of thin-wall tubing with a check valve at the bottom can be used.
3. A device to measure the depth of water in the cavity. Use a float attached to the end of a nonexpanding, flexible measuring tape or a lightly weighted, wooden rod graduated for easy measuring. The float must be smaller than the diameter of the hole and care must be taken

Table 29–1. Values of C for Eq. [4] for an auger hole underlain by an impermeable or infinitely permeable layer (after Boast & Kirkham, 1971).

		s/H for impermeable layer								s/H	s/H for infinitely permeable layer			
H/r	y/H	0	0.05	0.1	0.2	0.5	1	2	5	∞	5	2	1	0.5
1	1	447	423	404	375	323	286	264	255	254	252	241	213	166
	0.75	469	450	434	408	360	324	303	292	291	289	278	248	198
	0.5	555	537	522	497	449	411	386	380	379	377	359	324	264
2	1	186	176	167	154	134	123	118	116	115	115	113	106	91
	0.75	196	187	180	168	149	138	133	131	131	130	128	121	106
	0.5	234	225	218	207	138	175	169	167	167	166	164	156	139
5	1	51.9	48.6	46.2	42.8	38.7	36.9	36.1		35.8		35.5	34.6	32.4
	0.75	54.8	52.0	49.9	46.8	42.8	41.0	40.2		40.0		39.6	38.6	36.3
	0.5	66.1	63.4	61.3	58.1	53.9	51.9	51.0		50.7		50.3	49.2	46.6
10	1	18.1	16.9	16.1	15.1	14.1	13.6	13.4		13.4		13.3	13.1	12.6
	0.75	19.1	18.1	17.4	16.5	15.5	15.0	14.8		14.8		14.7	14.5	14.0
	0.5	23.3	22.3	21.5	20.6	19.5	19.0	18.8		18.7		18.6	18.4	17.8
20	1	5.91	5.53	5.30	5.06	4.81	4.70	4.66		4.64		4.62	4.58	4.46
	0.75	6.27	5.94	5.73	5.50	5.25	5.15	5.10		5.08		5.07	5.02	4.89
	0.5	7.67	7.34	7.12	6.88	6.60	6.48	6.43		6.41		6.39	6.34	6.19
50	1	1.25	1.18	1.14	1.11	1.07	1.05			1.04			1.03	1.02
	0.75	1.33	1.27	1.23	1.20	1.16	1.14			1.13			1.12	1.11
	0.5	1.64	1.57	1.54	1.50	1.46	1.44			1.43			1.42	1.39
100	1	0.37	0.35	0.34	0.34	0.33	0.32			0.32			0.32	0.31
	0.75	0.40	0.38	0.37	0.36	0.35	0.35			0.35			0.34	0.34
	0.5	0.49	0.47	0.46	0.45	0.44	0.44			0.44			0.43	0.43

so the tape or the rod does not touch the cavity wall. The floating device is best suited for shallow holes. As an alternative, an electric probe can be used to measure the depth of water in the well (see Luthin, 1949; Van Bavel & Kirkham, 1948). The electric probes are available commercially or can easily be constructed.
4. Stop watch or regular watch to measure time.
5. Data sheet. Prepare a set of data sheets for recording. A sample data sheet is shown in Fig. 29–2.

29–2.1.3 PROCEDURE

1. Clean plant materials, trash, and loose soil from the selected area.
2. Bore a hole with minimum disturbance to the wall, and extend the hole at least 30 cm below the water level. Check the soil texture while

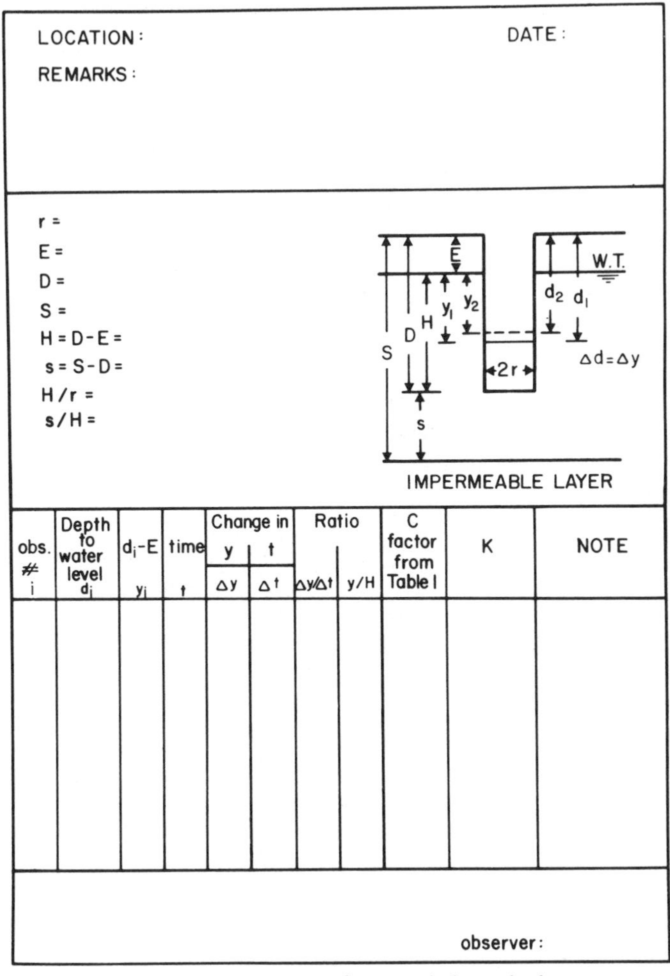
LOCATION: DATE:

REMARKS:

r =
E =
D =
S =
H = D - E =
s = S - D =
H / r =
s / H =

obs. # i	Depth to water level d_i	d_i-E y_i	time t	Change in y Δy	Change in t Δt	Ratio Δy/Δt	Ratio y/H	C factor from Table I	K	NOTE

observer:

Fig. 29–2. Data sheet for auger-hole method

digging the hole to determine any layering of the profile. If the soil is layered, an alternative method may be preferable. Make sure that the groundwater is not under artesian pressure. If the groundwater is under pressure, the water level in the hole will increase abnormally fast when the confining layer is penetrated.

3. To eliminate puddling effects, remove the water from the hole and allow the groundwater to fill the cavity. Repeat this step several times. Measurements can be made during this process to obtain an estimate for the rate of rise of water level in the hole. Dump the water away from the hole or keep it in a large bucket so that water flow into the hole will not be affected by the water dumped around the area.
4. After step *3*, allow equilibrium with the groundwater. Measure the diameter of the hole, $2r$, the depth of water in the hole, H, the distance between the bottom of the hole and the underlying impermeable layer s. (Or alternatively, estimate s.)
5. Remove the water from the hole and measure the rate of rise by measuring the change in the water level during a given period of time. Make more than one measurement of Δy and Δt before the depth of water in the hole, h, reaches about half of the depth of the groundwater in the hole at equilibrium, H.
6. Let the water in the hole come to equilibrium with the water table and repeat step *5*. If the results are not consistent with the results of the previous step, steps *5* and *6* must be repeated until consistent results are obtained for consecutive runs.

29–2.1.4 CALCULATIONS

Use Eq. [2] or [3] for an approximation of the saturated hydraulic conductivity. If a more accurate result is desired, find the shape factor from Table 29–1 and calculate K using Eq. [4]. To find the shape factor from Table 29–1, calculate the values of s/H, y/H, and H/r and find the value of C for the corresponding row and column. If the values of s/H, y/H and/or H/r fall between the given values in the table, the logarithm of H/r, y/H and C should be taken before doing an interpolation. For the case where an impermeable layer exists, the nomograph presented as Fig. 29–3 can be used as an alternative to Table 29–1 and Eq. [4]. The use of the nomograph will be demonstrated by an example: Assume an auger hole with $r = 6$ cm, $s = 30$ cm, $H = 60$ cm, $y = 45$ cm, and $\Delta y/\Delta t = 10$ cm/h. Find the value of y/H ($= 0.75$) on the curved scale (Point I). Connect Point I to Point O and find $H/r = 10$ on the line (Point II). At Point II, construct a perpendicular line until it intersects the curve on $s/H = 0.5$ (Point III). Draw a horizontal line to intersect the vertical scale at $C = 15.7$ (Point IV). Connect the Point IV to the value of $\Delta y/\Delta t$ (Point V) and continue to Point VI, reading the value of K as 4.4 cm/day or 0.18 cm/h.

If the soil below the water table is composed of two or more layers, the conductivity of each layer could be calculated separately. For a two-

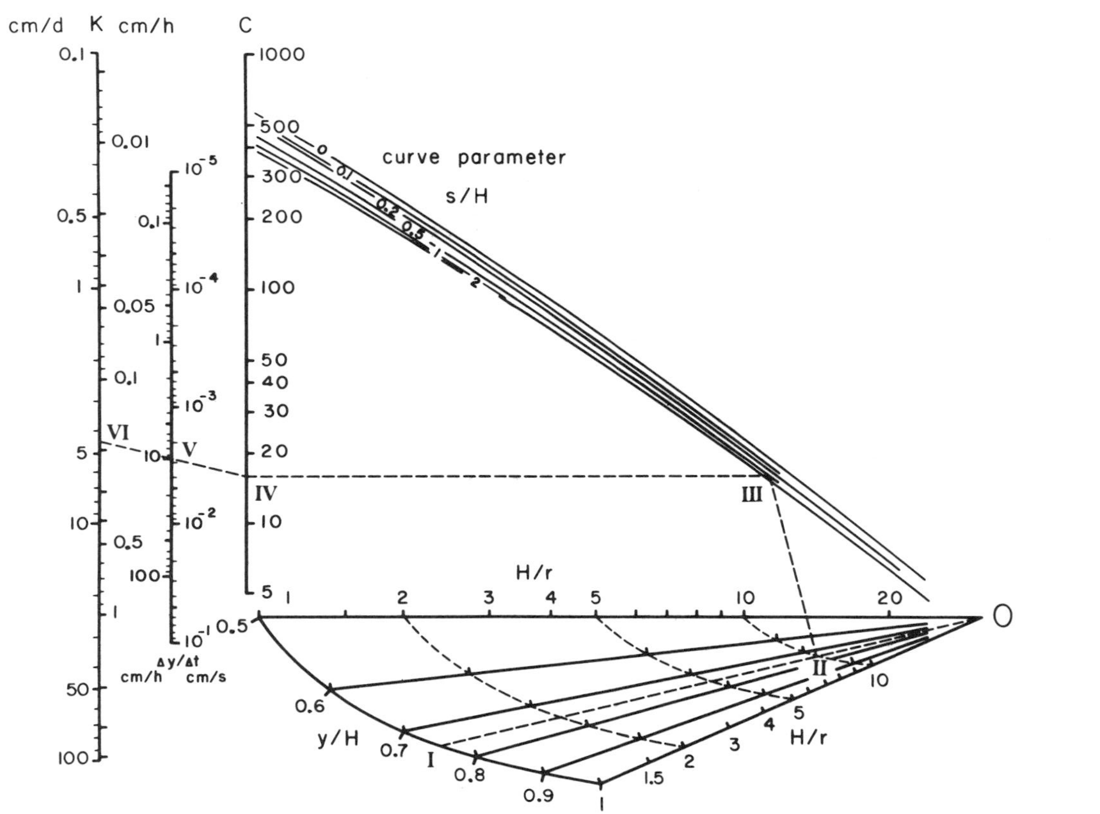

Fig. 29–3. Nomograph to determine the saturated hydraulic conductivity above an impermeable layer by the auger-hole method.

layered soil, bore out the soil through the upper layer at least 10 cm above the second layer (Fig. 29–4A). Determine the conductivity of the upper layer using the procedure described earlier. Deepen the hole through the second layer (Fig. 29–4B) and measure the conductivity again. The calculated value, K, is for the combined profile and the conductivity of the second layer may be calculated using the following equation (Luthin, 1957):

$$K_2 = (KH_2 - K_1H_1)/(H_2 - H_1) \quad [5]$$

For a third layer, the procedure could be extended by assuming the upper two layers as one layer. The error resulting from the assumptions made for the calculation of K for two-layered soil is $< 10\%$ (see Luthin, 1957). If an accurate value of K is desired for the second layer, possibly the measurements can be delayed until the water table falls below the first layer.

29–2.1.5 COMMENTS

The saturated hydraulic conductivity measured by the auger-hole method is dominated by the average value of the horizontal conductivity of the profile (Maasland, 1955). The volume of the soil where the conductivity is measured is about $10Hr^2$ to $40Hr^2$ (units L^3) for commonly used hole diameters (about 10–20 cm) (Bouwer & Jackson, 1974).

Maasland and Haskew (1957) indicated that auger-hole measurements are reproducible to within 10%. They also indicated that the best results are obtained if the amount of return flow into the hole is limited to 20% of the amount of water removed from the hole. Boersma (1965a) suggested that the measurement be completed before the height of the water in the hole, h, reaches to 20% of the depth of water in the hole at equilibrium, H (i.e., $h/H = 0.2$). Boast and Kirkham (1971), on the other hand, presented shape factor values, C, for auger holes up to one-half full. Therefore, it is advised not to make any measurements after the hole is one-half full.

Bouma (1983) found that the auger hole gave a much smaller saturated hydraulic conductivity than alternative methods for some Dutch

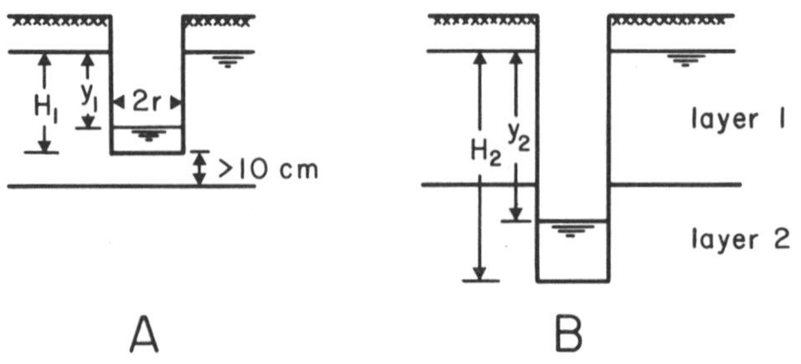

Fig. 29–4. Geometry of the auger hole for a two-layered soil.

clay soils (also see Bouma et al., 1979a). He indicated that smearing the wall of the auger hole during boring of the hole closes the planar pores which allow the water movement into the hole (Bouma et al., 1979b). Hoffman and Schwab (1964), however, reported that the saturated conductivity values obtained by the auger-hole procedures were generally greater than those obtained from drainage tile flow for a silty clay soil in a lakebed region in Ohio.

If smearing of the hole wall, and therefore clogging of the pores and planar voids is suspected, step *3* (as discussed in procedure) should be repeated until puddling effects are eliminated. If after step *3* the effects of smearing are still present, an alternative method, e.g., the column method using large undisturbed columns (Bouma et al., 1976, 1979a, 1981; Bouma and Dekker, 1981), must be used.

Serious errors might result if proper steps are not taken or if equations, tables, or nomographs are extrapolated beyond their range. For example, the water table level in the hole should equilibrate before measurements start (Kirkham, 1965). If measurements are made in units other than those suggested for the equations, tables, or nomographs, proper conversions must be performed so that all units are consistent.

Although it is difficult to use the auger-hole method in rocky or very gravelly soils, it could be performed if a power-assisted auger is available. For unstable soils, the cavity could be lined with a perforated pipe to avoid caving.

When the water level is above the soil surface or if artesian conditions exist, the results obtained by the auger-hole method are not reliable. Also, if the soil is layered extensively or small strata of high permeability occur, results may be inaccurate. According to Luthin (1957), Eq. [5] and the corresponding procedure can give reliable results if the conductivity of the lower layer in a two-layered soil is greater than the conductivity of the upper layer. If the conductivity for the lower layer is less, a negative value might result.

29–2.2. Piezometer Method

29–2.2.1 PRINCIPLES

The principle of the piezometer method for measuring the saturated hydraulic conductivity of soils was first presented by Kirkham (1946). Luthin and Kirkham (1949) developed a field procedure, which consists of installing a piezometer tube or pipe into an auger hole as big as the tube's diameter without disturbing the soil. A cavity is then provided at the bottom of the pipe (see Fig. 29–5) and after the puddling effect is eliminated, water is removed from the cavity. The rate of rise of water in the pipe then is measured and the conductivity is calculated with the help of nomographs or tables.

The conductivity is calculated by

Table 29–2. Values of C/r for Eq. [6] for piezometer method (after Youngs, 1968).

h_c/r	H/r	s/r for impermeable layer							s/r for infinitely permeable layer						
		∞	8.0	4.0	2.0	1.0	0.5	0	∞	8.0	4.0	2.0	1.0	0.5	0
0	20	5.6	5.5	5.3	5.0	4.4	3.6	0	5.6	5.6	5.8	6.3	7.4	10.2	∞
	16	5.6	5.5	5.3	5.0	4.4	3.6	0	5.6	5.6	5.8	6.4	7.5	10.3	∞
	12	5.6	5.5	5.4	5.1	4.5	3.7	0	5.6	5.7	5.9	6.5	7.6	10.4	∞
	8	5.7	5.6	5.5	5.2	4.6	3.8	0	5.7	5.7	5.9	6.6	7.7	10.5	∞
	4	5.8	5.7	5.6	5.4	4.8	3.9	0	5.8	5.8	6.0	6.7	7.9	10.7	∞
0.5	20	8.7	8.6	8.3	7.7	7.0	6.2	4.8	8.7	8.9	9.4	10.3	12.2	15.2	∞
	16	8.8	8.7	8.4	7.8	7.0	6.2	4.8	8.8	9.0	9.4	10.3	12.2	15.2	∞
	12	8.9	8.8	8.5	8.0	7.1	6.3	4.8	8.9	9.1	9.5	10.4	12.2	15.3	∞
	8	9.0	9.0	8.7	8.2	7.2	6.4	4.9	9.0	9.3	9.6	10.5	12.3	15.3	∞
	4	9.5	9.4	9.0	8.6	7.5	6.5	5.0	9.5	9.6	9.8	10.6	12.4	15.4	∞
1.0	20	10.6	10.4	10.0	9.3	8.4	7.6	6.3	10.6	11.0	11.6	12.8	14.9	19.0	∞
	16	10.7	10.5	10.1	9.4	8.5	7.7	6.4	10.7	11.0	11.6	12.8	14.9	19.0	∞
	12	10.8	10.6	10.2	9.5	8.6	7.8	6.5	10.8	11.1	11.7	12.8	14.9	19.0	∞
	8	11.0	10.9	10.5	9.8	8.9	8.0	6.7	11.0	11.2	11.8	12.9	14.9	19.0	∞
	4	11.5	11.4	11.2	10.5	9.7	8.8	7.3	11.5	11.6	12.1	13.1	15.0	19.0	∞
2.0	20	13.8	13.5	12.8	11.9	10.9	10.1	9.1	13.8	14.1	15.0	16.5	19.0	23.0	∞
	16	13.9	13.6	13.0	12.1	11.0	10.2	9.2	13.9	14.3	15.1	16.6	19.1	23.1	∞
	12	14.0	13.7	13.2	12.3	11.2	10.4	9.4	14.0	14.4	15.2	16.7	19.2	23.2	∞
	8	14.3	14.1	13.6	12.7	11.5	10.7	9.6	14.3	14.8	15.5	17.0	19.4	23.3	∞
	4	15.0	14.9	14.5	13.7	12.6	11.7	10.5	15.0	15.4	16.0	17.6	20.1	23.8	∞
4.0	20	18.6	18.0	17.3	16.3	15.3	14.6	13.6	18.6	19.8	20.8	22.7	25.5	29.9	∞
	16	19.0	18.4	17.6	16.6	15.6	14.8	13.8	19.0	20.0	20.9	22.8	25.6	29.9	∞
	12	19.4	18.8	18.0	17.1	16.0	15.1	14.1	19.4	20.3	21.2	23.0	25.8	30.0	∞
	8	19.8	19.4	18.7	17.6	16.4	15.5	14.5	19.8	20.6	21.4	23.3	26.0	30.2	∞
	4	21.0	20.5	20.0	19.1	17.8	17.0	15.8	21.0	21.5	22.2	24.1	26.8	31.5	∞
8.0	20	26.9	26.3	25.5	24.0	23.0	22.2	21.4	26.9	29.6	30.6	32.9	36.1	40.6	∞
	16	27.4	26.6	25.8	24.4	23.4	22.7	21.9	27.4	29.8	30.8	33.1	36.2	40.7	∞
	12	28.3	27.2	26.4	25.1	24.1	23.4	22.6	28.3	30.0	31.0	33.3	36.4	40.8	∞
	8	29.1	28.2	27.4	26.1	25.1	24.4	23.4	29.1	30.3	31.2	33.8	36.9	41.0	∞
	4	30.8	30.2	29.6	28.0	26.9	25.7	24.5	30.8	31.5	32.8	35.0	38.4	43.0	∞

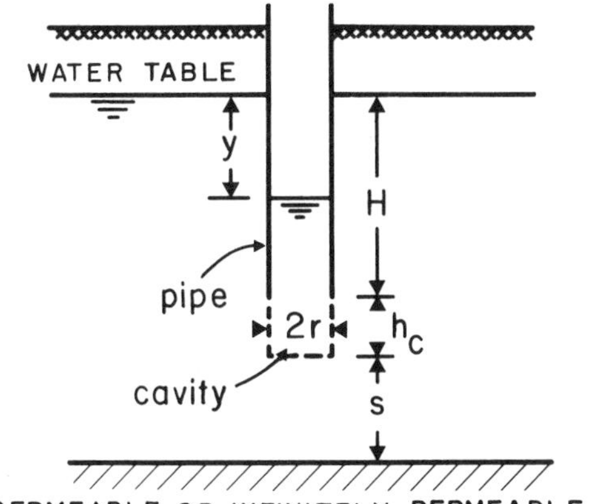

Fig. 29–5. Diagram of a piezometer hole.

$$K = \{\pi r^2/[C\,(t_{i+1} - t_i)]\}\ \ln(y_i/y_{i+1}) \qquad [6]$$

where

K = saturated hydraulic conductivity (units L/T),
r = radius of the cavity (units L),
y_i = the difference between the depth of groundwater and the depth of water in the pipe (units L) at time t_i (units T),
y_{i+1} = the difference between the depth of groundwater and the depth of water in pipe at time t_{i+1}, and
C = a shape factor (units L).

Luthin and Kirkham (1949) presented values of C for certain geometries. Johnson et al. (1952) presented a nomograph for a piezometer tube 4.9 cm in diameter and a cavity 10.9 cm long. Youngs (1968) determined values of C/r (dimensionless) for a variety of cases using an electrical analog. His results are presented in Table 29–2, where H is the length of the pipe extended in the groundwater, h_c is the length of cavity and s is the distance between the bottom of the cavity and the impermeable or infinitely permeable layer (see Fig. 29–5).

29–2.2.2 SPECIAL APPARATUS

1. Piezometer tube. Any kind of pipe, plastic or metal, with a diameter in the range such that Table 29–2 can be used. Aluminum pipes used for the neutron probe with a diameter of about 5 cm are suitable.
2. Auger. Use a bucket auger or screw type auger that fits inside the piezometer pipe.
3. Data sheet. Prepare a set of date sheets as shown in Fig. 29–6.
4. Items 2, 3, and 4 described in auger hole method.

29–2.2.3 PROCEDURE

1. Clear plant materials, trash, and loose soil from the area.
2. Bore a hole to a depth of about 10 cm. Remove the auger and insert

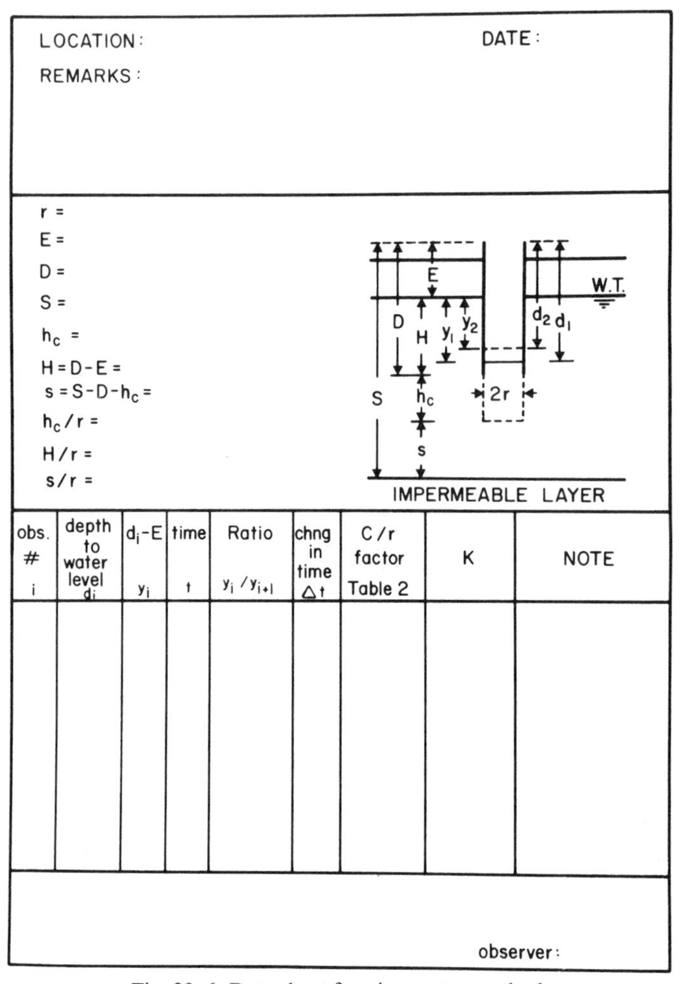

Fig. 29–6. Data sheet for piezometer method.

the piezometer pipe into the hole. Insert the auger into the pipe, excavate an additional 10 to 15 cm, and tap the pipe into the hole. Repeat the process until the bottom of the pipe is at the desired depth below the water table. Carefully bore a cavity at the bottom of the hole to a depth of h_c such that Table 29–2 may be used.

3. To eliminate the puddling effect, insert a tube down the pipe into the cavity and remove the water from the cavity. Let the water rise into the pipe and remove the water from the cavity again. Repeat this step until the rate of rise of the water reaches a constant value and the water from the cavity does not contain appreciable amounts of soil materials.
4. Allow the water level in the pipe to equilibrate with the groundwater. Measure and record H, r, h_c, and s (see Fig. 29–5).

5. Remove the water from the pipe to a distance y below the water table and record the rise of water in the pipe y_i and y_{i+1} in a specified length of time $\Delta t = t_{i+1} - t_i$. This measurement can be repeated a few times before the depth of water in the pipe reaches to within about 20 cm of the water table.
6. Let the water in the pipe come to equilibrium with the groundwater and repeat step *5*. If the rate of rise of water in the pipe is not consistent with the results of the previous measurement, repeat steps *4* and *5*. Continue the process until the results of two successive measurements are consistent.
7. If the hydraulic conductivity of soil at deeper depths is required, go back to step *2* and continue to extend the hole to the desired depth as in that step. Then, repeat steps *3* to *6* to find K.

29–2.2.4 CALCULATIONS

1. Calculate values of h_c/r and H/r. Determine if there is an impermeable or infinitely permeable layer at a distance s below the bottom of the cavity.
2. Find the value of C/r from the corresponding column and row of Table 29–2. Interpolate between the values if necessary.
3. Calculate the hydraulic conductivity using Eq. [6].

For the case where an impermeable layer exists at depth s, the nomograph in Fig. 29–7 can be used to find the value of C/r and the conductivity. As an example, assume $H = 80$ cm, $r = 5$ cm, $h_c = 30$ cm, and $s = 40$ cm. Find the value of h_c/r (=6) on the scale H/r (=16) (Point I), draw a vertical line to intersect the curve $s/r = 8$ (Point II), draw a horizontal line to intersect scale C/r (Point III), and read C/r (=22.5). Using Eq. [6] for $y_1/y_2 = 1.4$ and $t_2 - t_1 = 10$ min, the calculated value of K is 0.023 cm/min or 1.41 cm/h. To use the nomograph, connect Point III to the value of r (5 cm) (Point IV) and continue to intersect scale A at Point V. Connect the value of y_1/y_2 (Point VI) to $t_2 - t_1$ (Point VII) to intersect scale B at Point VIII. Connect the two points on scales A and B (i.e., Points V and VIII, respectively). Read or interpolate the value of $K = 1.5$ cm/h at the intersection of the latter line and scale K.

29–2.2.5 COMMENTS

The piezometer method, unlike the auger-hole method, can be used to measure either horizontal or vertical hydraulic conductivity. If the length of cavity h_c is great compared to the diameter of the cavity, the measurement is primarily the horizontal conductivity. If h_c is small compared to r, then the vertical conductivity is approached. A special case of the piezometer method, called the "tube method," was first suggested by Kirkham (1946) and developed by Frevert and Kirkham (1948). In the tube method, no cavity is provided at the bottom of the piezometer tube (i.e., $h_c = 0$). The conductivity determined by the tube method is the vertical conductivity, and the procedure is exactly the same as the

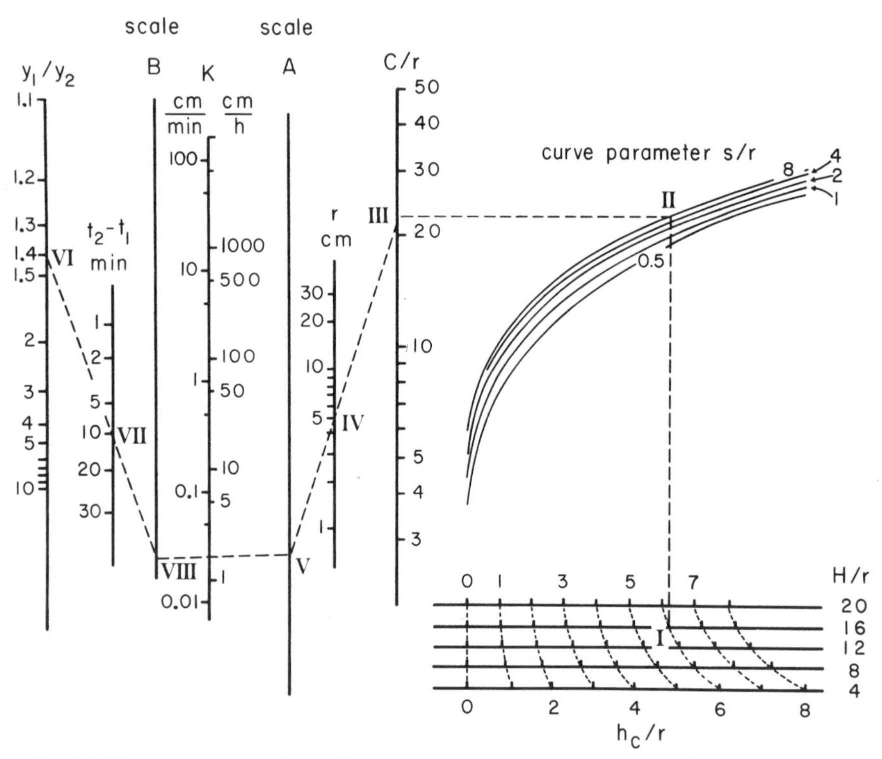

Fig. 29–7. Nomograph to determine the saturated hydraulic conductivity above an impermeable layer by piezometer procedure.

piezometer tube method. Values of C/r for the tube method are given in Table 29–2, and Eq. [6] is used to calculate the conductivity.

For stratified soils the piezometer method can be used to determine the conductivity of each individual layer. The piezometer method is not suitable for rocky and gravelly soils. In these soils, it will be difficult to bore the soil through the pipe, and the pipe might not fit tightly in the auger hole, causing channeling along the side of the pipe.

To measure the conductivity at deeper depths, King and Franzmeier (1981) bored a hole somewhat larger than diameter of the tube and then drove the tube to the desired depth. The large hole was backfilled and tamped to form a seal around the tube. The above procedure must be used with caution, to make sure that the soil will not cave in. Also, the cavity at the bottom of the pipe must be dug after the pipe is installed.

29–2.3 Other Methods

There are other methods to determine the saturated hydraulic conductivity in the presence of a shallow water table which will be discussed briefly. For additional information see Luthin (1957) and Bouwer and Jackson (1974).

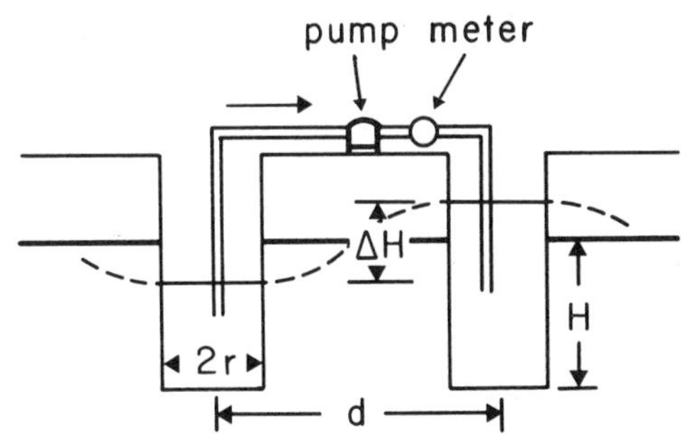

Fig. 29–8. Diagram of the geometry of two-well techniques.

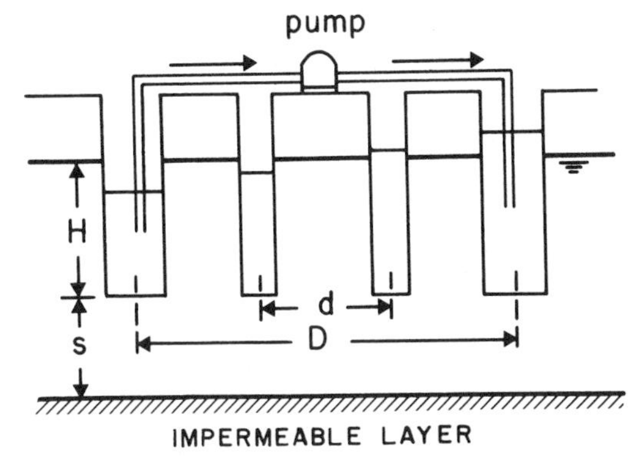

Fig. 29–9. Geometry of the four-well technique to determine the saturated hydraulic conductivity.

Childs (1952) and Childs et al. (1953) proposed a two-well technique which consists of two auger holes of equal diameter ($2r$) extended to the same depth, preferably to an impermeable layer (Fig. 29–8). The distance between the centers of the two holes, d, (units L) is about 1 m. Water is pumped from one hole and carried to the other hole at a constant rate Q (units L^3/T) until an equilibrium between the level of the water in the two holes is reached. The hydraulic conductivity can be calculated by the equation

$$K = Q \cosh^{-1}(d/2r)/[\pi \Delta H(H + L_f)] \qquad [7]$$

where H is the average depth of water in the holes (units L), ΔH is the equilibrium hydraulic head difference between the two holes, L_f is an end correction factor, (units L), and $\cosh^{-1}$ is the inverse hyperbolic cosine. The end correction factor is zero if the wells are extended to an impermeable layer and there is no capillary fringe.

To overcome the problem of clogging in the walls and bottom of the receiving wells (pump-in well) in the two-well technique, Kirkham (1954) proposed the four-well technique (Fig. 29–9). Two additional wells are located symmetrically between the discharge (pump-out) well and receiving well. The two center wells can be of smaller diameter and cased (piezometer-type wells). The distance between the two inner wells is d and the radius and the distance between the outer wells are r and D, respectively. The equations defining saturated hydraulic conductivity K where $D > 12r$ are

$$K = 0.221\ (GQ/H\Delta H) \qquad [8a]$$

for $D/d = 3$, and

$$K = 0.512\ (GQ/H\Delta H) \qquad [8b]$$

for $D/d = 1.5$ (Snell & van Schilfgaarde, 1964).

In the above equations, G is a geometrical factor (dimensionless), Q is the flow rate (units L^3/T), H is the average depth of water in the well (units L), and ΔH is the water level difference between the two inner wells. The coefficients (0.221 and 0.512) are dimensionless. Using an electric analog, Snell and van Schilfgaarde determined the geometrical factor which is available as Fig. 3 and 4 of Snell and van Schilfgaarde (1964) and also Fig. 23–9 and 23–10 of Bouwer and Jackson (1974).

Smiles and Youngs (1963) proposed a method known as the multiple-well technique, in which an array of wells is constructed on the circumference of a circle. The discharge and receiving wells are arranged in an alternate and symmetrical fashion (Fig. 29–10). The advantage of the multiple-well over the two-well technique is the larger soil volume sampled for K. However, the problem of clogging of receiving wells still exists. The equation for calculating K is

$$K = [2Q/n\pi\Delta H(H + L_f)]\ \ln(4a/nr) \qquad [9]$$

where n is the number of auger holes on the circle, a is the radius of the circle, ΔH is the average hydraulic head difference for each pair of ad-

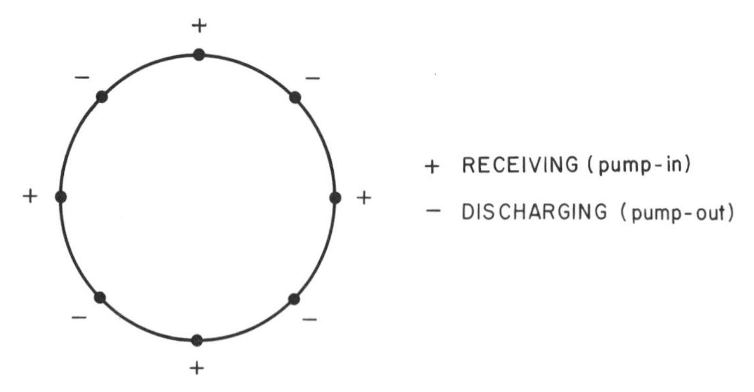

Fig. 29–10. Geometry of the multiple-well technique.

jacent wells, Q is the total flow rate into the system, and H, r, and L_f are the same as for the two-well method.

The pit-bailing method was developed by Healy and Laak (1973) to measure the in situ saturated hydraulic conductivity for septic-tank drain-field design. Recently, Bouwer and Rice (1983) extended the use of the pit-bailing technique to measure saturated conductivity of the soil below a water table. The procedure is useful particularly for stony soils where the auger-hole and other techniques are not practical.

In the pit-bailing method a large hole is dug extending below the water table (a backhoe can be used). After the water level in the pit comes to equilibrium with the water table, the water level in the pit is lowered rapidly. The rise of the water level in the hole is then measured for calculating the hydraulic conductivity.

Bouwer and Rice (1983) applied the piezometer theory to the pit-bailing technique. To find the shape factor for the pit hole, they extrapolated results of Youngs (1968) (also see Table 29–2). Recently, Boast and Langebartel (1984) have applied the auger-hole technique to the pit-bailing method and presented shape factors for pits with length/radius ratios ranging from 0.05 to 2, and for conditions ranging from an empty pit to 90% full. In their analysis, both impermeable barrier and very permeable lower boundary conditions were considered. The shape factors presented by Boast and Langebartel are an extension of the shape factors for auger holes by Boast and Kirkham (1971). Bouwer and Rice (1983) found that piezometer-derived geometry factors gave better results in a laboratory sand-tank than geometrical factors derived for an auger hole. Although there are discrepancies between the values of hydraulic conductivity calculated by the two methods, both procedures seem to be more reliable than the method originally suggested by Healy and Laak (1973) (see Boast & Langebartel, 1984).

Stibbe et al. (1970) described a method to evaluate saturated conductivity using large in situ monoliths. In principle, trenches are dug around a soil column (1.5-m^2 surface to a depth about 1.8 m, in their study). A series of piezometers are installed in the column and a plywood box is constructed around the column. The conductivity is measured by ponding water at the surface and evaluating the rate of water entry into the column. Bouma et al. (1976) described yet another procedure to isolate an undisturbed soil column in the field and used it to measure the vertical conductivity of some Dutch soils (also see Bouma & Dekker, 1981). Bouma et al. (1981) also described a field method to measure the saturated conductivity of the soil adjacent to the tile drains. The procedure is based on isolating an undisturbed soil volume around a tile drain and measuring the saturated K on the undisturbed sample. Although these procedures are alternative methods to auger-hole and piezometer-tube methods, their applicability is limited and the procedures resemble those of laboratory methods. For more detailed discussion of these procedures, the reader is referred to the above articles.

29-3 DEEP WATER TABLE METHODS

The methods to determine the saturated hydraulic conductivity in locations above a water table or in the absence of a water table are more elaborate and time consuming. In fact, a large quantity of water may be needed to saturate the nearby soil before the measurement. These methods include the double-tube method, shallow well pump-in method, cylindrical permeameter method, air entry permeameter and infiltration gradient technique.

29-3.1 Double-tube Method

29-3.1.1 PRINCIPLES

The double-tube method as proposed by Bouwer (1961, 1962, 1964) utilizes two concentric cylinders installed in an auger hole. An auger hole is dug to the desired depth, cleaned with equipment shown in Fig. 29-11, and the outer tube is installed in the hole and pushed about 5 cm into the hole bottom. The inner tube and the top-plate assembly (as shown in Fig. 29-12) are then installed in the outer tube and pushed about 2 cm into the bottom of the hole. With valve B open and valve C closed, water is supplied to both inner and outer tubes through valve A. After water reaches the top of the outer tube standpipe, valve A is adjusted so that water level in the inner tube standpipe stays at the zero mark (the end of the outer tube standpipe is at the same height as the zero mark on the inner tube standpipe). This can be achieved by allowing the outer standpipe to overflow slightly. Due to the large head of water, a wet zone with positive pressure is created at the bottom of the hole.

After saturation of the bottom of the hole is achieved (usually after 1 h for fine-textured soils), two sets of data are collected. First, valve B is closed and the water level in the inner tube standpipe decreases while the water level in the outer tube stays at the top of the outer tube standpipe. The fall of water in the inner tube standpipe, H, and time, t, are recorded; t starts with the closing of valve B. Due to the decrease in head

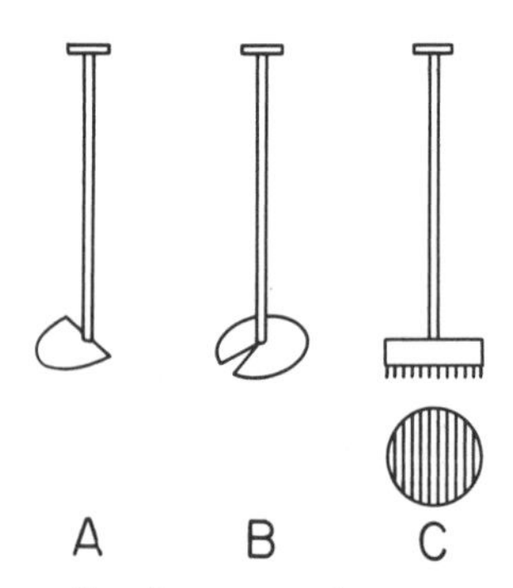

Fig. 29-11. Schematic diagram of equipment used to construct the auger hole for double-tube method. (A) spoon; (B) rotary or hole planer; and (C) hole cleaner.

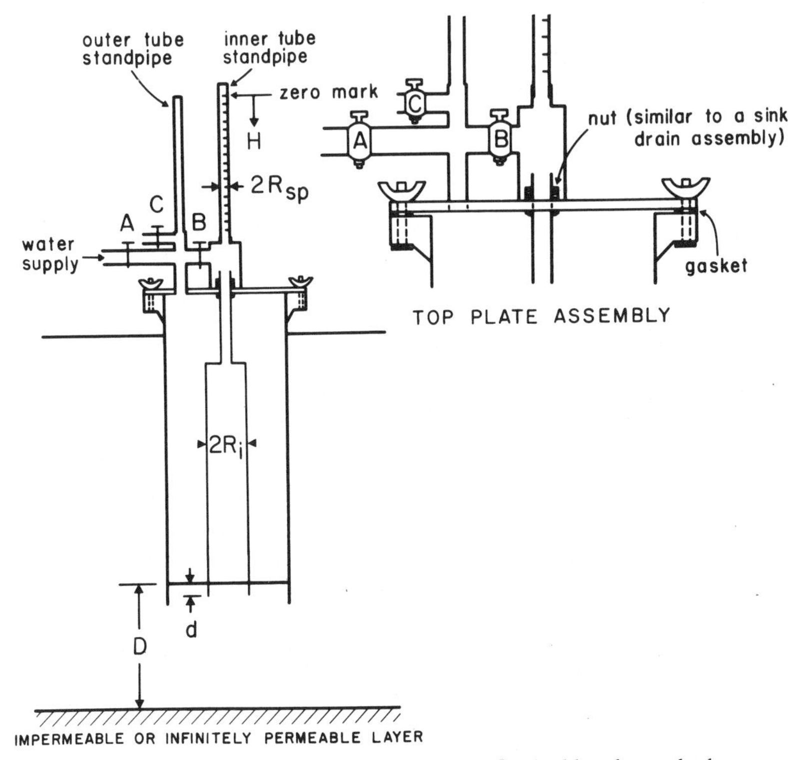

Fig. 29–12. Diagram of the equipment used for double-tube method.

in the inner tube and the greater head in the outer tube, the rate of fall of head in the inner tube decreases with time (see Fig. 29–13). After H vs. t is obtained, valve B is opened and water level in the inner tube standpipe is brought back to the zero mark. After the initial equilibrium is reached, valve B is closed and valves A and C are adjusted in such a way that the level of water in the outer tube standpipe stays at the same level as water in the inner tube standpipe. The rate of fall of water in the inner tube standpipe is recorded and the fall of water H is plotted vs.

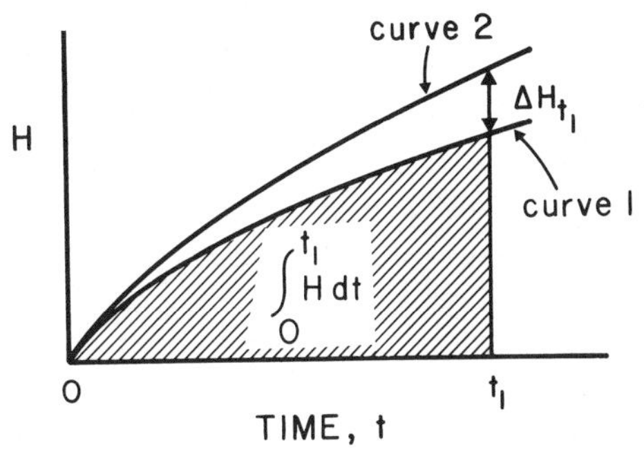

Fig. 29–13. Graph of H vs. t for double-tube procedure.

time on the same graph as data set 1 (Fig. 29–13). The second curve lies above the first curve because there is no difference between the head of water in the inner and outer tubes.

The saturated hydraulic conductivity is calculated using the H vs. t graphs and

$$K = R_{sp}^2 \, \Delta H_{t_1} / (FR_i \int_0^{t_1} H \, dt) \qquad [10]$$

where

R_{sp} = radius of the inner tube "standpipe",
R_i = radius of the inner tube,
ΔH_{t_1} = vertical distance between the two curves, (see Fig. 29–13) at a certain time $t = t_1$,
$\int_0^{t_1} H \, dt$ = area under the lower curve between $t = 0$ and $t = t_1$, and
F = a dimensionless quantity dependent on the geometry of the flow system. The value of F depends on the diameter of the inner tube R_i, depth of penetration of the inner tube d, and the depth and nature of the lower boundary. The nomograph in Fig. 29–14A gives values of F for an impermeable layer located below the bottom of the hole. If a highly permeable layer is located below the hole, the nomograph in Fig. 29–14B can be used.

29–3.1.2 APPARATUS

1. Double-tube apparatus. The double-tube apparatus is commercially available (from Soil Test, Inc., 2205 Lee Street, Evanston, IL 60202, USA) under the name "Tempe Double Tube Hydraulic Permeability Device." The apparatus can also be assembled as shown in Fig. 29–12. The materials needed are:
   a. Galvanized steel or aluminum pipe, approximately 25 cm in diameter.
   b. Galvanized steel or aluminum pipe, approximately 12 cm in diameter.
   c. Top assembly plate as shown in Fig. 29–12.
   d. Clear plastic or glass tubes with marks for the inner tube standpipe.
   e. Gasket and necessary hardware to assemble the equipment.
2. Auger. Use any auger that can make a round, clean hole with a diameter equal to that of the outer tube. (i.e., 25 cm).
3. Hole-cleaning equipment (Fig. 29–11):
   a. Spoon, to remove soil from the hole.
   b. Rotary or hole planer.
   c. Hole cleaner, consisting of thin steel plates, 2 cm wide, which are mounted 1 cm apart on a cylindrical block of the same diameter as the auger hole.
4. Auxiliary water reservoir: a 200-L drum is well suited for the purpose.

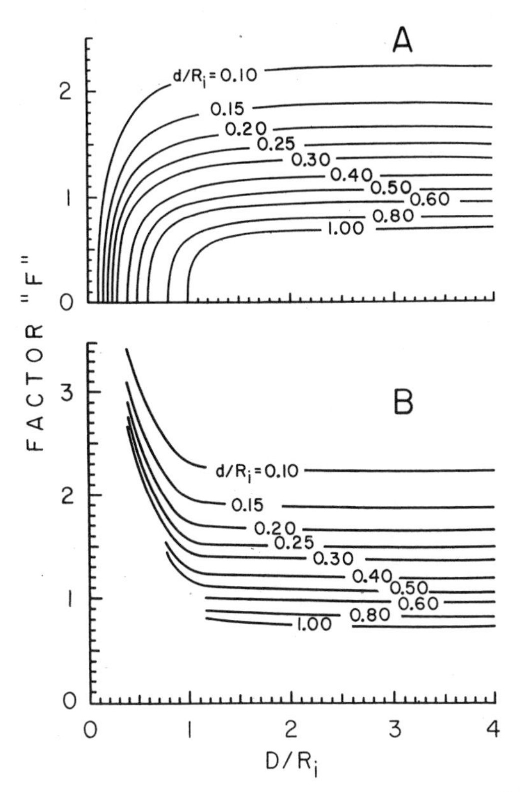

Fig. 29–14. Values of F for Eq. [10] for double-tube method. (A) An impermeable layer below the hole; (B) An infinitely permeable layer below the hole (Bouwer, 1961).

5. Data sheet such as the one shown in Fig. 29–15.
6. Graph paper.
7. Planimeter (if available) to measure the area under the curve.
8. Watch or timer.

29–3.1.3 PROCEDURE

1. Bore out the hole to the desired depth. Use the rotary planer (Fig. 29–11) to square the bottom of the hole. Insert the hole cleaner and push down so that the steel plates penetrate the hole bottom for about 1 cm. Apply a slight torque to the device, causing the soil to stick between the plates, and pull it up. If the soil is hard and/or very dry, add a small amount of water to the hole and wait for the water to soften and moisten the soil before proceeding with the rotary planer.
2. After the hole is squared and cleaned, apply a protective 1- to 2-cm layer of sand to the hole.
3. Insert the outer tube into the hole and push it down so it penetrates about 5 cm into the soil. Backfill any cavity that may be around the outer tube and tamp it to form a tight seal.

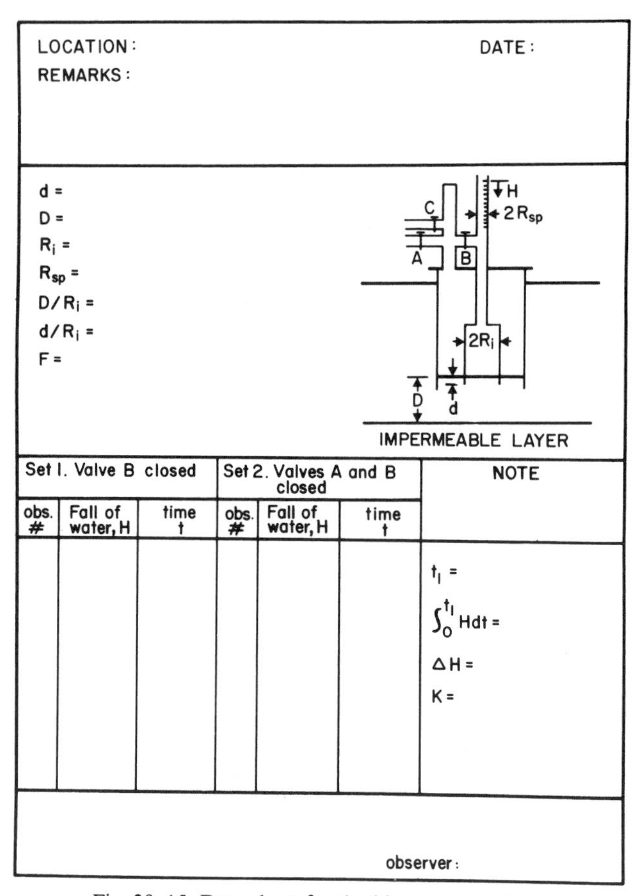
LOCATION: DATE:

REMARKS:

d =

D =

R_i =

R_{sp} =

D/R_i =

d/R_i =

F =

Set 1. Valve B closed			Set 2. Valves A and B closed			NOTE
obs. #	Fall of water, H	time t	obs. #	Fall of water, H	time t	
						t_1 = $\int_0^{t_1} H dt$ = ΔH = K =

observer:

Fig. 29–15. Data sheet for double-tube procedure.

4. Fill the outer tube with water and insert the inner tube into the outer tube. Locate the inner tube at the center of the outer tube and push it down so it penetrates the soil about 2 cm.
5. Assemble the top-plate assembly and standpipes. Supply water to both inner and outer tubes through valve A, keeping valve B open.
6. Adjust valve A so that the standpipe on the outer tube overflows slightly and keep valve B open so the water level in the inner tube standpipe is at the zero mark.
7. After saturation is achieved (about 1 h for fine-textured soils), close valve B and add a small amount of water to the inner tube standpipe so the water level is slightly above the zero mark. When the water level reaches the zero mark, start to record the fall of water in the inner tube standpipe, *H*, and time, *t*. Record the time for at least every 5-cm drop of *H* for a total of at least 30 cm.
8. After step *7* is completed, open valve B and allow the water in the inner tube standpipe to reach the zero mark and repeat step *6* until

initial equilibrium is reached. Wait for a period at least 10 times as long as the time required to collect the first data set as described in step *7*.

9. Close valve B and adjust valves A and C so that the water level in the standpipe for the outer tube is at the same level as water in the inner tube standpipe. Record the time and fall in the water level in the standpipe from the time that valve B is closed. Measurement of time can be made for every 5-cm drop of water head for a total of at least 30 cm.
10. Repeat steps *6* to *9* for another time. If the values obtained in the later run are not consistent with the results of the previous run, repeat the measurements (steps *6* to *9*) for another time. This must be repeated until the results of two consecutive runs are consistent.

29–3.1.4 CALCULATIONS

1. Plot the last results obtained from step *7* (H vs. t) curve 1 as shown in Fig. 29–13.
2. Plot the last results obtained from step *9* (H vs. t) curve 2 on the same graph.
3. Measure ΔH_{t1} and the area under curve 1 from $t = 0$ to $t = t_1$.
4. Find the value of F, using the nomograph in Fig. 29–14A or 29–14B depending on whether or not an impermeable layer or highly permeable layer is located below the auger hole.
5. Use Eq. [10] to calculate K.

29–3.1.5 COMMENTS

A simplified procedure for calculating K is proposed by Bouwer and Rice (1964), in which the curved relations between H and t are replaced with straight lines (Fig. 29–16) and Eq. [10] is modified to a simpler equation

$$K = 2R_{sp}^2\,\Delta t/(FR_i t_1^2) \qquad [11]$$

where Δt is the difference between the time required for the water level

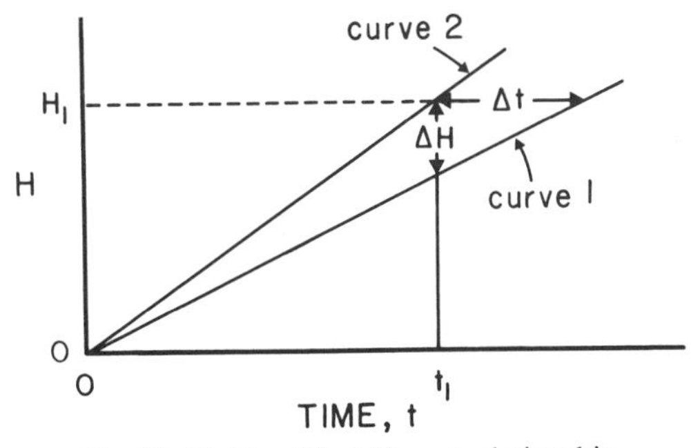

Fig. 29–16. Simplified H vs. t relationship.

in step *9* to reach H_1 (i.e., t_1) and for the water level in step *7* to reach H_1. Other parameters are the same as in Eq. [10]. For more information see Bouwer and Rice (1964).

Due to the soil disturbance around the inner tube, the diameter of the outer tube must be at least twice that of the inner tube (Bouwer & Rice, 1967). An inner tube with a diameter of <10 cm is not recommended for this procedure. Combinations of 10 and 20 or 12.5 and 25 cm for the inner and outer tube diameters have proved to be satisfactory.

The hydraulic conductivity value obtained by this method is affected by both vertical and horizontal conductivity of the soil. However, it is closer to the vertical direction conductivity for anisotropic soils. For more information see Bouma and Hole (1971).

Depending on the permeability of the soil, the double-tube method requires 2 to 6 h for completion. The method requires over 200 L of water for each site and is not suitable for rocky soils.

29–3.2 Shallow Well Pump-in Method

29–3.2.1 PRINCIPLES

To measure the saturated hydraulic conductivity by the shallow well pump-in technique (also referred to as constant head well permeameter or dry auger hole method), a hole is bored to the desired depth and a constant head of water is maintained in the hole (Fig. 29–17). When the water flow into the soil reaches a constant value, i.e., a steady-state condition, the flow is measured while the level is kept constant.

For an impermeable layer located at relatively great depth where $s > 2H$, the equation for K is due to Glover (Zangar, 1953, Eq. [8B]):

$$K = Q\,[\sinh^{-1}(H/r) - (r^2/H^2 + 1)^{1/2} + r/H]/(2\pi H^2)\,. \qquad [12]$$

When $H >> r$, Eq. [12] can be simplified to (Zangar 1953, Eq. [10B])

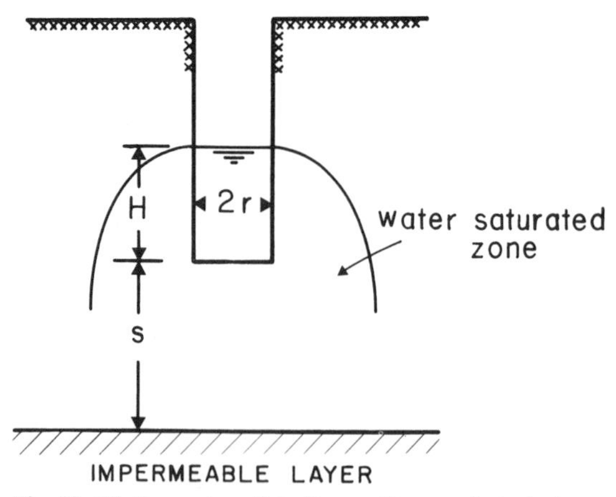

Fig. 29–17. Geometry of shallow well pump-in technique.

$$K = Q\,[\sinh^{-1}(H/r) - 1]/(2\pi H^2) \quad [13]$$

or equivalently to

$$K = Q\{\ln[(H/r) + (H^2/r^2 + 1)^{1/2}] - 1\}/(2\pi H^2) \quad [14]$$

where Q is the flow rate at equilibrium (units L^3/T), H is the depth of water in the auger hole (units L), r is the radius of the hole (units L), and s is the distance between the impermeable layer and the bottom of the hole (units L). For conditions where $0 < s < 2H$, the resulting equation is:

$$K = 3Q\,\ln(H/r)/[\pi H(3H + 2s)]\,. \quad [15]$$

Note that in shallow well pump-in, the auger hole should not penetrate any impermeable layer.

To prevent caving, a perforated pipe or commercial well screen may be used as casing. If casing materials are not available, the hole may be filled with coarse sand. When a casing pipe or screen is used, a layer of sand, gravel, or burlap must be placed at the bottom of the hole to prevent erosion caused by water application.

29–3.2.2 SPECIAL APPARATUS

1. Auger (use a bucket auger), with diameter on the order of 5 to 15 cm.
2. Constant head system (see Fig. 29–18) consists of
   a. A 200-L tank (a 54-gallon drum is suitable)
   b. A 250-ml flask.
   c. Tall cylinder, 1 m long, 10-cm diameter.
   d. Glass or metal tubing.
   e. Plastic tubing.
   f. Four rubber stoppers for cylinder, flask, and hole into drum.
3. Water supply tank with 1- to 2-m^3 (1000–2000 L) capacity.
4. Garden hose for rapid filling of the constant head reservoir.
5. Casing (perforated) pipe or commercial well screen, if available.
6. Sand or gravel, to be used when casing pipe or screen is not available.
7. Sand, gravel, or burlap to be placed on the bottom of the hole where casing is used.
8. Large round brush to clean the wall of the auger hole. As an alternative, narrow pieces of flat brush can be assembled around a stick.
9. Measuring tape or meter stick.
10. Watch or timer.
11. Data sheet, such as the one shown in Fig. 29–19.

29–3.2.3 PROCEDURE

1. Clean plant material, trash, and loose soil from the surface.
2. Bore a hole to the desired depth with minimum disturbance to the

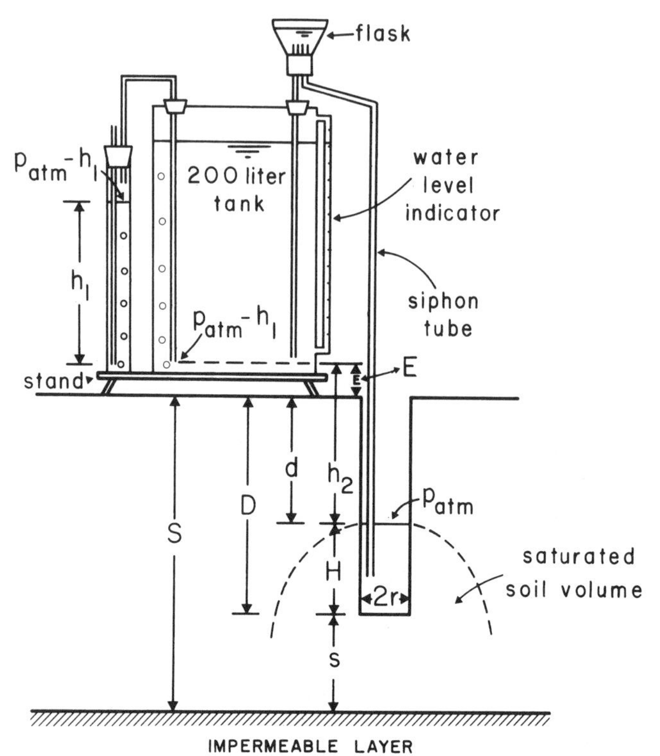

Fig. 29–18. Diagram of the constant head device and geometry of the shallow well pump-in set up.

wall of the hole. Check and record the texture of the profile as you bore out the hole. Make sure that you do not penetrate any impermeable layer.

3. After the hole is dug to the desired depth, brush the side of hole to remove any possible sealing and compaction caused by the auger. Use a metal brush with short and separate bristles for fine textures.
4. Measure the dimensions of the hole (r and D) and determine (or estimate) the depth of impermeable layer from the bottom of the hole s. Also determine the depth of water to be maintained in the hole (H).
5. Case the hole with perforated pipe or screen. The perforation should extend from the bottom of the hole to the predetermined water level. If coarse sand or gravel is used in place of casing, fill the hole to within 15 cm of the predetermined controlled water level.
6. Set up the constant head system as shown in Fig. 29–18. At equilibrium, the constant head device maintains the water level in the hole so that $h_l = h_2$. When the water level in the hole drops, a pressure gradient is created which causes the water to move to the hole through the siphon tube while air enters the water reservoir from the tall cylinder to replace the water. As a result, the air pressure in the tall

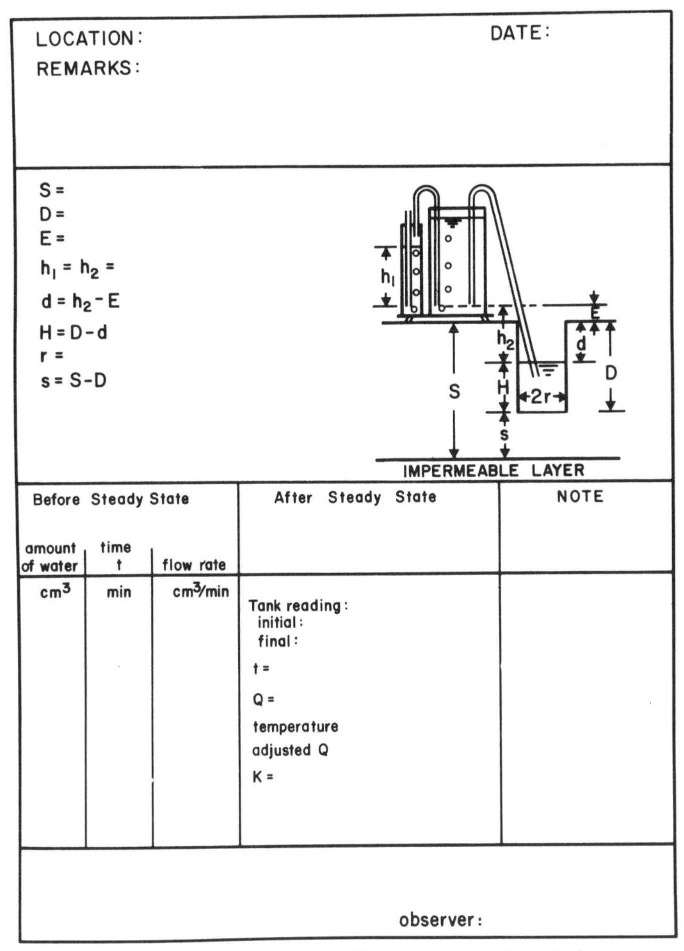

LOCATION: DATE:

REMARKS:

S =

D =

E =

$h_1 = h_2 =$

$d = h_2 - E$

$H = D - d$

r =

s = S - D

Before Steady State			After Steady State	NOTE
amount of water	time t	flow rate		
cm^3	min	cm^3/min	Tank reading: initial: final: t = Q = temperature adjusted Q K =	

observer:

Fig. 29–19. Data sheet for shallow well pump-in technique.

cylinder decreases, causing the air to enter the chamber through the regulator pipe. The difference between the air pressure inside the tall cylinder and atmospheric pressure corresponds to the length h_1. The air pressure head at the tip of the connecting tube inside the reservoir is therefore equal to $(P_{atm}/\rho g) - h_1$, which in turn must be equal to $(P_{atm}/\rho g) - h_2$. This maintains the water level inside the auger hole at a distance $h_2 = h_1$ below the tip of the connecting tube in the reservoir. A 250-mL flask filled with water will trap any air bubbles formed in the siphon system. The siphon system may be replaced by a spigot attached to the lower part of the reservoir and connected to a plastic tube extended below the water level in the hole. (As an alternative to the constant head system, a small constant-level float-valve can be used. The float-valve can be lowered in the hole, secured at the predetermined depth by a rod, and connected to the water reservoir by plastic or rubber tubing.)

7. Adjust the level regulator tube in the constant-head system to maintain water at the predetermined level (or install the float-valve at the desired depth).
8. Fill the hole to the water level and start the constant-head system to keep the water level steady.
9. Check the water reservoir tank regularly and refill the tank as needed. Check the water temperature and record the data.
10. Record the time and the amount of water moving from the tank regularly. Time intervals should be short enough so that the tank never runs out of water.
11. Compute the flow rate for each measurement. The flow rate can be corrected for any desired temperature by multiplying the rate by the ratio of water viscosity at the temperature during the measurement and the desired temperature (cf. viscosity for water at 10, 20, and 30°C are 1.3077, 1.002, and 0.7975 centipoise, respectively). A reference temperature may be chosen as the mean annual soil temperature at the depth of interest.
12. If the data for consecutive measurements are not consistent, repeat steps *8* to *11*. If the flow rate has become constant (over a 24-h period or for at least three consecutive measurements each for a few hours), proceed with calculations.

29–3.2.4 CALCULATIONS

Compare the depth of the impermeable layer from the bottom of the hole s and depth of water in the hole H, and select the appropriate equation (Eq. [12] to [14] or [15]) to calculate the hydraulic conductivity.

29–3.2.5 COMMENTS

The hydraulic conductivity determined by this procedure can be taken as an average hydraulic conductivity for the full depth of the hole being tested. However, in reality, the calculated K reflects the conductivity of the most permeable layers. If the soil is uniform, the measured value is dominated by the horizontal conductivity.

The volume of soil effective in testing is about the same as for the auger hole method [i.e., about 10 Hr^2 to 40 Hr^2 (units L^3)]. The hydraulic conductivity, however, might not be the same as by the auger hole method. In comparing the two methods, the measured values differed by as much as 100% (Boersma, 1965b).

The disadvantages of the method are the requirement of large quantities of water and a considerable amount of equipment, and the long duration of testing. The construction of the auger hole and installation of the equipment will take several hours, and the test itself might require several days. Water with chemical composition comparable to the natural soil or irrigation water is desirable.

Reynolds et al. (1983) presented new analytical and numerical solutions for the shallow well pump-in technique which result in considerably higher calculated saturated conductivity values. Compared to Eq. [14], their numerical solution produces increases in the saturated conductivity values of 68 and 65% for H/r ratios of 5 and 10, respectively. More recently Reynolds et al. (1985) and Philip (1985) have extended the theory of water flow from the auger hole to include the effects of unsaturated flow on the measured saturated conductivity.

In addition to the theory, Reynolds et al. (1983) presented a constant head apparatus (called a Guelph Permeameter, which is commercially available from Soilmoisture Equip. Corp., P. O. Box 30025, Santa Barbara, CA 93105) for maintaining a constant water level in a small auger hole. With their apparatus the conductivity can be measured in a small well of approximately 2 cm radius with a 10 to 20 cm depth of water in the hole (e.g., H/r ratios of 5–10). Such an auger hole can be constructed by a small screw soil auger or by a 4-cm diameter soil probe equipped with an inward-beveled cutting tip instead of the usual outward-beveled cutting tip. According to them, the measurement time and the amount of water needed for conductivity determination depend on the soil texture, initial water content of the soil, and H/r ratio. In their study, conductivity measurements on sandy loam soils near field capacity required < 0.5L of water and were completed in 5 min. Sandy soils below field capacity and clayey soils required measurement time of 15 to 30 min and up to 2 L of water. For detailed procedures with example calculations see Reynolds and Elrick (1986). Talsma and Hallam (1980) also used a simplified well permeameter assembly to determine K in an auger hole 6.4 cm diameter using 2 to 3 L of water. They reported that steady state flow was achieved after 20 min for initially dry soil, and after 8 min for moist soil.

The procedure used in the shallow well pump-in technique is similar to percolation tests used for septic tank suitability. However, in percolation tests, the rate of water flow into the soil (after saturation is achieved) is determined by measuring the rate of fall of water level in the hole rather than maintaining a constant head. For a more complete discussion of the percolation test, see Barbarick (1975) or Barbarick et al. (1976).

29–3.3 Other Methods

29–3.3.1 PERMEAMETER (OR CYLINDRICAL PERMEAMETER) METHOD

A large-diameter hole is dug to the desired depth and a cylinder (such as infiltrometer ring) $>$ 35 cm long, with a diameter of 45 to 50 cm, is placed in the center of the hole. The cylinder is driven about 15 cm into the soil with minimum disturbance to the natural profile. Four tensiometers are placed around the cylinder in a symmetrical fashion 10 cm from the cylinder and about 23 cm below the level surface inside the hole. Water is applied to the hole and inside the cylinder to a depth of about 15 cm and the tensiometers are monitored. When the tensiometers read zero, it is assumed that saturation is achieved and the rate of water flow into the soil through the cylinder is measured. Before a positive pressure builds up (i.e., tensiometers show < zero tension) the measurement is terminated and the conductivity is calculated using Darcy's equation.

The procedure is time-consuming and requires in excess of 100 L of water. It measures the conductivity in the vertical direction and is not suitable for rocky soils. For more information see Winger (1960) or Boersma (1965b).

29–3.3.2 INFILTRATION-GRADIENT METHOD

Bouwer (1964) proposed the infiltration-gradient technique to measure the vertical hydraulic conductivity. It is a modification of the cylinder-permeameter and double-tube method. Two concentric tubes are placed in an auger hole with small, fast-reacting piezometer tubes placed at different depth inside the inner cylinder. The piezometer tubes provide a complete vertical hydraulic gradient for the system when equal water depths, ranging from 20 to over 200 cm, is kept within the two cylinders. The method determines the *K* in the vertical direction. It takes about 3 h to complete, requires 100 L of water, and is not suitable for stony soils. For more information regarding the procedure and equipment see Bouwer (1978) and Bouwer and Jackson (1974). Both vertical and horizontal conductivities can be measured by combining the infiltration gradient technique with the double-tube method in the same hole (see Bouwer, 1964).

29–3.3.3 AIR-ENTRY PERMEAMETER METHOD

This method is a fast technique to determine the hydraulic conductivity *K* above a water table, with about 10 L of water. The method, as developed by Bouwer (1966), yields *K* from Darcy's law using the infiltration rate under a high head and gradient without using a tensiometer and/or piezometer installed in the soil. Fig. 29–20 shows the apparatus to measure the air entry value and hydraulic conductivity at the soil surface or at the bottom of a pit.

A cylinder, 20 to 30 cm in diameter and over 10 cm long, is driven about 10 cm into the soil with a minimum disturbance to the soil. A layer of sand is placed inside the cylinder with a disk on top to dissipate the energy and prevent the puddling effect of water application. The top-plate assembly is then secured and water is applied to the ring through the reservoir keeping the gage valve closed. The air escape valve is kept open to allow the air to escape and is shut when all the air is driven out of the system.

The infiltration rate is calculated by stopping the water application to the supply tank when the wetting front reaches the lower part of the cylinder. (The time for the wetting front to reach the lower part of the cylinder is estimated by a few trials before the actual procedure starts.) After the infiltration rate is measured, the supply valve is closed, halting

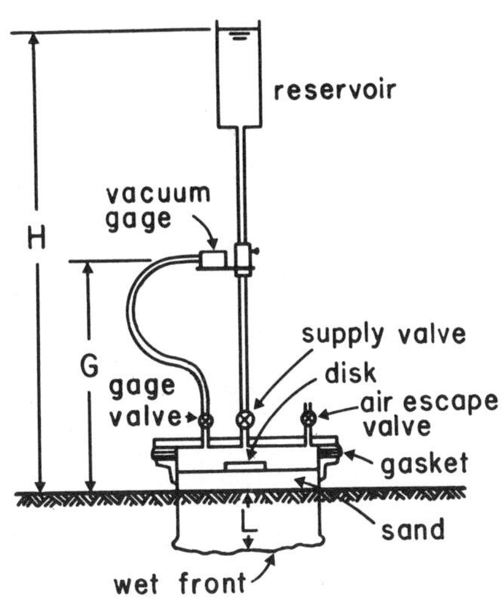

Fig. 29–20. Diagram of the equipment for the air-entry permeameter technique (after Bouwer, 1966).

the wetting front movement, and the gage valve is opened to measure the pressure inside the cylinder. The pressure inside the cylinder decreases to a minimum at which air begins to bubble up through the soil. As soon as the minimum pressure is reached, the equipment is removed and the depth of the wet front is determined by digging. The air entry value $P_a/\rho g$ (units L) is evaluated from the minimum pressure and the conductivity is calculated by

$$K = L(\Delta H/\Delta t)(R/R_c)^2/[H + L - (P_a/2\rho g)] \qquad [16]$$

where L is depth of the wet front when the supply valve was closed; H is the height of the water level above the soil; $\Delta H/\Delta t$ is the rate of fall of water level in the reservoir just before the supply valve was closed; and R and R_c are the radius of the reservoir and cylinder, respectively.

The conductivity calculated by this method is for the wetted zone, and according to Bouwer (1966), the saturated conductivity may be estimated as double the measured K. For heavy soils, however, measured K needs to be multiplied by a factor of 4. For more information see Bouwer (1966) and Bouwer and Jackson (1974).

Aldabagh and Beer (1971) evaluated the air entry permeameter technique by measuring the saturated conductivity at three different depths. They selected 30 sampling locations within a 150- by 60-m rectangular field. Their results indicate that the air-entry permeameter is a fast and reliable method for measuring K above a water table. The low manpower, low water-requirement and short time needed to make the measurement are among the advantages listed by them.

A modified version of the air-entry permeameter technique has been

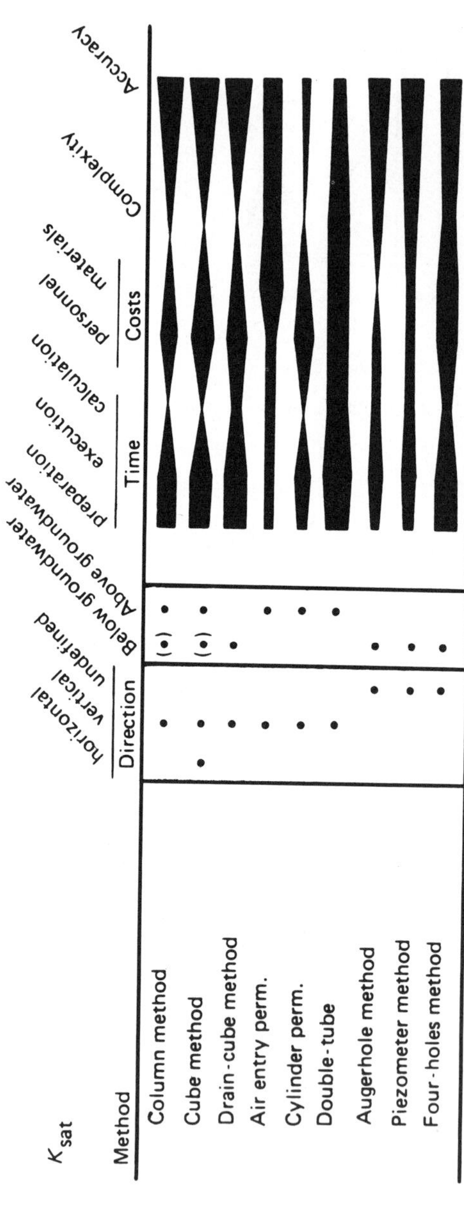

Fig. 29–21. Operational aspects of nine methods for measuring saturated hydraulic conductivity (K_{sat}) (after Bouma, 1983.).

presented by Topp and Binns (1976). In their procedure, the position of the wetting front can be detected by a fine tensiometer probe. These investigators also recommend the procedure for its reliability, low water requirement, and speed. McKeague et al. (1982) used the procedure to compare measured saturated conductivity with estimated values from soil morphology. They discussed some of the problems associated with using the procedure in the presence of biopores and cracks (i.e., macropores) in the soil.

29–4 COMMENTS

One of the important considerations in determining the saturated hydraulic conductivity in the field is the spatial variability of the soil hydraulic conductivity (Nielsen et al., 1973; Warrick & Nielsen, 1980). The conductivity values obtained from two locations only a few meters apart could be drastically different due to soil heterogeneity.

The presence of macropores can lead to unrealistic values of conductivity in the field (Bouma, 1983; Beven & Germann, 1982). Undoubtedly, even a few large pores will allow water to move at rates faster than the saturated conductivity of surrounding soil body. The amount of macropores and their contribution to the saturated conductivity of the soil depend on structure. Bouma et al. (1982), for example, described infiltration into a soil with vertical worm-channels. They state that on the average each channel occurred in a 200-cm^2 cross-sectional area. Therefore they suggested that, ideally, the area of infiltration in the infiltrometer should be a multiple of 200 cm^2 and should contain a corresponding number of channels for that particular site.

To obtain an unbiased estimate of the soil hydraulic conductivity, an optimal soil sample is needed for analysis. The sampling volume should contain a proportional amount of macropores compared to the main soil body under consideration; i.e., a minimum volume of soil is needed such that the variability of the macropores does not change with increasing sample volume. Such a sample size is referred to as the representative elementary volume (REV) (see Fig. 4 of Beven & Germann, 1982). Bouma (1983) presented REV values for four hypothetical classes of soil texture and structure. His reported values range from 10^2 cm^3 for a sandy soil with no peds to 10^5 cm^3 for a clayey soil with large peds and continuous macropores.

Care should be taken in selecting the appropriate number of replications and the site of each conductivity measurement within a field. Bouma (1983) evaluated the auger-hole method as well as other procedures, with respect to time required, cost, complexity, and accuracy of the measurements. His results are presented as Fig. 29–21, where the width of the band indicates favorability of the parameters on top. For example, the ease of preparation of the auger-hole method is indicated by a narrow band, whereas the difficulty of the double-tube method is shown as a wide band.

29-5 REFERENCES

Aldabagh, A. S. Y., and C. E. Beer. 1971. Field measurement of hydraulic conductivity above a water table with air-entry permeameter. Trans. Am. Soc. Agric. Eng. 14:29–31.

Barbarick, K. A., 1975. Percolation tests for septic tank suitability of typical southern Arizona soils. M.S. Thesis, Univ. of Ariz., Tucson.

Barbarick, K. A., A. W. Warrick, and D. F. Post. 1976. Percolation tests for septic tank suitability in southern Arizona soils. J. Soil Water Conserv. 31:110–112.

Beven, K., and P. Germann. 1982. Macropores and water flow in soils. Water Resour. Res. 18:1311–1325.

Boast, C. W., and D. Kirkham. 1971. Auger hole seepage theory. Soil Sci. Soc. Am. Proc. 35:365–373.

Boast, C. W., and R. G. Langebartel. 1984. Shape factors for seepage into pits. Soil Sci. Soc. Am. J. 48:10–15.

Boersma, L. 1965a. Field measurement of hydraulic conductivity below a water table. *In* C. A. Black et al. (ed.) Methods of soil analysis, part 1. Agronomy 9:222–233.

Boersma, L. 1965b. Field measurement of hydraulic conductivity above a water table. *In* C. A. Black et al. (ed.) Methods of soil analysis, part 1. Agronomy 9:234–252.

Bouma, J. 1983. Use of soil survey data to select measurement techniques for hydraulic conductivity. Agric. Water Manage. 6:177–190.

Bouma, J., C. F. M. Belmans, and L. W. Dekker. 1982. Water infiltration and redistribution in a silt loam soil with vertical worm-channels. Soil Sci. Soc. Am. J. 46:917–921.

Bouma, J., and L. W. Dekker. 1981. A method of measuring the vertical and horizontal K_{sat} of clay soils with macropores. Soil Sci. Soc. Am. J. 45:662–663.

Bouma, J., L. W. Dekker, and J. C. F. M. Haans. 1979a. Drainability of some Dutch clay soils: a case study of soil survey interpretation. Geoderma 22:193–203.

Bouma, J., L. W. Dekker, and H. L. Verlinden. 1976. Drainage and vertical hydraulic conductivity of some Dutch "knik" clay soils. Agric. Water Manage. 1:67–78.

Bouma, J., and F. D. Hole. 1971. Soil structure and hydraulic conductivity of adjacent virgin and cultivated pedons at two sites: A typic Arguidoll (silt loam) and a typic Entrochrept (clay). Soil Sci. Soc. Am. Proc. 35:316–319.

Bouma, J., A. Jongerius, and D. Schoonderbeck. 1979b. Calculation of saturated hydraulic conductivity of some pedal clay soils using micromorphic data. Soil Sci. Soc. Am. J. 43:261–264.

Bouma, J., J. W. Van Hoorn, and G. H. Stoffelsen. 1981. Measuring the hydraulic conductivity of soil adjacent to tile drains in a heavy clay soil in The Netherlands. J. Hydrol. 50:371–381.

Bouwer, H. 1961. A double tube method for measuring hydraulic conductivity of soil in situ above a water table. Soil Sci. Soc. Am. Proc. 25:334–339.

Bouwer, H. 1962. Field determination of hydraulic conductivity above a water table with the double-tube method. Soil Sci. Soc. Am. Proc. 26:330–335.

Bouwer, H. 1964. Measuring horizontal and vertical hydraulic conductivity of soil with the double-tube method. Soil Sci. Soc. Am. Proc. 28:19–23.

Bouwer, H. 1966. Rapid field measurement of air entry value and hydraulic conductivity of soil as significant parameters in flow system analysis. Water Resour. Res. 2:729–738.

Bouwer, H. 1978. Groundwater hydrology. McGraw-Hill Book Co., New York.

Bouwer, H., and R. D. Jackson. 1974. Determining soil properties. *In* J. van Schilfgaarde (ed.) Drainage for agriculture. Agronomy 17:611–672.

Bouwer, H. and R. C. Rice. 1964. Simplified procedure for calculation of hydraulic conductivity with the double-tube method. Soil Sci. Soc. Am. Proc. 28:133–134.

Bouwer, H., and R. C. Rice. 1967. Modified tube diameter for the double-tube apparatus. Soil Sci. Soc. Am. Proc. 31:437–439.

Bouwer H., and R. C. Rice. 1983. The pit bailing method for hydraulic conductivity measurement of isotropic or anisotropic soil. Trans Am. Soc. Agric. Eng. 26:1435–1439.

Childs, E. C., 1952. The measurement of the hydraulic permeability of saturated soil in situ. I. Principles of a proposed method. Proc. R. Soc. London, A. 215:525–535.

Childs, E. C., A. H. Cole, and D. H. Edwards. 1953. The measurement of the hydraulic permeability of saturated soil in situ. II. Proc. R. Soc. London, 216:72–89.

Freeze, R. A., and J. A. Cherry. 1979. Groundwater. Prentice-Hall, Englewood Cliffs, NJ.

Frevert, R. K., and D. Kirkham. 1948. A field method for measuring the permeability of soil below the water table. Highw. Res. Board. Proc. 28:433–442.

Healy, K. A., and R. Laak. 1973. Sanitary seepage fields—site evaluation and design. Civil Eng. Dep., Univ. of Connecticut, Storrs, CT.

Hoffman, G. J., and G. O. Schwab. 1964. Tile spacing prediction based on drain outflow. Trans. Am. Soc. Agric. Eng. 7:444–447.

Johnson, H. P., R. K. Frevert, and D. D. Evans. 1952. Simplified procedures for measurement and computation of soil permeability below the water table. Agric. Eng. 33:283–286.

King, J. J., and D. P. Franzmeier. 1981. Estimation of saturated hydraulic conductivity from soil morphological and genetic information. Soil Sci. Soc. Am. J. 45:1153–1156.

Kirkham, D., 1946. Proposed method for field measurement of permeability of soil below the water table. Soil Sci. Soc. Am. Proc. 10:58–68.

Kirkham, D. 1954. Measurement of the hydraulic conductivity in place. Symp. on permeability of soil. Am. Soc. Test Mater. Spec. Pub. 163:80–97.

Kirkham, D. 1958. Theory of seepage into an auger hole above an impermeable layer. Soil Sci. Soc. Am. Proc. 18:204–208.

Kirkham, D. 1965. Saturated conductivity as a characterizer of soil for drainage design. p. 24–31. *In* Drainage for efficient crop production Conf., Chicago, IL. 6–7 December. American Society of Agricultural Engineers, St. Joseph, MI.

Kirkham, D., and C. H. M. van Bavel. 1948. Theory of seepage into auger hole. Soil Sci. Soc. Am. Proc. 13:75–82.

Luthin, J. N. 1949. A reel-type electric probe for measuring water table elevations. Agron. J. 41:584.

Luthin, J. N. 1957. Measurement of hydraulic conductivity in situ. *In* J. N. Luthin (ed.) Drainage of agricultural lands. Agronomy 7:420–431.

Luthin, J. H., and D. Kirkham. 1949. A piezometer method for measuring permeability of soil in situ below a water table. Soil Sci. 68:349–358.

Maasland, M. 1955. Measurement of hydraulic conductivity by the auger hole method in anisotropic soil. Soil Sci. 81:379–388.

Maasland, M. 1957. Soil anisotrophy and land drainage. *In* J. H. Luthin (ed.) Drainage of agricultural lands. Agronomy 7:216–287.

Maasland, M., and H. C. Haskew. 1957. The auger hole method of measuring the hydraulic conductivity of soil and its application to the tile drainage problems. p. 8.69–8.114. *In* Trans. Int. Congr. Irrig. Drain. 3rd, New Delhi, India.

McKeague, J. A., C. Wang, and G. C. Topp. 1982. Estimating saturated hydraulic conductivity from soil morphology. Soil Sci. Soc. Am. J. 46:1239–1244.

Nielsen, D. R., J. W. Biggar, and K. T. Erh. 1973. Spatial variability of field measured soil-water properties. Hilgardia 42:215–259.

Philip, J. R. 1985. Approximate analysis of the bore hole permeameter in unsaturated soil. Water Resour. Res. 21:1025–1033.

Reynolds, W. D., and D. E. Elrick. 1986. A method for simultaneous in situ measurement in the vadose zone of field saturated hydraulic conductivity, sorptivity and the conductivity-pressure head relationship. Ground Water Monit. Rev. 6:84–95.

Reynolds, W. D., D. E. Elrick, and B. E. Clothier. 1985. The constant head well permeameter: effect of unsaturated flow. Soil Sci. 139:172–180.

Reynolds, W. D., D. E. Elrick, and G. C. Topp. 1983. A reexamination of the constant head well permeameter method for measuring saturated hydraulic conductivity above the water table. Soil Sci. 136:250–268.

Smiles, D. E., and E. G. Youngs. 1963. A multiple-well method for determining the hydraulic conductivity of a saturated soil in situ. J. Hydrol. 1:279–287.

Snell, A. W., and J. van Schilfgaarde. 1964. Four-well method of measuring hydraulic conductivity in saturated soils. Trans. Am. Soc. Agric. Eng. 7:83–87, 91.

Stibbe, E., T. J. Thiel, and G. S. Taylor. 1970. Soil hydraulic conductivity measurement by field monoliths. Soil Sci. Soc. Am. Proc. 34:952–954.

Talsma, T., and P. M. Hallam. 1980. Hydraulic conductivity measurement of forest catchments. Aust. J. Soil Res. 30:139–148.

Topp, G. C., and M. R. Binns. 1976. Field measurement of hydraulic conductivity with a modified air-entry permeameter. Can. J. Soil Sci. 56:139–147.

Topp, G. C., and S. Sattlecker. 1983. A rapid measurement of horizontal and vertical components of saturated hydraulic conductivity. Can. Agric. Eng. 25:193–197.

Van Bavel, C. H. M., and D. Kirkham. 1948. Field measurement of soil permeability using auger holes. Soil Sci. Soc. Am. Proc. 13:90–96.

van Beers, W. F. J. 1958. The auger hole method. Int. Inst. Land Reclam. and Improve. Bull. no. 1. Wageningen, The Netherlands.

Warrick, A. W., and D. R. Nielsen. 1980. Spatial variability of soil physical properties in the field. p. 319–344. *In* D. Hillel (ed.) Application of soil physics. Academic Press, New York.

Winger, R. J., Jr. 1960. In-place permeability tests and their use in subsurface drainage. p. 11.417–11.469. *In* Trans. Int. Congr. Comm. Irrig. Drain., 4th, Madrid, Spain.

Youngs, E. G. 1968. Shape factors for Kirkham's piezometer method for determining the hydraulic conductivity of soil in situ for soils overlaying an impermeable floor or infinitely permeable stratum. Soil Sci. 106:235–237.

Zangar, C. N. 1953. Theory and problems of water percolation. Engineering Monogr. No. 8, Bur. of Reclamation, U. S. Dep. of Interior.

30 Hydraulic Conductivity, Diffusivity, and Sorptivity of Unsaturated Soils: Field Methods

R. E. GREEN

University of Hawaii
Honolulu, Hawaii

L. R. AHUJA

Water Quality and Watershed Research Laboratory
Agricultural Research Service, USDA
Durant, Oklahoma

S. K. CHONG

Southern Illinois University
Carbondale, Illinois

30-1 INTRODUCTION

Application of soil water flow theory to many practical problems requires estimates or measurements of hydraulic conductivity or soil water diffusivity for unsaturated soil over the water content range of interest. In some cases, laboratory measurements (chapter 28,) on cores taken from the field, or calculation methods which utilize the water content-pressure head relationship (chaper 31 in this book) may provide useful conductivity or diffusivity data. In situ field measurements are generally preferred, however, if resources are available and the site is sufficiently accessible and physically amenable to field measurement—e.g., has reasonably level topography, is not too stony, and has predominantly vertical flow during drainage. A larger area of measurement and preservation of field structure are inherent advantages of field methods over laboratory methods.

Most of the methods described in this chapter have as their principal objective the measurement of hydraulic conductivity in relation to either water content or soil water pressure head. Diffusivity is calculated directly with some methods, but can be derived easily from measured conductivity data and the water content-pressure head relationship.

Four field methods of determining hydraulic conductivity in unsaturated soils are presented in sections 30-2 through 30-5. The unsteady drainage-flux approach (often called the instantaneous profile method, after Watson, 1966) is described in section 30–2, and is followed by a simplified version in section 30–3. Two steady-flux methods are described in sections 30–4 and 30–5; the steady flux is imposed by an artificial crust in the former method and by sprinkler irrigation in the latter.

Sorptivity, a hydraulic property of unsaturated soils which was defined by Philip (1957b) in his development of infiltration theory, is closely related to hydraulic conductivity and soil water diffusivity and is included in this chapter. Sorptivity measurements in the field under two types of water-entry conditions, slightly positive pressure (ponded) and slightly negative pressure (water entry through a porous plate), are described in sections 30–6 and 30–7, respectively. Field methods of measuring sorptivity will likely undergo continued development and testing as interest grows in the use of sorptivity data in infiltration prediction and for characterizing the spatial variability of hydraulic properties of surface soils. Sorptivity can also be obtained indirectly from other hydraulic functions (Philip, 1955; Brutsaert, 1976; Chong et al., 1982). Likewise, field-measured sorptivities can be used with supplementary measurements to calculate unsaturated hydraulic conductivity (Russo & Bresler, 1980) or diffusivity (Clothier & White, 1981). Only the direct field methods of determining sorptivity are presented in this chapter.

30–2 UNSTEADY DRAINAGE-FLUX METHOD

30–2.1 Principles

This method of determining unsaturated hydraulic conductivity and diffusivity in situ is based on Darcian analysis of transient soil water content and hydraulic head profiles during vertical drainage following a heavy irrigation or rainfall. The soil is wetted beyond the maximum depth for which the determinations are desired, and the groundwater table should be sufficiently below this depth in order to obtain maximum possible unsaturated drainage. The soil surface is generally covered to prevent evaporation, although with some modification the method can also be applied in the presence of evaporation (Arya et al., 1975). Isothermal conditions are assumed to exist in the soil profile during the course of drainage, neglecting the effect of any temperature changes that might occur. The drainage-flux method was first used in the field by Richards et al. (1956). It was developed further by Nielsen et al. (1964), Rose et al. (1965) and van Bavel et al. (1968). Watson (1966) improved upon the analysis of data by replacing the computation of differences in time and depth by the presumably more accurate instantaneous profile method. Since then, the method has been used by many other investigators to determine hydraulic conductivity of well-drained soils.

The equation describing one-dimensional, isothermal, nonhysteretic, unsaturated flow of water during drainage is

$$\partial\theta(z, t)/\partial t = \partial/\partial z \{K(\theta) [\partial H(z, t)/\partial z]\} \quad [1]$$

where $\theta(z, t)$ is the transient volumetric soil water content, $H(z, t)$ is the hydraulic head, $K(\theta)$ is the unsaturated hydraulic conductivity, z is the vertical distance coordinate, and t is the time. If z is measured positive downward with respect to the reference level (e.g., the soil surface), $H(z, t)$ at any point in the soil profile is given by

$$H(z, t) = h(z, t) - z \quad [2]$$

where $h(z, t)$ is the transient soil water pressure head. The initial condition for Eq. [1] is the soil moisture profile at the moment infiltration of water at the soil surface ceases. With the surface covered thereafter to prevent evaporation, the upper boundary condition for drainage is zero flux at $z = 0$. With this condition, Eq. [1] is integrated with respect to z, between the limits of $z = 0$ and any desired depth, say $z = z_1$, to obtain for a given time

$$\int_0^{z_1} \frac{\partial\theta(z,t)}{\partial t} dz = K(\theta) \frac{\partial H(z,t)}{\partial z}\bigg|_{z_1} \quad [3a]$$

or

$$\frac{\partial}{\partial t} \int_0^{z_1} \theta(z,t) dz = K(\theta) \frac{\partial H(z,t)}{\partial z}\bigg|_{z_1} . \quad [3b]$$

Equation [3b] is used to determine $K(\theta)$ at desired values of z_1 from analysis of θ and H profiles measured at frequent time intervals. The left- and right-hand sides of Eq. [3b] are evaluated at a given time.

The hydraulic head profiles for different times are measured with tensiometers installed to different depths. The measurements are thus restricted to a pressure head range of about 0.0 to -8.5 m of water (pressure range of 0.0 to -85 kPa), but this range includes pressures associated with significant liquid water movement. For measuring pressure heads less than -8.5 m of water, methods given in chapter 25 may be employed. The soil water content profiles should be measured in close proximity to the tensiometers. A neutron moisture probe provides a convenient way to do that. A field gamma probe may also be used for this purpose. An alternative approach is not to measure the water content profiles directly, but to obtain them from soil water pressure head profiles derived from tensiometers and water content-pressure relationships for different depths measured separately in the laboratory. Measurement of these relationships requires undisturbed soil cores from different soil horizons at the experimental site and the procedures of chapter 26.

The unsaturated soil water diffusivity, $D(\theta)$, for any soil depth and θ may best be determined by a method based on its definition

$$D(\theta) = K(\theta)\, dh/d\theta\,. \quad [4]$$

Use of this equation requires that the soil water content-pressure head relationship, $\theta(h)$, be established first for each soil depth. This is accomplished by synchronizing the data from in situ measurements of soil water content and pressure head profiles described above and fitting a curve through $\theta\ (h)$ data points for each depth. When the water content profiles are not measured directly in the field, the $\theta\ (h)$ relationships measured on soil cores that are used to obtain $\theta\ (z, t)$ from pressure head profiles $[h(z, t)]$ are also used to determine $dh/d\theta$ for Eq. [4].

When evaporation from the soil surface is not prevented during drainage measurements, the above procedures of determining $K(\theta)$ and $D(\theta)$ are applied separately to data from above and below a transient "plane of zero flux," which is taken as the reference limit in integrating Eq. [1] to obtain two modifications of Eq. [3] (Arya et al., 1975). This procedure has been found to be very useful for slow-draining soils.

30–2.2 Method

The equipment and procedure described in the following are for the commonly used field-plot method, with in situ measurement of soil water content profiles with a neutron probe. Measurement of steady-state infiltration rate before the start of the drainage process is included. A suggested layout of the field plot and location of the neutron-probe access-tube and tensiometers are shown in Fig. 30–1. This layout and the dimensions can be modified within reasonable limits depending upon the

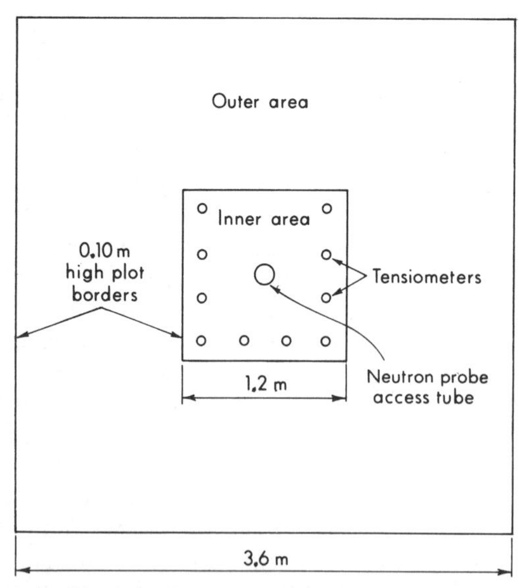

Fig. 30–1. Planar view of field plot for measurement of unsaturated hydraulic conductivity and diffusivity by the unsteady drainage-flux method.

field conditions. Some of the equipment described below can also be substituted with other suitable devices. An alternative to such a field plot would be a double-ring infiltrometer with multiple-depth tensiometers (Ahuja et al., 1976, 1980) and a neutron probe access tube, as shown in Fig. 30–2. In this set-up, the inner multiple-depth tensiometer, located at the vertical axis of the axisymmetric flow system, is used to obtain vertical hydraulic gradients, while the outer multiple-depth unit serves as a check on the inner tensiometer, and along with the adjoining neutron probe measurements, provides in situ soil water content-pressure head relationships.

30–2.2.1 EQUIPMENT

1. Replicate tensiometers, equipped with mercury manometers (e.g., Model 2600-A, Soilmoisture Equipment Corp., Santa Barbara, CA 93105) or other pressure measuring devices such as pressure transducers, to be installed at selected soil depths (e.g., 0.15, 0.45, 0.75, 1.05, and 1.35 m) or by soil horizon. Single-depth tensiometers may be replaced by more sensitive multiple-depth tensiometers which provide pressure-head measurements at various depths on the same vertical axis (e.g., Model 2510-A, Soilmoisture Equipment Corp., Santa Barbara, CA 93105).
2. A calibrated neutron moisture gauge (e.g., Model 3330, Troxler Elec-

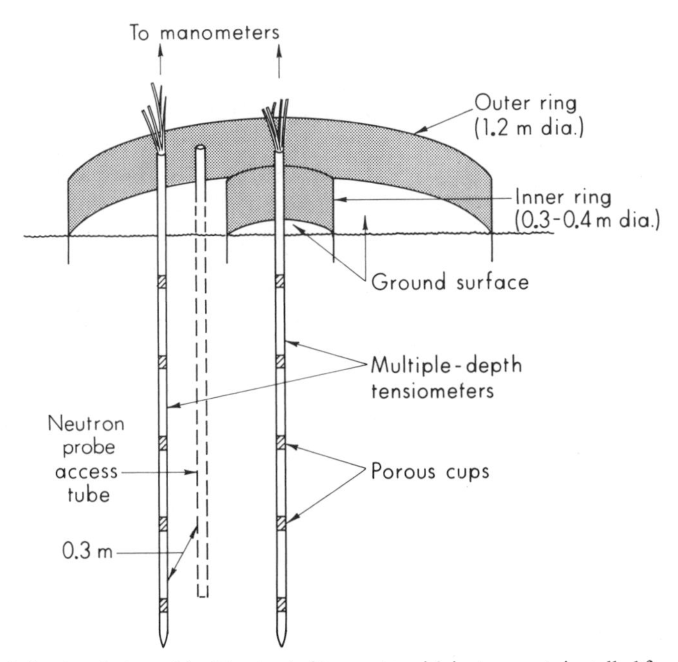

Fig. 30–2. Sectional view of double-ring infiltrometer with instruments installed for unsteady drainage-flux method.

tronic Laboratories, Inc., Research Triangle Park, NC 27709). A separate calibration is required for the 0- to 0.15-m depth interval (van Bavel & Stirk, 1967).

3. A neutron-probe access-tube of the type (aluminum or steel) and diameter that was used in calibration of the neutron moisture gauge. The tube length should be 0.3 m longer than the maximum depth of measurement.

   If the soil water content profiles are not measured directly, then replace items 2 and 3 above by equipment to obtain fairly large undisturbed soil cores: 76 to 100 mm diameter by 76 mm long brass or aluminum rings, a hydraulic jack and frame (Chong et al., 1982b), or other such devices to drive the rings into soil smoothly with minimum disturbance of soil structure.
4. Wooden boards (approximately 0.25 m high) to enclose a 3.6 by 3.6 m plot and a 6-mil plastic sheet to seal the enclosure from the inside.
5. Metal frame, 1.2 by 1.2 by 0.2 m, made of 3-mm thick sheeting, to enclose the 1.2 by 1.2 m inner area of the plot.
6. A source of water supply and a 200-L metered water reservoir (e.g., a modified oil barrel) and two float valves to control water level in inner and outer areas of the plot.
7. A 6-mil plastic sheet and preferably also some 30-mm thick styrofoam sheets to cover the plot surface during drainage.
8. A 5- by 5-m canvas canopy or fiberglass-roofing shelter to minimize the effect of rainfall.

30–2.2.2 PROCEDURE

1. Delineate a 3.6- by 3.6-m plot on a level area, dig a narrow trench 0.15-m deep around it, and install the wooden boards on edge to project 0.1 m aboveground. Line the inside of the boards and corners with the plastic sheet. Backfill gaps with loose soil and compact to prevent leakage of water (when ponded) through the boundaries.
2. Delineate a 1.2- by 1.2-m inner area in the middle of the plot as shown in Fig. 30–1. Install the metal frame around this area, 0.1 m into the ground and 0.1 m above, using a wood board and a sledge hammer to drive the frame into the soil.
3. Install a neutron-probe access-tube at the center of the inner area and the replicate sets of tensiometers to appropriate depths around it (Fig. 30–1). Methods of installing access tubes and tensiometers with manometers are given in other chapters in this monograph.
4. Clamp float valves to the plot borders, one for the inner area and another for the outer area. Connect them to a water source with hoses, and pond and maintain water in both areas to a depth of about 50 mm.
5. Erect the canopy or shelter over the plot. Also devise a temporary raised platform, with concrete blocks and wooden boards, for work in the inner area without stepping in the plot.

6. After enough water has infiltrated to wet the soil profile to the maximum depth of measurement (as estimated by a rough calculation or a quick measurement with the neutron probe), purge all the tensiometers and manometers with deaerated water. Let the ponded-water infiltration continue until all the tensiometers have become steady.
7. Connect the hose supplying water to the inner area to the metered water tank. Record the rate of infiltration.
8. Read the tensiometers and take neutron probe data in the profile at 0.1- to 0.2-m depth intervals, starting at the 0.1-m depth.
9. Disconnect water supply to both float valves. Record the time when ponded water just disappears in each area, and thus establish zero time for the drainage process. Cover the entire plot with the sheet of plastic, overlaid with the sheets of styrofoam or a layer of dry soil.
10. Read tensiometers and measure moisture profiles with the neutron probe frequently during the first 24 h of drainage, the time intervals depending upon the soil type. Increase the time interval between readings gradually to once a day and then every other day, according to the rate of changes in pressure heads and moisture contents. Continue measurements as long as appreciable changes are occurring.

    If the neutron probe is found not to be reliable enough for measuring soil water contents in the 0- to 0.15-m depth increment (even with a separate calibration), sample this layer of soil within the inner area to determine water contents for intermediate times.
11. If the transient soil water content profiles are not measured in situ (with the neutron probe or other means), dig a pit in the inner area after all tensiometric measurements have terminated, and obtain replicate undisturbed soil cores for each soil horizon or depth interval of interest. Determine water content–pressure head relationships (water characteristics) on these cores in the laboratory, using methods given in chapter 26.

30–2.2.3 ANALYSIS OF DATA

1. Calculate hydraulic heads (H) from tensiometric data for different positions and times (see chapter 23). Plot H vs. time for each position of measurement and draw a smooth curve through the data. Read values of H from the curves at some selected times (t_1, t_2, . . .) and tabulate them. For a given time, fit a curve through each set of H-vs.-z profile data. A cubic spline curve (Erh, 1972), preferably a least-squares cubic spline curve (de Boor and Rice, 1968; Ahuja et al., 1980), gives a flexible fitting. Using the equation of the fitted curve, determine the gradients $\partial H/\partial z$ at selected positions (z_1, z_2, . . .) where $K(\theta)$ and $D(\theta)$ are to be determined; results can be tabulated in a format such as is given in Table 30–1.For convenience in later analysis, choose z_1, z_2, . . . as multiples of the depth interval of neutron probe measurements, e.g., 0.1 m. Alternatively, $\partial H/\partial z$ may be determined by a finite difference procedure (see Fluhler et al., 1976). Determine (from Eq.

Table 30–1. Suggested format for tabulating results in calculation of $K(\theta)$, $K(h)$ and $D(\theta)$ with the unsteady drainage-flux method.

1 Depth, z_i	2 Time, t_j	3 $\partial H/\partial z$	4 $\int_0^{z_i} \theta dz$	5 $\frac{d}{dt}[\int \theta dz]$	6† K	7 θ	8 h	9 $dh/d\theta$	10‡ D
z_1	t_1								
	t_2								
	t_3								
	⋮								
z_2	t_1								
	t_2								
	t_3								
	⋮								
⋮									
z_n	t_1								
	t_2								
	t_3								
	⋮								

† Col. 6 = Col. 5 ÷ Col. 3. ‡ Col. 10 = Col. 6 × Col. 9.

[2]) soil water pressure heads, h, at positions z_1, z_2, . . . from the fitted curve of $H(z)$ and record these $h(z)$ values in column 8 of Table 30–1.

2. Calculate soil water contents, $\theta(z)$, from neutron probe data. Plot θ vs. time for each position of measurement and draw a smooth curve through the data. Read values of θ from the curves at the times selected in Step 1. Using the water content profile for a given time, estimate the integral $\int\theta(z, t)dz$ of Eq. [3b] by a trapezoidal approximation for each position z_1, z_2, . . ., taking water content in the 0- to 0.1-m soil layer to be the same as measured at 0.1-m depth. For example, for data points at 0.1-m depth interval and selected position z_1

$$\int_0^{z_1} \theta(z,t)dz = 0.1\theta_1 + \sum_{i=1}^{n_1-1} 0.1(\theta_i + \theta_{i+1})/2 \quad [5]$$

where θ_i is the soil water content measured at the ith point in the profile, starting from the top, and n_1 is the number of data points down to soil depth z_1. Record $\int\theta dz$ values calculated for each value of z in column 4 of Table 30–1. Also record (in column 7) the water content values determined for depths z_1, z_2, ...

If the soil water content profiles are not measured directly, (i) obtain the soil-water pressure head profiles from curves fitted to the hydraulic head profiles in Step 1 above using Eq. [2]; (ii) obtain $h(z)$ at small depth intervals (e.g., 0.1 m), for each time of measurement such that z_1, z_2 . . . are exact multiples of those intervals; (iii) convert these $h(z)$ data points to water content values using the water content–pressure relationships for different soil depth intervals determined on soil cores in the laboratory; (iv) then use the procedure of Eq. [5] to determine the integrals $\int\theta(z, t)dz$.

3. For each of the positions z_1, z_2 . . . fixed, fit a curve through the data of $\int\theta\ (z, t)dz$ vs. time, and evaluate the derivatives $\partial[\int\theta(z, t)dz]/\partial t$ at different times. Again, a least-squares cubic spline fit is recommended for its general applicability. The time derivatives determined here (see left-hand side of Eq. [3b]) are fluxes at fixed positions and times.
4. Obtain the unsaturated hydraulic conductivity values by dividing the fluxes calculated above (column 5, Table 30–1) with the hydraulic gradients at the same positions and times (column 3, Table 30–1) calculated in Step 1. Associate these conductivity values with soil water contents and pressure heads at the corresponding positions and times (Columns 7 and 8, Table 30–1).
5. Calculate the field saturated hydraulic conductivity $[K(\theta_s)_i]$ at each position z_i by dividing the steady-state infiltration rate (q) with the hydraulic head gradient (dH/dz) prevailing at that time and depth; that is, $K(\theta_s)_i = q/(dH/dz)_i$. Relate this K value to the corresponding soil water content and pressure head measured at steady state.
6. For each position z_1, z_2, . . ., fit a curve through each h-vs.-θ data set (columns 8 and 7) and evaluate the slopes $dh/d\theta$ at each value of the set. Calculate soil water diffusivities at different water contents at each position from synchronous $K(\theta)$ and $dh/d\theta$ data using Eq. [4].

The above procedure (with tabulation in a table such as Table 30–1) can be executed as a series of hand calculations but is easily adapted to computation by a programmable calculator or computer. An alternative calculation procedure by Hillel et al. (1972) requires the same field data and should yield similar results as the above procedure when soils are relatively uniform with depth, so that approximations of differential and integral quantities are satisfactory.

30–2.2.4 COMMENTS

1. A separately enclosed inner area in the field plot, as shown in Fig. 30–1, is not necessary if the measurement of infiltration rate and hydraulic conductivities in the profile under steady, ponded conditions are not desired. Some investigators measure the steady infiltration rate in the plot as a whole. Such a measurement may not involve much lateral-flow error if the plot is large enough. However, it is best to have the steady-state rate and conductivities be representative of the same (inner) area where unsaturated conductivities are determined.
2. When the inner and outer areas are not separated, it is not necessary to maintain a constant ponded-water level in the plot for measurement of infiltration rates; the rates can be determined more easily by periodically measuring the decline in water level with a hook gauge.
3. The field plot method as described above cannot be used on a sloping area or in a forested area where underground tree roots may extend laterally into the plot. In such cases, a large vertical column of soil

can be isolated by digging a trench around it, enclosing the column in an open-ended metal box, and sealing the gaps between the column and walls of the box with bentonite slurry or cement.

4. Similarly, for soils in which the horizons differ so greatly in hydraulic conductivity that significant lateral water flow occurs during drainage, the plot area should be isolated from the surrounding area to a depth exceeding the restricting horizon. Generally, the buffer area provided in the experiment (Fig. 30–1 and 30–2) should be adequate to prevent unsaturated lateral flow from the inner area.
5. Measurement of hydraulic heads with tensiometers under ponded-water infiltration conditions can be adversely affected if there is channeling of water around tensiometer stems. Proper tensiometer installation includes boring a hole of the same diameter as the tensiometer stem and packing soil around the stem near the soil surface to prevent rapid water movement down the hole. A soil slurry can be used to improve contact between the porous cup and the soil (Baker et al., 1974).
6. The tensiometers can be monitored automatically and continuously by using pressure transducers connected to a data acquisition system (Watson, 1967). These devices reduce the lag that may otherwise occur in the tensiometric response.
7. The errors associated with hydraulic conductivity values determined by this method were evaluated by Fluhler et al. (1976). In their analysis the relative errors in $K(\theta)$ were 20 to 30% of the measured K in the wet range (where K was large) but much greater in the drier range (where K was small). Errors in K values obtained during the early stages of drainage are due primarily to errors in determining the hydraulic gradient, while at later times (small K values) errors in water content measurement are more serious.
8. With certain simplifying assumptions about functional forms for $K(\theta)$ and $\theta(h)$ and the drainage process, $K(h)$ and $D(h)$ may be determined from analysis of tensiometric data alone, without any measurement of soil water content profiles (Ahuja et al., 1980). This analysis can also simultaneously determine $\theta(h)$ with measurement of the soil water content profile at just one time during drainage. The $K(\theta)$ and $D(\theta)$ can then be obtained from $K(h)$ and $D(h)$. Zachmann et al. (1982) and Dane and Hruska (1983) have used repeated numerical solutions of the flow equations as a means to identify parameters in $K(\theta)$ and $\theta(h)$ functions. These simplified approaches should be tested further.

30–3 SIMPLIFIED UNSTEADY DRAINAGE-FLUX METHOD

30–3.1 Principles

30–3.1.1 HYDRAULIC CONDUCTIVITY

The simplified method described in this section determines $K(\theta)$ using only the periodic measurements of $\theta(z, t)$ during redistribution of water in the soil profile following ponded infiltration. In addition to the assumption of negligible lateral flow in the soil layer for which $\partial\theta/\partial t$ is being evaluated, the simplified method assumes a unit hydraulic gradient, i.e., $\partial H/\partial z = -1$, during drainage.

The assumption of unit gradient during redistribution of soil water without evaporation following infiltration in a uniform soil profile was introduced by Black et al. (1969). This assumption was used in the determination of hydraulic conductivity of unsaturated soil by Nielsen et al. (1973). With $\partial H(z, t)/\partial z = -1$ at z_1 in Eq. [3a] and with the left-hand side of Eq. [3a] estimated by $z_1(\partial\theta^*/\partial t)$, in which θ^* is the average water content in the soil profile to depth z_1 at a given time, the resulting equation relating unsaturated hydraulic conductivity at $z=z_1$ to the transient water content profile is

$$K(\theta)_{z_1} = -z_1 \, (\partial\theta^*/\partial t) \qquad [6]$$

where

$$\theta^* = \frac{1}{z_1}\int_0^{z_1}\theta(z,t)\,dz.$$

While Eq. [6] may be used directly to calculate $K(\theta)$ over discrete time intervals in which measured changes in soil water content can be used to provide an estimate of $\partial\theta^*/\partial t$, further refinements have been developed. Chong et al. (1981) used the assumption of Richards et al. (1956) and Gardner et al. (1970) that water content in the soil profile during post-infiltration redistribution diminishes with time in a manner such that

$$\theta^* = at^b \qquad [7]$$

where a and b are constants. Substitution of the derivative of Eq. [7] into Eq. [6] gives K as a function of t, i.e.,

$$K(\theta)_{z_1} = -z_1 bat^{b-1}\,. \qquad [8]$$

Solution of Eq. [7] for t explicitly and substitution in Eq. [8] yields the working equation developed by Chong et al. (1981),

$$K(\theta)_{z_1} = -z_1 ba^{1/b}\,[\theta^*]^{(b-1)/b}\,. \qquad [9]$$

The above procedure yields $K(\theta)$ as a power function, a form widely used to represent conductivity as a function of water content (Brooks & Corey, 1964; Campbell, 1974; Clapp & Hornberger, 1978). Alternative simplified field procedures based on measured values of θ vs. t for a draining soil profile assume an exponential K-θ relationship (Libardi et al., 1980); these alternatives are discussed briefly in section 30–3.2.3.

Use of a single tensiometer at the depth z_1 along with $\theta(z, t)$ measurements provides a means of obtaining K as a function of soil water pressure head, h, as well. Often, the absolute value of pressure, $|h|$ (viz. suction), can be expressed as a power function of time (Chong et al., 1981); i.e.,

$$|h| = mt^n . \quad [10]$$

Solving Eq. [10] for t and substitution in Eq. [8] gives

$$K(h)_{z_1} = -z_1 abm^{-(b-1)/n} [|h|]^{(b-1)/n} . \quad [11]$$

To apply Eq. [11] at pressures near zero, the conductivity can be assumed to be constant between $h = 0$ and $h = h_a$, the air entry pressure. The air entry pressure to use in Eq. [11] corresponding to the saturated water content, θ_s, can be obtained by solving Eq. [10] for the value of t that gives θ_s when inserted in Eq. [7].

30–3.1.2 SOIL WATER DIFFUSIVITY

The field data required to determine $K(h)$ by Eq. [11], i.e., $\theta(z, t)$ and h vs. t (the latter only at the depth of interest, z_1), are sufficient to calculate $D(\theta)$ as proposed by Gardner (1970). Substitution of Eq. [6] into Eq. [4] and use of the chain rule results in

$$D(\theta)_{z_1} = -z_1(\partial h/\partial t)_{z_1} . \quad [12]$$

Further substitution of the derivative of Eq. [10] into [12] and use of Eq. [7] to change the dependent variable from t to θ^* (as was done previously in going from Eq. [8] to [9]) results in the power function

$$D(\theta)_{z_1} = -z_1 mna^{-(n-1)/b} [\theta^*]^{(n-1)/b} . \quad [13]$$

30–3.2 Method

30–3.2.1 EQUIPMENT AND PROCEDURE

The field set-up for the simplified method of determining $K(\theta)$ is basically the same as for the detailed drainage-flux method (see section 30–2.2). The double-ring infiltrometer is preferred over the larger plot with wood borders for mobility and ease of installation. The principal difference between the detailed and simplified methods is that the simple approach requires only the measurement of $\theta(z, t)$ for the depths of interest. This can be accomplished with a neutron moisture probe, as described in section 30–2.2, or by soil sampling at various times during drainage, with subsequent gravimetric determination of water content. The soil-sampling method has the disadvantage of requiring soil bulk density values for each horizon to allow conversion of gravimetric water contents to volumetric water contents. However, soil sampling will frequently be the most practical way to obtain profile water contents over time because: (i) soil variability may require calibration of a neutron probe at each field site, and (ii) neutron-probe results for soil water near the soil surface (top 15 cm) are frequently inaccurate. Since the simplified method will likely be used mostly for measuring $K(\theta)$ in the upper zone of the soil profile (e.g., 0–0.5 m), the requirement of bulk density data is not a serious problem, as undisturbed core samples can be taken easily near the surface after all water-content sampling is completed. Bulk density data (and an assumed or measured particle density) allow calculation of porosity, which in turn can be used to estimate the field saturated value of θ as porosity $\times$ 0.85 (Chong et al., 1981).

Following infiltration, measure the soil water content profiles to depth z_1 in increments of 0.1 to 0.15 m at appropriate times as suggested in section 30–2.2.2 (item 10). If soil sampling is used, samples should be taken in duplicate, preferably within the inner ring. If the inner ring is no larger than 0.3 m in diameter, it is probable that only two sets of samples can be taken within the inner ring, and it will be necessary to take subsequent soil samples in the buffer zone near the inner ring. If the depth z_1 (to which the average water content is to be determined) is sufficiently deep to require more than two increments of soil sampled from a given sampling hole, sampling at greater depths is facilitated by use of a 0.05-m diameter bucket auger for the upper depths and a smaller-diameter screw auger for greater depths. If several depth increments are required, the bucket auger can be used to enlarge the hole to the greatest depth yet sampled with the screw auger, to allow easy sampling of the subsequent depth interval with the smaller-diameter auger.

If $K(h)$ or $D(\theta)$ is to be determined, a tensiometer must be installed at each depth z_i for which calculations are to be accomplished. Tensiometers will normally be placed near the center of the inner ring. Tensiometers equipped with manometers or pressure transducers, and preferably with rapid-flow porous ceramic tips, are required. Tensiometer readings can be taken easily and more frequently than soil water content measurements (unless a neutron soil moisture probe is used). It is not necessary that soil water pressure and water content be measured at the same times.

30–3.2.2 ANALYSIS

When soil water content (volume basis) is determined for equal depth increments, either by neutron probe or soil sampling, the mean soil water content to depth z_1 for n increments is simply

$$\theta^* = 1/n \sum_{i=1}^{n} \theta_i \,.$$

If depth increments are of unequal length, then the water content for each increment must be weighted for the increment length, i.e.,

$$\theta^* = [1/\sum_{i=1}^{n} L_i)] \sum_{i=1}^{n} \theta_i L_i ,$$

in which L_i is the length of the ith depth increment.

The parameters a and b in Eq. [7] are determined by converting Eq. [7] to the linear form by logarithmic transformation, i.e., $\log \theta^* = \log a + b \log t$, and fitting the transformed equation to the experimental data by a least-squares procedure. The same procedure is followed in determining m and n in Eq. [10] when $K(h)$ or $D(\theta)$ is to be determined.

30–3.2.3 COMMENTS

1. The simplified method of Chong et al. (1981) will generally be used to determine hydraulic functions for the upper soil horizons in soils that are reasonably homogeneous with respect to hydraulic conductivity. Determinations of $K(h)$ and $D(\theta)$ require tensiometers at each depth for which these functions are desired, and this makes the methodology more complex and perhaps less useful when numerous measurements are required and resources are limited. In fact, one might as well use the detailed method of section 30–2 when tensiometers are used for two or more depths. Thus, the method is most useful when only $K(\theta)$ is required.
2. The assumption of unit hydraulic gradient is often not satisfied strictly in field soils because of nonuniformity of hydraulic properties in the profile caused by natural horizon differences and tillage effects. The relative hydraulic conductivities in the soil profile (in relation to the conductivities at a given reference depth) will normally vary with time. For example, a tilled surface horizon may have the highest conductivity in the profile at the beginning of profile drainage (near saturation), but the highest relative conductivity may be associated with a deeper horizon a day or two later. This change in relative conductivity between various depths over time results in a change in hydraulic head gradient over time. Chong et al. (1981) found that for well-drained Oxisols unit gradient was achieved at intermediate times (about 17 h), with gradient less than unity at earlier times and greater than unity at later times.
3. Two alternative methods of calculating $K(\theta)$ with field measurements of only $\theta(z,t)$ were developed by Libardi et al. (1980). These methods, called the "θ-method" and the "flux-method," utilize Eq. [6] assuming $\partial H/\partial z = -1$ and require that K be related to θ by an exponential relationship, viz., $K = K_0 \exp[\beta(\theta - \theta_0)]$, in which β is a constant and K_0 and θ_0 are the values of K and θ during steady infiltration. In a comparison of these two methods of calculating $K(\theta)$ with the method

of Chong et al. (1981) described above, designated as "CGA method" in the paper by Libardi et al. (1980), the latter authors found no significant difference in the results obtained by the three methods. If one desires to express the $K(\theta)$ relationship by an exponential function, then a method of Libardi et al. would be preferred over the method of Chong et al., which gives $K(\theta)$ in a power-function form. Dane (1980) reported that the method of Libardi et al. required modification for a coarse-textured soil.

30–4 CRUST-IMPOSED STEADY FLUX METHOD

30–4.1 Principles

In contrast to the unsteady flow methods described in the previous sections, in which the downward flux of water past a given depth in the soil profile is calculated by invoking the conservation of mass as drainage proceeds, this steady flux method involves the establishment of a known flux of water through an isolated pedestal of soil by means of a crust at the soil surface. When the water flux through a soil horizon is maintained at a value below the saturated conductivity under steady flow with unit hydraulic gradient, the hydraulic conductivity is equal to the imposed flux; this conductivity is associated with a soil-water pressure less than zero, which is conveniently measured by a single small tensiometer inserted into the soil pedestal, giving one point on a $K(h)$ curve. Successive measurements with different crusts, each with a different hydraulic conductivity, provide an in situ characterization of the $K(h)$ relationship over the range of pressures achieved.

The method was proposed by Hillel and Gardner (1970) and developed for field use by Bouma et al. (1971). Crusts made of puddled soil material in the first trials were later replaced by crusts composed of varying amounts of gypsum and silica sand (Bouma & Denning, 1972), and more recently by crusts made of mixtures of sand and quick-setting hydraulic cement (Bouma et al., 1983). Evaluation of the effect of the dimensions of the soil pedestal on the measured conductivities was accomplished by Baker (1977).

Some considerations that may influence the choice of this method by potential users are:

1. The method is applicable at relatively high soil water contents, corresponding to soil water pressures of 0 to −7 kPa.
2. The method yields K as a function of h and not as a function of θ, although $K(\theta)$ may be calculated from an independently determined $\theta(h)$ curve.
3. Measured $K(h)$ values will generally be those associated with the absorption hysteresis loop of $K(h)$, in contrast to the desorption loop associated with the unsteady drainage methods. The normal procedure for the crust method is to make a series of measurements with suc-

cessive crusts having increasing hydraulic conductivities, after starting with the least conductive crust over soil that is relatively dry. Steady flow of water with each crust is achieved faster in the absorption mode than in the desorption mode that would prevail if measurements were made first on wet soil with a highly conductive crust, then on soil with progressively decreasing water content associated with lower-conductivity crusts.

30–4.2 Method

The crust method presented here was proposed and tested by Bouma et al. (1971) with modifications by Bouma and Denning (1972). A schematic diagram of the experimental apparatus (from Baker, 1977) is shown in Fig. 30–3. Essential components include: (i) a carved pedestal of soil about 0.25 m in diameter and 0.3 m deep, the exterior of which is covered with aluminum foil; (ii) an infiltration ring (fitting the upper portion of the soil pedestal tightly) to which a lucite cover is bolted with a rubber seal to make the resulting chamber water-tight; (iii) a gypsum-sand or hydraulic-cement and sand crust on the soil surface; (iv) a constant-head water source device with which inflow volumes can be accurately mea-

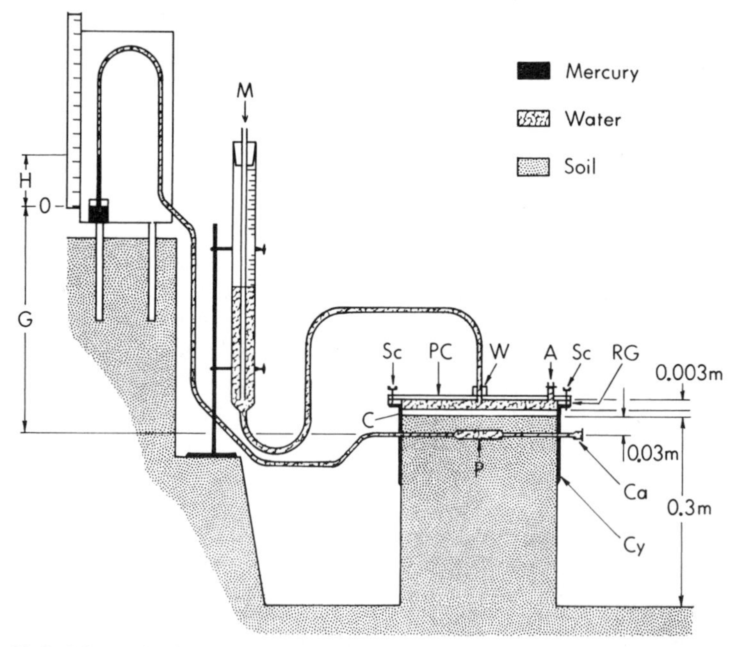

Fig. 30–3. Schematic diagram of a field installation of the measurement apparatus for the crust-imposed steady flux method (after Baker, 1977). M = constant-head device; Sc = wing nut; PC = plastic cover; W = water inlet; A = air outlet; RG = rubber gasket; C = gypsum-sand crust, Ca = tensiometer cap; Cy = metal cylinder with sharpened edge; H = height of mercury column above mercury pool; and G = height of mercury pool above tensiometer porous cup, P.

sured; and (v) a tensiometer to determine when steady flow has been achieved and also to give the pressure corresponding to the measured hydraulic conductivity.

30–4.2.1 EQUIPMENT

1. A metal ring infiltrometer, 0.25-m inside diameter and 0.1 m high, with a 0.025-m wide flange at the top with holes to accomodate bolts for securing the plastic cover. Two opposing holes in the cylinder wall, 0.015-m diameter and 0.045 m below the top, allow insertion of the tensiometer in the soil column.
2. Acrylic plastic cover, 0.013 m thick and 0.30-m diameter, to which a rubber gasket is glued to provide a seal when the cover is bolted to the infiltrometer flange. Cover is fitted with a water inlet and an air outlet.
3. Aluminum foil to cover the soil pedestal.
4. Anhydrous calcium sulfate (plaster of Paris) such as Ultracal-30, produced by U.S. Gypsum Co., or hydraulic cement, which is available in most hardware stores. Common gypsum ($CaSO_4 \cdot 2H_2O$) is not satisfactory as it does not harden irreversibly upon wetting.
5. Pencil-sized tensiometer cups with air-entry pressure head of about -1 m.
6. Clear plastic flexible tubing to connect water source to infiltration unit and tensiometer to mercury manometer.
7. Constant head device such as a graduated buret (0.1 to 0.5 L volume, depending on the crust conductivity).
8. Mercury manometer.
9. Small auger to bore hole for tensiometer cup.
10. A water source.

30–4.2.2 PROCEDURE

1. Determine the depth increment of the soil profile for which the $K(h)$ measurement is desired.
2. Prepare a horizontal plane at the desired level in the soil profile using appropriate cutting tools and a carpenter's level.
3. Carve a cylindrical column of soil, at least 0.3 m high with a diameter slightly larger than the inner diameter of the infiltration ring. Measurements at depths below 0.5 m will require that a large pit be dug in the process of isolating a soil pedestal, as indicated in Fig. 30–3.
4. Fit the metal cylinder (Cy in Fig. 30–3) on the soil column so that the soil surface inside the infiltrometer is about 0.015 m below the top end of the cylinder; the upper surface of the metal cylinder should be slightly sloped so that air can escape from the air vent (A) when the plastic cover (PC) is secured in place.
5. Wrap aluminum foil around the exposed lower portion of the soil column and shovel loose soil around the column to hold the foil in place and stabilize the column.

6. Mix the appropriate proportions of dry gypsum (or hydraulic cement) and silica sand to achieve the approximate crust conductivity desired. A common series of four crusts would be composed of 100, 50, 20, and 12% gypsum by volume. Continue mixing the crust material while adding water until a thick paste is obtained. Transfer the paste to the prepared soil surface and spread it over the surface to provide a crust of uniform thickness over the entire surface. Special care should be exercised to insure that the crust seals tightly with the inside wall of the metal cylinder to avoid flow at the boundary. The crust will harden irreversibly in 1200 to 1800 s. Bouma and Denning (1972) measured subcrust soil water pressures of −1.8, −3.0, and −5.2 kPa, respectively, with 5-mm thick crusts composed of 30, 50, and 100% gypsum. Install the tensiometer into the soil column through the access holes in the infiltration cylinder. One type of tensiometer installation is shown in Fig. 30–3, but other types might be equally satisfactory. The type in Fig. 30–3 has the advantage of being easily purged with deaerated water by injecting water through the capped end (Ca in Fig. 30–3). A conventional tensiometer cup, without the open-end extension, could be inserted into the soil column with a hole < 0.1 m, and thus would not interefere with water flow as much as the installation shown in Fig. 30–3. If the soil is very dry when the equipment is first installed, the tensiometer can be made operational only after the water application has been initiated so that soil water pressures are greater than the air entry pressure of the tensiometer cup.
8. Introduce water to the infiltrometer chamber from the constant-head device (M in Fig. 30–3) with the air vent (A) open until entering water has displaced all air in the chamber, then close the air vent. Adjust the bottom of the air inlet tube in the constant-head device so that it is at the same level as top of the cover on the infiltrometer ring.
9. Record water inflow volumes and tensiometer readings vs. time until steady flow has been achieved. Continue the water inflow measurements for about 1 h to get an average value of the steady flux, q, corresponding to the measured soil water pressure head, h. If the soil is sufficiently uniform with depth so that a unit hydraulic gradient can be assumed, the hydraulic conductivity is equal to the flux; thus $K(h) = q$.

30–4.2.3 COMMENTS

1. Soils having evident layering or tillage pans may not develop a unit hydraulic gradient over the depth interval being measured. Thus, it may be necessary to install a second tensiometer in the soil column to measure the gradient during steady flow. If the hydraulic head gradient is represented by dH/dz, then the hydraulic conductivity is given by $K(h) = q/(dH/dz)$, in which h is the average soil water pressure obtained with the two tensiometers during steady flow.
2. This method requires considerable effort and time to prepare the site,

and the measurement of K at four or more soil water pressures requires several days. The method has good precision (Baker, 1977) and thus the effort required may be warranted for certain applications. Also, while other methods are usually unsatisfactory on steeply sloping land, the crust method is well adapted to such conditions.

3. Optimum dimensions of the soil column may vary with the application, but a height/diameter ratio of 1.25 or greater has been recommended (Baker, 1977).
4. The experimental set-up in Fig. 30–3 (minus the crust) can be used to measure the hydraulic conductivity of saturated soil. The sides of the pedestal are covered with dental-grade plaster to prevent boundary flow (Baker, 1977).

30–5 SPRINKLER-IMPOSED STEADY FLUX METHOD

30–5.1 Principles

This method differs from the crust-imposed steady flux method (section 30–4) principally by the way in which the flux of water moving through the soil profile is maintained at a selected steady value below the infiltrability of the soil. Sprinkler application of water provides a means of controlling the flux of water entering the soil surface; given sufficient time, the flux approaches a constant value throughout the profile. A controlled-flux method was proposed by Youngs (1964), who demonstrated its use on laboratory columns of porous materials. Hillel and Benyamini (1974) used the sprinkler method in field measurements of hydraulic conductivity on a sandy loam soil. Tensiometers and a neutron moisture probe can be used to monitor soil water pressure and water content, respectively, at various depths in the soil profile until steady flow is achieved. As with the crust method, the hydraulic conductivity at a selected steady flux is given by the flux divided by the gradient in hydraulic head over the depth interval of interest. To obtain hydraulic conductivity as a function of water content and/or soil water pressure one must conduct a series of tests on the same plot, starting with a low application intensity on relatively dry soil until steady flow is achieved, and subsequently increasing the intensity step-wise for successive measurements. When steady flow is reached with each application intensity, the water content and pressure at each depth are measured so that K can be related to θ and h. A major limitation of the method is the difficulty of applying water at the low fluxes associated with low water contents, thus the method is useful for determining K only at relatively high water contents. Hillel and Benyamini (1974) applied the sprinkler method and the unsteady drainage-flux method (section 30–2) to the same field soil and found that the $K(\theta)$ values obtained with the two methods were consistent, even though the methods yielded conductivity data over different water content ranges and the measurements were for absorption and desorption, respectively.

30–5.2 Method

The method is straightforward, involving only the establishment of steady water flux by sprinkler application of water and the measurement of θ and h at the depths of interest for each flux. Uniform water distribution over the plot surface is required; this is a major requirement for the selection of a sprinkler system to establish the steady flux. Hillel and Benyamini (1974) used sprinkler equipment described by Morin et al. (1967) and Rawitz et al. (1972), which provided a range of intensities. Water was applied to a 4-m diameter plot at intensities of 2.78×10^{-7} to 2.78×10^{-5} m/s (1 to 100 mm/h). Sprinkler infiltrometers are discussed in detail in Chapter 33.

Other requirements include: (i) adequate plot size to minimize the effects of lateral movement at the boundaries on steady vertical flux in the vicinity of the neutron access tube and/or tensiometers and (ii) the absence of relatively impervious layers in the profile, which would impede vertical water movement and thus introduce error in the assumed vertical flux.

30–5.3 Comments

1. A change in the hydraulic properties of the surface soil over the period during which a sequence of measurements is made is a major disadvantage of this method for tilled or otherwise unstable soils. The surface should be protected from water-drop impact with chopped straw or other porous material, but even with this precaution many tilled soils will be subject to settling, which will likely affect the hydraulic conductivity at high water contents.
2. The difficulty of establishing a constant flux with low sprinkler intensities limits the utility of the method to measurement of K at fairly shallow depths, e.g., 0 to 0.5 m. Achievement of steady fluxes at greater depths requires too much time for the method to be of practical use.
3. The measured $K(h)$ values will be those associated with the absorption loop of $K(h)$, in contrast to the unsteady drainage-flux methods which give $K(h)$ and $K(\theta)$ from desorption. The common observation that there is little or no hysteresis in $K(\theta)$ for many soils was confirmed for Rehovot sandy loam by Hillel and Benyamini (1974) when the infiltration and drainage methods of measuring $K(\theta)$ gave consistent results (i.e., K values from both methods could be represented by a single $K(\theta)$ curve).
4. While sprinklers provide a means of applying water uniformly over an area, other application methods may suffice under some conditions. For example, a trickle irrigation system was used by van de Pol et al. (1977) to establish a steady flux in a field study of solute movement;

water emitters were arranged in a 0.3-m square grid system. This spacing of water sources would preclude determination of hydraulic conductivity near the soil surface, perhaps in the 0- to 0.3-m depth interval. Closer spacing of emitters would allow K determinations of soil nearer to the surface, but would not allow a measurement of K in the surface layer.

30–6 SORPTIVITY BY PONDED INFILTRATION

30–6.1 Principles

Cumulative infiltration, I, into a horizontal soil column, having uniform properties and water content, is proportional to the square root of time, t; i.e., $I = St^{1/2}$. The coefficient of proportionality, S, which varies with initial water content and differs between soils, was termed *sorptivity* by Philip (1957b). Downward one-dimensional cumulative infiltration for small times can be described by a rapidly converging power series in $t^{1/2}$ (Philip, 1957a), viz.,

$$I = St^{1/2} + At + Bt^{3/2} \dots . \quad [14]$$

For a brief time after initiation of infiltration, perhaps less than 120 s, the first term of this infiltration equation dominates the flow. Thus it is possible to estimate the value of the sorptivity, S, for a given initial water content by measuring the cumulative infiltration vs. time immediately following the application of water.

30–6.2 Method

This field sorptivity procedure involves measurement of cumulative infiltration as the head of ponded water in a single-ring infiltrometer falls over time. This technique was developed by Talsma (1969). It is a simple and rapid method with no specialized equipment required. Vertical flow is assured by insertion of the infiltration ring 0.10 to 0.15 m into the soil, which usually exceeds the depth of the wetting front during the brief period of infiltration. The effect of decreasing head on the measured sorptivity is minimized by starting with a small head of only 20 to 30 mm of water. Talsma (1969) assessed the effect of head change on S for several soils differing in soil texture and concluded that the error introduced by the drop in head during measurement is usually negligible compared with natural soil variability. Alternative constant-head, ponded infiltration techniques were found to be more difficult and less precise.

The ponding method is appropriate only for soils in which there is negligible flow through large cracks or channels (see section 30–6.2.3). The method is usually satisfactory for finely tilled surface soils or subsoils.

30–6.2.1 EQUIPMENT

1. Infiltrometer ring, 0.30 m in diameter and 0.15 to 0.20 m high.
2. Graduated (mm) metal scale (200 mm long) or graduated glass tube (about 5 mm inside diameter) for measurement of water surface level over time. A clamping device is needed to hold the scale at the proper angle; a thermometer clamp attached to a rod pushed vertically into the soil just outside the ring provides a good clamp for a capillary tube. The capillary action in glass tubing facilitates reading the level of the water in the inclined tube.
3. Can or plastic beaker, 2-L volume, to apply known volume of water to ring.
4. Porous material with low water-holding capacity, such as a plastic scrubbing pad, to break impact of water poured into the infiltration ring.
5. Stopwatch with provision for stopping and reading a sequence of event times ("lap time") while total elapsed time continues on the clock.
6. Tape recorder if the measurements are to be made by one person.

30–6.2.2 PROCEDURE

1. Prepare a level, plant-free surface about 0.5 by 0.5 m.
2. Install infiltration ring to a depth of 0.10 to 0.15 m. Compact soil next to the ring with a pencil-sized rod to prevent boundary flow.
3. Take a composite soil moisture sample to a depth of 50 to 100 mm, preferably outside the ring; a small-diameter core sampler is most appropriate.
4. Attach the measurement scale or capillary tube to the ring or auxiliary support rod at an angle α of 5 to 30° relative to the horizontal soil surface. The bottom of the scale should rest on the ground surface. The change in water level is amplified by a factor of $(\sin \alpha)^{-1}$. For example, when $\alpha = 30°$, a twofold amplification in the reading is achieved. If a glass tube is used to measure the water level, place a small piece of flexible plastic sheet on the soil surface underneath the glass tube; this will reduce the movement of soil particles into the tube.
5. Calibrate the beaker, which is to be used to apply a known volume of water into the ring. A volume of 1.6 L will provide a layer of water 20 mm deep in a 0.30-m diameter ring.
6. Quickly apply the water on the porous pad at the side of the ring opposite the measuring scale and start the clock simultaneously. A tape recorder may be used if a single operator has to record both time and water level during the brief measurement period.
7. Obtain as many water height vs. time readings as possible until the water surface is about 5 mm above the soil surface. Generally the best practice is to read the times corresponding to preselected indices on the measurement scale. Five or more readings are required for calculation of a reliable value of *S*.

8. Calculate cumulative infiltration as a function of time using the proper amplification factor (calculated from the scale angle).
9. Plot cumulative infiltration, I, vs. $t^{1/2}$ and determine the linear portion of the curve. Delete data for early times that may be in error due to water roughness; also delete data for later times that do not conform to the linear $I - t^{1/2}$ relationship.
10. Calculate the slope S by a least-squares fit.
11. Determine the initial soil water content associated with the measured S value, using samples taken in the field (item 3).

30–6.2.3 COMMENTS

1. Each sorptivity measurement, as described above, yields one point on the $S(\theta_n)$ curve, where θ_n is initial water content. A linear approximation of the $S(\theta_n)$ relationship for a range of water contents from saturation (θ_s) to air-dry can be obtained from the line plotted from $S = 0$ at $\theta = \theta_s$ through the measured value of S at $\theta = \theta_n$. Such an approximation was found to give reasonable estimates of infiltration at a number of field sites (Chong & Green, 1979).
2. Ponding of water on the soil surface may result in inaccurate estimates of sorptivity in soils that have worm holes, root channels, large cracks, or other large void spaces that are not representative of the soil matrix. In addition to the likely problem of inaccurate representation of larger areas with sorptivity measurements on such soils, sorptivities measured by this method are also subject to errors due to (i) the dependence of sorptivity on head and (ii) the contribution of terms containing powers of $t > 1/2$ in Eq. [14], so that the hydraulic conductivity at saturation contributes significantly to the measured cumulative infiltration. Talsma (1969) found that such soils had relatively large values of the ratio of hydraulic conductivity to sorptivity. The sorptivity procedure described in the next section (30–7) was designed to circumvent these problems.
3. Sorptivity values measured over a field or watershed are likely to be log-normally distributed; this conclusion results from both theoretical considerations (Brutsaert, 1976) and field measurements (Chong & Green, 1979; Sharma et al., 1980). Thus, a representative sorptivity value for a large area from many measurements may best be obtained from the geometric mean rather than the arithmetic mean.
4. Sorptivities measured by this method are representative of a thin layer of soil at the surface, generally < 50 mm.

30–7 SORPTIVITY BY INFILTRATION AT NEGATIVE MATRIC PRESSURE

30–7.1 Principles

This field method, suggested by Dirksen (1975) and utilized in the field by Russo and Bresler (1980) and by Clothier and White (1981), is a variant of the Talsma (1969) method presented in the previous section. It differs principally in the way in which water is applied to the soil surface. Instead of a ponded water source on the soil surface, a constant-head device delivers water to the soil through a porous plate in contact with the soil; a slight negative pressure at the bottom of the porous plate prevents water from entering large voids. Thus, the method obviates the problems caused by large voids when sorptivity is measured by ponded infiltration.

30- 7.2 Method

30-7.2.1 EQUIPMENT

The sorptivity-measuring apparatus illustrated in Fig. 30-4 (from Chong & Green, 1983) differs only slightly from that of Clothier and White (1981), the latter having a sintered glass plate (porosity no. 1) at the base instead of the drilled Plexiglas plate. Chong and Green (1983) found that the Plexiglas plate with about 15×10^4 holes per square meter (hole diameter, 1.05 mm) was less fragile and had less impedance to flow than the sintered glass plate. The upright Plexiglas tube (350 mm high, 22 mm inside diameter) serves both as water reservoir and sight tube for measuring cumulative infiltration into the soil through the 80-mm diameter porous plate. A millimeter scale is taped to the Plexiglas tube to

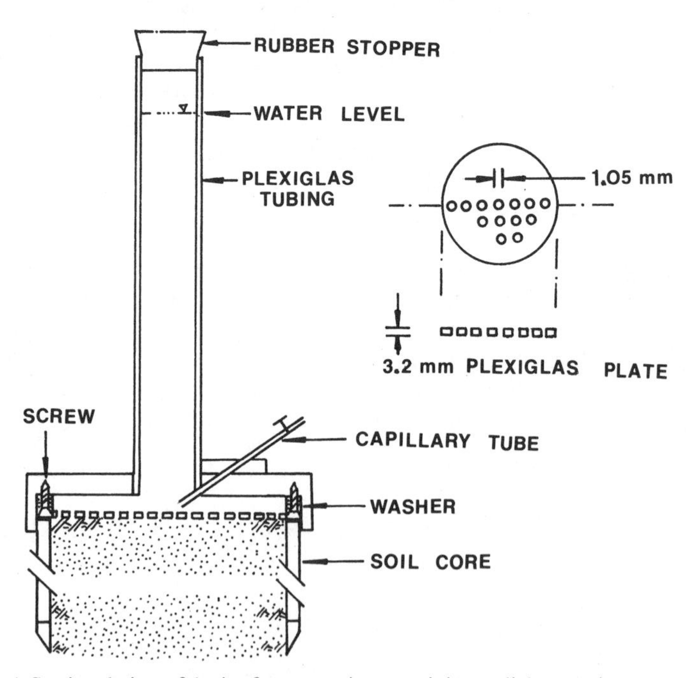

Fig. 30-4. Sectional view of device for measuring sorptivity at slight negative pressure (after Chong & Green, 1983).

facilitate measurement of the water level as infiltration proceeds. When the sorptivity device is filled with water and the stopper at the top is secured, water flows through the plate only if air enters the chamber through the capillary tube. Assuming no head loss through the porous plate, the pressure head, h, at the bottom of the plate, is given by

$$h = z - 2\sigma/r\rho g$$

where z is the elevation difference between the bottom of the plate and the lower end of the capillary tube having a radius r, σ is the surface tension of water, ρ is the density of water, and g is the acceleration due to gravity. The device used by Clothier and White (1981), with $r = 0.33$ mm and $z = 5$ mm, delivered water to the soil surface at a pressure head of -40 mm water. The value of h for the unit shown in Fig. 30–4, with $r = 2$ mm and $z = 3.5$ mm, was a little over -10 mm. The pressure head, h, can thus be varied as desired by adjusting the height of the capillary above the plate or by varying the inside diameter of the capillary tube. Clothier and White used a hypodermic syringe needle in place of the capillary tube.

Other essential equipment includes a water supply container (about 0.3 m diameter and 0.5 m high), a stopwatch, and a thin-walled Plexiglass cylinder (50 to 70 mm high) for enclosing a soil column.

30–7.2.2 PROCEDURE

The site is prepared and soil samples taken for soil moisture determination as described earlier (Section 30–6.2.2). The Plexiglas cylinder is inserted into the soil until the upper edge is flush with the soil surface. On compact soils it may be necessary to carve a soil column slightly larger than the cylinder, onto which the Plexiglas cylinder is pressed. Soil particles are removed from the upper edge of the cylinder to allow the porous plate of the sorptivity device to make good contact with the soil surface.

The sorptivity device is filled with water by submerging it in a large water container, the stopper at the top is secured, and the device is carefully removed from the container and placed on the soil column. A valve on the capillary tube is closed while the device is moved to avoid air entry into the chamber. The value is opened just before the device is placed on the soil column, at which time the clock is started. Several water height vs. time measurements can usually be made in the measurement period of 120 s or less. A tape recorder facilitates data recording so that the measurements can be accomplished by one operator.

30–7.2.3 COMMENTS

1. Although this method is quite new and has not been used widely, the published results of Clothier and White (1981) and Russo and Bresler (1980) plus our own recent experience on a variety of soils in Illinois

and Hawaii suggest that the method is simple, rapid and reliable. Clothier and White compared this method (using 86-mm diameter soil column) with the ponded infiltration method (0.3-m diameter ring) of Talsma (1969), presented in section 30–6, and found that the ponded method gave higher values of S (>2 times) and a slightly higher standard deviation. The two methods gave nearly the same results when the measured areas for the two methods were equal (86-mm diameter) and care was taken to avoid large voids in the soil surface when installing the cylinders. Clothier and White (1981) concluded that inclusion of large voids or channels in the larger infiltration cylinder gave higher sorptivities and a larger standard deviation.

2. Although the principal objective of this method is to obtain sorptivities at given initial water contents, the diffusivity vs. water content relationship can be obtained also by measuring sorptivity and the advance of the wet front concurrently when the soil is initially nearly air dry. Clothier and White (1981) present the details. A related method of deriving $K(h)$ and $\theta(h)$ from sorptivity and other field-measured hydraulic properties was used by Russo and Bresler (1980).

30–8 REFERENCES

Ahuja, L. R., S. A. El-Swaify, and A. Rahman. 1976. Measuring hydrologic properties of soil with a double-ring infiltrometer and multiple-depth tensiometers. Soil Sci. Soc. Am. Proc. 40:494–499.

Ahuja, L. R., R. E. Green, S. K. Chong, and D. R. Nielsen. 1980. A simplified functions approach for determining soil hydraulic conductivities and water characteristics in situ. Water Resour. Res. 16:947–953.

Arya, L. M., D. A. Farrell, and G. R. Blake. 1975. A field study of soil water depletion patterns in the presence of growing soybean roots: 1. Determination of hydraulic properties of the soil. Soil Sci. Soc. Am. Proc. 39:424–436.

Baker, F. H. 1977. Factors influencing the crust test for in situ measurement of hydraulic conductivity. Soil Sci. Soc. Am. J. 41:1029–1032.

Baker, F. G., P. L. M. Veneman, and J. Bouma. 1974. Limitations of the instantaneous profile method for field measurement of unsaturated hydraulic conductivity. Soil Sci. Soc. Am. Proc. 38:885–888.

Black, T. A., W. R. Gardner, and G. W. Thurtell. 1969. The prediction of evaporation, drainage and soil water storage for a bare soil. Soil Sci. Soc. Am. Proc. 33:655–660.

Bouma, J., C. Belmans, L. W. Dekker, and W. J. M. Jeurissen. 1983. Assessing the suitability of soils with macropores for subsurface liquid waste disposal. J. Environ. Qual. 12:305–311.

Bouma, J., and J. L. Denning. 1972. Field measurement of unsaturated hydraulic conductivity by infiltration through gypsum crusts. Soil Sci. Soc. Proc. 36:846–847.

Bouma, J., D. I. Hillel, F. D. Hole, and C. R. Amerman. 1971. Field measurement of hydraulic conductivity by infiltration through artificial crusts. Soil Sci. Soc. Proc. 35:362–364.

Brooks, R. H., and A. T. Corey. 1964. Hydraulic properties of porous media. Colorado State Univ. Hydrol. Paper no. 3, Fort Collins, CO.

Brutsaert, W. 1976. The concise formulation of diffusive sorption of water in a dry soil. Water Resour. Res. 12:1118–1124.

Campbell, G. S. 1974. A simple method for determining unsaturated conductivity from moisture retention data. Soil Sci. 117:311–314.

Chong, S. K., and R. E. Green. 1979. Application of field-measured sorptivity for simplified infiltration prediction. p. 88–96. *In* Proc. Symposium on Hydrologic Transport Modeling. ASAE Pub. 4-80. New Orleans, LA, 10–11 December. American Society of Agricultural Engineers.

Chong, S. K., and R. E. Green. 1983. Sorptivity measurement and its application. p. 82–91. *In* Proc. National Conference on Advances in Infiltration. ASAE Pub. 11-83. Chicago, IL, 12–13 December. American Society of Agricultural Engineers, St. Josephs, MI.

Chong, S. K., R. E. Green, and L. R. Ahuja. 1981. Simple in situ determination of hydraulic conductivity by power function descriptions of drainage. Water Resour. Res. 17:1109–1114.

Chong, S. K., R. E. Green, and L. R. Ahuja. 1982a. Determination of sorptivity based on in-situ soil water redistribution measurements. Soil Sci. Soc. Am. J. 46:228–230.

Chong, S. K., M. A. Khan, and R. E. Green. 1982b. A portable hand-operated soil core sampler. Soil Sci. Soc. Am. J. 46:433–434.

Clapp, R. B., and G. M. Hornberger. 1978. Empirical equations for some soil hydraulic properties. Water Resour. Res. 14:601–604.

Clothier, B. E., and I. White. 1981. Measurement of sorptivity and soil water diffusivity in the field. Soil Sci. Soc. Am. J. 45:241–245.

Dane, J. H. 1980. Comparison of field and laboratory determined hydraulic conductivity values. Soil Sci. Soc. Am. J. 44:228–231.

Dane, J. H., and S. Hruska. 1983. In-situ determination of soil hydraulic properties during drainage. Soil Sci. Soc. Am. J. 47:619–624.

de Boor, C., and J. R. Rice. 1968. Least-square cubic spline approximation, 1. Fixed knots. CSD TR 20, P.1–30. Purdue University, Lafayette, IN.

Dirksen, C. 1975. Determination of soil water diffusivity by sorptivity measurements. Soil Sci. Soc. Am. Proc. 39:22–27.

Erh, K. T. 1972. Application of spline function to soil science. Soil Sci. 114:333– 338.

Fluhler, H., M. S. Ardakani, and L. H. Stolzy. 1976. Error propagation in determining hydraulic conductivities from successive water content and pressure head profiles. Soil Sci. Soc. Am. J. 40:830–836.

Gardner, W. R. 1970. Field measurement of soil water diffusivity. Soil Sci. Soc. Am. Proc. 34:832– 833.

Gardner, W. R., D. Hillel, and Y. Benyamini. 1970. Post-irrigation movement of soil water. 1: Redistribution. Water Resour. Res. 6:851–861.

Hillel, D., and Y. Benyamini. 1974. Experimental comparison of infiltration and drainage methods for determining unsaturated hydraulic conductivity of a soil profile in situ. p. 271–275. *In* Isotope and Radiation Techniques in Soil Physics and Irrigation Studies 1973. International Atomic Energy Agency, Vienna.

Hillel, D., and W. R. Gardner. 1970. Measurement of unsaturated conductivity and diffusivity by infiltration through an impeding layer. Soil Sci. 109:149–153.

Hillel, D., V. D. Krentos, and Y. Stylianou. 1972. Procedure and test of an internal drainage method for measuring soil hydraulic characteristics in situ. Soil Sci. 114:395–400.

Libardi, P. L., K. Reichardt, D. R. Nielsen, and J. W. Bigger. 1980. Simple field methods for estimating soil hydraulic conductivity. Soil Sci. Soc. Am. J. 44:3–7.

Morin, J., D. Goldberg, and I. Seginer. 1967. A rainfall simulator with a rotating disk. Trans. ASAE 10:74–77.

Nielsen, D. R., J. W. Bigger, and K. T. Erh. 1973. Spatial variability of field-measured soil-water properties. Hilgardia 42:215–260.

Nielsen, D. R., J. M. Davidson, J. W. Biggar, and R. J. Miller. 1964. Water movement through Panoche clay loam soil. Hilgardia 35:491–506.

Philip, J. R. 1955. Numerical solution of equations of the diffusion type with diffusivity concentration-dependent. Trans. Faraday Soc. 51:885–892.

Philip, J. R. 1957a. The theory of infiltration. 1. The infiltration equation and its solution. Soil Sci. 83:345–357.

Philip, J. R. 1957b. The theory of infiltration. 4. Sorptivity and algebraic infiltration equations. Soil Sci. 84:257–264.

Rawitz, E., M. Margolin, and D. Hillel. 1972. An improved variable intensity sprinkling infiltrometer. Soil Sci. Soc. Am. Proc. 36:533–535.

Richards, L. A., W. R. Gardner, and G. Ogata. 1956. Physical processes determining water loss from soil. Soil Sci. Soc. Am. Proc. 20:310–314.

Rose, C. W., W. R. Stern, and J. E. Drummond. 1965. Determination of hydraulic conductivity as a function of depth and water content for soil in situ. Aust. J. Soil Res. 3:1–9.

Russo, D., and E. Bresler. 1980. Field determinations of soil hydraulic properties for statistical analyses. Soil Sci. Soc. Am. J. 44:697–702.

Sharma, M. L., G. A. Gander, and C. G. Hunt. 1980. Spatial variability of infiltration in a watershed. J. Hydrol. 45:101–122.

Talsma, T. 1969. In situ measurement of sorptivity. Aust. J. Soil Res. 7:269–276.

van Bavel, C. H. M., and G. B. Stirk. 1967. Soil water measurement with an ^{241}Am-Be neutron source and an application to evaporimetry. J. Hydrol. 5:40–46.

van Bavel, C. H. M., G. B. Stirk, and K. J. Brust. 1968. Hydraulic properties of a clay loam soil and the field measurement of water uptake by roots: 1. Interpretation of water content and pressure profiles. Soil Sci. Soc. Am. Proc. 32:310–317.

van de Pol, R. M., P. J. Wierenga, and D. R. Nielsen. 1977. Solute movement in a field soil. Soil Sci. Soc. Am. J. 41:10–13.

Watson, K. K. 1966. An instantaneous profile method for determining the hydraulic conductivity of unsaturated porous materials. Water Resour. Res. 2:709–715.

Watson, K. K. 1967. A recording field tensiometer with rapid response characteristics. J. Hydrol. 5:33–39.

Youngs, E. G. 1964. An infiltration method of measuring the hydraulic conductivity of unsaturated porous materials. Soil Sci. 97:307–311.

Zachmann, D. W., P. C. DuChateau, and A. Klute. 1982. Simultaneous approximation of water capacity and soil hydraulic conductivity by parameter identification. Soil Sci. 134:157–163.

31 Hydraulic Conductivity of Unsaturated Soils: Prediction and Formulas

YECHEZKEL MUALEM

The Hebrew University of Jerusalem
Rehovot, Israel

31-1 INTRODUCTION

Solution of unsaturated flow problems usually requires predetermination of the soil hydraulic properties, namely, (i) the relationship between the capillary head, ψ, and the moisture content, Θ and (ii) the dependence of the hydraulic conductivity, K, upon the moisture content (sometimes their derivatives are applied as a matter of convenience). It would be desirable to determine all necessary data by direct measurements, but often this is impossible for one or more of the following reasons:

1. The measurements are costly and time-consuming.
2. The hydraulic properties of soils are of a hysteretical nature. Different relationships prevail for wetting and drying processes, and the actual relationships between K, ψ, and Θ depend upon the preceding history.
3. The soil variability is such that the amount of data required to represent the hydraulic properties accurately is enormous.
4. The values of hydraulic conductivity of some soils may vary by several orders of magnitude within the water content range of interest. Most measurement systems can not efficiently cover such a wide range.
5. The available experimental data can not represent the complete relationships describing the hydraulic properties. To replace the missing information several methods have been applied to compute the hydraulic properties using either empirical, semi-empirical, or theoretical models. This chapter reviews the various approaches for estimating the $K(\Theta)$, and/or $K(\psi)$ relationships and provides guidance for selecting formula or calculation procedure appropriate for the available soil data and characteristics.

A considerable hysteresis loop is observed in the $K(\psi)$ plane. However, experimental work indicates that the $K(\Theta)$ hysteresis is much less significant than the $K(\psi)$ hysteresis, and therefore many models disregard hysteresis in $K(\Theta)$. These models are reviewed in detail, and their com-

 Methods of Soil Analysis, Part 1. Physical and Mineralogical Methods–Agronomy Monograph no. 9 (2nd Edition)

patibility with the hysteretic phenomenon is discussed. Recent theories allowing prediction of $K(\Theta)$ or the $K(\psi)$ hysteresis are briefly mentioned.

31–2 THEORY AND COMPUTATIONAL FORMULAS

31–2.1 General Concepts

The macroscopic definition of hydraulic conductivity stems from Darcy's law

$$\bar{q} = -K \nabla\phi \qquad [1]$$

where $\bar{q}$ is the specific discharge vector and $\nabla\phi$ is the gradient of the hydraulic head. The validity of Eq. [1] for saturated flows at sufficiently low Reynolds numbers has been confirmed by numerous theoretical and experimental studies, which also have led to relationships between the hydraulic conductivity and the physical parameters of the fluid and solid matrix (e.g., the Kozeny-Carman equation). Several studies deal with prediction of the saturated hydraulic conductivity (K_{sat}). They are not considered here unless they can be extended in a straight-forward manner to unsaturated soils.

For unsaturated soils ($\Theta < \Theta_{sat}$), Eq. [1] is of a more limited applicability (Braester et al., 1971). The dependence of K upon the fluid and matrix properties and the sequence of wetting and drying processes is more complex than in the saturated state. Darcy's law, however, is still generally used as a working tool which applies to a wide range of unsaturated flows. The assumption that K is a function of Θ or ψ, for a given soil and fluid, is adopted. Equation [1] takes the form

$$\bar{q} = -K(\Theta) \nabla\{\psi(\Theta) + Z + (P_a/\gamma)\} \qquad [2]$$

where Z is the elevation, P_a is the air pressure, and γ is the water specific weight. It is generally assumed that the relative hydraulic conductivity

$$K_r(\Theta) = K(\Theta)/K_{sat} \qquad [3]$$

is a scalar function. Often the degree of saturation (S), effective water content (θ), or effective saturation (S_e), defined by

$$S = \Theta/\Theta_{sat}; \qquad \theta = \Theta - \Theta_r; \qquad S_e = (\Theta - \Theta_r)/(\Theta_{sat} - \Theta_r) \qquad [4]$$

respectively, are used as independent variables instead of θ. θ_r is usually referred to as the residual moisture content.

31–2.2 Empirical Forms for $K(\psi)$ and $K(\Theta)$

This approach can be applied when some measured data for either $K(\psi)$ or $K(\Theta)$ are available. It may serve one or more of the following objectives:

1. To allow a closed-form analytical solution for some unsaturated flow problems.
2. To simplify the computational procedure of numerical solution, save computer time, and improve accuracy.
3. To systematically extrapolate the measured curve.
4. To minimize the measurements required for statistical representation of hydraulic conductivity distribution in the field.

For these purposes, simple mathematical formulas are used to represent the $K(\Theta)$ or the $K(\psi)$ relationship. Several formulas frequently used are given in Table 31–1. The coefficients in these formulas are determined by adjustment to measured data. The smallest number of measured points required is equal to the degrees of freedom of the adopted formula. When the number of measurements exceeds the number of coefficients, a curve-fitting procedure can be applied to minimize error. In the case of objective (4), comprehensive detailed measurements are carried out in a representative location in the field. These data are used to select the best-fitting formula. Assuming that the $K(\psi)$ or $K(\Theta)$ in any other point in the field displays the same features, only the varying coefficients have to be determined in other chosen locations. As mentioned above, the amount of necessary measurement in such cases is minimal, just equal to the number of unknown coefficients.

This approach should be used with regard to some restrictions. No

Table 31–1. Empirical formulas for the unsaturated hydraulic conductivity.

Reference	Wind (1955)	Gardner (1958)	Gardner's type
Function	$K = \alpha\lvert\psi\rvert^{-n}$	$K_r = \exp(\alpha\psi)$; $K = a/(\lvert\psi\rvert^n + b)$	$K_r = [(\psi/b)^n + 1]^{-1}$
Reference	Brooks & Corey (1964)		Averjanov (1950)
Function	$K = K_s$ for $\psi \geq \psi_{cr}$; $K_r = (\psi/\psi_{cr})^{-n}$ for $\psi \leq \psi_{cr}$		$K_r = S_e^{\,n}$; $S_e = (\Theta - \Theta_r)/(\Theta_s - \Theta_r)$; $n = 3.5$
Reference	Rijtema (1965)		King (1964)
Function	$K = K_s$ for $\psi \geq \psi_{cr}$; $K_r = \exp[\alpha(\psi - \psi_{cr})]$ for $\psi_1 \leq \psi \leq \psi_{cr}$; $K = K_1(\psi/\psi_1)^{-n}$ for $\psi < \psi_1$		

For more details, see references.

single relationship is valid for all types of soils. Even when a certain formula may be found suitable for a class of soil, coefficients may still vary considerably from soil to soil. Also, because of the hysteretic phenomenon all empirical formulas for $K(\psi)$ represent the soil characteristic for a continuous process of wetting or drying only. Different processes require at least adjustment of the coefficients, but sometimes may require fitting different formulas. As the $K(\Theta)$ function generally displays a considerably diminished hysteresis, it is generally preferable to use $K(\Theta)$ rather than $K(\psi)$ whenever successive wetting and drying processes are considered.

The mathematical formulas for $K(\Theta)$ or $K(\psi)$ also allow use of an "inverse approach" for determining soil water hydraulic characteristics. Instead of taking direct measurements of hydraulic conductivity, other flow variables are measured, such as water content distribution profiles, infiltration rate, or cumulative infiltration under prescribed boundary conditions. The best representative formula and its coefficients for the soil under consideration are determined by fitting the theoretical solution of the actual flow problem to corresponding measured data.

Gardner (1958) determined the values of coefficients of certain formulas for several soil classes. These values may serve as a guideline for rough estimation of hydraulic conductivity when no data are available, but the soil classification is.

31–2.3 Computation Based on Macroscopic Models

The objective of these models is to derive an analytical formula for the $K(\Theta)$ relationship. The following steps are common to all models in this group:

1. Assume an analogy between laminar flow, generally represented by the relationship between velocity vector $\overline{V}$ and potential field ϕ on a microscopic level

$$\nabla\phi = (\nu/g)\ \nabla^2\overline{V} \qquad [5]$$

   and flow in porous media, where ϕ and $\overline{V}$ can be defined only in a macroscopic sense.

2. Solve Eq. [5] analytically for an extremely simple laminar flow system to derive mathematical formulas interrelating macroscopic flow variables, such as average velocity, $\overline{u}$, and hydraulic gradient in the x direction, hydraulic conductivity, and hydraulic radius. The general form of these equations for flow in the x direction is

$$\overline{u} = -(R^2 g/C\ \nu)\ (d\phi/dx) \qquad [6]$$

   where $R = A/P$ is hydraulic radius (A is the fluid area in a cross section and P is the solid wetted perimeter), ν is liquid kinematic velocity, and C is a constant depending on the shape of the flow system.

3. Make a direct analogy between these variables and the corresponding macroscopic variables defined for the bulk of the soil-water-air system.

Averjanov (1950) considered unsaturated flow in a simple cylindrical capillary where the wetting fluid forms a homogeneously spread film along the tube wall, leaving the central part occupied by air. After a few simplifications, he derived an approximate relationship

$$K_r = S_e^{3.5} . \quad [7]$$

Solving the same case, Yuster (1951) obtained a similar relationship

$$K_r = S_e^{2.0} \quad [8]$$

by assuming that the nonwetting fluid in the center of the tube flows under the same gradient as the wetting fluid. The remarkable difference between the exponent in Eq. [7] and [8] may indicate that K depends not only on the soil matrix but also on flow conditions.

Kozeny (1927) determined the saturated hydraulic conductivity for a porous medium made of spherical particles

$$K_{sat} = (g/C\,\nu\,A_s^2)\,\Theta_{sat}^3 \quad [9]$$

where g is gravity acceleration, ν is kinematic viscosity, A_s is solid surface area, and C is a flow-configuration constant. Following Kozeny, Irmay (1954) generalized Eq. [9] for unsaturated porous media. Since C and A_s in the actual configuration are unknown, Irmay suggested the use of K_{sat} as a reference value, which yields:

$$K_r = K/K_{sat} = S_e^3 . \quad [10]$$

Irmay (1971) found experimental support for Eq. [10] using data of Topp and Miller (1966).

The advantage of Eq. [7], [8], and [10] over the empirical formulas listed in Table 31–1 is that they have a theoretical basis, are very convenient for practical use, and require a small amount of data for calibration (Θ_{sat}, Θ_r, and K_{sat}). However, Childs and Collis-George (1950) criticized these models for neglecting the effect of pore-size distribution on hydraulic conductivity. Brooks and Corey (1966) also questioned the assumption of constant A_s in Eq. [9]. They showed that, for tubes of extreme eccentricity, A_s would change with Θ. In such cases, changes in the degree of saturation may be reflected in variation of flow configuration parameter C.

With regard to the above observations, Mualem (1978) modified the macroscopic approach by interpreting Eq. [6] differently. R is considered to be a characteristic length scale

$$R = \Theta/A_w \quad [11]$$

where A_w is the area, per unit bulk volume, of an imaginary interface between the effective water content and the immobile water content θ_r. Upon this interpretation A_w may vary with the water content. Assuming a power function relation

$$A_w = A_{sat}\, S_e^m \qquad [12]$$

where m is a soil-water parameter, the relative hydraulic conductivity takes a general form

$$K_r(S_e) = S_e^{3-2m} = S_e^n\,. \qquad [13]$$

A qualitative analysis indicates that m can be positive for a granular porous medium and negative for unstructured soils of fine texture. These findings are verified by experimental data for 50 soils. The lower limit of n is found to be 2.5, while high values (up to $n = 24.5$) are found for fine-textured soils. Accordingly, m varies in the range of $+0.5$ to < -20 for this group of soils. Therefore, the power n can not be regarded as a universal constant for all porous media but should rather be considered a soil-water characteristic parameter. Using retention data for 50 soils, a good correlation is found between w, the energy per unit volume of soil required to drain the saturated soil down to the wilting point,

$$w = \int_{\Theta_{15\ \mathrm{atm}}}^{\Theta_{\mathrm{sat}}} \gamma_\omega \psi\, d\Theta \qquad [14]$$

and n, where γ_ω is the specific weight and w is measured in 10^2 Pa. An empirical linear formula which relates n to w is derived (Fig. 31–1)

$$n = 3.0 + 0.015\, w \qquad [15]$$

leading to

$$K_r\,(S_e) = S_e^{3.0+0.015\, w}\,. \qquad [16]$$

Comparison of predicted K_r using Eq. [7], [10], and [16] with experimental data for 50 soils shows that very good agreement is obtained between measured $K_r(\Theta)$ and theoretical curves computed by Eq. [16]. Better accuracy is achieved by the new model, Eq. [16], when soils of fine texture are considered. Typical results are shown in Fig. 31–2. On the overall basis of 50 soils, the mean square root deviation has been considerably reduced from 437% for Irmay's model and 380% for Averjanov's model to 151% by Eq. [16].

31–2.4 Computation of Relative Hydraulic Conductivity Based on Statistical Models

The objective of the "statistical models," including those of Purcell (1949), Childs and Collis-George (1950), Burdine (1953), Wyllie and

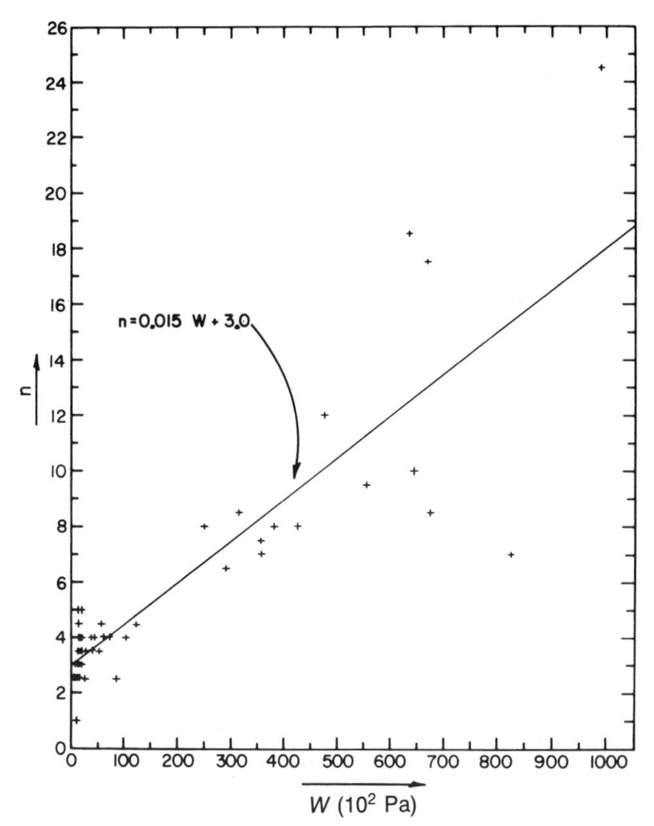

Fig. 31-1. Optimal values of the power n, computed by best-fitting procedure to experimental data, and the function $n = 3.0 + 0.015\ w$ (after Mualem, 1978).

Gardner (1958), Farrel and Larson (1972), Mualem (1976a, 1976b) and Mualem and Dagan (1978), is to compute K_r with regard to the soil physical properties as reflected by the measured retention curve. The methodology characterizing the statistical models is based on three common assumptions:

1. The porous medium may be regarded as a set of interconnected pores randomly distributed in the sample. The pores are characterized by their length scale, often called "the pore radius" (r), and described in statistical terms by $f(r)$, where $f(r)dr$ is the relative volume of pores of radius $r \rightarrow r + dr$. The areal pore distribution is the same for any cross section and equal to $f(r)$.
2. The Hagen-Poiseuille equation, Eq. [6], is assumed valid at the level of the single pore and thus used to estimate the hydraulic conductivity of the elementary pore unit. In this sense, it is a microscopic approach. The total hydraulic conductivity has to be determined by integration over the contributions of the filled pores.
3. The soil retention curve is considered analogous to the pore radii distribution function. Using the capillary law (for given fluids)

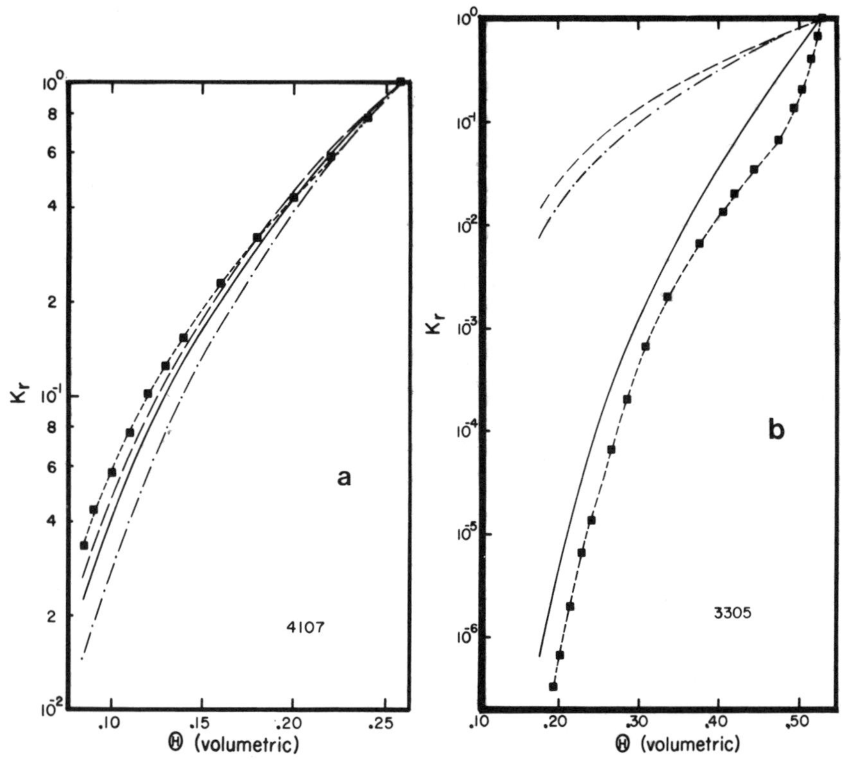

Fig. 31–2. The relative hydraulic conductivity for typical granular soil (a) and soil of fine texture (b), measured (solid squares) and computed using the models of Averjanov (dash-dot-line) and Irmay (1954). The solid line represents the power function derived by the generalized macroscopic approach (after Mualem, 1978).

$$r = C_1/\psi \qquad [17]$$

where C_1 is a constant coefficient, r is uniquely related to the capillary head (ψ) at which the pore is filled and drained. By definition $f(r)dr$ is the contribution of the filled pores of radius $r \rightarrow r + dr$ to the moisture content, namely

$$d\theta(r) = f(r)dr \qquad [18]$$

and thus

$$\theta(R) = \int_{R_{min}}^{R} f(r)dr\,. \qquad [19]$$

Models differ from each other in the interpretation of the geometrical configuration of the elementary pore and the estimate of its contribution to total permeability.

Gates and Lietz (1950), following Purcell (1949), computed the unsaturated hydraulic conductivity of a bundle of parallel capillaries having a radii distribution function identical to the measured $f(r)$ of the soil as defined by Eq. [18]. The unsaturated state is obtained by assuming that pores (capillaries) can be either totally filled or empty, and for a given ψ only tubes with radius smaller than $R = C_1/\psi$ are filled. Using Eq. [6] to estimate the elementary pore contribution to flux (dq), the total discharge of the model at a given moisture content is obtained by

$$q(\theta) = -\frac{g}{8\nu}\frac{d\phi}{dx}\int_{R_{\min}}^{R(\theta)} r^2 f(r)\,dr \qquad [20]$$

or as a direct function of ψ

$$q(\theta) = -\frac{C_1^2 g}{8\nu}\frac{d\phi}{dx}\int_0^{\theta}\frac{d\theta}{\psi^2}. \qquad [21]$$

The analogy between the Darcy equation, Eq. [2], and Eq. [21] indicates that the hydraulic conductivity is proportional to the integral in Eq. [21], leading to

$$K_r(\theta) = \int_0^{\theta}\frac{d\theta}{\psi^2}\Big/\int_0^{\theta_{\text{sat}}}\frac{d\theta}{\psi^2}. \qquad [22]$$

Fatt and Dykstra (1951) modified the model of straight capillaries to account for the tortuosity of the flow path. They assumed that tortuosity is inversely proportional to a power function of the pore radius, r^b, which led to

$$K_r(\theta) = \int_0^{\theta}\frac{d\theta}{\psi^{2+b}}\Big/\int_0^{\theta_{\text{sat}}}\frac{d\theta}{\psi^{2+b}}. \qquad [23]$$

Fatt and Dykstra found that b was approximately unity, based on examination of data for several soils. They indicated, however, that b may vary for different soil types.

Burdine (1953) accounted for the tortuosity using the square of the effective saturation as a correction factor for Eq. [22].

$$K_r(\theta) = S_e^2\int_0^{\theta}\frac{d\theta}{\psi^2}\Big/\int_0^{\theta_{\text{sat}}}\frac{d\theta}{\psi^2}. \qquad [24]$$

Comparison of Eq. [22] and [24] with experimental results of a series of soils showed that the Burdine equation is significantly more accurate than Eq. [22]. Burdine applied the same model to derive the relative hydraulic conductivity of the nonwetting fluid, K_{rn},

$$K_{rn}(\theta) = (1 - S_e)^2\int_{\theta}^{\theta_{\text{sat}}}\frac{d\theta}{\psi^2}\Big/\int_0^{\theta_{\text{sat}}}\frac{d\theta}{\psi^2}. \qquad [25]$$

He found very good agreement between Eq. [25] and observations.

Childs and Collis-George (1950) and Wyllie and Gardner (1958) analyzed the effect of random variation of the pore size on the hydraulic conductivity. To simulate this phenomenon, they sectioned a porous column normal to the flow direction and randomly rejoined together the opposite faces. In such a case, the probability that pores of radius (r) in one face should be connected to pores of radius (ρ) in the other is

$$a(r,\rho) = f(r)\, f(\rho)\, dr\, d\rho\,. \qquad [26]$$

To simplify computation, Childs and Collis-George made two assumptions: (i) that the resistance to flow is caused only by the smaller radius in the rejoined cross-section and (ii) that there is no more than one connection between pores. Accordingly, using Eq. [6] and [26], the contribution to the specific discharge of pores characterized by the pair r, ρ, is

$$d\bar{q} = M\, a(r, \rho)\, \rho^2\, \nabla\phi \qquad [27]$$

where ρ is the smaller radius and M is a constant which accounts for geometry and fluid properties. By integration of Eq. [27] over the filled pores and making an analogy between the resulting expression for q and the Darcy equation, Eq. [2], one obtains

$$K(\theta) = M \int_{\rho = R_{min}}^{\rho = R(\theta)} \int_{r=\rho}^{r = R(\theta)} \rho^2 f(\rho) f(r)\, dr\, d\rho + M \int_{\rho = R_{min}}^{\rho = R(\theta)} \int_{r = R_{min}}^{r=\rho} r^2 f(r) f(\rho)\, d\rho\, dr. \qquad [28]$$

With Eq. [27] and [28] valid for this model, K can be computed for any given θ using the measured soil water retention curve. Childs and Collis-George suggested [to transform the $\theta(\psi)$ to a $\theta(r)$ curve ($r \propto 1/\psi$)] dividing it into constant r intervals and carrying out the integration by Eq. [28]. This tedious computational procedure has been improved by Marshall (1958), who used equal water content intervals, and further modified by Kunze et al. (1968) to yield

$$K(\theta)_l = M_1 \Delta\theta^2 \sum_{i=1}^{\ell} \frac{2(\ell - i) + 1}{\psi_i^2}; \qquad M_1 = \frac{1}{2} \frac{\sigma^2}{\gamma\mu} \qquad [29]$$

where ℓ is the number of $\Delta\theta$ intervals, σ is the water surface tension, and μ is the dynamic viscosity. Using data on 10 soils, Marshall found fair agreement between $K(\theta)$ computed by Eq. [29] and the measured curves. However, Nielsen et al. (1960) have found that computation of the hydraulic conductivity is significantly improved when a correlation factor is applied to match the predicted K with the measured value at saturation.

This conclusion, strengthened by a series of later works, implies that the $K_r(\theta)$ equation

$$K_r(\theta) = \sum_{i=1}^{\ell} \frac{2(\ell - i) + 1}{\psi^2} \Big/ \sum_{j=1}^{m} \frac{2(m - 1) + 1}{\psi^2}\ ; \qquad m = \theta_{\text{sat}}/\Delta\theta \quad [30]$$

is a more accurate computational formula than Eq. [29]. Kunze et al. (1968) indicated that for $m > 32$, K is practically independent of m. Mualem (1974, 1976a) showed that the analytical solution of Eq. [28] is

$$K_r(\theta) = \int_0^{\theta} \frac{(\theta - \vartheta)}{\psi^2}\, d\vartheta \Big/ \int_0^{\theta_{\text{sat}}} \frac{(\theta_{\text{sat}} - \vartheta)}{\psi^2}\, d\vartheta \qquad [31]$$

where ϑ is a dummy variable of integration representing the effective water content as a function of ψ. This form allows the derivation of a closed-form solution for $K_r(\theta)$ when the soil water retention curve is given as a mathematical function.

Other investigators including Millington and Quirk (1961), Jackson et al. (1965), Brutsaert (1967), Green and Corey (1971), Bruce (1972), Jackson (1972), and Mualem and Dagan (1976) have discussed or experimentally tested this model. Millington and Quirk modified the Childs-Collis-George (CCG) model by multiplying the results with a correction factor S_e^n to account for correlation between pores of adjacent cross-section. They estimated n to be 4/3. Figure 31–3 shows a comparative test of various formulas carried out by Jackson et al. (1965). Kunze et al. (1968), followed by Jackson (1972), recommended $n = 1$ on the basis of

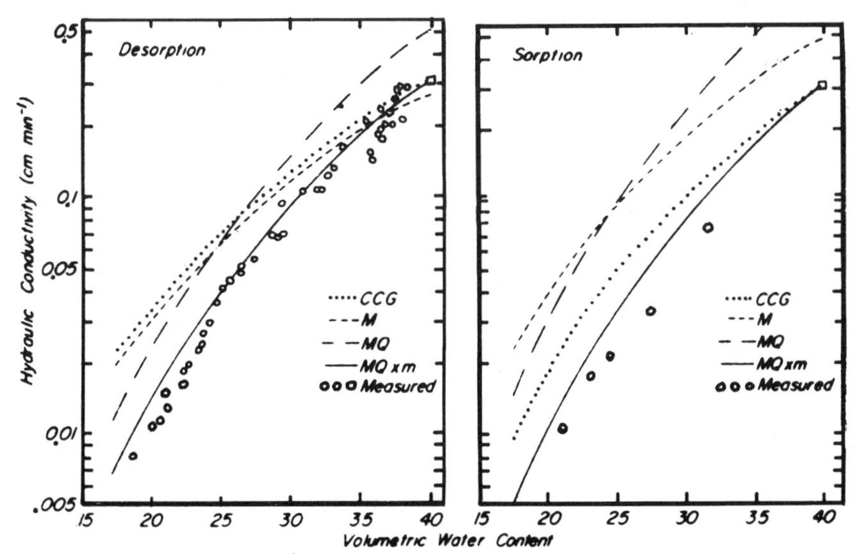

Fig. 31–3. Hydraulic conductivities measured and calculated in drainage and imbibition for 50- to 500-μm sand (after Jackson et al., 1965).

better agreement with measured data. The differences, however, between computed results with $n = 4/3$ and $n = 1$ are minor.

Wyllie and Gardner (1958) analyzed a simplified picture of the CCG model, made of randomly joined thin layers traversed by parallel capillaries. They assumed that flow is controlled mainly by the joint interface between the layers. As a result of random rejoining of the porous slab, the probability that a pore of radius r at one layer will encounter pores on the other is $a_e(r) = f(r)\,\theta/\lambda$, where λ is a constant correction factor. Accordingly, the porosity of the contact interface is interpreted as if the radius of each pore is reduced to $r_e = r(\theta/\lambda)^{1/2}$. Substitution of $a_e(r)$ for $f(r)$ and r_e for r in Eq. [20] yields Burdine's formula, Eq. [24]. It seems, however, that the physical interpretation of the CCG idea of random joining of identical cross-sections of porous medium has been given too much weight. In flow cases, where the Hagen-Poiseuille equation or the Darcy law is applicable, there is no resistance to flow at a cross section.

Mualem (1976a) analyzed a conceptual model of porous media similar to that of CCG. Consider a porous slab of thickness $\Delta x(x \rightarrow x + \Delta x)$ along the flow line. In homogeneous soil, the pore area distributions at the two slab sides are identical. For $\Delta x \gg R_{max}$, complete randomness of the relative position of the two slab faces can be assumed. The probability of pores of radii $r \rightarrow r + dr$ at x encountering pores of radii $\rho \rightarrow \rho + d\rho$ at $x + dx$ is $a(r, \rho) = f(r)f(\rho)\,drd\rho$. In this case, however, no direct connection between the pores r and ρ exists along the x axis. In the other extreme case, when $\Delta x \rightarrow$ o the correlation between the slab's two faces is complete and the pores at both faces are identical. In order to take into account the effect of changes in pore configuration on hydraulic conductivity, Δx should be same order of magnitude as the pore radii. In such a case, the probability of the pore domain characterized by (r, ρ) is

$$a(r, \rho) = G(R, r, \rho)f(r)f(\rho)drd\rho \qquad [32]$$

where $G(R, r, \rho)$ is a correction factor accounting for partial correlation between r and ρ at a given water content $\theta(R)$. The contribution of the actual flow configuration in pores (r, ρ) to the total hydraulic conductivity is assessed using two simplifying assumptions: (i) there is no bypass between the slab pores, and (ii) the pore configuration can be replaced by a pair of capillary elements whose lengths are proportional to its radii

$$l_1/l_2 = r/\rho\,. \qquad [33]$$

The hydraulic conductivity of the sequel is then found proportional to $r_e^2 = r\rho$. To account for the eccentricity of the flow path, a tortuosity factor $T(R, r, \rho)$ is applied. As there is no mechanism to estimate $T(R, r, \rho)$ and $G(R, r, \rho)$ independently, both factors are assumed to be a power function of the effective saturation. These simplifying assumptions lead to

$$K_r(\theta) = S_e^n \left[\int_0^{\theta} \frac{d\theta}{\psi} \bigg/ \int_0^{\theta_{sat}} \frac{d\theta}{\psi} \right]^2. \qquad [34]$$

The value of the power n depends upon the specific soil-fluid properties, and thus may vary considerably for different soils. However, using relative permeability data of 45 soils, Mualem found that the optimal value, which minimizes the overall discrepancies between measured and predicted K_r, is $n = 0.5$.

While all models of K_r require knowledge of the residual water content, soil-water retention curves are rarely measured as low as θ_r. Kunze et al. (1968) particularly investigated the effect of using a partial $\psi(\theta)$ curve on the computed hydraulic conductivity. They concluded that the accuracy of the computed K_r is significantly improved when a complete $\psi(\theta)$ curve is used—namely, when the condition $\theta = \theta_r$; $K_r = 0$ is realistically prescribed. Brooks and Corey (1964), Mualem (1976a, 1976b), and Van Genuchten (1980) have suggested different methods for extrapolation of the measured $\theta(\psi)$ and determination of the residual moisture content.

The relative efficiencies (with respect to prediction accuracy) of the models of Averjanov, Burdine, Childs and Collis-George (as modified by Mualem, and Millington and Quirk), and Mualem, Eq. [6], [24], [31] (multiplied by $S_e^{4/3}$) and [34] with ($n = 0.5$), respectively, have been tested against each other and on data for 45 soils (Mualem, 1976a). The results indicate that there is no single model that fits every soil. However, on an overall basis, $K_r(\theta)$ curves computed with Mualem's formula have been in better agreement with measured data than the other models. Averjanov's model yields good results for sand (Fig. 31–4b), while for soils of fine texture, agreement with experimental data is often unsatisfactory (Fig. 31–4a). Among the statistical models, Burdine's equation has been found the least accurate as a predictive tool, especially for soils in which $d\theta/d\psi \neq 0$ as $\psi \to 0$. In such cases, there is an abrupt fall of the computed hydraulic conductivity near saturation which is only rarely supported by experimental data.

Mualem and Dagan (1978), following the conceptual model of Mualem (1976a), presented a systematic analysis by which they could derive the models of Purcell (1949), Childs and Collis-George (1950), Burdine (1953) and Mualem (1976a), as well as other existing and new versions, from some common statistical principles.

When matching the computed K with measured values at saturation, the general form of the relative permeability is

$$K_r(\theta) = \int_{\substack{\text{over} \\ \text{filled pores}}} r_e^2 \, da_e \bigg/ \int_{\substack{\text{over} \\ \text{all pores}}} r_e^2 \, da_e . \qquad [35]$$

The differences between the various statistical models stem from the assumptions regarding the dependence of r_e and da_e upon r, ρ, and $R(\theta)$

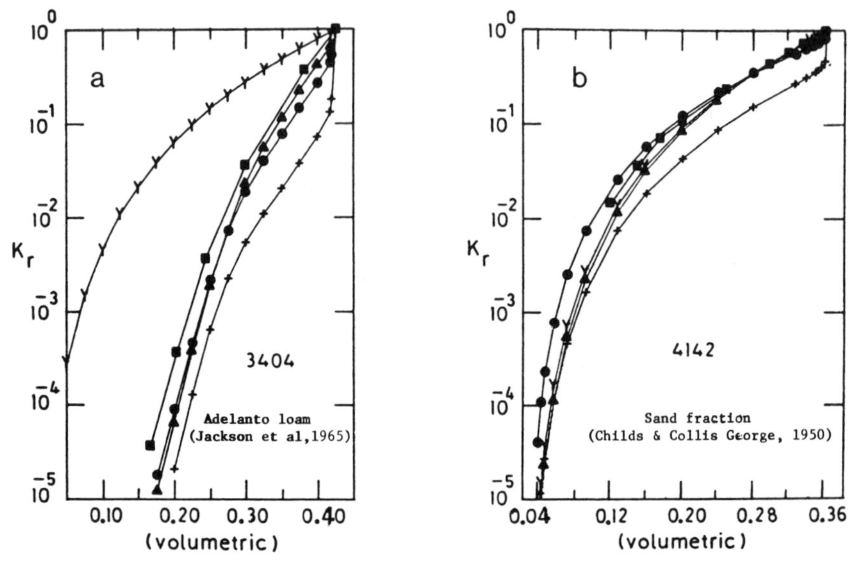

Fig. 31–4. Computed relative hydraulic conductivities using four models: Mualem *-*-*; WG +-+-+; Averjanov Y-Y-Y; MQ ▲-▲-▲, and the measured curve ■ ■ ■ (After Mualem, 1976a).

(R being the largest radius of the filled pores at water content θ). The "effective radius," r_e, and "area effective to flow," da_e, can be expressed in general terms by

$$r_e = J(r, \rho, R) \quad [36]$$

$$da_e = F(r, \rho, R)\, dr d\rho \quad [37]$$

leading to

$$K_r(\theta) = \frac{\int_{\text{over filled pores}} J(r,\rho,R)F(r,\rho,R)\,dr\,d\rho}{\int_{\text{over all pores}} J(r,\rho,R)F(r,\rho,R)\,dr\,d\rho} \quad [38]$$

where $F(r, \rho, R)$ can be statistically interpreted as the function which describes the probability of having connection between pores of radius r in one cross section of the soil at x, and pores of class ρ at the other cross section at $x + dx$, at water content $\theta(R)$. Accordingly, $J(r, \rho, R)$ is interpreted as the equivalent pore radius of the sequence of pores r,ρ, capable of conducting the same flux under the same gradient $d\phi/dx$. Table 31–2 shows the resulting formulas for K_r when correlation between pores and tortuosity effect are taken into account systematically. These expressions can be represented by three general formulas:

$$K_r(\theta) = S_e^n \int_0^\theta \frac{(\theta - \vartheta)}{\gamma^{2+b}} \, d\vartheta \bigg/ \int_0^{\theta_{sat}} \frac{(\theta_{sat-\vartheta})}{\gamma^{2+b}} \quad [39]$$

$$K_r(\theta) = S_e^n \int_0^\theta \frac{d\theta}{\psi^{2+b}} \bigg/ \int_0^{\theta_{sat}} \frac{d\theta}{\psi^{2+b}} \quad [40]$$

$$K_r(\theta) = S_e^n \left[\int_0^\theta \frac{d\theta}{\psi^{1+b}} \bigg/ \int_0^{\theta_{sat}} \frac{d\theta}{\psi^{1+b}} \right]^2 . \quad [41]$$

Equations [39] to [41] have two degrees of freedom and thus may adapt better to a wide spectrum of soils.

Based on the analogy between water, heat, and electrical conductivity, Farrel and Larson (1972) suggested using existing equations of electrical conductivity derived for heterogenous media made of spherical or cylindrical elements to determine the hydraulic conductivity of unsaturated soil. They also assumed that the soil retention curve represents the pore water distribution function. To simplify computation, they assumed that the measured $\psi(\theta)$ curve can be represented by an empirical function

$$\psi = \psi_{cr} \exp[\alpha(1 - S_e)].$$

However, the resulting equation for $K(\psi)$ is not explicit and requires an iterative computation. Comparison between this model and the Childs and Collis-George model, based on hypothetical soil-water retention curves, indicates a very significant difference between the predicted hydraulic conductivity curves, by the two models, for soils with a wide range of pore distribution. Unfortunately, neither the authors nor other investigators have checked the model against experimental results. It thus remains questionable whether the observed deviations from the CCG model represent better accuracy of the Farrel and Larson model or results from an unsound analogy.

To save computer time in numerical solutions and simplify computation of $K(\theta)$ for other practical uses, analytical $\psi(\theta)$ functions are sometimes fitted to the measured retention curves. These functions, when introduced in the integral formulas of K_r, yield closed-form analytical formulas of $K_r(\theta)$ or $K_r(\psi)$ (Brooks & Corey, 1964; Brutsaert, 1967; Mualem, 1974, 1976a; Van Genuchten, 1980). Table 31–3 summarizes some of these formulas obtained by using various models. It is interesting that the three statistical models yield the power function relationship $K_r = S_e^m$ for soil-water retention represented by $S_e = (\psi/\psi_{cr})^{-\lambda}$. Furthermore, the models of Childs and Collis-George and of Mualem lead to identical results. The soil-water function proposed by Van Genuchten

$$S_e = [1 + (\psi/\psi_1)^\alpha]^{-\beta} \quad [42]$$

does not allow a simple closed form integration when applied by various

Table 31–2. Formulae of $K_r(\theta)$ derived on the basis of different statistical models.

$K_r(\theta) = \iint J(r,\varrho,R)\,F(r,\varrho,R)\,dr\,d\varrho \,/\, \oint\!\!\oint J(r,\varrho,R)\,F(r,\varrho,R)\,dr\,d\varrho$			
Correlation	A-full Purcell	B-zero Childs & Collis-George and Wyllie & Gardner	C-partial MQ, Mualem
$F(r,\varrho,R)$	$f(r)\,\delta(r-\varrho)$	$f(r)\,f(\varrho)$	$g(R)\,f(r)\,f(\varrho)$
$\int_0^R\int_0^R F(r,\varrho,R)\,dr\,d\varrho$	$\theta(R)$ $f(r)\,dr = d\theta$	$\theta^2(R)$ $f(r)\,dr = d\theta$	$\theta^n(R);\ 1 \le n \le 2$ * $f(r)\,dr = d\theta$
Effective radius $J(r,\varrho,R)$	Relative hydraulic conductivity, $K_r(\theta)$		
r^2	$\int_0^\theta \frac{d\theta}{\psi^2} \Big/ \int_0^{\theta_{sat}} \frac{d\theta}{\psi^2}$	$S_e \int_0^\theta \frac{d\theta}{\psi^2} \Big/ \int_0^{\theta_{sat}} \frac{d\theta}{\psi^2}$	$S_e^{\,n-1} \int_0^\theta \frac{d\theta}{\psi^2} \Big/ \int_0^{\theta_{sat}} \frac{d\theta}{\psi^2}$
Childs & Collis-George $r^2;\ r < \varrho$ $\varrho^2;\ r > \varrho$		$\int_0^\theta \frac{(\theta-\vartheta)}{\psi^2(\vartheta)}\,d\vartheta \Big/ \int_0^{\theta_{sat}} \frac{(\theta_{sat}-\vartheta)}{\psi^2(\vartheta)}\,d\vartheta$	$S_e^{\,n-2} \int_0^\theta \frac{(\theta-\vartheta)}{\psi^2(\vartheta)}\,d\vartheta \Big/ \int_0^{\theta_{sat}} \frac{(\theta_{sat}-\vartheta)}{\psi^2(\vartheta)}\,d\vartheta$
Wyllie & Gardner $r^2\,\theta(R)$		$S_e^2 \int_0^\theta \frac{d\theta}{\psi^2} \Big/ \int_0^{\theta_{sat}} \frac{d\theta}{\psi^2}$	

(continued on next page)

Table 31-2. Continued.

$K_r(\theta) = \iint J(r,\varrho,R)\,F(r,\varrho,R)\,dr\,d\varrho / \oiint J(r,\varrho,R)\,F(r,\varrho,R)\,dr d\varrho$			
Correlation	A-full Purcell	B-zero Childs & Collis-George and Wyllie & Gardner	C-partial MQ, Mualem
Mualem $r\,\varrho$		$\left[\int_0^{\theta}\frac{d\theta}{\psi} / \int_0^{\theta_{sat}}\frac{d\theta}{\psi}\right]^2$	$S_e^{\,n-2}\left[\int_0^{\theta}\frac{d\theta}{\psi} / \int_0^{\theta_{sat}}\frac{d\theta}{\psi}\right]^2$
$r^2\,\theta^m$	$s_e^{\,m}\int_0^{\theta}\frac{d\theta}{\psi^2} / \int_0^{\theta_{sat}}\frac{d\theta}{\psi^2}$	$S_e^{\,m+1}\int_0^{\theta}\frac{d\theta}{\psi^2} / \int_0^{\theta_{sat}}\frac{d\theta}{\psi^2}$	$S_e^{\,n+m-1}\int_0^{\theta}\frac{d\theta}{\psi^2} / \int_0^{\theta_{sat}}\frac{d\theta}{\psi^2}$
$r^2\,\theta^m;\ r < \varrho$ $\varrho^2\,\theta^m;\ r > \varrho$		$S_e^{\,m}\int_0^{\theta}\frac{(\theta-\vartheta)}{\psi^2}\,d\vartheta / \int_0^{\theta_{sat}}\frac{(\theta_{sat}-\vartheta)}{\psi^2}\,d\vartheta$	$S_e^{\,n+m-2}\int_0^{\theta}\frac{(\theta-\vartheta)}{\psi^2}\,d\vartheta / \int_0^{\theta_{sat}}\frac{(\theta_{sat}-\vartheta)}{\psi^2}\,d\vartheta$
$r\varrho\,\theta^m$		$S_e^{\,m}\left[\int_0^{\theta}\frac{d\theta}{\psi} / \int_0^{\theta_{sat}}\frac{d\theta}{\psi}\right]^2$	$S_e^{\,n+m-2}\left[\int_0^{\theta_{sat}}\frac{d\theta}{\psi} / \int_0^{\theta_{sat}}\frac{d\theta}{\psi}\right]^2$
r^{2+b}	$\int_0^{\theta}\frac{d\theta}{\psi^{2+b}} / \int_0^{\theta_{sat}}\frac{d\theta}{\psi^{2+b}}$	$S_e\int_0^{\theta}\frac{d\theta}{\psi^{2+b}} / \int_0^{\theta_{sat}}\frac{d\theta}{\psi^{2+b}}$	$S_e^{\,n-1}\int_0^{\theta}\frac{d\theta}{\psi^{2+b}} / \int_0^{\theta_{sat}}\frac{d\theta}{\psi^{2+b}}$
$r^{2+b};\ r < \varrho$ $\varrho^{2+b};\ r > \varrho$		$\int_0^{\theta}\frac{(\theta-\vartheta)}{\psi^{2+b}}\,d\vartheta / \int_0^{\theta_{sat}}\frac{(\theta_{sat}-\vartheta)}{\psi^{2+b}}\,d\vartheta$	$S_e^{\,n-2}\int_0^{\theta}\frac{(\theta-\vartheta)}{\psi^{2+b}}\,d\vartheta / \int_0^{\theta_{sat}}\frac{(\theta_{sat}-\vartheta)}{\psi^{2+b}}\,d\vartheta$
$(r\varrho)^{1+b}$		$\left[\int_0^{\theta}\frac{d\theta}{\psi^{1+b}} / \int_0^{\theta_{sat}}\frac{d\theta}{\psi^{1+b}}\right]^2$	$S_e^{\,n-2}\left[\int_0^{\theta}\frac{d\theta}{\psi^{1+b}} / \int_0^{\theta_{sat}}\frac{d\theta}{\psi^{1+b}}\right]^2$

Table 31-3. Models of the relative hydraulic conductivity and the results for three analytic approximations of the $S_e(\psi)$ relationship.

	Reference	A	B	C	D
1	Reference	$g(K, S_e, \psi)$ \ $f(\psi, S_e)$	$S_e = (\psi/\psi_{cr})^{-\lambda}$	$\psi = \psi_{cr} \exp[\alpha(1 - S_e)]$	$S_e = [1 + (\psi/\psi_1)^{\alpha}]^{-\beta}$
2	Mualem form of Childs & Collis-George CCC, $n = 0$ MQ, $n = 4/3$ Kunze, $n = 1$	$K_r(S_e) = S_e^n \dfrac{\int_0^{S_e} (S_e - s)\psi^{-2}(s)\,ds}{\int_0^1 (1 - s)\psi^{-2}(s)\,ds}$	$K_r(S_e) = S_e^{(2+2\lambda)/\lambda+n}$ $K_r(\psi) = (\psi/\psi_{cr})^{-(2+2\lambda)-n\lambda}$	$K_r(S_e) = S_e^n \dfrac{[\exp(2\alpha S_e) - 2\alpha S_e - 1]}{[\exp(2\alpha) - 2\alpha - 1]}$	No simple closed form integral.
3	Burdine Wyllie & Gardner	$K_r(S_e) = S_e^2 \dfrac{\int_0^{S_e} \psi^{-2}(s)\,ds}{\int_0^1 \psi^{-2}(s)\,ds}$	$K_r(S_e) = S_e^{(2+3\lambda)\lambda}$ $K_r(\psi) = (\psi/\psi_{cr})^{-(2+3\lambda)}$	$K_r(S_e) = S_e^2 \dfrac{[\exp(2\alpha S_e) - 1]}{[\exp(2\alpha) - 1]}$	$K_r(S_e) = S_e^2[1 - (1 - S_e^{1/\beta})^{\beta}]$ $\beta = 1 - 2/\alpha$
4	Mualem $n = 0.5$	$K_r(S_e) = S_e^n \left[\dfrac{\int_0^{S_e} \psi^{-1}(s)\,ds}{\int_0^1 \psi^{-1}(s)\,ds}\right]^2$	$K_r(S_e) = S_e^{(2+2\lambda)/\lambda+n}$ $K_r(\psi) = (\psi/\psi_{cr})^{-(2+2\lambda)-n\lambda}$	$K_r(S_e) = S_e^n \dfrac{[\exp(2\alpha S_e) - 2\exp(\alpha S_e) + 1]}{[\exp(2\alpha) - 2\exp(\alpha) + 1]}$	$K_r(S_e) = S_e^n[1 - (1 - S_e^{1/\beta})^{\beta}]^2$ $\beta = 1 - 1/\alpha$
5	Yuster $n = 2.0$ Irmay $n = 3.0$ Averjanov $n = 3.5$ Corey $n = 4.0$ Mualem $n = 3.0 + 0.015w$	$K_r(S_e) = S_e^n$	$K_r(\psi) = (\psi/\psi_{cr})^{-n\lambda}$	$K_r(\psi) = [1 - (1/\alpha)\ln(\psi/\psi_{cr})]^n$	$K_r(\psi) = [1 + (\psi/\psi_1)^{\alpha}]^{-\beta n}$
6	Wind, Wesseling	$K_r = (\psi/\psi_{cr})^{-n}$; $\psi < \psi_{cr}$ $K_r = 1$; $\psi > \psi_{cr}$	$K_r(S_e) = S_e^{n/\lambda}$	$K_r(\psi) = \exp[-\alpha n(1 - S_e)]$	Incompatible models
7	Gardner	$K = \dfrac{a}{\lvert\psi\rvert^n + b}$; $K_r = \dfrac{1}{(\psi/\psi_1)^n + 1}$	Incompatible models	Incompatible models	$K_r(S_e) = \{[(1 - S_e^{1/\beta})/S_e^{1/\beta}]^{n/\alpha} + 1\}^{-1}$
8	Gardner	$K_r = \exp[a(\psi - \psi_{cr})]$; $\psi < \psi_{cr}$ $K_r = 1$; $\psi > \psi_{cr}$	$K_r = \exp[a\psi_{cr}^{-1}(S_e^{-1/\lambda} - 1)]$	$K_r(S_e) = \exp\{(a/\psi_{cr})\exp[\alpha(1 - S_e)]\}$	Incompatible models

models in its general form, namely, for independent α and β. However, models of Burdine and Mualem yield simple closed-form formulas when α and β are related by $\beta = 1 - 2/\alpha$ and $\beta = 1 - 1/\alpha$, respectively. It should be noted, therefore, that the resulting $K_r(S_e)$ formulas from Burdine and Mualem models correspond to different (S_e,ψ) functions.

31–2.5 Hysteresis of the Hydraulic Conductivity

The hydraulic conductivity, capillary head, and water-content relations are not unique functions. They depend on the history of wetting and drying processes to which the porous medium was subjected. Figure 31–5 shows typical hysteretic loops measured by Topp and Miller (1966). The $K(\psi)$ hysteresis is magnified by orders of magnitude relative to the $\theta(\psi)$ hysteresis, while the $K(\theta)$ hysteresis is much smaller. The maximum hysteretic difference (between points on the main drying and wetting curves having the same value of ψ) relative to the value on the main wetting curve is approximately 300% in the $S(\psi)$ plane (Fig. 31–5) and 20 thousand percent in $K(\psi)$, but only 40% in $K(\theta)$. This phenomenon has been observed by several investigators including Nielsen and Biggar (1961), Topp (1969), Poulovassilis (1970), Talsma (1970), and Vachaud and Thony (1971). However, Staple (1965) and Dane and Wierenga (1975) reported a considerable $K(\theta)$ hysteresis, with an order of magnitude deviation between corresponding values of K_r on the main wetting and the main drying curves. The experimental results indicate also that for some soils $K_r(\theta)$ in a drying process is higher than $K_r(\theta)$ in a wetting process, while for other soils the higher conductivity is observed in the wetting processes. Thus, it is preferable to use the $K(\theta)$ relationship rather than $K(\psi)$ for any practical use in which wetting and drying processes are involved, if hysteresis is to be neglected.

All quantitative models and methods of calculating $K(\theta)$ or $K(\psi)$ discussed above do not account for the soil-water hysteresis. The statistical models, for instance, are underlain by the assumption that each soil has a single pore-distribution function represented by the soil-water retention curve. This assumption is not consistent with the hysteretic nature of unsaturated soil. One may use, of course, one branch of the $\theta(\psi)$ hysteresis (either drying or wetting) to compute a single $K_r(\theta)$ function (Kunze et al., 1968), which would be acceptable from a practical point of view. Theoretically, however, it is not absolutely sound. One of the contradictory results is that the $K(\theta)$ formulas yield different K_{sat} values and require, therefore, different matching factors depending on whether the soil is brought to saturation on one branch of the $\theta(\psi)$ loop or another. Consequently, $K(\theta)$ curves computed from the two main branches of the hysteresis loop can neither duplicate nor yield a closed hysteretic loop which complies with the $\theta(\psi)$ loop. Mualem (1974, 1976b) proposed principles that allow prediction of the hysteresis in $K(\psi)$ as well as $K(\theta)$.

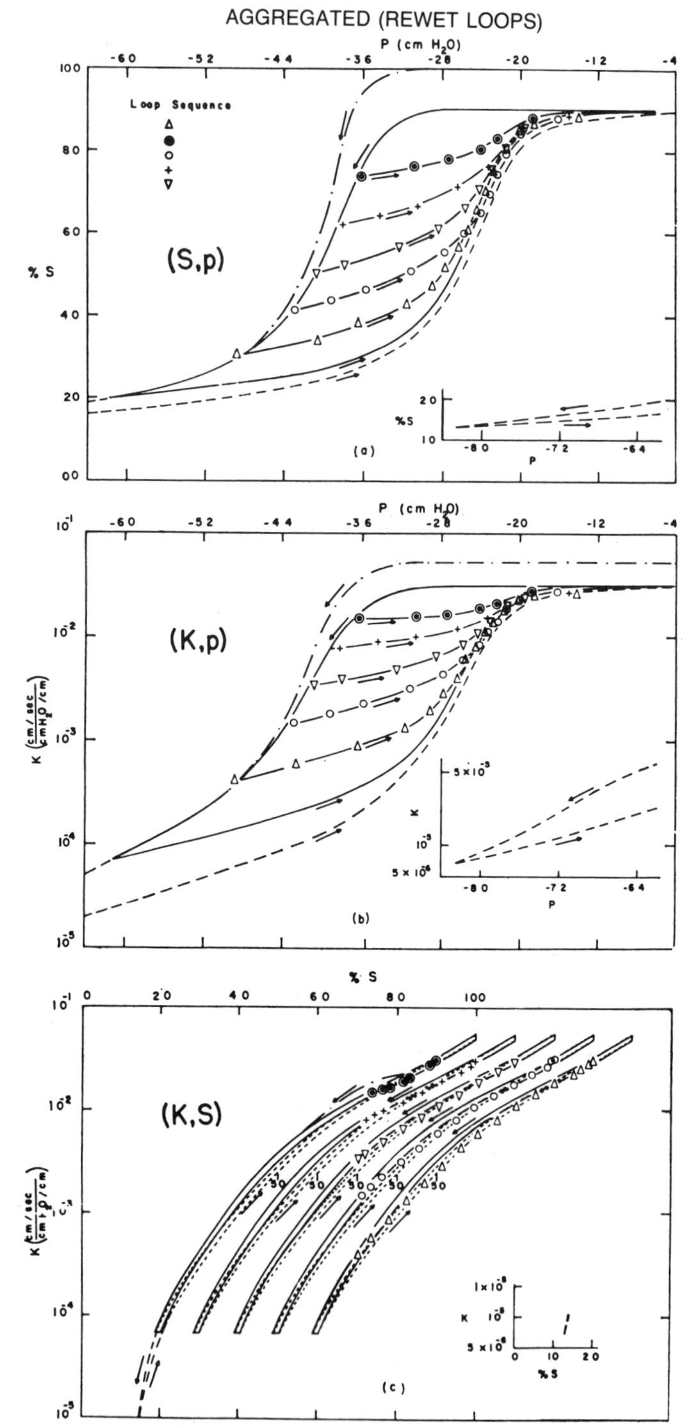

Fig. 31–5. Hysteresis of the $S(\psi)$, $K(\psi)$, and $K(S)$ measured on aggregated glass beads (after Topp & Miller, 1966).

31-3 RECOMMENDATIONS

In the above review of the state of the art of empirical and theoretical models for calculating the hydraulic conductivity functions, a rather wide range of possible approaches has been identified. The following comments and suggestions may help the reader select an appropriate approach to follow. In general, the method to be used will depend on the soil data and other information that may be available. Four cases are considered.

31-3.1 Case A.

No measured conductivity or retention data are available. Only the soil classification (texture and structure) is known.

1. Estimate the saturated hydraulic conductivity, the residual water content, and the saturated water content, based on general information and guidelines in textbooks (Baver et al., 1974; Van Wijk, 1966; Corey, 1977; Marshall & Holmes, 1979) and collections of soil-water data (Brooks & Corey, 1964; Holtan et al., 1968; Mualem, 1976a, 1976c).
2. Use a power function, Eq. [13], to calculate the relative hydraulic conductivity:

$$K_r = S_e^n$$

with n equal 3 to 3.5 for sands, 3.5 to 5 for loams, and 5 to 8 for clays. Other values can be adopted based on resemblance between the soil under consideration and soil data found in literature. Granular, well-structured soils tend to behave like coarse-textured soils, and an exponent of 3.5 to 4.0 may well represent their behavior.

3. As an alternative, $K(\psi)$ may be estimated using Gardner's type of equation (Table 31-1) and following his estimated values of a, b, and n for various soils (Gardner, 1958).

31-3.2 Case B.

The soil-water retention curve for a drying process and the saturated hydraulic conductivity are available.

1. If the minimum measured value of water content, Θ_{min}, is higher than the residual water content, extrapolate the measured retention curve into the dry range and estimate the residual water content. Extrapolation can be done either graphically by hand, based on experience and knowledge as mentioned in Case A, and following Corey (1977), or analytically. To apply the latter procedure an empirical function should be fitted to the measured data. The power function

$$S_e = S_{e\,min}\,(\psi/\psi_{min})^{-\lambda}; \qquad S_e = (\Theta - \Theta_r)/(\Theta_s - \Theta_r) \qquad [43]$$

is one of the relatively simple formulas that can be used. Since Eq.

[43] cannot produce the S-shape of the water retention curve, only data in the range between the inflection point and the minimum measured value, $\psi_{\min}$, should be used for the curve-fitting. Note that by Eq. [43] the extrapolated curve is anchored at $(\psi_{\min}, S_{e\,\min})$, and thus continuity of the first and the second order is achieved. For any assumption of Θ_r, the corresponding λ can be calculated by a regression procedure. The highest correlation factor should be the criterion for choosing the final values of Θ_r and λ.

2. Use Eq. [34] to compute the relative hydraulic conductivity in the extrapolated section by

$$K_r(S_e) = \frac{S_e^{1/2}}{M}\left[\frac{S_e}{(1 + 1/\lambda)\psi}\right]^2; \quad \psi < \psi_{\min} \qquad [44a]$$

and for $\psi > \psi_{\min}$ by

$$K_r(S_e) = \frac{S_e^{1/2}}{M}\left[\frac{S_{e\,\min}}{(1 + 1/\lambda)\psi_{\min}} + \int_{S_{e\,\min}}^{S_e}\frac{dS_e}{\psi}\right]^2 \qquad [44b]$$

where

$$M = \left[\frac{S_{e\,\min}}{(1 + 1/\lambda)\psi_{\min}} + \int_{S_{e\,\min}}^{1}\frac{dS_e}{\psi}\right]^2. \qquad [44c]$$

It is recommended that 30 or more intervals of water content (or S_e) be used for the numerical integration in Eq. [44b] and [44c]. On the average, Eq. [44] has been found more accurate than Eq. [24] or [31].

3. As an alternative way of calculating $K(\theta)$, with an overall expected accuracy similar to that obtained by Eq. [43] and [44], the macroscopic approach can be applied using Eq. [14] and [16]. Extrapolation of the measured retention curve may be required if the water content at wilting point (−1.5 MPa) is not already given. This method yields a continuous power function relationship between the relative hydraulic conductivity and the effective saturation. The method should not be used for soils with significant macroscopic porosity as a result of cracks, dead roots, and worm holes.
4. When closed-form functions for both $\Theta(\psi)$ and $K(\Theta)$ are desired as, for example, in a numerical solution of flow problem, one of the three possible $S_e(\psi)$ empirical formulas presented in the first row of Table 31–3, can be fitted to measured water-retention data. The corresponding formulas for $K_r(S_e)$ either in the second or the fourth rows are recommended for their better accuracy.

Equation [1B] of Table 31–3 has the advantage of being a very simple function. For sand and granular soils with a well defined air entry value,

Eq. [1B] can be fitted to the measured retention data in the range between ψ_{min} and $\psi = 0$, following a procedure similar to that discussed in Case B2.

Equation [1D] of Table 31–3 features the S-shape characteristic of the soil water retention curve. Therefore, it may better represent the measured retention curves in soils that do not display an abrupt increase in capillary head near saturation (i.e., $d\theta/d\psi > 0$ as $\psi \rightarrow 0$), as in the case of silt and clay soil. However, the analytical procedure of curve-fitting in this case would be more complicated than that required for Eq. [1B]. If a numerical program for curve-fitting (Van Genuchten, 1980) is available, Eq. [1D] and the corresponding $K_r(S_e)$ function, Eq. [4D], are recommended for use when unsaturated fine-textured soils are considered.

31.3.3 Case C.

A complete $\Theta(\psi)$ hysteresis loop and K_{sat} are available.

1. If a single $K(\Theta)$ curve is desired, use the main drying curve and follow guidelines presented in Case B.

31.3.4 Case D.

Data for $K(\Theta)$ and $\Theta(\psi)$ are available.

1. An empirical form (one of those presented in Table 31–1) can be fitted to the data to represent the hydraulic conductivity–water content relationships by a mathematical function, to facilitate use of the data in solutions of the soil water flow equation.

31–4 REFERENCES

Averjanov, S. F. 1950. About permeability of subsurface soils in case of incomplete saturation. p. 19–21. *In* English Collection, Vol. 7, (1950), as quoted by P. Ya Palubarinova, 1962, The theory of ground water movement. (English translation by I. M. Roger DeWiest, Princeton University Press).

Baver, L. D., Gardner, W. H., and W. R. Gardner. 1972. Soil physics. John Wiley and Sons, New York.

Braester, C., Dagan, G., Neuman, S., and D. Zaslavsky. 1971. A survey of equations and solutions of unsaturated flow in porous media. Technion Project no. A10-SWC-77, Haifa. Israel Institute of Technology.

Brooks, R. H., and A. T. Corey. 1964. Hydraulic properties of porous media. Hydrology Paper 3, Colorado State Univ., Fort Collins, CO.

Brooks, R. H., and A. T. Corey. 1966. Properties of porous media affecting fluid flow. J. Irrig. Drain. Div., Am. Soc. Civil Eng. 92(IR2):61–88.

Bruce, R. R. 1972. Hydraulic conductivity evaluation of the soil profile from soil water relations. Soil Sci. Soc. Am. Proc. 36:550–560.

Brutsaert, W. 1967. Some methods of calculating unsaturated permeability. Trans. ASAE 10:400–404.

Burdine, N. T. 1953. Relative permeability calculation size distribution data. Pet. Trans. Am. Inst. Min. Metal. Pet. Eng. 198:71–78.

Childs, E. C., and G. N. Collis-George. 1950. The permeability of porous materials. R. Soc. London, Proc., A 201:392–405.

Corey, A. T. 1977. Mechanics of heterogeneous fluids in porous media. Water Resources Publications, Ft. Collins, CO.

Dane, J. H., and P. J. Wierenga. 1975. Effect of hysteresis on the prediction of infiltration redistribution and drainage of water in a layered soil. J. Hydrol. 25:229–242.

Farrel, D. A., and W. E. Larson. 1972. Modeling the pore structure of porous media. Water Resour. Res. 3:699–706.

Fatt, I., and H. Dykstra. 1951. Relative permeability studies. Trans. Am. Inst. Min., Metall. Pet. Eng. 192:249–255.

Gardner, W. R. 1958. Some steady state solutions of the unsaturated moisture flow equation with application to evaporation from a water table. Soil Sci. 85:228–232.

Gates, J. I., and W. T. Lietz. 1950. Relative permeabilities of California cores by the capillary-pressure method, drilling and production practice. Am. Pet. Inst. Q:285–298.

Green, R. E., and J. C. Corey. 1971. Calculation of hydraulic conductivity: a further evaluation of some predictive methods. Soil Sci. Soc. Am. Proc. 35:3–8.

Holtan, H. N., England, C. B., Lawless, G. P., and G. A. Schumaker. 1968. Moisture-tension data for selected soils on experimental watersheds. USDA-ARS 41-144. U. S. Government Printing Office, Washington, DC.

Irmay, S. 1954. On the hydraulic conductivity of unsaturated soils. Trans. Am. Geophys. Union 35:463–467.

Irmay, S. 1971. A model of flow of liquid-gas mixtures in porous media and hysteresis of capillary potential. *In* Symposium on mathematical modeling of transport processes in porous media. Houston, TX. 28 February-4 March. AICHE, New York.

Jackson, R. D. 1972. On the calculation of hydraulic conductivity. Soil Sci. Soc. Am. Proc. 36:380–382.

Jackson, R. D., Reginato, R. J., and C. H. M. van Bavel. 1965. Comparison of measured and calculated hydraulic conductivities of unsaturated soils. Water Resour. Res. 1:375–380.

King, L. G. 1964. Imbibition of fluids by porous solids. Ph.D. thesis. Colorado State Univ., Ft. Collins.

Kozeny, J. 1927. Üeber kapillare Leitung des Wassers im Boden. Zitzungsber. Akad. Wiss. Wien. 136:271–306.

Kunze, R. J., Uehara, G., and K. Graham. 1968. Factors important in the calculation of hydraulic conductivity. Soil Sci. Soc. Am. Proc. 32:760–765.

Marshall, T. J. 1958. A relation between permeability and size distribution of pores. J. Soil Sci. 9:1–8.

Marshall, T. J., and J. W. Holmes. 1979. Soil physics. Cambridge University Press, New York.

Millington, R. J., and J. P. Quirk. 1961. Permeability of porous solids. Faraday Soc. Trans. 57:1200–1206.

Mualem, Y. 1974. Hydraulic properties of unsaturated porous media: a critical review and new models of hysteresis and prediction of the hydraulic conductivity. Technion, Proj. no. 38/74. Israel Institute of Technology, Haifa.

Mualem, Y. 1976a. A new model for predicting the hydraulic conductivity of unsaturated porous media. Water Resour. Res. 12:593–622.

Mualem, Y. 1976b. Hysteretical models for prediction of the hydraulic conductivity of unsaturated porous media. Water Resour. Res. 12:1248–1254.

Mualem, Y. 1976c. A catalogue of the hydraulic properties of soils. Proj. 442. Technion, Israel Institute of Technology, Haifa.

Mualem, Y. 1978. Hydraulic conductivity of unsaturated porous media: generalized macroscopic approach. Water Resour. Res. 14:325–334.

Mualem, Y., and G. Dagan. 1976. Methods of predicting the hydraulic conductivity of unsaturated soils. Proj. No. 442. Technion, Israel Institute of Technology, Haifa.

Mualem, Y., and G. Dagan. 1978. Hydraulic conductivity of soils: unified approach to the statistical models. Soil Sci. Soc. Am. J. 42:392–395.

Nielsen, D. R., and J. W. Biggar. 1961. Measuring capillary conductivity. Soil Sci. 92:192–193.

Nielsen, D. R., D. Kirkham, and E. R. Perrier. 1960. Soil capillary conductivity; comparison of measured and calculated values. Soil Sci. Soc. Am. Proc. 24:157–160.

Poulovassils, A. 1970. The effect of the entrapped air on the hysteresis curves of a porous body and its hydraulic conductivity. Soil Sci. 109:154–162.

Purcell, W. R. 1949. Capillary pressures—their measurement using mercury and the calculation of permeability therefrom. Pet. Trans. Am. Inst. Min., Metall. Pet. Eng. 186:39–48.

Rijtema, P. E. 1965. An analysis of actual evapotranspiration. Agric. Res. Rep. 659. Center for Agricultural Publications and Documentation, Wageningen, The Netherlands.

Staple, W. J. 1965. Moisture tension, diffusivity and conductivity of a loam soil during wetting and drying. Can. J. Soil Sci. 45:78–86.

Talsma, T. 1970. Hysteresis in two sands and the independent domain model. Water Resour. Res. 6:964–970.

Topp, G. C. 1969. Soil water hysteresis measured in a sandy loam compared with the hysteretic domain model. Soil Sci. Soc. Am. Proc. 33:645–651.

Topp, G. C., and E. E. Miller. 1966. Hysteretic moisture characteristics and hydraulic conductivities for glass-bead media. Soil Sci. Soc. Am. Proc. 30:156–162.

Vachaud, G., and J. L. Thony. 1971. Hysteresis during infiltration and redistribution in a soil column at different initial water content. Water Resour. Res. 7:111–127.

Van Genuchten, M. T. 1980. A closed form equation for predicting the hydraulic conductivity of unsaturated soils. Soil Sci. Soc. Am. J. 44:892–898.

Van Wijk, W. R. (ed.) 1966. Physics of plant environment. North-Holland Publishing Co., Amsterdam.

Wind, G. P. 1955. Field experiment concerning capillary rise of moisture in heavy clay soil. Neth. J. Agric. Sci. 3:60–69.

Wyllie, M. R. J., and G. H. F. Gardner. 1958. The generalized Kozeny-Carman equation 11. A novel approach to problems of fluid flow. World Oil Prod. Sect. 146:210–228.

Yuster, S. T. 1951. Theoretical considerations of multiphase flow in idealized capillary systems. Proc. World Pet. Congr., 3rd 2:437–445.

32 Intake Rate: Cylinder Infiltrometer

HERMAN BOUWER
U.S. Water Conservation Laboratory, ARS, USDA
Phoenix, Arizona

32-1 INTRODUCTION

Cylinder infiltrometers are metal cylinders that are pushed or driven a small distance into the soil. The area inside the cylinder is flooded with water and the rate at which water moves into the ground is measured. The flooding and the measurements normally are continued until the infiltration rate has become essentially constant. Cylinder infiltrometers frequently have a diameter of about 30 cm and a height of 20 cm, and they penetrate the soil about 5 cm. Unfortunately, the cylinder diameters commonly used are too small for accurate measurement and much larger diameters (as large as practicable but certainly not < 1 m) are necessary if meaningful data are to be obtained (see section 32–2.3). Cylinder infiltrometers are used to determine infiltration rates for inundated soils, for example, in connection with (i) surface or flood-type irrigation systems, (ii) infiltration basins for groundwater recharge, (iii) seepage from streams, canals, reservoirs, or wastewater lagoons, (iv) movement of leachate into the ground below sanitary landfills or other garbage disposal facilities, and (v) effectiveness of clay linings, soil compaction, or other treatment to reduce infiltration or seepage.

When using infiltrometers to predict infiltration rates for a given system, one must be careful that the conditions of that system (water quality, temperature, soil conditions, surface conditions, lengths of inundation, infiltration flow system in soil, etc.) are duplicated as much as possible with the infiltrometer so that the measurement is realistic. This is not always readily done (see sections 32–2.3 and 32–2.4). Thus, while cylinder infiltrometers in principle are simple devices and the necessary measurements can be taken with great precision, the results can be grossly in error if the conditions of the system for which they are to predict infiltration are not exactly duplicated in the infiltrometer test.

Cylinder infiltrometers should not be used where the geometry of the inundated surface drastically differs from that simulated by a circular device, such as flooded furrows between row crops. The relation between infiltration rates from furrows and cylinders is quite complex (Bouwer, 1961), so that infiltration behavior in furrows can best be measured in

 Methods of Soil Analysis, Part 1. Physical and Mineralogical Methods—Agronomy Monograph no. 9 (2nd Edition)

furrows (see chapter 34). By the same token, cylinder infiltrometers should not be used to predict infiltration where the water is in the form of falling drops (rainfall, sprinkler irrigation). Such conditions are much better simulated by sprinkler-type infiltrometers (see chapter 33).

32–2 PRINCIPLES

32–2.1 Time and Depth Effects

Infiltration of water into soil from surface inundations can be described by the Green-Ampt theory (Green & Ampt, 1911), which considers the infiltrated water to advance downward in the soil as piston flow (uniform hydraulic conductivity in wetted zone and constant pressure head of water at wetting front; see also Bouwer, 1978). Applying Darcy's equation to this flow system (Fig. 32–1) yields

$$v_i = K_w (H_w + L_f - h_f)/L_f \quad [1]$$

where v_i = infiltration rate (length/time), K_w = hydraulic conductivity of wetted zone (length/time), H_w = water depth above soil, h_f = pressure head of water at wetting front, and L_f = depth of wetting front.

Since the wetted zone seldom if ever is completely saturated, K_w of the wetted zone is less than the saturated hydraulic conductivity K_s. Limited experience indicates that K_w is about 0.5 K_s (Bouwer, 1966). The pressure head of the water at the wetting front can be taken as the critical pressure-head h_{cr} for wetting (Bouwer, 1964; Mein & Farrell, 1974; Morel-Seytoux & Khanji, 1974). The critical pressure head is defined as $\int K_h dh/K_0$ and is obtained from the relation between the unsaturated hydraulic conductivity K_h and the (negative) pressure head h of the water. K_0 refers to the hydraulic conductivity at zero pressure head and is equal to K_s when there is no entrapped air. For soils with a relatively uniform particle size, h_{cr} can be taken as the water-entry value of the soil, which in turn is equal to about one-half the air-entry value (Bouwer, 1966). Both h_{cr}

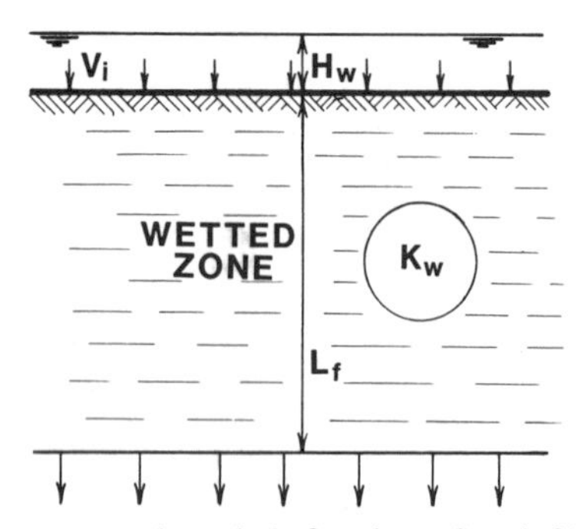

Fig. 32–1. Geometry and symbols for piston-flow infiltration system.

and K_w can be measured in situ with the air-entry permeameter (Bouwer, 1966; Aldabagh & Beer, 1971; Topp & Binns, 1976). Values of h_{cr} may range from -10 cm of water for medium to coarse sand to -100 cm and less (more negative) for fine-textured soils without structure (Bouwer, 1964,1966). The validity of the critical-pressure head and of other parameters used to estimate the term h_f in Eq. [1] was investigated by Freyberg et al. (1980).

Equation [1] shows that water depth affects infiltration rate most at the beginning of infiltration while L_f is still small. As infiltration progresses and L_f increases, the effect of water depth on infiltration rate decreases and eventually becomes negligible. The equation also shows that the infiltration rate at the beginning of the infiltration (small L_f) is much larger than K_w, but that it approaches K_w as infiltration continues and L_f increases. Eventually, the infiltration rate becomes essentially constant and equal to K_w, regardless of H_w. In this chapter, "the" infiltration rate implies the final infiltration rate, unless otherwise defined.

Where the soil is nonuniform and the profile contains a number of layers of different hydraulic conductivity, the final infiltration rate is equal to the harmonic mean of the hydraulic conductivities of the various layers (see section 32–4) if the layers become less permeable with depth (Bouwer, 1969). If the soil becomes more permeable with depth, there is a point in the soil profile below which the soil will not be completely wetted by the advancing wet front (Bouwer, 1976). The final infiltration rate in that case can still be calculated with Eq. [1], but the harmonic mean must be calculated only for the K_w values above that point. If the point at which the soil is no longer completely wetted is not deep compared to H_w, the water depth will then have a significant effect on final infiltration rate.

32–2.2 Restricting Surface Layers

The extreme case of a soil becoming more permeable with depth is where the soil is covered by a crust or other layer of low permeability that restricts the infiltration rate (Fig. 32–2). These conditions can develop where the soil surface is exposed to the impact of falling water drops, where the surface soil is compacted by traffic or other means, where the infiltrating water contains suspended solids (organic or inorganic) that accumulate on the surface or in surface-connected pores, where bacteria and other organisms grow on the surface and produce slimes, polysaccharides, and other metabolic products that clog the soil, etc. Whether or not infiltration is controlled by a thin surface layer of low permeability is readily ascertained by determining the infiltration rate first for the natural surface condition and then after the surface layer has been carefully removed. A drastic increase in infiltration rate indicates that there was a restricting layer at the surface.

The restricting surface layer normally is < 1 cm thick and its hydraulic properties are expressed in terms of the hydraulic impedance R_r, which is the thickness L_r of the layer divided by its hydraulic conductivity

K_r (Bouwer, 1964), or $R_r = L_r/K_r$. Since the flow through the surface layer numerically will be considerably less than the saturated hydraulic conductivity K_s of the underlying soil, the latter will not be completely wetted by the infiltrated water. Hence, the underlying soil will remain unsaturated and eventually reach a water content at which the corresponding unsaturated hydraulic conductivity is numerically equal to the infiltration rate. This is because the hydraulic gradient in the unsaturated material will approach 1 (Bouwer, 1964). Since the critical pressure head h_{cr} is essentially the center of the h-range where most of the reduction in K_h occurs when h decreases (becomes more negative), the pressure head in the unsaturated soil below the restricting layer can thus be approximated as h_{cr} for a range in v_i (or K_h). Thus, after infiltration has progressed for a long enough time to produce steady flow conditions in the upper portion of the unsaturated soil, the infiltration rate can be calculated by applying Darcy's equation to the restricting layer, yielding

$$v_i = (H_w - h_{cr})/R_r \qquad [2]$$

where R_r is the hydraulic impedence (L_r/K_r) of the restricting layer (Fig. 32–2). R_r can be determined in situ from the relation between infiltration rate and H_w. This relation should be measured for a wide range in H_w (for example, 0.5 to 1 m), as can be done with tall cylinder infiltrometers or by equipping the cylinder with a top plate and plastic standpipe as used in the air-entry permeameter (Bouwer, 1966). A plot of v_i against H_w then should yield a straight line with slope $1/R_r$ and intercept on the v_i-axis of $(L_r - h_{cr})/R_r$ (Bouwer, 1964).

Where R_r of the top layer is very large, as in severely clogged soils or artificial clay linings, h below the restricting layer may be considerably less (more negative) than h_{cr}. In that case, h below the restricting layer should be evaluated from the relation between unsaturated hydraulic conductivity and (negative) pressure head of the underlying material (Bouwer, 1982).

For coarse-textured materials and structured clays with well-developed macropores, h_{cr} is relatively close to 0. Equation [2] then shows that v_i is significantly affected by water depth, especially if H_w is relatively large. On the other hand, if h_{cr} is -30 cm of water, a change in H_w from 5 cm to 10 cm will increase v_i by only 14%. Increasing H_w from 50 to 100 cm would almost double v_i in this case, however.

Equations [1] and [2] and the discussions in the preceding paragraphs should enable the investigator to decide what water depth to use in the

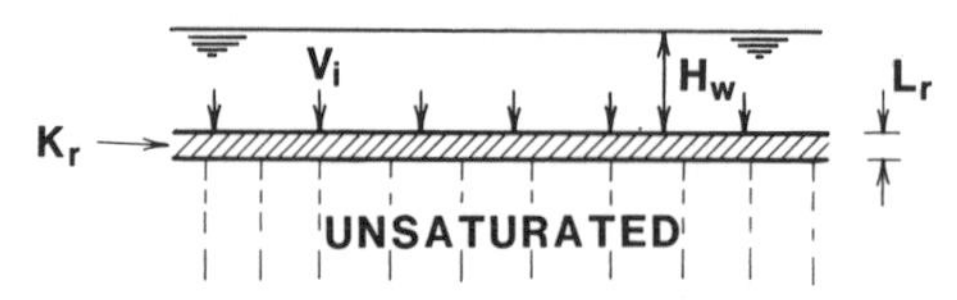

Fig. 32–2. Geometry and symbols for infiltration into soil with restricting surface layer.

infiltrometer, how important it is to maintain constant water depth during the measurements, how the infiltration rate can be extrapolated to different values of H_w, how important it is to have the same H_w in the infiltrometer as in the system for which the infiltration rate is to be predicted, and how long to continue the measurements. Measured relations between H_w and v_i can also be used to determine what soil conditions prevail at a given site. If, for example, v_i rapidly becomes constant and appears to vary significantly with H_w, it is likely that infiltration is restricted by a layer of low permeability at the surface and that h_{cr} of the underlying material is relatively close to zero.

32–2.3 Errors Due to Lateral Divergence of Flow

In most instances, one wishes to know the "vertical" infiltration rate—the rate that occurs when the infiltrated water moves vertically downward through the entire soil profile, as will be the case below relatively large inundated areas. Frequently, however, the flow below cylinder infiltrometers is not straight down but diverges laterally. The infiltration rate inside the infiltrometer will then be higher than the true infiltration rate for vertical flow. Lateral divergence can be caused by capillary forces, by layers of reduced hydraulic conductivity deeper in the profile, and by the water depth in the infiltrometer.

Lateral divergence by capillary forces is due to the fact that pressure heads in the unsaturated soil adjacent to the infiltrometer are less (more negative) than those in the wetted zone directly below the infiltrometer, thus creating a hydraulic gradient radially away from the wetted zone (Fig. 32–3). This can occur in uniform as well as in layered soils. The overestimation of the vertical infiltration rate due to lateral capillary forces depends on the ratio between the cylinder diameter d and the unsaturated-flow capability of the soil as reflected in its h_{cr}. Analyses with a resistance network (Bouwer, 1961) showed that the measured final in-

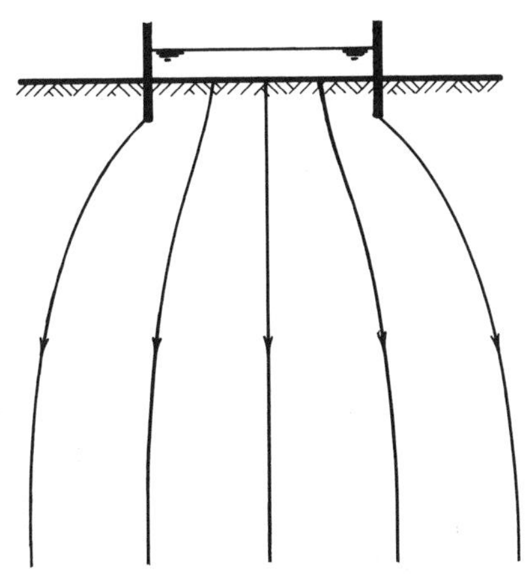

Fig. 32–3. Divergence of streamlines below cylinder infiltrometer.

filtration rate gives the true vertical infiltration rate (which theoretically is equal to K_w) correctly only if $h_{cr}/d = 0$ (Fig. 32–4). If $h_{cr}/d = -1$, which, for example, would be true for a 30-cm-diameter cylinder in a soil with $h_{cr} = -30$ cm of water, the final infiltration rate from the cylinder would be about 3.5 times the true vertical infiltration rate. If the cylinder diameter had been 5 cm, the final infiltration rate for that particular soil would have been overestimated by more than 11 times. This shows that old beer cans do not make good infiltrometers!

The distribution of the infiltration rates within the cylinder infiltrometer (Fig. 32–5), as obtained by resistance network analog (Bouwer, 1961), shows that infiltration rates at all points inside the cylinder are much greater than the true vertical infiltration rate for which $v_i/K_w = 1$. Thus, when lateral capillary gradients below the cylinder infiltrometer cause the flow to diverge, it does not help to put a smaller cylinder

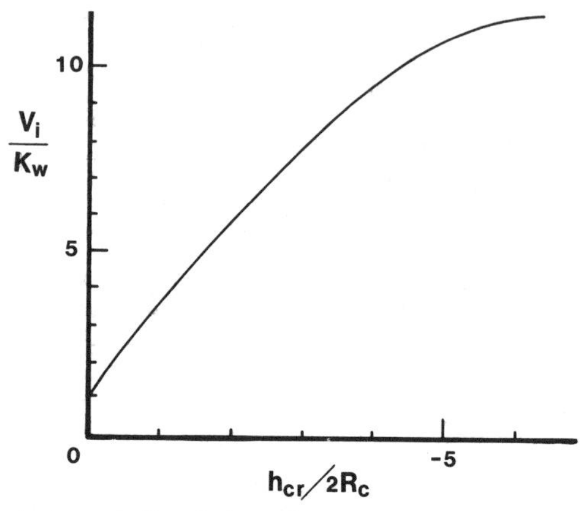

Fig. 32–4. Relation between infiltration rate from cylinder (expressed as v_i/K_w) and ratio of critical pressure head h_{cr} to cylinder diameter $2R_c$ (from Bouwer, 1961).

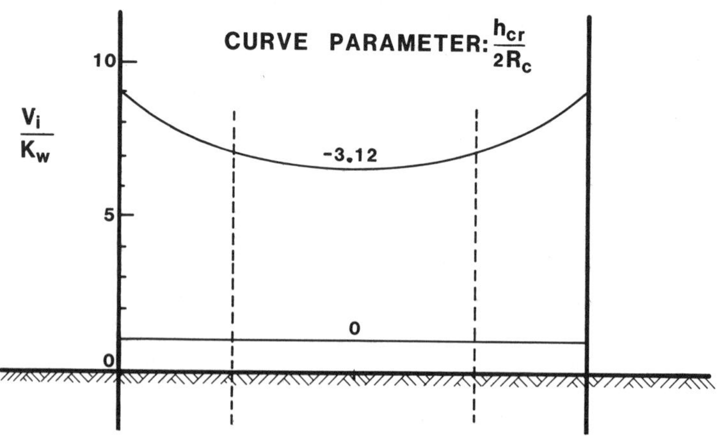

Fig. 32–5. Distribution of infiltration rates inside cylinder for two values of $h_{cr}/2R_c$ (from Bouwer, 1961).

concentrically inside the big cylinder and measure the infiltration rate in it, in hopes of getting a measure of the true vertical infiltration rate (see also Swartzendruber & Olson, 1961a, 1961b). This demonstrates the fallacy of buffered cylinder infiltrometers to "take care of" lateral divergence and "edge effects." The only way to measure the true vertical infiltration rate of the soil is to use cylinders with such large diameters that the ratio h_{cr}/d is essentially zero. At this ratio, all point infiltration rates inside the cylinder will be close to the infiltration rate for vertical flow and there is no need to put a smaller infiltrometer inside the main cylinder to get a "buffered" system. For the sand in the model studies by Swartzendruber and Olson (1961b), this situation was reached when the cylinder diameter was 1.2 m.

Restricting layers deeper in the profile also can cause lateral flow below cylinder infiltrometers (Evans et al., 1950). If under those conditions only a small surface area is inundated, as in a cylinder infiltrometer, a perching groundwater mound can establish itself above the restricting layer (Fig. 32–6). This mound can spread laterally and the infiltrated water can move through the restricting layer over a horizontal area larger than that covered by the infiltrometer. The infiltration rate in the infiltrometer will then exceed the rate that will occur when the entire soil surface is inundated and all water has to move straight down through the soil and the restricting layer (Fig. 32–7). The difference between the final infiltration rate from the infiltrometer and that for true, vertical infiltration depends on the hydraulic conductivity profile and on the depth of the restricting layer(s). Errors of several hundred percent are quite possible. Again, to minimize this error, the infiltrometer should be as large as practicable (see also section 32–4).

A third cause of lateral divergence of flow in the soil below cylinder infiltrometers is the pressure head of the water at the soil surface due to the water depth inside the cylinder. The larger this pressure head in relation to the cylinder diameter, the more the flow will diverge and the

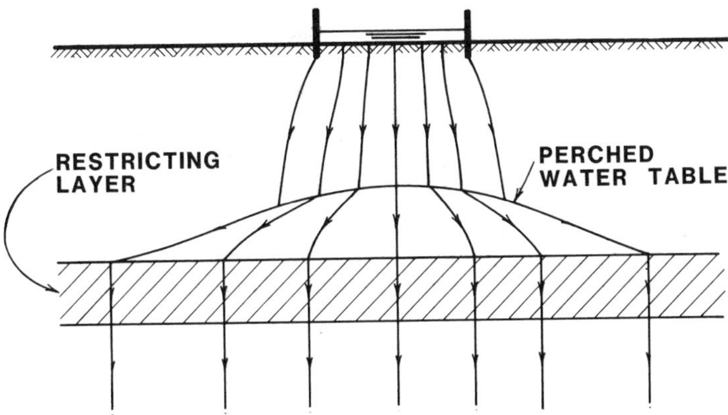

Fig. 32–6. Infiltration from cylinder into soil with restricting layer (hatched area) and perched groundwater mound.

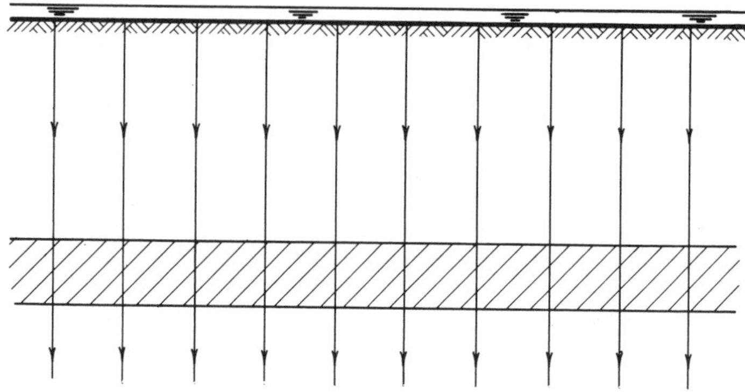

Fig. 32–7. Infiltration from large inundated area into soil with restricting layer (hatched area).

more the cylinder will overestimate the vertical infiltration rate of the soil. Thus, the water depth inside the infiltrometer should always be shallow to minimize this type of divergence. If it is desirable to know the infiltration rate for larger values of H_w, v_i should still be measured for small H_w in the infiltrometer. Infiltration rates for larger H_w can then be calculated with Eq. [1] or with other infiltration equations.

True vertical flow below infiltrometers of relatively small diameter can be expected only where the infiltration is controlled by a crust or other restricting layer at the surface of the soil. The flow through the restricting layer will then be vertically downward regardless of the diameter of the infiltrometer. There may be some flow divergence in the underlying unsaturated material that is affected by the diameter of the infiltrometer, but for normal size infiltrometers this will probably have only a minor effect on the pressure head below the restricting layer and hence on the infiltration rate inside the cylinder.

32–2.4 Other Sources of Error

In addition to the systematic error in infiltration measurement due to lateral divergence of the flow below the infiltrometer, a number of other factors can affect the results. One of these is soil disturbance by the insertion of the cylinder into the soil. If this disturbance is a soil compaction, the measured infiltration rate will be too low. If the soil has a surface crust or other restricting layer at or near the surface that is disrupted by the installation of the infiltrometer, or there is imperfect contact between the restricting layer and the inside cylinder wall, the measured infiltration rate will be too high. Effects of soil disturbance can be minimized by using large-diameter, thin-walled cylinders with beveled edges, and careful installation techniques (see section 32–3).

Soil can also be disturbed by careless application of water to the cylinder after it has been installed (erosion and puddling of surface). Clays and other fine particles brought temporarily in suspension in the water inside the cylinder can settle out on the soil again during the rest of the test and can create a restricting layer on the surface. The resulting infiltration rates will then be less than the actual infiltration rate of the soil. To minimize this error, direct impact of water on unprotected soil surfaces inside the cylinder should be avoided (see section 32–3.4).

Sometimes, infiltrometers are placed in larger inundated areas to determine local infiltration rates. This could be done to measure "point" infiltration rates in flood-irrigated fields (for example, to determine spatial variability of irrigation) and to measure local infiltration or seepage rates in irrigation canals or reservoirs (for example, to determine distribution of seepage rates and to delineate areas most in need of lining). For these measurements, the water level inside the cylinder infiltrometer should be maintained at the same elevation as that outside the infiltrometer. Otherwise, errors due to "leakage" flow between the two bodies of water through the soil and below the cylinder will result (Bouwer, 1963). Where the water is so deep that open cylinders become awkward to use, as in seepage measurement of deep channels or reservoirs, a cylinder of normal height is used, but it is covered by a top plate and connected with tubing to a higher reservoir to allow measurement of the outflow from the cylinder while the pressure head inside the cylinder is the same as that corresponding to the water surface in the canal or reservoir (Fig. 32–8). In this construction and application, the cylinder infiltrometer is called the seepage meter (see section 32–3.5).

The chemical composition (including type and concentration of sediment and other suspended solids) of the water used in the infiltrometer should be the same as that for which the infiltration rate is to be predicted. This is because the ionic composition of the water affects the flocculation-deflocculation status of the clay and, hence, the soil structure and hydraulic conductivity. Sediment and other suspended solids in the infiltrating water will form restricting layers at the surface, which can greatly reduce the measured infiltration rate.

The temperature of the infiltrating water should be the same as that of the soil to avoid soil air being dissolved in the infiltrating water (which

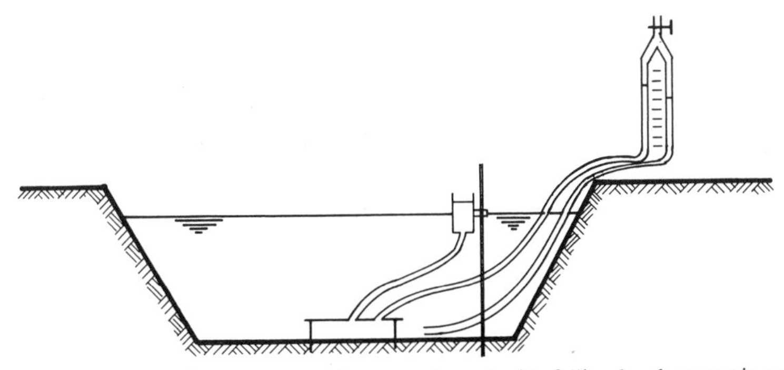

Fig. 32–8. Schematic of seepage meter in open channel with falling-level reservoir and U-tube manometer.

would increase K_w and hence v_i). If the infiltrating water is colder than the soil, dissolved air from the infiltrating water will go out of solution and increase the entrapped air content of the wetted zone. This is called air binding, and it reduces K_w and hence v_i. Another "air effect" is restriction of movement of air in the soil beneath the advancing wet front. This produces a "backing-up" effect on the wetted zone and causes an increase in the pressure head at the wet front (h_f in Eq. [1]), which in turn decreases the infiltration rate (Wilson & Luthin, 1963; Peck, 1965). Large inundated areas and restricting layers at some depth in the profile contribute to restricted air movement below the wetted zone. Thus, restricted air movement may not affect the infiltration from a cylinder infiltrometer, but it could reduce infiltration rates initially when the inundated area is much larger.

Biological effects on infiltration rate include growth of bacteria and algae on the soil surface, which can reduce infiltration rates through the accumulation of biomass and metabolic products. Heavy algae growth can also increase the pH of the water through uptake of CO_2 for photosynthesis, which in turn can lead to precipitation of $CaCO_3$ on the soil and further reduction in infiltration rates. Bacterial growth in the soil itself reduces infiltration rates by the physical presence of the bacteria, by their metabolic products, and by entrapment of the gases they produce. Worms tend to move upwards and crawl out of the soil when the surface is covered with water. The resulting open worm holes can greatly increase infiltration rates, particularly if there are many worms in the soil and the test is carried out for a long time.

The preceding discussions clearly show that predicting infiltration rates with cylinder infiltrometers is fraught with errors and uncertainties. In the hands of careful operators, infiltrometers can be useful for predictive or comparative purposes. Precise data will almost never be attainable, however, and expression of processed cylinder infiltrometer data in more than one or two significant figures creates a false impression of the accuracy of the technique.

32–3 METHOD

32–3.1 Construction

Cylinder infiltrometers typically are constructed from thin-walled steel pipe. Diameters commonly range from 20 to 30 cm and heights from 10 to 20 cm. The edge of the cylinder should be beveled to reduce soil disturbance when the cyliinder is pushed or driven into the soil. As discussed in section 32–2.3, the diameter should be as large as practicable to get a true measure of the vertical infiltration rate. Thus, cylinder infiltrometers with a diameter of 1 m or more are much better than those with smaller diameters. Larger cylinders can be rolled from sheet metal strips. The joint can be riveted, welded, or soldered; if riveted, caulking

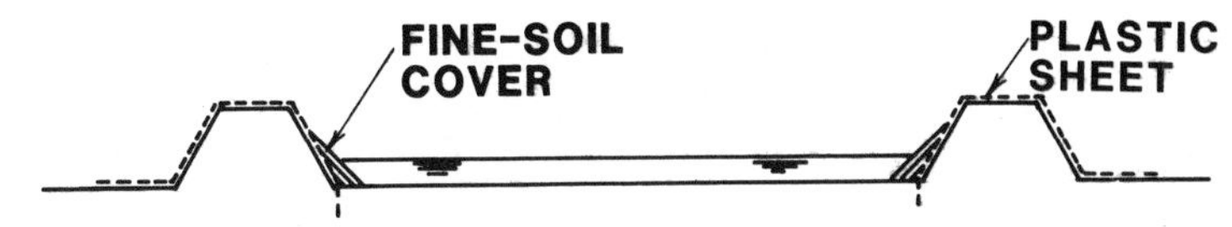

Fig. 32–9. Diked-area infiltrometer with plastic sheet on dikes and fine soil at inside toe to cover area disturbed by placing plastic into soil.

compound or other sealer must be used to obtain a leakproof joint. Metal lawn-edging and strips of linoleum or tar paper with adequate overlap at the joint are another possibility. Large square infiltrometers made from sheet metal or wooden planks forced into the soil are also suitable.

Large "infiltrometers" can also be constructed by forming a low ridge or dike with the soil. Seepage through the dike can be prevented by covering it with a plastic sheet that extends a few centimeters into the ground at the inside toe of the ridge (Fig. 32–9). This causes some soil disturbance, but if the resulting infiltrometer is relatively large (for example, several square meters) the effect on infiltration will probably be minor and will be outweighed by the benefits obtained from the large size. The area where the plastic sheet has been inserted into the soil can be covered with a thin layer of bentonite or other fine soil material to avoid excessive infiltration due to soil disturbance (see Fig. 32–9).

Infiltrometers should be kept in good condition. They should be cleaned after each use to prevent soil from drying and sticking to the metal. Such soil could later increase the soil disturbance when the meter is installed again. Rusting should also be prevented. Nicks and dents in the beveled edge, which are readily formed if the cylinder is used in gravelly soil, should be smoothed out with a ball peen hammer before the cylinder is installed again.

32–3.2 Single- and Double-Ring Infiltrometers

With double-cylinder infiltrometers, a smaller cylinder is placed concentrically inside another cylinder. Diameters typically may be 20 cm for the inner cylinder and 30 cm for the outer cylinder. Equal water levels must be maintained in both cylinders (Bouwer, 1963), and the infiltration rate is measured on the inner cylinder only. The thought behind the double-cylinder system was to let the outer, annular space between the two rings "absorb" all the edge and divergence effects, so that the infiltration from the inner ring would be a true measure of the vertical infiltration rate of the soil. As discussed in section 32–2.3, this thought is erroneous and the infiltration from the inner cylinder is also very much affected by divergence and edge effects. Field studies also have shown that there is little difference between the infiltration rate from the inner cylinder and that of the larger cylinder alone (Burgy & Luthin, 1956). Increasing the size of the infiltrometer is the only way to reduce the effect of lateral divergence of the flow below the cylinder (see section 32–2.3).

Double-cylinder infiltrometers could be effective only where there is a surface crust, an impeding layer on the surface, or other soil condition that makes it difficult to get a good bond between the undisturbed soil and the cylinder wall. When the area inside the cylinder is then covered with water, flow along the inner edge of the cylinder may cause the measured infiltration rate to be too high. Leakage along the cylinder wall can be minimized by placing another cylinder around the infiltrometer and maintaining the water level in there at exactly the same height as that in the inner cylinder.

32–3.3 Installation

Cylinder infiltrometers should be installed with as little disturbance of the soil as possible. Leave the soil surface in its natural condition and remove only woody stems, rocks, or other items that may get caught under the cylinder edge when the device is inserted. In soft soils, it may be possible to push the cylinder into the soil. In harder soils (e.g., dry clays), it must be driven in. Pushing or driving should be done uniformly and straight down, without rocking the cylinder or pushing first one side and then the other. Where driving is needed, a special driver consisting of a piece of the same metal pipe as that from which the cylinder is made is useful. Guides are attached to the inside of the driving device so that it snugly rests on the infiltrometer (Fig. 32–10). A crossbrace of angle iron is welded to the inside of the driving device and a rod with sliding weight is mounted in the center of the crossbrace (Fig. 32–10). Raising and dropping the weight then drives the infiltrometer into the soil. This technique makes it possible to keep the infiltrometer straight and to give it more uniform penetration than can be obtained by laying a piece of wood on the infiltrometer and pounding it with a hammer.

The depth of penetration of the cylinder should be as small as possible to minimize soil disturbance, but large enough to prevent water from blowing out a hole under the cylinder wall. Penetrations of about 5 cm are usually sufficient, unless H_w is relatively large. In some studies, penetration depths of 30 cm or more have been used, probably as an effort to "force" one-dimensional infiltration flow in the upper part of the soil

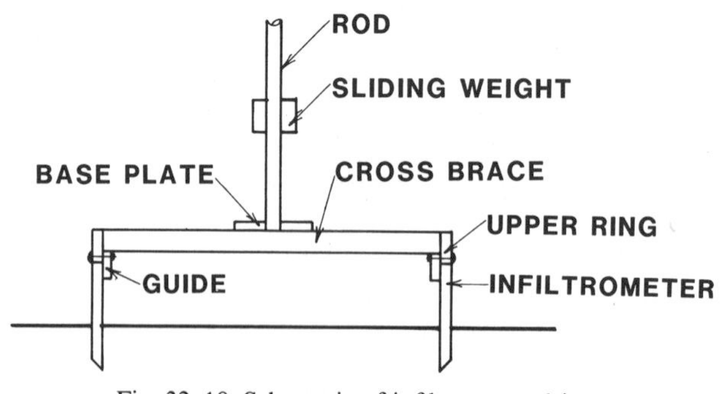

Fig. 32–10. Schematic of infiltrometer driver.

profile. This, however, could increase soil disturbance. A better way to obtain more one-dimensional flow would be to increase the diameter of the infiltrometer. If the soil surface is cracked, the cylinder should penetrate the soil at least as deep as the cracks, to prevent water from escaping laterally through cracks below the cylinder. If after installing the cylinder there is some separation between the soil inside the cylinder and the cylinder wall, the soil should be pushed or packed back against the cylinder. Likewise, if a surface crust is damaged by the cylinder penetration and there is no good contact between the crust and the cylinder wall, bentonite or other fine-powdered clay or soil should be placed along the inside of the wall to prevent excessive infiltration.

32–3.4. Water Supply, Constant Water Depth, and Infiltration Measurement

The water used in infiltrometer tests must be of the same quality and composition as the water in the real system for which the infiltration rate is to be predicted. Also, the water must be at about the same temperature as the soil (see section 32–2.4). The water for the test can be stored in buckets, bottles, drums, tanks, or other reservoirs. Water can be delivered to the infiltrometer by gravity, using a flexible tube. The end of the tube should be placed on a small inverted lid or similar device to break up the stream and to protect the soil surface against erosion or puddling. Covering the entire soil surface inside the infiltrometer with a cloth or a layer of coarse sand or fine gravel effectively protects the surface against erosion. However, the resulting infiltration rate may be too low because infiltration is blocked at the contact areas between the cover material and the soil surface itself.

Constant water depth in the infiltrometer can be maintained manually (for example, by frequently adding a small amount of water or adjusting a valve in the supply line), or automatically with a float valve inside the cylinder or a Mariotte-syphon arrangement (Fig. 32–11). The latter requires a stoppered bottle as a water reservoir. Two tubes are inserted through the stopper: one for siphoning the water to the infiltrometer and the other for letting air into the bottle. The bottom of this "bubble" tube is set at the level at which the water surface in the infiltrometer is to be maintained. The principle of the Mariotte syphon is

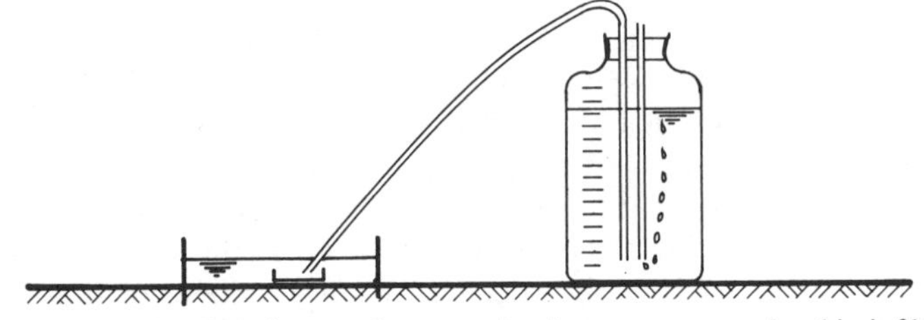

Fig. 32–11. Schematic of Mariotte syphon to maintain constant water level in infiltrometer.

that the pressure inside the bottle at the level of the end of the bubble tube is at atmospheric value, which then maintains the water surface in the infiltrometer at the same height as the end of the bubble tube. As mentioned in section 32–2.3., water depths in cylinder infiltrometers generally should be as small as possible.

The infiltration rate can be calculated from the rate of fall of the water level in the reservoir. Also, the water supply to the infiltrometer can be periodically stopped and the infiltration rate can be determined with hook or point gages by measuring the time it takes for the water surface to drop a small distance (a few millimeters, for example). The latter technique is especially adaptable where the infiltrometer is already surrounded by water, as in a flooded field or in a pond or stream. Water can then be admitted to the infiltrometer by connecting it to the water outside with a siphon or a valved tube through the wall. The water inflow is then periodically stopped to measure the rate of fall of the water level inside the infiltrometer.

Depending on the purpose of the infiltrometer test, the measurements should be continued until a certain amount of water has moved into the ground or until the infiltration rate has become essentially constant.

32–3.5 Seepage Meters

Where the infiltrometer is used to measure seepage from deeper inundations (e.g., open channels or ponds), the water inside the covered infiltrometer, called a "seepage meter," is connected to a reservoir (Fig. 32–8) that can be raised or lowered on a rod driven into the bottom (Bouwer & Rice, 1963). If no measurement is taken, the reservoir is positioned on the rod so that it is completely submerged and the pressure head inside the seepage meter is equal to that due to the water surface in the channel or pond. When a seepage measurement is to be taken, the reservoir is raised and positioned a few centimeters above the water surface in the channel or reservoir. Time and water-level measurements are then taken to determine the rate of fall of the water level in the reservoir at the instant that this water level is at the same height as that in the channel or pond. This can be done with a hook gage inside the reservoir. However, a more accurate and convenient way is to connect the water inside the seepage meter and that in the channel or pond outside the seepage meter with flexible tubes to an inverted U-tube manometer placed on the bank of the channel or pond (Fig. 32–8; Bouwer & Rice, 1963). The waters are then drawn into the U-tubes by applying a suction at the top of the U-tube and closing the top valve (Fig. 32–8) when the water levels have moved about half way up the manometer tubes. When the reservoir is raised for a measurement, the water level in the leg of the U-tube connected to the seepage meter will go up, and that in the other U-tube will go down. These movements will then reverse when the reservoir is positioned on the rod and the water level in the reservoir starts to fall due to the seepage from the seepage meter, yielding curves

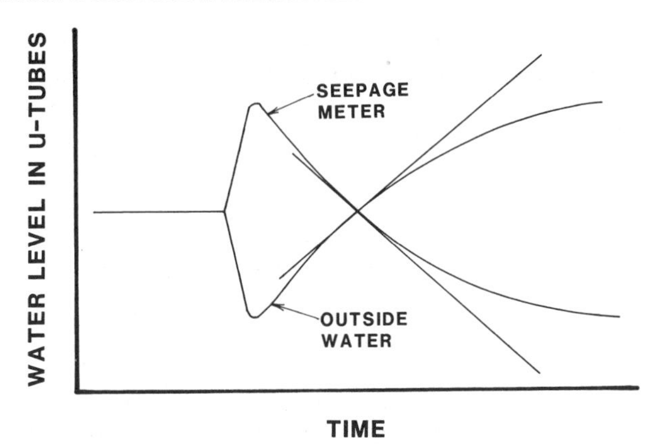

Fig. 32–12. Response of water levels in U-tube manometer after raising reservoir for seepage measurement.

as in Fig. 32–12. Where the curves cross, the water level in the reservoir is at the same height as that in the channel or pond. The angle between the tangents to the curves at the point of intersection then yields the rate of fall of the water surface in the reservoir when the pressure head inside the seepage meter is the same as that due to the water surface in the channel or pond (Bouwer & Rice, 1963). Multiplying this rate by the ratio of the cross-sectional area of the reservoir to that of the seepage meter then yields the seepage rate. In practice, a cluster of two or three reservoirs of different diameters (e.g., 0.02, 0.1, and 0.3 times the diameter of the seepage meter) connected with a two- or three-way valve enables rapid and accurate measurement of a wide range of seepage rates.

32–3.6 Variability and Number of Measurements

Because of soil variability, a number of infiltrometer tests must be made to obtain an accurate value for "the" infiltration rate (final or intermediate) of an entire field, infiltration basin, irrigation border, lagoon, or other system. The data can also be used to estimate variances, confidence limits, and other statistical aspects. Even for the theoretical case of a uniform soil, replicate tests would still be necessary because of the various sources of error in the measurement technique itself (section 32–2.4.).

Few systematic studies have been made on spatial variability of infiltration rates and variation in infiltrometer data. In a recent study, Sharma et al. (1980) performed 26 tests on a 9.6-ha watershed in Oklahoma that consisted of three different types of silt loam. Infiltrometers were of the double-cylinder type with cylinder diameters of 76.2 and 45.7 cm. The results were expressed in terms of the two-parameter Philip equation (see Eq. [4] and [5] in section 32–4). The resulting values of the sorptivity (S_i) in these equations ranged within one order of magnitude, and those of the factor A in these equations within two orders of magnitude. The frequency distribution of the accumulated infiltration I_t and of the parameters S_i and A was more log-normal than normal.

Where there is such great spatial variability, the infiltration behavior of the entire watershed could not be accurately described by an infiltration curve based on average values of S_i and A. Also, while point infiltration fluxes may be initially vertical, and hence independent of each other, multidimensional infiltration flow in the soil may develop later and infiltration rates may become interdependent. In that case, the infiltration rate for a large area may be between the arithmetic and harmonic means of the individual values and the geometric mean may be appropriate (Sharma et al., 1980; Bouwer, 1978, p. 132–133). Application of the similar-media concept showed that the geometric and harmonic means of the scaling factors were better than the arithmetic mean (Sharma et al., 1980).

Some indication of spatial variability of infiltration rates can also be gleaned from studies on the variation of hydraulic conductivity. For example, measurement of the hydraulic conductivity of the upper part of a loam-soil profile with the air-entry permeameter showed that about 20 replicate measurements were necessary to get the standard deviation down to about 10% of the mean (Bouwer, 1966). Aldabagh and Beer (1971) applied the air-entry permeameter to a field containing four soil types (three loams, one silty-clay loam) to measure the mean vertical hydraulic conductivity (calculated as the harmonic mean, see section 32–4) of the top 0.9 m of the soil profile. They concluded that 11 such measurements randomly spaced would give an average mean vertical hydraulic conductivity that is within 20% of the true value. Since the final infiltration rate theoretically equals the mean vertical hydraulic conductivity of the soil profile, a similar number of measurements could be required for determining the vertical infiltration rate. No hard and fast rules, however, can be given for the number of infiltrometer measurements that are necessary for a given situation. This number can best be determined on the basis of the measurements as they are taken, starting, for example, with five tests and adding more if the results vary too much to get an acceptable average infiltration rate. Soil surveys should precede a program of infiltrometer measurements so that average values can be determined by soil type. Infiltrometers with large diameters, as needed to minimize effects of lateral divergence due to capillary gradients (see section 32–2.3), also should be more effective in masking point variations in infiltration rates, and hence should produce more uniform values than infiltrometers with small diameters.

32–4 COMMENTS

An alternative to the direct measurement of infiltration is the measurement of the input parameters for infiltration equations, such as K_w and h_f in the Green-Ampt equation (Eq. [1]). Values of K_w in the vertical or near-vertical direction can be determined in situ with the air-entry permeameter for surface soils and with the air-entry permeameter, infiltration-gradient technique, and double-tube method for layers deeper in the profile (Bouwer, 1978, and references therein; chapter 30 in this book). The value of h_f can be measured directly with the air-entry permeameter, or it can be evaluated from the relation between unsaturated hydraulic conductivity K_h and pressure head for the soil in question. Relations between water content and pressure head may also be suitable, at least to get an estimate of h_f.

The accumulated infiltration I_t is equal to fL_f, where f is the fillable porosity of the soil (difference between volumetric water content before and after wetting). Since the rate of advance of the wet front dL_f/dt is equal to v_i/f, Eq. [1] can be integrated to obtain a relation between I_t (as fL_f) and t (see, for example, Eq. [8.2] on p. 254 of Bouwer, 1978). This equation then makes it possible to calculate the accumulated infiltration as a function of time. The necessary value of f can be determined by water content measurements of the soil before and after wetting.

The Green-Ampt equations can be applied to homogeneous as well as to layered soils. In the latter case, which includes soil profiles that become less permeable as well as more permeable with depth, the calculation of v_i and I_t consists of a step procedure (Bouwer, 1969, 1976). These articles also describe how the procedure can be applied to profiles where f is not constant (f tends to decrease with depth because antecedent water contents tend to increase with depth).

Calculation of v_i from measured values of K_w and h_f may be the best method for evaluating vertical infiltration rates of soils that contain restricting layers at some depth. For such profiles, it may be difficult to get infiltrometers that are large enough to measure the true vertical infiltration rate or the infiltration rate from large irrigation borders or infiltration basins. The final vertical infiltration rate can then be calculated on the basis of the mean vertical hydraulic conductivity of the soil profile. Ignoring H_w and h_f, the final infiltration rate will be equal to this mean hydraulic conductivity. For a layered profile, the mean vertical hydraulic conductivity is the harmonic mean (Bouwer, 1978, and references therein), defined as

$$\overline{K}_z = Z/(z_1/K_1 + z_2/K_2 + \dots z_n/K_n) \qquad [3]$$

where

$\overline{K}_z$ = mean hydraulic conductivity of layered soil in vertical direction,
Z = total thickness of packet of layers or of wetted zone,
z = thickness of each layer, and
K = hydraulic conductivity (rewet value) of each layer.

If the soil profile is divided into n layers of equal thickness, Eq. [3] becomes $\overline{K}_z = n/(1/K_1 + 1/K_2 + \dots 1/K_n)$. If K increases with depth, the soil profile is not completely rewetted for the entire depth, because the soil below the point where the infiltration rate is numerically equal to the rewetted hydraulic conductivity remains unsaturated. In that case, Eq. [3] should be applied only to the part of the profile that is completely rewetted (Bouwer, 1976).

Vertical infiltration can also be calculated with the two-parameter Philip equation

$$v_i = \tfrac{1}{2} S_i t^{1/2} + A \qquad [4]$$

where S_i is the sorptivity and A is a factor related to K of the soil (Philip, 1969, and references therein). Equation [4] can be integrated to obtain the following equation for the accumulated infiltration:

$$I_t = S_i t^{1/2} + At\,. \qquad [5]$$

The term S_i depends on the pore configuration of the soil, the initial water content of the soil, and H_w. Values of S_i can be determined from infiltrometer measurements by plotting I_t vs. $t^{1/2}$ for the first few minutes of the test, where I_t increases essentially linearly with $t^{1/2}$ and S_i is evaluated as the slope of the straight-line portion of the curve (Talsma, 1969). In terms of the parameters of the Green-Ampt equation, S_i can also be approximated as $(2f\,K_w\,h_f)^{1/2}$ (Youngs, 1968). The A term in Eq. [4] and [5] can be taken as K_w for long-term infiltration, and as $\tfrac{1}{2}K_w$ for short-term infiltration (Youngs, 1968; Talsma & Parlange, 1972). Equations [4] and [5] are not directly applicable to layered soils.

The infiltration rates calculated from the hydraulic conductivity profile of the soil are for pure water. If the infiltrating water in the actual system contains sediments or other suspended solids, the reduction in infiltration rate due to the accumulation of solids on the soil surface must be estimated. Mathematical expression of the clogging process generally yields equations with an exponential decay in infiltration rate (Behnke, 1969). The coefficients in these equations must be experimentally determined for the sediment concentration of the water, type of sediment, type of bottom material, and infiltration rates of the particular installation. Berend (1967) developed the empirical equation

$$v_i = v_o - \alpha C_s I_t \qquad [6]$$

where

v_i = infiltration rates at time t (m/day),
v_0 = initial infiltration rate (m/day),
α = coefficient describing clogging properties of system ($L\,g^{-1}\,day^{-1}$),
C_s = concentration of suspended material in infiltrating water that will be retained on the soil surface ($g\,L^{-1}$), and
I_t = accumulated infiltration at time t (m).

Reported values for α include a range of 0.1 to 5.5 $L\,g^{-1}\,day^{-1}$ for floodwaters with relatively high sediment contents of 0.14 to 2 g L^{-1} (Berend, 1967), and a range of 1.2 to 6.5 L g^{-1} day^{-1} for secondary sewage effluent with a high algae content in a system of rapid infiltration basins west of Phoenix, AZ (Bouwer & Rice, 1976).

Because of soil clogging and resulting reduction in infiltration rate, infiltration basins for groundwater recharge normally are regularly dried to allow shrinking and cracking of the clogged layer and decomposition of biomass and organic compounds that may have accumulated on the soil surface during flooding. Cleaning the bottom by scraping the surface layer off may also be needed periodically. The optimum sequence of flooding and drying periods and the optimum cleaning frequency must be experimentally determined for the local conditions of infiltrating water, soil, and climate.

The accumulated infiltration over a long period (1 year, for example) in infiltration systems for groundwater recharge is called the hydraulic loading rate. For secondary sewage effluent infiltrating in a loamy sand in the Salt River bed west of Phoenix, Arizona, maximum hydraulic loading was obtained with flooding periods of 2 weeks alternated with drying periods of 10 days in summer and 20 days in winter. This schedule produced a hydraulic loading rate of 120 m $year^{-1}$. The infiltration rate for pure water for this system was 1.2 m day^{-1} or 430 m $year^{-1}$. Thus, the maximum hydraulic loading rate for intermittent flooding with secondary sewage effluent was 27% of the continuous infiltration rate for pure water. Percentages like these are often used to convert infiltration rates for pure water to hydraulic loading rates for water of less quality in intermittent-inundation systems. Lower percentages can be expected in areas with more rain and cooler climates, because drying and decomposition of the clogging materials then takes longer, or where the effluent or other wastewater contains more suspended solids and organics. Thus, under less favorable conditions, hydraulic loading rates for water of impaired quality in intermittently flooded systems may be as low as 5% of the continuous infiltration rate for pure water.

Prediction of hydraulic loading rates for water of impaired quality from measured or calculated infiltration rates for pure water should be done only for exploratory or preliminary planning purposes, or for comparisons. The resulting values will seldom if ever be sufficiently reliable for final planning or design of systems. Pilot projects and infiltration measurements for the actual water and type of infiltration facilities involved are required for such purposes.

32–5 REFERENCES

Aldabagh, A. S. Y., and C. E. Beer. 1971. Field measurement of hydraulic conductivity above a water table with air-entry permeameter. Trans. ASAE 14:29–31.

Behnke, J. J. 1969. Clogging in surface spreading operations for artificial ground-water recharge. Water Resour. Res. 5:870–876.

Berend, J. E. 1967. An analytical approach to the clogging effect of suspended matter. Bull. Int. Assoc. Sci. Hydrol. 12:42–55.

Bouwer, H. 1961. A study of final infiltration rates from cylinder infiltrometers and irrigation furrows with an electrical resistance network. Trans. Int. Congr. Soil Sci., 7th 6:448–456.

Bouwer, H. 1963. Theoretical effect of unequal water levels on the infiltration rate determined with buffered cylinder infiltrometers. J. Hydrol. 1:29–34.

Bouwer, H. 1964. Unsaturated flow in ground-water hydraulics. Hydraul. Div., Am. Soc. Civ. Eng. 90(HY5):121–144.

Bouwer, H. 1966. Rapid field measurement of air entry value and hydraulic conductivity as significant parameters in flow system analysis. Water Resour. Res. 1:729–738.

Bouwer, H. 1969. Infiltration of water into nonuniform soil. J. Irrig. Drain. Div., Am. Soc. Civ. Eng. 95(IR4):451–462.

Bouwer, H. 1976. Infiltration into increasingly permeable soils. J. Irrig. Drain. Div., Am. Soc. Civ. Eng. 102(IR1):127–136.

Bouwer, H. 1978. Groundwater hydrology. McGraw-Hill Book Co., New York.

Bouwer, H. 1982. Design considerations for earth linings for seepage control. Ground Water 20:531–537.

Bouwer, H. and R. C. Rice. 1963. Seepage meters in seepage and recharge studies. J. Irrig. Drain. Div., Am. Soc. Civ. Eng. Proc. 89(IR1):17–42.

Bouwer, H., and R. C. Rice. 1976. Wastewater renovation by spreading treated sewage for groundwater recharge. Annu. Rep. U. S. Water Conservation Laboratory, Phoenix, AZ, p. 20-1 to 20-41.

Burgy, R. H., and J. N. Luthin. 1956. A test of the single and double ring types of infiltrometers. Trans. Am. Geophys. Union 37:189–191.

Evans, D. D., D. Kirkham, and R. K. Frevert. 1950. Infiltration and permeability in soil overlying an impermeable layer. Soil Sci. Soc. Am. Proc. 15:50–54.

Freyberg, D. L., J. W. Reeder, J. B. Franzini, and I. Remson. 1980. Application of the Green-Ampt model to infiltration under time-dependent surface water depths. Water Resour. Res. 16:517–528.

Green, W. H., and G. A. Ampt. 1911. Studies on soil physics. I. The flow of air and water through soils. J. Agric. Sci. 4:1–24.

Mein, R. G., and D. A. Farrell. 1974. Determination of wetting front suction in the Green-Ampt equation. Soil Sci. Am. Proc. 38:872–876.

Morel-Seytoux, H. J., and J. Khanji. 1974. Derivation of an equation of infiltration. Water Resour. Res. 9:795–800.

Peck, A. J. 1965. Moisture profile development and air compression during water uptake by bounded porous bodies: 3. Vertical columns. Soil Sci. 100:44–51.

Philip, J. R. 1969. Theory of infiltration. p. 216–296. *In* V. T. Chow (ed.) Advances in hydroscience. Academic Press, New York.

Sharma, M. L., G. A. Gander, and C. G. Hunt. 1980. Spatial variability of infiltration in a watershed. J. Hydrol. 45:101–122.

Swartzendruber, Dale, and T. C. Olson. 1961a. Sand-model study of buffer effects in the double-ring infiltrometer. Soil Sci. Soc. Am. Proc. 25:5–8.

Swartzendruber, Dale, and T. C. Olson. 1961b. Model study of the double ring infiltrometer as affected by depth of wetting and particle size. Soil Sci. 92:219–225.

Talsma, T. 1969. In situ measurement of sorptivity. Aust. J. Soil Res. 7:269–276.

Talsma, T., and J.-Y. Parlange. 1972. One-dimensional vertical infiltration. Aust. J. Soil Res. 10:143–150.

Topp, G. C., and M. R. Binns, 1976. Field measurements of hydraulic conductivity with a modified air-entry permeameter. Can. J. Soil Sci. 56:139–147.

Wilson, L. G., and J. N. Luthin. 1963. Effect of air flow ahead of the wetting front on infiltration. Soil Sci. 96:136–143.

Youngs, E. G. 1968. An estimation of sorptivity for infiltration studies from moisture moment considerations. Soil Sci. 106:157–163.

33 Intake Rate: Sprinkler Infiltrometer[1]

ARTHUR E. PETERSON AND GARY D. BUBENZER
University of Wisconsin
Madison, Wisconsin

33-1 INTRODUCTION

For several decades researchers have tried to develop sprinkling infiltrometers that duplicate the rainfall characteristics of a natural storm. Such duplication is extremely difficult. In some instances the rainfall characteristics of goegraphical regions are not known in sufficient detail to allow accurate descriptions of the storms, or if the rainfall characteristics are known the researcher may not be able to find a device capable of reproducing all desired characteristics. Given the fact that the nozzle or drop-forming device will not give complete simulation, choices must be made among the rainfall parameters to simulate. Such decisions must be made even though research has not yet clearly established the relative importance of each parameter. These choices must then depend upon the nature of the project and criteria developed by the investigator.

Attempts to simulate natural rainfall which began in the early 1930s and continued through the 1940s were reviewed in an excellent article by Mutchler and Hermsmeier (1965). Two basic types of simulators were developed. One group of simulators produced rainfall by forming drops on the tips of yarn or small-diameter glass, stainless steel, brass, or polyethylene tubes (Ekern, 1950; Barnes & Costel, 1957; Adams et al., 1957; Mutchler & Moldenhauer, 1963; Chow & Harbaugh, 1965). The second group of simulators used nozzles to form drops. Initially, nozzles were usually designed to spray upward or horizontally from a fixed position to increase wetted area and reduce the application intensity (Wilm, 1943). Impact velocity of drops produced in this way was low compared to that of normal raindrops.

In the late 1950s, Meyer and McClune (1958) developed the Rainulator, which produced rainfall characteristics more nearly approximating the kinetic energy levels of natural storms. Bubenzer and Meyer (1965) used oscillating nozzle systems to reduce the period of intermittency of the Rainulator. In the late 1950s a sprinkling infiltrometer was developed

[1]Contribution from Departments of Soil Science and Agricultural Engineering, College of Agricultural and Life Sciences, University of Wisconsin-Madison, Madison, WI 53706.

by Bertrand and Parr (1961) for the study of infiltration. Amerman et al. (1970) developed a variable intensity sprinkling infiltrometer by intercepting any desired portion of the spray from the nozzle. Recently, Foster et al., (1982) developed a simulator that can be used to vary the rainfall intensity during the test. To obtain higher impact velocities, spray nozzles in these systems were pointed downward to take advantage of the initial drop velocity.

Much of the recent emphasis has been on developing sprinkling infiltrometers that will have characteristics of natural rainfall. One of the first steps in the design of a sprinkling infiltrometer involves the development of a list of criteria to be met. Storm characteristics make up a portion of this list. A review of literature shows that six criteria related to storm characteristics recur:

1. Drop-size distribution similar to that of natural rainfall (Borst & Woodburn, 1940; Meyer & McCune, 1958; Bertrand & Parr, 1961; Chow & Harbaugh, 1965; Nassif & Wilson, 1975; Shriner et al., 1977).
2. Drop velocity at impact near terminal velocity (Meyer & McCune, 1958; Bertrand & Parr, 1961; Nassif & Wilson, 1975).
3. Rainfall intensity corresponding to natural conditions (Meyer & McCune, 1958; Bertrand & Parr, 1961; Chow & Harbaugh, 1965; Shriner et al., 1977).
4. Uniform rainfall and random drop-size distribution (Borst & Woodburn, 1940; Meyer & McCune, 1958; Bertrand & Parr, 1961; Chow & Harbaugh, 1965; Shriner et al., 1977).
5. Total energy approaching that of natural rainfall (Bertrand & Parr, 1961; Munn & Huntington, 1976).
6. Reproductive storm patterns (Meyer & McCune, 1958; Bertrand & Parr, 1961; Shriner et al., 1977).

In a survey of rainfall criteria for 28 developers or users of rainfall simulators, Bubenzer (1979) found that two-thirds of the respondents using nozzle-type simulators considered all of the first four criteria. More than 90% of all responses indicated that mean drop size, intensity, and uniformity of coverage were criteria used in selecting a simulator for their research.

Meyer and Harmon (1979) also identified additional important characteristics of rainfall simulators, including: (i) research area of sufficient size to satisfactorily represent the treatments and conditions to be evaluated, (ii) raindrop application nearly continuous throughout the study area, (iii) angle of impact not greatly different from vertical for most drops, (iv) satisfactory characteristics when using during common field conditions, such as temperature and moderate winds, and (v) portability for movement from research site to research site.

33–2 DESIGN PARAMETERS

Impact velocity, drop-size distribution, and intensity level are the basic design parameters which must be considered in the design or selection of a rainfall simulator. Several parameters, such as kinetic energy per unit of rainfall or per unit drop area, and momentum per unit rainfall or drop impact area, have been developed using these basic parameters. However, no single combined parameter has been conclusively related to rainfall erosivity over a wide range of conditions (Meyer et al., 1970). Therefore, close reproduction of rainfall drop size and impact velocity seems essential for many infiltration studies.

Extensive studies have been conducted on the fall velocity of various sized water drops falling through still air (Laws, 1941; Gunn & Kinzer, 1949). Rainstorms are often accompanied by significant winds. The actual impact velocity of the raindrop is, therefore, a function of wind speed and turbulence as well as drop size (Van Heerden, 1964). Such effects are usually neglected in simulator design. Simulators based on Laws' data (1941) or other similar studies will represent minimum design impact levels.

The median drop size and the drop-size distribution of natural rainfall are dependent upon rainfall intensity. Most researchers have used an exponential equation to express the relationship between size and intensity (Laws & Parsons, 1943; Chapman, 1948; Rogers et al., 1967). Carter et al. (1974) found a cubic equation best described this relationship for storms with intensities up to 254 mm/h for the south-central USA. McGregor and Mutchler (1976) developed a three-term exponential relationship to relate the median drop size to intensity for the Holly Springs data. Much of the reported differences in median drop size reported in the literature may be attributed to wide random variation within and among storms, to geographical factors influencing storm type, and to a lack of data at the high intensity levels. Despite this variation, the results indicate that there is a rapid increase in mean drop diameter with intensity for rainfall rates up to about 50 mm/h. There is also strong evidence that at higher intensities the mean drop diameter tends to remain nearly constant or decrease slightly.

For many types of research, knowledge of the median drop size is not sufficient for the design of rainfall simulators. In cases where drop impact plays an important role in the infiltration or erosion process, it is important that the simulator's drop-size distribution be similar to that of natural storms. Typical drop-size distributions for various locations throughout the USA are presented in Fig. 33–1 and 33-2. Because of differences in methods of reporting, it is difficult to compare drop-size distributions between regions. A portion of the apparent regional variation is probably due to random variation and differences in methods of grouping the data.

Selection of a design intensity for simulator development must depend upon the objectives of the investigator. Since the most runoff is usually associated with high-intensity storms, most simulators have been designed to apply water at relatively high intensities. However, regional differences in intensity and storm characteristics must be considered. In

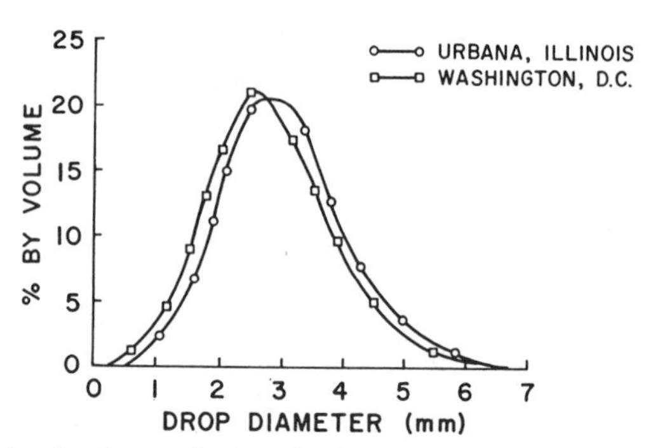

Fig. 33–1. Drop-size distribution for two sites in the USA at an intensity of approximately 50 mm/h (Bubenzer, 1979).

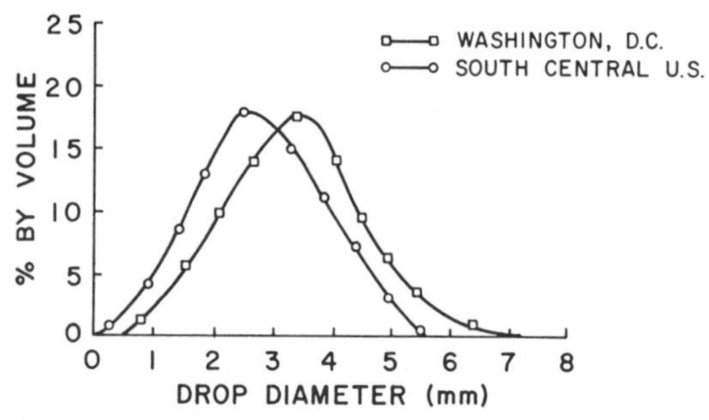

Fig. 33–2. Drop-size distribution for two sites in the USA at an intensity of approximately 150 mm/h (Bubenzer, 1979).

some areas a major portion of the total annual runoff may be associated with low-intensity rain on thawing or snow-covered fields. Such storms might contribute only a small portion of the runoff and annual soil loss in other parts of the USA, where major runoff and soil loss are associated with intense thunderstorm activity.

Most simulators in use today do not allow researchers to vary storm characteristics during a rainfall event. Temporal variations in intensity are known to have an effect upon both the amount and peak rate of runoff. The effect of this variation on hydrologic processes is not well documented. Recent advances in simulator design have given researchers the opportunity to consider temporal intensity variations (Foster et al., 1982).

Impact velocity, drop size and intensity are not independent of each other. In nature the three are interrelated in a complex manner. Most rainfall simulators are not designed to consider the dynamic nature of the rainfall process. For example, intensity reduction on most simulators is controlled by reducing the percentage of the time the spray is striking the plot (Bubenzer & Meyer, 1965; Amerman et al., 1970) or by reducing the number of nozzles spraying onto the plot (Swanson, 1965). Average intensity is, therefore, reduced through intermittent or reduced application; however, neither the impact velocity nor the drop-size distribution is reduced at the low intensities, as would occur in natural rainfall. This would appear not to be a serious problem until the intensity is reduced below 50 mm/h.

33–2.1 Rainfall Intermittency

Major advances have been made in the development of sprinkling infiltrometers which more nearly simulate natural storm conditions. The simplest of the nozzles that have been used for rainfall simulation were a sprinkling can, rose flares, and shower heads that are common in garden and household use. The spray in these methods obviously left much to be desired. Great improvement has been made in developing nozzles that will approximate normal rainfall in intensities. However, the intermittency of some of the sprinkling infiltrometers has been subject to question. In these simulators rainfall intensity is controlled by varying the on-off time. Sloneker and Moldenhauer (1974) showed that the energy to initiate runoff may be increased as the nozzle off-time is increased. Water intake rate is increased as application is increased, even though runoff is taking place. Thus some of the basic factors must continue to be studied to better understand how relative comparisons can be made with natural storms.

33–3 SIMULATOR SYSTEMS

In general the sprinkling rainfall simulators have been divided into two groups. The first group, described in Table 33–1, includes simulators that produce rainfall by means of a nozzle. The second category of rainfall simulators are those which use drop formers to produce rainfall.

33–3.1 Nozzle Systems.

Several different nozzles with widely varying drop characteristics have been used on modern rainfall simulators. From this group four nozzles seem to predominate. The Spray Engineering Company's 7LA nozzle was first used on the Purdue Sprinkling Infiltrometer by Bertrand and Parr (1961).

The Rainulator (Meyer & McCune, 1958) used the Spraying Systems 80100 Veejet nozzle. This nozzle was selected because it closely approxiamted the drop-size distribution of erosive storms in the Midwest. The Rainulator has also been modified to meet special research needs. Seimens and Oschwald (1978) constructed a self-propelled unit. Swanson (1965) used the same nozzle on a trailer-mounted, rotating boom simulator. Bubenzer and Meyer (1965) also used the 80100 Veejet nozzle to develop an oscillating laboratory simulator. The oscillating simulator greatly reduced the period of intermittency of the Rainulator. The oscillating nozzle concept has also been incorporated into the interrill sim-

Table 33–1. Rainfall simulators and infiltrometers using nozzles to simulate rainfall.

Rainfall simulator location	Nozzle	Pressure	Nozzle movement and spray pattern	Drop size D_{10}	Drop size D_{50}	Drop size D_{90}	Intensity	Plot size per unit	Use	Reference
		kPa		mm	mm	mm	mm/h	m		
Rainulator USDA-SEA-AR Lafayette, IN	Spraying Systems 80100 Veejet	41.4	Lateral Intermittent	1.0	2.1	3.0	64 & 127	4 by 11.5	Erosion	Meyer & McCune, 1958 Hermsmeier et al., 1963
Rainulator USDA-SEA-AR Watkinsville, GA	Spraying Systems 80100 Veejet	41.4	Lateral Intermittent	1.0	2.1	3.0	64 & 127	4 by 11.5	Erosion	Hermsmeier et al., 1963
Rainulator USDA-SEA-AR Morris, MN	Spraying Systems 80100 Veejet	41.4	Lateral Intermittent	1.0	2.1	3.0	64 & 127	4 by 11.5	Erosion Infiltration	Hermsmeier et al., 1963
Rainulator Univ. of Illinois Urbana, IL	Spraying Systems 80100 Veejet	41.4	Lateral Intermittent	1.0	2.1	3.0	64 & 127	4 by 11.5	Erosion	Seimens & Oschwald, 1978
Rainulator Dep. of Prim. Ind. Toowooba, Queensland, Australia	Spraying Systems 80100 Veejet	41.4	Lateral Intermittent	1.0	2.2	3.2	30–200	4 by 22.5	Erosion Infiltration	McKay & Loch, 1978
Modified Rainulator New Mexico State Univ. Las Cruces, NM	Spraying Systems 80100 Veejet	41.4	Lateral Intermittent	1.0	2.1	3.0	101 & 203	5.0 by 6.5	Erosion	Anderson et al., 1968
New Rainulator USDA-SEA-AR Lafayette, IN	Spraying Systems 80100 Veejet 80150 Veejet	41.4	Oscillating Intermittent	1.0 1.1	2.1 2.5	3.2 4.2	2–127	4 by 11.5	Erosion Infiltration Runoff	
Rotating Boom USDA-SEA-AR Lincoln, NE	Spraying Systems 80100 Veejet	41.4	Rotating Intermittent	1.0	2.1	3.0	64 & 127	4 by 11	Erosion Infiltration	Swanson, 1965
Rotating Boom USDA-SEA-AR Ames, IA	Spraying Systems 80100 Veejet	41.4	Rotating Intermittent	1.0	2.1	3.0	64 & 127	4 by 11	Erosion	Swanson, 1965
Laboratory Simulator Univ. of Wisconsin Madison, WI	Spraying Systems 80100 Veejet	41.4	Oscillating Intermittent	1.0	2.1	3.0	38	1 by 5	Erosion Runoff	Bubenzer & Meyer, 1965

(continued on next page)

Table 33–1. Continued.

Rainfall simulator location	Nozzle	Pressure	Nozzle movement and spray pattern	Drop size D_{10}	Drop size D_{50}	Drop size D_{90}	Intensity	Plot size per unit	Use	Reference
		kPa		mm	mm	mm	mm/h	m		
Inter-rill Simulator USDA-SEA-AR Oxford, MS	Spraying Systems 80100 Veejet 80150 Veejet	41.4	Oscillating Intermittent	0.7 1.1	1.6 2.5	3.2 4.2	10–127	0.7 by 0.9	Erosion	Meyer & Harmon, 1979
Inter-rill Simulator Michigan Tech Univ. Houghton-Hancock, MI	Spraying Systems 80100 Veejet 80150 Veejet	41.4	Oscillating Intermittent	0.7 1.1	1.6 2.5	3.2 4.2	10–127	0.7 by 0.9		Meyer & Harmon, 1979
USGS USDA-SEA-AR Sidney, MT	Rainjet 78C	207	Stationary Continuous	0.7	1.5	2.4	64		Erosion Infiltration	Lusby, 1977
Rotating Disk Simulator Soil Erosion Res. Stn. Emek, Hefer, Israel	Spraying Systems 1HH12 Fulljet 1.5 H30 Fulljet	60	Stationary Intermittent				9–74 15–143	1.0 by 1.5	Erosion	Morin et al., 1967
Rotadisk Rainulator Univ. of Arizona Tuscon, AZ	Spraying Systems 1.5 H30 Fulljet		Stationary Intermittent				17–1520	1.5 by 1.5	Erosion Infiltration	Cluff, 1971
Morin and Goldberg Type Gunnedah Soil Conserv. Res. Ctr. Gunnedah, Australia	Spraying Systems 1.5 H30 Fulljet	70	Stationary Intermittent	1.9	2.6	4.3	58–115	1.0 by 1.5	Erosion	Marston, 1978
Waite Institute Waite Agric. Res. Inst. South Australia	Spraying Systems 1.5 H30 Fulljet	69	Stationary Intermittent		2.4		10–150	1.0 by 1.0	Erosion	Grierson, 1975
Rainfall Simulator Cornell University Ithaca, NY	Spraying Systems 7309 Flat Teejet 8015 Flat Teejet	137–275	Rotating Intermittent				17–282		Pesticide movement	Brockmen et al., 1975
Portable Simulator Commonwealth Atherton, Queensland, Australia	Rose Sprayhead		Lateral Intermittent		1.3		80	2.0 by 3.3	Erosion Infiltration Nutrient movement	Costin & Gilmour, 1970
Raintower USDA-SEA-AR Manhattan, KS	Spraying Systems 14WSQ & 35WSQ		Stationary Continuous	1.0	2.1	3.9	18	1.5 by 3.1	Erosion	Lyles et al., 1969

(continued on next page)

Table 33–1. Continued.

Rainfall simulator location	Nozzle	Pressure	Nozzle movement and spray pattern	Drop size D_{10}	Drop size D_{50}	Drop size D_{90}	Intensity	Plot size per unit	Use	Reference
		kPa		mm	mm	mm	mm/h	m		
Laboratory Simulator Univ. of Salford Lancashire, U.K.	Childs (PVC)	45	Stationary				0–300	6.2 by 4.1	Infiltration Runoff	Nassif & Wilson, 1975
RAINS Oak Ridge National Lab Oak Ridge, TN	Beta Fog SRN303		Stationary Continuous	0.4		1.2	5–27	1.0 by 1.0	Infiltration Nutrient transport	Shriner et al., 1977
Type F Infiltrometer USDA-SEA-AR Beltsville, MD	Type F	193–248	Stationary Continuous				46–64	2.0 by 3.9	Erosion Infiltration	Wilm, 1943
Rocky Mountain Infilt. USDA-Forest Service Ft. Collins, CO	Type F	138–206	Stationary Continuous				127	0.3 by 0.8	Erosion Infiltration Runoff	Dortignac, 1951
Intermountain Infilt. USDA-Forest Service Ogden, UT	Type F	241	Stationary Continuous				25–152	0.6 by 1.8	Erosion Infiltration Runoff	Packer, 1957
Rocky Mountain Infilt. Utah State Univ. Logan, UT	Type F	138–206	Stationary Continuous				127	0.5 by 0.7	Erosion Infiltration	Neeuwig, 1969
Rainfall Simulator Australia	Spraying Systems 8070 Veejet	41.4	Lateral Intermittent					4.6 by 4.6	Erosion	Turner & Langlord, 1969
Palouse Field Station USDA-SEA-AR Pullman, WA			Stationary Continuous				2–2000	2.6 by 13.1	Erosion	
Palouse Infiltrometer Univ. of Idaho Moscow, ID	Spraying Systems 14WSQ Fulljet	41.4	Stationary Intermittent	0.8	1.7	2.6	1–50	2 by 2	Infiltration	
Purdue Sprinkling Infiltrometer Purdue Univ. Lafayette, IN	Spray Engr Co 7LA	41.4	Stationary Continuous	0.1	1.2	2.4	119	1.2 by 1.2	Infiltration	Bertrand & Parr, 1961
	5B			0.1	0.8	1.5	64			
	5D			0.1	0.6	1.5	82			

(continued on next page)

Table 33-1. Continued.

Rainfall simulator location	Nozzle	Pressure	Nozzle movement and spray pattern	Drop size D_{10}	Drop size D_{50}	Drop size D_{90}	Intensity	Plot size per unit	Use	Reference
		kPa		mm	mm	mm	mm/h	m		
Modified Purdue Type Univ. of Wisconsin Madison, WI	Spray Engr Co 7LA	41.4	Stationary Intermittent	0.1	1.2	2.4	2-111		Infiltration Nutrient transport	Dixon & Peterson, 1964 Amerman et al., 1970
Modified Purdue Type USDA-SEA-AR Tuscon, AZ	Spray Engr Co 7LA	41.4	Stationary Continuous	0.1	1.2	2.4	119	1.0 by 1.0	Infiltration Erosion Soil management	Dixon & Peterson, 1968
Modified Purdue Type Univ. of Missouri Columbia, MO	Spray Engr Co 7LA	41.4	Stationary	0.1	1.2	2.4	2-111			
Modified Purdue Type USDA-SEA-AR Columbia, MO	Spray Engr Co 7LA	41.4	Stationary Intermittent	0.1	1.2	2.4	2-111			
Variable Intensity Inf. Hebrew Univ. Rehovot, Israel	Spray Engr Co 7LA	41.4	Stationary Intermittent	0.1	1.2	2.4	2-111	1.2 by 1.2	Infiltration	Rawitz et al., 1972
RFER Colorado State Univ. Fort Collins, CO	Rainjet 78C	193	Stationary	0.5	1.2	3.0	12-100		Runoff	Holland, 1969
Sprinkler Head Grid North Dakota State Univ. Mandan, ND	Rainjet 78C	193	Stationary Continuous	0.7	1.4	2.8	36 & 58	13 by 26	Erosion Infiltration	Holland, 1969
USGS USGS Lakewood, CO	Rainjet 78C	193	Stationary Continuous	0.6	1.4	2.8	50		Erosion Infiltration Runoff	Lusby, 1977
USGS Bureau of Land Management Denver, CO	Rainjet 78C	193	Stationary Continuous	0.6	1.4	2.8	50		Erosion Infiltration	Lusby, 1977
USGS USDA-SEA-AR Tuscon, AZ	Rainjet 78C	193	Stationary Continuous	0.6	1.4	2.8	50		Erosion Infiltration	Lusby, 1977

ulator (Meyer & Harmon, 1979) and the new Rainulator (Foster et al., 1982). The interrill simulator and the new Rainulator use the 80150 Veejet nozzle as well as the 80100 Veejet. The drop size distribution and the kinetic energy level obtained with the 80150 Veejet are somewhat greater than those for the 80100.

Holland (1969) used the Rainjet 78C nozzle on a 13- by 26-m plot simulator. Lusby (1977) modified this simulator for field work. The primary advantages of the simulator are the ease of assembly and the flexibility of plot size and shape. Drop sizes and energy levels are, however, lower than from those simulators using the Veejet 80100 and 80150 nozzles.

The Spraying Systems 1.5H30 Fulljet nozzle has also been used on several simulators. This nozzle is used in connection with the rotating disk (Morin et al., 1967) to reduce rainfall intensity. The median drop size of 2.6 mm compares well with erosive natural storms.

Several other nozzles are described in Table 33–1. Included in this group is the Type F nozzle described by Wilm (1943), Dortignac (1951), and Packer (1957). While drop sizes from the Type F nozzle are large, the kinetic energy is low due to the short fall-height of the simulator. The Spraying Systems 14WSQ Fulljet nozzle is being used on the Palouse Infiltrometer to simulate the small drop size and low-intensity events of the Pacific Northwest. The Beta Fog SRN 303 nozzle, used by Shriner et al. (1977) on the RAINS simulator, produces a drop size much less than those of natural rainfall.

33–3.1.1 PURDUE SPRINKLING INFILTROMETER

The sprinkling infiltrometer, often called the Purdue Sprinkling Infiltrometer, (Bertrand & Parr, 1961), was developed in conjunction with a North Central Regional Technical Committee 40(NC-40) and was used to identify the water infiltration rates into soils of the Upper Midwestern Region. The drop size distribution, intensity, spray diameter, energy, and variation in spray pattern for three different nozzles at different heights and pressures are given in Table 33–2. Infiltration rates for 150 sod and bare soils throughout the North Central Region were determined using this system (Jones, 1979). These basic nozzle data have been used on most of the sprinkling infiltrometers in recent years.

33–3.2 Drop-Former Systems

The second group of rainfall simulators included in the inventory use drop-formers to produce rainfall. Early simulators used small pieces of yarn to form drops (Ekern, 1950). Recent simulators have used glass capillary tubes, hypodermic needles, polyethylene tubing, and brass or stainless tubes as drop formers. Reported drop diameters range from 2.2 (Bubenzer & Jones, 1971) to 5.6 mm (Adams et al., 1957). Most simulators produce drops at a constant size for a given rainfall module. Brak-

Table 33-2. Drop size distribution, intensity, spray diameter, energy and variation in spray pattern for three nozzles at different heights and pressures (Bertrand & Parr, 1961).

Nozzle	Pressure	Height	Variation over 1.22 by 1.22 m plot	Proportion of all drops in each drop size range (mm)									Intensity per hour	Spray diameter	Energy per ha/cm
				3.327 or larger	3.326–2.794	2.793–2.362	2.361–1.651	1.650–1.410	1.409–1.168	1.167–0.833	0.832–0.147	0.146 or smaller			
	kPa	m	%										mm	m	MT m
7LA	20.7	2.7	7.12	22.4	6.9	8.7	19.3	6.5	7.1	10.5	13.1	5.5	112	3.35	2509
7LA	27.6	2.7	5.56	17.1	12.0	9.1	18.3	7.2	6.6	10.2	12.6	6.9	108	3.54	2528
7LA	34.5	2.7	5.31	4.2	6.6	8.9	19.7	7.9	8.4	15.1	19.6	10.4	110	3.84	2305
7LA	41.4	2.7	4.04	1.0	5.1	6.7	19.3	10.3	8.3	13.9	20.1	14.8	119	4.15	2249
7LA	62.0	2.7	5.01										138		
7LA	20.7	2.4	8.33	9.7	8.2	10.6	23.2	8.9	8.9	12.4	12.0	5.9	140	3.23	2275
7LA	27.6	2.4	5.75	4.5	8.9	10.8	23.5	9.8	8.8	12.4	13.3	7.6	135	3.35	2275
7LA	34.5	2.4	6.35	3.3	10.3	8.0	24.8	10.6	8.9	13.2	13.6	7.3	129	3.66	2301
7LA	41.4	2.4	6.99	3.1	3.2	7.9	22.8	10.7	9.9	15.9	15.6	10.8	128	3.84	2182
7LA	41.4	2.3	5.62												
7LA	41.4	1.8	8.36												
5B	17.6	2.7	15.39												
5B	34.5	2.7	6.64			1.4	7.8	7.2	10.7	24.7	31.1	16.9	70	2.44	1587
5B	41.4	2.7	6.54				5.8	5.7	10.2	20.4	36.6	21.3	64	2.93	1476
5B	48.3	2.7	9.92				8.4	5.0	7.8	21.4	40.9	16.0	69	3.05	1516
5B	34.5	2.4	5.20			4.0	9.9	8.6	11.5	25.6	31.3	12.6	78	2.44	1624
5B	41.4	2.4	7.59				5.8	6.6	8.5	22.1	36.3	20.8	75	2.74	1457
5B	48.3	2.4	7.56			0.4	4.2	6.0	7.7	20.2	36.6	24.8	75	3.05	1394
5B	20.7	2.4	20.0												
5B	27.6	1.8	23.6												
5B	41.4	1.8	8.94												
5B	37.9	1.8	10.7												
5D	41.4	1.8	21.77												
5D	27.6	1.8	31.82												
5D	62.0	2.4	6.61				3.9	8.5	10.8	24.5	38.6	13.4	90	3.05	1777
5D	62.0	2.7	7.28			0.3	3.0	4.8	8.5	21.6	40.2	21.8	82	3.05	1844
5D	48.3	2.7	13.65												
5D	41.4	2.7	9.91												
5D	34.5	2.7	13.49												
5D	27.6	2.7	17.66												
5D	20.7	2.7	43.87												

ensiek et al. (1979) used compressed air blowing around the drop-former to produce drops of varying sizes from the same module. Most of these simulators cover relatively small plots. The most notable exceptions are the simulators developed by Chow and Harbaugh (1965) and Peterson (1977) and the laboratory simulators located at Purdue and Utah State Universities.

33–3.3 System Modifications

Various states of the North Central Region modified the Purdue unit, with a major modification being by Dixon and Peterson (1964). Their vacuum accumulation systems (Dixon & Peterson, 1968) simplified the operation while increasing accuracy. Amerman et al. (1970) and Rawitz et al. (1972) described slotted rotating disk units for the Purdue Infiltrometer to reduce rainfall intensity. There have been no major changes in the nozzle.

Interest also developed on theoretical investigations of one-dimensional infiltration and redistribution situations. Examples are papers by Hanks and Bowers (1962), Gardner et al. (1970), and Hillel and Gardner (1970). These investigations all dealt with homogeneous, uniform soils. In order to use the sprinkling infiltrometer to validate this multidimensional flow theory in the field it was necessary to modify the sprinkling unit to provide a wide range of intensities. To provide the desired impact energy at very low intensities, the drop size should reproduce natural rainfall. The low intensities have small drop size (Table 33–2); thus, the varying-intensity sprinkling infiltrometer was developed by Amerman et al. (1970) and later modified by Rawitz et al. (1972). This involved developing a rotating disk unit with pie-slice openings, which intercepted

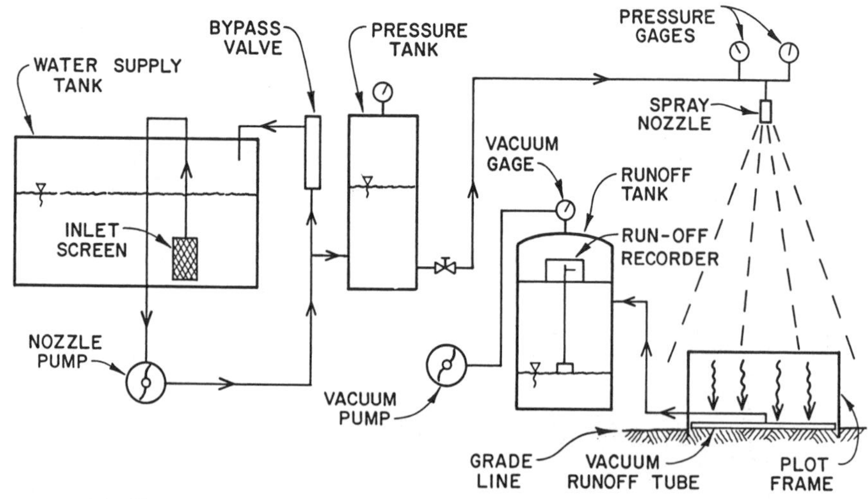

Fig. 33–3. Water flow chart for spray infiltrometer (Dixon & Peterson, 1968).

and recirculated from 50 to 99% of the nozzle output. Thus, the nozzle intensity could be reduced by anything > 50% down to even 1%. This provided much greater latitude for use of sprinkling infiltrometers. Figure 33–3 gives a view of the modified unit and the vacuum accumulation. Dixon and Peterson (1968) developed a shunt system that permitted sampling at any time during runoff, since the shunt was simply hooked into the vacuum system between the collection frame and the vacuum accumulating tank. This permitted characterization of the runoff at any time.

33–4 A RECOMMENDED SPRINKLING INFILTRATION SYSTEM

The Purdue Sprinkling Infiltrometer (Bertrand & Parr, 1961) as modified by others (Dixon & Peterson, 1964) and modified by Dixon (R. M. Dixon, 1983, personal comunication) to electric power and metric units has proven very satisfactory for infiltration research.

33–4.1 Apparatus

33–4.1.1 PUMPING UNIT

To reduce mechanical troubles and labor requirements, the spray nozzle pump and runoff vacuum pump are driven by electric motors. A lower capacity nozzle pump may be used (3785 L/h instead of 15 140 L/h) to take some of the load off the pressure control system. Flow rate of the 7LA nozzle at 41.4 kPa is about 1890 L/h; thus about one-half of the water is by-passed, even with the lower capacity pump.

A large supply of water is required to simulate rain for a 2-h run; therefore, a supply tank of about 3785 L is desirable. To transport this quantity of water requires a heavy-duty carrier. An auxiliary tank and carrier are desirable for extended runs. Satisfactory tanks with suction pipe and foot valve (such as discarded underground fuel tanks) can frequently be obtained at little cost from excavating contractors.

33–4.1.2 NOZZLE PRESSURE-CONTROL SYSTEM

For control of nozzle pressure, a low-pressure spring-loaded by-pass valve is used. A by-pass valve tends to compensate for variations in engine or pump speed, thereby reducing variations in nozzle pressure.

A pressure gauge was installed at the top of the pressure tank so that pressure adjustments can be made before water is released to the nozzle. To provide a safeguard against water-logging the system, a water level tube was installed on the side of the pressure tank.

33–4.1.3 RUNOFF COLLECTION SYSTEM

The runoff collection system includes the plot frame, runoff "wand", and runoff collector. The plot frame is fabricated from four steel plates welded to form a square (Fig. 33–4). Runoff is collected with a 9.5-mm diameter brass tube placed on the soil surface parallel and adjacent to the lower side of the plot frame and connected at the center to the vacuum line. Runoff is picked up via 3.2-mm holes drilled at 6.4-mm intervals along the underneath side. This allows the "wand" to follow a settling plow layer downward and thus prevents ponding and sedimentation by removing runoff at the level of the soil surface. It also permits installation of the 30.5-cm plot frames to the depth required to block lateral water flow, since depth is not dictated by runoff holes or slots. It also permits elimination of the surface-storage variable of contour-furrow treatments, and extends the sprinkler infiltrometer research to nearly level topography.

33–4.1.4 RUNOFF MEASURING SYSTEM

The runoff measuring system consists of a runoff accumulation tank, a leveling stand, and a portable water stage recorder. Tank dimensions were chosen such that 10 cm of runoff could be accumulated, with 12.5 cm of vertical pen movement (from bottom to top of chart) equivalent to 5 cm of runoff. The leveling stand is designed for ease of leveling (when tank is full or empty), to resist settling, and for low cost of construction.

33–4.1.5 TOWER AND COVER

The tower for supporting the nozzle and wind shield consists of a top plate with three 2.92-cm pipe stubs welded at the proper angle to permit the three 5-cm thin-walled steel tubing legs of the tripod to slip over and be secured. The tower is easily erected by one person in 5 to 10 min. Once erected, the tower (without removal of cover) can be trans-

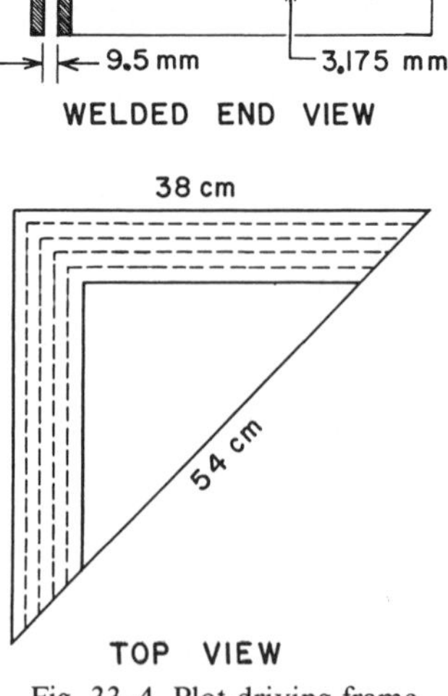

Fig. 33–4. Plot driving frame.

ported to new plot sites by slipping small wheels onto the axles located on two of the tower base brackets. One person can move a covered tower; however, two people are recommended, especially in high winds.

A modified parachute canopy provides a suitable tower cover for complete protection against wind and rain. The nylon canopy is light weight, strong, mildew resistant, water-proof, semitransparent, and inexpensive.

33–4.1.6 SPRAY NOZZLE ASSEMBLY

Using a modified trailer hitch, a ball-and-socket arrangement was devised for suspension of the nozzle. With the aid of a bull's eye level and a plumb line, the nozzle can be quickly pointed vertically downward over the plot center and then locked into position with a set screw. Rough adjustment is made by altering the length of the carriage bolt which passes through the ball.

33–4.1.7 VARIABLE-INTENSITY SPRINKLING INFILTROMETER

The 7LA nozzle provides an intensity of 63.5 mm/h. The infiltrometer must be modified to produce low intensities. Amerman et al. (1970) developed a variable-intensity unit. It consisted of a rotating disk with four open, adjustable slots mounted on a portable, single-nozzle sprinkling infiltrometer. When rotated at about 60 rpm, essentially uninterrupted artificial precipitation reached the ground surface. At intensities ranging from 1 mm/h to more than 150 mm/h uniformity coefficients were esentially equal to those achieved by the nozzle itself.

33–4.2 Equipment Suppliers

The names and addresses of suppliers that have provided satisfactory materials are listed. They are identified only as a convenience, not as a recommendation. Local suppliers may be preferred.

1. Pressure regulator for controlling nozzle pressure Dresser Industrial Valve & Instrument Div.
   Dresser Industries, Inc.
   7332 E. Adams St.
   Paramount, CA 90723
   Reverse acting pressure regulator having specifications
   a. 2760R series for regulating back pressure;
   b. Size—1-1/2 inch;
   c. Range—7 to 15 lb.;
   d. Single seated with 1-1/2-inch orifice;
   e. Bronze body;
   f. Union ends of malleable iron 1 each
2. Vacuum regulator for controlling runoff tank vacuum

Gast Manufacturing Corp.
Box 117
Benton Harbor, MI 49022
AA204, Vacuum Relief Valve — 1 each
AA840A, Vacuum Relief Valve — 1 each

3. Sprinkler nozzle
   Spray Engineering Company
   100 Camridge St.
   Burlington, MA 01803
   7LA-Full Cone Nozzle; 303 stainless steel; 3/4 inch female connection — 1 each
   Model 156 Strainer with: (i) 100-mesh manel liner; (ii) 3/4-inch connections — 1 each
   Flushing Valve & Nipple for 3/4-inch strainer — 1 each
   Cleaning Brush for Strainer Cylinder for 3/4-inch strainer — 1 each

4. Pressure and vacuum gauges for water pressure system and air vacuum system
   Marsh Instrument Co.
   Div. Colorado Mfg. Corp.
   Skokie, IL 60076
   Type 1 Pressure, Marsh Quality Pressure Gauge with: (i) 15-lb dial range; (ii) 3 1/2-inch dial size; (iii) CP case pattern; (iv) Marsh recalibrator — 3 each
   Type 2 Vacuum Marsh Quality Vacuum Gauge with: (i) 30-inch dial range; (ii) 3-1/2-inch dial size; (iii) CP case pattern; (iv) Marsh recalibrator — 1 each

5. Electrical vacuum pump for runoff tank; vacuum and runoff conduits
   Van Waters and Rogers, Inc.
   850 South River Road
   Sacramento, CA 95691
   54908, Air Pressure–Vacuum Pump, 3.8 ft^3 min^{-1}; Serial no. 67 362697, Model no. 5KH39KG 416T — 1 each
   56421, Rubber Tubing, Garlock, black 3/8-inch inside diameter — 200 feet

6. Waterstage recorder
   Belfort Instrument Co.
   1600 S. Clinton St.
   Baltimore, MD 21224
   a. 5-FW-1 Portable Liquid Level Recorder with: (a) gears for 5:12 and 1:12 stage ratio; (ii) 6-inch float and counter weight; (iii) 30 foot stainless steel tape; (iv) 6-h clock gears — 2 each
   b. 5573, Pt. 6 & 5572, Pt. 6; extra 6-h clock gears — 4 set
   c. 5574, Pt. 144 & 5572, Pt. 144; extra 144-h clock gears — 4 set

d. 695-30, Graduated perforated tape, 30-ft length 1 set
e. (v) 5-1940-AB, charts for liquid level recorder 5 set
f. (vi) 5-744, hook gauge with metric graduation 1 each
g. (vii) 5-745, stilling well 1 each
h. (viii) P/N 9664 and P/N 9665; float wheel gear, cam shaft gear-conversion for 5:6 ratio 4 set
i. (ix) 5592; recorder ink no. 10 black, 1-oz. bottle with dropper cap 2 btl
j. (x) Part no. 56,00, Pt. 2; lubricating oil, general purpose in 2-oz. bottles 3 btl

7. Chart drive for recorder
   Belfort Instrument Company
   1600 S. Clinton St.
   Baltimore, MD 21224
   18–29, Chart Drive, fully jeweled anchor escapement, complete with cylinder, chart clip, 6-h gears and mounting nut. This part is for a portable water recorder Belfort Cat. no. 5-FW-1 1 each

8. Power source for driving electric vacuum and water pumps and water pump for spray nozzle
   Sears, Roebuck & Co.
   2650 E. Olympic Blvd., 10th Floor
   Los Angeles, CA 90054
   32AF32033N, Alternator, 3650 W, 4500 W surge, rope start, continuous duty. Ser. no. 120029 1 each
   42K2532N, Electric water pump, centrifugal, 63 gpm at 10 psi, 1/2 HP, self-priming 1 each

9. Voltage tester for alternator
   Sears, Roebuck & Co.
   2650 E. Olympic Blvd., 10th Floor
   Los Angeles, CA 90054
   32AF32231, Voltage tester 1 each

10. Vacuum tank for accumulating runoff
    Nevada Sheet Metal
    2101 Timber Way
    Reno, NV 89502
    Galvanized steel cylinders having: (i) one end open; (ii) one end closed with 14 gauge galvanized steel reinforced by cross bending; (iii) dimensions of 32.5 cm i.d. by 150 cm long; and (iv) welded construction 5 each

11. Various tanks — local supplier
    a. 3785-L supply tank 1 each
    b. 200-L pressure tank 1 each

12. Plot frames — (as many as desired) — local supplier
    Steel plate, 3.17 mm by 30.5 by 100 cm

Cut and welded to form a square 100 cm on the inside lower side of frame V sharpened slightly to facilitate driving

13. Plot-driving frame — four required (each corner)
    Steel—3.175 mm — right triangle — 38 by 38 by 54 cm
    Steel—9.5 mm — 6 — 38 mm pieces

14. Plot drivers
    Galvanized pipe, 20 mm by 120 cm
    Steel plate, 50 by 152 by 152 mm
    Pipe reducing breaking, 50 by 38 mm
    Rubber stopper no. 11, 19-mm hole
    Pipe coupling, galvanized, 50 mm — welded to steel plate

15. Runoff (calibration) pan
    15 ga. galvanized steel sheet, corners riveted and soldered, 6.35 mm larger than plot frame

16. Tower for suspending spray nozzle pipe
    Thin-wall steel tubing, 31.75 by 365 cm (three or four required); tripod preferred

17. Parachute for windshield (personal type) — panels removed to fit tower

The availability of above items 10, 12, 13, 14, 15, 16, and 17 locally is preferred, since several must fit together (items 12, 13, and 15). Specifications for the variable intensity arrangement are given by Amerman et al. (1970).

33–4.3 Procedure

33–4.3.1 INSTALLATION OF RUNOFF COLLECTION SYSTEM

Duplicate plot sites should be selected which will minimize differences due to soil micro-relief, cover (including living vegetation, plant residue, stones, and clods), and compaction. Before installation, plot frames are oriented with the lower side exactly perpendicular to the slope. The angle-iron driving frame is installed and the plot frame is then driven uniformly into the soil.

33–4.3.2 SETTING UP RUNOFF MEASURING SYSTEM

The runoff measuring system, consisting of a recorder, runoff tank, and leveling stand, is set up midway between the pumping unit and the runoff collection system and then leveled. The recorder chart should bear such annotations as date, time, 15-min time intervals, treatment, and replicate. Hoses are connected so that runoff flow is cycled from the runoff collector to the runoff tank.

33–4.3.3 SPRAY-NOZZLE INSTALLATION AND ORIENTATION

The nozzle assembly is attached to the terminal bracket of the tower with a single machine bolt. This bolt is initially tightened with the nozzle in the center of the horizontal adjustment. The nozzle is then roughly oriented over the plot with a plumb line, after which the tower is staked down using the holes provided in the base brackets. A 1.9-cm garden hose is connected between the pressure tank and the nozzle. The parachute canopy is slipped on the tower and anchored at the periphery via shroud lines and steel stakes. Final orientation of the nozzles is made at the nozzle assembly. The orientation is correct when the nozzle is (i) located directly over plot center, (ii) located 275 cm above plot center, and (iii) pointing vertically downward. A separate adjustment is provided to satisfy each of the above conditions.

33–4.3.4 INITIATION OF RUN

The spray is turned on after the recorder pen tracing horizontally along the chart zero line reaches a heavy time line, and the vacuum has reached 34.5 kPa in the runoff (accumulation) tank. Spring tension of the by-pass valve is adjusted to give 41.4 kPa at the spray nozzle or about 65.6 kPa at the pressure tank. Nozzle pressure and runoff-pump pressure should be checked frequently. The time at which the entire plot area begins contributing to runoff should be noted. Periodically, the recorder clock and pressure gauges should be checked for accuracy.

33–4.3.5 CALIBRATION OF NOZZLE

At the beginning or end of each run the application rate of the nozzle is determined as follows:

1. Put the runoff pan on the plot frame and place collection "wand" in the pan.
2. Turn on spray after the recorder pen has been allowed to trace horizontally along the chart zero line for a few minutes to a heavy time line, and 34.5 kPa vacuum has developed in the runoff tank.
3. Spray for 20 min at a nozzle pressure of 41.4 kPa.
4. Determine application rate on the basis of the last 15 min of the run.

33–4.4 Data Processing

When 1-L samples are collected via the shunt system (direct from runoff line), the sediment concentration is calculated as follows: Evaporate a 25-mL sample of water applied (your blank); subtract this weight from the sediment weight remaining after evaporating a 25-mL aliquot of the runoff sample. Multiply the net sediment weight by 40 to obtain the grams per liter concentration in the runoff.

When sampling from the accumulation tank it is always necessary to add water to zero the float in the waterstage recorder at the beginning of each run. Thus, the following step *must* be added to obtain the final concentration in the runoff. Determine liters of water required to zero the float. After the run, determine the sediment concentration in the diluted runoff by the previous technique, then multiply the liters of runoff plus the liters necessary to zero the waterstage recorder by the concentration (g/L) of the diluted runoff. Divide this number by liters of runoff to obtain the concentration (g/L) of the runoff (based on the undiluted concentration).

To obtain the sediment loss from the plot fram area during the applied "rain," multiply the concentration (g/L) by total runoff (liters). Plot-frame areas of 1 m^2 are convenient for work and calculations, but any area of approximately that size is satisfactory.

33–5 COMMENTS

33–5.1 North Central Regional Project

Nearly three dozen of the important agricultural soils of the North Central Region were characterized using the Purdue Infiltrometer (Jones, 1979). The committee agreed on the soils to be tested and the standard field procedures. This wealth of data is available in a 93-p appendix to the Illinois Agricultural Experiment Station Bull. 760. Most states made measurements on clipped bromegrass as well as on corn seedbed. Some made measurements on corn after cultivation. All sites used both a dry and a wet run, which means that 24 h after the dry run was completed, another run so-labeled "wet" was also made. Data collected at each site included the antecedent moisture content from the 0- to 15-mm depth, water application rate, the time from start of a test until the start of runoff, the average infiltration rate at selected time intervals, and the equilibrium infiltration rate. The Fayette silt loam (fine-silty, mixed, mesic Typic Hapludalfs) infiltration rates for both dry and wet runs on clipped bromegrass and bare soil are given in Tables 33–3 and 33–4.

The regional study (Jones, 1979) found, in ten of the thirteen cases, that there was no significant difference in the equilibrium infiltration rates for corn seedbed compared to those for cultivated corn. In the three cases where there were differences, special physical properties of the soils affected cracking of the profile or crusting of the surface. This shows that the changes in soil conditions related to the growth of the corn plant (*Zea mays* L.) and one additional trip through the field with machinery are insufficient to justify infiltration measurements under both field conditions. The chance that a 25- to 50-year storm will occur during the one to three weeks of the year when the soil is fully exposed, or even during a 10-year study, is remote. However, new cultural practices need to be evaluated. The sprinkling infiltrometer presents a method for rapid screening of these new techniques.

Table 33–3. Moisture infiltration rates on soils with corn seedbed treatment (Jones, 1979).†

Identification		Percent antecedent moisture, 0–15 cm deep	Rainfall intensity, cm/h	Minutes to initial runoff	Mean rate, cm/h								
Rep.	Dup.				15 min	30 min	45 min	60 min	75 min	90 min	105 min	120 min	Equil. rate
				Initial run (July 1965)									
1	1	13.4	9.9	12	9.9	9.1	6.6	5.3	3.6	2.8	2.5	2.0	2.0
1	2		9.9	3	9.9	7.9	5.3	3.6	2.3	2.8	2.3	0.5	1.3
2	1		10.4	15	10.4	6.4	3.3	3.3	1.3	0.2	2.0	1.8	1.8
2	2		10.4	5	10.4	9.9	7.9	6.1	4.6	3.8	4.3	4.1	4.1
				Wet run (July 1965)									
1	1	16.2	9.6	7	2.3	1.5	1.0	0.2	0.0	0.0	0.0	--	0.2
1	2		10.4	2	4.3	2.3	0.8	1.3	1.5	1.0	1.5	2.3	2.0
2	1		9.6	2	5.6	1.0	0.2	0.0	0.0	0.0	0.0	--	0.0
2	2		9.6	2	5.3	1.3	1.0	1.0	--	0.5	--	0.5	0.5

† Soil: Fayette loam (Wisconsin). Crop stage: No crop.

Table 33–4. Moisture infiltration rates on soils with clipped bromegrass treatment (Jones, 1979).†

Identification		Percent antecedent moisture, 0–15 cm deep	Rainfall intensity, cm/h	Minutes to initial runoff	Mean rate, cm/h								
Rep.	Dup.				15 min	30 min	45 min	60 min	75 min	90 min	105 min	120 min	Equil. rate
				Initial run (July 1965)									
1	1	12.3	9.6	10	9.6	7.6	5.1	3.6	3.1	2.8	2.5	2.7	2.7
1	2		10.7	8	9.6	8.1	6.4	6.6	5.8	5.6	5.1	4.3	4.3
2	1		9.9	12	9.4	6.1	5.1	4.3	1.0	2.0	1.8	1.8	1.8
2	2		9.9	10	8.6	6.6	5.1	4.1	3.6	3.3	2.8	1.0	1.0
				Wet run (July 1965)									
1	1	14.4	9.6	8	7.6	4.1	3.0	3.6	3.3	3.6	2.3	3.3	2.8
1	2		10.7	5	7.9	3.3	4.7	3.6	4.8	2.0	3.0	1.3	1.3
2	1		9.9	10	7.9	3.6	2.5	3.1	0.8	3.0	3.0	1.3	1.3
2	2		9.9	8	7.9	3.6	1.5	2.5	2.0	1.5	2.0	1.0	1.0

† Soil: Fayette loam (Wisconsin).

Table 33-5. Ratios for the comparison of runoff and erosion from natural rainfall plots and from spray-infiltrometer plots, Miami silt loam soil (Madison, WI, 1964; Dixon, 1966).

Rainfall type	Runoff ratio†	Erosion ratio
Simulated	0.0719	0.0256
Natural	0.0877	0.0215

† Ratio = $\frac{\text{Losses from minimum-tilled corn land with stover cover}}{\text{Losses from conventional-tilled corn land}}$

Ratios are based on 60-min losses under simulated rainfall and 7-year losses under natural rainfall. Simulated rainfall ratios are based on the mean of duplicate runs.

33-5.2 Evaluation of C-factor in USLE

The prediction accuracy of the water erosion equation which was developed by Wischmeier and Smith (1965) depends in part upon the evaluation of the cropping management factor (C-factor). Since the C-factor ranges from 0 to 1, the utility of having the erosion equation depends on the reasonable accurate estimate of this factor. Dixon (1966) used the normal runoff plots at Madison, WI, for a comparison of this cropping factor to estimate the water erosion. The 7LA nozzle supplies 1.9×10^7 J/ha per hour (2813 foot tons per acre per hour) corresponding to the energy of a natural storm with an intensity of 67.5 mm/h (Wischmeier & Smith, 1958). This is the energy equivalent to a 50-year, 1-hour storm at Madison, WI. To provide this energy the 7LA nozzle provides a sprinkling rainfall with an application of 120 mm/h; consequently the energy per unit of rainfall is low compared to natural rainfall. The additional water provides more overland flow and may therefore help to compensate for the small plot size. Comparisons of tillage practices and crop residue management systems are possible over a wide range of soil types and slopes using this system. The effectiveness of various covers of erosion on highway back slopes was studied by Meyer and Mannering (1963) and on construction sites (Meyer et al., 1971) as well as urban lawn infiltration rates and fertilizer runoff losses (Kelling & Peterson, 1975).

Sprinkling infiltrometer and natural runoff and erosion data have been compared at the University of Wisconsin Experimental Farm, Madison. A set of long-time natural rain runoff plots were available for comparisons with spray infiltrometer measurements made on each of these plots on the Miami silt loam soil. The comparisons are given in Table 33-5. The data in Table 33-5 suggest that there is approximately a 1:1 relationship between the simulated-rainfall from the sprinkling infiltrometer and the natural-rainfall ratios. Thus if 60-min losses can be determined by infiltrometer and losses from the standard practice can be determined or are known for natural rainfall, then natural rainfall losses for the new soil management practices may be able to be estimated (Dixon & Peterson, 1971). Extreme care must always be taken when trying to extrapolate from a small runoff plot or measurement to a larger area, especially on a watershed basis. Decades of research have shown that erosion loss or sediment movement will be much greater on small plots than on larger areas. The integration effects of surface roughness, slope, vegetation, and many other factors frequently do not apply to the same degree on the small plot as they would on any larger area.

33–6 CONCLUSIONS

Although there are some obvious limitations to the use of sprinkling infiltrometers for field studies, infiltrometers have proved their value as a research tool (Moldenhauer, 1979). While sprinkling infiltrometers can be somewhat costly in terms of their operation, much valuable data can be collected in the course of a single growing season. Natural-rainfall plots are extremely costly to operate and data may take years to accumulate. It is doubtful that we ever again will be content to depend on natural-rainfall plots as a sole source of data. It appears that rainfall simulators are here to stay (Moldenhauer, 1979).

Because of the complexity of the erosion process and the interaction among rainfall, runoff, and soil erodibility, no single, simple parameter covers the entire range of conditions (Peterson & Dixon, 1971). Until such a parameter is identified, researchers must determine the importance of the raindrop in the erosion or infiltration process being investigated. Where raindrop action is not of great importance to the process being studied, such as erosion or nutrient movement beneath an agronomic crop canopy or infiltration beneath a mulched surface, drop characteristics of the simulator may be of minor importance. However, when raindrop action is critical in the process, as in splash erosion or infiltration into an exposed soil, every attempt must be made to accurately simulate the drop-size distribution, impact velocity, and intensity of the natural erosive storms of the area.

33–7 REFERENCES

Adams, J. E., D. Kirkham, and P. P. Nielsen. 1957. A portable rainfall simulator infiltrometer and physical measurements of soil in place. Soil Sci. Soc. Am. Proc. 21:473–477.

Amerman, C. R., D. I. Hillel, and A. E. Peterson. 1970. A variable intensity sprinkling infiltrometer. Soil Sci. Soc. Am. Proc. 34:830–832.

Anderson, J. U., A. E. Stewart, and P. C. Gregory. 1968. A portable rainfall simulator and runoff sampler. New Mexico State Univ. Agric. Exp. Stn. Rep. 143.

Barnes, O. F., and G. Costel. 1957. A mobile infiltrometer. Agron. J. 49:105–107.

Bertrand, A. R., and J. F. Parr. 1961. Design and operation of the Purdue sprinkling infiltrometer. Purdue Univ. Res. Bull. 723.

Borst, H. L., and R. Woodburn. 1940. Rainfall simulator studies of the effect of slope erosion and run-off. USDA-SCS SCS-TP-36. U.S. Government Printing Office, Washington, DC.

Brankensiek, D. L., W. J. Rawis, and W. R. Harmon. 1977. Application of an infiltrometer system for describing infiltration into soils. Paper no. 77-2553. ASAE winter meeting, Chicago. American Society of Agricultural Engineers, St. Joseph, MI

Brockman, F. E., W. B. Duke, and J. F. Hunt. 1975. A rainfall simulator for pesticide leaching studies. Weed Sci. 23:533–535.

Bubenzer, G. D. 1979. Rainfall characteristics important for simulation. USDA Rainfall Simulator Workshop. Tucson, AZ. p. 22–34. ARM-W-10. U.S. Government Printing Office, Washington, DC.

Bubenzer, G. D., and B. A. Jones, Jr. 1971. Drop size and impact velocity effects on the detachment of soil under simulated rainfall. Trans. ASAE 14:625–628.

Bubenzer, G. D., and L. D. Meyer. 1965. Simulation of rainfall and soils for laboratory research. Trans. ASAE 8:73, 75.

Carter, C. E., J. D. Greer, H. J. Braud, and J. M. Floyd. 1974. Raindrop characteristics in south central United States. Trans. ASAE 17:1033–1037.

Chapman, G. 1948. Size of raindrops and their striking force at the soil surface in a red pine plantation. Trans. Am. Geophys. Union 29:664–670.

Chow, V. T., and T. E. Harbaugh. 1965. Raindrop production for laboratory watershed experimentation. J. Geophys. Res. 70:6111–6119.

Cluff, C. B., 1971. the use of a realistic rainfall simulator to determine relative infiltration rates of contributing watershed to the Lower Gila below Point Rock Dam. Univ. of Ariz., Water Resources Center. Tucson, AZ.

Costin, A. B., and D. A. Gilmour. 1970. Portable rainfall simulator and plot unit for use in field studies of infiltration, runoff, and erosion. J. Appl. Ecol. 7:193–200.

Dixon, R. M. 1966. Water infiltration responses to soil management practices. Ph.D. thesis. Univ. of Wisconsin, (order no. 66-5903; Diss. Abstr. 27:4.)

Dixon, R. M., and A. E. Peterson. 1964. Construction and operation of a modified spray infiltrometer. Univ. of Wisconsin-Madison Res. Rep. 15.

Dixon, R. M., and A. E. Peterson. 1968. A vacuum system for accumulating runoff from infiltrometers. Soil Sci. Soc. Am. Proc. 32:123–125.

Dixon, R. M. and A. E. Peterson. 1971. Water infiltration control: a channel system concept. Soil Sci. Soc. Am. Proc. 35:968–973.

Dortignac, E. J. 1951. Design and operation of Rocky Mountain infiltrometer. USDA-FS Stn. Paper no. 5. Rock Mountain Forest Range Experiment Station, Ft. Collins, CO.

Ekern, P. C. 1950. Raindrop impact as the force initiating soil erosion. Soil Sci. Soc. Proc. 15:7–11.

Engel, O. G. 1955. Waterdrop collisions with solid surfaces. J. Res. Natl. Bur. Stand. 54:281–298.

Foster, G. R., W. H. Neibling, and R. A. Natterman. 1982. A programmable rainfall simulator. ASAE Paper no. 82-2570. ASAE winter meetings. Chicago. American Society of Agricultural Engineers, St. Joseph, MI.

Gardner, W. R., D. Hillel, and Y. Benyamini. 1970. Post-irrigation movement of soil water: I. Redistribution. Water Resour. Res. 6:851–861.

Grierson, I. T. 1975. The effect of gypsum on the chemical and physical properties of a range of red brown earths. Unpublished M. Agric. Sci. thesis. University of Adelaide, South Australia.

Gunn, R., and G. D. Kinzer. 1949. Terminal velocity of water droplets in stagnant air. J. Meteorol 6:243–248.

Hanks, R. J., and S. A. Bowers. 1962. Numerical solution of the moisture flow equation for infiltration into layered soils. Soil Sci. Soc. Am. Proc. 26:530–534.

Hermseier, L. F., L. D. Meyer, A. P. Barnett, and R. A. Young. 1963. Construction and operation of a 16-unit Rainulator. USDA-ARS 41-62.

Hillel, D., and W.R. Gardner. 1970. Measurement of unsaturated conductivity and diffusivity by infiltration through an impeding layer. Soil Sci. 109:149–153.

Holland, M.E. 1969. Design and testing of a rainfall system. Colorado State Univ. Exp. Stn. CER 69–70 MEH 21. Ft. Collins, CO.

Jones, B. A. 1979. Water infiltration into representative soils of North Central Region. Univ. of Illinois Agric. Exp. Stn. Bull. 760 (with appendix).

Kelling, K. A. and A. E. Peterson. 1975. Urban lawn infiltration rates and fertilizer runoff under simulated rainfall. Soil Sci. Soc. Proc. 39:348–352.

Laws, J. O. 1941. Measurement of the fall velocity of waterdrops and raindrops. Trans. Am. Geophys. Union. 22:709–721.

Laws, J. O., and D. A. Parsons. 1943. Relation of raindrop size to intensity. Trans. Am. Geophys. Union 24:452–460.

Lusby, G. G. 1977. Determination of runoff and sediment by rainfall simulation. In erosion: research techniques, erodibility and sediment delivery. Geological Abstracts, Norwich, England.

Lyles, L., L. A. Disrud, and W. P. Woodruff. 1969. Effects of soil physical properties, rainfall characteristics and wind velocity on clod disintegration by simulated rainfall. Soil Sci. Am. Proc. 33:302–306.

Marston, D. 1978. The use of simulated rainfall in assessing the erodibility of various stubble management practices. Proc. Conf. on Agric. Eng., 1978. Toowomba, Australia.

McGregor, J. C., and C. K. Mutchler. 1976. Status of the R factor in northern Mississippi. p. 135–142. *In* G. R. Foster (ed.) Soil Erosion Prediction and Control. Soil Conservation Society of America, Ankeny, IA.

McKay, M. E., and R. J. Loch. 1978. A modified Meyer rainfall simulator. Proc. Conf. on Agric. Eng., 1978. Toowomba, Australia. p. 78–81.

Meeuwig, R. O. 1969. Infiltration and soil erosion as influenced by vegetation and soil in northern Utah. J. Range Manag. 23:185–188.

Meyer, L. D., and W. C. Harmon. 1979. Rainfall simulator for evaluating erosion rates and sediment sizes for row sideslopes. Trans. ASAE 22:100–103.

Meyer, L. D., and J. V. Mannering. 1963. Crop residues as surface mulches for controlling erosion on sloping land under intensive cropping. Trans. ASAE 6:322, 323, 327.

Meyer, L.D., and D.L. McCune. 1958. Rainfall simulator for runoff plots. Agric. Eng. 39:644–648.

Meyer, L.D., W.H. Wischmeier, and W.H. Daniel. 1971. Erosion, runoff and revegetation of denuded construction sites. Trans. ASAE 14:138–141.

Meyer, L.D., W.H. Wischmeier, and G.R. Foster. 1970. Mulch rates required for erosion control on steep slopes. Soil Sci. Soc. Am. Proc. 34:928–931.

Moldenhauer, W. C. 1979. Rainfall simulation as a research tool. USDA Rainfall Simulator Workshop. Tucson, ZA. p. 90–95. ARM-W-10.

Moldenhauer, W. C., W. G. Lovely, N. P. Swanson, and H. D. Currence. 1971. Effect of row grades and tillage systems on soil and water losses. J. Soil Water Conserv. 26:193–195.

Morin, J., D. Goldberg, and I. Seginer. 1967. A rainfall simulator with a rotating disk. Trans. ASAE 10:74–77.

Munn, J.R., and G. L. Huntington. 1976. A portable simulator for erodibility and infiltration measurements on rugged terrain. Soil Sci. Soc. Am. J. 40:622–624.

Mutchler, C. K., and L. L. Hermsmeier. 1965. A review of rainfall simulators. Trans. ASAE 8:67–68.

Mutchler, C. K., and W. C. Moldenhauer. 1963. Applicator for laboratory rainfall simulator. Trans. ASAE 6:220–222.

Nassif, S. H., and E.M. Wilson. 1975. The influence of slope and rain intensity on runoff and infiltration. Hydrol. Sci. Bull. 20:539–553.

Packer, P. E. 1957. Intermountain infiltrometer. Misc. Pub. 14. USDA-FS, Intermountain Forest and Range Exp. Stn., Ogden, UT.

Peterson, A. E., and R. M. Dixon. 1971. Water movement in large soil pores. Res. Rep. 75. College of Agric. and Life Sciences, University of Wisconsin, Madison, WI.

Peterson, R. J. 1977. Laboratory simulation of soil erosion. Unpublished MS thesis. Agronomy Department, College of Agric. Colorado State University, Ft. Collins, CO.

Rawitz, E., M. Margolin, and D. I. Hillel. 1972. An improved variable-intensity sprinkling infiltrometer. Soil Sci. Soc. Am. Proc. 36:533–535.

Rogers, J. S., L. C. Johnson, D. M. A. Jones, and B. A. Jones, Jr. 1967. Sources of error in calculating the kinetic energy of rainfall. J. Soil Water Conserv. 22:140–142.

Seimens, J. C., and W. R. Oschwald. 1978. Corn-soybeans tillage systems: erosion control, effects on crop production, costs. Trans. ASAE 21:293–302.

Shriner, D. S., C. H. Abner, and L. K. Mann. 1977. Rainfall simulation for environmental application. Oak Ridge National Lab. Environmental Sciences Div. Pub. 1067.

Sloneker, L. L., and W. C. Moldenhayer. 1974. Effect of varying the on-off time of rainfall simulator nozzles on surface sealing and intake rate. Soil Sci. Soc. Am. Proc. 38:157–158.

Swanson, N. P. 1965. Rotating boom rainfall simulator. Trans. ASAE 8:71–72.

Turner, A. K., and K. A. Langlord. 1969. A rainfall simulator and associated facilities for hydrologic studies. J. Aust. Inst. Agric. Sci. 35:115–199.

Van Heerden, W. M. 1964. Splash erosion as affected by the angle of incidence of raindrop impact. Ph.D. Thesis. Purdue University, West Lafayette, IN (Diss. Abstr. 65-2653).

Wilm, H. G. 1943. The application and measurement of artificial rainfall on types FA and F infiltrometers. Trans. Am. Geophys. Union 24:480–484.

Wischmeier, W. H. and D. D. Smith. 1958. Rainfall energy and its relationship to soil loss. Trans. Am. Geophys. Union 39:284–291.

Wischmeier, W. H., and D. D. Smith. 1965. Predicting rainfall erosion loss from cropland east of the Rocky Mountains. USDA Agric. Handb. 202, U.S. Government Printing Office, Washington, DC.

34 Intake Rate: Border and Furrow[1]

D. C. KINCAID

Snake River Conservation Research Center, ARS, USDA
Kimberly, Idaho

34-1 INTRODUCTION

Water infiltration data which are to be used for evaluation, planning, or management of surface irrigation systems should be obtained by flooding or furrow-flow methods. This chapter describes methods that can be used for determining infiltration rates under actual operating conditions of border or furrow systems.

Volume-balance methods are most often used on low gradient borders and also can be used on level basins and level furrows. The type of irrigation system and slope are the main factors which determine whether the border or furrow method should be used. To achieve high application efficiencies, borders are generally constructed with slopes of 0.005 (m/m) or less. On slopes less than about 0.005, the average depth of water in surface storage is continually changing and can be a significant portion of the applied volume. On furrows having slopes greater than 0.005, the surface storage depth can usually be totally neglected or assumed to be constant. The border method consists of measuring surface storage and inflow volumes, whereas the furrow method consists of measuring inflow and outflow rates. The border method yields intake data applicable to the initial or advance portion of an irrigation, whereas the furrow inflow-outflow method yields data applicable to periods after water has advanced to the outflow station. The recirculating flow method is a modified inflow-outflow method that can evaluate intake rates for a relatively short time span.

The use of laser-controlled scraping equipment to construct level and graded borders has made the surface storage-advance method more practical because soil surface elevations and field slopes are more uniform and fewer measurements are needed to accurately determine surface storage volumes.

Davis and Fry (1963) compared four methods of determining intake

[1]Contribution from the Western Region, Agricultural Research Service, U. S. Department of Agriculture; University of Idaho College of Agriculture Research and Extension Center, Kimberly, cooperating.

in furrows. They stated that intake in furrows should be measured with flowing water when possible and that volume balance and inflow-outflow methods were the most accurate. They found that cylinder infiltrometers (Haise et al., 1956) and furrow infiltrometers (Bondurant, 1957; Shull, 1961) tend to overestimate intake rates in coarse-textured soils and underestimate furrow intake rates in fine-textured soils. Gilley (1968) and Kincaid (1979) also found that cylinder infiltrometers tend to underestimate intake rates on borders on fine-textured soils. Fangmeier and Ramsey (1978) found that ponded infiltration tests underestimate infiltration in furrows and that equations developed from advance data underestimate infiltration during the continuing and recession phases. They also found that intake volumes were linearly dependent on the furrow wetted perimeter. Merriam and Keller (1978) described field procedures for evaluating irrigation systems, including measurement of intake rates in furrows and flooded basins.

Erie (1962) gave a review of the factors affecting intake rates under gravity-irrigation conditions. Some of these factors are soil texture and water content, compaction due to tractor traffic, prior tillage, surface soil conditions, cracking of soils, crop cover, and hydrostatic head. Intake rates are often high for the first irrigation and tend to decrease as the growing season advances, but may increase late in the season due to high water use rates or blocking of the furrows by foliage and resultant increases in wetted perimeters of furrows. Intake rates decrease with time when the whole surface is wet, as in border irrigation. Intake from furrows tends to remain more constant with time until wetting zones from adjacent furrows begin to overlap. Under some conditions on medium-to-coarse-textured soils, constant rate furrow streams have been observed to "back up," indicating increased intake rates. Kemper et al. (1982) presented a possible explanation. Furrow streams are often set in the early morning hours when the soil and water temperatures are low, and the backing-up occurs in the latter part of the day when temperatures are high. The increased infiltration rate may be a result of increasing water temperatures and decreased viscosity of the water. This phenomenon deserves further study in the future.

34–2 THE VOLUME BALANCE ADVANCE METHOD

34–2.1 Principles

The volume balance-advance method involves determining of the coefficients for an empirical time-based infiltration function by utilizing data from the advance phase of an irrigation. Authors who have used this type of analysis include Davis and Fry (1963), Christiansen et al. (1966), Gilley (1968), Norum and Gray (1970), Wu (1971), Lal and Pandya (1972), Singh and Chauhan (1973), Kincaid (1979), and Elliot and Walker (1982). All of these methods are similar in principle and differ only in the assumptions made concerning: (i) the shape of the intake function, (ii) the rate of advance, or (iii) the depth of surface storage, i.e, constant or variable.

This section describes the mathematical relationships under one set of assumptions and the procedures used in collecting field data and calculating the resultant values of the intake constants.

Several empirical volume or depth infiltration functions have been used. Perhaps the most widely used is the power function

$$Z = k\tau^{a} \quad [1]$$

where Z is depth of intake, m; τ is intake opportunity time, min; and k and a are coefficients related to soil properties.

More recently, Philip (1954) proposed the equation

$$Z = S\tau^{1/2} + A\tau \quad [2]$$

where S is a constant related to capillarity and A is a constant related to the gravitational effect on infiltration. The USDA-SCS (1974), has adopted the equation

$$Z = k\tau^{a} + c \quad [3]$$

where k, a, and c are empirical constants. Equation [3] reduces to Eq. [1] when $c = 0$.

The basic continuity relationship for border irrigation is

$$Qt = \int_{0}^{s} y\mathrm{dx} + \int_{0}^{s} Zdx \quad [4]$$

where Q is inflow rate per unit width, m^2/min; t is elapsed time, min; x is distance from the inflow end, m: y is depth of surface flow, m; s is distance of water front advance, m. The first term in Eq. [4] is the total volume applied, the second term is surface storage volume, and the third term is total infiltrated volume at any time t. Figure 34–1 shows surface storage and infiltrated depth profiles at two different times.

In the following analysis, Eq. [1] will be used to describe infiltration. The intake opportunity time is $\tau = t - t_s$, where t_s is the time necessary for the flow to advance a distance, s. The advance distance can be represented by the power function

$$s = bt^{h} \quad [5]$$

where b and h are empirical constants. Elliot and Walker (1982) recommend the two-point method for determining the advance constants. Advance times to two known distances are substituted into Eq. [5] separately to solve for b and h.

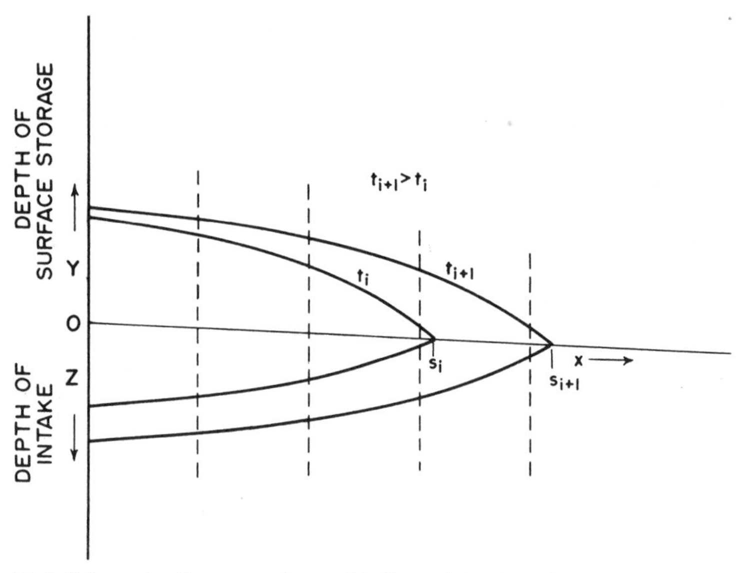

Fig. 34–1. Schematic of water surface and infiltrated depth profiles on a sloping border.

The infiltrated volume integral (last right-hand term of Eq. [4]) can be evaluated by combining Eq. [1] and [5], expanding the result in a binomial series, and integrating term by term from 0 to t. This procedure results in the following relationship for infiltrated volume as a function of time:

$$V_z = f(a,h)\ Pkbht^{a+h} \qquad [6]$$

where V_z is total infiltrated volume, m^3; P is wetted perimeter, m; t is elapsed time, min; b, h, k, and a are constants; and the function $f(a, h)$ is a binomial series as follows:

$$f(a, h) = \frac{1}{h} - \frac{a}{h+1} + \frac{a(a-1)}{2!(h+2)} - \frac{a(a-1)(a-2)}{3!(h+3)} + \ldots . \qquad [7]$$

Equation [6] was developed by Gilley (1968), and also by Christiansen et al. (1966), in a different form. This equation shows that if the intake and advance can be described by power functions, the infiltrated volume will also follow a function of this form. Experimentally, these relationships have been found to work well. The advance constants b and h are evaluated from field advance data. Infiltrated volume is calculated from measurements of inflow and surface storage volume. Values of the intake constants are then determined by Eq. [5] and [6], with known values of b and h. Gilley (1968) used this method on borders having slopes between 0.0002 and 0.005. Kincaid (1979) found the method could be used on zero-slope borders, and extended the method for level furrows by modifying the advance function to describe the total wetted area rather than advance distance.

For borders and furrows of about 0.005 slope or greater, the surface storage depth can be assumed to be constant. The advantage of assuming a constant surface depth is that intake functions other than Eq. [1] may be evaluated. Singh and Chauhan (1973), Christiansen et al. (1966), and others have used this assumption to develop various methods of computing advance or evaluating the intake function. Norum and Gray (1970) proposed a curve matching technique to evaluate the constants of Eq. [1] or [2].

In general, it is best to measure surface storage depths in the field, determine whether or not they can be considered constant, and select an appropriate method of analysis based on the intake function desired. The field procedures described here will provide data necessary for use of any of the analytical methods.

34–2.2 Equipment

1. Flumes (described in detail in the next section) or weirs, orifice plates, or pipeline meters (obtainable in all sizes) for inflow measurement.
2. A surveyor's level, rod, and tape measure.
3. Steel rods approximately 1 m in length for bench-mark stakes.
4. Staff gauges or hook gauges for water surface elevation measurements. Figure 34–2 describes the construction and use of an inexpensive hook gauge.

34–2.3 Procedures

1. Select borders (or furrows) to be tested, with uniform soils and slope if possible.
2. Install inflow measuring equipment.
3. Set six or more bench-mark stakes along one side of the border at about 30-m spacing. It is desirable to space the bench marks closer together near the inflow end. Set the stakes so they can be easily reached from the border dike. Set the tops of the stakes 100 to 150 mm above the border surface.
4. Measure the border bottom width and side slopes of the dike or berm at each bench-mark station. (A method of measuring a furrow profile is given in the next section.)
5. Take level rod readings on the top of each bench-mark stake. Take at least six rod readings on the soil surface in a line across the border at each station. On level borders, bench marks may be set to a constant elevation to reduce error and facilitate data reduction.
6. Determine the required inflow rate. (Furrow inflow rates are described in the next section.) Border flows range from 0.05 to 0.2 m^3/min per meter of width, depending upon soil, slope, and surface conditions. Initiate flow.

7. Maintain constant flow rate until water passes the last station.
8. Estimate the advance time to each station as the time when about one-half of the soil surface is covered by water at the station. Record time to the nearest minute.
9. Measure the water surface elevation relative to the top of the stake (bench mark) using a hook gauge and scale preferably reading in millimeters. An accuracy of ±2 mm in water surface measurements is sufficient. Record water surface measurements 1, 2, 5, and 10 min after water reaches the station and at 10- to 30-min intervals thereafter.
10. Shut off the water when water has passed the last station. If recession

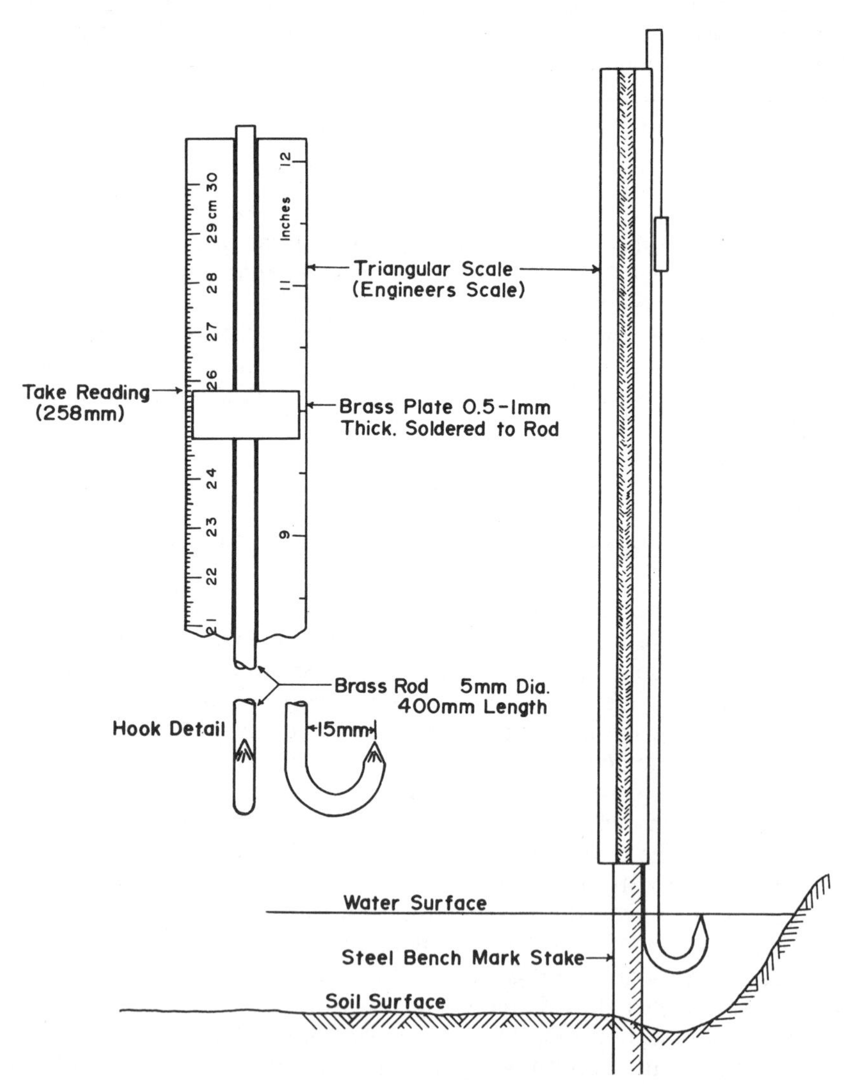

Fig. 34–2. Hook gauge for water surface elevation measurements.

data are desired, record water surface readings immediately before and several times immediately after water is shut off, and then less frequently until all water has infiltrated.

34–2.4 Data Analysis

Tables 34–1 to 34–3 list data from a representative level border advance test (Kincaid, 1979). Table 34–1 gives locations of the bench mark stakes, average soil surface elevations and bottom widths. Table 34–2 lists hydrograph data in the form of water surface elevations computed from hook gauge readings. The actual field readings and bench-mark elevations are not shown. A computer program was written to convert field readings to elevations at the time intervals shown, and compute the total volume in surface storage at any time. The advance distance at any time was computed by log-log interpolation between hydrograph stations. The water surface profile was assumed to be linear between stations. Figure 34–3 shows plotted hydrograph data. A plot of the data is useful in finding and correcting errors and determining whether enough points have been included to accurately describe the hydrographs.

Table 34–1. Dimensions and bench mark stations, Border 5.

Bench mark	Distance	Soil surface elevation	Width
	m	mm	m
1	0.6	165	2.99
2	31.4	156	2.96
3	57.3	147	3.11
4	66.8	130	3.08
5	92.1	160	3.02
6	123.8	165	2.96
Average			3.02

Table 34–2. Hydrograph data—time (min), water surface elevation (mm).

Bench mark											
1		2		3		4		5		6	
min	mm	min	mm	min	mm	min	mm	min	mm	min	mm
1	178	22	174	51	166	62	138	98	179	149	193
3	198	24	185	58	183	65	168	100	183	151	198
6	203	27	189	71	192	69	180	103	185	157	206
9	206	47	199	96	201	98	201	137	203	165	212
20	212	72	208	141	208	139	208	156	210	200	206
43	218	95	210	153	211	153	211	164	212	265	196
73	222	142	217	162	213	163	213	199	206	311	188
94	222	158	217	195	206	198	206	264	196	0	0
143	226	161	216	261	194	262	197	310	188	0	0
158	227	194	207	307	188	309	190	0	0	0	0
160	220	261	197	0	0	0	0	0	0	0	0
193	210	306	188	0	0	0	0	0	0	0	0
260	197	0	0	0	0	0	0	0	0	0	0
305	192	0	0	0	0	0	0	0	0	0	0

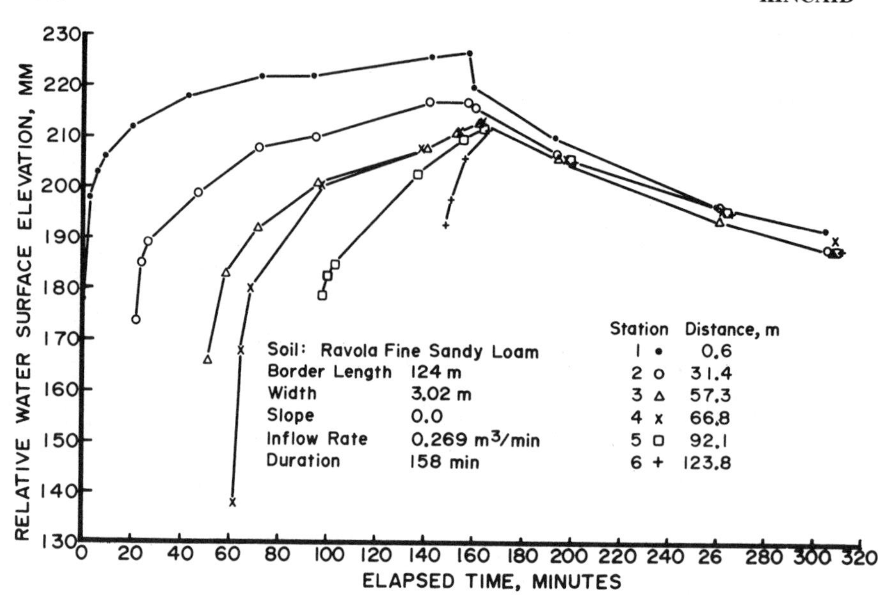

Fig. 34–3. Water surface hydrographs (CSU Farm, Fruita, CO. 5 Sept. 1978).

Table 34–3. Calculated volumes, advance distance, and wetted perimeter.

Time	Inflow	Surface storage	Infiltrated volume	Advance distance	Wetted perimeter
min		m^3		m	
10	2.92	0.91	2.01	11.5	1.98
20	5.75	2.60	3.15	27.8	2.06
30	8.57	4.51	4.06	39.2	2.89
40	11.40	5.68	5.71	48.2	2.75
50	14.18	6.89	7.29	56.5	2.66
60	16.86	8.91	7.95	65.1	3.06
70	19.54	11.06	8.48	72.8	3.13
80	22.19	12.50	9.69	80.0	3.05
90	24.82	14.04	10.78	86.9	2.99
100	27.45	15.52	11.92	93.7	3.22
110	30.07	16.56	13.51	100.1	3.16
120	32.70	17.69	15.01	106.4	3.11
130	35.33	18.94	16.38	112.5	3.06
140	37.95	20.27	17.69	118.5	3.02
150	40.51	22.11	18.41	123.7	3.25
160	42.51	22.89	19.62	123.7	3.26
190	42.51	20.70	21.81	123.7	3.24
220	42.51	18.66	23.85	123.7	3.21
250	42.51	16.77	25.73	123.7	3.19
280	42.51	14.92	27.58	123.7	3.17
310	42.51	13.08	29.42	123.7	3.15

Table 34–3 shows computed volumes, advance distance, and wetted perimeter at 10-min time intervals. The time interval should be small enough to provide at least six points on the infiltrated volume curve. Figure 34–4 shows a plot of the advance and infiltrated volume curves

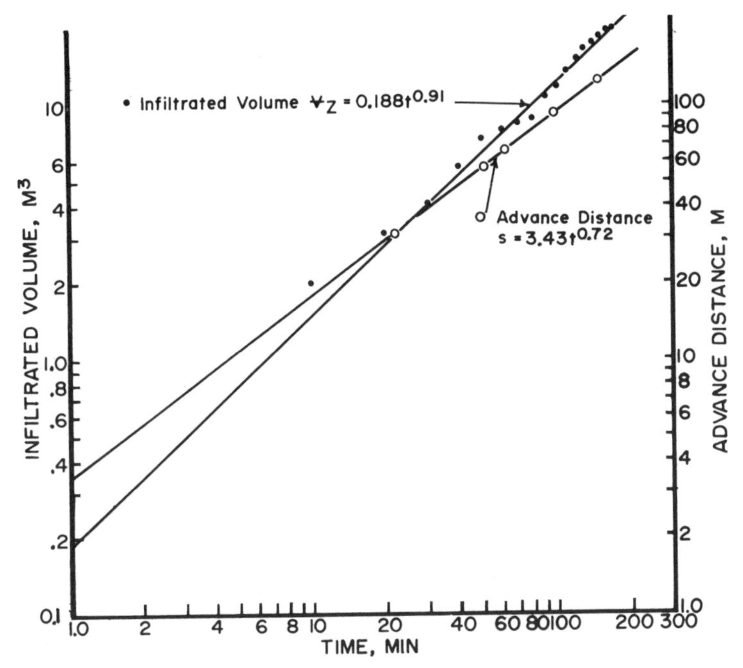

Fig. 34–4. Advance and infiltrated volume curves.

on log-log paper. The advance time to each hydrograph station is plotted and a straight line is fitted to determine the advance constants b and h (h is the slope of the line on the log-log plot, and $b = s$ when $t = 1.0$). The infiltrated volumes are often somewhat erratic at small times, due to the errors in determining surface storage volume for the initial advance. The infiltrated volume data prior to 30 min and after 158 min (the time at which water reached the end of the border) was ignored in fitting a straight line as shown in Fig. 34–4.

The border used in this test was relatively narrow, and the average wetted perimeter was used to determine the intake constants rather than the average bottom width. The surface storage volume and wetted perimeter were computed by assuming a trapezoidal cross section having side slopes of 0.5, vertical/horizontal. The average wetted perimeter was P = 3.14 m.

The value of the constants $b = 3.43$, $h = 0.72$, $a + h = 0.91$ and $Pkbh\, f(a, h) = 0.188$ are obtained from Fig. 34–4. The value of a is $a + h - h = 0.91 - 0.72 = 0.19$, and the value of $f(a, h)$ is calculated by Eq. [6]. By substitution, $k = 0.188/[Pbh\, f(a,h)] = 0.020$ m.

On sloping borders and furrows, the surface flow depths approach a constant normal flow depth, the determination of which is beyond the scope of this chapter. However, if field measurements indicate that a constant flow depth is adequate, the infiltrated volume curve can be constructed by using the assumed surface depth, the advance curve, and inflow rate. The intake constants can then be computed as described above.

34–2.5 Comments

The preceding section describes a method of determining coefficients for equations describing rate of water advance in border ($s = bt^h$) and infiltration ($Z = k\tau^a$). The method assumes that the intake and advance rates can be described by these power functions of time. If the advance or infiltration data cannot be adequately described by a power function, then different intake functions should be used. The reader is referred to the papers previously mentioned which describe other volume balance-advance techniques.

The power function describes intake best on medium- to fine-textured soils where the basic intake rates are relatively small. The intake constants derived from a test may vary with the total time of a test. Intake constants from a relatively short test should not be used to make predictions for longer time spans.

A more general method of using volume balance-advance data for determining infiltration rates is as follows. The measured infiltrated volume curve is contructed as described previously from field data. A predicted infiltrated volume curve is then constructed using the advance data and an assumed infiltration curve. A comparison of the measured and predicted infiltrated volume curves will indicate whether or not the assumed infiltration curve is reasonable. The infiltration curve can be adjusted and a new predicted volume curve calculated. By trial and error, a best-fit infiltration curve can be estimated, or alternative infiltration models can be compared.

34–3 THE INFLOW-OUTFLOW METHOD

34–3.1 Principles

The rate of inflow into an irrigation furrow minus the rate of outflow, at any time, is equal to the furrow intake rate plus the rate at which channel storage is changing. Flow depth is proportional to flow rate at a particular point in a furrow, and since intake rates generally decrease with time, channel storage usually increases with time. However, on furrow slopes greater than about 0.005, the rate of change in surface storage is small (after advance) and may be neglected. The average intake rate can then be taken as the inflow minus outflow rate, and the corresponding time is the average intake opportunity time for the entire furrow. The advance time should be limited to $< 25\%$ of the total time span for the test.

To obtain data for the initial part of the intake curve, it would be desirable to test shorter lengths. However, the accuracy of the flow measurements is limited to about ± 3% and sufficient length of furrow must be used to obtain an appreciable flow reduction between the inflow and outflow stations to achieve reasonably good estimates of infiltration. Shockley et al. (1959) gave the following recommendations for furrow test length:

Fine-textured soils	100–300 m
Medium-textured soils	60–150 m
Coarse-textured soils	30–60 m

They also recommend measuring inflow and outflow from two or more adjacent and relatively short furrows to gain a measure of variability. This method reduces the advance time for a given furrow stream size. The maximum recommended nonerosive stream size for each furrow is given by

$$Q = 38/s \quad [8]$$

where Q is the furrow inflow rate, L/min, and s is the furrow slope in percent. The recommended stream size is one-half to three-fourths of the maximum nonerosive stream.

The flow of water in furrows can be measured either volumetrically or by any of a number of flow-rate measuring devices. Shockley et al. (1959) recommended the volumetric method for flow less than about 80 L/min. Volumetric measurements can be made by collecting the entire flow in a calibrated container and measuring the time to fill the container with a stopwatch. For accurate measurements (± 5%), the container should require at least 4 s to fill. The container is placed in a hole dug in the furrow and the water is run through a tube (approximately 75 mm in diameter) placed in the bottom of the furrow and cantilevered over the container. Inflow measurements can also be made in this manner by using ditch spiles or siphon tubes.

Two types of flumes have been used successfully for direct furrow-flow rate measurement. Small Parshall flumes were described by Robinson (1957), and more recently, small trapezoidal flumes were described by Robinson and Chamberlain (1960). The trapezoidal flumes were also described in ASAE Standard S359.1 (adopted 1975), which gives details on accuracy, construction, and calibration.

The trapezoidal flumes fit furrow channel shapes better than Parshall flumes and can be installed with very little excavation. The V-notch trapezoidal flumes are suitable for flows up to about 150 L/min, and the 50.8-mm (2-inch) throat trapezoidal flumes can be used for larger flows. Fiberglass V-notch trapezoidal flumes can be obtained from the Powlus Manufacturing Co., Inc., Twin Falls, ID 83341.[2]

---

[2]Mention of trade products or companies in this chapter does not imply that they are recommended or endorsed by the Department of Agriculture over similar products of other companies not mentioned. Trade names are used for convenience in reference only.

The discharge relationship for the V-notch flume of Robinson and Chamberlain (1960) is

$$Q = 0.001281\ h^{2.58} \quad [9]$$

where Q is flow rate in L/min and h is water depth in the flume inlet in millimeters. The calibration for the Powlus flume is

$$Q = 0.00169\ h^{2.46} \quad [10]$$

The accuracy of the flow measurement depends on the accuracy of the flume throat dimensions and the accuracy of the stage measurement. To obtain 5% accuracy using the above relationship for the V-notch flume, the stage measurement should be accurate to within ± 1 mm. Materials of construction can have a slight affect on the calibration of a flume and it is recommended that calibration checks be made on new flumes. The errors due to calibration can be reduced by using identical devices to measure both inflow and outflow from furrows and by exchanging the devices between replicated tests.

Flumes should be installed so that the bottom of the flume is flush with or slightly above the furrow bottom. Trapezoidal flumes require very little head loss for free flow conditions and can be used in furrows with slopes as low as 0.002.

Orifice meters are used extensively for flow measurement and are inexpensive to construct. Trout (1983) stated that well-made orifices can measure flows to within $\pm 3\%$ in the field. The basic equation for an orifice is

$$Q = 0.0105\ C\ A\ H^{0.5} \quad [11]$$

where Q is flow rate in L/min, H is head measured from the water surface to the center of the orifice, mm, A is area of the orifice, mm^2, and C is discharge coefficient which has a value of about 0.65 for freeflow and about 0.61 for submerged flow. Calibration tests should be run to determine the value of C for particular orifices.

For submerged orifices, the head is the difference in elevation between the water surfaces. Submerged orifices measure flow more accurately than the other devices because $H \propto Q^2$ for orifices and the exponent on Q is much lower on the other devices, going down to 0.4 for flumes. This same factor causes more head loss at the orifice, which can increase upstream water levels and cause more infiltration than would occur without the orifice plate in place. This effect on infiltration will be practically negligible if head loss at the orifice is less than 0.05 times the elevation difference between the inflow and outflow orifices. A metal box or large pipe with several orifices can distribute water equally to several adjacent furrows when the head is the same on all orifices. Miller and Rasmussen

(1978) used a 100-mm diameter orifice pipe with constant head control for inflow measurement and regulation. A chassis punch can be used to punch uniform size holes in sheet metal.

Weirs are free overfall devices having characteristics intermediate between flumes and orifices. Commercial propeller-type pipe flow meters are available and are accurate to about ± 3%.

34–3.2 Procedures

1. Select furrows to be tested and determine the locations for measuring devices. At least four furrows or groups of furrows should be tested at a site. Test adjacent furrows and supply water to a buffer furrow on each side of the test furrows if tests are to be of long duration so wetting zones will overlap. Determine, if possible, which of the furrows are traffic furrows.
2. Install the measuring devices and water control facilities.
3. Measure the exact furrow length between inflow and outflow measuring devices.
4. Set stakes at three or more intermediate points equally spaced. If wetted perimeter measurements are to be obtained, at each intermediate point drive a stake in the ridge on each side of the furrow so that a straightedge laid across them will be level.
5. Measure the furrow profile at each intermediate stake. Furrow profiles can be measured by using a graduated straightedge placed on the stakes in a level position across the furrow and measuring from this datum line to the soil surface with an adjustable square or point gauge. The stakes are left in place for later measurement of water flow depth. Take rod readings with a surveyor's level to determine the elevation of the furrow invert and top of the stakes at each intermediate point.
6. Select the furrow stream size, select orifice size or tube size, adjust head controls to maintain constant flow, start the flow, and record the start time.
7. Record the time when the furrow stream reaches each intermediate staked point and the outflow point.
8. Record inflow and outflow rates at 15- to 30-min intervals for the duration of the test. The duration of the test should be sufficient to define the shape of the intake curve, which may be 1 to 2 h on coarse-textured soils and up to 10 h on fine-textured soils. For best results, the test duration should be three or four times the advance time from the inflow to the outflow station.
9. If infiltration per unit area of wetted perimeter is desired, measure water surface levels at the intermediate staked points using the straightedge and point gauge or hook gauge.

34–3.3 Data Analysis

1. Compute inflow and outflow rates and loss rates for each time when outflow measurements were made. Also, compute average elapsed intake opportunity time for each of the loss rates.

2. If rate of loss per unit area of wetted perimeter is desired, compute average total wetted area from the furrow profile and water surface data.
3. Convert loss rates to intake rate per unit length of furrow (L h^{-1} m^{-1}) or infiltration rate (mm h^{-1}) based on wetted furrow area. An equivalent field intake rate (mm h^{-1}) can also be calculated by dividing the intake rate per unit length by the furrow spacing.
4. Calculate furrow slope from furrow invert elevations.

34–3.4 Comments

Intake rates determined by the inflow-outflow method will generally produce a more gradually changing intake rate curve than the volume balance method. The results are most applicable to large intake times and are somewhat dependent on the furrow length and advance rate.

34–4 RECIRCULATING FLOW METHOD

34–4.1 Principles

The recirculation method for measuring furrow infiltration combines some of the advantages of the inflow-outflow method and the furrow blocking method, while maintaining field flow conditions. Tests can be run in off-season and on shorter furrow lengths than is practical with the inflow-outflow method. Basically, the method involves introducing a constant inflow to a furrow from a supply reservoir, collecting the runoff at an outflow weir, and pumping the runoff back to the supply reservoir. The accuracy of this method is potentially high, since the total intake is measured volumetrically, avoiding the errors of inflow-outflow rate measurements. Nance and Lambert (1970) used this method to test 4.5-m (15-ft) furrow lengths. Walker and Willardson (1983) and Wallender and Bautista (1983) described improved versions of the recirculating infiltrometer. In principle, the length of furrow is limited only by the size of the supply reservoir and the total intake volume per unit length. The test section can be long enough to avoid local variations and minimize end effects, and short enough to keep filling time to $< 5\%$ of irrigation time. A length of 5 to 50 m would be desirable, requiring approximately 0.5 to 4 m^3 of water. Two or more adjacent furrows can be run simultaneously by dividing the inflow and combining the runoff in one sump. This tends to average the effect of tractor tire compaction in alternate furrows. If the irrigation run is to continue after the wetting fronts meet, water should be maintained in buffer furrows adjacent to the test furrows. The buffer furrows may be ponded rather than flowing, to reduce the water requirement.

34–4.2 Equipment

A cylindrical tank equipped with a float-operated water stage recorder may be used as a supply reservoir. A precision volumetric water meter could be used to refill the reservoir if necessary during a test.

A small reservoir equipped with a float valve serves as a constant-head device for use with a calibrated orifice for flow measurement. A large-diameter pipe with uniform orifices can be used to distribute water to several furrows. Metal plates or boxes are used to block the end of the furrow sections. The inflow boxes should be constructed so that water entering the furrow will not erode the soil. The outflow boxes should be equipped with adjustable weirs and a sump for collecting runoff. A gasoline-engine-powered self-priming pump with a float-controlled throttle provides automatic pump-back regulation.

34–4.3 Procedure

1. Select a test site and construct furrows, if necessary, to the desired length, spacing, and depth.
2. Install the inflow and outflow boxes, being careful not to disturb the furrows in the test sections.
3. Measure the average furrow slope, cross sections, and soil water content if desired.
4. Determine the inflow rate desired and initiate the flow.
5. Record the time of advance and the time at which water begins to flow back into the supply reservoir. Record water levels in the supply reservoir and any additional volumes of water added.
6. Adjust the outflow weir so that the flow depth is nearly constant throughout the test section, and so that the flow near the downstream end neither backs up nor erodes the soil.
7. Measure flow depths in the furrows at several times during the test.

34–4.4 Data Analysis

Calculate and plot cumulative intake volume per unit furrow length vs. average time of intake. Convert data to intake rates if desired.

34–4.5 Comments

The recirculation method can be used to determine infiltration rates for smaller time periods than the inflow-outflow method. It is also applicable to studying the effects of flow rate and depth on infiltration rates. Buffer furrows may not be needed if alternate furrows are tested, if furrow spacing is large, or if relatively short-duration tests are made. Soil probings can be used to determine the extent of the wetting pattern. The wetting pattern can also be observed by cutting a trench across the furrow. Erosion and sediment content of the runoff water can cause problems with recirculation. If sediment concentration is a factor, special sediment handling equipment may be necessary.

34-5 REFERENCES

American Society of Agricultural Engineers Standard S359.1. 1975. Trapezoidal flumes for irrigation flow measurement. American Society of Agricultural Engineers, St. Joseph, MI.

Bondurant, J. A. 1957. Developing a furrow infiltrometer. Agric. Eng. 38:602–604.

Christiansen, J. E., A. A. Bishop, F. W. Kiefer, Jr., and Y. S. Fok. 1966. Evaluation of intake rate constants as related to advance of water in surface irrigation. Trans. ASAE 9:671–674.

Davis, J. R., and A. W. Fry. 1963. Measurement of infiltration rates in irrigated furrows. Trans. ASAE 6:318–319.

Elliot, R. L., and W. R. Walker. 1982. Field evaluation of furrow infiltration and advance functions. Trans. ASAE 15:396–400.

Erie, Leonard. 1962. Evaluation of infiltration measurements. Trans. ASAE 5(1):11–13.

Fangmeier, D. D., and M. K. Ramsey. 1978. Intake characteristics of irrigation furrows. Trans. ASAE 21(SW4)696–700.

Gilley, J. R. 1968. Intake function and border irrigation. Unpublished Master's thesis, Dep. of Agricultural Engineering, Colorado State Univ., Fort Collins.

Haise, H. R., W. W. Donnan, J. T. Phelan, L. F. Lawhon, and D. G. Shockley. 1956. The use of cylinder infiltrometers to determine the intake characteristics of irrigated soils. USDA-ARS and SCS. ARS Handb. 41-7. U.S. Government Printing Office, Washington, DC.

Kemper, W. D., B. J. Ruffing, and J. A. Bondurant. 1982. Furrow intake rates and water management. Trans. ASAE 25:333–339, 343.

Kincaid, D. C. 1979. Infiltration on flat and furrowed level basins. Paper no. 79-2109. ASAE, summer meeting, Winnipeg, Canada. 1979. American Society of Agricultural Engineers, St. Joseph, MI.

Lal, R., and A. C. Pandya. 1972. Volume balance method for computing infiltration rates in surface irrigation. Trans. ASAE 15:69–72.

Merriam, J. L., and J. Keller. 1978. Farm irrigation system evaluation: A guide for management. Utah State University, Logan, UT.

Miller, D. E., and W. W. Rasmussen. 1978. Measurement of furrow infiltration rates made easy. Soil Sci. Soc. Am. J. 42:838–839.

Nance, L. A., Jr., and J. R. Lambert. 1970. A modified inflow-outflow method of measuring infiltration in furrow irrigation. Trans. ASAE 13:792–794, 798.

Norum, D. I., and D. M. Gray. 1970. Infiltration equation from rate of advance data. J. Irrig. Drain. Div. ASCE 96(IR2):111–119.

Philip, J. R. 1954. An infiltration equation with physical significance. Soil Sci. 77:153–157.

Robinson, A. R. 1957. Parshall measuring flumes of small sizes. Agric. Exp. Stn., Colorado State Univ. Tech. Bull. 61.

Robinson, A. R., and A. R. Chamberlain. 1960. Trapezoidal flumes for open channel flow measurement. Trans. ASAE 3:120–124, 128.

Shockley, D. G., J. T. Phelan, L. F. Lawhon, H. R. Haise, W. W. Donnan, and L. E. Meyers. 1959. A method for determining intake characteristics of irrigation furrows. USDA-ARS and SCS, ARS Handb. 41–31. U.S. Government Printing Office, Washington, DC.

Shull, Hollis. 1961. A by-pass furrow infiltrometer. Trans. ASAE 4:15–17.

Singh, Pratap, and H. S. Chauhan. 1973. Determination of water intake rate from rate of advance. Trans. ASAE 16:1081–1084.

Trout, T. J. 1983. Orifice plates for furrow flow measurement. Paper no. 83-2573. ASAE winter meeting, Chicago. American Society Agricultural Engineers, St. Joseph, MI.

U. S. Department of Agriculture–Soil Conservation Service. 1974. Soil Conservation Service National Engineering Handbook, Sec. 15, chapter 4, Border Irrigation.

Walker, W. R., and L. S. Willardson. 1983. Infiltration measurements for simulating furrow irrigation. Proc. Natl. Conference on Advances in Infiltration. 12–13 December, Chicago. ASAE Pub. 11-83. American Society of Agricultural Engineers, St. Joseph, MI.

Wallender, W. W., and E. Bautista. 1983. Spatial variability of water distribution under furrow irrigation. Paper no. 83-2574. ASAE winter meeting, Chicago. American Society of Agricultural Engineers, St. Joseph, MI.

Wu, I-pai. 1971. Overland flow hydrograph analysis to determine infiltration function. Trans. ASAE 14:294–300.

35 Evaporation from Bare Soil Measured with High Spatial Resolution

C. W. BOAST
University of Illinois
Urbana, Illinois

35-1 INTRODUCTION

Measurement of water evaporation from soil in fallow fields and in fields with partial or complete vegetative cover is necessary for the analysis of a number of agronomic and environmental questions. The ability to quantify evaporation from bare soil and to partition evapotranspiration from vegetated soil into its two components—evaporation (from soil) and transpiration (from plants)—is critical to such areas as irrigation scheduling, yield prediction, evaluation of water extraction by roots, and modeling of the movement of chemicals in soil.

Methods for measuring evaporation of water from *medium* or *large* areas of uncropped or cropped soil are discussed in the earlier Methods of Soil Analysis monograph by Robins (1965) and are discussed elsewhere by Van Bavel (1961), Tanner (1967), Black et al. (1969), Ritchie (1972), Aston and van Bavel (1972), Jackson et al. (1977), Hanks and Ashcroft (1980), Aboukhaled et al. (1982), and Hillel (1982). The methods include:

1. *Meteorological* methods, where readings of such variables as incoming or net radiation, air temperature, relative humidity, and wind speed are taken above a bare soil surface or above or within a crop and are interpreted to give estimates of evaporation;
2. *Remote sensing* methods, where evaporation is evaluated (usually in conjunction with meteorological methods) by determining certain radiative or reflective properties of the soil or crop as viewed from a great distance;
3. *Water balance* approaches, where a measurement or estimate of the soil water flux at one depth is combined with measurements of water content changes between that depth and the soil surface; and
4. *Lysimetric* methods, where a body of soil is hydrologically isolated from the surrounding soil, and water loss is determined occasionally or continuously (usually by weighing).

The methods discussed in this chapter are lysimetric. Most weighing lysimeters range from 0.5 to 2.0 m deep and their areal extent is on the

 Methods of Soil Analysis, Part 1. Physical and Mineralogical Methods—Agronomy Monograph no. 9 (2nd Edition)

order of 0.1 to 10 m². Because of their large mass they generally are weighed in situ, although a few examples of lysimeters small enough to be removed temporarily from container holes for mass determination have been described (Miller, 1878; Rykachev, 1898; Popov, 1928; Stanhill, 1956; Abramova, 1968; Staple, 1974; Al-Khafaf et al., 1978; Boast & Robertson, 1982; Walker, 1983; Shawcroft & Gardner, 1983).

In some cases lysimeters include a supply of water at the bottom. This type of lysimeter is limited to measuring evaporation during the first stage of evaporation in the surrounding soil, that is, while actual evaporation from this soil equals potential evaporation. After the end of the first stage, the rate of evaporation from the lysimeter may exceed that from the surrounding soil.

Two methods for measuring evaporation of water from *small* areas of bare soil are described in this chapter. In the first method (Arkin et al., 1974), a flat soil-covered "atmometer" tray is connected to a constant-suction water supply (Fig. 35-1). The rate of evaporation is determined by measuring the rate of water loss from the supply. Because the supply suction is close to zero, the soil surface is constantly wet. Therefore the method is designed to measure only first-stage evaporation from soil. The surface area of the device described is 0.1 m², although the method ap-

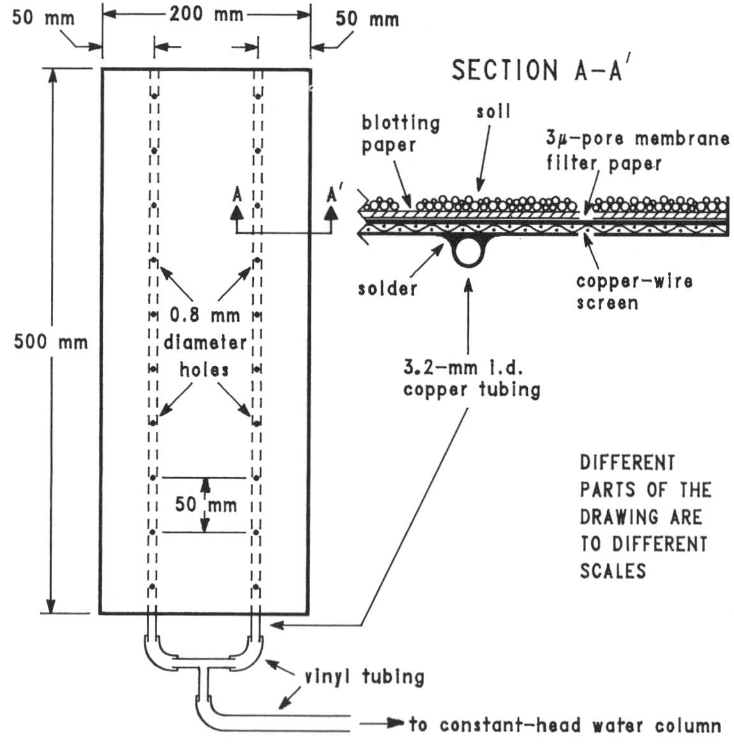

Fig. 35–1. Top view and cross section of the first-stage evaporimeter tray (after Arkin et al., 1974).

plies, with a decrease in sensitivity, to smaller areas if greater spatial resolution is needed.

The second method involves a small "microlysimeter," which is produced by pushing a thin-walled cylinder into the soil, removing it, sealing the bottom, weighing it, subjecting it to the same evaporative conditions as the surrounding soil for a time, and reweighing it. This method is applicable to both wet and dry soil surface conditions. Boast and Robertson (1982) evaluated the method, as applied to fallow soil, to determine how long a microlysimeter can be used before its isolation from the surrounding soil causes it to deviate enough from reality that the soil in the microlysimeter must be discarded (and a new sample taken if subsequent readings are desired).

Neither of the two methods requires much equipment, and evaporation can be determined in circumstances where traditional methods are impractical or impossible. For instance, the methods can be used at a large number of locations where the cost of larger lysimeters is prohibitive. In addition, with these two methods evaporation can be measured in situations for which the spatial resolution of traditional lysimeters is too large. For example, evaporation can be measured as a function of distance from a crop row under conditions of partial canopy cover. Finally, the methods make possible an estimate of the partitioning of evaporation from a cropped field into the two components: transpiration from plants and evaporation from soil.

35-2 FIRST-STAGE EVAPORIMETER METHOD

35-2.1 Principles

The rate of evaporation from a wet soil surface is controlled by the energy balance at the surface. The net radiation energy (incoming solar and indirect radiation minus outgoing reflected and emitted radiation) can do three things: enter the soil, enter the air, and cause water to evaporate. In order to match the energy balance of the surrounding soil a device for measuring evaporation should be similar to the soil in a number of ways: (i) be situated so that it is subjected to the same incoming radiation, (ii) be of the same color and temperature so that reflection and emission of radiation are the same, (iii) have the same roughness and aerodynamic properties so that transfer of heat into the air is the same, and (iv) have similar thermal properties and be in good thermal contact with the underlying soil so that transfer of heat into the soil is the same. If all of these criteria are satisfied, then the energy available for evaporating water is the same in the evaporimeter as in the surrounding soil, and as long as the soil surface is wet in both, the rate of evaporation is the same. As the natural soil surface dries out, the rate of evaporation from the soil becomes less than that from the first-stage evaporimeter, and evaporation measurement requires a different method (see section 35–3).

35–2.2 Method

35–2.2.1 APPARATUS

The evaporation tray, shown in top view and in cross section in Fig. 35–1, consists of a thin sheet-metal tray 3.2 mm deep, to which water is supplied by two copper tubes (soldered along the under side of the tray) via a series of holes drilled through the tray and the upper wall of the copper tubing. To construct the evaporimeter, a screen is first placed on the tray to provide space for the water to be distributed over the whole surface of the tray. Then a membrane filter paper is placed over the screen and bonded around the edges of the tray with contact cement to make an airtight seal. A sheet of blotting paper is placed on top of the membrane to keep it clean, and a thin layer (ca. 1 mm) of sieved soil is placed on the blotting paper to simulate the soil surface.

The water supply is a constant-head "Mariotte" device (See Chapter 28), which maintains a constant water suction in the evaporation tray as the volume of water in the supply changes. For this, Arkin et al. (1974) used a 0.76-m tall clear plastic cylinder with a 51-mm inside diameter and an air inlet about 5 mm above the bottom of the cylinder.

35–2.2.2 PROCEDURE

Install the evaporation tray with its upper surface even with the soil surface. To minimize temperature sensitivity, bury the tubing that connects the evaporation tray to the water supply about 0.1 m below the soil surface. Measure the water loss either visually by use of a graduated Mariotte cylinder or by recording with a pressure transducer the air pressure in the air-filled chamber above the water in the Mariotte device. For the latter option, Arkin et al. (1974) used a scanning valve to record water loss from a number of evaporimeters using just one pressure transducer. To minimize temperature sensitivity, they housed the pressure transducer and the recording equipment, as well as the Mariotte water column, in an insulated box.

35–2.2.3 COMMENTS

The area of the evaporation tray and the volume of the Mariotte device should be tailored to the intended application. The values described here were chosen because they are small enough for measuring evaporation below a row-crop canopy and large enough to give evaporation rates of roughly the same precision as traditional weighing lysimeters.

Arkin et al. (1974) found no difference in evaporation rate when suctions of 0.05 m and 0.2 m were applied to the evaporimeters in field tests. They adopted a 0.05-m suction because it minimized air leaks.

The first-stage evaporimeter is designed to match an undisturbed wet soil surface as closely as possible in the ways listed in section 35–2.1. Incoming radiation is controlled by the placement of the evaporimeter, and outgoing radiation is controlled by the use of soil on the surface of the evaporimeter to match the natural soil color. Matching of surface roughness is not always practical. While it might in principle be possible to use a thick, rough soil layer, short-term storage or loss of water in this soil would result in a discrepancy between the actual evaporation rate and the readings taken on the Mariotte cylinder. Arkin et al. (1974) investigated whether the heat flux through the evaporimeter matches that of undisturbed soil by measuring differences between undisturbed soil surface temperature and the evaporimeter soil surface temperature (by infrared thermometer) and differences between temperatures at 20, 40, 60, and 80 mm below these surfaces (by thermocouples). They concluded that "the thermal environment of the soil was not appreciably altered by the evaporation plates."

It is essential in the use of a first-stage evaporimeter to know when the first stage is over, i.e., when the rate of evaporation from the evaporimeter starts to deviate (due to the supply of nearly suction-free water in the evaporimeter) from that of the undisturbed soil. Arkin et al. (1974) suggest that this time could be determined by comparing the soil surface temperature (Aston & Van Bavel, 1972) of the evaporimeter with that of undisturbed soil. Arkin et al. (1974) found that evaporation rates from first-stage evaporimeters and rates from a traditional weighing lysimeter agreed until the lysimeter soil entered "second-stage (declining rate) drying after 30 hours and 5 mm of cumulative evaporation."

They also demonstrated the value of the evaporimeter in monitoring diurnal variations in evaporation rate, and Adams et al. (1976) evaluated the effect on soil water evaporation of row spacing and of straw mulches. Although the evaporimeters have not been evaluated for use in long-term studies, clearly in such use the thin soil layer has to be maintained from time to time when it is disturbed by precipitation, irrigation, or runoff.

35–3 MICROLYSIMETER METHOD

35–3.1 Principles

The validity of any lysimetric method for determining evaporation hinges on whether the evaporation from the isolated body of soil is essentially the same as from a comparable nonisolated body. A number of factors can cause conditions in a lysimeter to deviate from reality: imposition of a plane of zero flow or a water table at the bottom of the lysimeter, cutting of roots by the lysimeter walls, disturbance of the soil inside the lysimeter during construction, and conduction of heat by the lateral walls.

A common approach to these problems is to employ lysimeters that are large in depth (so that the effect on the evaporation rate of the unnatural hydrological condition at the bottom of the lysimeter is minimized) and in lateral extent (so that the relative importance of lateral boundary effects is minimized). A solution to the disturbance problem is to construct the walls of the lysimeter around the body of soil to be studied or to mechanically push a lysimeter cylinder into undisturbed soil.

An approach that addresses both the depth problem and the disturbance problem is to employ lysimeters of any size, including very small ones which can be pushed into undisturbed soil, and to use them only as long as the rate of evaporation from them equals (to a desired degree of accuracy) that from a comparable body of soil which has not been isolated from its surroundings.

A microlysimeter's eventual deviation from reality is caused by its isolation. However, the "lifetime" of a microlysimeter cannot be prolonged by establishing good contact between the soil in the microlysimeter and the surrounding soil. If there is flow between the surrounding soil and the lysimeter soil, the measured change of mass is not indicative of the amount of evaporation occuring either in the lysimeter soil or in undisturbed soil. Rather, it is the amount of evaporation minus the amount of water entering the body from the surrounding soil (or plus the amount leaving). Ironically, although it is the microlysimeter's isolation that ultimately causes it to deviate from reality, it is also this isolation which allows it to measure evaporation.

35–3.2 Method

35–3.2.1 APPARATUS

Metal or plastic cylinders, generally at least 76 mm long and 76 mm in diameter, are used. The walls are tapered at the bottom to 0.5 mm (Fig. 35–2a) to ease insertion into the soil. The bottom edge is left 0.5 mm thick to minimize damage to the cylinders when hard dry soil or small rocks are encountered. (In dry soil, aluminum cylinders are damaged even when rocks are not present. Boast and Robertson (1982) used brass cylinders with very little damage, and Walker (1983) and Shawcroft and Gardner (1983) used plastic cylinders successfully.)

35–3.2.2 PROCEDURE

Push the cylinders into the soil either to within about 6 mm of the rim as shown in Fig. 35–2b or all the way to the rim, depending on the bottom sealing method adopted (see below). Push the cylinders either by hand or by a hydraulic device. In either case it helps to place an "extension" cylinder (20–30 mm tall for hydraulic pushing, 70–80 mm tall for

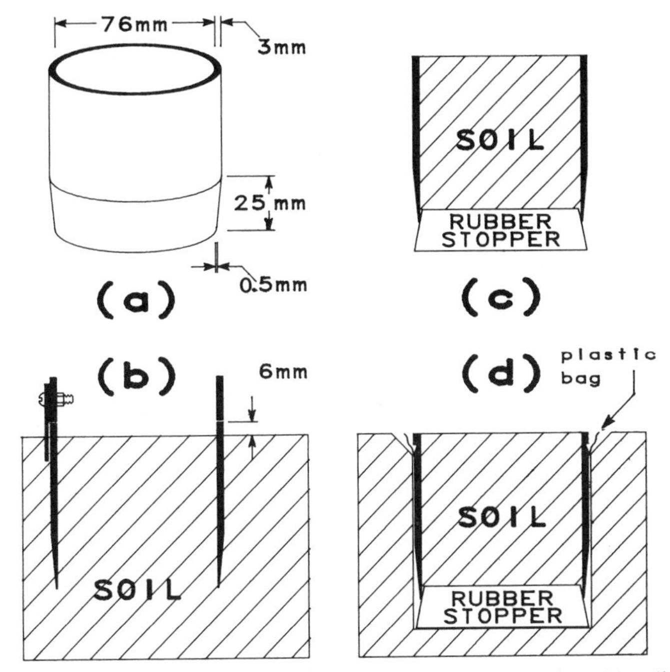

Fig. 35–2. Procedure for microlysimeter determination of evaporation: (a) cylinder, (b) cylinder pushed into soil using extension cylinder, (c) microlysimeter at time of mass determination, and (d) microlysimeter in field between mass determinations (after Boast & Robertson, 1982).

manual pushing) on the upper rim of the microlysimeter cylinder. The extension is held in place by three or four tabs (only one is shown in Fig. 35–2b) which project downward on the outside of the microlysimeter cylinder.

For fallow soil, if possible, push the cylinders into the soil long before they are to be removed (preferably when the soil surface is moist, but not so wet that compression occurs). This practice avoids the disturbance which occurs if the cylinders are pushed into an already-cracked soil surface and allows the cylinder material to reach thermal equilibrium with the soil before the test begins.

Remove the cylinder from the field, clean soil from the outside, trim the soil even with the bottom, and seal the bottom of the cylinder. The sealing method used by Boast and Robertson (1982) is to push a no. 14 rubber stopper, cut to a thickness of 20 mm, up into the cylinder. This method, which is shown in Fig. 35–2c, raises the soil surface to the level of the rim and seals the bottom of the sample against water movement.

Alternatively, push the cylinder into the soil all the way to the rim instead of the way it is shown in Fig. 35–2b and seal the microlysimeter by using a sheet metal bottom (Shawcroft & Gardner, 1983) or by taping a circle of thin plastic sheeting, cut with slits so that it folds easily, to form a cap on the bottom of the microlysimeter (Walker, 1983). One disadvantage of the rubber-stopper technique is that there is a danger of damaging the soil surface when the stopper is pushed into the bottom of the cylinder.

After the bottom of the cylinder is sealed, determine the mass of the microlysimeter. There are two practical advantages to using cylinders which are small in diameter and in length. They are easier to initially insert into the soil, and their mass can be determined on a precision balance. Hand insertion of cylinders 76 mm in diameter and 76 mm long is feasible under many soil moisture conditions. The mass of such a cylinder, filled with soil and stoppered, is typically < 1100 g, so a 1200-g balance can be used.

After the microlysimeter is weighed, put it in a thin plastic bag (to protect the outside from contamination) and place it in a pre-formed hole in the soil (Fig. 35–2d) at a location with the same aspect and shading as the sampling location. Then trim the bag so it is level with the soil surface.

To produce the preformed hole, push a thin-walled metal cylinder, slightly larger in diameter than the microlysimeter, into the soil and remove the cylinder and the soil contained in it. To refine the size of the hole, place an empty microlysimeter cylinder, sealed at both ends, in the hole for a few days or weeks to allow rain and drying to form the soil closely around it. For manipulation, make small notches on each side of the cylinders to serve as handles and make two indentations in the soil to expose the notches (Fig. 35–2d).

After the microlysimeter has been exposed to environmental conditions for a time, remove it from the hole and plastic bag, and determine its mass again. The difference between the two masses divided by the circular cross-sectional area of the cylinder is the cumulative evaporative flux density. With a surface area of 4500 mm^2, cylinders 76 mm in diameter lose 4.5 g water for each millimeter of water evaporated. Hence, if a balance which is precise to 0.01 g is used, the cumulative evaporative flux density is known to better than 0.005 mm.

The microlysimeter can be repeatedly weighed and replaced in the preformed hole as long as it is not kept too long in the weighing environment, where evaporation may not occur at the same rate as in the field. Discard the soil in the microlysimeter when the no-flow bottom boundary condition or lateral root cutting by a microlysimeter located close to plants causes evaporation to deviate too much from that of undisturbed soil. A procedure for correction of small deviations for a medium-textured fallow surface soil is described in the next section.

35–3.2.3 COMMENTS

Boast and Robertson (1982) examined, for fallow Drummer silty clay loam (Typic Haplaquoll), the questions: (i) how long does the loss of water from a microlysimeter equal (to a desired degree of accuracy) the loss from undisturbed soil, and (ii) how is this time affected by the length of the microlysimeter, the evaporativity, and the wetness of the soil at the beginning of the measurement period. Using microlysimeters containing 106 and 146 mm of soil as standards, they determined the time, denoted t_d, when microlysimeters containing 70 mm or less of soil deviated by 0.5 mm from "infinitely long" behavior.

An effort was made to obtain samples at times of typical "wet" and "dry" soil conditions: 0.26 and 0.13 g water/g dry soil (corresponding to about 20 and 2000 kPa) in the surface 20 mm. (The soil surface appeared wet (black) and dry (gray) for the two groups of samples.)

In the experiment, the microlysimeters were not placed in holes in the field (Fig. 35–2d), as is the usual practice, but rather were placed in a constant evaporativity environment for a period of up to seven days. During this time deviations of the short microlysimeters (20, 44, and 70 mm) from the long ones (106 and 146 mm) were determined. For evaporativity ranging from 2 to 9 mm/day, the microlysimeters containing 70 mm of soil were found to be accurate to within 0.5 mm cumulative evaporation for at least 1 or 2 days, depending on initial soil wetness (Fig. 35–3 and Table 35–1).

Reynolds and Walker (1984) numerically simulated evaporation from cores which were of length 50, 100, and 200 mm, which were initially "wet" and which were subjected to an imposed evaporativity of 6.4 mm/day (see their Fig. 1a, or Eq. [3] with $h_v = 0.39\ \mathrm{cm_w^3\ cm_{pm}^{-2}\ min^{-1}\ MPa^{-1}}$, not $h_v = 3.9 \times 10^{-9}$ as given in Table 2 (personal communication, 1985, W. D. Reynolds). As shown in their Fig. 4 the cumulative evaporation from a 50-mm core deviated from that for a 200-mm core by 0.5 mm after approximately 2 days for Pachappa sandy loam (Mollic Haploxeralf) and after about 10 days for Yolo light clay (Typic Xerorthent). These numerically simulated deviation times are, if anything, larger than the values of t_d given in Fig. 35–3 or given for "wet" soil in Table 35–1.

For 20-, 44-, and 70-mm long microlysimeters, according to Table 1 and Fig. 5 of Boast and Robertson (1982), the value of t_d is proportional

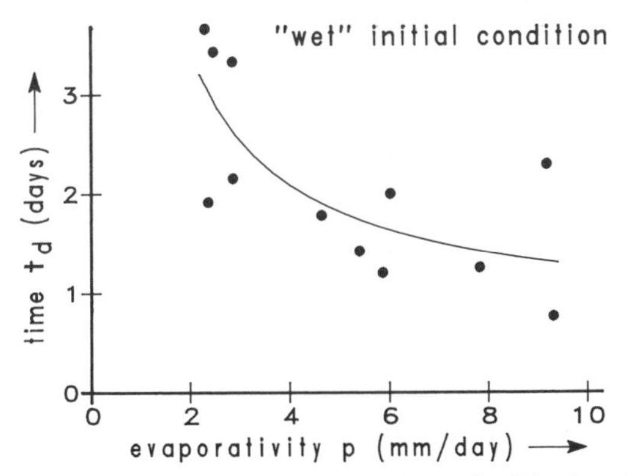

Fig. 35–3. Time t_d at which microlysimeters with 70 mm of soil in them have evaporated 0.5 mm less than microlysimeters with 106 and 146 mm of soil, for a fallow silty clay loam soil starting with a "wet" initial condition.

Table 35–1. Means and standard deviations, for a fallow silty clay loam soil, of time t_d and soil water loss, when the cumulative evaporation from microlysimeters with 70 mm soil in them deviates by 0.5 mm from ones with 106 and 146 mm of soil.

Initial condition	Evaporativity	Number of runs	Time t_d	Water loss from soil in the long microlysimeters at time t_d
	mm/day		days	mm
Wet	8.8 ± 0.8	3	1.4 ± 0.8	7.7 ± 0.9
Wet	5.5 ± 0.6	4	1.6 ± 0.4	6.0 ± 0.7
Wet	2.6 ± 0.3	5	2.9 ± 0.8	7.6 ± 0.9
Dry	6.2 ± 0.9	5	3.4 ± 1.3	3.7 ± 0.7
Dry	3.7 ± 1.0	5	4.0 ± 2.0	3.8 ± 0.7

to the length of the microlysimeter raised to at least the first power. If an assumption of proportionality between length and t_d is valid for microlysimeters longer than 70 mm, then the value of t_d for microlysimeters 200 mm long is at least 4 days (values in Table 35–1 multiplied by 200/70).

Other researchers have used cores of length > 70 mm: Walker (1983), 120 mm; Shawcroft and Gardner (1983), 50, 100, and 200 mm. Walker (1983) suggests that "evaporimeters should be replaced if surface drying ... determined by eye in this experiment ... does not simultaneously occur in the soil and evaporimeters." He replaced the evaporimeter soil typically every 8 to 10 days (G.K. Walker, 1984, personal communication). Shawcroft and Gardner (1983) used one set of microlysimeters for an entire summer, adding water to them at about 14-day intervals (R.W. Shawcroft, 1984, personal communication) such that the total soil water content inside a lysimeter matched that of an equivalent depth of soil outside the lysimeter. They suggest that agreement between water loss from their lysimeters and water loss from the surrounding soil depends "on a series of compensating errors in order to simulate actual conditions." For example, in one case where they reported good agreement between cumulative water loss from lysimeters and the surrounding soil, the volumetric soil water content in the lysimeter initially happened to be 0.09 larger than in the surrounding soil.

Based on the time-course of the deviation of 70-mm microlysimeters from longer ones, Boast and Robertson (1982) developed a procedure for estimating and correcting for the systematic errors that occur when the microlysimeter method is used in fallow soil:

1. Estimate t_d either using the curve in Fig. 35–3 (for wet initial conditions, where t_d is strongly dependent on the evaporativity), or using Table 35–1 (for dry initial conditions, where it is not).
2. If the time t that the microlysimeter has been exposed to environmental conditions is greater than t_d, then discard the soil in the microlysimeter and start over with new soil.
3. If $t \leq t_d$, calculate t/t_d.

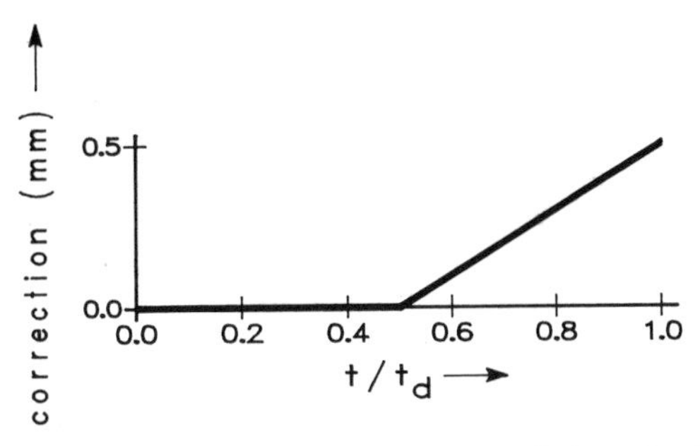

Fig. 35–4. Estimated error, for a fallow silty clay loam soil, for microlysimeters with 70 mm of soil in them, as a function of the fraction t/t_d of the time t_d when the cumulative evaporation deviates by 0.5 mm from infinite-depth behavior.

4. Find this value on the horizontal axis of Fig. 35–4 and estimate the correction (in mm) that is to be added to the measured value from the curve.

For a greater margin of safety the soil can be discarded (step 2) when the time t equals $t_d/2$.

The value of the proposed procedure for error estimation is restricted by limitations on the data from which it is developed. It is based solely on data from a fallow, fine-textured surface soil, high in organic matter. Determination of the applicability of the microlysimeter method to other soils and to cropped soils, and quantification of t_d for other soils and other microlysimeter lengths and materials requires further work.

In the meantime, as the microlysimeter method is used under conditions for which it has not been evaluated, quantitative checks on the method should be made. For example, soil can be left in some microlysimeter cylinders for long periods and the rate of evaporation from these microlysimeters on later days can be compared to the rates of evaporation from microlysimeters taken "fresh" on these same days. Or, for fallow soil, results from microlysimeters of various lengths can be compared. In addition, qualitative checks such as measurement of soil surface temperature by infrared thermometer and visual evaluation of the soil surface may be of some use.

The microlysimeter method is valid for use both during first-stage evaporation and later. However, it is time-consuming, so the most practical strategy may be to use the method of section 35–2 during the first stage of evaporation and to switch to the microlysimeter method only when necessary.

35–4 REFERENCES

Aboukhaled, A., A. Alfaro, and M. Smith. 1982. Lysimeters. Irrigation and Drainage Paper 39. Food and Agriculture Organization of the United Nations, Rome.

Abramova, M. M. 1968. Evaporation of soil water under drought conditions. Sov. Soil Sci. 1968:1151–1158.

Adams, J. E., G. F. Arkin, and J. T. Ritchie. 1976. Influence of row spacing and straw mulch on first stage drying. Soil Sci. Soc. Am. J. 40:436–442.

Al-Khafaf, S., P. J. Wierenga, and B. C. Williams. 1978. Evaporative flux from irrigated cotton as related to leaf area index, soil water, and evaporative demand. Agron. J. 70:912–917.

Arkin, G. F., J. T. Ritchie, and J. E. Adams. 1974. A method for measuring first-stage soil water evaporation in the field. Soil Sci. Soc. Am. Proc. 38:951–954.

Aston, A. R., and C. H. M. van Bavel. 1972. Soil surface water depletion and leaf temperature. Agron. J. 64:368–373.

Black, T. A., W. R. Gardner, and G. W. Thurtell. 1969. The prediction of evaporation, drainage, and soil water storage for a bare soil. Soil Sci. Soc. Am. Proc. 33:655–660.

Boast, C. W., and T. M. Robertson. 1982. A "micro-lysimeter" method for determining evaporation from bare soil: description and laboratory evaluation. Soil Sci. Soc. Am. J. 46:689–696.

Hanks, R. J., and G. L. Ashcroft. 1980. Applied soil physics. Springer-Verlag New York, New York.

Hillel, Daniel. 1982. Introduction to soil physics. Academic Press, New York.

Jackson, R. D., R. J. Reginato, and S. B. Idso. 1977. Wheat canopy temperature: a practical tool for evaluating water requirements. Water Resour. Res. 13:651–656.

Miller, S. H. 1878. Prize essay on evaporation. Utrecht society of arts and sciences. J. W. Leeflang, Utrecht.

Popov, V. P. 1928. Pochvennaia vlaga i metody ee izucheniia (Soil moisture and methods for its study). (In Russian.) Trudy Mleevskoi sodovo-ogorodnoi opytnoi stantsii, Otdel s.-kh. meteorologii no. 16 (Bull. Mleev Horticultural Exp. Stn., Section of rural meteorology no. 16).

Reynolds, W. D., and G. K. Walker. 1984. Development and validation of a numerical model simulating evaporation from short cores. Soil Sci. Soc. Am. J. 48:960–969.

Ritchie, J. T. 1972. Model for predicting evaporation from a row crop with incomplete cover. Water Resour. Res. 8:1204–1213.

Robins, J. S. 1965. Evapotranspiration. *In* C. A. Black et al. (ed.) Methods of soil analysis. 1st ed. Agronomy 9:286–298.

Rykachev, M. A. 1898. Novyi isparitel' dlya nablyudeniya nad ispareniem travy. (New evaporator for observing evaporation by grasses.) (In Russian.) Zapiski Akademii Nauk 7(3) (Memoires de l'Acad. Imp. Sciences de St. Petersbourg. Serie 8, Classe. Phys.-Mathem. 7, no. 3).

Shawcroft, R. W., and H. R. Gardner. 1983. Direct evaporation from soil under a row crop canopy. Agric. Meteorol. 28:229–238.

Stanhill, G. 1956. Further studies on the loss of water from plant/soil systems in the field. p. 41–42. *In* Sixth (1954–1955) Annual Report of the National Vegetable Research Station. British Society for the Promotion of Vegetable Research, Wellesbourne, Warwick, England.

Staple, W. J. 1974. Modified Penman equation to provide the upper boundary condition in computing evaporation from soil. Soil Sci. Soc. Am. Proc. 38:837–839.

Tanner, C. B. 1967. Measurement of evaporation. *In* R. M. Hagen et al. (ed.) Irrigation of agricultural lands. Agronomy 11:534–574.

Van Bavel, C. H. M. 1961. Lysimetric measurements of evapotranspiration rates in the eastern United States. Soil Sci. Soc. Am. Proc. 25:138–141.

Walker, G. K. 1983. Measurement of evaporation from soil beneath crop canopies. Can. J. Soil Sci. 63:137–141.

36 Field Capacity and Available Water Capacity

D. K. CASSEL
North Carolina State University
Raleigh, North Carolina

D. R. NIELSEN
University of California
Davis, California

36–1 GENERAL INFORMATION

The soil serves as a leaky reservoir, which holds water that may be withdrawn by plants. The amount of water held in this reservoir varies from soil to soil—indeed, from soil horizon to soil horizon—and is related to the pore size distribution of the soil. At any given moment the amount of water held by the reservoir is dependent upon factors such as the type of plant cover, plant density, stage of plant growth, rooting depth, evaporation and transpiration rates, amount of water infiltrated, rate of wetting, nature of the soil horizonation or layering, and the length of time since the last rainfall or irrigation event. For many agronomic applications, it is of value to quantify the amount of water held in this reservoir that can be used by plants or that affects the extent to which deep percolation may occur. The amount of water retained by the reservoir at the upper or "full" end is referred to as field capacity; the amount of water retained at the lower or "dry" end is the permanent wilting point. The available water capacity is the difference between these two limits and is defined as the quantity of water held by a soil at the upper or "full" end that is available for plant use. The mere presence of water held between field capacity and the permanent wilting point is not sufficient; the soil must be interlaced with plant roots at a high enough density to allow the soil water to be extracted down to the permanent wilting point. Soils still retain water when the soil is at the permanent wilting point. This water is generally considered unavailable to plants in quantities large enough to sustain life. After or during a rainfall or irrigation event, water is usually present in the soil at water contents greater than field capacity. For soils without shallow water tables, most of this water is generally considered unavailable to plants because it drains out of the soil rapidly. This unavailable water will be discussed in more detail later.

It is important to realize that the limits of plant-available water are those for a soil profile from the soil surface to the bottom of the rooting zone. The rooting depth is time variant, thus causing field capacity and permanent wilting points, each represented by mean soil water contents over the entire rooting zone, to change with time, especially early in the growing season. Because various crops have different rooting depths, the limits of available water for a given soil may vary with the kind of crop. In addition, soil profiles are heterogeneous owing to textural, mineralogical, and structural changes with depth, which in turn affect the water retention properties. Other factors which cause field capacity to vary are compacted layers and pans, textural discontinuities, and fluctuating water tables.

To illustrate the concepts discussed above, consider the plot of soil water content versus soil depth for each 0.15-m increment for the upper and lower limits of soil water availability, for the soil profile shown in Fig. 36–1A. If the rooting zone for this soil is 0.60 m, the field capacity and permanent wilting point for the soil are 0.21 and 0.05 cm/cm, respectively, and are shown in Fig. 36–1B as average water contents over the entire 0.60-m depth. The available water summed over the 0.60-m depth, i.e., the shaded area in Fig. 36–1B, is the available water capacity. The field capacity, permanent wilting point, and available water capacity values for the entire 1.50-m profile, i.e., for rooting systems which permeate the entire profile, are represented by the values shown in Fig. 36–1C.

36–2 FIELD CAPACITY

The field capacity (FC) concept has widespread use and is commonly used by agronomists, extension specialists, soil scientists and conserva-

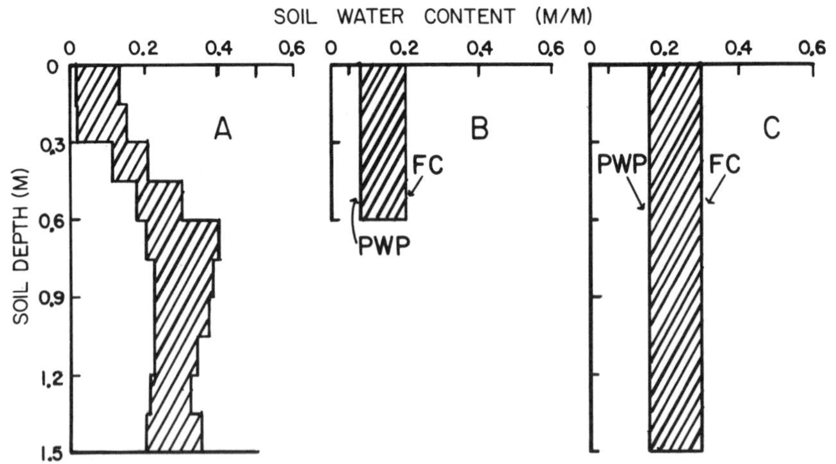

Fig. 36–1. Upper and lower limits of available water measured in 0.15-m increments (A) and these limits expressed for this profile for a 0.6-m rooting depth (B) and for a 1.5-m rooting depth (C).

tionists, agricultural engineers, and others to refer to the water content of a recently irrigated soil. Unfortunately, the FC concept is commonly misunderstood. Typically, FC for a given soil is treated as a "constant," i.e., as a specific value which, once measured, never changes. As shown in Fig. 36–1, rooting depth changes occur with time, thus causing FC to vary with time.

Veihmeyer and Hendrickson (1931) introduced the field capacity concept and defined it as "the amount of water held in the soil after the excess gravitational water has drained away and after the rate of downward movement of water has materially decreased." Their intention was twofold: (i) to define the upper limit of the plant available water retained by the soil, and (ii) for irrigated regions, to provide a concept to encourage farmers not to irrigate excessively, but to take advantage of the soil reservoir and irrigate less frequently. Several assumptions not stated in this FC definition are that the soil is deep and permeable, no evaporation occurs from the soil surface, and no water table or slowly permeable barriers occur at shallow depths in the profile. Under these conditions, if the well-drained soil receives sufficient water to wet the rooting zone and to allow drainage of the excess water into the soil below the rooting zone, then the drainage rate decreases rapidly after water is no longer applied to the soil surface and becomes negligible within a period often said to be from 24 to 72 h.

An inherent problem with the FC concept is the difficulty in defining when the drainage rate becomes negligible. A drainage rate of 1 mm/day compared with an evapotranspiration rate of 5 mm/day may seem negligible under rainfed or irrigated conditions that provide as much as 40 mm of water per week. On the other hand, a drainage rate of only 0.1 mm/day results in > 30 mm of water leaching past the root zone per year in the Northern Great Plains of the USA. Over several years this has allowed the development of undesirable saline seeps on land which has been alternatively cropped and fallowed. The downward flux of water out of the rooting zone may become negligible in less than a day for coarser-textured soils, while certain finer-textured soils may require a week or more. It is important to stress that the water content measured at FC is not exactly constant but continues to decrease, albeit at a slower and slower rate. For example, the average soil water content of a 1-m deep soil profile might well decrease after FC has been declared owing to a flux of 1 mm/day at the bottom of the profile; a few days later the soil water flux may have decreased to only 0.1 mm/day. Ogata and Richards (1957) reported a logarithmic rate of decrease in water content for certain soils.

The ideal soil conditions listed above are not found for many field soils. *In situ field capacity* is defined in the *Glossary of Soil Science Terms* (Soil Sci. Soc. Am., 1984) as the amount of water remaining in a soil 2 or 3 days after having been wetted and after free drainage is negligible. The amount may be expressed on the basis of weight or volume. Salter and Williams (1965) have justified this definition assuming that water that is slowly draining after 48 h is subject to interception by plant roots, thus making the water available.

Water movement out of the initially wetted portion of the soil profile is described by the following equation:

$$\partial\theta/\partial t = (\partial/\partial z)\left[K(\theta)\frac{\partial H}{\partial z}\right] \quad [1]$$

where

θ = volumetric water content,
t = time,
z = depth, measured positive downward,
K = hydraulic conductivity which is a function of θ or h,
H = hydraulic head ($h - z$), and
h = soil water pressure.

After infiltration, the wetted part of a homogeneous soil, in the absence of a shallow water table, has a soil water pressure head gradient ($\partial h/\partial z$) approaching zero (Gardner, 1970). Hence, the hydraulic gradient is approximated by the gravity head gradient, which is equal to -1. As water drains out of the wetted part of the profile, $\partial\theta/\partial t$ is assumed to be independent of z, as supported by observation, thus allowing Eq. [1] to be integrated with respect to z to obtain

$$L\,(d\theta/dt) = -K(\theta) \quad [2]$$

where L is the depth of initial wetting. The left-hand side of Eq. [2] is the rate of drainage from the initially wetted part of the profile, assuming evaporation at the soil surface is nil.

If a soil is not homogeneous, but contains layers differing in their soil water properties, it may still be assumed that within each layer, the value of $\partial h/\partial z$ approaches zero; hence, integration of Eq. [1] yields

$$(\partial/\partial t)\int_0^L \theta dz = K(\theta)(\partial H/\partial z)|_{z=L} - K(\theta)(\partial H/\partial z)|_{z=0} \quad [3]$$

$$\mathrm{L}\,(\partial\bar{\theta}/\partial t) = -K(\theta) \quad [4]$$

where $\bar{\theta}$ is the mean soil water content from the soil surface to depth L,

$$\bar{\theta} = (1/L)\int_0^L \theta\, dz\,. \quad [5]$$

The drainage rate at depth L can be obtained from observations of the rate of change of θ or a knowledge of the value of the hydraulic conductivity K at depth L. Unfortunately, at the present time, there is no established "standard" drainage rate that is considered negligible. However, once the user establishes the drainage rate considered to be negligible, Eq. [4] gives the value of $K(\theta)$ corresponding to that negligible drainage rate. If either the $K(h)$ or $K(\theta)$ relationships is known, the value of h or θ that corresponds to FC can be found.

Even if $\partial h/\partial z$ fails to approach zero, as it may in some finer-textured soils, the value of $K(\theta)$ will decrease to a small value. Gardner (1960) defined FC as that water content below which the hydraulic conductivity is sufficiently small that redistribution of moisture in the soil profile due to a hydraulic head gradient can usually be neglected. For some coarser-textured soils this reduction in hydraulic conductivity may occur within 6 to 24 h (Cassel & Sweeney, 1974) whereas some finer-textured soils continue to drain for several weeks (Davidson et al., 1969). Hence, it is emphasized that FC is not truly an equilibrium water content but instead is that water content at which the soil water flux out of the rooting zone becomes negligible and no significant change in water content occurs with time.

One example of a soil that exhibits a definite FC is a medium-textured soil material underlain by coarse sand. In this case $\partial H/\partial z$ may approach 0 at the interface, but $\partial H/\partial z \cong -1$ and $\partial z/\partial z \cong 1$ in the upper layer. Hence as Eq. [1] shows, it is possible for $\partial \theta/\partial t$ to approach zero even though $K(\theta)$ remains high, but flow becomes nil because $\partial H/\partial z$ approaches zero. Field capacity in the overlying medium-textured material therefore occurs because $\partial H/\partial z$ approaches zero, while in the lower layer, FC occurs because $K(\theta)$ approaches zero.

The presence of a shallow water table in the soil profile reduces the amount and rate of drainage and redistribution as compared to that in a deep, well-drained soil because the hydraulic head gradient in Eq. [1] approaches zero. Field conditions with permanent or temporary perched water tables are common, especially in humid and subhumid climates. If FC is determined when the water table is near the soil surface, it is important to realize that FC, and thus the amount of water available to the plant, will decrease as the water table recedes. For a hydraulic head gradient of zero, hydraulic equilibrium is attained throughout the profile above the water table and the amount of water retained by the soil is dependent upon the pore size distribution and depth to the water table (see chapters 26 and 27). Field capacity of this soil is not the limit of available water because large amounts of water can be obtained from upward flow from the water table under low upward hydraulic gradients. A special case of the shallow water table as it affects the upper limit of plant-available water in soils or other growth media in containers is discussed in section 36–3.

The degree of nonuniformity of a field soil is important. For example, uniform, deep permeable soils such as those studied by Veihmeyer and Hendrickson (1931) to define FC are seldom found in nature. Instead most soils have distinct horizons, which may have either diffuse or abrupt boundaries, and which generally differ in texture and structure. Each layer has its own physical properties. The presence and arrangement of alternate coarse- and fine-textured soil horizons and their effect upon FC are discussed in detail by Miller (1969, 1973), Miller and Bunger (1963), Clothier et al. (1977), and Unger (1971). The presence of coarse-textured material underlying fine-textured material increases the water content of the upper layer when the drainage rate becomes negligible, whereas fine-textured underlying coarse-textured soil may, in some cases, reduce the water content of the upper layer. Changes in soil structure with depth also affect FC, as does the presence of tillage pans and compacted layers (Eagleman & Jamison, 1962).

Several other factors are important in affecting the FC of soils. The water redistribution rate following irrigation depends upon soil texture and depth of wetting. For a sandy soil, the amount of water retained per unit depth at a given time decreased as wetting depth increased (Carbon, 1975); these results were opposite for clay loam and silt loam textured soils (Biswas et al., 1966). The relationship between FC and the initial soil water content and depth of wetting was discussed by Coleman (1944) and Richards and Moore (1952).

Even though attempts have been made to remove the term "field capacity" from technical usage (Sykes & Loomis, 1967; Richards, 1960), its use persists in both technical and practical applications. To date, no alternative concept or term has been advanced to identify the upper limit of available water. It is imperative from a practical standpoint that the FC concept be clarified and maintained until a viable alternative is advanced.

36–2.1 In Situ Field Capacity

36–2.1.1 PRINCIPLE

Water is added to a field soil to wet the soil profile to the desired depth. After the water has redistributed into the drier underlying soil and drainage from the initially wetted zone becomes negligible, the water content is taken as in situ FC.

36–2.1.2 APPARATUS

1. Shovel.
2. Plastic sheeting.
3. Water tank, pipe line, or sprinkler.
4. Straw or mulching material.
5. Soil auger or soil sampling probe.
6. Moisture cans.
7. Balance.
8. Drying oven.
9. Bulk density sampler.
10. Trencher.

36–2.1.3 PROCEDURE

Select appropriate field sites. The number of sites chosen should be based upon the degree of soil variability within the field being studied. Soil variability arises because more than one soil type may be present or because soil properties within a given soil type vary laterally as well as vertically. The number and potential spacing of sites should be based upon the variance structure of the FC observations. Size of the experimental plot area at each site should be at least 3 m on a side and preferably as large as 4 m or greater.

Install a water-tight dike around the perimeter of the plot. An earthen dike can be used for fine-textured soils or a wooden framework dike can be installed. For soils having abrupt textural changes in the soil profile, e.g., a sandy loam A horizon overlying a clay loam B horizon, it is necessary to install a dike around the plot perimeter deep enough to prevent lateral movement of water in the rapidly permeable A horizon at the expense of infiltration into the B horizon (Miller, 1963). Lateral water movement in the above case is controlled by extending the trench down into the less permeable material, wrapping the walls of the trench with plastic sheeting, and compacting the backfill in the trench.

The in situ FC determination should be initiated when the soil is at its normal driest condition, e.g., in early fall when roots have reduced water content to a low level in certain climatic regions of the USA. Sufficient water is applied to nearly level plots by flooding the soil surface to allow water to infiltrate to the desired depth (at least 75 cm). The amount of water to apply may be estimated from the initial soil water content distribution within the profile or by using one or more tensiometers installed at the desired depth. Water is applied to sloping plots with sprinklers. Dispersion of the soil surface resulting from water drop impact is minimized by applying a thick layer of straw or other convenient mulching material over the soil surface.

Immediately after infiltration ceases, cover the site with an evaporation barrier such as polyethylene sheeting. After redistribution of soil water in the profile has proceeded for 48 h (see section 36–2.1.4), remove the evaporation barrier and take soil samples to determine water content. Soil samples are taken to the depth of initial wetting and are collected in convenient depth increments based upon soil morphology. Sampling must be confined to the area near the center of the plot to avoid boundary effects. A neutron meter or gamma ray attenuation device could be substituted for gravimetric sampling. In all cases, a sufficient number of measurements per depth are made to provide estimates of the average water content to within prescribed limits of reliability. No fewer than four samples per depth are required.

The soil water content at in situ FC is calculated by

$$\mathrm{FC_w} = M_w/M_s \qquad [6]$$

or

$$FC_v = FC_w \times \rho_b/\rho_w = M_w/V_a\, \rho_w \quad [7]$$

where

FC_w = in situ FC (g water/g soil),
FC_v = in situ FC (cm^3 water/cm^3 soil),
M_w = mass of water,
M_s = oven-dried mass of soil,
ρ_b = soil bulk density,
ρ_w = water density, and
V_a = bulk soil volume.

36–2.1.4 COMMENTS

Site selection is an extremely important consideration owing to the natural and induced soil variability in field soils. The number of sites required to obtain a reliable estimate of in situ FC in a field depends upon the field size, the distribution of soil types within it, and the spatial variability of each soil type. Because this procedure is so laborious, time-consuming, and expensive, especially for soils requiring trenching, the tendency is to measure in situ FC at only one site. Such a practice may be acceptable when the site is that of a modal mapping unit consistent with soil surveys. The trenching requirement for soils having abrupt textural changes can be circumvented by greatly increasing the surface area of the plot. However, the amount of water applied to each plot increases. If trenching is performed, it is pertinent to compact the fill material back into the trench to prevent the short-circuiting of water downward between the soil material and the plastic sheeting barrier.

At some locations, it may be possible to determine in situ FC of soils wetted by natural rainfall. The optimum time of year to employ this approach is when the crop rooting zone is relatively dry. Wetting by natural rainfall is perhaps the most desirable approach because no trenching is necessary; indeed, dikes are not even required. Moreover, due to the savings in labor associated with this procedure, effort can be directed toward sampling more sites, to reduce the variance. A distinct disadvantage is that the investigator is subject to the vagaries of weather.

A properly calibrated neutron meter or gamma ray attenuation device (see chapter 21) can be substituted for gravimetric sampling. The depth of infiltration can be monitored while water is being applied, thus establishing the time when water reaches the bottom of the rooting zone. The number of access tubes per plot should be commensurate with the precision and/or accuracy required.

A variation of this field procedure is described below. The representative site is selected, wetted, and drained as above. Tensiometers are installed at selected depths in and at the base of the desired depth and are monitored periodically during the ensuing drainage period. The mean tensiometer readings at the desired depths after 48 h of drainage are considered to be the soil water pressure values at in situ FC. The in situ FC is derived from soil water characteristic curves made using undisturbed soil samples taken from the corresponding soil depth (chapter 26). There exists some doubt about the reliability of this method because the laboratory-measured soil water characteristic data do not correlate well with the water contents measured at the corresponding soil water pressure values in the field.

For practical reasons, the arbitrary drainage time of 48 h has been used in the field capacity definition. However, it is once again important to emphasize that, in reality, the hydraulic conductivity of coarser-textured soils may become "negligible" in < 24 h, while some finer-textured soils may continue to drain at a "nonnegligible rate" for periods exceeding 1 week or more. Investigators are encouraged to examine more thoroughly the rates of redistribution within each soil, and perhaps establish a more appropriate drainage time at which to sample for FC, depending upon the individual requirements.

36–2.2 Field Capacity Approximations

There is *no* good alternative for measuring FC other than the in situ field method described in section 36–2.2.1. It is realized, however, that many times a crude estimate of FC is desired and that the effort for in situ field measurement will not or cannot be expended. For this reason, several procedures for obtaining FC approximations are presented. These procedures should be used only when it is impossible to use the in situ FC procedure described above. Hence, the procedures below are presented and labeled as approximations because, in general, they do not give the same value of FC as determined by the in situ method (section 36–2.1).

The literature contains many examples showing that the equilibrium water content of disturbed soil samples subjected to a given soil water pressure correlates with in situ field capacity measurements. In preparation, the disturbed soil sample is ground, passed through a 2-mm sieve, and placed in small rings on a porous plate using the procedure discussed in chapter 26. In general, the lower positive pressures applied to the pressure chamber (5 and 10 kPa) are used for coarser-textured soils, the 33 kPa value for medium-textured soils, and the 50 kPa pressure for finer-textured soils (Rivers & Shipp, 1977; Jamison & Kroth, 1958; Coleman, 1947). Jamison and Kroth (1958) selected the equilibrium water content at 33 kPa to estimate FC for their soils even though the pressure values ranged from > 10 kPa to > 100 kPa. Over time, the use of the 33-kPa water content value has been adopted as field capacity by persons who do not fully understand the FC concept as presented in section 36–2. Selection of the appropriate pressure for desorption of a particular soil sample is not straightforward (Cassel & Sweeney, 1974). For example, the in situ FC for a particular coarse-textured soil may correlate better with the 33-kPa water content rather than the 10-kPa water content; for coarse-textured soils as a group, however, the 10-kPa water content is usually considered to be the better choice. Some reasons for this observed deviation are differences in organic matter content, soil structure, degree of compaction, degree of wetting, and percent sand, silt, and clay (Reeve et al., 1973). Because of the confusion surrounding this desorption procedure, it is recommended that sieved samples not be used to estimate FC, but rather soil cores be used as described in the following two methods.

36–2.2.1 LARGE SOIL CORE METHOD (APPROXIMATE)

36–2.2.1.1 Principles (see section 36–2.1.1).

36–2.2.1.2 Apparatus.

1. Open-ended 16-guage galvanized steel cylinder, 1 m long and 0.3 m diameter; accompanying base plate.
2. Tractor-mounted hydraulic press or other device for pushing the galvanized cylinder into the soil.
3. Shovel.
4. Soil auger.
5. Drying oven.
6. Balance.
7. Moisture cans.

36–2.2.1.3 Procedure. This procedure is a modification of one proposed by Shaw (1927). Select a representative site. Prior to taking soil cores, coat each open-ended cylinder both inside and outside with a light coat of mineral- or petroleum-based oil. Edges of the steel cylinder are beveled to displace soil outward, thus minimizing compaction of the soil core. Obtain each undisturbed soil core by forcing the 16-gauge steel cylinder down into moist soil with a truck- or tractor-mounted hydraulic press. Push the container into the soil until the rim of the cylinder is 0.05 m above the soil surface. Excavate soil from around the cylinder and carefully remove the soil core from its natural position. Attach a steel or wood cap to the base of the cylinder to prevent soil loss and disturbance of its structure. Pack the soil cores in straw and transport them to the laboratory. The soil cores can be stored indefinitely.

Prior to determining the FC approximation, the base plate is removed from each soil core, and the core is placed on top of dry soil having the same or finer texture as the soil at the base of the soil core. Water is added to the soil surface until it infiltrates to the desired depth and then allowed to redistribute for 48 h. Soil samples are obtained with a soil auger in convenient increments based upon soil morphology to the depth of initial wetting. Field capacity approximation values are computed according to equations presented in section 36–2.1.3 except that FC_w and

FC_v are replaced by FC_w^a and FC_v^a, respectively, where the superscript a indicates that the values are approximations.

36–2.2.1.4 Comment. This procedure appears more laborious, tedious, time-consuming, and expensive than the in situ procedure described in section 36–2.1. If, for example, the investigator desires FC information for only one modal profile of a given soil series, then this method does require as much effort as the in situ procedure. Yet if the purpose is to assess the variability of FC within a given soil series or within a field or across a toposequence, this core procedure may well require less effort. Moreover, the soil cores can be collected at one's convenience. The cores can then be used for further experimentation if desired (Cassel et al., 1974).

The diameter of the soil core should be large enough to encompass the significant features of the soil structure as it occurs in the field. It is possible that a core diameter < 0.30 m may be applicable for soils having fine to medium class structure. Cores with diameters exceeding 0.30 m are heavy and difficult to handle; however, a neutron meter can be used to measure water content in large cores. A double-source gamma radiation attentuation system could be substituted for gravimetric soil sampling. A single-source gamma system can be used if independent measurements of bulk density with depth are obtained (Cassel et al., 1974). Instead of standing the open-ended soil on top of dry soil, an alternative procedure is to install a porous plate or a series of porous cups in the soil at the base of the core and maintain a constant suction (Cassel et al., 1974). Some limitations of this method are discussed in section 36–2.2.2.4.

36–2.2.2 SMALL SOIL CORE METHOD (APPROXIMATE)

36–2.2.2.1 Principle As indicated in section 36–1, FC is reached when drainage from the initially wetted portion of the soil profile become negligible. The soil water pressure associated with in situ field capacity has been measured for a broad range of soils and was summarized by McIntyre (1974a). Reported soil water pressure values range from -2.5 kPa (Smith & Browning, 1947) to values approaching -50 kPa (Jamison, 1956). Because many of the reported soil water pressure values at in situ field capacity are near the -10 kPa value (Marshall & Stirk, 1949), this value is arbitrarily chosen for the procedure but may be adjusted to that commensurate with that pressure manifested in situ. Water is expelled from an undisturbed soil core in contact with a saturated porous plate by applying a differential pressure of 10 kPa (see chapter 26). This procedure is similar to that proposed by Hanks et al. (1954).

36–2.2.2.2 Apparatus.

1. Soil sampler and core rings for obtaining undisturbed soil cores of 0.100 to 0.350 dm^3.
2. Shovel.

3. Pressure plate apparatus or similar device.
4. Moisture cans.
5. Balance.
6. Drying oven.
7. Spatula.
8. Knife.

36–2.2.2.3 Procedure. After selecting a representative site, a number of soil cores consistent with statistical procedures are taken from each horizon or soil depth using a soil core sampler. The integrity of the soil structure as it existed in the field must be maintained in the soil cores.

In the laboratory, trim the ends of the soil core even with the sample retaining-ring and carefully place the soil core plus retaining ring on the ceramic plate. Rotate the core slightly to assure good contact. The soil is wetted by capillary rise over a 12-h period by maintaining the water table at the base of the core. Completely saturate the soil core by slowly raising the water level until the entire soil core is submerged.

Swelling soils, which are collected in the field when the soil is wet, must be removed from the retaining ring prior to placing the soil on the ceramic plate. Allow the soil core to begin wetting by maintaining a pressure < -5 kPa on the porous plate. After several hours, increase the pressure head to zero. Finally, slowly raise the water table level until the entire sample is submerged. If it is necessary to obtain the swelling soil cores from the field in a relatively dry state, the soil cores should be wetted in the same manner except that an initial pressure of -50 kPa should be maintained on the porous plate (McIntyre, 1974b).

Remove the excess water. Place the ceramic plate in the pressure plate device, apply a pressure equivalent to -10 kPa, and allow water to drain from the soil until equilibrium is attained—i.e., until water no longer flows from the pressure plate. Remove the soil sample from the pressure chamber with a wide spatula, weigh, and oven-dry it for 24 h at 105°C. Calculate water content on either a weight or volume basis.

In lieu of collecting soil cores, it may be desirable or even necessary for some soils to obtain a series of clods (or peds) that have soil structure representative of selected depths of the soil in question. When individual peds are collected, the resulting FC approximation is biased toward a higher water content because the "cracks" between the peds which occupy space and drain at soil water pressures > -10 kPa are ignored. Obtain the desired number of clods or peds from each depth; the oven-dry soil mass should range between 0.1 and 0.3 kg. The soil water content at the time the samples are collected from the field must be dry enough to prevent shattering. Individually wrap each sample with soft packing material and place it in a box, using packing material to separate the individual samples. Prevent sudden bumps or jolts during transportation to the laboratory.

The samples are wetted using the procedure for swelling soils. A flat surface on one side of the clod or ped should be cut with a knife to promote good soil to ceramic plate contact.

36–2.2.2.4 Comments. It is emphasized that this procedure is a field capacity approximation. The assumption of a −10 kPa soil water pressure at in situ field capacity does not apply to all soils; in general, the soil water pressure at in situ FC for coarser-textured soils is > -10 kPa, while the pressure head for finer-textured soils is less. Consequently, the −10 kPa soil water pressure yields data somewhat more reliable for medium-textured soils and perhaps should be adjusted for a particular condition or known pressure existing under field conditions for a particular soil.

The time required to reach equilibrium may be slow because of poor soil sample–ceramic plate contact. A contact material placed between the soil core and the porous plate may increase the rate at which equilibrium is attained. Contact materials include soil material similar to that of the soil core or clod, very fine glass beads, and grinding powders (McIntyre & Watson, 1975). A suitable contact material must remain saturated at the pressure applied to the ceramic plate and have a high hydraulic conductivity. Use of a contact material which drains at high soil water pressures yields erroneously high approximations of FC.

36–2.2.3 CENTRIFUGE METHOD (APPROXIMATE)

36–2.2.3.1 Principle. For laboratory approximations of FC, the use of soil cores is desirable because field structure is preserved. This is especially important for FC estimates because water retention is dependent on the shape and size of pores, rather than upon total soil particle surface area. Unfortunately, the costs associated with collecting, trimming, storing, and determining the water content at a known pressure (see section 36–2.2.2.3.) are high, and in some cases variability among cores may also be high. Often estimates of FC are desired when no soil cores are available, particularly from greater soil depths, while disturbed, sieved samples are available. Clearly a choice exists between (i) FC estimates with higher probable accuracy and higher cost for a few observations and (ii) FC estimates in disturbed soils which may have greater precision, albeit their mean may not be accurate owing to loss of field structure.

Beginning with the work of Briggs and McLane (1907), a voluminous amount of field-measured and laboratory-estimated FC data using the centrifuge were published prior to the 1950s (e.g., Israelsen, 1918; Veihmeyer & Hendrickson, 1931; Russell & Richards, 1938; Schaffer et al., 1937). Results indicated that the following parameters can be varied to obtain the desired result: soil sample thickness, angular velocity, centrifuge radius, and time of centrifugation. The moisture equivalent was introduced and defined as the amount of water retained by a soil when a layer of soil material 10 mm thick is centrifuged for 40 min in a gravitational field equal to $1000 \times g$ (Briggs & McLane, 1907).

Consider a "water saturated sample" having inner and outer radii equal to r_1 and r_2, respectively, revolving at a constant angular velocity ω. When the water in the sample reaches hydraulic equilibrium with the soil, the force F acting on the water of mass m is

$$F/m = (-\partial\Psi/\partial r) = -\omega^2 r\,. \qquad [8]$$

Integrating Eq. [8] and applying the boundary condition that $\Psi = 0$ at r_2 we get

$$\Psi = (\omega^2/2)\,(r_1{}^2 - r_2{}^2)\,. \qquad [9]$$

Observe that the maximum force occurs at r_2 for $\Psi = 0$, while the greatest $|\Psi|$ in the sample occurs at r_1.

For the 10-mm thick sample having outer radius r_2 as defined for the moisture equivalent,

$$F/m = -1000\ g = -\omega^2 r \qquad [10]$$

at $r_2 = 15$ cm giving $\omega = 2.56 \times 10^2\ s^{-1}$ or

$$\omega = \frac{256\ \text{rad/s}}{2\pi\ \text{rad/rev}} \times \frac{60\ \text{s}}{\text{min}} = 2445\ \text{rev/min}. \qquad [11]$$

The soil water pressure head at boundary r_1, derived from Eq. [9], is

$$h = \Psi/g = (\omega^2/2g)\,(r_1{}^2 - r_2{}^2)\,. \qquad [12]$$

The average soil water pressure head $\bar{h}$ throughout the soil sample is

$$\bar{h} = \int_{r1}^{r_2} h\ dr/\int_{r1}^{r_2} dr = (-\omega^2/6g)(2r_2 + r_1)(r_2 - r_1)\,. \qquad [13]$$

For a 10-mm thick soil sample having outer radius of 0.15 m, Table 36–1 shows the pressure head distribution at hydraulic equilibrium, the angular velocity, and revolutions per second required to obtain $\bar{h}$ values of 0.50, 1.00, 1.50, 2.00, 3.00, 4.00, and 5.00 m.

36–2.2.3.2. Apparatus.

1. Centrifuge with drum head having a radius of at least 0.15 m.
2. Brass sample cups with screen bottoms.
3. Whatman no. 2 filter paper.
4. Moisture cans.
5. Sieve, 2 mm.
6. Balance.

36–2.2.3.3. Procedure. Perform all determinations in duplicate,

Table 36-1. The pressure head distribution in a 10-mm thick soil sample with outer radius 0.15 m for selected values of mean pressure head for the entire sample. The angular velocity and rpm for each case are also given.

	Mean pressure head (m)						
Radius	−0.50	−1.00	−1.50	−2.00	−3.00	−4.00	−5.00
	m						
0.140	−0.99	−1.98	−2.97	−3.95	−5.93	−7.91	−9.89
0.142	−0.80	−1.59	−2.39	−3.19	−4.98	−6.37	−7.96
0.144	−0.60	−1.20	−1.80	−2.41	−3.61	−4.81	−6.01
0.146	−0.40	−0.81	−1.21	−1.61	−2.42	−3.23	−4.04
0.148	−0.20	−0.41	−0.61	−0.81	−1.22	−1.63	−2.03
0.150	0	0	0	0	0	0	0
ω	82	116	142	164	200	231	259
rpm	781	1104	1352	1516	1912	2208	2468

placing the duplicate samples opposite each other in the centrifuge head to keep it balanced. Cover the screen bottoms of the brass centrifuge cups with Whatman no. 2 filter paper. Add 30 g of air-dry soil, previously screened through a 2-mm sieve, to each cup. Wet the samples from the bottom by standing them in a tray, adding water to the tray, and allow them to soak for several hours. Load the wetted samples into the drum head of the centrifuge, and, as quickly as feasible, bring them to a speed of 2468 revolutions per minute and maintain for 30 min. After stopping the centrifuge, quickly remove the brass cups, discard the filter paper, and transfer the soil to tared moisture cans. Obtain the appropriate weighings to calculate the water content as given in Eq. [6] and [7].

36–2.2.3.4 Comments. The method presented above has been shown to provide good estimates of in situ FC for finer-textured soils. For medium- and coarse-textured soils the force applied, i.e., the revolutions per minute, would have to be altered to values which correlate more closely with field-measured values of FC.

36–3 CONTAINER CAPACITY

36–3.1. Principle

Containerized plants are plants grown in greenhouse pots or containers. Many commercial potting media have been developed to grow containerized plants. These media are mixtures of such materials as vermiculite, bark, peat moss, coarse sand, and soil. Several components are mixed in appropriate amounts to provide a medium with the desired balance of smaller pores which retain water and larger pores which drain freely and provide adequate aeration. Soil material may also be used as a medium to grow plants in containers. Usually the soil material is sieved and sometimes it is ground prior to its use.

Container capacity (CC) is defined as the water content of the potting medium, initially thoroughly wetted, after free drainage from holes in the base of the container ceases. The CC concept was proposed by White (1964) and discussed by Spomer (1974, 1975). In reaching CC, water held in the larger pores freely drains into air (atmospheric pressure) from the soil at the base of the container where the soil water pressure is zero.

Probably the most blatant misuse of the FC concept has been its use to represent the equilibrium water content of field soil packed in small containers after drainage from the thoroughly wetted soil ceases. A field soil typically does not have a large volume of large pores which freely drain at soil water pressures near zero; instead, these soil pores retain water thus giving rise to poor aeration. Figure 36–2 shows the volume fractions of water- and air-filled pores versus container height for several media at CC, i.e., after drainage from the base of a 0.15-m high container of medium has ceased and the system has reached equilibrium. For the manufactured potting material (medium A) the volume of large pores provides adequate aeration. For the medium sand (medium B) and the clay loam (medium C), the pore sizes are too small to drain at soil water pressures greater than the −1.5 kPa value that occurs at the top of the container and these two media remain in a nearly saturated condition. Such poorly aerated conditions do not provide a suitable soil atmosphere for most plants.

It is clear that CC and FC are not identical. The term FC, therefore, is invalid for describing the equilibrium water content of a soil or other porous medium held in a container and which is initially wetted and allowed to drain into the atmosphere through openings in the base of the container. Hence CC of the containerized soil is greater than the FC value of the same well drained soil in the field. However, if a shallow water table is present in the field soil, an additional amount of water is retained by the soil (compared with that held at FC when the assumptions in section 36–2 are met), and the soil water content determined after drainage ceases corresponds to the CC. If the bases of finer-textured containerized media remain in contact with water, large amounts of water can be obtained by the plant due to upward flow from the water table under low upward hydraulic gradients.

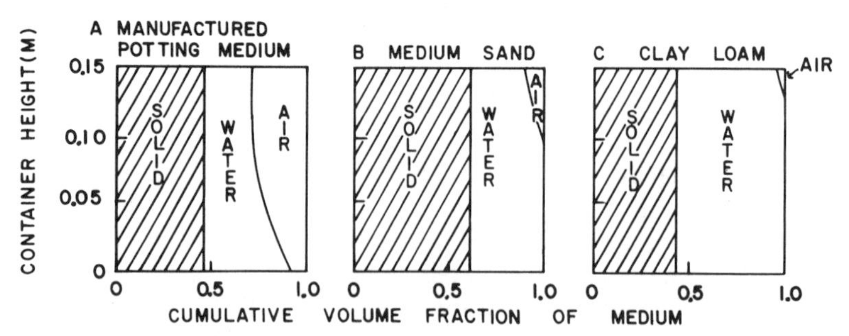

Fig. 36–2. Volume fractions of solids, water, and air at CC as a function of container height for a manufactured potting medium (A) and two soil materials (B and C).

36–3.2. Apparatus

1. Drying oven.
2. Moisture cans.
3. Balance.
4. Soil sampling tube such as an Oakfield probe.

36–3.3. Procedure

Wet the soil or other porous medium held in the container according to the wetting procedure for nonswelling soils given in section 36–2.2.2.3. For very porous potting mixes made from ingredients such as sand, bark, peat, or soil, water can be ponded on the medium surface until water begins to flow out of the holes in the base of the container. Care must be taken that excess flow does not occur down the edges or through large continuous pores, thus allowing excessive drainage from the bottom before the entire volume of medium has become wet. Cover the top of the container to prevent evaporation. Punch a small hole in the cover. Position the container so that water drips freely into the air from holes in the base of the container. Allow the container to drain for 6 h. Drainage will cease in < 1 h for many potting media. The whole container can be weighed, dried at 105°C for 24 h, and reweighed as representative samples of the medium can be taken and oven-dried. For either method, container capacity is calculated by

$$\mathrm{CC_w} = M_w/M_s \quad [14]$$

or

$$\mathrm{CC_v} = \mathrm{CC_w} \times \rho_b/\rho_w = M_w/V_a\rho_w \quad [15]$$

where $\mathrm{CC_w}$ = CC (kg water/kg medium) and $\mathrm{CC_v}$ = CC (m^3 water/m^3 medium). The other symbols are defined in section 36–2.1.3.

36–3.4. Comments

The container capacity is the mean water content of the soil in the container. An important difference exists with regard to the physics associated with FC and CC. As noted in section 36–2.1, FC is reached when drainage becomes negligible due to a reduction in the unsaturated hydraulic conductivity or hydraulic gradient. Container capacity in shallow containers is reached when the hydraulic head becomes constant at each elevation in the container. Container capacity, then, is an equilibrium water content value, whereas FC is not. Thus, the time required for a soil to drain to CC is less than the time required for the same soil to drain to FC. The water content in the container soil at CC increases with depth as dictated by the equilibrium soil water pressure, which decreases with elevation to a value of zero at the base of the container. For some finer-textured materials, the soil will remain at apparent saturation if the container height is not tall enough to create a soil water pressure small enough to get out of the capillary fringe. For the clay loam in Fig. 36–2, a soil water pressure of −1.3 kPa is required to overcome the capillary fringe. On the other hand, water content in a uniform soil at FC is nearly constant with depth in the soil zone that was initially wetted.

36–4 PERMANENT WILTING POINT

The establishment of the lower limit of available water retained by the soil reservoir is of considerable practical significance. Two lower limits of available water have been defined. Briggs and Shantz (1912) found that the water content at the onset of wilting was reproducible for a given soil and that a plant failed to make additional growth when this soil water content was maintained. Furr and Reeve (1945) later defined this water content to be the incipient wilting point and as "the soil water content at which the lowest pair of leaves of a particular kind of plant at a particular growth stage wilts and fails to recover in a water-saturated atmosphere." The second wilting point is the *permanent wilting point* (PWP), which is defined (Soil Sci. Soc. Am., 1984) as "the water content of a soil when indicator plants growing in that soil wilt and fail to recover when placed in a humid chamber." In general, there is a considerable range in water content between incipient and permanent wilting (Gardner & Nieman, 1964).

The PWP concept has encountered considerable criticism but its use continues. The PWP is determined using an indicator plant, usually a dwarf sunflower (*Helianthus annuus* L.). The soil water pressure at this PWP was shown to be approximately −1500 kPa (Richards & Weaver, 1943). Evidence has since been presented showing that the lower limit of available water varies with the plant species and corresponds to different soil water pressures. Sykes and Loomis (1967) reported soil water pressures at PWP ranging from approximately −500 kPa for sunflower to less than −3000 kPa for intermediate wheatgrass [*Agropyron intermedium* (Host) Beaux]). For most soils, except for some fine-textured ones, the change in soil water content between pressures of −800 and −3000 kPa is negligible (McIntyre, 1974a).

The environmental conditions under which a particular plant, e.g., sunflower, is grown is important because it affects evaporative demand and the rate of water uptake required to maintain turgor. Thus, soil water contents at incipient and permanent wilting become functions of atmospheric demand as well as the stage of plant growth and the rooting characteristics, which are affected by soil structure and compaction. Moreover, in the field, the top soil may dry to water contents less than the PWP due to evaporation, while roots as deep as 2 or 3 m or even deeper are actively taking up water.

Wilting is a diurnal cyclic phenomenon which attains a maximum near the period of maximum evaporative demand. The plant may recover overnight and the cycle may be repeated until permanent wilting occurs. Gardner and Nieman (1964) found that the water content at permanent wilting was too low and did not represent a true lower limit for any of the following plant processes: transpiration, cell division, or cell enlargement. It is therefore desirable to establish a wilting point based upon plant responses. Although this is a good idea, it is pragmatically impossible with the present level of knowledge.

Based upon the above discussion, it is recognized that the incipient and permanent wilting points under field conditions are not constants for any given soil, but that the values, in reality, are determined by the integrated effect of the plant, soil, and atmospheric conditions. The extent and depth to which roots pervade the soil profile, the distribution of plant cover, the relative proportion of the soil surface allowed to evaporate to soil water contents less than the PWP, and the microclimate within the crop canopy all preclude an exact definition of the PWP under field conditions.

36–4.1. Sunflower Method for Permanent Wilting Point (PWP)

36–4.1.1. PRINCIPLE

A well watered plant is allowed to become established. Water is then withheld from the soil, gradually subjecting the plant to water stress. The soil water content at which the plant wilts and remains permanently wilted is the PWP.

36–4.1.2. APPARATUS AND MATERIALS

1. Moisture cans.
2. Cardboard containers.
3. Polyethylene bags.
4. Balance.
5. Drying oven.
6. Cotton balls.
7. Dwarf sunflower seeds (*Helianthus annuus* L.).

36–4.1.3. PROCEDURE

The procedure is based upon one proposed by Furr and Reeve (1945): the PWP is determined based upon the soil water content at which a dwarf sunflower permanently wilts when grown under conditions of low evaporative demand. Air-dry the soil and force it through a 2-mm sieve. Perform the following procedures in triplicate. Weigh 0.600 kg of soil into a tared no. 2 friction-top moisture can or a cylindrical cardboard freezer carton of similar size lined with a polyethylene bag. Add lime and nutrients as needed. Add sufficient water to bring the soil to one-half container capacity. With a lead pencil, punch three equally spaced holes to a depth of 12 mm; place one dwarf sunflower seed into each hole and cover. Weigh the container of wetted soil and lid and record as "initial weight."

After the seeds germinate, select the healthiest seedling and cut off the remaining two at soil level (do not pull). Cut a 10-mm diameter hole at the appropriate location on the lid and place the lid on the can, allowing the seedling to extend upward through the hole. Continue to water the sunflower seedling periodically and grow it until the third set of leaves appears. The can lid can be partially removed to facilitate watering.

After the third pair of leaves begins to develop, once again adjust the soil water content to its initial value. Place a wad of cotton between the sunflower stem and the side of the hole in the lid, to allow air exchange but prevent evaporation from the soil. If a plastic-lined cardboard container is used, wrap a wad of cotton, approximately 15 mm in diameter, around the base of the sunflower stem, and tie the plastic bag snugly shut at the point where the bag contacts the cotton.

Continue to grow the sunflower in an atmosphere of low evaporative demand until all three pairs of leaves first appear to be wilted. Transfer the container holding the wilted plant to a dark, humid chamber and leave overnight (usually 14 to 16 h). If one or more sets of leaves regain turgidity by the following morning, return the sunflower to its growing environment for one more day. Repeat the process of alternate wilting and humidification until all three pairs of leaves fail to recover when left overnight in the dark, humid chamber.

After permanent wilting is attained, cut off the sunflower at soil level with a razor blade and discard the plant. Weigh the soil, dry at 105 °C for 24 h, and reweigh. If the plastic-lined cardboard container is used, place a representative subsample of soil, or preferably the entire soil sample, into a tared moisture can and weigh. Reweigh the subsample after oven-drying.

Determination of the PWP is most precise if the sunflower roots are removed from the soil prior to its initial weighing. In most cases, however, the error introduced by including the roots with the soil mass is negligible.

The water content at the PWP is calculated by

$$\mathrm{PWP_w} = M_w/M_s \qquad [16]$$

or

$$\mathrm{PWP_v} = \mathrm{PWP_w} \times \rho_b/\rho_w = M_w/V_a\,\rho_w \qquad [17]$$

where $\mathrm{PWP_w}$ = PWP (kg water/kg soil), $\mathrm{PWP_v}$ = PWP (m^3 water/m^3 soil), and the other symbols are as defined in section 36–2.1.3.

36–4.1.4. COMMENTS

For some soils, it is difficult to decide by visual observation if a plant has recovered from wilting after remaining overnight in a dark, humid chamber. If this situation arises, weigh the container containing the wilted sunflower in question and return the plant to its growing environment for one more day. At the end of the day, again place the plant in the dark, humid chamber overnight. If it is still difficult to determine whether the plant is permanently wilted the following morning, reweigh the container again. If the water loss between consecutive weighings is < 0.001 to 0.002 kg, the plant is considered permanently wilted.

The incipient wilting point (Furr & Reeve, 1945) may also be determined using the procedure of section 36–4.1.2. The water content at which the lowest pair of leaves wilts and recovers when the plant is placed in the dark, humid chamber is the incipient wilting point. As indicated above, the incipient wilting point is more closely associated with cessation of growth than the PWP (Gardner & Nieman, 1964).

36–4.2 Pressure Outflow Apparatus PWP Approximation

36–4.2.1. PRINCIPLE

The principle of the pressure plate and pressure membrane apparatus is discussed in chapter 26. By statistical correlation procedures, it has been observed that the PWP measured by the sunflower method (section 36–4.2.1) is approximately equal to the soil water content of a disturbed soil sample placed on a permeable membrane or porous plate and equilibrated with an applied pressure of 1500 kPa (Richards & Weaver, 1943; Lehane & Staple, 1960; Cassel & Sweeney, 1974).

36–4.2.2. APPARATUS

1. Mortar and pestle or soil grinder.
2. Sieve, 2 mm.
3. Pressure plate or pressure membrane apparatus.
4. Rubber sample rings to retain soil samples.
5. Regulated air pressure system.
6. Moisture cans.
7. Spatula.
8. Balance.
9. Drying oven.

36–4.2.3. PROCEDURE

Air-dry the soil, crush with a mortar and pestle or soil grinder, and pass through a 2-mm sieve. Discard the material retained by the 2-mm sieve. If the soil is stony, the percent by weight of coarse fragments must be determined on a subsample for use in the PWP computations (Reinhart, 1961). Place the soil samples on the pressure membrane or porous plate apparatus and continue using the procedure outline in chapter 26. A positive pressure of 1500 kPa is applied in the pressure chamber.

The permanent wilting point approximation on a weight basis (PWP^a_w) and on a volume basis (PWP^a_v) are given as

$$PWP^a_w = M_w/M_s \qquad [18]$$

and

$$PWP^a_v = PWP^a \times \rho_b/\rho_w = M_w/V_a\,\rho_w\,. \qquad [19]$$

Symbols are defined in section 36–2.1.3.

For soils having > 2% by weight coarse fragments

$$PWP^a_w = (M_w/M_s)\,(100 - P_r)/100 \qquad [20]$$

where P_r = % gravel on a weight basis.

36–4.2.4. COMMENTS

The 1500-kPa pressure plate results correlate so well with the PWP measured by the sunflower method for nonsaline soils that the former is usually used in lieu of the time-consuming sunflower method.

36–5 AVAILABLE WATER CAPACITY

36–5.1. Principle

The available water capacity (AWC) of a soil is the amount of water retained in the soil reservoir that can be removed by plants. For field soils, the AWC is estimated by the difference in soil water content between FC and PWP; for coarse potting mixtures used in containers for plant propagation in greenhouses, the AWC is estimated by the difference between CC and PWP. The water content at CC for natural field soils, which are usually sieved and placed in containers, is too wet for growth of most plants. Because of the basic difficulties in establishing an exact value for FC and PWP, an uncertainty exists in the AWC computation. Personal experiences suggest that the uncertainty in FC is greater than that for PWP. The reader is referred to the sections on FC (section 36–2) and PWP (section 36–4) to review the limitations and uncertainties associated with making each measurement.

36–5.2. Procedure

The available water capacity is calculated by

$$AWC_w = FC_w - PWP_w \qquad [21]$$

or

$$AWC_v = FC_v - PWP_v \quad [22]$$

where AWC_w and AWC_v are reported in kg/kg and m^3/m^3, respectively. Sometimes AWC_v is reported on a volumetric basis per unit area or as m/m.

36–5.3. Comments

The AWC concept and the "water availability" concept (Gardner, 1960) are related but are not the same. A specified amount of water is retained by the soil reservoir; the upper limit of the AWC is defined as the water content in the initially wetted zone after drainage into an initially drier zone below is reduced to a negligible level. The lower limit of the AWC is the water content at which plants, growing in an environment of low evaporative demand, permanently wilt. The concept of "water availability" (Gardner, 1960) involves the interaction of three major factors: (i) the kind of plant, physical condition, stage of maturity, root distribution, etc.; (ii) the potential gradient existing at the root-soil water interface; and (iii) the unsaturated hydraulic conductivity near the water-absorbing portions of the root. For example, at the incipient wilting point, water availability for a particular plant is reduced to the point where growth no longer occurs, but the plant remains alive. If the plant is placed under conditions of lower evaporative demand, water availability to the plant for this soil-plant system increases and the plant once again begins to grow. If the plant is now placed under conditions of higher evaporative demand, the roots continue to take up water, but the degree of wilting increases. If the plant is kept in the high evaporative demand environment, it cannot take up all of the water retained by the AWC because the plant permanently wilts under the high evaporative demand conditions. On the other hand, under conditions of low evaporative demand, the plant may extract all of the water held by the AWC. Hence, AWC represents the *maximum* amount of water that is available to the plant under *low* evaporative demand conditions. Most crop plants will be injured if the soil is depleted into the wilting range. Hence there exists a difference between AWC as defined and that which is practically "useable." If one compares AWC for a particular crop computed from in situ observations of FC_v and PWP_v with AWC_v^a computed from FC_v^a and PWP_v^a for the same soil, the results may be quite different. The laboratory methods do not consider root distribution. Recent work by Ratliff et al. (1983) and Cassel et al. (1983), has shown that in general, the in situ and approximate methods give water content values which are within ±0.02 to 0.03 m/m of each other.

Under field conditions, 48 h may be insufficient time for drainage to become negligible; i.e., FC is not attained and the upper level of the AWC is not well defined. For such cases, water remaining in the soil after 48 h is draining slowly and may be utilized by the plant (Wilcox, 1960;

Winter, 1975). The AWC of soil in this case might more appropriately be called the effective available water capacity (Miller and Aarstad, 1973).

A problem exists when one attempts to calculate the AWC for swelling soils because bulk density increases as the soil water content decreases from FC to PWP. Hence, confusion exists with respect to which value of ρ_b to use in Eq. [7], [11], and [13]. McIntyre (1974a) discussed this soil shrinkage problem and presented methodology for dealing with it. He reported that the error introduced by not correcting for shrinkage probably does not exceed 5% of the measured AWC. This level of error lies within the limits of accuracy of the AWC determination itself. For further discussion on computation of AWC of swelling soils, the reader is referred to the treatment by McIntyre (1974a).

36-6 REFERENCES

Biswas, T. D., D. R. Nielsen, and J. W. Biggar. 1966. Redistribution of soil water after infiltration. Water Resour. Res. 2:513–514.

Briggs, L. J., and J. W. McLane. 1907. The moisture equivalent of soils. USDA Bureau of Soils Bull. 45.

Briggs, L. J., and H. L. Shantz. 1912. The wilting coefficient for different plants and its indirect determination. USDA BPI Bull. 230.

Carbon, B. A. 1975. Redistribution of water following precipitation on previously dry sandy soils. Aust. J. Soil Res. 13:13–19.

Cassel, D. K., T. H. Krueger, F. W. Schroer, and E. B. Norum. 1974. Solute movement through disturbed and undisturbed soil cores. Soil Sci. Soc. Am. Proc. 38:36–40.

Cassel, D. K., L. F. Ratliff, and J. T. Ritchie. 1983. Models for estimating in-situ potential extractable water using soil physical and chemical properties. Soil Sci. Soc. Am. J. 47:764–769.

Cassel, D. K., and M. D. Sweeney. [1974 sic, undated.] In situ soil water holding capacities of selected North Dakota soils. North Dakota Agric. Exp. Stn. Bull. 495.

Clothier, B. E., D. R. Scotter, and J. P. Ken. 1977. Water retention in soil underlain by a coarse-textured layer. Theory and a field application. Soil Sci. 123:392–399.

Coleman, E. A. 1944. The dependence of field capacity upon the depth of wetting of field soils. Soil Sci. 58:43–50.

Coleman, E. A. 1947. A laboratory procedure for determining the field capacity of soils. Soil Sci. 63:277–283.

Davidson, J. M., L. R. Stone, D. R. Nielsen, and M. E. LaRue. 1969. Field measurement and use of soil-water properties. Water Resour. Res. 5:1312–1321.

Eagleman, J. R., and V. C. Jamison, 1962. Soil layering and compaction effects on unsaturated moisture movement. Soil Sci. Soc. Am. Proc. 26:519–522.

Furr, J. A., and J. O. Reeve. 1945. The range of soil moisture percentages through which plants undergo permanent wilting in some soils from semi-arid, irrigated areas. J. Agric. Res. 71:149–170.

Gardner, W. R. 1960. Dynamic aspects of water availability to plants. Soil Sci. 89:63–73.

Gardner, W. R. 1970. Field measurement of soil water diffusivity. Soil Sci. Soc. Am. Proc. 34:832–833.

Gardner, W. R., and R. H. Nieman. 1964. Lower limit of water availability to plants. Science 143:1460–1462.

Hanks, R. J., W. E. Holmes, and C. B. Tanner. 1954. Field capacity approximation based on the moisture-transmitting properties of the soil. Soil Sci. Soc. Am. Proc. 18:252–254.

Israelsen, O. W. 1918. Studies on capacities of soils for irrigation water, and on a new method of determining volume weight. J. Agric. Res. 13:1–36.

Jamison, V. C. 1956. Pertinent factors governing the availability of soil moisture to plants. Soil Sci. 81:459–471.

Jamison, V. C., and E. M. Kroth. 1958. Available moisture storage capacity in relation to textural composition and organic matter content of several Missouri soils. Soil Sci. Soc. Am. Proc. 22:189–192.

Lehane, J. J., and W. J. Staple. 1960. Relationship of the permanent wilting percentage and the soil moisture content at harvest to the 15-atmosphere percentage. Can. J. Soil Sci. 40:264–269.

Marshall, T. J., and G. B. Stirk. 1949. Pressure potential of water moving downward into soil. Soil Sci. 68:359–370.

McIntyre, D. S. 1974a. Water retention and the moisture characteristic. p. 43–62. *In* J. Loveday (ed.) Methods for analysis of irrigated soils. Tech. Commun. Commonw. Bur. Soils no. 54. CAB: Fornham Royal, England.

McIntyre, D. S. 1974b. Sample preparation. p. 21–37. *In* J. Loveday (ed.) Methods for analysis of irrigated soils. Tech. Commun. Commonw. Bur. Soils no. 54. CAB: Fornham Royal, England.

McIntyre, D. S., and C. L. Watson. 1975. Contact materials for laboratory soil water studies. p. 1–11. *In* CSIRO Aust. Div. Soils Tech. Paper no. 26. Commonwealth Scientific and Industrial Research Organization, Melbourne, Australia.

Miller, D. E. 1963. Lateral moisture flow as a source of error in moisture retention studies. Soil Sci. Soc. Am. Proc. 27:716–717.

Miller, D. E. 1969. Flow and retention of water in layered soils. USDA Conservation Res. Rep. 13. U.S. Government Printing Office, Washington, DC.

Miller, D. E. 1973. Water retention and flow in layered soil profiles. p. 107–117. *In* R. R. Bruce et al. (ed.) Field soil water regime. Spec. Pub. 5. Soil Science Society of America, Madison, WI.

Miller, D. E., and J. S. Aarstad. 1973. Effective available water and its relation to evapotranspiration rate, depth of wetting, and soil texture. Soil Sci. Soc. Am. Proc. 37:763–766.

Miller, D. E., and W. C. Bunger. 1963. Moisture retention by soil with coarse layers in the profile. Soil Sci. Soc. Am. Proc. 27:586–589.

Ogata, G., and L. A. Richards, 1957. Water content changes following irrigation of bare-field soil that is protected from evaporation. Soil Sci. Soc. Am. Proc. 21:355–356.

Ratliff, L. F., J. T. Ritchie, and D. K. Cassel. 1983. Field-measured limits of soil water availability as related to laboratory measured properties. Soil Sci. Soc. Am. J. 47:770–775.

Reeve, M. J., P. D. Smith, and A. J. Thomasson. 1973. The effect of density on water retention properties of field soils. Soil Sci. 24:355–367.

Reinhart, K. G. 1961. The problem of stones in soil moisture measurement. Soil Sci. Soc. Am. Proc. 25:866–870.

Richards, L. A. 1960. Advances in soil physics. Trans. Int. Congr. Soil Sci., 7th. 1:67–69.

Richards, L. A., and D. C. Moore. 1952. Influence of capillary conductivity and depth of wetting on moisture retention in soil. Trans. Am. Geophys. Union 33:531–540.

Richards, L. A., and L. R. Weaver. 1943. Fifteen-atmosphere percentages as related to the permanent wilting percentage. Soil Sci. 56:331–339.

Rivers, E. D., and R. F. Shipp. 1977. Soil water retention as related to particle size in selected sands and loamy sands. Soil Sci. 126:94–100.

Russell, M. B., and L. A. Richards. 1938. The determination of soil moisture energy relations by centrifugation. Soil Sci. Soc. Am. Proc. 3:65–69.

Salter, P. J., and J. B. Williams. 1965. The influence of texture on the moisture characteristics of soils determining the available water capacity and moisture characteristic curve of a soil. Soil Sci. 16:1–15.

Schaffer, R. J., J. Wallace, and F. Garwood. 1937. The centrifuge method of investigating the variation of hydrostatic pressure with water content in porous materials. Trans. Faraday Soc. 33:723–734.

Shaw, C. F. 1927. The normal moisture capacity of soils. Soil Sci. 23:303–317.

Smith, R. M., and D. R. Browning. 1947. Soil moisture retention and pore space relations for several soils in the range of field capacity. Soil Sci. Soc. Am. Proc. 12:17–21.

SSSA. 1984. Glossary of soil science terms. Soil Science Society of America, Madison, WI.

Spomer, L. A. 1974. Two classroom exercises demonstrating the pattern of container soil water distribution. Hortic. Sci. 9:152–153.

Spomer, L. A. 1975. Small soil containers as experimental tools: soil water relations. Commun. Soil Sci. Plant Anal. 6:21–26.

Sykes, D. J., and W. E. Loomis. 1967. Plant and soil factors in permanent wilting percentages and field capacity storage. Soil Sci. 104:162–173.

Unger, P. W. 1971. Soil profile gravel layers: I. Effect on water storage, distribution, evaporation. Soil Sci. Soc. Am. Proc. 35:631–634.

Veihmeyer, F. J., and A. H. Hendrickson. 1931. The moisture equivalent as a measure of the field capacity of soils. Soil Sci. 32:181–194.

White, J. W. 1964. The concept of "container capacity" and its application to soil-moisture-fertility regimes in the production of container-grown crops. Ph.D. diss. Pennsylvania State Univ., University Park. (Diss. Abstr. 25:64–13, 425).

Wilcox, J. C. 1960. Rate of soil drainage following irrigation. II. Effects on determination of rate of consumptive use. Can. J. Soil Sci. 40:15–27.

Winter, E. J. 1975. Measurement of soil moisture properties. Minist. Agric. Fisheries, Forestry Tech. Bull. 29., p. 240–248.

37 Temperature

STERLING A. TAYLOR[1]
Utah State University
Logan, Utah

RAY D. JACKSON
United States Water Conservation Laboratory
Agricultural Research Service, USDA
Phoenix, Arizona

37-1 GENERAL INTRODUCTION

Temperature is a concept that is widely used to characterize the thermal properties of a system. The growth of biological systems is optimal within certain ranges of temperature and inhibited or prevented beyond such boundaries. The agricultural significance of temperature is readily ascertained when one considers that, in addition to plant growth, physical, chemical, and microbiological processes occurring in soil are strongly influenced by temperature. The influence of temperature on such processes has been, and will probably continue to be, intensively studied. Such studies require the measurement of temperature. Several means of making this measurement are described in this chapter.

Temperature cannot be measured directly. It can only be estimated by its influence on some property of matter that responds to variation in the intensity of heat in the body of matter. Changes in properties of matter listed in Table 37-1 have been found most useful for practical temperature measurements. Instruments built to take advantage of any of these properties of matter are called thermometers. In order that all of these properties will give the same indication of the temperature of a

Table 37-1. Thermometric substances and their thermometric properties.

Thermometric substances	Thermometric property
Mercury or liquid in a gllass capillary	Volume
Bimetal strip	Length
Platinum or other wires and thermistors	Electrical resistance
Thermocouples and thermels	Thermal emf
Gas or vapor at constant volume	Pressure

[1] Deceased 8 June 1967.

given body, calibration techniques have been established, and certain standard references have been agreed upon.

37–2 KINDS OF THERMOMETERS USED IN SOILS WORK

37–2.1 Introduction

There are numerous kinds of thermometers that can be used for measuring soil temperatures. Some of the commonly used types are mercury or liquid in glass, biometallic, bourdon, and electrical resistance thermometers. The choice of a thermometer for a given application depends, among other things, upon availability, degree of required precision, accessibility to location of sensing element, and physical size of element. Proper calibration and installation are necessary if reliable results are to be expected.

37–2.2 Mercury or Liquid-in-Glass Thermometers

Instruments of this kind should be checked against a standard thermometer or at a reference point, such as the ice point, once each year to adjust for any change in calibration that might result from irreversible changes in volume of the glass bulb.

A correction must be made if the stem is subjected to a temperature different from that which the instrument was intended to indicate. If n degrees of the mercury column are out of a medium that is at temperature T, and if the mean temperature of the emergent stem is T_s, then the necessary correction to the reading of an immersion thermometer is

$$\Delta T = n(\gamma_h - \gamma_g)(T - T_s) \qquad [1]$$

where γ_h and γ_g are the cubical expansion coefficients of mercury and glass. For most glass, $\gamma_h - \gamma_g = 1/6200$ to a sufficient degree of accuracy if n is in centigrade degrees. It is difficult to measure T_s precisely.

Care must be exercised in reading any mercury or liquid-in-glass thermometer to see that the line of sight is in the plane through the end of the mercury and perpendicular to the plane of the thermometer.

37–2.3 Bimetallic Thermometers

These thermometers are made commercially by welding together two bars of different metals and rolling the resulting compound bar into a strip. The metals generally used are invar and brass or invar and steel. Because of the difference in linear expansion of the two metals, the strip bends in response to temperature changes. An indicating arm or pointer to indicate the amount of angular deformation is attached to a helical coil of the bimetallic strip. The angular deflection is calibrated in terms of temperature. The instrument can be made with an adjustable zero by providing an adjusting screw at the place where the helix is attached to the frame. A pen can be attached to the indicating arm to make the instrument self-recording.

This kind of thermo-sensing element is used extensively in soils laboratories for regulating constant-temperature baths, environment chambers, and constant-temperature rooms. Dial-type thermometers based on this principle are used both in the laboratory and in the field.

The bimetal thermometers are generally less precise than the mercury-in-glass thermometers. The advantages responsible for their use are their lower lag and increased durability and their mechanical advantage, which makes it possible for them to be made self-recording.

Precautions to avoid exposure to direct radiation must be taken if these thermometers are to give reliable measurements of temperature.

37–2.4 Bourdon Thermometers

This type of thermometer consists of a curved tube of elliptical cross section connected through a capillary to a bulb that is inserted into the soil. The system is completely filled with some organic liquid. The bulb is usually long and about 1 cm in diameter. An increase in temperature causes the organic liquid to expand and increase the pressure inside the soil bulb and the curved capillary tube. This causes the curved tube to become slightly less curved, thus changing the position of the pointer attached to it. If a pen is attached to the pointer on the end of the bourdon element, a continuous record of the temperature can be made on a chart that is attached to a clock-operated drum. When properly placed horizontally in the soil, it will give a good indication of the temperature at that depth. If the bulb is placed vertically in the soil, an integrated or average temperature over a depth interval can be obtained.

The most troublesome problem in using this type of instrument is that the instrument and capillary tube are relatively sensitive to changes in temperature and are responsive to direct radiation heating. Radiant heating can be reduced by properly shielding the bourdon element from direct radiation and by burying at least 3 m of the protected capillary tube at the same depth as the bulb.

37–2.5 Electrical Resistance Thermometers

There are two types of resistance thermometers in general use. One type depends upon the increase in resistance with an increase in temperature of a wire such as nichrome, copper, silver, or platinum. The temperature coefficient of resistance of platinum wire at 0°C is about 0.35% per degree. Resistance changes are measured with a bridge, or in some cases a potentiometer, and are related to temperature changes. Recording bridges or potentiometers yield a continuous record of resistance and hence of temperature.

The second type of resistance thermometer is a semiconductor called a *thermistor.* Thermistors have a high negative temperature coefficient of resistance, on the order of 4% per centigrade degree. This coefficient is opposite in sign to and about 10 times larger than that of platinum resistance thermometers. Thermistors are available in various shapes, e.g., spheres, discs, and rods, all in various sizes, thus permitting great flexibility in application. Thermistors are, however, not all uniform; therefore, each unit must be separately calibrated.

Some thermistors have been found to change resistance with time; thus recalibration is required. In discussing this problem, Friedberg (1955) says:

> Thermistors undergo systematic resistance changes with time, generally becoming more stable after aging at elevated temperatures (100°C.) for several days or weeks, making preaging essential in most thermometric applications. In some instances, similar stabilization may be achieved by the passage of currents much larger than the usual measuring current through the material for shorter periods. Properly aged and electrically formed thermistors have been found to have resistance at 100°C. reproducible to within ±0.01°C. over periods of several months. Stability is less satisfactory when these thermometers are used up to 300°C.

Small bead-type thermistors mounted in glass probes, similar to VECO 32A1 (Western Electric Type 14-B), of the Victory Engineering Co., Union, NJ, are suitable for measurements of freezing-point depression, heat of wetting, heat capacity, and thermal conductivity. The thermistors may be used as glass rods or may be mounted in metallic shields as desired. If the thermistors are mounted in metallic shields, the heat capacity is increased, and the response time is modified accordingly. A direct-current Wheatstone bridge, capable of recording 9999 ohms to the nearest ohm, and a galvanometer to indicate the null point are satisfactory for reading temperatures to about 0.003°C (Richards & Campbell, 1948). More sensitive instruments and more precise laboratory techniques are necessary to achieve greater accuracy and precision. Care must also be taken to minimize self-heating of the thermistor.

For some applications, thermistors are purposely self-heated. However, for precise temperature measurements, self-heating must be controlled. This is attained by applying a low voltage to the thermistor. To estimate the maximum voltage allowable, the dissipation constant for each type of thermistor must be known. This constant is usually given by the manufacturer for the thermistor immersed in a particular fluid, e.g., still air, moving air, still oil, or water. The dissipation constant is the power required to raise the temperature of the thermistor 1 centigrade degree when immersed in a specified fluid. For example, a VECO 32A8 bead thermistor has a dissipation constant given by the manufacturer of 0.7 mW per centigrade degree in still air and a nominal resistance of 2000 ohms at 25°C. If the required precision of the air temperature measurement is 0.01C°, the allowable power is (0.7) (0.01) = 0.007 mW (7

$\times 10^{-6}$ W). The maximum voltage that can be applied with negligible self-heating is $E = (7 \times 10^{-6} \times 2 \times 10^{3})^{1/2} = 0.118$ V.

For measuring heat conductivity where the precise location of the temperature determination must be known, the VECO 34A1 thermistor of the Victory Engineering Company is satisfactory. It is the size of a very small sand grain and can be used to measure temperature with about the same precision as the 32 A1, which has similar characteristics. It can be placed at a point in the soil with little or no disturbance to the soil. With a sufficiently sensitive bridge and good circuits, temperature can be measured with a precision of 0.003C° or better. This precision greatly exceeds the accuracy of most laboratory standard thermometers, but this fact does not preclude their use in measuring differences or variations in temperature with greater precision than standard thermometers, provided that the exact temperature on the international standard scale is not important.

Temperatures indicated by thermistors may be continuously recorded with a recording potentiometer by placing a small (1.3 V) constant-voltage battery in series with a protective resistance large enough to reduce the voltage across the thermistor, low enough to prevent self-heating. A recording potentiometer is placed across the thermistor to measure the voltage drop across it. The voltage drop varies with temperature.

The magnitude of the protective resistance R_p necessary to reduce the voltage drop V_T across the VECO 32A8 thermistor of the last example to 0.118 V, if a 1.3-V mercury battery B_B is used and if the resistance of the thermistor R_T is 2000 ohms, is $R_p = (V_B/V_T)R_T = (1.3/0.118)\,(2000) = 22000$ ohms.

Bridge circuits can also be conveniently used for recording temperatures measured with thermistors. An appropriate recorder can be placed in the circuit in place of the galvanometer; and, with proper choice of resistance ratios and recorder sensitivity, a satisfactory range of temperatures can be recorded. Some suppliers of thermistors will provide tested circuit diagrams for various applications.

37–2.6 Thermocouples

Thermoelectric junctions, called *thermocouples*, are made by joining two dissimilar metals at two different places, to form two junctions. If the entire circuit is composed of only two metals, the total electromotive force in the circuit is proportional to the difference in temperature of the two junctions. One junction is called the *measuring* (or hot) *junction*, and the other, the *reference* (or cold) *junction*. For measurement of temperature, the reference junction is kept at a constant temperature, for example, in melting ice. The electrical potential produced is usually measured with a potentiometer when precise measurements are desired. Also, a galvanometer or millivoltmeter can be used. Some commercially available potentiometers have a built-in correction for the temperature of the reference junction, thus eliminating the need of an ice bath or other constant-temperature source. These methods are also adaptable to recording equipment.

If several thermocouples are wired in series so that their hot and cold junctions are composed of several different couples, the emf is increased and greater precision is possible. Such units are called *thermels* or *thermopiles.*

One great advantage to thermocouples and thermistors for measuring soil temperature, in addition to their adaptability to automatic recording, is their small size and almost instantaneous response to temperature changes. They have a low heat capacity and the lead wires may be very small, so that there is little influence of the temperature outside the soil on the reading. They may be placed precisely, and therefore indicate the temperature at a given point.

Thermocouples with an appropriate potentiometer and a suitable reference have numerous uses for measuring temperature in the soils laboratory. Differential thermal analysis of soil minerals, differences in temperature between moist soil and dry soil, the temperature rise that occurs upon wetting of soil, and the evaporative cooling are some of the applications. Properly calibrated galvanometers or recording potentiometers may be used for detecting and recording the difference in temperature between the two junctions.

Thermocouples have been used to measure the vapor pressure of soil water. These methods, however, require special precautions to remove all contact and junction potentials and require thermally shielded and guarded circuits and switches. At the time that this chapter was first written (1961), methods were not sufficiently developed to allow their adoption as routine laboratory procedures. Twenty years later, the procedures have been refined to the stage that field soils are routinely monitored. Two general methods have been developed (see chapter 24). One is based upon a measure of the temperature difference between a reference, which is usually the constant temperature bath, and the wet-bulb temperature that is obtained by inserting a tiny thermal junction, containing a droplet of water, into the soil environment (Richards & Ogata, 1958). The other method is based upon cooling one junction to the dew point while retaining the other at the temperature of the bath. The Peltier effect is utilized to cool the junction to the dew point (Korven & Taylor, 1959; Monteith & Owens, 1958).

37-3 CALIBRATION OF THERMOMETERS

37-3.1 Introduction

Temperature is a fundamental physical property that can be measured using any method that has been correctly related to the accepted standard scale. A set of fundamental fixed points to which specified values have been assigned by international agreement is used as a basis for calibrating thermometers. These fixed points are the freezing and boiling points of the materials given in Table 37–2 (Hall, 1955).

37–3.2 Principles

In addition to the fixed points defined in Table 37–2, a continuous temperature scale requires that the means which are to be used for interpolation between the fixed points be specified. For this purpose, the scale has been divided into three regions. From −182.97 to 630.5°C, the scale is defined by the electrical resistance of a standard platinum resistance thermometer. For purposes of interpolation in the region of the scale between 0 and 630.5°C, use is made of the equation

$$R_T = R_0(1 + AT + BT^2)\,, \qquad [2]$$

where R_T is the resistance at temperature T, R_0 is the resistance of 0°C, and A and B are constants that are determined by measurements at the melting point of ice, the boiling point of water, and the boiling point of sulfur. Below 0°C a different equation,

$$R_T = R_0[1 + AT + BT^2 + C'(T - 100)T^3] \qquad [3]$$

is used for interpolation purposes. The additional constant is determined by measurement of the boiling point of liquid oxygen (−182.97°C).

From 630.5 to 1063°C, the scale is defined by means of a 10% rhodium-platinum against pure platinum thermocouple using the equation

Table 37–2. Fundamental fixed points and primary fixed points of the International Temperature Scale under the standard pressure of 1 013 250 dynes/cm².

Fixed point	Value adopted 1927	Value adopted 1948
	°C	°C
Temperature of equilibrium between liquid oxygen and its vapor (boiling point of oxygen)	−182.97	−182.970
Temperature of equilibrium between ice and air-saturated water (melting point of ice) (fundamental fixed point)†	0.000	0
Temperature of equilibrium between liquid water and its vapor (boiling point of water) (fundamental fixed point)†	100.00	100
Temperature of equilibrium between liquid sulfur and its vapor (boiling point of sulfur)	444.60	444.600
Temperature of equilibrium between solid and liquid silver (freezing point of silver)	960.5	960.8
Temperature of equilibrium between solid and liquid gold (freezing point of gold)	1063	1063.0

† The freezing and boiling points of water under standard conditions are fundamental fixed points. The other points are given values as close to the Celsius temperature as possible and fixed by definition to the precision listed; these points are called primary fixed points and may change slightly as more precise methods are developed.

$$E = a + bt + ct^2 . \quad [4]$$

The constants are determined by measurements at the freezing point of antimony (630.5°C), silver, and gold. So that the scale will be continuous, it is prescribed that the freezing point of the actual sample of antimony which is used shall be determined by means of a platinum resistance thermometer, and that the value so determined shall be used in calculation of the thermocouple calibration. Alternatively, the thermocouple may be compared directly with the resistance thermometer at a temperature close to 630.5°C.

In soils work, it is unlikely that there will be concern with temperatures above 1063°C; consequently, that region is not considered.

In most soils laboratories, the standard instruments prescribed are not available for the purpose of comparing and checking instruments used for measuring soil temperature; consequently, secondary standards of sufficient accuracy for the intended purposes are used. A thermocouple of platinum with an alloy of platinum containing 10% rhodium can be read to ±0.01°C with a precise potentiometer. This instrument can be calibrated either by the supplier or by the National Bureau of Standards to give a good secondary standard over the range of temperatures generally covered in soils work. Mercury-in-glass thermometers may also be calibrated to give a satisfactory secondary standard for most purposes.

37–3.3 Method

37–3.3.1 SPECIAL APPARATUS

1. Primary standard platinum resistance thermometer or a secondary standard thermometer.
2. Dewar flask or thermos bottle.
3. Stirrer.

37–3.3.2 PROCEDURE

Begin by checking the ice point. [If the necessary equipment is available, a more stable point is the triple point of water, which is 0.0100°C above the ice point (Stimpson, 1955)]. Suspend the standard thermometer and the thermometer to be checked near the center of a Dewar flask. Fill the Dewar flask with crushed ice made from distilled water (if ice from distilled water is not available, it may be made from regular tap water in a domestic freezing unit with errors <0.01°C). Add liquid water to the crushed ice to displace all the air, and allow the system to establish temperature equilibrium. This temperature will give the ice point (0°C or 32°F).

Gradually add heat until all the ice melts. Place the stirrer in the Dewar flask, and stir the solution vigorously. Add hot water to the system to raise the temperature. When stable readings are attained, compare the reading of the standard thermometer to that of the instrument being checked or calibrated. Maintain the water at the proper level near the top of the Dewar flask by siphoning off the excess.

37–3.3.3 COMMENTS

If mercury-in-glass thermometers are being used or calibrated, care should be taken to see that the thermometer is immersed to the proper depth, depending on whether the thermometer is intended for total immersion or partial immersion. If electrical resistance or thermoelectric thermometers are being used, care should be taken to see that the leads do not conduct heat into the bath and to the sensing unit, thereby changing the temperature reading. This can be avoided by immersing 20 to 25 cm of electrically insulated lead in the same bath with the thermometer. Simultaneous readings should be taken on both the standard thermometer and the instrument being calibrated at approximately 10-degree intervals over the range from the freezing point to the boiling point of water.

A pressure correction for the melting point of ice will probably be unnecessary except for the most exacting work. If it is necessary to make a correction, it can be done by applying the Clapeyron equation

$$dT/dP = T(V_l - V_s)/\Delta H_s$$

where T is temperature, P is pressure, and V_l and V_s are the specific or molar volume of liquid water and ice, respectively, at the freezing point. For ice at 0°C, $T = 273.2\text{K}$, $(V_l - V_s) = -0.0906\ \text{cm}^3\ \text{g}^{-1}$, and $\Delta H_v =$ 333.6 J g^{-1}; hence $dT/dP = -0.742\text{K/kPa}$.

The solubility of air in water at the freezing point under standard conditions is given by Dorsey (1940, p. 605) as 1.29×10^{-3} g formula weight of air dissolved in 1 kg of water at equilibrium. This causes a depression of the freezing point of 2.4×10^{-3} degrees. The solubility of gas increases by about 0.4% per atmosphere (101.3 kPa), which is entirely negligible for normal variations in atmospheric pressure, so that the correction of 2.4×10^{-3} degrees for solubility of air is adequate.

This boiling point of water is a fundamental fixed point at a standard pressure of 1.01325 bars (the standard atmosphere). If the pressure is different from the standard reference pressure, a correction must be made for the temperature of the boiling point. For precise work, the correct boiling point should be taken from tables [for example, Dorsey (1940, p. 580)]. However, the simple linear average of 0.39°C reduction in boiling point per 10 mm decrease in barometric pressure below the standard atmosphere is accurate to the nearest 0.1°C.

To read the correct temperature at the boiling point of water, the thermometer should be placed at a point far from the surface through which heat is supplied, the rate of heating should be low, mixing should be thorough, and the thermometer should be screened from direct radiation. The observed temperature will then be near to that of the true boiling point.

Precise calibration can only be justified for use in the laboratory where conditions can be carefully controlled. Under these conditions, the use of a "triple-point" apparatus would remove some of the sources of variation discussed above (Stimpson, 1955). In the field, installation errors and the continual fluctuation in temperatures with depth and time make precise determinations difficult; hence, any calibration needs to be only as accurate as the level of experimental error encountered in the field measurements.

37–4 FIELD MEASUREMENTS

37–4.1 Introduction

There are no single methods for measuring soil temperature that are recommended to the exclusion of others. The user should always be sure that the methods and instruments he is using are properly related to the standard scale.

The more commonly used methods for measuring the soil temperatures will now be described. There may be other methods and equipment that can be used to achieve the same purpose.

37–4.2 Methods

37–4.2.1 ONE-POINT

The thermometer most appropriate for the measurement desired should be selected. The sensing element should then be placed at the point (location and depth) for which the measurements of temperature are desired. Care should be taken to ensure that the sensing element or thermometer bulb is in good thermal contact with the soil. The apparent temperature of the soil air is likely to be somewhat different from the temperature of the solid and liquid portions because of the low thermal conductivity of the air (Onchukov, 1957). Good thermal contact is particularly important if the thermometer has a large heat capacity.

Mercury-in-glass thermometers have a high heat capacity. At depths >30 cm in the field, specially constructed thermometers are used. One method is to hang the thermometers in a tube or pipe and pull them out for reading. In order that its temperature shall not change appreciably during the process, the bulb is surrounded by a mass of paraffin wax to give it a high lag coefficient. This is permissible only because the temperature at depths of more than a few inches changes very slowly; and in most applications, it is probably not serious if the observed temperature represents the temperature that actually occurred in the soil a few minutes earlier. It has been observed, however, that in some of the thermometers at temperatures <5°C, appreciable error is caused by the thermal contraction of the wax and resultant thermometer bulb deformation

(Garvitch & Probine, 1956). Consequently, thermometers of this kind should be checked frequently to see that they are indicating the correct temperature. It is possible to make a suitable wax that will not cause this contraction.

A more serious error is probably attributable to the thermal conductivity of the steel pipe that is usually used as a well. A mild steel pipe terminated in a sharp cone that can be driven into the soil without serious disturbance is usually provided. The pipe should be heavily painted to minimize its effect on the measured temperature. Convection of heat within the tube when the surface is colder than the deeper soil would also be expected to have an influence on the observed temperature.

Thermometers should be shielded from direct radiation of the sun. Electric thermometers should have 60 to 90 cm of lead wires buried at the same depth as the sensing element to avoid errors in reading that result from direct radiation.

When precise measurements of temperatures and temperature gradients are desired, one needs to use a thermometer with a very low heat capacity that will respond almost instantaneously to temperature changes. The location of the unit may also need to be known with accuracy. Thermistor units and thermocouples are ideally suited for these measurements since they are small and quick to respond. It is very important that they be installed properly in the soil. With sensitive units of this kind and with precision measuring equipment, one can readily detect a difference in temperature between the air space and the soil particles with the associated liquid if the temperature is changing rapidly, as it does during summer in the surface 5- to 10-cm depths of soil (Onchukov, 1957).

37–4.2.2 TEMPERATURE DISTRIBUTION

In the past little attention has been given to depths of temperature measurement, and the data are reported for many different depths. It is desirable to have measurements at the same depths so that the data from place to place can be directly compared. It is suggested that, whenever possible, temperatures be measured at 10, 20, 50, 100, 150, and 300 cm (Richards et al., 1952; Blanc, 1958). Temperatures may also be measured at 5 cm and at the surface of the soil, but special precautions are necessary to achieve proper thermal contact and to avoid errors resulting from direct radiation when surface measurements are made.

Surface temperatures can be measured by using a sheet of conducting metal and a thermometer firmly attached to a piece of insulating board 1 to 2 cm in thickness. Good thermal contact is made between the metal and the thermometer, and the two are attached to the insulating board with an epoxy resin or other suitable cement. Electrical thermometers are best suited for this kind of measurement, but mercury thermometers may also be used successfully. In use, the metal is placed in direct contact with the soil surface, and the 1- to 2-cm board to which it is attached shields it from direct radiation. The board shield may be eliminated if the metal is covered with about 1 cm or more of soil. A convenient size for the metallic conductor, and one that appears to give satisfactory readings of surface temperature, is 20 by 20 cm. Many other sizes have been used with satisfactory results. Whenever this method is used, one should report the size and dimensions of the metallic conductor and shield used. For semipermanent or seasonal installations, copper screens covered with a thin layer of soil might be substituted for metallic plates.

The type and amount of ground cover, as well as the soil moisture status, have a marked effect on soil temperatures. For this reason, it is recommended that, whenever possible, measurements be made under a uniform sod or in a bare soil plot (Newman et al., 1959). The sod should be clipped regularly so that the grass remains between 5 and 7.5 cm tall. The sod should not be irrigated. In arid regions where grass will not grow, measurements should be made under a vegetative cover that is as near to that of the native vegetation as possible. In irrigated regions, it may be necessary to irrigate the sod in a manner similar to that used in adjacent areas. Whenever it is convenient, it is suggested that temperatures be measured under bare plots maintained free of vegetation by herbicides or by scraping the soil surface with a hoe in order that there may be a universal comparison for all kinds of climatic zones.

37–4.2.3 INTEGRATED VALUES

Many applications of soil temperature data to biology require an integrated or average soil temperature. The simplest and probably the least desirable method for doing this is to take readings once or twice a day at various depths and make a simple average of all observations during the time interval and over the depth interval for which information is desired.

A somewhat better method, but still not completely satisfactory, is to average the maximum and minimum temperatures that are observed each day. This information can be obtained either from maximum- and minimum-registering thermometers or from thermographs of the temperature at the desired point. If results obtained by this method are compared with those calculated by hourly average temperatures, it is found that in the winter the median calculated from daily maximum and minimum temperatures is above the hourly average, while in the summer it is below. These variations, however, are often $<1°C$ and may be negligible.

There are several methods of continuously integrating the temperature variations so that one reading gives the integrated value. MacFayden (1956) has proposed a method for obtaining the weekly integrated temperature at a point in the soil. His method consists of using a thermistor connected in series with a silver volt-ampmeter. A voltage of 1.3 volts was supplied by means of a dry cell. A separate thermistor and volt-ampmeter is required for each position to be measured. The battery and any measuring or metering circuit can be common for all positions at a site. An electrochemical device known as a solion two-terminal integrator type SV 150 is available from Self Organizing Systems, Inc., 6612 Denton Drive, Dallas, TX. This device can be used in place of the silver voltameter in a circuit similar to the one described. The unit gives a voltage which can be calibrated to give the integrated temperature. The unit is disconnected from the thermistor circuit and taken into the laboratory where the voltage is measured with a potentiometer. A rough estimate of the reading may also be obtained by noting the color of the integrator solution through a window provided in the chamber.

Thermistors with well-insulated leads are placed in the soil at the desired position. Since the resistance of the thermistor has a very marked temperature coefficient, the amount of current that will flow through it (at constant voltage) is a function of the temperature and time. The amount of current that flows through the volt-ampmeter in a week is determined by accurately weighing the amount of silver that is transferred from one electrode of the volt-ampmeter to the other during the period of observation. The amount of silver transferred must be calibrated with average temperature using a particular thermistor, since thermistors are not sufficiently uniform to permit use of a universal calibration curve.

A resistance thermometer for integrating temperatures over several soil locations and depth intervals has been proposed by Tanner (1958). Several resistance thermometers are wired in series and are attached to a specially designed Wheatstone bridge. For rapid measurements, a galvanometer is used. For the continuous record that is necessary for a time integration, a strip-chart recorder is used. Since the method depends on an unbalanced circuit, the battery must be of constant voltage such as provided by mercury or alkali cells. Precautions must be taken to see that the battery current is either very small or is applied for only short time periods to avoid undesirable self-heating of the thermometers.

If the units are placed vertically in the soil, they will integrate the temperature over the depth of the soil in which the unit is buried as well as over the several locations in which the units are placed. If placed horizontally in the soil, they will integrate temperatures over the several locations but at only one depth. In either method of installation, the leads should be buried for several feet to reduce the effect of thermal conductivity of the lead wires that are exposed to different temperatures.

37–5 REFERENCES

Blanc, M. L. 1958. The climatological investigation of soil temperature. World Meteorological Organization, Geneva, Switzerland, Tech. Note 20.

Dorsey, N. E. 1940. Properties of ordinary water-substance. American Chemical Society Monogr. Ser. no. 81. Reinhold Publishing Corp., New York.

Friedberg, S. A. 1955. Semiconductors as thermometers. p. 359–382. *In* H. C. Wolfe (ed.) Temperature, its measurement and control in science and industry. Vol. 2. Reinhold Publishing Corp., New York.

Garvitch, Z. S., and M. C. Probine. 1956. Soil thermometers. Nature (London) 177:1245.

Hall, J. A. 1955. The international temperature scale. p. 115–139. *In* H. C. Wolfe (ed.) Temperature, its measurement and control in science and industry. Vol. 2. Reinhold Publishing Corp., New York.

Korven, H. C., and S. A. Taylor. 1959. The Peltier effect and its use for determining relative activity of soil water. Can. J. Soil Sci. 39:76–85.

MacDowell, J. 1957. Soil thermometers. Nature (London) 179:328.

MacFayden, Amyan. 1956. The use of a temperature integrator in the study of soil temperature. Oikos 7:56–81.

Monteith, J. L., and P. C. Owen. 1958. A thermocouple method for measuring relative humidity in the range 95–100 percent. J. Sci. Inst. 35:443–446.

Newman, J. E., R. H. Shaw, and V. E. Suomi. 1959. The agricultural weather station. Wisconsin Agric. Exp. Stn. Bull. 537.

Onchukov, D. N. 1957. The phenomenon of heat and moisture transmission in soils and subsoil (In Russian). Mosk. Tekhnol. Inst. Pisheh. Promgsch. Tr. 1957 (8):55–63.

Richards, L. A., and R. B. Campbell. 1948. Use of thermistors for measuring the freezing point of solutions and soils. Soil Sci. 65:429–436.

Richards, L. A., and G. Ogata. 1958. Thermocouple for vapor pressure measurement in biological and soil systems at high humidity. Science 128:1089.

Richards, S. J., R. M. Hagan, and T. M. McCalla. 1952. Soil temperature and plant growth. *In* B. T. Shaw (ed.) Soil physical conditions and plant growth. Agronomy 2:304–336.

Stimpson, H. F. 1955. Precision resistance thermometry and fixed points. p. 141–168. *In* H. C. Wolfe (ed.) Temperature, its measurement and control in science and industry. Vol. 2. Reinhold Publishing Corp., New York.

Tanner, C. B. 1958. Soil thermometers giving the average temperature of several locations in a single reading. Agron. J. 50:384–387.

38 Heat Capacity and Specific Heat

STERLING A. TAYLOR[1]

Utah State University
Logan, Utah

RAY D. JACKSON

United States Water Conservation Laboratory
Agricultural Research Service, USDA
Phoenix, Arizona

38-1 INTRODUCTION

The rates of biological and chemical reactions and hence the rate of crop growth are influenced by the temperature of the soil. The temperature, in turn, depends directly upon the specific heat and heat capacity of the soil. The amount of temperature change in response to the absorption or release of heat is governed by the heat capacity. Temperature will increase faster in the spring, and the magnitude of the diurnal temperature changes will be greater in soil with lower heat capacity. In addition, heat transfer through soil depends upon the specific heat.

The heat capacity of natural soil is strongly dependent upon the soil porosity and water content, both of which are subject to rapid fluctuations. If the specific heat and amounts of each soil constituent and the water content are known, the heat capacity of the soil-water system can be calculated; hence it is desirable to have a reliable method for determining the specific heat of soil constituents. Baver (1956, p. 370) gives the specific heat for some soil constituents and indicates the procedure used for calculating the heat capacity of a soil-water system.

38-2 PRINCIPLES

The heat capacity of a system at Kelvin temperature T is the limit of the ratio $\delta Q/\delta T$ as δT approaches zero, where δQ is the amount of heat that must be introduced into the system to increase its temperature from T to $T + \delta T$. There are two heat capacities that are used in thermodynamic applications. One is the heat capacity at constant volume C_v, which is given by the equation

[1]Deceased 8 June 1967.

 Methods of Soil Analysis, Part 1. Physical and Mineralogical Methods—Agronomy Monograph no. 9 (2nd Edition)

$$C_v = (\partial U/\partial T)_v \qquad [1]$$

where U is the total (sometimes called *internal*) energy of the system. The energy U is related to the heat content Q by the relation $Q = \Delta U + W$, where W is the work performed by the system. The work term can usually be expressed as the work of expansion or shrinking at some constant pressure P. That is, $W = P\Delta V$, where ΔV is the change in volume of the system. The heat content or the enthalpy H of the system at constant pressure is $U + PV$, and so the equation for the heat capacity of a system at constant pressure is

$$C_p = [\partial(U + PV)/\partial T]_p = (\partial H/\partial T)_p\,. \qquad [2]$$

The heat capacity, as defined above, is an extensive property and varies with the amount of material in the system. When expressed on the basis of a unit mass of substance, it becomes an intensive property that is independent of the size of the system. The term is then called the *specific heat*. If the size of the unit is taken as a mole, rather than unit mass, the term is called the *molar heat*.

The most useful term for soil science is the specific heat at constant pressure, since it is possible to calculate the enthalpy, entropy, and free energy of the systems from such data.

The heat capacity at constant pressure can be measured with sufficient accuracy for most purposes by calorimetric means. That is,

$$\tilde{C}_p = [(H_2 - H_1)/(T_2 - T_1)](\Delta Q/\Delta T)\,. \qquad [3]$$

If the curve of C_p vs. T is nearly linear between T_1 and T_2, and if ΔT is sufficiently small, the value of $\tilde{C}_p$ can be identified with C_p at $(T_1 + T_2)/2$. This approximation can be safely made in most cases if $T_2 - T_1$ does not exceed 5°C.

38–3 METHOD

The origin of the method described here has been lost; however, it is an adaptation of the method of mixtures (Estermann, 1959, p. 260–269) modified to eliminate the heat of wetting.

38–3.1 Special Apparatus

1. Calorimeter: Use a commercial calorimeter. Alternatively, improvise one by immersing a 500-mL Dewar flask in a constant-temperature bath or by placing it in an insulated box. Fit the flask with an insulated cover, and equip it with a stirrer and a thermometer (such as a Beckman differential thermometer, a thermoelectric thermometer, or a resistance thermometer) capable of indicating a temperature difference of 0.01°C or less.

2. Accessory vessel: Use a commercial calorimeter. Alternatively, improvise an accessory vessel as described above (1.), omitting the constant-temperature bath or insulated box if desired. Calibrate the thermometer in the same vessel with and at the same time as the thermometer in the calorimeter, so that the two thermometers give the same temperature readings.

38–3.2 Procedure

Add a known quantity of soil to the calorimeter, along with a measured amount of water, sufficient to form a dilute suspension. Stir the suspension slowly until thermal equilibrium is established. Record the temperature to the nearest 0.01°C.

From the accessory vessel, add a measured quantity of water at a higher known temperature such that the final temperature will be about 1 degree, but not more than 5 degrees, higher than the initial temperature in the calorimeter. Record the initial temperature of the calorimeter and the accessory water and the final temperature of the calorimeter to the nearest 0.01°C. Weigh or measure all soil samples and water additions to at least three significant figures.

Evaluate the heat capacity $\tilde{C}_p$ of the soil sample from the difference T_c between the initial and final temperatures of the calorimeter, the temperature T_a of the added water, the water equivalent C_c (or heat capacity) of the calorimeter, the mass M_s of soil, the mass M_c of water initially in the calorimeter, and the mass M_a of water added from the accessory vessel.

Determine the water equivalent or heat capacity of the calorimeter, including the thermometer, stirrer, and vessel, by measuring the temperature of the calorimeter containing a known amount of water, placing a measured amount of water at a higher known temperature in the calorimeter, and observing the final temperature. The equation for this calculation is

$$C_c = M_{wc}c_w - M_{wa}c_w(\Delta T_a/\Delta T_c) \quad [4]$$

where M_{wc} and M_{wa} are the mass of water initially in the calorimeter and the mass added in grams, respectively, c_w is the specific heat of water at the mean temperature of the determination (4184 at 15°C.) in joules kg^{-1} K^{-1}, and ΔTa and ΔTc are the temperature changes in degrees for the water added and that already in the calorimeter. Since temperature differences appear in Eq. [4], either Celsius or Kelvin degrees can be used. The heat capacity of the calorimeter is given in J kg^{-1} K^{-1}, and C_c is the mean heat capacity of the calorimeter over the temperature interval of the experiment.

When the heat capacity of the calorimeter is known, the average specific heat of the soil sample (cal per g C°) may be determined by the formula

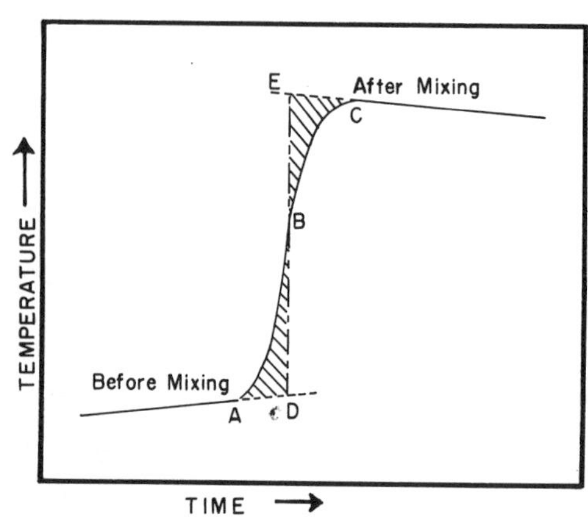

Fig. 38–1. Graphical correction for heat loss from the calorimeter during measurement of heat capacity and specific heat.

$$c_s = c_w(M_{wa}/M_s)(\Delta T_a/\Delta T_c) - (M_{wc}c_w + C_c)/M_s \quad [5]$$

where M_s is the mass of soil used in the experiment and the other symbols are as previously defined. Example: Assume that 0.05 kg of soil were mixed with 0.100 kg of water and brought to a temperature of 14.00°C. Then 0.0775 kg of water at 20.00°C was introduced from the auxiliary vessel. The final temperature was 16.00°C. The heat capacity of the calorimeter at this temperature had been determined to be 175.7 J K^{-1}. Near 15°C the specific heat of water (c_w) is 4186 J kg^{-1} K^{-1}. From the data M_{wa} = 0.0775 kg, M_s = 0.050 kg, M_{wc} = 0.100 kg, and C_c = 175.7 J K^{-1}, c_s = 4184 (0.0775/0.050) (4.00/2.00) – (0.100 × 4184 + 175.7)/(0.050 = 1096 J K^{-1} of soil.

38–3.3 Comments

It is frequently necessary to correct for the thermal leakage of the calorimeter and to compensate for energy added by stirring. This is accomplished by correcting ΔT_c from a temperature vs. time plot as shown in Fig. 38–1. In rough work, it is sometimes sufficient to extrapolate the initial and final linear portions of the temperature-time curve to a time midway between the end of the fore period and the maximum and to take the temperature difference at the point. A better approximation is to take the length of the vertical line *DE* so constructed as to make the areas *ABD* and *BCE* equal.

38–4 REFERENCES

Baver, L. D. 1956. Soil physics. 3rd ed. John Wiley and Sons, New York.

Estermann, I. 1959. Methods of experimental physics. Vol. 1, Classical methods. Academic Press, New York.

39 Thermal Conductivity and Diffusivity

RAY D. JACKSON

United States Water Conservation Laboratory
Agricultural Research Service, USDA
Phoenix, Arizona

STERLING A. TAYLOR[1]

Utah State University
Logan, Utah

39-1 GENERAL INTRODUCTION

The properties of soil which govern the flow of heat through it are of interest to several disciplines. Soil scientists, meteorologists, and agronomists are concerned with such problems as the temperature distribution in soil and lower air layers, the energy balance at the surface of the earth, the influence of temperature on water movement, and the measurement of soil water content by thermal methods. Engineers require information on heat conductance in soil in connection with buried cables, road construction, and heat transfer from heat pump coils.

The flow of heat through soil involves the simultaneous operation of several different transport mechanisms. Conduction is responsible for the flow of heat through the solid materials, while across the pores, three mechanisms, conduction, convection, and radiation, act in parallel. When water is present, latent heat of distillation is an additional factor involved in heat transfer.

Calculations of temperatures and quantities of heat flow within soil are usually made using equations derived to describe the conduction of heat in homogeneous, isotropic solids. The assumptions under which these equations were derived are not always met in soils. As an example, when heat is applied to a moist soil, mass transfer of water from hotter to colder areas carries heat with it. Within a pore, water may evaporate from the hot side and condense on the cold side, with the simultaneous transfer of a quantity of heat equal to the mass of water moved times the heat of vaporization of water.

The mathematics describing heat flow in porous materials such as soils which account for these additional factors are complicated and not

[1] Deceased 8 June 1967.

fully developed. In many cases, if certain conditions are observed, mechanisms of heat transfer other than conduction may be neglected and the well-known, well-developed mathematics of heat conduction in homogeneous isotropic solids can be applied to the description of heat flow in soils. The assumptions underlying the mathematics of conduction will apply to the following discussion. The term *heat transfer* will refer only to heat transfer by conduction.

Conductive heat flow may be compared to the flow of electricity. Thermal resistivity is analogous to electrical resistivity. The reciprocal of the resistivity is the conductivity. The thermal conductivity is defined as the quantity of heat that flows through a unit area in a unit time under unit temperature gradient.

Another useful heat-transfer coefficient is the thermal diffusivity. Mathematically, the diffusivity is the ratio of the conductivity to the product of the specific heat and density. The diffusivity is a measure of the change of temperature which would be produced in a unit volume by the quantity of heat which flows through the volume in a unit time, when a unit temperature gradient is imposed across two opposite sides of the volume. The term *thermal diffusivity* was first used by Lord Kelvin.

There are many ways to measure the thermal diffusivity and conductivity. In some instances a method must be selected to meet special conditions. All methods are either a steady-state heat flow (temperature not a function of time) or transient heat flow (temperature changes with time). When a steady-state method is used to study heat transfer in moist, unsaturated soils, a moisture gradient as well as a temperature gradient results. When a constant temperature difference is applied across a moist soil column, the soil near the hot face becomes drier; that near the cold face becomes wetter. Smith (1940) measured the thermal conductivity of moist soils using a steady-state method and observed the moisture gradients. The meaningfulness of the thermal conductivity coefficient measured in such systems is questionable. Steady-state methods are, however, sufficiently accurate for measuring the diffusivity and conductivity of dry soils.

Methods using transient heat flow are generally considered to be most useful for soils work. The advantages of transient methods are that water movement in response to temperature gradients is minimized, and a long wait for thermal gradients to become constant is not required. Patten (1909) and De Vries (1950) discussed these points in detail and recommended using transient systems with small temperature gradients for measurements of thermal conductivity and diffusivity of soil.

A review of published works on thermal conductivity and diffusivity measurements in soil is given by Baver (1956), Crawford (1952), and Richards et al. (1952). A method of calculating theoretically the conductivity of granular materials of known composition is given by De Vries (1952a).

39–2 GENERAL PRINCIPLES

The general equation of heat conduction, which describes both transient and steady-state heat flow, is

$$\partial T/\partial t = k\nabla^2 T \qquad [1]$$

where T is temperature, t is time, k is the thermal diffusivity, and ∇^2 is the Laplacian operator.

For one-dimensional flow of heat in the direction of distance, x, Eq. [1] becomes

$$\partial T/\partial t = k\partial^2 T/\partial x^2 \qquad [2]$$

and for radial flow of heat in the direction of distance r from a line source

$$\partial T/\partial t = k[(\partial^2 T/\partial r^2) + (1/r)(\partial T/\partial r)] \,. \qquad [3]$$

Derivation of the above equations is given in standard references on heat conduction, such as Carslaw and Jaeger (1959) and Jakob (1949). These references discuss the assumptions upon which the equations are based. Hence, the applicability of the equations for describing heat flow in soils may be ascertained.

The relation between the thermal conductivity K and the thermal diffusivity k is

$$K = kc_{sw} \,. \qquad [4]$$

The term c_{sw} is the volumetric heat capacity. It is the product of the specific heat and the density. The SI unit of volumetric heat capacity is joules per kilogram Kelvin.

The specific heat and density of both solids and water must be considered when calculating the volumetric heat capacity. For soils, the volumetric heat capacity is given by

$$c_{sw} = \rho_s(c_s + c_w\theta) \qquad [5]$$

where ρ_s is the density of dry soil, c_s the specific heat of dry soil, c_w the specific heat of water, and θ the ratio of the mass of water to the mass of dry soil.

39–3 THERMAL CONDUCTIVITY

39–3.1 Introduction

Classically, the thermal conductivity has been determined using steady-state methods. As discussed at the beginning of this chapter, the results obtained by the use of steady-state methods on unsaturated soils are questionable because of the temperature-induced water movement. In addition, steady-state methods are almost entirely limited to the laboratory, whereas some transient-state methods are readily adaptable to both the laboratory and field.

The transient-state cylindrical-probe method at present offers one of the better means of measuring the thermal conductivity of soil in situ. The probe can be placed horizontally or vertically at various depths in the soil, and the method can be adapted for laboratory use.

The probe method, when used in unsaturated soils, is also affected by temperature-induced water movement, but to a lesser degree than steady-state methods. The over-all effect of the water movement on the measured conductivity is determined by the probe size, type of soil, water content of the soil, and many other factors. De Vries and Peck (1958b) analyzed theoretically the effects of water movement on thermal conductivity measurements and concluded that this effect is small. Their theory can be used to account for the water movement if a high degree of accuracy is desired.

The construction of the probe requires careful workmanship, especially if the diameter is small. A probe larger than the one discussed herein is described by Lachenbruch (1957). Woodside (1958) discussed the chronological development of the cylindrical probe method. His references to this method are fairly complete. Information on steady-state methods can be found in papers by Kersten (1949), Smith and Byers (1939), and Waddams (1944).

39–3.2 Principles

If an infinitely long line source of heat is embedded in an infinite, homogeneous, isotropic medium, then the flow of heat away from the source is described by Eq. [3]. For thermal conductivity measurements, the infinite line source is approximated by a long electrically heated wire enclosed in a cylindrical probe. The probe is introduced into the material, heating current is supplied to the wire, and the temperature rise is measured with a thermocouple placed next to the wire.

The temperature rise $(T - T_0)$ at a radial distance r from the source is represented by

$$(T - T_0) = [q/(4\pi K)][-Ei/(-r^2/4kt)] \qquad [6]$$

where q is the heat produced per unit time and unit length of the source, K the conductivity, k the diffusivity, t the time, T_0 the temperature at time $t = 0$, and

$$-Ei(-r^2/4kt) = \int_{r2/4kt}^{\infty} (1/u) \exp(-u)\, du$$
$$= -\gamma - \ln(r^2/4kt) + (r^2/4kt) - (r^2/4kt)^2/4 + \ldots$$

is the exponential integral. (For numerical values see Jahnke and Emde [1945].) In the exponential integral, γ is Euler's constant (0.5772 · · ·), and u is a variable of integration.

For values of $r^2/4kt \ll 1$, all terms after the logarithmic term may be neglected. Thus

$$T - T_0 = q/(4\pi K)[-\gamma - \ln (r^2/4kt) \\ = q/(4\pi K)[c + \ln t\,. \qquad [7]$$

Errors caused by neglecting terms after the second in the exponential integral and those caused by the finite radius of the heat source are accounted for by introducing a time correction factor t_0 into Eq. [7]. For the heating period, the temperature rise is, to a good approximation

$$T - T_0 = q/(4\pi K)[d + \ln (t + t_0)], \qquad \text{for } t < t_1 \qquad [8]$$

where d is a constant and t_1 is the time at the end of the heating period.

The temperature difference can also be obtained during the time of cooling. Thus for $t > t_1$

$$T - T_0 = q/(4\pi K)[d + \ln (t + t_0)] \\ - q/(4\pi K)[d + \ln (t - t_1 + t_0)]. \qquad [9]$$

In Eq. [8] and [9], t_0 is a constant which depends upon the dimensions of the probe and the thermal properties of the probe and the medium. For probes of 0.1 cm diameter or less and for $t > 60$ s, t_0 may be neglected. For larger probes or smaller times, t_0 may be evaluated graphically. A plot is made of $T - T_0$ vs. ln t. For large values of t, a straight line results. This straight line is extrapolated to smaller values of t. A value of t_0 is chosen such that the adjusted experimental points fall on the extrapolated line. Thus a straight line ensues, the slope of which is used to determine the thermal conductivity K.

The conductivity K is calculated by equating the measured slope S to the theoretical slope $q/4\pi K$ as given by Eq. [8] and [9]. It is usually more convenient to use common instead of Naperian logarithms for plotting the temperature rise vs. The logarithm of time data; consequently, the measured slope is

$$S = 2.303q/4\pi K. \qquad [10]$$

The heat produced q is obtained from current and resistance measurements; or, using Ohm's law, the voltage may be used in conjunction with the current or resistance. If I is the current in amperes and R the resistance in ohms per centimeter of probe, then I^2R is the watts of heat produced per centimeter ($1\text{W} = 1\ \text{J s}^{-1} = 1/4.186\ \text{cal s}^{-1}$).

Substituting I^2R for q and rearranging, Eq. [10] yields

$$K = 18.34 I^2 R/S \text{ W m}^{-1} \text{ K}^{-1}. \quad [11]$$

39–3.3 Method (De Vries, 1952b)

39–3.3.1 SPECIAL APPARATUS

1. Galvanometer: A galvanometer of sensitivity of 2 μV per mm deflection on a galvanometer scale at 1 m, is a suitable instrument. Thermocouple potentiometers that indicate temperatures to 0.02°C are suitable.
2. Dewar flask.
3. Sensitive Wheatstone bridge.
4. Stopwatch.
5. Six-volt storage battery.
6. Cylindrical probe: The components to be used in the construction of the probe are as follows: enameled constantan heating wire, 0.01 cm in diameter, of specific resistance about 0.63 ohms per cm, and about 20 cm long; glass capillary tube of about 0.04 cm outside diameter, 10 cm long; fine monel gauze; enameled copper and constantan thermocouple wire of 0.01 cm diameter; collodion for insulating; and paraffin wax.

   Fold the heating wire once, and introduce it into the glass capillary. Roll one layer of gauze along a metal wire so that it forms a cylinder of about 0.14 cm outer diameter. Because the individual wires of the gauze interlock closely at the seam, no soldering is necessary.

   Make a thermocouple by soldering the copper and constantan wires, and insulate the junction with a thin layer of collodion. Place the thermocouple junction and the glass capillary containing the heating wire into the monel gauze cylinder. Locate the thermocouple junction adjacent to the glass capillary midway between the ends of the probe. Fill the space remaining in the gauze cylinder with paraffin wax.

   Solder both ends of the heating wire to copper wires which lead to the power source. Place a plastic jacket over the end of the probe to protect the wires extending from the element.

   Make a second thermocouple, insulate it, and connect the constantan lead to the constantan lead of the couple in the probe. Connect the copper thermocouple leads to a galvanometer. Make the leads of the second couple of sufficient length that it can be placed in the soil at the same depth as the probe, or in an insulated flask located on the soil surface.

39–3.3.2 PROCEDURE

Excavate the soil to the desired depth of measurement. For horizontal positioning, bore a hole of the same diameter as the probe into the side of the excavation. Insert the probe. Locate the thermocouple cold junction at the same depth but at a sufficient horizontal distance ($\approx$15 cm) away from the probe so that it will not be influenced by the heating of the probe. Alternatively, place the cold junction above the soil in a Dewar flask filled with melting ice. Replace the excavated soil.

Measure the resistance R of the heating wire with a sensitive Wheatstone bridge. Connect the heater lead wires through a switch to a 6-V storage battery. Switch on the current, and measure the temperature rise $(T - T_0)$ with the galvanometer or portable thermocouple potentiometer. Read the time of each temperature measurement with a stop watch. After 3 min, switch off the heating current. Note the temperature decline as a function of time.

Plot the temperature rise $(T - T_0)$ as a function of time, and determine t_0 as previously described in section 39–3.2. Plot $(T - T_0)$ as a function of $\ln(t + t_0)$, and determine the slope of the line. For the cooling curve, plot $q/(4\pi K)[d + \ln(t + t_0)] - (T - T_0)$ as a function of $\ln(t - t_1 + t_0)$. Find the required value of $q/4\pi K)[d + \ln(t + t_0)]$ for this plot by extrapolating the line obtained for $t \leq t_1$. Again determine the slope. For both the heating and the cooling periods, the plots should yield straight lines of slope $q/4\pi K$.

39–3.3.3 CALCULATIONS

Average the measured slopes for the heating and cooling curves. Calculate the thermal conductivity using Eq. [11]. As an example, if the measured resistance of the probe is 1.26 ohms per cm, the current 0.03 A, and the slope of the temperature rise vs. the common logarithm of time is 0.065°C, then

$$K = 18.34(0.03)^2(1.26)/0.065 = 0.32 \text{ W m}^{-1} \text{ K}^{-1}.$$

39–3.3.4 COMMENTS

The construction of the probe requires careful workmanship. Although the monel gauze and the paraffin wax construction offer some rigidity, the probe is somewhat fragile. The advantages of having such a small probe are that it gives a good approximation of a line source of heat required by the theory, and that the dissipation of the heat is controlled largely by the soil and little by the probe. The metal of the gauze aids in providing a good thermal contact between the probe and the soil; however, care must be taken that the boring for the probe is not too large.

The possible contact resistance and the effect of the conductivity and heat capacity of the probe material were treated theoretically by De Vries and Peck (1958a). They concluded that in many cases the simple theory is sufficient.

The observed value of the time correction t_0 for this probe is about

−2 s. Accuracy of the conductivity values obtained by this method is ±5%.

39–4 THERMAL DIFFUSIVITY

39–4.1 Principles

Consider a sample of soil in a container having uniform cross section and insulated sides. The cross section is usually taken as a cylindrical or rectangular, but it may be of irregular shape. The temperature of the sample is initially uniform at ambient temperature T_0. At time $t = 0$, one end of the sample is brought into contact with a heat source of constant temperature T_s. A solution of Eq. [1], which relates the temperature T occurring in a plane a distance x from the heat source to the time t, is (Carslaw & Jaeger, 1959)

$$T - T_0 = (T_s - T_0)[1 - \operatorname{erf} x/(4kt)^{1/2}] \qquad [12]$$

where k is the thermal diffusivity of the soil and

$$\operatorname{erf} x/(4kt)^{1/2} = 2\pi^{-1/2} \int_0^{x/(4kt)1/2} \exp(-u^2) du.$$

The term erf is called the error function or the probability integral.

Equation [12] is applicable only if the soil sample is sufficiently long or if the time is sufficiently short that the temperature at the end of the sample opposite from the heated face is not changed during the experiment. This restriction is necessary because a semi-infinite medium is assumed in the derivation of the equation.

In practice T_s, T_0, and x are measured, heat is applied to one face of the sample, and the temperature T a distance x from the source is measured at one or more times. The term $1 - (T - T_0)/(T_s - T_0)$ is calculated for each time. The resulting numerical value is equal to the error function of argument $x/(4kt)^{1/2}$. A value of $x/(4kt)^{1/2}$ is obtained from tables of the error function in a manner similar to obtaining an antilogarithm from tables of logarithms. The diffusivity k can be calculated from this value, since x and t are known. A table of the error function (probability integral), sufficiently complete for this method, is given by Larsen (1958). Tables of the *normal* probability integral are not the same as tables of the probability integral; hence, it is necessary to use a conversion factor if these are to be used.

39–4.2 Method (Jackson, 1960)

39–4.2.1 SPECIAL APPARATUS

1. Stopwatch.
2. Sample container and heat exchanger: Exact dimensions of component

parts of the sample container and heat exchanger are not critical. The dimensions given below have proven satisfactory.

Construct the sample containers of plastic (a suitable plastic is polymethylmethacrylate, which is available under the tradenames of Plexiglas and Lucite) or other material having a low thermal conductivity. Cut cylindrical containers at least 10 cm long from plastic tubing 10 cm in diameter. If rectangular containers are desired, make them at least 10 by 10 by 10 cm. Cut a plastic plate, fit a gasket to it, and bolt it on one end of the tubing. The gasket makes the joint watertight. Bolt the plate on the tubing in such a manner that it can be removed to replace the soil sample.

Construct the heat exchanger on the other end of the tubing in the following manner. Cut a 1.3-cm piece of the 10-cm tubing. Cement six plastic lugs to the outside of the 1.3-cm piece and six lugs on the sample container. Align the lugs on the two pieces before cementing them in place. After the cement has hardened, drill a hole through the aligned lugs. This will allow the two pieces to be bolted together. Cut a thin copper plate and a plastic plate the same size as the container cross section. Install the copper plate between the sample container and the 1.3-cm piece of tubing. Bolt the pieces together. Make the joints water-tight by applying paraffin wax.

Drill two 1.3-cm holes in the plastic plate—one in the center, the other near the outside edge. Cement two pieces of 1.3-cm outside diameter plastic tubing about 5 cm long in the holes. These form the inlet and outlet for the heat source (water at constant temperature). On the inside edge of the plate, construct a baffle using strips of 1.3-cm plastic to form a spiral path from the center to the outside edge. Fit the completed plate on the 1.3-cm piece of 10-cm diameter tubing with the baffle on the inside touching the copper plate and with the small tubes protruding.

For the temperature-sensing device, drill two 0.25-cm holes in the side of the sample container at distances 1 and 2 cm from the copper plate. Install glass-coated thermistor rods 5 cm long in the holes. Veco 32A1 thermistors, available from the Victory Engineering Co., Union, NJ 07083, are suitable for this application. Seal the thermistors in the holes with paraffin wax. This construction lends rigidity to the thermistors and makes a watertight seal. The sensing points of the thermistors thus installed are located approximately in the center of the cross section of the container and about 1 and 2 cm from the copper plate.

3. Temperature-controlled water bath: Obtain a well-stirred, large-capacity (30- to 40-L) water bath. Install a pump in the bath to provide a means of circulating water through the heat exchanger. Control the water-bath temperature by use of a power-proportioning precision temperature controller. The "Electron-O-Therm Senior," available from the Bayley Instrument Co., P.O. Box 538, Danville, CA 94526, is a suitable instrument.

4. Temperature recorder: Record temperature with a recording bridge-type potentiometer. Endeavor to obtain precision of 0.01 °C.

39–4.2.2 PROCEDURE

Measure the distance from the copper plate to each thermistor with a cathetometer. Remove the bottom of the sample container and fill the container with soil. Take care to avoid disturbance of the thermistors, since the distance from the copper plate to the thermistors is critical. This distance may be checked upon completion of the measurement if care is exercised in removal of the soil.

After the soil has been packed into the container, insulate the sides with Styrofoam or other insulating material. (Styrofoam is a trade name for expanded polystyrene produced by the Dow Chemical Co. The thermal conductivity of Styrofoam is about 3.6×10^{-2} W m^{-1} K^{-1}.) Connect 1-cm inside diameter rubber tubing to the inlet and outlet of the heat exchanger. Connect the inlet tubing to the pump in the water bath, and put the outlet tubing in the bath for the return flow. Measure the initial temperature T_0 of the sample and the temperature T_s of the water bath. Refer to chapter 37 for a procedure for using and calibrating thermistors.

Begin a run by switching on the pump motor. Start a stopwatch the instant the heat exchanger is filled with water. Let the water circulate through the heat exchanger continuously during the measurement. Record the temperature T of each thermistor, and note the time the temperature is measured. Take the first temperature and time readings about 1 min after the pump motor is switched on. Take five readings of temperature and time for each thermistor at intervals of approximately 1 min.

39–4.2.3 CALCULATIONS

Use the measured values of T_0, T_s, x, and T at a specific t, and calculate $1 - (T - T_0)/(T_s - T_0)$. This is the numerical value of erf $x/(4kt)^{1/2}$. Consult a table of the error function to obtain the numerical value of $x/(4kt)^{1/2}$. Denote this numerical value by y, and calculate the diffusivity using the expression

$$k = x^2/(4ty^2). \qquad [13]$$

Calculate k for each set of measurements of T and t for each thermistor. Average the values of k obtained.

As an example, let $T_s = 34.00°C$, $T_0 = 24.00°C$, $x = 1.00$ cm, $t = 300$ s, and $T = 28.5°C$. From Eq. [12]

$$\text{erf}\, x/(4kt)^{1/2} = 1.00 - (28.50 - 24.00)/(34.00 - 24.00) = 0.5500.$$

In tables of the error function, the number 0.5500 lies between 0.54987 and 0.55071. Values of y corresponding to the latter two numbers are 0.534 and 0.535, respectively. Linear interpolation yields a value of y = 0.5342. The thermal diffusivity is, using Eq. [13],

$$k = (1)^2/[(4)(300)(0.5342)^2] = 0.00292 \text{ cm}^2 \text{ s}^{-1} = 2.92 \times 10^{-7} \text{m}^2 \text{ s}^{-1}.$$

39–4.2.4 COMMENTS

The sample container should be positioned in such a way that the best possible contact is maintained between the soil and the copper plate. This may be accomplished by placing the container so that heat flow is either horizontal or vertical with the heat exchanger on the bottom. If the heat flow is vertical, the heat flow direction should be downward; that is, the water bath temperature T_s is less than ambient temperature T_0. The terms $(T - T_0)$ and $(T_s - T_0)$ in Eq. [12] are therefore negative, but their ratio is positive, and the equation is not changed. Heat flow downward is preferred to reduce the effects of convection. In general, it is best to make the temperature-time determinations when the ratio $(T - T_0)/(T_s - T_0)$ is within the range of 0.1 to 0.6.

With adequate temperature-indicating equipment, using two thermistors, and making five determinations per thermistor, results can be obtained with a coefficient of variability <3% for dry soils. Greater variabilities are to be expected for moist soils.

The size of the sample is not critical. The cross section should be large enough that edge effects are negligible. Even with insulated sides, it is desirable to have the temperature-sensing devices at least twice as far from the edge as from the hot face. The container needs only to be sufficiently long that the temperature at the end does not change during the experiment.

If the copper plate used in the construction of the heat exchanger and the container is very thin, it may bend when water is turned into the heat exchanger. The bending can be reduced by soldering a bolt to the plate near its center and bringing the bolt out through the plastic cover plate.

For best results, the apparatus should be kept in a constant-temperature room.

39–5 REFERENCES

Bayer, L. D. 1956. Soil physics. 3rd ed. p. 370–379. John Wiley and Sons, New York.

Carslaw, H. S., and J. C. Jaeger. 1959. Conduction of heat in solids. 2nd ed. Oxford University Press, London.

Crawford, C. B. 1952. Soil temperature, a review of published reports. Highway Res. Board Spec. Rep. 2:19–41.

De Vries, D. A. 1950. Some remarks on heat transfer by vapour movement in soils. Trans. Int. Congr. Soil Sci., 4th 2:38–41.

De Vries, D. A. 1952a. The thermal conductivity of granular materials. Inst. Int. du Froid, Bul., Annexe 1952-1. p. 115-131.

De Vries, D. A. 1952b. A nonstationary method for determining thermal conductivity of soil *in situ*. Soil Sci. 73:83-89.

De Vries, D. A., and A. J. Peck. 1958a. On the cylindrical probe method of measuring thermal conductivity with special references to soils: I. Extension of theory and discussion of probe characteristics. Aust. J. Phy. 11:255-271.

De Vries, D. A., and A. J. Peck. 1958b. On the cylindrical probe method of measuring thermal conductivity with special reference to soils: II. Analysis of moisture effects. Aust. J. Phys. 11:409-423.

Jackson, R. D. 1960. The importance of convection as a heat transfer mechanism in two-phase porous materials. Ph.D. diss, Colorado State Univ., Fort Collins, CO.

Jahnke, E., and F. Emde. 1945. Tables of functions. p. 6-8. Dover Publications, New York.

Jakob, M. 1949. Heat transfer. John Wiley and Sons, New York.

Kersten, M. S. 1949. Thermal properties of soils. Univ. of Minnesota Eng. Exp. Stn. Bull. 28.

Lachenbruch, A. H. 1957. A probe for measurement of thermal conductivity of frozen soils in place. Trans. Am. Geophys. Union 38:691-697.

Larsen, H. D. 1958. Rinehart mathematical tables, formulas, and curves. p. 162-165. Rinehart and Co., New York.

Patten, H. E. 1909. Heat transference in soils. USDA. Bur. Soils Bull. 59. U. S. Government Printing Office, Washington, DC.

Richards, S. J., R. M. Hagan, and T. M. McCalla. 1952. Soil temperature and plant growth. *In* B. T. Shaw (ed). Soil physical conditions and plant growth. Agronomy 2:304-336.

Smith, W. O. 1940. Thermal conductivities in moist soils. Soil Sci. Soc. Am. Proc. (1939) 4:32-40.

Smith, W. O., and H. G. Byers. 1939. The thermal conductivity of dry soils of certain of the great soil groups. Soil Sci. Soc. Am. Proc. 3:13-19.

Waddams, A. L. 1944. The flow of heat through granular material. J. Soc. Chem. Ind. 63:337-340.

Woodside, W. 1958. Probe for thermal conductivity measurement of dry and moist materials. Am. Soc. Heating and Air-Conditioning Eng. J. Sect., Heating, Piping and Air Conditioning. p. 163-170. September.

40 Heat Flux

M. FUCHS

Agricultural Research Organization
Institute of Soils and Water
Bet Dagan, Israel

40-1 INTRODUCTION

The soil heat-flux density, i.e., the amount of heat flowing in the soil per unit area per unit time, determines the thermal environment of the soil. As a component of the ground energy budget it depends mainly on the exposure of the soil surface to solar radiation and its radiative properties. The partition of net radiation between heat flow into the air and into the soil is governed by the energy transport properties at the soil-atmosphere interface (Lettau, 1951). Strong gradients and considerable heterogeneity of the thermal properties of the soil and air layer near the ground complicate the description of the physical mechanisms ruling this partition and make it difficult to predict soil heat flow accurately from Fourier's law of heat conduction, the equation of continuity, and the forcing function of the radiant energy input. For many studies, therefore, soil heat flux must be determined by measurement.

The porous structure of the soil allows heat to move through a multiple network of pathways. Each pathway has its own mode of transport and its own thermal properties. Heat is conducted through the solid phase at a rate depending on the mineralogical and organic composition of the soil. Conduction occurs in the liquid water films surrounding the soil particles. The low thermal conductivity of air impedes heat transport through the air-filled pores, but here conduction is paralleled by radiative exchange and free convection.

Variation in time and space of the water content results in the most significant changes of the thermal properties of any given soil (de Vries, 1963). As water flows through nonisothermal soil, the equilibrium between vapor, liquid, and ice phases is modified along the temperature gradient. The equilibrium water vapor pressure is higher in the warmer regions of the soil, allowing evaporation and melting, which absorb latent heat. As water flows to the colder regions where the equilibrium vapor pressure is lower, condensation and freezing release latent heat. This mechanism parallels the conduction of sensible heat (de Vries, 1958). The magnitude of latent heat transport depends upon the absolute temperature, the temperature gradient, and the hydraulic properties of the soil. In addition, water flow contributes a convective component to heat transport.

Climate-driven soil heat flux is cyclic. The diurnal and annual periods are dominant, but other fluctuations caused by irregularities of the weather may modify heat propagation in the soil. During days when incoming solar radiation reaches 1000 W m^{-2}, Fuchs and Hadas (1972) have measured maximum soil heat flux densities in excess of 300 W m^{-2} on a bare soil surface. But the amplitude of diurnal soil heat flux fluctuations decreases sharply with depth. At 0.6 m in a moist sandy loam, it is attenuated to 1% of the surface value.

The multiplicity of methods for measuring soil heat flux density is dictated by the multiple forms of the process. The choice of a method should therefore be determined by the nature of the heat transport, the conditions of the soil, and the accuracy required. Some measurement procedures make use of the cyclic nature of soil heat flux. Clearly such measurements will ignore the steady-state component of heat flux, which occurs often in controlled environments or when the soil is used as a heat exchanger. Convenience of work and ease of processing should also be considered for long-term and routine data collection.

40–2 CALORIMETRIC METHOD

40–2.1 Principles

The calorimetric evaluation of the soil heat flux density is based on the heat balance of a finite layer of soil:

$$G_{i-1} = \delta z_i C_i \,(\partial T_i/\partial t) + G_i \qquad [1]$$

this equation being written for the ith layer of soil beneath the surface. The thickness of the layer, δz, should be selected to be sufficiently small so that C, the heat capacity of the soil, is constant across it. $\partial T/\partial t$ is the rate of change of the mean temperature of the layer and G_i is the heat flux density at the bottom of the layer.

If we can ascertain the presence of an nth layer in the profile for which the temperature gradient, $\partial T/\partial z$, is zero, i.e.,

$$G_n = 0 \qquad [2]$$

then the soil heat flux density at level j is given by

$$G_{j-1} = \sum_{i=j}^{n} \delta z_i C_i \;\; (\partial T_i/\partial t)\,. \qquad [3]$$

when $j = 1$, Eq. [3] yields G_0, the soil heat flux density at the surface.

Steady-state heat flux density occurs without temperature change, and therefore cannot be measured by this method. Furthermore, as energy is conserved throughout the profile, $\partial T/\partial z$ is never zero.

40–2.2 Procedure

The data needed for computing the soil heat flux density using Eq. [3] are soil temperature and heat capacity, whose measurements are explained in chapters 37 and 38. The temperature of each layer has to be measured as a function of time in order to determine $\partial T/\partial t$. There must also be enough layers (sufficient depth) to find the level n where $\partial T/\partial z = 0$. The heat capacity depends upon the composition of the soil, the bulk density, and the water content. Its value must be determined for each layer and updated to account for the change of water content with time. The largest vertical variation of the soil heat capacity occurs near the surface and necessitates temperature sampling in thin layers. In homogeneous soils, the thickness of the layers may increase in arbitrary increments proportional to depth. In layered soils, the thicknesses should be adjusted according to the distribution of horizons.

40–2.3 Apparatus

The instrumentation to measure soil temperature is simple. Nevertheless, data for heat flux calculation must satisfy certain conditions. Temperature gradients are steepest near the surface, where changes with time and depth are most rapid. As mean layer temperature change is required in Eq. [3], Suomi (1957) and Tanner (1958) proposed the use of large thermometers to directly average the temperature across the layer. However, this averaging procedure distorts considerably non-linear soil temperature profiles, and causes very serious errors in the location of the depth where $\partial T/\partial z = 0$. It is preferable to measure temperature with miniature sensors at precisely located depths. Thermometers with diameter 1 mm or less should be used above 100-mm depth.

Fine-gauge thermocouple wires and microbead thermistors have suitable dimensions. Both types of sensors have low precision, thermocouples because of microvolt range sensitivity, and thermistors because of curvilinear and dispersed electrical properties. Typical maximum rates of temperature changes in the upper layers are of the order of 0.001 K s^{-1}, so that a precision of 0.1 K determines the mean hourly value of heat storage to within 3%. Sensing the temperature with a thermopile consisting of 5 to 10 thermocouples connected in series provides the desired precision on standard recording equipment. As thermopile output indicates the temperature difference between hot and cold junctions, one group of junctions is used as sensor, the other serves as reference. It is recommended to bury the reference junctions at a depth between 100 and 200 mm, where a large precision thermometer can be installed to measure the reference temperature. Deeper placement of the reference junctions is not desirable because temperature differences > 20 K may occur, in which case resolution of the recording equipment must exceed 0.5% to yield 0.1 K precision. Distribution of the measuring junctions in a horizontal plane provides a means of averaging out the small-scale spatial variability of temperature.

Microbead thermistors have small size and high sensitivity. They provide absolute temperature readings. But the precision of 0.1 K can be achieved only through individual calibration of each thermistor used, over the entire range of expected soil temperatures.

In the deeper layers of the soil, temperature gradients are weak, allowing the use of larger sensing elements. The rate of temperature change also decreases with depth. The attentuation of the diurnal fluctuation is typically by a factor of 10 per layer of 0.3 m. Consequently, the maximum rate of temperature change to be expected at a depth of 0.3 m is approximately 0.4 K/hour. At this depth the amplitude of the diurnal heat flux density is 1/10 of the value at the surface. Hence, a 10% error in the heat storage of the layers below 0.3 m will contribute only a 1% error in the surface soil heat flux density. The resulting thermometric precision required for hourly determinations of the soil heat flux density is 0.04 K. Metal resistance thermometers with adjustable resistance bridges are recommended for this application.

It should be emphasized that the exact soil temperature with reference to the absolute temperature scale is of little importance. The accuracy requirement is for the difference between consecutive hourly measurements of the temperature. Therefore, resolution and stability are the important specifications in the choice of the thermometers.

40–2.4 Comments

The calorimetric method uses heat capacity and temperature change to measure sensible heat flux. Latent heat flow can theoretically be accounted for by using an apparent heat capacity which has a temperature-dependent term proportional to the latent heat change. Heat of vaporization can be neglected because of the small mass of water in vapor phase (Fuchs et al., 1978). When latent heat of fusion is involved, temperature changes become very slow and cannot be detected with usual field instrumentation.

The depth of the layer for which $\partial T/\partial z = 0$ sets the limit of summation in Eq. [3]. This layer moves during the day from the surface down to approximately 0.7 m deep. Soil temperature profile measurements must extend to that depth. As the maximum contribution of each layer of soil to the heat flux density decreases nearly logarithmically with depth, it is recommended to place thermometers at 5, 10, 20, 40, 80, 160, 320, and 640 mm below the soil surface.

In order to evaluate the heat storage with the same error everywhere in the profile, thermometric precision must be proportional to layer thickness. Consider, for example, a thermometer distribution in the soil profile as given above, and a uniform heat capacity of 2×10^6 J m^{-3} K^{-1}. A precision of 0.1 K down to a depth of 80 mm confines errors of the average hourly soil heat flux density to $<$ 2 W m^{-2}. Below 80 mm, measurements to within 0.04 K at 160 mm and 0.02 K for the deeper layers are required to obtain 2 W m^{-2} accuracy.

The stringent requirements of precision and the large number of measurements needed impose practical limitations on the use of the calorimetric method. Its main purpose is to serve as a reference for checking other methods. Even so, it is restricted to diurnal and higher-frequency heat flux density variations. Variations with longer periodicities penetrate deeper in the soil. The depth of 99% attentuation of the annual cycle is around 12 m, where temperature changes are of the order of 0.001 K per day. Application of the calorimetric method to measure the annual cycle is therefore prohibitively expensive in equipment and data processing.

40–3 THE GRADIENT METHOD

40–3.1 Principles

The gradient method for measuring the heat flux density, G, is a direct application of Fourier's law for heat conduction:

$$G = \lambda \, \partial T/\partial z \qquad [4]$$

where λ is the thermal conductivity and $\partial T/\partial z$ is the temperature gradient in the soil.

Soil temperature gradients can be accurately measured especially in the upper layers of the soil. But the thermal conductivity of any particular soil varies considerably as a function of moisture content and temperature. De Vries (1963) has shown experimentally and by computation that the thermal conductivity of a quartz sand could vary from 0.25 to 1.9 W m^{-1} K^{-1} at 20°C, to 2.5 W m^{-1} K^{-1} at 60°C, for changes of volumetric water fractions from 0 to 0.1. Riha et al. (1980) reported similar data for sand and a thermal conductivity of silt loam from 0.1 to 1.0 W m^{-1} K^{-1} for moisture content between 0 and 0.55.

The effect of water content is due to the large difference between the thermal conduction of air and of water, which are 0.025 and 0.57 W m^{-1} K^{-1}, respectively. Furthermore, the apparent thermal conductivity of air-filled pores increases as a result of latent heat transport in water vapor flow driven by the temperature gradient. The contribution of latent heat increases nearly exponentially with temperature and soil water potential. Soil moisture gradients (of opposite sign to temperature gradients) in drying soils reduce or cancel the contribution of latent heat to conduction (De Vries, 1958; Kimball et al., 1976b). Latent heat of fusion transport also enhances thermal conduction in soils. But it is significant over a temperature band $<$ 0.1 K wide near 273 K (Fuchs et al., 1978).

40–3.2 Procedure

The main difficulty associated with the gradient method for determining soil heat flux density is the variability of thermal conductivity. The method relies very heavily on the feasibility of in situ determinations of thermal conductivity. Measurements of this property are reviewed in chapter 39. The procedure most relevant to soil heat flux measurements is based on the transient dissipation of a heat source inserted in the soil (van Wijk & Belghith, 1966; Fritton et al., 1974; Riha et al., 1980).

The temperature gradient may be determined by differentiating a smooth curve fitted to a temperature profile with respect to depth (Kimball, 1976). A simpler alternative is to measure the average temperature difference across a layer of soil at the desired depth.

40–3.3 Apparatus

Specifications for thermometer size and accuracy are identical to those described in the calorimetric method. Precision requirements are more easily met because all temperature measurements can be made truly differential. Thermocouples and thermopiles are very well suited for this purpose.

40–3.4 Comments

The gradient method is applicable to transient and steady-state heat flow. It can determine instantaneous heat flux density. Time averages can be obtained by time-scaled integration of temperature gradients over intervals for which the thermal conductivity is constant.

The accuracy of the method depends largely on the reliability of the thermal conductivity data. Because of the large variations in conductivity caused by water content and temperature changes, the application of the method near the soil surface is not practical. For example, the thermal conductivity of a silt loam at a depth of 40 mm and an average moisture content of 0.15 has a diurnal cyclic variation between 0.59 and 0.31 W m^{-1} K^{-1} (Hadas, 1977). The minimum depth at which the method can be reliably used will vary from soil to soil, but in general it will be around 200 mm.

40–4 THE COMBINATION METHOD

40–4.1 Principles

Diurnal variation of heat storage in the layer above 200 mm may amount to approximately 75% of the soil heat flux density. As heat storage is measured with best accuracy in the upper layers of the soil, and the gradient method is reliable below 200 mm, the combination of the calorimetric and gradient measurement is recommended for determining heat flux at the soil surface.

The combination method derives directly from Eq. [1] with $i = 1$ and $\delta z = 200$ mm.

40–4.2 Apparatus and Procedure

Specifications of soil thermometers used in the calorimetric and gradient methods also apply for the combination method. As heat storage is measured only in the upper 200-mm layer, a resolution of 0.1 K on the soil temperature reading is sufficient.

The temperature gradient at 200 mm can be evaluated by a direct temperature difference measurement between thermometers located at 150- and 250-mm depths. An error of 0.1 K on the temperature difference measurement does not contribute an error of > 1 W m^{-2} on the evaluation of the heat flux density.

40–4.3 Comments

An interesting variant of the combined calorimetric-gradient approach is the so-called null-alignment method (Kimball & Jackson, 1975), which uses the fact that the temperature gradient is zero somewhere above 200 mm in the early morning and in the late afternoon. The change of heat storage in the soil layer between the depth at which the temperature gradient is zero and the 200-mm depth is then used to determine the heat flux density at 200-mm depth according to Eq. [3]. For example, consider a soil in the warming phase of the diurnal cycle. Assume that at 0800 h the temperature gradient is zero at a depth of 70 mm, and that at 0830 h the zero gradient is found at 80 mm. By linear interpolation, the zero gradient at time 0815 h is 75 mm below the surface. The change in temperature that has occurred from 0800 to 0830 h in the layer between 75 and 200 mm multiplied by the heat capacity of the layer, divided by the time elapsed, 1800 s, is the heat flux density at the 200-mm depth. This datum is combined with a concurrent measurement of the temperature gradient across the 200-mm depth to find the thermal conductivity which, at this depth, remains fairly constant throughout the day. Measurements of the temperature gradient can now be used to determine the heat flux density at 200 mm any time of the day. The soil heat flux density at the surface is obtained by summing Eq. [1] for the soil layers above 200 mm. The procedure simplifies considerably the instrumentation because heat storage needs to be determined in the upper layers of the soil only. The null-alignment technique relies for its accuracy on the exact positioning of the level where $\partial T/\partial z$ is zero. This level is found by numerical smoothing of the soil temperature profile (Kimball, 1976) and iterative interpolation.

40–5 SOIL HEAT FLUX PLATE METHOD

40–5.1 Theoretical Basis

A heat-flux plate is a device that senses the flow of heat. It is made of a flat piece of solid material with rigid shape and constant thermal conductivity. The two large sides of the plate are fitted with temperature sensors in a configuration suitable for differential readings. The plate is inserted in the soil with its large sides in the plane perpendicular to the direction of heat flow.

The flow of heat in the soil produces a temperature gradient across the plate. The output so generated is directly proportional to the heat flux density through the plate, G_m which is also related to the heat flux density in the soil, G. Philip (1961) has shown that the ratio of heat flow in the soil to that through the plate is given as

$$G/G_m = 1 - \alpha r (1 - \lambda/\lambda_m) \quad [5]$$

where λ is the thermal conductivity of the soil, λ_m is the thermal conductivity of the plate, r is the ratio of plate thickness to square root of the area facing the heat flow, α is an empirical factor depending on the shape of the plate (approximately 1.7 for a rectangular plate). Variations of the soil thermal conductivity modify the relation between the output of the plate and the heat flux density in the soil. The design of the heat-flux plate should seek to reduce this effect. This can be achieved by diminishing the relative thickness of the plate. Increasing the thermal conductivity of the plate also lessens the effects of fluctuations in soil thermal conductivity. In practice, heat conduction through the plate is impeded by the contact between the soil and the plate, because heat conduction is only through the areas of the plate directly touching the soil matrix (Fuchs & Hadas, 1973). The contact resistance decreases the effective thermal conductivity of the plate. It depends upon the texture and structure of the soil, but can be reduced if the thermal admittance of the plate is large. The thermal admittance is defined as $(\lambda_m C_m)^{1/2}$, where C_m is the heat capacity of the plate. A high thermal admittance of the plate increases heat flow across contact areas and evens the temperature field on the face of the plate.

40–5.2 Apparatus

Thermopiles are most often used to generate the signal of heat-flux plates. They can be made very thin by vacuum deposition of metallic films, or by a combination of wire winding and electroplating. The junctions are mounted symmetrically on the two sides of a flat support of rigid material with high dielectric constant (plastic polymer, ceramic, or glass). An outer shield protects the junctions against corrosion and provides electrical insulation. Contact with the soil occurs through the shield, which should have large thermal admittance. For this reason, several designs of heat-flux plates have a metallic outer shield. Tanner (1963) recommends the use of anodized aluminum sheets. Aluminum has very high thermal conductivity. Anodization produces an aluminum oxide layer < 10 μm thick, which provides adequate electrical insulation and excellent resistance to the corrosive agents of the soil. Tanner (1963) has given a detailed description of the construction of a heat-flux plate using a glass microscope slide as a support for a copper-constantan thermopile.

Heat-flux plates are impervious to water flow. Water vapor may form or condense on them and thereby modify the heat flux. This source of error can be made small by reducing the size of the plate. Concurrently, the plate must be very thin to conform with the requirement that the ratio, *r*, be small. Plates 3 to 4 mm thick, with a face area around 2500 mm^2—*i.e.*, with $r < 0.1$—provide adequate accuracy for most applications.

Manufacturers of meteorological instruments offer a variety of soil heat-flux plates. The performance of these plates can be evaluated on the basis of Eq. [5]. However, it is recommended to check their calibration in the soil in which they will be used.

40–5.3 Calibration Procedure

The uncertainty regarding the thermal resistance at the contact between heat-flux plate and soil, and the empirical parameter in Eq. [5] necessitate experimental determination of the relation between plate signal and soil flux density. As thermal contact depends on texture and compaction of the soil, plates should be calibrated in a medium similar to the soil in which they will be used. Calibration procedures which do not take into account the problem of heat transfer between the soil and the plate are usually inadequate.

Ideally, heat flux plates should be calibrated in situ. For field conditions this can be done by comparison with the calorimetric method (Hogstrom, 1974) or the gradient method (Kimball et al., 1976a). It has also been done by placing a large plane heater of known power output over the soil surface (Blackwell & Tyldesley, 1965).

Laboratory calibration procedure using soil as the heat transfer medium requires a regulated heat supply and a controllable heat sink to produce a known steady-state heat flux density. The soil used must be dry. Otherwise the temperature gradient induces water flow and stratifies the thermal properties of the calibration medium.

The most suitable heat source is a flat electrical heater connected to a stabilized power supply. The current and voltage are measured to determine the power output. All the energy generated by the heater should dissipate through the soil in the direction perpendicular to the heat-flux plate. Heat losses in the other directions must be prevented. Hatfield and Wilkins (1950) used thermostated auxiliary heaters to maintain isotherm conditions through the bottom and side walls of a soil container. In this fashion all the energy produced by the main heater is channeled through the top of the container. Biscoe et al. (1977) considerably simplified the procedure by using the zero output at zero heat flux of an uncalibrated plate to control the operation of the auxiliary heaters. An even simpler calibration device consists of a large flat heater in the central large plane of symmetry of a flat soil container (Fuchs & Tanner, 1968; Fuchs & Hadas, 1973). Both sides of the heater are packed evenly with soil. The heat-flux plates are inserted in the middle of the soil layers near the center of the heater. Symmetry assures that heat dissipates equally through both sides of the heater. Thermal insulation reduces heat losses through the small edges of the container.

40–5.4 Comments

The theory of heat flow disturbance caused by local heterogeneity (Philip, 1961) indicates how design and calibration can improve the performance of heat flux plates. However, as the variability of the soil thermal properties remains the major source of error, the validity of soil heat-flux plate measurements must be checked in the field. Biscoe et al. (1977) compared calibrated plate measurements with heat flux densities obtained as the residual term of the energy balance. The mean relative error was approximately 5%, indicating that the calibration procedure of the plates was correct. Individual comparisons, however, were widely dispersed, because of inaccuracies of the energy balance measurements and variability of plate response.

Fuchs and Hadas (1972) verified heat-flux plate calibration by measuring the heat flux profile in the soil. The resulting heat flux divergence matched closely the heat storage in the soil layer between the plates, in both wet and dry soil. Kimball et al. (1976a) compared surface soil heat flux densities measured by combined calorimetric and various conduction methods, including heat-flux plates at a depth of 50 mm. Heat-flux plate and null alignment differed on the average by 10% at midday, but were within the range of values obtained by the other methods.

Field and laboratory tests confirm that correct design can improve the reliability of heat-flux plates (Fuchs & Hadas, 1973). The calibration of a plate making contact with the soil through a sheet of anodized aluminum changes by $< 8\%$ over a wide range of soil texture and moisture content. For the same range of soil conditions, the calibration of a poorly conductive plate varies by 22%.

Heat-flux plates respond only to sensible heat flow. In drying soils they have to be placed below the level at which liquid water is vaporized. Otherwise they measure the total downward heat flowing through the upper dry layer, but do not sense the latent heat flow transported upward by the water vapor. Burying the plate at a depth of 50 mm is satisfactory because evaporation from a soil covered by a dry layer 50 mm thick is negligible. The contribution of this upper layer to the surface soil heat flux density can be measured using the calorimetric method.

40–6 FURTHER COMMENTS

Concepts underlying the measurement of soil heat flux density are simple. The basic instrument used is a thermometer whose specifications depend upon the method of measurement and the nature of the soil.

The calorimetric method requires intensive soil temperature profile sampling and miniature sensors in the upper layers of the soil. Deeper in the soil, absolute temperature must be measured with high precision. Large amounts of data have to be collected and processed. Concurrent measurements of the moisture content are also needed to compute heat capacity. Steady-state heat flow cannot be detected.

The gradient method can be used below 200 mm, where variations of moisture content and temperature are sufficiently small not to affect the soil thermal conductivity which can either be measured in situ or evaluated from theoretical considerations. The method is most readily applied to slowly varying or steady-state heat flow. In the case of fluctuating heat flow, it must be combined with the calorimetric method to account for heat storage in the top soil layer. As the calorimetric measurement of the upper layers of the soil can be simplified, the combined method is often favored.

The heat-flux plate method is the easiest to use in the field, as it provides a direct reading of the heat flux density. Plates can be installed near the surface, leaving heat storage evaluation to the thin upper soil layer only. But heat flux plates are specialized sensors subject to errors of utilization. Calibration must be performed in a medium with physical and thermal properties similar to those of the soil in which the plates are used. As few auxiliary measurements are required, the method is well suited for long-term determinations of the soil heat flux density.

40–7 REFERENCES

Biscoe, P. V., R. A. Saffell, and P. D. Smith. 1977. An apparatus for calibrating soil heat flux plates. Agric. Meteorol. 18:49–54.

Blackwell, M. J., and J. B. Tyldesley. 1965. Measurements of natural evaporation: comparison of gravimetric and aerodynamic methods. p. 141–148. *In* F. E. Eckardt (ed.) Methodology of plant eco-physiology. Proc. Montpellier Symp. UNESCO, Paris.

De Vries, D. A. 1958. Simultaneous transfer of heat and moisture in porous media. Trans. Am. Geophys. Union 39:909–916.

De Vries, D. A. 1963. Thermal properties of soils. p. 210–235. *In* W. R. van Wijk (ed.) Physics of plant environment. North-Holland Publishing Co., Amsterdam.

Fritton, D. D., W. J. Busscher, and J. E. Alpert. 1974. An inexpensive but durable thermal conductivity probe for field use. Soil Sci. Soc. Am. Proc. 38:854–855.

Fuchs, M., G. S. Campbell, and R. I. Papendick. 1978. An analysis of sensible and latent heat flow in a partially frozen unsaturated soil. Soil Sci. Soc. Am. J. 42:379–385.

Fuchs, M., and A. Hadas. 1972. The heat flux density in a non-homogeneous bare loessial soil. Boundary-layer Meteorol. 3:191–200.

Fuchs, M., and A. Hadas. 1973. Analysis of the performance of an improved soil heat flux transducer. Soil Sci. Soc. Am. Proc. 37:173–175.

Fuchs, M., and C. B. Tanner. 1968. Calibration and field test of soil heat flux plates. Soil Sci. Soc. Am. Proc. 32:326–328.

Hadas, A. 1977. Evaluation of theoretically predicted thermal conductivities of soils under field and laboratory conditions. Soil Sci. Soc. Am. J. 41:460–466.

Hatfield, H. S., and F. J. Wilkins. 1950. A new heat flow meter. J. Sci. Instr. 27:1–3.

Hogstrom, U. 1974. *In situ* calibration of ground heat flux plates. Agric. Meteorol. 13:161–168.

Kimball, B. A. 1976. Smoothing data with cubic splines. Agron. J. 68:126–129.

Kimball, B. A., and R. D. Jackson. 1975. Soil heat flux determination: a null-alignment method. Agric. Meteorol. 15:1–9.

Kimball, B. A., R. D. Jackson, F. S. Nakayama, S. B. Idso, and R. J. Reginato. 1976a. Soil heat flux determinations: temperature gradient method with computed thermal conductivity. Soil Sci. Soc. Am. J. 40:25–28.

Kimball, B. A., R. D. Jackson, R. J. Reginato, F. S. Nakayama, and S. B. Idso. 1976b. Comparison of field-measured and calculated soil-heat fluxes. Soil Sci. Soc. Am. J. 40:18–25.

Lettau, H. H. 1951. Theory of surface-temperature and heat-transfer oscillations near a level ground surface. Trans. Am. Geophys. Union 32:189–200.

Philip, J. R. 1961. The theory of heat flux meters. J. Geophys. Res. 66:571–579.

Riha, S. J., K. J. McInnes, S. W. Childs, and G. S. Campbell. 1980. A finite element calculation for determining thermal conductivity. Soil Sci. Soc. Am. J. 44:1323–1325.

Suomi, V. E. 1957. Soil temperature integrators. p. 1–24. *In* H. H. Lettau and B. Davidson (ed.) Exploring the atmosphere's first mile. Pergamon Press, New York.

Tanner, C. B. 1958. Soil thermometer giving the average temperature of several locations in a single reading. Agron. J. 50:384–387.

Tanner, C. B. 1963. Basic instrumentation and measurements for plant environment and micrometeorology. Univ. of Wisconsin, Dep. of Soil Sci. Soils Bull. 6.

van Wijk, W. R., and A. Belghith. 1966. Determination of thermal conductivity, heat capacity and heat flux density in soils by non-stationary methods. p. 335–361. *In* W. E. Sopper and H. W. Lull (ed.) Int. Symp. forest hydrology. Pergamon Press, New York.

41 Heat of Immersion

D. M. ANDERSON
Texas A&M University
College Station, Texas

41-1 INTRODUCTION

Measurement of the heat of wetting (immersion) was first applied to the study of soils by Mitscherlich (1899), who apparently hoped by this means to characterize the colloidal fraction. Janert (1934) later offered an improved apparatus and procedure, and contributed a number of measurements. Since that time many others have measured heats of wetting of soils and soil constituents, but the characterization of soils on this basis has not been widely adopted, principally because it offers no clear-cut advantage over other more firmly established criteria.

In contrast with other fields in which calorimetry has been fruitful, the necessary supplemental and complementary work has not been advanced sufficiently in soil science to permit realization of the full potential of calorimetric studies of soils. This situation is changing rapidly, however. Among the areas where the determination of heats of immersion may be expected to be particularly revealing are in studies of the following: energetics of ion-exchange phenomena, thermodynamics of soil-water systems, wettability and surface reactivity of soils, energetics of soil-water-air interfaces, and polarizability and topography of surface-active sites of the individual soil constituents. This is in addition to the tabulation of the integral and differential heats of immersion of various soils and their constituents.

Since the heat of immersion is only one of several possible calorimetric measurements in the thermochemical study of soils, it is necessary to distinguish among several heat effects. The most common is the integral heat of immersion, more widely known as the heat of wetting. It is simply the heat evolved on immersing a soil, at a known initial water content (usually oven-dry), in a large volume of water. The differential heat of immersion is more useful in the detailed study of surfaces and interfaces than is the integral heat of immersion. The differential heat of immersion is defined as the ratio of the increment of heat evolved ∂q to the increment of water ∂m added and uniformly distributed at constant temperature throughout an amount of soil sufficiently large that the added water does not change the soil water content appreciably. This concept is identical with that of partial molar and partial specific quantities developed in chemical thermodynamics. Therefore, the differential heat of immersion may also be called the partial molar or partial specific heat of immersion; it is symbolized by $(\partial q/\partial m)_T$ and is expressed in units of energy per unit mass of water. The physical significance of the differential heat of immersion is best comprehended by studying the diagram in Fig. 41–1. From the definition and the diagram, it is apparent that the differential heat of immersion aι any given water content is simply the slope, at that point, of the line resulting from a plot of the integral heat of immersion as a function of the initial water content of the sample. A fuller discussion has been given by Edlefsen and Anderson (1943).

There are two other heat effects analogous to the integral and differential heats of immersion; they are the integral and differential heats of adsorption. In principle, they may be regarded simply as the sum of the heat of condensation and the appropriate heat of immersion (Edlefsen & Anderson, 1943; Brunauer, 1945). The integral heat of adsorption usually is obtained, by means of the Clausius-Clapeyron equation, from adsorption isotherms. When obtained in this way, it is called the *isosteric heat of adsorption*. In this context, the term isosteric means equal surface coverage and refers to the fact that the isosteric heat of adsorption is evaluated by comparing points at equal amounts of water adsorbed on two isotherms obtained at different temperatures. Several different definitions have been adopted for the differential heat of adsorption, however; and since there is not unanimity on this point, each investigator should give due consideration to possible choices (Brunauer, 1945; Hill, 1948; Adamson, 1960).

Sample preparation is of critical importance in determining the heat

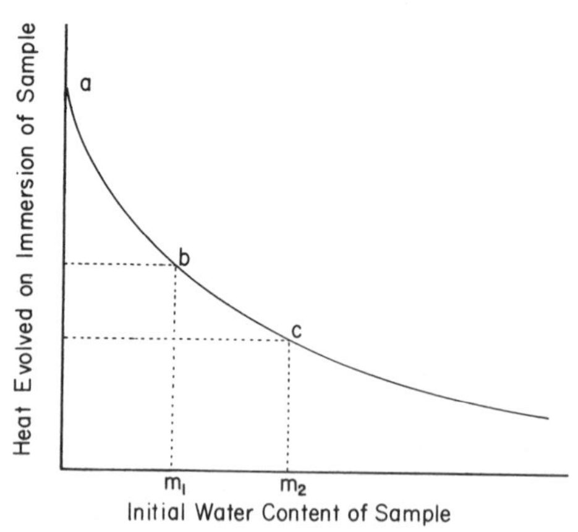

Fig. 41–1. Integral heat of immersion of soil in water as a function of initial water content. The ordinate at a gives the heat evolved on immersing and thoroughly dispersing a perfectly dry sample in a large container of water. Points b and c on the ordinate give the heat evolved when samples having initial water contents m_1 and m_2, respectively, are similarly immersed. The difference in ordinates corresponding to b and c gives the heat evolved in changing the water content of the soil from m_1 to m_2.

of immersion. Attempts to utilize the heat of immersion as a general criterion for characterizing soils have been complicated greatly by this fact. But dependence of heats of immersion on sample condition and preparation should be regarded as a hitherto neglected asset that can be of value in obtaining information more specific than a mere characterization. The heat of immersion evidently is sensitive to, and therefore reveals, properties of the sample which are not as yet understood or are not completely interpretable, but may be of considerable importance. Strict attention to the chemical and physical state of the sample, as it is prepared and introduced into the calorimeter, is therefore necessary if the measurement is to be utilized fully. Details of sample preparation must, however, be left to the individual investigator since they depend largely upon the purpose of the investigation.

41-2 PRINCIPLES

Most calorimetric studies require special procedures and techniques determined largely by the purpose of a particular investigation, but three features are common to all. To be meaningful, any thermochemical study must include (i) a determination of an amount of energy, usually but not always inferred from an observed heat effect; (ii) a specification of the process to which the energy is attributed; and (iii) a determination of the extent of the process associated with the measured energy (Rossini, 1956). In a heat-of-immersion measurement, the energy is determined by inference from the measurement of a heat effect. The second requirement is met by so arranging the experiment that the heat of immersion can be distinguished from any other heat effects present. The last requirement is met by controlling the weight, water content, and the physical and chemical condition of the soil sample before and during the process. In determining the heat of immersion, all this may conveniently be done in a water-filled calorimeter surrounded by a jacket of constant, uniform temperature.

The calorimetric part of the investigation consists of measuring the temperature of the calorimeter and its contents before and after immersing the soil. The amount of electrical energy which, when transformed into heat inside the calorimeter, produces exactly the same temperature change is then determined and is regarded as equivalent to the energy released on immersion of the sample.

The essential parts of the calorimeter are: (i) The calorimeter vessel; (ii) a thermometer; (iii) an electrical resistance heater; (iv) a thermostated surrounding jacket; and (v) a stirrer for the calorimeter contents.

The calorimeter must contain the water in which the soil is immersed; it must provide a means for introducing the sample, and in addition, must thermally insulate its contents. Provision must be made for stirring the contents to ensure thorough dispersion and wetting of the sample and to maintain a uniform temperature distribution. And last, means must be provided for the determination of an energy input by electrical resistance heating to determine the "energy equivalent" of the calorimeter. A wide-mouth, silvered Dewar flask has been found satisfactorily adapatable as a calorimeter vessel in fulfilling all these requirements.

The thermometer must be sensitive and stable. The development of thermistors has recently led to their use as calorimeter thermometers, since they possess the required sensitivity and have been proven to be sufficiently stable for this purpose. As a result, thermistor thermometers may in time replace the conventional thermopile and resistance thermometers in most ordinary calorimeters.

The energy equivalent of the calorimeter may be determined by passing an electrical current through a small coil of constantan, manganin, or other resistance wire having a low temperature coefficient and suitable diameter and length. The resistance of this coil must be accurately known. Then when an accurately known, constant potential drop is maintained across the coil for an accurately measured interval of time, the energy input, E, is given by

$$E = V^2t/R \quad [1]$$

where V is the electric potential drop across the heater coil, R is the resistance of the heater coil, and t is the time interval during which the potential is maintained across the heater.

The calorimeter vessel should be surrounded by a thermostated fluid, so that all thermal leaks in the calorimeter can be reliably controlled. Immersion of the calorimeter and attendant apparatus in a water bath often has been employed with success but with considerable inconvenience. A satisfactory thermostated air-bath is almost as easy to assemble and is certainly to be preferred because of the convenience afforded.

The calorimeter contents must be stirred to effect rapid and complete wetting of the sample and to ensure a uniform temperature throughout. Stirring, however, unavoidably introduces energy into the calorimeter that must be accounted for. The best stirring device, therefore, will efficiently mix the calorimeter contents with the least uncertainty in energy input. A three- or four-bladed screw propeller turning inside a tube mounted vertically in the calorimeter vessel is good stirring device (White, 1928). The tube length and diameter should be adjusted to provide the most effective stirring, and the propeller should be powered by a constant-speed motor external to the calorimeter.

The calorimeter gains heat because of stirring and will either gain or lose heat, as the case may be, because of thermal leaks. For this reason a correction usually must be made to the temperature rises accompanying both the heat of immersion and the input of electrical energy used to calibrate the calorimeter. The corrected temperature rise may be obtained either graphically or analytically. Most soil scientists, in determining heats of immersion, seem to have preferred to obtain it graphically from a plot of the temperature-time data. The plot consists of an initial portion and a final portion, during which the rate of temperature change is nearly constant, separated by the portion during which the heat of immersion is evolved. Lines extending the initial and final portions of the curve may be drawn, and the vertical distance separating them may be measured at a point on the abscissa judged to yield the best estimate for the corrected temperature rise. Usually the point chosen is where the calorimeter temperature has reached about two-thirds of the temperature change attributable to the heat of immersion. This method of correcting for extraneous heat effects often may be judged to be adequate, but for the most accurate determinations an analytical method of correction is to be preferred. Obtaining the corrected temperature rise analytically and formally not only is more accurate, but serves also to accent the operational features of each individual calorimetric study and, therefore, forces a consideration of possible errors which might otherwise be overlooked. The general availability of programmable bench-top computers now makes this approach even more desirable.

Figures 41–2 and 41–3 show two typical temperature-time plots of data obtained on immersion of a clay. They help clarify a discussion of the corrections for the heat of stirring and thermal leakage referred to in section 41–3.2. In both figures the symbols have identical meanings; θ_i and θ_f are the temperatures of the calorimeter contents when the sample is immersed and when all the heat of immersion has been evolved, respectively. θ_j is the temperature of the thermostated enclosure containing the calorimeter, and θ_∞ is the temperature the calorimeter contents would eventually reach if stirring were continued indefinitely and all other conditions were kept constant. The plots are conveniently divided into three

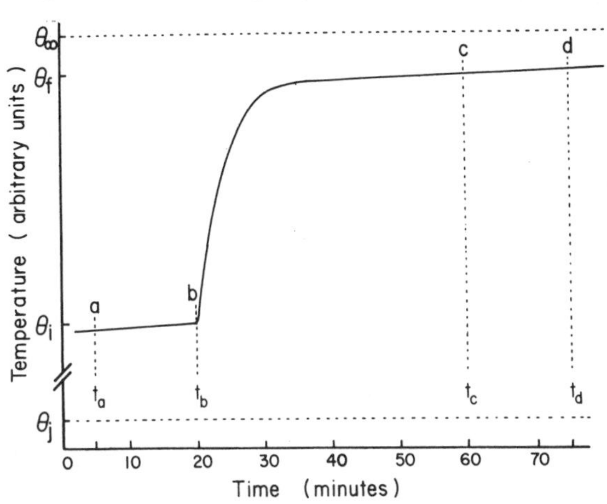

Fig. 41–2. A typical temperature–time curve for the heat of immersion of a clay soil. In this case, the soil sample was immersed at a temperature θ_i which is below θ_∞, the temperature which the calorimeter contents would reach in infinite time if stirring and all other conditions were kept constant; θ_f, the temperature after evolution of the heat of immersion, was also lower than θ_∞.

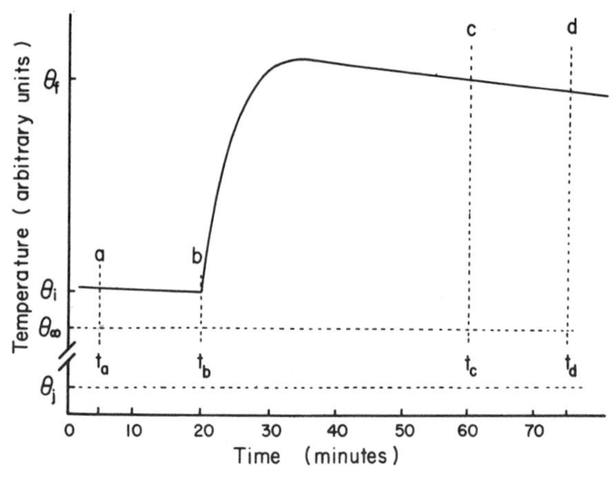

Fig. 41–3. A typical temperature–time curve for the heat of immersion of a clay soil. In this case, the soil was immersed at a temperature θ_i which was higher than θ_∞, the temperature which the calorimeter contents would reach in infinite time if stirring and all other conditions were kept constant. θ_f, the temperature after evolution of the heat of immersion, was higher than θ_∞.

portions. Portions *ab* and *cd* are characterized by a rate of temperature change that is nearly constant. During the intervening period, *bc*, when the heat of immersion is being evolved, θ is seen to be a more complicated function of time. In Fig. 41–2, because the determination was begun at a temperature sufficiently lower than θ_∞, it is seen that during all three periods the calorimeter was heating. In Fig. 41–3, however, the temperature at the beginning of the determination was slightly above θ_∞, so that during periods *ab* and *cd* the calorimeter was cooling and only during a part of period *bc* was it heating. In the discussions that follow, one should keep in mind that it is usually convenient to express temperature in terms of an increase or decrease in signal thermistor resistance or out-of-balance bridge voltage. And since the calorimeter energy-equivalent is determined under conditions as nearly identical as possible to those existing when the heat of immersion is determined and in identical terms, conversion of the data to temperatures in degrees is not necessary.

As was mentioned earlier, the rate of temperature rise observed during the immersion of a soil in water is affected not only by the heat of immersion but also by the heat of stirring the calorimeter contents and the heat transfer between the calorimeter and its surroundings. This can be expressed mathematically as

$$d\theta_{obs}/dt = u + k(\theta_j - \theta) + (d\theta_{imm}/dt) \qquad [2]$$

where $d\theta_{obs}/dt$ is the observed rate of temperature rise, u is the constant rate of temperature rise from the heat of stirring, k is a constant called the *leakage modulus*, θ_j is the constant temperature of the calorimeter jacket, θ is the variable temperature inside the calorimeter, and $d\theta_{imm}/dt$ is the rate of temperature rise from the heat of immersion. The second term on the right-hand side of Eq. [2] represents the rate of heat transfer between the calorimeter interior and its surroundings according to Newton's law of cooling. This term is positive or negative, depending upon whether heat is transferred into or out of the calorimeter.

When an "infinite" amount of time has elapsed after a heat-of-immersion measurement, it is to be expected that

$$u = k(\theta_{\infty} - \theta_j) \quad [3]$$

where θ_{∞} is the temperature inside the calorimeter after "infinite" elapsed time. Substituting this result into Eq. [2], one obtains

$$d\theta_{\mathrm{obs}}/dt = k(\theta_{\infty} - \theta) + (d\theta_{\mathrm{imm}}/dt)\,. \quad [4]$$

The above equation can be rearranged to give an expression for the temperature rise from the heat of immersion:

$$d\theta_{\mathrm{imm}} = d\theta_{\mathrm{obs}} + k\theta\, dt - k\theta_{\infty}\, dt \quad [5]$$

which on integration gives

$$\Delta\theta_{\mathrm{imm}} = \Delta\theta_{\mathrm{obs}} + k \int_{t_b}^{t_c} \theta\, dt - k\theta_{\infty}(t_c - t_b) \quad [6]$$

where t_b and t_c are, respectively, the time when the heat-of-immersion measurement begins and when it ends, giving an observed temperature rise $\Delta\theta_{\mathrm{obs}}$. The integral term in Eq. [6] usually cannot be evaluated analytically, but must be evaluated by graphically integrating the observed heat-of-immersion curve.

An expression for k, the leakage modulus, may be derived by considering the heat-balance equation (Eq. [4]) before the heat-of-immersion measurement is made:

$$d\theta_{\mathrm{obs}}/dt = k(\theta_{\infty} - \theta)\,. \quad [7]$$

The solution of this differential equation is

$$\ln\,[(\theta_{\infty} - \theta)/(\theta_{\infty} - \theta_i)] = -kt \quad [8]$$

where θ equals θ_i when t equals zero. Since this equation is linear in t, a plot of the left-hand side against time enables one to calculate k.

Each determination of the heat of immersion should be accompanied by a measurement of the energy equivalent of the calorimeter found by introducing a known amount of heat into the calorimeter by an electrical resistance heater and observing the resultant temperature rise. The corrected temperature rise for this determination is calculated by means of an equation exactly like Eq. [6] and then is divided into the electrical input, to get the energy equivalent of the calorimeter. All the information required is now obtainable and may be introduced into the relation

$$Q_{imm} = \Delta\theta_{imm}(Q_E/\Delta\theta_E) \quad [9]$$

to actually compute the heat of immersion Q_{imm} for a particular sample. In Eq. [9], $\Delta\theta_E$ is the corrected temperature rise corresponding to a measured input of electrical energy, Q_E.

To obtain and evaluate Eq. [6], it was first necessary to regard the rate of temperature change due to thermal leaks as a linear function of the calorimeter temperature. Each experimenter is obliged to verify the validity of this assumption for his calorimeter whenever this method of correction is used; it will usually be found valid over the small temperature range normally encountered, provided the heat of stirring is kept constant. Perhaps the best way to determine the extent to which a given stirring arrangement meets this requirement is, first, to check the constancy of the stirring speed. But because of the change in viscosity of the calorimeter liquid on immersing the sample, and because of the numerous collisions between the propeller of the stirrer and large soil particles and glass from the broken sample vial, a constant rate of stirring may not result in a sufficiently constant heat of stirring during the determination. Therefore, it may be deemed desirable in some instances to try to evaluate the extent to which the heat of stirring is changed on immersion of the sample. Perhaps a good way to assess it is to make several determinations in which k and θ_∞ are carefully measured both before and after the heat of immersion is obtained. The difference in k and θ_∞ before and after immersing the sample, in conjunction with Eq. [3], will provide a means of evaluating the change in the heat of stirring occurring during the determination. If an appreciable difference is found, a method of correcting for it should be sought or else the sample size or other parameter(s) should be adjusted so that the difference is reduced to the point where it may be neglected.

41–3 METHOD

This procedure is patterned after that of Zettlemoyer et al. (1953), although it corresponds in most respects to that of many other investigators before and since. It has been modified on the basis of the author's experience to fit what he judges to be the general circumstances of soils laboratories where this measurement might be utilized.

41–3.1 Special Apparatus

41–3.1.1 CALORIMETER

A relatively simple and inexpensive calorimeter adequate for measuring heats of immersion of soils and soil constituents is shown in Fig.

41–4. It consists of a wide-mouth, 500-mL, silvered Dewar flask. A cover cut from a machinable plastic such as melamine, bakelite, or plexiglass is provided to prevent evaporation of the calorimeter contents. A ring cut from the same material can be cemented to the Dewar flask to provide a means of fastening on the cover with bolts or screws. A good adhesive for this purpose is one of the epoxy adhesives. It should be applied so that there are only three or four points of contact between ring and flask; otherwise excessive strains may develop during the curing of the adhesive, which would result in an easily broken calorimeter. An O—ring groove should be cut in the cover to coincide with the lip of the flask, to provide a means of ensuring a vapor-proof calorimeter cover.

The cover should contain openings and be provided with fittings to accommodate the stirring shaft, the sample-holder supports, a sample-breaking rod, the heater, and the thermistor leads. The stirring shaft may be made of bakelite, fiberglas, or some other poorly conducting, machinable material that is, at the same time, rigid and strong. A stirrer

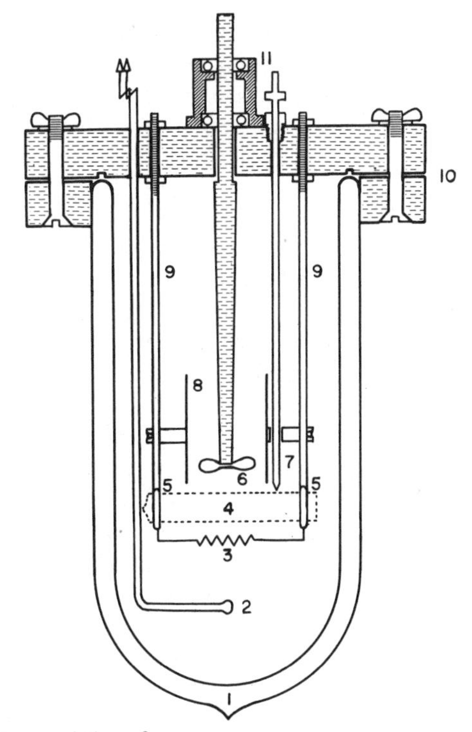

Fig. 41–4. Calorimeter consisting of
1. 500-mL silvered Dewar flask.
2. Thermistor thermometer.
3. Resistance-wire heater.
4. Glass sample vial.
5. Sample vial holders.
6. Stirrer.
7. Sample-breaker rod.
8. Stirrer tube, fastened by two insulated "ears" to sample-holder supports.
9. Supports for sample holder and electrical conductors for the resistance heater.
10. Plexiglass top.
11. Bearing assembly for stirrer.

tube made from thin copper or stainless steel tubing can be supported from the calorimeter cover by the two brass rods shown in Fig. 41–4. If the stirrer tube is fastened to the rods by an epoxy resin or some other suitable nonconducting substance in such a way that no electrical connection exists between them, the two brass rods can also serve as electrical leads to the resistance heater. The stirrer tube should have a diameter of about 3 cm and a length of about 5 cm. For a tube of this size, a three-bladed propeller 1.8 to 2.0 cm in diameter located near the center of the stirring tube will provide nearly optimum stirring, provided that the water level and position of the stirring tube and propeller are properly adjusted in the calorimeter. This adjustment is best made by trial until it is satisfactory for the particular Dewar flask used.

The stirrer shaft must be brought through the calorimeter cover in such a way that the shaft will turn easily without permitting evaporation of the calorimeter contents. Zettlemoyer et al. (1953) used a vacuum-seal stirrer. Another good solution offered by White (1928) consists of an oil seal made by an upside-down cup soldered or glued to the shaft; the upside-down cup is positioned so that it rotates in an oil-filled, circular moat in the calorimeter cover. Figure 41–4 shows another arrangement that has proved satisfactory. The two bearings are press-fitted or glued into the housing and to the stirrer shaft. (Two FaFrir FS160B-C1-Sw ball bearings carefully cleaned and reoiled with a light, varnish-free oil have proved satisfactory.) The stirrer shaft may be connected to a constant-speed motor either directly or by a belt or pulley. A direct coupling, consisting of a short length of thick-walled rubber tubing having an inside diameter slightly smaller than both the motor and the stirrer shafts, has proved satisfactory; it permits mounting the motor outside the thermostated enclosure housing the calorimeter, and it can be connected easily and simply. A similar approach has been recommended by Featherstone and Dickenson (1976).

Electrical connections can be made in many ways. A good method is to drill the necessary holes in the cover, pass the leads through, and seal them in the cover with epoxy resin. Of course, sealing glands, plugs, and other devices may be used, but with hardly any advantage. Both thermistor and resistance heater should be located near the bottom of the propeller tube and positioned securely. This, too, may be accomplished in various ways. The brass rods that support the stirring tube may be used to position the thermistor and also may be used as leads and supports for the heater. A clamping device to secure the glass sample vial at each end may also be attached to these rods as shown in Fig. 41–4. A third rod is brought into the calorimeter through a packing gland positioned so that by dropping or tapping it sharply it can be used to break the glass vial containing the sample. Provision should be made to limit its excursion, so that no danger to the calorimeter will exist.

The heater may be made by (i) wrapping bare resistance wire of about 24 gauge around a glass tube 10 to 12 mm in diameter, (ii) heating it to incandescence to anneal it, by passing an electric current through it while it is still tightly wrapped around the tube, (iii) cutting it to a length that will have a resistance of about 10 ohms, and (iv) connecting it to the brass rods, previously referred to, with solder or screws. It should be positioned so that it does not interfere with the sample-breaking rod and so that relatively little abrasion from soil and broken glass will occur.

The thermistor which is to serve as an electrical thermometer may be any one of a large variety now on the market. Because such a large choice is possible and since final performance is the best indication of a good choice, only the essential requirements will be mentioned here. The thermistor should have a high temperature coefficient in the 10 to 50°C range for maximum sensitivity. It should be small in size to keep its thermal lag low; it and its leads should be well insulated; and it should be stable and dependable. Finally, the resistance of the thermistor at its normal operating temperature should be chosen to fit the resistance bridge, potentiometer, or other measuring device to be used in conjunction with it. Resistance ranging from 100 ohms to about 50 000 ohms may be appropriate for various situations, so that a considerable latitude of choice exists. Thermistors sealed in plastic may be satisfactory, but those hermetically sealed in glass or metal are preferred. It is also possible to employ thermistor assemblies which consist of a thermistor sealed in a tiny metal probe with its insulated leads brought out through a packing gland. These are available from Fenwall Electronics, Inc., Framingham, MA; Gulton Industries, Inc., Metuchen, NJ; Cole-Parmer Instrument and Equipment Co., 224 W. Illinois Street, Chicago, IL; and many other venders. An assembly such as this offers a convenient means of positioning and housing the thermistor.

41–3.1.2 AUXILIARY ELECTRICAL DEVICES

The precision and sensitivity of the calorimeter depend to a great extent on the instrument used to measure the change in resistance of the thermistor; the highest sensitivity, therefore, will be achieved with a resistance bridge of high sensitivity. A Mueller bridge capable of detecting a 0.0001-ohm resistance change was used by Zettlemoyer et al. (1953), and they reported being able to detect temperature changes of 0.00002°C reliably. An ordinary potentiometer with a good galvanometer or an electronic null-point indicator has proved satisfactory in detecting temperature changes of 0.0001°C. A calorimeter sensitivity of about 0.1% is thus attainable with relative ease.

The resistance heater is also a most vital part of the calorimeter. With it, one determines the absolute amount of heat energy corresponding to a measured temperature rise. This quantity of heat is calculated from a knowledge of the electrical current passed through the heater, the electrical potential drop across that portion of the heater circuit inside the calorimeter, and the length of time this potential is applied. All these must therefore be measured accurately.

A suitable circuit for supplying and measuring electrical energy added to the calorimeter and its contents is shown in Fig. 41–5 (Coops et al., 1956, p. 48–49; White, 1928, p. 135). It consists of a standard resistor, a voltage divider, two double-pole, double-throw switches, a ballast resistor, a storage battery, and a potentiometer with standard cell. Current from the storage battery flows through the ballast resistor, R_1, or the heater, R_2, depending upon the position of switch A. The resistance of R_1 should be made as nearly equal to that of R_2 as possible, so that the current drawn from the storage battery will be nearly the same when switch A is in either position. The ballast resistor is provided to stabilize the storage battery output by allowing current to be drawn from the storage battery for an hour or so before switching to the heater. In utilizing the circuit shown in Fig. 41–5, V_2, the potential drop across the calorimeter heater, is compared to V_3, the potential drop across the standard resistor R_3. The electrical power input to the calorimeter is then V_2V_3/R_3. To facilitate making alternate measurements of V_2 and V_3, the voltage-divider ratio and the resistances of the heater coil and the standard resistor should be adjusted so that the potentiometer reading is nearly the same in each case. Otherwise, to make the measurements as rapidly as is required, it may be necessary to employ two potentiometers.

When the investigation warrants it, the electrical power source may be arranged to minimize the effect of the rapid change in the heater resistance at the beginning of the heating period; a method has been suggested by Skinner et al. (1962, p. 211–214). Care should be taken in wiring the circuit to ensure that the leads from the heater are in intimate thermal contact with the calorimeter and its thermostated surroundings. Whenever possible, wires should be arranged so as to reduce heat flux by conduction. In some applications, simpler, less sensitive circuits and procedures may be more appropriate; this the experimenter himself should determine.

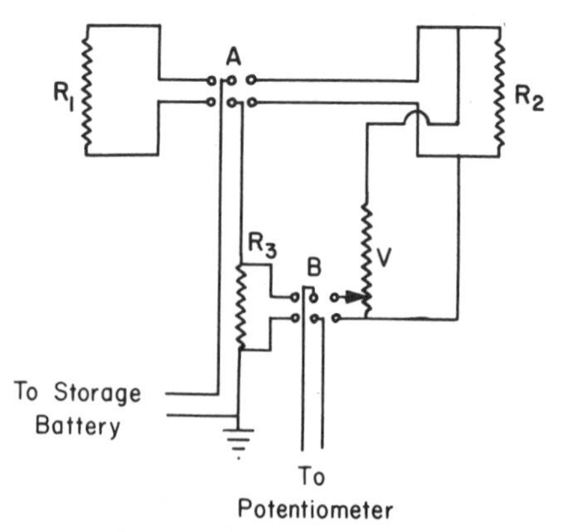

Fig. 41–5. Circuit for determining the electrical energy equivalent of the calorimeter. The circuit elements are: R_1, ballast resistor; R_2, calorimeter heater; R_3, standard resistor; V, voltage divider; A and B, double-pole, double-throw switches.

The measurement of time during calibration of the calorimeter may be accomplished satisfactorily with a stopwatch and manual operation of the heater switch. If one aims for an accuracy of better than 1 part in 1000, however, it becomes necessary to use some means of coupling between switch and timer. Skinner described one way in which this may be done and referred to other methods (Skinner et al., 1962, p. 181).

41–3.1.3 THERMOSTATED ENCLOSURE

A satisfactory constant-temperature air bath can be constructed easily from an electronic proportional thermoregulator (for example, see Anderson and Jones, 1961), a 10- to 50-W resistance-wire heater, a small fan, and an enclosure lined with alternating layers of aluminum foil and styrofoam or other suitable insulation. The motor powering the fan is best mounted outside the box with its shaft passing into the box through a small opening or bearing. Temperature control with fluctuations and drift of the order of ± 0.01°C is not difficult to achieve.

41–3.2 Procedure

Prepare an accurately weighed 4- to 5-g soil sample, and seal it watertight in a glass ampule of a size and shape determined by the construction of the calorimeter. Fasten the filled sample vial into place in the calorimeter beneath the breaker rod. Fill the calorimeter with a measured amount (about 400 mL) of distilled water at about the temperature that is to be maintained in the thermostated enclosure surrounding the calorimeter. Attach the calorimeter cover, engage the stirrer, connect the electrical leads, mount the assembly in a thermostated enclosure, and begin stirring.

Begin taking readings of the thermistor resistance, noting accurately the time each reading was taken. By means of the electric heater add energy in small increments, raising the calorimeter temperature until the rate of temperature change in the calorimeter is constant and small enough (a few ten-thousandths of a degree centigrade per minute) to permit subsequent correction of the results for the heat of stirring and thermal leakage. Break the sample vial with the breaker rod, and continue recording the thermistor resistance every 10 to 15 s. When it becomes evident that the heat of immersion has been evolved completely, introduce an accurately measured amount of electrical energy through the resistance heater; continue recording the thermistor resistance every 10 to 15 s until it is evident that the thermistor response to the heat evolved has been completed.

Plot the thermistor resistance vs. time on a large sheet of coordinate paper. The plots will have characteristic shapes such as those of Fig. 41–2 and 41–3. From a comparison of the resistance–time data accompanying the sample immersion with that accompanying the input of electrical energy, determine the uncorrected temperature rises resulting from the heat of immersion and the electrical heating. Correct the observed temperature rises for the heat flux due to thermal leakage and the heat of stirring.

When the corrected temperature rise accompanying the input of a measured quantity of electrical energy has been determined, it is referred to as the electrical equivalent of the calorimeter. From the electrical energy equivalent of the calorimeter and a knowledge of the weight of the sample immersed, calculate the heat of immersion from Eq. [9] as described in section 41–2. Subtract the temperature rise due to breaking the glass sample vial. (Several blank determinations using a sealed vial containing 4 to 5 g of large glass beads but otherwise identical with that normally employed will yield a relationship between sample weight and the heat of breaking, from which an acceptable correction for each determination can be obtained.) Express the results as joules per gram of sample (dry-weight basis). In addition, record clearly the initial water content of the sample and other details of its physical and chemical state as well as the temperature of the determination.

41–3.3 Comments

In selecting a calorimeter design, the rate of the process, the size of the anticipated heat effect, the accuracy of the result desired, and whether or not other measurements or operations are to be made must be considered. For soils, one may expect the heat effect to be observed in from 2 to 15 min after the sample is immersed and to amount to from 8 to 120 J per g of soil. An accuracy of measurement of 0.1% is more than sufficient for all but the most elaborate and demanding studies.

Neutraglas Ampuls No. 12011-L (Kimball Glass Co., a subsidiary of Owens-Illinois, Toledo, OH) or their equivalent have been found to be ideal for sample containers. The water-tight seal is easily made by fusing the prepared neck in a flame.

The amount of water used in the calorimeter is not actually critical, but it should be known and reported, since the observed integral heat of immersion depends in part on the final state of dilution of the soil-water system. Whereas the heat evolved on dilution of a soil suspension may be extremely small, it is nonetheless real and in some rare cases may have to be taken into account in comparing data from different investigations.

Stirring at about 3500 revolutions per minute with a constant-speed motor provides sufficient agitation to break up and uniformly distribute the soil sample on immersion. This is a high stirring rate and produces enough heat in the calorimeter to maintain its temperature several degrees above that of the thermostated enclosure. According to White (1928), a high heat of stirring is very undesirable because it adds considerable uncertainty to the result. Unfortunately, because of the difficulty in thoroughly dispersing some soils, there seems to be no way to avoid a rather high heat of stirring in this determination. The heat of stirring is most conveniently accounted for if it is constant. It is necessary, therefore, that the stirring rate be controlled and held constant. When it is possible to reduce the stirring rate, as in the case of easily dispersed soils, this certainly should be done. The heat of immersion usually will be evolved within 2 to 15 min after breaking the sample free, depending upon the texture and consistency of the soil, unless the sample is not adequately dispersed by the stirrer.

The temperature of the sample and water must be known at the time the heat of immersion is measured. It may be obtained from the thermistor calibration when the data of the determination are being analyzed. The total temperature rise usually will be only a few hundredths of a degree, and at the most a few tenths of a degree, so that there is no serious problem in assigning a temperature to the process. For a discussion of this point, however, see Rossini (1956, p. 16–20).

Care should be taken to eliminate all sources of error or uncertainty in determining the exact amount of heat imparted to the calorimeter and its contents. This requires a careful analysis of the heating and measurement circuit as it actually exists when all connections are finally made. Alternate measurements of the potential drop across the heater and the standard resistor permit calculation and a check, by Ohm's law, of the electrical energy input, provided that all the circuit elements are known. By adjusting the heater current and the heating period, one should try to obtain nearly the same total temperature rise and nearly the same rate of change of temperature as when the heat of immersion was evolved. Errors then tend to become self-compensating.

The temperature rise due to breaking the sample free is an important correction in precision calorimetry. On breaking the vial, the calorimeter liquid does PV work that is converted into heat by the viscous or locally turbulent collapse of liquid into the void space. It has been found (Guderjahn et al. 1958) that for empty glass and metal bulbs, the heat evolved on breaking the bulb varies linearly with the void volume according to the equation

$$H = PV + 0.03 \pm 0.07 \text{ J.} \qquad [10]$$

This result is regarded as justification for the procedure given in the foregoing section. In the more precise investigations this correction should be measured with accuracy for each determination.

In addition to reporting all relevant details of sample preparation and experimental procedure, the accuracy and precision of the results should be estimated. The former must be estimated from the principles of measurement and the accuracy with which the measurements were made. Investigators are obliged to furnish such an estimate when calorimetric data are reported. The latter may be best expressed by reporting at least (i) the number of observations made under supposedly identical circumstances, (ii) the mean of the observed values, and (iii) the standard deviation, when the complete data are found to be too bulky for publication (Rossini, 1956).

41-4 REFERENCES

Adamson, A. W. 1960. Physical chemistry of surfaces. Interscience Publishers, New York.

Anderson, D. M., and R. C. Jones. 1961. An inexpensive control circuit for mercury thermoregulators. Soil Sci. Soc. Am. Proc. 25:416–417.

Brunauer, Stephen. 1945. The adsorption of gases and vapors: I. Physical adsorption. Princeton University Press, Princeton, NJ.

Coops, J., R. S. Jessup, and K. van Ness. 1956. Reactions in a bomb at constant volume. p. 27–58 *In* F. D. Rossini (ed). Experimental thermochemistry, chapter 3. Interscience Publishers, New York.

Edlefsen, N. E., and A. B. C. Anderson. 1943. Thermodynamics of soil moisture. Hilgardia 15:31–298.

Featherstone, J. D. B., and N. A. Dickenson. 1976. An isothermal dilution calorimeter for enthalpy of mixing determinations in the region of 288 to 343 K. J. Chem. Thermodynamics. 8:955–992.

Guderjahn, C. A., D. A. Paynter, P. E. Berghausen, and R. J. Good. 1958. Heat of bulb breaking in heat of immersion calorimetry. J. Chem. Phys. 28:520–521.

Hill, T. L. 1948. Statistical mechanics of adsorption: V. Thermodynamics and heat of adsorption. J. Chem. Phys. 17:520–535.

Janert, H. 1934. The application of heat of wetting measurements to soil research problems. J. Agric. Sci. 24:136–150.

Mitscherlich, E. A. 1899. Dissertation, Kiel. (cited by H. Janert. J. Agric. Sci. 24:136).

Rossini, F. D. 1956. Experimental thermochemistry. Interscience Publishers, New York.

Rossini, F. D. 1956. Assignment of uncertainties. p. 59–74. *In* F. D. Rossini (ed). Experimental thermochemistry, chapter 4. Interscience Publishers, New York.

Skinner, H. A., J. M. Sturtevant, and S. Sunner. 1962. The design and operation of reaction calorimeters. p. 157–219. *In* H. A. Skinner (ed). Experimental thermochemistry II, chapter 9. Interscience Publishers, New York.

White, W. P. 1928. The modern calorimeter. Am. Chem. Soc. Monogr. 42.

Zettlemoyer, A. C., G. J. Young, J. J. Chessick, and F. H. Healey. 1953. A thermistor calorimeter for heats of wetting: entropies from heats of wetting and adsorption data. J. Am. Chem. Soc. 57:649–652.

42 Solute Content

J. D. RHOADES

U. S. Salinity Laboratory, Agricultural Research Service, USDA
Riverside, California

J. D. OSTER

University of California
Riverside, California

42-1 INTRODUCTION

The term *solute content* as applied to soils, refers to the major dissolved inorganic solutes. Soluble solutes in soils can be determined or estimated from measurements made: (i) on aqueous extracts of soil samples, (ii) on samples of soil water per se, obtained from the soil, (iii) in situ, using either buried porous salinity sensors that imbibe and equilibrate with soil water or four-electrode probes or electrode systems, and (iv) remotely, using electromagnetic-induction electrical-conductivity sensors. The choice of method depends upon the purpose of the determination and the accuracy required. Methods of collecting water samples using in situ samplers and methods of measuring soluble salts with in situ or remote monitors are dealt with in this chapter. Chemical methods of determining solutes and extraction procedures are dealt with in Part 2 of Agronomy 9 (Page et al., 1982).

Ideally, it would be desirable to know the individual solute concentrations in the soil water over the entire range of field water contents and to obtain this information immediately in the field. No practical methods are available, at present, to permit such determinations, although determinations of *total* solute concentration can be made in situ or remotely using electrical signals from sensors. Such determinations are valuable for survey, monitoring, and irrigation and drainage management needs, and in many cases are supplanting the need for more conventional analytical procedures. If a particular solute concentration is needed, then a sample either of soil or of the soil water is required. Relatively wet soil conditions are required to obtain soil water samples. Concentrations of solutes in extracts obtained from soil samples give relative comparisons only, since the soils are adjusted to unnaturally high water contents during extraction. Methods for extraction of soil samples are given in Part 2 of

 Methods of Soil Analysis, Part 1. Physical and Mineralogical Methods—Agronomy Monograph no. 9 (2nd Edition)

Agronomy 9. A combination of the various methods minimizes the need for sample collection and chemical analysis, especially when monitoring solute changes with time and characterizing large field or project situations.

42–2 COLLECTION OF SOIL WATER SAMPLES USING IN SITU SAMPLERS

42–2.1 Principles

Soil water samples are useful for assessment of solute transport as functions of depth and time, and of the in situ concentrations of plant essential nutrients and pesticides. The major task is to obtain the sample without unduly altering its composition in the process. Methods of soil water sampling may be classified as follows: (i) displacement, (ii) compaction, (iii) centrifugation, (iv) molecular adsorption, (v) suction, and (vi) pressure membrane extraction. Of these, only suction is commonly used in the field. Sufficient sample for chemical analysis can be obtained in most cases. For these reasons, this method will be described herein, with a discussion of its limitations. For some needs, methods of obtaining soil water samples from soil samples may be desired. Such methods will not be reviewed here nor will a specific method be described, because none has been sufficiently adopted in practice, to date, to warrant its general recommendation. The most recent discussions and descriptions of these methods may be found in the following references: (i) displacement methods are described in Adams (1974), (ii) a combination displacement/centrifugation method has been developed by Mubarak and Olsen (1976, 1977), (iii) adsorption techniques are described in Shimshi (1966) and Tadros and McGarity (1976), and (iv) centrifugation techniques have been developed by Davies and Davies (1963), Yamasaki and Kishita (1972), Gilman (1976), Dao and Lavy (1978), and Kinniburgh and Miles (1983).

Soil water samples are collected in situ with vacuum extractors. The suction method, first proposed by Briggs and McCall (1904), is useful for extracting water from the soil when the soil-water matric potential is more than about −30 kPa. Water movement in soils becomes very slow at matric potentials < −30 kPa. While different porous-tube devices have been used (Krone et al., 1952; Krugel et al., 1935; Jackson et al., 1976), the most common is the porous ceramic cup (Krugel et al., 1935; Brooks et al., 1958; Wagner, 1962; Reeve & Doering, 1965). Kohnke et al. (1940) assembled a bibliography on early vintage extractor construction and performance. Chow (1977) designed a vacuum sampler that shuts off automatically when the desired volume of sample is collected. Improved and specialized versions have been more recently developed for different purposes, including a miniaturized sampler which eliminates sample transfer in the field (Harris & Hansen, 1975) and samplers which function at depths greater than the 10-m suction lift of water (Parizek & Lane, 1970; Wood, 1973). Recently soil water has been extracted using cellulose-acetate hollow fibers (Jackson et al., 1976; Levin & Jackson, 1977), which are thin-walled and flexible. Claimed advantages include flexibility, small diameter, minimal chemical interaction of solute with the tube matrix, and sample compositions similar to those obtained from ceramic extraction cups. Large-scale vacuum extractors (15 cm wide by 3.29 m long) have been used to assess deep percolation losses and chemical composition of soil water (Duke & Haise, 1973). Collection "pan"-type collectors have also been used to collect soil percolate (Jordan, 1978). Because of their extensive use, only the use of porous ceramic cups and suction apparatus will be described in detail.

As demonstrated by Hansen and Harris (1975) and Alberts et al. (1977), various errors in sampling soil water can occur with the use of porous ceramic-cup extractors. These include errors related to the rate of extraction, the chemical composition of the cup, the rate of soil water movement, and the nonhomogeneity of the soil solution composition. Nitrate contents of soil water were influenced by sampler intake rate, plugging, and sampler depth and size. To reduce sample variability, samplers with uniform permeabilities and size should be used in conjunction with uniform sampling intervals and vacuum. Cups must be cleaned with dilute acid before use, as they can release some solutes to solution (Wolff, 1967). Soil water samples are representative of the soil solution surrounding the cup at the time of sampling. If water is flowing in the soil profile and the objective is to obtain a representative sample of all the water passing the sampler, then a number of samples should be collected over time in proportion to the soil-water flux. If the samples are not, a bias will be produced in the data (Hansen & Harris, 1975). Alberts et al. (1977) compared soil-core sampling with ceramic-cup extractions for determining nitrate in the soil profile and concluded that spatial variability makes it difficult to interpret nitrate data collected by either sampling technique. Nielsen et al. (1973), Biggar and Nielsen (1976), and van De Pol et al. (1977) have aptly used soil water extractors to determine salt flux in fields and have demonstrated that field variability is very large. They concluded that soil water samples, being "point samples," can provide good indications of relative changes in the amount of solute flux, but not quantitative amounts, unless the variability of such measurements is properly established. As pointed out by England (1974), serious doubts remain about the representativeness of water samples collected by ceramic extraction cups. Because the composition and concentration of soil water are not homogeneous throughout its mass, water extracted from large pores at low suctions (as collected by vacuum extractors) may have compositions very different from that extracted from micropores. A point source of suction (such as a porous cup) samples a sphere, draining different-sized pores differently dependent upon distance from the point, the amount of applied suction, the hydraulic conductivity of the medium, and the soil water content. Another concern is the problem of adsorption of ions by the ceramic cup itself. Although in situ soil water samples can be easily obtained from wet soils by vacuum extraction, the user needs to be aware of the limitations.

42–2.2 Apparatus

The apparatus for sampling the soil solution in the field is shown schematically in Fig. 42–1. The ceramic cups should have a minimum conductance of 1.4×10^{-5} cm^2/s (0.50 mL h^{-1} kPa^{-1}). The ceramic cups are attached to plastic tubes (sampling tubes) of such length to locate the cup at the desired depth in the soil. Suction is applied to the sampling tubes from a portable pre-evacuated vacuum tank A through a manifold B to which sample bottles C are attached. Although three sampling tubes are shown, any number of tubes may be used with appropriate manifold and sample-bottle connections. Neoprene tubing is durable for outside use, and convenient to use for all connections from tank to bottles and to sampling tubes. The valve on the vacuum tank is opened to apply suction to the sampling tube through the interior copper extraction tube, which extends to the bottom of the ceramic cup. In response to a pressure gradient across the ceramic wall, the soil solution moves through the ceramic cup into the sampling tube. The soil solution is withdrawn from the sampling tube by releasing the clamp on the air-inlet tube at D. The admission of air at C increases the pressure within the sampling tube and forces the solution from the sampling tube into the sample bottle at C.

The portable vacuum tank may be evacuated by a laboratory line or

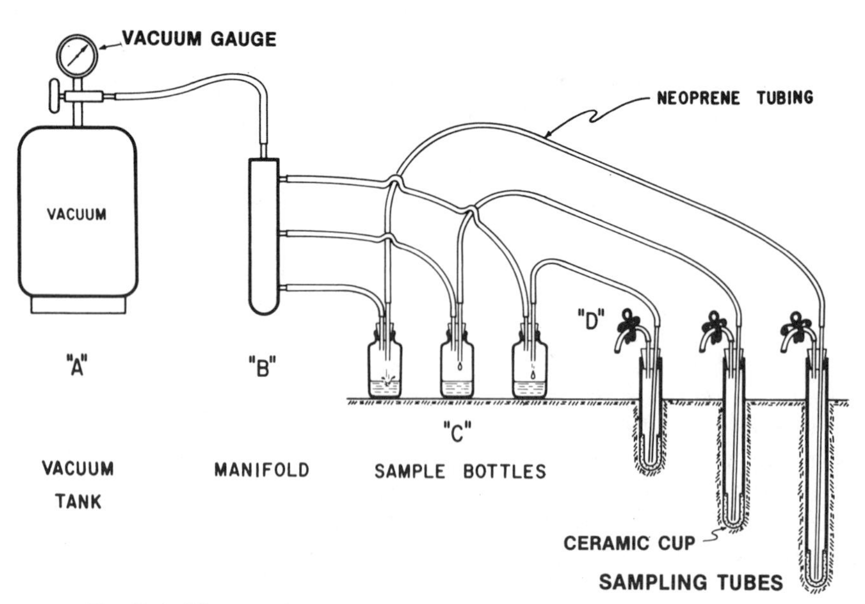

Fig. 42–1. Diagram of vacuum extractor apparatus for sampling soil water.

some other source of vacuum. It may be replaced by either a portable motor-driven vacuum pump or a hand pump capable of producing >75 kPa of suction.

42–2.3 Procedure

42–2.3.1 INSTALLATION

Core a hole using a soil sampling tube that is slightly larger than the diameter of the soil water sampler. Install the sampler in the hole at the desired depth. Depending on the soil properties and purpose of sampling, one of several methods can be used to install the sampler. Local soil (usually sieved through a 6-mm sieve) may be used to backfill around the emplaced sampler. The soil should be tamped thoroughly using a long rod after increments of backfill have been added, to ensure good soil contact with the porous ceramic cup and to complete sealing of the cored hole. Alternatively, the sampler can be inserted into a cored hole in which a soil slurry has been poured to a depth that will cover the ceramic cup. This procedure embeds the cup in the soil slurry and insures good soil-cup contact (though it may also result in a slowly permeable zone adjacent to the cup if the soil material puddles easily). The remainder of the cored hole is backfilled as described above. A third method consists of embedding the cup in fine (<0.074 mm) silica sand by first pouring in some of the sand, then inserting the sampler, and then adding more sand to cover the cup. The remainder of the cored hole is then backfilled with soil as before. To better isolate the ceramic cup from soil above and below, and particularly to prevent excessive percolation through the back-filled hole, wet bentonite can be poured into the bottom of the cored hole, followed by a small amount of fine silica sand, then the sampler, then more sand to cover the cup, and then more wet bentonite. The remainder of the hole is then backfilled as before.

42–2.3.2 SAMPLE COLLECTION

Various techniques can be used to apply vacuum to the sampler and to collect the water (Reeve & Doering, 1965; Wagner, 1962). In general, a vacuum of 50 to 85 kPa is normally applied to the sampler via vacuum tubing with collection bottles and over-flow water traps inline. The time required to collect a water sample varies with the suction applied and the hydraulic conductivity and water content of the soil. For soils near field capacity with good conductivity, enough sample for most analytical needs can be collected in several hours. Under less ideal conditions, several days or more may be needed to collect an adequate sample.

42–2.3.3 SAMPLE HANDLING

Collected samples should be preserved as described by Rhoades (1982). Analytical procedures are also given in this chapter.

42–2.3.4 COMMENTS

For more detail on other types of specialized extraction equipment, including a miniaturized sampler which eliminates sample transfer in the field and samplers which function at depths greater than the suction lift of water (~10 m), see the review of Rhoades (1978).

42–3 MEASURING SOLUBLE SALTS WITH IN SITU OR REMOTE MONITORS

For many purposes, the total salt concentration of the soil water is all that is needed. In such cases, in situ or remote devices capable of measuring total salt concentration of the soil water (or a related parameter) can be used advantageously. Four kinds of sensors are now available, each with its own advantages and limitations: (i) porous matrix sensors, which imbibe and maintain diffusional equilibrium with soil water and measure its electrical conductivity directly, (ii) four-electrode soil electrical conductivity sensors, (iii) electromagnetic-induction soil electrical-conductivity sensors, and (iv) time-domain reflectometry (TDR) and insertion parallel-guide electrodes. The last three all measure the electrical conductivity of bulk soil, which depends upon the soil water content and salt concentration. The use of such sensors has steadily increased in agricultural research where continuous monitoring of soil salinity in soil columns, lysimeters, and field experiments is often required (Oster & Ingvalson, 1967; Todd & Kemper, 1972; Rhoades, 1972, Oster et al., 1973; Wierenga & Patterson, 1974; Hoffman et al., 1978.

42–3.1 Porous Matrix Sensors

42–3.1.1 PRINCIPLES

The electrical conductivity of soil-water, κ_w, can be measured in situ with a buried electrical conductivity cell made from ceramic (Kemper, 1959). The basic principle underlying the measurement is that spaced electrodes embedded in porous ceramic can be used to measure its electrical conductance, *L*. The measurement assumes diffusional equilibrium between the ceramic and soil solution, and constant ceramic water content as the soil wets and dries.

42–3.1.2 APPARATUS

The salinity sensor sold by Soilmoisture Equipment Corp.[1] (Fig. 42–2) is based on the design of Richards (1966). It consists of a 1.5-mm thick by 6-mm diameter electrolytic element made from 1500-kPa ceramic with

[1]Mention of a trademark or proprietary product in this manuscript does not constitute a guarantee or warranty of the product by the U. S. Department of Agriculture, and does not imply its approval to the exclusion of other products that may also be suitable.

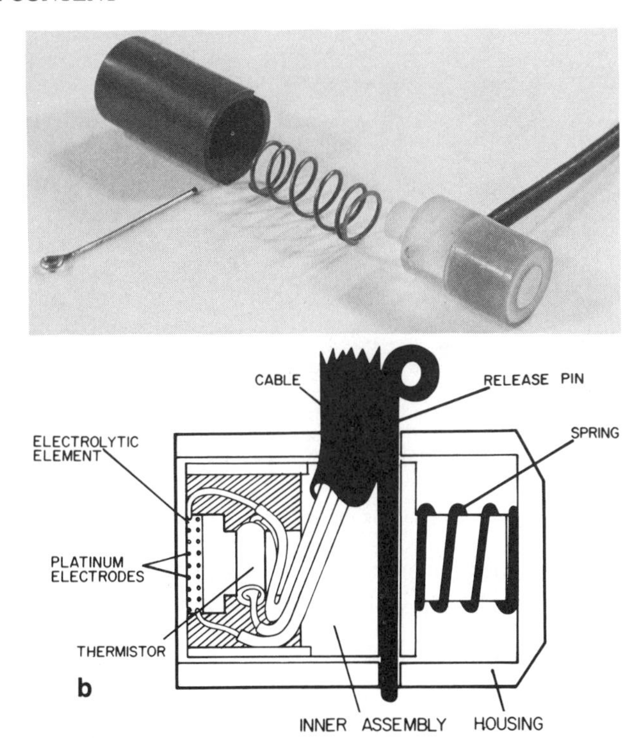

Fig. 42–2. (a) A porous matrix salinity sensor with spring, housing, and pin in disassembly. (b) Schematic of porous-matrix salinity-sensor.

platinum screen electrodes, which is cast into epoxy so that only one surface of the disk can contact the soil. The epoxy insulates the electrodes from the soil and provides a structure upon which the rest of the sensor is built. The sensor also includes a thermistor to measure temperature, so that the conductance of the ceramic electrical conductivity cell may be referenced to a standard temperature. The housing and spring provide a mount to ensure good contact between the electrolytic element and the soil when the release pin is pulled. When the sensor is used in conjunction with the salinity bridge designed and sold by the same company (Fig. 42–3), the readout is in decisiemens per meter (dS m^{-1} or mmhos cm^{-1}) automatically corrected to 25 °C.

42–3.1.3 INSTALLATION

These sensors can be installed using the same methods as described for the soil water extractors in section 42–2.3.1. They should be soaked in water for at least 24 h before installation. If possible, the soaking time should be increased to several days, because the initial wetting process is slow and can take from 3 to 5 days before the ceramic element is fully wetted. The commercial units are spring-mounted; after the sensor is placed in the soil, with backfilling and tamping, the pin is withdrawn by

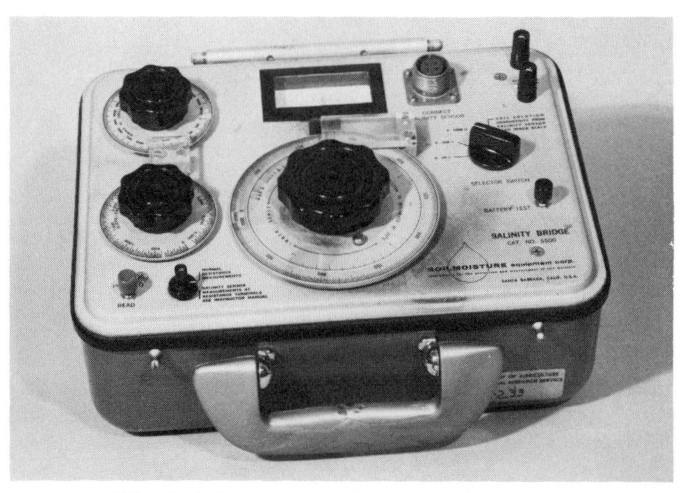

Fig. 42–3. A porous-matrix salinity-sensor meter.

a pull wire. The spring then exerts a steady force to maintain good contact between the porous sensing element and the soil.

42–3.1.4 PROCEDURE

Proper sensor operation depends on the water retentivity characteristics of the ceramic, stable calibration curves for both the electrolytic element and thermistor, and salt diffusional equilibrium between the soil and ceramic. Although the electrolytic element is made of fine-pored ceramic with a bubble pressure of 1500 kPa, L decreases with matric potential (Ingvalson et al., 1970; Aragues, 1982). At −100 kPa the decrease is about 10%. Consequently, sensor readings obtained in drier soils are not accurate. Upon rewetting, sensor conductance returns to normal. Sensors have been made from fine-pored glass which remains saturated to matric potentials of −2000 kPa (Enfield & Evans, 1969). However, the attachment of lead wires to the electrodes stresses the glass in a manner which often results in cracks.

The conductance of the electrolytic element, L, increases linearly with increasing electrical conductivity, κ, of the ceramic water solution. A typical calibration relationship is

$$L = 0.2 + 0.1\kappa, \quad 1 < \kappa < 40 \text{ dS m}^{-1}. \qquad [1]$$

When κ is < 1 dS m^{-1}, the relationship between L and κ is curvilinear, and L approaches zero as κ approaches zero.

Measurement of L in three or four mixed Na/Ca chloride solutions of known κ ($1 < \kappa < 40$ dS m^{-1}) is sufficient for calibration. Additional calibration points are required if measurements of $\kappa < 1$ dS m^{-1} are contemplated. Several precautions need to be taken. Sensors should be allowed to equilibrate with the calibration solution for at least 8 h before

L is measured. If the sensor is initially dry, equilibrate for several days until two succeeding readings, separated by 24 h, differ by $< 2\%$. Three to five days may be required. Although wetting the sensors under vacuum hastens this process, it is not recommended (Aragues, 1982). This wetting procedure causes all ceramic pores to be water-filled, whereas under normal operating conditions this may not occur. Conductance measurements of one sensor by two AC-resistance Wheatstone bridges that are otherwise identical can differ by as much as 4%. This difference is a result of differences between the capacitance of the sensor and the electrical circuitry of the resistance bridge. The resistance readings of a sensor can be grossly different if measured by AC-resistance Wheatstone bridges made by different manufacturers. Consequently, the best procedure is to use the same instrument to calibrate and read the sensor.

The thermistors used in the commercial sensors all have a common resistance–temperature relationship. The following equation, obtained using polynomial regression techniques, relates the temperature, T (°C), to the corresponding thermistor resistance, R (ohms), and to its resistance at 25 °C, R_{25}:

$$T = 7.38\ (R^2/R_{25}) - 39.2\ (R/R_{25}) + 56.9 \qquad [2]$$

for $7 < T < 30$°C. To determine R_{25}, equilibrate the salinity sensor in a stirred temperature bath at 25 ± 0.1°C for at least 5 min and measure its resistance. Thermal equilibrium occurs within 3 to 4 min (Aragues, 1982).

The calibration characteristics of the electrolytic element and thermistor can change with time (Wood, 1978). For the electrolytic elements, both coefficients in Eq. [1] can change. For 205 commercial sensors used for a period of 3 to 5 years, without an intervening period where the electrolytic element was dried, 60% of the calibration slopes (Eq. [1]) increased and 16% decreased. The remainder did not change. Based on the criterion that errors in measured κ ($1 < \kappa < 16$ dS m^{-1}) of 15 to 20% are tolerable, about 70% of these sensors [and other commercial sensors tested by Wood (1978)] retained stable calibrations over a period of 3 to 5 years. Using similar criteria, Aragues (1982) reported that, of 22 commercial sensors tested after 1.2 years of use, only 30% were stable. Sensors should be recalibrated whenever possible, and if they cannot be recalibrated for a period of several years because of experimental design requirements, replication of sensor placement should be used to reduce the potential error.

Thermistor calibration characteristics also change (Wood, 1978). The resistance at 25°C increases with time. After 5 years the average increase was 234 ohms or about 12% of the nominal value of 2000 ohms. This is equivalent to an error in measured temperature of -4°C. The corresponding error in temperature-compensated κ is $+ 9\%$.

Sensor response time is dependant on ion diffusion between the solution in the ceramic and that in the soil (Wesseling & Oster, 1973). Response time depends on thickness of the ceramic conductivity cell, the diffusion coefficients in soil and ceramic, and the fraction of the ceramic surface in contact with soil. The total response time in aqueous solutions is about 10 h. However, for soils, response times can be considerably longer (Wood, 1978), and they increase with decreasing water content (Aragues, 1982). At volumetric soil water contents between 0.3 and 0.45, the response time is increased by a factor of about 3; at water content of 0.05 to 0.2, the factor is about 10. At the lower water contents, the response is still sufficiently fast to monitor changes in salinity which occur over a time interval of at least 5 days.

42–3.1.5 COMMENTS

The primary use of salinity sensors is to continuously monitor κ of soil water at a selected location over a relatively long time. Sensor readings can be made as often as needed (limited only by their response time). Only a few minutes are required to obtain a reading. Similar data cannot be obtained using soil samples because of spatial variation in κ and changes caused by sample removal and dilution with water to obtain an extract. Although similar data can be obtained by vacuum extractors, the use of salinity sensors permits measurement in drier soils and without restrictions with regard to the frequency of sampling and consequent changes in soil solutions around the measuring device. If properly calibrated, buried four-electrode sensors can be used in place of salinity sensors, as discussed later.

Although salinity sensors are useful to monitor salinity at a given location, they are not as useful to monitor field salinity changes (because of the small volume sampled and appreciable soil heterogeneity), nor are they well suited for diagnosis or inventory needs (because of their lack of portability and relatively long response time). Four-electrode and electromagnetic induction devices are recommended for such purposes.

An oscillator circuit system has been developed for automated salinity sensor measurements and data logging (Austin & Oster, 1973), which permits linear readings to be obtained with sensor lead lengths up to several hundred meters.

42–3.2 Bulk Soil Electrical Conductivity Sensors

Soil salinity and soil water electrical conductivity can also be determined from measurements of bulk soil electrical conductivity using four-electrode, electromagnetic induction, and TDR methods and equipment (Rhoades & Ingvalson, 1971; Rhoades & van Schilfgaarde, 1976; Corwin & Rhoades, 1982, 1984; Dalton et al., 1984; Rhoades & Corwin, 1980). Such equipment is now commercially available. With the four-electrode method, the resistance to current flow is measured between one pair of electrodes inserted in the soil while electrical current is caused to flow through the soil between another pair of electrodes. With an appropriate constant, varying with electrode configuration, soil electrical conductivity (κ_a) may be determined from the resistance measurement. With the electromagnetic-induction method, induced current flow in the soil is caused by imposition of a primary electromagnetic field. An induced and measurable secondary electromagnetic field is developed in proportion to κ_a, permitting the latter to be determined. With the TDR method, the dielectric constant, ϵ, and κ_a of the soil are determined from the attenuation of a voltage pulse and the time it takes to pass through the soil, as guided by two parallel rods inserted in the soil. Soil salinity and soil water κ values can be determined from κ_a as described below.

42–3.2.1 PRINCIPLES

Because most soil minerals are insulators, electrical conduction in saline soils is primarily through the pore water, which contains dissolved solutes. The electrical conductivity of the soil is also affected by the number, size, and continuity of soil pores and by the salt and water content of the soil. The contribution of the surface conductance, κ_s, of exchangeable cations to electrical conduction is relatively small and constant, because their mobility is less than that of solutes and little affected by salt concentration and water content.

The dependence of κ_a on the electrical conductivity of the soil water (κ_w), on volumetric water content (θ), on soil pore geometry (T), and on surface conductance (κ_s) is given by

$$\kappa_a = T\,(\kappa_w\theta) + \kappa_s \qquad [3]$$

where T is an empirically determined "transmission" coefficient dependent on θ as

$$T = a\theta + b \qquad [4]$$

with the constants a and b determined by linear regression (Rhoades et al., 1976). Both T and κ_s are properties of the solid soil phase, whereas κ_w and θ are properties of the soil liquid phase. The values of T and κ_s are related to soil type. For a given soil type

$$\kappa_a = A_1\,(\kappa_w\theta) + B \qquad [5]$$

where $A_1 = T$ and $B = \kappa_s$. If κ_a measurements are made at a reference, e.g., calibration water content, then

$$\kappa_a = A_2\kappa_w + B\,. \qquad [6]$$

The conductivity of the soil water, κ_w, can be determined from κ_a and θ measurements using Eq. [6], once calibration factors (A,B) for the soil of interest have been determined (Rhoades et al., 1976; Rhoades, 1980). For any given soil, the conductivity of a saturation extract, κ_e, is uniquely related to κ_w, so that one may also write

$$\kappa_a = A_3\kappa_e + B\,. \qquad [7]$$

Thus, soil salinity can be determined by making κ_a measurements at a reference soil water content using a calibration in the form of Eq. [7] for a given soil type. For irrigated soils, κ_a measurements should be made after irrigation when the soil water content is at field capacity. This water content is reproducible enough to establish the necessary calibrations. Under dryland conditions, κ_a values should be measured in early spring, preferably in fallowed land, to take advantage of relatively uniform soil water conditions when such soils are also near field capacity. Calibrations between κ_e and κ_a have been successfully determined for many soils in this manner and used to diagnose soil salinity (Rhoades, 1976, 1978).

42–3.2.2 APPARATUS

42–3.2.2.1 Four-electrode Sensors. A combination electric current source and resistance meter, four metal electrodes, and connecting wire are needed for large soil volume (array) measurements (Fig. 42–4). A four-electrode salinity probe, in which the electrodes are built into the probe (Rhoades & van Schilfgaarde, 1976) is needed for small soil volume measurements (Fig. 42–5). The current source-meter unit may be either a hand-cranked generator or a battery-powered type. Units designed for geophysical purposes generally read in ohms and should measure from 0.1 to 1000 ohms for soil salinity measurement needs. Units specifically designed for use with the four-electrode salinity probe are much smaller

Fig. 42–4. Photograph of four electrodes positioned in a surface array and a combination electric current generator and resistance meter.

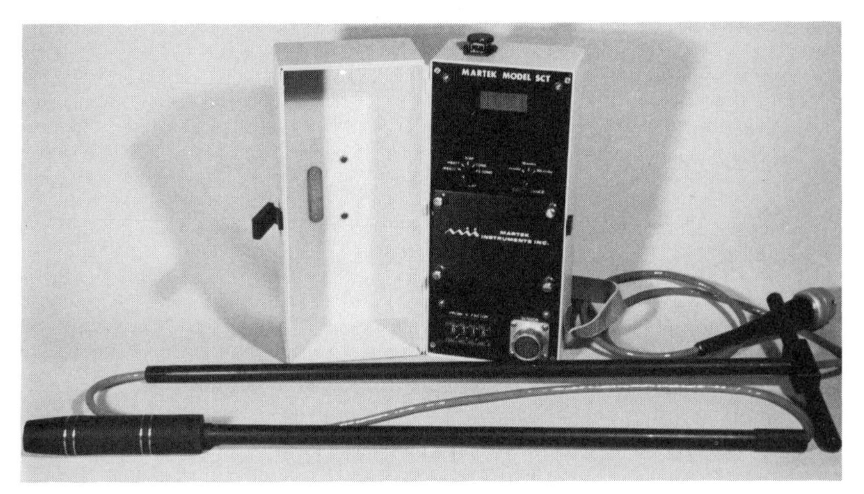

Fig. 42–5. Photograph of commercial four-electrode conductivity probe and generator-meter.

and more convenient. A commercial unit, Martek SCT[1], reads directly in κ_a values corrected to 25°C (Fig. 42–5).

Electrodes may be made of stainless steel, copper, brass, or almost any other noncorrosive conductive metals. Array electrode size is not critical, except that the electrode must be small enough to support its own weight while maintaining firm contact with the soil when inserted to a 5-cm depth or less. Electrodes 1.0 to 1.25 cm in diameter by 45 cm long are convenient for most array purposes, although smaller electrodes are preferred for determination of κ_a within shallow depths (< 30 cm).

Any flexible, well-insulated, multistranded, 12- to 18-gauge wire is suitable for connecting the array electrodes to the meter.

For survey or traverse work, the array electrodes may be mounted in a board with a handle so that soil resistance measurements can be made quickly for a given interelectrode spacing (Fig. 42–6). These "fixed-array" units save the time involved in positioning the electrodes. For most purposes, an interelectrode spacing of 30 or 60 cm is adequate and convenient (wider spacings require lengthy, cumbersome units).

42–3.2.2.2 Electromagnetic Induction Sensors. Figure 42–7 shows the commercially available EM soil salinity sensor (Geonics EM-38)[1] being held in the vertical (coils) position. This device has an intercoil spacing of 1 m, operates at a frequency of 13.2 kHz, is powered by 9-V transistor batteries, and reads κ_a directly. The coil configuration and intercoil spacing were chosen to permit measurement to effective depths of approximately 1 and 2 m when placed at ground level in a horizontal and vertical configuration, respectively. The device contains appropriate circuitry to minimize instrument response to the magnetic susceptibility of the soil and maximize response to κ_a.

The basic principle of operation of the EM soil electrical conductivity

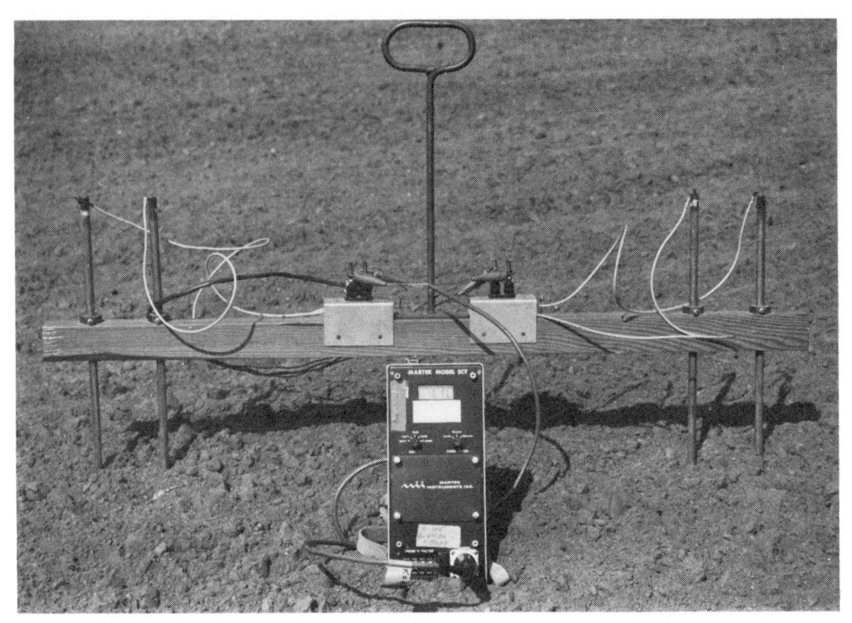

Fig. 42–6. A "fixed-array" four-electrode apparatus and generator-meter.

Fig. 42–7. An electromagnetic soil conductivity sensor.

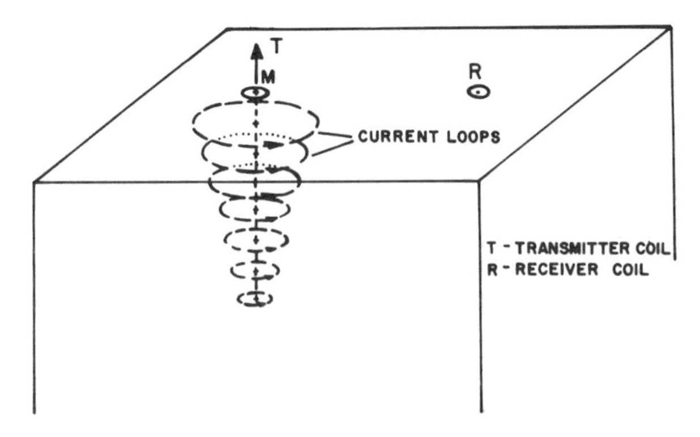

Fig. 42–8. Diagram showing the principle of operation of EM conductivity sensor.

meter is shown schematically in Fig. 42–8. A transmitter coil located in one end of the instrument induces circular eddy current loops in the soil. The magnitude of these loops is directly proportional to the conductivity of the soil in the vicinity of that loop. The current loops generate a secondary electromagnetic field that is proportional to the value of the current flowing within the loop. A fraction of the induced electromagnetic field from the loops is intercepted by the receiver coil, and the signal is amplified and formed into an output voltage that is also linearly related to soil κ.

42–3.2.2.3 Time-domain Reflectometry Sensors. Figure 42–9 shows a TDR insertion sensor (homemade unit) and a commercially available TDR tester (Tektronix 7603 oscilloscope-7512 TDR sampler)[1]. With this equipment the apparent dielectric constant of the soil, ϵ, is obtained by measuring the transit time, t, of a voltage pulse applied to the parallel transmission line (dual-rod probe sensor) of length L embedded in the soil of electrical conductivity κ_a, and applying the relation

$$\epsilon = (ct/2L)^2 \qquad [8]$$

where c is the velocity of light in vacuum. The signal is attenuated in proportion to κ_a so that the transmitted voltage, V_T, is reduced according to

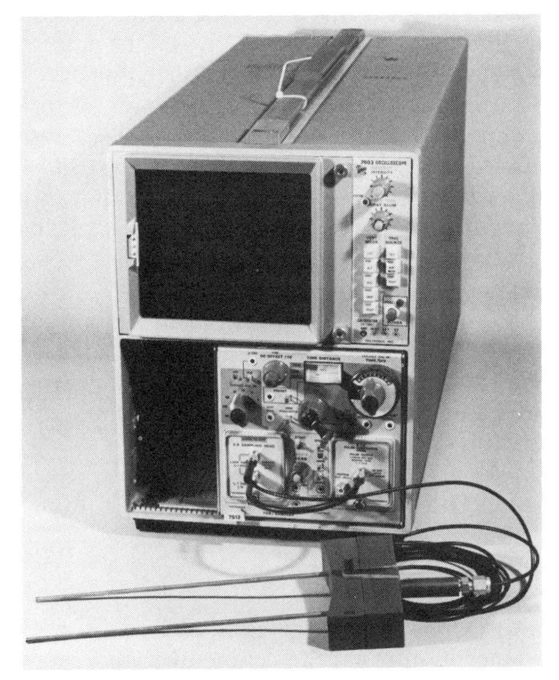

Fig. 42–9. A TDR probe and meter.

$$V_R = V_T \exp(-2\alpha L) . \quad [9]$$

The attenuation coefficient, α, increases linearly with κ_a as

$$\alpha = 60\pi\kappa_a/(\epsilon)^{1/2} . \quad [10]$$

Recent laboratory studies have shown that κ_a and κ_w are well correlated with V_R/V_T (Dalton et al., 1984). Since the practical attributes of this method have not been evaluated at this time, procedures for its use for field salinity measurements cannot be given. The method offers the distinct advantage of measuring both water content and soil electrical conductivity simultaneously.

42–3.2.3 PROCEDURES

42–3.2.3.1 Large Volume Measurements. For the purpose of determining soil salinity of entire root zones or some fraction thereof, it is desirable to make the measurement when the current flow is concentrated within this soil depth. This can be accomplished with the four-electrode equipment by selecting the appropriate spacing between the two current (outer) electrodes, which are placed in the soil surface. In this arrangement, four electrodes are placed in a straight line. With conventional geophysical resistivity measurements the electrodes are equally spaced in the so-called Wenner array. With the Martek SCT meter each of the inner pair of electrodes is placed a distance inward from its closest outer-pair counterpart equal to 10% of the spacing between the outer pair. The inner pair is used to measure the potential while current is passed between the outer pair. The effective depth of current penetration for either configuration (in the absence of appreciable soil layering) is about equal to one-third the outer-electrode spacing, y, and the average soil salinity to approximately this depth is measured (Rhoades & Ingvalson, 1971; Halvorson & Rhoades, 1974). Thus, by varying the spacing between current electrodes, one can measure average soil salinity or soil-water salinity to different depths and within different volumes of soil. One of the major advantages of this method is the much larger volume of soil brought under measurement than with soil samples, soil water extractors, or salinity sensors. The volume of measurement is about $\pi(y/3)^3$. Hence, effects of small-scale variations in field-soil salinity on sampling requirements can be minimized by these large-volume measurements.

For measurements taken in the Wenner array (electrodes equally spaced), the soil electrical conductivity is calculated from

$$\kappa_a = f_t/a\ R_t \quad [11]$$

where a is the distance between the electrodes in centimeters, R_t is the measured resistance in ohms at the field temperature t, and f_t is a factor to adjust the reading to a reference temperature of 25°C (see Table 15,

U.S. Salinity Laboratory Staff, 1954). For measurements made using the Martek SCT meter, a factor is supplied in chart form for each spacing of outer electrodes; this factor is dialed into the meter and the correct soil κ reading is displayed in the meter readout.

Large volumes of soil can also be measured with the electromagnetic induction technique. The volume and depth of measurement can be increased by increasing the spacing between coils, by reducing the current frequency, and by positioning the coils so that their axis is vertical to the soil surface plane. The effective depths of measurement of the Geonics EM-38 device are about 1 and 2 m when it is placed on the ground and the coils are positioned horizontally and vertically, respectively. The EM-38 device does not integrate soil κ_a linearly with depth. The 0 to 0.305, 0.305 to 0.61, 0.61 to 0.915, and 0.915 to 1.22 m depth intervals contribute about 43, 21, 10, and 6%, respectively, to the κ_a reading of the EM unit when positioned on the ground in the horizontal position (Rhoades & Corwin, 1980). Thus, the conductivity read by the EM device is given by

$$\kappa_{EM} = 0.43\kappa_{0-0.3} + 0.21\kappa_{0.3-0.6} + 0.10\kappa_{0.6-0.9} + 0.06\kappa_{0.09-1.2} + 0.1\kappa_{>1.2} \quad [12]$$

where the κ subscript designates the depth interval in the soil. While this response of the EM-38 device does not permit easy determinations of average soil κ_a (or salinity), it does give a weighted reading of κ_a (or salinity) that is close to that to which crops seem to respond under certain circumstances (Bernstein & Francois, 1973; Rhoades & Merrill, 1976). Since irrigated crops tend to remove the water approximately in the proportions 40:30:20:10 by successively deeper quarter-fractions of their root zone, the water-uptake-weighted conductivity for the profile, κ_{wu}, is

$$\kappa_{wu} = 0.4\ \kappa_1 + 0.3\ \kappa_2 + 0.2\ \kappa_3 + 0.1\ \kappa_4$$

where κ_1, κ_2, κ_3, and κ_4 are the conductivities of the successively deeper quarter-fractions of the root zone. This integrated conductivity is similar to that measured by the EM unit, and the device appears to generate a conductivity value that gives a measurement of soil κ_a close to that to which crop response is related.

In spite of the interesting feature of the EM-38 device described above, it is desirable to be able to determine soil κ_a-depth distributions for measuring average soil salinity and making management assessments. Since the proportional contribution of each soil depth interval to κ_a as measured by the EM unit can be varied by raising it aboveground to higher and higher heights, it is possible to calculate the κ_a-depth relation from a succession of EM measurements made at various heights aboveground (Rhoades & Corwin, 1980). The κ_a values of each soil depth interval are simply correlated with the succession of EM readings as

$$\kappa_{0-0.3} = \beta_0 EM_0 + \beta_1 EM_1 + \beta_2 EM_2 + \beta_3 EM_3 + \beta_4 EM_4; \quad [13a]$$

$$\kappa_{0.03\text{-}0.6} = \gamma_0\text{EM}_0 + \gamma_1\text{EM}_1 + \gamma_2\text{EM}_2 + \gamma_3\text{EM}_3 + \gamma_4\text{EM}_4;\ \text{etc.} \qquad [13b]$$

The coefficients in Eq. [13], as reported in Rhoades and Corwin (1980) are generally applicable, though exceptions have been found. Another series of equations has been derived to obtain the actual κ_a within a given soil depth interval from measurements made with the magnetic coils of the EM instrument positioned at ground level, first horizontally and then vertically (Corwin & Rhoades, 1982). For the depth increment $x1$–$x2$, the equations are of the form

$$\kappa_{(x1-x2)} = (k_1\ \text{EM}_H - k_2\ \text{EM}_V)/k_3 \qquad [14]$$

where EM_V and EM_H are the apparent bulk soil electrical conductivities measured electromagnetically at the soil surface in the vertical and horizontal positions, respectively; $x1$–$x2$ is the soil depth increment; and k_1, k_2, and k_3 are empirically determined coefficients for the depth increment.

42–3.2.3.2 Small Volume Measurements. Sometimes information on salinity distribution within a specified fraction of the whole root zone is desired, or permanently installed devices are needed. For such conditions, the four-electrode salinity probe (Rhoades & van Schilfgaarde, 1976) and burial-type probe (Rhoades, 1979) are recommended, respectively (although the EM-38 device can be used as described above and a new EM-probe is under development for this purpose). In these devices (see Fig. 42–5 and 42–10), four annular rings are molded in a plastic probe that is slightly tapered so it can be inserted into the soil to a desired depth via a hole made with a coring tube. In the portable version, the

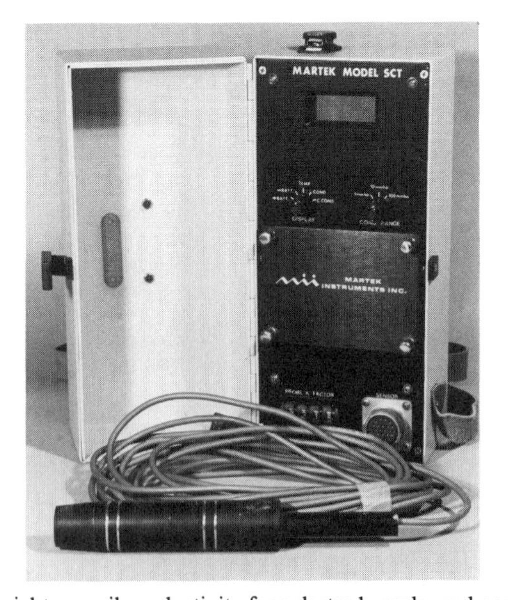

Fig. 42–10. A burial-type soil conductivity four-electrode probe and generator-meter.

probe is attached to a shaft through which the electrical leads are passed and connected to a meter. In the burial unit (Fig. 42–10), the leads from the probe are brought to the soil surface. The volume of sample under measurement can be varied by changing the spacing between the current electrodes. The standard commercial unit, Martek SCT, measures a soil volume of about 2350 cm^3.

To determine soil κ with the four-electrode probe, core a hole in the soil to the desired depth of measurement using an Oakfield soil sampling tube (or sampler of similar diameter). Insert the four-electrode sensor into the soil and record the resistance or the displayed κ, depending on the meter being used. When using meters which display resistance, κ is calculated as

$$\kappa_a = \mathrm{k}\, f_t / R_t \qquad [15]$$

where k is an empirically determined geometry constant (cell constant) for the probe in units of 1000 cm^{-1}, R_t is the resistance in ohms at the field temperature, and f_t is a factor to adjust the reading to a reference temperature of 25°C (see Table 15, U.S. Salinity Laboratory Staff, 1954).

42–3.2.4 COMMENTS

Average soil, κ_e, or soil water, κ_w, conductivity may be determined from the κ_a measurements using calibrations established between κ_e and κ_a or κ_w and κ_a, respectively, for the soil type(s) in question, if the κ_a determinations are made at approximately the same water content as that at which the calibrations were made. Similar determinations may also be made of more limited and discrete soil volumes with the four-electrode salinity probe.

A separate calibration for each individual soil of interest is not usually necessary. Calibrations are similar enough for soils of similar water-holding capacities and textures so that suitable salinity appraisals can frequently be made using generalized calibrations (Rhoades, 1978, 1980). If specific calibrations are required, however, they are simply established using one of three methods (Rhoades, 1976; Rhoades & Halvorson, 1977).

The depth to which average soil κ_a is measured with the horizontal array method may be varied by varying the spacing between the electrodes. The effective depth of measurement of soil κ_a (and hence soil salinity) is about equal to one-third the outer electrode spacing, provided the soil is essentially uniform in soil physical properties to this depth. Thus, by varying the spacing between electrodes, the average soil salinity can be determined to different depths, provided the soil has a relatively uniform texture through this depth. A single calibration, such as that appropriate to the surface soil, cannot be applied to the subsoil if the subsoil's texture is appreciably different. For such cases, the four-electrode salinity probe should be used to determine the soil κ_a in each discrete stratum or depth interval. Salinity can then be interpreted from calibrations appropriate to the soil type in each stratum.

As shown in Eq. [5], water content, as well as salinity, affects soil electrical conductivity. To remove this influence, determinations are made when the soil is near calibration water content. Deviations of water content from calibration levels are not as troublesome as might be assumed at first glance, because of compensating factors operative during the soil drying process. This may be explained with reference to Eq. [5]. After water infiltration and rapid drainage following an irrigation, most of the remaining loss of soil water occurs by transpiration. In this process the salts are essentially excluded during uptake by the crop and remain behind in the water, thereby proportionately increasing its concentration. Hence, the product ($\kappa_w \cdot \theta$) tends to remain the same, i.e., the total amount of soluble electrolyte in the soil water is not greatly decreased during water loss. Thus, the prime factor producing changes in measurable κ_a with changes in θ is the degree to which T, the tortuosity-related term, is affected. The degree of this effect is not great (Rhoades et al., 1976); it produces a percentage error in κ_a about equivalent to the percentage deviation in water content at time of measurement from that at calibration. The extreme limit of deviation in water content in irrigated soils following drainage is 50%, i.e., field capacity to permanent wilting percentage. One would normally make κ_a determinations following an irrigation, when the soil would not differ much from the field-capacity water content. A deviation in water content at time of measurement from that at calibration equal to 20% will cause an error of about 20% in κ_a. For more discussion in this regard, see the review by Rhoades (1984).

42-4 REFERENCES

Adams, Fred. 1974. Soil solution. p. 441–481. *In* E. W. Carson (ed.) The plant root and its environment. University Press, Virginia, Charlottesville.

Alberts, E. E., R. E. Burwell, and G. E. Schuman. 1977. Soil nitrate-nitrogen determined by coring and solution extraction techniques. Soil Sci. Soc. Am. J. 41:90–92.

Aragues, R. 1982. Factors affecting the behavior of soil salinity sensors. Irrig. Sci. 3:133–147.

Austin, R. S., and J. D. Oster. 1973. An oscillator circuit for automated salinity sensor measurements. Soil Sci. Soc. Am. Proc. 37:327–329.

Bernstein, L., and L. E. Francois. 1973. Leaching requirement studies: sensitivity of alfalfa to salinity of irrigation and drainage waters. Soil Sci. Soc. Am. Proc. 37:931–943.

Biggar, J. W., and D. R. Nielsen. 1976. Spatial variability of the leaching characteristics of a field soil. Water Resour. Res. 12:78–84.

Briggs, L. J., and A. G. McCall. 1904. An artificial root for inducing capillary movement on soil moisture. Science 20:566–569.

Brooks, R. H., J. O. Goertzen, and C. A. Bower. 1958. Prediction of changes in the compositions of the dissolved and exchangeable cations in soils upon irrigation with high-sodium waters. Soil Sci. Soc. Am. Proc. 22:122–124.

Chow, T. L. 1977. A porous cup soil-water sampler with volume control. Soil Sci. 124:173–176.

Corwin, D. L., and J. D. Rhoades. 1982. An improved technique for determining soil electrical conductivity depth relations from above ground electromagnetic measurements. Soil Sci. Soc. Am. J. 46:517–520.

Corwin, D. L., and J. D. Rhoades. 1984. Measurement of inverted electrical conductivity profiles using electromagnetic induction. Soil Sci. Soc. Am. J. 48:288–291.

Dalton, F. N., W. N. Herklerath, D. S. Rawlins, and J. D. Rhoades. 1984. Time domain reflectrometry: Simultaneous measurement of the soil water content and electrical conductivity with a single probe. Science 224:989–990.

Dao, T. H., and T. L. Lavy. 1978. Extraction of soil solution using a simple centrifugation method for pesticide adsorption-desorption studies. Soil Sci. Soc. Am. J. 42:375–377.

Davies, B. E., and R. J. Davies. 1963. A simple centrifugation method for obtaining small samples of soil solution. Nature (London) 198:216–217.

Duke, H. R., and H. R. Haise. 1973. Vacuum extractors to assess deep percolation losses and chemical constituents of soil water. Soil Sci. Soc. Am. Proc. 37:963–964.

Enfield, C. C., and D. D. Evans. 1969. Conductivity instrumentation for in situ measurement of soil salinity. Soil Sci. Soc. Am. Proc. 33:787–789.

England, C. B. 1974. Comments on "A technique using porous cups for water sampling at any depth in the unsaturated zone," by W. W. Wood. Water Resour. Res. 10:1049.

Gilman, G. P. 1976. A centrifuge method for obtaining soil solution. Div. Soils Rep. 16, CSIRO, Melbourne, Australia.

Halvorson, A. D., and J. D. Rhoades. 1974. Assessing soil salinity and identifying potential saline-seep areas with field soil resistance measurements. Soil Sci. Soc. Am. Proc. 38:576–581.

Hansen, E. A., and A. R. Harris. 1975. Validity of soil-water samples collected with porous ceramic cups. Soil Sci. Soc. Am. Proc. 39:528–536.

Harris, A. R. and E. A. Hansen. 1975. A new ceramic cup soil-water sampler. Soil Sci. Soc. Am. Proc. 39:157–158.

Hoffman, G. J., C. Dirksen, R. D. Ingvalson, E. V. Maas, J. D. Oster, S. L. Rawlins, J. D. Rhoades, and J. van Schilfgaarde. 1978. Minimizing salt by irrigation management—design and initial results of Arizona field studies. Agric. Water Manage. 1(3):233–252.

Ingvalson, R. D., J. D. Oster, S. L. Rawlins, and G. J. Hoffman. 1970. Measurement of water potential and osmotic potential in soil with a combined thermocouple psychrometer and salinity sensor. Soil Sci. Soc. Am. Proc. 34:570–574.

Jackson, D. R., F. S. Brinkley, and E. A. Bendetti. 1976. Extraction of soil water using cellulose-acetate hollow fibers. Soil Sci. Soc. Am. J. 40:327–329.

Jordan, C. F. 1978. A simple, tension-free lysimeter. Soil Sci. 105:81–86.

Kemper, W. D. 1959. Estimation of osmotic stress in soil water from the electrical resistance of finely porous ceramic units. Soil Sci. 87:345–349.

Kinniburgh, D. G., and D. L. Miles. 1983. Extraction and chemical analysis of water from rocks. Environ. Sci. Technol. 17:362–368.

Kohnke, H., F. R. Feibelbis, and J. M. Davidson. 1940. A survey and discussion of lysimeters and a bibliography on their construction and performance. USDA Misc. Pub. 372. U.S. Government Printing Office, Washington, DC.

Krone, R. B., H. F. Ludwig, and J. F. Thomas. 1952. Porous tube device for sampling soil solutions during water-spreading operations. Soil Sci. 73:211–219.

Krugel, C., C. Dreyspring, and W. Heins. 1935. A new suction apparatus for the complete separation of the soil solution from the soil itself. Superphosphate 8:101–108.

Levin, M. J., and D. R. Jackson. 1977. A comparison of *in situ* extractors for sampling soil water. Soil Sci. Soc. Am. J. 41:535–536.

Mubarak, A., and R. A. Olsen. 1976. Immiscible displacement of the soil solution by centrifugation. Soil Sci. Soc. Am. J. 40:329–331.

Mubarak, A., and R. A. Olsen. 1977. A laboratory technique for appraising in situ salinity of soil. Soil Sci. Soc. Am. J. 41:1018–1020.

Nielsen, D. R., J. W. Biggar, and K. T. Erh. 1973. Spatial variability of field-measured soil-water properties. Hilgardia 42:215–259.

Oster, J. D., and R. D. Ingvalson. 1967. In situ measurement of soil salinity with a sensor. Soil Sci. Soc. Am. Proc. 32:572–574.

Oster, J. D., L. S. Willardson, and G. J. Hoffman. 1973. Sprinkling and ponding techniques for reclaiming saline soils. Trans. ASAE 16:89–91.

Page, A. L., R. H. Miller, and D. R. Keeney (ed.) 1982. Methods of soil analysis, Part 2. Agronomy 9.

Parizek, R. R., and B. E. Lane. 1970. Soil-water sampling using pan and deep pressure-vacuum lysimeters. J. Hydrol. 11:1–21.

Reeve, R. C., and E. J. Doering. 1965. Sampling the soil solution for salinity appraisal. Soil Sci. 99:339–344.

Rhoades, J. D. 1972. Quality of water for irrigation. Soil Sci. 113:277–284.

Rhoades, J. D. 1976. Measuring, mapping, and monitoring field salinity and water table depths with soil resistance measurements. FAO Soils Bull. 31:159–186.

Rhoades, J. D. 1978. Monitoring soil salinity: A review of methods. p. 150–165. *In* L. G. Everett and K. D. Schmidt (ed.) Establishment of water quality monitoring programs. American Water Resource Association, St. Paul, MN.

Rhoades, J. D. 1979. Inexpensive four-electrode probe for monitoring soil salinity. Soil Sci. Soc. Am. J. 43:817–818.

Rhoades, J. D. 1980. Predicting bulk soil electrical conductivity versus saturation paste electrical conductivity calibrations from soil properties. Soil Sci. Soc. Am. J. 45:42–44.

Rhoades, J. D. 1982. Soluble salts. *In* A. L. Page et al. (ed.) Methods of soil analysis, Part 2. Agronomy 9:167–178.

Rhoades, J. D. 1984. Principles and methods of monitoring soil salinity. p. 130–142. *In* I. Shainberg and J. Shalhevet (ed.) Soil salinity under irrigation—processes and management. Springer-Verlag, New York, New York.

Rhoades, J. D., and D. L. Corwin. 1980. Determining soil electrical conductivity–depth relations using an inductive electromagnetic soil conductivity meter. Soil Sci. Soc. Am. J. 45:255–260.

Rhoades, J. D., and A. D. Halvorson. 1977. Electrical conductivity methods for detecting and delineating saline seeps and measuring salinity in Northern Great Plains soils. USDA-ARS-42. U.S. Government Printing Office, Washington, DC.

Rhoades, J. D., and R. D. Ingvalson. 1971. Determining salinity in field soils with soil resistance measurements. Soil Sci. Soc. Am. Proc. 35:54–60.

Rhoades, J. D., and S. D. Merrill. 1976. Assessing the suitability for irrigation: theoretical and empirical approaches. FAO Soil Bull. 31:69–109.

Rhoades, J. D., P. A. C. Raats, and R. J. Prather. 1976. Effects of liquid-phase electrical conductivity, water content, and surface conductivity on bulk soil electrical conductivity. Soil Sci. Soc. Am. J. 40:651–655.

Rhoades, J. D., and J. van Schilfgaarde. 1976. An electrical conductivity probe for determining soil salinity. Soil Sci. Soc. Am. J. 40:647–651.

Richards, L. A. 1966. A soil salinity sensor of improved design. Soil Sci. Soc. Am. Proc. 30:333–337.

Shimshi, Daniel. 1966. Use of ceramic points for sampling of soil solution. Soil Sci. 101:98–103.

Tadros, V. T., and J. W. McGarity. 1976. A method for collecting soil percolate and soil solution in the field. Plant Soil 44:655–667.

Todd, R. M., and W. D. Kemper. 1972. Salt dispersion coefficients near an evaporating surface. Soil Sci. Am. Proc. 36:539–543.

U.S. Salinity Laboratory Staff. 1954. L. A. Richards (ed.) Diagnosis and improvement of saline and alkali soils. USDA Agric. Handb. 60. U.S. Government Printing Office, Washington, DC.

van De Pol, R. M., P. J. Wierenga, and D. R. Nielsen. 1977. Solute movement in a field. Soil Sci. Soc. Am. J. 41:10–13.

Wagner, G. H. 1962. Use of porous ceramic cups to sample soil water within the profile. Soil Sci. 94:379–386.

Wesseling, J., and J. D. Oster. 1973. Response of salinity sensors to rapidly changing salinity. Soil Sci. Soc. Am. Proc. 37:553–557.

Wierenga, P. J., and T. C. Patterson. 1974. Quality of irrigation return flow in the Mesilla Valley. Trans. Int. Congr. Soil Sci., 10th 10:216–222.

Wolff, R. G. 1967. Weathering Woodstock granite near Baltimore, Maryland. Am. J. Sci. 265:106–117.

Wood, J. D. 1978. Calibration, stability, and response time for salinity sensors. Soil Sci. Soc. Am. J. 42:248–250.

Wood, W. W. 1973. A technique using porous cups for water sampling at any depth in the unsaturated zone. Water Resour. Res. 9:486–488.

Yamasaki, A., and A. Kishita. 1972. Studies on soil solutions with reference to nutrient availability. I. Effect of various potassium fertilizer on its behavior in the soil solution. Soil Sci. Plant Nutr. 18:1–6.

43 Solute Diffusivity[1]

W. D. KEMPER

Snake River Conservation Research Center, ARS, USDA
Kimberly, Idaho

43–1 INTRODUCTION: SOLUTES IN WATER

Molecules and ions in soil water are commonly bound to portions of the semicrystalline water lattice by dipole-dipole or charge-dipole electrical attractions. Consequently, when the semicrystalline water lattice moves in response to body forces exerted by gravity or to electrical potential gradients, the solute molecules and ions are carried or convected with the lattice (as outlined in chapter 45 in this book). When water moves, portions of the water lattice nearest the solid surfaces generally move relatively slowly and those further from solid surfaces move more quickly. This range of lattice velocities results in dispersion or mixing of solution components, which is discussed in chapter 45.

Bonding energy of solutes to the water lattice is similar in magnitude to the kinetic energy of unbound water molecules. Consequently, impacts by thermally agitated molecules often dislodge solute molecules or ions from the lattice and they move through the lattice colliding occasionally with water and other solute molecules, imparting their kinetic energy to them and eventually becoming adsorbed in new positions on the water lattice. This movement of solutes with respect to the water lattice is random for individual ions or molecules in an isothermal system. Consequently, when the initial concentration of a solute is higher in one zone than in an adjacent zone, a net flux or diffusion from the zone of higher to the zone of lower concentration will occur. Diffusion of solutes with respect to the water lattice is often slower than convection. Consequently, when measuring diffusion coefficients, elimination of convection is a primary requisite.

In porous media and particularly near solid surfaces, the velocity of the water lattice (convection) is relatively slow and diffusion is often the dominant means by which fertilizer ions move away from solution in the vicinity of solid fertilizer particles, where their concentration is high, to other portions of the soil water where their concentration is lower.

---

[1]Contribution from USDA-ARS, Snake River Conservation Research Center, Kimberly, ID 83341.

Similarly, there is a net diffusive flux toward the root surfaces, which have reduced the nutrient ion concentration in their vicinity. The rates of these and other diffusive transport phenomena in soil can be predicted if water and soil characteristics are known and diffusion coefficients have been determined as a function of those characteristics.

43–2 DIFFUSION CONCEPTS

43–2.1 Fick's First Law (for Steady-state Diffusion)

43–2.1.1 DIFFUSION OF SOLUTES IN WATER

When steady gradients of concentration are maintained (i.e., dc/dx does not change with time) and convective movement of the solution does not occur, kinetic theory predicts and experiments have shown that

$$J = -D\,\partial C/\partial x\,. \qquad [1]$$

Since Fick's time, this law has also been formulated in terms of chemical potential which allows consideration of activity coefficients (Jost 1952). If a scientist is dealing with systems where expected concentration ranges are large enough to encompass large differences in activity coefficients, utilization of the chemical potential form avoids the tendency of D to vary slightly with concentration, which is inherent in Eq. [1] under these conditions.

If the flux J of the diffusing component is expressed in g/cm^2 s, C in g/cm^3, and x in cm, then D is in cm^2/s, which is the accepted dimensional unit for diffusion coefficients. Other sets of units may be used for J and C as long as their quantity factors are consistent. For instance, chemists generally describe concentration in terms of moles (or equivalents) per liter, which is identical to millimoles (or milliequivalents) per milliliter. To keep D in cm^2/s, the flux J must be expressed in millimoles (or milliequivalents) per cm^2/s.

There are many advantages in working with units used by chemists, since they have accumulated extensive literature, determined the bulk solution diffusion coefficients of most ions (see Table 43–1), demonstrated that diffusion coefficients are inversely proportional to the viscosity of water, and shown that diffusion coefficients are equal to the electrically determined ion mobility times its average thermal energy, kT, (where k is Boltzmann's constant and T is the temperature in Kelvin). Since the viscosity of water in soils (other than the first two molecular layers adsorbed on mineral surfaces) is not greatly different from its viscosity in bulk solution (Kemper et al., 1964), diffusion coefficients from the chemistry literature provide good estimates of ion mobility in soil solution.

Convective movement of bulk solution has been a continuing prob-

Table 43–1. Diffusion coefficients of ions, D, calculated from their limiting equivalent conductivities in water at 25 °C (Robinson & Stokes, 1959).†

Cation	D	Anion	D
	cm²/s		cm²/s
H^+	9.30×10^{-5}	OH^-	5.26×10^{-5}
Li^+	1.03×10^{-5}	F^-	1.48×10^{-5}
Na^+	1.33×10^{-5}	Cl^-	2.03×10^{-5}
K^+	1.96×10^{-5}	Br^-	2.08×10^{-5}
Rb^+	2.07×10^{-5}	I^-	2.05×10^{-5}
Cs^+	2.06×10^{-5}	NO_3^-	1.90×10^{-5}
Ag^+	1.65×10^{-5}	ClO_3^-	1.72×10^{-5}
NH_4^+	1.96×10^{-5}	BrO_3^-	1.49×10^{-5}
$CH_3NH_3^+$	1.56×10^{-5}	IO_3^-	1.08×10^{-5}
Mg^{2+}	0.71×10^{-5}	IO_4^-	1.45×10^{-5}
Ca^{2+}	0.79×10^{-5}	HCO_3^-	1.19×10^{-5}
Sr^{2+}	0.79×10^{-5}	Formate	1.45×10^{-5}
Ba^{2+}	0.85×10^{-5}	Acetate	1.09×10^{-5}
Cu^{2+}	0.72×10^{-5}	SO_4^{2-}	1.07×10^{-5}
Zn^{2+}	0.70×10^{-5}	$C_2O_4^{2-}$	0.99×10^{-5}
Co^{2+}	0.73×10^{-5}	CO_3^{2-}	0.92×10^{-5}
Pb^{2+}	0.93×10^{-5}	$P_3O_9^{3-}$	0.74×10^{-5}
La^{3+}	0.62×10^{-5}	$P_2O_7^{4-}$	0.64×10^{-5}

† Diffusion coefficients of salts can be calculated from these coefficients for ions using Eq. 43–3. Diffusion coefficients in water at any other temperature, T, can be calculated from $D_T = (M_{25}/M_T)\, D_{25}$ where M_{25} and M_T are the viscosities at the respective temperatures.

lem in measurement of diffusion coefficients in bulk water. Consequently, since the relation between electrically induced mobility of ions and their diffusion coefficients has become well established, most published ionic diffusion coefficients (see Table 43–1) have been derived from electrical mobility determinations.

43–2.1.2 EFFECTS ON ION DIFFUSION OF SOLUTION VISCOSITY AND MOBILITIES OF CO-DIFFUSING OR COUNTER-DIFFUSING IONS

Unless there are electrode reactions taking place in the system that can add and remove electrons from solution, a net transfer of electric charge by ions builds up electrical potential gradients which quickly cause net electron flux carried by the ions to be zero, i.e.,

$$\sum_{i=1}^{n} Z_i\, J_i = 0 \qquad [2]$$

where Z_i is the valence of the ion species (negative for anions) and J_i is the flux of that species in a specific direction. As has been pointed out by several investigators (Low, 1981; Kemper & Quirk, 1972; Kemper et al., 1972), these electrical potential gradients, resulting from different mobilities of diffusing ions as a salt diffuses through a system, cause many interesting phenomena which are beyond the scope of this discussion. However, a directly pertinent electrical interaction is the role that a co-diffusing or counterdiffusing ion species has on the diffusion coefficient of the first species.

Codiffusion takes place when a cation species and an anion species diffuse in the same direction. Table 43–1 shows that the two ion species of any particular salt probably have different mobilities. If the anion is the faster-diffusing ion, it creates a negative charge in the direction of flux, which retards the flux of the anions and increases the flux of the cations. The electrical potential gradient increases until the flux of the two species is the same, and it can be shown (Low, 1981) that this causes the diffusion coefficient of the salt D_{+-} to be related to the diffusion coefficients of the anion D_- and the cation D_+ by

$$D_{+-} = [D_+D_-(Z_+C_+ + Z_-C_-)]/(D_+Z_+C_+ + D_-Z_-C_-) \qquad [3]$$

where the valences Z are absolute values and C_+ and C_- are the cation and anion concentrations, respectively. For instance, if the salt diffusing were $LiNO_3$ and concentrations of Li^+ and NO_3^- were the same, the diffusion coefficients of Li^+ and NO_3^- in Table 43–1 and Eq. [3] tell us that the diffusion coefficient for $LiNO_3$ at 25°C in water would be 1.34 $\times$ 10^{-5}cm^2/s, which is identical to the value given for $LiNO_3$ in the *American Institute of Physics Handbook*. Consequently, the diffusion coefficients of most salts in water at 25°C can be calculated from data on ionic diffusion coefficients in Table 43–1 and Eq. [3].

Values of the diffusion coefficients at 25°C can be multiplied by the viscosity of water at 25°C and divided by the viscosity of water at any other temperature to obtain diffusion coefficients at that temperature.

Counterdiffusion takes place when two ionic species with charge of the same sign are diffusing in opposite directions. In this case, an equation identical to Eq. [3], except defining the ions as species 1 and 2 rather than as + and −, can be used to estimate their combined counter-diffusion coefficient. The components of the salt are diffusing in the same direction and will commonly be equal to each other so that D_{+-} will be relatively constant. When ions are counterdiffusing, their concentrations gradients are in opposite directions, and consequently a diffusion coefficient which varies with distance is probable. The codiffusion or counterdiffusion coefficients can be neither lower than the lowest nor higher than the highest coefficient of the diffusing ions.

43–2.1.3 APPLICATION OF FICK'S FIRST LAW TO POROUS MEDIA

Diffusion of mineral solutes is practically limited to the liquid phase, which occupies only a fraction of the soil volume, and the cross-sectional area available for diffusion is decreased by this same fraction. Moreover, the diffusion pathway L_e between two points is tortuous and longer in soils than the straight distance L between those points, and for reasons discussed by Porter et al. (1960) this tortuosity reduces the diffusion by the factor $(L/L_e)^2$. Electrical interactions, such as negative adsorption of cations and anions from constrictions in the flow paths (e.g., Van Schaik & Kemper 1966), can reduce the ion mobility by another factor, γ. If most of the ions are diffusing in the first two molecular water layers adsorbed on mineral surfaces, they will be hindered by the low mobility of water, and the fraction α representing the average mobility, divided by the mobility in bulk solution, can be appreciably < 1.0.

Consequently, the diffusion coefficient in porous media, D_p, is lower than the diffusion coefficient D in bulk solution. The relation of these coefficients to each other may be expressed as

$$D_p = \theta \, (L/L_e)^2 \, \gamma \, \alpha \, D \, . \qquad [4]$$

Some investigators find it convenient to express solute concentration in terms of quantity per cubic centimeter of bulk soil volume, C_b, rather than in terms of quantity per cubic centimeter of solution. In essence, they move the term θ (i.e., $cm^3 H_2O/cm^3$ soil) in Eq. [4] into the concentration term to make this change (i.e., $C_b = \theta \, C$). When they substitute C_b for C in Eq. [1], their diffusion coefficient becomes

$$D_b = (L/L_e)^2 \, \gamma \, \alpha \, D \, . \qquad [5]$$

Thus, in porous media when the bulk volume concentration C_b is used, Fick's first law becomes

$$J = D_b \, (\partial C_b/\partial x) \, . \qquad [6]$$

Arnold Klute (personal correspondence) pointed out that Eq. [6] is valid only when θ is constant. Otherwise, since by general definition

$$\partial C_b \theta/\partial x = \theta \, (\partial C/\partial x) + C(\partial \theta/\partial x) \, ,$$

unqualified use of Eq. [6] implies that $C(\partial \theta/\partial x)$ is a driving force for diffusion, and it is easily demonstrated that this is not true.

The necessity for this qualification is avoided if solute concentrations are expressed in terms of quantity of solute per unit volume of solution, C, and θ is left in the diffusion coefficient; i.e.,

$$J = D_p \, (\partial C/\partial x) \, . \qquad [7]$$

When comparing data obtained from studies which have used Eq. [6] and [7], it is helpful to remember that $D_b \theta = D_p$.

In steady-state measurement of diffusion coefficients, a solution-phase concentration gradient is often imposed (see section 43–3.1) across a soil that contains more cations adsorbed on mineral surfaces than are in solution. While the mobility of these adsorbed cations is less than in bulk solution, the energies of bonds holding monovalent cations to isomorphous substitution exchange spots on clays are of the same order of magnitude as energies of those holding cations to the water lattice and as the thermal energy kT (Shainberg & Kemper, 1966). Consequently, mobilities of sodium ions adsorbed on smectite clays can be as high as 60% of their mobilities in bulk solution. Relative mobilities of 30% for Ca^{2+} adsorbed on smectite clays were also measured (Van Schaik et al., 1966). Adsorbed ions may also attach directly to clay mineral lattice edges in covalent or coordinate covalent linkages. Energy of these bonds is often four or more times kT, and consequently their relative mobility is $< 2\%$. Concentration of adsorbed cations can be much greater than of solution ions when solution concentrations are in the range commonly encountered in soils. Consequently, adsorbed cations often play a significant role in diffusion.

When diffusion involves a cation and an anion species moving in the same direction, Eq. [3] expresses the relation of the diffusion coefficient of the salt to that of the ions where the "concentrations" of the ions are calculated from the amounts in the solution plus the amounts adsorbed times their relative mobilities. From Eq. [3] it can be deduced that when a salt is diffusing, its diffusion coefficient approaches that of the anion if the system contains large amounts of adsorbed cations.

If the diffusion involves two cations diffusing in opposite directions, the coefficients defining the interdiffusion at any plane normal to the diffusion will be dominated by the diffusion coefficient of the ion having the lowest concentration at that plane.

In the interdiffusion case, the effective concentration of mobile cation (solution cation plus adsorbed cation times its relative mobility) is generally large compared to that of the solution cation. The effective concentration gradient is larger than the solution concentration gradient by this same ratio. Consequently, a diffusion coefficient calculated by multiplying the flux times the diffusion distance and dividing by the solution concentration difference between the two boundaries (as in section 43–3) will include the contribution of adsorbed ions. In some cases (e.g., Van Schaik et al., 1966), this will cause such measured porous system interdiffusion coefficients to exceed the diffusion coefficients of either of the interdiffusing cation species in water.

43–2.2 Fick's Second Law (Transient State)

The mass balance principle for one-dimensional diffusion of a solute which may be involved in adsorption and desorption may be written

$$\theta \partial C/\partial t = \partial J/\partial x - \partial S_a/\partial t \quad [8]$$

where θ is the volumetric water content, C is the amount of the species per unit volume of water, J is the diffusion flux, and S_a is the amount of the species adsorbed per unit of bulk soil volume. The relation between the concentration of the diffusing species in solution and the amount, S_a, adsorbed per unit volume of the medium can often be expressed as a reasonably linear isotherm of the type

$$S_a = bC + b_o \qquad [9]$$

where b and b_o are constants. Then, since

$$\partial S_a/\partial C = b, \qquad [10]$$

$$\partial S_a/\partial t = (\partial S_a/\partial C)\,(\partial C/\partial t) = b\,(\partial C/\partial t)\,. \qquad [11]$$

Substituting this relation for $\partial S_a/\partial t$ and using Fick's first law,

$$J = -\,D_p\,(\partial C/\partial x)\,, \qquad [7]$$

in Eq. [8] yields

$$\theta\partial C/\partial t = \partial(D_p\partial C/\partial x)/\partial x - b\partial C/\partial t\,. \qquad [12]$$

If we make θ and the bulk density of the porous medium constant, then D_p is constant and Eq. [12] may be rearranged to

$$\partial C/\partial t = [D_p/(\theta + b)]\,(\partial^2 C/\partial x^2)\,. \qquad [13]$$

In this equation $\theta + b$ is the capacity of the water and solid per unit volume of soil to hold more of the diffusing species as the concentration in the solution phase increases by one unit. This capacity function is similar to the heat capacity in the one-dimensional transient-state heat flow equation, the form of which is essentially the same as Eq. [13]. Solutions for many boundary conditions which have been developed for the transient heat flow equations (Carslaw & Jaeger, 1959) can be utilized to obtain measurements of $D_p/(\theta + b)$.

If the diffusing species is not adsorbed on the solid phase, then $b = 0$, and since θ is easily determined, D_p can be calculated directly from D_p/θ. If the diffusing species is adsorbed, determination of the adsorption isotherm, Eq. [9], is generally necessary to allow calculation of D_p from $D_p(\theta + b)$.

If the total isotherm is nonlinear, a concentration range can often be selected for measurement purposes in which the adsorption isotherm is reasonably linear and b can be assumed to be essentially constant.

43–3 STEADY-STATE METHOD

43–3.1 Principles

Equation [7] is the differential form of the basic steady-state equation (Fick's first law) for porous media. This equation can also be written in a finite difference form

$$\Delta Q/\Delta T = A\ D_p\ (\Delta C/\Delta x) \qquad [7a]$$

in which $(\Delta Q/\Delta T)/A$ is equivalent to the flux, J, and $\Delta C/\Delta x$ is equivalent to the concentration gradient $\partial C/\partial x$, ΔQ is the amount of the component that has diffused in the time ΔT, A, and Δx are the cross-sectional area and thickness of the diffusion media, and $\Delta C = C_H - C_L$ is the concentration difference maintained between the high- and low-concentration sides of the diffusion medium.

The desired porous system diffusion coefficient, D_p, is equal to the measured flux divided by the maintained concentration gradient and A.

43–3.2 Apparatus

Cells of the type outlined in Fig. 43–1A can provide the boundary conditions needed for measurement of diffusion coefficients. Good circulation of solution in the two end compartments is needed to keep concentrations next to the porous plates at the desired C_H and C_L levels. Convective flow of solution through the soil can move large amounts of the diffusing solute unless care is taken to avoid hydraulic gradients in the cell. Means for detecting and preventing convective flow should be part of the measurement system. For instance, in Fig. 43–1B the height of the meniscus between air and water phases in the tube on the left side can be monitored to detect solution flux. The pressure in the air-filled flask can then be adjusted to prevent the convective flux.

In Fig. 43–1B, the tube leading into the compartment containing soil provides a means for allowing air phase in at the desired pressure when measurements on unsaturated soil systems are desired. Desaturation to the desired level by increasing air pressure through this tube should precede measurements of the diffusive flux.

43–3.3 Procedure

Soil or clays in the center section are retained by porous barriers. A large volume of solution at a relatively high and essentially constant concentration is circulated through one side of the cell. The initial solute concentration on the other side is zero or relatively low and must also be circulated so that solutes coming through the barriers are mixed quickly throughout the solution in that compartment.

After an appropriate time interval, solution on the low-concentration

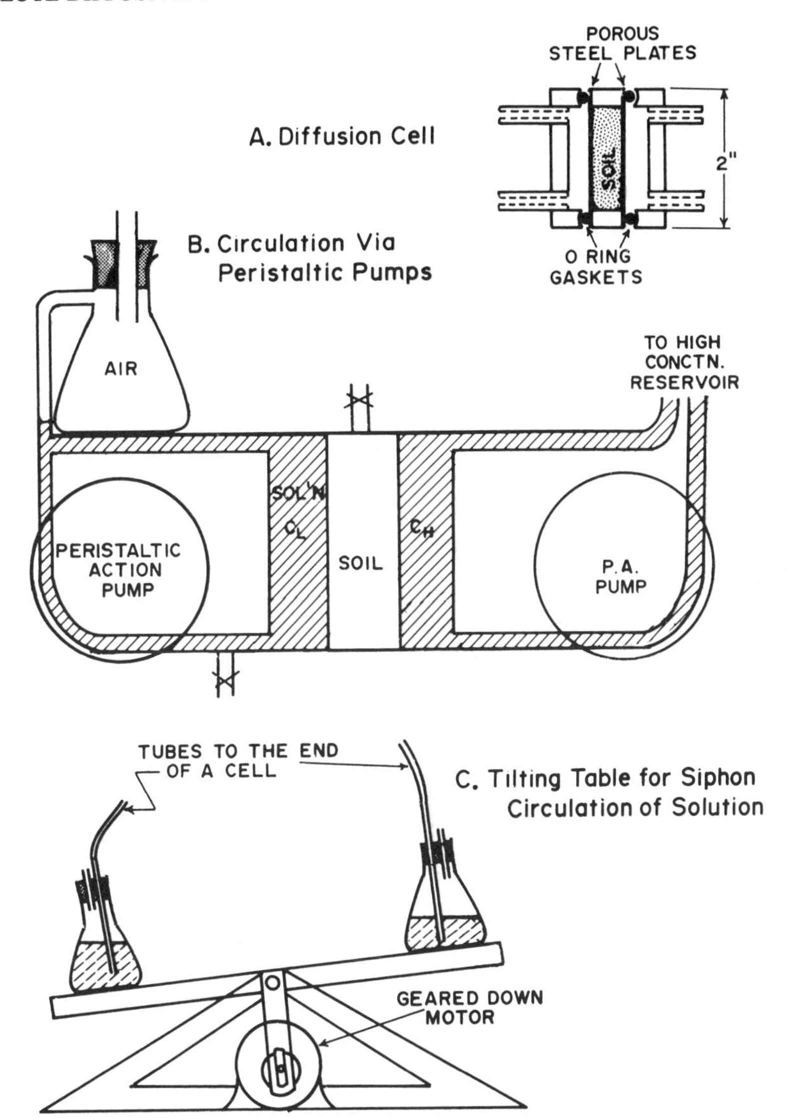

Fig. 43–1. Diffusion cell and circulation systems for steady-state method. A. Diffusion cell. B. Circulation via peristaltic action pumps. C. Tilting table for siphoning solution through the ends of cells.

side is replaced and the concentration of the solution circulated for that time interval is determined. If the final low-side concentration C_{Lf} minus the initial concentration C_{Li} is more than 2/10 of the concentration difference between the high and low sides the next time interval should be decreased. When $(C_{Lf} - C_{Li})$ is less than 2/10 of the difference between the high and low sides and has the same value for successive equal time periods, the flux is calculated as

$$J = (\Delta Q/\Delta t)/A = (C_{Lf} - C_{Li})\ V/(A\ \Delta\ t) \qquad [14]$$

where V is the volume of solution on the low-concentration side. Increasing C_L on the low side causes the average concentration difference $\overline{\Delta C}$ across the cell to be

$$\overline{\Delta C} = C_H - \overline{C_L} = C_H - (C_{Lf} + C_{Li})/2\ .$$

The concentration gradient is thus $\overline{\Delta C}/\Delta x$, where Δx is the thickness of soil in the cell. The form of Eq. [7] used for determining D_p is

$$D_p = J/(\overline{\Delta C}/\Delta x). \qquad [15]$$

43–3.4 Comments

For this method to give results within ± 5% accuracy, the thickness of the two retaining plates must be kept to < 5% of the soil thickness, or the diffusion coefficient through the plates must be determined separately and used with the flux to determine the portion of the concentration difference occurring through the retaining plates. This portion is then subtracted from the total concentration difference, to obtain a more accurate assessment of concentration difference across the soil.

The solution on the low-concentration side should be measured and replaced at regular intervals until successively measured flux rates are the same. Regularity of replacing the solution is particularly essential when the diffusing component is adsorbed by the porous medium.

If the soil thickness is about 1 cm and the solute is not adsorbed, steady-state flux rates may be achieved within 2 or 3 days. If the diffusing solute is strongly adsorbed by the medium (e.g., P in soil), months may be required to attain the steady-state condition. To reduce the time required to attain such steady-state conditions, the soil sample can be assembled in layers which have been equilibrated with solution concentrations ranging from those planned on the high- to those planned on the low-concentration side (Olsen et al., 1965).

An alternative way of achieving circulation of the solutions in the two end compartments is to provide a slowly oscillating (e.g., one cycle/5 min) tilting table-top such as that outlined in Fig. 43–1C. Solution is siphoned from one flask to the other through one end of the cell shown in Fig. 43–1A. Elevation of the cell above the table can provide suctions on the soil water in the 0.2- to 2-m range. To minimize convective flow from one end to the other in the diffusion cell, an identical pair of flasks, filled to the same levels and set at exactly the same distance from the pivot line of the table, are attached to the tubes on the other end of the cell. Convective flow can be detected by determining the weights of the solution from the pair of flasks before and after the time period during which the fluid has been circulated. If the pairs of flasks attached to the two ends contain the same amounts of solution, the flasks are at the same distances from the pivot line of the table, and soil water suction is > 100 cm, appreciable convective flow through the cell does not generally occur. However, solution weights should always be checked to make sure the system has not malfunctioned. When this siphoning method of circulating solution in the two end compartments is used, concentration changes in both the high and low concentration compartments should be determined, and the average concentration on the high side must be used in determining the average concentration difference.

Water contents at which this method has been used have generally been at suctions <20 kPa because thin porous barriers, such as the stainless steel plates indicated in Fig. 43–1, have bubbling pressures in the range of only 10 to 50 kPa. Pressure cells could be assembled with the ends at atmospheric pressure, the barriers of high bubbling pressure materials, and positive pressures inside the cell. However, the equipment would be expensive and cumbersome. To avoid this machinery, most scientists in the past have used the transient state methods discussed below when working at low water contents.

43–4 TRANSIENT-STATE METHODS

43–4.1 Principles

Combining Fick's first law and the principle of conservation of mass in one dimension leads to Eq. [13], which is Fick's second law for linear diffusion.

Many solutions for Eq. [13] have been developed that allow determining diffusion coefficients from simple experimental setups. One of the most useful is the solution for two blocks of soil containing different solute concentrations, which are brought together at time = 0 as indicated in Fig. 43–2. The blocks on the left and right have initial solute concentrations of C_o and C_s, respectively. After contact times t_1, t_2, t_3, and t_4, the solute concentrations on the two sides change as indicated. Given

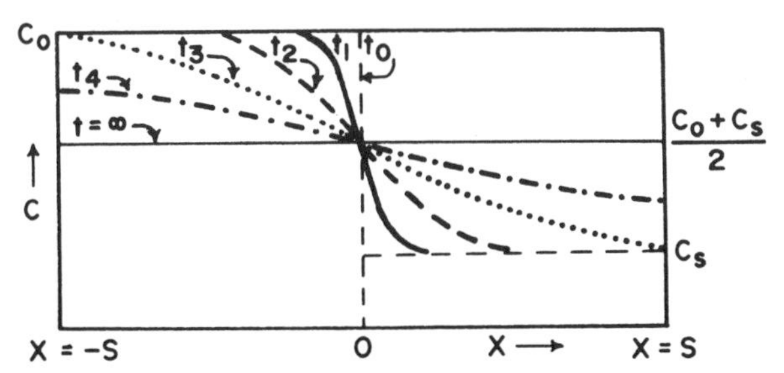

Fig. 43–2. Concentration gradient after bringing two blocks of soil with initial concentrations of C_o and C_s together at $t = 0$.

infinite time, the concentrations on both sides become equal to $(C_o + C_s)/2$, and the amount of material that has moved across, q_∞, is equal to half of the difference in the amounts initially in the two blocks. Crank (1956, p. 15–17) provided a solution for this boundary condition applied to Eq. [13] which can be manipulated to give the quantity, q, that has diffused from one block to the other as

$$q/q_\infty = 1 - (8/\pi^2) \sum_{i=0}^{\infty} \left\{ \frac{\exp[-(2i + 1)^2 (\pi/2)^2 \; D_p t/Bs^2]}{(2i + 1)^2} \right\} \quad [16]$$

where s is the cylinder length and $B = \theta + b$.

Equation [16] is plotted in Fig. 43–3, where it can be compared to the simpler equation

$$q/q_\infty = (2/\sqrt{\pi}) \, (D_p t/Bs^2)^{1/2} \, , \quad [17]$$

which as indicated in Fig. 43–3 is adequately accurate for these boundary conditions as long as $q/q_\infty < 0.6$.

43–4.2 Apparatus for Linear Transient Measurements

Apparatus necessary for measurements of $D_p/(\theta + b)$ is relatively simple if Eq. [16] or [17] is used to relate diffusion from one block of soil to another. Essentially, it consists of two "half cells" of known cross-

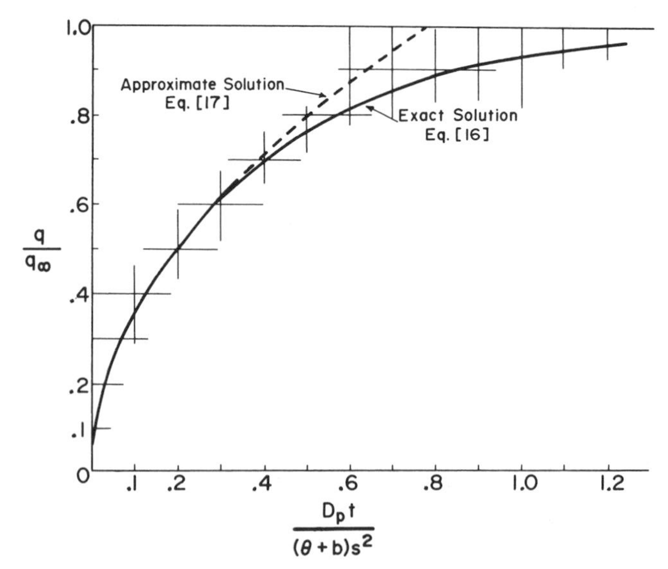

Fig. 43–3. Comparison of approximate (Eq. [17]) and exact (Eq. [16]) relations between q/q_∞ and $D_p t/Bs^2$ where $B = \theta + b$.

sectional area and length, open on one end, which will hold two blocks of the porous medium in the desired configuration and will allow the operator to establish firm contact between the blocks of soil when they are brought together (Brown et al., 1964). Since diffusion coefficients are highly dependent on water content, means must also be provided to prevent evaporation of water from the system. Electricians tape around the joint of two cells usually restrains water loss sufficiently if the time required for good measurement is less than 3 or 4 days. Longer diffusion times may require a more complete seal against continued vapor loss—for example, placing the joined cells in a sealable container such as a wide mouthed fruit jar.

If a complete profile of the solution concentration in the blocks is desired (Phillips & Brown, 1964), equipment for slicing the porous medium is needed. However, such detailed concentration data are generally integrated to obtain the total amount, q, of the component that has diffused into the opposite block, and are then used in equations such as Eq. [16] and [17] to determine $D_p(\theta + b)$. Consequently, for simply determining $D_p/(\theta + b)$, equipment for slicing the sample is not necessary. Selection of an optimum length, s, of the half-cells requires some knowledge or estimates of the value of $D_p/(\theta + b)$ and the time that can be allowed for the measurement. If the operator has ball-park estimates of these factors and wants to keep $q/q_\infty < 0.6$, he can use Eq. [17] to calculate the desired lengths, s, of the half-cells.

43–4.3 Procedure for Linear Transient-state Measurements

Two samples of the soil or clay are brought to the desired water content. Each sample is then brought to the desired solute concentration (or isotope activity level). This may be done by mixing a known amount of solute or isotope thoroughly with the soil prior to packing disturbed soils into the half-cells, or if undisturbed cores and nonadsorbed solutes are being used, leaching may be used to move the solute into the cores. In all cases it is advisable to allow time for equilibration so that concentration within the half-cell will be uniform when the two half-cells are brought together. Time required for such equilibration will be proportional to s^2 and $(\theta + b)D_p$, as indicated by Eq. [17] and Fig. 43–3.

When the ratio $D_p/(\theta + b)$ is small, s should be small, since the diffusing component does not move far into the opposite block. In such cases, it is particularly critical to separate the half-cells at precisely the plane at which they were joined together. In some cases (Olsen et al., 1965) flat, fine-threaded open-weave nylon net has been placed between the two half-cells when they were assembled. Then when they are to be separated, pulling the net slightly cuts the soil at the desired plane. If netting is to be used in this manner, its strands should be sufficiently widely spaced and small in size that they block no more than a few percent of the cross section for diffusion at the junction plane.

Care should be taken in this and other measurements of diffusion to avoid convective flow. Using paired blocks of soil with the same water potential or content will generally avoid appreciable convective transfer of solute from one block to the other.

After the half-cells are separated, counts of radioactivity in each of the cells are made, or the amount of the diffusing solute or ion can be measured in the two half-cells by appropriate extraction and analytical procedures. The ratios of these counts or quantities (i.e., q/q_{∞}) are then used in Eq. [17] or Fig. 43–3, with known s and t, to determine $D_p/(\theta + b)$.

43–4.4 Comments on Linear Transient-state Measurements

Measurements of this type provide the ratio of the porous system diffusion coefficient divided by the capacity factor $[D_p/(\theta + b)]$. Often it is the ratio that is needed in equations to predict diffusion. However, in some equations, the value of D_p alone is also needed. To obtain values of D_p an independent measure of the capacity factor is needed. In other words, the adsorption isotherm (Eq. [9]) of that component in the porous medium must be determined. An example of such adsorption isotherms is given in Fig. 43–4 (from Olsen et al., 1962), in which the dashed portions of the isotherms near the origin are obviously not linear. However, in the solution concentration range from 0.05 to 0.3 μg/mL (which ranges from deficient to adequate for most crops), the isotherm is represented adequately by the indicated straight lines, and $\theta + b$ is essentially constant. Examples of isotherms for Na, K, Mg, and Ca are given by Olsen and Kemper (1968).

In some cases where it is difficult to define which portion of a mineral constituent is "labile" (i.e., is released from the solid phase and moves as concentration decreases), this could be determined by measuring D_p

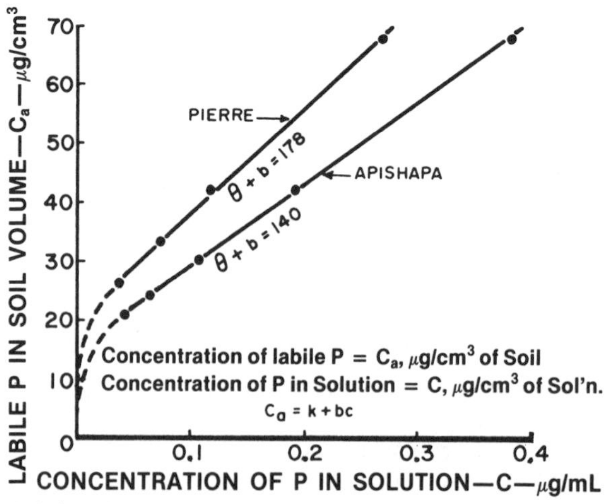

Fig. 43–4. Relations between labile P and solution P on Apishapa and Pierre soils.

using a steady-state method and $D_p/(\theta + b)$ using the transient state method and calculating $\theta + b$ from their quotient.

A somewhat complicating factor is that adsorption and desorption isotherms often appear to be different. Nye (1979) pointed out that part of these differences may be due to the time required to achieve complete equilibrium between the solid and solution phases. Consequently, significantly different b values for the adsorption and desorption isotherms are less likely to be encountered when the diffusion rates are slow (small concentration gradients) than when the diffusion rates are fast.

43–5 RADIAL TRANSIENT CONDITIONS

The diffusion coefficient D_p of a solute in a soil can also be estimated from the rate at which solute diffuses from a porous tube into soil (initially devoid of the solute) as indicated in Fig. 43–5. Knowing the concentration of solution entering the porous tube C_o, the concentrations of solution exiting from the porous tube, C_e, the rate at which the solution is flowing through the tube and the length of the tube allows calculation of the rate, J_i, per unit length at which the solute is diffusing out of the tube. When these measurements are made at time t following initiation of the solution flow, the tube radius, a, is known and the capacity factor $B = \theta + b$ is known, the value of D_p can be calculated.

From Carslaw and Jaeger (1959, Fig. 42, p. 336, Eq. [8]) a dimensionless variables graph can be drawn for these boundary conditions,

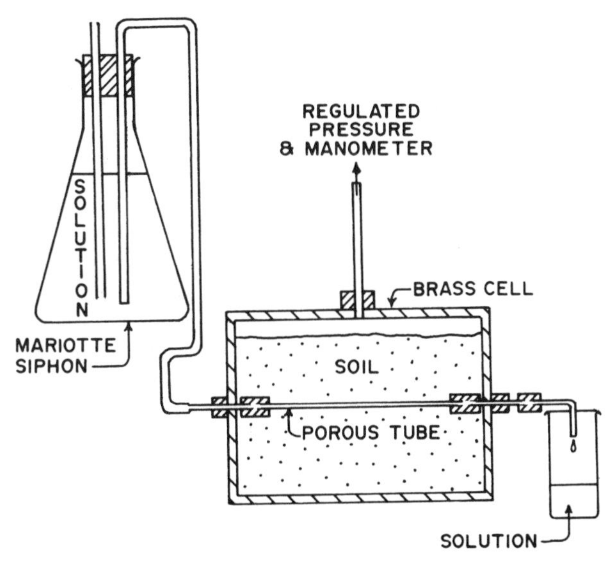

Fig. 43–5. Apparatus used for measuring the diffusion coefficient to a porous tube under transient-state radial conditions when the desorption isotherm is known or the diffusing component is not adsorbed. (from Kemper et al., 1971)

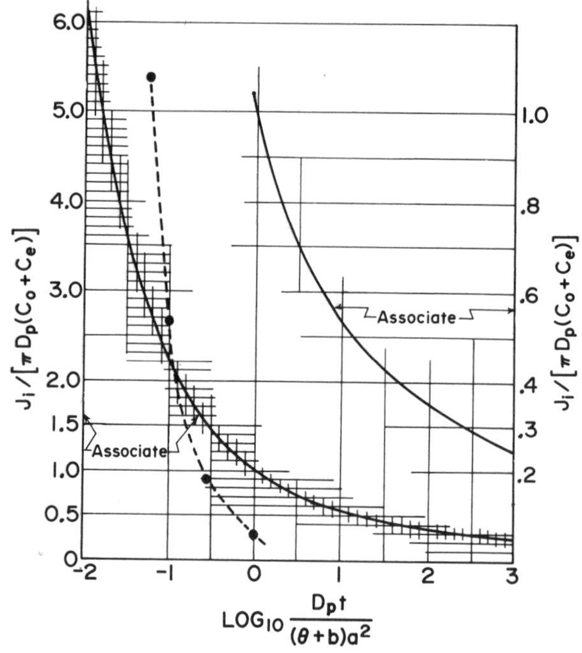

Fig. 43–6. General solution for the boundary conditions at $t = 0$, $C = C_o$, and $t > 0$, when $r = a$, $C = C_a$.

which when translated into the units defined above, becomes the solid curves in Fig. 43–6. The porous system diffusion coefficient D_p appears in the terms on both the ordinate and the abscissa in this figure and cannot be determined directly. However, when the capacity factor $B = \theta + b$ is known along with J_i, C_o, C_e, a, and t, about four tentative values of D_p, spanning the possible range, can be selected and the dimensionless variables containing D_p on the ordinate and abscissa can be calculated and plotted for each tentative D_p. The correct value of D_p can be calculated from the point at which this line intersects the solid line in Fig. 43–6.

An example of this procedure has been drawn as the dashed line in Fig. 43–6 for which it was assumed that $C_o = 0.007N$ and $C_e = 0.005N$, $J_i = 10^{-8}$ meq/cm² s, $B = 100$ and $T = 10^5$, $a = 0.0316$ cm. Tentative values for D_p of 5×10^{-8}, 10^{-7}, 3×10^{-7}, and 10^{-6} cm²/s were used to calculate the four points connected by the dashed line in Fig. 43–6. The correct value is then calculated from where this dashed line intersects the solid line. For this data set the intersection is where $J_i/[\pi D_p(C_o + C_e)] = 2.15$. Interjecting the values of J_i, C_o, and C_e given above, $D_p = 1.3 \times 10^{-7}$ cm²/s.

There is a finite concentration drop from the inside to the outside of the porus tube (Fig. 43–5) and the pertinent radius for diffusion into the soil to be used in the abscissa of Fig. 43–6 is the outer radius of the tube. Concentration at the outside radius, m, of the tube is $(C_o + C_e)/2$ minus the concentration drop that occurs as the solute diffuses through the wall of the tube. That concentration drop can be estimated from the inner and outer radii (a and m, respectively), the flux rate ($J \times L$), and the coefficient of diffusion of that solute in the porus walls of the tube, D_{pt}. The latter may be estimated by filling the cell (Fig. 43–5) with solution of concentration C_m and passing distilled water slowly through the tube. Then the average concentration on the inner wall of the tube is half the outgoing concentration (i.e., C'_e). From the flow rate and low concentration of the solution emerging from the tube, the diffusion flux J_2 can be determined and then the value of D_{pt} in the walls of the porous tube can be calculated from the steady-state equation

$$J_2 = 2\ D_{pt}L(C_m - C'_e/2)/\ln(m/a) \qquad [18]$$

where a and m are the inner and outer radii of the porous tube and L is its length. Then, knowing a, m, and D_{pt} for these walls and the flux rate $J_i \times L$ occurring at any time in the transient measurement of diffusion in soils, the value of $C_m - C'_e/2$ (the concentration drop across the wall of the porous tube) can be estimated from Eq. [18]. This value is subtracted from $(C_o + C_e)/2$ and the remainder is multiplied by two and used in place of $C_o + C_e$ in the denominator of the ordinate in Fig. 43–6, along with the value of m in place of a in the abscissa in the process of calculating a more accurate value of D_p from measured values of J_i.

43–6 GENERAL COMMENTS

Except for H^+ and OH^-, the diffusion coefficients of the ions in water as listed in Table 43–1 fall in the range from 0.62×10^{-5} to $2.08 \times 10^{-5} cm^2/s$. This is a relatively small range compared to the differences in D_p that occur as a result of differences in water content in soil and which are as great as 50-fold (Olsen et al., 1965). The extreme sensitivity of the porous system diffusion coefficient to water content occurs because decreasing water content decreases the cross-sectional area of flow, increases the tortuosity, increases the average water viscosity, and reduces the average pore size, thus reducing all four of the factors, θ, L/L_e, α, and γ, in Eq. [4].

The value of D_p/B will often be from one to three orders of magnitude smaller than D_p when the ion is adsorbed on the mineral surfaces, particularly when the solution concentration is low, and therefore, the amount in solution is small compared to the amount adsorbed. Space limitations prevent full discussion of all the factors pertaining to measurement of diffusion coefficients and to the significance of those coefficients. The reader is referred for further details to the review papers by Nye (1979); Low (1981); Jackson et al. (1963); Olsen and Kemper (1968), and the other papers cited.

43-7 REFERENCES

Brown, D. A., B. E. Fulton, and R. E. Phillips. 1964. Ion diffusion: I. A quick freeze method for measurement of ion diffusion in soil and clay systems. Soil Sci. Soc. Am. Proc. 28:628–631.

Carslaw, H. S., and J. C. Jaeger. 1959. Conduction of heat in solids. Clarendon Press, Oxford, England.

Crank, J. 1956. The mathematics of diffusion. Oxford University Press, Oxford, England.

Jackson, R. D., D. R. Nielsen, and F. S. Nakayama. 1963. On diffusion equations applied to porous materials. USDA Publ. ARS 41–186. U.S. Government Printing Office, Washington, DC.

Jost, W. 1952. Diffusion in solids, liquids, and gases. Academic Press, New York.

Kemper, W. D., B. A. Stewart, and L. K. Porter. 1971. Effects of compaction on soil nutrient status. p. 178–189. *In* W. M. Carleton et al. (ed.) Compaction of agricultural soils. ASAE Monogr. St. Joseph, MI.

Kemper, W. D., D. E. L. Maasland, and L. K. Porter. 1964. Mobility of water adjacent to mineral surfaces. Soil Sci. Soc. Am. Proc. 28:164–167.

Kemper, W. D., and J. P. Quirk. 1972. Ion mobilities and electric charge of external clay surfaces inferred from potential differences and osmotic flow. Soil Sci. Soc. Am. Proc. 36:426–433.

Kemper, W. D., Isaac Shainberg, and J. P. Quirk. 1972. Swelling pressures, electric potentials, and ion concentrations and their role in hydraulic and osmotic flow through clays. Soil Sci. Soc. Am. Proc. 36:229–236.

Low, P. F. 1981. Principles of ion diffusion in clays. p. 31–45. *In* D. E. Baker (ed.) Chemistry in the soil environment. Spec. Pub. 40. American Society of Agronomy, Madison, WI.

Nye, P. H. 1979. Diffusion of ions and uncharged solutes in soils and clays. Adv. Agron. 31:225–272.

Olsen, S. R., W. D. Kemper, and R. D. Jackson. 1962. Phosphate diffusion to plant roots. Soil Sci. Soc. Am. Proc. 26:222–227.

Olsen, S. R., W. D. Kemper, and J. C. Van Schaik. 1965. Self diffusion coefficients of phosphorus in soil measured by transient and steady state methods. Soil Sci. Soc. Am. Proc. 28:154–158.

Olsen, S. R., and W. D. Kemper. 1968. Movement of nutrients to plant roots. Adv. Agron. 20:91–151.

Phillips, R. E., and D. A. Brown. 1964. Ion diffusion: comparison of apparent salt and counter diffusion coefficients. Soil Sci. Soc. Am. Proc. 28:758–763.

Porter, L. K., W. D. Kemper, R. D. Jackson, and B. A. Stewart. 1960. Chloride diffusion in soils as influenced by moisture content. Soil Sci. Soc. Am. Proc. 24:450–463.

Robinson, R. A., and R. H. Stokes. 1959. Electrolytic solutions. Butterworths, London.

Shainberg, Isaac, and W. D. Kemper. 1966. Conductance of alkali cations in aqueous and alcoholic bentonite pastes. Soil Sci. Soc. Am. Proc. 30:700–706.

Van Schaik, J. C., and W. D. Kemper. 1966. Chloride diffusion in clay water systems. Soil Sci. Soc. Am. Proc. 30:22–25.

Van Schaik, J. C., W. D. Kemper, and S. R. Olsen. 1966. Contribution of adsorbed cations to diffusion in clay water systems. Soil Sci. Soc. Am. Proc. 30:17–22.

44 Solute Dispersion Coefficients and Retardation Factors

M. Th. VAN GENUCHTEN

U.S. Salinity Laboratory, Agricultural Research Service, USDA
Riverside, California

P. J. WIERENGA

New Mexico State University
Las Cruces, New Mexico

44-1 INTRODUCTION

Modern agriculture uses substantial quantities of fertilizers, pesticides, and other chemicals that are beneficial only in the upper part of the soil profile. Translocation of these chemicals to the subsoil makes them not only unavailable for plant uptake, but also poses a threat to the quality of underlying groundwater systems. Chemicals dumped in waste disposal sites are also subject to translocation to groundwater, drains, or even surface streams. The same is true for radioactive materials and other chemicals spilled from waste storage reservoirs.

Various theoretical models have been developed over the years to describe chemical transport in soils. Success of these models depends to a large degree on our ability to quantify the transport parameters that enter into these models. Important parameters are the fluid flux, the dispersion coefficient, and the adsorption or exchange coefficients in case of interactions between the chemical and the solid phase. Simple linear adsorption or exchange can be accounted for by introducing a retardation factor in the transport equation. Various zero- or first-order production or decay coefficients may be required also (for example, when predicting transport of certain organic compounds, N species, or radionuclides).

A large number of methods are available for determining the dispersion coefficient and the retardation factor from observed solute concentration distributions. This chapter describes five different techniques that are applicable to both laboratory and field displacement experiments. A few other methods exist that, in addition to those discussed here, can also be used to obtain estimates for the dispersion coefficient and the retardation factor. They include the method of moments (Aris, 1958;

Agneessens et al., 1978; Skopp, 1985; Valocchi, 1985; Jury & Sposito, 1985; among others), and methods to determine the two coefficients from the location and peak concentration of a short or instantaneous surface-applied tracer pulse (Kirkham & Powers, 1972; Saxena et al., 1974; Yu et al., 1984). The reader is referred to the original studies for a discussion of these additional methods.

44–2 THEORETICAL PRINCIPLES

All available methods for determining the required transport parameters from observed concentration distributions are based on analytical solutions of the solute transport equation. Consequently, we will first give a brief discussion of the transport equation and of various boundary conditions that have been used to derive analytical solutions of that equation.

44–2.1 Transport Equation

Consider the situation where water containing a dissolved tracer is applied to a tracer-free soil profile. As more of the solution is added, the initially very sharp tracer front near the soil surface becomes more and more spread out (dispersed) due to the combined effects of diffusion and convection. Transport of the dissolved tracer consists of three components:

Convective or Mass Transport (J_m). Convective (or advective) transport refers to the passive movement of the dissolved tracer with flowing soil water. In the absence of diffusion, water and the dissolved tracer move at the same average rate

$$J_m = q\,C \qquad [1]$$

where q is the volumetric fluid flux density and C is the volume-averaged solute concentration.

Diffusive Transport (J_D). Diffusion is a spontaneous process that results from the natural thermal motion of dissolved ions and molecules. Diffusive transport in soils tends to decrease existing concentration gradients, and in analogy to Fick's law, can be described by

$$J_D = -\theta D_m \frac{\partial C}{\partial x} \qquad [2]$$

where θ is the volumetric water content, D_m is the porous medium ionic or molecular diffusion coefficient, and x is distance. Because of a tortuous flow path, D_m in soils is somewhat less than the diffusion coefficient in pure water (D_o):

$$D_m = D_o\tau \qquad [3]$$

where τ is a dimensionless tortuosity factor, ranging roughly from 0.3 to about 0.7 for most soils.

Dispersive Transport (J_h). Dispersive transport results from the fact that local fluid velocities inside individual pores and between pores of different shapes, sizes, and directions, deviate from the average pore-water velocity. Such velocity variations cause the solute to be transported down-gradient at different rates, thus leading to a mixing process that is macroscopically similar to mixing caused by molecular diffusion. Dispersion is a passive process that, unlike diffusion, occurs only during water movement. On the other hand, diffusion always forms an integral part of the overall dispersion process by reducing flow-induced concentration gradients within and between pores. Because of the passive nature of the dispersion process, the term *mechanical dispersion* is often used to describe mixing caused by local velocity variations (Fried & Combarnous, 1971; Bear, 1972; Freeze & Cherry, 1979). Laboratory and field experiments have shown that dispersive transport can be described by an equation similar to Eq. [2] for diffusion

$$J_h = -\theta D_h \frac{\partial C}{\partial x} \qquad [4]$$

where D_h is the mechanical dispersion coefficient (Bear, 1972). This coefficient is generally assumed to be a function of the fluid velocity

$$D_h = \lambda v^n \qquad [5]$$

where λ is the dispersivity and v the average interstitial or pore-water velocity, approximated by the ratio q/θ. The exponent n in Eq. [5] is an empirical constant, roughly equal to 1.0. For most laboratory displacement experiments involving disturbed (repacked) soils and for certain uniform field soils, λ is on the order of about 1 cm or less (assuming that $n = 1$). For transport problems involving undisturbed field soils, especially when aggregated, λ is usually about one or two orders of magnitude larger.

Because of the macroscopic similarity between molecular diffusion and mechanical dispersion, the coefficients D_m and D_h are often considered additive

$$D = D_m + D_h \qquad [6]$$

where D is the longitudinal hydrodynamic dispersion coefficient (Bear, 1972), further referred to simply as the dispersion coefficient. Other terms frequently used for D are the *apparent diffusion coefficient* (Nielsen et al., 1972; Boast, 1973), and the *diffusion-dispersion coefficient* (Hillel, 1980), while the term *hydrodynamic dispersion coefficient* sometimes has been reserved for D_h only (Shamir & Harleman, 1966; Nielsen et al., 1972; Boast, 1973).

Combining Eq. [1], [2], [4], and [6] leads to the following expression for the solute flux, J_s:

$$J_s = -\theta D \frac{\partial C}{\partial x} + qC. \qquad [7]$$

Substituting Eq. [7] into the equation of continuity

$$\frac{\partial}{\partial t}(\theta C + \rho S) = -\frac{\partial J_s}{\partial x} \qquad [8]$$

yields the transport equation

$$\frac{\partial}{\partial t}(\theta C + \rho S) = \frac{\partial}{\partial x}\left(\theta D \frac{\partial C}{\partial x} - qC\right) \qquad [9]$$

where S is the adsorbed concentration (mass of solute per unit mass of soil), ρ is the soil bulk density, and t is time. The two terms on the left side of Eq. [9] account for changes in solute concentrations associated with the liquid and solid phases, respectively.

We assume here that S and C can be related by a linear or linearized equilibrium isotherm of the form

$$S = kC \qquad [10]$$

where k is an empirical distribution coefficient. The assumption of linear adsorption generally is valid only at low concentrations. Note that Eq. [9] assumes that the chemical is not subject to any production or decay processes. A few comments about the determination of zero- and first-order production or decay coefficients from observed displacement experiments are given in section 44–9.

If, in addition to linearized equilibrium adsorption, steady water flow in a homogeneous soil profile is assumed (θ and q are constant in time and space), Eq. [9] reduces to

$$R \frac{\partial C}{\partial t} = D \frac{\partial^2 C}{\partial x^2} - v \frac{\partial C}{\partial x} \qquad [11]$$

where R is the retardation factor

$$R = 1 + \rho k/\theta. \qquad [12]$$

If there are no interactions between the chemical and the soil, k becomes zero and R reduces to one. In some cases R may become less than one, indicating that only a fraction of the liquid phase participates in the transport process. This may be the case when the chemical is subject to anion exclusion or when relatively immobile liquid regions are present, for example inside dense aggregates, that do not contribute to convective transport. In case of anion exclusion, $(1-R)$ may be viewed as the relative anion exclusion volume.

44–2.2 Boundary Conditions and Analytical Solutions

To complete the mathematical description of transport through semi-infinite field profiles ($0 \leqslant x < \infty$) or finite laboratory columns of length L ($0 \leqslant x \leqslant L$), Eq. [11] must be augmented with auxiliary conditions describing the initial concentration of the system and the boundary conditions. Proper formulation of the boundary conditions is important when analyzing laboratory displacement experiments involving relatively short columns, as well as for interpreting tracer data from laboratory or field profiles exhibiting large dispersivities λ. Also, incorrect use of boundary conditions for laboratory tracer experiments can lead to serious errors when the experimental results subsequently are extrapolated to field situations.

Table 44–1 summarizes four available analytical solutions of Eq. [11] for both semi-infinite (A-1, A-2) and finite systems (A-3, A-4). The analytical expressions hold for the relative concentration c, which is defined as

$$c(x,t) = [C(x,t) - C_i]/(C_o - C_i) \qquad [13]$$

where C_o is the concentration of the applied solution and C_i the initial concentration. Both C_i and C_o are assumed to be constant.

When a tracer solution is applied at a specified rate from a perfectly mixed inlet reservoir to the surface of a finite or semi-infinite soil profile, continuity of the solute flux across the inlet boundary leads directly to a third-type or flux-type boundary condition of the form

$$\left(-D\frac{\partial C}{\partial x} + vC\right)\bigg|_{x=0+} = vC_o \qquad [14]$$

where $0+$ indicates evaluation at the inlet boundary just inside the medium.

For semi-infinite systems in the field we also need a boundary condition that specifies the behavior of $C(x,t)$ when $x \rightarrow \infty$. For our discussion it is sufficient to require that

$$\frac{\partial C}{\partial x}(\infty, t) = 0. \qquad [15]$$

The analytical solution for boundary conditions [14] and [15] is given by case A-2 in Table 44–1. This solution correctly evaluates volume-averaged, in situ or resident concentrations in semi-infinite field profiles. For example, one may verify that solution A-2 satisfies the mass balance requirement

Table 44–1. Analytical solutions of Eq. [11] for various boundary conditions.

Case	Inlet boundary condition	Exit boundary condition	Analytical solution
A-1	$C(0, t) = C_o$	$\frac{\partial C}{\partial x}(\infty, t) = 0$	$c = \frac{1}{2}\,\mathrm{erfc}\left[\frac{Rx - vt}{2(DRt)^{1/2}}\right] + \frac{1}{2}\exp\left(\frac{vx}{D}\right)\mathrm{erfc}\left[\frac{Rx - vt}{2(DRt)^{1/2}}\right]$
A-2	$\left(-D\frac{\partial C}{\partial x} + vC\right)\Big\vert_{x=0} = vC_o$	$\frac{\partial C}{\partial x}(\infty, t) = 0$	$c = \frac{1}{2}\,\mathrm{erfc}\left[\frac{Rx - vt}{2(DRt)^{1/2}}\right] + \left(\frac{v^2 t}{\pi DR}\right)^{1/2}\exp\left[-\frac{(Rx - vt)^2}{4DRt}\right] - \frac{1}{2}\left(1 + \frac{vx}{D} + \frac{v^2 t}{DR}\right)\exp\left(\frac{vx}{D}\right)\mathrm{erfc}\left[\frac{Rx + vt}{2(DRt)^{1/2}}\right]$
A-3	$C(0, t) = C_o$	$\frac{\partial C}{\partial x}(L, t) = 0$	$c = 1 - \sum_{m=1}^{\infty}\frac{2\beta_m \sin\left(\frac{\beta_m x}{L}\right)\exp\left[\frac{vx}{2D} - \frac{v^2 t}{4DR} - \frac{\beta_m^2 Dt}{L^2 R}\right]}{\left[\beta_m^2 + \left(\frac{vL}{2D}\right)^2 + \frac{vL}{2D}\right]}$ $\beta_m \cot(\beta_m) + \frac{vL}{2D} = 0$
A-4	$\left(-D\frac{\partial C}{\partial x} + vC\right)\Big\vert_{x=0} = vC_o$	$\frac{\partial C}{\partial x}(L, t) = 0$	$c = 1 - \sum_{m=1}^{\infty}\frac{\frac{2vL}{D}\beta_m\left[\beta_m \cos\left(\frac{\beta_m x}{L}\right) + \frac{vL}{2D}\sin\left(\frac{\beta_m x}{L}\right)\right]\exp\left[\frac{vx}{2D} - \frac{v^2 t}{4DR} - \frac{\beta_m^2 Dt}{L^2 R}\right]}{\left[\beta_m^2 + \left(\frac{vL}{2D}\right)^2 + \frac{vL}{D}\right]\left[\beta_m^2 + \left(\frac{vL}{2D}\right)^2\right]}$ $\beta_m \cot(\beta_m) - \frac{\beta_m^2 D}{vL} + \frac{vL}{4D} = 0$

References: A-1: Lapidus and Amundson (1952) A-3: Cleary and Adrian (1973)
A-2: Lindstrom et al. (1967) A-4: Brenner (1962)

$$vC_o t = R\int_0^\infty [C(x,t) - C_i]\,dx. \qquad [16]$$

In other words, whatever material is added at the surface (term on the left) should be present in the soil profile (term on the right).

Instead of Eq. [14], a first- or concentration-type input boundary condition has also been used

$$C(0,t) = C_o\,. \qquad [17]$$

This equation assumes that the concentration itself can be specified at the inlet boundary, a situation that usually is not possible in practice. Analytical solution A-1 for this condition (see Table 44–1) fails to satisfy mass balance Eq. [16], the largest errors occurring at relatively small values of the dimensionless group v^2t/DR (van Genuchten & Parker, 1984). Hence, this solution should not be used for evaluating volume-averaged concentrations in semi-infinite field profiles. On the other hand, solution A-1 is useful for estimating solute fluxes at any point in the profile. As shown by Brigham (1974), Kreft and Zuber (1978), and Parker and van Genuchten (1984a), among others, solutions A-1 and A-2 are related through the transformation

$$C_f = C - \frac{D}{v}\frac{\partial C}{\partial x} \qquad [18]$$

where C_f, to be identified with analytical solution A-1, represents flux-averaged or flowing concentrations, in contrast to C (solution A-2), which represents volume-averaged or resident concentrations. In other words, the solute flux J_s (Eq. [7]) at any point in the profile is directly specified by $J_s = qC_f$ with C_f given by solution A-1.

Proper formulation of the exit boundary condition for displacement through finite laboratory columns is considerably more difficult than for semi-infinite field profiles. In analogy with Eq. [14], mass conservation requires the solute velocity to be continuous across the exit boundary

$$\left(-D\frac{\partial C}{\partial x} + vC\right)\bigg|_{x=L-} = vC_e \qquad [19]$$

where $L-$ indicates evaluation just inside the column, and where C_e is the effluent concentration. This equation assumes negligible dispersion in the exit reservoir (or after-section). Because of an extra unknown (C_e), Eq. [19] leads to an indeterminate system of equations; hence, an additional relation is needed to fully describe the system. One such equation is based on the intuitive assumption that the concentration should be continuous at $x = L$:

$$C(L-,\,t) = C_e(t)\,. \qquad [20]$$

Substitution of this equation into Eq. [19] leads to the frequently used boundary condition (Danckwerts, 1953)

$$\frac{\partial C}{\partial x}(L,t) = 0. \quad [21]$$

The analytical solution for boundary conditions [14] and [21], derived by Brenner (1962), is given by case A-4 in Table 44–1. Brenner's solution describes volume-averaged concentrations inside the column. Because of the zero concentration gradient at $x = L$, this solution also defines a flux concentration (Eq. [18]) at the lower boundary. Hence, Brenner's solution correctly interprets effluent concentrations as representing flux-averaged concentrations. Table 44–2 (case A-4) shows the resulting expression for the relative effluent concentration in terms of the number of pore volumes, T, leached through the column and the column Peclet number, P

$$T = vt/L \quad [22a]$$

$$P = vL/D. \quad [22b]$$

Computational programs for evaluation of Brenner's series solution are readily available (van Genuchten & Alves, 1982). The series solution converges only for relatively small values of P. For large P-values, computationally efficient and very accurate approximate solutions have also been obtained (van Genuchten & Alves, 1982).

Analogous to Eq. [16] for semi-infinite systems, the mass balance requirement for finite columns is

$$v\int_0^t [C_o - C_e(\tau)]\,d\tau = R\int_0^L [C(x,t) - C_i]\,dx. \quad [23]$$

This equation states that whatever is added to the column minus whatever is leaving that column (left side) must be stored in the column (right side). Upon substitution, it is easily demonstrated that solution A-4 satisfies Eq. [23]. Tables 44–1 and 44–2 also list analytical solution A-3 for boundary conditions [17] and [21]. Contrary to A-1 and A-2, solutions A-3 and A-4 are not related through the transformation given by Eq. [18]. Moreover, solution A-3 fails the mass-balance requirement (Eq. [23]), and as will be shown later, also violates a mass balance for the effluent curve. Hence, this solution should never be applied to displacement experiments.

The analysis above shows that Brenner's solution A-4 is based on the assumption that the concentration is continuous at $x = L$. An alternative formulation is possible by assuming that solute distributions inside the column are not affected by an outflow boundary or effluent collection system, thus considering the column to be part of an effectively

Table 44–2. Expressions for the relative effluent concentration, $c_e(T)$, in terms of the column Peclet number (P) and pore volume (T) for the four analytical solutions listed in Table 44–1.

Case	Relative effluent concentration	
A-1	$c_e(T) = \frac{1}{2}\,\mathrm{erfc}\left[\left(\frac{P}{4RT}\right)^{1/2}(R-T)\right] + \frac{1}{2}\exp(P)\,\mathrm{erfc}\left[\left(\frac{P}{4RT}\right)^{1/2}(R+T)\right]$	
A-2	$c_e(T) = \frac{1}{2}\,\mathrm{erfc}\left[\left(\frac{P}{4RT}\right)^{1/2}(R-T)\right] + \left(\frac{PT}{\pi R}\right)^{1/2}\exp\left[-\frac{P}{4RT}(R-T)^2\right] - \frac{1}{2}\left(1 + P + \frac{PT}{R}\right)\mathrm{erfc}\left[\left(\frac{P}{4RT}\right)^{1/2}(R+T)\right]$	
A-3	$c_e(T) = 1 - \sum_{m=1}^{\infty} \frac{2\,\beta_m \sin(\beta_m)\exp\left[\frac{P}{2} - \frac{PT}{4R} - \frac{\beta_m^2 T}{PR}\right]}{\beta_m^2 + \frac{P^2}{4} + \frac{P}{2}}$	$\beta_m \cot(\beta_m) + \frac{P}{2} = 0$
A-4	$c_e(T) = 1 - \sum_{m=1}^{\infty} \frac{2\,\beta_m \sin(\beta_m)\exp\left[\frac{P}{2} - \frac{PT}{4R} - \frac{\beta_m^2 T}{PR}\right]}{\beta_m^2 + \frac{P^2}{4} + P}$	$P\beta_m \cot(\beta_m) - \beta_m^2 + \frac{P^2}{4} = 0$

semi-infinite system. This in turn suggests that analytical solution A-2 of Table 44–1 adequately describes volume-averaged concentrations inside the column. Substituting this solution into Eq. [19] leads then to a different expression for the relative effluent concentration

$$c_e(t) = \frac{1}{2}\,\mathrm{erfc}\left[\frac{RL - vt}{2(DRt)^{1/2}}\right] + \frac{1}{2}\exp\left(\frac{vL}{D}\right)\mathrm{erfc}\left[\frac{RL + vt}{2(DRt)^{1/2}}\right] \quad [24]$$

which is the same as solution A-1 of Table 44–1, evaluated at $x = L$. This result is consistent with the notion that effluent concentrations are flux-averaged concentrations (Van Genuchten & Parker, 1984; Parker & Van Genuchten, 1984a). Hence, solution A-1 of Table 44–2 correctly predicts effluent curves from finite columns, given the assumption that the exit boundary does not affect transport inside the column.

To appreciate the effects of various boundary conditions, Fig. 44–1 shows, for three different values of P, calculated effluent curves based on the four analytical solutions listed in Table 44–2. Note that the solutions deviate drastically when $P = 1$, but slowly converge when P increases. Solutions A-1 and A-4 are similar when $P = 5$ and essentially identical when $P = 20$.

An important attribute of a breakthrough curve from a finite column is the area above each curve. This area, sometimes referred to as the

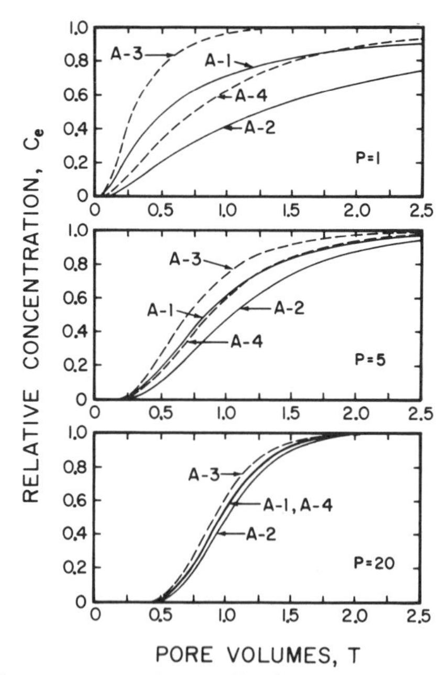

Fig. 44–1. Relative effluent concentration profiles for analytical solutions A-1 (Lapidus & Amundson, 1952), A-2 (Lindstrom et al., 1967), A-3 (Cleary & Adrian, 1973) and A-4 (Brenner, 1962). The effluent curves are plotted for three values of the column Peclet number, P.

holdup (H), represents the amount of material that can be stored in the column. Mathematically, H is given by

$$H = \int_0^{\infty} [1 - c_e(T)]\, dT. \qquad [25]$$

One may verify that $H = R$ for solutions A-1 and A-4. For solution A-2, H is given by (van Genuchten & Parker, 1984)

$$H = R[1 + (1/P)] \qquad [26]$$

while for solution A-3

$$H = R[1 - (1/P) - (e^{-P}/P)]\,. \qquad [27]$$

Figure 44–2 gives a plot of the relative holdup (H/R) vs. P for the four analytical solutions. Note that H/R deviates substantially from unity for solutions A-2 and A-3. Hence, highly inaccurate estimates for R can be obtained when these two solutions are fitted to column effluent data, especially when P is small. For example, solution A-2 overestimates R by about 50% when $P = 2$.

In conclusion, only solutions A-1 and A-4 for the effluent curve lead to correct estimates for R, irrespective of the value of P. However, these two solutions may yield slightly different estimates for P, especially when P is less than about 5 (Fig. 44–1). As pointed out before, solutions A-1 and A-4 are based on different assumptions regarding the physics of flow and transport at or near the lower boundary. Because of some inconsistencies in the stipulation of concentration continuities at the inlet and outlet boundaries for Brenner's solution (van Genuchten & Parker, 1984) and because of its numerically more tedious form compared to A-1, we recommend that solution A-1 always be used to calculate flux-averaged concentrations, whether they pertain to finite systems (effluent curves) or semi-infinite field profiles. We similarly recommend that solution A-2 always be used for volume-averaged resident concentrations.

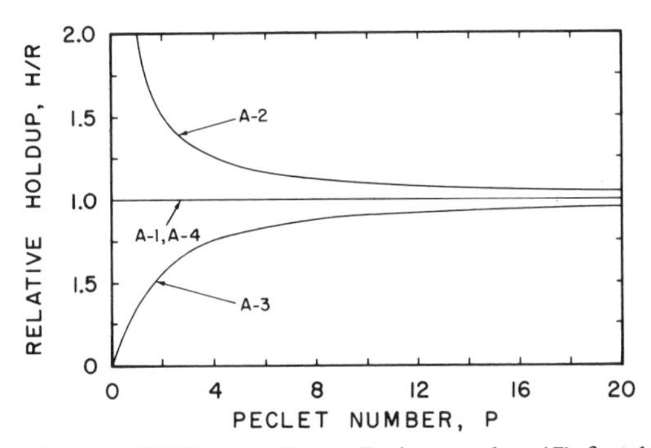

Fig. 44–2. Relative holdup (H/R) vs. column Peclet number (P) for the four analytical solutions of Table 44–2 (see also Fig. 44–1).

In some cases, breakthrough curves are obtained within laboratory columns by means of suction cups or other extraction devices. In that case, it is difficult to reason exactly which concentration mode will be observed. Because the soil solution is extracted at a fixed location in the soil, it is unlikely that the observed concentrations are flux concentrations. However, because of the transient behavior of displacement experiments and the uncertainty of how exactly flow lines are disrupted by the installation and performance of suction cups, observed concentrations are probably not exactly those of resident concentrations either. This problem is not only pertinent for laboratory experiments but also for in situ field measurements.

An expression not listed in Tables 44–1 or 44–2 but frequently used to describe displacement experiments is (Danckwerts, 1953; Rifai et al., 1956)

$$c(x,t) = \frac{1}{2}\,\mathrm{erfc}\left[\frac{Rx - vt}{2(DRt)^{1/2}}\right] \qquad [28]$$

This equation provides a close approximation of the four analytical solutions in Table 44–1 for relatively large values of P. For example, Eq. [28] follows from solutions A-1 and A-2 by retaining only the first term of the analytical expressions. Equation [28] can be derived also from Eq. [11] by assuming either an infinite system ($-\infty<x<\infty$) or a purely dispersive system in which molecular diffusion is neglected (Rifai et al., 1956; Kirkham & Powers, 1972). Even though Eq. [28] is formally not applicable to either laboratory or field experiments, its simple form and the fact that the equation provides a close approximation of the analytical solutions when P is large makes it an attractive tool for deriving simple and approximate expressions for D in terms of measurable parameters.

44–3 EXPERIMENTAL PRINCIPLES

44–3.1 Special Apparatus

The apparatus for determining D and R consists of a precision constant-volume pump, a soil column, and a fraction collector (Wierenga et al., 1975) (Fig. 44–3). For unsaturated flow experiments, controlled vacuum is necessary, as well as a vacuum container to house the fraction collector. Columns for holding the soil are most conveniently constructed from plexiglass. Construction details are available from the second author.

The lower end of the plexiglass column has a bottom plate with an O-ring closely fitted inside the column. The top part of the bottom plate contains a fritted glass porous plate with a slightly smaller diameter. Space below the fritted glass plate and a hole in the center of the bottom plate

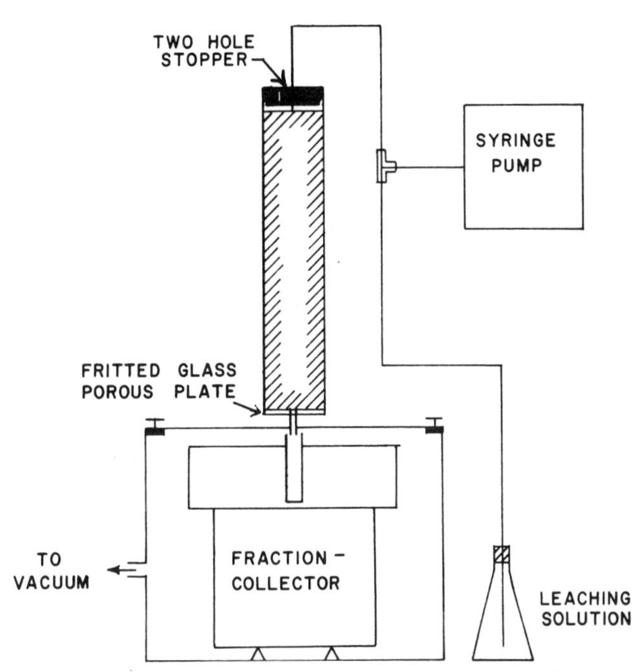

Fig. 44–3. Schematic diagram of experimental apparatus for column displacement experiments.

allow for drainage from the column. To reduce mixing in the end-plate assembly, it is important to keep the pore space of the end assembly as small as possible. For ions that react with fritted glass, stainless steel porous plates (Michigan Dynamics, Garden City, MI 48135) may be used. Unfortunately, these plates generally have a lower air-entry value than fritted glass plates. Alternatively, silver membranes (Selas Flotronics, Huntingdon Valley, PA 19006) may be used. They should be placed on top of the stainless steel plate or other coarse, porous material for support. Silver membranes have a high air-entry value and a low resistance to water flow. They are also very thin (0.5 mm) and thus contain minimal pore-space volume. At the upper end of the vertically placed soil column, the tracer solution can be applied through a porous plate or by means of an assembly of hypodermic needles (Gaudet et al., 1977) that spreads the solution evenly over the soil surface.

The column is placed above the fraction collector to collect the effluent in small-volume fractions. For unsaturated flow experiments, the fraction collector is placed inside a vacuum container which is connected to a vacuum supply by means of a vacuum regulator (Moore Products Co., Spring House, PA 19477). By locating the column outlet through a hole in the plexiglass cover of the container, a constant vacuum will be maintained at the lower end of the column. The magnitude of the required vacuum depends on the rate at which the solution is applied to the top of the column, and the hydraulic properties of the soil inside the column.

Ideally, the vacuum should be such that unit gradient flow conditions and a uniform water content distribution with depth exist within the column during the experiment. Constant flow conditions are best maintained by applying the solution with a precision constant-volume pump. Maintaining constant flow conditions is particularly important for experiments of long duration. When several columns are leached simultaneously, a multichannel syringe pump as described by Wierenga et al. (1973) may be considered. Where maintaining constant-flow conditions is less critical, or when the soil is to be maintained close to saturation, burettes may be used to apply the solution to the columns (Nielsen & Biggar, 1961). Although burettes make it more difficult to maintain constant-flux conditions, they do allow for a much better control of the pressure-head gradient inside the soil columns. With either pumps or burettes, it is important to be able to switch rapidly from one solution to another, with minimal mixing between the applied solutions in the inlet assembly.

44–3.2 Experimental Procedure

Upon collecting effluent samples, determining their concentration by standard chemical procedures, and plotting these concentrations vs. either time, volume of effluent ($V = Aqt$), or pore volume (T), an effluent curve is obtained. The number of pore volumes is calculated by dividing the amount of water leached through the column (V) by the liquid capacity ($V_o = A\theta L$) of the column

$$T = V/V_o = vt/L \qquad [29]$$

where L is the length and A the cross-sectional area of the column. The analysis of effluent curves is greatly facilitated by plotting relative concentrations (Eq. [13]) vs. pore volumes. Figure 44–4 shows typical effluent curves for the movement of tritiated water (^3H_2O) through a 30-cm long soil column and for chromium (Cr^{6+}) transport through a 5-cm long column of sand. Relevant data for these and two other experiments are listed in the appendix. The ^3H_2O-data are hypothetical insofar as they were calculated with solution A-1 (Table 44–2), with $R = 1$ and $P = 30$; the chromium data were actually measured. Both sets of data will be used in later sections to check the accuracy of the different methods.

We emphasize here that use of the dimensionless parameters P and T does not suggest that the different methods below are restricted only to finite laboratory soil columns. The parameter L in case of field-measured concentration-time curves simply refers to the soil depth at which the concentrations were observed. Hence, the methods below apply also to semi-infinite field profiles, provided that only methods based on solution A-2 (Tables 44–1 and 44–2) be used to estimate P and R from

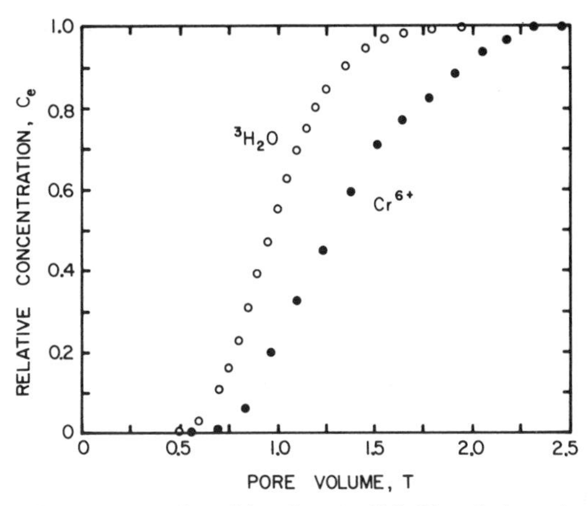

Fig. 44–4. Effluent curves for tritiated water (^3H_2O) and chromium (Cr^{6+}).

volume-averaged field concentration data. Once P is determined with whatever method, the value of D follows immediately from Eq. [22b].

44–4 METHOD I: TRIAL AND ERROR

44–4.1 Principles, Procedure, and Example

Estimates for P and R can be obtained by comparing the experimental curve directly with a series of calculated distributions and selecting those values of P and R that provide the best fit with one of the theoretical curves. An approximate estimate for R can be obtained first by locating the number of pore volumes ($T = R$) at which the relative concentration of the observed curve reaches 0.5. This property is based on Eq. [28], which at $x = L$ and in terms of P and T reduces to

$$c_e(T) = \tfrac{1}{2}\,\mathrm{erfc}[(P/4RT)^{1/2}\,(R - T)]\,. \qquad [30]$$

Because erfc(0) = 1, it follows from Eq. [30] that $c_e(R) = 0.5$. As an example, Fig. 44–5 compares the ^3H_2O-curve of Fig. 44–4 with theoretical curves based on solution A-1 of Table 44–2, using $R = 1$ and various values of P. Inspection of this figure shows that the correct value of P should be about 30 to 40.

44–4.2 Comments

From Fig. 44–5, it must be clear that trial-and-error methods can be cumbersome and time-consuming, and that they yield estimates for P and R that are not necessarily reproducible by two different investigators or by the same investigator on two different occasions.

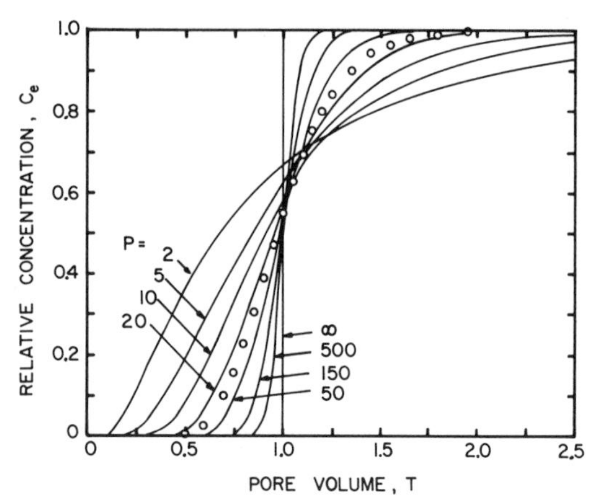

Fig. 44–5. Calculated effluent curves based on analytical solution A-1 (Table 44–2) and for various values of the column Peclet number, P. The "observed" curve (open circles) was obtained with $R = 1$ and $P = 30$.

44–5 METHOD II: FROM THE SLOPE OF AN EFFLUENT CURVE

44–5.1 Principles

Consider first the approximation given by Eq. [30] for the relative effluent concentration. Differentiation of this equation with respect to T, evaluating the resulting equation at $T = R$, and solving for P yields (see also Rifai et al., 1956)

$$P = 4\pi\, R^2\, S_T^2 \qquad [31]$$

where S_T is the slope of the effluent curve after exactly R pore volumes. The retardation factor, R, again can be estimated by locating the value of T at which the relative concentration equals 0.5.

The method above is based on Eq. [30] and hence can give only approximate estimates for P and R. However, a similar method can also be derived from analytical solution A-1 of Table 44–2. Differentiation of that solution with respect to T and solving for P at $T = R$ also leads to Eq. [31]. Contrary to Eq. [30], however, solution A-1 does not yield a value of 0.5 after R pore volumes. This is shown in Fig. 44–6, where relative effluent concentrations at $T = R$, i.e., $c_e(R)$, are plotted vs. P for the four analytical solutions of Table 44–2. Note that $c_e(R)$ for case A-1 is always > 0.5, especially when P is small. Once approximate values of P and R are obtained by means of Eq. [31] and the initial assumption that $c_e(R) = 0.5$, curve A-1 of Fig. 44–6 can be used to obtain a better estimate for $c_e(R)$. By locating that estimate on the measured curve and reading the associated value of T, an improved estimate for R results. If $c_e(R)$ differs greatly from 0.5, it may be necessary to graphically recalculate

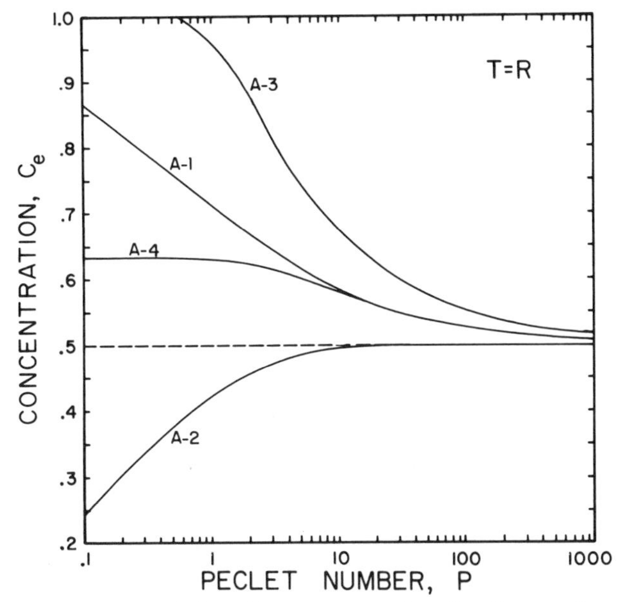

Fig. 44–6. Effect of P on the relative effluent concentration after R pore volumes ($T = R$) for the four analytical solutions of Table 44–2.

the slope S_T at $T = R$. Once improved estimates for R and S_T are available, the final value for P is obtained by again using Eq. [31].

44–5.2 Procedure

After plotting relative concentrations vs. pore volumes, $c_e(T)$, determine graphically the slope S_T of that curve at a relative concentration of 0.5. The value of T at $c_e = 0.5$ gives an initial approximation for R. Given the initial estimates for S_T and R, use Eq. [31] to obtain a first approximation for P. Use curve A-1 of Fig. 44–6 to obtain an improved estimate for the relative concentration after R pore volumes. Locate this concentration on the experimental effluent curve; the value of T at that point gives an improved estimate for R. If needed, graphically recalculate the slope S_T at the point. Application of Eq. [31] leads to an improved value for P.

44–5.3 Example

Using the tritiated water effluent curve as an example, it follows from Fig. 44–7 that the slope S_T at $c_e = 0.5$ equals $1/(1.29 - 0.65) = 1.56$, while $R = T = 0.96$ at that point. Substitution of these values into Eq. [31] yields $P = 28.2$. For this P-value, c_e at $T = R$ will be about 0.55 (curve A-1 of Fig. 44–6). The value of T at $c_e = 0.55$ in Fig. 44–7 is about 1.00, which is our improved estimate for R. Assuming that the dimensionless slope S_T of the observed curve at a relative concentration

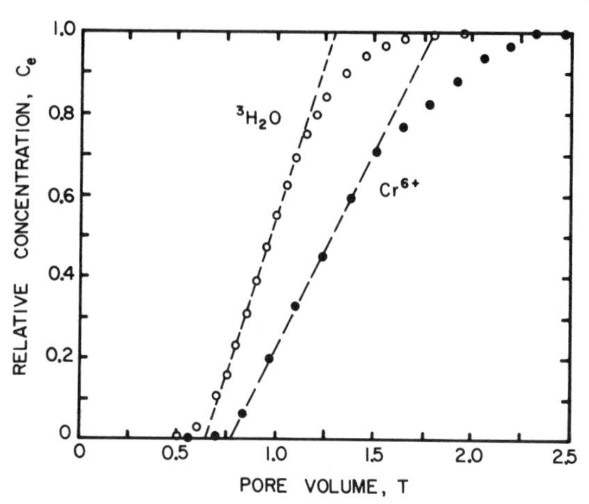

Fig. 44–7. Calculation of the dispersion coefficient from the slope of the effluent curve at $T = R$ (Method II; see text).

of 0.55 is the same as at 0.5, application of Eq. [31] with $R = 1.00$ and $S_T = 1.56$ leads to a final estimate of 30.6 for P. In actuality, the slope at $c_e = 0.55$ should be slightly smaller than at 0.5, thus causing P to be somewhat overestimated. Ignoring this small effect, we conclude that P and R are 30.6 and 1.00, respectively. Results obtained with this and the next two methods are summarized in Table 44–3. Figure 44–8 shows graphically the final results for both ^3H_2O and Cr^{6+}.

44–5.4 Comments

Method II is relatively easy to apply, requires only a minimum number of calculations, and still gives fairly accurate answers. This method is based on analytical solution A-1 of Table 44–2, and hence can be applied only to flux-averaged concentrations (effluent curves). However, for not too small values of P the method should give fairly accurate answers also for observed laboratory and field in situ measurements. Because Eq. [30] as compared to A-1 is a more accurate approximation of solution A-2 in Table 44–2, we recommend that the initial estimates for P and R be used directly in that case, i.e., without using the iteration involving curve A-1 of Fig. 44–6.

44–6 METHOD III: FROM A LOG-NORMAL PLOT OF THE EFFLUENT CURVE

44–6.1 Principles

Consider again the approximation, Eq. [30], for the relative effluent concentration. Inverting this equation yields

Table 44-3. Estimates for P and R based on several methods and for different analytical solutions. For comparison, the estimates based on Eq. [30] are also included.

Exp. no.	Tracer	Method		Analytical solution A-1	A-2	A-3	A-4	Eq. [30]
1	^3H_2O	II	P	30.6	--	--	--	28.2
			R	1.00	--	--	--	0.96
		III	P	30.3	29.9	29.4	29.2	30.7
			R	1.000	0.975	1.036	1.000	0.975
		IV	P	30.00	29.54	29.37	28.96	30.49
			R	1.000	0.967	1.035	1.000	0.968
2	Cr^{6+}	II	P	21.6	--	--	--	19.4
			R	1.34	--	--	--	1.28
		III	P	18.7	18.3	17.9	17.6	19.1
			R	1.350	1.284	1.419	1.350	1.284
		IV	P	19.65	19.19	18.95	18.59	20.11
			R	1.349	1.280	1.424	1.349	1.284
3	Cl^-	II	P	266.0	--	--	--	262.0
			R	0.92	--	--	--	0.914
		III	P	252.0	251.6	251.1	251.9	252.4
			R	0.922	0.919	0.925	0.922	0.919
		IV	P	253.6	253.1	253.1	253.0	254.1
			R	0.921	0.918	0.925	0.921	0.918
4	^3H_2O	II	P	32.4	--	--	--	30.4
			R	0.97	--	--	--	0.940
		III	P	26.8	26.4	26.0	25.7	27.1
			R	0.987	0.948	1.020	0.987	0.948
		IV	P	26.76	26.31	26.10	25.72	27.26
			R	0.973	0.937	1.012	0.973	0.938

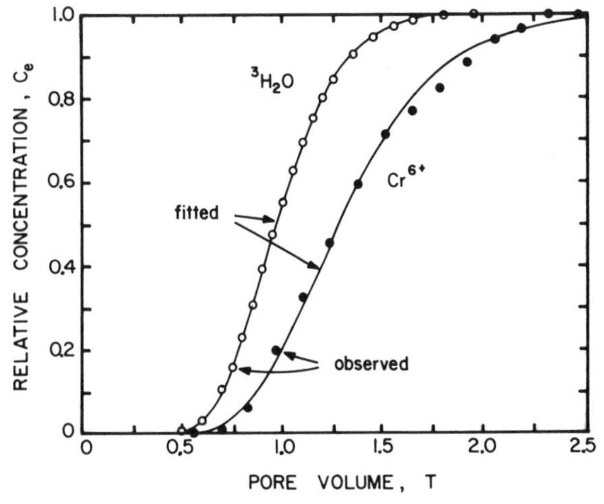

Fig. 44-8. Observed and fitted effluent curves for tritiated water and chromium (Exp. no. 1 and 2).

$$\text{inverfc}(2c_e) = (P^{1/2}/2)(1 - T/R)/(T/R)^{1/2} \quad [32]$$

where inverfc is the inverse complementary error function. For values of y close to one, the following approximation is very accurate

$$(1 - y)/y^{1/2} \sim -\ln(y)\,. \quad [33]$$

Applying Eq. [33] with $y = T/R$ to Eq. [32] yields

$$\text{inverfc}(2c_e) = (P^{1/2}/2)\ln(T/R) = (P^{1/2}/2)[\ln(R) - \ln(T)]\,. \quad [34]$$

This equation shows that P can be obtained from the slope of the curve inverfc($2c_e$) vs. ln(T). Calling this slope α, we have

$$P = 4\alpha^2 - \Delta \quad [35]$$

where a correction factor Δ is introduced to account for the approximate nature of Eq. [33]. Because of this approximation, the value of Δ depends on the range of T-values used in the correlation of Eq. [35], and consequently also on the concentration range of that correlation. Equations [34] and [35] can be applied also to the four analytical solutions of Table 44–2, provided the value of Δ is properly adjusted for each case. Table 44–4 gives approximate values of Δ for the four solutions and for different concentration ranges. An alternative and more complicated expression for P as a function of α was given earlier by Rose and Passioura (1971). Their expression, which holds only for solution A-4 and for relative concentrations between 0.05 and 0.95, does not yield better results than the much simpler Eq. [35].

Once α is derived graphically from a plot of inverfc($2c_e$) vs. ln(T), the value of P is calculated with Eq. [35] using a suitable value for Δ taken from Table 44–4. R again is estimated by first using Fig. 44–6 to find $c_e(R)$ and subsequently locating that concentration on the experimental curve. The value of T associated with this point provides an estimate for R.

Table 44–4. Values for Δ in Eq. [35] associated with the four analytical solutions of Table 44–2 and for different concentration ranges.†

Relative concentration range	Analytical solution				
	A-1 $P > 4$	A-2 $P > 4$	A-3 $P > 9$	A-4 $P > 4$	Eq. [30] $P > 2$
0.20–0.80	0.5	0.9	1.3	1.6	0.07
0.10–0.90	0.6	1.0	1.4	1.7	0.17
0.05–0.95	0.7	1.1	1.5	1.8	0.27
0.02–0.98	0.8	1.2	1.7	1.9	0.42
0.05–0.80	0.7	1.0	1.6	1.8	0.2

† For comparison, values of Δ for approximation in Eq. [30] are also included. The values for Δ are only valid for the indicated P-values.

The method above is fairly straightforward, the main challenge being an accurate inversion of the complementary error function. One can do this by using available tables of the erfc-function (e.g., chapter 7 of Abramowitz and Stegun, 1965), or more conveniently by making use of the following inversion formula:

$$\text{inverfc}(y) = \begin{cases} f(p) & p = [-\ln(y/2)]^{1/2} \quad (0 < y < 1) \\ -f(p) & p = [-\ln(1 - y/2)]^{1/2} \quad (1 \leqslant y < 2) \end{cases} \quad [36]$$

where

$$f(p) = p - \frac{1.881796 + 0.9425908p + 0.0546028p^3}{1 + 2.356868p + 0.3087091p^2 + 0.0937563p^3 + 0.0219104p^4}. \quad [37]$$

This approximation, which we obtained by improving Eq. [26.2.23] of Abramowitz and Stegun (1965), has an absolute error in inverfc of less than 4.10^{-5} over the range $-4 < \text{inverfc} < 4$.

Because of the relation between the complementary error function and the normal probability function, Eq. [34] also implies that a straight line emerges when c_e is plotted vs. $\ln(T)$ on probability paper. This property obviates the need to invert erfc for each new experiment. Figure 44–9 gives an example of the type of plot that must be prepared for this purpose. The vertical axis at the left, to be identified with the observed concentrations, shows the regular probability scale between 0.0001 and 0.9999. The vertical axis on the right shows equally spaced intervals for inverfc($2c_e$). For easy reference, the plot also includes for selected values of c_e several horizontal lines exhibiting the exact values of inverfc($2c_e$). The horizontal axis of the log-normal plot (Fig. 44–9) of course must be logarithmic.

44–6.2 Procedure

Prepare a log-normal plot as shown in Fig. 44–9. If no probability paper is available, a log-normal plot can be constructed either by using published tables of the erfc-function, or by making use of Eq. [36] and [37]. Once this plot is constructed, graphically calculate the slope α of the curve inverfc($2c_e$) vs. $\ln(T)$. Use Table 44–4 to find the value of Δ applicable for a given analytical solution and a given concentration range judged to be appropriate for the analysis. Use Eq. [35] to calculate P from α and Δ. Given P, use one of the curves in Fig. 44–6 to estimate the concentration c_e after R pore volumes. Use this value of $c_e(R)$ to find $\ln(T) \equiv \ln(R)$, and hence R, from the experimental curve.

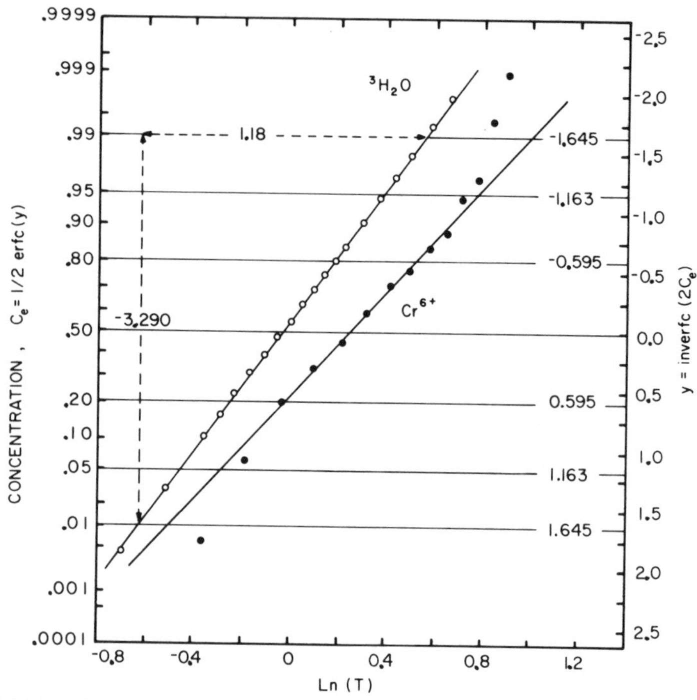

Fig. 44–9. Logarithmic-normal plots of the effluent curves shown in Fig. 44–3 (Method III).

44–6.3 Example

Figure 44–9 shows the same experimental data of Fig. 44–4 plotted vs. $\ln(T)$ on probability paper. The solid lines were eye-fitted through the data points. For the ^3H_2O curve all data points between relative concentrations of 0.02 and 0.98 were used, while for Cr^{6+} only data between 0.10 and 0.90 were considered. When drawing the curves, we recommend using more weight for observed data points close to $c_e = 0.5$; this procedure avoids too much influence being given to data points involving very low and very high concentrations (Fig. 44–9). Although deviations from the straight-line behavior at these end points of the fitted curve are important both from an experimental and conceptual point of view, these deviations are visually greatly enhanced with a probability scale. The fitted curves in Fig. 44–9 can be obtained also by applying least-squares techniques (Rose & Passioura, 1971; Passioura et al., 1970). Unfortunately, such an approach again will put too much emphasis on the extreme low and high concentrations.

The slope α of the ^3H_2O-curve (see Fig. 44–9) equals $(-1.18/3.29)$ or 2.79. Equation [35] with $\Delta = 0.8$ (Case A-1 of Table 44–3) gives $P = 30.3$. The relative concentration after R pore volumes for this P-value is about 0.54 (Fig. 44–6), which leads to an estimate of 0.0 for $\ln(R)$. Hence $R = 1.00$. Estimates of P and R for solution A-1 and the other three analytical solutions are listed in Table 44–3.

The procedure for chromium is exactly the same. The slope α is determined graphically from Fig. 44–6: $\alpha \sim 2.326/1.06 = 2.194$. With $\Delta = 0.6$ (Case A-1 for the concentration range 0.1 to 0.9 in Table 44–4), this leads to $P = 18.7$. Since $c_e = 0.56$ at $T = R$ for this P-value (Fig. 44–6), it follows from Fig. 44–9 that $\ln(T)$ is about 0.30. This in turn gives $R = 1.35$ for the chromium curve.

44–6.4 Comments

While a log-normal plot of experimental data is not expected to yield exactly a straight line because of the approximations discussed above, deviations of the magnitude of those shown for the chromium curve in Fig. 44–6 cannot be explained on that basis. Several situations will lead to these types of deviations, such as the use of undisturbed, aggregated, or otherwise structured soils, soils containing large macropores that induce preferential solute transport, channeling of water along the walls of a poorly packed soil column during saturated flow, or fingering caused by density differences between the displacing and displaced solution. Also chemical effects, such as nonlinear adsorption or cation exchange in general, can lead to serious deviations from the log-normal distribution. Figure 44–10 shows two additional experimental curves, one for chloride transport through a loam soil that yields a straight line and one for tritiated water movement through an aggregated clay loam that quite severely deviates from the straight log-normal line. Analysis of the chloride data is straightforward, yielding P and R-values as shown in Table 44–3. Unfortunately, some judgment is needed when analyzing the ^3H_2O curve. In this case, a straight line was drawn through the data points located between relative concentrations of 0.2 and 0.8. The fitted effluent curves for Cl and ^3H_2O are both shown in Fig. 44–11. Note that the observed tritiated water curve deviates considerably from the straight log-normal plot in Fig. 44–10, but that the fitted curve in Fig. 44–11 is still quite reasonable; serious deviations occur only at the higher concentrations. The fitted curve in this case was based on analytical solution A-2 (Table 44–2). The other solutions generated essentially the same fitted curve, albeit with slightly different values for P and R (Table 44–3).

Method III involves a few more calculations than the somewhat simpler Method II, notably the cumbersome inversion of the complementary error function. Fortunately, the number of calculations can be reduced drastically by first preparing special logarithmic-normal paper as shown in Fig. 44–6. Once this special paper is available, the analysis can be carried out quickly and easily. Results obtained with Method III are generally more accurate than those based on Method II. This is a direct consequence of the fact that the slope S_T for Method II must be determined after exactly R pore volumes. Because this slope changes from point to point on the curve, its value is not easy to pinpoint exactly and in a reproducible manner. On the other hand, the approximately straight log-normal plot of Method III allows a much broader concentration range

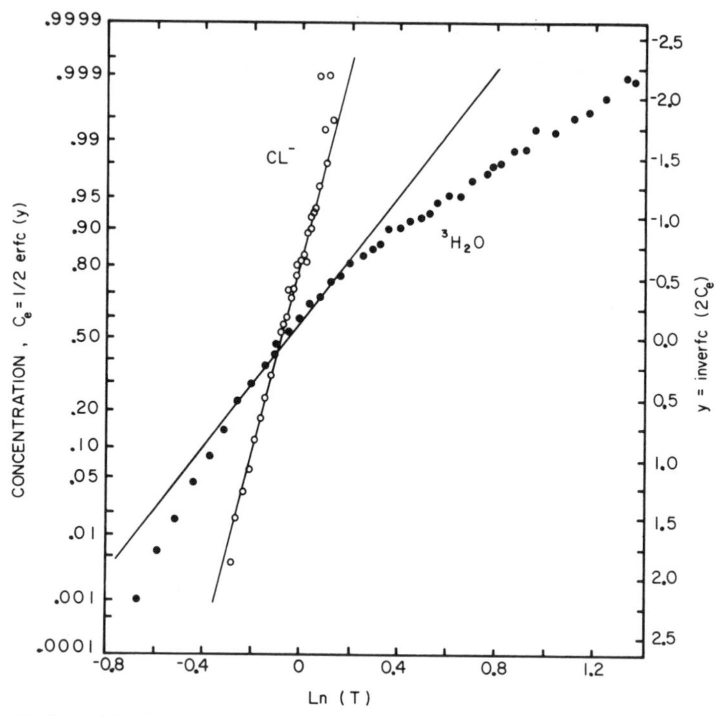

Fig. 44–10. Logarithmic-normal plots of effluent curves for Exp. no. 3 and 4 (Table 44–3).

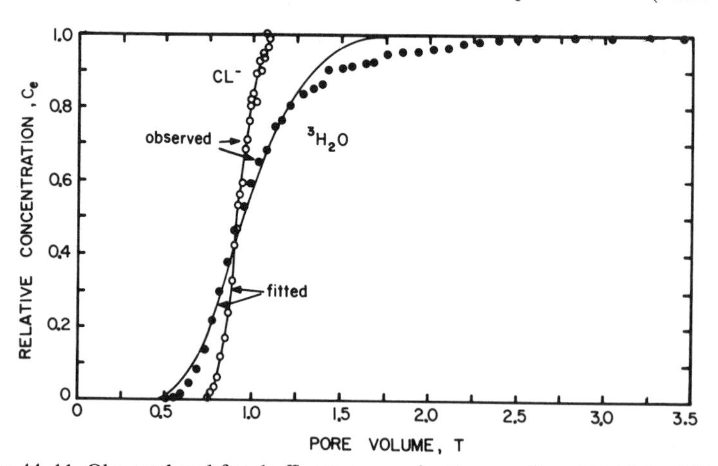

Fig. 44–11. Observed and fitted effluent curves for Exp. no. 3 and 4 (Method III).

to be used when determining the slope α, which in turn will lead to a more accurate estimate for P. An additional advantage of Method III is its applicability to all four analytical solutions, whereas Method II holds only for solution A-1. A disadvantage of Method III is that results based on this method can become quite inaccurate when P becomes less than about 5 (see Table 44–4).

44–7 METHOD IV: LEAST-SQUARES ANALYSIS OF THE EFFLUENT CURVE

44–7.1 Principles

The trial-and-error method discussed earlier (Method I) is based on a series of comparisons between observed and calculated effluent curves until a satisfactory visual fit of the observed curve is obtained. This method can be expanded into a more formal approach by continuously adjusting P and R until a least-squares fit of the observed data is obtained. This is done by minimizing the residual sum of squares (R_s)

$$R_s = \sum_{i=1}^{n} \left[c_e(L, T_i) - c(L, T_i) \right]^2 \qquad [38]$$

where $c_e(L,T_i)$ and $c(L,T_i)$ are the observed and calculated data points at pore volume T_i, and n is the number of observed data points. Several methods for minimizing R_s are available, ranging from relatively simple graphical methods to computerized solutions using existing least-squares inversion methods (Elprince & Day, 1977; Laudelout & Dufey, 1977; Le Renard, 1979; van Genuchten, 1980, 1981; Parker & van Genuchten, 1984b). Table 44–3 gives for all four experiments the fitted values of P and R as obtained with the least-squares program of van Genuchten (1980). This program is available upon request from the senior author.

44–7.2 Comments

Results based on least-squares optimization methods are the most accurate, are valid for all values of P and R, and are reproducible on different occasions and by different investigators. Once programmed, computerized least-squares methods are also by far the most convenient ones to use. In addition, they can be extended easily to situations where pulse-type effluent curves are present, i.e., to cases where the tracer is in the feed solution for only a relatively short period of time. Parker and van Genuchten (1984b) recently published a least-squares optimization method that can be used to estimate D and R from observed breakthrough data obtained at more than one distance from the soil surface. Clearly, such an approach extracts the most information from available experimental data.

44–8 METHOD V: FROM CONCENTRATION-DISTANCE CURVES

In some cases dispersion coefficients must be estimated from concentration-distance curves. Such curves can be obtained by sectioning laboratory soil columns or by determining solute concentration in situ, either in laboratory soil columns or in field soils. Two methods for these situations are discussed briefly below. Both methods are based on Eq. [28], and consequently, can yield only approximate estimates for D and R. Not mentioned any more are trial-and-error and least-squares methods; application of these methods to concentration-distance curves is, at least in principle, straightforward. For the documentation of a least-squares optimization program applicable to concentration distributions vs. distance, see Parker and van Genuchten (1984b).

Differentiating Eq. [28] with respect to x and evaluating the resulting expression at $Rx = vt$ leads to the following equation for D:

$$D = R/4\pi t\, S_x^2 \quad [39]$$

where S_x is the slope of the experimental curve at $c = 0.5$. Since at this point $Rx = vt$, Eq. [39] can be written also in the form

$$D = v/4\pi x_o\, S_x^2 \quad [40]$$

where x_o is the value of x where the relative concentration attains a value of 0.5. Equations [39] and [40] are evaluated in a similar way as Eq. [31] for Method II. Contrary to Method II, however, the present equations are applicable only to Eq. [28]. Hence, they cannot be applied to any of the analytical solutions in Table 44–1.

A procedure somewhat similar to Method III follows by inverting Eq. [28]:

$$\mathrm{inverfc}(2c) = (Rx - vt)/2(DRt)^{1/2}\,. \quad [41]$$

This equation shows that D can be obtained also from the slope, β, of the curve inverfc($2c$) vs. x. The predictive equation for D in this case is

$$D = R/4\beta^2 t \qquad (R = vt/x_o)\,. \quad [42]$$

44–9 OTHER TRANSPORT MODELS

The different methods described above all deal with transport Eq. [11], a model that considers linear equilibrium adsorption but ignores production or decay. For many organics, N species, or radionuclides, additional terms are usually needed that describe zero- and/or first-order production or decay processes. A general transport model for that case is

$$R\frac{\partial C}{\partial t} = D\frac{\partial^2 C}{\partial x^2} - v\frac{\partial C}{\partial x} - \mu C + \gamma \quad [43]$$

where μ and γ are first- and zero-order rate coefficients. Because of the production and decay terms, simple methods analogous to Methods II, III, and V are not available for estimating D, R, μ, and/or γ. However, an easy-to-use computer program that couples a least-squares optimization method with various analytical solutions of Eq. [43] recently has been made available by Parker and van Genuchten (1984b). Their program can be applied to observed concentration distributions vs. time and/or distance.

Least-squares computerized methods can be conveniently used also to estimate parameters that appear in more complicated transport models. For example, van Genuchten (1981) published a program that is applicable to various physical and chemical nonequilibrium models. His program is limited to breakthrough curves in time. The program of Parker and van Genuchten (1984b) considers similar nonequilibrium models applicable to concentration-distance curves as well.

44–10 GENERAL COMMENTS

As outlined above, several methods are available for determining D (or P) and R. The two simplest methods (I and II) are probably most appropriate when the effluent curve is poorly defined. This may be due to a limited number of (inaccurate) data, considerable scatter between the data points, or serious deviations from the mostly sigmoidal-type curves shown in this study. Methods based on least-squares techniques or on log-normal plots of the data are best reserved for cases where the experimental curve is well-defined. Of all existing methods, least-squares computer methods are the most accurate and at the same time the most convenient ones to use. Provided that the necessary computer facilities are available, we strongly recommend their use.

As was pointed out already in section 44–2, we emphasize again that the fitted values of D and R ultimately are associated with one of the analytical solutions in Table 44–1. Methods based on different analytical solutions but applied to the same experimental data will lead to different values for D and R. These differences can become especially noticeable for relatively small values of the dimensionless group (vx/D), or in the case of effluent curves of the column Peclet number, P. Figure 44–1 suggests that for P-values greater than about 20, the fitted value of the dispersion coefficient becomes more or less independent of the method used to determine that value. In practice, this means that experiments should be carried out on columns of at least 10 to 15 cm length in case of homogeneous soils with relatively narrow pore-size distributions, and on columns of at least 30 to 50 cm length if undisturbed, aggregated, or other soils are present that have a relatively broad pore-size distribution. If column experiments are used primarily to determine the retardation factor, R (or the linearized distribution coefficient, k), then a shorter column is allowed, provided that only methods based on analytical solutions A-1 or A-4 of Table 44–2 are used.

44–11 APPENDIX

Data for four column displacement experiments.

Example	1		2		3		4	
Tracer:	^3H_2O		Cr^{6+}		Cl^-		^3H_2O	
θ (cm^3 cm^{-3})	0.400		0.184		0.363		0.453	
ϱ (g cm^{-3})	1.400		1.68		1.53		1.22	
q (cm day^{-1})	10.0		3.61		5.16		17.0	
L (cm)	30.0		5.0		30.0		30.0	
	T	c_e	T	c_e	T	c_e	T	c_e
	0.50	0.0042	0.558	0.000	0.749	0.004	0.512	0.001
	0.60	0.0294	0.695	0.006	0.768	0.017	0.556	0.006
	0.70	0.1015	0.831	0.061	0.787	0.035	0.599	0.016
	0.75	0.1586	0.967	0.198	0.806	0.059	0.643	0.045
	0.80	0.2279	1.103	0.325	0.825	0.114	0.686	0.082
	0.85	0.3057	1.236	0.450	0.845	0.169	0.730	0.138
	0.90	0.3881	1.375	0.592	0.864	0.240	0.774	0.216
	0.95	0.4709	1.511	0.705	0.883	0.326	0.817	0.296
	1.00	0.5507	1.647	0.768	0.902	0.421	0.861	0.376
	1.05	0.6247	1.783	0.841	0.911	0.467	0.904	0.465
	1.10	0.6913	1.919	0.881	0.921	0.531	0.948	0.529
	1.15	0.7497	2.055	0.944	0.930	0.562	0.992	0.593
	1.20	0.7996	2.191	0.966	0.940	0.594	1.035	0.655
	1.25	0.8414	2.327	0.994	0.949	0.709	1.079	0.685
	1.35	0.9037	2.463	0.999	0.959	0.688	1.122	0.745
	1.45	0.9436			0.968	0.712	1.166	0.764
	1.55	0.9679			0.978	0.767	1.218	0.806
	1.65	0.9822			0.987	0.806	1.284	0.834
	1.80	0.9929			0.997	0.822	1.340	0.850
	1.95	0.9973			1.006	0.838	1.384	0.866
					1.016	0.814	1.428	0.901
					1.025	0.893	1.501	0.905
					1.035	0.900	1.558	0.915
					1.044	0.924	1.646	0.923
					1.054	0.935	1.689	0.928
					1.063	0.940	1.754	0.947
					1.073	0.963	1.848	0.953
					1.082	0.999	1.935	0.953
					1.092	0.993	2.022	0.968
							2.110	0.974
							2.196	0.979
							2.283	0.981
							2.392	0.986
							2.517	0.987
							2.608	0.993
							2.824	0.994
							3.040	0.995
							3.257	0.996
							3.473	0.998

44-12 REFERENCES

Abramowitz, M., and I. A. Stegun. 1965. Handbook of mathematical functions. Fourth printing. Applied Math. Ser. 55, U. S. Government Printing Office, Washington, DC.

Agneessens, J. P., P. Dreze, and L. Sine. 1978. Modélisation de la migration d'éléments dans les sols. II. Détermination du coefficient de dispersion et de la porosité efficace. Pedologie 27:373–388.

Aris, R. 1958. On the dispersion of linear kinematic waves. Proc. R. Soc. London, Ser. A 245:268–277.

Bear, J. 1972. Dynamics of fluids in porous media. American Elsevier Publishing Co., New York.

Boast, C. W. 1973. Modeling the movement of chemicals in soils by water. Soil Sci. 115:224–230.

Brenner, H. 1962. The diffusion model of longitudinal mixing in beds of finite length. Numerical values. Chem. Eng. Sci. 17:229–243.

Brigham, W. E. 1974. Mixing equations in short laboratory cores. Soc. Pet. Eng. J. 14:91–99.

Cleary, R. W., and D. D. Adrian. 1973. Analytical solution of the convective-dispersive equation for cation adsorption in soils. Soil Sci. Soc. Am. Proc. 37:197–199.

Danckwerts, P. V. 1953. Continuous flow systems. Chem. Eng. Sci. 2:1–13.

Elprince, A. M., and P. R. Day. 1977. Fitting solute breakthrough equations to data using two adjustable parameters. Soil Sci. Soc. Am. J. 41:39–41.

Freeze, R. A., and J. A. Cherry. 1979. Groundwater. Prentice Hall, Englewood Cliffs, NJ.

Fried, J. J., and M. A. Combarnous. 1971. Dispersion in porous media. Adv. Hydrosci. 7:169–282.

Gaudet, J. P., H. Jegat, G. Vachaud, and P. J. Wierenga. 1977. Solute transfer, with exchange between mobile and stagnant water, through unsaturated sand. Soil Sci. Soc. Am. J. 41:665–671.

Hillel, D. 1980. Fundamentals of soil physics. Academic Press, New York.

Jury, W. A., and G. Sposito. 1985. Field calibration and validation of solute transport models for the unsaturated zone. Soil Sci. Soc. Am. J. 49:1331–1341.

Kirkham, D., and W. L. Powers. 1972. Advanced soil physics. Wiley-Interscience, New York.

Kreft, A., and A. Zuber. 1978. On the physical meaning of the dispersion equation and its solutions for different initial and boundary conditions. Chem. Eng. Sci. 33:1471–1480.

Lapidus, L., and N. R. Amundson. 1952. Mathematics of adsorption in beds. IV. The effect of longitudinal diffusion in ion exchange chromatographic columns. J. Phys. Chem. 56:984–988.

Laudelout, H., and J. E. Dufey, 1977. Analyse numérique des expériences de lixiviation en sol homogène. Ann. Agron. 28:65–73.

Le Renard, J. 1979. Description d'une méthode de détermination du coefficient de dispersion hydrodynamique longitudinal en milieu poreux homogène saturé en eau. II. Utilisation d'un modèle mathematique a trois paramètres dont un phénomenologique. Ann. Agron. 30:109–119.

Lindstrom, F. T., R. Hague, V. H. Freed, and L. Boersma. 1967. Theory on the movement of some herbicides in soils: linear diffusion and convection of chemicals in soils. Environ. Sci. Technol. 1:561–565.

Nielsen, D. R., and J. W. Biggar. 1961. Miscible displacement in soils: I. Experimental information. Soil Sci. Soc. Am. Proc. 25:1–5.

Nielsen, D. R., R. D. Jackson, J. W. Cary, and D. D. Evans (ed.) 1972. Soil water. American Society of Agronomy and Soil Science Society of America, Madison, WI.

Parker, J. C., and M. Th. van Genuchten. 1984a. Flux-averaged and volume-averaged concentrations in continuum approaches to solute transport. Water Resour. Res. 20:866–872.

Parker, J. C., and M. Th. van Genuchten. 1984b. Determining transport parameters from laboratory and field tracer experiments. Virginia Agric. Exp. Stn. Bull. 84.

Passioura, J. B., D. A. Rose, and K. Haszler. 1970. Lognorm: a program for analysing experiments on hydrodynamic dispersion. Tech. Memorandum 70/6. CSIRO, Div. of Land Res., Canberra, Australia.

Rifai, M. N. E., W. J. Kaufman, and D. K. Todd. 1956. Dispersion phenomena in laminar flow through porous media. Sanitary Eng. Res. Lab., Univ. of California, Berkeley. Inst. Eng. Res. Ser. no. 93(2).

Rose, D. A., and J. B. Passioura. 1971. The analysis of experiments on hydrodynamic dispersion. Soil Sci. 111:252–257.

Saxena, S. K., F. T. Lindstrom, and L. Boersma. 1974. Experimental evaluation of chemical transport in water-saturated porous media: 1. Nonsorbing media. Soil Sci. 118:120–126.

Shamir, U. Y., and D. R. F. Harleman. 1966. Numerical and analytical solutions of dispersion problems in homogeneous and layered aquifers. Rep. no. 89. Hydrodynamics Lab. Dep. of Civil Engineering, Massachusetts Institute of Technology.

Skopp, J. 1985. Analysis of solute movement in structured soils. p. 220–228. *In* J. Bouma and P.A.C. Raats (ed.) Proc. ISSS Symp. Water and Solute Movement in Heavy Clay Soils. International Institute for Land Reclamation and Improvement, Wageningen, Netherlands.

Valocchi, A. J. 1985. Validity of the local equilibrium assumption for modeling sorbing solute transport through homogeneous soils. Water Resour. Res. 21:808–820.

van Genuchten, M. Th. 1980. Determining transport parameters from solute displacement experiments. Res. Rep. 118. U.S. Salinity Lab., Riverside, CA.

van Genuchten, M. Th. 1981. Non-equilibrium transport parameters from miscible displacement experiments. Res. Rep. 119. U. S. Salinity Lab., Riverside, CA.

van Genuchten, M. Th., and W. J. Alves. 1982. Analytical solutions of the one-dimensional convective-dispersive solute transport equation. USDA-ARS Tech. Bull. 1661. U.S. Government Printing Office, Washington, DC.

van Genuchten, M. Th., and J. C. Parker. 1984. Boundary conditions for displacement experiments through short laboratory soil columns. Soil Sci. Soc. Am. J. 48:703–708.

Wierenga, P. J., R. J. Black, and P. Manz. 1973. A multichannel syringe pump for steady state flow in soil columns. Soil Sci. Soc. Am. Proc. 37:133–134.

Wierenga, P. J., M. Th. van Genuchten, and F. W. Boyle. 1975. Transfer of boron and tritiated water through sandstone. J. Environ. Qual. 4:83–87.

Yu, C., W. A. Yester, and A. R. Jarrett. 1984. Simultaneous determination of dispersion coefficients and retardation factors for a low level radioactive waste burial site. Radioact. Waste Manage. Nucl. Fuel Cycle 4:401–420.

45 Water and Solute Flux

R. J. WAGENET
Cornell University, Ithaca, NY

45-1 GENERAL INFORMATION

The movement of water and dissolved materials through the soil profile must often be predicted. Whether the specific issue focuses on irrigation water, rainfall, agricultural chemicals, salts, or waste materials, it is often necessary to estimate the displacement (flux) of water and solutes that will occur during a given time period. Only a few approaches have been developed to measure water and solute flux in the field, most efforts having been centered upon the development of relatively indirect methods of calculating water and solute movement. These latter approaches can be mathematically complicated and are relatively untested under field situations, so that their ability to predict fluxes on a field basis remains unresolved. As more is learned of the nature of transport processes under field conditions, it is becoming increasingly clear that only approximate prediction of water or solute flux can presently be made. These approximate predictions can be made using very simplified conceptualizations of the soil-water-solute system. The reliability of simplified models as well as much more mechanistic and data-intensive approaches remains obscure.

The flux density of water or solute, hereafter referred to as flux, is usually defined as a mass of material moving through a unit cross-sectional area of soil in a unit time period. Flux must be distinguished from a water content or solute concentration. The latter is an instantaneous capacity and not a rate measurement of water or solute behavior. Water content (chapter 21 in this book) and solute content (chapter 42 in this book) can be measured by a variety of techniques, and these measurements can be mathematically manipulated to provide estimations of flux. In laboratory experiments, water and solute flux can be directly measured as the rate of effluent flow from a soil column experiment or, in very limited field cases, as drainage rate from a lysimeter. However, there currently does not exist a direct measurement technique for water or solute flux on a field basis. All techniques for estimating flux on a field basis, whether direct or indirect, are limited by the localized nature of any individual measurement, since these measurements have to be extrapolated to a larger area that may not be of the same character as the region in which the measurement was originally made. This problem of spatial variability of the processes that determine water and solute fluxes has only recently been approached by soil scientists interested in assessing variability and its implications in water and solute movement (chapter 3 in this book). These studies are demonstrating that whatever method is used to estimate field fluxes of water or solute, the use of that method should be subject to considerable caution, with some restraint necessary in extrapolation of localized measurements to large areas.

45–2 WATER FLUX

The quantitative description of water flow through porous media has been relatively recently developed. The basic relationship describing water flow through saturated materials was postulated by Henry Darcy (Darcy, 1856), and understanding of the forces of soil water retention was first provided by Buckingham (1907) in the early 20th Century. The relationship of water content and the potential energy status of soil water followed about 15 years later (Gardner, 1920), with the basic equations describing water flow in unsaturated soils being presented in 1931 (Richards, 1931). The formulation of these concepts had produced by the early 1950s the basis for description of water flux in soil. The period that immediately followed focused mainly on the testing, evaluation, and refinement of these basic principles under carefully controlled laboratory conditions. A substantial technology developed to measure and predict water flux under such experimental regimes. Since 1970, the focus of attention has shifted to description of water and solute movement under field conditions using as starting points the basic relationships developed in the laboratory. It is becoming increasingly apparent that problems of spatial variability of flow properties in field soils will require development of new, complementary approaches (e.g., Nielsen et al., 1981b) if the fundamental relationships developed in the laboratory are to be used to describe fluxes on a field basis.

The water flux in a vertical one-dimensional soil system is expressed according to Darcy's law as

$$q_w = -K(dH/dz) \qquad [1]$$

where q_w is the soil water flux (L/T), H is the hydraulic head (L), K is a proportionality factor called the hydraulic conductivity (L/T), and z is vertical distance or depth (L). The terms *water* and *soil water* will be interchangeably used in this discussion to represent the solution phase of soil. When the water flow regime is not constant with space and time, Eq. [1] is combined with the continuity equation to become the Richards equation

$$\frac{\partial \theta}{\partial t} = \frac{\partial}{\partial z}\left[K(\theta)\frac{\partial H}{\partial z}\right] \qquad [2]$$

where θ is the volumetric soil water content (L^3/L^3), $K(\theta)$ now represents the dependence of hydraulic conductivity upon the water content, and t is time. The flux of water at depth z during nonsteady flow can be generally expressed as

$$q_w(z) = q_w(z_0) - (1/\Delta t)\int_{z_0}^{z} \Delta\theta \, dz \qquad [3]$$

where q_w is the volumetric flux of water (L/T) and it is assumed that no sources or sinks of water exist in the interval Δz.

Equation [3] is difficult to solve analytically for most initial and boundary conditions that apply to field cases. This has resulted in several types of approaches to estimation of water flux under field conditions. These include:

1. Direct measurement with a **soil water flux meter**

Soil water flux meters that measure the flux past a point in the profile have been described by Ivie and Richards, 1937; Richards et al., 1937; Byrne et al., 1967, 1968; Byrne, 1971; Cary, 1968, 1970; and Dirksen, 1972, 1974. These meters are subject to problems including the localized nature of the measurement, the disruption of the soil during installation, and interruption of normal soil water flow patterns, but do provide reasonable direct water flux measurements if properly used.

2. Estimation from **continuity principles**

This approach utilizes measurement of the water content as a function of depth and time in the profile to calculate the flux from the continuity equation for soil water. Several variations of the approach are possible. In the *water balance* method the continuity equation is applied in an integrated form to a profile. All the components of the water balance except the flux across the lower boundary of the profile are measured or estimated, and the flux (drainage) is then calculated from the balance equation. In another variation, *water-content–depth–time* relations are measured and Eq. [3] is applied to estimate the flux at depth z (Rose & Stern, 1965; Nielsen et al., 1973).

3. Prediction with **numerical models**

Finite-difference and finite-element methods have been used to provide approximate solutions of Eq. [2] (e.g., Hanks & Bowers, 1962; Bresler, 1973). These models generally assume some unique relationship between K, θ, and matric potential energy as a means of calculating q_w under transient conditions. These models have been extended to description of solute movement through the use of the dispersion-convection equation of solute transport. Flux is usually calculated by integrating predicted profiles of water content or solute concentrations.

4. Estimation using **scaling techniques**

These approaches utilize scaling methodologies developed from similitude anaysis in an attempt to represent in Eq. [2] the spatial changes in $K(\theta)$ relationships that are obvious from field measurements, with water flux thereby estimated as a function of spatially variable soil hydraulic properties (Warrick et al., 1977a, 1977b; Simmons et al., 1979). Such methods are most often used in analytical solutions, but have been applied in at least two cases (Bresler et al., 1979; Wagenet & Rao, 1983) to estimate both water and solute flux using a numerical model.

Approaches 3 and 4, though often useful at a research level, will not be presented here and the reader is referred to the cited papers for description of the approaches.

From a practical standpoint, the water flux probably can be estimated just as accurately by a simple, relatively empirical technique (e.g., method 2 above) as it can be estimated from the much more tedious and data-intensive modeling techniques (i.e., methods 3 and 4 above). The assumptions made in generating simple descriptions of water flux are no less limiting than are the current assumptions contained in the numerical models relative to spatial and temporal behavior of hydraulic conductivity, water content, and matric potential relationships. There is substantial opportunity for improvement of these latter approaches, but until this is accomplished, there is no reason to abandon, for general purposes, the semi-empirical techniques that require only moderate training and minimal data collection.

45–2.1 Soil Water Flux Meter

45–2.1.1 PRINCIPLES

The volumetric water flux is measured by intercepting soil water flow with an in situ soil water flux meter. The conductance of the meter and hydraulic head loss across the meter are used to calculate water flux through the meter, which is interpreted as a measure of the soil water flux. The meter described here was developed by Dirksen (1972, 1974) and consists of two thin filter plates located at the top and bottom of a cylindrical chamber (Fig. 45–1). These plates intercept and redistribute the soil water flow. The flow intercepted in one porous filter plate is

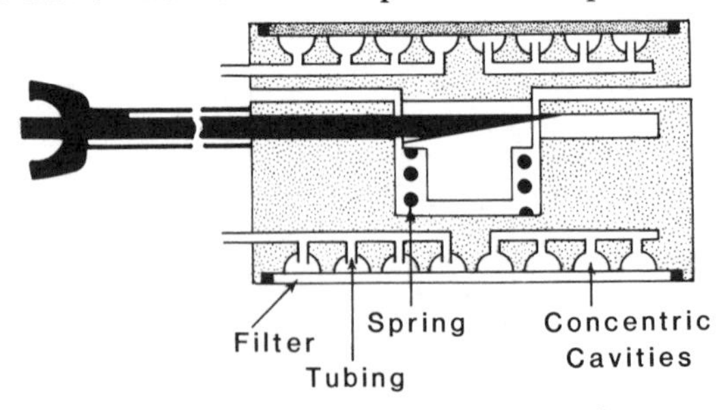

Fig. 45–1. Cross section of cylindrical soil water flux meter.

diverted through a fine-metering valve (Fig. 45–2) and distributed back into the soil with the other filter. The hydraulic head loss across the valve is measured and the hydraulic resistance of the valve adjusted so that the total hydraulic head loss across the meter remains approximately equal to the head loss measured with tensiometers inserted into undisturbed soil at some adjacent location. In this way, the water flux through the meter is adjusted so that it will represent the soil water flux, which can then be determined using the equations presented below. The tensiometers are located close enough to the meter to represent local flow conditions, but far enough away that their presence does not disrupt the flow regime. According to Cary (1968), two diameters of the meter appears to be a safe distance. The specific principles and procedures described below are those of Dirksen (1972, 1974), and unless identified otherwise, will pertain to his method. For details of meter design, calibration, and operation the reader is referred to the references, as there are many subtleties associated with the technique.

The flux of water intercepted by a soil water flux meter (q_m, cm/day) is

$$q_m = K_m(\Delta H_m/L) \qquad [4]$$

where K_m (cm/day) is the equivalent hydraulic conductivity of the meter, ΔH_m is the head loss (cm) across the meter, and L is the height of the meter (cm). The equivalent conductivity of the meter must be approximately equal to the conductivity of the bulk soil solution; otherwise, divergence or convergence of water flow will result in the vicinity of the meter. That is, if $K_m < K(\theta)$, flow will diverge away from the meter with measured fluxes too small to correctly represent the soil water flux; conversely, if $K_m > K(\theta)$, flow will converge to the meter and water flux through the bulk soil will be overestimated. It is upon this issue that most design differences exist among water flux meters.

If the value of K_m can be adjusted with changes in $K(\theta)$ such that

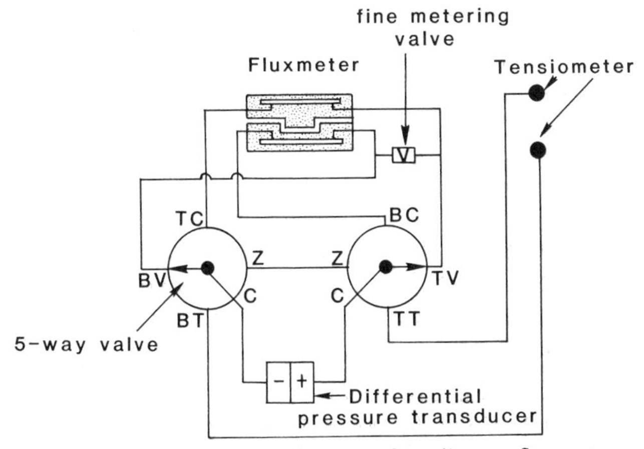

Fig. 45–2. Diagram of hydraulic system for soil water flux meter.

ΔH_m remains equal to ΔH (hydraulic head difference in bulk soil), then the continuity of the flow system is never interrupted by the presence of the meter and accurate soil water fluxes can be measured. This is accomplished by the meter described above, which varies the hydraulic resistance through the meter and adjusts this resistance until $\Delta H_m = \Delta H$. The head loss ΔH in bulk soil is equal to that measured with the tensiometers over the same distance L in undisturbed soil nearby. If the resistance is calibrated, values of q_m and q_w can be determined.

If the flow rate through the meter is led to a fine-metering valve (where the variable resistance is applied), then the flow rate Q_m (cm^3/day) through the meter is:

$$Q_m = \alpha_v \Delta H_v \quad [5]$$

where α_v is a calibration factor (cm^3/cm-day) of the value that depends upon the valve setting (Dirksen, 1972), and ΔH_v is the total head loss across the valve. Equation [5] can be expressed as

$$Q_m/A = q_m = K_v\ (\Delta H_v/L) \quad [6]$$

where

$$K_v = \alpha_v L/A\,. \quad [7]$$

A major consideration in the field is the ability to keep ΔH and ΔH_m perfectly matched, since this requires continuous monitoring of the system. However, if ΔH and ΔH_m are not matched perfectly, the remaining convergence or divergence of flow can be calculated from

$$(q_m - q_w)/q_w = -1.7\ (\Delta H_m - \Delta H)/\Delta H\,. \quad [8]$$

This relationship was found (Dirksen, 1972) empirically for a 13-fold range of fluxes in two different soils for values of $(\Delta H_m - \Delta H)/\Delta H$ ranging between -0.54 and $+0.31$. The numerical coefficient in Eq. [8] is a ratio between two dimensionless quantities and will depend partially on the diameter-to-height ratio of the meter. Solving Eq. [8] for q_w gives:

$$q_w = q_m/(2.7 - 1.7\ \Delta H_m/\Delta H)\,. \quad [9]$$

This equation makes use of the flux meter more practical, as there is no need to match ΔH_m and ΔH at all times, but simply to keep them within the range for which Eq. [8] is valid.

Once q_w and ΔH_s are obtained, the hydraulic conductivity of the soil can be calculated directly from Darcy's law as

$$K(\theta) = q_w L/\Delta H\,. \quad [10]$$

The soil water flux meter is thereby also a "soil hydraulic conductivity meter."

A basic problem associated with all flux measurements made with water flux meters is that the fluxes will be different than those in the surrounding soil. The meter described above attempts to minimize these effects by installing the meters from a horizontal direction into undisturbed soil. Such installation is made possible by incorporating a compression spring into the meter that pushes the top and bottom filter plates against the top and bottom of a cavity excavated to the size of the meter. A slanted pin makes it possible to reverse the process and remove the meter from the access hole. In this way, the flux of water both above and below the meter occurs under relatively undisturbed conditions. Meters proposed by other researchers do not include this design, but instead require excavation and backfilling of vertical holes, which will substantially alter the water flow properties in the vertical axis of the meter. The effect of such disturbance is difficult to predict, and is best avoided by horizontal insertion.

45–2.1.2 MATERIALS

1. Soil water flux meter (Fig. 45–1 and 45–2)
2. Two tensiometers, size 8 cm long by 0.5 cm thick
3. Shovel or backhoe to dig a soil pit to the desired depth
4. Access hole cutter (Dirksen, 1974) for flux meter and tensiometers.

45–2.1.3 PROCEDURE

The soil water flux meter should first be calibrated by methods described by Dirksen (1972). In the field, excavate a soil pit sufficiently deep that the desired depth of water flux measurement is located on an accessible face of the pit. The pit need be only large enough to allow convenient work space for horizontal insertion of the flux meter. Prepare an access hole for the insertion of the meter using the cutter described by Dirksen (1974). Before removing the plate and guide for the access cutter, drill also the access holes for the tensiometers. Then remove the cutter. Insert the flux meter and the tensiometers. Use the remainder of the access hole as a location to set the fine-metering valve, pressure transducer, and other small temperature-sensitive devices. Seal the access holes, preferably with insulating material, to allow easy retrieval of the flux meter and other instruments at a later date. Backfill the soil pit, marking carefully the location of the flux meter. Measure total head loss across the fine-metering valve, and calculate flux (q_w) as described above in Eq. [9].

45–2.1.4 COMMENTS

Several soil water flux meters other than Dirksen's have been developed (Ivie & Richards, 1937; Richards et al., 1937; Byrne et al., 1967, 1968; Cary, 1968, 1970, 1971; Byrne, 1971). All require considerable technique for proper calibration, operation and installation and some appreciation of the theory associated with the measurement, if reliable soil water flux estimates are to be gathered. For this reason the method is not often used. It does, however, represent a direct measurement technology that could be used by a properly trained professional.

In Dirksen's meter, temperature variations have been found to impose unsatisfactory effects upon the tensiometers, the fine metering valve, and pressure transducers. Diurnal fluctuations in temperature at the soil surface influence tensiometer readings, but are not matched by similar diurnal fluctuations at soil depths of measurement. The temperature of the fine-metering valve must be monitored, as the hydraulic head loss across the valve is directly proportional to the viscosity of water. These instruments therefore need to be insulated from temperature change, whether it be diurnal in nature or produced from exposure during equipment servicing and data collection.

Water flux meters represent a direct measurement technique, but require a comparatively high degree of training and sophisticated equipment. The estimate of water flux obtained can be biased by installation, calibration, and environmental factors. Additionally, the very localized nature of the measured water flux is difficult to extrapolate to a field basis, given the spatial variability of $K(\theta)$ relationships that have been measured in the field (Nielsen et al., 1973). When the expense of each water flux meter and the need to locate at least several of these meters at each depth of interest are combined, the method is rather unattractive in terms of time and money for the collection of a large number of flux measurements in the field. Therefore, although the technique has worked well for those researchers who have perfected its subtleties, it does not appear to be pragmatic on a large-area basis and should be used only for very localized and specialized studies.

45–2.2 Water Content Method

45–2.2.1 PRINCIPLES

The water flux at the bottom depth of a soil profile is estimated by measuring in one time period all other components of the water balance for that profile, and then either calculating the flux by difference, or in the case where all other components are zero, from the change in soil water content.

The equation of continuity for the soil water can be integrated with respect to time over an arbitrary time interval Δt, and with respect to position over an arbitrary depth of soil profile to obtain the water balance for the profile:

$$P + I = \mathrm{ET} + R + \Delta\theta + D \qquad [11]$$

where P = precipitation (L), I = irrigation (L), ET = evapotranspiration (L), R = runoff (L), $\Delta\theta$ = change in soil water storage (depletion), and D = drainage (L). The values of all components of this equation are associated with the time interval Δt. Equation [11] can be solved for the drainage, D:

$$D = P + I - \Delta\theta - R - \mathrm{ET}\,. \qquad [12]$$

If the terms on the right-hand side of Eq. [12] can be measured or otherwise evaluated, the drainage can be calculated. The average soil water flux at the bottom of the profile for the time interval Δt is given by

$$q_w = D/\Delta t\,. \qquad [13]$$

The value of Δt is usually determined by the frequency of observation of the water content of the profile.

In specific cases, some of the terms in Eq. [12] may be zero. For example, if there is no irrigation water applied during the interval Δt, I is zero. If no plants are present and there is negligible soil surface evaporation, ET is zero.

If D is positive, the flux q_w will be positive. Values of the flux greater than zero represent a downward flow across the bottom of the profile, and vice versa.

When plants are present, the actual evapotranspiration, ET, must be measured or estimated by some means. Some alternatives for estimating ET are given in section 45–2.2.3.

Several variations of treatment of the water balance are possible. For example, the continuity equation for soil water can be integrated with respect to depth, from the soil surface ($z = 0$) to a depth L to obtain:

$$\int_0^L (\partial\theta/\partial t)\, dz = q_w\,(L, t) - q_w(0, t) \qquad [14]$$

where it has been assumed that no sources or sinks of water exist in the depth L. If the flux at the surface, $q_w\,(0, t)$, is assumed to be zero, Eq. [14] can be used to estimate the flux at the depth L from measurements of the soil water content as a function of depth and time. If the data for $\theta(z, t)$ can be fitted with an analytic function that can be integrated and differentiated, the integral can be evaluated by analytic means. Alternatively, numerical methods may be applied to the measured water content data to evaluate the integral, and hence the flux. If the flux at the surface is not zero, but is known, Eq. [14] can still be used to evaluate the flux at the depth L.

Another formulation of the balance equation may be obtained by introducing the average water content at time t, $\langle\theta\rangle$, of the soil profile to the depth L, viz:

$$\langle\theta\rangle = (1/L) \int_0^L \theta(z, t)\, dz \qquad [15]$$

into the left-hand side of Eq. [14] to obtain

$$L\, \partial\langle\theta\rangle/\partial t = q_w(L, t) - q_w(0, t)\,. \qquad [16]$$

If the flux at the surface is known, Eq. [16] can be used to calculate the flux at depth L from the change in the average water content of the profile to depth L.

45–2.2.2 MATERIALS

The following data are required:

1. Precipitation (cm), measured on a daily basis using a gauge located on or close to the site.
2. Evapotranspiration (cm), measured on a daily basis using a weighing lysimeter, or estimated using any of a variety of theoretical methods presented by e.g., Rosenberg et al. (1968), Tanner (1967, 1968) and Jensen (1974).
3. Irrigation (cm), measured on a daily basis by, for example, metering the quantity of water applied to the field.
4. Runoff (cm), measured on a daily basis using, for example, a weir and recording hydrograph as explained by Brakensiek et al. (1979).
5. Soil water content with depth as a function of time, measured gravimetrically with a neutron probe or tensiometers as explained in chapter 21.

45–2.2.3 PROCEDURE

Only general instructions are given, as specific techniques will vary according to the equipment used. Measure irrigation, rainfall, and surface runoff on a daily basis. Calculate evapotranspiration on a daily basis. Measure soil water content with depth (through the root zone if a plant is present) at regular time intervals. A 5-to 10-day interval is suggested.

45–2.2.4 CALCULATIONS

In the presence of measured surface fluxes of water and calculated evapotranspiration, Eq. [12] is used to estimate D. Calculate the change in soil water storage ($\Delta\theta$) from

$$\Delta\theta = \sum_{z=0}^{L} \langle\theta_j\rangle\Delta z - \sum_{z=0}^{L} \langle\theta_{j+1}\rangle\Delta z \qquad [17]$$

where $0 \leqslant z \leqslant$ L, and $\langle\theta_j\rangle$ and $\langle\theta_{j+1}\rangle$ are the mean water contents over depth interval Δz ($0 < \Delta z \leqslant L$) at times j and $j+1$. Then calculate D using Eq. [12] and estimate the flux with Eq. [13].

If water flux is to be estimated over shorter time intervals, then it is necessary to use the $\theta(z,t)$ information to calculate the average volumetric water content for each time interval over the depth L ($0\leqslant z \leqslant L$) using Eq. [15]. The integral in that equation can be evaluated by numerical means from a tabulation of θ vs. depth at a given time. The average water content of the profile versus time could be obtained by repeated application of this process at a series of times.

A sample calculation is given for a soil with a 100-cm rooting depth.

Table 45-1. Water input and losses and water content changes for sample calculation.

Day	I	P	ET	R	W_B	W_E	$\Delta\theta$	D
	cm							
1	--	--	0.6	--	24.4	23.7	−0.7	0.1
2	--	0.2	0.3	--	23.7	23.4	−0.3	0.2
3	--	--	0.8	--	23.4	22.6	−0.8	0.0
4	--	--	0.5	--	22.6	22.1	−0.5	0.0
5	10.0	--	0.5	--	22.1	31.6	+9.5	0.0
6	--	12.0	0.2	0.5	31.6	37.0	+5.4	5.9
7	--	--	0.2	--	37.0	36.2	−0.8	0.6
8	--	--	0.5	--	36.2	35.2	−1.0	0.5
9	--	--	0.5	--	35.2	34.3	−0.9	0.4
10	--	--	0.5	--	34.3	33.4	−0.9	0.4

The average volumetric soil water content at the beginning ($t = 0$) is 0.25. Measurements of P, I, and $\Delta\theta$ are made daily, and ET is calculated daily by one of the methods referenced in section 45–2.2.5. Since it is necessary in this example to keep the time increment of calculation short, one of the more fundamental relationships, such as the Bowen ratio method, aerodynamic-profile method, or combined transport-energy balance method (Penman) would have been used to generate the ET data in Table 45–1. W_B and W_E are total water (cm) in the 100-cm profile at the beginning and end of each day, respectively.

The drainage, D, for each time increment is calculated from

$$D = P + I - \Delta\theta - R - \mathrm{ET}\,.$$

On a daily basis, $q_w = D$.

45–2.2.5 COMMENTS

The water balance method is intended to provide, with as few measurements as possible, an estimate of water flux at a given depth on a short-term basis. Water flux over longer times can be calculated by summing the short-term estimates. In the presence of plants, the method requires calculation of actual ET over short time intervals. Errors in estimating water flux can result if too large a time increment is used. For example, if a rainfall event is concentrated in a short time period, substantial downward water flux can result. This flux will be relatively independent of plant extraction and will perhaps occur relatively quickly, depending on soil properties. The effects of such an event upon the water balance are obscured or even lost when measurements of $\Delta\theta$ and ET are calculated over long-time increments. This consideration becomes more important as the permeability of the soil increases.

The major limitation of this method is the need to estimate actual ET, a process which begins by calculating potential ET ($\mathrm{ET_p}$). The measurements required for calculating $\mathrm{ET_p}$ depend on the specific relationships used. Once $\mathrm{ET_p}$ has been obtained, it must be converted to actual

ET for the time period of interest, due to limitations imposed upon ET by soil water content. The simplest feedback models are of the form

$$ET = \kappa_s ET_p \quad [18]$$

where κ_s is a soil coefficient ranging between 0 and 1. The value of κ_s has been studied in a range of situations (summarized by Lomas & Levin, 1979), with attempts made to establish the relationship between κ_s and soil water content (Veihmeyer, 1927; Penman, 1948; Thornthwaite & Mather, 1957; Denmead & Shaw, 1962). This relationship is generally not simple, due to climatological effects, variations with plant age and amount of ground cover, and rooting patterns. This problem has been considered by Hanks (1974) in a simple model that attempts to incorporate these dynamic aspects. If there are no plants present in the system, it is necessary to consider soil-limited evaporation (Black et al., 1969; Ritchie, 1972; Tanner & Jury, 1976) as a means of modifying potential evaporation rates.

The water balance method is useful in describing the irrigation return flow from an agricultural field or leaching from a landfill or waste disposal site. It can also be used to generalize potential contributions to groundwater from the leaching of soluble materials. This last purpose can be accomplished by combining the estimated water flux with consideration of solute concentrations as discussed below. It is important to note that small errors in estimating ET will result in large errors in estimation of water flux, particularly as the magnitude of the flux decreases in proportion to the amount of ET.

Equation [15] reduces during dry periods or when $q_w(0,t) = 0$ to

$$L\ \partial\langle\theta\rangle/\partial t = K(\theta)[(\partial h/\partial z) - 1]_{z=L} \quad [19]$$

where h is the soil water matric potential and $K(\theta)$ is the hydraulic conductivity at depth L. A further simplification has been proposed (Rose & Stern, 1965), which assumes that in the absence of plant adsorption of water and in all except wet cases, the slope $\partial h/\partial z$ near the soil surface is much greater than unity (Rose & Stern, 1965). The flux q_w can thereby be approximated as

$$q_w = K(\theta)[(\partial h/\partial z)]_{z=L} \quad [20]$$

where the second term of Eq. [19] becomes of much less importance. Notice that Eq. [19] and [20] require knowledge of $h(z, t)$, which can be provided either by tensiometer measurements or by a soil-water characteristic curve and measurements of $\theta(z, t)$. Deeper within the profile and without plant extraction, $\partial h/\partial z > 0$, and Eq. [20] can be written as

$$q_w = K(\theta) \quad [21]$$

which allows the water flux to be estimated from measurement of water content, given knowledge of the $K(\theta)$ relationship, which can be estimated by various means (e.g., chapters 28–31 in this book).

If the component terms of the water balance equation are expressed as equivalent depths of water, then Eq. [19], [20], or [21] can be integrated to give the drainage component (D) as

$$D = \int_0^t q_w \, dt \; . \qquad [22]$$

Measurements of matric potential with depth or the use of a soil-water characteristic curve in conjunction with measurements of h provide all the information required to estimate drainage by any of these methods, so long as the other components of the water balance are known.

45–3 SOLUTE FLUX

As irrigation and rainwater infiltrate and redistribute within the soil profile, solute materials are displaced to deeper depths, out of the plant root zone and perhaps into groundwater. Similarly, the rise of water from water tables can carry dissolved materials upward, where they may influence soil properties or plant growth. The amount of solute displaced is influenced by physical and chemical processes related to both water and solute. Generally, the nature of these processes has been studied in detail in the laboratory, with only recent extension of the basic relationships to field description of solute transport.

Solute concentrations can be measured directly in the field, either by soil sampling or with porous ceramic extraction devices (chapter 42 in this book). These concentrations are instantaneous representations of solute presence in the profile and must be combined with a representation of water flux if the solute flux, here defined as a mass of solute moving through a unit cross-sectional area in a unit time, is to be calculated. This solute flux is the combination of both diffusive and convective fluxes (Wagenet, 1983), where the latter is defined as the product of the solute concentration and the volumetric solution phase (water) flux. All the inherent problems of spatially averaging water flux measurements are also germaine to the issue of calculating solute flux in the field. As with water, this has led to two general approaches to estimation of solute flux. First, simplified representations of the interaction of water and solute movement have been developed that can provide gross estimates of the leaching of soluble materials. These methods range from simply multiplying the water flux by a time-averaged concentration to slightly more fundamental models that describe drainage and wetting front penetration (Rao et al., 1976; DeSmedt & Wierenga, 1978a, 1978b; Rose et al., 1982a, 1982b). At a second level, more complicated models have been developed that attempt to integrate the physical and chemical mechanisms influencing solute movement. These models are generally based on numerical differencing techniques and use the prediction of water flux as an input necessary for the prediction of solute flux (Childs & Hanks, 1975; Bresler,

1973; van Genuchten & Wierenga, 1974; Wood & Davidson, 1975; van Genuchten, 1978a, 1978b; Gureghian et al., 1979; Robbins et al., 1980). Recently, the stochastic nature in the field of many of the parameters used in these models has been recognized by increased sampling and study of field cases (Nielsen et al., 1981b). This has led to a shift from the use of strictly deterministic models that use single values of input parameters to models that use a mixed deterministic-stochastic approach. These models (Gelhar, 1976; Tang & Pinder, 1977; Warrick et al., 1977a, 1977b; Bresler et al., 1979; Bresler & Dagan, 1979; Dagan & Bresler, 1979; Simmons et al., 1979; Smith & Hebbert, 1979) have not often been used for most management purposes and are not discussed here. This presentation will focus on methods that may be used by a majority of soil scientists or others interested in estimating solute flux.

A wide variety of mechanisms must be considered in predicting solute flux. The most often used of the several theoretical approaches that have been developed is miscible displacement theory. This representation of solute displacement as the combined processes of chemical diffusion and mass flow can be traced to origins in the chemical engineering literature. In the early 1960s the theory was extended to questions of solute displacement in soils (Nielsen & Biggar, 1961, 1963; Biggar & Nielsen, 1962) and was found reasonable in several soil types under a range of laboratory conditions. The approach was later expanded to include nonsteady flows of water and solute and to descriptions of sources or sinks of solute that act during transport. The usefulness of the basic approach, particularly for carefully controlled laboratory situations of solute transport, has been reinforced by a large number of subsequent investigations (summarized by van Genuchten & Cleary, 1979; Nielsen et al., 1981a). The equations arising from this work are interchangeably known as miscible displacement, dispersion-convection, or solute transport equations.

Miscible displacement theory assumes that a solute being transported through a volume increment of soil is subject to mixing processes within the soil pores. There are two such processes. One is chemical in nature, resulting from the diffusion of solute in response to concentration gradients existing in the soil solution. The other is physical in nature resulting from variations in water flow velocity within each pore and between pores. This latter phenomenon is often termed *hydrodynamic dispersion* (Day, 1956). These two mixing processes occur simultaneously during the imposed mass flow of solute in response to the movement of the bulk soil water. With these fundamental assumptions, derivations of miscible displacement equations have been developed (Bear, 1972; Kirkham & Powers, 1972), generally as statements of the law of conservation of mass. In the equations, the soil is assumed to be homogeneous and isotropic. The most simple case has a steady-state flow of water into and out of a small elemental volume of soil in the one-dimensional z-direction. The water content is also constant with z. Considering a solute that does not interact with the soil matrix, for which diffusion-dispersion and mass flow are the two mechanisms of transport, and conserving mass during the distance and time of transport, solute displacement can be represented as

$$\partial c/\partial t = D\,(\partial^2 c/\partial z^2) - v\,(\partial c/\partial z) \qquad [23]$$

where c is solute concentration (M/L^3), D is the apparent diffusion coefficient (L^2/T), v is the pore water velocity (L/T), and z and t are space (L) and time (T), respectively. When water flow is transient, Eq. [23] must be revised to include consideration of changing water content and water flux. That is

$$\frac{\partial(\theta c)}{\partial t} = \frac{\partial}{\partial z}\left(\theta D\,\frac{\partial c}{\partial z}\right) - \frac{\partial}{\partial z}\,(q_w c) \qquad [24]$$

where θ is the volumetric soil water content (L^3/L^3) and q_w is the volumetric flux of water (L/T).

These two equations have provided the basis for a wide range of studies on solute flux. The value of the apparent diffusion coefficient, D can be measured under laboratory conditions (chapter 43 in this book). The volumetric flux of water, particularly in laboratory and lysimeter experiments, can be measured directly, and the average pore water velocity can be calculated using some estimate or measure of the soil water content along the distance of flow. With these parameters in hand, both analytic and numeric solutions to Eq. [23] and [24] are useful in predicting solute concentration as functions of depth and time. This prediction of concentration is then used to estimate flux, generally by integrating concentration changes over time with reference to a cross-sectional area through which transport is occurring. For laboratory experiments and simple field cases, such an approach is acceptable and provides a reasonable estimate of the flux. When prediction of solute flux over large areas in the field becomes the focus, this approach becomes more approximate, as will be seen later.

Equations [23] and [24] represent the case of a solute that does not experience any substantial interaction physically or chemically with the soil solids and is not chemically or biologically altered during transport. The best example of such a solute, though not widespread in nature, is tritium (^3H_2O). In most temperate soils, the case of a noninteracting solute is closely approximated by chloride ion, although chloride may at times be influenced by the process of anion exclusion, a consideration often ignored given the degree of resolution available in determining values of D and v. To varying degrees, all cations and anions commonly found in soils are potentially subject to chemical precipitation or dissolution reactions, adsorption-desorption or cation exchange interactions with the soil matrix, and biological transformation and assimilation processes. It is therefore necessary to alter the basic miscible displacement equations if the flux of one of these reactive solutes is to be estimated.

If the objective is the description of a reactive solute that can undergo source- or sink-type reactions, then conceptually, Eq. [23] and [24] can be modified for steady-state water flow to

$$\rho \frac{\partial s}{\partial t} + \theta \frac{\partial c}{\partial t} = \theta D \frac{\partial^2 c}{\partial z^2} - q_w \frac{\partial c}{\partial z} \pm \sum_{i=1}^{n} \phi_i \qquad [25]$$

and for transient water flow to

$$\frac{\partial(\rho s)}{\partial t} + \frac{\partial(\theta c)}{\partial t} = \frac{\partial}{\partial z}\left(\theta D \frac{\partial c}{\partial z}\right) - \frac{\partial}{\partial z}(q_w c) \pm \sum_{i=1}^{n} \theta \phi_i \qquad [26]$$

where s represents the mass of solute associated with the solid phase (M/M) such as in adsorption-desorption, ϕ_i represents the rate of removal or resupply (M/L^3-T) of solute, and ρ is soil bulk density (M/L^3), usually assumed constant with time. Deviation of these equations is presented in more detail by Wagenet (1983) or van Genuchten and Cleary (1979).

A variety of analytical and numerical solutions to forms of Eq. [25] and [26] have been developed (Lapidus & Amundson, 1952; Lindstrom et al., 1967, 1971; Lindstrom & Boersma, 1970, 1971; van Genuchten & Wierenga, 1976; Wagenet et al., 1977; Selim et al., 1977; Cameron & Klute, 1977; Rao et al., 1979; Tillotson et al., 1980). These are generally not useful under field conditions due primarily to (i) the large amount of input data and (ii) assumptions that the basic physical, chemical, and biological processes modeled are deterministic under field conditions. In light of these limitations it appears, as was the case with the different approaches used to calculate water flux, that solute flux in the field can presently be as reliably calculated by using a simple model as by using much more complicated and data-intensive models. Several of these simplified approaches are presented below.

It is important to note that experimental techniques for directly measuring solute flux under laboratory conditions are well identified, centering upon the use of soil columns and leaching lysimeters (Elrick et al., 1966; Mansell et al., 1968; Kirda et al., 1973; Misra et al., 1974; Starr et al., 1974; Wagenet & Starr, 1977; Rao et al., 1979). The techniques and procedures required to successfully conduct laboratory solute transport experiments are explained in chapter 44, in which methodology is described to measure the apparent diffusion coefficient. This methodology, although it requires some practiced experimental technique, is straightforward. However, the extrapolation of laboratory results to field cases requires caution. In general, parameters that can be measured and controlled quite accurately in the laboratory, such as the water content, water flux, and the apparent diffusion coefficient, are often subject to wide ranges of spatial and temporal variation in the field.

45–3.1 Average Concentration Method

45–3.1.1 PRINCIPLES

Measurement of soil–water solute concentration is made using soil sampling or ceramic cup extractors. Using an estimate of the water flux, the measured solute concentration is used to provide an estimated solute flux. It is assumed that the convective flux, which is the product of the soil water flux and the solute concentration, is the dominant component of the solute flux. The values of the soil water flux and solute concentrations used are average values for the time interval selected.

45–3.1.2 MATERIALS

The following are required:

1. An in situ sampling device for withdrawal of soil solution, or equipment for soil sampling, or one of the other methods discussed in chapter 42 for measurement of solute concentration at depths of interest in the soil profile.
2. An estimate of the water flux obtained from one of the methods described in section 45–2.

45–3.1.3 PROCEDURE

Estimate the water flux. Measure solute concentration in the soil profile (chapter 42 in this book) and express the answer as g-solute/cm^3 soil solution.

45–3.1.4 CALCULATIONS

Calculate the solute flux (q_s) from

$$q_s = (q_w)\,(c_j + c_{j+1})/2 \qquad [27]$$

where q_w = average water flux density for the time interval with limits j and $j + 1$, respectively. The time interval selected is usually based upon the method used to estimate q_w.

A sample calculation begins by assuming the water flux, q_w, for a 10-day period to be 1.5 cm/day, c = 10 meq/L (NaCl) at the 200-cm depth at t = 0, and c = 8 meq/L (NaCl) at the 200-cm depth at t = 10 days. Calculation of the average solute flux for the 10-day period involves first converting the concentration in meq/L to g/cm^3, where

$$(10 \text{ meq NaCl/L})\,(58 \text{ mg/meq})\,(\text{g}/1000 \text{ mg})\,(\text{L}/1000 \text{ cm}^3) = 5.80 \times 10^{-4}\,(\text{g NaCl/cm}^2)\,.$$

The solute flux is then calculated, using Eq. [36], to be

$$q_s = (1.5 \text{ cm/day})(10 \text{ day})\left(\frac{5.8 \times 10^{-4} + 4.64 \times 10^{-4}}{2}\right)$$

$$= 7.83 \times 10^{-4}(\text{gNaCl})/\text{cm}^2 \text{ per day}).$$

45–3.1.5 COMMENTS

This method is relatively straightforward once a reasonable estimate of the water flux is obtained. However, inadequate representation of the spatial variability of solute in the measurement of solute concentrations can impose serious errors on estimation of solute flux. In highly permeable situations with substantial downward movement of water and solute, variations in solute concentration at a selected depth within a 1-ha area can be large enough that errors in the solute flux of orders of magnitude can result. Presently, the only way to overcome such errors is to take enough samples that the variability of the field is somewhat characterized, with this characterization used to provide confidence limits on the estimate of solute flux.

The estimate of solute concentrations used in Eq. [27] could be better than the estimate of water flux, if the latter is obtained by the approximate techniques of section 45–2.2 or 45–2.3 and the former is gained from a large number of field observations. In such a case, it should be obvious that better resolution in estimating solute flux may not actually be achieved by an intensive field sampling program for solute distribution, if that program is not accompanied by an equal effort to identify spatially variable water fluxes.

45–3.2 Approximate Analytic Solution Method

45–3.2.1 PRINCIPLES

Solute displacement under transient conditions is predicted from an approximate analytic solution of the miscible displacement equation, Eq. [23]. Solute flux is calculated by Eq. [27]. This method is similar to the one presented in section 45–3.1.1, in that convective flux is considered to be the main transport process. However, in this case an approximate analytic solution that considers diffusion-dispersion is used to estimate solute concentrations, rather than measuring them as in section 45–3.1.1.

It has been shown (e.g., DeSmedt & Wierenga (1978a, 1978b) that solute breakthrough curves (effluent contributions) resulting from steady flow through profiles nonuniform in water content may be described with good accuracy by assuming average values of water content (θ), water flux (q_w), and dispersion (D). These observations have led to the development of approximate analytical solutions (Rose et al., 1982b) useful in estimating solute movement in the field. If Eq. [23] is solved according to the following boundary conditions

$$c = 0 \quad x > 0 \quad t = 0 \qquad [28a]$$

$$c = c_0 \quad x = 0 \quad 0 < t \leqslant t_1 \qquad [28b]$$

$$c = 0 \quad x = 0 \quad t > t_1 \qquad [28c]$$

which describe the displacement of a pulse of solute through the soil, and the solution is intended to apply only for cases where $t \gg D/v^2$, then the approximate solution

$$c = c_0/2 \,\{\mathrm{erf}\,[(z + z_0 - vt)/(4DT)^{1/2}] - \mathrm{erf}\,[(z - vt)/(4DT)^{1/2}]\} \qquad [29]$$

can be used to calculate solute concentration, where c_0 = concentration of solute in the applied pulse (M/L); z = depth (L); t = time; D = apparent diffusion coefficient (L^2/T), which may be estimated from $D = 0.6 + 4.65\ v^{1.14}$ (Biggar & Nielsen, 1976); and erf (α) denotes the error function of α. The value of z_0 is calculated from

$$z_0 = I/\langle\theta\rangle \qquad [30]$$

where I = depth of added pulse of water (L) and $\langle\theta\rangle$ = average volumetric water content over the depth and time of interest. The average pore water velocity, v, over time t is calculated from

$$v = \langle q_w\rangle/\langle\theta\rangle \qquad [31]$$

where $\langle q_w\rangle$ is the average water flux at the depth of interest over the time considered, as estimated by the methods presented earlier in this chapter.

The solute flux at any depth z can now be calculated from Eq. [27], which is rewritten in the form

$$q_s = 1/t\ \Sigma_{j=1}^{n}\ \langle q_w\rangle_j\ [(c_j + c_{j+1})/2]\ \Delta t \qquad [32]$$

where $\langle q_w\rangle_j$ = average water flux at depth z during the time interval with limits j and $j + 1$, t = total time over which the summation is calculated, and $\Delta t = t_{j+1} - t_j$.

45–3.2.2 MATERIALS

The following are required:

1. Neutron probe, tensiometers, or soil sampling equipment to measure the average soil water content in the depth of interest. See chapter 21.
2. Laboratory facilities, such as colorimetric or spectrophotometric methods, for determination of the concentration of solute in added water.
3. An estimate of the average water flux at the depth of interest, obtained by one of the methods described in section 45–2.
4. A metering or gauging system to measure the depth of water (containing solute) added to the field.

45–3.2.3 PROCEDURE

Measure water content through the depth of interest at 30-cm intervals or less, the depth of added water that contains solute, and the concentration of the added solute. The time at which the solute is added to the soil is denoted as $t = 0$, and water contents and flux are measured only after that time. Measurements should be collected relatively frequently so that a good estimate of $\langle q_w\rangle$ is obtained. This would be from intervals of approximately 1 day or less in coarse-textured soils, up to perhaps weekly in fine-textured soils.

45–3.2.4 CALCULATIONS

Calculate q_w and $\langle\theta\rangle$ from methods presented in section 45–2. Calculate z_0 and v using Eq. [30] and [31]. Estimate D from, for example, $D = 0.6 + 4.65\ v^{1.14}$. Use z_0, v, D, and c_0 to calculate $c(z,t)$ from Eq. [29]. Estimate the solute flux q_s at depth z from Eq. [32].

A sample calculation first assumes that (i) the average volumetric soil water content over a 30-day period is 0.25 to the 100-cm depth, (ii) a uniform 5-cm application of water (I = 5 cm) containing 100 mg/L NO_3^- is made to 1.0 ha at time = 0, and (iii) the average water flux (q_w) over 30 days is 1.0 cm/day. Then

$v = q_w/\langle\theta\rangle = (1.0 \text{ cm/day})/0.25 = 4.0 \text{ cm/day}$,

$z_0 = I/\langle\theta\rangle = 5 \text{ cm}/0.25 = 20 \text{ cm}$, and

$D = 0.6 + 4.65\ (4.0)^{1.14} = 23 \text{ cm}^2\text{/day}$.

To calculate the flux, values of c should be calculated at short intervals. Here we choose 3 days. Using Eq. [29], a curve was calculated for $c(t)$ at z = 90 cm (Fig. 45–3).

The flux of salt is found from

$$q_s = \langle q_w\rangle/t\ [\Sigma_{j=1}^{n}(c_j + c_{j+1})/2]\ \Delta t = 3.7 \times 10^{-3} \text{ mg } NO_3^- \text{ cm}^{-2} \text{ soil day}^{-1}$$

Over a 28-day period this represents a transport of 1.05×10^7 mg NO_3^-/ha. The total NO_3^- applied was 1.25×10^7 mg/ha. This demonstrates that about 83% of the applied NO_3^- had passed the 90-cm depth after 28 days, at which time the peak concentration was at about 100 cm.

45–3.2.5 COMMENTS

The approximate analytical solution presented in Eq. [29] was developed from well-defined initial and boundary conditions which, in a strict sense, limit the application of the solution to those cases with steady-state water flow both before and after the pulse application of solute. According to these criteria, Eq. [29] cannot be used to describe field conditions of transient solute flux under nonsteady water flow regimes. However, given the great variability of measured field water fluxes, it is

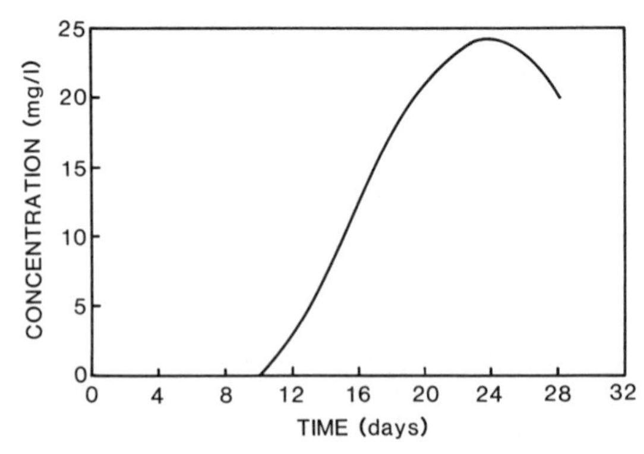

Fig. 45–3. Calculated solute concentration at the 90-cm depth using Eq. [29].

possible that use of Eq. [29] or similarly constructed equations to describe transient cases will not, in fact, grossly misrepresent either water or solute fluxes in the field.

The use of Eq. [29] to describe transient cases presumes that field variability of q_w and θ create a situation in which the resolution of solute flux estimations can only be expected to fall within broad limits. More fundamentally correct mathematical approaches, such as numerical or stochastic models or other types of simple models (i.e., see section 45–3.4) probably include assumptions or measurement requirements which prevent the statement that they are either more accurate, more reliable, or easier to use than the approximate analytical solution. This issue is not yet resolved in the literature and will probably be addressed in the next few years. At present, it has not been proved that use of Eq. [29], as proposed here, represents any more or less accurate a method for predicting field solute fluxes than other approaches with their inherent weaknesses.

Given Eq. [29], the time (t_m) at which a maximum solute concentration (peak concentration, c_{max}) will appear at a given depth, L, can be estimated from

$$t_m = 1/v\ [L + (z_0/2)] \qquad [33]$$

and the value of the maximum solute concentration at that depth and time will be

$$c_{max} = c_0\ \mathrm{erf}\ [\mathrm{I}/4\theta(\mathrm{Dt})^{1/2}]\ . \qquad [34]$$

The above theoretical development (FAO/IAEA, 1976) assumes no solute interaction with the soil matrix during transport, and no plants present in the system. As such, it is limited to description of anion movement in a noncropped soil. The latter assumption will not be addressed here, but Eq. [29] can be modified to consider, again in approximate terms, the interaction of a cation with the soil surface during transport. To do this, Eq. [25], can be revised as (Wagenet, 1983)

$$\partial c/\partial t = D^*\ (\partial^2 c/\partial z^2) - v^*\ (\partial c/\partial z) \qquad [35]$$

where

$$D^* = D/[1 + \rho\ (R/\theta)]$$

$$v^* = v/[1 + \rho\ (R/\theta)]$$

and R is defined by a linear Langmiur isotherm describing adsorption-desorption of the solution according to

$$s = Rc \qquad [36]$$

where s is defined below Eq. [26].

Equations [29] and [34] are now altered by substituting D^* for D and v^* for v. A substantial body of literature exists on the use of Eq. [34] and [35], including summary discussions by van Genuchten and Cleary (1979), Nielsen et al. (1981a), and van Genuchten and Alves (1981). If the solute is monotonically decreasing during transport due to biological conversion, additional alteration of Eq. [30] or [34] is possible (Misra et al., 1974; Wagenet et al., 1977; Davidson et al., 1978) and simplified analytical solutions can be developed.

45–3.3 A Model for Long-term Estimation of Solute Flux

45–3.3.1 PRINCIPLES

Long-term trends in solute concentration and water flux are predicted and used to provide an estimate of the solute flux.

This model, presented by Rose et al. (1979), is not designed to predict the solute concentration profile as is the case in more detailed models of soil solute movement. The primary purpose is to estimate long-term trends in water and solute flux using observations made in the short term. It is assumed that the water balance presented in Eq. [11] can be simplified to

$$I + P - ET = D \qquad [37]$$

where there is assumed to be no runoff, and over the long term $\Delta\theta$ is zero.

Applying the principles of conservation of mass to solute in the layer of soil of arbitrary depth z, in cases where $D > 0$, the mean concentration in the top z cm of the soil profile at some time t can be shown to be (Rose et al., 1979)

$$\langle s(t)\rangle = \langle s_0\rangle + [(Ic/D\lambda) - \langle s_0\rangle]\,[1 - \exp(-D\lambda t/z\langle\theta_s\rangle)] \qquad [38]$$

where $\langle s(t)\rangle$ = spatial mean solute concentration in the top z cm of soil at time = t (M/L^3), $\langle s_0\rangle$ = value of $\langle s\rangle$ at $t = 0$, $\langle\theta\rangle_s$ = average saturated volumetric water content (L^3/L^3) over depth z (L), I = irrigation rate (L/T), c = solute concentration in irrigation water (M/L^3), D = drainage (or leaching) flux (L/T), and t = time measured from an initial time when $\langle s\rangle = \langle s_0\rangle$. The dimensionless parameter λ is defined by

$$\lambda = s(z)/\langle s\rangle \qquad [39]$$

which relates the solute concentration at depth z, $s(z)$, to the spatial mean solute concentration in the top z depth of soil at time t. It is preferable to determine λ for z greater than the depth of rooting, where the value of λ is relatively independent of z. The value of λ is assumed constant for all t, an assumption discussed by Rose et al. (1979).

If $D = 0$, s increases linearly with time as given by

$$\langle s(t)\rangle = s_0 + (Ic/z\langle\theta\rangle_s)t\,. \quad [40]$$

If $D > 0$, s approaches a final steady-state value (s_f) given by

$$s_f = Ic/D\lambda\,. \quad [41]$$

The solute flux q_s at time t can be calculated from

$$q_s = D\langle s\rangle \quad [42]$$

where the sign of q_s indicates the direction of salt movement (upward is negative, downward is positive).

45–3.3.2 MATERIALS

The following are required:

1. Measurement with a weighing lysimeter located in the field, or estimates using environmental data (e.g. Jensen, 1974) of evaporation (E, cm).
2. A metering or gauging system to measure irrigation (I, cm).
3. A gauging system to measure precipitation (P, cm).
4. Soil sampling equipment for measurement of solute concentration in the soil profile (chapter 42 in this book) at selected depth intervals at two different times (g salt/g soil solution).
5. Laboratory facilities, such as colorimetric or spectrophotometric methods, for determination of solute concentration in added irrigation water.

Values of evaporation, irrigation, or rainfall are used on an annual basis in the calculations below.

45–3.3.3 PROCEDURE

Two sets of soil samplings are required, one preferably before the start of solute leaching ($t = 0$) and another at a later date, after some solute has been redistributed through the soil profile. If this is not possible, the theory presented above can be generalized as explained under section 45–3.3.5 below.

Collect soil samples at 30-cm depth intervals throughout the soil region in which solute flux is to be estimated. Analyze for total soluble salt. Repeat the sampling at a later date. Measure or estimate the values of E, I, and P to be used in the calculations below.

45–3.3.4 CALCULATIONS

Express the soil solute concentration, $s(t)$, as g salt/g dry soil. Calculate $\langle s_0\rangle$ as the mean of those samples taken at time zero. The value of λ is determined from Eq. [39] using $\langle s \rangle$ and $s(z)$ gained from the second sampling. Convert both $\langle s_0 \rangle$ and c to units of g salt/cm^3 soil solution, where $\langle s_0 \rangle$ in units of g/g should be multiplied by soil bulk density (g/cm^3) and divided by $\langle \theta_s \rangle$. The value of D is calculated from Eq. [37] where E, I, and P are known. The value of $\langle s \rangle$ at some future time t at depth z can then be estimated from Eq. [38] using known values of s_0, I, c, D, and λ. The solute flux is then calculated from Eq. [42].

A sample calculation follows (from Rose et al., 1979). Given that $\lambda = 1.42$, $D = 8.47$ cm/year, $I = 142$ cm/year, $\langle s_0 \rangle = 1.62 \times 10^{-4}$ g NaCl/g dry soil, $c = 10$ meq NaCl/L, and $\langle \theta_s \rangle = 0.43$, it is desired to calculate q_s. Assuming all salt is present as NaCl, it is necessary to convert s_0 and c to g/cm^3 of soil solution. Assuming the soil to be water saturated ($\langle \theta_s \rangle = 0.43$) and the bulk density to be 1.54 g/cm^3 gives

$$\langle s_0 \rangle = (1.62 \times 10^{-4})\left(\frac{\text{g NaCl}}{\text{g dry soil}}\right)\left(\frac{1.54\ \text{g dry soil}}{\text{cm}^3\ \text{vol}}\right)\left(\frac{\text{cm}^3\ \text{vol}}{0.43\ \text{cm}^3\ \text{H}_2\text{O}}\right)$$

$$= 5.80 \times 10^{-4}\ \frac{\text{g NaCl}}{\text{cm}^3\text{H}_2\text{O}}$$

and

$$c = 8.70 \times 10^{-4}\ \frac{\text{g NaCl}}{\text{cm}^3\ \text{H}_2\text{O}}.$$

Therefore, from Eq. [38]

$$\langle s \rangle = \langle s_0 \rangle + \left(\frac{Ic}{D\lambda} - \langle s_0 \rangle\right)\left[1 - \exp\left(-\frac{D\lambda t}{z\langle \theta_s \rangle}\right)\right] = 7.67 \times 10^{-3}\ \frac{\text{g NaCl}}{\text{cm}^3\ \text{H}_2\text{O}}$$

and

$$q_s = D\langle s \qquad 6.50 \times 10^{-2}\ \frac{\text{g NaCl}}{\text{cm}^2/\text{year}}$$

45-3.3.5 COMMENTS

This method is primarily intended to describe those cases in which low hydraulic conductivity of the soil restricts drainage, and is appropriate for interpreting long-term trends from observations made in the shorter term (Rose et al., 1979). It does not describe solute profiles resulting from transient water and solute flow regimes, but leaves these descriptions to other theoretical approaches (e.g., Raats, 1975; Robbins et al., 1980). The strength of the approach is that a relatively limited data base, consisting only of field data on the amount and quality of irrigation water applied and profile information on the distribution of an assumed noninteractive solute, can be used to predict long-term trends in a manner useful for irrigation management decisions.

The value of λ must be estimated from the measured solute profile at any time $t = t_1$, and is assumed to hold for all $t > 0$. Replicate soil samples each subdivided into depth intervals can be used to determine $s(z)$ and $\langle s \rangle$. Given one value of $\langle s \rangle$ and no information on D ($D = q_w$ at bottom of root zone), known values of I, C, $\langle s_0 \rangle$, $\langle \theta_s \rangle$, λ, and t can be used in Eq. [38] to yield D. The value of $\langle s \rangle$ can then be computed for all t assuming the irrigation regime remains as it was for the data that produced D.

The behavior of $\langle s \rangle$ is quite dependent on whether the value of D is positive (downward) or negative. If D is negative, Eq. [38] predicts s will increase for all times with no final limiting steady-state value, a prediction consistent with the assumed situation for this model of an infinitely soluble salt. If D is positive, s approaches a final steady-state value of $\langle s_f \rangle$, given by Eq. [41], and if $D = 0$, then $\langle s \rangle$ increases with time (Eq. [40]).

A steady-state approach to description of the leaching of solutes has long been promoted through the "leaching requirement" concept (U.S. Salinity Laboratory, 1954). The present model reduces to this form if $\lambda = 1$ in Eq. [41], but is obviously of wider versatility in describing transient cases. If it is not possible to determine salinity conditions prior to initiating irrigation, Eq. [38] can be generalized by replacing $\langle s_0 \rangle$ with $\langle s(t_1) \rangle$ and $\langle s(t) \rangle$ with $\langle s(t_2) \rangle$, and letting t in Eq. [38] represent $(t_2 - t_1)$. Two sequential observations of $\langle s \rangle$ thus provide an estimate of D. These observations should be chosen with $(t_2 - t_1)$ large enough for a significant change in s.

The assumption is made that salts are nonreactive, both in terms of precipitation-dissolution type reactions and cation exchange. The former assumption is reasonable for systems dominated by chloride or sodium sulfate salts, but is not adequate for gypsiferous or calcareous soils. Further refinement of the approach is needed for these cases. The assumption that the effect of cation exchange is negligible becomes increasingly acceptable as time progresses, as little change in the composition of the exchanger phase will be expected once the chemical regime is fairly constant in terms of ionic composition.

The method of determining D (water flux) appears to be quite an efficient one for soils of low permeability, where measurement techniques are often difficult and changes are rather slow. Since D is small relative to I or E, it is difficult to accurately calculate it using most classical methods (i.e., those in section 45–2.2, 45–2.3, 45–2.4 above). The technique used here seems adequate to complement other more traditional methods that are of use in more permeable soils.

45–3.4 Model for Short-term Estimation of Water and Solute Movement

45–3.4.1 PRINCIPLES

The solute peak location is estimated after infiltration and redistribution of the soil water to field capacity. The model is constructed upon a "piston displacement" concept of water flow, which assumes that (i) all soil pores participate in water and solute transport, and (ii) the soil water initially present in the soil profile is displaced ahead of the water entering the soil. Additionally, it is assumed that the solute is not subject to any interaction with the soil matrix (the case of adsorption is covered by Rao et al., 1976) and undergoes no resupply to or removal from solution. If a reactive solute is to be considered, the depth of penetration of the solute peak can be amended by including a retardation factor that will represent ion-soil interaction. Similarly, plant extraction of solute and chemical or biological transformation of solute can be incorporated to increase the utility of this approach. Such was accomplished for N transport and transformation by Davidson et al. (1978).

45–3.3.2 MATERIALS

The following are required:

1. Neutron probe, tensiometers, or soil sampling equipment to measure the initial soil water content (before first water application). See chapter 21 for details.
2. An estimate of the field capacity water content, measured as described in chapter 36.
3. A metering or gauging system for measuring the depth of each increment of water added as rainfall or precipitation.
4. A weighing lysimeter to measure evapotranspiration on a daily basis, or an estimate of evapotranspiration using any of a variety of theoretical methods presented, for example, by Rosenberg et al. (1968) or Jensen (1974).
5. An estimate of the density distribution of roots within the root zone, obtained by excavating plants at several times from planting until maturity.
6. Laboratory facilities to analytically measure the concentration of solute in the applied water.

45–3.4.3 PROCEDURE

Measure the initial water content in the soil profile to the depth of interest. Use depth increments no larger than 15 cm, and preferably 10 cm in size. Measure the irrigation or rainfall and evapotranspiration on a daily basis. Measure the concentration of solute in each increment of applied water. Measure the distribution of roots to the maximum depth of rooting and for each depth increment used in water content determination, express the quantity of roots in that increment, R_i, as a proportion of the total roots ($0 < R_i \leq 1$). Measure the field capacity water content.

45–3.4.4 CALCULATIONS

Calculation of solute flux involves the proper construction of a model using the operational steps listed, executed with the required data. The soil is divided into depth increments, preferably smaller than 15 cm.

For the first wetting of a soil assumed to be at uniform initial volumetric water content, θ_i, calculate the position of the wetting front, z_w, resulting from the infiltration of I cm of water. This is given by

$$z_w = I/(\theta_{fc} - \theta_i) \; ; \qquad \theta_{fc} > \theta_i \qquad [43]$$

where θ_{fc} is the field capacity volumetric water content.

Calculate the position of the solute peak, z_s, resulting from this first wetting, where

$$z_s = I/\theta_{fc} \, . \qquad [44]$$

If plant roots are present in the system, calculate the average daily soil water content decrease in the root zone resulting from plant extraction, $\Delta\theta_z$, as

$$\Delta\theta_z = (ET/z_i) \, R_i \qquad [45]$$

where ET = daily evapotranspiration (cm) and R_i = the proportion of roots ($0 \leqslant R_i \leqslant 1.0$) contained in depth interval z_i (cm).

Calculate the soil water deficit, I_d (cm) (the difference between the field capacity water content and the soil water content established as the result of depletion) from

$$I_d = z_s \sum_{t=0}^{t=t_1} \Delta\theta . \qquad [46]$$

The value of I_d is calculated for the time interval t_1 between additions of irrigation or rain water.

Calculate the depth of effective water (I_e) added in the second irrigation. This is the water available to move the solute peak from position z_s, and is given by

$$I_e = I_n - I_d \qquad [47]$$

where the subscript n ($n \geqslant 2$) refers to any water addition after the first increment.

The water content profile resulting from the I_nth addition of water can be calculated by considering for each depth increment of soil, the depth of water (proportion of I_e) needed to return that increment to field capacity. This water is subtracted from I_e, with the remainder of I_e available to wet deeper into the profile. Wetting will continue until all I_e cm of water have been distributed in the profile.

The new position of the solute peak, z_s^*, can be calculated from

$$z_s^* = z_s + (I_e/\theta_{fc}) \, . \qquad [48]$$

Calculation of any succeeding solute displacement is made by substituting

z_s^* for z_s in Eq. [48] and calculating a new value of z_s^* from the appropriate value of I_e.

The following is a sample calculation of solute flux using the computational procedure described in Eq. [43–48]. Assume the following conditions:

1. Irrigation: Day 1 (2 cm).
2. Rainfall: Day 2 (3 cm).
3. ET: 0.5 cm/day.
4. The solute is located at $z = 0$ at $t = 0$.
5. $\theta_{fc} = 0.280$.
6. Evapotranspiration occurs on days with irrigation or rainfall, but only *after* these events.
7. Initial conditions of water content, θ_i, and rooting distribution, R_i, as presented in Table 45–2.

The water content profile and the position of the peak front on Days 1 and 2 can be calculated using this information. Recognizing that a nonuniform water content profile exists at $t = 0$, it is necessary to calculate the depth of water required to wet each layer to field capacity. On Day 1, for the first layer (0–10 cm), the depth of water required (I_1) is

$$I_1 = (10 \text{ cm})\,(0.280 - 0.200) = 0.80 \text{ cm}.$$

Similarly for 10 to 20 cm: $I_2 = (10 \text{ cm})\,(0.280 - 0.200) = 0.60$ cm; and for 20 to 30 cm: $I_3 = 0.60$ cm. The total $= 2.00$ cm. The first rainfall therefore wets the top 30 cm to θ_{fc}. The position of the solute peak is given by

$$z_s = I/\theta_{fc} = 2.00 \text{ cm}/0.28 = 7.14 \text{ cm}.$$

The value of $\Delta\theta$ is constant each day, since R_i and ET are constants (in this example). A sample calculation of $\Delta\theta$ for the 0- to 10-cm depth is

$$\Delta\bar{\theta}_z = (\text{ET}/z_1)R_1 = \frac{0.5 \text{ cm/day}}{10 \text{ cm}}(0.4) = 0.020 \text{ day}^{-1}.$$

The water content profile at the end of Day 1, after ET, is denoted as θ_1 in Table 45–2.

Table 45–2. Water content (θ) profiles, root density distribution (R_i), and calculated depletion ($\Delta\bar{\theta}_z$) for example case of short-term estimation of water and solute movement.

Depth	θ_i	R_i	θ_1	θ_2	$\Delta\bar{\theta}_z$
cm					
0–10	0.200	0.40	0.260	0.260	0.020
10–20	0.220	0.25	0.267	0.267	0.013
20–30	0.220	0.15	0.272	0.272	0.008
30–40	0.200	0.10	0.195	0.275	0.005
40–50	0.180	0.10	0.175	0.275	0.005
50–55	0.150	0.00	0.150	0.280	0
55	0.150	0.00	0.150	0.150	0

On Day 2, it is again necessary to calculate the depth of water required to bring each layer to field capacity. By layer (I_i), this is

$$I_1 = (10 \text{ cm})(0.280 - 0.260) = 0.20 \text{ cm}$$
$$I_2 = (10 \text{ cm})(0.280 - 0.267) = 0.13 \text{ cm}$$
$$I_3 = (10 \text{ cm})(0.280 - 0.272) = 0.08 \text{ cm}$$
$$I_4 = (10 \text{ cm})(0.280 - 0.195) = 0.85 \text{ cm}$$
$$I_5 = (10 \text{ cm})(0.280 - 0.175) = \underline{1.05 \text{ cm}}$$
$$\text{Total} = 2.31 \text{ cm}$$

A total of 3 cm of rainfall fell on Day 2, with 2.31 cm distributed within the root zone to recharge the profile to field capacity to the 50-cm depth. The balance of the water, 0.69 cm, was lost to the plant system as drainage. The depth that it wetted can be calculated as

$$z_w = 50 + \frac{0.69}{(0.280 - 0.150)} = 55.3 \text{ cm} .$$

The soil profile will therefore be at field capacity water content to approximately the 55 cm depth, with $\theta = 0.150$ below this depth.

The new position of the solute peak can be calculated using Eq. [48], where the effective irrigation I_e is

$$I_e = 3 - [7.14\ (0.280 - 0.260)] = 2.48$$

from which

$$z_s^* = 7.14 + (2.48/0.28) = 16.00 \text{ cm} .$$

Therefore, although there has been some drainage past 50 cm resulting from the two water applications, the solute has been displaced only to approximately the 16-cm depth.

The water content profile after ET on Day 2 is given by subtracting $\Delta\bar{\theta}_z$ from 0.280 for each depth in the root zone. This gives θ_2 in Table 45–2.

45–3.4.4 COMMENTS

The model presented here (Rao et al., 1976) is intended to be used for management purposes in which estimates of solute transit time are required, and can be used to estimate the time of arrival of a solute peak at some soil depth. The approach is limited to the case of homogeneous, well-drained profiles, and will not in its present form describe the presence of a water table or flow in layered situations. The approach is limited to identifying the position of a solute peak, and does not include representation of spreading about the peak due to dispersion-diffusion effects.

Recently, DeSmedt and Wierenga (1978b) have presented some approximate methods that can be used to include description of solute dispersion during transport, and these can be employed to estimate solute concentration once the value of z_s^* has been identified. Equations [43] and [44] show that a nonreactive solute will, during infiltration, move through the soil on the wetting front if the soil is initially dry ($\theta_i = 0$) or follow behind the wetting front if the soil is initially wet ($\theta_i > 0$). This concept holds in soils that do not experience macropore transport of water and solutes, and is important because it demonstrates that water leaving a field as leachate during a water application event does not generally contain solute present at the soil surface prior to the water application. The leachate is, rather, that water already present in the soil profile prior to the water application.

Models of the type presented here are most useful in situations in which only an approximate description of water and solute fluxes is needed. Situations where such general guidance is useful are found in the screening of fertilizer or pesticide management programs as a function of soil type, climatic regime, or chemical properties. A broad estimate of solute fate across a range of conditions can be efficiently and approximately provided by a relatively simple model such as the one presented here. These models also can serve as educational tools, useful in condensing the wide range of physical, chemical, and biological mechanisms that affect fluxes into a form useful to educate nonprofessionals. More research is needed to develop appropriate simplified models that best represent the complicated natural system that produces a particular water or solute flux.

45–4 SUMMARY

Describing solute flux on a field basis is presently one of the most challenging problems facing soil scientists. One hundred years after the first development of the Darcy equation for water flow and 75 years after the first recognition that water and solutes do not always move together (Schlichter, 1905), it is still questionable whether or not soil scientists can predict with a reasonable degree of certainty on a mechanistic basis what the water and solute flux are in the field at a particular depth. Only within the last 10 years has this problem been approached directly by measuring, on a large scale, water and solute distribution in field soil profiles. This first series of investigations has demonstrated that the manner in which most processes operate spatially and temporarily in the field is not compatible with the structure of most models used to predict water contents and solute concentrations and calculate fluxes.

The emphasis in predicting solute flux is shifting from the use of strictly deterministic models that use single values of input parameters to models that use a mixed deterministic-stochastic approach. These models will hopefully allow better representation of the variability inherent in field-determined values of solute transport parameters and transformation processes. However, the best-constructed representations of water and solute flux in the field are useful only if the construction, validation, and application are based on data truly representative of the field situation. This prerequisite has rarely been met by sampling programs in the field. It is only reasonable to assume that as new mathematical means to predict water and solute flux are developed, the technology of field sampling should also advance. In any field study of water and solute flux, sampling programs should consider the number, location, and frequency of samples to be collected, the physical scale of the individual observations compared to the scale over which the process to be inferred from the observation is occurring, and the variance structure of the collected samples. The majority of the data presently available relative to water and solute flux were not collected under these conditions, primarily because soil scientists have only recently recognized the need to employ new techniques to cope with heterogeneous field situations (chapter 3 in this book).

45-5 REFERENCES

Bear, J. 1972. Dynamics of fluids in porous media. Elsevier Science Publishing Co., New York.

Biggar, J. W., and D. R. Nielsen. 1962. Miscible displacement: II. Behavior of tracers. Soil Sci. Soc. Am. Proc. 26:125–128.

Biggar, J. W., and D. R. Nielsen. 1976. Spatial variability of the leaching characteristics of a field soil. Water Resour. Res. 12:78–84.

Black, T. A., W. R. Gardner, and G. W. Thurtell. 1969. The prediction of evaporation, drainage and soil water storage for a bare soil. Soil Sci. Soc. Am. Proc. 33:655–660.

Brakensiek, D. L., H. B. Osborn, and W. J. Rawls. 1979. Field manual for research in agricultural hydrology. USDA Agric. Handb. 224, U.S. Government Printing Office, Washington, DC.

Bresler, E. 1973. Simultaneous transport of solutes and water under transient unsaturated flow conditions. Water Resour. Res. 9:975–986.

Bresler, E., H. Bielorai, and A. Laufer. 1979. Field test of solution flow models in a heterogeneous irrigated cropped soil. Water Resour. Res. 15:645–652.

Bresler, E., and G. Dagan. 1979. Solute dispersion in unsaturated heterogeneous soil at field scale: II. Applications. Soil Sci. Soc. Am. J. 43:467–472.

Buckingham, E. 1907. USDA Bur. of Soils Bull. 38. U.S. Government Printing Office, Washington, DC.

Byrne, G. F. 1971. An improved soil water flux sensor. Agric. Meteorol. 9:101–104.

Byrne, G. F., J. E. Drummond, and C. W. Rose. 1967. A sensor for water flux in soil. "Point source" instrument. Water Resour. Res. 3:1073–1078.

Byrne, G. F., J. E. Drummond, and C. W. Rose. 1968. A sensor for water flux in soil. 2. "Line source" instrument. Water Resour. Res. 4:607–611.

Cameron, D. R., and A. Klute. 1977. Convective-dispersive solute transport with a combined equilibrium and kinetic adsorption model. Water Resour. Res. 13:183–188.

Cary, J. W. 1968. An instrument for in situ measurements of soil moisture flow and suction. Soil Sci. Soc. Am. Proc. 32:3–5.

Cary, J. W. 1970. Measuring unsaturated soil moisture flow with a meter. Soil Sci. Soc. Am. Proc. 34:24–27.

Cary, J. W. 1971. Calibration of soil heat and water flux meters. Soil Sci. 111:399–400.

Childs, S. W., and R. J. Hanks. 1975. Model of soil salinity effects on crop growth. Soil Sci. Soc. Am. Proc. 39:617–622.

Dagan, G., and E. Bresler. 1979. Solute dispersion in unsaturated heterogeneous soil at field scale: I. Theory. Soil Sci. Soc. Am. J. 43:461–467.

Darcy, H. 1856. Les fountaines publique de Ville de Dijon. Dalmont, Paris.

Davidson, J. M., D. A. Graetz, P. S. C. Rao, and H. M. Selim. 1978. Simulation of nitrogen movement, transformation and uptake in plant root zone. USEPA Ecological Res. Ser. EPA-600/3-78-029. U.S. Government Printing Office, Washington, DC.

Day, P. R. 1956. Dispersion of a moving salt-water boundary advancing through saturated sand. Trans. Am. Geophys. Union 37:595–601.

Denmead, O. T., and R. W. Shaw. 1962. Availability of soil water to plants as affected by soil moisture content and meteorological conditions. Agron. J. 54:385–390.

DeSmedt, F., and P. J. Wierenga. 1978a. Approximate analytical solution for solute flow during infiltration and redistribution. Soil Sci. Soc. Am. J. 42:407–412.

DeSmedt, F., and P. J. Wierenga. 1978b. Solute transport through soil with nonuniform water content. Soil Sci. Soc. Am. J. 42:7–10.

Dirksen, C. 1972. A versatile soil water flux meter. p. 425–442. *In* Proc. 2nd Symp. on Fundamentals of Transport Phenomena in Porous Media, Vol. 2. IAHR, ISSS. Guelph, Ontario, Canada.

Dirksen, C. 1974. Field test of soil water flux meters. Trans. ASAE 17:1038–1042.

Elrick, D. E., K. T. Erh, and H. K. Krupp. 1966. Application of miscible displacement techniques to soils. Water Resour. Res. 2:717–728.

FAO/IAEA. 1976. Tracer manual on crops and soils. Tech. Rep. Ser. no. 171 (STI/DOC/10/171). International Atomic Energy Agency, Vienna.

Gardner, W. 1920. The capillary potential and its relation to soil moisture constants. Soil Sci. 10:357–359.

Gelhar, L. W. 1976. Effects of hydraulic conductivity variations in groundwater flows. p. 21/1–21/3. *In* L. Gottschalk et al. (ed.) Proc. 2nd Int. Symp. Stochastic Hydraulics. International Association of Hydraulics Research, Lund, Sweden.

Gureghian, A. B., D. S. Ward, and R. W. Cleary. 1979. Simultaneous transport of water and reacting solutes through multilayered soils under transient unsaturated flow conditions. J. Hydrol. 41:253–278.

Hanks, R. J. 1974. Model for predicting plant yield as influenced by water use. Agron. J. 66:660–665.

Hanks, R. J., and S. A. Bowers. 1962. Numerical solution of the moisture flow equations for infiltration into layered soils. Soil Sci. Soc. Am. Proc. 26:530–534.

Ivie, J. O., and L. A. Richards. 1937. A meter for recording slow liquid flow. Rev. Sci. Instr. 8:86–89.

Jensen, M. E. (ed.) 1974. Consumptive use of water and irrigation requirements. American Society of Civil Engineering, New York.

Kirda, C., D. R. Nielsen, and J. W. Biggar. 1973. Simultaneous transport of chloride and water during infiltration. Soil Sci. Soc. Am. Proc. 37:339–345.

Kirkham, D., and W. L. Powers. 1972. Advanced soil physics. Wiley-Interscience, New York.

Lapidus, L., and N. R. Amundson. 1952. Mathematics of adsorption in beds. 6: The effect of longitudinal diffusion in ion-exchange and chromatographic columns. J. Phys. Chem. 56:984–988.

Lindstrom, F. T., and L. Boersma. 1970. Theory of chemical transport with simultaneous sorption in a water saturated porous medium. Soil Sci. 110:1–9.

Lindstrom, F. T., and L. Boersma. 1971. A theory on the mass transport of previously distributed chemicals in a water saturated sorbing porous medium. Soil Sci. 111:192–199.

Lindstrom, F. T., L. Boersma, and D. Stockard. 1971. A theory on the mass transport of previously distributed chemicals in a water saturated sorbing porous medium: isothermal cases. Soil Sci. 112:291–300.

Lindstrom, F. T., R. Haque, V. H. Fried, and L. Boersma. 1967. Theory on the movement of some herbicides in soils. Environ. Sci. Tech. 1:561–565.

Lomas, J., and J. Levin. 1979. Irrigation. p. 217–264. *In* J. Seeman et al. (ed.) Agrometeorology. Springer-Verlag New York, New York.

Mansell, R. S., D. R. Nielsen, and D. Kirkham. 1968. A method for the simultaneous control of aeration and unsaturated water movement in laboratory soil columns. Soil Sci. 106:114–121.

Misra, C., D. R. Nielsen, and J. W. Biggar. 1974. Nitrogen transformations in soil during leaching: 2. Steady-state nitrification and nitrate reduction. Soil Sci. Soc. Am. Proc. 38:294–299.

Nielsen, D. R., and J. W. Biggar. 1961. Miscible displacement: I. Experimental information. Soil Sci. Soc. Am. Proc. 25:1–5.

Nielsen, D. R., and J. W. Biggar. 1963. Miscible displacement: III. Theoretical considerations. Soil Sci. Soc. Am. Proc. 26:216–221.

Nielsen, D. R., J. W. Biggar, and K. T. Erh. 1973. Spatial variability of field measured soil-water properties. Hilgardia 42:215–259.

Nielsen, D. R., J. W. Biggar, and C. S. Simmons. 1981a. Mechanisms of solute transport in soils. p. 115–135. *In* I. K. Iskander (ed.) Modeling wastewater renovation. John Wiley and Sons, New York.

Nielsen, D. R., J. Metthey, and J. W. Biggar. 1981b. Soil hydraulic properties, spatial variability and soil water movement. p. 47–68. *In* I. K. Iskander (ed.) Modeling wastewater renovation. John Wiley and Sons, New York.

Penman, H. L. 1948. Natural evaporation from open water, bare soil, and grass. Proc. R. Soc. Ser. A. 193:120–145.

Raats, P. A. C. 1975. Distribution of salts in the root zone. J. Hydrol. 27:237–248.

Rao, P. S. C., J. M. Davidson, and L. C. Hammond. 1976. Estimation of nonreactive and reactive solute front locations in soils. p. 235–242. *In* Proc. Hazardous Waste Res. Symp., July. (EPA-600/9-76-015). US Environmental Protection Agency, Cincinnati, OH.

Rao, P. S. C., J. M. Davidson, R. E. Jessup, and H. M. Selim. 1979. Evaluation of conceptual models for describing nonequilibrium adsorption-desorption of pesticides during steady flow in soils. Soil Sci. Soc. Am. J. 43:22–28.

Richards, L. A. 1931. Capillary conduction of liquids in porous mediums. Physics 1:318–333.

Richards, L. A., M. B. Russel, and O. R. Neal. 1937. Further developments on apparatus for field moisture studies. Soil Sci. Soc. Am. Proc. 1:55–63.

Ritchie, J. T. 1972. Model for predicting evaporation from a row crop with incomplete cover. Water Resour. Res. 8:1204–1213.

Robbins, C. W., R. J. Wagenet, and J. J. Jurinak. 1980. A combined salt transport-chemical equilibrium model for calcareous and gypsiferous soils. Soil Sci. Am. J. 44:1191–1194.

Rose, C. W., F. W. Chicester, J. R. Williams, and J. T. Ritchie. 1982a. A contribution to simplified models of field solute transport. J. Environ. Qual. 11:146–150.

Rose, C. W., F. W. Chichester, J. R. Williams, and J. T. Ritchie. 1982b. Application of an approximate analytic method of computing solute profiles with dispersion in soils. J. Environ. Qual. 11:151–155.

Rose, C. W., P. W. A. Dayananda, D. R. Nielsen, and J. W. Biggar. 1979. Long-term solute dynamics and hydrology in irrigated slowly peremable soils. Irrig. Sci. 1:77–87.

Rose, C. W., and W. R. Stern. 1965. The drainage component of the water balance equation. Aust. J. Soil Res. 3:95–100.

Rosenberg, N. J., H. E. Hart, and K. W. Brown. 1968. Evaporation-review of research. Nebraska Agric. Exp. Stn. Bull. MP20.

Schlicter, C. S. 1905. The rate of movement of underground waters. U.S.G.S. Water Supply and Irrigation Paper no. 140. Washington, DC.

Selim, H. M., J. M. Davidson, and P. S. C. Rao. 1977. Transport of reactive solutes through multilayered soils. Soil Sci. Soc. Am. J. 41:3–10.

Simmons, C. S., D. R. Nielsen, and J. W. Biggar. 1979. Scaling of field-measured soil water properties. I. Methodology. II. Hydraulic Conductivity and Flux. Hilgardia 47:77–174.

Smith, R. E., and R. H. B. Hebbert. 1979. A Monte Carlo analysis of the hydrologic effects of spatial variability of infiltration. Water Resour. Res. 15:419–429.

Starr, J. L., F. E. Broadbent, and D. R. Nielsen. 1974. Nitrogen transformations during continuous leaching. Soil Sci. Soc. Am. Proc. 38:283–289.

Tang, D. H., and G. F. Pinder. 1977. Simulation of groundwater flow and mass transport under uncertainty. Adv. Water Res. 1:25–30.

Tanner, C. B. 1967. Measurement of evapotranspiration. *In* R. M. Hagan et al. (ed.) Irrigation of agricultural lands. Agronomy 11:534–574.

Tanner, C. B. 1968. Evaporation of water from plants and soil. p. 74–106. *In* T. T. Kozlowski (ed). Water deficits and plant growth. Academic Press, New York.

Tanner, C. B., and W. A. Jury. 1976. Estimating evaporation and transpiration from a row crop during incomplete cover. Agron. J. 68:239–243.

Thornthwaite, C. W., and J. R. Mather. 1957. Instructions and tables for computing potential evapotranspiration and the water balance. Publications in climatology, Vol. 10, no. 3. Drexel Institute of Technology, Centerton, PA.

Tillotson, W. R., C. W. Robbins, R. J. Wagenet, and R. J. Hanks. 1980. Soil water, solute, and plant growth simulation. Utah Agric. Exp. Stn. Bull. 502.

U.S. Salinity Laboratory Staff. 1954. Diagnosis and improvement of saline and alkali soils. USDA Agric. Handb. 60. U.S. Government Printing Office, Washington, DC.

Veihmeyer, F. J. 1927. Some factors affecting the irrigation requirements on deciduous orchards. Hilgardia 2:125–288.

van Genuchten, M. Th. 1978a. Numerical solutions of the one-dimensional saturated-unsaturated flow equation. Princeton Univ. Water Resour. Program. Rep. no. 78-WR-09.

van Genuchten, M. Th. 1978b. Mass transport in saturated-unsaturated media: One-dimensional solutions. Princeton Univ. Water Resour. Program. Rep. no. 78:WR-11.

van Genuchten, M. Th., and W. J. Alves. 1981. A compendium of available analytical solutions of the one-dimensional convective-dispersive transport equation. U.S. Salinity Lab., Riverside, CA.

van Genuchten, M. Th., and R. W. Cleary. 1979. Movement of solutes in soil: computer simulated and laboratory results. p. 349–386. *In* G. H. Bolt (ed.) Soil chemistry. B. Physico-chemical models. Elsevier Science Publishing Co., New York.

van Genuchten, M. Th., and P. J. Wierenga. 1974. Simulation of one-dimensional solute transfer in porous media. New Mexico Agric. Exp. Stn. Bull. 628.

van Genuchten, M. Th., and P. J. Wierenga. 1976. Mass transfer studies in sorbing porous media. I. Analytical solutions. Soil Sci. Soc. Am. J. 40:473–480.

Wagenet, R. J. 1983. Principles of salt movement in soil. p. 123–140. *In* D. W. Nelson et al. (ed.) Chemical mobility and reactivity in soil systems. Spec. Pub. 11. American Society of Agronomy, Madison, WI.

Wagenet, R. J., J. W. Biggar, and D. R. Nielsen. 1977. Traciing the transformations of urea fertilizer during leaching. Soil Sci. Soc. Am. J. 41:896–902.

Wagenet, R. J., and B. K. Rao. 1983. Description of nitrogen movement in the presence of spatially variable soil hydraulic properties. Agric. Water Manage. 6:227–242.

Wagenet, R. J., and J. L. Starr. 1977. A method for the simultaneous control of the water regime and gaseous atmosphere in soil columns. Soil Sci. Soc. Am. J. 41:658–659.

Warrick, A. W., G. J. Mullen, and D. R. Nielsen. 1977a. Scaling of field-measured soil hydraulic properties using a similar media concept. Water Resour. Res. 13:355–362.

Warrick, A. W., G. J. Mullen, and D. R. Nielsen. 1977b. Predictions of soil water flux based upon field measured soil water properties. Soil Sci. Soc. Am. J. 41:14–19.

Wood, A. L., and J. M. Davidson. 1975. Fluometron and water content distributions during infiltration: Measured and calculated. Soil Sci. Soc. Am. Proc. 38:820–825.

46 Gas Diffusivity

D. E. ROLSTON
University of California
Davis, California

46-1 INTRODUCTION

Diffusion is the principal mechanism in the interchange of gases between the soil and atmosphere. The interchange results from concentration gradients established within the soil by the respiration of microorganisms and plant roots, by production of gases associated with biological reactions such as fermentation, nitrification, and denitrification, and by soil incorporation of materials such as fumigants and anhydrous ammonia. The diffusion of water vapor within the soil also occurs due to differences in vapor pressure gradients induced by temperature differences or by evaporative conditions at the soil surface.

Several gases are of particular interest in relation to gaseous diffusion in soils. The two gases which have received the most attention are oxygen (O_2) and carbon dioxide (CO_2) (Bremner & Blackmer, 1982). Other soil gases of interest are nitrous oxide (N_2O) and nitrogen (N_2), which result from the biological reduction of nitrate in soils. The process of transforming ammonium to nitrate may also result in production of N_2O. Other gases such as methane and ethylene can be produced in soil under certain conditions. The diffusion of gases such as methyl bromide, a fumigant, is also of interest.

The diffusion of gases in soil can be described by Fick's law

$$F/(At) = f = -D_p\,(\partial C/\partial x) \qquad [1]$$

where F is the amount of gas diffusing (g gas), A is the cross-sectional area of the soil (m^2 soil), t is time (s), f is the gas flux density (g gas m^{-2} soil s^{-1}), C is concentration in the gaseous phase (g gas m^{-3} soil air), x is distance (m soil), and D_p is the soil-gas diffusivity (m^3 soil air m^{-1} soil s^{-1}).

The soil-gas diffusivity has been related both empirically and theoretically to the soil-air content. Currie (1960b) found that the following empirical relationship fits many dry materials:

$$D_p = \gamma\, \epsilon^{\mu} D_o \qquad [2]$$

where γ and μ represent measures of pore shape, D_o (m^2 air s^{-1}) is the gas diffusivity in a region containing only gas (no soil), and ϵ is the soil air content (m^3 air m^{-3} soil). Values of γ generally lie between 0.8 and 1.0 and increase with the mean porosity of each type of material. Other researchers have used similar forms of Eq. [2] with $\gamma = 0.66$, $\mu = 1.0$ (Penman, 1940a, 1940b); $\gamma = 1.0$, $\mu = 3/2$ (Marshall, 1959); and $\gamma = 1.0$, $\mu = 4/3$ (Millington, 1959). Millington (1959) modified the equation to include the effect of water-filled pores to give

$$D_p = (\epsilon^{10/3}/E^2)\, D_o \qquad [3]$$

where E is the total void ratio (m^3 voids m^{-3} soil), which is the sum of the soil air content and the soil water content on a volume basis. Equations [2] and [3] are not always valid for very wet soil and strongly aggregated soil (Millington & Shearer, 1971).

Since discrepancies between measured soil-gas diffusivities stemming from Eq. [1] and those calculated from Eq. [2] and [3] occur, it is often desirable to measure the diffusivity for particular situations. A laboratory method using soil cores and a proposed field method for measuring soil-gas diffusivities will be described.

46–2 LABORATORY METHOD

46–2.1 Principles

The laboratory method is based upon establishing gas of concentration C_0 within a chamber as shown by the diagram of Fig. 46–1. One end of a soil core of concentration C_s is placed in contact with the gas within the chamber. The other end of the soil core is maintained at concentration C_s. The gas of interest diffuses either into or out of the chamber depending upon the concentration C_0 with respect to that outside the core. Obviously, the other gases making up the atmosphere will diffuse in an opposite direction to that of the gas of interest. The time rate of change of concentration in the chamber is related to the soil-gas diffusivity.

The laboratory method described here is based upon solutions given by Currie (1960a). Similar procedures have been used by several investigators (Rust et al., 1957; Gradwell, 1961; McIntyre & Philip, 1964; Shearer et al., 1966; Flühler, 1973).

The unsteady diffusion of a gas which is nonreactive (physically, biologically, and chemically) is described by the combination of Fick's first law (Eq. [1]) and the continuity equation

$$\epsilon(\partial C/\partial t) = D_p(\partial^2 C/\partial x^2)\,. \qquad [4]$$

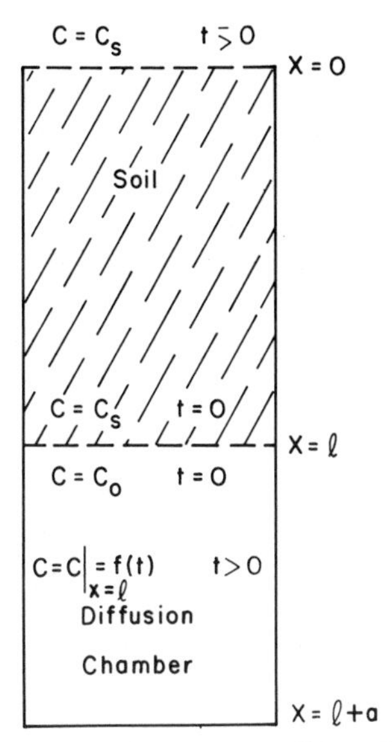

Fig. 46–1. Diagram giving initial and boundary conditions for the laboratory method of measuring the soil-gas diffusivity.

In developing Eq. [4] it has been assumed that the soil is uniform with respect to diffusivity and that ϵ is constant in space and time. Equation [4] may be solved subject to the boundary and initial conditions given in Fig. 46–1. The solution (Carslaw & Jaeger, 1959, p. 128) for the concentration in the chamber C is

$$\frac{C - C_s}{C_o - C_s} = \sum_{n=1}^{\infty} \frac{2h \exp(-D_p \alpha_n^2 t/\epsilon)}{l(\alpha_n^2 + h^2) + h} \qquad [5]$$

where $h = \epsilon/(a\epsilon_c)$, ϵ_c is the air content of the chamber (1.0 m^3 air m^{-3} chamber), a is the length of the chamber or the volume, V, of the chamber per area, A, of soil, and α_n with $n = 1,2, \ldots$ are the positive roots of $\alpha l \tan \alpha l = hl$. Equation [5] was used by Currie (1960a) for the case where $C_s = 0$. At some time greater than zero, the terms for $n \geq 2$ are negligible with respect to the first term, and Eq. [5] reduces to

$$\frac{C - C_s}{C_o - C_s} = \frac{2h \exp(-D_p \alpha_1^2 t/\epsilon)}{l(\alpha_1^2 + h^2) + h} . \qquad [6]$$

Thus, a plot of ln $[(C - C_s)/(C_o - C_s)]$ vs. time, t, becomes linear with slope $-D_p \, \alpha_1^2/\epsilon$ for sufficiently large t.

46–2.2 Apparatus and Materials

The apparatus and materials will be described for measuring the mutual diffusivity of argon (Ar) and N_2, but also apply to other gases of interest. Argon and N_2 are used because they are relatively unreactive biologically, chemically, and physically. Argon has approximately the same values for gas diffusivity and solubility in water as does O_2. Thus, the Ar and N_2 mixture is expected to have a similar soil-gas diffusivity to that of air.

1. Soil cores. Undisturbed soil cores can be taken from the field with samplers described in the literature. The typical double-cylinder, hammer-driven core sampler for obtaining soil samples for bulk density (chapter 13 in this book) is appropriate. Sieved, packed samples may also be used.
2. Argon gas. Commercially available compressed Ar gas in metal tanks is convenient and satisfactory.
3. Diffusion apparatus. A diagram of the diffusion vessel including the attached soil core is given by Fig. 46–2. The apparatus can be constructed with several different variations. A commonly available vacuum desiccator (without the lid) can be used for the diffusion vessel. A base plate constructed of material such as acrylic plastic can then make an airtight seal with the ground glass portion of the desiccator. The base plate should be constructed in such a way that a sliding plate containing a hole of the same diameter as the soil core can be made

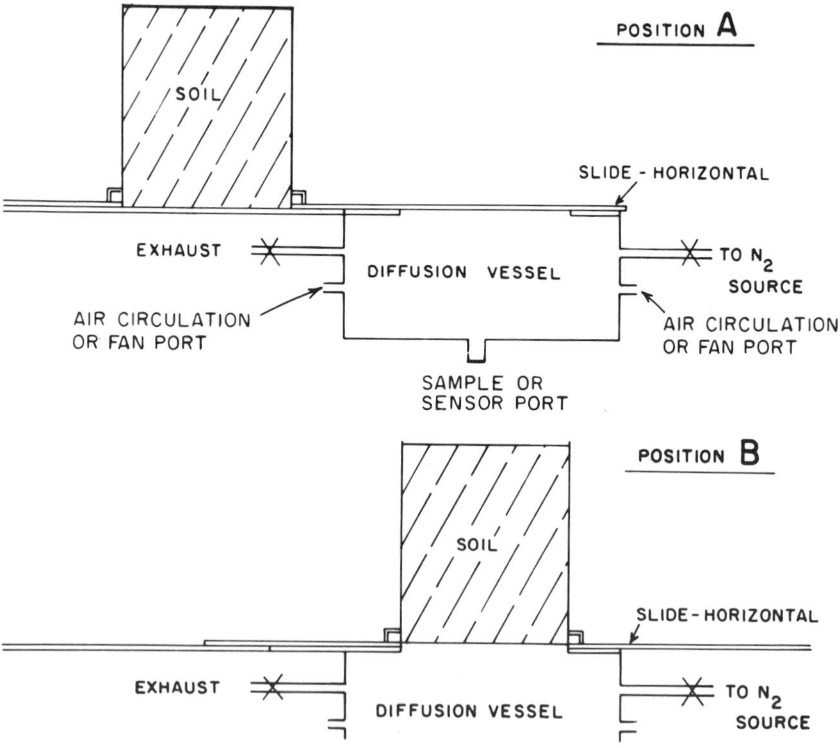

Fig. 46–2. Diagram of a typical laboratory diffusion apparatus. Redrawn from Evans (1965).

to slide over a similar hole in the base plate. Airtight seals can be made between the sliding surface of the sliding plate and the base plate using a lubricant or material such as stopcock grease. The sliding plate should also have an attachment which will hold the undisturbed soil core. The base plate should .'so contain four or five ports through which tubing can be connected for flushing, for removal of gas samples, or for circulation of air within the vessel with a fan or pump. If the volume of the diffusion chamber is relatively small ($\simeq$100 mL), Currie (1960a) found that stirring of the air in the diffusion chamber was not required. The fifth hole may be needed as a sample port. The size and shape of the diffusion vessel, as well as the length and diameter of the soil core, are flexible and can be adapted to various situations.

4. Gas chromatograph. It is convenient to measure N_2 gas with a common gas chromatograph available in many laboratories. See Bremner and Blackmer (1982) for details of gas chromatography.

46–2.3 Procedure

Assemble the diffusion apparatus as shown by Fig. 46–2, and accurately measure the volume of the diffusion chamber by filling with water. Obtain undisturbed soil cores (or packed, sieved samples) in the sample cylinders as described in chapter 13. Place soil core and sample cylinder on the sliding plate of the diffusion apparatus and seal the sample container to the sliding plate using paraffin, silicone sealant, epoxy, rubber O-ring, or any other suitable material to ensure an airtight seal.

Flow Ar through the diffusion vessel with the soil core in position A until an N_2 concentration near 10% is obtained within the diffusion vessel. An initial concentration of 10% N_2 within the diffusion chamber is desirable in order to have a large enough peak height on a chromatogram to insure satisfactory separation from the Ar peak. Stop the flow of Ar and close the valves on the inlet and outlet purging ports. If a fan or air circulation pump is required, these should be operated continuously. The constant concentration of N_2 attained within the diffusion vessel is the initial concentration, C_o.

Slide the sample to position B so that the soil core is located directly over the hole in the base plate of the diffusion chamber. Record the time or start a timer immediately. This time will be $t = 0$. The concentration of N_2 within the diffusion vessel should be measured as frequently as practical. For fairly dry soils, with expected large diffusivities, the concentration should be measured on a time scale of minutes. For fairly wet soils, with small diffusivities, the concentration within the diffusion vessel should be measured on a time scale of hours.

Plot ln $[(C - C_s)/(C_o - C_s)]$ vs. t. For the Ar–N_2 system described above, C_s is the concentration of N_2 in the atmosphere (888.8 g m^{-3}). Sampling should be continued long enough so that several data points can be obtained on a linear portion of the plot.

Draw a straight line through the data points or run a linear regression

Table 46–1. The first six roots, $\alpha_n l$, of $\alpha l \tan \alpha l = hl$. The roots of this equation are all real if $hl > 0$.

hl	$\alpha_1 l$	$\alpha_2 l$	$\alpha_3 l$	$\alpha_4 l$	$\alpha_5 l$	$\alpha_6 l$
0	0	3.1416	6.2832	9.4248	12.5664	15.7080
0.001	0.0316	3.1419	6.2833	9.4249	12.5665	15.7080
0.002	0.0447	3.1422	6.2835	9.4250	12.5665	15.7081
0.004	0.0632	3.1429	6.2838	9.4252	12.5667	15.7082
0.006	0.0774	3.1435	6.2841	9.4254	12.5668	15.7083
0.008	0.0893	3.1441	6.2845	9.4256	12.5670	15.7085
0.01	0.0998	3.1448	6.2848	9.4258	12.5672	15.7086
0.02	0.1410	3.1479	6.2864	9.4269	12.5680	15.7092
0.04	0.1987	3.1543	6.2895	9.4290	12.5696	15.7105
0.06	0.2425	3.1606	6.2927	9.4311	12.5711	15.7118
0.08	0.2791	3.1668	6.2959	9.4333	12.5727	15.7131
0.1	0.3111	3.1731	6.2991	9.4354	12.5743	15.7143
0.2	0.4328	3.2039	6.3148	9.4459	12.5823	15.7207
0.3	0.5218	3.2341	6.3305	9.4565	12.5902	15.7270
0.4	0.5932	3.2636	6.3461	9.4670	12.5981	15.7334
0.5	0.6533	3.2923	6.3616	9.4775	12.6060	15.7397
0.6	0.7051	3.3204	6.3770	9.4879	12.6139	15.7360
0.7	0.7506	3.3477	6.3923	9.4983	12.6218	15.7524
0.8	0.7910	3.3744	6.4074	9.5087	12.6296	15.7587
0.9	0.8274	3.4003	6.4224	9.5190	12.6375	15.7650
1.0	0.8603	3.4256	6.4373	9.5293	12.6453	15.7713
1.5	0.9882	3.5422	6.5097	9.5801	12.6841	15.8026
2.0	1.0769	3.6436	6.5783	9.6296	12.7223	15.8336
3.0	1.1925	3.8088	6.7040	9.7240	12.7966	15.8945
4.0	1.2646	3.9352	6.8140	9.8119	12.8678	15.9536
5.0	1.3138	4.0336	6.9096	9.8928	12.9352	16.0107
6.0	1.3496	4.1116	6.9924	9.9667	12.9988	16.0654
7.0	1.3766	4.1746	7.0640	10.0339	13.0584	16.1177
8.0	1.3978	4.2264	7.1263	10.0949	13.1141	16.1675
9.0	1.4149	4.2694	7.1806	10.1502	13.1660	16.2147
10.0	1.4289	4.3058	7.2281	10.2003	13.2142	16.2594
15.0	1.4729	4.4255	7.3959	10.3898	13.4078	16.4474
20.0	1.4961	4.4915	7.4954	10.5117	13.5420	16.5864
30.0	1.5202	4.5615	7.6057	10.6543	13.7085	16.7691
40.0	1.5325	4.5979	7.6647	10.7334	13.8048	16.8794
50.0	1.5400	4.6202	7.7012	10.7832	13.8666	16.9519
60.0	1.5451	4.6353	7.7259	10.8172	13.9094	17.0026
80.0	1.5514	4.6543	7.7573	10.8606	13.9644	17.0686
100.0	1.5552	4.6658	7.7764	10.8871	13.9981	17.1093
∞	1.5708	4.7124	7.8540	10.9956	14.1372	17.2788

on those data points within the linear part of the plot. The slope of the line is equal to $-D_p\,\alpha_1^2/\epsilon$. The value of α_1 can be determined from Table 46–1. An independent determination of ϵ must be made from the soil bulk density and water content. The value of D_p can then be determined from the slope. In the example of Fig. 46–3, the slope of the linear portion of the curve is equal to -0.058, to yield a value of D_p equal to 0.0005 m^3 soil air m^{-1} soil h^{-1} ($\epsilon = 0.1$, $\alpha_1 = 3.41$).

46–2.4 Comments

It is important to have precise temperature control when making measurements, since the diffusivity is strongly dependent upon temper-

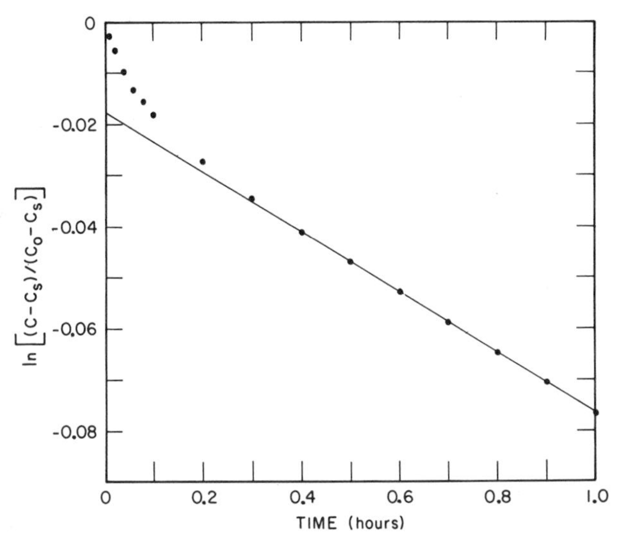

Fig. 46–3. A plot of ln $[(C - C_s)/(C_o - C_s)]$ vs. time using hypothetical data from a soil core with values of $\epsilon = 0.1$ m^3 m^{-3}, $l = 76$ mm, $A = 4540$ mm^2, and $V = 0.5$ L.

ature. When results are reported, the temperature should be stated. The following equation from Currie (1960a) allows the soil-gas diffusivity measured at any temperature to be calculated for any other temperature:

$$D_{T2} = D_{T1}\ (T_2/T_1)^{1.72}$$

where D_{T2} and D_{T1} are the diffusivities at temperatures T_2 and T_1 (K), respectively.

The time required before a straight line is obtained can be estimated from the following equation:

$$t = \frac{\epsilon}{(\alpha_2^2 - \alpha_1^2)D_p} \ln\left\{ \frac{B\left[\alpha_1^2 + (\epsilon^2 A^2/V^2) + (\epsilon A/Vl)\right]}{\alpha_2^2 + (\epsilon^2 A^2/V^2) + (\epsilon A/Vl)} \right\} \qquad [7]$$

where B is the ratio between the first and second terms of the infinite sum of Eq. [5]. For instance, B would be equal to 100 when the second term is 1% of the value of the first term and would be 1000 when the second term is 0.1% of the value of the first term.

Equation [7] shows that the time required before a plot of ln $[(C - C_s)/(C_o - C_s)]$ vs. t is linear is dependent upon the volume (V) of the diffusion chamber, the length (l) and cross-sectional area (A) of the soil core, the soil-air content (ϵ), and the soil-gas diffusivity (D_p). For $V = 2$ L, $A = 5000$ mm^2, $l = 0.1$ m, $B = 1000$, and fairly dry soil with high soil-air content (0.5) and soil-gas diffusivity (0.03 m^2 h^{-1}), the plot of ln $[(C - C_s)/(C_o - C_s)]$ vs. t gives a straight line after approximately 3 min. However, for the same values of V, A, l, and B, but with a wet soil with a small soil-air content (0.05) and a small soil–gas diffusivity (0.000015

$m^2\ h^{-1}$), the time required before a linear plot can be obtained is approximately 3 h. Therefore, for dry soil, it would be convenient to determine D_p from the slope $(-D_p\ \alpha_1^2/\epsilon)$ of a plot of ln $[(C - C_s)/(C_o - C_s)]$ vs. t. However, for a wet soil, it may be desirable to use all the measured values even for short times. In that case, Eq. [5], using several terms, would have to be curve-fitted to the measured values of $(C - C_s)/(C_o - C_s)$ in order to determine the best-fitted values of D_p. Values of $\alpha_n l$ are given in Table 46–1 (Carslaw & Jaeger, 1959).

Several alterations of the apparatus design and procedures are possible dependent upon particular requirements and gases of interest. For instance, rather than flushing the diffusion chamber to establish an initial concentration near zero of a particular gas, it may be desirable to add gas or to flush the chamber to maintain some higher level of gas and maintain gas of zero concentration at the end of the soil core not in contact with the diffusion chamber. Also, rather than flushing the diffusion chamber with Ar, helium (He) could be used. Since the mutual diffusivity of He and N_2 is approximately twice that of Ar and N_2, a correction would be required. Several variations on the size of the soil core and the size and shape of the diffusion apparatus are also feasible. However, the size of the diffusion vessel and the soil core will have an effect on the time required before a linear plot of ln $[(C - C_s)/(C_o - C_s)]$ vs. time would be attained.

An alternate approach for handling the data is based upon the method proposed by Taylor (1949). Equation [1] can be rearranged according to

$$dF/dt = -D_p\ A(dC/dx)\ . \qquad [8]$$

The amount of gas in the diffusion chamber is simply the concentration of gas at any time multiplied by the volume of air in the chamber, V(m^3 air), given by

$$F = C\ V \qquad [9]$$

where C (g gas m^{-3} air) is the concentration in the chamber. Substituting Eq. [9] into Eq. [8] gives

$$dC/dt = -(D_p A/V)\ (dC/dx)\ . \qquad [10]$$

The concentration gradient within the soil core (dC/dx) can be approximated by $(C - C_s)/l$ where C_s is the concentration at the upper end of soil core in contact with the atmosphere and l (m soil) is the length of the soil core. Making this substitution in Eq. [10] and integrating gives

$$\ln\ [(C - C_s)/(C_o - C_s)] = -D_p\ At/(Vl)\ . \qquad [11]$$

By plotting the left-hand side of Eq. [11] vs. t, a straight line should result with a slope equal to $-D_p\ A/(Vl)$. From the slope, the value of D_p can be calculated. This equation indicates that a plot of ln $[(C - C_s)/(C_o - C_s)]$ vs. t should give a straight line for all values of t. However, Eq. [11] is not valid for small times due to the fact that the concentration gradient within the soil during early phases of the experiment cannot be approximated by $(C - C_s)/l$.

Equation [11] also does not take into account a change in storage of gas within the soil as concentration changes. The error incurred by neglecting the change in storage can be determined by setting the slope from Eq. [6] $(-D_{p1}\alpha_1^2/\epsilon)$ equal to the slope from Eq. [11] $(-D_{p2}A/Vl)$. Solving for D_{p1} (from Currie's method) in terms of D_{p2} (from Taylor's method) allows a correction factor to be determined for various diffusion apparatus volumes. Figure 46–4 gives the correction factor, K, $(D_{p1} = K\ D_{p2})$ as a function of the ratio of the volume of soil air to the volume of chamber air $(\epsilon Al/V) = hl)$. It is obvious from Fig. 46–4 that K is nearly 1.0 for small values of the ratio but increases rapidly when the volume of soil air is > 10% of the volume of chamber air. Thus, the diffusivity (D_{p2}) calculated according to Taylor's method will require a correction for certain diffusion chamber and soil core sizes. The parameter which is most easily manipulated is the volume of the diffusion vessel, V. If V is sufficiently large, the correction required for change in storage is negligible. For example, the diffusion vessel used by Taylor with $V = 2$ L, $A = 1900$ mm^2, $l = 35.5$ mm, and $\epsilon = 0.5$ (assumed) would result in a correction factor of 1.005 (a 0.5% underestimate). Bakker and Hidding (1970) also used a similar approach for correcting the soil-gas diffusivity

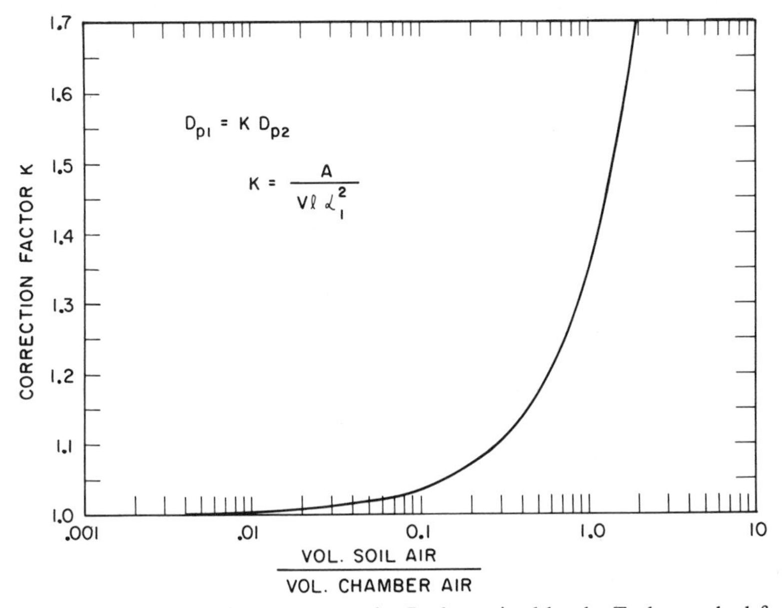

Fig. 46–4. The factor required to correct the D_p determined by the Taylor method for a change in storage within the soil. The abscissa is a log scale.

due to a change in storage for their particular apparatus. The equations of Currie and Taylor will give the same values of the soil-gas diffusivity for $t \gg 0$ when the linear part of the curve is attained if the correction due to change in storage is also made.

Potential errors associated with this method are: (i) drying of a wet soil core during measurements, (ii) mass flow, (iii) reaction of gas with soil or chamber, and (iv) diffusion of gas through chamber parts. For very wet soil cores, water-saturated air should be slowly flowed across the top of the soil core exposed to the atmosphere in order to prevent evaporation. This can be accomplished by adding a small chamber to the top of the soil core in which water-saturated air is slowly flowed through the chamber. Mass flow through the soil core may be induced by pressure fluctuations, high air movement across the top of the core, or removal of large volumes of gas from the diffusion chamber through sampling. If the gas of interest strongly reacts with the soil or the gas is consumed or produced within the soil, the soil-gas diffusivity will be in error. This error could be substantial for very wet soil, requiring fairly long times to conduct the diffusion experiment. Diffusion of gas through tubing connections, leaks, rubber septa, etc. may result in error, especially if the experiment is conducted over long time periods.

46–3 FIELD METHOD

Several field methods for measuring the soil-gas diffusivity have been proposed (McIntyre & Philip, 1964; Lai et al., 1976; Rolston & Brown, 1977). These methods offer the possibility of directly measuring the diffusivity in situ. However, none of these methods has been adequately investigated and tested. The method by McIntyre and Philip (1964) will be proposed as a means of measuring the diffusivity in the field. However, the method also requires further evaluation.

46–3.1 Principles

The method involves insertion of a tube into the surface of the field soil and supplying the gas of interest to the confined soil core from a well-stirred reservoir or chamber of finite volume placed over the inserted tube. A solution of Eq. [4] for a uniform, semi-infinite porous medium of initially constant gas concentration and appropriate boundary conditions is available (Carslaw & Jaeger, 1959, p. 306). The problem is similar to that depicted in Fig. 46–1 except that the diffusion chamber is above the soil and the soil extends semi-infinitely downward. The adaption of the solution to measurement of the soil-gas diffusivity has been made by McIntyre and Philip (1964). The solution of this problem for the concentration at the soil surface and in the well-stirred diffusion chamber is given by

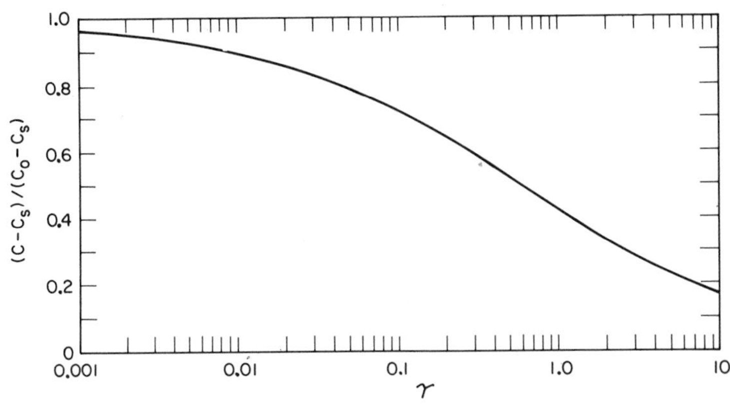

Fig. 46–5. Functional relationship between gas concentration within the diffusion chamber and the reduced variable τ (Eq. [12]). The abscissa is a log scale. Redrawn from McIntyre and Philip (1964).

$$(C - C_s)/(C_o - C_s) = \exp(\tau) \operatorname{erfc}(\tau^{1/2}) \quad [12]$$

where C is the concentration of the gas of interest in the diffusion chamber at any t, C_s is the concentration of the gas in the soil at $t = 0$, C_o is the concentration of the gas in the chamber at $t = 0$, erfc is the complimentary error function, and τ is the reduced variable for time given by

$$\tau = \epsilon D_p t/a^2 \quad [13]$$

where $a = V/A$ is the volume of the chamber per unit area of the soil tube (m^3 air m^{-2} soil) and the other variables were defined previously.

The functional relationship between $(C - C_s)/(C_o - C_s)$ and τ is shown in Fig. 46–5. The concentration of gas in the diffusion chamber is measured as a function of time. The value of $(C - C_s)/(C_o - C_s)$ can then be calculated for any value of time t. For this particular value of $(C - C_s)/(C_o - C_s)$ a value of τ can be determined from Fig. 46–5. The value of D_p can then be determined from Eq. [13], since ϵ (determined independently), t, and a are known.

46–3.2 Apparatus and Materials

The apparatus and materials will be described for measuring the mutual diffusivity of O_2 diffusing into air as given by McIntyre and Philip (1964). The same equipment is applicable to other gases of interest.

1. Diffusion apparatus. A sharpened metal cylinder with a flange soldered to the top of the cylinder is constructed so that it can be pushed or driven into the surface of a field soil. The cylinder used by McIntyre and Philip was 76 mm long and 76 mm in diameter. The metal cylinder and diffusion chamber are depicted in Fig. 46–6 (redrawn from McIntyre & Philip, 1964). The diffusion chamber can be constructed

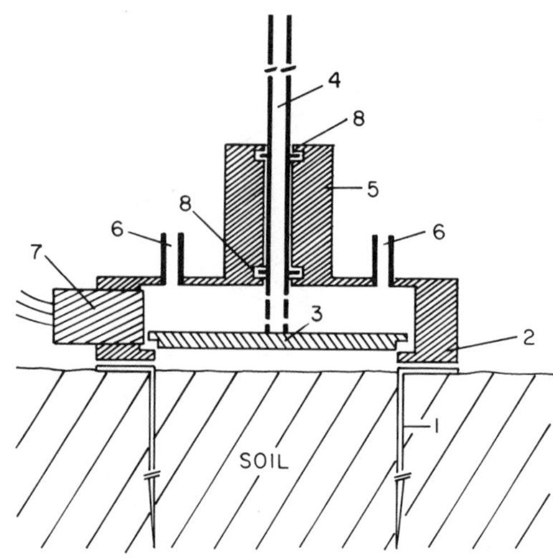

Fig. 46–6. Diagram of the field diffusion apparatus. The various components of the apparatus are: 1—Metal cylinder, 2—Reservoir, 3—Gate, 4—Tube attached to gate provides gas inlet, 5—Housing for tubing slide, 6—Gas outlet tubes, 7—Gas sensor, and 8—Rubber O-rings. Redrawn from McIntyre and Philip (1964).

of acrylic plastic or metal. The chamber should be constructed so that the lower part of the chamber forms a tight seal with the flanged cylinder placed in the soil. The volume of the diffusion chamber used by McIntyre and Philip (1964) was 0.025 m^3 air m^{-2} soil within the cylinder. The "gate" must be shaped so that when closed it will be flush with the base of the diffusion chamber. When fully opened, the gate should seal off the chamber flushing tubes.

2. Oxygen gas. Commercially available compressed O_2 gas in metal tanks is convenient and satisfactory.
3. Equipment for gas analysis. An O_2 analyzer based upon the polarographic O_2 sensor (Ecolyzer Model 60-600, Teledyne Model 320B/RC, Beckman Model 715) is convenient when the diffusivity of O_2 is being measured. Many of these devices are portable. The sensor should be inserted into the diffusion chamber (Fig. 46–6). A gas chromatograph must be used for other gases (see section 46–2.2 and Bremner & Blackmer, 1982), and the diffusion apparatus of Fig. 46–6 must be modified to allow for gas sampling.

46–3.3 Procedure

Push or drive the sharpened metal cylinder into the surface of the field soil until the flange soldered to the top of the cylinder is seated firmly on the soil surface. Slowly flush the soil in the cylinder with air in order to establish a constant and known concentration, C_s, of O_2 in the soil. The flow rate must be small enough so as to not alter the water content and distribution. While the soil cylinder is being flushed with air, close the gate of the diffusion chamber, and flush with 100% O_2.

After flushing sufficiently, stop the flow, and add several drops of water through the outlet ports to ensure saturation of the chamber air with water vapor. Place the diffusion chamber near the soil cylinder for enough time to ensure temperature equilibrium of the diffusion chamber. Measure the concentration of O_2 within the diffusion chamber. This concentration is C_o.

Place the diffusion chamber on the metal cylinder ensuring that an air-tight seal is made between the bottom of the diffusion chamber and the flange of the cylinder by the use of stopcock grease or other means. Open the gate of the diffusion chamber and begin making concentration measurements at regular intervals of up to 2 or 3 min.

Take a sample of the soil within the cylinder or the entire soil cylinder in order to determine the soil-air content ϵ.

The theory requires that soil within the cylinder be uniform and semi-infinite. However, since a finite cylinder length is used in practice, care must be taken that the depth of penetration of the O_2 from the diffusion chamber does not exceed the length of the cylinder or the depth where the soil ceases to be uniform. McIntyre and Philip (1964) derived a relationship to estimate the time that the depth of penetration would exceed a certain depth. Rearranging their equation and solving for time gives

$$t' < [0.054\, x'^2/(D_p/\epsilon)_{max}] \quad [14]$$

where t' is the maximum time beyond which the theoretical analyses may no longer be valid, x' is the depth of the soil cylinder or the depth where the soil ceases to be uniform, and $(D_p/\epsilon)_{max}$ is the maximum expected value of the ratio of the diffusivity and soil-air content. McIntyre and Philip estimated that for a 76-mm-long soil cylinder and values of $(D_p/\epsilon)_{max}$ of approximately 0.015 $m^2\ h^{-1}$, t' should be no greater than 1.5 min.

Thus the procedure for determining D_p is to measure the value of $(C - C_s)/(C_o - C_s)$ for some time t, say 1.5 min, determine the corresponding value of τ from Fig. 46–5, and calculate D_p from Eq. [13]. This procedure should be repeated for several different times between 0 and t' (Eq. [14]). The several different values of D_p should be approximately equal if the soil is relatively uniform and the conditions of the theory are satisfied.

46–3.4 Comments

The theoretical analysis for this method is not valid if there is diffusion of the gas of interest below the soil cylinder depth or below the depth where the soil ceases to be uniform, whichever is smaller. The theory is also based on the assumption that the gas within the diffusion chamber is well mixed. This assumption is reasonably good for small (few hundred milliliters) chamber volumes, but may result in error for larger volumes unless artificial mixing by fans or pumps is undertaken.

Theory also requires that the diffusing gas have no sinks within the soil. Obviously, this assumption is violated for O_2 due to consumption by soil organisms. However, since experimental times may be small, O_2 consumption may be considered to be negligible with respect to diffusion. Possible errors due to sampling are similar to those discussed in section 46–2.4. Since experimental times are small, instrument response times must also be small.

46–4 REFERENCES

Bakker, J., and A. P. Hidding. 1970. The influence of soil structure and air content on gas diffusion in soils. Neth. J. Agric. Sci. 18:37–48.

Bremner, J. M. and A. M. Blackmer. 1982. Composition of soil atmospheres. *In* A.L. Page et al. (ed.) Methods of soil analysis, part 2. 2nd ed. Agronomy 9:873–902.

Carslaw, H. S., and J. C. Jaeger. 1959. Conduction of heat in solids. 2nd ed. Claredon Press, Oxford.

Currie, J. A. 1960a. Gaseous diffusion in porous media. Part 1. A non-steady state method. Br. J. Appl. Phys. 11:314–317.

Currie, J. A. 1960b. Gaseous diffusion in porous media. Part 2. Dry granular materials. Br. J. Appl. Phys. 11:318–324.

Evans, D. 1965. Gas movement. *In* C.A. Black et al. (ed.) Methods of soil analysis, Part 1. Agronomy 9:319–330.

Flühler, H. 1973. Sauerstaffdiffusion in Boden. Mitt. Schweiz. Anst. Forstl. Versuchs wes. 49:125–250.

Gradwell, M. W. 1961. A laboratory study of the diffusion of oxygen through pasture topsoils. N.Z. J. Sci. 4:250–270.

Lai, S.-H., J. M. Tiedje, and A. E. Erickson. 1976. In situ measurement of gas diffusion coefficient in soils. Soil Sci. Soc. Am. J. 40:3–6.

Marshall, T. J. 1959. The diffusion of gas through porous media. J. Soil Sci. 10:79–82.

McIntyre, D. S., and J. R. Philip. 1964. A field method for measurement of gas diffusion into soils. Aust. J. Soil Res. 2:133–145.

Millington, R. J. 1959. Gas diffusion in porous media. Science 130:100–102.

Millington, R. J., and R. C. Shearer. 1971. Diffusion in aggregated porous media. Soil Sci. 111:372–378.

Penman, H. L. 1940a. Gas and vapor movements in the soil. I. The diffusion of vapors through porous solids. J. Agric. Sci. 30:437–462.

Penman, H. L. 1940b. Gas and vapor movements in the soil. II. The diffusion of carbon dioxide through porous solids. J. Agric. Sci. 30:570–581.

Rolston, D. E., and B. D. Brown. 1977. Measurement of soil gaseous diffusion coefficients by a transient-state method with a time-dependent surface condition. Soil Sci. Soc. Am. J. 41:499–505.

Rust, R. H., A. Klute, and J. E. Gieseking. 1957. Diffusion-porosity measurements using a non-steady state system. Soil Sci. 84:453–463.

Shearer, R. C., R. J. Millington, and J. P. Quirk. 1966. Oxygen diffusion through sands in relation to capillary hysteresis: I. Calibration of oxygen cathode for use in diffusion studies. Soil Sci. 101:361–365.

Taylor, S. A. 1949. Oxygen diffusion in porous media as a measure of soil aeration. Soil Sci. Soc. Am. Proc. 14:55–61.

47 Gas Flux

D. E. ROLSTON
University of California
Davis, California

47-1 INTRODUCTION

The need for measuring the soil-gas flux, especially at the soil surface, arises from interest in quantifying sources and sinks of gases within the soil. Knowledge of the amount of water vapor transport across the soil-atmosphere interface is important for irrigation management. The flux of O_2 and CO_2 is of interest in evaluating the respiration of microorganisms and roots in soil. The importance of the nitrogen cycle in agricultural and natural ecosystems requires that the volatile losses of nitrogen compounds such as NH_3, N_2O, N_2, and the trace amounts of NO and NO_2 be quantified in order to evaluate losses and efficiencies of fertilization. Recently, much effort has been directed at measuring the N_2O flux from soils in order to evaluate the potential impact of nitrogen fertilization on N_2O transport to the atmosphere, resulting in potential depletion of the ozone layer of the lower stratosphere. The flux of other gases such as fumigants is also of interest in order to quantify effectiveness of organism control and potential health hazards. The flux of hydrocarbons from soil can be of importance in some situations such as methane gas diffusing from landfill sites.

This chapter will discuss three basic methods for determining the flux of gas within the soil and specifically at the soil surface. One method is based upon independent measurements of the soil-gas diffusivity and the gas concentration gradient, and the use of Fick's law to calculate the flux within the profile or at the soil surface. A second method of directly measuring the soil-gas flux is to place closed chambers over the soil surface and measure the accumulation of the particular gas of interest within the chamber. A third approach, similar to the second method, is to place a chamber over the soil surface and continually flow gas through the chamber and either sample or trap the gas of interest in the effluent stream. A possible fourth category of methods (not discussed here) is micrometeorological approaches based on making various measurements in the atmosphere above soil. Although these methods have been used to determine flux of H_2O, CO_2, and NH_3 from soils, they measure flux in the air and only indirectly the soil-gas flux. For some gases, the presence

 Methods of Soil Analysis, Part 1. Physical and Mineralogical Methods—Agronomy Monograph no. 9 (2nd Edition)

of a crop may provide an additional source or sink for the gas of interest. Readers should refer to Denmead et al. (1978) for more information on micrometeorological methods.

47–2 FLUX CALCULATED FROM FICK'S LAW

47–2.1 Principles

The steady flux of gas in soil can be described by Fick's law of diffusion

$$(F/At) = f = -D_p\,(dc/dx) \qquad [1]$$

where F is the amount of gas diffusing (g gas), A is the cross-sectional area of the soil (m^2 soil), t is time (s), f is the gas flux density (g gas m^{-2} soil s^{-1}), c is concentration in the gaseous phase (g gas m^{-3} soil air), x is distance (m soil), and D_p is the soil-gas diffusivity (m^3 soil air m^{-1} soil s^{-1}). The total amount F diffusing from the soil or across any particular depth in the soil over some time period t_1 can be calculated from

$$F = -AD_p \int_0^{t_1} (dc/dx)dt\,. \qquad [2]$$

In order to apply either Eq. [1] or [2], the soil-gas diffusivity, D_p, and the concentration gradient, dc/dx, must be known at the depth for which the flux is to be calculated. Values of D_p can be measured or calculated by the methods discussed in chapter 46. In addition to the possible errors associated with the methods discussed in that chapter, one must also be concerned with the variability of the values of D_p in field soil or laboratory soil columns. Rolston (1978) provided estimates of the possible uncertainties or variabilities involved in measuring the soil-gas diffusivity within a relatively small soil area.

The concentration gradient is the change in concentration of gas over an infinitesimally small distance dx within the soil. In practice, the concentration gradient is measured by sampling the soil-gas concentration at a minimum of two finite depths, x_1 and x_2, within the soil. In theory, these two depths should be as close to each other as possible yet be far enough apart that a measurable difference in concentration can be determined. If one wished to measure the soil-gas flux at the soil surface, the concentration should be measured at $x = 0$ and at $x = 20$ to 50 mm below the soil surface. It is also important to measure the concentration at several lateral locations of the same soil depth in order to determine an "average" concentration at a particular depth. Since in practice the concentration can be measured only at finite points within the soil, it is important that the concentration as a function of depth have a reasonably smooth gradient. If the gradient is not smooth and in fact reversed in slope at some point in the soil, it becomes difficult to estimate the gradient accurately. Such a case has been demonstrated by Rolston et al. (1978) with measurements of N_2O and N_2 within a field soil actively undergoing denitrification. The largest concentrations of gas occurred at a shallow depth of only 50 mm below the soil surface. Thus, it was difficult to accurately evaluate or estimate the concentration gradient near the soil surface. Calculations of the soil-gas flux from Eq. [1] and [2] are limited to those cases where the concentration gradient at a specific depth can be adequately characterized. The method can also only be used under those conditions where the sources or sinks of the particular gas are negligible or zero over the depth where the concentration gradient is measured. This again points out the necessity of establishing small distances between sampling points so that the amount of gas produced or consumed over that depth increment can be considered to be small in relation to the amount of gas diffusing.

Concentrations must be measured in situ with sensors or by removal of small samples of gas at the depths of interest with subsequent analysis in the laboratory. Measuring concentration in situ is presently limited to O_2. For other gases, samples must be removed and analyzed with laboratory instrumentation.

47–2.2 Apparatus and Materials.

1. Samplers for gas withdrawal. A diagram of a representative sampler is given by Fig. 47–1. The internal gas volume of the sampler should be very small, inasmuch as the sampler must be purged before sample removal. The internal air volume of the sampler described by Fig. 47–1 is 0.8 mL/m of sampler length.
2. Gas syringes for removal of samples from soil. Disposable syringes, 1- or 2-mL, with a glass barrel and rubber plunger (Glaspak-Becton, Dickinson, and Co.). These syringes will hold samples for up to 12 h with negligible loss of gas. An alternative means of sampling the gases is with vacutainers (evacuated blood collection tubes—Sherwood Medical Industries). Vacutainers can hold samples for 7 days or more with negligible loss of gas. Syringes or vacutainers should be tested to determine if they contain contaminants such as organic chemicals from plastic parts, which may interfere with analyses.
3. Platinum electrode for in situ O_2 measurement. A description of the electrode and measurement system is given by McIntyre (1970) and also in chapter 49.
4. Gas chromatograph. Samples of gas removed from field samplers are transported to the laboratory in syringes or sample tubes. Gases such as O_2, CO_2, and N_2 can be analyzed with a thermal conductivity detector (Varian Model 920) or an ultrasonic detector (Tracor Model 150G). Nitrous oxide can be analyzed with a hot ^{63}Ni electron-capture

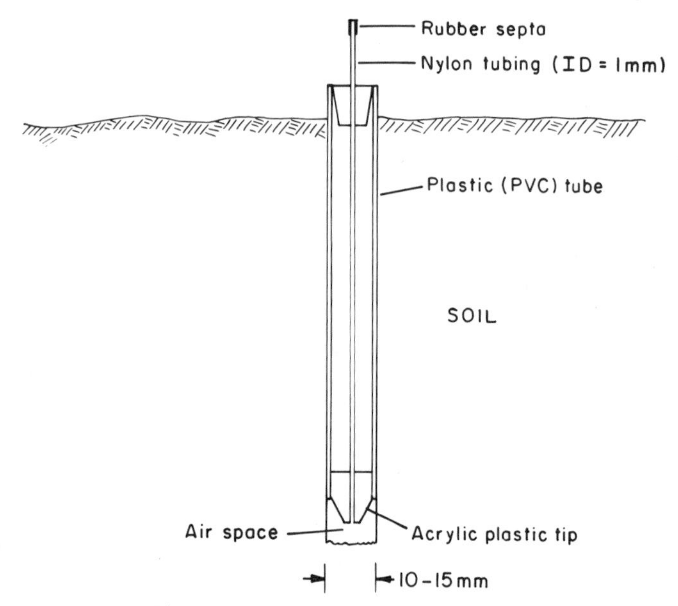

Fig. 47–1. Diagram of soil gas sampler.

detector (Packard Model 427). More detail on gas chromatography is presented by Bremner and Blackmer (1982).

47–2.3 Procedure

Install samplers in the field soil by augering (wood bit welded on pipe) a hole slightly smaller in diameter than that of the outside of the soil air sampler (Fig. 47–1). The depth of the hole should be slightly deeper than the desired depth of the sampler tip in order to provide an air space around the tip. Care should be taken that the tube of the sampler provides a tight fit with the soil so that leakage of gas does not occur around the outside of the sampler. The small inner tubing of the sampler should be slightly longer than the sampler and be sealed at the end with a rubber septum. If the flux is to be measured at the soil surface, the shallowest sampling depth should be near the soil surface (20 or 50 mm). It is advantageous to measure concentrations at several depths in order to improve the estimate of the concentration gradient at particular depths or at the soil surface.

Remove samples of gas by inserting the needle of the syringe through the rubber septum and withdrawing at least 1.5 times the volume of the sampler and discarding that amount from the syringe. It is advisable to then pull approximately another 0.5 times the volume of the sampler in order to clear the needle and any residual gas left in the syringe, and then discard the second small sample. The volume of the sample required for analysis as well as any extra that might be needed to purge the sample loop of the gas chromatograph should then be taken. The needle of the syringe should be inserted into a solid rubber stopper in order to prevent leakage of gas from the syringe. Alternatively, the final sample can be removed with a vacutainer.

Transport the gas samples to the laboratory and analyze for the concentration of the particular gas of interest. See Bremner and Blackmer (1982) for methods of analyses. Analyses should be conducted within a few hours after sampling.

Plot the concentration as a function of soil depth if concentrations at several depths have been measured. For several samples taken at any one particular depth, an arithmetic mean should be determined. If enough samples are available to determine the frequency distribution (normal, log normal, etc.) the mean should be calculated using the correct frequency distribution.

From the means of concentration at particular depths, determine the concentration gradient from the slope of a plot of concentration vs. x if a smooth curve has been drawn or smoothed mathematically (Kimball, 1976; Synder, 1976; Erh, 1972) through the averages of several depths. If only two depths are available, draw a straight line through the concentrations at the two depths and calculate the slope ($\Delta c/\Delta x$).

Calculate the flux at several different times by substitution of the measured concentration gradients for those times into Eq. [1]. In order to determine the total amount of gas diffusing from the soil over some time period t_1, Eq. [2] is used with measured concentration gradients at several different times. If it is reasonable to assume that the soil-gas diffusivity remains constant over the time period t_1, then D_p can remain outside the integral. However, if D_p is varying as a function of time, it must also be included within the integral. The value of the integral is approximated by the area beneath a plot of flux vs. time.

47–2.4 Comments

The number of samplers placed at a particular depth will be dependent upon the variability of the gaseous concentration and the area over which the flux is estimated. Rolston (1978) presented data giving the uncertainty in measuring the concentration at a particular depth from five samplers placed within an area of 1 m^2. He measured concentrations of N_2O in a wet soil profile with five samplers at each depth. The 95% confidence interval was approximately 100% of the mean N_2O concentration. For drier soils, one would expect that the variability of a gaseous constituent would be less than that in a wet soil. The number of samplers needed will be dependent upon the gas of interest and soil-water content at the time of analysis.

The major errors and uncertainties associated with calculating flux or total amount from Eq. [1] and [2] occur due to the variability of the soil-gas diffusivity and concentration gradients within the soil, inability to accurately determine the concentation gradient near the soil surface for cases where surface flux is required, and production or consumption of the gas of interest over the depth interval for which the concentration gradient is measured. Other errors may also result from changes in the soil-gas diffusivity over the time period of the measurement due to drying of the soil or redistribution of the soil water; changes in temperature over the time period of the flux measurement; errors in removing gas samples, especially in very wet soils, resulting in mass flow of gas; and additional gas exchange resulting from pressure fluctuations, which result in greater than normal diffusion. The greatest uncertainties in calculating the soil-gas flux using Eq. [1] occur due to the uncertainties and variability of the soil-gas diffusivity and concentration gradients within field soils. This uncertainty can be especially large for soils close to saturation, which are of interest in trying to calculate flux of denitrification gases occurring under anaerobic or near-anaerobic conditions. Associated with the variability in the concentration gradient is the inability to adequately determine the slope and shape of the concentration gradient very near the soil surface.

The primary advantage of this method is that it does not greatly disturb the soil environment as does the placement of chambers over the soil surface. If samples are taken at several depths, the method also gives information about concentrations within the soil and zones of sources and sinks. The major disadvantage of the method is due to the uncertainty associated with measuring the concentration gradients and the soil-gas diffusivities in spatially and temporally variable soil profiles. The method also requires a large number of samples to adequately determine soil-gas diffusivities and concentration gradients.

The variability of gas fluxes in field soils can be expected to be different by possibly orders of magnitude, due to local differences in porosity, texture, and water content. All methods will result in uncertainties and variability in sampling field soils, although the variability using the method of calculating flux from the diffusion coefficient and concentration gradient will inherently be greater than from other methods due to the scale of measurement.

47–3 CLOSED CHAMBER METHOD

47–3.1 Principles

In this method, a closed chamber is placed over the soil surface and the increase in concentration of gas within the chamber is measured as a function of time. The gas within the chamber must be well mixed so that concentration gradients do not exist. For small chambers, this mixing is usually accomplished by diffusion. For large chambers, a small fan within the chamber may be needed. The flux of gas at the soil surface is calculated from

$$f = (V/A)\,(\Delta c/\Delta t) \qquad [3]$$

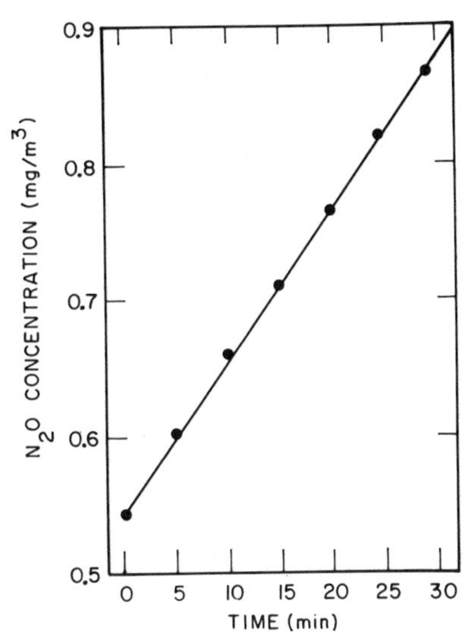

Fig. 47–2. Concentration of gas within closed chamber as a function of time. Redrawn from Matthias et al. (1980).

where f is the flux density of gas (g gas m^{-2} soil s^{-1}), V is the volume of air within the chamber (m^3 air), A is the area of the soil within the chamber (m^2 soil), and $(\Delta c/\Delta t)$ is the time rate of change of gas concentration in the air within the chamber (slope of line of Fig. 47–2) with units of g gas m^{-3} air s^{-1}.

This method measures the surface flux directly and is not dependent upon measuring individual components to calculate the flux as is the previous method (section 47–2). However, this method can only be used to determine the flux at the soil surface.

In theory, the method can only approximate the real soil-gas flux for measurements taken over short time periods after placing the cover over the soil surface. As soon as a small amount of gas diffuses into the chamber, the concentration at the soil surface increases, the concentration gradient decreases, and the flux must also decrease. For large times, with a constant soil-gas diffusivity and gas production, it is expected that the concentration of gas within the chamber would become constant with time. Thus, it is critical that the cover be placed over the soil surface for only small enough times (15 min to 3 h) to obtain a few data points. In practice a closed chamber is placed over the soil surface, the increase in the gas of interest within the chamber is measured over a short time period, and the chamber removed. This procedure would be repeated several times in order to estimate fluxes during a daily or longer time period.

47–3.2 Apparatus and Materials

1. Closed chamber. A chamber developed by Matthias et al. (1980) is illustrated diagrammatically in Fig. 47–3. It consists of an insulated,

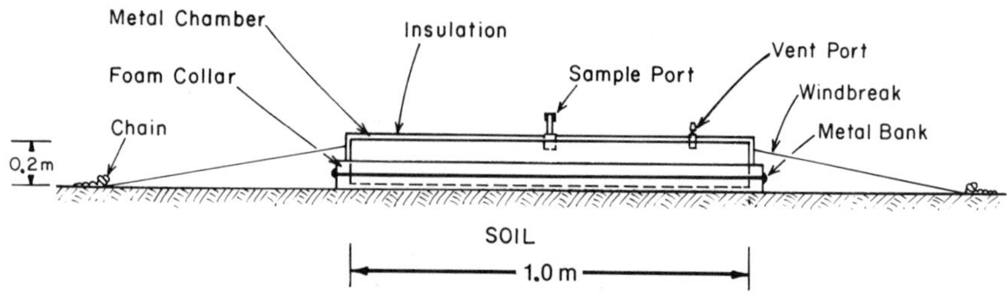

Fig. 47–3. Diagram of closed chamber for directly measuring gas flux at soil surface. Redrawn from Matthias et al. (1980).

cylindrical, open-bottom chamber (diameter 0.88 m, height 0.17 m) fabricated from 16-gauge galvanized sheet metal. The bottom of their chamber circumscribes an area of 0.6 m^2. The chamber is insulated with white styrofoam 20 mm thick and is fitted with a high-density polyurethane foam collar (50 mm by 0.1 m) covered with 6-mil polyethylene film. This collar is attached to the chamber by means of a tight-fitting metal band that encircles the collar and chamber. The polyurethane foam collar insures a good seal between the chamber and the soil surface. The chamber is fitted with a windbreak fabricated from 6-mil polyethylene film, in order to minimize wind-induced movement of air into and out of the chamber. This windbreak extends from the top of the chamber to the soil surface, about 0.8 m outside the metal chamber. It is held to the soil surface by means of a 9-mm link chain placed along the perimeter of the windbreak. The top of the chamber is fitted with an air sample port and a vent port. The air sample port either can be connected to air sample bottles as described by Matthias et al. (1980) or can be used to obtain one or several samples of the gas within the chamber using syringes or vacutainers. The vent port was used by Matthias et al. (1980) to minimize mass flow of soil air during removal of air samples from the chamber, inasmuch as they removed 1.0 L of gas at each sampling. The chamber design of Matthias et al. (1980) may possibly be improved by incorporating the vent design of Hutchinson and Mosier (1981), which allows equilibration of pressure fluctuations yet minimizes contamination from ambient air.

Other types of closed chambers may be more appropriate for specific applications. For instance, Rolston et al. (1978) used a chamber consisting of a thick sheet of acrylic plastic with rubber tubing on the lower edge to make an airtight seal with the top of a wood border placed into the soil to a depth of 0.6 m. Other chamber designs were described by Findlay and McKenney (1979), Terry et al., (1981), and Seiler and Conrad (1981).

2. Gas sampling devices. Sampling and storage devices can be either syringes or vacutainers as described in section 47–2.2, or glass bottles fitted with stopcocks as described by Matthias et al. (1980).
3. Instrumentation for measuring gas concentrations. See section 47–2.2 and Bremner and Blackmer (1982).

47–3.3 Procedure

Assemble the chamber as shown in Fig. 47–3 and calculate the volume of the chamber.

Place the chamber on the soil surface. Start a timer and place the windbreak on the chamber. The temperature of the air within the chamber should also be measured in order to calculate mass per unit volume of chamber air.

Remove samples of gas from the chamber using either syringes, vacutainers, or evacuated glass bottles (Matthias et al., 1980). For sampling by syringes or vacutainers, it is desirable to remove three or four samples from the chamber in order to insure repeatability of the analyses.

The time interval for removal of samples from the chamber is dependent upon the flux of gas from the soil and the sensitivity in measuring the particular gas of interest. The gas within the chamber should be sampled several times (3–6) in order to plot concentration as a function of time, to be used in calculating the gas flux. The chamber should be left on the soil surface long enough to obtain a few data points as a function of time, yet not long enough to result in a flattening of the curve and a decrease in flux.

Transport the air sample bottles, vacutainers, or gas syringes to the laboratory. Determine gas concentrations in samples by methods given in section 47–2.2 and Bremner and Blackmer (1982).

Plot data with concentration on the ordinate and time on the abscissa according to the example given in Fig. 47–2. Only those data should be used which can be approximated by a straight line. If later data points systematically deviate from the straight line due to the theoretical flattening of the curve for large times, an error will result if a straight line is drawn through all data points. Calculate the flux from Eq. [3].

47–3.4 Comments

Potential major errors associated with the closed chamber technique are (i) underestimation of the gaseous flux due to the increase in concentration within the cover with time, (ii) changes in the soil environment (decreased evaporation, changes in temperature, and lack of exchange of gases), and (iii) insufficient mixing of gas within the chamber, which would result in gradients and errors in sampling. The most serious error probably results from leaving the chamber on the soil surface too long, resulting in changes in the soil environment and an increase of gas concentration within the chamber, which causes a significant decrease in the flux. Rolston et al. (1978) proposed a technique for correcting the flux due to the concentration increase that involved sampling at one depth within the soil. Hutchinson and Mosier (1981) developed a correction method that involved sampling the concentration within the chamber at three times separated by equal time intervals. If the plot of concentration within the chamber as a function of time is not linear, these correction techniques should be employed.

If the chamber of Mathias et al. (1980) is used, another possible error may occur if the windbreak is not used around the chamber. This is especially serious when wind is blowing across the soil surface, resulting in pressure fluctuations and turbulence around the chamber and causing mass flow of gas beneath the chamber. Matthias et al. (1980) also point out that errors could result if the chamber is not insulated. The temperature within an uninsulated chamber can change over a short time period and cause changes in pressure, biological activity, and the flux of gas from the soil. Errors would also occur if chamber materials adsorbed any of the soil gases of interest.

Alternate chamber designs are feasible. For instance, the volume of the chamber and area of soil covered are flexible, although it is important to choose the area large enough to encompass the soil structural units and to make the chamber height as small as possible to maximize sensitivity. The sensitivity for measuring the flux will be greater for a small chamber volume than for a large volume. Matthias et al. (1980) were able to use a fairly large chamber volume for measuring N_2O flux, since precision of measuring N_2O in air is great. However, precision for measuring other gases may not be as great and would require keeping the chamber height small in order to increase sensitivity. Chambers can be constructed from various materials as long as the chamber is insulated and is designed so that a tight seal can be attained with the soil surface.

The time that the chamber must be left on the soil surface will be dependent upon specific gases and situations. For measuring the flux of N_2O, Matthias et al. (1980) left the chamber on the soil surface for 20 min with samples removed at 5- min intervals. However, analytical techniques for measuring N_2O are very sensitive. Such sensitivity may not exist for other gases. For instance, if one is interested in measuring N_2 tagged with ^{15}N, the required time period that the cover would remain in place would be 1.0 to 4.0 h even with high ^{15}N enrichments of applied fertilizer (Rolston et al., 1978).

The advantages of this particular closed chamber technique are (i) it does not cause significant disturbance of the soil texture, structure, or bulk density by driving chambers or tubes into the soil, (ii) it is not limited to sites where electricity or special equipment are available, (iii) the chamber is inexpensive to fabricate, transport, and use, and (iv) the method measures the gaseous flux directly, without the need to sample the gases within the soil or measure transport parameters such as the soil-gas diffusivity.

47–4 FLOW-THROUGH CHAMBER METHOD

47–4.1 Principles

In this method a chamber is formed over the soil by driving a cylinder (closed at one end) a few millimeters into the soil surface. Air is caused to flow through the chamber, and the increase in concentration of the gas in the effluent stream is measured. The flux of gas at the soil surface is calculated from the increase in concentration of the particular gas in the effluent stream over that in the influent stream. In addition to the concentration increase of gas within the effluent stream, the flow rate of gas through the chamber must also be accurately measured.

The gas flux from the soil within the chamber is calculated from

$$f = \delta v / A \quad [4]$$

where f is the soil gas flux density (g gas m^{-2} soil s^{-1}), A is the cross-sectional area of the soil (m^2 soil), δ is the difference in concentration of gas between the ambient air and the effluent flow stream (g gas m^{-3} air), and v is the volume flow rate of air through the chamber (m^3 air s^{-1}). The value of δ should be taken after equilibration has been attained.

47–4.2 Apparatus and Materials

1. Flow-through chamber. The suggested chamber (Fig. 47–4) is similar to that described by Denmead (1979). It consists of an open-ended cylinder of steel or material strong enough to be forced into the soil. The cylinder has a sharpened cutting edge on the bottom and a reinforced flange on the top. The flange is used for driving the cylinder approximately 0.1 m into the soil so as to isolate a small soil column. The chamber has a Plexiglas lid, which is supported 0.05 mm above the cylinder flange by metal shims at eight points around its circumference. A cylindrical baffle extending to 0.5 mm above the ground surface is cemented inside the lid, leaving 2.5 mm clearance between the baffle and the cylinder wall. When outside air is drawn continuously through the chamber, air enters through the gap between the lid and the flange and is drawn under the baffle and through the chamber to an exit port in the lid.

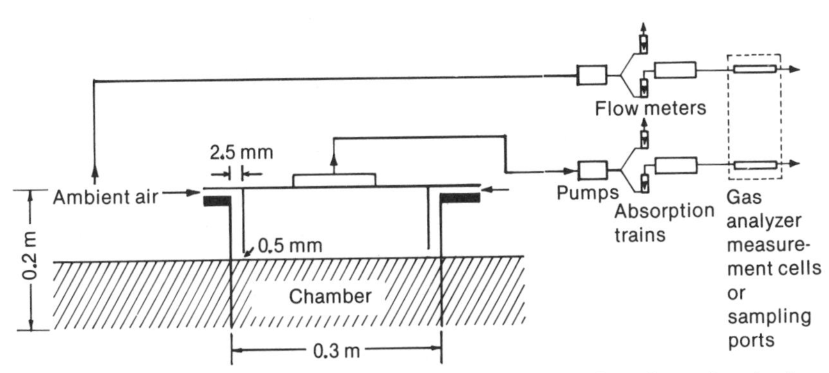

Fig. 47–4. Diagram of installation for measuring gas flux at soil surface using the flow-through system. Redrawn from Denmead (1979).

2. Pumps. Tubing is connected from the exit port of the chamber to a pump (Fig. 47–4). Various kinds of pumps may be used for drawing air through the chamber, such as Masterflex tubing pumps.
3. Flow meters. The tubing on the exit side of the pump is split, dividing the flow to two flow meters. One meter is vented to the atmosphere. Air flowing through the other meter passes to a gas analyzer or sampling ports. The flow meter arrangement allows the choice of air flow rate through the chamber while maintaining a constant flow of air to the analyzer or sampling points. Flow meters can be of two types, rotameters (Matheson) or linear mass flowmeters (Matheson). The mass flowmeters are more accurate and more expensive than the rotameter type.
4. Adsorption trains. Depending upon the gas of interest and the analysis technique, it may be necessary to place adsorption trains between the flow meters and the gas analyzer in order to eliminate gases that would interfere with the analysis of a particular gas. For an infrared analyzer for instance, and with the gas of interest being N_2O, it would be necessary to remove H_2O vapor, CO_2, and CO from the effluent stream, since those gases would interfere with the N_2O analysis. Denmead (1979) used a train of absorbers consisting of tubes of magnesium perchlorate (to remove H_2O vapor) and ascarite (to remove CO_2), a silver oxide catalyst furnace (LEDCO, Model 507-100) operating at 250°C (to convert CO to CO_2), and a further tube of ascarite.
5. Gas analysis system. After the gas from the chamber flows through the flow meters (and possibly absorption trains), its concentration is measured by a gas analyzer or samples of gas are removed with syringes or vacutainers from the flow stream in order to be analyzed later by gas chromatography. The concentration of the ambient air entering the chamber must also be measured either by alternately analyzing ambient air with a similar flow system or by sampling the ambient air concentration of the gas of interest. An infrared gas analyzer (Denmead, 1979) can be used for analyzing CO_2 or N_2O. Infrared gas analyzers can be purchased from several companies (Beckman, Infrared Industries Inc., Horiba Instruments, Inc.). In general, other gases would be measured by gas chromatography, which would require removal of samples of gas from the flow stream with syringes or vacutainers, transporting samples to the laboratory, and analyzing the gas by injection into a gas chromatograph (see section 47–2.2 and Bremner & Blackmer, 1982). The volume of syringes or vacutainers needed for removing samples from the effluent flow stream will depend upon the size of sample required to provide an amount of gas within the range of sensitivity of the gas chromatographic analysis.

47–4.3 Procedure

Drive the metal cylinder approximately 0.1 m into the soil. Place the cover on the chamber and connect tubing to the outlet port, pump, flow

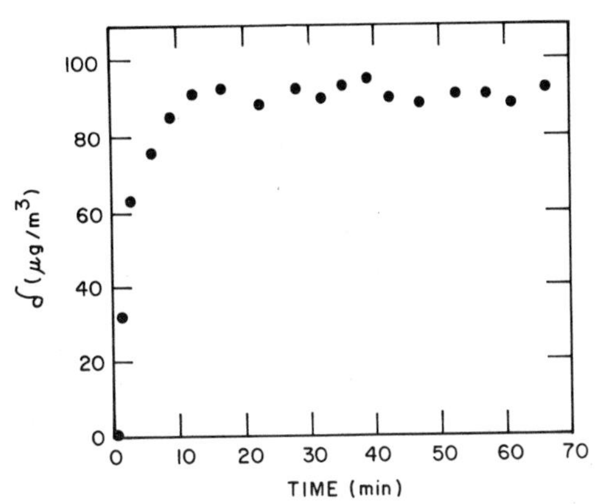

Fig. 47–5. Difference in gas concentration between the influent and effluent flow streams as a function of time for a flow-through system. Redrawn from Denmead (1979).

meters, and analyzers or sampling ports according to the diagram of Fig. 47–4.

Begin flowing air through the chamber and flow system. Measure the flow rate of the effluent stream accurately. When the gas flow through the chamber is initiated, there will be a period during which the concentration of gas in the soil and in the chamber air adjust to new equilibrium values different from those before the chamber was placed over the soil surface. Estimates of the steady flux during this time period before equilibration will be erroneous. The time required to attain equilibrium will be dependent upon the chamber volume and the flow rate of gas through the system. For a chamber volume of 0.007 m^3 and a flow rate of 17 and 50 mL s^{-1} through the chamber, the equilibration time was about 35 and 12 min, respectively, for the design of Denmead (1979). To determine the appropriate flow rate for the particular chamber volume, plot the change in concentration of the gas in the effluent stream as a function of time and determine the amount of time before equilibration is reached. An example of the change in concentration of the effluent stream as a function of time is given by Fig. 47–5 (Denmead, 1979).

Analyze the concentration of gas in the effluent stream with an infrared analyzer, or sample the gas stream with syringes or vacutainers. Also measure the gas concentration in the ambient air at points near where the ambient air enters the chamber.

Plot the concentration of gas in the effluent stream or the increase or decrease in concentration above ambient concentration in the effluent stream as a function of time.

The gas flux from the soil within the chamber is calculated from Eq. [4]. The value of δ should be taken from that part of the curve of Fig. 47–5 after equilibration has been attained.

47–4.4 Comments

The potential errors that may result from the flow-through chamber technique are (i) those due to mass flow of gas to or from the soil caused by pressure differences between the chamber and the soil gas phase, (ii) those incurred from measuring the change in concentration of gas within the effluent stream before equilibration has been attained, (iii) those caused by changes in soil temperature and other soil environmental conditions within the chamber, which result from leaving the chamber over the soil surface for long time periods, and (iv) those caused from inaccurate mesurement of gas flow rate. The error due to pressure differences within the chamber and mass flow of gas can be minimized by insuring that the area for the entry of ambient air into the chamber is relatively large. The system designed by Denmead (1979) results in a large area for the influent gas to enter the chamber, which results in small pressure differences between the chamber air and the soil air. If one is unsure whether significant pressure differences may result for a particular chamber design, measurements of pressure differences should be made between the chamber air and soil air or the possible pressure change calculated from Poiseuille's equation for flow through an orifice. Kanemasu et al. (1974) and Kimball and Lemon (1971) showed that pressure deficits of 1 Pa resulted in significant mass flow. Thus, an allowable pressure deficit must be much smaller than 1 Pa. The importance of mass flow induced by pressure deficits may also be evaluated by determining if the flux from the soil is indeed different for different flow rates of gas through the chamber.

The error due to nonattainment of equilibrium can be minimized by insuring that equilibration within the chamber has occurred by plotting effluent concentration vs. time and using only the value of δ after equilibration occurs. Changes in soil environmental conditions can be minimized by leaving the chamber on the soil surface for the minimum time required to attain a measurement of flux.

In practice, the choice of air flow rate will be a compromise between the need for rapid equilibrium and the sensitivity of the gas concentration measurements. The amount of gas evolving from the soil will determine the flow rate required for the particular analysis technique. For small rates of evolution of gas from the soil, small flow rates would be required in order to obtain concentrations in the effluent stream large enough to measure. However, for small flow rates, the equilibration time would be large. Some trial and error may be required to determine the best compromise between flow rate, chamber volume, and sensitivity of the flux measurement.

Several alternate system designs are feasible, such as chamber volume and area of soil covered. Ryden et al. (1978) used a large, single orifice for the entry of gas into the chamber. Nitrous oxide gas in the effluent stream was trapped on molecular sieve 5A and the trap transported to the laboratory for analyses. The methods for sampling or analyzing the gas within the effluent stream are many and are dependent upon the particular gases of interest.

An advantage of the open flow-through chamber technique is the maintenance, within the chamber and the soil, of environmental conditions nearly the same as those in the field. Since ambient air is flowing into the chamber, temperature, O_2, and H_2O vapor can be maintained at levels approximately equal to those of the air outside the chamber. The problems associated with gas build-up beneath a closed chamber do not occur with the open flow-through system.

A disadvantage of the system is that it requires a pump to cause flow of gas through the chamber. This requirement limits the technique to areas where electricity is available or to situations where small battery-operated pumps can be obtained.

Another disadvantage of the flow-through chamber technique is that by flowing gas through the chamber, dilution of the particular gas of interest occurs and may reduce concentration levels in the effluent stream below measurable levels. Thus, the ability to measure small fluxes is decreased. This becomes particularly important in attempting to measure the flux of gases where the analytical capability is not great enough to detect small concentrations of gas above or below ambient levels. Trial calculations using Eq. [4] will give an estimate of the analytical capability or flow rate required to measure a minimum gas flux.

47-5 COMMENTS ON MECHANISMS CAUSING MASS FLOW

The three methods discussed above are based upon the assumption that interchange of gas from the soil to the atmosphere can be attributed to diffusion processes. Other mechanisms, however, may cause flow or transport of gas within or across the soil surface to be greater than that due to diffusion alone. Pressure fluctuations due to barometric pressure changes and air turbulence effects due to wind can result in mass flow and dispersion of soil gases. The infiltration of rain and irrigation water into soil results in mass flow of gas ahead of the wetting front. The redistribution and drainage of soil water results in mass flow of gas causing transport greater than that from molecular diffusion alone. Extreme differences in temperature would also be expected to result in pressure differences, resulting in mass flow. A review of mechanisms causing mass flow has been made by Kimball (1983).

For high wind speeds across the soil surface, Kimball and Lemon (1971) determined that the effects of air turbulence on gas exchange increased the transport of gas over that due to diffusion alone, especially for coarse-textured surface materials such as straw and coarse sand. Kimball (1973) found that the average effective diffusion coefficient for water vapor during afternoon periods was 1.26 greater than the molecular diffusion coefficient.

Farrell et al. (1966) also evaluated vapor transport in soil due to air turbulence and concluded that air turbulence effects could lead to gas exchange greater than that due to diffusion alone. Scotter et al. (1967) evaluated effects of cyclic flow on transport in addition to that transport arising from diffusion. Whenever mass flow occurs due to pressure fluctuations, wind turbulence, thermal gradients, and water movement, the increased mixing that occurs over that due to molecular diffusion is dispersion. Dispersion coefficients of Scotter et al. (1967) increased by as much as 50% over molecular diffusion coefficients for porous surface soils. Simple methods are not yet available to quantitatively predict the effect that air turbulence may have on gas exchange. However, data indicate that for soils moist enough to support plants, winds will normally have little effect on soil-gas movement (Kimball, 1983).

Kimball (1983) concludes that barometric pressure fluctuations may cause changes in the diffusive flux of up to 60% in soils with no impermeable barriers at a depth of < 100 m. For soils with < 15 m to a water table or other barrier, barometric pressure changes probably have little effect on gas exchange rates.

47-6 REFERENCES

Bremner, J. M., and A. M. Blackmer. Composition of soil atmospheres. *In* A. L. Page (ed.) Methods of soil analysis, Part 2. Agronomy 9:873–902.

Denmead, O. T. 1979. Chamber systems for measuring nitrous oxide emission from soils in the field. Soil Sci. Soc. Am. J. 43:89–95.

Denmead, O. T., R. Nulsen, and G. W. Thurtell. 1978. Ammonia exchange over a corn crop. Soil Sci. Soc. Am. J. 42:840–842.

Erh, K. T. 1972. Application of spline function to soil science. Soil Sci. 114:333–338.

Farrell, D. A., E. L. Greacen, and C. G. Gurr. 1966. Vapor transfer in soil due to air turbulence. Soil Sci. 102:305–313.

Findlay, W. I., and D. J. McKenney. 1979. Direct measurement of nitrous oxide flux from soil. Can. J. Soil Sci. 59:413–421.

Hutchinson, G. D., and A. R. Mosier. 1981. Improved soil cover method for field measurement of nitrous oxide fluxes. Soil Sci. Soc. Am. J. 45:311–316.

Kanemasu, E. T., W. L. Powers, and J. W. Sij. 1974. Field chamber measurements of CO_2 flux from soil surface. Soil Sci. 118:233–237.

Kimball, B. A. 1973. Water vapor movement through mulches under field conditions. Soil Sci. Soc. Am. Proc. 37:813–818.

Kimball, B. A. 1976. Smoothing data with cubic splines. Agron. J. 68:126–129.

Kimball, B. A. 1983. Canopy gas exchange: gas exchange with soil. p. 215–226. *In* H. M. Taylor et al. (ed.) Limitations to efficient water use in crop production. American Society of Agronomy, Madison, WI.

Kimball, B. A., and E. R. Lemon. 1971. Air turbulence effects upon soil gas exchange. Soil Sci. Soc. Am. Proc. 35:16–21.

Matthias, A. D., A. M. Blackmer, and J. M. Bremner. 1980. A simple chamber technique for field measurement of emissions of nitrous oxide from soils. J. Environ. Qual. 9:251–256.

McIntyre, D. S. 1970. The platinum microelectrode method for soil aeration measurement. Adv. Agron. 22:235–283.

Rolston, D. E. 1978. Application of gaseous-diffusion theory to measurement of denitrification. p. 309–335. *In* D. R. Nielsen and J. G. MacDonald (ed.) Nitrogen in the environment, Vol. 1. Nitrogen behavior in field soil. Academic Press, New York.

Rolston, D. E., D. L. Hoffman, and D. W. Toy. 1978. Field measurement of denitrification: I. Flux of N_2 and N_2O. Soil Sci. Soc. Am. J. 42:863–869.

Ryden, J. C., L. J. Lund, and D. D. Focht. 1978. Direct in-field measurement of nitrous oxide flux from soils. Soil Sci. Soc. Am. J. 42:731–737.

Scotter, D. R., G. W. Thurtell, and P. A. C. Raats. 1967. Dispersion resulting from sinusoidal gas flow in porous materials. Soil Sci. 104:306–308.

Seiler, W., and R. Conrad. 1981. Field measurement of natural and fertilizer-induced N_2O release rates from soils. J. Air Pollut. Control Assoc. 31:767–772.

Synder, W. M. 1976. Interpolation and smoothing of experimental data with sliding polynomials. USDA-ARS, Bull. ARS-S-83. U.S. Government Printing Office, Washington, DC.

Terry, R. E., R. L. Tate, and J. M. Duxbury. 1981. Nitrous oxide emissions from drained, cultivated organic soils of south Florida. J. Air Pollut. Control Assoc. 31:1173–1176.

48 Air Permeability

A. T. COREY
Colorado State University
Fort Collins, Colorado

48-1 INTRODUCTION

Two fluid phases, i.e., a water solution and a gaseous mixture, normally exist within the root zone of soils. In any dynamic process involving a change in water content, a portion of one phase is displaced by the other. The relationships between water content and the permeabilities for water and air are factors that should be considered in a complete analysis of fluid displacements in soils.

Soil scientists frequently analyze bulk water transport without reference to the simultaneous bulk transport of soil air. This is because the viscosity of air is small compared to that of water, permitting air to move into or out of soil with a negligible pressure gradient. Consequently, air often remains at nearly atmospheric pressure everywhere within an interconnected gas phase so that the flow of air has a negligible effect on the flow of water. For this reason, permeability to water as a function of water content has been studied much more often than has permeability to air.

Although pressure gradients in soil air undoubtedly have little effect on fluid displacements in many cases, it is not always possible to ignore this factor. Resistance to air flow is sometimes significant during infiltration. Dixon and Linden (1972), for example, have reported the effect of air compression ahead of a wetting front during infiltration under border irrigation. Youngs and Peck (1964), Peck (1965), McWhorter (1971), and Morel-Seytoux and Khanji (1974) also have investigated the effect of air resistance on infiltration.

The effect of air resistance during drainage is probably less obvious than during infiltration. However, Corey and Brooks (1975) have found that water in draining soil columns develops pressure surges that can be associated with air entry through restrictions within the soil pore space.

In addition to its significance in the analysis of fluid displacements, the relationship between water content and air permeability may have indirect significance in relation to the transport of soil gases by molecular diffusion. Diffusion of gases is affected by several of the factors that affect permeability to air, e.g., soil porosity, water content, and the continuity and tortuosity of the air-filled pore space.

Measurement of air permeability k_a as a function of water content Θ is, therefore, a promising tool for obtaining new information useful in analyzing problems of aeration as well as infiltration and drainage in soils. However, it is a tool which soil scientists seldom use and has never been developed for routine use.

48-2 PRINCIPLES

Air permeability k_a is defined as the coefficient in the air-flux equation in the form

$$q_a = -(k_a/\eta_a) \nabla (p_a + \rho_a g h)$$

in which

q_a = volume flux per unit area, dimensions Lt^{-1},
η_a = viscosity of air, dimensions FtL^{-2},
p_a = pressure of air, dimensions FL^{-2},
ρ_a = density of air, dimensions ML^{-3},
g = acceleration of gravity, dimensions Lt^{-2} or FM^{-1}, and
h = elevation, dimension L.

Because of the small value of ρ_a, the term $\rho_a g h$ is usually neglected so that the flux equation is written as

$$q_a = -(k_a/\eta_a) \nabla p_a .$$

The coefficient k_a has dimensions of L^2, the usual SI unit being the square micrometer. It is a function of water content and geometry of the pore space. Fluid properties theoretically affect k_a only indirectly as they affect the geometry of the pore space. Geometric factors include the size-distribution of air-filled pore space as well as the continuity and tortuosity of this space.

Theoretically, k_a for a dry, stable, porous medium should be equal to the intrinsic permeability k_w with a liquid at the maximum liquid content. Actually, air at atmospheric pressure does not act as a true fluid continuum in soils, so that the fluid velocity is not zero at solid boundaries as is the case with liquids. Consequently, k_a for a dry soil is always larger than k_w at the maximum liquid content. This fact has been pointed out by Klinkenberg (1941), and the phenomenon is known as "gas slippage" or the "Klinkenberg effect."

Klinkenberg (1941) has shown how the effect of gas slippage can be evaluated by measuring k_a for a range of mean pressures and extrapolating the curve of the results to an infinite mean pressure. The value of k_a found by extrapolation equals the value of k_w obtained with liquids that do not react with the solid matrix. The extrapolated value of k_a is often smaller than k_a found at atmospheric pressure by 20 to 30%. However, with soils having large specific surfaces, the difference may be much greater than 20 to 30%.

When water is used to determine the intrinsic permeability of soils, the value obtained is usually substantially smaller than even the value of k_a corrected for gas slippage. This is because of the interaction between water and soil solids, mainly the clay. Reeve (1953) has employed the ratio k_a/k_w to evaluate the stability of soils in the presence of water.

The theory of how transport mechanisms of two fluid phases are interrelated in displacement processes was presented by Buckley and Leverett (1942). Their analysis implies that the relative importance of air resistance in a water-air displacement is indicated by the ratio

$$\eta_a k_w/\eta_w k_a$$

in which η_a and η_w are the viscosities of air and water, and k_a and k_w are the permeabilities of air and water, respectively.

The fact that k_w increases and k_a decreases as the water content, Θ, increases was first demonstrated experimentally by Wyckoff and Botset (1936). Because of this, the ratio k_w/k_a becomes large as Θ approaches its maximum field value. Consequently, soil air may develop a significant pressure gradient in spite of the small viscosity of air when air is displaced through regions of high water content. Therefore, the relationship $k_a(\Theta)$, particularly in the range of large Θ, is a factor of importance in the analysis of fluid displacements in soils. In particular, the minimum value of Θ at which $k_a = 0$ is often a significant factor in a displacement process. This value of Θ, i.e., the value corresponding to the "air-entry pressure," is often about 80 to 90% of the maximum value of Θ.

48–3 METHODS

The first experiments in which $k_a(\Theta)$ was measured were conducted by Wyckoff and Botset (1936). The measurements were made by forcing the flow of gas-liquid mixtures through long tubes of unconsolidated sands, with varying ratios of flow rates for the two fluids. In this way, both $k_a(\Theta)$ and $k_w(\Theta)$ were obtained simultaneously. The original procedures, however, were for samples much larger than those that can usually be removed from a well bore or from the field.

Methods for measuring $k_a(\Theta)$ can be classified broadly as unsteady-state methods or steady-state methods. The unsteady-state methods referred to here are to be distinguished from the method of Kirkham (1947), which employed a falling-pressure permeameter to measure k_a at an existing value of Θ in the field. The latter was not intended for measuring k_a with Θ as an explicit variable.

Unsteady-state methods for obtaining $k_a(\Theta)$ involve calculations based on measurements made during a fluid displacement process, i.e., during a change in Θ. Steady-state methods involve direct measurements of k_a at particular values of Θ while Θ is held constant and uniform along the axis of flow in the sample.

48–3.1 Unsteady-state Methods

The best known of the unsteady-state methods is the Welge (1952) technique, which depends upon an integration of the Buckley-Leverett displacement equation (1942). With this method, a wetting fluid is displaced from a sample by a nonwetting fluid under an extremely large pressure gradient. The pressure gradient is large enough to permit the assumption that the two fluids flow through the sample under the same pressure gradient, even though the fluid content is nonuniform.

Values of k for each fluid are determined from the instantaneous rates at which each of the two fluids emerges from the outflow end of the sample. The corresponding "wetting-phase" contents are those calculated for the outflow face of the sample by the Welge integration.

More recently, Jones and Roszelle (1978) have employed an analysis which they called the "principle of linear scalability," to simplify the mathematics and expedite the calculations involved in the determination of permeabilities from displacement experiments on laboratory samples. The analyses of Welge and Jones and of Roszelle have been described in the text *Mechanics of Heterogeneous Fluids in Porous Media* by Corey (1977).

Unsteady-state methods have important advantages in that permeabilities for both fluid phases as functions of liquid content can be determined in a short time. However, because the procedure developed by Welge requires extremely large pressure gradients, it is suitable for displacing fluids of high viscosity only. When air is used to displace water, the fluids emerge at high velocity and difficulties are encountered in making the necessary measurements with the required speed and precision. However, it is conceivable that some other unsteady-state procedure could be devised that would be more satisfactory for measuring $k_a(\Theta)$ of soils.

48–3.2 Steady-state Methods

An important requirement for the direct steady-state measurement of $k_a(\Theta)$ is to maintain a uniform Θ along the axis of flow. Since Θ is a function of the pressure difference between air and water, the pressure gradient in the two fluids must be equal in magnitude and direction.

One method, called the "stationary-liquid method" involves air flowing upward through a sample in response to a pressure gradient equal to that in the static liquid. The latter method was pioneered by Osoba et al. (1951) for consolidated porous rock samples. It was first used for disturbed soil samples by Brooks and Corey (1964).

Another method involves both fluids flowing with equal pressure gradients. It was pioneered in the petroleum industry by Hassler (1944) for measurement of both k_a (Θ) and K_w (Θ) on small cores of porous rocks. The method has been described by Osaba et al. (1951). It was first used on a soil sample by Corey (1957) and later used in a modified form by Brooks and Corey (1966).

48-3.2.1 APPARATUS FOR STATIONARY-LIQUID METHOD

The apparatus described below was originally designed by Brooks and Corey (1964) for disturbed soil samples using a hydrocarbon liquid as the wetting fluid. The equipment is a modification of that used earlier by Corey (1954) for measurement of k_a (Θ) for porous rock samples and by Corey (1957) for obtaining k_a (Θ) on a soil sample molded in an acrylic plastic cylinder.

A schematic illustration of the overall flow system is shown in Fig. 48-1. Details of the sample holder and flow meter are shown in Fig. 48-2 and 48-3, respectively. The flow meter is made of glass, except for the teflon plug in the lower stopcock. Materials used for the construction of the sample holder are indicated on Fig. 48-2.

A suitable flow meter is a critical part of the entire apparatus, especially for determining the minimum value of Θ at which k_a exists. A version of a soap-film flow meter designed by Corey (1954) specifically for this purpose is illustrated in Fig. 48-3. It employs two graduated tubes of contrasting sizes to permit precise measurement of flow rates < 0.1 mL/min, as well as > 1 mL/s.

No calibration of this flow meter is required, other than the initial calibration of the tubes at the time of fabrication. The graduated tubes

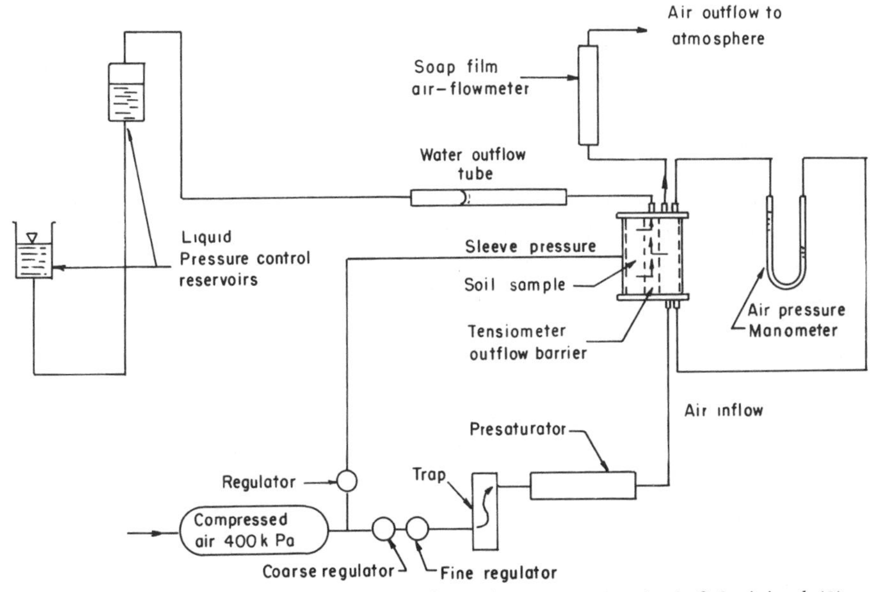

Fig. 48-1. Flow and pressure control system for stationary-water method of obtaining k_a(Θ).

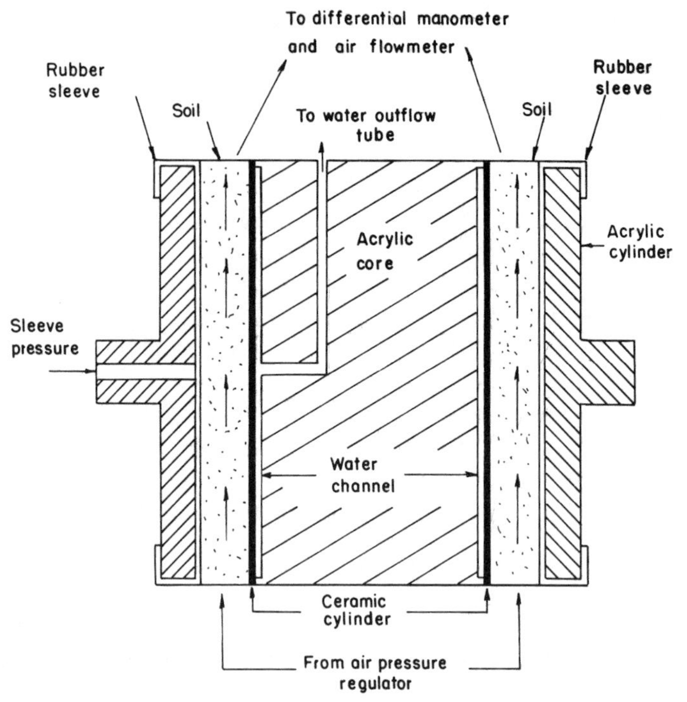

Fig. 48–2. Sample holder for stationary-water method of obtaining $k_a(\Theta)$.

are placed inside of an evacuated glass jacket so that they are insulated from fluctuations in the ambient temperature which might otherwise affect measurements of small flow rates. The four-way stopcock at the top permits the direction of flow through the tubes to be reversed without reversing the direction of flow through the sample. This permits a single stable soap film to be used for repeated measurements, sometimes over a period of hours.

Although flow meters of the type illustrated are not stock items, they can be fabricated from stock parts by a skilled glass blower. The graduated tubes, for example, are fabricated from chemical pipettes calibrated at the factory. Care must be taken to *completely* evacuate the insulator jacket because even a small concentration of gas molecules greatly increases the rate of heat conduction. A clear soap solution, available from any toy shop in soap-bubble kits, is excellent for creating stable soap films. Colored solutions containing dyes or other additives are not suitable. It is *essential* to avoid silicone-based lubricants for the stopcocks because an extremely small contamination with silicone will prevent the formation of stable soap films and is extremely difficult to remove.

The sample holder shown in Fig. 48–2 retains the soil in an annular space between a water-filled ceramic cylinder and a rubber tube which lines the inner wall of the sleeve cylinder. Bypassing of air along the

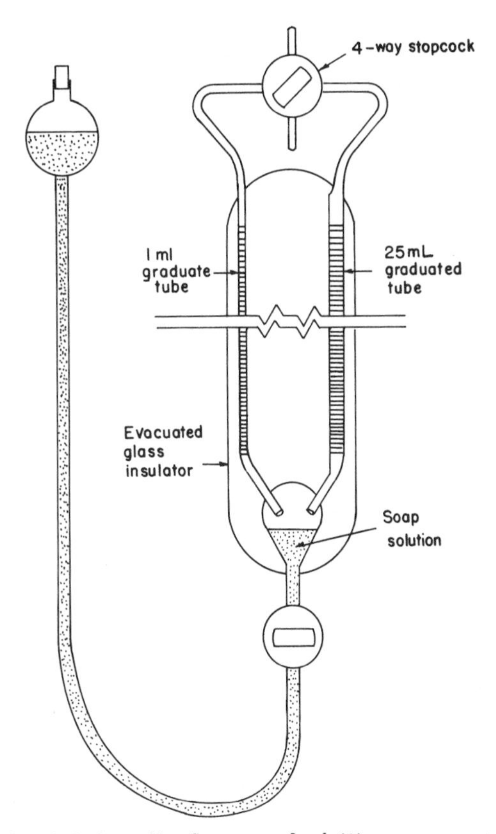

Fig. 48–3. Soap-film flowmeter for $k_a(\Theta)$ measurements.

boundaries is prevented by a pneumatic pressure of about 50 to 100 kPa, which holds the sleeve firmly against the outer circumference of the sample. This pressure also keeps the sample in firm contact with the ceramic cylinder through which the sample is desaturated.

The pressurized sleeve is an essential part of the sample holder, since soils tend to shrink away from a rigid boundary as they desaturate. Other features of the system include a vacuum control device for desaturating the sample in increments, a calibrated tube for measuring water removed from the sample during desaturation, a U-tube manometer for measuring the air pressure difference, and a presaturator to ensure that air flowing through the sample does not remove water from the sample as vapor.

Air enters and leaves the sample through grooves in the end plates (not shown in the figure). Soil is prevented from clogging the grooves by fiber glass disks at the top and bottom of the sample. These must have a negligible resistance to air flow relative to the resistance of the dry soil. Consequently, the fiber glass must desaturate at a very small negative water pressure and yet be sufficiently closely woven to retain the soil.

48–3.2.2 PROCEDURES

48–3.2.2.1 Packing the Sample Holder. The first step in operating the equipment shown in Fig. 48–1 is to pack air-dry soil in the sample holder. This must be done before the top end plate is bolted in place. A variety of packing techniques may be employed, including that described by Reeve and Brooks (1953). It is desirable to pack the soil so that it is as nearly homogeneous as possible throughout the annular space. Any segregation of solid particles or aggregates according to size will greatly affect the $k_a(\Theta)$ relationship.

A fiberglass disk is first placed over the grooves in the bottom end plate, and after packing a similar disk is placed on top of the sample before bolting the top end plate in place.

48–3.2.2.2 Saturating the Sample. The soil, ceramic cylinder, and internal water channels are saturated with water by placing the sample holder assembly in a vacuum tank. The tank is then evacuated, after which deaerated water is admitted to immerse the entire assembly. The tank is then returned to atmospheric pressure. Leads to the water outflow tube are filled with the same deaerated water and connected to the sample holder.

It is often desirable to use a saturating solution with an electrolyte concentration similar to that of the soil solution in the field because $k_a(\Theta)$ is sensitive to the kind and concentration of salts in the solution, especially in the case of soils containing swelling clays (Jamil, 1976).

48–3.2.2.3 Connecting the Flow System. A slight negative gauge pressure is established at the water outflow tube before pressurizing the sleeve and connecting the air flow leads and manometer. This small suction should be sufficient to remove excess water from the air flow passages and fiberglass disks, which may have entered due to the slight compression resulting from the sleeve pressure. The initial suction should not be enough to remove water from the soil sample.

Before connecting the air flow leads, the inflow air regulators are adjusted to produce a null output from the fine regulator or, at least, no more than a small positive gauge pressure corresponding to a few centimeters of water head. The flow leads (into and from the sample holder) may then be connected. The manometer leads are the last to be connected, and should not be connected until after a soap film has been established across the smaller graduated tube as explained below.

48–3.2.2.4 Preparing the Flow Meter for Operation. The first step in operating the flow meter is to block the lead from the sample holder (by clamping or by turning a three-way stopcock). The soap solution stopcock (at the bottom of the meter) is then opened and the soap solution reservoir is raised to fill the smaller graduated tube with the solution. The larger tube will not be filled because the inflow lead is blocked. The reservoir is then lowered slowly to drain the smaller tube. When the soap solution reaches the bottom of the tubes, the four-way stopcock is turned abruptly so that the smaller tube is blocked and the larger tube is connected to the outlet. The soap solution continues to drain, with air entering through the larger tube which empties completely. The soap solution is allowed to drain until its level is slightly below the lower end of the two tubes, at which time the lower stopcock is closed.

A soap film will now be stretched across the base of the smaller tube. Because the walls of this tube have been moistened with the soap solution, the film can move upward and downward without breaking—often for a period of several hours. The next step is to move the film to a position near the lower graduations on the tube. This is done by again reversing the four-way stopcock and opening the lower stopcock slightly and briefly. A small inflow of soap solution from the reservoir causes the film to move upward from its position at the base of the tube. The lower stopcock is closed again when the film reaches the first graduations. The water manometer is then connected across the sample holder and the apparatus is ready for operation.

48–3.2.2.5 Desaturating the Sample. The fine regulator is adjusted so that the head difference indicated by the U-tube manometer equals the height of the soil column. The pressure difference between air and water should then be equal at the top and bottom of the sample. The air pressure at the top is atmospheric at all times since the outlet lead is connected to the atmosphere. The air pressure at the bottom of the sample is held constant by adjustment of the fine regulator. It is essential that the fine regulator be capable of precise pressure control so that the input air pressure does not fluctuate. Any such fluctuation is transmitted to the manometer fluid causing a displacement that is reflected in the movement of the soap film, producing an error in any flow rate measurement being made at the time of the fluctuation.

The pressure of the soil water relative to the soil air pressure, before desaturation begins, can be calculated from the elevation difference of the levels in the two pressure control reservoirs. A correction is made for the elevation of the outflow water tube relative to the soil surface and for the capillarity of the outflow tube.

The initial position of the air-water interface in the outflow tube is recorded, and desaturation is initiated by increasing the elevation difference of the control reservoirs. This adjustment is made in small increments of about 2 cm during the earlier stages of desaturation. A significant but small desaturation usually occurs with each increment of water pressure decrease. At this stage, only a few minutes are required for equilibrium at each incremental pressure.

During each desaturation the soap film moves slightly, normally in a direction opposite to the eventual flow direction. This is because air displaces water being removed from the soil. In any case, displacement of the soap film is small and stops when the water outflow ceases. About 10 to 15% of the total pore volume of water typically is removed before a steady and continuous movement of the soap film is observed. The film then moves in a direction corresponding to upward flow through the sample.

Initially, the movement of the soap film in the direction of flow may be erratic. An explanation for this is as follows: Before the last water film in an otherwise interconnected air channel is removed, there is a pressure difference as indicated by the manometer head across the immediate region of the obstruction. The instant this restriction is removed, the controlled pressure drop is spread more or less evenly over the height of the sample. Apparently, the obstructing water film sometimes reforms after it is broken, due to a local sudden decrease in air pressure. Therefore, an intermittent flow may take place for a short period. This difficulty is temporary, however, and the flow quickly becomes very steady.

The first steady flow occurs at a water content called the "air-entry water content," corresponding to a water pressure equal to the "air-entry pressure." The movement of the soap film at this time is typically very slow, even in the smaller tube. If the soap film had been placed across the larger tube, the movement might not be observed.

Desaturation, however, tends to proceed at a maximum rate at the air-entry water content (White et al., 1970). The operator may elect to slow or stop the desaturation at this time before measuring the flow rate. This is done by increasing the water pressure slightly. The operator may also decide to permit desaturation to continue until equilibrium (at the applied pressure increment) is reached. In the latter case, an air flow rate is measured while desaturation is proceeding.

In either case, the position of the outflow meniscus is recorded at the time air flow first becomes stable. This information is later used to calculate the air-entry water content. The air-entry pressure can be recorded only if equilibrium at the controlled water pressure is achieved. An unknown pressure drop exists across the ceramic cylinder during desaturation of the sample. Consequently, the soil water pressure is unknown until desaturation stops. This difficulty could be overcome by adding a tensiometer ring, isolated from the desaturation cylinder, and connecting this to a null pressure transducer.

Desaturation is continued, either by increments or in an uninterrupted fashion, by periodically lowering the pressure at the liquid outflow meniscus. Air flow rates are measured after suitable increments of desaturation. A measurement of the flow rate after desaturation increments of 5% of the pore volume will usually define the $k_a(\Theta)$ relationship adequately. The process is usually continued until the value of k_a increases by only a small percentage after each additional desaturation. At this time, desaturation proceeds slowly regardless of the increment of water pressure, and the operator may elect to decrease the increments of desaturation before ending the experiment.

To choose a suitable increment of desaturation between flow measurements, it is necessary to make an estimate of the pore volume. This is done by making a reasonable estimate of the porosity and multiplying this times the volume of the annular space occupied by soil.

48–3.2.2.6 Measuring the Flow Rate. The direction of air flow through the meter is upward in the tube connected to the atmosphere through the four-way stopcock and downward in the tube connected to the sample holder. The direction of flow through the meter tubes is reversed by turning the stopcock 90°. This reverses the direction of soap film travel while the direction of air flow in the sample remains constant in an upward direction.

In measuring the flow rate, a stopwatch is used to obtain the time of travel of the soap film between two graduations corresponding to increments of flow of perhaps 0.1 to 10 mL. In the early stages, with water contents at or near the air-entry value, an increment of 0.01 mL may be justified. A soap film in the smaller tube, capacity 1 mL, is used for flow increments up to about 0.5 mL. When the flow rate increases enough that larger increments are more suitable, a soap film is established across the larger tube.

When a new soap film needs to be established during the desaturation process, it is necessary to shunt the flow from the sample holder directly to the atmosphere and to close the inflow opening to the flow meter. This may be accomplished by providing a three-way stopcock in the lead from the sample holder which is turned so that air discharges directly to the atmosphere and the path to the flow meter is blocked. It is then possible to establish a soap film in either of the tubes desired (without interrupting flow through the sample) by employing the procedure previously described. Flow from the sample may then be redirected through the meter and additional measurements made.

In order to avoid the inconvenience of establishing new soap films more often than necessary, the operator should reverse the travel direction before the films move past either the bottom or top of the graduated tube.

48–3.2.2.7 Completing the Experiment. After the final flow measurement has been made, the last position of the water outflow meniscus is recorded. All of the soil is quickly removed from the sample holder while maintaining the final water pressure in the ceramic cylinder. The soil is placed in a soil moisture sample container. The weight of the container plus the moist soil is obtained. The soil is then oven-dried in the container, and afterwards the dry weight of container plus the soil is obtained.

From these measurements, the value of Θ at the end of the experiment can be calculated as explained in chapter 22.

Finally, the particle density of the soil solids is determined using a pycnometer as described in chapter 14.

48–3.2.3 CALCULATIONS

Air permeability in square micrometers is calculated from the relationship

$$k_a = \eta_a VL/tA\rho_1 gh) \times 10^8$$

in which

η_a = viscosity of air at the room temperature (poises),
V = increment of volume displacement of soap film (cm^3),
L = height of sample (cm),
t = time for increment of displacement (s),
A = area of soil cross section perpendicular to axis of flow (cm^2),
ρ_1 = density of manometer fluid (g cm^{-3}),
g = acceleration of gravity (dynes $g^{-1} \simeq 981$), and
h = manometer displacement (cm).

In this equation, the factor 10^8 converts k_a in cm^2 to k_a in μm^2.

Water contents corresponding to each k_a value measured are obtained from material balance calculations. The amount of water removed from the sample between flow measurements is calculated from the corresponding distance of advance of the meniscus in the outflow tube and the cross-sectional area of the tube. The volume increments are added to the volume of water remaining at the end of the experiment to obtain the intermediate water contents. The total volume of water held in the sample at the beginning of desaturation is used to calculate the pore volume V_p of the sample.

A check on the value of V_p calculated by the procedure described above is made from the dry mass M_s of the soil sample and the particle density ρ_s. The volume of dry soil V_s is given by M_s/ρ_s. This volume is subtracted from the total volume of the annular space in which the soil is packed to give V_p. The two values of V_p should provide porosity values that differ by $< 0.5\%$.

48–3.3 Other Steady-state Methods

48–3.3.1 SIMULTANEOUS WATER-FLOW METHOD

With the simultaneous water-flow method, both fluids flow in response to equal pressure gradients. Theoretically, the flow could be in any direction. A procedure of this type employed by Brooks and Corey (1966) created downward flow through the sample. Figure 48–4 shows the distinguishing details of the sample holder used for this method. In some respects this sample holder is similar to that used for the stationary-liquid method, i.e., a pressurized rubber sleeve seals the boundaries and maintains firm soil contact with ceramic porous cylinders used to desaturate the sample.

However, in this case the ceramic barriers are used to conduct water into as well as out of the sample. Separate ceramic cylinders are provided for inflow and outflow. The flow, as it leaves the inflow cylinder and enters the outflow cylinder, is not one-dimensional. However, two separate ceramic rings are used as tensiometers to measure the piezometric head at two horizontal planes between which the flow should be nearly

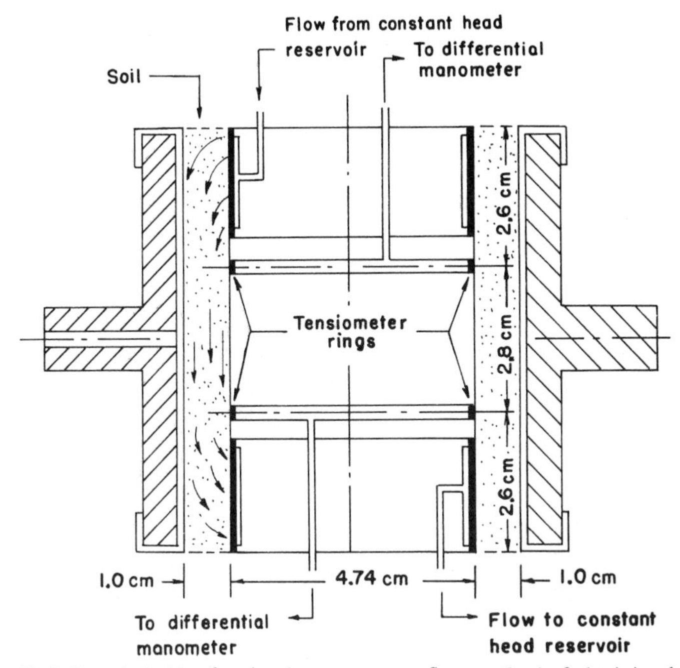

Fig. 48–4. Sample holder for simultaneous water-flow method of obtaining $k_a(\Theta)$.

one-dimensional. The piezometer rings are connected to null pressure transducers to permit the soil water pressure as well as piezometric head difference to be measured directly while desaturation continues.

Unlike the stationary-liquid method, this procedure does not permit the operator to stop the water flow while air permeabilities are being measured. However, it is possible to manipulate the pressure gradients in both the air and water phases so that inflow of water equals the outflow and a steady state is reached.

Practically the same results are obtained in a shorter time by allowing the water outflow to exceed inflow by a small percentage at all times. In this case, the flow rate of the water is taken as the average of the inflow and outflow rate. Furthermore, the latter procedure is not strictly a steady-state method. It is more accurately called a quasi steady-state method.

When employing the simultaneous water-flow method, Brooks and Corey (1966) mounted the entire sample holder (with its flexible leads attached) on a sensitive balance. The water content of the sample was calculated from the change in weight of the assembly as the sample was desaturated. The latter procedure was used because of the difficulties of making accurate material balance calculations when water moves into as well as out of the sample. However, material balance calculations were made as a check on the weight calculations.

The pressure control apparatus used with the simultaneous water flow method by Brooks and Corey is very complex. Furthermore the operation procedures required much patience, skill, and experience for satisfactory results. It is not likely that the latter method will ever become widely used. Consequently, details of the apparatus and its operation are not presented here.

48–3.3.2 HASSLER METHOD

The simultaneous liquid-flow method pioneered by Hassler et al. (1944) was used for a soil sample by Corey (1957). The principle of this method is similar to that described in section 48–3.3. In this case, however, a more or less undisturbed sample was forced into a tube made from acrylic plastic. Afterwards the sample was air-dried and plastic end plates were attached to both ends of the tube to seal the sample. It was then placed in a heated oil-filled tank to soften the plastic while pressure was applied, molding the plastic to the soil sample.

Afterwards the contents of the tank were cooled, the pressure was returned to atmospheric, and the sample with its plastic jacket was removed. The end plates were then removed to expose the end faces of the otherwise confined sample.

Porous ceramic pads were pressed into contact with the end faces of the soil sample (after it was saturated with water) to conduct water into or out of the sample. These pads contained grooves to conduct the air. In addition, smaller concentric pads were used as tensiometers on each end of the sample.

The methods of pressure control, measurement of flow rates, and measurement of the water content were accomplished by procedures similar to those previously described. However, an important disadvantage of this method is that the passage of air into and out of the sample is affected by conditions imposed at the end boundaries. This is unavoidable because of the necessarily complex end pads that must be in firm contact with the soil, causing some local compression of the soil and preventing air passage from at least half of the end area.

Furthermore, although the plastic was firmly molded to the dry sample, it is possible that air bypassing did occur when the sample was resaturated and then desaturated during the experiment. For these reasons, the method is not recommended for use with soil samples, especially those containing swelling clays.

48–3.3.3 OSOBA METHOD

The stationary-liquid method of obtaining $k_a(\Theta)$ was pioneered by Osoba et al. (1951). When used for undisturbed porous rock samples, it can be precise, rapid, and convenient and provides highly reproducible results. This is why Brooks and Corey (1966), White et al. (1970), and Jamil (1976) have used porous rock samples for obtaining basic information concerning air permeability as a function of liquid content.

The only difference in the apparatus used for measuring $k_a(\Theta)$ with rock samples (from that described in section 48–3.2.1) is the design of the sample holder. A convenient core holder for rock samples has been described by Corey (1977). With this core holder, the sample is easily removed after each k_a measurement and the corresponding value of Θ is obtained gravimetrically. The sample is desaturated by periodic blotting or by evaporation.

Unfortunately, a procedure of comparable convenience and precision for measuring $k_a(\Theta)$ with undisturbed soil samples is not yet available.

48-4 LIMITATIONS

None of the methods described in the foregoing sections of this chapter is suitable for measurement of $k_a(\Theta)$ for undisturbed soils. This is unfortunate because $k_a(\Theta)$ is sensitive to the bulk density and structure of the soil. The structure of the soil, particularly as it exists in the field, is difficult if not impossible to reproduce in disturbed laboratory samples.

The basic difficulty is due to the fragility of undisturbed soil samples, which makes it practically impossible to remove them from a permeameter to determine or change their water content. The samples must remain in the permeameter during the entire experiment, and air bypassing at their boundaries must be prevented despite shrinkage during desaturation. A method of desaturating the samples and measuring the resulting Θ values without removing the samples from the permeameter also must be devised.

Any procedure which involves desaturating samples through the exposed ends is likely to result in interference with the free flow of air and may cause a local disturbance of the soil structure. Air permeability is very sensitive to soil structure.

The problem of air bypassing at sample boundaries is far more critical than is the problem of water bypassing during the measurement of $k_w(\Theta)$. This is because air, being a nonwetting fluid, is not confined to the smaller pore channels as Θ is reduced. Water bypassing is a problem only when the sample is fully saturated, at which time shrinkage has not yet taken place. The opposite situation exists for air bypassing. A small shrinkage away from the confining boundary results in a large error in the k_a measurement.

The problem of providing an exit for water during desaturation (while maintaining a flexible confinement at the boundaries of an undisturbed soil sample) provides a formidable experimental challenge. This problem apparently has not yet been solved.

It would be desirable also to have a method for measuring $k_a(\Theta)$ in the field, eliminating the necessity of removing soil samples. A procedure for measuring k_a in the field has been proposed by Kirkham (1947). However, the method of Kirkham involves the establishment of air flow in the soil in response to a falling air pressure. If Θ is large at the time k_a is measured, the falling pressure results in a changing value of Θ which is not measured or accounted for in the theory. Consequently, Kirkham's method is applicable primarily for water contents approaching that at field capacity or less. The same limitation applies to several other methods that have been reviewed by Grover (1955). It is the value of k_a at larger values of Θ, however, which is likely to be related to soil aeration problems or to affect the dynamics of fluid displacements in soils.

48-5 REFERENCES

Brooks, R. H., and A. T. Corey. 1964. Hydraulic properties of porous media. Hydrology Paper no. 3, Colorado State Univ., Fort Collins.

Brooks, R. H., and A. T. Corey. 1966. Properties of porous media affecting fluid flow. J. Irrig. Drain. Div., Am. Soc. Civ. Eng. 92:455–467.

Buckley, S. E., and M. C. Leverett. 1942. Mechanism of fluid displacement in sands. Trans. Am. Inst. Min., Metall. Pet. Eng. 146:107–116.

Corey, A. T. 1954. The interrelation between gas and oil relative permeabilities. Prod. Mon. 19:32–41.

Corey, A. T. 1957. Measurement of water and air permeability in unsaturated soil. Soil Sci. Soc. Am. Proc. 21:7–10.

Corey, A. T. 1977. Mechanics of heterogeneous fluids in porous media. Water Resources Publications, Fort Collins, CO.

Corey, A. T., and R. H. Brooks. 1975. Drainage characteristics of soils. Soil Sci. Soc. Am. Proc. 39:251–255.

Dixon, R. M., and D. R. Linden. 1972. Soil-air pressure and water infiltration under border irrigation. Soil Sci. Soc. Am. Proc. 36:948–953.

Grover, B. L. 1955. Simplified air permeameters for soil in place. Soil Sci. Soc. Am. Proc. 19:414–418.

Hassler, G. L. 1944. U.S. Patent 2, 345, 935.

Jamil, A. 1976. Effect of clay swelling on the hydraulic parameters of porous sandstones. Ph.D. thesis. Dep. of Agricultural Engineering, Colorado State Univ., Fort Collins, CO (Diss. Abstr. 77–12041).

Jones, S. C., and W. O. Roszell. 1978. Graphical techniques for determining relative permeability from displacement experiments. J. Pet. Technol. (May) 1978:807–817.

Kirkham, D. 1947. Field method for determination of air permeability of soil in its undisturbed state. Soil Sci. Soc. Am. Proc. 11:93–99.

Klinkenberg, L. J. 1941. The permeability of porous media to liquid and gases. Am. Pet. Inst. Drill. Prod. Pract. 1941:200–213.

McWhorter, D. B. 1971. Infiltration affected by flow of air. Hydrology Paper no. 49, Colorado State Univ. Fort Collins, CO.

Morel-Seytoux, H. J., and J. Khanji. 1974. Derivation of an equation of infiltration. Water Resour. Res. J. 10:795–800.

Osoba, J. S., G. G. Richardson, J. K. Kerver, J. A. Hafford, and P. M. Blair. 1951. Laboratory measurements of relative permeability. Trans. Am. Inst. Min., Metall. Pet. Eng. 192:47–56.

Peck, A. J. 1965. Moisture profile development and air compression during water uptake by bounded porous bodies, 3: Vertical columns: Soil Sci. 100:44–51.

Reeve, R. C. 1953. A method of determining the stability of soil structure based upon air and water permeability measurements. Soil Sci. Soc. Am. Proc. 17:324–329.

Reeve, R. C., and R. H. Brooks. 1953. Equipment for subsampling and packing fragmented soil samples for air and water permeability tests. Soil Sci. Soc. Am. Proc. 17:333–336.

Welge, H. J. 1952. A simplified method for computing oil recovery by gas or water drive. Trans. Am. Inst. Min., Metall. Pet. Eng. 195:91–98.

White, N. F., H. F. Duke, D. K. Sunada, and A. T. Corey. 1970. Physics of desaturation in porous materials. J. Irrig. Drain. Div., Am. Soc. Civ. Eng. 96:165–191.

Wyckoff, R. D., and H. G. Botset. 1936. The flow of gas-liquid mixtures through unconsolidated sands. Physics 7:325–345.

Youngs, E. G., and A. J. Peck. 1964. Moisture profile development and air compression during water uptake by bounded porous bodies: 1. Theoretical introduction. Soil Sci. 98:280–294.

49 Oxygen Electrode Measurement

C. J. PHENE
Water Management Research Laboratory
Agricultural Research Service, USDA
Fresno, California

49-1 INTRODUCTION

Soil aeration is the process by which gases consumed or produced under the soil surface are exchanged for gases in the aerial atmosphere. Aeration in soil may result from changes in soil temperature, barometric pressure, wind, rain, and evaporation, but the major continuously operating mechanism of gas exchange is provided by gaseous diffusion through the pore space of the soil (Penman, 1940). A plant root environment in which the O_2 supply is ample for respiration and exchange of CO_2 and other noxious gases produced by microorganisms is a requisite for vigorous plant growth.

Indices that describe the soil aeration process include: (i) percent air-filled porosity, (ii) air permeability, (iii) gas composition in open pores, (iv) diffusion in the gas phase, and (v) diffusion through the gas-liquid-solid medium surrounding a cylindrical Pt microelectrode (Lemon & Erickson, 1952; Letey, 1965). In this chapter, we deal with the O_2 diffusion rate (ODR) measurable with the Pt microelectrode and the O_2 concentration in the gaseous and liquid phases measurable with the membrane-covered electrode.

Russell (1950) stated that the evaluation of the aeration conditions at the interface between the root and the soil system presents the greatest possibility of defining the influence of aeration on plant growth. In the process of respiration, plants quickly take up the environmental O_2 surrounding the roots in the rhizosphere, and an increasing oxygen concentration gradient develops between the soil atmosphere and the atmosphere next to the root surrounded by the water film. Thus, movement of O_2 from the atmosphere to the actively respiring rhizosphere involves diffusion through the gaseous phase of the soil, and diffusion through the gas-liquid phase boundary and liquid phase of the water film (Lemon & Erickson, 1952). Since the diffusion coefficient of O_2 is approximately 2.4×10^{-5} cm^2 s^{-1} in water and 1.8×10^{-1} cm^2 s^{-1} in air, the limiting factor for this transport is usually the diffusion rate through the water film rather than through the gas-filled pore space.

In view of the above concepts, a method to measure the ODR through the liquid phase to a reducing surface approximating that of the plant root should be useful for assessment of soil aeration. Since the Pt microelectrode method was developed by Lemon and Erickson in 1952, it has been the subject of considerable use and discussion (Lemon & Erickson, 1955; Lemon, 1962; Stolzy & Letey, 1964; Letey, 1965; McIntyre, 1966, 1971; and many others). McIntyre (1971) listed several objections concerning its theoretical basis, but stated that the method can be useful for correlation purposes as reported by McIntyre (1966) and Phene et al. (1976). McIntyre (1971) concluded that the "established correlation between root growth and ODR" as reported by Valoras and Letey (1966) is only valid for their specific soil and experimental conditions. Sojka and Stolzy (1980) reviewed greenhouse studies of the effect of low soil O_2 on stomatal closure and reported that when the ODR dropped below 0.2 $\mu g\ cm^{-2}\ min^{-1}$, the leaf diffusion resistance increased as an exponential function for a wide range of plant species. They concluded that "elimination of adequate O_2 from the soil profile can cause stomatal closure, which is not fully explained by increased root resistances as some researchers have suggested."

Hutchins (1926), Raney (1949), and Van Bavel (1954) developed methods to measure the ODR in the gaseous phase in soil. Davies (1962) discussed the theory of the membrane-covered electrode (O_2 cathode) adapted from biological techniques for tissues and blood O_2 content measurements. Neville (1962) and Willey and Tanner (1963) described instruments capable of measuring partial pressures of O_2 in gases and in soil, respectively. The instrument developed by Willey and Tanner is equipped with temperature compensation and can provide precise, continuous in situ measurement of soil O_2.

49–2 PLATINUM ELECTRODE METHOD

49–2.1 Principles

The model which is assumed for the microelectrode-soil system is shown in Fig. 49–1. The oxygen diffusion to the electrode is assumed to be described by Fick's (first) law. Because of the geometry of the electrode, an axisymmetrical radial coordinate system is used

$$q = -D\,(\partial c/\partial r) \quad [1]$$

where q is the oxygen diffusion flux in the r direction, D is the effective diffusion coefficient in the soil-water film around the electrode, and $\partial c/\partial r$ is the concentration gradient in the radial direction. The continuity equation

$$\partial c/\partial t = -(1/r)\,[\partial(rq)/\partial r] \quad [2]$$

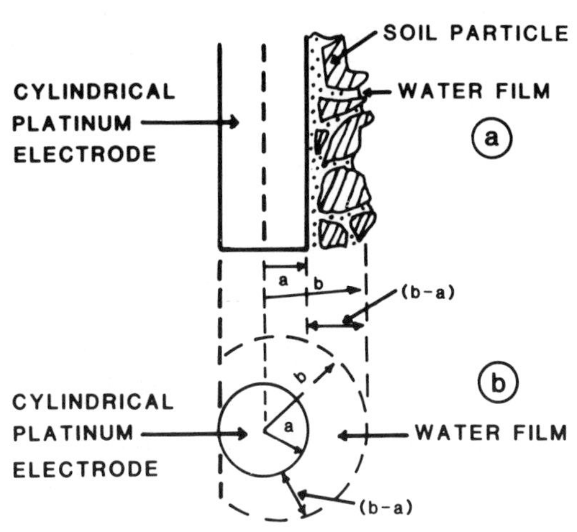

Fig. 49–1. The model which is assumed in order to explain microelectrode behavior. (a) Particles and solution separating the electrode from gas-filled pores (after Kristensen & Lemon, 1964). (b) Coaxial cylindrical model with water film of mean thickness (b-a) or separating the electrode from gas-filled pores (from McIntyre, 1971).

in combination with Eq. [1] and the assumption of steady-state diffusion yields

$$\partial^2 c/\partial r^2 = -(1/r)(\partial c/\partial r) \qquad [3]$$

where D has been considered constant.

Letey (1965) assumed the following concentration boundary conditions for the bare Pt microelectrode (Fig. 49–1):

$$c = 0 \text{ when } r = a \text{ (electrode radius)}$$

$$c = c_2 \text{ when } r = b \text{ (equilibrium } c \text{ in solution at the liquid-gas interface)}.$$

Letey (1965) stated that the condition $c = 0$ at the microelectrode surface is valid only when the electrode is placed in an environment causing low flux of O_2. McIntyre (1971) indicated that the electric current is limited by the ODR to give a current-voltage function of the type shown in Fig. 49–2. Integrating Eq. [2] and inserting the assumed boundary conditions gives

$$\left.\frac{\partial c}{\partial r}\right|_{r=a} = \frac{c_2}{a(\ln b - \ln a)}\,. \qquad [4]$$

The diffusion rate is equal to the product of the effective diffusion coefficient (D_e) and the concentration gradient as stated by Eq. [4]:

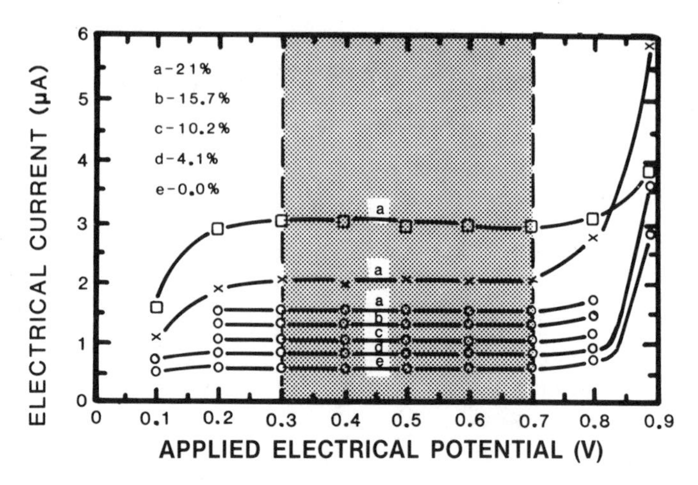

Fig. 49–2. Electrical current-voltage relations for water-saturated media: □, sand; X, clay suspension, o, glass beads 18 μm median diameter. Letters a-e refer to concentration of O_2 in equilibrium with saturating solution (after McIntyre, 1971).

$$\frac{Q}{At} = \frac{D_e c_2}{a(\ln b - \ln a)} \qquad [5]$$

where Q is the amount of O_2, A is the surface area of electrode, t is the time, and D_e is the effective diffusion coefficient of O_2 through the medium surrounding the electrode. The diffusion coefficient D_e is dependent on the properties of the soil medium surrounding the electrode:

$$D_e = D_o\, \theta\, [L/L_e]^2 \qquad [6]$$

where D_o is the diffusion coefficient of O_2 through pure water, θ is the fraction of the surface area of the electrode covered with water as opposed to solid, and $[L/L_e]$ is the tortuosity factor of the diffusing path (Letey, 1965).

In the shaded portion of Fig. 49–2, the current is independent of voltage and a function only of the diffusion rate of O_2 to the microelectrode surface. Hence, if the electrical potential of the Pt microelectrode is lowered sufficiently with respect to a reference electrode, the O_2 at the microelectrode surface is reduced electrolytically until the O_2 concentration at the surface is zero. The reduction and diffusion rates of O_2 are equal and independent of the voltage, and the resulting electrical current (i in microamperes) is expressed as

$$i = n\, FAf_{a,t} \qquad [7]$$

where n is the number of electrons required to reduce one molecule of O_2 ($n = 4$), F is the Faraday constant ($F = 96\ 500$ coulombs/mol of O_2), A is the surface area of the electrode, and $f_{a,t}$ is the flux of O_2 at the surface of the microelectrode of radius a at time t.

The O_2 flux ($f_{a,t}$) is calculated by measuring the steady-state current (i) after 4 or 5 min, assuming that the rate of O_2 reduction is limited by the rate of O_2 diffusion and equal to it

$$f_{a,t} = \frac{iM}{nFA} = \frac{D_o\theta(L/L_e)^2 C_2}{a(\ln b - \ln a)} \quad [8]$$

where M is the molecular weight of O_2 (M = 32 g/mol) and is inserted to convert the units from mol to g. The O_2 flux ($f_{a,t}$) is limited by and is equal to the ODR in solution as shown in Eq. [8], and the transport mechanism is strictly a diffusion process. Substitution of the values for M, n, and F into Eq. [8] gives

$$\text{ODR} = C\,(i/A)\ \mu\text{g cm}^{-2}\ \text{min}^{-1} \quad [9]$$

with $C = M60/nF = 0.00497\ \mu\text{g}\ \mu A^{-1}\ \text{min}^{-1}$. This equation includes all physical factors which affect the ODR to a single root of constant dimensions similar to that of the Pt microelectrode: the apparent mean liquid path length (b-a), the diffusion coefficient of O_2 through water, the ratio of solid to liquid in contact with the microelectrode, and the O_2 concentration in the gas-filled pore space (Letey, 1965).

49–2.2 Equipment

49–2.2.1 BASIC OXYGEN DIFFUSION RATE (ODR) MEASUREMENT SYSTEM

The ODR measurement system consists of the following components, indicated by the numbers in Fig. 49–3:

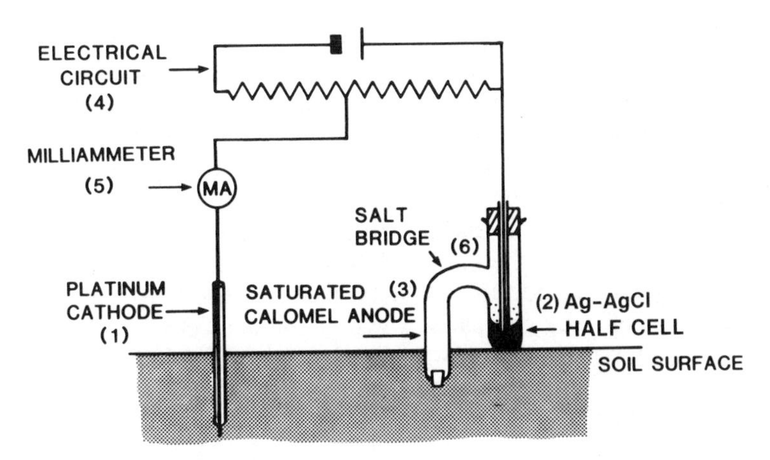

Fig. 49–3. Diagram of apparatus used to make in situ soil measurement of O_2 diffusion (from Lemon & Erickson, 1955).

1. The platinum microelectrode (cathode) (1).
2. The Ag–AgCl half-cell (2).
3. The reference electrode (anode) (3) and KCl-saturated agar salt bridge (6).
4. The electrical circuit required to apply the electrical potential (4).
5. A milliammeter to measure the output current (5).

49–2.2.2 CONSTRUCTION OF PT MICROELECTRODE AND MEASUREMENT SYSTEM

A method for constructing and testing high quality Pt microelectrodes and an electrical bridge for making the ODR measurement were developed by Letey and Stolzy (1964). This instrumentation, designed to measure sequentially 10 Pt microelectrodes, was subsequently improved by Dick's Machine Shop, with commercially available equipment (Dick's Machine Shop, Lansing, MI 48901)[1] to provide three functions: (i) an external electrical potential source, (ii) a current measurement in the external circuit between the Pt and the reference electrodes, and (iii) application of a holding voltage to the other nine electrodes.

Plastic-insulated 12-gauge Cu wire (U.S. wire 12, solid TW, 600 V) is used to make ODR microelectrodes as follows:

1. Cut the Cu wire to desired length (this can vary based on the requirements, usually 15 to 25 cm) and slip off the insulation from the Cu wire.
2. Weld Pt wire (22-gauge) to the Cu wire and cut the Pt wire to a length of 8 mm.
3. After welding the Pt wire to the Cu wire, polish the Cu wire and slip the insulation back onto the Cu wire down to the Cu-Pt weld.
4. Apply epoxy adhesive (Epoxi-Patch 0150 or 0151, Hysol Division, Dexter Corporation, Olean, NY 14760) to the edge of the insulation and taper to a point half-way down the Pt wire.
5. Cut the insulation at the end of the wire and solder an electrical gold-plated pin connector (Winchester 2520S) to the Cu wire of each microelectrode to facilitate the installation and exchange of the Pt microelectrodes (Phene et al., 1976).

Examine the electrodes weekly during use to make sure that the Cu wire does not become exposed.

Equation [9] indicates that the surface area (A) of the electrode must be known to calculate the ODR. Methods for measurement of A were developed by Letey and Stolzy (1964), who showed statistically that the measurement of A by comparison with a standard electrode immersed in a 3% bentonite suspension was significantly better than the measurement of A by the direct method.

Measure the surface area (A) of the Pt electrode as follows: prepare

[1]Mention of trade names in this chapter does not constitute a recommendation for use by the U.S. Department of Agriculture.

a 3% bentonite suspension and thoroughly mix with a stirrer to equilibrate the suspension with atmospheric O_2. Allow the suspension to stand for 5 min; then place a standard electrode and the nine electrodes to be calibrated in the suspension. Apply the potential, and after another 10 min record the readings. Calculate the unknown length (X) of the microelectrode by the equivalent ratios

$$\frac{\text{Unknown length } (X)}{\text{Standard length}} = \frac{\text{Current of unknown}}{\text{Current of standard}} . \qquad [10]$$

Then multiply the length by the circumference of the Pt wire and add this lateral area to the area of the base of the Pt wire.

Polish new electrodes partially oxidized by the welding process with a precious metal cleaner (any jeweler's precious metal cleaner is adequate) before calibration and/or use for measurement; failure to clean the electrodes may result in erratic measurements.

The Ag–AgCl reference half-cell consists of a Lucite container, which holds the salt solution, and approximately 2000 cm^2 of rolled Ag foil (0.0127 cm thick). Insert a strip of Ag foil through a slit in the Lucite container, and seal with epoxy cement to provide electrical contact with the anode. Insert a Lucite tube into the container and connect it to the reference electrode with a Tygon tube (Stolzy & Letey, 1964).

The reference electrode (3) (numbers in parentheses refer to numbers in Fig. 49–3) consists of a long, cylindrical, porous ceramic cup (100 kPa bubbling pressure) filled with KCl-saturated agar-agar gel. Connect the reference electrode (3) to the Ag–AgCl half-cell (2) with a Tygon tube (6) filled with the same gel to the reference electrode (3). Insert the reference electrode (3) in the soil near the Pt electrode (1) to be measured. (See section 49–2.4 for a more permanent system.)

The basic electrical circuit to apply the electrical potential and measure the output current is shown in Fig. 49–4. A more detailed version for automatic measurements is shown in Fig. 49–6.

49–2.3 Procedures

49–2.3.1 INSTALLATION OF ELECTRODES

Reduction of O_2 at the microelectrode surface (cathode) is obtained by lowering the electrical potential of the microelectrode with respect to a large, nonpolarizable, saturated calomel electrode (anode), as shown in Fig. 49–3 (Lemon & Erickson, 1955).

Push the Pt microelectrodes into the soil for measurements at shallow depths. At greater depths, the problem of electrode installation becomes more complex (see section 49–2.4 for suggestion of permanent installation in a cylinder). Irrespective of installation technique, it is imperative that the soil-O_2 status at the point of measurement is not altered and that a good contact be established between the reference electrode and the soil.

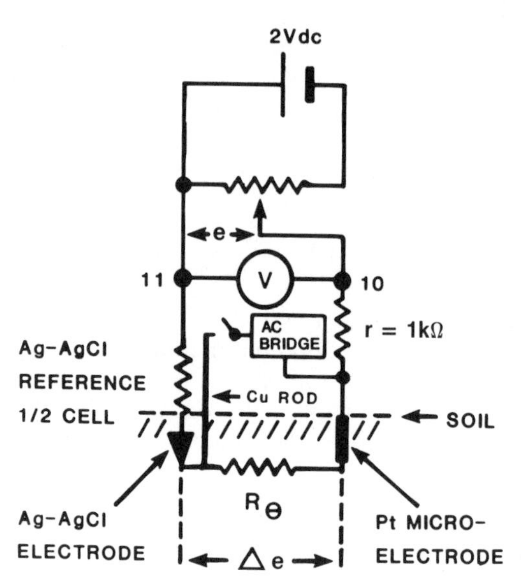

Fig. 49–4. Electrical circuit for ODR measurement (after Kristensen, 1966).

49–2.3.2 MEASUREMENTS OF ELECTRICAL CURRENT AND SOIL RESISTANCE, AND CALCULATION OF ODR

Equation [9] states that the ODR is proportional to the ratio of the steady-state current, i, to the surface area of the Pt microelectrode, A. However, Kristensen (1966) demonstrated that the soil electrical resistance, R_θ, decreases as the water content of the soil increases. The electrical current flow is dependent on the flux of O_2 to the electrode and on the actual potential voltage between the electrodes, so that as the soil dries out, the electrical current should rise unless a simultaneous increase in electrical resistance occurs. Hence, the actual O_2 reduction measurement will partly depend on the electrical resistance according to Ohm's law. Kristensen suggested installing a Cu rod in the soil next to the Ag–AgCl electrode to measure the electrical current flow equal to the ODR, and to correct the current for quantitative measurements.

Apply a potential of −0.65 V and wait 4 or 5 min for steady-state current to be achieved. Measure the steady-state current (i) between the reference cell and the Pt microelectrode (Fig. 49–4).

Measure R_θ with an AC bridge between the Cu rod and each Pt microelectrode. Kristensen (1966) gave an example of the correction method and provided a nomogram (Fig. 49–5) for correction to a standard applied voltage e'. Apply a voltage (e.g., −0.7 V; a potential difference somewhat higher than the standard potential is suggested), measure i and R_θ, and calculate the actual voltage, e' (voltage corrected for the effect of soil moisture on R_θ), using Eq. [11]. Using the nomogram and the measured value of the electrical current, i, correct to a standard e' by following the dotted line in the positive Y-direction until it intersects with the

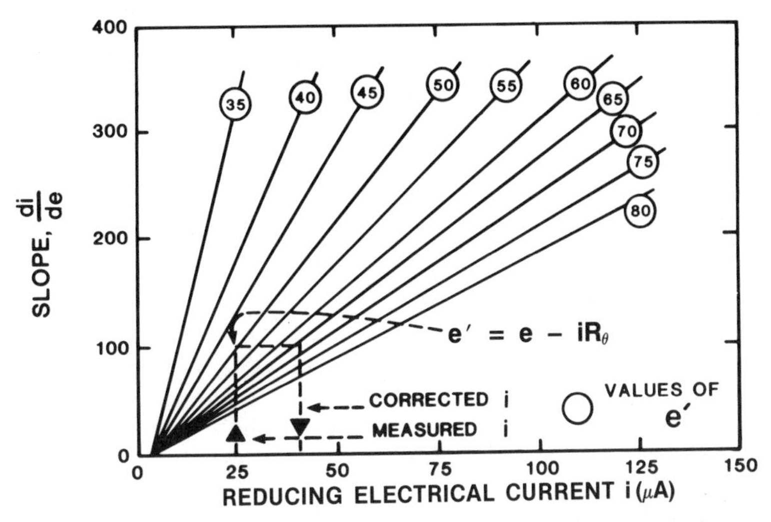

Fig. 49–5. Nomogram for correcting the reducing current i to a standard electrical potential e'. Numbers in circles on lines refer to different values of e' (after Kristensen, 1966).

calculated e' value, and from this intersection in the positive X-direction until it intersects with the line for the standard e' (e.g., −0.65 V). The corresponding electrical current, i, on the X axis is the corrected or true current.

Calculate the true voltage e' from the relation

$$e' = e - \Delta e = e - iR_{\Theta} \times 10^{-6} \qquad [11]$$

where e is the voltage measured between terminals 10 and 11 (Fig. 49–4), Δe is the voltage drop across R_θ, and i is the current (μA) flowing between the electrodes and measured across r.

49–2.4 Comments

Lemon and Erickson (1955) discussed the limitation of the Pt microelectrode for ODR measurements, Birkle et al. (1964) analyzed the various factors affecting the measurements, and McIntyre (1971) presented a detailed review and analysis of the electrochemistry of the methods, the physical effects of the electrode insertion in soil, and the electrical requirements for steady-state operation of the Pt microelectrode. McIntyre questioned the universal validity and applicability of the Pt microelectrode for providing meaningful measurements of soil aeration other than for correlation purposes. On the other hand, Phene et al. (1976) and Campbell and Phene (1977) independently determined similar soil matric potential threshold values for aeration by respectively measuring ODR with the Pt microelectrode and O_2 concentration with a membrane electrode. Until a better method is established, measurements of soil aeration with the Pt microelectrode can provide valuable information as to the relative dynamics of soil and plant root aeration. However, users of the method should observe and consider the limitations and requirements of the method and the need for standardization of the procedure so that the results can be comparable.

Birkle et al. (1964) outlined some major factors influencing the ODR measurements:

1. Establishing steady-state current conditions. A high current occurs immediately after the potential is applied. After 4 or 5 min under the applied potential, the reducing current is essentially constant and can be used for calculating the ODR.
2. The choice of applied potential. Any potential between −0.55 and −0.75 V can be used with the 22-gauge microelectrode. Many investigators have published results obtained with an applied potential of −0.65 V, and this voltage has been recommended for standardization purposes.
3. Installation of electrodes. Good contact between the reference electrode and the soil is necessary, and a hole slightly smaller than the porous cup of the reference electrode should be used for easy but firm insertion in the soil. Installation of the microelectrodes at great depth would be more meaningful if the electrode could be buried permanently, but the occurrence of "poisoning" requires that the electrodes be accessible for cleaning and recalibration.
4. Poisoning. This expression is used to indicate almost anything that can happen (except breakage) and may influence ODR measurements (Rickman et al., 1968). Most often, this effect results from a chemical deposit which changes the characteristics of the Pt surface. Rickman et al. (1968) studied two of the phenomena that cause electrode poisoning: (i) oxide plating of the Pt tip and (ii) the magnitude and nature of the poisoning affecting electrodes installed for periods of 4 weeks or more in different soils. No consistent influence of electrode oxidation conditions on ODR measurements was found in a clay suspension, probably because all measurements were made in a standard medium which restricted the current density on the electrodes to low values (40 $\mu A/cm^2$ or less).

   Results from the study of the second phenomenon showed that salts (principally calcium bicarbonate) and clay particles (principally biotite) were deposited on Pt microelectrodes left in place in a loamy sand for 2 months, and reduced the ODR by an average of 50% when compared to ODRs from periodically reinserted electrodes in the same soil. However, this phenomenon did not occur in clay soils when the electrodes were handled in a similar manner. The procedure used to determine poisoning is questionable, since it requires removing the electrode from the soil (thus changing the contact geometry with the soil-water system) and also measures a different O_2 environment. Generally, poisoning may have occurred if electrodes have been in the soil several days and ODR measurements are continuously decreasing when soil water measurements indicate that they should be increasing.

5. Soil factors affecting ODR measurements.
   a. *Soil moisture.* ODR measurements increase with decreasing soil water until reduction of electrode wetting causes an increase in R_θ and an associated lower reduction current. Birkle et al. (1964) emphasized the fact that this ODR decrease is in no way associated with low O_2 diffusion, but is an artifact associated with the method. The correction suggested by Kristensen (1966) that was described above could be applied.
   b. *Salt content of the soil.* The electrical conductivity of the soil may affect ODR, since the O_2 reduction reaction is electrochemical; however, Birkle et al. (1964) found that for KCl and K_2SO_4, oxygen solubility is the only important factor in determining the diffusion rate in salts. They concluded that enough salt is present in soil to provide sufficient electrical conductivity for maximum ODR.
   c. *Oxygen concentration.* Lemon and Erickson (1955) reported that the reducing current increased linearly with O_2 concentration, with a slope dependent upon the electrode's equilibration time with the soil suspension. Although their method allowed a residual current at zero O_2 concentration, Birkle et al. (1964), using an improved procedure, showed that the linear relationship extends through the origin, indicating that no substance other than O_2 is reduced.
   d. *Temperature.* Oxygen diffusion rate depends upon temperature-controlled factors such as concentration, diffusion coefficient, reaction rate, and others less important. Letey et al. (1962) showed that an increase in ODR of 1.8% $°C^{-1}$ was in agreement with theoretical calculations based on the diffusion coefficient and the concentration of O_2 in water in the temperature range typical of agricultural soil systems.

Birkle et al. (1964) and later McIntyre (1971) suggested that in unsaturated soils the current-voltage functional relationships do not exhibit the zero slope patterns shown in Fig. 49–2, but rather indicate a dependence on the applied voltage or the effective voltage. The applied voltage dependence of the measurement suggests that the limiting process is not ODR through the water film, but instead some electrochemical process. However, the whole theoretical derivation of the established model was based on a boundary condition which required that O_2 reduction be limited by the ODR through the water film. Hence, after this prerequisite is established, one can test for the non-voltage-dependence and based on these results either accept or reject the ODR measurement.

McIntyre (1971) stated in the conclusions of his review that the established diffusion model is not applicable and that the microelectrode used by Lemon and Erickson (1952) and others in no way simulates a root. He reviewed the electrochemistry of the system, electrode kinetics, and transport processes, discussed the physical effects of insertion of the electrode on the environment, and developed theories for new models. The most obvious discrepancy between the established and proposed models and reality results from the basic assumption that the system is strictly physical and static. The root system is not limited to a static, physical system but involves the interaction of complex, dynamic, biochemical and physiological systems subjected to complex and variable external environmental conditions. Consequently, considerations should be given to the case of an elongating root, diurnally expanding and contracting in response to various soil and climatic stresses, so that the system is not strictly dependent on ODR for its O_2 supply but also on the dynamics of the roots; however, this is not the objective of this chapter and here we will be concerned with describing the established method, its limitations, and its uses.

Phene et al. (1976) transformed and adapted the basic bridge for rapid sequential measurements of 72 electrodes in sets of nine with an automatic data acquisition system (DAS) (Fig. 49–6). The basic electrical circuit designed for automatically measuring ODR is similar to that described by Letey et al. (1964). At 6-h intervals the solenoid relay S_1 is closed, and after 4 min the voltages across resistors no. 1 through 9 are measured by scanning between terminal no. 10 and each one of the nine channels for each set of electrodes. When the relay S_1 is closed, the electrical current flowing between the two electrodes is equal to the voltage drop measured across the resistors (no. 1 through 9) divided by their resistance, and is used to calculate the ODR. The reference voltage between terminals no. 10 and 11, initially adjusted to -0.65 V, is measured and recorded every 6 h. The electrical current (i) flowing in the circuit is directly proportional to the voltage potential between the reference and microelectrode and the rate of O_2 diffusing to the microelectrode, and it is inversely proportional to the internal resistance of the circuit (Kristensen, 1966). The internal resistance is the sum of the half-cell resistance (approximately 300 ohms), the resistance used to measure the voltage drop (1000 ohms), and the soil resistance (R_θ), which varies with soil water, texture, temperature, and salt content.

Bornstein and McGuirk (1978) suggested minor electronic modifications for a commercial system.[2] Although there are some disagreements regarding the value of the suggested modifications (Jensen & Stolzy, 1979; McGuirk & Bornstein, 1979), McGuirk and Bornstein (1979) concluded that "with these modifications, the Jensen Model B ratemeter is a reliable and convenient research tool."

Mann and Stolzy (1972) suggested an improved construction method for Pt microelectrodes which essentially provides leak-proof electrodes for use in water-logged soils. Brass alloy brazing rods (approximately 2 mm OD) are cut to the desired length and both ends are filed flat. A 1.0-mm hole, 3 to 4 mm deep, is drilled in one end, and 22-gauge Pt wire

[2]This equipment can be purchased from Jensen Instruments, 3612 6th Avenue, Tacoma, WA 98406. Model "B" Soil ODR Meter, now available as Model "C".

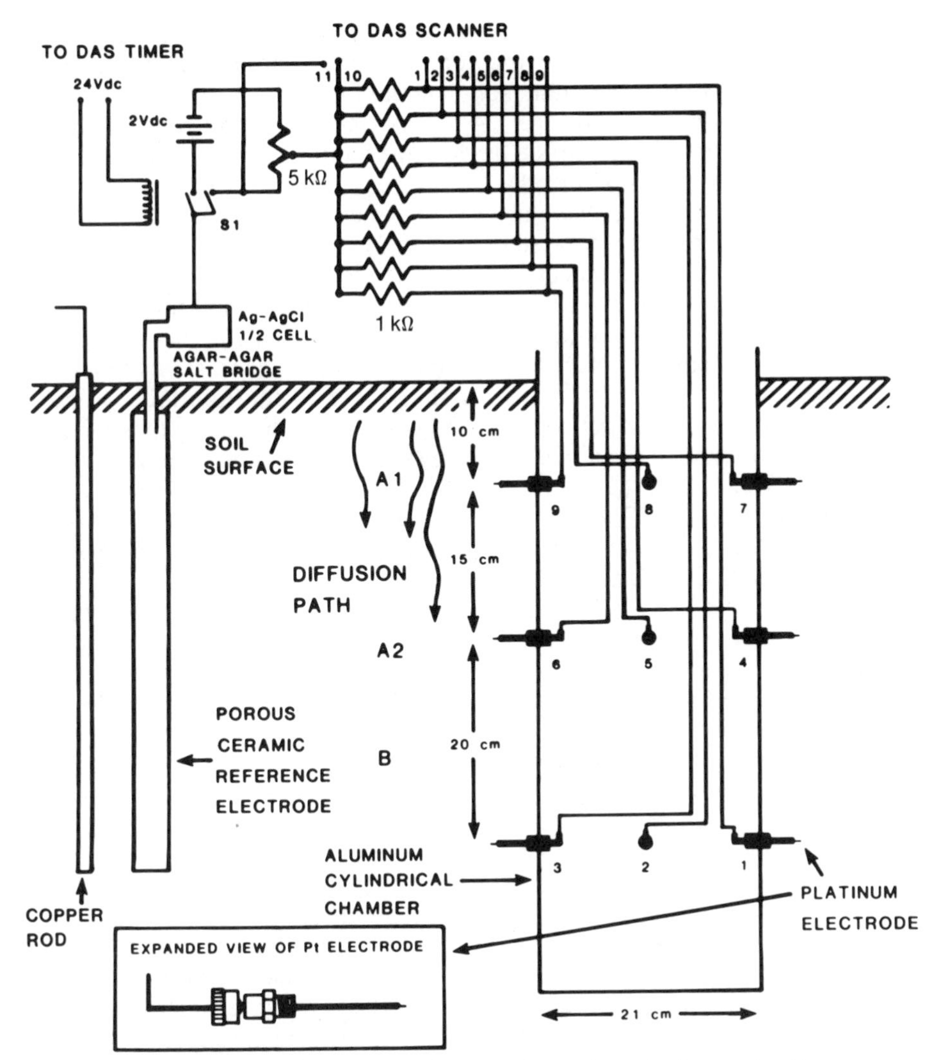

Fig. 49–6. Platinum microelectrode installation and electrical circuitry for in situ measurement of ODR with automated Data Acquisition System (Phene et al., 1976).

is soldered into the hole leaving approximately 1 cm of Pt wire protruding. Excess solder is carefully removed by filing and sanding with emery cloth. A 1-cm length of Tygon tubing (6 mm OD) is slipped on the other end of the brass rod, which is placed in a glass tube (8 mm OD, 6 mm ID) treated with a releasing agent. Approximately 4 to 6 mm of Pt wire should be left protruding from the end of the glass tube. Polyester casting resin, prepared by adding several drops of hardener and stirring slowly to prevent formation of air bubbles, is poured into the glass tube. The soldered Pt-brass junction is completely encapsulated in resin. The electrodes are cured overnight before removing the hardened electrodes from the glass mold. A sudden twist on the rod is usually sufficient to loosen the electrode from the tube. After extraction, the resin surrounding the Pt wire is filed to a point to facilitate soil penetration, and the completed electrodes are cured an additional 24 h. Tests by Mann and Stolzy (1972) have shown that this electrode has a significantly improved performance over commonly used electrodes.

Phene et al. (1976) have developed a method for in situ installation of the Pt microelectrode system that seals the soil around the Pt microelectrode and permits easy access to the Pt electrodes for cleaning or replacement without disturbing the diffusion path to the soil surface. Four holes are drilled and tapped at equidistant points along the circumference of a 21-cm diameter aluminum or PVC cylinder. Each set of holes is located respectively at a distance of 15, 30, and 50 cm from the top edge of the cylinder. The lower edges of the cylinders are sharpened to facilitate installation, and the cylinders are pressed into the soil. The soil is removed from within the cylinder and a bottom plate is installed and sealed with silicone rubber cement. Each of the tapped holes has an O-ring vacuum fitting, 0.95 cm in diameter (Cajon Company, 32550 Old South Miles Road, Cleveland, OH 44139; Cajon Ultra-torr male connector), screwed into it. The soil surrounding the cylinder is saturated with water and the Pt microelectrodes are pushed manually through the O-ring vacuum fittings into the saturated soil, which insures thorough electrode wetting. The O-ring vacuum fitting is tightened to seal the electrodes in place and prevents air exchange between the chamber and the soil. Each electrode is connected to the measuring bridge (Fig. 49–6). The measuring bridge, the half-cell, and the electrical connections can be stored inside the cylinder, which is closed with a small wooden cover. Data are inspected and partially evaluated daily to determine normal electrode functioning or possibility of electrode poisoning. Installation of the Pt microelectrodes in the cylinder prevents loosening of the electrode in the soil during the experimental period, and possibly, the ensuing diffusion of O_2 along the body of the electrode. By using this method of installation, the soil above the electrode is not disturbed, which provides a more natural field environment for the electrodes and the plants. If poisoning is suspected, the Pt electrodes can be pushed further into the soil with a rotating motion so that the abrasive property of the soil can clean the Pt tip (Stolzy & Letey, 1964).

49–3 MEMBRANE ELECTRODE METHOD

49–3.1 Principles

Davies (1962) has reviewed the theory, developmental methods, nature of the electrode reactions, and characteristics of the O_2 cathode and its applications. When a voltage is applied between the cathode and a nonpolarizable reference anode, the diffusion current reaches a plateau (Fig. 49–2). In this voltage range (0.3–0.7 V), all of the O_2 that diffuses to the polarographic electrode is reduced, and the O_2 concentration at the electrode surface is near zero. The electrode current is proportional to the rate of reduction of O_2, is limited by the rate of diffusion of O_2 to the electrode, and is expressed by Eq. [7] (Letey, 1965).

When a membrane permeable to O_2 is installed over the face of the cathode, the physical operation of the system is not changed (Willey & Tanner, 1961, 1963; Neville, 1962; Davies, 1962), because under constant pressure and temperature, the diffusion impedance of the membrane and electrolyte layers is constant and the O_2 "diffusion current" of the covered electrode is directly proportional to the O_2 concentration (partial pressure) at the outer surface of the membrane. When the electrode is covered, it can be used in liquids (Carritt & Kanwisher, 1959; Krog & Johansen, 1959; Reeves et al., 1957; Sawyer et al., 1959; Willey & Tanner, 1961) and in gases (Sawyer & Interrante, 1961; Sawyer et al., 1959); Willey & Tanner, 1961). The membrane can be used to prevent the cathode from reacting with chemicals in solution, thus preventing "poisoning," and/ or to provide a constant diffusion layer in contact with the cathode. Figure 49–7 illustrates the boundary conditions required when measuring in air under steady-state conditions with the membrane-covered cathode. The steady-state rate of oxygen diffusion is expressed as

$$Q/At = D_m (C_1/L) \qquad [12]$$

where Q is the amount of O_2, A is the area of the electrode, t is the time, D_m is the effective diffusion coefficient of O_2 through the membrane and the medium surrounding the cathode, L is the thickness of the membrane, and C_1 is the O_2 concentration in air.

The electric current is given by

$$i = n\, FAD_m (C_1/L) \qquad [13]$$

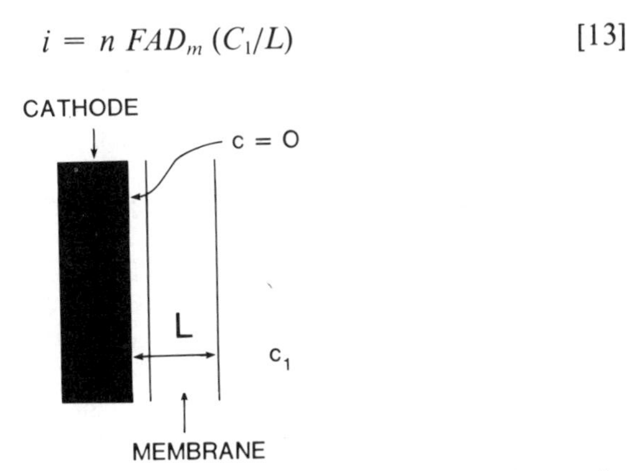

Fig. 49–7. Steady-state boundary conditions for the membrane-covered cathode operating in air (Letey, 1965).

using the terms defined in Eq. [7] and [12]. Equation [13] indicates that the current i is directly proportional to the concentration of O_2 in air, C_1.

Willey and Tanner (1963) stated that although there is some question about the real nature of the O_2 reduction reaction, the reaction probably should be written as a direct reduction of O_2:

$$O_2 + 2H_2O + 4e^- \rightarrow 4\ OH^- . \qquad [14]$$

Willey and Tanner (1963) found that most commercially available membrane-covered electrodes are designed primarily for laboratory use and do not offer the temperature compensation or the ruggedness needed for agricultural field operation. Because an increase of temperature decreases the solubility of O_2 in aqueous solution and increases the diffusivity of the membrane and solution layers, it is found that the electrode current increases with temperature at about the same rate as the diffusivity of the membrane, so that a linear relationship exists between reciprocal temperature and the logarithm of the current. Because of this temperature dependence, readings either must be corrected for temperature or some form of compensation must be provided.

49–3.2 Equipment

49–3.2.1 MEASUREMENT SYSTEM

1. Membrane-covered electrode.
2. Portable DC voltmeter.
3. 1.35 VDC source (Mallory[1] Mercury Battery RM-42R), depending on the electrode being used.

49–3.2.2 CONSTRUCTION OF ELECTRODE

Willey and Tanner (1961, 1963) and Neville (1962) have designed, constructed, and tested membrane-covered O_2 electrodes for various applications. The probe of Willey and Tanner (1963) has several advantages for application in soil, since it was designed to have: "(i) temperature compensation, (ii) a rugged construction suuitable for inserting in an access tube in the soil, (iii) a construction that requires only machine shop techniques, and (iv) a flexible design that can be modified easily to accommodate different electrode sizes."

The temperature-compensated Willey and Tanner probe is designed for low O_2 consumption (Fig. 49–8). The oxygen consumption by the probe is about 1.3×10^{-10} g s^{-1} in air at standard conditions. The probe is easily constructed and the calibration is stable. The current from the electrode is only 1.5 μA in air, which can cause some recording problems. Nevertheless, the probe appears to be well adapted for routine soil measurements.

The O_2 electrode consists of a 0.5-mm thick, 3.2-mm diameter Pt

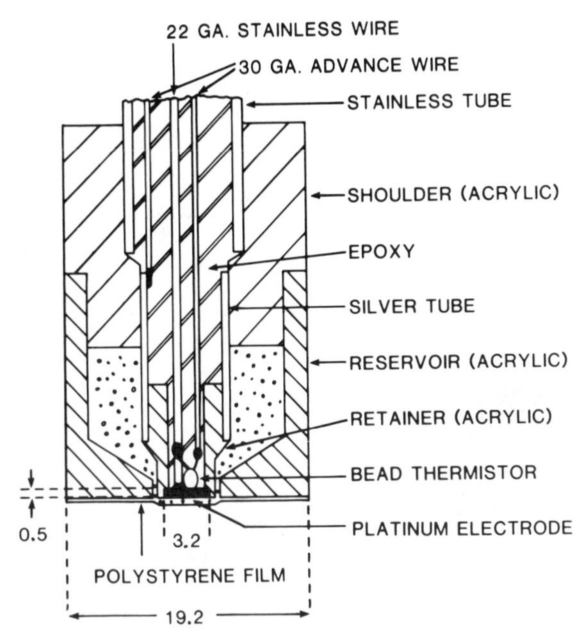

Fig. 49–8. Willey and Tanner membrane-covered, temperature compensated, 3.2 mm diameter O_2 electrode (all dimensions are in millimeters) (Willey & Tanner, 1963).

disk sealed into an acrylic retainer with epoxy cement. The electrode face is covered with thick polystyrene film that is cemented to the end of the reservoir tube. The Ag tube extending through the center of the probe supports the retainer and Pt electrode.

The probe can be sealed in a 19.0-mm inside diameter access tube with an expandable rubber plug. Because of the temperature dependence of the measurement, Willey and Tanner (1961, 1963) considered using polystyrene film instead of polyethylene- or Teflon-film-covered electrode as used in earlier probes. The probe with the polystyrene-membrane-covered electrode has a temperature coefficient of $< 2\%\ °C^{-1}$, as opposed to temperature coefficients in the range of 5% for polyethylene or Teflon. This lower temperature coefficient permits easy temperature compensation with a thermistor circuit. In their application, Willey and Tanner cemented bead thermistors to the back of the electrodes to provide good thermal contact between the electrode and thermistor, thus minimizing thermal time constants and the effects of any external temperature gradients.

When temperature compensation is used, the O_2 electrode current (which increases with temperature) is measured as a voltage drop across a thermistor circuit, shown diagrammatically in Fig. 49–9, that decreases in resistance with increasing temperature. The thermistor circuit (Willey & Tanner, 1963) is designed to have a temperature coefficient resistance equal (except for sign) to the O_2 electrode temperature coefficient of current. With this arrangement, the voltage drop across the thermistor circuit

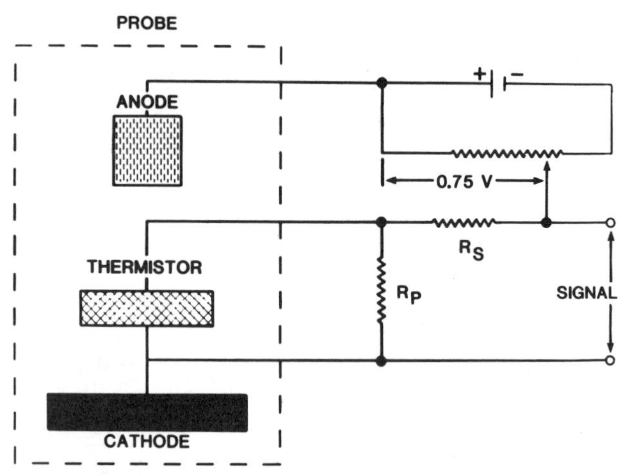

Fig. 49–9. Power supply and temperature compensation circuit for the polystyrene membrane-covered electrode (Willey & Tanner, 1963).

does not change with temperature, but remains a linear function of O_2 concentration. Electrical components for the circuit shown in Fig. 49–9 can be selected for adequate temperature compensation in the range of operation. The resistances R_p and R_s are selected by satisfying the following conditions:

$$d(\ln R)/d(1/T) = \alpha T_0^2 \quad [15]$$

$$d^2(\ln R)/d(1/T)^2 = 0 \quad [16]$$

where R is the effective resistance of the thermistor circuit ($R = 1/R_n + 1/R_p + R_s$ and R_n is the nominal resistance of the thermistor), α is the temperature coefficient of the oxygen electrode at temperature T_0, and T is the absolute temperature (in K). The error of this circuit is about 1% in a temperature range of ± 15°C about the temperature for which R_p and R_s are selected. For example, if the resistances R_p and R_s are selected for a temperature of 20°C, then the temperature compensation will be within ± 1% for a temperature range from 5 to 35°C, a normal range of ambient temperatures. If the thermistor shown in Fig. 49–9 has a nominal resistance equal to 5 kΩ and the ambient temperature is 20°C, the output signal will be about 11 mV (R = 7.5 kΩ at 20°C). Larger output signals can be obtained by selecting thermistors with larger resistance, but a maximum of 30 mV should be maintained to avoid variations in electrode voltages.

Commercially available membrane-covered electrodes using similar theory have been designed and marketed and can be used for measurement of O_2 in a gas and liquid phases. They are available from

Beckman Instruments, Inc., P.O. Box C-1900, Irvine, CA 92713;

Chemtronics, Inc., P.O. Box 6996, San Antonio, TX 78209;
Yellow Springs Instruments Co., Inc., Box 279, Yellow Springs, OH 45387; and
Jensen Instruments, 3612 6th Avenue, Tacoma, WA 98406.

49–3.3 Procedures

49–3.3.1 OPERATION AND CALIBRATION OF THE ELECTRODE

Connect the power leads to a 1.35-VDC mercury battery (Mallory, RM-42 R) and the output signal wires to the voltmeter, as shown in Fig. 49–9. The membrane-covered electrode can normally be operated for 1 year by applying 0.75 VDC from a single 1.35 VDC mercury battery or from similar voltage sources (Fig. 49–9). The electrolyte should be changed monthly for reliable, continuous operation (Willey & Tanner, 1963).

The probe responds rapidly to changes in O_2 concentration. The (1/*e*) time constant is about 15 s, and 135 s are required to reach equilibrium. Thinner membranes are necessary for more rapid response, but are more fragile. If the probe is in an access well, the response depends mainly upon the volume of the cavity, the geometry of the connection to the soil, and soil porosity and diffusivity.

The electrode output is linear with O_2 concentration and the current is nearly zero at zero O_2 concentration; thus, the probe requires calibration at only one concentration point—atmospheric O_2 (20.7%). The accuracy of this calibration procedure is about 2% (relative error) and should be adequate for most measurement purposes. If an expanded scale is used, an occasional calibration of the electrode against a high-purity nitrogen gas to test the zero O_2 concentration and against known O_2 gas mixtures to expand the scale is desirable. When this method is used the voltage measurement is equal to the O_2 concentration in the atmosphere sampled.

Calibration drifts are not a usual problem. Willey and Tanner (1963) report that the small electrode probe has been operated in air for as long as 45 days with 2% drift. In soil, the probes have been operated continuously only for periods of up to 2 weeks, but satisfactory operation over longer periods should be possible. Huck (M. G. Huck, 1975 personal communication; 1971 annual report, USDA-ARS, Auburn, AL,) reported that in situ operation requires recalibration or calibration adjustments to be performed at least monthly for routine measurements and more often for precise studies.

49–3.3.2 INSTALLATION AND MAINTENANCE

The membrane-covered O_2 electrode can be used to measure O_2 in a gas phase; measurement in a liquid phase can also be made, providing liquids are kept flowing to prevent establishment of a liquid diffusion layer at the membrane surface. Oxygen measurements can be made in situ with the probe inserted in a soil cavity or by taking soil gas samples with syringes and making measurements with the electrode system in the laboratory. However, if measurements are to be made in situ, the electrode must be calibrated and serviced regularly. Therefore, the installation must provide easy access for performing these tasks.

Campbell and Phene (1977) used a commercial membrane-covered electrode (Chemtronics, Model LP-10) to measure the O_2 concentration of 5-mL gas samples. These samples were obtained with a gas-tight syringe through a septum covering a 0.1-cm (ID) tube, which led to an inverted gas diffusion chamber buried in soil. The electrode was mounted in a cuvette and sealed with a septum. The voltage output of the probe was measured with a portable DC voltmeter in a temperature-controlled environment. This method allows processing of 20 to 30 samples per hour. With new process-control technology, this process could be fully automated.

Letey (1965) cautioned that water should be prevented from getting on the membrane. An artificial cavitation around the electrode or a double-membrane electrode system, as described by Willey (1974), can be used to avoid this difficulty. Since the measurement process consumes O_2 and thereby modifies the environment to be measured, periodic rather than continuous measurements are recommended.

M. G. Huck (1975, personal communication; 1971 annual report, USDA-ARS, Auburn, AL) described a method for long-term installation and computerized measurement of membrane-covered O_2 electrodes (Chemtronics, Model GP-10) in the Rhyzotron soil bins. Each probe was installed inside a 100-cm^3 glass diffusion chamber having an opening large enough to admit the probe in an inverted position (membrane up). Thus it was protected from the mechanical pressure exerted by the soil on the membrane. The gas inside the chamber was in diffusion equilibrium with the gas in the surrounding soil. During flooding periods a gas bubble was always in contact with the electrode membrane and prevented water from filling the cavity. At monthly intervals, the gas in the cavity was flushed with N_2 gas to obtain a zero-value, and with ambient air to obtain a 20.7% value.

Willey and Tanner (1963) designed the probe so that it can be sealed in an access tube with an expandable rubber plug (Fig. 49–10). A stainless steel wire is connected at the terminal end of the probe and is tightened against a fiber tube to expand the plug against the access tube wall. The probes fit a 1.9-cm i.d. access tube. An O-ring can be placed in a groove on the shoulder to provide a seal. However, the probe may act as a piston when inserted in the access tube and a transient pressure and flushing of soil gases will occur.

Standard preparation, use, and calibration of commercial electrodes are specified in manufacturer's operation manuals for each specific system.

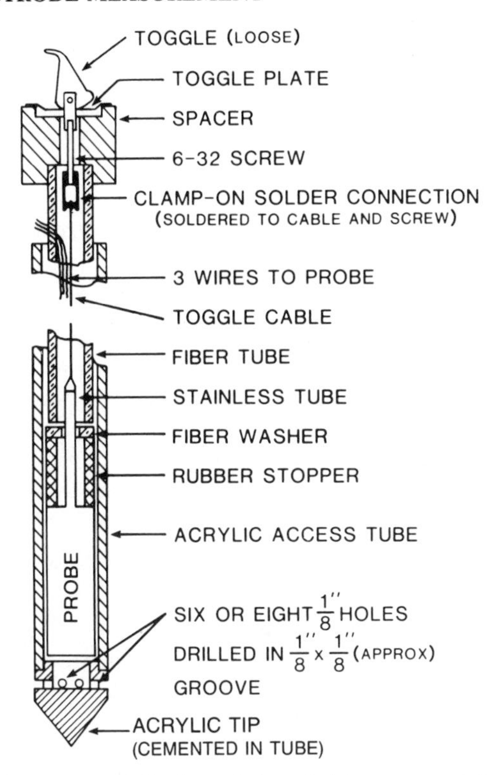

Fig. 49–10. Membrane-covered O_2 electrode installed in an access tube for long-term use (after Willey & Tanner, 1963).

49–3.4 Comments

Letey (1965) emphasized that when using the membrane-covered electrode to measure O_2 concentration in solution, the concentration at the outer boundary of the membrane be kept constant by stirring or flowing the solution. If the solution is stationary, the diffusion layer will extend beyond the membrane into the solution and change the diffusion coefficient and path length in Eq. [14]. The calibration factor will no longer be applicable and the concentration cannot be calculated from the electrical current.

The temperature-compensated probe (Willey & Tanner, 1963) can be used to provide precise (2–5%) in situ measurements of soil O_2 concentration, but long-term use will require frequent calibration of the probe, change of electrolyte, and possibly change of the membrane. Operation of the electrode below the freezing point of the electrolyte (−5°C) will not be possible unless special additives have been added to the solution.

Activation of commercial electrodes, calibration, and measurements of O_2 concentration should follow the specific instructions furnished with the equipment.

49-4 REFERENCES

Birkle, D. E., J. Letey, L. H. Stolzy, and T. E. Szuszkiewicz. 1964. Measurement of oxygen diffusion rates with the platinum microelectrode. II. Factors influencing the measurements. Hilgardia 35:555–566.

Bornstein, J., and M. McGuirk. 1978. Modifications to a soil oxygen diffusion ratemeter. Soil Sci. 126:280–284.

Campbell, R. B., and C. J. Phene. 1977. Tillage, matric potential, oxygen and millet yield relations in a layered soil. Trans. ASAE 20:271–275.

Carritt, D. E., and J. W. Kanwisher. 1959. An electrode system for measuring dissolved oxygen. Anal. Chem. 31:5–9.

Davies, P. W. 1962. The oxygen cathode. p. 137–179. *In* W. L. Nastuk (ed.) Physical techniques in biological research. Academic Press, New York.

Hutchins, L. M. 1926. Oxygen supplying power of the soil. Plant Physiol. 1:95–150.

Jensen, D. R., and L. H. Stolzy. 1979. Comments on the paper "Modifications to a soil oxygen diffusion ratemeter." Soil Sci. 128:61–62.

Kristensen, K. J. 1966. Factors affecting measurements of oxygen diffusion rate (ODR) with bare platinum microelectrodes. Agron. J. 58:351–354.

Kristensen, K. J., and E. R. Lemon. 1964. Soil aeration and plant root respiration. III. Physical aspects of oxygen diffusion in the liquid phase of the soil. Agron. J. 56:295–301.

Krog, J., and K. Johansen. 1959. Construction and characteristics of Teflon-covered polarographic electrode for intravascular oxygen determination. Rev. Sci. Inst. 30:108–109.

Lemon, E. R. 1962. Soil aeration and plant root relations. I. Theory. Agron. J. 54:167–170.

Lemon, E. R., and Erickson, A. E. 1952. The measurement of oxygen diffusion in the soil with a platinum microelectrode. Soil Sci. Soc. Am. Proc. 16:160–163.

Lemon, E. R., and A. E. Erickson. 1955. Principle of the platinum microelectrode as a method of characterizing soil aeration. Soil Sci. 79:383–392.

Letey, J. 1965. Measuring aeration. p. 6–10. *In* Drainage for Efficient Crop Production, Conf. Proc. ASAE. Chicago, IL. American Society of Agricultural Engineers, St. Joseph, MI.

Letey, J., and L. H. Stolzy. 1964. Measurement of oxygen diffusion rates with platinum microelectrode. I. Theory and equipment. Hilgardia 35:555–566.

Letey, J., L. H. Stolzy, N. Valoras, and T. E. Szuszkiewicz. 1962. Influence of soil oxygen on growth and mineral concentration of barley. Agron. J. 54:538–540.

Mann, L. D., and L. H. Stolzy. 1972. An improved construction method for platinum microelectrodes. Soil Sci. Soc. Am. Proc. 33:853–854.

McGuirk, M., and J. Bornstein. 1979. Response to the letter to the editor. Soil Sci. 128:63.

McIntyre, D. S. 1966. Characterizing soil aeration with a platinum microelectrode. I. Response in relation to field moisture conditions and electrode diameter. Aust. J. Soil Res. 4:95–102.

McIntyre, D. S. 1971. The platinum microelectrode method for soil aeration measurements. Adv. Agron. 22:235–283.

Neville, J. R. 1962. Electrochemical device for measuring oxygen. Rev. Sci. Inst. 33:51–55.

Penman, H. L. 1940. Gas and vapor movements in the soil. I. The diffusion of vapours through porous solids. J. Agric. Sci. 30:437–461.

Phene, C. J., R. B. Campbell, and C. W. Doty. 1976. Characterization of soil aeration in situ with automated oxygen diffusion measurements. Soil Sci. 122:271–281.

Raney, W. A. 1949. Field measurement of oxygen diffusion through soil. Soil Sci. Soc. Am. Proc. 14:61–65.

Reeves, R. B., D. W. Rennie, and J. R. Pappenheimer. 1957. Oxygen tension of urine and its significance. Fed. Proc. Fed. Am. Soc. Exp. Biol. 16:693–696.

Rickman, R. W., J. Letey, G. M. Aubertin, and L. H. Stolzy. 1968. Platinum microelectrode poisoning factors. Soil Sci. Am. Proc. 32:204–208.

Russell, M. B. 1950. Soil aeration and plant growth. Monogr. Agronomy Dep., Cornell University, Ithaca, NY.

Sawyer, D. T., R. S. George, and R. C. Rhodes. 1959. Polarography of gases: Quantitative studies of oxygen and sulfur dioxide. Anal. Chem. 31:2–5.

Sawyer, D. T., and L. V. Interrante. 1961. Electrochemistry of dissolved gases: II. Reduction of oxygen at platinum, palladium, nickel and other metal electrodes. J. Electroanal. Chem. 2:310–327.

Sojka, R. E., and L. H. Stolzy. 1980. Soil-oxygen effects on stomatal response. Soil Sci. 130:350–358.

Stolzy, L. H., and J. Letey. 1964. Measurements of oxygen diffusion rates with the platinum microelectrodes. III. Correlation of plant response to soil oxygen diffusion rates. Hilgardia 35:567–576.

Valoras, N., and J. Letey. 1966. Soil oxygen and water relationships to rice growth. Soil Sci. 101:210–215.

Van Bavel, C. H. M. 1954. Simple diffusion well for measuring soil specific diffusion impedance and soil air composition. Soil Sci. Soc. Am. Proc. 18:229–234.

Willey, C. R. 1974. Elimination of errors caused by condensation of water on membrane-covered oxygen sensors. Soil Sci. 117:343–346.

Willey, C. R., and C. B. Tanner. 1961. Membrane-covered electrode for oxygen measurements in soils. Univ. of Wisconsin Soils Bull. 3.

Willey, C. R., and C. B. Tanner. 1963. Membrane-covered electrode for measurement of oxygen concentration in soil. Soil Sci. Soc. Am. Proc. 27:511–515.

50 Air Pressure Measurement

H. FLÜHLER
Swiss Federal Institute of Technology
Zurich, Switzerland

A. J. PECK
CSIRO, Division of Groundwater Research
Wembley, Western Australia

L. H. STOLZY
University of California
Riverside, California

50-1 INTRODUCTION

The flow of air which results from pressure gradients in soils is often neglected. Yet this process can play an important role in transient soil water flow and soil aeration. Measurements of the air pressure in soils are essential to improve our understanding of these phenomena. Indirectly, these measurements provide evidence of the existence of soil macrostructure.

Soil air pressures have been measured in laboratory columns and in the field. Within soil columns the air pressure ahead of a wetting front can exceed atmospheric pressure by 5 kPa (50 mbar) and occasionally as much as 11 kPa (Wilson & Luthin, 1963; Peck, 1965a, 1965b; Adrian & Franzini, 1966; McWhorter, 1971; Vachaud et al., 1973; Raj Pal & Stroosnijder, 1976). In open-bottom or otherwise ventilated columns, the air pressure increase is negligibly small. Flood irrigation or intense rainfall may raise the air pressure in field soils by approximately 1 to 3 kPa (Dixon & Linden, 1972; Linden et al., 1977). As a consequence, large pores and cracks in the soil remain air-filled and do not contribute to water movement (Dixon & Peterson, 1971; Dixon, 1975; Linden & Dixon, 1976). When air is confined between a wetting front and an air-impervious layer, the resulting pressure increase induces a lateral air flow over considerable distances and may cause a significant water table depression (Linden & Dixon, 1973, 1975). Under such conditions the hydraulic gradients and the water-conducting cross section are diminished and the tortuosity of the flow paths increased, leading to a sharp decline of the

 Methods of Soil Analysis, Part 1. Physical and Mineralogical Methods—Agronomy Monograph no. 9 (2nd Edition)

infiltration rate which was observed in all studies quoted above. When there is a reversal of soil air pressure changes, the relationship between capillary pressure and soil water content is shifted to another hysteretic loop (Youngs & Peck, 1964). Subsequent drainage is, therefore, affected by the preceding air pressure history.

The volumetric water and air contents of a soil are parameters affecting the transient development of soil air pressures when water displaces air, or vice versa. The analytical treatments of two-phase flow presented by Brustkern and Morel-Seytoux (1970) and Parlange and Hill (1979) demonstrate that infiltration models should include the soil air pressure unless air pressure invariance is explicitly proven. Since the viscosity of air is about 50 times less than that of water, the pressure gradients in the air phase become important only in soils close to saturation. Thus, it is often a good approximation to neglect the effects of transient air pressure on water movement in well-drained soils and ventilated laboratory columns. Hence, effects of air pressure on infiltration are most likely to be important in poorly drained field soils under flood irrigation or intense rainfall.

Soil air can be isolated from the atmosphere as minute bubbles in individual pores, regions of soil which are bypassed during wetting, or partially air-filled pockets which occupy many pores in a soil structural unit. Large regions of soil air can be trapped between a wetting front and an air-impervious stratum. Hence, these discrete and isolated volumes of soil air may range in size from fractions of a microliter to thousands of liters or even more. Air bubble entrapment is a common occurrence in wetted soil samples. Smith and Browning (1942) measured an average volumetric air content of 9% and a maximum of 22% in soil samples saturated under laboratory conditions.

A pressure buildup in the soil air signals that air displacement is severely restricted. This in turn indicates that gaseous diffusion and hence soil aeration in regions of occluded air are also impeded. The response of soil aeration to gas phase discontinuities was demonstrated in the field by ponding a large loess soil monolith with water (Flühler & Läser, 1975). When the surface layer was sealed off by the infiltration water, an immediate drop of oxygen partial pressure was observed with a simultaneous increase of the soil air pressure.

It might also be useful to consider soil air pressure as an indirect expression of structural properties, because an air pressure buildup usually affects the relative magnitude of the air permeability and hydraulic conductivity, which are both indices of soil structure. Furthermore, an increased air pressure usually exerts a mechanical stress on the soil matrix. Immediate wetting of soil crumb surfaces can raise the internal air pressure to more than 10 kPa (Stroosnyder & Koorevaar, 1972).

Usually, the most important factors contributing to soil air pressure changes are air displacement by water infiltration and high air-flow resistance. Temperature changes modify the soil air pressure according to

Boyle's law, but this effect is probably more important in the laboratory than in the field. The relatively small amplitudes of short-term temperature variations, especially in wet field soil profiles, are not likely to affect air pressure significantly. Some evidence presented by Anderson and Linville (1962) and Wilson and Luthin (1963) showed that the heat of wetting slightly increases the gaseous volume to be displaced. Microorganisms can increase or decrease the amount, and hence the pressure, of entrapped air by consuming relatively insoluble O_2 and producing highly soluble CO_2. In anaerobic environments the O_2/CO_2 counterdiffusion is not equimolar, since there is more CO_2 produced than O_2 respired. Under such conditions, nongaseous substrates are transformed into gaseous constituents such as N_2, N_2O, H_2S, hydrocarbons, and others. In waterlogged soils "air" bubble or "air" pocket formation should therefore be expected. As a result of air entrapment, atmospheric pressure fluctuations can change the water table position (Peck, 1960; Stevenson & Van Schaik, 1967; Norum & Luthin, 1968). Solution or dissolution of entrapped gases reduces this effect when the barometric pressure changes slowly (Peck, 1969).

The existing data on soil air pressure and the understanding of its role in transient water flow and soil aeration suggest that this variable is more often relevant than one is inclined to assume. It is probably one of the most frequently neglected variables in soil physics research.

50-2 PRINCIPLES

50-2.1 General

In the simplest measurements of soil air pressure, an air-filled access tube connects the soil atmosphere to a manometer (Fig. 50-1) that has one arm open to the ambient atmosphere. Thus, the soil air pressure is measured relative to the barometric pressure. Some calculations require the absolute pressure, which is the sum of the gauge and barometric pressures. In order to measure the soil air pressure, this (nonwetting) phase must be continuous between the soil and the measuring system. In this sense, the function of a soil air pressure gauge is basically analogous to that of a tensiometer. In tensiometry, liquid phase continuity is established between the soil water and a device for recording fluid pressure. This is made possible by the inclusion of a porous barrier which prevents the entry of air into the pressure measuring system, although the pressure in this system is usually less than atmospheric.

The water contents of soil regions with an air pressure greater than atmospheric are often close to saturation. At locations where the water pressure exceeds the barometric pressure, the water tends to intrude into the air-filled access tube. This can be prevented by using a hydrophobic membrane permeable to air but impermeable to water (Flühler & Läser, 1975). A pressure of > 10 kPa could be applied without forcing water through such water-repellent barriers. The use of an access tube covered

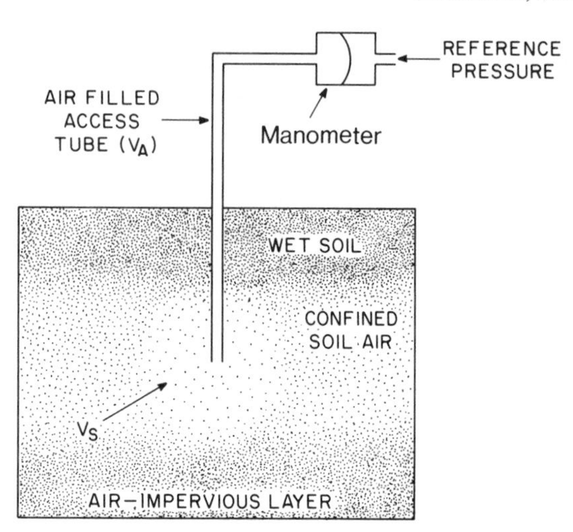

Fig. 50–1 Diagram of a soil air pressure gauge. Various types of manometers such as pressure transducers, fluid manometers, and anaeroid barometers can be used depending upon internal air volume of the access system, V_A, and the soil air volume, V_S, surrounding the orifices of the tube.

with a hydrophobic membrane is fully analogous to the common (liquid phase) tensiometer, except that the gaseous phase is kept continuous instead of the liquid phase.

The design of soil air pressure gauges (access tube + manometer) described in the literature varies with respect to (i) volume of the access tube system, (ii) size of the opening in the soil, and (iii) type of manometer. Dixon and Linden (1972) clearly distinguished between point measurements and measurements within a large soil air mass trapped in the soil profile. Local measurements were made using small-diameter access tubes open only at the tip, whereas the soil air pressure within the subsoil was tapped using a large-diameter access tube perforated over several 0.1-m increments of depth. Point measurements of air pressure within soil aggregates, such as those reported by Stroosnyder and Koorevaar (1972), require an access system with an internal air volume which is small compared with that entrapped within the crumb. These authors used a water-filled hypodermic needle, which was connected with a sensitive pressure transducer. This system ceased to function when the wetting front reached the tip of the needle.

Soil air and soil water pressures (P_{nw} and P_w, respectively) are related to the capillary potential Ψ:

$$\Psi = P_w - P_{nw} . \qquad [1]$$

Both P_w and P_{nw} must be measured relative to a common datum. Interfacial tension and the curvature of air-water interfaces in the soil result in the pressure differential between the two phases. As long as the interfacial curvature remains constant, any change in P_{nw} (soil air pressure) will be compensated by an equal adjustment in P_w (soil water pressure). Attempts have been made to measure the pressure difference across the air-water interfaces directly according to Eq. [1]. Vachaud et al. (1973) used tensiometers equipped with a pressure transducer. An air-filled hypodermic needle connected the soil atmosphere in the vicinity of the tensiometer cup to the reference side of the transducer membrane. Using a fluid switch, the soil water pressure could be measured relative to barometric pressure or pressure of the soil air.

50–2.2 Measurement System Parameters

As will be shown below, three parameters determine the response of a soil air pressure gauge to soil air pressure changes: (i) the volume of soil air being tapped, V_S, (ii) the gaseous volume within the access system, V_A, and (iii) the sensitivity of the manometer, S_M (Eq.[2]). Unfortunately, the soil air pressure gauges mentioned in the previous sections were not explicitly defined in these terms. However, the relevant differences in design and response of these systems are obvious from the published descriptions.

As pointed out earlier, air pressure gauges are, in principle, nonwetting phase tensiometers. In contrast to tensiometry, the tapped phase is extremely compressible. Hence, more care must be taken to minimize the influence of the access system upon the measured property. The sensitivity of a soil air pressure gauge (access tube + manometer) can be defined similarly to that of tensiometers (Richards, 1949) [cf.chapter 23 in this book]:

$$S = dP/dV \text{ (kPa cm}^{-3}) \qquad [2]$$

where dV is the gas volume which must enter the access system for a pressure increase dP. This volume increment depends upon compression of the air within the access tube (dV_A) and the volume change within the manometer (dV_M):

$$dV = dV_M - dV_A \text{ (cm}^3) \,. \qquad [3]$$

Note that dV_A is negative when dP is positive. The subscripts A and M refer to the access system and manometer, respectively, as shown in Fig. 50–2. In pressure transducers the volume change dV_M is caused by membrane deflection. In capillary manometers it corresponds to the displacement of the manometer fluid. The manometer sensitivity is defined by

$$S_M = dP/dV_M \text{ (kPa cm}^{-3}) \,. \qquad [4]$$

To express dV_A in terms of the initial conditions P and V_A, imagine a

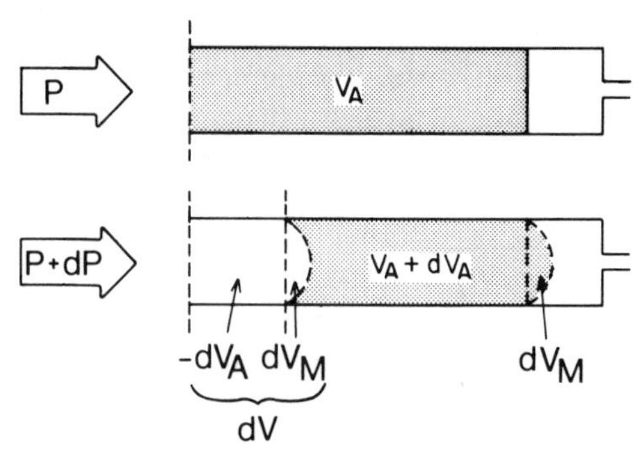

Fig. 50–2 When soil air pressure increases, a gaseous volume dV is vented into the access system. For a positive dP the internal air volume change dV_A is negative.

membrane confining a volume V_A of air in the gauge. A pressure change dP will alter the volume V_A according to

$$P\,dV_A + V_A\,dP = 0\,. \qquad [5]$$

Since this equation is an expression of Boyle's law, P is the absolute pressure. Substitution of Eq. [4] and [5] into Eq. [3] and Eq. [3] into Eq. [2] defines the overall gauge sensitivity S (access system + manometer)

$$S = S_M\,P/(V_A S_M + P)\,. \qquad [6]$$

In terms of a dimensionless expression which we call the gauge characteristic ϵ

$$\epsilon = S_M\,V_A/P \qquad [7]$$

the relative sensitivity S/S_M (Fig. 50–3) becomes

$$S/S_M = 1/\epsilon + 1\,. \qquad [8]$$

Since ϵ is intrinsically positive, the gauge sensitivity, S, is less than that of the manometer. It indicates to what degree the manometer itself is limiting the response of the entire measuring system. If $V_A\,S_M$ equals P (absolute pressure of soil air) the gauge sensitivity, S, is 50% of S_M. The sensitivity of pressure transducers and also their reading precision is usually quite high ($S_M > 10$ kPa cm^{-3}), although nonlinear response and calibration instability occasionally cause problems. Capillary manometers filled with a low-density fluid such as water allow precise readings but their sensitivity is low ($S_M \sim 3$ kPa cm^{-3}). For two-arm manometers (U-tube) the sensitivity is twice that of one-arm manometers,

$$S_M = 2\,\rho\,g/\pi\,d^2\,, \qquad [9]$$

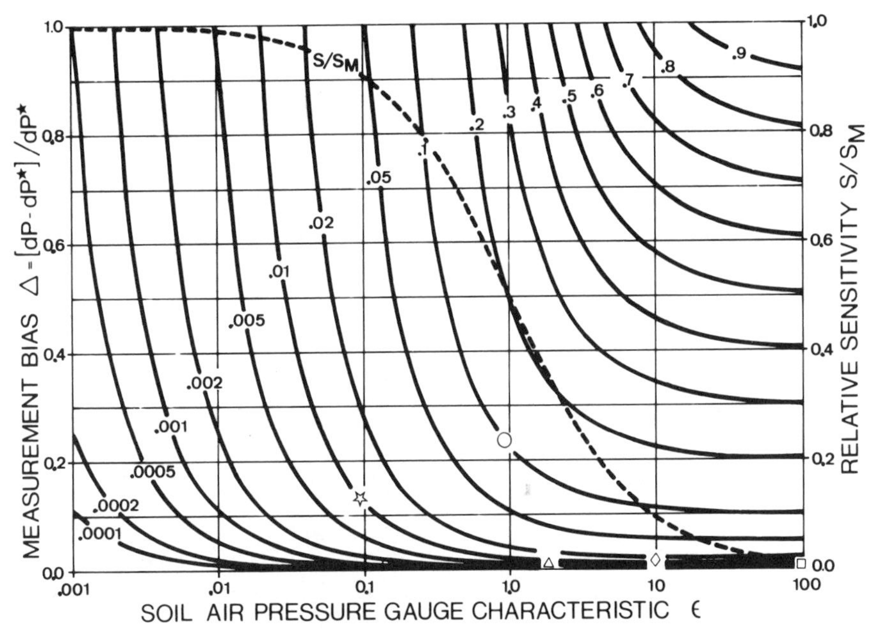

Fig. 50–3 The measurement bias Δ (solid lines) and the relative sensitivity S/S_M (broken line) depend upon the pressure gauge characteristic ϵ (Eq.[7],[8], and [17]). The numbers on the curves refer to the volume ratio κ. The symbols indicate the values estimated in Table 50–1.

where ρ denotes fluid density, g the gravity constant, and d the capillary i.d. With mercury instead of water as manometer fluid, the sensitivity S_M is higher at the expense of a lower reading precision ($S_M \sim 80$ kPa cm^{-3}).

The change of soil air pressure measured with any gauge is less than it would be without such an artificial extension of the soil atmosphere, since a volume element dV (Eq.[3]) must pass to or from the gauge to change the indicated pressure. The difference between the undisturbed and measured pressure changes dP and dP^*, respectively, may be expressed as a measurement bias Δ defined by

$$\Delta = (dP - dP^*)/dP^* \, . \qquad [10]$$

This bias should be both minimized and estimated.

The behavior of the entire system (soil atmosphere and gauge together) is defined by the variables V_S (soil air volume), P, S_M, and V_A. In order to derive an expression for Δ in terms of these four parameters, we assume that intruding or draining water displaces a given volume dV_S. Applying Boyle's law (Eq.[5]) to the soil atmosphere, excluding the influence of the soil air pressure gauge, yields

$$dP = -\,(dV_S/V_S)\,P \, . \qquad [11]$$

Considering the entire system, the soil air pressure gauge included, with a total volume $V_t = V_S + V_A$, the measured pressure change is

$$dP^* = -(dV_t/V_t)\,P\,. \qquad [12]$$

Since the access tube is assumed to be rigid, the volume change of the entire system is

$$dV_t = dV_S + dV_M \qquad [13]$$

where dV_S and dV_M have opposite signs.

Substituting Eq.[4] and [11] into Eq.[13] and Eq.[13] and [12] into Eq.[10] leads to

$$\Delta = (V_A\,S_M + P)/(V_S\,S_M - P)\,. \qquad [14]$$

Knowing the magnitude of S_M, V_A, V_S, and P, one can therefore calculate the bias Δ and adjust observed pressure changes to determine the changes one would expect in a soil not disturbed by the soil air pressure gauge

$$dP = dP^*\,(1 + \Delta)\,. \qquad [15]$$

From Eq.[14] it follows that Δ can be minimized more effectively by reducing the gaseous volume V_A than by increasing the manometer sensitivity S_M.

The dimensionless gauge characteristic ϵ (Eq.[7]) and the ratio κ defined by

$$\kappa = V_A/V_S \qquad [16]$$

may be used to illustrate relationships between ϵ and Δ (Fig.50–3)

$$\Delta = \kappa\,(\epsilon + 1)/(\epsilon - \kappa)\,. \qquad [17]$$

The characteristics of several air pressure gauges are listed in Table 50–1. The three parameters, S_M, V_A, and V_S are estimated based upon the description given in the quoted publications. The volume V_A and sensitivity S_M of different manometers vary by several orders of magnitude. The soil air volume V_S cannot be accurately determined because it is somewhat arbitrary to define a sphere of influence around the soil end opening of the access tube. In addition, V_S varies with moisture content. The values of V_S in Table 50–1 are minimum estimates. The pressure gauge characteristics, ϵ, and the volume ratios, κ, shown in Table 50–1 are also indicated in the nomogram (Fig. 50–3). In two of the field experiments, ϵ was large and the volume ratio κ small, allowing the use of Hg manometers instead of highly sensitive pressure transducers. From

Table 50–1. Characteristics of soil air pressure gauges.

Observed system	Air volume†		Volume ratio	Manometer	Sensitivity	Gauge characteristic	Authors/access system
	Soil V_S (radius‡)	Access tube V_A	$\varkappa = V_A/V_S$		S_M†	ϵ	
	cm³				kPa cm⁻³		
Confined air, field soils	50 000 (~50 cm)	500	0.01	Anaeroid barometer	20	100	Dixon & Linden, 1972 Perforated access tube
Entrapped air, field soils	500 (~5 cm)	5	0.01	Water manometer	2	0.1	Dixon & Linden, 1972 Capillary access tube
Confined and entrapped air, field soils	500 (~11 cm)	0.5	0.001	Pressure transducer	2000	10	Flühler & Läser, 1975 Hydrophobic membrane probe
Confined air, soil columns	10 (~6 cm)	0.1	0.01	Pressure transducer	2000	2	Vachaud et al., 1973 Hypodermic needle
Entrapped air, aggregates	0.5 (~1 cm)	0.05	0.1	Pressure transducer	2000	1	Stroosnyder & Koorevaar, 1972 Water-filled hypodermic needle with air bubble

† V_S, V_A, and S_M are estimated based upon experimental description.
‡ The radius of the sphere of influence is estimated assuming an air-filled porosity of 10%.

this we may conclude that the design of soil air pressure gauges can and should be adapted to the particular experimental dimensions. In terms of the nomogram, ϵ should be maximized and κ minimized.

50–3 METHODS

50–3.1 Instrument Construction

The measured pressure changes will always be less than those occurring in the absence of a measuring system (cf.Eq.[15]). Hence, the mere presence of the measuring device, especially that of the access tube, changes the soil air continuum. Knowing the pertinent parameters of the system described in this chapter enables the researcher to optimize the dimensions of the soil air pressure gauge and to take the unavoidable errors into account.

For any application, the soil air pressure gauge should be designed and built according to the specifics of the medium under consideration. Since the volume of soil air considered may vary by several orders of magnitude, it is not possible to recommend a particular design of equipment, but rather a procedure for optimum design and construction. An example is given in parenthesis to illustrate each step.

1. Describe the system to be measured in terms of its spatial dimensions such as volume, shape, and minimum volumetric air content of the soil region with a confined gas phase. Estimate the smallest expected soil air volume, V_S, within the "sphere of influence" of the access tube opening(s) (example : V_S = 1000 cm^3).
2. Design the access tube system with a minimum internal volume V_A (example : V_A = 50 cm^3). The estimate of V_S dictates the maximum size and the location of the access tube opening(s) (cf.Linden & Dixon, 1972) For local soil air pressure measurements the size of the opening should be small relative to V_S. In order to tap the gas phase of an entire soil horizon the access tube may be perforated over the required depth interval. In soils with a significant air pressure buildup the opening(s) are most likely exposed to positive water pressures, at least for short periods of time. To avoid plugging by water menisci, the opening diameter and the I.D. of the access tube should be either large (i.e., $>$ 0.5 cm) or protected by a porous hydrophobic membrane. Most filter-producing companies offer such material with air entry values $>$ 10 kPa (For example, Millipore Filter Corp., Bedford, MA 01730 or Gelman Sciences, Inc., Ann Arbor, MI 48106). For the access system, it is also advantageous to use water-repellent tubing material (i.e., nylon or teflon), which drains more easily after being plugged with water.
3. Determine the volume ratio $\kappa = V_A / V_S$ (Eq.[16]) (example: κ = 0.05).
4. Define a maximum acceptable bias Δ (Eq.[10]) (example : Δ = 0.10). Note that any measurement of a soil air pressure change dP^* is less

than the value dP which would occur in absence of the additional internal air volume V_A of the access system.

5. Calculate the soil air pressure gauge characteristic ϵ from Eq. [18]

$$\epsilon = \kappa (\Delta + 1)/(\Delta - \kappa) \quad [18]$$

or obtain it from Fig. 50–3 for $\Delta = 0.10$ and $\kappa = 0.05$ (example: $\epsilon = 1.1$).

6. The required manometer sensitivity S_M is obtained from Eq.[7] (example: for $\epsilon = 1.1$, $V_A = 50$ cm^3, and $P = 100$ kPa, the sensitivity is $S_M = 2.2$ kPa cm^{-3}; hence, a U-tube water manometer with an i.d. of 0.17 cm would be adequate). To calculate d in meters, use S_M in Pa m^{-3}, $g = 9.81$ m s^{-2}, and ρ in kg m^{-3} (cf. Eq. [9]).

50–3.2 Installation and Operation

The access tube must tightly fit into the surrounding soil. The technique used for installation of air sampling probes by Roulier et al. (1974) is recommended for small access tubes (i.d. < 1 cm). A stainless steel access tube is pushed into the ground using a guiding clamp with handles. After installation the access tube is cleared by gently blowing air through it. Those parts protruding aboveground should be protected from excessive temperature changes. When pressure transducers are used, periodic recalibration is advisable. Therefore, the transducers must be either detachable from the access tube or connected to a fluid switch that allows calibration with respect to atmospheric pressure.

50–4 REFERENCES

Adrian, D. D., and J. B. Franzini. 1966. Impedance to infiltration by pressure build-up ahead of the wetting front. J. Geophys. Res. 71:5857–5862.

Anderson, D. M., and L. Linville. 1962. Temperature fluctuations accompanying water movement through porous media. Science 131:1370–1371.

Brustkern, R. L., and H. J. Morel-Seytoux. 1970. Analytical treatment of two-phase infiltration. J. Hydraul. Div., Am. Soc. Civ. Eng. 96 (HY12):2535–2548.

Dixon, R. M. 1975. Design and use of closed-top infiltrometers. Soil Sci. Soc. Am. Proc. 39:755–763.

Dixon, R. M., and D. R. Linden. 1972. Soil air pressure and water infiltration under border irrigation. Soil Sci. Soc. Am. Proc. 36:948–953.

Dixon, R. M., and A. E. Peterson. 1971. Water infiltration control: a channel system concept. Soil Sci. Soc. Am. Proc. 35:968–973.

Flühler, H., and H. P. Läser. 1975. A hydrophobic membrane probe for total pressure and partial pressure measurements in the soil atmosphere. Soil Sci. 120:85–91.

Linden, D. R., and R. M. Dixon. 1973. Infiltration and water table effects of soil air pressure under border irrigation. Soil Sci. Soc. Am. Proc. 37:94–98.

Linden, D. R., and R. M. Dixon. 1975. Water table position as affected by soil air pressure. Water Resour. Res. 11:139–143.

Linden, D. R., and R. M. Dixon. 1976. Soil air pressure effects on route and rate of infiltration. Soil Sci. Soc. Am. J. 40:963–965.

Linden, D. R., R. M. Dixon, and J. C. Guitjens. 1977. Soil air pressure under successive border irrigations and simulated rain. Soil Sci. 124:135–139.

McWhorter, D. B. 1971. Infiltration affected by flow of air. Hydrology Paper 49. Colorado State Univ., Fort Collins.

Norum, D. I., and J. N. Luthin. 1968. The effects of entrapped air and barometric fluctuations on the drainage of porous mediums. Water Resour. Res. 4:417–424.

Parlange, J-Y., and D. E. Hill. 1979. Air and water movement in porous media: compressibility effects. Soil Sci. 127:257–263.

Peck, A. J. 1960. The water table as affected by atmospheric pressure. J. Geophys. Res. 65:2383–2388.

Peck, A. J. 1965a. Moisture profile development and air compression during water uptake by bounded porous bodies. 2. Horizontal columns. Soil Sci. 99:327–334.

Peck, A. J. 1965b. Moisture profile development and air compression during water uptake by bounded porous bodies. 3. Vertical columns. Soil Sci. 100:44–51.

Peck, A. J. 1969. Entrapment, stability and persistence of air bubbles in soil water. Aust. J. Soil Res. 7:79–90.

Raj Pal, and L. Stroosnijder. 1976. Experimental verification of two calculation methods for relating horizontal infiltration to soil-water diffusivity. Geoderma 15:413–424.

Richards, L. A. 1949. Methods of measuring soil moisture tension. Soil Sci. 68:95–112.

Roulier, M. H., L. H. Stolzy, and T. E. Szuszkiewicz. 1974. An improved procedure for sampling the atmosphere of field soils. Soil Sci. Soc. Am. Proc. 38:687–689.

Smith, R. M., and D. R. Browning. 1942. Persistent water unsaturation of natural soil in relation to various soil and plant factors. Soil Sci. Soc. Am. Proc. 7:114–118.

Stevenson, D. S., and J. C. Van Schaik. 1967. Some relations between changing barometric pressure and water movement into lysimeters having controlled water tables. J. Hydrol. (Amsterdam) 5:187–196.

Stroosnyder, L., and P. Koorevaar. 1972. Air pressure within soil aggregates during quick wetting and subsequent "explosion." I. Preliminary experimental results. Mededel. Fac. Landbouwwet., Rijksuniv. Ghent 37:1095–1106.

Vachaud, G., M. Vauclin, D. Khanji, and M. Wakil. 1973. Effects of air pressure on water flow in an unsaturated stratified vertical column of sand. Water Resour. Res. 9:160–173.

Wilson, L. G., and J. N. Luthin. 1963. Effect of air flow ahead of the wetting front on infiltration. Soil Sci. 96:136–143.

Youngs, E. G., and A. J. Peck. 1964. Moisture profile development and air compression during water uptake by bounded porous bodies. I. Theoretical introduction. Soil Sci. 98:290–294.

SUBJECT INDEX